Magnesium

Proceedings of the 7th International Conference on
Magnesium Alloys and Their Applications

Edited by
K. U. Kainer

DGM

WILEY-
VCH

7th International Conference on Magnesium Alloys and Their Applications

Organiser: DGM – Deutsche Gesellschaft für Materialkunde e. V.

Organisational Committees

International Committee

Prof. Dr. E. Aghion, Ben Gurion University, Beer Sheva,, Israel
Dr. J. Carpenter, U.S. Department of Energy, Washington, USA
Prof. Dr. E.-H. Han, Chinese Academy of Science, Shenyang, China
Prof. Dr. E. Ghali, Laval University, Quebec, Canada
Dr. J. Jackman, CANMET Materials Technology, Laboratory, Ottawa, Canada
Prof. S. Kamado, Nagaoka University of Technology, Japan
Dr. H. Kaufmann, Leichtmetall-Kompetenzzentrum, Ranshofen GmbH, Austria
Dr. N. Li, Ford Motor Company, Dearborn, USA
Dr. A. Luo, General Motors Corporation, Warren, USA
Prof. Dr. J. F. Nie, Monash University, Victoria, Melbourne, Australia
Dr. E. A. Nyberg, Pacific Northwest National, Laboratory, Richland, USA
Prof. Dr. F. Pan, Chongqing University, Chongqing, China
G. Patzer, International Magnesium, Association, Wauconda, USA
Prof. Dr. M. Pekguleryuz, McGill University, Montreal, Canada
Prof. Dr. L. L. Rokhlin, Baikov Institute of Metallurgy and Materials Science, Moscow, Russia
Dr. T. Wilks, Magnesium Elektron, Manchester, UK
Dr. H. Westengen, Hydro Aluminium, Porsgrunn, Norway

National Committee

Chairman
Prof. Dr. K.U. Kainer, GKSS-Forschungszentrum, Geesthacht GmbH

Prof. Dr. F.-W. Bach, Universität Hannover
Prof. Dr. A. Bührig-Polaczek, Gießerei-Institut der RWTH Aachen
Dr. B. Engl, MgF Magnesium Flachprodukte GmbH, Freiberg
Dr. P. Gregg, Franz Oberflächentechnik, Geretsried
Dipl.-Ing. Elke Hombergsmeier, EADS Deutschland GmbH, München
Dr. N. Hort, GKSS-Forschungszentrum, Geesthacht GmbH
Dr. P. Juchmann, Salzgitter Magnesium-Technologie GmbH
Dr. C. Kettler, Advanced Magnesium Technology, Heidelberg
Dipl.-Ing. G. Rienaß, Hydro Magnesium Marketing GmbH, Bottrop
Dipl.-Ing. C. Schendera, Hydro Magnesium, Bottrop
Prof. Dr. R. Schmid-Fetzer, TU Clausthal-Zellerfeld
Dr. S. Schumann, Volkswagen AG, Wolfsburg
Prof. Dr. R. F. Singer, Universität Erlangen-Nürnberg
Dr. G. Terlinde, Otto Fuchs KG, Meinerzhagen
Dr. M. Wappelhorst, Honsel GmbH & Co. KG, Meschede
Dr. J. Wesemann, Ford Forschungszentrum Aachen

Sponsor:

Co-Sponsors:

Magnesium

Proceedings of the 7th International Conference on
Magnesium Alloys and Their Applications

Edited by
K. U. Kainer

WILEY-
VCH

WILEY-VCH Verlag GmbH & Co. KGaA

Editor:
Prof. Dr. K. U. Kainer
GKSS-forschungszentrum
Institut für Werkstoffforschung
Max-Planck-Straße
21502 Geesthacht
Germany

7th International Conference on Magnesium Alloys
and Their Applications

from 06-09 November 2006 in Dresden
Organized byDGM (Deutsche Gesellschaft für Materialkunde e. V.)

All books published by Wiley-VCH are carefully produced. Nevertheless, editor, authors, and publisher do not warrant the information contained in these books, including this book, to be free of errors. Readers are advised to keep in mind that statements, data, illustrations, procedural details or other items may inadvertently be inaccurate.

Library of Congress Card No.: applied for

British Library Cataloguing-in-Publication Data:
A catalogue record for this book is aailable from the British Library

Bibliografic information published by the Deutsche Nationalbibliothek
The Deutsche Nationalbibliothek lists this publication in the Deutsche Nationalbibliografie;
detailed bibliografic data is available in the Internet at <http://dnb.d-nb.de>.

© 2007 WILEY-VCH Verlag GmbH & Co. KGaA, Weinheim

Printed on acid-free paper

Printed in the Federal Republic of Germany

All rights reserved (including those of translation in other languages). No part of this book may be reproduced in any form – by photoprinting, microfilm, or any other means – nor transmitted or translated into machine language without written permission from the publishers. Registered names, trademarks, etc. used in this book, even when not specifically marked as such, are not to be considered unprotected by law.

Composition: W.G.V. Verlagsdienstleistungen GmbH, Weinheim
Printing: betz-Druck GmbH, Darmstadt
Bookbinding: Litges & Dopf Buchbinderei GmbH, Heppenheim

ISBN: 978-3-527-31764-6

Preface

Magnesium is the most promising light metal for the universal use in the transportation industry. The intrinsic characteristics of light weight include promising mechanical properties, high recyclability and low primary material cost. Therefore, the interest for the implementation of magnesium alloys has increased both as a substitute and as a strategic material for new design.

Magnesium alloys have to compete in terms of economical aspects and design requirements with various light materials such as aluminium and polymers, but also with light weight steel based applications. After intensive research during the last decade, a number of new research programmes in Europe, China and North America have generated new ideas. The implementation of new alloys in power trains, the use of wrought alloys in 3C applications, and cast alloys for parts with decorative surfaces stand for sustainable research and development in the past.

Nevertheless, the production of parts by high pressure die casting is still the predominant technology. Wrought alloys have been niche products so far. The lack of knowledge on their behaviour, as well as to the limited availability of alloys for extrusion, rolled products and forgings are most obvious. New developments will open new opportunities for the increased use of wrought alloys. This was one focus of the conference. The other major focus addressed the corrosion behaviour and hence the development of reliable surface coatings.

The conference programme included the presentation of the research results of the Priority Programme "Extending the Range of Applications of Magnesium Alloys" (InnoMagTec) funded by the German Research Foundation (Deutsche Forschungsgemeinschaft, DFG.

The organisation of the conference required massive support by both the National and the International Committees. For turning the contribution of the committees into shape, the Conference Secretariat provided by the Deutsche Gesellschaft für Materialkunde did an excellent job. My special thanks go to Ms. A. Mangold, Ms. Y. Koall, Ms. C. Wendt.

These proceedings cover 168 papers originating from invited plenary lectures, as well as from contributed oral and poster presentations. I would like to extend my sincere thanks to the authors for their contribution. Finally I thank the publisher for excellent support.

November 2006,
Karl U. Kainer, Chairman

Contents

Alloy Development — 1

The Viability of Mg Alloy with Nano/Sub-micron Structure as a new Material for Practical Applications — 3
E. Aghion, A. Arnon

A Comparative Study of Carbon Additives on Grain Refinement of Magnesium Alloys — 8
P. Cao, Ma Qian, K. Kondoh, D.H. St.John

Advanced Gravity Casting Magnesium Alloys for the Aircraft Industry — 14
B. Bronfin, N. Moscovitch, A. Ben-Dov, J. Townsend, S. Mahmood, J. Vainola, S. Deveneyi

Elektron 21 – An Aerospace Magnesium Alloy for Sand Cast & Investment Cast Applications — 20
P. Lyon, I. Syed, S. Heaney

Microstructures and Mechanical Properties of Mg-Zn-Al Alloys Produced by High Pressure Die Casting and Sand Casting — 26
J.F. Nie, M. Easton, T.B. Abbott

Phase Formation in Mg-Sn Alloys Modified by Ca and Ce — 32
J. Gröbner, A. Kozlov, M. Hampl, R. Schmid-Fetzer

Microstructure and Micromechanical Properties of as-cast Mg-Sn-Ca and Mg-Sn-Mn Alloys — 37
J. Buršík, Y. Jirásková, V. Buršíková, T. Abu Leil, C. Blawert, W. Dietzel, N. Hort, K.U. Kainer

Microstructure and High Temperature Mechanical Properties of Mg-Al-Sn Alloys — 43
Yeon Jun Chung, Kwang Seon Shin, Woo Chul Cho

Effect of Heat-Treatment on the Microstructure, Microhardness and Corrosion of Cast Mg-3Sn-2Ca Alloy — 49
J. Buršík, K. Zábranský, Y. Jirásková, V. Buršíková, T. Abu Leil, C. Blawert, Y. Huang, W. Dietzel, N. Hort, K. U. Kainer, K.P. Rao

Elevated Temperature Alloys – Paths for Further Performance Gains in AE44 — 55
P. Bakke, A.L. Bowles, H. Westengen

Phase Composition and Creep Behavior of Mg-Rare Earth-Mn Alloys with Zn Addition — 67
B. Smola, I. Stulíková, J. Pelcová, N. Zaludová

Investigation of Mg-Al-Sr Phase Equilibria — 74
A. Janz, J. Gröbner, D. Mirković, R. Schmid-Fetzer, M. Medraj, Zhu Jun, Y.A. Chang

Hot Chamber Die-Casting of Magnesium Alloys *R. Cibis, A. Kiełbus*	81
Microstructural characterization of Mg-Al-Sr alloy *T. Rzychoń, A. Kiełbus, R. Cibis*	87
A new Look at the Mg-Al-Zn Family – Focus on Tailoring Properties through Composition Control *A.L. Bowles, P. Bakke, H. Westengen*	93
Mg-4Al-4RE Alloy for Die-Casting *T. Rzychoń, A. Kiełbus, R. Cibis*	100
Effective Alloy Development using Thixomolding® Focussing on Improved Creep Strength of AZ91 *H. Eibisch, A. Lohmüller, M. Scharrer, R.F. Singer*	106
Microstructures and Mechanical Properties of Extruded AZ31 Mg Alloy Processed by Roll-compaction Process	113
K. Kawabata, H. Oginuma, K. Kondoh, Microstructural Evolution of Mg-Al-Ca-Sn Alloy during Extrusion *J. Dzwonczyk, M.A. Leeflang, J. Zhou, J. Duszczyk*	119
Microstructure and Mechanical Properties of Strip Cast Mg-Zn-Al Alloys with Ca by Twin Roll Caster *K. Matsuzaki, K. Hatsukano, M. Kato, K. Hanada, T. Shimizu, Y. Murakoshi*	125
Production of Mg-Ni Alloys under High Pressure and High Temperature for use as Catalysts in Diamond Synthesis *A. J. Sideris, G.S. Bobrovnitchii*	131
Investigation of Structure and Properties of the Age-Hardenable Mg-Sm Alloys after Sever Plastic Deformation *L.L. Rokhlin, T.V. Dobatkina, N.I. Nikitina, V.N. Timofeev, S.V. Dobatkin, M.V. Popov*	137
Simulation and Modelling	**143**
Yielding of Magnesium: from Single Crystal to Structural Behaviour *S. Graff, D. Steglich, W. Brocks*	145
Microstructure Formation in Magnesium Alloys Investigated by Phase-Field Simulation *J. Eiken, I. Steinbach, B. Böttger*	151
Effects of Calcium on Texture Evolution and Plastic Anisotropy of the Magnesium Alloy AZ31 *T. Ebeling, Ch. Hartig, R. Bormann, J. Bohlen, D. Letzig*	158

Modeling Microstructure and Mechanical Properties of thin-walled Magnesium Castings *K. Weiss, A. Wendt, Ch. Honsel, G. Arruebarrena, I. Hurtado, J. Väinölä, S. Mahmood, J. Townsend, A. Ben-Dov, A. Schröder, A. Pinkernelle*	165

Casting and Recycling Melting — 173

Advancements in High Pressure Die Casting of Magnesium *H. Westengen, H. Gjestland*	175
A New Analytical Methodology for the Assessment of Grain Refinement in Magnesium Alloys *D. H. StJohn, Peng Cao, Ma Qian, M. A. Easton*	189
Creation of Fine and Spherical Magnesium Crystals via Control of Nucleation and Growth Instabilities *Ma Qian*	200
The Role of Fluorine Solubility in the Protection of Molten Magnesium *M. Syvertsen, G. Tranell, K. Aarstad, T.A. Engh*	207
Reduction and Avoidance of Ecologically Harmful Cover Gases for Handling Molten Magnesium *Fr.-W. Bach, M. Schaper, M. Hepke, A. Karger, J. Werner*	215
Ceramic Foam Filtration of Magnesium Alloy Melt *Bong Sun You, Chang Dong Yim, G.H. Wu*	221
Development of an Investment-Casting Process of Mg-Alloys for Aerospace Applications *G. Arruebarrena, I. Hurtado, J. Väinölä, S. Dévényi, J. Townsend, S. Mahmood, A. Wendt, K. Weiss, A. Ben-Dov, A. Schröder, A. Pinkernelle*	228
Evaluation of the Hot Tearing Susceptibility of Selected Magnesium Casting Alloys in Permanent Molds *J.P. Thomson, M. Sadayappan, M. Sahoo, D.J. Weiss*	234
Temperature Dependence of the Viscosity of Magnesium Alloy AZ91D with Additions of Calcium in Liquid State and Effect of Thermal Treatment of these Melts on the Structure that Forms after their Crystallization *I. S. Abaturov, P. S. Popel, I. G. Brodova, V. V. Astafiev, Peijie Li*	240
Advances in Magnesium Injection Molding (Thixomolding®) *M. Scharrer, A. Lohmüller, R. M. Hilbinger, H. Eibisch, R. Jenning, M. Hartmann, R. F. Singer*	248
Microstructure and Mechanical Properties of Rheo-diecast Mg-alloys *S. Liu, G. Ji, Y. Wang, Z. Fan*	256
Development of an AM50 Based Magnesium Recycling Alloy *D. Fechner, C. Blawert, P. Maier, K. U. Kainer, N. Hort*	262
The Properties of Magnesium Alloy Strip Produced by Melt Drag Process *M. Motomura, S. Nishida*	268

Experimental and Numerical Investigation of Texture Development During
Hot Rolling of Magnesium Alloys   274
R. Kawalla, C. Schmidt, H. Riedel, A. Prakash

Basic Research in the Contacting Condition between the Molten Magnesium Alloy
and the Roll on the Melt Drag Process   281
*Mitsugu Motomura, Takeshi Yamazaki, Yoshitaka Ando, Yuki Takano,
Shinichi Nishida*

Method for Horizontal Continuous Casting of Magnesium Alloy Slab   288
K. Tada

Wrought Alloys   **295**

Development and Application of Wrought Magnesium Alloys in China   297
Fusheng Pan, Aitao Tang, Siyuan Long, Mingbo Yang

Wrought Mg Alloy for Civil Aircraft Application – A Process Chain Approach   305
*M. Kettner, U. Noster, H. Kilian, R. Gradinger, W. Kühlein, A. Drevenstedt, F.
Stadler, E. Ladstaetter, A. Lutz*

The Dynamic Recrystallization of Commercially Pure Mg and Three Mg-Y
Alloys   318
R. Cottam, J. Robson, G. Lorimer, B. Davis

Data Correction for Constitutive Analysis of Wrought Magnesium Alloys Based
on Compression Tests   324
F.A. Slooff, J.P. Boomsma, J. Zhou, J. Duszczyk, L. Katgerman

Change in Formability of AZ31 and AZ61 from Room to Elevated Temperatures   330
H. Takuda, H. Fujimoto, T. Hama, Y. Sakurada

Examination of the Forming Properties of Magnesium Wrought Alloys MRI 301F
and AZ80A   336
A. Löffler, S. Schumann, K. U. Kainer, B. Bronfin

Die Forging of Commercial and Modified Magnesium Alloys   344
J. Swiostek, C. Blawert, D. Letzig, K. U. Kainer, J. Bober, W. Hintze

Die-Forged Disks for Automobile Wheels in Magnesium Alloys   352
B. V. Ovsyannikov

A Study on the Microstructure of Strip Cast Magnesium   357
H. Palkowski, P.M. Rucki, H.-G. Brokmeier, S.-B. Yi

Properties of Strips and Sheets of Magnesium Alloy Produced by Casting-
Rolling Technology   364
R. Kawalla, M. Ullmann, M. Oswald, C. Schmidt

Single Pass Large Draught Rolling of Some Magnesium Alloys below 473K
by High Speed Rolling   370
T. Sakai, H. Utsunomiya, S. Minamiguchi, H. Koh

Evaluation of Forming Behaviour and Tribological Properties of Magnesium Sheet-metal Directly Rolled From Semi-continuously Casted Feedstock *B. Viehweger, M. Düring, A. Karabet, H. Hartmann, U. Richter, G. Pieper, F. Newiak*	377
Investigations on Formability and Tribology of Magnesium Sheets with Regard to Warm Forming Processes *J. Hecht, M. Merklein, M. Geiger, A. Stich*	386
Superplastic Forming of Magnesium Alloys: Composition and Microstructure Effects *R. Boissière, J.J. Blandin*	393
On the Superplastic Forming of the AZ31 Magnesium Alloy *F. Abu-Farha, M. Khraisheh*	399
Extrusion of Different AZ Magnesium Alloys *S. Mueller, K. Mueller, T. Huichang, P. Wolter, W. Reimers*	406
Effect of billet temperature and ram speed on the behavior of AZ31 during extrusion *M.A. Leeflang, J. Zhou, J. Duszczyk*	413
Influence of Extrusion Conditions on the Superplastic Properties of a Mg-8%Li Alloy Processed at Room Temperature by Equal-Channel Angular Pressing *M. Furui, H. Kitamura, H. Anada, T.G. Langdon*	419
Severe Plastic Deformation of AZ31 *S. Mueller, K. Mueller, W. Reimers*	425
Production and Properties of Small Tubes Made from MgCa0,8 for Application as Stent in Biomedical Science *Th. Hassel, Fr.-W. Bach, A.N. Golovko*	432
Microstructural Evolution	439
The Effect of Zinc and Gadolinium on the Precipitation Sequence and Quench Sensitivity of Four Mg-Nd-Gd alloys *L.R. Gill, G. W. Lorimer, P. Lyon*	441
Secondary Precipitation in the Magnesium Alloy WE54 *Yuchang Su, B. M. Gable, B. C. Muddle, J.-F. Nie*	447
Influence of Production Technology and Heat Treatment on the Precipitation Processes in Mg-Y-Nd-Zn-Zr Alloy *J. Pelcová, B. Smola, I. Stulíková*	453
Microstructure of Elektron 21 Magnesium Alloy After Heat treatment *A. Kielbus*	459
Structural Evolution on Thermal Treatments of EV31 Alloy *M. Massazza, G. Riontino, D. Lussana, A. Iozzia, P. Mengucci, G. Barucca, A. Di Cristoforo§, R. Ferragut, R. Doglione*	466

Microstructural Stability of Magnesium Alloys for Aerospace Applications S. Avraham, M. Bamberger, B. Bronfin, G. Arruebarena	473
The Refinement of Precipitate Distributions in an Age Hardenable Mg-Sn Alloy Through Microalloying C. L. Mendis, C. R. R. Hutchinson, C. J. Bettles, M. A. Gibson, S. Gorsse	480
Microstructural Evolution and Phase Formation in Novel Mg-Zn Based Alloys B. Rashkova, G. Dehm, J. Keckés, G. Levi, A. Gorny, M. Bamberger	486
Zirconium - rich Coring Structure in WE54 Alloy after Heat Treatment A. Kiełbus	492
Microstructural Characterization of Die Cast Mg-Al-Sr (AJ) Alloy M. Kunst, A. Fischersworring-Bunk, C. Liebscher, U. Glatzel, G. l'Esperance, P. Plamondon, E. Baril, P. Labelle	498
Procedure of Quantitative Description of Mg-Al Alloy Structure After Heat Treatment J. Adamiec, A. Kiełbus, J. Cwajna, J. Paśko	506
The Evolution of Dislocation Structure as a Function of Deformation Temperature in UFG Magnesium Determined by X-ray Diffraction T. Fabián, Z. Trojanová, R. Ku_el	512
Microstructure and Thermal Stability of Ultra Fine Grained Mg-4Tb-2Nd Alloy Prepared by High Pressure Torsion J. Cizek, I. Prochazka, B. Smola, I. Stulikova, M. Vlach, R. K. Islamgaliev, O. Kulyasova	517
Amorphous Precipitates Sandwiched by Long Period Stacking Structures in the Melt-quenched Mg98Cu1R1 (R=Y and Gd) K. Konno, M. Matsuura, M. Yoshida, M. Nishijima, K. Hiraga	523
Texture Development in Different Routes ECAP Processed Mg-Si Alloys by Neutron Diffraction Mingyi Zheng, Xiaoguang Qiao, Shiwei Xu, Xiaoshi Hu, Kun Wu, Weimin Gan, A. Schreyer, K. U. Kainer, H.-G. Brokmeier, B. Schwebke	529
Effect of Ca Addition on the Microstructure Evolution of AZ31 Alloy during Thermomechanical Processing E. Essadiqui, Jian Li, C. Galvani, P. Liu, F. Zarandi, S. Yue	535
Flow Instabilities of Mg-Al4-Zn1 Alloy during High Strain-rate Deformation J. Dzwonczyk, F. Slooff, J. Zhou, J. Duszczyk	541
Microstructure and Mechanical Properties of Rolled Magnesium Sheets for Deep-drawing Applications Fr.-W. Bach, M. Hepke, M. Rodman, A. Rossberg	547
The Influence of Strain Path on Texture Evolution in Magnesium Alloy AZ31 T. Al-Samman, G. Gottstein	553

Microstructural Evolution by Hot-Working of Extruded and Continuously
Casted Mg-Wrought-Alloys AZ31 and AZ80 — 560
B. Viehweger, M. Düring, L. Schaeffer

Mechanical Properties — **569**

Ductile Failure of Magnesium AZ31: Twinning when the c-axis is Compressed — 571
M. R. Barnett

Dynamic Strain Ageing in Magnesium Alloys — 575
P. Lukáč, Z. Trojanová

Behaviors of High Temperature Deformation and Texture Development of
AZ31 and AZ61 under Uniaxial Compression — 581
L. Helis, K. Okayasu, H. Fukutomi

Slip during Tension Testing of Mg-3Al-1Zn Sheet — 588
Z. Keshavarz, M.R. Barnett

Microstructure and Deformation Behaviour of Mg-4Li Alloy — 594
Z. Drozd, Z. Trojanová

Method or Determination of Specific Damping Capacity Based on Relaxation
Oscillations of Cantilever Beam in Magnesium Alloys — 601
B. Landkof
A. Kotler, H. Abramovich

Acoustic Emission Study of the Mechanical Anisotropy of AZ31 Magnesium
Alloy Sheet — 607
P. Dobroň, F. Chmelík, P. Lukáč, J. Bohlen, D. Letzig, K. U. Kainer

Surface Rolling on the High-Strength Wrought Magnesium Alloy AZ80 — 613
P. Zhang, J. Lindemann, C. Leyens

Rolling of AM50 Magnesium Cast Alloy — 620
J. Göken, K. Steinhoff, I. Stulíková, B. Smola, N. Hort, V. Očenášek

Static and Dynamic Dent Resistance of AZ31 Magnesium Alloy Sheet — 627
J. Kaneko, M. Sugamata, E. Yukutake

Elaboration and Mechanical Behaviour of a Mg-based Bulk Metallic Glass — 633
S. Puech, J. J. Blandin, J. L. Soubeyroux

Fracture Toughness Behavior of Pressure Die Cast Magnesium Alloys Under
Static and Impact Loading — 639
G. Dietze, D. Regener

Influence of the Microstructure and Local Mechanical Properties of Magnesium
Die Cast Components Made of MgAl6Mn on the Quality of Numerical Simulations — 645
E. Lieven, R. Koch, W. Böhme, S. Schwarz

Effect of Anodized Layer on Fatigue Behavior of AM60 Magnesium Alloy — 651
Y. Miyashita, S. A. Khan, Y. Mutoh, T. Koike

Structural Durability of MRI 153M Die-Cast Components *A. Esderts, P. David, C. Berger, M. Gugau, T. Troßmann, J. Grimm*	658
Fatigue Crack Propagation Behavior of Mg-Al-Zn Alloys *E. H. Han, W. Ke, W. Dietzel, K. U. Kainer, R. C. Zeng*	666
Microstructure-based Characterization of the Cyclic Deformation Behavior of the Magnesium Die-cast Alloys MRI 153M and MRI 230D *B. Ebel-Wolf, F. Walther, D. Eifler*	673
Tensile-Compressive Creep Asymmetry of Die Cast Magnesium Alloys AM50, AE44 and AJ62A *S. Xu, M.A. Gharghouri, M. Sahoo*	680
Mechanical Strength and Creep Properties of Heat Treated AZ Alloys *S. Mueller, B. Camin, W. Reimers*	687
Creep Behavior of Squeeze Cast and T5 Heat Treated Mg4Y2Nd1Sc1Mn *F. Hnilica, V. Očenášek, B. Smola, I. Stulíková*	693
Influence of Specimen Orientation on Creep of Mg-10 vol.% Ti PM-Composite *F. Dobeš, K. Milička, P. Perez, G. Garces, G. Garces*	699
Effect of Finely Dispersed Particles on Mechanical Properties of Magnesium Alloys *S. Mizuno, Y. Tamura, H. Tamehiro, K. Funami, M. Takaya*	704
Corrosion and Surface Treatment	**708**
Advances in the Active Environment Impact on the Viscoelastic Behavior of Magnesium Alloys *Y. B. Unigovski, E. M. Gutman, Z. Koren*	709
Evaluation of Mg SCC Using LIST and SSRT *N. Winzer, G. Song, A. Atrens, W. Dietzel, C. Blawert, K. U. Kainer*	715
Corrosion Behaviour of Magnesium Alloys: Material Specific Corrosion Testing *S. Bender, E. Boese, A. Heyn, J. Goellner*	721
Corrosion-Fatigue of AZ31 Wrought Magnesium Weldments *C.E. Cross, P. Xu, G. Ben-Hamu, D. Eliezer*	727
Investigation of the Corrosion of Welded Magnesium Joints *C. Berger, K. Eppel, J. Ellermeier, T. Troßmann, U. Dilthey, H. Masny, K. Woeste*	734
Effect of Salt Spray Corrosion on the Tensile Behavior of Wrought Magnesium Alloy AZ31 *Sp.G. Pantelakis, N.D. Alexopoulos, A.N. Chamos*	743
Characterization of Corrosion Interfaces in Extruded Mg-Al-Zn Alloys *I. Apachitei, F. Andreatta, L. E. Fratila-Apachitei, J. Dzwonczyk, A. Berkani, J. Duszczyk*	749

Influence of the Casting Method on Microstructure and Corrosion of AZ91 and AM50 *D. Zander, C. Pieper, U. Köster*	757
Effect of Second Phase on Corrosion Behavior of AZ31-xCa Magnesium Alloys *Chang Dong Yim, Young Min Kim, Bong Sun You, Rang Su Jang, Su Geun Lim*	763
The Role of Si and Ca in New Wrought Mg-Zn-Mn Based Alloys *G. Ben-Hamu, D. Eliezer, K. S. Shin*	769
Corrosion Behavior of Die-Cast MRI 153M and AZ91D Alloys in Alkaline and Neutral Chloride Media *S. Amira, D. Dubé, R.A. Tremblay, E. Ghali*	775
Study of the Corrosion Behaviour of New Mg Based Alloys *C. Juers, C.-E. Barchiche, E. Rocca, J. Hazan, J. Steinmetz*	782
Influence of the Alloy Composition on the Mechanical and Electrochemical Properties of Binary Mg – Ca – Alloys and its Corrosion Behaviour in Solutions at Different Chloride Concentrations *T. Hassel, F.-W. Bach, C. Krause*	789
Study on Chloridic Corrosion of Two Magnesium Alloys WE43 and WE54 *R.A. Khosroshahi, N. Parvini Ahmadi, A. Torabadi*	796
Investigation of the Passivation Behavior of Magnesium Alloys by Means of Cyclic Current-Potential-Curves *T. Troßmann, K. Eppel, M. Gugau, C. Berger*	802
Corrosion Resistance of an Electrodeposited Magnesium Layer from Grignard Reagents on Carbon Steel in Aqueous Chloride Media *S. Ben Hassen, E. Triki, L. Bousselmi, M. Razrazi, P. Berçot*	809
Comparative Corrosion And Electrochemical Properties Of Mg-Ni And Mg-Cu Alloy Systems After Cathodic Polarization *V. Shvets, V. Lavrenko, V. Talash, T. Khomko*	816
The Role of Biological Environments on Magnesium Alloys as Biomaterials *A. Eliezer, F. Witte*	822
Nanoparticle-based Inorganic Coatings for the Corrosion Protection of Magnesium Alloys *W. Fürbeth, F. Feil, M. Schütze*	828
Composite Coatings – The Newest Surface Treatment Technology For Magnesium *I. Ostrovsky,*	834
Coating and Surface Treatment of Magnesium-Based Alloys and Composites *V. Neubert, A. Bakkar, Ching-An Huang*	842
Anodizing of Magnesium Alloys *A. Berkani, P. Skeldon, G.E. Thompson, L.E. Fratila-Apachitei, I. Apachitei, J. Duszczyk*	849

Corrosion Resistance of Plasma-Anodized AZ91D Alloy: Effect of Additive C.-E. Barchiche, C. Juers, E. Rocca, J. Hazan, J. Steinmetz	856
Electroplating Processes with New Magnesium Alloys A. Dietz, G. Klumpp	862

Post Processing — **869**

Effects of Strain Rate and Filler Alloy on the Properties of Laser Welded AZ31 Alloy Sheet Y. Durandet, W. Song, M. Brandt, P. Cordini, A. Ostendorf	871
Microstructure Features of Hot Rolled AZ31 Magnesium Alloy for Friction Stir Welding L. Commin, J.-E. Masse, L. Barrallier, M. Dumont	877
Rational Friction Welding of High Creep Resistance Magnesium Alloy G. A. Pinheiro, C. G. Pankiewicz, J. Fernandez dos Santos, K. U. Kainer	883
Development of Face Milling Process for Mg-hybrid (Mg-Al, Mg-sintered steel) Materials P. Ozsváth, A. Szmejkál, J. Takács, M. Eidenhammer, F. Obermair	894
Arc Brazing of Magnesium Alloy AZ31 with Steel (DX 53 Z) G. Garg, J. Zschetzsche, E. Simmchen, M. Schaper, U. Fuessel, S. Pandey	901
Electromagnetic Compression of Magnesium Tubes and Process-related Improvement of Wrought Alloys by Micro-alloying V. Psyk, A. Brosius, M. Kleiner, C. Broer, M. Bosse, M. Schaper, Fr.-W. Bach	909
Efficient and Ecological Machining of Magnesium Hybrid Parts C. Sanz, E. Fuentes, F. Obermair, L. Muntada	916
Experiences with the Machining of Magnesium R. Schwerin, S. Joksch	922
Magnesium Matrix Composites	927
In Mould Heating of Continuous Carbon Fibre Preforms for Reinforced High Pressure Die Casting Magnesium Parts C. Oberschelp, G. Klaus, A. Bührig-Polaczek, J. Werner, A. Langkamp, W. Hufenbach, N. Hort, H. Dieringa, K. Kainer	929
Mechanical Behaviour of an Mg-8Li Alloy Reinforced With Sic Particles Z. Száraz, Z. Trojanová	935
Production of AZ91D/CNF by Compocasting/Ultrasonic Agitation Conjugated Process R. G. da Silveira Mussi, T. Motegi, F. Tanabe, H. Kawamura, K. Anzai, D. Shiba, M. Suganuma	941
Thresholds in Creep Behaviour of Magnesium Alloy Matrix Composites with Short Alumina Fibres K. Milička, F. Dobeš	947

Anisotropy of Dilatation Characteristics in Mg and Mg4Li Fibre Composites A. Rudajevová, P. Lukáč	953
Magnesium-Hydroxyapatite Composites as an Approach to Degradable Biomaterials F. Witte, J. Fischer, P. Maier, C. Blawert, M. Störmer, N. Hort	958

Applications   965

AM-lite®: an Innovative New Alloy Opens Fresh Opportunities for Magnesium Applications C. Kettler, T. Abbott, M. Murray, G. Dunlop, G. Stoesser, T. Fuest	967
Aluminum Bolts in Magnesium Engine Components I. Bertilsson, Zheng Tan, F. Bergman, A. Fischersworring-Bunk, M. Kunst, J. Thiele, T. Marx	976
Experimental Warm Hydroforming System for Mg AZ31 Alloy Sheet Using a Low Melting Point Alloy as Forming Medium M. Steffensen, J. Danckert	985
Magnesium with Magnetic Properties for Sensory Use Fr.-W. Bach, M. Schaper, M. Rodman, M. Nowak	991

Research Programs   997

The MagForge Project: European Community Research on Forging of Magnesium Alloys W.H. Sillekens, D. Letzig	999

Reports of the Priority Programme 1168 of the Deutsche Forschungsgemeinschaft (German Research counsil) InnoMagTec   1007

Detecting the Solid Fraction of Commercial Mg Alloys by Heat-Transfer Modeling: Comparison of DTA and DSC Experimental Basis Data D. Mirković, R. Schmid-Fetzer	1009
Thermodynamic Simulation of Blending Mg-Alloys by Thixomolding A. KozlovFetzer, M. B. Djurdjevic, R. Schmid-Fetzer	1015
Quantitative Understanding and Modeling of Microstructural Formation by Solidification Technique of Magnesium Alloys G. Klaus, B. Böttger, J. Eiken, A. Bührig-Polaczek, I. Steinbach, M. Ohno, J. Gröbner, R. Schmid-Fetzer	1021
Production of a Fine-grained Mg Alloy AZ31 with SiC Particles G. Vidrich, Y. Estrin, A. Schiffl, F. Pravdic, H. Kaufmann, H. Ferkel, N. Hort	1027
Grain Refinement of AZ31 by (SiC)P: Theoretical Calculation and Experiment R. Günther, Ch. Hartig, R. Bormann	1033
Creep Resistance of Highly Calcium-alloyed AZ91 D. Amberger, M. Göken, P. Eisenlohr	1042

Experimental and Numerical Investigation of Texture Development During Hot Rolling of Magnesium Alloys *R. Kawalla, C. Schmidt, H. Riedel, A. Prakash*	1048
The Influence of Calcium and Cerium Mischmetal on the Deformation Behavior of Mg-3Al-1Zn *T. Laser, Ch. Hartig, R. Bormann, M. R. Nürnberg, D. Letzig*	1055
Corrosion Fatigue Behaviour of Newly Developed Mg Wrought Alloys *C. Fleck, A. Schildknecht, K.A. Weidenmann, A. Wanner, D. Löhe*	1062
Comparison of Dieless Clinching and Dieless Rivet Clinching of Magnesium *R. Neugebauer, S. Dietrich, C. Kraus*	1069
Influence of Cutting and Non-Cutting Processes on the Corrosion Behavior and the Mechanical Properties of Magnesium Alloys *F.-W. Bach, B. Denkena, P. Alpers, M. Bosse, K. Weinert, N. Hammer*	1076
Development of Innovative Technologies for Joining Magnesium Alloys and Dissimilar Materials Related to Components *S. Mücklich, B. Wielage, M. Horstmann, O. Hahn*	1085
Monotonic Properties and Cyclic Deformation Behavior of Friction Stir Welded Mg/Mg-, Mg/Al- And Al/Al-Joints *G. Wagner, M. Gutensohn, D. Eifler*	1092
Author Index	1099
Subject Index	1105

Alloy Development

The Viability of Mg Alloy with Nano/Sub-micron Structure as a new Material for Practical Applications

E. Aghion and A. Arnon
Department of Materials Engineering, Ben-Gurion University of the Negev, Beer-Sheva, Israel

1 Abstract

The development of new magnesium alloys with consolidated nano/sub-micron structure may introduce a new generation of super-light alloys that possess significantly higher specific strength. Consolidated nano/sub-micron structure relates to alloys with the combined microstructure of the nano-crystalline phase and the sub-micron phase.

The aim of the present study was to evaluate the viability of a consolidated nano/sub-micron magnesium alloy having the composition of AZ31, as a new structural material for practical applications. The viability evaluation was based on metallurgical examination using SEM, HRSEM, TEM, mechanical testing, and electrochemical corrosion measurements.

The results show that the nano/sub-micron structured alloy has nearly more than twice the hardness and strength of the conventional reference alloy. However, in terms of ductility and corrosion resistance, the nano/sub-micron structured alloy showed significantly inferior properties which limits its potential use in practical applications.

2 Introduction

The large spectrum of new magnesium alloys developed in the last few years highlights the increasing demand of the transportation and electronic industries for advanced structural material with improved properties [1–8]. The requirements become even more particular when new magnesium alloys are developed to address specific and critical applications [9,10]. One of the most important advantages of magnesium based alloys is their specific strength (strength to density ratio). This property can be improved if the microstructure is composed of relatively small grains, as in the case of nano-structured material.

The aim of the present paper is to examine the viability of Mg alloys having a consolidated nano/sub-micron structure as a new structural material for practical applications. The magnesium alloy selected for this study was based on the composition of AZ31 (2.5–3.5 % Al, 0.6–1.4 % Zn, 0.2 %Mn, max. 0.3 % others, and Mg – balance). The alloy was produced by „MBN Nanomaterialia" using mechanical alloying synthesis technology [11]. This was followed by direct extrusion of the nano-structured powder to obtain consolidated rods for metallurgical assessment.

3 Experimental procedure

The chemical composition of the consolidated nano/sub-micron magnesium alloy was within the following range: 2.5–3.5 % Al, 0.8–1.2 % Zn, 0.2–0.4 % Mn, <1 % others, with Magnesium making the balance. This composition basically complies with the composition of AZ31 magnesium alloy which was used in this study as a reference material in the form of regular extruded rods.

The microstructure examination was carried out using optical and electron microscopy (SEM with EDS, HRSEM and TEM). TEM analysis was obtained after developing a special method for sample preparation which incorporated an electro-polishing process with nitric acid solution.

The mechanical properties were determined using standard round specimens according to E8-95 ASTM spec. Hardness measurements were carried out using the standard range of Rockwell B.

Potentiodynamic polarization analysis in 3.5 % NaCl solution was carried out to evaluate the environmental behavior of the tested materials. The scanning rate of the potentiodynamic measurements was 5 mV/sec and the potential range was nearly 2000 mV. The consolidated nano/sub-micron structured alloy was tested in as-received condition and after annealing at 345 °C for 1 hr. The results obtained were introduced in terms of corrosion current which was converted into standard corrosion rate expressions (mpy).

4   Results and Discussion

The microstructure of the nano/sub-micron structured alloy obtained by SEM and HRSEM is shown in Figures 1 and 2, respectively. The SEM analysis revealed the presence of pure magnesium stringers along the extrusion direction. In addition, the microstructure contained micro-porosity and Fe-Mn base inclusions.

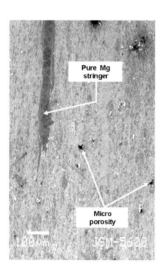
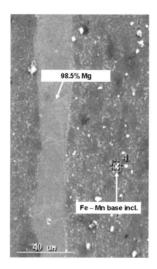

Figure 1: Microstructure of nano/sub-micron structured alloy obtained by SEM and EDS analysis

HRSEM revealed a combined microstructure of nano-crystalline and sub-micron phases. The typical crystal size of the nano phase was up to 40 nm, while the crystal size of the sub-micron phase was up to 150 nm. It is believed that the sub-micron phase developed from the nano-structured powder during the extruding process at a relatively high temperature.

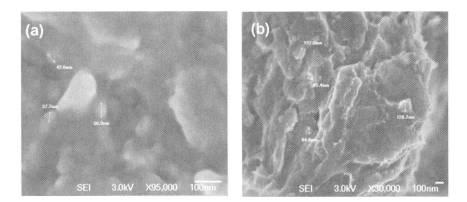

Figure 2: Microstructure of nano/sub-micron structured alloy obtained by HRSEM

TEM analysis showing the dislocation morphology of the nano/sub-micron structured alloy, along with its selected area diffraction pattern, is shown in Figure 3. This reveals the presence of high dislocation density which was the result of plastic deformation of the nano-structured powder during the direct extrusion process. The non-cyclic pattern diffraction is indicative of the fact that the micro-structure was not made of pure nano-scale phase.

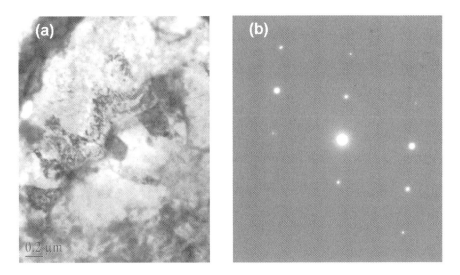

Figure 3: TEM analysis of nano/sub-micron structured alloy (a) Typical microstructure (b) Selected area diffraction pattern

In terms of the mechanical properties, it was evident the hardness and tensile strength of the nano/sub-micron structured alloy was 104 Rockwell B and 440 MPa, compared to 44 Rockwell

B and 240 MPa for the regular extruded AZ31 alloy. This result was supported by the fact that the average elongation of the nano/sub-micron structured alloy was only 1 % compared to more than 20 % for the regular reference alloy. The results obtained by the hardness and elongation measurements are clearly indicative of the brittle nature of the nano/sub-micron structured alloy.

The electrochemical behavior of the nano/sub-micron structured alloy and conventional AZ31 reference alloy in 3.5 % NaCl solution is shown in Figure 4 in as-received condition and after annealing. The results obtained show that the corrosion currents of the nano/sub-micron alloy in as-received condition and after annealing were $5 \cdot 10^{-4}$ A / cm^2 (443 mpy) and $1.6 \cdot 10^{-4}$ A / cm^2 (142 mpy), respectively. This clearly reveals that the annealing process has no significant effect on the magnitude of the corrosion rate. The corrosion current of the conventional AZ31 reference alloy was significantly lower, $1.6 \cdot 10^{-5}$ A / cm^2 (14 mpy).

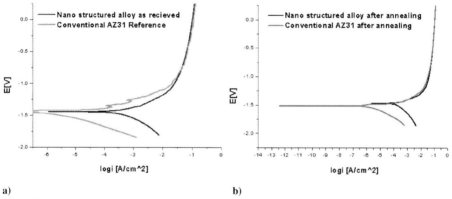

Figure 4: Electrochemical behavior of nano/sub-micron structured alloy and conventional AZ31 reference alloy obtained by potentiodynamic polarization in 3.5 % NaCl solution (a) In as-received condition (b) After annealing at 345 °C for 1 hr

Comparing the corrosion resistance of the nano/sub-micron alloy with the conventional alloy, it is evident that the corrosion rate of the nano/sub-micron alloy is one order of magnitude higher, indicative of extremely low corrosion resistance. The significantly reduced corrosion resistance of the nano/sub-micron structured alloy is mainly attributed to presence of residual pure magnesium stringers and cathodic phases such as Fe base inclusions. The interaction between the two phases in a corrosive environment creates a significant galvanic effect and, consequently, detrimental corrosion degradation. This degradation mechanism is amplified by the presence of micro-porosity.

5 Conclusions

The results obtained by the present study clearly indicate that the newly developed nano/sub-micron structured magnesium alloy in its present form and quality can not be used as a structural material for practical applications. This is mainly attributed to the inherent defects in terms of pure magnesium residue, heavy element inclusions, and micro-porosity that have a significant detrimental effect on ductility and corrosion performance. The inherent defects are generated during the mechanical alloying synthesis process. Hence, it is believed that the status of the nano/sub-micron magnesium alloy as a new structural material can be improved only if the inherent defects are adequately addressed in terms of the mechanical alloying process.

6 References

[1] K. Jereza, R. Brindle, S. Robison, J. N. Hryn, D. J. Weiss, B. M. Cox, „The road to 2020: Overview of the Magnesium casting industry technology roadmap," Magnesium technology, TMS San Antonio, Texas, USA, March 2006, pp. 89–94.

[2] B. Bronfin, E. Aghion, V. Von Buch, S. Schumann, H. Friedrich, M. Katzer, „High temperature resistant magnesium alloys", Patent numbers: EP 1329530A1 & EP 1329530B1 (European patent office), DE 60200928C0 (Germany), AT 0273400E (Austria), CA2415729AA (Canada), US 6767506 (USA) 2004.

[3] E. Aghion, B. Bronfin, F. von Buch, S. Schumman and H. Friedrich, „Newly developed Magnesium alloys for power train applications", JOM, Nov. 2003, pp. 30–33.

[4] M.O Pekguleryuz, A. A. Kaya, „Magnesium die-casting alloys for high temperature applications," Magnesium technology, TMS Charlotte, N. Carolina, USA, March 2004, pp. 281–287.

[5] M. Kettner, F. Pravdic, W. Fragner, K.U. Kainer, „Vertical Direct Chill (VDC) Casting of a Novel Magnesium Wrought Alloy with Zr and Re Additions (ZK10): Alloying Issues," TMS San Antonio, Texas, USA, March 2006, pp. 133–138.

[6] A. Luo, „Research and development challenges for Magnesium applications in Automotive structures," Second International Conference on Magnesium, Beijing, China, June 2006, paper No, A–038.

[7] S. Kamado, Y. Kojima, „Development of Magnesium alloys with high performance," Second International Conference on Magnesium, Beijing, China, June 2006, paper No. A–308.

[8] D. St. John, „Overview of CAST and Australian Magnesium research," Second International Conference on Magnesium, Beijing, China, June 2006, paper No. A–302.

[9] H. Westengen, P. Bakke, J. I. Skar, H. Gjestland, „New Magnesium die casting alloys: Driving developement of critical automotive applications," 63rd annual World Magnesium Conference IMA, Beijing, China, May 2006.

[10] T. Abbot, M. Murray, G. Dunlop, „AM-lite: A new Magnesium diecasting alloy for decorative applications," 63rd annual World Magnesium Conference IMA, Beijing, China, May 2006.

[11] P Matteazzi, G. Le Caer, A. Mocallin, „Synthesis of Nanostructured materials by mechanical alloying," Ceramics International 1997, pp. 39–44.

A Comparative Study of Carbon Additives on Grain Refinement of Magnesium Alloys

P. Cao[1], Ma Qian[2], K. Kondoh[3] and D. H. StJohn[4]
[1] Department of Engineering, University of Waikato, Private Bag 3105, Hamilton, New Zealand
[2] BCAST (Brunel Centre for Advanced Solidification Technology), Brunel University, Uxbridge, UB8 3PH, UK
[3] JWRI (Joining and Welding Research Institute), Osaka University, Japan
[4] CAST CRC, School of Engineering, University of Queensland, Brisbane, Queensland 4072, Australia

1   Introduction

To overcome the shortcomings of the superheating process (e.g. the high operating temperature and the applicability to relatively small volumes of melt only), post-1940s research on the grain refinement of magnesium alloys focused on developing an alternative method applicable to large volumes of melt at low operating temperatures. These efforts led to the discovery of the carbon inoculation process [1, 2]. A number of carbon based additives were tested for their grain refining potential, for example, carbonaceous gases, solid carbon or carbon-bearing powders and silicon carbides [3–7].

Following the discovery of the grain-refining effect of carbon additions, considerable efforts were made to develop a commercially reliable grain refiner for Mg-Al alloys. From an operational aspect, the carbonaceous gases and organic chlorides can be more readily introduced to magnesium melts than the solid carbon additives. In addition, bubbling a melt also helps to degas the melt. The addition of carbon in the form of C_2Cl_6 or Cl_2-CCl_4 blend was widely used before the 1970s because of the combined action of grain refining and degassing. The major problem of using carbonaceous gases, particularly chlorine-based gases, is the emission of chlorinated hydrocarbons (CHC). Other developments avoided gaseous additions. For example, Renger and Simon [8] reported an effective grain refiner named Nucleant 5000, which appeared to work satisfactorily on AZ91 alloys. Wax-fluorspar-carbon grain refiners were developed early in England [9] and recently in Canada [10]. Liu et al. [11] reported an Al_4C_3-SiC-Al alloy suitable for grain refinement of Mg-Al alloys such as AZ31 and AZ63. Motegi et al. [12] recently disclosed a carbon grain refiner that contains either pure carbon powder or a mixture of carbon and Nb_2O_5 or V_2O_5 powder. However, to date no carbon-based grain refiners have proved reliable in delivering consistent grain refinement of Mg-Al alloys. This study compares the grain-refining effect of five carbon-based grain refiners provided by different suppliers (part of the experimental results were reported previously [13]). The mechanism of grain refinement by carbon addition is discussed.

2   Experimental Procedure

Two series of grain refinement tests were conducted. The first series were conducted to assess the effectiveness of several carbon grain refiners that are available from suppliers. Melts of the Mg-3%Al-1%Zn (AZ31) alloy were prepared from commercial purity (99.7 %) magnesium (Norsk Hydro), aluminium (99.7 %), and zinc (99.95 %) ingots. In total, four carbon additives were tested, three from Japan and one from China (see Table 1). Two addition levels of each refiner, i.e. 0.1 % and 0.5 %, were made. Melting was conducted in an electrical resistance furnace under a protective cover gas (SF6 in 49 % dry air and 50 % CO2). Boron nitride (BN) coated mild steel crucibles were employed. The weight of each melt is ~ 1 kg. Following the addition of the refiner, the melts were held for 10 minutes and then a conical sample in the dimensions of ϕ 20 mm × ϕ 30 mm × 25 mm was taken from the top of the melt using a BN-coated mild steel conical ladle. Both melting and sampling were performed at 730 °C.

The second series were designed to determine which types of magnesium alloys can be grain refined by carbon and to further clarify the grain refinement mechanism associated with carbon in Mg-Al alloys. Grain refinement tests of pure Mg, Mg-3%Zn and Mg-3%Al alloys were conducted on the 1 kg scale of melt using a commercially available carbon grain refiner, Nucleant 5000 (FG1 in Table 1). Sublimed magnesium (99.98 %), high purity aluminium (99.99 %) and commercial purity zinc (99.95 %) were used as raw materials. The amount of carbon addition was fixed at 0.6% in the form of Nucleant 5000. A similar melting and sampling procedure was followed as used in the first series of tests. Each conical sample was sectioned transversely approximately 15 mm above the bottom of the sample and polished for optical examination. The average grain size was measured in the central region of the polished surface, in accordance with the linear intercept method described by ASTM standard E112-96.

Table 1. Carbon-based grain refiners assessed in this study

Carbon additive	Supplier	Form	Alloys tested	Effectiveness
Nucleant 5000	Foseco, Germany	Powder	Pure Mg	Negligible
			Mg-3%Zn	Negligible
			Mg-3%Al	Very marked
SD1	Shangdong, China	Ingot	AZ31	Unsatisfactory
KK1	Tokyo, Japan	Pellet	AZ31	Marked
KK2	Tokyo, Japan	Powder	AZ31	Very marked
KK3	Tokyo, Japan	Pellet	AZ31	Mild

3   Results

3.1   Tests of Carbon Refiners on Commerical Purity AZ31 Alloys

Figure 1 compares the efficiency of various sources of the grain refiners. The carbon refiner SD1 did not show a grain refining effect on AZ31, while the other refiners, KK1, KK2 and KK3, more or less resulted in grain refinement. The refiner KK2 showed very marked grain refining ability: the grain size of AZ31 decreased significantly from 495 μm to 170 μm. The observations of the SD1 refiner is unfortunately inconsistent with the reported observations at the addition level tested [14].

3.2 Pure Mg, Mg-3%Zn and Mg-3%Al using Nucleant 5000

Figures 2 through 4 show the results obtained from the grain refining tests of pure magnesium, Mg-3%Zn and Mg-3%Al alloys using Nucleant 5000. As expected, little or negligible grain refinement was observed in pure Mg and Mg-3%Zn at the addition level of 0.6 % of carbon refiner (Figs. 2 and 3). In contrast, obvious grain refinement was observed when an addition of 0.6 % of carbon refiner Nucleant 5000 was introduced to the melt of a Mg-3%Al alloy under the same conditions, as shown in Fig. 4.

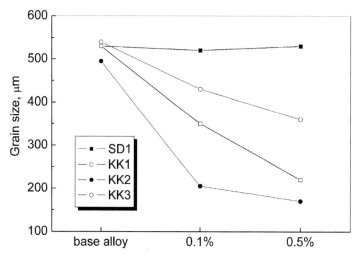

Figure 1: Grain refining effectiveness of the addition of 0.1% and 0.5 wt% of various carbon additives to AZ31 alloys

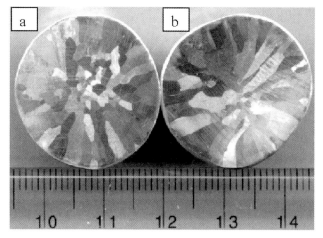

Figure 2: Carbon inoculation of pure Mg: (a) base metal and (b) after treatment with an addition of 0.6 % of carbon in the form of Nucleant 5000. No grain refinement was observed

Figure 3: Carbon inoculation of Mg-3%Zn: (a) base alloy and (b) after treatment with an addition of 0.6 % of carbon in the form of Nucleant 5000. No grain refinement was observed

Figure 4: Carbon inoculation of Mg-3%Al: (a) base alloy and (b) after treatment with an addition of 0.6 % of carbon in the form of Nucleant 5000. The average grain size was reduced from 400 μm to 170 μm after carbon grain refinement

3.3 Nucleant Particles in Carbon Refined Mg-Al Samples

The grain-refined Mg-3%Al samples were examined using scanning electron microscopy (SEM). Unfortunately no carbon-containing nucleant particles were detected at the centre of magnesium grains. The reason is unclear. However, if the generally accepted Al_4C_3 theory [13,15] is assumed to be correct, then the failure to observe nucleant particles is probably because Al_4C_3 can be hydrolysed by water during sample preparation [16], as Al_4C_3 is reactive to water by the following reaction:

$$Al_4C_3 + 12\ H_2O \rightarrow 4\ Al(OH)_3 + 3\ CH_4.$$

4 Discussion

Table 2 summarizes the results of carbon grain refinement of magnesium alloys reported by different researchers. As was shown previously by Davis et al. [5] and further confirmed by the present work, carbon is not a grain refiner for magnesium alloys that do not contain aluminium. Instead, it is a good grain refiner for Mg-Al based alloys, e.g. Mg-3%Al, AZ31 and AZ91.

Table 2: Comparison of the grain size of various magnesium alloys inoculated by carbon

	Before carbon inoculation	After carbon inoculation	Reference
Pure Mg	Columnar	Columnar	[5]
Mg-1.5%Mn	1524 µm	2540 µm	[5]
Mg-3 %Zn	890 µm	890 µm	[5]
Pure Mg	Columnar	Columnar	This work
Mg-3%Zn	450 µm	500 µm	This work
Mg-3%Al	400 µm	170 µm	This work
Mg-3%Al	380 µm	180 µm	[16]
Mg-3%Al	400 µm	120 µm	[17]
AZ31	500 µm	170 µm	Fig.1, KK2
AZ31	530 µm	220 µm	Fig.1, KK1
AZ31	550 µm	370 µm	Fig.1, KK3
AZ31	530 µm	550 µm	Fig.1, SD1
AZ91	650 µm	190 µm	[8]
AZ91	150 µm	70 µm	[18]

Although Al_4C_3 particles were not observed in this study, the Al_4C_3 nuclei theory still appears to be the most reasonable mechanism. The work by Yano et al. [18,19], Lu et al. [16,20] and the present study generally support the Al_4C_3 theory. Jin et al. [17] proposed another grain refining mechanism which attributes the grain refinement to the segregation of carbon. If this were true, carbon grain refinement would have been observed in pure Mg and Mg-Zn alloys. Apparently, this is not the case. This segregation mechanism has been discussed previously by Qian and Cao [13].

Motegi's group once proposed a hypothesis on the basis of the formation of Al_2OC [21], which has very good crystallographic matching with magnesium [22]. Recently, it appears that Motegi has returned to the Al_4C_3 hypothesis [23].

5 Summary

Carbon inoculation has no effect on magnesium alloys that do not contain aluminium. Grain refinement of Mg-Al based alloys by carbon inoculation stems from the interaction between carbon and aluminium. The Al_4C_3 nuclei theory is the most reasonable mechanism proposed for carbon inoculation of Mg-Al type alloys. The tests of several grain refiners supplied by different companies showed that it is still a long term task for metallurgists to develop a commercially reliable grain refiner for Mg-Al alloys.

6 Acknowledgements

The CAST Cooperative Research Centre was established under and is supported in part by the Australian Government's Cooperative Research Centre Programme. PC would like to thank Mr. Y. Liu of Shandong University, China for providing carbon refiner SD1 used in this study. MQ gratefully acknowledges the support from EPSRC, UK.

7 References

[1] Hultgren, R., Mitchell, D.W., Trans. AIME **1945**, 161, 323–327.
[2] Holm, V.C.F., Krynitsky, A.I., The Foundry **1947**, 75, 81,228–240.
[3] Nelson, C.E., Holdeman, G.E., US Patent 2380863 (1945)
[4] Nelson, C.E. Trans. AFS **1948**, 56, 1–23.
[5] Davis, J.A., Eastwood, L.W., DeHaven, J. Trans. AFS **1945**, 53, 352–362.
[6] Mahoney, C.H., Tarr, A.L., LeGrand, P.E. Trans. AIME **1945**, 161, 328–350.
[7] Easton, M.A, Schiffl, A., Yao, J.-Y., Kaufmann, H. Scripta Mater. **2006**;55: 379–382.
[8] Renger, K., Simon, R., in *Proceedings 1st International Light Metals Technology Conference*, CAST Centre Pty Ltd **2003**, p.231-234.
[9] Fox, F.A. British Patent GB608941 (1948).
[10] Karlsen, D.O., Oymo, D., Westengen, H., Pinfold, P., Stromhaug, S.I., in *Proceedings of the international symposium on light metals processing and applications,* Canadian Institute of Mining, Metallurgy and Petroleum **1993**, p.397–408.
[11] Liu, Y., Liu, X., Xiufang, B. Mater Lett **2004**, 58, 1282–1287.
[12] Motegi, T., Miyazaki, K., Tezuka, Y., Yoshibara, K., Yano, E. US Patent 6616729 (2003).
[13] Qian, M., Cao, P. Scripta Mater. **2005**;52:415–419.
[14] Liu, Y., Liu, X, Li, T., Bian, X. Chinese J. Nonferrous Metals **2003**, 13, 622–625.
[15] Emley, E.F. Principles of Magnesium Technology. Pergamon, Oxford, 1966, p.208.
[16] Lu, L. Dahle, A.K., StJohn, D.H. Scripta Mater. **2006**;54:2197–2201.
[17] Jin, Q., Eom, J.P., Lim, S.G., Park, W.W., You, B.S. Scripta Mater **2003**, 49, 1129–1132.
[18] Yano, E., Tamura, Y., Motegi, T., Sato, E. Journal of Japan Inst Light Metals **2001**, 51, 599-603.
[19] Yano, E., Tamura, Y., Motegi, T., Sato, E. Mater Trans **2003**, 44, 107–110.
[20] Lu, L. Dahle, A.K., StJohn, D.H. Scripta Mater **2005**, 53, 517–522.
[21] Tamura, Y., Kono, N., Motegi, T., Sato, E. Journal of Japan Inst. Light Metals **1998**; 48:395.
[22] Zhang, M.X., Kelly, P.M., Qian, M., Taylor, J.A. Acta Mater. **2005**; 53:3261–3270.
[23] Motegi, T. Mater. Sci. Eng. A **2005**; 413–414: 408–411.

Advanced Gravity Casting Magnesium Alloys for the Aircraft Industry

B. Bronfin[1], A. Ben-Dov[2], J.Townsend[3], S. Mahmood[3], J.Vainola[4], S.Deveneyi[5] and N. Moscovitch[1]

[1]Dead Sea Magnesium Ltd, Beer Sheva, [2] IAI, Tel-Aviv, [3] Stone Foundries, London, [4] VTT, Helsinki, [5] Specialvalimo, Helsinki,

1   Introduction

Selection of structural materials for aircraft applications is dictated by a number of factors including mechanical, corrosion and physical properties (particularly density) as well as availability and cost. Magnesium alloys being the lightest of all of the commonly used structural metals can be considered as one of very attractive candidates. Furthermore, magnesium alloys have other desirable properties such higher than for other alloys strength-to–density ratio combined with excellent machinability and damping capacity as well as good castabilty. On the other hand, magnesium alloys exhibit a number of negative features including inferior strength –ductility relationship, less than desirable fatigue strength, fracture toughness and creep resistance. Although general corrosion behaviour of high purity magnesium alloys is considered as acceptable, galvanic corrosion is regarded as very poor and special coating and fasteners are mandatory for most applications [1, 2].

Magnesium alloys have been known to aeronautical engineers for a long, long time. However, they have always suffered prejudice on the twin scores of fire hazard and corrosion. Nevertheless there are some aircraft applications where existing magnesium alloys are being currently used. Thus, magnesium alloys ZE41 and WE43 are used for a range of helicopter transmission castings [3]. In the aero engine industry, the above alloys are also being used along with EZ33 and EQ21 alloys both in civil and military aircrafts [4–7]. However, it should be recognized that such applications are very limited due to the reasons being partly technical and to great extent "emotional" [1]. On the other hand, according to Work program of the thematic priority ' Aeronautics and Space " which is the part of 6th EU Framework programme "World Aeronautics is entering a new age of aviation- the age of sustainable growth- characterized by the need of more affordable, cleaner, quieter, safer and more secure air travel." Thus, it is evident that magnesium alloys may and have to play a crucial role in achieving the above targets.

In respond to the growing demand for weight reduction, the consortium consisting of 14 partners initiated a comprehensive research project titled IDEA (Integrated Design and Product Development for the Eco-efficient Production of Low-weight Airplane Equipment). The development of new alloys is essential part of this project. The requirements to new alloys set by end-user are summarized in Table 1.

The alloys are designated for room temperature applications. A secondary structural part - motion transfer housing shown in Fig.1 was selected as one of reference castings.

Table 1: Technical requirements to properties of sand casting alloys

- TYS -220 MPa, UTS -290 MPa, Elongation- 3%
- Axial fatigue strength for 10^6 cycles:
  0.28 UTS at R = –1
  0.50 UTS at R = 0.2
  Strict requirements to general and galvanic corrosion

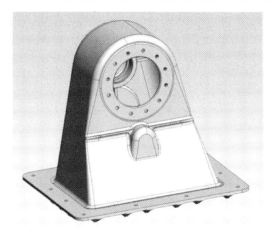

Figure1: Motion transfer housing

2 Metallurgical Background to the Development of Gravity Casting Alloys

The development of new gravity casting alloys was based on implementation of major mechanisms affecting properties of those alloys. These mechanisms are precipitation hardening, solid solution strengthening and grain boundary strengthening. In general, alloying principles for the development of high strength magnesium alloys for gravity casting applications after T6 heat treatment can be reduced to the following rules [8, 9]. Grain refinement is considered as a very important tool at the development of gravity casting alloys. It is well known and documented that Zr has a grain refining effect when added to magnesium leading to the greater casting integrity and improved mechanical properties [10]. The soluble Zr fraction mostly provides the fine grain size obtained in Zr contained alloys. Mg-Zr alloys have more consistent properties through thin and thick sections and are not prone to through-wall porosity, which can cause lubricant leakage.

The main alloying elements should have a wide range of solid solubility in magnesium, which decreases sharply at room temperature. The most desirable alloying systems are those in which extensive solubility at solid solution treatment temperature can be retained with industrially adequate cooling rate in supersaturated solid solution at room temperature. This factor is very important to provide a marked response to aging.

Solutes should also have a low diffusion coefficient, to provide strong interatomic bonds and to form solid solution, which has no response to aging over service conditions. For enhanced elevated temperature properties, alloying elements should form thermally stable intermetallic compounds that have the coherency with the matrix and are effective obstacles against deformation and strengthen grain boundaries. The melting point of the precipitate is a good indication of its thermal stability.

The first precipitates to nucleate are very often metastable and coherent with the matrix providing excellent precipitation hardening. As aging progresses, metastable precipitates are transformed into stable equilibrium phases. The morphology of the precipitates including their shape and interparticle spacing affect both ambient strength and creep resistance. Based on the strict requirements listed in Table 1 and taken into consideration the availability and commercial attractiveness of the alloying elements, some of them such as Nd, Y, Gd and Ag seem to be most promising for the development of new alloys. It is believed that Nd should be considered as the main alloying element due to optimal combination of enhanced solid solubility, availability and cost. However, Nd content higher than 3 % results in significant embrittlement. Yttrium and particularly gadolinium have higher solubility in magnesium than neodymium On the other hand, they are relatively expensive and Gd has also a great atomic weight that requires its higher weight additions in order to obtain the same atomic percent of Gd compared to Y and Nd. Therefore, the most challenge is to find optimal combination in concentrations of the above three elements.

Zinc content should be limited in most cases to 0.3–0.8 % as zinc combines with Nd, Y and Gd to form stable eutectic intermetallics thereby nullifying the contribution of the above elements to precipitation hardening. However, higher Zn content can be used in some cases when solid solution strengthening can be considered as alternative strengthening mechanism.

Heat treatment is a very important factor for obtaining a required combination of service properties. It should be selected a based on compromise between mechanical properties requirements and commercially acceptable holding time at solid solution treatment and particularly at aging. Solid solution treatment should be performed at the highest practicable temperature to dissolve coarse eutectic intermetallic phases formed during casting process. Practically the solid solution treatment is conducted at the temperatures about 20–30 °C below the solidus temperature of the alloy. The most challenge is usually associated with selection of the temperature and time of aging because these variables significantly affect the final properties.

In addition to their influence on mechanical properties and creep behaviour, alloying elements should provide good castability (increased fluidity, low susceptibility to cracking, reduced porosity and greater casting integrity) combined with improved corrosion resistance and affordable cost. With regard to the last factor it is believed that it is not so critical as for the automotive industry because, for example, in the civil aircraft the value of a pound in weight saved is equal to $300 US [2].

Based on the above principles three new alloys designated MRI 204, MRI 205 and MRI 207 have been developed in the framework of the IDEA project. The alloys MRI 204 and MRI 207 are based on Mg-Zn-Zr-Nd-Y (Gd) alloying systems and are in fact also creep resistant alloys. Precipitation hardening and grain boundary strengthening are major mechanisms contributing to the strength of the above alloys. On the other hand, solid solution and grain boundary strengthening underlying unique properties of MRI 205, which is designated for room temperature applications that require high strength, combined with increased ductility. The present paper addresses mechanical properties of newly developed gravity casting alloys in comparison with commercial magnesium alloy ZE41-T5 and aluminium alloy A357-T6, which are used in some aeronautic applications. In addition the effect of casting technology is also demonstrated and discussed.

3 Experimental Procedure

The study was carried out on tensile and fatigue test bars that were produced by sand, investment and gravity die casting technologies. The quality of specimens was evaluated by X-Ray radiography, using Seifert Eresco 200 MF constant potential X-Ray tube.

Tensile specimens' configuration was in accordance with BS 2970 (Form F). The room temperature tensile properties were determined in accordance with ASTM E8M. A minimum fifteen tensile specimens of each alloy were tested and the results averaged.

Fatigue specimens were machined from separately cast cylindrical bars with a 16 mm diameter. The specimens' configuration was in accordance with ASTM E466 (specimens with a continues radius between ends). The specimens had the radius of curvature 120 mm, the minimum diameter of 6.35 mm and the ends diameter of 14 mm. The reduced section was polished to 0.1 mm roughness. All specimens were tested in T6 in conditions that were preliminarily optimised for each alloy based on numerous experiments. Creep tests were performed in the temperature range 175–250 °C according to ASTM E139 standard.

4 Results and Discussion

The results of tensile and compression tests of new sand casting alloys are summarized in Table 2 in comparison with the benchmark commercial alloys ZE41-T5 and A357-T6.

Table 2: Mechanical properties of MRI sand casting alloys in T6 condition compared to commercial alloys

Alloy	Casting method	TYS [MPa]	UTS [MPa]	E [%]	CYS [MPa]
MRI 201	Sand casting	194±9	272±9	6±1	207±6
MRI 204	Sand casting	206±4	306±8	4±1	245±8
	Investment casting	210±3	302±2	3±1	234±6
	Gravity die casting	212±4	305±2	3±1	238±6
MRI 205	Sand casting	220±4	325±4	10±2	227±6
MRI 207	Sand casting	220±8	304±6	4±1	259±7
	Investment casting	215±3	285±3	3±0	249±9
	Gravity die casting	215±2	280±15	3±1	237±9
ZE41-T5*	Sand casting	140	220	5	140
WE43-T6*	Sand casting	180	260	6	190
A357-T6*	Sand casting	230	290	3	250

* Metals Handbook Data

It is evident that MRI 201-T6 alloy significantly outperforms commercial magnesium alloy and has tensile and compressive properties similar to those of significantly more expensive WE43 alloy. With regard to MRI 204, MRI 205 and MRI 207 alloys, table 2 demonstrates that their tensile and compressive properties in T6 condition are comparable with those of aluminium alloy A357-T6. In addition, it should be noted that unique combination of strength and ductility was obtained on MRI 205-T6 alloy. Furthermore, table1 distinctly illustrates that casting method practically does not affect tensile and compressive properties of MRI 204 and MRI 207 alloys

One of the most important requirements to new alloys set by end user is improved fatigue behaviour. The results of axial fatigue tests (R = –1) performed on MRI alloys are given in Figure 2 in comparison with the data obtained for aluminium alloy A357-T6(shown as a lower limit and mean values lines). As can be distinctly seen MRI 205-T6 is superior to A357-T6 in axial fatigue behaviour both at high and low stresses. Other two new magnesium alloys exhibit fatigue performance similar to that A357-T6 alloy.

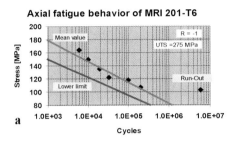
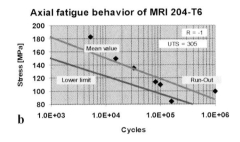
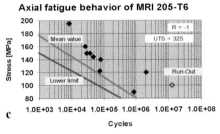

Figure 2: Axial fatigue behavior of MRI 201-T6 (a), MRI 204 -T6 (b) and MRI 205-T6 (c) in comparison with aluminum alloy A357 - T6 (represented by lower limit and mean value lines)

New alloys MRI 201, MRI 204 and MRI 207 are also designated for elevated temperature applications. For example, creep properties of MRI 201S and MRI 207 alloys are listed in Tables 2 and 3, respectively. It is evident that both MRI 201 and MRI 207 exhibit excellent creep resistance and may operate at 200-250°C under high stresses. This performance makes new alloys good candidates for high-temperature applications in automotive, aircraft, motorsport and defense industries such as engine block, cylinder head cover, helicopter gearbox housings etc.

Table 3: Creep properties of MRI201– T6 sand castings (stress [MPa] to produce certain creep strain at different temperatures)

Time under load, h	0.1%	0.2%	0.5%
At 200°C			
10	172	180	190
100	154	170	176
1000	88	102	143
at 250 °C			
10	74	84	96
100	49	75	80
1000	34	48	55

Table 4: Creep properties of MRI207– T6 sand castings (stress [MPa] to produce certain creep strain at different temperatures)

Time under load, h	0.1%	0.2%	0.5%
At 200 °C			
10	175	183	195
100	156	173	180
1000	92	105	145
At 250 °C			
10	70	80	92
100	45	70	80
1000	30	45	50

In conclusion, it should be noted that the results obtained so far are very promising and can serve as a background for further research. It is evident that the additional data including corrosion behaviour should be obtained prior to some practical decisions can be drawn.

5   References

[1]   F.H. Froes et al, Proceedings of Annual World Magnesium Conference, IMA 2000, Vancouver, 2000, 56–63.
[2]   E .Aghion , B. Bronfin and I. Schwartz , Proceedings of the 36th Israel Annual Conference on Aerospace Sciences, Technion, Israel, 1996, 353–362.
[3]   J.M. Arlhac and J.C.Chaize, Proceedings of the Third International Magnesium Conference,(Ed: G.W.Lorimer), Manchester, 1966, 213–229.
[4]   B. Geary , Proceedings of the Third International Magnesium Conference, (Ed: G.W.Lorimer), Manchester, 1966, 565–574.
[5]   F.de Mestral and M.Brun, Magnesium Alloys and their Applications, (Eds.; B.Mordike and F.Hehmann), 1992, 389–396.
[6]   N. Zeumer, Magnesium Alloys and their Applications, (Eds.: B.Mordike and K.Kainer),1998, 125–132.
[7]   L. Duffy, Materials World, March 1996, 127–130.
[8]   B.Bronfin et al, EP Patent 1,329,530, Aug.11, 2004.
[9]   B.Bronfin et al, Magnesium Technology 2005, (ed. H.I. Kaplan),Warrendale, PA: TMS, 2005, 395–401.
[10] E.F. Emley, Principles of Magnesium Technology, Pergamon, Oxford, 1966.

Elektron 21 – An Aerospace Magnesium Alloy for Sand Cast & Investment Cast Applications

Paul Lyon, Ismet Syed, Steve Heaney
Magnesium Elektron, Manchester, England

1 Introduction

Magnesium alloys are used primarily because of their lightweight, being only 2/3 the density of Aluminium. A range of alloys are commercially available from Magnesium–Aluminium alloy AZ91, to the highest strength, high temperature Magnesium – Zirconium alloy Elektron WE43. These alloys, and those with properties between, are serving the aerospace market well. There is however, a need for an alternative alloy with similar properties to Elektron WE43, but with foundry handling and associated costs that are more similar to the simpler alloys. Of course any new alloy must also be corrosion resistant.

Elektron 21[1] was developed by Magnesium Elektron to fulfil these needs of the aerospace and speciality industry. Primarily this is used in the form of sand castings; Elektron 21 has, however, also proved to be suitable for investment casting. Investment casting, offers further weight saving, by the ability to reduce casting section thickness.

This paper aims to illustrate the effect of some of the main alloying additions and their effect on castability and properties.

2 Effect of Alloy Constituents on Elektron 21 Properties

Elektron 21 contains Neodymium, Heavy Rare Earth (Gadolinium), Zinc, and is grain refined by the addition of Zirconium.

The alloy is heat treated to the fully solution treated and aged condition (T6).

This combination of elements and heat treatment generates a combination of good castability, mechanical properties and corrosion resistance.

2.1 Neodymium

Neodymium is known to improve the strength of Magnesium. Early work [1] indicated the benefits of Neodymium compared with other Rare Earth alloys, namely improved strength, particularly at temperature. These benefits relate to both the solubility limits of Neodymium and the potential to generate stable precipitates within the grain structure and at grain boundaries.

Neodymium, combined with other elements for additional benefit, is used in several commercially available alloys. For example QE22 (Mg, 2 %Ag, 2 % Nd, 0.5 % Zr) and Elektron WE43 (Mg 4 %Y, 2.3 % Nd, 1 % HRE, 0.5 % Zr).

---

1. Elektron 21 is a Magnesium Elektron patented alloy

Whilst Neodymium is beneficial to properties, actual content is restricted by the solid solubility limit (Approximately 3.6 % at 550 °C) [2,3].

2.2 Gadolinium

Gadolinium is classed as a Heavy Rare Earth (atomic number 64). Like Neodymium, Gadolinium shows a decreasing solid solubility as temperature falls, indicating potential for precipitation strengthening. Unlike Neodymium, however, the maximum solid solubility [2, 3] is high (24 %) and solid solubility at 200 °C is still approximately 4 %. This would suggest that high levels of Gd would need to be present to achieve precipitation hardening (>4 %). This is commercially unattractive.

2.3 Neodymium/Gadolinium Combined

Addition of Neodymium to Gadolinium containing alloys reduces the solid solubility of Gadolinium [4], shifting the solvus toward lower Gadolinium levels. This is potentially attractive since this offers the possibility of improved precipitation hardening response, at lower levels of Gadolinium than the binary system offers. Work in the 1990's by Magnesium Elektron and others [5, 6] pursued this aspect, but continued to employ comparatively high levels of Gadolinium (5–10 %).

Further work [7] illustrated the benefit of lower Gadolinium additions in terms of age hardening response, demonstrating some improvements in ageing response and peak hardness below 2 % Gadolinium.

A further benefit of the Gadolinium/Neodymium combination is improved castability in terms of microshrinkage [8]. This can be explained by considering the Neodymium/Gadolinium relationship in more detail.

Increases in Neodymium improve castability. As Neodymium content increases in Mg-Nd-Zr alloys, a peak in Ultimate tensile strength is reached, after which, a reduction in strength occurs. This reduction approximately coincides with the solid solubility of Neodymium in Magnesium, leading to excess (brittle) grain boundary phase after heat treatment.

When the Gadolinium component of the TRE (Total Rare Earth) is increased in Elektron 21, properties are not reduced. This effect can be explained by the higher solid solubility (than Neodymium) of Gadolinium. This allows dissolution of some of the Gadolinium rich grain boundary phase into the matrix during heat treatment. This in turn, allows the TRE level to be increased (tests indicated a reduction in microshrinkage) without detriment to tensile properties [8,9].

2.4 Zinc

Zinc is often added to Magnesium in sufficient quantities to achieve precipitation strengthening. Commercially available alloys employing Zinc include ZE41 (Mg 4 % Zn, 1.5 % RE, 0.5 % Zr). Lower levels of Zinc are also used in Mg-Al alloys (AZ91 Mg, 9 % Al, 1 % Zn, 0.3 % Mn) where Zinc improves strength without reducing ductility. This attractive benefit would not occur if Aluminium content alone were to be increased to achieve the same level of strength [9].

Use of Zinc at low levels (0–1.3 %) in Magnesium-Neodymium alloys can alter the ageing response of the alloy [10], high Zinc levels change the form and location of Zinc containing precipitate, which can reduce the hardness.

Zinc can also influence the corrosion behaviour of Magnesium alloys.

Zinc content is carefully controlled in Elektron 21. Further details of the effect of Zinc on ageing performance of Elektron 21 are provided in the next sections.

2.5   Zinc, Gadolinium and Neodymium Combined

Characterisation of ageing behaviour during the development of Elektron 21 in the range 0–1.3 % Zn, showed that increasing Zinc content delayed the onset of over ageing but, if levels were too high (1.3 %), this also resulted in a lower peak hardness and tensile properties [12, 13]. Meanwhile, Gadolinium resulted in an increased age hardening response, which was also maintained as time at temperature increased.

The precipitation sequence in Elektron 21 has now been defined by Gill [14] as follows:

$$Mg_{ssss} \rightarrow \beta''_{(DO_{19})} \rightarrow \beta'_{(FCC)}$$

Gill found that peak hardness, at the ageing temperature of 200 °C, is obtained from the B" precipitate. The main constituent of the strengthening precipitates is Neodymium. Zinc is associated and Gadolinium has also been detected. Gadolinium contributes to solid solution hardening. Zinc can modify the form of precipitates developed and can also influence quench sensitivity.

2.6   Zirconium

The most significant advantage of Zirconium additions to Magnesium is the potent grain refining effect. This can result in typical grain sizes within Mg-Zr alloy castings in the range 30- 80 microns. This is of benefit not only in terms of improved mechanical properties and consistency, but also castability and corrosion performance.

Fine-grained structures have improved hot strength, reducing the opportunity for cracking during solidification. The form of any micro shrinkage is modified by bulk feeding during solidification, whereby a „slurry" of grains will infill during solidification [10]. A consequence of this is that if microshrinkage is present in a Magnesium – Zirconium alloy, it is far less likely to be outcropping and results in pressure tight castings.

2.7   Mechanical Property Comparison

Using the optimised mixture of elements described, Elektron 21 generates a useful range of mechanical properties which can best be summarised by comparison with other Magnesium alloys currently available (see Figure 1).

Up to approximately 200 °C (392 °F) strength of Elektron 21 is most similar to Elektron WE43. Properties are superior to ZE41 and AZ91. The difference is emphasised as temperature increases.

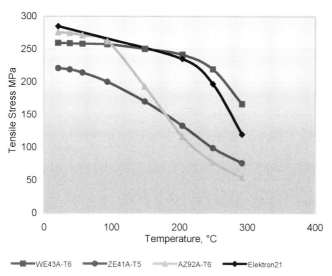

Figure 1: Effect of Elevated Temperature on the Tensile Strength of various Magnesium Alloys

3 Elektron 21 Specification

Using some of the data described in this paper, compositional ranges were finalised. This combined with over 200 mechanical test results from Magnesium Elektron and foundries through out the world, helped to achieve AMS (Aerospace Material Specification) approval.

AMS tensile property minimums are shown in Table 1 and compared with some existing AMS specification minimum values.

The ASTM designation for Elektron 21 is EV31A.

Table 1: AMS minimum properties in separately cast bars & samples cut from castings

	AMS No.	Separately Cast Bars MPa (KSI)			AMS No.	Cut from Casting MPa (KSI)		
		0.2% PS	UTS	Elong %		0.2% PS	UTS	Elong %
Elektron 21 (EV31a)	4429	145(21)	248(36)	2	4429	145(21)	248(36)	2
WE43	4427	172(25)	221(32)	2	4427	152(22)	217(31½)	2
ZE41	4439	134(19.5)	200(29)	2½	4439	121(17.5)	179(26)	2
AZ91	4446	110(16)	235(34)	3	4446	85(12)	115(17)	

It will be noted that Elektron 21 (EV31A) shows excellent consistency, with specification minima the same irrespective of whether samples are cut from castings or separately cast bars.

4 Investment Cast Elektron 21

To date investment casting of Magnesium has been comparatively limited because of the potential reaction between Magnesium and the Silica containing moulds. Unlike sand moulds, investment moulds are difficult to inhibit due to the high firing temperatures required.

The low reactivity of Elektron 21 compared to many other Magnesium alloys, spawned the idea that this alloy could be attractive for investment casting [15].

Casting trials with eight investment casting companies who were familiar with Magnesium proved Elektron 21 to be suitable for investment casting. No reaction was observed in any of the castings produced. An example is shown in figure 2.

Subsequent work included activities with Cast Technology International (CTi). The objective here was to increase the size of castings, which could be made in Elektron 21 by improving mould preparation technology. This work allowed the production of large casting sections (see figure 3) without reaction with the mould. This is important not because one would wish to produce thick sections by investment casting, but because this indicates that larger volumes of metal can now be poured without mould reaction; hence large thin wall castings can be produced.

Evaluation of mechanical properties, showed investment cast Elektron 21 to be very similar to sand cast material. Importantly, the results were consistent showing little difference with foundry or section thickness.

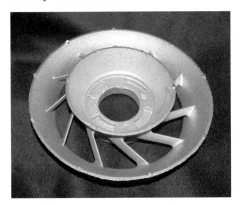

Figure 2: Elektron 21 Investment Casting

Figure 3: Elektron 21 Investment Casting. Section Thickness is up to 75mm. (3") Height of Component is 350mm. (14")

Success with investment casting means that the property benefits of sand cast Elektron 21 are now also available in the form of light weight investment castings.

5 Use and Applications

Elektron 21 (EV31A) has already been specified for motorsport and US military applications; in the latter case, replacing Aluminium.

For Aerospace applications, Elektron 21 is currently under successful evaluation on numerous programmes, where lightweight and corrosion resistance are required. To serve these evaluations, castings produced range from small investment castings to large sand castings of 365 Kgs poured weight.

Compared to other applications where Magnesium is already used, the drive is to replace less corrosion resistant alloys. Elektron 21 fulfils this desire for applications, which do not require the full temperature requirements available from Elektron WE43 (250 °C), but do require a corrosion resistant alternative for use at up to approximately 200 °C with lower overall component cost.

Outside of the applications currently using Magnesium, Elektron 21 offers a solution to weight reduction compared to Aluminium and other heavier materials.

6   References

[1] Leontis, Feisel, Sand Cast Mg-RE-Zr Alloys **1957,** Trans Aime, 209, 1245
[2] Nayeb-Nashemi, Clark, Phase Diagrams of Binary Magnesium Alloys **1988**
[3] Rokhlin, Magnesium Alloys containing RE Metals **2003,** Taylor & Francis
[4] Negishi, Nishimura et al, **1995.** Journal of Japan Institute of Light Metals. Vol 45 pt 2
[5] P. Daly **1999**. Evaluation of Mg-Nd-Gd/Dy-Zr Alloys, Magnesium Elektron MR10/843
[6] Kamoda, Kojima **1996**. Ageing Characteristics & High Temperature Properties of Mg-HRE Alloys. Third International Magnesium Conference, Manchester
[7] Frankel **2003**. Elektron 21 Evaluation – Materials, Science Centre Investigation for Magnesium Elektron
[8] Lyon, Conroy **2003**. Internal Chicago Magnesium/Magnesium Elektron Internal Report
[9] Lyon, Wilks, Syed **2005**. TMS Magnesium Technology. P303-308
[10] Emley. Principles of Magnesium Technology Pergamon Press. **1966**
[11] Wilson, Bettles et al **2003**. Precipitation Hardening in Mg-Nd-Zn Alloys. Materials Science Forum pp 267–272
[12] Gill, Lorimer, Lyon **2003**. Microstructure & Property Relationships of Mg-RE-Zn-Zr Alloys. DGM Conference, Germany
[13] Gill **2004**. Internal Progress Report for Magnesium Elektron
[14] Gill **2006**.PhD Microstructure/Property relationship of Mg-RE-Zn-Zr Alloys.
[15] Lyon, Thompson, Rowett **2005** Precision Casting of Magnesium- A Light Weight Solution. 53rd Investment Casting Conference. November **2005** Dearborn USA.

Microstructures and Mechanical Properties of Mg-Zn-Al Alloys Produced by High Pressure Die Casting and Sand Casting

J.F. Nie[1], M. Easton[1], and T.B. Abbott[2]

[1] CAST CRC, Department of Materials Engineering, Monash University, Victoria 3800, Australia
[2] Advanced Magnesium Technologies Pty. Ltd., Milton BC, Queensland 4064, Australia

1 Introduction

Magnesium die casting alloys based on the Mg-Al system have been studied extensively for use in vehicles due to the weight savings they provide and also for their excellent castability. These alloys include those designated AM50, AM60, AZ91, AS21, AS41 and AE42. Though AZ91 and AM alloys exhibit good tensile yield strengths (100–200 MPa) at ambient and elevated temperatures, they are all prone to substantial creep deformation when exposed to moderate loads at temperatures close to or above 150 °C, presumably as a result of discontinuous precipitation of $Mg_{17}Al_{12}$ phase [1]. Alloys AE42 and AS21 that have been considered for elevated temperature applications, are expensive and/or difficult to cast.

Since it is difficult to eliminate discontinuous precipitation via microalloying additions [2], the present investigation has been focused on design of alloy microstructures that do not contain any particles of the equilibrium $Mg_{17}Al_{12}$ phase. The approach used involves ternary additions of large amounts of Zn to the Mg-Al alloys so that the composition of the ternary alloy is not in the (α(Mg) + $Mg_{17}Al_{12}$), but is in the (α + Φ) two phase field of the Mg-Al-Zn ternary phase diagram. The equilibrium Φ phase has a composition of $Mg_{21}(ZnAl)_{17}$ and a primitive orthorhombic structure [3].

While it has been known for some time that the creep properties of some Mg-Zn-Al alloys are superior to AM60 and AZ91[4–6] these alloys have not been subjected to detailed metallurgical study. This work involves characterization of microstructures and evaluation of mechanical (tensile and creep) properties of a range of such Mg-Zn-Al alloys and alloy AM60 that are produced by high-pressure die casting (HPDC) and sand casting (SC). The compositions of the Mg-Zn-Al alloys were selected to allow evaluation of the variation in tensile yield strength and creep resistance as a function of the weight ratio of Zn:Al and the total amount of intermetallic particles at grain boundaries.

2 Experimental Procedures

The die-cast Mg-Zn-Al (ZA) alloys were produced from AM60, commercial purity zinc and aluminium. The ZA alloys were die cast at a constant superheat of 100 °C. The die contained two round samples for tensile and creep testing, and its temperature was kept more or less constant during casting. The die-cast samples had a cross section of 5 mm in diameter, a gauge length of 30 mm and a total length of 66 mm. ZA alloys were also produced by sand casting, and tensile and creep properties of these alloys in the T6 condition were measured and compared these with those obtained from die-cast alloys of similar compositions. The sand-cast alloys were produced from commercially pure (99.7 %) Mg and Al ingots, high purity (99.9 %) Zn ingots, and Al-10wt%Mn master alloy. The melts were prepared in a coated iron crucible and protected under an atmosphere of 0.5%SF_6 in dry air. The melts were heated to 750 °C and held there for 30 minutes to ensure sufficient dissolution. These four alloys were homogenised at 325 °C for 72 hours, water quenched and then aged for 8h at 200 °C.

Table 1: The compositions of the HPDC and SC alloys tested in this work in weight percent. Approximately 0.3 wt%Mn was present in each of the alloys

Alloy (HPDC)	Zn	Al	Alloy (SC)	Zn	Al
AM60	0.01	5.8	ZA84	8.0	3.88
ZA85	8.73	5.22	ZA85	7.74	4.88
ZA86	8.35	5.78	ZA105	10.16	4.7
ZA105	10.16	5.37	ZA107	9.57	6.47
ZA107	10.32	6.81			

Tensile tests were performed at both room temperature and 150 °C using a screw-driven Instron 4505 machine equipped with a heating chamber. For tensile tests at 150 °C, it usually took about 20 minutes for specimens to reach the selected temperature. Tensile creep tests were carried out at 150 °C using constant load machines. Creep samples were immersed for 10 minutes in oil baths of 150±1 °C before they were loaded for creep tests. Creep strains were measured using extensometers that were connected directly onto the gauge length of the specimens. Specimens for metallography were etched in a solution of 6 g picric acid, 40 ml water, 40 ml acetic acid and 100 ml ethanol. Precipitate microstructures were characterised using a Philips CM20 transmission electron microscope operating at 200 kV.

3 Experimental Results

3.1 Microstructures of HPDC and SC Alloys

From optical observations, die-cast ZA alloys appeared to have very similar microstructures. There was a distinct bi-modal grain size distribution, Fig. 1. A significant amount of large dendrites (>20µm) were observed which probably formed in the shot sleeve before filling. Surrounding these was a much finer structure with a cell size less than 5 µm in diameter. This structure was finer near the wall of the casting. Evidence of coring was found in each cell where the dark region near the edge of the grains probably contained higher levels of solute elements, due to the eutectic reactions. Intermetallic particles were found at the cell edge in the centre of the darkened matrix phase and their volume fraction increased with the concentration of alloying elements in the alloy. Figure 2 shows the die-cast microstructure of alloy ZA107, recorded from regions close to the surface of the runner section of the casting. The size of dendrites was approximately 2 µm in diameter. A significant number of intermetallic particles were visible in the micrograph. These particles formed an almost continuous network along dendritic boundaries. Electron diffraction patterns recorded from such particles indicated that they had a quasi-crystalline structure, Fig. 2(b). The as-cast microstructures of HPDC alloys did not appear to contain any particles or precipitates of the equilibrium precipitate phase $Mg_{17}Al_{12}$. There was also no evidence of discontinuous precipitation of $Mg_{17}Al_{12}$ in these alloys.

The grain sizes of sand-cast alloys are typically of 200 μm. The fraction of retained intermetallics in T6 samples of the ZA84, ZA85, ZA105 and ZA107 alloys increased with alloying content. The primary intermetallic particles retained in the T6 samples of the sand-cast alloys were found to be the equilibrium φ phase, Fig. 3(a). The peak-aged microstructures had a relatively fine-scale distribution of icosahedral precipitates [7]. The distribution of such precipitate varied from grain to grain for each alloy, Fig. 3(b). Examinations of peak-aged microstructures of all four alloys did not detect any particles or precipitates of $Mg_{17}Al_{12}$ phase.

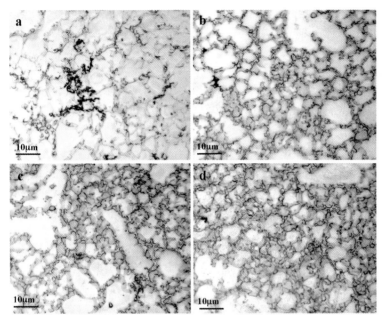

Figure 1: Optical micrographs showing microstructures of HPDC alloys of (a) AM60, (b) ZA85, (c) ZA105, and (d) ZA107

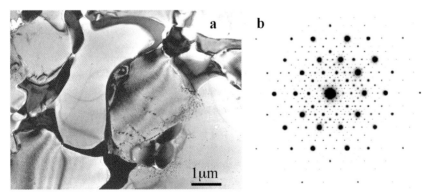

Figure 2: (a) Transmission electron micrograph showing microstructure of die-cast ZA107, and (b) electron diffraction pattern recorded from an intermetallic particle in (a)

3.2 Tensile and Creep Properties of HPDC and SC Alloys

The tensile properties of HPDC alloys are provided in Table 2. Many of these alloys exhibited limited ductility due to cracks in the shoulder region resulting from hot tearing. Therefore, it is difficult to assess the true ductility of ZA alloys in tensile samples. When tested at 150 °C, the yield strengths of the die-cast ZA alloys were typically about 105 MPa and were substantially superior to AM60 (~85 MPa). Figure 4 shows creep curves obtained at 60 MPa at 150 °C. The die-cast ZA alloys all had substantially superior creep properties compared to AM60, even though the creep rates obtained in the present work were not as good as those previously published for ZA85 [6], Fig. 5(b). It was further observed that the die-cast ZA alloys performed extremely well at low stresses (for example, 35 MPa), where the minimum creep rate is comparable with AE42. However, at the higher stress of 60 MPa the creep performance was not as good as die-cast AS21 or AE42, Fig. 5.

Figure 3: Transmission electron micrographs showing (a) a primary intermetallic particle of equilibrium φ phase, and (b) solid-state precipitates in T6 samples of ZA85 (8 h at 200 °C)

Table 2: Comparison of tensile properties of die-cast and sand-cast ZA alloys at room temperature and 150 °C

Alloy	R.T.			150 °C		
	0.2 % Proof (MPa)	UTS (MPa)	Elongation (%)	0.2 % Proof (MPa)	UTS (MPa)	Elongation (%)
AM60-HPDC	119	173	4.5	85	136	12.3
ZA85-HPDC	–*	106	0.3	106	118	0.6
ZA86-HPDC	105	177	1.6	104	154	4.5
ZA105-HPDC	–*	121	0.4	104	140	1.8
ZA107-HPDC	159	154	0.5	108	163	5.8
ZA84-T6	137	180	1.2	93	172	8.7
ZA85-T6	138	197	1.8	119	154	10.8
ZA105-T6	154	235	3.0	123	169	15.6
ZA107-T6	147	170	0.8	188	173	9.3
AZ91-T6	130	235	5.0	98	139	5.8

*These samples failed before significant yielding due to hot tearing in the shoulder region

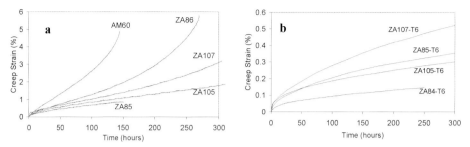

Figure 4: Creep curves of (a) HPDC, and (b) SC ZA alloys tested at 150 °C and 60 Mpa

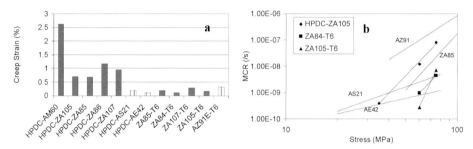

Figure 5: (a) Comparison of creep strains of ZA alloys and other alloys in the literature at 150 °C, 60 MPa and 100 hours. The data are from [8] for die-cast AS21 and AE42, and [9] for sand-cast AZ91E-T6. (b) Minimum creep rate (MCR) against applied stress at 150 °C for selected ZA alloys compared with data from die-cast ZA85 [6], AZ91 [10], and AS21 and AE42 [8].

Tensile properties obtained from T6 samples of sand-cast alloys at ambient and elevated temperatures are also provided in Table 2. The data for AZ91-T6 was obtained from an ingot cast alloy (solution treated for 24 h at 425 °C and aged for 24 h at 200 °C) [9]. The ZA alloys had comparable or better properties than AZ91. However, it needs to be realised that defects were often found on the fracture surface of the samples with low elongations, and therefore the low elongation (and tensile strength) of, eg. ZA107, may be due to a poor casting rather than being inherent in the alloy.

All four sand-cast alloys were creep tested at 60 MPa and 75 MPa at 150 °C. It was found that ZA84 performed better than other ZA alloys, Figs. 4 and 5. The better performance of ZA84 compared to ZA105, and ZA85 compared to ZA107, suggested that increasing the intermetallic content did not improve the creep rate and for some reason the higher intermetallic containing alloys actually had a higher creep rate. Comparisons of ZA alloys with other alloys in the literature, Fig. 5, indicated that: (i) the creep strains of ZA84-T6 and ZA85-T6 obtained at 100h at 150 °C and 60 MPa were lower than that of sand-cast AZ91-T6, and were comparable to that of die-cast AS21 and AE42, (ii) the minimum creep rate of ZA84-T6 was substantially lower than that of die-cast AZ91 and was also lower than that of die-cast AS21, and (iii) the creep resistance of sand-cast ZA alloys was much better than that of ZA alloys produced by high-pressure die casting.

4 Conclusions

The microstructures of die-cast ZA alloys do not appear to contain particles of the equilibrium precipitate phase $Mg_{17}Al_{12}$. There is also no evidence of discontinuous precipitation of $Mg_{17}Al_{12}$ in these alloys. Tensile yield strengths of the die-cast ZA alloys are higher than those of die-cast alloy AM60 at both ambient and elevated temperatures. Under the condition of 60 MPa at 150 °C, the creep resistance of die-cast ZA alloys is significantly better than that of die-cast AM60, but is worse than that of die-cast AS21 and AE42. However, at lower stress levels at 150 °C the creep resistance of die-cast ZA alloys becomes better than AS21 and is competitive to AE42.

T6 samples of sand-cast ZA alloys have relatively large volume fractions of solid-state precipitates within individual magnesium grains. Tensile yield strengths of these sand-cast alloys are in the range 135–150 MPa. These strength values are higher than that of AZ91-T6 samples. Under testing conditions of 60 MPa at 150 °C, the creep resistance of the ZA-T6 samples is significantly better than that of AZ91-T6, and is comparable to that of die-cast AS21. Tensile yield strengths and creep resistance of sand-cast ZA alloys are significantly better than those of die-cast ZA alloys of similar compositions.

5 Acknowledgements

The authors wish to thank Andrew Yob, Gary Savage, Qian Ma, and Graham Prior for their kind assistance in this project. The CAST CRC was established under and is supported in part by the Australian Federal Government's Co-operative Research Centre Scheme.

6 References

[1] Dargusch M., Dunlop G.L., in Proc. of Magnesium Alloys and Their Applications, eds. Mordike B.L. and Kainer K.U., Werkstoff-Informationsgesellschaft 1998, 277–282.
[2] Bettles C.J., Humble P., Nie J.F., in Proc. of 3rd International Magnesium Conference, ed. Lorimer G.W., The Institute of Materials 1996, 403–17.
[3] Bourgeois L., Muddle B.C., Nie J.F., Acta Mater, 2001, 49, 2701–2711.
[4] Vogel M., Kraft O., Dehm G., Arzt E., Scripta Mater., 2001, 45, 517–524.
[5] Zhang Z., Couture A., Luo A., in Proc. of Magnesium Alloys and Their Applications, eds. Mordike B.L. and Kainer K.U., Werkstoff-Informationsgesellschaft 1998, 289–294.
[6] Vogel M., Kraft O., Arzt E., in Proc. of Magnesium Alloys and Their Applications, ed. Kainer K.U., Wiley-VCH 2000, 693–698.
[7] Mendis C.L., Muddle B.C., Nie J.F., in Proc. of 4th Pacific Rim International Conference on Advanced Materials and Processing (PRICM-4), eds. Hanada S., Zhong Z., Nam S.W. and Wright R.N., Japan Institute of Metals 2001, 1207–1210.
[8] Dargusch M., PhD Thesis, University of Queensland, 1999.
[9] Bettles C.J., Forwood C.T., Internal Report, CAST, 2000.
[10] Regev M., Aghion E., Rosen A., Bamberger M., Mater. Sci. Eng. A, 1998, 252, 6–16.

Phase Formation in Mg-Sn Alloys Modified by Ca and Ce

J. Gröbner, A. Kozlov, M. Hampl, R. Schmid-Fetzer
Clausthal University of Technology, Clausthal-Zellerfeld, Germany

1 Introduction

The development of Al-free Mg-alloys is a promising innovation. These new alloys enable the use of Zr as grain refiner, which is not possible in alloys of the AZ and AM group because of detrimental Al-Zr compound formation. Base alloying systems for this purpose are XT (Mg-Ca-Sn) and ZK (Mg-Zn-Zr). Additional alloying elements for these systems could be Ce, Ca, and other elements. Final goal of this long-term investigation is to study the phase formation as prerequisite to focused alloy development and analysis of grain refinement and microstructure design.

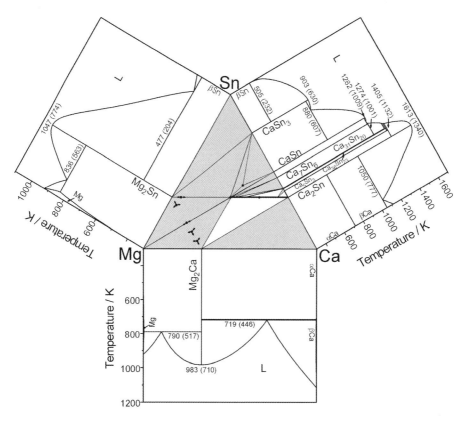

Figure 1: The calculated isothermal section of the Mg-Sn-Ca system at 298 K. Dots represent the investigated samples compositions; arrows point to the identified phases. Three-phase triangles are shaded.

Emley suggests [1] that Sn offers strengthening at the expense of ductility in Mg-alloys. Combined with Ca additional precipitates can be expected. In the present work first results on the ternary base system Mg-Sn-Ca are shown [2]. The second part is related to the Mg-Ca-Ce base system. Preliminary calculations of phase formation in quaternary Mg-Sn-Ca-Ce alloys are performed.

2   The Ternary Mg-Sn-Ca System

The present experimental investigation of the Mg-Sn-Ca ternary system was carried out with X-ray powder diffraction (XRD) analysis, metallographic analysis, scanning electron microscopy with energy dispersive X-ray microanalysis (SEM/EDX), and differential thermal analysis (DTA). Within this work, key samples were selected on the basis of preliminary thermodynamic calculations to provide relevant missing information on the Mg-Sn-Ca phase equilibria. Samples were placed on the section Mg_2Ca-Mg_2Sn and also on the section Mg_2Sn-Ca_2Sn to investigate three-phase equilibria and the ternary solubilities of the involved phases and to get additional information like liquidus data and the primary crystallizing phase. Another sample was prepared for confirmation of $Ca_{6.2}Mg_{3.8}Sn_7$ ternary compound. Starting materials were Ca granules (99.99 mass% Alfa, Karlsruhe), Mg pieces (99.98 mass% Alfa, Karlsruhe) Sn bar (99.999 mass%, Alfa, Karlsruhe). All purity designations are related to the metal basis.

A consistent thermodynamic model was developed using the Calphad method incorporating all experimental data. The resulting ternary phase diagram is quite complex because of substitution-type solid solution $Ca_{2-x}Mg_xSn$ (0<x<1) based on binary Ca_2Sn compound and associate formation in liquid phase in the binaries Ca-Sn and Mg-Sn. In the ternary liquid, five species are present: three pure components: Ca, Mg and Sn, and two associated complexes, Ca_2Sn and Mg_2Sn. The ternary phase CaMgSn was modeled as a three-sublattice line compound CaSn(Ca,Mg) to reflect the experimentally observed ternary solubility starting from the binary phase Ca_2Sn and ending at the composition CaMgSn. The calculated isothermal section of the Mg-Sn-Ca phase diagram at 298 K is given in Figure 1, together with the edge binary phase diagrams.

The investigated sample compositions are shown in different symbols according to the phase analysis. The arrows point to the measured phase compositions. The earlier reported phase $Ca_{6.2}Mg_{3.8}Sn_7$ has been confirmed. Large ternary solubilities were confirmed for the binary phase Ca_2Sn. The solubility of third elements in other binary compounds was not found.

3   The Ternary Mg-Ca-Ce System

Further addition of Ce involves possible formation of intermetallic phases at high temperature in the melt and the interaction with other alloying elements. The Mg-Ca-Ce system is one of the pertinent key systems, the phase diagram of which was previously unknown. The Mg-rich corner of this ternary system is studied by same methods described above. The liquidus surface of the Mg-rich corner and some vertical phase diagram sections are constructed using the results of DSC measurements and investigations of the microstructures of these samples. Details of this investigation will be given in [3]. The calculated isothermal section of the Mg-Ca-Ce system at 400 K is given in Figure 2, together with the edge binary phase diagrams. Varying the cerium composition, several possible precipitation phases can be formed. The most important of them is the Mg-rich intermetallic phase $CeMg_{12}$. This phase and Mg_2Ca are the two intermetallics in equilibrium with the (Mg) phase at 400 K.

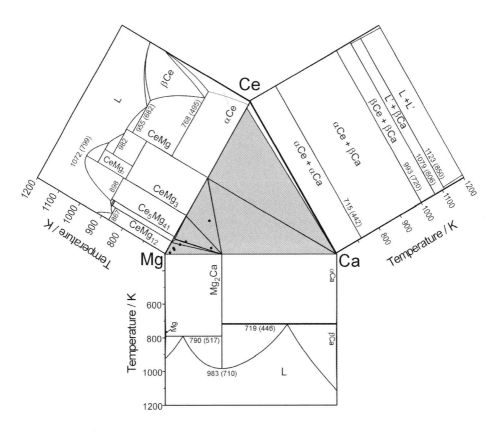

Figure 2: The calculated isothermal section of the Mg-Ca-Ce system at 400 K. Dots represent the investigated samples compositions. Three-phase triangles are shaded.

4 Quaternary Mg-Sn-Ca-Ce Alloys

Using these thermodynamic data sets an extrapolation in the quaternary system Mg-Ca-Ce-Sn is possible. The mathematical background was described earlier [4]. It is emphasized that the current calculations are based on the tentative thermodynamic models developed; further work is in progress to confirm and refine these data.

The phase formation during solidification of potential new Mg-alloys in this system can be calculated. The results for equilibrium solidification of the alloys TXE311 (Mg-3Sn-1Ca-1Ce wt.%) and TXE310 (Mg-3Sn-1Ca-0.5Ce wt.%) are shown in Figs. 3 a and b.

Ca_2Sn (with composition close to the limit of CaMgSn) and $CeMg_{12}$ were predicted as precipitations during solidification in equilibrium state in both of the calculated alloys. Ca_2Sn is formed from the melt after the primary crystallization of (Mg). $CeMg_{12}$ is formed in a invariant reaction at 590°C where the alloys were totally solidified. In the Ce-rich TXE311 alloy the amount of $CeMg_{12}$ is larger than that of Ca_2Sn, whereas it is opposite in the TXE310 alloy with half the content of Ce. In both alloys the equilibrium solidification interval is quite large, about 50 °C.

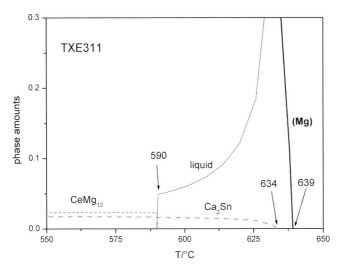

Figure 3a: Calculated equilibrium phase amounts of TXE311 alloy

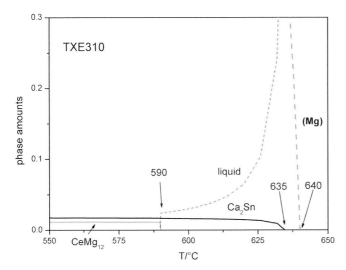

Figure 3b: Calculated equilibrium phase amounts of TXE310 alloy

5 Acknowledgement

This study is supported by the German Research Foundation (DFG) in the Priority Programme „DFG-SPP 1168: InnoMagTec" under grant no. Schm 588/26.

6 References

[1] Emley, E.F., Pergamon Press, Oxford **1966**, 232.
[2] Kozlov, A., Ohno, M., Arroyave, R., Liu, Z. K., Schmid-Fetzer, R., to be published.
[3] Hampl, M., Gröbner J., Schmid-Fetzer, R., to be published.
[4] Schmid-Fetzer, R., Gröbner, J., Adv. Eng. Mater., **2001**, *3*, 947–961.

Microstructure and Micromechanical Properties of As-cast Mg-Sn-Ca and Mg-Sn-Mn Alloys

J. Buršík[1], V. Buršíková[2], Y. Jirásková[1], T. Abu Leil[3], C. Blawert[3], W. Dietzel[3], N. Hort[3], K.U. Kainer[3]

[1] Institute of Physics of Materials, Academy of Sciences of the Czech Republic, Brno, Czech Republic
[2] Masaryk University, Faculty of Science, Dept. of Physical Electronics, Brno, Czech Republic
[3] Institut für Werkstoffforschung, GKSS Forschungszentrum Geesthacht GmbH, Geesthacht, Germany

1 Abstract

To launch a new class of magnesium alloys and to make a selection of suitable candidates for further studies, series of Mg-Sn-Ca and Mg-Sn-Mn samples were prepared with two levels of Sn (3 and 5 wt.%) and with third element ranging from 0 to 2 wt.%. The samples were studied in the as-cast condition by means of scanning electron microscopy (SEM) with energy dispersive X-ray analysis (EDX). In most cases existing secondary phases could be clearly visualized and analyzed by EDX. Image analysis on SEM micrographs was used in several cases to estimate the volume fraction of secondary phases. Mechanical properties were studied on polished surfaces using a depth sensing indentation test with Berkovich indenter and loads ranging from 10 to 1000 mN. This method enabled to obtain and compare material parameters like universal hardness, plastic hardness, elastic modulus, elastic and plastic deformation work and indentation creep resistance.

2 Introduction

Magnesium alloys represent a broad class of materials with high application potential due to their low density (in general) and e.g. good creep and/or corrosion resistance (in case of particular alloys) [1,2]. Knowledge of many Mg-based systems is still at its beginning. To make a selection of suitable candidates for practical application, series of multicomponent alloys are cast and their microstructural, mechanical, corrosion and other characteristics are studied, depending on intended usage of particular alloy.

Magnesium alloys with an addition of tin and calcium or manganese are developed as a low cost alternative for creep resistant magnesium alloys. In this work the effect of additions of Ca and Mn as ternary elements in Mg-Sn system is investigated. As a first step of characterization of new trial melts, the microstructure of cast alloys was studied from the viewpoint of phase composition and precipitates morphology. Mechanical properties were studied as well, using a depth sensing indentation (DSI) test with Berkovich indenter to obtain and compare material parameters like universal hardness, plastic hardness, elastic modulus, elastic and plastic deformation work and indentation creep resistance.

3 Experimental

Series of Mg-Sn-Ca and Mg-Sn-Mn alloys were selected for this study with two levels of Sn (3 and 5 wt. %) and with third element ranging from 0 to 2 wt.% (see Table 1 for nominal compositions). Alloys were prepared using high purity metals melted at 740 °C under Ar+SF$_6$ protective atmosphere. Large cylindrical billets of 100 mm in diameter and 500 mm in length were cast into permanent steel moulds preheated to 200 °C. The samples were studied in the as-cast condition. Small coin-shaped pieces (height 4 mm, diameter 18 mm) were cut from die cast rods of 14 magnesium alloys. Both planar surfaces were grinded with emery paper up to grit 1200 and one side was finally polished using OPS solution by Struers. The microstructure and phase composition were studied using a Jeol JSM 6460 scanning electron microscope (SEM) equipped with Oxford Instruments energy dispersive X-ray analyzer INCA Energy (EDX). In most cases existing secondary phases could be clearly visualized and analyzed by EDX. Image analysis on SEM micrographs was used in several cases to estimate the volume fraction of secondary phases.

Mechanical properties of as-cast alloys were studied on polished surfaces using a Fischerscope H100 XYP depth sensing indentation tester with Berkovich indenter and loads ranging from 10 to 1000 mN.

Table 1: Nominal compositions of studied alloys (wt %)

Alloy No.	1	2	3	4	5	6	7	8	9	10	11	12	13	14
Mg	Bal	bal	bal	bal	bal	bal	bal	bal	bal	bal	bal	bal	bal	bal
Sn	5.0	5.0	5.0	5.0	5.0	5.0	5.0	3.0	3.0	3.0	3.0	3.0	3.0	3.0
Ca		0.5	1.0	1.5	2.0				0.5	1.0	1.5	2.0		
Mn						0.5	1.0						0.5	1.0

4 Results

4.1 Microstructure

SEM micrographs were taken preferably using a signal of backscattered electrons to visualize compositional changes and hence better distinguish different phases. In the binary alloys Mg-5Sn and Mg-3Sn a small volume fraction of precipitates is present (Fig. 1). Precipitates decorate the continuous linear network of Sn-enriched regions developed due to segregation in course of solidification. Identification of precipitates as Mg$_2$Sn was supported by EDX analysis and by previous XRD experiments [3].

Adding Ca to binary Mg-Sn alloys causes the precipitation of additional phase in the form of thin plates or needles. According to EDX analyses this phase contains all three alloying elements. Plates form isolated islands in the alloy Mg-5Sn-0.5Ca (Fig. 2a) and they interconnect at higher Ca contents (Fig. 2b). Coarser precipitates in Fig. 2b are of the same type as fine ones. Mg$_2$Sn precipitates are in alloys No. 4 and 5 with higher Ca content very rare.

Microstructure of the alloy Mg-3Sn-0.5Ca is similar to that of Mg-5Sn-0.5Ca (Fig. 3a). However, with increasing Ca content alloys 10, 11 and 12 contain besides plates of MgCaSn also increasing volume fraction of precipitates of binary phase Mg$_2$Ca (Fig. 3b).

Mn addition to Mg-Sn alloys (alloys No. 6, 7, 13 and 14) causes only a slight increase of volume fraction of Mg_2Sn phase. This effect is most pronounced in alloy Mg-5Sn-1Mn (Fig. 4), where in some areas the Mg_2Sn phase along Sn-rich regions is continuous.

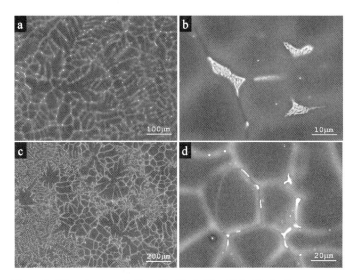

Figure 1: Microstructure of alloys Mg-5Sn (a,b) and Mg-3Sn (c,d)

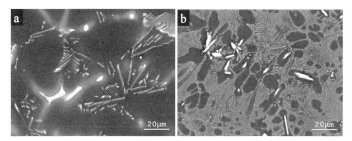

Figure 2: Microstructure of alloys Mg-5Sn-0.5Ca (a) and Mg-5Sn-1.5Ca (b)

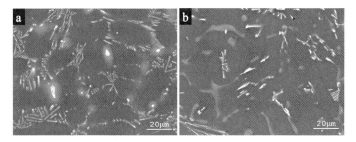

Figure 3: Microstructure of alloys Mg-3Sn-0.5Ca (a) and Mg-3Sn-2Ca (b)

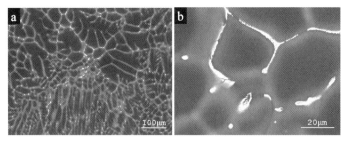

Figure 4: Microstructure of alloy Mg-5Sn-1Mn (a) and its detail at higher magnification (b)

4.2 Mechanical Properties

Fischerscope H100 DSI tester enables to register the indentation depth as a function of the applied load during both the loading and unloading part of the indentation test. On the basis of this method material parameters such as universal hardness HU (resistance against elastic and plastic deformation according to DIN50359-1), elastic part and irreversibly dissipated part of the indentation work (W_e and W_{irr}), plastic hardness HU_{pl} (i.e. resistance against plastic deformation) and effective elastic modulus Y may be determined. The accuracy of the depth measurement is about ±1nm.

Mechanical parameters of 14 alloys reported in Table 2 were carried out at maximum load of 100 mN. Here W_{tot} is the total indentation work, h_{max} is the maximum indentation depth at given maximum load L_{max} = 100 mN, effective elastic modulus $Y = E / (1 - v^2)$ (E is the Young's modulus and v is the Poisson's ratio). Universal hardness HU is calculated as $L_{max} / (26.43\ h_{max}^2)$. Plastic hardness HU_{pl} evaluated from loading-unloading curves recorded in course of DSI test is directly comparable with standard (e.g. Vickers) microhardness. To obtain older HV units, HU_{pl} in MPa should be divided by 10.

The maximum load may be kept constant for certain period of time before unloading and from the time dependence of the indentation depth the indentation creep resistance CR of the tested material can be evaluated. We used the regime with a loading period of 20s followed by a hold time of 5s, followed by an unloading period of 20s and finished by holding the minimum load for 5s. CR in Table 2 is the depth change during 5s hold at L_{max}, expressed as a percentage of h_{max}.

Plots in Fig. 5 show the dependence of universal hardness and effective elastic modulus on Ca content in Mg-3Sn-xCa and Mg-5Sn-xCa alloys.

5 Discussion

According to the results of image analysis of selected SEM micrographs the volume fraction of MgCaSn phase increases with increasing content of Ca in Mg-3Sn-xCa and Mg-5Sn-xCa alloys. In Mg-3Sn-xCa system we could observe also an increasing amount of Mg_2Ca phase. The increase of volume fraction of secondary phases is accompanied by increasing universal hardness and effective elastic modulus of the alloys (see Tab. 2 and Fig. 5). The measured values of effective elastic moduli are quite low compared to known data on other magnesium alloys [4]. This should be verified in next work using other experimental methods.

Table 2: Mechanical parameters measured by depth sensing indentation tester

Alloy	HU [MPa]	W_e/W_{tot} [%]	HU_{pl} [MPa]	Y [GPa]	h_{max} [μm]	CR [%]
1	440 ± 20	17 ± 1	520 ± 30	23 ± 1	2.93 ± 0.06	1.07 ± 0.12
2	430 ± 25	14 ± 2	490 ± 34	27 ± 1	2.97 ± 0.09	1.05 ± 0.23
3	437 ± 17	13 ± 1	487 ± 23	31 ± 1	2.94 ± 0.06	1.11 ± 0.09
4	528 ± 23	16 ± 1	612 ± 24	29 ± 1	2.68 ± 0.06	1.39 ± 0.20
5	512 ± 33	12 ± 1	562 ± 40	38 ± 1	2.72 ± 0.09	1.34 ± 0.08
6	462 ± 23	15 ± 1	533 ± 28	28 ± 1	2.86 ± 0.07	0.77 ± 0.01
7	368 ± 5	11 ± 1	409 ± 5	29 ± 1	3.21 ± 0.02	0.64 ± 0.13
8	327 ± 7	10 ± 1	356 ± 8	29 ± 1	3.40 ± 0.04	0.93 ± 0.08
9	410 ± 6	15 ± 2	477 ± 16	28 ± 2	3.04 ± 0.02	0.87 ± 0.22
10	476 ± 28	13 ± 1	526 ± 32	34 ± 1	2.82 ± 0.08	1.35 ± 0.26
11	497 ± 6	12 ± 1	543 ± 14	37 ± 1	2.76 ± 0.02	1.32 ± 0.18
12	575 ± 27	13 ± 1	633 ± 45	41 ± 1	2.57 ± 0.06	1.24 ± 0.27
13	446 ± 16	14 ± 1	503 ± 23	29 ± 1	2.91 ± 0.05	0.97 ± 0.17
14	508 ± 29	13 ± 1	575 ± 40	36 ± 1	2.73 ± 0.02	0.65 ± 0.18

Alloys with Mn addition (namely alloys 7 and 14 with 1 wt.% Mn) have the lowest CR parameter, i.e. the highest creep resistivity at selected testing regime.

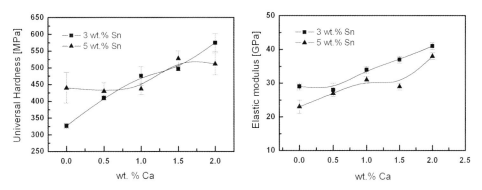

Figure 5: Dependence of mechanical properties on Ca content in Mg-3Sn-xCa and Mg-5Sn-xCa alloys

6 Summary

Fourteen ternary alloys of Mg-Sn-Ca and Mg-Sn-Mn ternary systems were prepared and studied in the as-cast condition by means of analytical scanning electron microscopy. Existing secondary phases were identified according to their chemical composition measured by energy dispersive X-ray analysis. The results of microstructure study were complemented with mechanical properties measured by means of depth sensing indentation test. Increasing amount of Ca in Mg-5Sn-xCa and Mg-3Sn-xCa content was correlated with increasing universal hardness and effective elastic modulus. Results of short creep tests performed by applying a static load between loading and unloading curve of DSI tests showed, that alloys Mg-5Sn-1Mn and Mg-3Sn-1Mn have the highest creep resistance.

7   Acknowledgements

This work was supported by travel grant received under DAAD-AS CR (D-CZ 13/05-06) and by the IPM Research Plan (AV0Z20410507) that enabled the joint research work presented in this paper.

8   References

[1]   Mordike, B.L.; Materials Science and Engineering A **2002**, *324*, 103–112.
[2]   Vogel, M., Kraft O., Arzt, E., Scripta Materialia **2003**, *48*, 985–990.
[3]   Bowles, A.L., Blawert, C., Hort, N., Kainer, K.U., in *Magnesium Technology 2004, Editor A. Luo*, TMS **2004**, p. 307–310.
[4]   Magnesium, Proceedings of the 6th International Conference Magnesium Alloys and Their Applications, Edited by K.U. Kainer, DGM, **2003**, p. 97, p. 419, p. 451.

Microstructure and High Temperature Mechanical Properties of Mg-Al-Sn Alloys

Woo Chul Cho, Yeon Jun Chung and Kwang Seon Shin
Department of Materials Science and Engineering
Seoul National University, Seoul 151–744, Korea

1   Introduction

Magnesium alloys are increasingly used in the automobile industry because of their low density, superior die castability and good balance of strength and ductility. The most widely used magnesium alloys are in the Mg-Al series, such as AM50, AM60 and AZ91. However, the applications of these alloys have been limited to temperatures around 120ºC because of their poor high temperature mechanical properties [1–3]. In Mg-Al alloys, the main deformation mechanisms at elevated temperatures are the rapid softening of the $Mg_{17}Al_{12}$ phase and grain boundary sliding. Therefore, it is important to reduce the amount of the $Mg_{17}Al_{12}$ phase and introduce thermally stable precipitates at grain boundaries as well as in the grain interior by adding proper alloying elements [4–7]. However, the addition of Al is necessary to improve die castability and room temperature strength. Recently developed heat resistant magnesium alloys use Ca, Sr, and rare-earth as alloying elements [8–16]. In this study Sn, Sr, Ca and misch metal (50wt.% Ce-20wt.% Nd-20wt.% La-10wt.% Pm) were selected as alloying elements that form thermally stable intermetallic compounds. In particular, the effects of Sn on the microstructure and high temperature mechanical properties were examined. Sn has several suitable characteristics as an alloying element for elevated temperature applications such as considerable solid solubility and low diffusivity in Mg. Moreover, the addition of Sn introduces the thermally stable Mg_2Sn precipitate, which effectively controls grain boundary sliding and dislocation motion at elevated temperatures.

2   Experimental Procedures

All test specimens were die-cast on a 320 ton cold chamber high-pressure die casting machine at an injection pressure of 260 kgf/cm^2 in this study. Melt temperature was maintained at 680 °C in an atmosphere of SF_6 and CO_2. The die temperature was fixed at 200 °C using a die temperature controller before undertaking the die casting operation. Table 1 summarizes the alloy designations and chemical analysis results of these specimens as examined by spectrometer. The microstructure was examined by optical microscopy and scanning electron microscopy with EDS.

For microstructure observation, die-cast specimens were mechanically polished with sand paper (#1200) and then with 0.3 μm alumina powders. These specimens were then etched with 0.5% HNO_3 + 99.5% distilled water for 2 to 5 seconds depending upon alloy composition. The volume fraction of the second phase was analyzed using an image analyzer. Tensile tests were conducted at room temperature, 150 °C and 200°C at an initial strain rate of $2 \cdot 10^{-4}$/sec. The specimens were heated to the desired temperature and held at that temperature for 20 minutes before the tensile tests. High temperature creep tests were carried out at 150 °C with an applied stress of 70 MPa.

Table 1: Chemical compositions of the die-cast specimens

Alloy	Composition (wt.%)					
	Al	Sn	Sr	Ca	Mm	Mg
A5	4.9	-	-	-	-	Bal.
AT54	5.1	4.3	-	-	-	Bal.
AT58	5.1	8.1	-	-	-	Bal.
AT58JXE	5.2	7.8	1.1	1.1	0.7	Bal.

3 Results and Discussion

3.1 Microstructure Analyses

Figure 1(a)~(d) show the effects of alloying elements on microstructure of Mg-5wt.% Al alloys. The results show that an increase in Sn decreases grain size and increases the volume fraction of the precipitates. In the A5 alloy, a continuous network of the $Mg_{17}Al_{12}$ phase is observed along the grain boundaries. In contrast, semi-continuous grain boundary particles are observed in the AT54 and AT58 alloys. Figure 1(d) shows the microstructure of the AT58JXE alloy with Sr, Ca and Mm (Ce-rich) addition. As seen in this figure, the grain size of the AT58JXE alloy is much smaller than those of other alloys. Most of the grain boundaries show that the proeutectic α is surrounded by the eutectic structures of $Mg_{17}Al_{12}$ or Mg_2Sn.

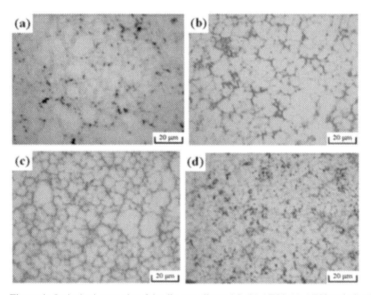

Figure 1: Optical micrographs of the die-cast alloys; A5, (b) AT54, (c) AT58 and (d) AT58JXE

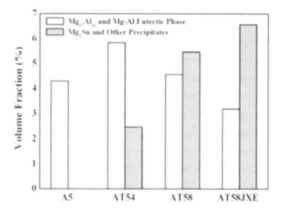

Figure 2: Volume fraction of the second phase in the die-cast alloys

Figure 2 shows the volume fraction of the second phase present in the investigated alloys. With increase in Sn content, the volume fraction of the thermally stable Mg_2Sn phase increased, while the volume fraction of the thermally unstable $Mg_{17}Al_{12}$ and Mg-Al eutectic phase decreased. These results clearly indicate that the addition of Sn can improve the high temperature mechanical properties of the Mg-5wt.% Al alloys. Figure 3 shows the scanning electron micrographs of the AT58 and AT58JXE alloys displaying the second phase particles. It was confirmed by the SEM/EDS analyses that the second phases were Mg_2Sn, $Al_{11}Ce_3$, MgCaSn and MgAlSr compounds. Since these precipitates are more thermally stable than the $Mg_{17}Al_{12}$ phase at elevated temperatures, they are expected to stabilize the microstructures and improve the high temperature mechanical properties of Mg-Al alloys. As shown in Figure 3(b), the size of the Mg_2Sn precipitates decreased in the case of the AT58JXE alloy. It is considered that the addition of Ca, Sr and Mm decreased the size of the Mg_2Sn phase. Figure 4 shows the result of an EDS line scan along the line marked in Figure 3(a). From the EDS line scan, it is evident that the matrix (black phase) is α-Mg, the gray precipitates are $Mg_{17}Al_{12}$ and the white precipitates are Mg_2Sn. Additionally, the high concentration of Al around the Mg_2Sn phase suggests that the Mg_2Sn phase which has a high melting point (770.5 °C) precipitated first and the $Mg_{17}Al_{12}$ phase precipitated around the Mg_2Sn precipitates. Thus, it can be observed that the Mg_2Sn phase is sometimes surrounded by the $Mg_{17}Al_{12}$ phase.

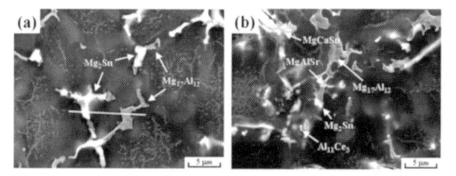

Figure 3: Results of phase identification with SEM/EDS; (a) AT58 and (b) AT58JXE

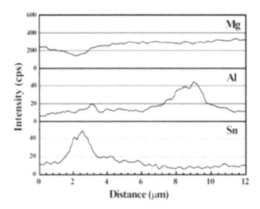

Figure 4: EDS line scan of the AT58 alloy

3.2 Mechanical Properties

Figure 5 compares the stress-strain curves of the die-cast specimens at room temperature, 150 °C and 200 °C, respectively. It was found that the yield strength and ultimate tensile strength increased and elongation decreased at all temperatures with increase in Sn. The yield

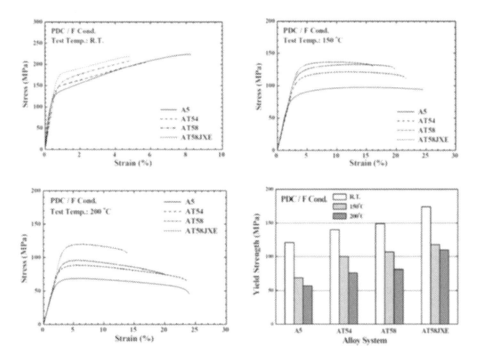

Figure 5: Stress-strain curves of the die-cast alloys and comparisons of the yield strengths at different temperatures

strength and ultimate tensile strength of the A5 alloy rapidly drops with increase in temperature. However, in the case of the AT54 and AT58 alloys, the magnitudes of strength decrease are relatively smaller than that of the A5 alloy. It is known that Sn atoms existinside the matrix as solute atoms and the Mg_2Sn precipitates at grain boundaries contributes to the increase in strength at high temperatures. In the case of the AT58JXE alloy, the yield strength and ultimate tensile strength are 105 MPa and 119 MPa at 200 °C, respectively, which are the highest values found in this study. These results show that the effects of adding small amounts of Sr, Ca and Mm to the Mg-Al-Sn alloys greatly improve the high temperature mechanical properties. Thus, it is concluded that the thermally stable precipitates Mg_2Sn, Al_4Sr, Mg_2Ca, $Al_{11}Ce_3$, MgCaSn and MgAlSr compounds act as the strengthening phases by blocking the movement of dislocations effectively and suppressing grain boundary sliding at high temperatures. Table 2 shows the results of creep tests conducted at 150°C with an applied stress of 70 MPa. The minimum creep rate decreased as the Sn content increased. As mentioned previously, the presence of the thermally stable Mg_2Sn phase improved the creep resistance of the Mg-Al alloys. The AT58JXE alloy shows good creep resistance since the smaller Mg_2Sn particles are more evenly dispersed by adding Sr, Ca and Mm. The stress exponent for the AT58JXE alloy is found to be 3.5. Comparisons with the values of the stress exponent reported in the literature suggest that the dominant creep mechanism for the present AT58JXE alloy is the dislocation climb controlled creep [1]. The presence of the thermally stable Mg_2Sn, Al_4Sr, Mg_2Ca, $Al_{11}Ce_3$, MgCaSn and MgAlSr compounds contribute to the improved creep resistance by impeding the dislocation movement.

Table 2: Minimum creep rate of the die-cast alloys at 150 °C with a stress of 70 MPa

Alloy	Minimum creep rate (s^{-1})
A5	$1.26 \cdot 10^{-6}$
AT54	$7.07 \cdot 10^{-7}$
AT58	$3.14 \cdot 10^{-7}$
AT58JXE	$1.26 \cdot 10^{-8}$

4 Conclusions

The addition of Sn to the Mg-5wt.% Al alloys causes the formation of the Mg_2Sn phase at grain boundaries, suppresses the formation of the $Mg_{17}Al_{12}$ phase and reduces the grain size. Due to the high temperature stability of the Mg_2Sn phase, the addition of Sn improves the high temperature tensile and creep properties of the Mg-5wt.% Al alloys. The additions of Ca, Sr and Misch metal significantly improve the creep resistance of the Mg-Al-Sn alloy.

5 Acknowledgements

This work was financially supported by the Korean Ministry of Commerce, Industry and Energy.

6 References

[1] A.A. Luo, Mater. Sci. Forum, 419-422, **2003**, 57–66.
[2] S. Schumann and H. Friedrich, Mater. Sci. Forum, 419–422, **2003**, 51–56.
[3] Y.G. Na, C.D. Yim, S.C. Park and K.S. Shin, Mater. Sci. Forum, 419–422, **2003**, 285–290.
[4] M.O. Pekguleryuz, Mater. Sci. Forum, 350-351, **2000**, 131–140.
[5] Y.J. Chung and K.S. Shin, Mater. Sci. Forum, 475-479, **2005**, 537–540.
[6] M.S. Yoo, Y.C. Kim, S.H. Ahn and N.J. Kim, Mater. Sci. Forum, 419–422, **2003**, 419–424.
[7] M.S. Yoo, K.S. Shin, and N.J. Kim, Metall. Mater. Trans. A, 35A, **2004**, 1629–1632.
[8] B.L. Mordike, Mater. Sci. & Eng., A324, **2002**, 103–112.
[9] I.P. Moreno et al., Scripta Mater., 48, **2003**, 1029–1034.
[10] B.L. Mordike, J. Mater. Proc. Tech., 117, **2001**, 391–394.
[11] Y.J. Chung and K.S. Shin, Magnesium Technology 2005 (TMS), **2005**, 425–428.
[12] E. Brail, Pierre Labelle and M.O. Pekguleryuz, JOM Nov., **2003**, 34–39.
[13] K. Ozturk, Y. Zhong, A. Luo, and Z. Liu, JOM Nov., **2003**, 40–44.
[14] C.D. Lee and K.S. Shin, Met. Mater. Int., 9(1), **2003**, 21–27.
[15] W.C. Cho, H.C. Jung, K.S. Shin, in Proceedings of the 1st Asian Symposium on Magnesium Alloys, Jeju, Korea, **2005**, 105–108.
[16] S. Cohen, G.R. Goren-Muginstein, S. Avraham, G. Dehm and M. Bambeger, Magnesium Technology 2004, **2004**, 301–305.

Effect of Heat-Treatment on the Microstructure, Microhardness and Corrosion of Cast Mg-3Sn-2Ca Alloy

J. Buršík[a], K. Zábranský[a], Y. Jirásková[a], V. Buršíková[b], T. Abu Leil[c], C. Blawert[c], Y. Huang[c], W. Dietzel[c], N. Hort[c], K.U. Kainer[c], K.P. Rao[d]

[a] Institute of Physics of Materials, Academy of Science of the Czech Republic, Brno
[b] Masaryk University, Faculty of Science, Dept. of Physical Electronics, Brno
[c] Institut für Werkstoffforschung, GKSS Forschungszentrum Geesthacht GmbH, Geesthacht
[d] City University Hong Kong, 83 Tat Chee Avenue, Kowloon

1 Introduction

Magnesium alloys are among the very good lightweight structural materials with relatively high-strength-to-weight ratio and excellent technological properties. Therefore they attract special attention of electronic, automotive and aircraft industries. On the other hand a drawback of magnesium as a structural material is its high chemical activity leading often to a low corrosion resistance. Since many different mechanically loaded components are subjected to prolonged cyclic stresses in an active environment, it is of significant and practical interest to look for new Mg-based alloys. They should be up to standard of the economical production process and easy workability. Of importance is also surface stability and eventual possibility of modification in terms of the topography and chemistry.

One possible solution is to add Ca to Mg-Sn alloys. Sn as an alloying element is cheap, has a low melting point, and is known to improve castability [1, 2]. It forms with magnesium an intermetallic Mg_2Sn phase precipitating in Mg-matrix and contributing to strengthening of this alloy [3]. The first published results show a positive effect of Ca in the technological processing of molten magnesium [4], the oxidation and corrosion properties [5-7] and the creep resistance [8].

In this paper the structural and physical properties of the as-cast Mg-3 wt.%Sn-2 wt.%Ca (Mg-3Sn-2Ca) alloy (I) are compared with those obtained for the alloy after solution heating followed by furnace cooling (II), solution heating with following quenching into water (III), and solution heating followed by aging (IV).

2 Experimental

The Mg-3Sn-2Ca alloy used for the present study, has the nominal composition of 3 wt.% Sn, 2 wt.% Ca and rest Mg. It was prepared using high purity Mg (99.99 %), Sn (99.96 %) and Ca (98.5 %). The alloy was melted at 740 °C under the protection of SF_6-argon mix cover gas, held for 10 minutes before casting it into permanent steel moulds with a mould temperature of 200 °C to obtain large cylindrical billets of 100 mm diameter and 350 mm length. For the microstructure examinations round discs of about 4 mm thickness and 18 mm diameter were prepared. The specimens were ground from both planar sides with a 1200 grade silicone carbide paper, cleaned in ultrasonic bath using alcohol and dried. One side of all specimens was finally polished using OPS solution by Struers. A part of specimens was left in the initial as-cast state (I), second part was solution treated at 500 °C for 6 hours followed by cooling inside the furnace

(II), third part was exposed to solution treatment at 500 °C for 6 hours followed by quenching into water (III) and the last part was solution treated at 500 °C for 6 hours followed by ageing at 300 °C for 6 hours (IV).

The microstructure of specimens was investigated using a Jeol JSM 6460 Scanning Electron Microscope (SEM) equipped with Oxford Instruments energy dispersive X-ray analyzer INCA Energy (EDX) facility. The phases were recognized also by X-Ray Diffraction (XRD) using Siemens diffractometer operating at 40 kV and 40 mA with Cu K_α radiation. The measurements were done for $20° < 2\theta < 120°$ with a step of 0.02°. A count time of 3 seconds per step was applied.

The corrosion behavior of the specimens has been studied by means of salt spray test and potentio-dynamic (polarization) measurements. For the polarization test the specimens were ground on one side, while for the salt spray test the specimens were ground on both sides. The potentio-dynamic measurements were conducted for the specimens in states I ÷ IV at room temperature, in 5% NaCl solution with a pH value adjusted to 11 using NaOH. The set-up has used specimens of 18 mm diameter as working electrodes with an exposed area of 1.03 cm^2 that was in contact with the electrolyte, in conjunction with an Ag/AgCl reference electrode and a platinum counter electrode. After recording the free corrosion potential for 30 minutes, the polarization scan was started at −250 mV relative to the free corrosion potential with a scan rate of 0.2 mV/s. The corrosion rate was calculated using Tafel slope from the cathodic branch of the polarization curve. The specimens of the same dimension were also used in salt spray corrosion tests. The specimens were immersed in 5 % NaCl solution of pH value 7 at 35 °C for 48 hours as per ASTM B177-85 standard. The corroded specimens were washed off by salt fog followed by washing with water and chromic acid to remove the oxides and dried subsequently. The weight loss of the specimens was used to calculate the average corrosion rate in millimeters per year as practiced by others.

A Fischerscope H100 depth sensing indentation tester equipped with Vickers indenter was used to study the indentation response of the alloy. The tester enables to register the indentation depth as a function of the applied load during loading and unloading parts of the test. On the basis of this method universal hardness *HU* (resistance against elastic and plastic deformation according to DIN50359-1) can be determined. The accuracy of the depth measurement is about ±1 nm. Several tests were made at different maximum indentation loads (i.e. several different indentation depths) in order to study the load dependence of the mechanical properties of the alloy in the as-cast (I) and heat-treated (II, III, IV) states at room temperature. The loading period of 20 s was followed by a hold time of 5 s, an unloading period of 20 s, and the test concluded after holding the minimum load for 5 s. At least nine different maximum loads were chosen in the range 10 to 1000 mN. Indentation test at each selected condition was repeated 9 to 16 times in order to minimize the experimental errors. The microhardness values were averaged and the 95 % confidence level was determined.

3 Results and Discussion

Scanning electron micrographs of the investigated Mg-3Sn-2Ca alloy in the as-cast (I) and treated (II, III, IV) states are shown in Figure 1. The matrix in the as-cast (I, Figure 1a) state is formed of Mg in which precipitates of Mg$_2$Ca are visible as drab grey plates. They are also found in the heat treated states of the alloy, but they are of irregular shape and less distinct in the furnace cooling case (II, Figure 1b). Another type of precipitates, appearing mostly needle type and bright, were analyzed to be CaMgSn. This phase was found in all the heat treated states of the alloy. In the microstructure of specimen III (Figure 1c) it appears in a finer form compared to the other three cases. This probably contributes to the hardening of the material as documented by microhardness measurements presented later in the paper.

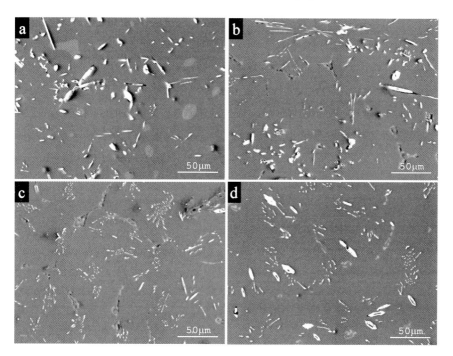

Figure 1: Scanning electron micrographs of Mg-3Sn-2Ca alloy in the as-cast I (a) and heat treated II (b), III (c) and IV (d) states

The annealing at 500 °C for 6 hrs followed by further steps of treatments (states II, III, IV) has influenced only slightly the composition of Mg matrix in which always about 0.3–0.5 wt.% of Ca was detected. The identification of the two types of precipitating particles described above has been confirmed by XRD patterns shown in Figure 2, which clearly indicate the presence of Mg_2Ca and CaMgSn phases in all specimens.

The results obtained from the two types of corrosion tests for the alloy in as-cast (I) and heat-treated (II–IV) states are summarized in Figure 3. The corrosion rates measured by the polarization test fall within a small range for the specimens in all states. Slightly higher differences in corrosion rates are observed from salt spray test whereas the specimen III yields the lowest corrosion rate. The slightly higher corrosion rates obtained from salt spray test could be ascribed to lower pH value of the solution used.

The average values of calculated microhardness as a function of the applied load are shown in Figure 4. Higher microhardness values were obtained for specimen III. This is in good correlation with the results obtained on microstructure by SEM and by EDX analysis (Figure 1c). The microhardness values obtained for specimen III at maximum load of 50 mN had substantially higher scatter (~15 %) on the contrary to the rest of the data with relative scatter under 10 %. In this case discontinuities (pop-ins) were observed on the loading curves indicating that the indentation has induced changes around the indentation point (cracking, twinning) in the material. With loads of about 5 mN the results are highly localized and hence sensitive to presence of either solid solution effect or secondary phases. Further investigations are needed to compare these hardness values with those obtained with conventional hardness measurements.

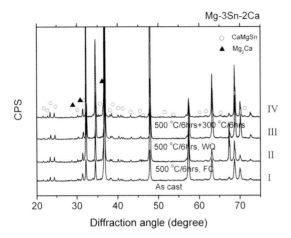

Figure 2: XRD patterns of the Mg-3Sn-2Ca alloy in the as-cast (I) and heat treated (II, III, IV) states

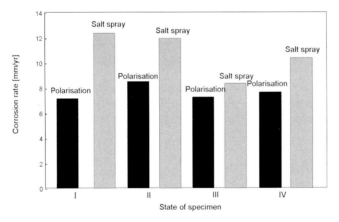

Figure 3: The corrosion rates of Mg-3Sn-2Ca specimens in the as-cast (I) and heat treated (II, III, IV) states in dependence on type of corrosion test

4 Conclusions

The phase composition, microstructure and corrosion resistance completed by microhardness measurements of a new magnesium based cast alloy were investigated with respect to various heat-treatment conditions.

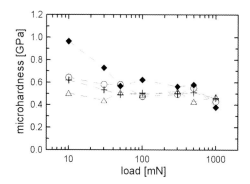

Figure 4: Dependence of microhardness on the applied load for Mg-3Sn-2Ca alloy in the as-cast (I-o) and different heat-treated (II-△, III-♦, IV-+) states

The Mg-3Sn-2Ca alloy was studied in the as-cast (I) state and in the states after solution heating at 500 °C / 6 hrs followed by furnace cooling (II), solution heating at 500 °C / 6 hrs followed by quenching in water (III) and solution heating at 500 °C / 6 hrs followed by aging at 300 °C / 6 hrs (IV).

Scanning electron microscopy has revealed drab grey precipitates of Mg_2Ca and needle type bright precipitates of CaMgSn changing slightly in their shape and size in dependence on heat treatment. The solution heating at 500 °C / 6 hrs followed by quenching in water (III) has yielded the microstructure with the finest form of both these precipitates. This has contributed to hardening of the material as documented by measurement of the higher microhardness in comparison to other states (I, II, IV). The X-ray diffraction has confirmed the phase analysis of SEM of specimens in all states investigated. The corrosion rates obtained with the salt spray and polarization tests have fallen within a narrow range.

5   Acknowledgment

This work was supported by travel grant received under DAAD-AS CR (D-CZ 13/05-06) and the IPM Research Plan (AV0Z20410507) that enabled the joint research work presented in this paper.

6   References

[1]   Dead Sea Magnesium Ltd., High Strength and Creep Resistant Magnesium Alloys, European Patent, EP 1 308 531 Al, 2003.
[2]   Hitachi Ltd., High Strength Mg Based Alloy and its Uses, EP1 108799, European Patent, 2001.
[3]   A.L. Bowles, C. Blawert, N. Hort, K.U. Kainer, Magnesium Technology 2004, (Ed.: A.A. Luo), TMS, 2004, 307–310.
[4]   M. Sakamoto at al., J. Mat. Sci. Let. 1997, 16, 1048–1050.
[5]   B.S. You, W.-W. Park, I.-S. Chung, Scripta Mat. 2000, 42, 1089–1094.

[6] Y. Fan, at al., Mat. Sci. Forum 2005, 488–489, 869–872.
[7] T. Abu Leil, K.P. Rao, N. Hort, C. Blawert and K.U. Kainer, in: Magnesium Technology 2006 (Ed.: A. Luo, N.R. Neelameggham, R.S. Beals), TMS, 2006, San Antonio, Texas, USA, 281–286.
[8] T. Abu Leil, Y. Huang, H. Dieringa, N. Hort, K.U. Kainer, J. Buršík, Y. Jirásková and K.P. Rao, IMA's 63rd Annual World Magnesium Conference, Beijing, China, June 2006, send for publication.

Elevated Temperature Alloys – Paths for Further Performance Gains in AE44

P. Bakke, A.L. Bowles, H. Westengen
Competence Centre Magnesium, Hydro Magnesium, Porsgrunn, Norway

1 Introduction

The beneficial effect of rare earth elements (Ce, La, Nd and Pr) on Mg-alloys was discovered in the 1930's [1,2]. A series of alloys containing rare earth elements was established, following the successful development of zirconium induced grain refinement. These alloys include ZE41, EZ33, QE21 and WE43. Common to these alloys is their good castability in sand moulds, and the excellent elevated temperature properties [3,4].

In high pressure die casting, grain refining results from the rapid cooling of alloys of the AE-family were studied in some details in the 1970's [5–7]. Based on the early studies AE42 was introduced as a candidate alloy for automotive engine and transmission parts in the early 1990's [8-10]. Since then AE42 has been the benchmark Mg alloy with respect to elevated temperature properties and corrosion [11,12]. Unlike today, when the price of rare earth elements is lower than that of Ca and Sr, the rare earth element additions made AE42 unattractive from a cost point of view. The new creep resistant die casting alloy AE44 patented by Hydro Magnesium consists of 4% Al, which provides solid solution strengthening, and 4% mischmetal (MM) which combines with Al, forming dispersoid phases, identified mainly as Al_2RE, in the grain boundary regions. The continuous 3D network of dispersoids effectively prevents creep arising from thermal activation and grain boundary sliding without affecting the ductility [13]. The AE44 die casting alloy possesses a unique combination of excellent creep properties at elevated temperatures, outstanding ability to absorb energy under crash loads and superior corrosion performance, providing an opening to unexplored avenues in automotive component design [14,15]. The Corvette Z06 front cross member is an excellent example of a component that probably could not be realized in any other magnesium die casting alloy [16,17].

It is well known that alloys of the Mg-Al-Ca [18,19] and Mg-Al-Sr systems [20,21] may provide significant improvement in elevated temperature properties. Investigations of alloys of the Mg-Al-RE-Ca and Mg-Al-RE-Sr systems have mainly been restricted to RE < 2 wt% [22,23]. In the Mg-Al-RE-Ca system alloys with extremely interesting high temperature properties have been identified [22].

In the commercially available AE alloys, rare earth elements are present in the form of mischmetal that typically consists of Ce (>45 %), La (20–35 %), Nd (10–20 %) and Pr (4–10 %). Modification of AE44 by increasing Al and/or RE is possible if the processing temperature is increased significantly. For example, an alloy of the type AE45 would partly overlay the liquidus line at 680 °C, thereby requiring higher processing and casting temperature compared to standard alloys [24]. Paths for modification of AE44 are thus restricted to changes in the composition of RE, and the addition of elements that dissolve in the liquid AE44 alloy. This paper discusses potential improvements of AE44 by additions of Ca and Sr to commercially available MM based AE44.

2 Experimental

Two types of alloys were prepared:
- Commercially available MM based AE44 with additions of Ca up to 2 wt% (1.2 at%)
- Commercially available MM based AE44 with additions of Sr up to 4 wt% (1.2 at%)

Specimens were high pressure die cast on a 420 ton Bühler SC42D Evolution machine with a five cavity plate/bar die as indicated in Fig. 1. This die contains two 140 × 100 × 3 mm corrosion test plates (1), a 6 mm cylindrical tensile creep bar (2), a full size 6 mm cylindrical tensile bar (ASTM B557) (3), and a 10 × 10 × 50 mm charpy impact bar (4).

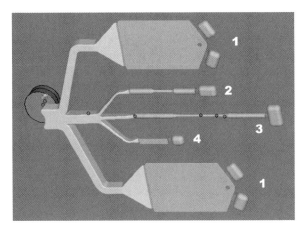

Figure 1: The five cavity plate/bar die for the Bühler SC42D Evolution die casting machine at CC Magnesium, Porsgrunn.

Permanently mould cast medallions (disk samples) were withdrawn from the alloy melts just prior to die casting. These were analysed by ICP-AES and XRF-UNIQUANT for chemical composition.

Standard 6 mm test bars from each composition were tensile tested in a 10 kN Instron machine at room temperature and elevated temperatures in accordance with ISO 6892. In all tests a strain rate of 1.5 mm/min (0.03 min^{-1}) was used until 0.5% strain, after which the strain rate was increased to 10 mm/min (0.2 per min^{-1}). From each alloy, 10 specimens were tested at room temperature while only three samples were tested at each of the elevated temperatures.

Die cast tensile creep specimens were subjected to single level creep testing (ASTM E139) for a minimum of 250 hours at 150 °C/90 MPa, 175 °C/75 MPa and 200 °C/60 MPa.

Stress relaxation compression testing was performed at 175 °C in accordance with ASTM E328-86 and established practice [25]. The initial load was 12 kN corresponding to 106 MPa. The cylindrical specimens (12 mm dia., 5 mm length) were cut from one arbitrary end of a die cast tensile creep bar. For the experimental alloys, two parallels were tested at each condition. The stress relaxation compression test is a strongly simplified but relevant substitute for a full bolt load retention test.

Un-notched charpy test bars were subjected to Vickers hardness testing using the 0,5 mm indentation pyramid (HV 5) according to ISO standard 6507/1.

Impact strength testing with a 50 J pendulum was performed on 10 un-notched charpy test bars per alloy in accordance with the ISO 148 standard.

Salt spray corrosion tests were performed on the standard 3 mm thickness die cast test plates in accordance with ASTM B117. Five parallels were tested.

Specimens for microstructural studies were cut and prepared from one corner (near the gate) of an arbitrary test plate in the as cast condition. The specimens were examined on a JSM5600 Pioneer Scanning Electron Microscope (SEM) system. Energy Dispersive Spectroscopy (EDS analysis) was performed with an accelerating voltage of 15 kEV.

3 Results

3.1 Chemical Analyses

The chemical analyses of the experimental alloys are shown in Table 1. The full chemical analyses of alloys AEX440 and AEX441 were not undertaken. Except for Ca, the analyses of these alloys are expected to be the same as for reference alloy AE44.

Table 1: Chemical analyses of the alloys

	Al	Mn	Fe	Ce	La	Nd	Pr	ΣRE	Ca	Sr
AE44	4,1	0,24	0,0035	2,18	1,34	0,5	0,24	4,3	0,007	
AEX 440									0,50	
AEX 441									0,98	
AEX 442	4,2	0,25	0,0023	2,21	1,34	0,52	0,24	4,3	1,88	
AE44	4,0	0,27	0,0014	2,13	1,08	0,54	0,19	3,9		
AEX 440	3,8	0,25	0,0009	2,01	1,1	0,51	0,09	3,7		0,60
AEX 441	3,9	0,22	0,0007	2,06	1,21	0,53	0,17	4,0		1,07
AEX 442	3,8	0,13	0,0029	1,91	1,12	0,49	0,15	3,7		1,50
AEX 444	3,8	0,22	0,0009	2,43	1,4	0,62	0,25	4,7		4,26

3.2 Tensile Properties and Hardness

Figure 2 shows TYS and UTS at room temperature as functions of the Ca and Sr content of the alloys. It is seen that both Ca and Sr have a positive effect on TYS. In the case of Ca, TYS increased from 143 MPa in the reference AE44 to 190 MPa with 2 wt% Ca added (20 % increase). This is partly due to Ca entering solid solution (solid solution hardening) [26,27] as also indicated by the Vickers hardness measurements, Fig. 3.

For AE44 + 4wt% Sr the TYS has risen to 175 MPa compared to 143 MPa for the reference AE44 (25 % increase). Sr is not expected to provide a significant contribution to the yield strength through solid solutions strengthening as the maximum solid solubility of Sr in Mg is 0.03 at% [27], the increase in strength is most likely a result of strengthening due to the precipitate dispersion. The Vickers hardness is not significantly influenced by strontium until a more or less continuous grain boundary phase adds considerable strength at high Sr additions. This hardening is accompanied by a dramatic fall in ductility as shown in Fig. 4. This may also explain the lower slope for the increase in strength (TYS and hardness) for Sr additions compared to Ca additions as strengthening from relatively large precipitates is not very effective.

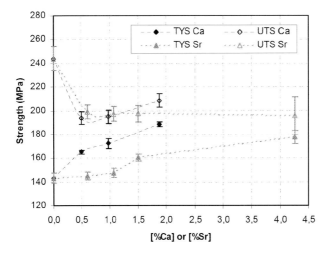

Figure 2: Tensile properties of the AE44+Ca and AE44+Sr alloys at RT as a function of wt% Ca or Sr. Error bars indicate ±1 standard deviation

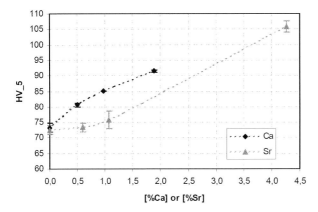

Figure 3: Vickers hardness (HV5) of the AE44+Ca and AE44+ Sr alloys as a function of wt% Ca or Sr. Error bars indicate ±1 standard deviation

Additions of Ca and Sr result in a noticeable loss in the ductility, Fig. 4. With the addition of as little as 0.5 wt% of Sr or Ca, only 2–3 % elongation at fracture is measured. Larger additions decrease the ductility even further. As will be discussed later, it is likely that phases deleterious to the ductility (grain boundary phases) are formed after only minor additions (0.5 wt%) resulting in the dramatic loss in ductility.

Figures 5 and 6 shows tensile yield strength as a function of temperature for the AEX and AEJ alloys, respectively. The tendency is quite clear, the higher the content of alloying elements the higher TYS at all temperatures. It is also noticed that the TYS vs. T curves are almost parallel for all alloys, including the AE44 reference alloys.

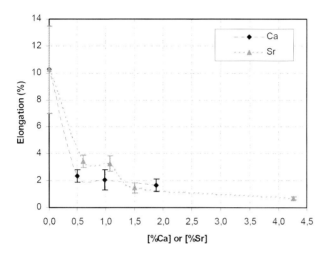

Figure 4: Elongation at fracture for the AE44+Ca and AE44+Sr alloys at RT as a function of wt% Ca or Sr. Error bars indicate ±1 standard deviation

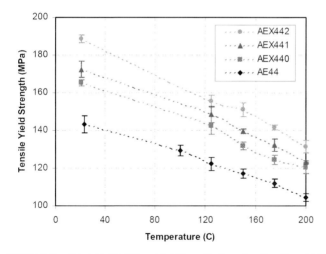

Figure 5: Tensile yield strength as a function of temperature for alloys AE44, AEX440, AEX441 and AEX442. Error bars indicate ±1 standard deviation

3.3 Impact

Figure 7 gives the impact properties of the AE44 alloy series with Sr and Ca additions, compared to AZ91D, AM60 and AE42. While typical impact energy for AE44 is 17.5 J, the addition of just 0.5wt% Sr or Ca reduces the impact energy absorption of the alloys to 4–6 J. Further additions of Sr and Ca result in even lower impact energies. The data indicate that, for equivalent content of alloying elements, the impact absorption energy of the AEJ alloys is slightly lower than that of the AEX alloys.

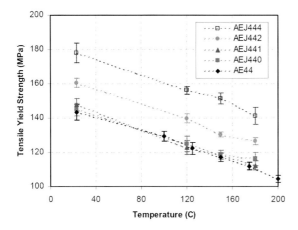

Figure 6: Tensile yield strength as a function of temperature for alloys AE44, AEJ440, AEJ441, AEJ442 and AEJ444. Error bars indicate ±1 standard deviation

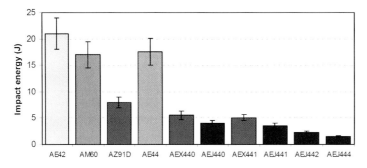

Figure 7: Impact properties of the AE44+Ca and AE44+ Sr alloys. Alloys AE42, AM50 and AZ91D are included for comparison. Error bars indicate ±1 standard deviation

3.4 Tensile Creep

Results from tensile creep tests at 150 °C/90 MPa, 175 °C/75 MPa and 200 °C/60 MPa are shown in Figures 8, 9 and 10, respectively. From Fig. 8 it is seen that at 150 °C/90 MPa all alloys obtain less than 0.1% creep strain within the testing period (350–500 h). The alloys are in practice indistinguishable, as the differences in both creep strain and creep rates are within the experimental accuracy of the method. The tests at 175 °C/75 MPa, Fig. 9, show no significant improvement due to additions of Sr to AE44, while the results at 200 °C/60 MPa, Fig. 10, indicate higher creep rates of the AEX alloys compared to standard AE44.

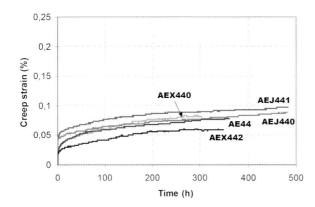

Figure 8: Creep strain at 150 °C/90 MPa for AE44, AEX440, AEX442, AEJ440 and AEJ441

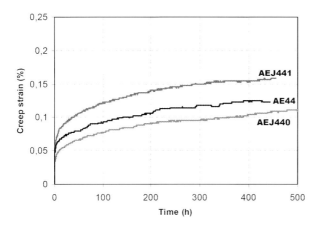

Figure 9: Creep strain at 175 °C/75 MPa for AE44, AEJ440 and AEJ441

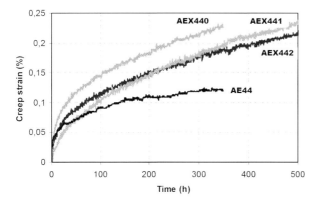

Figure 10: Creep strain at 200 °C/60 MPa for AE44, AEX440, AEX441 and AEX442

3.5 Stress Relaxation Compression

Figure 11 shows the remaining load after 100 h at 175 °C and 12 kN initial load for the AE44-type alloys, AEX alloys, AE42 and aluminium alloy A380. The addition of Sr or Ca to AE44-type alloys has a positive effect on the stress relaxation under compression. AEX441 has the best performance (72 %), significantly better than AE44 (55 %). This is still considerably less than A380 (90 %). The data indicate that Sr and Ca have a relatively similar effect on the stress relaxation behaviour. It is also interesting to notice that additions in excess of about 1 wt% Sr or Ca give no further improvements.

3.6 Corrosion

From Fig. 12 it is seen that Ca and Sr additions to AE44 up to 2 and 4%, respectively, do not have any significant effect on the corrosion properties as measured in the ASTM B117 salt spray test. In particular, it seems that Ca does not have the same adverse effect on corrosion properties of AE44 as it has on other alloys such as AM50. Also, the corrosion properties of AE44 appear to be unaffected by additions of Sr.

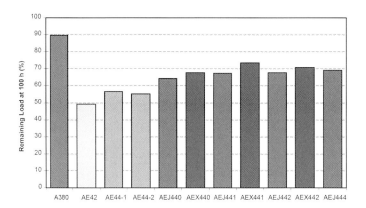

Figure 11: Remaining load after 100 h at 175 °C/12 kN initial load. Alloys AE42 and aluminium alloy A380 are shown for comparison

3.7 Microstructure

Typical optical micrographs for the alloys AE44, AEJ440 and AEJ444 are shown in Fig 13. For AE44, there is an almost continuous network of grain boundary phase. The grain boundary phase has previously been reported to be Al-RE types phases (eg. Al_3RE, $Al_{11}RE_3$, Al_2RE) [13]. This grain boundary phase has a lamellar-type morphology and there is also a small number of polygonal intermetallics. In AE44 there appears to be only three phases: the matrix, the grain boundary lamellar phase and a small number of polygonal precipitates. With the addition of 0.5 wt% strontium (AEJ440) there is a significant change of the microstructure. A new grain boundary phase appears (circled in Figure 13: Typical optical micrographs of (a) AE44, (b) AEJ440 and (c) AEJ44413 b), and additionally there is a larger number of polygonal precipitates. The microstructure has coarsened considerably. After the addition of 4 wt% strontium (AEJ444), the microstructure shows another significant change. The light coloured grain boundary phase present in AEJ440 is now the most dominating grain boundary phase and almost no lamellar phase can be seen. The number of polygonal shaped intermetallics has also increased and some of these particles are relatively large (circled particle in Figure 13c). The particles appear mostly in, or near the grain boundaries. The change in microstructure to large, coarse phases with the addition of Sr will result in the significant loss of impact strength and ductility.

From the EDS analysis, Fig. 14, the light grey phase seen in AEJ444 is likely to be AlSr (containing RE).

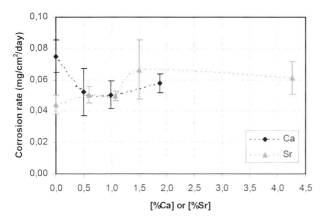

Figure 12: Salt spray corrosion rates of the AE44+Ca and AE44+Sr alloys as a function of wt% Ca or Sr. Error bars indicate ±1.0 standard deviation

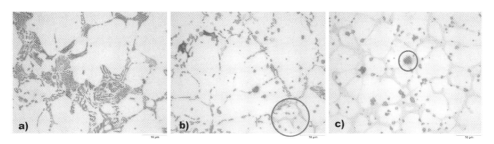

Figure 13: Typical optical micrographs of (a) AE44, (b) AEJ440 and (c) AEJ444

Figure 15 illustrates the change of microstructure from AE44 to AEX441 containing 1 wt% Ca. Although the continuous 3D network is still present, the lamellar structure of AE44 is partly replaced by polygonal shaped particles with relatively well-defined geometry. Since the ductility of AE44 is attributed to the lamellar structure of the grain boundary phases [13], it may be expected that this partial change to the more sharply defined precipitates is responsible for the drastic reduction in ductility. However, the polygonal shaped dispersoids are, as demonstrated, at least as effective in preventing creep arising from grain boundary sliding.

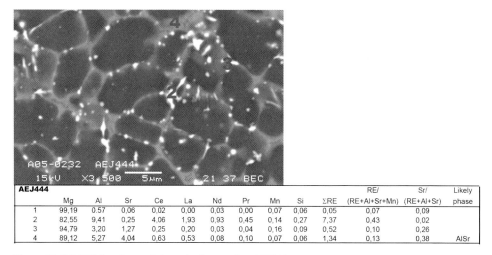

AEJ444	Mg	Al	Sr	Ce	La	Nd	Pr	Mn	Si	ΣRE	RE/(RE+Al+Sr+Mn)	Sr/(RE+Al+Sr)	Likely phase
1	99,19	0.57	0,06	0,02	0,00	0,03	0,00	0,07	0,06	0,05	0,07	0,09	
2	82,55	9,41	0,25	4,06	1,93	0,93	0,45	0,14	0,27	7,37	0,43	0,02	
3	94,79	3,20	1,27	0,25	0,20	0,03	0,04	0,16	0,09	0,52	0,10	0,26	
4	89,12	5,27	4,04	0,63	0,53	0,08	0,10	0,07	0,06	1,34	0,13	0,38	AlSr

Figure 14: SEM-EDS analyses of phases in die cast alloy AEJ444

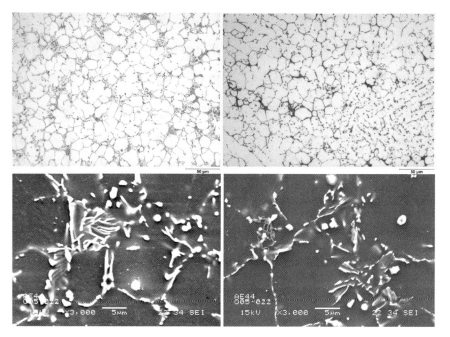

Figure 15: Optical micrographs and SEM-BEC micrographs of AE44 (left) and AEX441 (right)

The SEM-EDS analyses, Fig. 16, indicate at least two different phases in the grain boundary regions. These are Al_2RE or AlRE, possibly with tiny amounts of Ca (6,8,9,10), and one or more Ca-rich phase (1,2,3,4,5) also containing RE and Al. It is also interesting to notice that there seems to be an enrichment of La in the Ca-rich phases (1,2,4,5). Although the bulk alloy contains approximately 60% more Ce than La, there is more La than Ce in the Ca rich phases.

Also it is noticed that relative to the other RE elements, the fraction of Nd in the particles is considerably lower than in the mischmetal added.

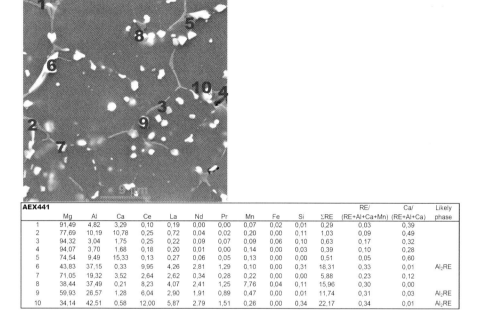

AEX441	Mg	Al	Ca	Ce	La	Nd	Pr	Mn	Fe	Si	ΣRE	RE/ (RE+Al+Ca+Mn)	Ca/ (RE+Al+Ca)	Likely phase
1	91,49	4,82	3,29	0,10	0,19	0,00	0,00	0,07	0,02	0,01	0,29	0,03	0,39	
2	77,69	10,19	10,78	0,25	0,72	0,04	0,02	0,20	0,00	0,11	1,03	0,09	0,49	
3	94,32	3,04	1,75	0,25	0,22	0,09	0,07	0,09	0,06	0,10	0,63	0,17	0,32	
4	94,07	3,70	1,68	0,18	0,20	0,01	0,00	0,14	0,00	0,03	0,39	0,10	0,28	
5	74,54	9,49	15,33	0,13	0,27	0,06	0,05	0,13	0,00	0,00	0,51	0,05	0,60	
6	43,83	37,15	0,33	9,95	4,26	2,81	1,29	0,10	0,00	0,31	18,31	0,33	0,01	Al$_2$RE
7	71,05	19,32	3,52	2,64	2,62	0,34	0,28	0,22	0,00	0,00	5,88	0,23	0,12	
8	38,44	37,49	0,21	8,23	4,07	2,41	1,25	7,76	0,04	0,11	15,96	0,30	0,00	
9	59,93	26,57	1,28	6,04	2,90	1,91	0,89	0,47	0,00	0,01	11,74	0,31	0,03	Al$_2$RE
10	34,14	42,51	0,58	12,00	5,87	2,79	1,51	0,26	0,00	0,34	22,17	0,34	0,01	Al$_2$RE

Figure 16: SEM-EDS analyses of phases in die cast alloy AEX441

4 Conclusions

In general, it is shown that modifications of alloy chemistry can lead to improvements of certain properties at the expense of others. Specifically, the results of this study indicate that additions of Ca or Sr to AE44 cause noticeable improvements of tensile yield strength and hardness. The improvements of remaining load in stress relaxation compression tests are significant, but these seem not to be fully reproduced in the tensile creep tests. The corrosion properties seem not to be affected by the additions of Ca or Sr.

On the negative side, the ductility and impact energy absorption are drastically reduced due to even 0.5wt% of either Sr or Ca. While there are basically two different phases in AE44, the matrix and the lamellar type grain boundary phase consisting of Al$_2$RE, Al$_3$RE and Al$_{11}$RE$_3$, there are additional Ca and Sr-rich phases formed in the grain boundaries of the AEX and AEJ alloys. Thus overall, the Hydro Magnesium-patented, elevated temperature alloy AE44 remains an excellent choice for achieving an outstanding balance of properties.

5 Acknowledgements

The following persons are acknowledged for contributing to this study: Per R. Gjertsen, Vidar Sjøberg, Sten Bjørneboe, Jan Ivar Skar, Eva Thorjussen, Arve Wilhelmsen, Kari S. Andersen.

6 References

[1] Beck, A., *The Technology of Magnesium and its Alloys*, F.A. Hughes & Co. Ltd, London UK, 1940. p. 250–252
[2] Beck, A., Ibid. p. 286
[3] Emley, E.F., *Principles of Magnesium Technology*, Pergamon Press, London UK, 1966. p. 308–341
[4] Unsworth, W., Intl. J. of Materials and Product Technology, 1989, *4(4)*, p. 359–378
[5] Nelson, K.E., *Proc. 6th SDCE, Paper no. 13*, 1971
[6] Foerster, G.S., *Proc. 7th SDCE, Paper no. 9372*, 1973
[7] Foerster, G.S., *8th SDCE Intl. Die Casting Exp. & Congress, Paper no G-T75-112*, 1975
[8] Mercer II, W.E., *SAE Tech. Paper 900788*, 1990
[9] Waltrip, J.S., *Proc. 47th IMA Conf., Cannes, France*, 1990, p. 124–129
[10] Aune, T; Westengen, H; Ruden, T., *SAE Techn. Paper 94077*, 1994
[11] Druschitz, A.P; Showalter, E.R; McNeill, J.B; White, D.L., *Magnesium Technology 2002, Ed. H.I.Kaplan, TMS-AIME, Warrendale, PA*, 2002, p. 117–122
[12] Berkmortel, J; Hu, H; Kearns, J.E; Allison, J.E., *SAE Techn. Paper 2000-01-1119*, 2000
[13] Bakke, P; Westengen, H., *Proc. 2nd Intl. Light Metals Technology Conference 2005, June 8-10, St. Wolfgang, Austria, Ed: H. Kaufmann, ISBN-3-902092-03-3*, 2005, p. 57–62
[14] Bakke, P; Westengen, H; Wang, G; Jekl, J; Berkmortel, R., *13th Magnesium Abnehmer Seminar, Aalen, Germany, Sept.* 2005
[15] Westengen, H; Bakke, P; Skar, J.I; Gjestland, H., *Proc. 63th IMA Conf., Beijing*, 2006
[16] Li, N; Osborne, R; Cox, B; Penrod, D., *SAE Techn. Paper 2005-01-0337*, 2005
[17] Aragones, J; Goundan, K; Kolp, S; Osborne, R; Ouimet, L; Pinch, W., *SAE Techn. Paper 2005-01-0340*, 2005
[18] Norsk Hydro, British Patent GB1,163,200, Filed 30-01-1967
[19] Luo, A; Balogh, M.P; Powell, B.R., *SAE Techn. Paper 2001-01-0423*, 2001
[20] Pekguleryuz, M; Baril, E., *Magnesium Technology 2001, Ed.: J. Hryn, TMS-AIME, Warrendale PA*, 2001, p.119–125
[21] Labelle, P; Fischersworring-Bunk, A; Baril, E., *SAE Techn. Paper 2005-01-0729*, 2005
[22] Koike, S; Washizu, K; Tanaka, S; Baba, T; Kikawa, K., *SAE Techn. Paper 2000-01-1117*, 2000
[23] Dargusch, M.S; Pettersen, K; Bakke, P; Nogita, K; Bowles, A.L; Dunlop, G.L., Intl. Journal of Cast Metals Research, 2004. *17(3)*, p. 170–173
[24] Bakke, P; Westengen, H; Brassard, C., *Proc. Intl. Symposium on Magnesium Technology in the Global Age, Oct. 1-4, 2006, Montréal, Canada, Ed: M. Pekguleryuz, CIM*, 2006
[25] Aune, T.K; Ruden, T., *SAE Tech. Paper 920070*, 1992
[26] Beck, A., Ibid. p. 137–138
[27] Nayeb-Hashemi, A.A; Clark, J.B., *Phase Diagrams of Binary Magnesium Alloys*, ASM International, Metals Park, OH, USA, 1988

Phase Composition and Creep Behavior of Mg-Rare Earth-Mn Alloys with Zn Addition

B. Smola, I. Stulíková, J. Pelcová, N. Žaludová
Faculty of Mathematics and Physics, Charles University, Prague, Czech Republic

1   Introduction

Creep resistant magnesium alloys with Y, rare earths (RE) and the small addition of Zr (WE54 and especially WE43) are successfully used in aeronautical and automobile industry for the applications up to about 250 °C [1]. Recently developed Mg alloys with ~ 1 wt.% Sc, ~ 4 wt.% RE (Y, Gd, Nd) and ~1 wt.% Mn [2–8] are superior to the WE alloys and exhibit high creep resistance at temperatures over 300 °C. It is due to the high thermal stability of tiny Mn_2Sc phase discs parallel to basal plane of the α-Mg matrix and precipitation of very thin hexagonal basal plates (THBP) of the RE and Mn containing phase at elevated temperatures (~ 250–275 °C) [3, 9]. Both these basal precipitates are very effective inhibitors of cross slip of basal dislocations and non-basal slip. Thin basal plates in a dense arrangement can be also effective obstacles for climb of basal dislocations [6]. Less expensive alternative to Sc in Mg-RE based alloys can be Zn, which also forms intermetallic phases with Mg and/or RE as plates on basal planes of α-Mg matrix [10]. Recently THBPs with the same structure as in Mg-Sc-RE-Mn alloys were reported to form in Mg-Zn-Ca [10] and Mg-RE-Zn-Zr alloys [11]. Structure and properties of binary Mg-Zn and ternary Mg-Zn-RE (inclusive Y) alloys (mainly with higher Zn and RE content) have been extensively investigated for a long time; see e.g. [12–15].

In this contribution we report the results on the investigation of creep behavior, mechanical properties and phase development during isochronal and isothermal annealing of Mg alloys with relatively low RE, Zn and Mn content, namely Mg4Y1Zn1Mn, Mg3Y2Nd1Zn1Mn and Mg4Ce1Zn1Mn (nominal wt.%).

2   Experimental Details

Alloys investigated were squeeze cast under a protective gas atmosphere (Ar + 1%SF_6). The composition of alloys is listed in Table 1.

The creep tests (mostly dip tests) were carried out in the temperature range 200–350 °C at the constant load 30 and 40 MPa. The tensile tests were undertaken at temperatures from room temperature to 400 °C at a constant strain rate $\approx 10^{-4}$ s^{-1}. The specimens were mounted in the preheated deformation machine furnace 15 minutes before the test started. This time was proved to be sufficient for specimen heating to deformation temperature (up to 400 °C).

The isochronal annealing response of relative electrical resistivity changes was determined in the range 20 to 510 °C on the as cast material. Isochronal annealing was carried out in steps of 30 K/30 min followed by quenching into liquid N_2. Heat treatment was performed in a stirred silicon oil bath up to 240 °C or in a furnace with argon protective atmosphere at higher temperatures. The resistivity was measured at 77 K after each heating step on the H-shaped specimens.

Relative electrical resistivity changes $\Delta\rho/\rho$ were obtained within an accuracy of 10^{-4}. The resistivity was measured by means of the dc four-point method with a dummy specimen in series. The influence of parasitic thermoelectromotive force was suppressed by polarity reversal.

Table 1: Composition of alloys studied (in wt.%)

Alloy	Y	Nd	Ce	Zn	Mn
Mg4Y1Zn1Mn	3.88	-	-	1.19	1.13
Mg3Y2Nd1Zn1Mn	3.22	2.12	-	1.22	1.45
Mg4Ce1Zn1Mn	-	-	3.72	0.92	1.16

The development of microstructure was followed using transmission electron microscopy (TEM) and electron diffraction (ED) (JEOL JEM 2000FX electron microscope). Corresponding mechanical properties were characterized by Vickers hardness HV3 measured at room temperature. The phase composition was qualitatively determined using Link AN 10000 microanalyser. The specimens prepared for TEM and hardness measurement were heat-treated using the same annealing procedure as for electrical resistivity.

3 Results and Discussion

The as cast alloys investigated exhibit typical dendritic structure with the mean grain size about 80 μm. The grains coarsen only slightly after the annealing at 500 °C for 8 h, see Fig. 1. The Mg4Y1Zn1Mn and Mg3Y2Nd1Zn1Mn alloys show very high creep resistance up to 350 °C comparable or better than that of high creep resistant MgScREMn alloys, see Figs 2 and 3. They are superior to the WE alloys concerning minimum creep rates. Time to rupture of the Y and Zn containing alloys is considerably longer than that of Y and Sc containing ones, cf. Fig. 3. Contrary to this, the minimum creep rate of Mg4Ce1Zn1Mn alloy increases rapidly with increasing temperature, cf. Fig. 2.

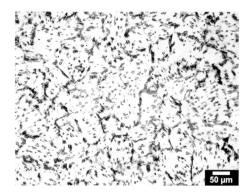

Figure 1: Structure of Mg4Y1Zn1Mn alloy after annealing at 500 °C for 16 h

Mechanical properties of the alloys studied, characterized by yield tensile stress (YTS) $R_{p0.2}$ and tensile strengths (UTS) R_m, are compared in Figs 4 and 5 respectively. Both YTS and UTS of Y containing alloys are stable up to 300 °C, contrary to the behavior of these parameters of the Mg4Ce1Zn1Mn alloy, which decrease with increasing temperature in the whole temperature range.

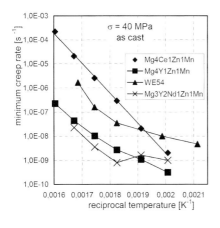

Figure 2: Temperature dependence of minimum creep rate of as cast alloys at 40 MPa compared to WE54 alloy

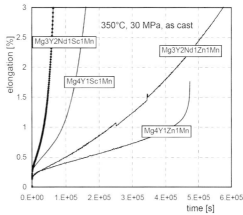

Figure 3: Creep curves at 350 °C and 30 MPa of MgYZnMn and MgYNdZnMn alloys. Note a comparison to MgYScMn and MgYNdScMn alloys

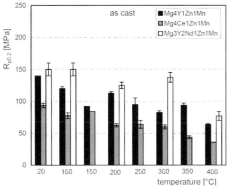

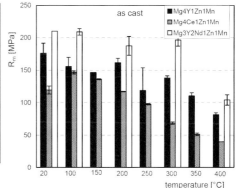

Figure 4: Temperature dependence of the yield tensile stress $R_{p0.2}$

Figure 5: Temperature dependence of the ultimate tensile stress R_m

Relative resistivity changes response to the isochronal annealing is plotted in Fig. 6 for the as cast alloys studied. Both Mg4Y1Zn1Mn and Mg3Y2Nd1Zn1Mn alloys exhibit slight increase of the resistivity in the range 150–270 °C and 150–240 °C respectively followed by decrease to the main minima (at 390 °C and 330 °C respectively). Also Mg4Y1Sc1Mn [5] and Mg4Y2Nd1Sc1Mn [8] exhibited similar slight resistivity increase in the same temperature range. The resistivity of the Mg4Ce1Zn1Mn alloy decreases slowly from 150 °C to 270 °C and then drops markedly to the minimum at 360 °C. Similar dramatic resistivity response to the isochronal annealing in the same temperature region was observed in Mg4Ce1Sc1Mn alloy [5] and was also reported for Mg-1.3 wt.% Ce in temperature range 200–325 °C [16].

The resistivity increases from the main minima for all alloys. It exceeds the initial value on annealing at temperatures over 450 °C in Y containing alloys. Contrary to this the value of the resistivity of the Mg4Ce1Zn1Mn stays under initial value. The main resistivity decrease and following increase can be ascribed to the precipitation of the equilibrium $Mg_{12}Ce$ phase and its subsequent dissolution at higher temperatures. Reconstruction of grain boundary (GB) eutectic, which consists of this phase and α-Mg matrix and is present in as cast state as the consequence of relatively low Ce solubility, can contribute to the resistivity increase observed.

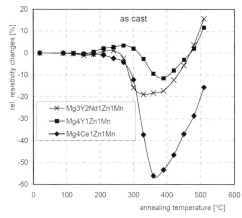

Figure 6: Isochronal annealing curves of relative resistivity changes measured at 77K

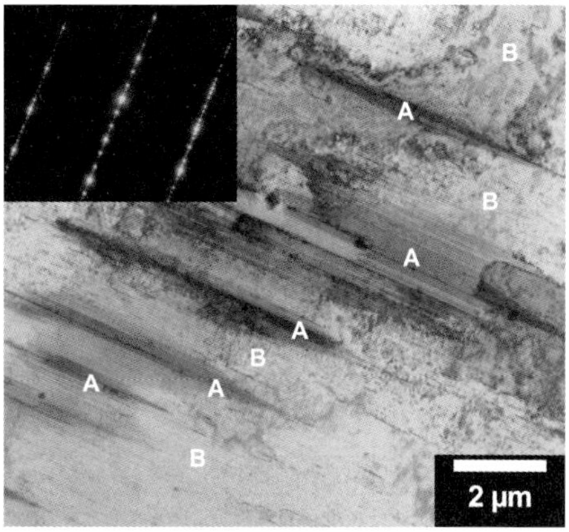

Figure 7: Microstructure of as cast Mg4Y1Zn1Mn alloy. Inset is electron diffraction pattern of 18R - A. B - α-Mg matrix with hexagonal basal platelets

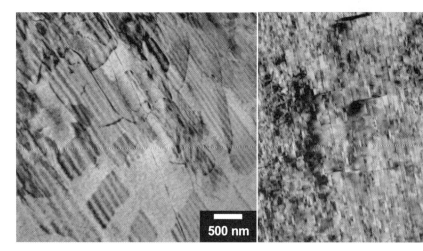

Figure 8: Split basal dislocations with SFs in the as cast Mg3Y2Nd1Zn1Mn alloy

Figure 9: THBPs and prismatic plates of stable phase in Mg3Y2Nd1Zn1Mn alloy annealed up to 330 °C

The microstructure development was monitored by TEM in both creep resistant alloys. The structure of as cast Mg4Y1Zn1Mn alloy, shown in Fig. 7, consists of two phases (A and B), namely α-Mg matrix (B) with the dense arrangement of THBPs (the same structure as reported for MgREScMn [3, 5, 8] and MgZnRE [11, 17] alloys) and plate-like 18R long period stacking (LPS) structure, e.g. [15, 18]. Only relatively high density of split basal dislocations was found in the as cast Mg3Y2Nd1Zn1Mn alloy, cf. Fig. 8.

The main resistivity decrease in the Mg3Y2Nd1Zn1Mn alloy can be attributed to: (i) the increasing content and/or ordering of Zn and RE atoms in stacking faults (SF) that leads to the transformation of SF into THBPs [17] and (ii) precipitation of prismatic plates of the phase isomorphic to the stable phase observed in WE alloys; see Fig. 9 and cf. [8]. The latter precipitation is responsible for a slowing down resistivity increase up to 390 °C. Prismatic plates were still observed after annealing up to 420 °C together with small oval particles of Mn and Y containing phase, which persist up to 480 °C. The density of THBP decreases; images of planar defects parallel to basal plane can be mostly interpreted as those of SFs.

The resistivity decrease to the minimum at 390 °C and subsequent increase can be ascribed to the similar processes in Mg4Y1Zn1Mn alloy, namely transformation of SFs to THBPs, enhanced formation of 18R LPS phase and precipitation of the stable phase isomorphic with the $Mg_{24}Y_5$ phase. These processes are not so pronounced in this alloy as the phases involved are present already after squeeze casting (the stable phase in the form of GB eutectic). The precipitation of small oval particles (similar as in the Mg3Y2Nd1Zn1Mn alloy) containing Mn and Y slows down the resistivity increase from the minimum, cf. Fig. 10.

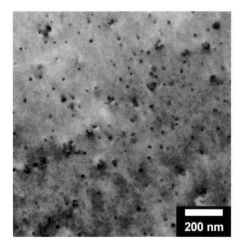

Figure 10: Mn and Y containing oval particles in Mg4Y1Zn1Mn alloy annealed up to 480 °C

The structure development observed in the Y containing alloys manifest clearly that the excellent high temperature creep resistance and stability of YTS and UTS up to 300 °C is due to both the low SF energy (which inhibits cross slip of basal dislocations) and the dense arrangement of THBPs (which are very effective obstacles to the non-basal slip). Hardness HV3 response of Mg3Y2Nd1Zn1Mn alloy to the same isochronal annealing as in resistivity measurement was insignificant. This reflects very low effect of basal plates with the high aspect ratio on the flow stress of basal dislocations at room temperature [19].

4 Conclusions

The results of the present investigation can be summarized as follows:

The addition of ~ 1 wt. % Zn to the Mg-Y(Nd)-Mn alloys lowers significantly stacking fault energy, enhances the formation of thin hexagonal basal platelets and leads to the formation of 18R long period stacking structure phase.

This does not happen in the Ce containing alloy, most probably due to the low solubility of Ce in Mg and weak affinity of Ce and Mn.

The low stacking fault energy and precipitation of very thin hexagonal plates parallel to basal plane of the α-Mg matrix secures outstanding high temperature creep resistance and stability of mechanical properties up to 300 °C.

5 Acknowledgement

Financial supports by Czech Science Foundation (project 106/06/0252) and by Ministry of Education of Czech Republic (the research program MSM 0021620834) are gratefully acknowledged.

6 References

[1] Polmear, I.J., Mater. Sci. Technol. 1994, *10*, 1–16.
[2] Stulíková, I., Smola, B., von Buch, F., Mordike, B.L., Mat.-wiss. u. Werkstofftech. 2001, *32*, 20–24.
[3] Smola, B., Stulíková, I., von Buch, F., Mordike, B.L, Mat. Sci.Eng. A 2002, *324*, 113–117.
[4] Stulíková, I., Smola, B., von Buch, F., Mordike, B.L., Mat.-wiss. u. Werkstofftech. 2003, *34*, 102–108.
[5] Smola, B., Stulíková, I., Pelcová, J., von Buch, F., Mordike, B.L., Z. Metallkde. 2003, *94*, 553–558.
[6] Smola, B., Stulíková, I., Pelcová, J., Mordike, B.L, J. Alloys and Comp. 2004, *378*, 196–201.
[7] Mordike, B.L., Stulíková, I., Smola, B., Metall. Mater. Trans. A 2005, *36A*, 1729–1736.
[8] Stuliková, I., Smola, B., Pelcová, J., Vlach, M., Mordike, B.L., Z. Metallkde. 2005, *96*, 821–825.
[9] Smola, B., Stuliková, I., Pelcová, J., von Buch, F., Mordike, B.L., in: *Magnesium Alloys and their Applications* K.U. Kainer (Ed.), Wiley-VCH Verlag, Weinheim 2000, p. 92–97.
[10] Nie, J. F., Muddle, B.C. Scripta Mater. 1997, *37*, 1475–1481.
[11] Ping, D.-H., Hono, K., Nie, J.F., Scripta Mater. 2003, *48*, 1017–1022.
[12] Padezhnova, E.M., Mel'nik,E.V., Miliyevskiy, R.A.,Dobatkina, T.V., Kinzhibalo, V.V., Russian Metallurgy (Metally) (Engl. Transl.) 1982, *4*, 185.
[13] Luo, Z.P., Zhang, S.Q., J. Mater. Sci. Lett. 1993, *12*, 1490–1492.
[14] Wei, L.Y., Dunlop, G.L., Westengen, H., Metall. Mater. Trans. A 1995, *26A*, 1947–1955.
[15] Itoi, T., Seimiya, T. Kawamura, Y., Hirohashi, M., Scripta Mater. 2004, *51*, 107–111.
[16] Omori, G., Matsuo, S., Asada, H., J. Japan Inst. Metals 1973, *37*, 677–682.

[17] Suzuki, M., Kimura, T., Koike, J., Maruyama, K., Mater. Sci. Eng. A 2004, *387-389*, 706–709.
[18] Luo, Z.P., Zhang, S.Q., J. Mater. Sci. Lett. 2000, *19*, 813–815.
[19] Nie, J.F., Scripta Mater. 2003, *48*, 1009–1015.

Investigation of Mg-Al-Sr Phase Equilibria

A. Janz[1], J. Gröbner[1], D. Mirković[1], M. Medraj[2], Jun Zhu[3], Y. A. Chang[3] and R. Schmid-Fetzer[1]
[1]Clausthal University of Technology, Institute of Metallurgy, Clausthal-Zellerfeld, Germany
[2] Mechanical Engineering Department, Concordia University, Montreal, Canada
[3]Department of Materials Science and Engineering, University of Wisconsin, Madison, USA

1 Introduction

Special automotive applications, such as powertrain components or engine blocks, require sufficient creep resistance at elevated temperatures. For these elevated temperature applications new alloys were developed by using additions of rare earth (RE) elements or Ca and Sr.

The purpose of this work is to generate a comprehensive and consistent thermodynamic description of the phase equilibria based on the new experimental data obtained in this work together with the bulk of, partly re-assessed, experimental literature data. This is an important basis for a purposeful alloy development, enabling solidification calculations and the understanding of microstructures of promising alloys for various applications.

2 Experimental Data

2.1 Experimental Data from the Literature

Early experimental information of the ternary Al-Mg-Sr system is from several publications of the group of Makhmudov and coworkers [1–5]. Although these papers on the ternary Al-Mg-Sr system all originate from the same group, several discrepancies can be observed.

A critical evaluation and thermodynamic calculation of the related binary subsystems was given by [6]. The ternary literature was mentioned, but the ternary phase diagram was only extrapolated from binary data using the quasi-chemical model for the liquid phase. The key feature of this ternary system, that is the formation of substantial ternary solid solutions or compounds, was not taken into account. The calculated phase equilibria and triangulation of the ternary system does not follow those determined experimentally by Makhmudov et al. [1–5].

In one of the commercial Mg-Al-Sr alloys, AJ52x, investigated by Baril et al. [7] precipitates were found and claimed to be a ternary phase with unclear stoichiometry, tentatively named Al3Mg13Sr or Mg68.3-Al17-Sr14.7 (wt.%).

More recent experimental results are presented by Parvez et al. [8]. These data are carefully re-evaluated in cooperation with that group, based on the complete raw experimental information, and included in the present work. Part of these data required a substantial re-interpretation as detailed later.

In an investigation of phase equilibria in the Mg-rich corner the results of five samples are presented by Cao et al. [9]. Information was given on the primary phases observed and on the results of three samples annealed at 400 °C and investigated with SEM/BSE and EPMA.

2.2 Experimental Work

Within this work, seven new key samples were selected on the basis of preliminary thermodynamic calculations to provide relevant missing information on the Mg-Al-Sr phase equilibria. One sample should represent a commercial alloy („AJ62") in the Mg-rich corner, the other six samples were placed on the section Mg65Al35-Sr (wt.%) to investigate three-phase equilibria and the ternary solubilities of the involved phases and to get additional information like liquidus data and the primary crystallizing phase.

Table 1: List of Samples prepared in this work

sample no.	sample composition (wt. %)
C1	Mg92Al6Sr2
C2	Mg58.5Al31.5Sr10
C4	Mg45.5Al24.5Sr30
C5	Mg39Al21Sr40
C6	Mg26Al14Sr609
C7	Mg18.2Al9.8Sr72

After thermal analysis the samples were prepared metallographically. Care must be taken of the reactivity of the samples with oxygen which increased drastically with Sr-content. Therefore ethanol had to be used for grinding and polishing with strictly limited time. As a consequence, not all scratches in the micrographs could be removed, as will be shown later.

The microstructure was investigated with scanning electron microscopy mostly using back scattered electrons (SEM/BSE) and local compositions were analyzed with electron probe microanalysis (EPMA) using a CAMECA SX100 (Cameca, France).

Phase identification was based on local chemical composition as measured by EPMA. Detailed analysis and interpretation of the solidification sequence of these microstructures will be given in the discussion section, considering also the results of the thermodynamic calculations.

The ternary phase τ was clearly identified with an approximate content of ~13.5 wt.% Sr based on EPMA data, thus differing significantly from the γ-$Mg_{17}Al_{12}$ phase. The EPMA measurement of τ gives 46.5 wt.% Mg and 40 wt.% Al. This measured Mg/Al ratio of ô is very similar to that of γ-$Mg_{17}Al_{12}$. The fact that τ is a distinct ternary phase and not the suspected ternary solubility of Sr in γ-$Mg_{17}Al_{12}$ is further compounded by the marked contrast in the SEM/BSE micrographs in Fig 4, indicating a two-phase structure of τ + γ-$Mg_{17}Al_{12}$.

All of the original experimental data from samples prepared in Montreal, including unpublished SEM and EPMA data, were used for this work. The data which were partly previously published [8] were critically reassessed in detail. This included both the raw XRD-patterns as well as the raw thermal analysis DSC signal curves. The entity of this DSC data was reassessed with the objective to obtain both higher accuracy and proper information on the type of reaction. Some samples had to be dismissed due to inconsistencies between heating and cooling cycles.

3 Results and Discussion

The good agreement between the observed primary crystallizing phases and the calculation is noted. Generally, Al_4Sr is the dominating phase in the liquidus projection in this system, as shown in Fig. 1 . For Mg-rich alloys the primary fields of (Mg) and $Mg_{17}Sr_2$ are also supported by experimental data.

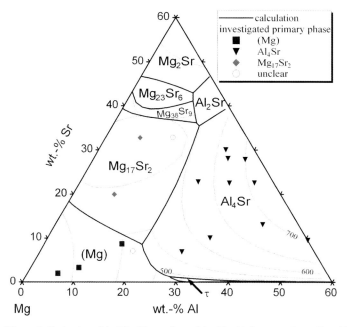

Figure 1: Projection of the Liquidus surface of the Mg-Al-Sr system from 60 to 100 wt.% Mg. Superimposed is experimental information on the primary phase

The isothermal section at 400 °C in conjunction with the complete phase diagrams of the edge binary systems is presented in Fig. 2. Three-phase fields are highlighted in grey shade and the substantial ternary solubilities of the binary phases are displayed with thick black lines. Calculated equilibria > 70 wt.% Sr are not supported by ternary experimental data. At 400 °C the τ phase, which forms below 477 °C, is present.

The samples C2-C7 in this work (Table 1) have been placed on the section Mg65Al35-Sr. The corresponding partial vertical section up to 50 wt.% Sr is given in Fig. 3. Superimposed are also the DSC results [8] of samples with compositions close to this section. A good agreement of the calculation with the measured liquidus data up to 50 wt.% Sr is noted. Also the invariant reactions are well reproduced by the thermodynamic modeling, such as the peritectic formation of the τ phase at 477 °C.

The thermodynamic modeling at more than 70 wt.% Sr is only partly supported by the experimental data. The measured ternary solubilities of the binary phases, especially along the Mg_2Sr-Al_2Sr section, and the accepted Sr-Al and Sr-Mg binary descriptions essentially determine the thermodynamic calculation in the Sr-rich corner. There is not much room to shift the calculated reaction temperatures by variations in the modeling. This disagreement above

70 wt.% Sr might be due to the existence of a ternary phase in this Sr-rich area as mentioned in [5]. Since the Sr-corner is not in the main focus of this work, this problem will not be further addressed.

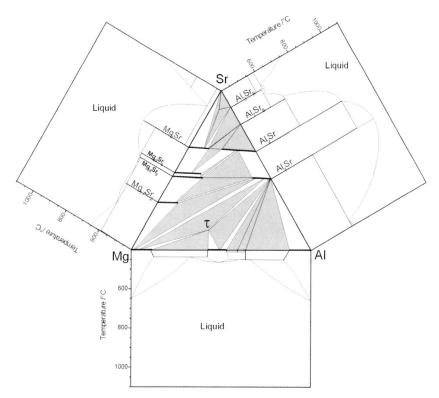

Figure 2: The calculated isothermal section of the Mg-Al-Sr system at 400 °C and the edge binary phase diagrams

The distinction between the phases τ and γ-$Mg_{17}Al_{12}$ by mass contrast is demonstrated in sample 08 (Fig. 4). After the primary solidification of the Al_4Sr-bars a secondary monovariant peritectic reaction, L + Al_4Sr = τ is indicated by the thermodynamic calculation. This equilibrium reaction occurs over the narrow temperature range of 473 to 460 °C, producing about 36 % of τ phase while nibbling off Al_4Sr from the primary amount of 42 % down to 35 %. This peritectic reaction type is in perfect agreement with Fig. 4, showing a peripheral rim of τ around all Al_4Sr crystals. This peritectic reaction ends at U6, L + Al_4Sr = τ + γ, where the equilibrium solidification ends in a production of a large amount of γ-$Mg_{17}Al_{12}$ and some τ.

This calculation is in excellent agreement with the microstructure in Fig. 4. Only a very small amount of Al_4Sr needs to be consumed in this reaction at 460 °C, L + Al4Sr = τ + γ, and this is probably the reason for the near-equilibrium completion of this reaction. By contrast, a Scheil calculation for that alloy predicts roaming of a residual liquid down to E4, where 7 % of (Mg) should form, in addition to the large amount of γ. A small amount of such eutectic (Mg)-precipitates might be present in the sample, e.g. in the area between the two marks of γ; the black spots above that area in Fig. 4 are holes.

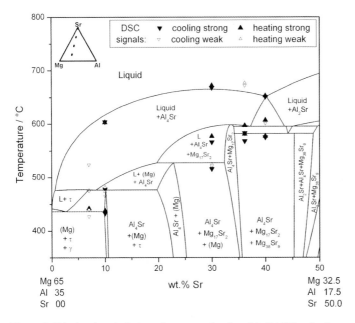

Figure 3: Calculated vertical phase diagram section from Mg65Al35 to the Sr corner compared with experimental data from DSC analysis

Table 2: Invariant four-phase reactions involving liquid phase in the ternary Al-Mg-Sr system.

Type	T/°C calculated	Reaction	T/°C measured	
			This work	[2, 4]
U1	611	L + Mg2Sr = Al2Sr + Mg23Sr6	619	610 ?
U2	606	Mg23Sr6 + Al2Sr = Mg38Sr9 + L		
U3	595.4	L + Mg38Sr9 = Mg17Sr2 + Al2Sr		
E1	595.3	L = Mg17Sr2 + Al2Sr + Al4Sr	607/597	
D	555	Al3Sr8 + â Sr = áSr + L		
U4	527	L + Mg17Sr2 = (Mg) + Al4Sr	521/519/527/524/523	
U5	477	L + Al4Sr = τ + (Mg)	473/476	
U6	460	L + Al4Sr = τ + γ	457/453	
E2	450	L = Al4Sr + (Al) + â	454/452	445
E3	449	L = Al4Sr + β + γ		
E4	436	L = (Mg) + γ + τ	438/442	438
U7	427	L + Al2Sr = Mg2Sr + Al7Sr8		
U8	423	L + Al7Sr8 = Mg2Sr + Al3Sr8		
E5	399	L = Al3Sr8 + Mg2Sr + áSr		

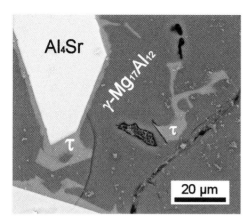

Figure 4: Scanning electron micrograph (SEM/BSE) of Mg30Al46Sr24 (sample 08) after slow cooling in DSC

4   Conclusion

The discrepancies in the experimental data in the ternary system Al-Mg-Sr concerning the existence of ternary phases or solubilities reported by Makhmudov et al. [1–5] and Parvez et al. [8] could be solved by investigating six key samples, combined with a more detailed evaluation of the complete raw experimental information of [8]. Substantial mutual solid solubilities of Al in binary Mg-Sr compounds and of Mg in binary Al-Sr compounds exist. The suspected Sr solubility in γ-$Mg_{17}Al_{12}$ is negligible. Only one distinct ternary phase τ, $Al_{38}Mg_{58}Sr_4$, exists; its composition in wt.% is Al37Mg50Sr13.

The thermodynamic modeling of the ternary phase equilibria is well supported by experimental data in the partial system Mg-Al-Al_2Sr-Mg_2Sr. A comprehensive and concluding experimental study and thermodynamic calculation of Al-Mg-Sr phase equilibria by our group is in preparation [10], with a meticulous study of the most complex equilibria involving the (Mg) phase.

5   Acknowledgements

The authors are grateful to Elhachmi Essadiqi and Jian Li of CANMET, Ottawa, Canada, for providing additional EPMA and SEM data. YAC wishes to thank NSF through the FRG Grant No. DMR-0309468 and Wisconsin Distinguished Professorship for financial support. This study is supported by the German Research Foundation (DFG) in the Priority Programme „DFG-SPP 1168: InnoMagTec" under grant no. Schm 588/27(29).

6   References

[1]   M.M. Makhmudov et al., Dokl. Akad. Nauk Tadzh. SSR, 1980, 23(1), 25-28
[2]   M.M. Makhmudov et al., Dokl. Akad. Nauk Tadzh. SSR, 1981, 24(7), 435-438
[3]   M.M. Makhmudov et al., Izv. Akad. Nauk SSSR, Met., 1981 , (6), 209-212

[4] M.M. Makhmudov et al., Russ. Metall., 1982 , (1), 122-124
[5] M.M. Makhmudov et al., Zavods. Lab., 1982, 48, 61-62
[6] P. Chartrand and A.D. Pelton, J. Phase Equilibria, 1994, 15 (6), 591-605
[7] E. Baril et al., JOM, 2003, 55, 34-39
[8] M.A. Parvez, M. Medraj, E. Essadiqi, G. Dénès, J. Alloys Compounds, 2005, 402, 170–185
[9] H. Cao, J. Zhu, C. Zhang, K. Wu, N.D. Saddock, J.W. Jones, T.M. Pollock, R. Schmid-Fetzer and Y. A. Chang, Z. Metallkunde, 2006, 97, 422-428
[10] Janz, J. Gröbner, D. Mirkoviæ, M. Medraj, Jun Zhu, Y. A. Chang, R. Schmid-Fetzer: To be published.

Hot Chamber Die-Casting of Magnesium Alloys

R. Cibis, A. Kielbus
NTP Cibis Ltd., Kedzierzyn-Kozle, Poland
Silesian University of Technology, Katowice, Poland

1 Abstract

A reduction of fuel consumption by decreasing the weight of vehicles is one of the factors which contribute to the development of new light materials. The most popular process of fabrication of casts from magnesium alloys is high-pressure die casting with the application of cold and hot chamber machines. The technology of die casting by means of a hot chamber machine used in NTP Cibis Ltd. in Kędzierzyn-Koźle is described in the paper. It is an efficient and cost-effective process ensuring high quality and dimensional stability of casts made of magnesium alloys. A complete magnesium melting and casting system consists of a hot chamber die-casting machine, melting and casting furnaces, a liquid metal transfer tube and cover gas supply.

2 Introduction

High-pressure die casting (with the application of cold and hot chamber machines) is a popular production processes applied for magnesium alloys. The process, due to economic reasons, should provide for automatic delivery of a liquid material to a mould. It is much easier in the case of hot chamber casting [1, 2]. The cold chamber process consists of two stages. At the first stage, liquid metal is fed into an injection chamber and then (stage two) it is injected into a mould. Moreover, during this process, neither the chamber nor the piston are heated, which makes it difficult to ensure an appropriate casting temperature. Such problems are not experienced while applying the hot chamber method. During the process, the siphon is immersed in a liquid metal bath, as the result of which it is constantly filled with metal, which ensures the obtaining of the required casting temperature. Due to placing the injection mechanism directly in the bath, the casting process is faster and, consequently, more efficient. Hot chamber machines are more competitive than cold chamber machines for small-size details (weight from 2 to 3 kg) due to a shorter time of the cycle, during which the capacity of up to 400 injections per hour may be obtained [1, 2].

3 Magnesium Melting System for Hot Chamber Die-casting

The magnesium melting system for hot chamber die casting is composed of a heating-melting furnace, two transfer tubes, two foundry furnaces, a protective gas generation unit and two hot chamber die casting machines [3]. A diagram of the system is provided in Figure 1. The system of furnaces is presented in fig. 1.

For the sake of ensuring safety and reliability of the furnaces' operation, their ceramic filling is made of new, aluminium oxides based material which does react with melted magnesium.

Furthermore, the melting pots are made of two steel layers of which the internal one is resistant to the impact of molten magnesium and the external one carries and distributes heat very well. A modern system of heaters (owing to an appropriate arrangement of heating components) ensures uniform and quick heating of the melting pot and the metal held inside, with minimum heat losses. The multiplied system of temperature measurement and the doubled control system prevent the magnesium alloy from its overheating. The differences between a melting furnace and a foundry furnace are related to their shape and power. Temperatures of the furnace interior and the liquid metal are measured by means of two thermocouples connected to independent measuring lines. The temperature of metal is additionally measured with a third thermocouple connected to an independent control system. In the case of the main control unit failure, this additional system enables disabling of the furnace and prevents it from exceeding maximum temperatures. In the lower part of the furnace, there are electrodes signalling the melting pot's crack and leakage of liquid metal. Their activation causes disabling of the furnace and switching on an alarm.

The applied system of furnaces for melting magnesium alloys is presented in Fig. 2.

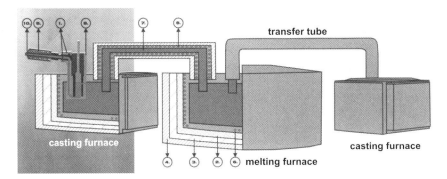

Figure 1: Magnesium melting system for hot-chamber die-casting

Figure 2: Furnace system for melting magnesium alloys

The casting machine can operate in three modes: manual, semi-automatic and automatic. The mould closure system consists of a closing mechanism, an ejecting mechanism, a mould height adjustment mechanism, cores mechanism and an elevating mechanism. The injection process in the Power Shot system enables flexible acceleration and deceleration of the shot piston movement as well as its complete halting. A view of a hot chamber die casting machine of the closing power of 260 tons is presented in fig. 3.

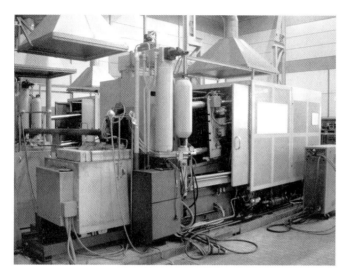

Figure 3: Hot-chamber die-casting machine

Due to the fact that the process takes place at the temperature of 630–690 °C, thus far exceeding the ignition temperature of magnesium, the whole liquid magnesium system is completely tight and the melting process takes place in the shielding gas atmosphere. The system generating shielding gases is composed of a mixer, a dosing unit and a distributor provided with a flow measuring system enabling precise dosing of gases to the furnace. Currently, as shielding gas, a 0.6 % mixture of SO_2 with dry air is applied.

4 Technology of Hot Chambers Die-casting of Magnesium Alloys

The process of manufacturing of components from magnesium alloys by means of the hot chamber method is based on the following pattern:
- pig sows of magnesium alloys after seasoning in a production hall are placed in a holding furnace; after attaining a temperature of ~150 °C and evaporating the whole moisture, they are placed in a melting furnace;
- in the melting furnace, the pig sows are melted and initial stabilisation of the molten metal temperature takes place; from the melting furnace, the liquid metal is delivered through electrically heated transfer tubes (based on the principle of inverted communicating vessels) to the foundry furnaces;
- in the foundry furnace, a siphon together with a shot piston is fully immersed in the liquid metal;

- the set values of injection parameters are entered;
- injection of liquid metal takes place – the shot piston forces the liquid metal through a channel in the siphon and through a nozzle to a mould;
- finished cast;
- in the course of the whole production cycle, permanent monitoring is conducted for temperatures in the individual zones as well as parameters of the casting machine operation (Fig. 4).

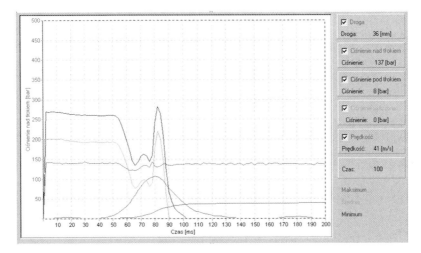

Figure 4: Monitoring of hot-chamber die casting parameters

5 Methodology of Cast Quality Tests in Industrial Condition

The quality of casts depends on numerous factors. Not all of them can be controlled, therefore, it is often difficult fully stabilize the entire process. In die casts of magnesium alloys, gas and shrinkage pores may appear, as well as other casting defects [4].

The methodology of cast's quality assessment in the conditions present at NTP Cibis Ltd. encompasses the following stages:

Figure 5: Cast flatness testing station

5.1 Preliminary Tests of the Cast Quality Conducted by the Machine Operator

Tests are conducted at two stages. Immediately after removing the cast from the mould and after trimming. In order to identify the defects, special masters are used.

5.2 Cast Flatness Tests

The casts flatness testing station is provided with six sensors responsible for measurements of the differences between the individual measuring points.

5.3 X-ray Cast Quality Tests

Tests are conducted on the MU 2000 X-ray defectoscope manufactured by YXLON International provided with a computer aided control system PXM 2500 in accordance with the ASTM E505 standard. Examples of the X-ray pictures are provided in Fig. 6.

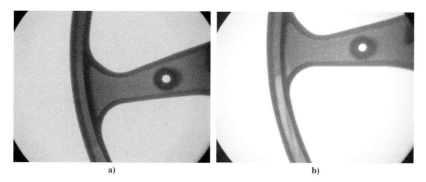

Figure 6: X-ray picture of a cast, a) good quality, b) with defects

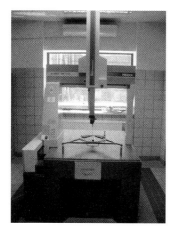
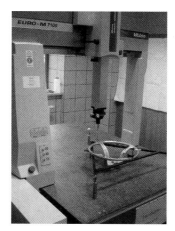

Figure 7: Three-dimensional measurement machine SPECTRUM 700

5.4 Measurement of Geometric Properties of the Cast

The measurements are conducted in a three-dimensional measurement machine SPECTRUM 700 of ZEISS (fig.7). The so-called critical dimensions settled with the customer are subject to control.

At customer's request, destructive examinations are conducted, including the measurement of porosity in sections, assessment of the microstructure and mechanical properties of casts. The tests are conducted at the Department of Material Sciences of the Silesian University of Technology.

6 Summary

The technology of hot chamber pressure die casting of magnesium alloys developed and applied in the NTP Sp. z o.o. company enables casting of a wide range of casts. The casts comply with all requirements specified by their recipients. As the charge material, alloys of the Mg-Al group are used, mostly AM50 , AM60, AM-lite and AZ91. Currently, research works are being carried out focused on the hot chamber pressure die casting of the Mg-Al alloys with an addition of rare-earth elements. The research methodology applied for the purpose of casts' assessment, as described in the paper, enables a precise evaluation of the quality of fabricated components. A continuous analysis (during casting) of the results obtained reduces defectiveness and enables an appropriate selection and subsequent modification of the casting process parameters.

7 Acknowledgements

The present work was supported by the Polish Ministry of Science and Higher Education under the research project No PBZ-KBN-114/T08/2004.

8 References

[1] Magnesium Die Casting Handbook (201), NADCA, Illinois, 1998.
[2] Avedesian M. M. and Baker H.: Magnesium and Magnesium Alloys. Metals Park, OH: American Society for Metals, 1999.
[3] NTP Cibis Ltd.- folders 2004.
[4] Walkington W.G.: "Die Casting Defects. Causes and solutions". North American Die Casting Association, Illinois U.S.A., 1997.

Microstructural Caracterization of Mg-Al-Sr Alloy

T. Rzychoń*, A. Kiełbus*, R. Cibis**
* Department of Materials Science, Silesian University of Technology, 40-019 Katowice, Poland
**NTP Sp. z o.o., 47-225 Kędzierzyn-Koźle, Poland

1 Introduction

Magnesium alloys with their weight advantage have unique application opportunities in the automotive industry. High-pressure die-casting is preferred over sand casting for its high productivity of complex near net shape. Although the Mg alloys used for die-casting, such as the AM (Mg-Al-Mn) and AZ (Mg-Al-Zn) series, posses excellent castability, their low creep strength restricts their application in components working above 120 °C [1,2]. Research and development in the 1980s led to the development of Mg-Al-RE (AE) alloy system, such as AE42, AE21 and AE52 [3]. Microstructure of AE42 die-casting alloy is composed of the lamellar phase $Al_{11}RE_3$, which dominates in the interdendritic eutectic regions of the alloy and small amount Al_2RE compounds. The $Al_{11}RE_3$ phase is unstable above 150 °C and decomposes into Al_2RE and $Mg_{17}Al_{12}$ phase. The creep strength decreases sharply with these phase changes [4]. There is need, therefore, for the development of new magnesium casting alloys with acceptable creep resistance and castability. Initial work in replacing RE additions had led to development of Mg-Al-Sr system [3]. Microstructure of these alloys consists of primary Mg dendrites and second phases in the interdednritic regions and depends on the amount of Al and Sr. The Al_4Sr morphology is lamellar or divorced-eutectic, $Al_3Mg_{13}Sr$ morphology is massive. The ratio of $Al_3Mg_{13}Sr$ phase to Al_4Sr is higher in AJ52x than in AJ62x. Trace quantities of $Mg_{17}Al_{12}$ particles are observed in Mg-Al-Sr alloys in AJ52 alloy. Czerwinski and Zielinska-Lipiec observed small amounts fine $Mg_{17}Sr_2$ plates [3,5-7].

2 Experimental

Ingots of AJ62 magnesium alloy containing 6 wt. % Al, 2 wt. % Sr and 0.3 wt. % Mn were investigated. Specimens for microstructure studies were mechanically polished using standard methods, etched with solution containing 0.7 ml H_3PO_4, 4g picric acid and 100 ml ethanol. The area fraction of phases were measured using the quantitative image analyzer „Metilo". The microstructure was characterized by optical microscopy (Olympus GX-70) and a scanning electron microscopy (Hitachi S3400N) equipped with an X-radiation detector EDS (VOYAGER of NORAN INSTRUMENTS). EDS analysis were performed with an accelerating voltage of 15 keV. X-ray diffractions (JEOL JDX-7S) with graphite monochromator were performed on the investigated material. The Cu-Kα X-ray radiation was used. TEM thin foils were obtained in the conventional way. Two final thinning steps were used; ion beam milling with argon or electropolishing using an electrolyte an electrolyte of 20 ml $HClO_4$ in 1000 ml ethanol. For microstructural characterization a JEM 2010 ARP transmission electron microscope equipped with an energy dispersive X-ray spectrometer (EDS) detector were used.

3 Results and Discussion

3.1 Microstructure of AJ62 Ingot

Typical microstructure of studied alloy is characterized by primary α-Mg dendrites, two phases in the interdendritic regions and globular particles, which are distributed inside the grains (Fig. 1). SEM observations indicates that interdendritic phase morphologies are lamellar eutectic and massive type (Fig. 2). The amount of lamellar phase was 3.7%, massive phase was 4.4% and globular particles below 1%, which were measured by quantitative metallography.

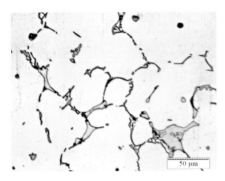

Figure 1: Optical micrograph of AJ62

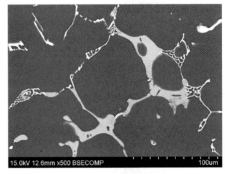

Figure 2: BSE image of AJ62

EDS analysis of the lamellar phase, which has the brightest contrast of the BSE image (Fig.3), shows enrichment of Al and Sr elements. The high amount of magnesium was probably caused by overlapping of the magnesium matrix with the particles. The chemical composition of massive phase indicates (Fig. 4), that it is ternary Mg-Al-Sr compound. Baril et al. reported this phase as $Al_3Mg_{13}Sr$ [3]. Globular particles are mainly composed of Al and Mn elements (Fig. 5).

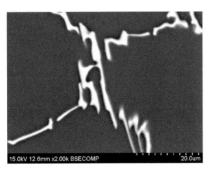 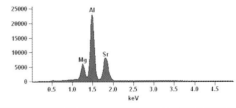

Chemical composition, at. %		
Mg	Al	Sr
15.59	63.43	20.98

Figure 3: The results of EDS analysis of lamellar phase in AJ62 alloy

TEM examinations provided the characteristic of precipitates which were detected by optical and scanning microscopy. The first, lamellar compounds formed eutectic with solid solution α-Mg were identified as Al_4Sr phase but chemical composition of this phase suggested that magnesium atoms can be present in this phase.

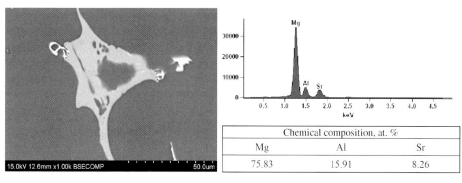

Figure 4: The results of EDS analysis of massive phase in AJ62 alloy

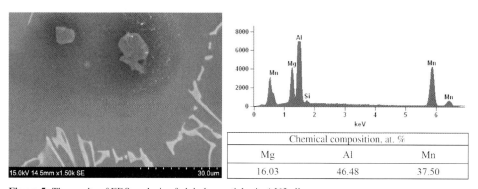

Figure 5: The results of EDS analysis of globular particles in AJ62 alloy

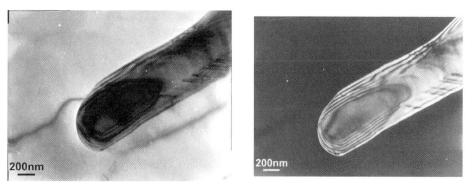

Figure 6: TEM image showing of lamellar Al_4Sr phase, bright and dark field, respectively

The massive phase formed partially divorced eutectic with solid solution α-Mg were not identified. Chemical composition of this phase suggested that it is probably $Al_3Mg_{13}Sr$ phase (Fig.10). The third type of precipitates was identified as Mn_5Al_8 phase (Fig.11).

In order to identify the existing phases in the alloys, XRD analysis was performed, the results of which are shown in Fig. 11. Two phases were positively identified in the pattern and as-

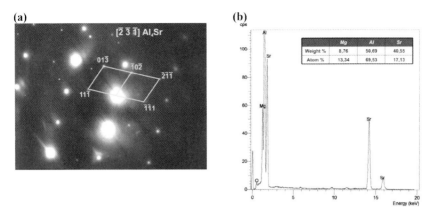

Figure 7: Diffraction pattern from both the matrix and lamellar phase (a) and EDS analysis from this phase (b)

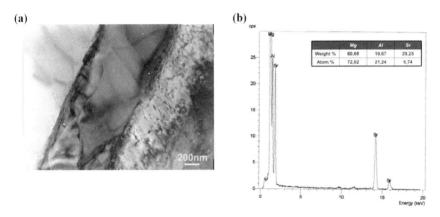

Figure 8: TEM image showing of massive phase (a) and EDS analysis from this phase (b)

sociated with the phases observed in microstructure: α-Mg and Al$_4$Sr. The reflections positions for magnesium, the major phase, are shifted to higher angles, a shift that is consistent with dissolved aluminum in this phase. The relative intensities of the different peaks of Mg in the XRD are different to the relative intensities obtainable from Mg powders. Due to this fact, analyzed sample has a textured effect and the observed intensities cannot be the only crystal structure function. The diffraction lines for Al$_4$Sr compound are shifted to lower angles. The distinctions between values of d_{hkl} observed in material and those found in ICDD files (07-0375) were 0.04 Å. It is consistent with increase the lattice parameters of this tetragonal phase. According to Zhang and Akiba [8], this change of lattice parameters is caused by the occupation of the Al sites by Mg atoms because the size of magnesium atoms is larger than that of aluminum atoms. Based on the EDS results of microanalysis of Al$_4$Sr phase it is possible. The Mg-Al-Sr phase, which has the largest area fraction determined by quantitative metallography, was not positively identified. The diffraction lines of Al$_8$Mn$_5$ phase was not observed in the pattern, due to small amount in alloy.

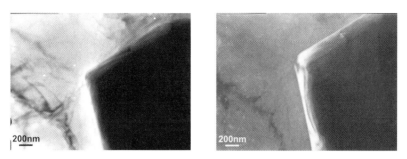

Figure 9: TEM image showing of Al_8Mn_5 phase, bright and dark field, respectively

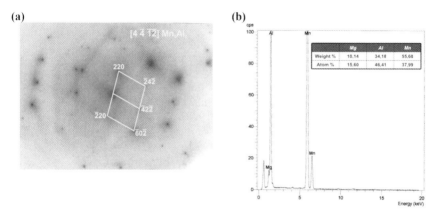

Figure 10: Diffraction pattern from both the matrix and Al_8Mn_5 particles (a) and EDS analysis from this phase (b)

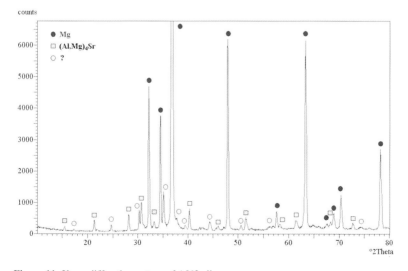

Figure 11: X-ray diffraction pattern of AJ62 alloy

4 Summary

The microstructure of AJ62 ingot consists of primary equiaxed α-Mg dendrites with the lamellar and massive phase formed partially divorced eutectic at the boundaries and globular particles Al_8Mn_5 phase, which are distributed inside the grain. The XRD and TEM results indicates, that the lamellar phase is rather $(Al,Mg)_4Sr$ than Al_4Sr. The dominates phase in this alloy is massive compound of $Al_3Mg_{13}Sr$, which consists islands of α-Mg, characteristic of a divorced eutectic reaction.

5 Acknowledgements

The present work was supported by the Polish Ministry of Education and Science under the research project No PBZ-KBN-114/T08/2004.

6 References

[1] E. Evangelista, S. Spigarelli, M. Cabibo, C. Scalabroni, O. Lohne, P. Ulseth, Materials Science and Engineering, A 410-411, 2005, 62–66.
[2] D. Wenwen, S. Yangshan, M. Xuegang, X. Feng, Z. Min, W. Dengyun, Materials Science and Engineering, A356, 2003, 1–7.
[3] E. Baril, P. Labelle, M.O. Pekguleryuz, JOM, November 2003, 34–39.
[4] B.R. Powell, V. Rezhets, M.P. Balogh, R.A. Waldo, JOM, August 2002, 34–38.
[5] M.O. Pekguleryuz, P. Labelle, D. Argo, E. Baril, Magnesium Technology 2003, 2003, 201–206.
[6] F. Czerwinski, A. Zielinska-Lipiec, Acta Materialia 53, 2005, 3433–3444.
[7] B. Jing, S. Yangshan, X. Shan, X. Feng, Z. Tianbai, Materials Science and Engineering A 419, 2006, 181–188.
[8] Q.A. Zhang, E. Akiba, Journal of Alloys and Compounds 360, 2003, 143–150.

A new Look at the Mg-Al-Zn Family – Focus on Tailoring Properties through Composition Control

A.L. Bowles, P. Bakke, H. Westengen
Competence Center Magnesium, Hydro Magnesium, Porsgrunn, Norway.

1 Introduction

The Mg-Al-Zn alloy system is well known, and while Mg-Al-Zn alloys have existed for many years, limited information is available for die casting alloys containing more than 1wt% Zn. Alloys from the Mg-Al-Zn system span a large range of properties from high ductility to high strength and high castability. With knowledge of the mechanical properties, microstructure, corrosion and castability of alloys from this system in relation to the chemical composition, new and improved alloys can be applied to a wide range of applications.

At moderate addition levels, the effect of both Al and Zn on the mechanical properties is quite similar. In general, the addition of either or both Al and Zn results in an increase in the strength and decrease in the ductility in Mg alloys [1]. Al and Zn both act as solid solution strengtheners, and Al and Zn form precipitates with Mg, giving the strengthening effect. In the as-high pressure die cast condition these precipitates form on the grain boundaries which results in the loss of ductility with the addition or either or both elements.

The addition of Al and/or Zn is beneficial for an alloys castability, with both elements improving the fluidity of alloys [2], leading to better mould filling). Another positive effect of additions of both Al and Zn is an improvement in the corrosion resistance [1]. In Al rich Mg-Al-Zn alloys, Zn segregates to the β phase and the size of the two-phase ($\alpha + \beta$) field increases (increasing the fraction of β phase). Alloys containing moderate levels of Zn are reported to have some castability problems, notably hot cracking [3]. High Zn (and binary Zn) alloys are amenable to age hardening (due to the formation of intermediate precipitates) and show improved creep properties over Mg-Al alloys [4].

To date the properties and applications of Mg-Al-Zn high pressure die cast alloys with more than 1wt% Zn have not been exploited. The present work forms the background for the study and mapping of the chemical composition, castability and mechanical properties of die cast Mg-Al-Zn alloys. It is the aim of this work to open the system for the exploitation of new alloys and applications.

2 Experimental

To produce alloys of the nominal compositions given in Table 1, ingots of AM20 and AM60 and pure Zn where alloyed together. As such, it should be noted that the Mn content is not optimized for the alloys. The alloys were high pressure die cast in a 420 ton Bühler SC42D Evolution machine in two different dies.

For standard tensile testing, a five-cavity plate/bar die (round tensile bar and creep bar, charpy bar and two corrosion plates) was used. To examine the cracking tendency and surface quality samples were cast in a ribbed box die [5].

Table 1: Nominal composition of alloys produced in wt%.

Zinc	Aluminium						
	2	4	6	8	10	12	14
0	X	X	X	X	X	X	X
1	X	X	X	X	X	X	X
2	X	X	X	X	X	X	X
3.5	X	X	X	X	X	X	X
5	X	X	X	X	X	X	X

Standard ⌀ 6 mm tensile bars were tensile tested according to ASTM B557. The tests were conducted with a strain rate of 0.03 min^{-1} until 0.5% strain, after which the strain rate was increased to 0.2 min^{-1}. Un-notched Charpy bars were tested according to ASTM E23. Salt spray corrosion testing was performed on the flat plate samples according to ASTM B117. Standard metallography was performed on the tensile bar samples.

Casting the alloys in a ribbed box die and counting the number of cracks determined the cracking tendency of an alloy. To measure the surface quality, a rating/ranking system was devised with the grade 5 given to the highest surface quality alloys (when cast in the ribbed box die) and 1 given to alloys of the poorest surface quality.

3 Results and Discussion

3.1 Mechanical Properties

Figure 1 presents an overview of the effect of changes in the alloy composition (total alloy content in at% of Al and Zn) on the tensile yield strength and ductility. In general the yield stress increases and ductility decreases with increasing alloy content. This is not a surprising result, as the volume fraction of intermetallics increases with addition of both Al and Zn; additionally, both elements have a solution hardening effect. These two mechanisms will result in both a higher yield stress and lower ductility (due to grain boundary intermetallics) [6]. At low alloy contents (less than 4 at%) the ductility results show a wide spread in the data. Relatively dilute casting alloys are more susceptible to the casting quality, this is typically seen in AM20 where a wide range of values for the ductility are commonly measured [7]. In such dilute alloys the casting parameters and die design are very important factors affecting the ductility.

Contour maps for the individual mechanical properties as a function of both the Al and Zn contents are shown in Figure 2. The yield stress increases towards high Al and Zn contents, this is expected due to the increasing volume fraction of intermetallic phases and an increase in the contribution to the strength from solid solution strengthening. From the figure Zn and Al contribute similarly to the increase in yield stress. Conversely to the yield stress, the maximum ductility occurs at low alloy contents. The maximum average ductility for binary Mg-Al alloys occurs at around 3–4wt% Al, as opposed to at 2wt%, the lowest alloy content tested (corresponding to alloy AM20); this is a result of the large scatter in the ductility for AM20 due the alloy's higher sensitivity to casting defects [7]. In terms of the decrease in ductility and impact, Zn has a slightly stronger effect than Al.

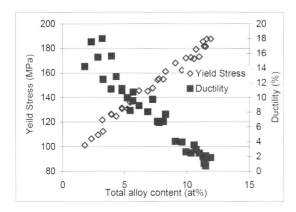

Figure 1: The yield stress (MPa) and ductility (%) of the alloys as a function of the total alloy content in atom%

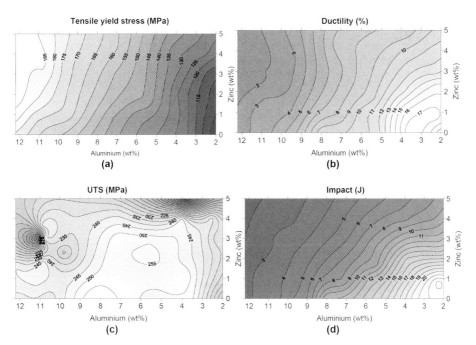

Figure 2: Contour plots of (a) tensile yield stress (MPa), (b) ductility (%), (c) ultimate tensile stress (MPa) and (d) Impact energy (J), versus the Al and Zn contents

To compare the variation in yield stress and ductility to the variation in chemical composition, a surface map and contour map can be overlaid as shown in Figure 3. The area centred around 1–3 wt% zinc and between 2 and 4 wt% aluminium (circled) has a maximum in the ductility. From the data it is seen that it is possible to make adjustments to the composition to increase the yield stress with little change in the ductility. For 3C components (small electronics – computers, cameras and cell phones, etc) the possibility to increase the ductility/impact properties while retaining the yield stress is an important consideration. By using such a mapping procedure the full potential of the alloy system is brought into consideration.

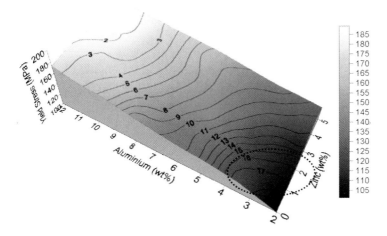

Figure 3: Overlay map of the yield stress (surface, scale in MPa shown on the right) and ductility (%) (contours) for the alloys investigated

3.2 Microstructure

A series of micrographs is shown in Figure 4; the increasing volume fraction of intermetallics with increasing alloy content is clearly seen as the alloy content increases from left to right. The castings showed microstructures typical of high pressure die cast magnesium.

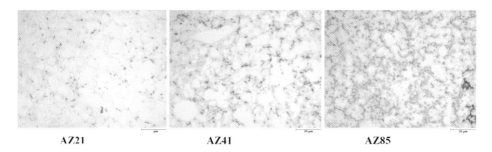

Figure 4: Optical micrograph series (x50) of selected alloys

From the optical micrographs it appears that only two phases are present in all the alloys investigated: the primary α-phase and the β-phase ($Mg_{17}Al_{12}$). Further examination by SEM revealed at least one additional phase (Figure 5), a high contrast phase. The corresponding EDS-spot analysis results are also recorded in Figure 5. The high contrast phase is highly zinc rich. There are reported to be two ternary intermetallics in the Mg-Al-Zn system, the τ-phase ($Mg_{32}(Al,Zn)_{49}$) and the φ-phase ($Mg_5Zn_2Al_2$) [8, 9] in addition the β phase $Mg_{17}Al_{12}$ ($Mg_{17}(Al,$

Zn)$_{12}$) can contain a large fraction of Zn. Shepeleva *et al.* [10] investigated die cast alloy Mg-9wt%Al-5.8wt%Zn (AZ96) and reported the Zn-containing phases Al$_2$Mg$_5$Zn$_3$ and Mg$_2$Zn from SEM investigations, while Vogel *et al.* [9] report a quasi-crystalline phase (I-phase) at the grain boundaries of ZA85 (AZ58). In the present study the high contrast phase stoichiometry has not been uniquely identified.

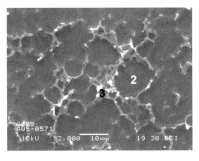

wt%	beta-phase Point 1	matrix Point 2	bright-phase Point 3
Mg	57.9	94.8	52.5
Al	23.1	3.5	15.2
Zn	19	1.6	31.4
Si	0.07	0.06	0.9

Figure 5: BS-SEM images of AZ85. There are at least 3 major phases occurring in the alloy: the matrix (α-Mg) the β-phase (light grey) and a third unidentified phase (bright particles in the β-phase). The corresponding EDS results from the points indicated in the micrograph are tabulated.

3.3 Corrosion

Salt spray corrosion testing was performed on the alloys and the results are given in Figure 6. The initial additions of Al and Zn result in significant improvements in the corrosion resistance (note that the Mn content has not been optimised for these experimental alloys). Readers should note that for alloys with moderate to high Al and Zn contents (higher than those reported herein), the salt spray corrosion behaviour does not show a linear decrease with alloy content as might be inferred from the data in Figure 5.

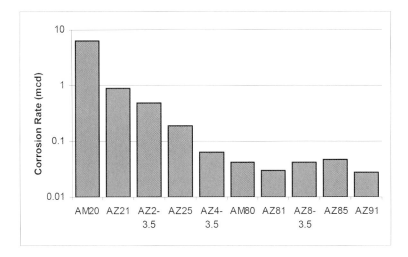

Figure 6: Salt spray corrosion results

3.4 Casting Quality

In alloy development it is necessary not only to develop an alloy with the required mechanical and/or chemical properties, but also one that can be easily produced by the desired process, in this case die casting. Contour maps of the average number of cracks recorded, and the surface finish grade are shown in Figure 7. From this information, castings showing the least tendency for cracking arise from both very low alloy contents and from binary alloys with approximately 9wt% Al. In terms of surface finish of the alloys tested, the best alloys are those from the centre of the compositions examined (around 7wt% Al and 2.5wt% Zn). The mapping of the casting properties (such as cracking, surface finish and sticking) against the alloy composition is an important tool for further alloy development, enhancing the ability to produce good quality castings.

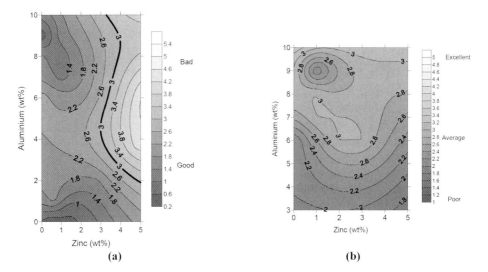

Figure 7: (a) Average number of cracks on a casting (average of 10 castings) and (b) Surface finish grade (1 = poor, 5 = excellent)

4 Conclusions

- Moderate additions of both Al and Zn results in an increase in the yield stress and decrease in the ductility and impact properties.
- For moderate additions of both Al and Zn the corrosion resistance increases, with increasing additions.
- For some alloy compositions an additional Zn-rich phase(s) is observed.
- There exists a region in the low alloy corner of the Mg-Al-Zn system that shows a broad ductility maximum and a range of yield strengths.
- The outlined procedure for mapping properties (mechanical, physical, chemical, castability and microstructure) against chemical composition allows for tailoring of specific properties through composition control.

5 References

[1] Emley, E.F., Principles of Magnesium Technology. 1966, London: Pergamon Press.
[2] Lü, Y., Dong, Q., et al., Fluidity of Mg-Al alloys and the effect of alloying elements. Z. Metallk., 2000. 91(6): p. 477–482.
[3] Foerster, G.S. New Developments in Mg Die Casting. in International Magnesium Assoc., 33rd Annual Meeting. 1976. Ohio, USA: International Magnesium Assoc.
[4] Zhang, Z. and Couture, A. The creep properties of Mg-Zn-Al alloys. in Magnesium Alloys and their Applications. 1998. Wolfsburg, Germany: Werkstoff-Informations GmbH.
[5] Sannes, S., Gjestland, H., et al. The Use of Quality Mapping to Predict Performance of Thin-Walled Magnesium Die Castings. in SAE 2005 World Congress & Exhibition. 2005. Detroit, USA: SAE.
[6] Bowles, A.L., Griffiths, J.R., et al. Ductility and the Skin Effect in High Pressure Die Cast Mg-Al Alloys. in Magnesium Technology 2001. 2001. New Orleans, USA: The Minerals, Metals and Materials Society.
[7] Bakke, P., Pettersen, K., et al. Enhanced ductility and strength through RE addition to magnesium die casting alloys. in Magnesium Technology 2003. 2003. San Diego, USA: TMS.
[8] Donnadieu, P., Quivy, A., et al., On the crystal structure and solubility range of the ternary f phase in the Mg-Al-Zn system. Z. Metallkd., 1997. 85(12): p. 911–916.
[9] Vogel, M., Kraft, O., et al., Quasi-crystalline grain-boundary ohase in the magnesium die-cast alloy ZA85. Scripta Mater., 2001. 45: p. 517–524.
[10] Shepeleva, L., Manov, E., et al. TEM study of the as-cst and aged microstruc-ture of Mg-Al-Zn alloys and the influence of Zn content on precipitation. in Magnesi-um Technology 2001. 2001: TMS.

Mg-4Al-4RE Alloy for Die-Casting

T. Rzychoń*, A. Kiełbus*, R. Cibis**
*Department of Materials Science, Silesian University of Technology, 40-019 Katowice, Poland
**NTP Sp. z o.o., 47-225 Kędzierzyn-Koźle, Poland

1 Introduction

Magnesium alloys with their weight advantage have unique application opportunities in the automotive industry. AZ91D alloy has been used to fabricate a variety of automobile parts, such as cam covers, oil adapters, steering wheels and so on. AM60 and AM50 alloys are frequently employed to manufacture instrument panels, steering wheel armatures and seat risers, with a view to its good toughness [1–3]. Typical microstructure of these alloys is composed of α-Mg matrix, $Mg_{17}Al_{12}$ precipitations and small volume of Al_8Mn_5 phase [4]. However, their applications are restricted when the temperature surpasses 120 °C, due to instability of β-$Mg_{17}Al_{12}$ phase [2-3]. A casting alloy that was developed for high temperature applications is AE42 (Mg-4Al-2RE). Aluminum is added to improve castability and room temperature mechanical properties and rare earth elements for creep resistance. However, the properties of AE42 deteriorate rapidly when the temperature is above 150 °C [5]. In this temperature the partially decomposition of $Al_{11}RE_3$ had been reported [7]. $Al_{11}RE_3$ is unstable and decomposes according to the reaction: $Al_{11}RE_3 \rightarrow 3Al_2RE + 5Al$. The released aluminum subsequently react with magnesium and this leads to the emergence of β phase, which attributes to the deterioration of creep resistance. Therefore a limitation to the use of AE42 magnesium alloys at high temperature still remains. The increase RE content can cause the improvement of creep resistance. With an increasing RE/Al ratio, Al_2RE phase gradually becomes the more dominant phase relative to $Al_{11}RE_3$. A higher RE content than that of AE42, possibly producing Al_2RE phase during solidification, this unwanted phase transformation may no longer occur [8–16].

2 Experimental

Ingots for die casting and high-pressure die cast of AE44 alloy with the chemical composition given in Table 1 were investigated. The rare earth additions were made as mischmetal with the approximate compositions: 50 % Ce, 25 % La, 20 % Nd, 3 % Pr. Die casting was carried out on 420 tone locking force cold-chamber die casting machine. Specimens for microstructure studies were mechanically polished using standard methods, etched with 5 % acetic acid. The microstructure was characterized by optical microscopy (Olympus GX-70) and a scanning electron microscopy (Hitachi S3400) equipped with an X-radiation detector EDS (VOYAGER of NORAN INSTRUMENTS). EDS analysis were performed with an accelerating voltage of 15 keV. X-ray diffractions (JDX-75 and X-Pert Philips diffractometer with graphite monochromator) were performed on the as cast and on the die cast samples. The Cu-Kα X-ray radiation was used.

Table 1: Chemical composition (wt-%) of experimental alloys

Alloy	Al	Mn	RE	Zn	Si
AE44	3.6-4.4	0.2-0.5	3.6-4.6	0.2	0.05

3 Results and Discussion

3.2 Microstrucutre of AE44 ingot

Microstructure of as-cast AE44 ingot under scanning electron microscope is showed in Fig. 1 and 2. There was observed of α-Mg matrix, acicular, globular and irregular precipitations. The brighter contrast in these precipitations is reflective of the presence of heavier RE elements. Further investigations of the rare earth content of the Al-RE compounds show that globular particles (point A, Fig. 1, Tab. 2) have higher content of cerium than other constituents of the Ce-rich mischmetal, while the acicular compounds (point B, Fig. 1, Tab. 2) have higher content lanthanum and neodymium than the others. Moreover, in microstructure are Mn-rich particles (point C, Fig. 2, Tab. 2). The aluminum dissolved in α-Mg is lower than its maximal solid solubility at room temperature (point D).

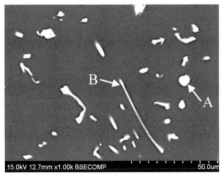
Figure 1: BSE image of AE44

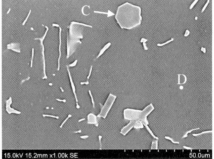
Figure 2: SE image of AE44 alloy

Table 2: Element distribution in different areas of AE44 as-cast structure from Fig. 1 and 2 (at.%)

Point	Mg-K	Al-K	Ce-L	La-L	Nd-L	Mn-K
A	11.12	71.26	15.54	2.08	–	–
B	32.63	49.27	11.91	3.19	2.99	–
C	5.65	55.22	4.16	–	–	34.97
D	99.57	0.43	–	–	–	–

Results of XRD analysis (Fig. 3) show the main intermetallic phase in AE44 alloys is $Al_{11}La_3$. Probably, microstructure consists of Al_3Ce and Al_2Ce phases, but their diffraction lines are very weak. The relative intensities of the different peaks of Mg in the XRD are almost equal to the relative intensities obtainable from Mg powders, the presence of preferred crystallographic orientation can be excluded once more. The amount Mn-rich phase is to small to be detected by X-ray diffraction. Based on the XRD pattern and results of microanalysis particles in this alloy can be identified as $Al_{11}La_3$, with some cerium substituting lanthanum. The molecular formula of these phases is written as $Al_{11}RE_3$.

According to Wei and Dunlop [16], because of the high chemical stability of $Al_{11}RE_3$, rare earth elements are combined with aluminum and form these phases until all the available RE were used without any formation of pseudobinary Mg-RE or pseudoternary Mg-Al-RE phases. It is interesting that no $Mg_{17}Al_{12}$ phase form in this alloy during slow solidification of standard size ingot. The $Mg_{17}Al_{12}$ has a softening effect at high temperatures and small volume fraction of this phase needs to be restricted if good creep resistance is to be obtained [16]. The crystallization of $Al_{11}RE_3$, Al_3RE and Al_2RE compounds reduces the amount of aluminum atoms in solid solution and diffusion of solute atoms of aluminum at elevated temperature could be minimize if the amount of aluminum in solid solution is reduced.

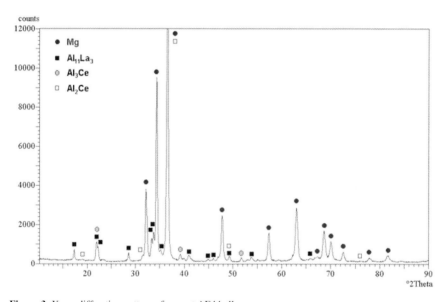

Figure 3: X-ray diffraction pattern of as-cast AE44 alloy

4   Microstructure of Die-cast AE44 Alloy

SEM micrographs taken from die-cast AE44 alloy are shown in Fig. 4 and 5, from which it can be seen that the die-cast microstructure of this alloy consists of light dendritic α-Mg matrix and a dark interdendritic network. The microstructure of the as-cast alloy was significantly coarser than those of the die casting because of the slower rate of cooling that was experienced. Considerable porosity was present in the die casting. Observations of the core regions of die-cast specimen reveal the grain boundary second-phases show a kind of lamellar divorced eutectic morphology with bright contrast (Fig. 4). The microstructure of the skin regions in the same sample is shown in Fig. 5. In these areas the interdendritic phase had a much finer microstructure than in core regions. It is consistent with higher rate of cooling these regions during die-casting. Qualitative analysis of lamellar divorced eutectic by energy dispersive X-ray analysis in a scanning electron microscope, indicated that this eutectic contained magnesium, aluminum and rare earth elements (Fig. 6). The bright precipitates are too fine to accurately determine chemical composition. EDS analysis also shows enrichment of RE elements in these areas. The results of microanalysis showed a relatively high aluminum content in the matrix of die-cast alloy compare to standard ingot. From the initial 3,6 at. % of aluminum, approximately 2 at. % is left in the matrix when one assumes that all the RE atoms bind to aluminum in Al-RE phases. The solubility of Al in α-Mg is known to decrease from about 11,5at. % at the eutectic temperature to 1at. % at room temperature [3]. The higher Al concentrations in solid solution than its maximal solubility can cause precipitation of $Mg_{17}Al_{12}$ phase during exploitation at elevated temperature.

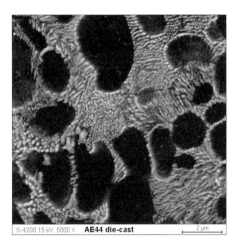

Figure 4: SE image of the core regions of die-cast AE44 alloy

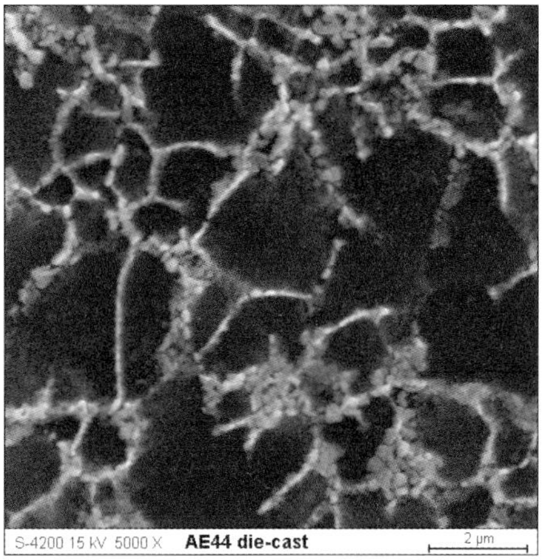

Figure 5: SE image of the skin regions of die-cast AE44 alloy

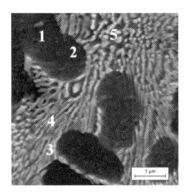

Point	Chemical composition, % at.			
	Mg	Al	La	Ce
1	97,04	1,99	0,35	0,62
2	95,98	2,56	0,6	0,85
3	83,39	11,59	1,81	3,21
4	85,8	9,24	1,8	3,16
5	84,13	11,86	1,28	2,73

Figure 6: EDS analysis of die-cast AE44 alloy

In order to identify the existing phases in the alloys, XRD analysis was performed, the results of which are shown in Fig. 7. The XRD indicates the die-cast microstructure of AE44 alloy is mainly composed of α-Mg phase and $Al_{11}RE_3$ (Al_4RE) and probably Al_3RE phase. The peak positions for magnesium, the major phase, are shifted to higher angles, a shift that is consistent with dissolved aluminum in this phase. During fast solidification Al_2RE phase was not observed. The dominant intermetallic phase in microstructure is $Al_{11}RE_3$. Powell et al. [7] suggested that $Al_{11}RE_3$ phase in die-cast AE42 alloy is unstable and susceptible to decompose according to the follow formula when the temperature surpasses 150 °C:

$$Al_{11}RE_3 \rightarrow 3Al_2RE + 5Al$$

This leads to the formation of β phase ($Mg_{17}Al_{12}$) resulting in the deterioration of creep properties.

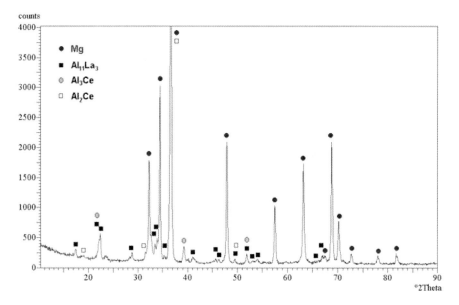

Figure 7: X-ray diffraction pattern of die-cast AE44 alloy

5 Conclusions

Based on the research results obtained, it has been found that:

1. The microstructure of ingot of AE44 alloy consists of α-Mg matrix with acicular, globular and irregular precipitations of $Al_{11}RE_3$, Al-RE-Mn, and probably Al_3RE and Al_2RE phases.
2. Globular particles of Al-RE phases have higher Ce/La ratio, while acicular compound of these phases have lower Ce/La ratio.
3. The aluminum concentrations in α-Mg matrix of ingot is lower than its maximal solid solubility at room temperature.
4. The microstructure of high-pressure die-cast AE44 alloy consists of primary equiaxed α-Mg dendrites with a lamellar eutectic consists of α-Mg, $Al_{11}RE_3$ and probably Al_3RE phases.

5. The microstructure of ingot is significantly coarser than that of the die casting.
6. In skin regions of die-cast alloy microstructure is a much finer microstructure than in core regions.
7. The aluminum content in α-Mg matrix of die-cast alloy is higher than its maximal solubility at room temperature.

6 Acknowledgements

The present work was supported by the Polish Ministry of Education and Science under the research project No PBZ-KBN-114/T08/2004.

7 References

[1] B.L. Mordike, Journal of Material Processing Technology, 117, 2001, 391–394.
[2] G. Pettersen, H. Westengen, R. Hoier, O. Lohne, Materials Science and Engineering A207, 1996, 115–120.
[3] H. Friedrich, B. Mordike, Magnesium Technology. Springer-Verlag Berlin Heidelberg 2006.
[4] A. Kiełbus, T. Rzychoń, in *Proceedings of the 13th scientific conference „New Technologies and Materials in Metallurgy and Materials Science and Engineering"*, Katowice, 2005, p.117–122 (in Polish)
[5] M.O. Pekguleryuz, E. Baril, Materials Transactions, Vol. 42, No. 7, 2001, 1258–1267.
[6] S. Xue, Y.S. Sun, S.S. Ding, Q. Bai, J. Bai, Materials Science and Technology, Vol. 21, No. 7, 2005, 847–853.
[7] B.R. Powell, V. Rezhets, M.P. Balogh, R.A. Waldo, JOM, August 2002, 34–38.
[8] AE alloys – the new family of creep resistant die casting alloys, Diecaster Bulletin, No. 11, October 2004.
[9] P. Bakke, H. Westengen, TMS, Magnesium Technology 2005, 2005 p.291–296.
[10] T. Rzychoń, A. Kiełbus, in *Proceedings of the 23rd Conference „Materials Science and Engineering School"*, Krakow-Ustron, 2005, p.149–154.
[11] A.A. Luo, International Materials Reviews, Vol. 49, No. 1, 2004, 13–32.
[12] I.P. Moreno, T.K. Nandy, J.W. Jones, J.E. Allison, T.M. Pollock, Scripta Materialia 45, 2001, 1423–1429.
[13] D. Wenwen, S. Yangshan, M. Xuegang, X. Feng, Z. Min, W. Dengyun, Materials Science and Engineering, A356, 2003, p. 1–7.
[14] M.M. Avedesian, H. Baker, ASM Spiecalty Handbook, Magnesium and Magnesium Alloys, 1999.
[15] Y. Lü, Q. Wang, X. Zeng, W. Ding, C. Zhai, Y. Zhu, Materials Science and Engineering A278, 2000, 66–76.
[16] L.Y. Wei, G.L. Dunlop, H. Westengen, Materials Science and Technology, Vol. 12, 1996, p.741–750.

Effective Alloy Development using Thixomolding® Focussing on Improved Creep Strength of AZ91

H. Eibisch, A. Lohmüller, M. Scharrer, R.F. Singer
Neue Materialien Fürth GmbH, Fürth

1 Introduction

In search of new Magnesium cast alloys for high pressure die castings, engineers have to accept a lot of time consuming melt preparations in adequate volumes and scores of crucible replacements for adjusting element additions. Moreover alloying in die casting is limited by critical factors like melt dross, oxidation, evaporation, segregation or dissolution of the alloying elements.

In contrast to this conventional alloy development technique the Thixomolding® (TM) process enables an easy way to screen a wide range of alloy compositions with much less effort in a much shorter time period. As the Thixomolding process uses Magnesium granules as feedstock material, additional alloy components can be dosed parallel to the main feedstock directly into the barrel by a second feeding system as separate granules. In addition the Thixomolding process enables a unique possibility to modify the microstructure of Magnesium cast alloys by generating a user defined solid phase fraction f_s between 0 % and approximately 50 %. The solid fraction is controlled by the barrel temperature of the TM machine.

Nowadays the majority of Magnesium alloy developments intend to improve the creep resistance – one of the most critical properties of Magnesium for power train applications. In order to strengthen Magnesium cast alloys for high temperature applications, alloying seems to be one of the most promising methods. Hence an adequate manipulation of the microstructure is indispensable for a reduction of creep dislocation movement in Magnesium solid solution.

In this study an alloy development survey based on AZ91 with the main intention to improve the creep resistance is presented. Starting with a series of pure AZ91 with varying solid phase content (2–45 %) the influence of the changing microstructure on compression creep testing was investigated. Furthermore the feasibility of separate Ca addition to AZ91 by an additional dosing system for the Thixomolding process was analysed and the potential of creep improvement with varying Ca content was quantified.

2 Experimental

The casting experiments were performed on a Japan Steel Works Thixomolding (TM) machine with a clamping force of 220 t. Simple plates with a thickness of 6 mm and a dimension of 120 × 120 mm were cast. Compression creep samples with a diameter of 5 mm and a length of 7 mm as well as flat tensile specimens were machined out of these casting plates. Process parameters were: die temperature 150 °C, injection speed 2 m/s, holding pressure 300 bar, shot weight 280 g, cycle time 50 s. As feedstock AZ91D granules supplied by Ecka Granules were used. The solid phase content f_s was adjusted to 0–45 % by reducing the barrel temperature starting from 605 °C down to 585 °C.

For adding Ca to the main feedstock a separate feeding system was installed (Figure 1). By controlling the mass flow of either dosing screws the desired Ca content can be adjusted. Since pure Ca has a high affinity to oxygen, MgCa30 granules (30 % Ca + 70 % Mg) were used for the alloying element additions.

Compression creep measurements were carried out until a steady state creep rate minimum or a maximum testing time of 200 h was reached. The setup and parameters for mechanical testing and creep experiments as well as the microstructural investigation methods are explained in detail in [1].

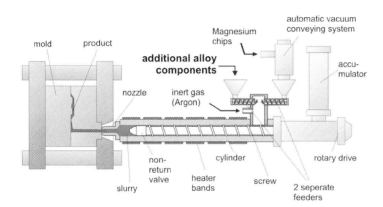

Figure 1: Schematic setup of a Thixomolding machine with an additional feeding system for alloying components (patented [2])

3 Results and Discussion

3.1 Influence of Semi-Solid Phases on Creep for AZ91

In general the dependence of the steady state creep rate $\dot{\varepsilon}$ on stress and temperature can be expressed for most metals as

$$\dot{\varepsilon} = A\sigma^n \exp\left(-Q_c/RT\right) \tag{1}$$

where A is a constant, σ the applied true stress, n the stress exponent, Q_c the activation energy for creep, R the gas constant and T the absolute temperature. The stress exponent is determined from a powerlaw plot (Figure 2a) and the activation energy can be calculated from the Arrheniusplot (Figure 2b). In both diagrams $\dot{\varepsilon}$ of AZ91 samples with varying solid phase content is significantly reduced with increasing f_s. The stress exponent as well as the activation energy does not change significantly with varying solid fraction.

Since all process and creep testing parameters remained unchanged except the process temperature of the TM castings, it is clear that the microstructure is the crucial parameter, which has to explain the observed behavior.

Figure 3: Microstructure of AZ91 with (a) 0 % and (b) 45 % solid fraction shows the microstructure of AZ91 with 0 % (a) and 45 % (b) solid fraction. The solid fraction, consisting of primary solidified α-solid solution in the barrel, contains only 3 % Al and slightly increases with raising f_s. This is consistent with the equilibrium phase diagram, since the Al solubility in α-Mg raises with decreasing process temperature. As a result the Al content in the residual liquid melt increases with rising f_s. After injection of the slurry into the mold the residual liquid melt itself solidifies as secondary α-grains and finally as a fully divorced eutectic consisting of eutectic α and precipitated β-phases (mainly $Mg_{17}Al_{12}$).

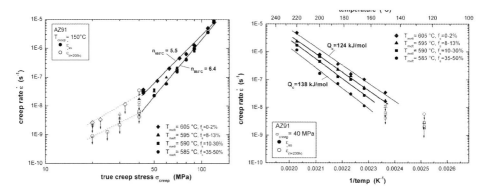

Figure 2: Steady state creep rates of AZ91 with varying solid fraction versus (a) applied true stress and (b) testing temperature

Moreover the β-phase content in the residual liquid melt increases with raising f_s. Thermodynamic calculations using Computherm with the Scheil solidification model verify that the β-phase fraction in the residual liquid melt equals 10.0 % for $f_s = 0$ % and 17.1 % for $f_s = 40$ %.

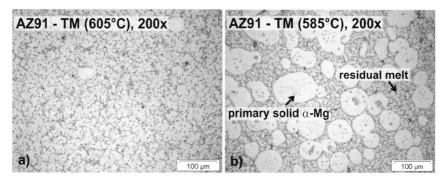

Figure 3: Microstructure of AZ91 with (a) 0 % and (b) 45 % solid fraction

The increasing Al content in the primary solid phases, the secondary α-grains and in the overall residual liquid melt was quantitatively proven by microprobe element mapping (Figure 4). Although the Al content rises in both α-phases, the overall Al content keeps constant at about 9 %, because the overall β-phase content decreases with raising f_s.

3.1.1 Creep Mechanisms

It is obvious, that the main creep mechanism of AZ91 in the tested temperature and stress range is dominated by dislocation creep, since the stress exponent varies between 5.5 and 6.4, which

is in accordance with literature data of numerous Mg alloys [3]. Nevertheless a significant reduction of creep deformation with increasing solid fraction is noticed. While the absolute alloy composition of AZ91 with and without primary solid phases keeps unchanged, the microstructural investigations show three main distinguishing characteristics with increasing solid fraction:
- an increase of the average α-Mg grain size due to the bigger primary solid phases and
- an Al enrichment of the residual melt associated with
- an increase of β-phase in the residual liquid melt

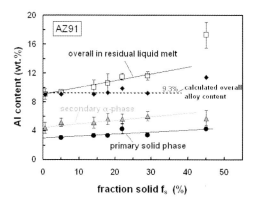

Figure 4: Al content in different phases of AZ91 versus f_s, measured by microprobe element mapping

The *grain size* is an important influencing factor for creep of most metals [4]. By reaching the micro scale a contribution of boundary sliding gains more and more importance. Thus the very fine grained microstructure of fully liquid cast AZ91 leads to an amplified creep deformation. To consider this effect the general equation (1) is modified to

$$\dot{\varepsilon}_{ss} = A \frac{1}{\overline{d}} \sigma^n \exp\left(-Q_c/RT\right) \qquad (2)$$

with $\overline{d}$ as average α-grain size calculated by a simple rule of mixture from the bimodal α-grain size distribution of the TM samples, namely the medium primary solid α-grain size $\overline{d}_{a,prim}$ and the medium secondary α-grain size $\overline{d}_{a,sec}$ of the residual liquid melt, as follows:

$$\overline{d} = f_s \cdot \overline{d}_{a,prim} + (1-f_s) \cdot \overline{d}_{a,sec} \qquad (3).$$

With the measured mean values of $\overline{d}_{a,prim} = 10\,\mu m$ and $\overline{d}_{a,sec} = 45\,\mu m$ the steady state creep rates of the AZ91 with varying solid fraction can be plotted versus the calculated inverse average α-grain size $1/\overline{d}$ (Figure 5). The dotted line in Figure 5 has a slope of 1 and demonstrates clearly an indirect proportional dependency of the creep rate from the average α-grain size.

As a result one explanation for the improved creep resistance of AZ91 with raising f_s can be ascribed to a reduced boundary sliding effect due to the increased grain size.

A second contribution to the improved creep deformation behavior might be attributed to the higher Al and therefore higher β-phase content in the residual liquid melt. In consequence of the increasing solid fraction, the structure of the accumulated β-phase precipitations become more contiguous and the dislocation movement might therefore be interfered with the more and more linked β-phase network structure.

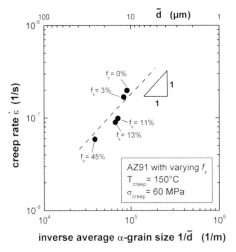

Figure 5: Creep rates of AZ91 with varying solid phase content versus the calculated inverse average α-grain size

A similar correlation between the morphology of the precipitates and the creep behavior can be observed by Ca addition to AZ91, as discussed in the following chapter.

3.2 Improved Creep Resistance of Ca Additions to TM AZ91

Unlike before, the experiments with Ca addition to AZ91 were performed at a constant process temperature of 605 °C, i.e. without any primary solid phase formation. The content of Ca was increased by 1 wt.% steps up to 5 wt.% to the main feeding control of AZ91. Glow discharge optical emission spectrometry (GDOES) as well as scanning electron microanalysis measurements (EPMA) of the actual Ca content of the castings are in quite good accordance with the intended Ca addition (Figure 6).

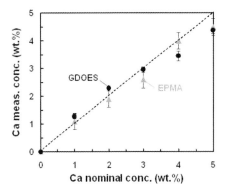

Figure 6: Correlation between the intended and measured Ca content in AZ91

The feasibility of this new alloying technique by adding separate alloying granules directly into the TM machine could therefore be evaluated successfully.

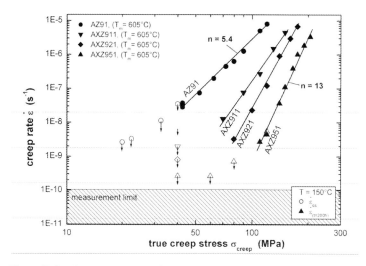

Figure 7: Steady state creep rates of AZ91 with varying Ca addition versus true creep stress

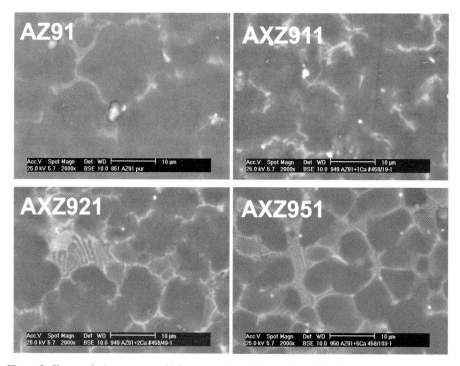

Figure 8: Change of microstructure with increasing Ca addition to AZ91 (SEM in as cast condition)

Concerning the creep properties of AZ91 alloyed by adding Ca, a remarkable creep rate reduction of up to 3 magnitudes for 5 % Ca can be observed. Figure 7 shows the creep rate versus the applied stress of AZ91 samples with varying Ca content. According to the ASTM convention the Ca containing alloys are labeled with „X" in the following (e.g. AXZ951 for AZ91 with 4.5–5.5 % Ca). The stress exponent increases from $n = 5.5$ for AZ91 to $n = 13$ for AXZ951. This is an obvious indication of a fundamental change of the creep mechanism, thus the dislocation movement is not anymore the dominant mechanism.

Associated with the remarkable increase of creep strength, a continuous change in microstructure with increasing Ca content can be observed (Figure 8). While for pure AZ91 $Mg_{17}Al_{12}$ is the predominantly precipitation, additional Al_2Ca is precipitated with raising Ca content. For AXZ921 no β-phase can be detected any more. Further Ca addition leads consequently to an increase of the Al_2Ca volume fraction and therefore to a more and more continuous precipitation network.

On the one hand the positive effect of Ca addition to Mg-Al alloys is most likely attributed to the hindrance of dislocation movements from one α-grain to another due to that continuous precipitation network, which act as a sustaining framework against plastic deformation. On the other hand a fine dispersed Ca precipitations within the α-grains may also be a possible contribution to reduce dislocation movements, which is discussed in detail in [1].

4   References

[1]   H. Eibisch, A. Lohmüller, R.F. Singer, to be published in *Zeitschrift für Metallkunde*, 2006
[2]   R.F. Singer, M. Hartmann, A. Lohmüller, P. Hutmann, R. Treidler, Patent DE 10303363 A1, 2003
[3]   K. Hirai, H. Somekawa, Y. Takigawa, K. Higashi: Materials Science and Engineering A 2005, *403*, 276-280
[4]   B. Ilschner, *Hochtemperaturplastizität* 1973, Springer Verlag, Berlin

Microstructures and Mechanical Properties of Extruded AZ31 Mg Alloy Processed by Roll-compaction Process

K. Kawabata, H. Oginuma and K. Kondoh
JWRI (Joining and Welding Research Institute), Osaka University, Ibaraki, Osaka 567-0047, Japan

1 Introduction

Magnesium-based alloys have been expected as structural materials for aerospace and aircraft applications, because of their lowest density among the elements widely used as industrial materials. Since grain refinement is especially one of the effective method enhancing mechanical properties of magnesium alloys due to their higher Hall-Petch factor than the other metallic materials, many investigators have attempted to refine the grains into sub-micrometer size through sever plastic deformation process, i.e., Equal-Channel-Angular-Extrusion [1], Accumulative Roll-Bonding [2] and Repeated Plastic Working [3]. Although these attempts have satisfactorily improved the strength and ductility compared to the input raw materials, the productivity and product cost are obstacles to apply to industrial process. In the present study, Roll-Compaction Process, RCP, which is prefer to the other sever plastic working process from the view point of industrial process is applied to refine the grains of AZ31 Mg alloy. The microstructures and mechanical properties of RCPed AZ31 are reported.

2 Experimental Procedure

2.1 RCP

The details of RCP were described in the previous paper [4]. RCP is the continuously direct plastic working on coarse powder as input raw material by employing twin rolls pressed each other. First, the raw powder is put into the rolls with the tolerance of zero in initial state. Band-like materials mechanically bonded by severe working on raw powder are output from the rolls, and fractured to arrange their shape and size by the roller ganulator. The plastic strain accumulates in the material by returning the powder with severe working to the rolls.

In this study, AZ31 alloy powder, having average length of 1.7 mm, was employed as input raw material. The rotating speed, the load and temperature of the rolls were 313 mm/s, 4.4 kN/mm and room temperature, respectively. The size of powder granulated from band-like materials through the rolls was under 7.5 mm. The number of cycles, N, from the roll compaction to the granulation was 10, 30, 50 and 80. The RCP equipment employed in this work have sufficient productivity and yield of 200 kg/h in one cycle.

2.2 Hot Extrusion and Evaluation

Each raw and RCPed AZ31 powders with N = 10, 30, 50 and 80 was compacted in a pressure of 600 MPa at room temperature. It was consolidated by indirect hot extrusion at 673 K. The preheating temperature of the green body before the extrusion was 673 K for 300 s in nitrogen gas atmosphere, and the heating rate was 1 K/s. The green compact was immediately extruded with an extrusion ratio of 37 after pre-heating.

Tensile specimens with a gage length of 10 mm and diameter of 3 mm were machined from the extruded rods with the tensile axis parallel to the extrusion direction. Tensile tests were carried out at room temperature under an initial strain rate of $3.0 \cdot 10^{-2}$ s^{-1}. The microstructures of as-RCPed powder and extrusion rods were observed by optical microscopy.

3 Results and Discussion

3.1 As-RCPed Powder

Figure 1 shows the optical microstructures of raw ($N = 0$) and RCPed AZ31 powders with N = 10, 30, 50 and 80. The microstructure of raw powder indicates coarse grain size of 200 – 500 μm with twins induced by machining from ingots. On the other hand, the microstructures of RCPed powder cannot be clearly observed grain boundaries by optical microscope, because high density of twins and/or dislocations is introduced through the process in the condition of only N = 10. Plastic flow of the powder becomes finer and more homogeneous with increasing the number of cycles.

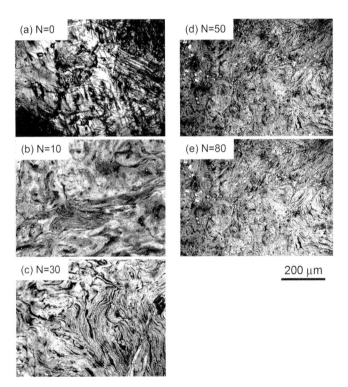

Figure 1: Optical microstructures of RCPed powders (a) N = 0, (b) N = 10, (c) N = 30, (d) N = 50 and (e) N = 80

Micro-Viker's hardness tests were performed in order to evaluate the accumulate strain through RCP. Figure 2 shows the hardening of the RCPed powder as a function of N. Though the hardness of the powder significantly increase with increasing N until $N = 10$, the hardness gradually increase with increasing N over 10 cycles. It means that small number of the cycles until $N = 10$ introduce a great part of whole strain until $N = 80$ and corresponds to the microstructure of the RCPed powder showing high density of twins and/or dislocations at $N = 10$.

3.2 Extruded AZ31 RCPed Powder

Figure 3 shows the microstructures of extruded AZ31 through RCP. All microstructures of the extruded rods indicate equiaxed grains without twins generated by dynamic recrystallization during hot extrusion. The microstructure of $N = 0$ without RCP exhibits coarse grain size of 21.4 µm uniformly, while that of $N = 10$ and 20 consists of coarse and fine grains of 20-30 µm and 1–2 µm, respectively. In the case above $N = 50$, the microstructures reveal homogeneous fine grain size of 1 µm. The mean grain size dependence on N is shown as Fig. 4. The mean grain size was obtained from image analysis used by Image Pro Plus 4.0. The mean grain size decrease with increasing N up to $N = 50$, while the grain size on the further cycles of RCP almost remain constant grain size of 1 µm. Although the sever plastic strain by RCP is almost saturated up to $N = 10$ as a consequence of the results as shown in Fig. 2, the grain size decrease until $N = 50$. The refinement effect of RCP is expected not only to provide sever plastic deformation but also to refine β phases and introduce the fine oxides retarding grain growth during hot extrusion.

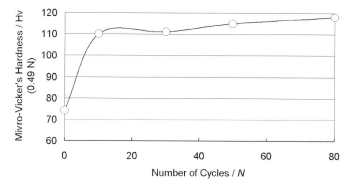

Figure 2: Dependence of hardness of RCPed AZ31 powder on number of cycles.

Figure 5 shows tensile properties of extruded AZ31 through RCP as function of N. Both tensile strength and 0.2 % proof strength increase with increasing N, while elongation gradually decrease with increasing N up to $N = 30$ and remain constant value of 18 % above $N = 30$. In general, magnesium alloys with coarse grains have poor ductility because the deformation mechanisms are limited by basal slip and twins. However, according to Koike et al. [5] in magnesium alloys with fine grains below 10 µm, non-basal slip can be activated by plastic compatibility stress associated with grain boundaries. As a result of that, magnesium alloys with the fine grains exhibit good ductility. Therefore, in spite of increasing tensile strength, the extruded AZ31 through RCP above N = 30, whose grain size below 10 µm, remain a good elongation.

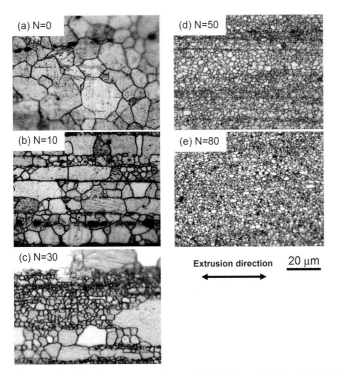

Figure 3: Optical microstructures of extruded AZ31 through RCP at 673 K (a) N = 0, (b) N = 10, (c) N = 30, (d) N = 50 and (e) N = 80

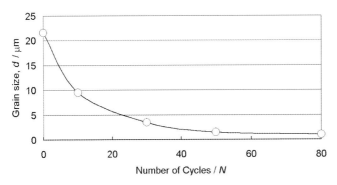

Figure 4: Dependence of mean grain size of extruded AZ31 through RCP on number of cycles

The yield stress, σ_y, of polycrystalline metals usually depends on the grain size according to the Hall-Petch relation as described below,

$$\sigma_y = \sigma_0 + kd^{-1/2} \tag{1}$$

where σ_0, k and d are yield strength of single crystalline material, Hall-Petch factor and grain size, respectively. Figure 6 shows the correlation between the grain size and the 0.2 % proof strength of extruded AZ31 through RCP. k value in the present study is 0.175 MPa/m$^{-1/2}$ and good agreement with that of 0.17 MPa/m$^{-1/2}$ in the other study [1].

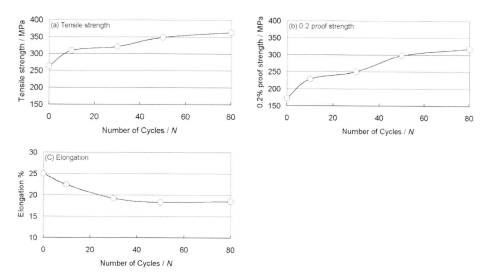

Figure 5: Tensile properties of extruded AZ31 trough RCP. (a) Tensile strength, (b) 0.2 % proof strength and (c) elongation

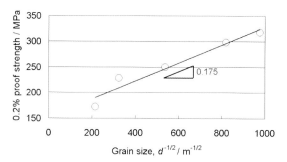

Figure 6: Hall-Petch relation of extruded AZ31through RCP

4 Summary

In the present study, Roll-compaction process, which is prefer to the other sever plastic deformation process from the view point of industrial process is applied to AZ31 alloy. AZ31 powders performed the process up to 80 cycles are compacted and extruded into rods. Tensile strength of extruded AZ31 through RCP increases with increasing the cycles of the process without significant degradation of ductility. Extruded AZ31 with 80 cycles have fine grains under 1 μm and show the excellent tensile strength and moderate elongation of 363 MPa and 18.5 %, respectively.

5 Acknowledgment

This work is financially supported by NEDO (New Energy and Industrial Technology Development Organization).

6 References

[1]   Y. Yoshida, L. Cisar, S. Kamado, Y. Kojima, Mater. Trans. 2003, 44, 468-475.
[2]   Y. Saito, H. Utsunomiya, N. Tsuji, T. Sakai, Acta Mater. 1999, 47, 579-583
[3]   K. Kondoh, T. Aizawa, Mater. Trans. 2003, 44, 1276-1283
[4]   K. Kondoh, Magnesium Technology 2005, TMS, 2005, p. 77.
[5]   J. Koike, T. Kobayashi, T. Mukai, H. Watanabe, M Suzuki, K. Maruyama, K. Higashi, Acta Mater. 2003, 51, 2055-2065.

Microstructural Evolution of Mg-Al-Ca-Sn Alloy during Extrusion

J. Dzwonczyk, M.A. Leeflang, J. Zhou, J. Duszczyk
Delft University of Technology, Delft

1 Introduction

Magnesium alloys are the lightest structural materials and hence suitable for applications in the automotive industry, which has increased attention to the vehicle weight and fuel economy [1, 2]. Many efforts have in recent years been made to develop new magnesium alloys and optimize processing technologies in order to obtain high quality products, which could compete with other construction materials for automotive applications. The development necessary in Mg alloys is frequently re-actualised for trends [3, 4] suggesting still unexplored opportunities in Mg alloy design (Figure 1).

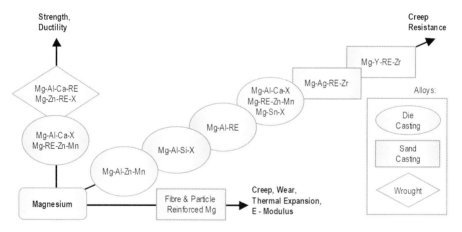

Figure 1: Development necessary in Mg alloys [3, 4]

One of the emerging groups in Mg alloy systems is Mg-Al-Ca-(X) with or without rare earth elements. Alloys in this group gain significant interest due to their excellent high temperature properties in the as-cast condition. Moreover, Mg-Al-Ca-(X) alloys can be successfully subjected to wrought processing, which makes this group even more attractive to the end-users.

In the present study, a Mg-Al-Ca-Sn alloy intended principally for powertrain applications at operating temperatures up to 190 °C [5] was investigated. Since this alloy is new, no comprehensive literature data about its microstructure after hot deformation are available. This research was performed on the alloy prepared in two metallurgical conditions and then extruded at different temperatures. It concerned the characterization of second-phase particles, analysis of microstructure evolution and material flow during extrusion.

2 Experimental Details

The material was provided in the as-cast and pre-extruded conditions. The chemical composition of the alloy is given in Table 1. The as-cast alloy was used for fundamental microstructure characterization, while the pre-extruded alloy (280 °C / 0.5–1.0 m/min) was subjected to further deformation at 400 °C and 500 °C using a 250 ton extrusion press. The details of extrusion runs are shown in Figure 2. In both cases, the extrusion process was stopped before completion and the discards were removed together with the last part of extrudate and then air-cooled. The extrusion reminders were subjected to microstructural analysis using an optical microscope, polarized light microscope, SEM/EDX and X-ray diffraction. Samples were prepared in accordance with the preparation routine for magnesium alloys [6]. To reveal grain structure, a picric-based etchant was applied.

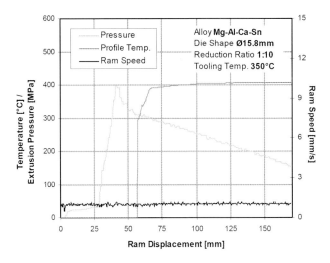

Figure 2: Extrusion of the Mg-Al-Ca-Sn alloy at 400°C. Parameter profilogram

Table 1: Alloy composition

Element	Al	Ca	Sn	Mn	Sr	Si	Fe	Cu	Mg
Wt. %	7.0	2.4	0.9	0.35	0.3	0.01	0.004	0.002	In bal.

3 Results and Discussion

3.1 Microstructure and Second-Phase Particles in the As-Received Alloy

The as-pre-extruded Mg-Al-Ca-Sn microstructure differed considerably from the as-cast one. It was observed that well-developed, recrystallized grains replaced completely the primary dendrites observed in the as-cast condition – compare Figure 3 (a) and (b). In both conditions, the material contained a large volume fraction of particles, which were formed during billet casting and retained after hot deformation. The pre-extrusion deformation caused intensive particle refinement and led to significant changes in their morphology.

The particles revealed in the Mg-Al-Ca-Sn alloy are shown in Figure 4. Basically, most of the particles contained aluminium.

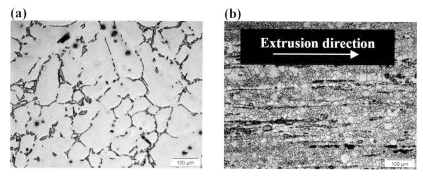

Figure 3: Microstructures of Mg-Al-Ca-Sn alloy in (a) the as-cast and (b) the as-pre-extruded condition

It is known that Al as an alloying element in Mg improves castability. However, its higher content will lead to the formation of the $Mg_{17}Al_{12}$ intermetallic compound, which due to its low melting temperature can be detrimental to the deformability of the alloy. In the alloy investigated, the added alloying elements appeared to successfully suppress the formation of the eutectic phase ($Mg_{17}Al_{12}$), so that this kind of precipitates was not observed in the alloy.

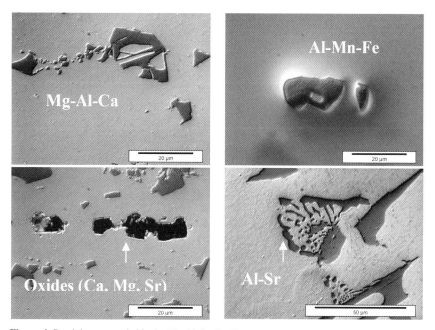

Figure 4: Precipitates revealed in the Mg-Al-Ca-Sn alloy

In Ca-containing Mg alloy, Al reacted with Ca, forming a high volume fraction of Al_2Ca intermetallic particles. In the as-cast condition, these particles were found in combination with Mg_2Ca (as a C36 phase reported also by Amerioun [7]), while extrusion led to their transformation into single Al_2Ca particles with a negligible amount of Mg_2Ca (as revealed by the SEM/EDX analysis). A small amount of Mg_2Ca in comparison with a large volume fraction of Al_2Ca can be explained by the fact that Al_2Ca is much more stable and forms easily due to the large cohesion forces between Al and Ca, i.e. high bonding energy [8].

Moreover, Al merged with strontium (approx. 0.25 wt.%) into an Al-Sr intermetallic phase with some addition of Mg (most likely Al_4Sr). These panel-like, very small particles were observed under the optical microscope only in the as-cast condition in the neighborhood of Al_2Ca, whereas in the as-pre-extruded condition their presence would have to be confirmed using X-ray diffraction, since pre-extrusion led to their refinement. It should be noted that strontium is highly reactive with oxygen, nitrogen and moisture and its use as a modifying agent is limited. As such, Sr may influence the amount of oxides. A large volume fraction of oxides were indeed observed in the pre-extruded condition and, as a matter of fact, EDX analysis revealed Sr as one of the oxide constituents.

Another group of precipitates, which contained Al, was recognized as an Al-Mn-Fe intermetallic with the stoichiometric formula of Al_8Mn_5, which is commonly present in Mg alloys. This compound was well described in the literature [9, 10]. The Al_8Mn_5 phase with a remarkable volume fraction in the material was easily recognizable due to its rounded shape, dark brown color and features similar to those reported in [11].

Regarding tin added to the alloy by 0.77 wt.%, it was expected that the compounds containing Sn, such as Mg_2Sn, Ca_2Sn or $CaSn_3$, would be present in the material. However, none of these phases were perceived. Instead, a large concentration of tin was found in oxides together with Ca, Mg and Sr. Analysis of oxides suggested that the major triggering element for their presence was actually calcium, which due to its high reactivity possessed a high capability of bonding with oxygen. Ca would bond with Mg and Sn to form a ternary MgCaSn compound, which was reported by Cohen [12] as a stable phase at elevated temperatures. With oxygen in the present material, however, it could be picked up by a Ca-containing compound and result in oxide formation.

3.2 As-Extruded Microstructure and Flow Analysis

Figure 5 shows a macrostructural overview of the Mg-Al-Ca-Sn alloy obtained after extrusion at 400 °C and 500 °C.

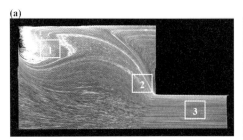
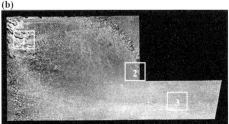

Figure 5: Mg-Al-Ca-Sn extrusion reminders obtained at (a) 400 °C and (b) 500 °C

As it can be seen, both extrusion conditions resulted in a parabolic pattern C typical of direct extrusion. However, the different temperatures applied led to significant differences in microstructure: large grains marking the peripheral region and the dead zone are visible only at 500 °C. On the other hand, compared with the 500 °C extrusion, 400 °C extrusion resulted in a somehow more extended dead zone.

From the macrostructures, it can also be seen that the pattern of inclusion distribution within the material was more expanded at 400 °C (represented by white lines) than at 500 °C. At 400 °C the origin of the material in the extrudate from the billet is more clearly visible. These morphological differences observed on a macro scale are a direct result of the response to deformation at those temperatures. Regarding the material flow, it was found that 500 °C extrusion caused far more advanced uniformity than 400 °C extrusion. Figure 6 shows the microstructures revealed in the regions marked on the extrusion reminders in Figure 5. It should be mentioned that the peripheral region in 500 °C extrusion contained significant amounts of inclusions and in this region, the grain size was widely distributed from a few microns to tens of microns. It is a result of intensive DRX suppression by the inclusions.

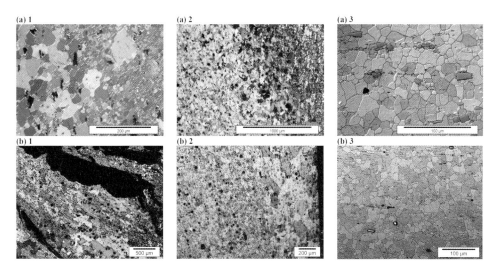

Figure 6: Microstructures of the Mg-Al-Ca-Sn alloy extruded at (a) 400 °C and (b) 500 °C. Numbers indicate the regions delineated in Figure 5.

It is important to mention that the extrusion under both of the conditions resulted in homogeneously distributed, well-developed equiaxed grains. Within the extruded parts an average grain size was calculated. The results are as follows: for 400 °C extrusion $d_{aver.} = 11$ μm $\pm$ 0.6, while for 500 °C extrusion $d_{aver.} = 17.2$ μm $\pm$ 1.3. Obviously, a higher extrusion temperature led to a larger average grain size.

Moreover, it was found that the particles present were involved in grain evolution. Figure 7 shows the microstructures revealed close to the extrusion die. It can be seen that local retardation of DRX took place. Such regions with particles situated between fine grains were only observed in the material extruded at 400°C, indicating the sensitivity of DRX to particle presence. Additionally, among the particles present, oxides appeared to have pronounced influence on recrystallized structure, while the other particles underwent intensive refinement. Also, the refined particles were found to shape the grain structure, since grain boundaries were often shifted to the regions maked by the traces of the refined particles.

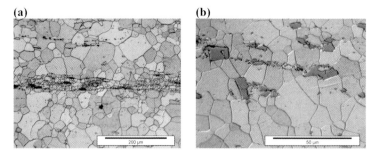

Figure 7: Particle behavior during extrusion at 400 °C (a) oxides retarding DRX and (b) Al$_2$Ca refinement – no influence on DRX appearance

4   Conclusions

The as-received Mg-Al-Ca-Sn alloy (in the as-cast and as-pre-extruded conditions) contained a high volume fraction of particles identified as: Mg-Al-Ca (Al$_2$Ca and Mg$_2$Ca), Al-Mn-Fe (Al$_8$Mn$_5$), Al-Sr (Al$_4$Sr) and complex oxides containing Ca, Mg and Sr as the main constituents. Pre-extrusion caused dynamic recrystallization, which removed the dendritic structure observed in the as-cast condition.

The reminders of the Mg-Al-Ca-Sn extrudate were subjected to microstructural examinations after extrusion performed at 400 °C and 500 °C. The results obtained led to the following conclusions:
- Extrusion resulted in a homogeneously distributed, equiaxed grain structure, although the extrusion at the higher temperature led to grain coarsening.
- The different temperatures applied caused remarkable variations in microstructures. The extrusion at 400 °C resulted in a more pronounced dead zone and relatively unstable flow within the non-extruded part, while 500 °C extrusion led to significant grain coarsening marking a narrow dead zone and the material flow in the non-extruded part was much more uniform.
- 400 °C extrusion exhibited morphological dependence on the particles present, while at 500 °C grains were recrystallized, which was not influenced much by the particles present.

5   References

[1]   H. Friedrich, S. Schumann, in Proceddings of the 2nd Israeli International Conference on Mg Science and Technology **2000**, 9.
[2]   R. R. Braeutigam, R. G. Noll, Report of Economic Statistics **1984**, *66(1)*, 80.
[3]   P. Maier, K. U. Kainer, in Proceedings of the 2nd International Light Metals Technology Conference (Ed. H. Kaufmann) **2005**, 242.
[4]   H. E. Friedrich, B. L. Mordike, Magnesium Technology, Springer-Verlag Berlin Heidelberg **2006**, 97.

[5] E. Aghion, D. Eliezer, Magnesium Alloys, Avi Moshe Bregman Ltd **2004**, 438–152.
[6] V. Kree, J. Bohlen, D. Letzig, K. U. Kainer; Practical Metallography **2004**, *41(5)*, 233.
[7] S. Amerioun, S. I. Simak, U. Haeussermann; Inorganic Chemistry **2003**, *42*, 1467–1474.
[8] X. Min, et al., Materials Chemistry and Physics **2002**, *78*, 88–93.
[9] R. M. Wang, A. Eliezer, E. M. Gutman, Material Science and Engineering **2003**, *A335*, 201–207.
[10] S. Barbagallo, et al., Journal of Alloys and Compounds **2004**, *378*, 226–232.
[11] R. M. Wang, A. Eliezer, E. Gutman, Materials Science and Engineering **2002**, *A344*, 281.
[12] S. Cohen, Zeitschrift Metallkunde **2005**, *96*, 1081–1087.

Microstructure and Mechanical Properties of Strip Cast Mg-Zn-Al Alloys with Ca by Twin Roll Caster

K. Matsuzaki, K. Hatsukano, M. Kato, K. Hanada, T. Shimizu and Y. Murakoshi
National Institute of Advanced Industrial Science and Technology (AIST), Tsukuba 305-8564, Japan

1 Introduction

Mg alloy is the lightest metallic constructional material with a density of 1.78 mg/m^3 and have an attractive possibility for weight saving the vehicle. Most of Mg components are produced by using die cast process because of its good castability. On the other hand, Mg wrought alloys have superior mechanical properties compared to cast alloys. In conventional Mg sheet production process, a thin sheets is produced through many steps, because of poor formability of Mg due to hcp structure. This might lead to an increase in the cost of Mg wrought alloys. The strip casting process makes it possible to directly obtain the sheets with thickness less than one or two mm from molten, leading to the reduction in cost and energy consumption. Furthermore, the strip casting is an effective process for refining the microstructure and reducing the segregation, leading to the improvement of mechanical properties. The strip cast Mg alloy is expected to have a high strength and good ductility. Therefore, strip casting is useful process for developing new wrought Mg alloy as well as manufacturing thin Mg alloy sheets. It is reported that[1] the addition of Ca to Mg-Al alloys is useful for preventing the oxidation or combustion of Mg molten alloy and make the casting process easier. Therefore, Ca added Mg alloys are expected to be candidate for new Mg wrought alloys.

In this paper, Mg-Zn based sheets with Ca are produced by strip casting, and their microstructure and mechanical properties are clarified.

2 Experimental Procedure

In this study, $Mg_{97}Zn_{1.5}Al_1Ca0.5$(ZA21Ca), $Mg_{95}Zn_{3.5}Al_1Ca_{0.5}$(ZA41Ca), $Mg_{93}Mg_{5.5}AlCa0.5$(ZA61Ca) and $Mg_{91}Zn_{7.5}Al_1Ca_{0.5}$(ZA81Ca) alloys were used. The mother alloy is prepared as follows. At beginning, the Ca rich Mg alloy is prepared by melting of pure Mg, Zn and Ca in a carbon crucible under an inert gas atmosphere. The resultant alloy ingots was melted with Mg, Zn and Al into a desired composition in a steel crucible under a protective atmosphere of Ar/SF6 gas mixture. After melting, the crucible was moved to the twin roll caster and the melt was poured between rotating rollers through the tundish made of Boron Nitride. The roller was made of steel, which diameter and width are 300 mm and 50 mm, respectively. The experimental condition is as follows; the roll gap was 1.0 mm and the roll speed was 16m/min. The casting temperature was in the range of 620 to 650 °C. The casting was carried out without cover gas. The cast strips were subjected to annealing. The microstructure was examined by an optical microscope. Hardness was measured by a micro Vickers hardness tester and tensile properties were measured by a Shimazu Autograh material testing machine.

3 Results and Discussion

Figure 1 shows an appearance of Mg-Zn-Al with 0.5%Ca alloy sheets with different Zn content produced by strip casting. The as-cast sheets have a thickness of about 1.4mm and show a smooth surface. Furthermore, no oxidation is observed although the casting is carried out without cover gas. This indicate that the addition of small amount of Ca is effective for preventing the oxidation for the Mg-Zn base alloy. This effect is confirmed for Mg-Al base alloy[1]. The addition of Ca also makes the strip casting process simple in Mg-Zn base alloy. One can see that the generation of cracks at edge occurs more notably with increasing Zn content. The substitution of Zn by Al reduces the generation of cracks. For example, strip cast AZ31 and AZ61 alloy containing 0.5 Ca show no cracks[2]. On the other hand, cracks are observed in the strip cast AZ91Ca alloy. The generation of cracks for strip cast AZ91 alloy has also been reported by Watari et al.[3]. This reason is not clear but may be due to the solidification behavior. It is reported[4] that the large freezing temperature range makes it difficult to control casting strip qualities. The temperature range between the solid line and liquidus line in Mg-Zn base alloy increase with Zn content. In the case of Mg-Al base alloy, the freezing rage also increases with Al content. The SEM observation reveals that the surface of cracks is irregular and quite different from that caused by conventional hot-rolling, which is smooth surface. Moreover, the distribution of melt along the roll length plays an important role in the forming of sound sheets by strip casting[4]. In the preset study, the distribution is not homogenous, and this maybe lead to the formation of cracks. In some case, however, cracks are hardly observed for the ZA61Ca, shown in Fig1(d). This suggest that the appropriate casting condition enable us to obtain sound sheet of Mg-Zn base alloy. This is under investigation.

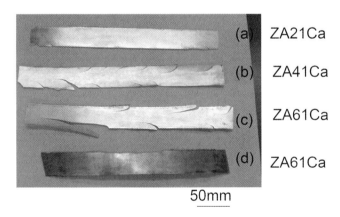

Figure 1: Strip cast ZA alloy with 0.5%Ca

Figure 2 shows X ray diffraction patterns of strip cast alloy sheets. This result indicates that the strip cast alloys consist of hcp-Mg and other peaks corresponding to the compound such as Al2Ca and MgZn are not observed. The intensity of 00l peaks are rather strongly observed, indicating the slight development of texture with a c-plane of hcp-Mg parallel to the roll surface. There is significant difference between alloys. The microstructure of as-cast alloy sheets are shown in Figure 3. As-cast alloys are composed of equiaxed grain and the fine microstructure with an average grain size of 10 to 20 µm is obtained. The grain size decreases with increasing

Zn content. This fine microstructure results from the rather high cooling rate by strip casting and is expected to improve the mechanical properties.

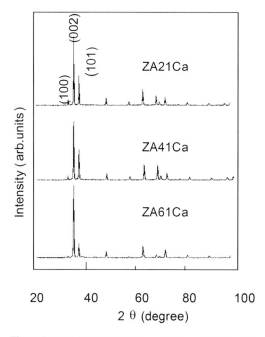

Figure 2: X ray diffraction patterns of strip cast ZACa alloys

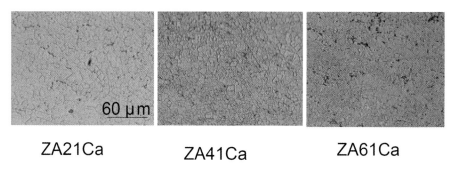

Figure 3: Optical microstructure of as-cast ZACa alloy sheets

The tensile mechanical properties of as-cast ZACa alloy sheets are shown in Figure 4. The yield strength (Y.S.) and ultimate tensile strength (UTS) of as-cast ZM21Ca alloy are 150 MPa and 200 MPa respectively, and the strength increases with Zn content. The ZA61Ca shows the highest UTS of 330 MPa with a Y.S. of 265 MPa. On other hand, the elongation decreases with Zn content. The elongation of 23% is obtained for the as-cast ZM21Ca and decrease to 11 % for ZA61Ca. The good ductility is obtained for the ZA21Ca alloy. It is notable that the strip cast ZA41alloy sheet has a high ductility with an elongation of 18 % while it shows large cracks at edge in as-cast sheet, shown Fig.1(b). This means also that the generation of cracks is due to the nature of molten alloy. The good ductility as well as high strength is obtained by grain refinement.

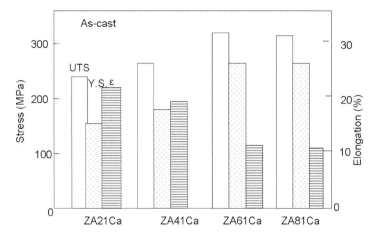

Figure 4: Tensile mechanical properties of strip cast ZACa alloy sheets

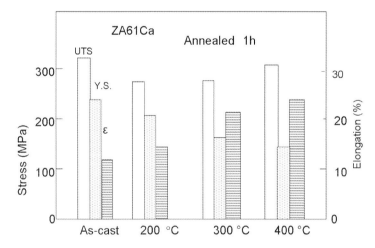

Figure 5: Change in mechanical properties of strip cast ZA61Ca alloy sheets with annealing temperature

The mechanical properties are improved by annealing. Figure 5 shows a change in the mechanical properties of strip cast ZA61Ca alloy sheet as a function of annealing temperature. Y.S decreases with increasing annealing temperature and reaches 140 MPa at 400 °C. UTS decreases by annealing at 200 °C but increases with further increase of annealing temperature. The UTS of 305 MPa is obtained for the sample annealed at 400 °C. The elongation increases monotonically with annealing temperature and reaches 23% at 400 °C . The increase of UTS at above 300 °C is caused by work hardening. It is said that ZA61Ca alloy sheets annealed at 400 °C shows a good ductility as well as high strength. In the other alloys, the elongation is improved by annealing. The ZA41CA annealed at 400 °C shows high ductility with an elongation of 32 %, but UTS and Y.S decrease to 210 and 110 MPa, respectively. Further improvement of mechanical properties is expected by optimization of annealing condition or thermomechanical treatment

4 Conclusion

Mg-Zn-Al-0.5%Ca alloys with deferent Zn content were strip cast into a sheet with a thickness of about 1.4mm. The conclusions are as follows:

1. The oxidation on the surface was not observed in spite of no cover gas during casting, It is found that the addition of Ca is useful for preventing the oxidation or combustion in Mg-Zn base alloy.
2. ZA21Ca alloy sheet is soundly formed by strip casting but the generation of cracks occurs at edge in sheet more markedly with increasing Zn content.
3. The as-cast alloy sheets consists of fine equiaxed grain with a grain size of 10 to 20 μm and the grain size decreases with increasing Zn.
4. The UTS and Y.S. of as-cast sheet increase with Zn content and the ZA61Ca shows a UTS of 330MPa and Y.S.of 265 MPa. ZA 21Ca and ZA41Ca show a good ductility. The elongation is 23 % for ZA21Ca and 18 % for Za41Ca and decreases to 11% for ZA61Ca.
5. The annealing at 400 °C improves the ductility and the elongation of 23 % with a strength of 305 MPa is obtained for ZA61Ca alloy.
6. It is found that strip cast Mg-Zn-Al alloy with 0.5%Ca show high strength with good ductility. The process is useful for producing thin Mg sheets with good mechanical properties.

5 References

[1] K. Matsuzaki, K. Hatsukano, Y. Torisaka, K. Hanada and T. Shimizu, Continuous Casting (ed. H. R. Müller), WILEY-VCH, Weinheim,2005, p. 77.
[2] K. Matsuzaki, K. Hatsukano, Y. Torisak, K. Hanada.T.Shimizu and M.Kato, Proceedings Materials Processing for Properties and Performance,Vol4, Tsukuba, Japan,2005,p. 225.
[3] H. Watari, T. Haga, K. Davey, H. Ona, S. Izawa. H.Hamano,T. Iwashita and M.Nakayama, Materials Science Forum 2003, 426–432,617–622.
[4] C. Yang, P. Ding, D. Zhang, F. Pan, Materials Science Forum 2005, 488-489, 427–430.

Production of Mg-Ni Alloys under High Pressure and High Temperature for use as Catalysts in Diamond Synthesis

A. J. Sideris, G.S. Bobrovnitchii
North Fluminense State University, Campos dos Goytacazes, RJ – Brazil

1 Introduction

Nowadays, diamond crystals (powder) for industrial applications are manufactured by processing graphite under high pressure and high temperature (HPHT) conditions. A catalyst/solvent metallic alloy contributes to an efficient graphite to diamond transformation at an equilibrium transition interface in the diamond-graphite system. The graphite is dissolved in the solvent metal to reduce the energetic threshold of transformation of graphite into diamond. Currently, the most widely-used metallic alloys are Ni-Mn, Fe-Co and Fe-Ni [1-3]. The use of these alloys allows synthesization at pressures from 4.5 GPa and at temperatures from 1500 K, resulting in diamonds with varied morphologies and properties.

The search for new catalysts that improve productivity and that can supply new properties to crystals has led to studies of synthesis processes with elemental catalysts; very interesting results have been achieved with magnesium (Mg). Recently, it was found that Mg, when used as a catalyst/solvent for the synthesis of diamonds, results in crystals with different morphologies when compared to those obtained with other catalysts/solvents [4, 5]. Synthetic diamonds produced with Mg have the characteristics required for semi-conductors [5]. However, the use of Mg presents practical problems, due to the elevated HPHT conditions that are needed for the synthesis processes. In addition to its instability, synthesization occurs only when the pressure and temperature in the reaction cell are relatively high, i.e., more than 7.0 GPa and over 2000 K, resulting in economically non-competitive diamonds.

A possible alternative to decrease the values of the HPHT parameters would be to consider the use of Mg not as single metal but as an alloy associated with other metals. One possible candidate is Ni, which is already used in catalyst/solvent alloys [1, 3].

Numerous processes have been developed over the years for the production of magnesium alloys. The production of Mg-Ni alloys is, however, a complex and expensive operation, due to the very high reactivity of Mg, particularly with oxygen. Therefore, the methods demand vacuum-producing devices, protective scums or inert atmospheres. In industry, metallurgist processes are used, though on a larger scale [6]. Special atmospheric conditions are needed to avoid intense oxidation of Mg in contact with air at high temperatures.

In an attempt to minimize practical problems associated with traditional methods for the fabrication of Mg-Ni alloys, we developed a new method based on high pressure (1.0 to 3.0 GPa) and high temperature (1400 to 2000 K) conditions. The application of high pressure conditions inside an isolated chamber, not only intensifies the interaction between the Mg and Ni powders, but also restricts interaction with oxygen. Furthermore, the high pressure inhibits the diffusion of the metallic atoms, resulting in a more homogeneous and smaller grain-size alloy [5]. Our objective was to develop intermetallic Mg_2Ni and $MgNi_2$ compositions, based on the diagram of the state of the Mg–Ni system under atmospheric pressure [7], and to demonstrate the success of this new method of production of Mg alloys. We investigated the fusion process of the Mg–Ni system under high pressure conditions, in order to produce Mg–Ni alloys with pre-established compositions for subsequent application in the diamond-synthesis process.

2 Experimental Procedures

The raw materials that were employed were powdered Mg and Ni of analytical purity (higher than 99.9%) with a grain size of 50 to 100 µm. Stoichiometric mixtures (0.667 at.% Mg + 0.333 at.% Ni and 0.333 at.% Mg + 0.667 at.% Ni) of the intermetallic alloys were prepared under atmospheric pressure. The mixtures were then compacted into 7.0 mm diameter, 8.0 mm thick cylinders under a pressure of 60 MPa during 5 seconds. The compositions (0.667 at.% Mg + 0.333 at.% Ni and 0.333 at.% Mg + 0.667 at.% Ni) were selected based on the diagram of the state of the Mg–Ni system under atmospheric pressure, with the purpose of obtaining the intermetallic Mg_2Ni and $MgNi_2$ compositions.

A cylindrical support made of graphite (grain size of 100 µm and purity 100%) was developed with four blind axial holes (Fig. 1), which can be filled with the cylinders of the Mg-Ni mixture. A cylindrical cover (10 mm thick and 30 mm in diameter), also made of graphite, was used to cover the mixture and assure electric conduction through the material. This four-hole support served as an oven, heated by electric current. The four holes of the support allow us to produce four different alloy compositions in one process, with four different samples being sub mitted to the same pressure, temperature and time parameters.

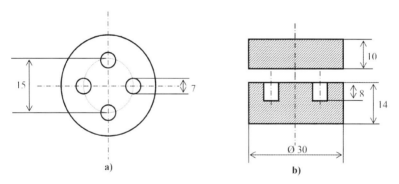

Figure 1: Graphite heater scheme: (a) Top view of graphite support; (b) Cross-section of graphite support and cover

The dimensions of the graphite cylinder were adapted to fit within the high pressure cell (capsule) of a high-pressure device (DPA) in a 2500 ton press; this capsule acts as a compressible environment. Four cylinders with different compositions of Mg-Ni were placed in the graphite heater and the entire assembly was then inserted into the central hole of the high pressure cell, resulting in the cell shown in Figure 2. Each capsule was placed in a high-pressure anvil-type device with a 55 mm diameter concavity [8]. The temperature and pressure inside the reaction volume were measured with methods described elsewhere [1]. Figure 2 schematically depicts a cross-section of the high pressure cell assembly used in the production of the Mg-Ni alloys under high pressure and temperature conditions.

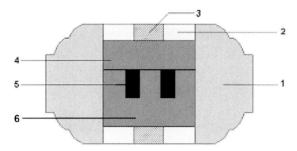

Figure 2: Cross-section of the high pressure cell assembly: (1) Calcite capsule; (2) Thermo-insulation ring; (3) Graphite disk conductor; (4) Graphite cover; (5) Micropowder mixture; (6) Graphite support

Based on the Mg and Ni concentrations (0.667 at.% Mg + 0.333 at.% Ni e 0.333 at.% Mg + 0.667 at.% Ni) and considering that the thermodynamic $\partial T_m/\partial P$ ratio for Ni is 33 K/GPa, and 70K/GPa for Mg, it was determined that the heating temperature of the mixture should not exceed 2000 K. This would ensure efficient fusion of the mixture's components. This temperature was induced in the high-pressure cell. The experiments were performed under a pressure of 1.0 to 3.0 GPa. The sample treatment time was set for 1 to 3 min and the heating/cooling rate was maintained at 20 to 30 K/s. At the end of each treatment, the device was disassembled and the samples removed. The sample surfaces were cleaned by sandpapering and polishing. The samples were prepared for optical microscopy analysis by chemical attack with an aqueous solution of H_2SO_4. Their microstructures were analyzed using a Neophot optical microscope, and the phases were identified based on X-ray diffraction, XRD, and analyses using K_α radiation of Cu.

The alloy samples of the Mg-Ni system were ground to 0.5 to 1.5 mm particles. These alloy particles were then used for investigation of spontaneous diamond crystallization from the Mg-Ni-C system with high pressure (5.0 to 7.0 GPa) and high temperature (1300 to 1700 K) treatments, using a toroidal anvil device [9].

3 Results and Discussion

The Mg-Ni alloys were produced with two distinct compositions: (a) 0.667 at.% Mg + 0.333 at.% Ni and (b) 0.333 at.% Mg + 0.667 at.% Ni. The idea was to produce the stoichiometric Mg_2Ni and $MgNi_2$ intermetallic compounds based on the diagram of the state of the Mg–Ni system under atmospheric pressure. Formation of MgC and MgC_2 due to HPHT reaction between magnesium and graphite can also occur [10]. However, the precise combination of stoichiometric amounts did not result in exactly corresponding intermetallic compounds after HPHT processing, due to the influence of the high pressures and high temperatures.

Figure 3 shows the XRD patterns for both alloys. In Figure 3 (a), the major XRD peak for alloy with 0.667 at.% Mg + 0.333 at.% Ni corresponds, as expected, to Mg_2Ni. However, lower intensity peaks were found for $MgNi_2$, and a significant peak indicated the presence of pure Mg. Quantitative phase analyses revealed that the amount of pure Mg was about as that of Mg_2Ni. By contrast, the XRD pattern for the 0.333 at.% Mg + 0.667 at.% Ni (Fig. 3 (b)) did not show any peak for pure Mg. As was also expected, the major peaks in Figure 3 (b) correspond to the $MgNi_2$ compound, but other peaks associated with Mg_2Ni were also observed. We were unable to identify the smaller peaks in the two diffractograms in Figure 3.

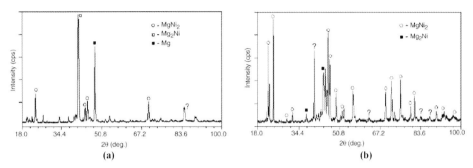

Figure 3: XRD patterns for: (a) 66.7%Mg+33.3%Ni alloy; (b) 33.3%Mg+66.7%Ni alloy

Marked differences were observed in the microstructure of the investigated alloys. Figure 4 shows a typical optical micrograph of the 0.667 at.% Mg + 0.333 at.% Ni. In addition to a relatively heterogeneous microstructure, Figure 4 shows the clear presence of grain boundaries associated with isolated inclusions, which were identified as pure Mg.

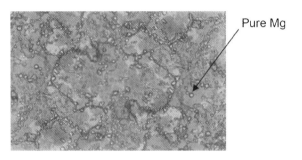

Figure 4: Optical micrograph of the 0.667 at.% Mg + 0.333 at.% Ni alloy

On the other hand, there were no grain boundaries in the microstructure of the 0.333 at.% Mg + 0.667 at.% Ni alloy (Fig. 5). The microstructure was formed by dendrites of $MgNi_2$, with interdendritic segregation but no sign of inclusions.

Based on the striking differences between the two alloys, one may infer that the 0.333 at.% Mg + 0.667 at.% Ni alloy (Fig. 5) has a more homogeneous structure, in which a crystallization mechanism probably superimposes on a diffusion process.

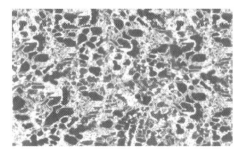

Figure 5: Optical micrograph of the 0.333 at.% Mg + 0.667 at.% Ni alloy

In principle, this alloy without pure Mg associated with grain boundaries, should be a better catalyst/solvent for the diamond synthesis process. A preliminary evaluation of the use of both alloys as catalyst/solvents in diamond synthesis under high pressure (5.0 to 7.0 GPa) and temperature (1300 to 1700 K) demonstrated the efficiency of the 0.333 at.% Mg + 0.667 at.% Ni alloy. Figure 6 illustrates the diamond crystals obtained from a system composed of a mixture of the 0.333 at.% Mg + 0.667 at.% Ni alloy with graphite powder subjected to the HPHT processing for 50 sec. Conversely, a similar operation using the 0.667 at.% Mg + 0.333 at.% Ni alloy as catalyst/solvent yielded no diamond crystals. Figure 6 (b) shows a diamond crystal with cubic morphology, due to the participation of Mg as catalyst/solvent.

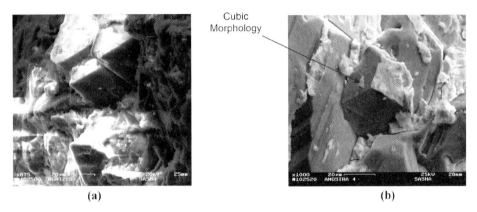

Figure 6: Diamond crystals obtained from graphite plus 0.333 at.% Mg + 0.667 at.% Ni alloy.

4   Conclusions

Mg-Ni alloys with variable stoichiometric compositions were produced under high-pressure conditions. The HPHT parameters of formation of the alloys were from 1.0 to 3.0 GPa and 1400 to 2000 K. The Mg-Ni alloys obtained with the stoichiometric compositions 0.667 at.% Mg + 0.333 at.% Ni gave a significantly different structure. In addition to the expected Mg_2Ni intermetallic composition, pure magnesium and $MgNi_2$ were also found.

Our studies using high pressure may serve as a basis for subsequent experiments aimed at studying the influence of heating kinetics and cooling of the metals on the production of alloys with well-defined compositions.

We propose that it is possible to produce other types of alloys under high pressures and temperatures, starting with light elements that possess low sublimation energy and react with gaseous mediums.

The heterogeneous microstructure of the Mg-Ni alloys with grain boundaries associated with isolated inclusions of pure Mg did not favor the synthesis of diamonds. On the other hand, spontaneous synthesis of diamonds took place in the presence of the $Mg-Ni_2$ alloy without inclusions of pure Mg, resulting in a reduction of HPHT parameters for synthesis in comparison with the Mg-C system. The morphology of the crystals that were formed leads us to speculate on the possibility of obtaining diamond crystals with new characteristics.

5   References

[1] Novikov, N.V.; Ivakhnenko, S. A.; and Katsay, M.Ya.: Kinetics of diamond crystals growth at high static pressure. MRS Int. Conf. Proc., 1999, pp.71–80.
[2] Sugano, T. et al. Pressure and temperature region of diamond formation in system graphite and Fe containing alloys. Diam. and Rel. Mat., no 5, 1996, pp.29–33.
[3] Borimskiy, A. I.; Delevi, V.G.; and Nagorniy, P. A.: Kinetics of formation and growth of diamonds in the Fe-Si-C system. J. Superhard Mat., vol. 21, no 3, 1999, pp. 6–11.
[4] Shulzhenko, A.A.; Novikov, N.V.; and Chipenko, G.V.: Peculiarities of diamond growth in the system based on Mg. J. of Superhard Mat., no 3, 1988, pp. 10–11.
[5] Shulzhenko, A.A. et al. State diagram of Mg-C system at 7.7 GPa pressure. J. of Superhard Mat., no 4, 1988, pp. 17–21.
[6] Berlonis L.E.A., Cabrera E., Hall-Barientos E. and Hall P.J., J. Mat. Res., 1 (2001) 45.
[7] Feufel H.; Sommer F., J. Alloys Comp. 224 (1995) 42.
[8] Novikov N.V., Prikhna A.I., Borimsky A.I., in: High Pr. Res. Ind. 8 216 AIRAPT Conference, vol. 2, 1982, p. 790. 217.
[9] Verestshagin L. F., Khvostantsev L. G., US Patent n. 3854854 – High Pressure Producing Apparatus, 1974.
[10] Novikov, N. V.: New trends in high-pressure synthesis of diamond. Diam. and Rel. Mat., no 8, 1999, pp.1427–1432.

Investigation of Structure and Properties of the Age-Hardenable Mg-Sm Alloys after Severe Plastic Deformation

L.L. Rokhlin[1], S.V. Dobatkin[1,2], T.V. Dobatkina[1], N.I. Nikitina[1], V.N. Timofeev[1], M.V.Popov[2]

[1]Baikov Institute of Metallurgy and Materials Science, Leninsky pr. 49, 119991 Moscow, Russia,
[2]Moscow State Institute of Steel and Alloys (Technological University), Leninsky pr. 4, 119049 Moscow, Russia

1 Introduction

Severe plastic deformation (SPD) is known to increase significantly strength properties of metals and alloys [1]. This method was used successfully for many metallic materials of various compositions and structure, including Mg alloys [2–5]. Magnesium-base alloys are used mostly as light structural materials and, therefore, their strength properties are of a great importance. In this work a possibility of strengthening of magnesium-base alloys by SPD together with aging was studied.

The work was performed using the perspective Mg-Sm system alloys characterized by high strength at room and elevated temperatures. The properties of this system alloys to a great extent are determined by their high strengthening effect during the decomposition of supersaturated magnesium solid solution. So, the investigations of the Mg-Sm alloys allow one to establish the joint effect of SPD and solid solution decomposition on strength properties.

2 Experimental Procedure

The phase diagram of the Mg-Sm system on the Mg side is presented in Fig.1 [6]. It shows relatively small solubility of Sm in solid Mg, decreasing with lowering temperature. Three binary alloys were used for the experiments. They contained 2.8, 4.5 and 5.5 mass% Sm. As the initial materials for the alloy preparation magnesium of the 99.96 %[1] purity and samarium of the 99.83 % purity (MG-96 and SmM-1 of Russian Grades) were used. The alloys were obtained by melting in electrical resistance furnace in steel crucibles under flux VI2 (Russian designation) containing 38–46 % $MgCl_2$, 32–40 % KCl, 3–5 % CaF_2, 5–8 % $BaCl_2$, 1.5 % MgO, <8 % ($NaCl+CaCl_2$). The flux preserved the melts from intensive oxidation and burning. The melts were poured into steel mould with three round cavities. Small ingots of 15 mm in diameter and of about 90 mm in length were obtained. The ingots were cut into cylindrical pieces which were solution treated at 510 °C for 5 hours and quenched in cold water. After quenching, the pieces were aged at 200 °C for 8 hours (the alloy Mg-4.5% Sm) and at 300 °C for 6 hours (the alloys Mg-2.8% Sm, Mg-4.5% Sm and Mg-5.5% Sm).The aging at 200 °C, 8 hours corresponded approximately to maximum strengthening during Mg solid solution decomposition and the aging at 300 °C, 6 hours corresponded to actually full Mg solid solution decomposition and softening [6]. Then the samples of 10 mm in diameter and 0.6 mm in thickness were subjected to SPD at room temperature and 200 °C by torsion under a pressure of 4 GPa to $\varepsilon \approx 6$ (5 revolutions).

Samples of the alloys without SPD and after SPD were studied by specific electrical resistance measurements, microhardness measurements and microscopy examination using optical

---
1. Here and further concentrations are presented in mass %.

and transmission electron microscopes. The specific electrical resistance was measured by the compensation method on the device based on the low-ohm potentiometer. The calculated error of the specific electrical resistance measurements was ±7 %. To conduct the measurements, the strips ~6 mm wide and ~10 mm long were cut from the round samples ~0.30–0.34 mm thick. The microhardness was measured using diamond pyramid at a load of 20 g for 10 seconds at the middle of the radius of the sample. Transmission electron microscopy was performed using the JEM-1000 (JEOL, Japan) apparatus with accelerating voltage of 750 kV.

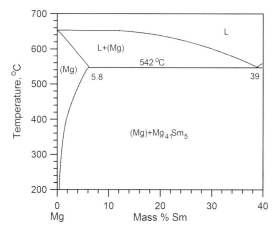

Figure 1: Mg side of the Mg-Sm phase diagram [6]

3 Results and Discussion

3.1 Specific Electrical Resistance and Microhardness of Mg-Sm Alloys after SPD

Results of the specific electrical resistance and microhardness measurements of the alloys after various treatments are given in Tables 1 and 2.

As one can see from the Tables, aging at 300 °C of the quenched Mg-2.8%Sm, Mg-4.5%Sm and Mg-5.5%Sm alloys decreases the electrical resistively due to the depletion of Mg solid solution by Sm. Microhardness measurements show that as compared with aging at 300 °C aging at 200 °C causes a higher hardening of the alloy Mg-4.5%Sm as a result of finer Sm-rich precipitates formed upon Mg solid solution decomposition. After aging at 300 °C, the microhardness is about the same that as in the as-quenched condition.

SPD at both room temperature and 200 °C results in significant strengthening of the alloys. In the Mg-Sm alloys, the microhardness after SPD at 200 °C is higher and the specific electrical resistance is lower than those after SPD at room temperature (Tabl.1,2). These facts can be explained by partial decomposition of Mg solid solution during deformation at 200 °C, This fact correlates with almost full decomposition of magnesium solid solution expected for preliminary aging at 300 °C, 6 h. and only partial depletion of it for preliminary aging at 200 °C, 8 h. On the other hand, in the case of SPD after aging at 300 °C, additional aging at 200 °C results in certain decrease in specific electrical resistance showing an additional depletion of Mg solid solution which confirms an additional Mg solid solution decomposition. Thus, SPD intensifies Mg solid solution decomposition and increases its strengthening effect.

Table 1: Specific electrical resistance of the Mg-Sm alloys after the various treatments

Alloy composition, %	Initial heat treatment	ρ (μOhm·cm)		
		Without SPD	SPD at 20 °C	SPD at 200 °C
Mg-2.8%Sm	Quenching from 510 °C, 5h	14.9	15.1	14.2
	Quenching + aging at 300 °C	10.7	8.9	9.2
Mg-4.5%Sm	Quenching	16.2	15.6	14.8
	Quenching + aging at 300 °C	11.3	12.1	–
	Quenching + aging at 200 °C	12.0	13.3	–
Mg-5.5%Sm	Quenching	19.3	18.2	16.8
	Quenching + aging at 300 °C	10.6	9.8	–

The highest strengthening was reached after aging at 200 °C when SPD was applied to the samples pre-aged at 200 °C.

i.e. strain aging. Some increase of specific electrical resistance after SPD of the alloy Mg-4.5%Sm aged preliminary at 200 and 300 °C suggests a partial dissolution of the Sm-rich precipitates. For the same heat treatment, the microhardness of the alloys increases with increasing Sm content.

3.2 Effect of SPD on Behavior during Aging of the Mg-Sm Alloys

Effect of SPD on behavior during aging of the Mg-Sm alloys was studied on the alloy with 4.5%Sm. After SPD the Mg-4.5%Sm alloy was aged isothermally at 200 °C for up to 64 hours. At this aging temperature, the decomposition of the Mg-Sm supersaturated solid solution is accompanied without SPD by strengthening and then softening [6]. The solid solution decomposition rate in this work was checked by specific electrical resistance and microhardness measurements. Results of the measurements are shown in Figs. 2 and 3. Decrease of the specific electrical resistance with increasing aging time (Fig. 2) shows, that during aging at 200 °C after only quenching + aging at 200 °C, 8 hours and quenching + aging at 300 °C, 6 hours without SPD decomposition of Mg-Sm supersaturated solid solution continues. However, for preliminary aging at 300 °C, 6 h. decomposition rate is significantly less because of higher depletion of the solid solution before aging at 200 °C. SPD results in the additional solid solution decomposition at 200°C that is revealed in the faster and higher specific electrical resistance decrease during aging for both initial states. Aging at 200 °C after SPD results in typical change of microhardness with appearance of maximum (Fig. 3). In the case of preliminary aging at 300 °C, 6 hours additional aging at 200 °C after SPD results in significantly lower microhardness maximum because of more depletion of Mg-Sm solid solution in initial state as compared with the case when aging 200 °C, 8 h is used before SPD.

Table 2: Microhardness of the Mg-Sm alloys after the various treatments

Alloy composition, %	Initial heat treatment	HV, kg/mm²		
		Without SPD	SPD at 20 °C	SPD at 200 °C
Mg-2.8%Sm	Quenching 510 °C, 5h	76.8	90.6	100.7
	Quenching + aging at 300 °C	78.8	86.1	95.8
Mg-4.5%Sm	Quenching	78.4	100,6	104,3
	Quenching + aging at 300 °C	81.5	99.6	–
	Quenching + aging at 200 °C	98.5	107.7	–
Mg-5.5%Sm	Quenching	88.4	107.7	109.4
	Quenching + aging at 300 °C	85.8	101.8	–

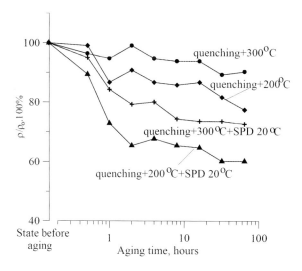

Figure 2: Specific electrical resistance of the Mg-4.5% Sm alloy vs. aging time at 200 °C (ρ_o is specific electrical resistance before aging)

3.3 Effect of Annealing on Strengthening after SPD

Investigations show, that strengthening effect caused by SPD in Mg-Sm alloys is enough thermal stable and is retained after annealing up to enough high temperatures. In Fig. 4 results of the microhardness measurements of the Mg-4.5%Sm alloy annealed at different temperatures after quenching + aging 200 °C, 8 h with/without additional SPD are shown. Annealing was performed successively increasing temperature by 50 °C from 100 °C up to 400 °C with exposure at every temperature for 1 h. As one can see, microhardness of the alloy after SPD surpasses that without SPD at all annealing temperatures.

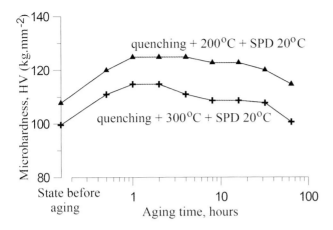

Figure 3: Microhardness of the Mg-4.5% Sm alloy vs. aging time at 200 °C

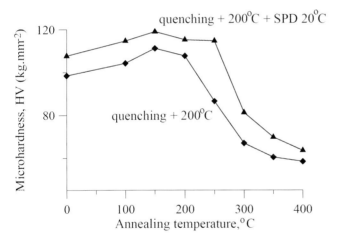

Figure 4: Microhardness change of the Mg-4.5%Sm alloy with increasing annealing temperature

3.4 Structure of the Mg-Sm Alloys after SPD

Transmission electron microscopy reveals formation of submicrocrystalline structure in the Mg alloys after SPD with quite fine grains. The remarkable feature of the structure formed is high dislocation density and not distinct boundaries between grains. The typical structure is shown in Fig. 5a. Annealing of the alloys after SPD results in formation of the more coarse grains. Simultaneously precipitations of the fine Sm-rich particles can be seen. The Sm-rich precipitates are seen on the grain boundaries and within grains (Fig. 4b). Obviously, Sm-rich precipitates restrain dislocation mobility and, as a result, strengthening effect caused by SPD is retained up to high annealing temperatures.

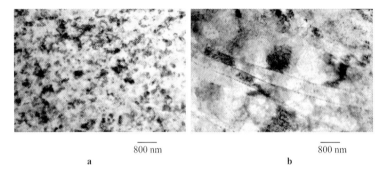

Figure 4: Structure of the Mg-Sm alloys. (a) – Mg-2.8%Sm, quenching + SPD at 20 °C, (b) – quenching + SPD at 200 °C + annealing 250 °C, 1 h

4 Conclusions

1. SPD of the Mg-Sm age-hardenable alloys results in significant strengthening.
2. Compared with aging at 300 °C, 6 h previous aging at 200 °C, 8 h results in higher strengthening of the alloys due to finer Sm-rich precipitates.
3. SPD intensifies the decomposition of Mg supersaturated solid solution upon aging.
4. Strengthening effect caused by SPD is retained after annealing at high temperatures.

5 References

[1] Lowe, T.C., R.Z. Valiev, R.Z., eds., *Investigations and Applications of Severe Plastic Deformation,* Kluwer Academic Publishing, Dordrecht, The Netherlands, **2000**.
[2] Mabuchi, M., Nakamura, M., Ameyama, K., Iwassaki, H., Higashi, K., Materials Science Forum. **1999,** 304–306, p. 67–72.
[3] Galiev, A., Kaibyshev, R., Scripta Materialia. **2004**, 51, p. 89–93.
[4] Chuvil'deev, V.N., Nieh, T.G., Gryaznov, M.Yu., Sysoev, A.N., Kopylov, V.I., Scripta Materialia. **2004**, 50, p. 861–865.
[5] R. Lapovok, P.F. Thomson, R. Cottam and Y. Estrin: Materials Science & Engineering, A, **2005,** 410-411, p. 390–393.
[6] Rokhlin, L.L., *Magnesium Alloys Containing Rare Earth Metals*, Taylor and Francis. London – New York. **2003,** p. 245.

Simulation and Modelling

Yielding of Magnesium: from Single Crystal to Structural Behaviour

S. Graff, D. Steglich, W. Brocks
GKSS Research Centre, Institute for Materials Research, Materials Mechanics, Geesthacht, Germany

1 Introduction

Thorough understanding of the relation between microstructural features and mechanical properties becomes a key competence in material science. This holds in particular for metals used in modern vehicles, where structural components with tailored properties are requested. Magnesium and its alloys are attractive for use in lightweight constructions because of their ratio of strength and density which may allow a significant weight reduction for structures. Due to their hexagonal crystallographic (hcp) structure and specific twinning mechanisms, magnesium alloys show a strong tension/compression asymmetry, which has to be accounted for in materials processing, construction design and structural assessment. Advanced material models, therefore, play a key role for a broad application of magnesium alloys in engineering. Deformation mechanisms, dislocation motion on specific slip systems and activation of twinning must be understood at the micromechanical level. The finite element method provides a helpful tool for the modeling of the mechanical response of single- and polycrystals. Thereby it is assumed that inelastic deformations and plastic flow are caused by slip on a limited number of slip systems and mechanical twinning.

In this work a model for crystal plasticity is used to identify the active deformation mechanisms of magnesium and to generate the resulting yield surface of magnesium rolled plates. The latter is than taken as reference for fitting model parameters of a phenomenological yield criterion and a hardening rule. This yield criterion allows for efficient simulation of mechanical deformation, e.g. the deep drawing process of a cup made out of a magnesium plate material.

The identification of deformation modes and fit of material parameters is realised using the experiments of Kelley and Hosford [1,2]. The yield criterion proposed by Cazacu and Barlat [3] and accounting for anisotropy and unlike tension and compression yielding is adopted in simulating the deep drawing process.

2 The Model for Crystal Plasticity

The kinematical theory for the mechanics of crystals has been established in the pioneering work of Taylor [4] and the theory by Hill [5], Rice [6], Hill and Rice [7]. The model of crystal plasticity used here employs the framework of Peirce et al. [8] and Asaro [9,10]. The implementation in the commercial finite element code ABAQUS used here is based on the user-material routine of Huang [11]. For the sake of brevity, only a summary of the relevant mechanisms and parameters is given here.

Slip in this framework is assumed to obey Schmid's law, the slipping rate $\dot{\gamma}^{(\alpha)}$ depends on the stress tensor $\boldsymbol{\sigma}$ solely through Schmid's resolved shear stresses,

$$\tau^{(\alpha)} = \mathbf{n}^{*(\alpha)} \cdot \frac{\rho_0}{\rho} \boldsymbol{\sigma} \cdot \mathbf{m}^{*(\alpha)}, \qquad (3)$$

where ρ_0 and ρ are the mass densities in the reference and current states. According to Peirce et al. [5], the constitutive equation of slip is assumed as a viscoplastic power law,

$$\frac{\dot{\gamma}^{(\alpha)}}{\dot{\gamma}_0^{(\alpha)}} = \left|\frac{\tau^{(\alpha)}}{\tau_Y^{(\alpha)}}\right|^n \mathrm{sign}\left(\frac{\tau^{(\alpha)}}{\tau_Y^{(\alpha)}}\right), \qquad (4)$$

where $\dot{\gamma}_0^{(\alpha)}$ is a reference strain rate and $\tau_Y^{(\alpha)}$ characterizes the current strength of the α slip system.

Three different hardening laws are applied in the following, namely linear hardening,

$$h(\overline{\gamma}) = h_0, \qquad (5)$$

Voce hardening (Agnew et al. [12]; Yi et al.[13]),

$$h(\overline{\gamma}) = h_0\left(1 - \frac{\tau_0}{\tau_\infty}\right)\exp\left(-\frac{h_0\overline{\gamma}}{\tau_\infty}\right), \qquad (6)$$

with $\tau\infty$ being the saturation stress, and particularly for deformation twinning,

$$h(\overline{\gamma}) = \begin{cases} h_0 & \text{for } \overline{\gamma} \leq \gamma_{ref} \\ h_0\left(\dfrac{\overline{\gamma}}{\gamma_{ref}}\right)^{m-1} & \text{for } \overline{\gamma} > \gamma_{ref} \end{cases} \qquad (7)$$

The latter law consists in linear hardening as $\overline{\gamma} \leq \gamma_{ref}$ and power law hardening as $\overline{\gamma} > \gamma_{ref}$. In this work, twinning is handled as additional slip mechanisms following Schmid's law, eqn. (3), and reorientation of crystallographic planes due to lattice rotation is not taken into account. The critical resolved and saturation shear stresses, τ_0 and τ_∞, the hardening parameters, h_0 and m the reference shear strain rate, $\dot{\gamma}_0$, and Norton exponent, n, as well as the coefficients describing latent hardening are model parameters to be identified. Basal $\langle\mathbf{a}\rangle$, prismatic $\langle\mathbf{a}\rangle$, pyramidal $\langle\mathbf{a}+\mathbf{c}\rangle$ plus tensile twinning on $\{10\overline{1}2\}$ are chosen for being the slip mechanisms [14]. These families include three basal, three prismatic, six pyramidal and six twinning systems.

3   Plane-Strain Compression Tests and Biaxial Tests

Plane-strain compression tests (channel die experiments) of Mg single crystals and polycrystalline rolled plate samples are used here in order to calibrate parameters of the crystal plasticity model. In case of the single crystal tests, the letters A to G correspond to the different initial orientations of the crystallographic planes with respect to compression and constraint directions. Each of the single crystals is taken as one finite element in the simulations. For the polycrystalline samples, the initial sample orientations are named by a combination of two of the letters R (rolling direction), T (transverse direction) and S (thickness direction). The first letter signifies the loading direction and the second letter the extension direction in the die. In magnesium rolled plate the $\vec{c}$ axis of the hexagonal structure is oriented in the plate thickness direction with a certain misorientation in rolling and in transverse direction. The specific rolled plate texture presented in [2] is mapped qualitatively to polycrystalline aggregates, discretized by a finite element mesh such that each grain is modeled as one finite element. In order to save computational time, however, each polycrystalline aggregate is modeled by only 8*8*8 grains, described by a single 8-nodes brick element each.

In order to retrieve information on the behaviour of polycrystalline representative volume elements (RVEs), the mechanical behaviour under plane stress conditions was investigated. Radial in-plane loading paths with different ratios of $\sigma_{Transverse}/\sigma_{Rolling}$ have been prescribed and the resulting mesoscopic strain was monitored. The polycrystalline aggregate was again discretized by 8*8*8 grains. The loaded surfaces have been subjected to periodic boundary conditions to ensure compatibility between the RVEs. By mapping the respective mesoscopic stress with fixed values for the plastic strain, yield surfaces can be constructed and the hardening behavior can be quantified.

Fig. 1a and Fig. 1b show the stress-strain curves obtained numerically, for the seven different orientations of a single crystal in the channel die experiment compared with the experimental data [1], and for the six initial orientations of polycrystalline samples compared with the experimental data [2], respectively.

Both simulations of single crystal and polycrystalline plane-strain compression tests give satisfactory results, beside general agreement between experimental and simulated results, in Fig. 1a and 1b, the following specific features are captured by the simulations:

- the high yield stress and the strong hardening of orientations A and B,
- the high yield stress but a lower hardening of curves C and D, as well as SL, TL, and LT,
- the unusual hardening behavior of curves E and F, LS and TS, with low yield stress and almost no hardening, at strains smaller than 6%, followed by a sudden increase in stress,
- orientation G with its very low stress level,
- orientation TL with relatively high yield stress and hardening.

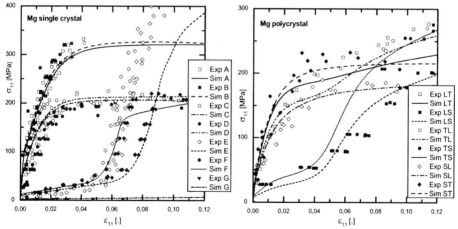

Figure 1: Experimental and simulation results of plane-strain compression test of Mg single crystals (left) and Mg polycrystalline rolled plate samples (right) for different initial orientations of the sample

As the micromechanical mechanisms of deformation are correctly mapped, plane stress simulations of a textured sample are conducted. The "mesoscopic" response of the RVE is taken to determine yield loci of the material. Fig. 2a displays yield loci for plastic strains from 0.01 to

0.10. They are not symmetric with respect to the coordinate origin, with the consequence that the yielding behavior is different in tension and compression and tensile strength is different in rolling and transverse direction. This simulated behavior is in good agreement with the observation reported by Kelley and Hosford [2]. Furthermore, hardening is not following an isotropic hardening law, but depends on the loading direction. It appears that for higher strains, the asymmetry is reduced, which can be explained by the completion of $\{10\bar{1}2\}$ twinning and "handover" of pyramidal slip systems. Fig. 2b illustrates the relative accumulated shear strain of the considered deformation modes at 0.01, 0.05 and 0.10 equivalent plastic strain as a function of the angle ρ representing the load path of constant ration $\sigma_T : \sigma_L$ to length direction of the plate. Basal is clearly the most active deformation mode over the whole range of biaxial tests. Moreover, this graph emphasises the competition between basal and prismatic glide which allow for deformation in a-direction. Twinning is very much active in the first percent of deformation strain and is therefore identified as being responsible for the reduced yield stresses in compression. Pyramidal slip plays an important role in case of biaxial tension, $\rho \in [30,60]$ and in compression after twinning saturates.

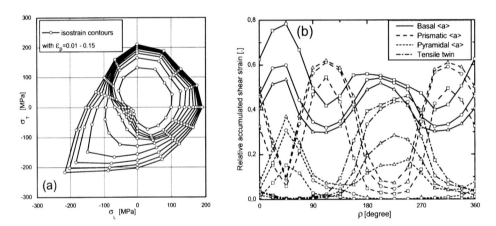

Figure 2: Yield loci and isocontours of plastic equivalent strain of an Mg rolled plate for plane stress loading in (L,T)-plane (a) and relative accumulated shear strain of the considered deformation mechanisms for plastic strain 0.01 (□)- 0.05 (○) - 0.10 (△) (b)

The presented methodology to simulate the plastic deformation behaviour of textured metals on the grain level is very much computational time consuming. This can be overcome by using phenomenological constitutive laws, which mask out the micromechanical mechanisms of deformation but describe the material as a continuum, using plastic potentials and associated flow rules. In this framework, materials symmetries can be exploited.

4 Phenomenological Modeling

The phenomenological yield surface proposed by Cazacu and Barlat [3] is a possible candidate to describe the deformation behaviour of magnesium alloys on the phenomenological level. The plastic potential includes the third invariant of the stress tensor and hence accounts for anisotropy and the strength differential effect. Its numerous model parameters (seven in the case of plane stress) are identified with the help of the generated yield surfaces of Fig. 2a. Since the yield surface shape changes with increasing deformation, the model parameters are defined as functions of the plastic strain and are thus not constant. An optimization procedure is used to fit the parameter functions and to guarantee convexity of the resulting surface.

Using this yield surface and its parameters, any mechanical process causing rate-independent isothermal plastic deformation can be predicted. Verification and validation of the model and the methodology of parameter determination has to be carried out. Lacking comprehensive experimental data, the authors present in the following numerical simulations of a simple deep drawing process based on different sets of model parameters.

Fig. 4 displays the simulation results considering four sets of model parameters, namely:

- **CaBaInd**, identified in [3] for 1% plastic strain (independent on plastic strain)
- **FitDep**, corresponding to yield surfaces of Fig. 3 (dependent on plastic strain)
- **FitInd**, initial values of set **FitDep** (independent on plastic strain)
- **MisesInd**, corresponding to an isotropic non-hardening von Mises material.

Fig. 3a shows the resulting cups for sets **MisesInd** (top) and **CaBaInd** (bottom). The cup is deformed homogeneously in case of an isotopic material, and a strong earing profile is obtained in case of the yield surface with pronounced anisotropy. Fig 4b shows the cup height after deep drawing along the cup circumference. The cup height is normalised with its initial radius. The earing profile obtained with sets **FitDep** and **FitInd** is moderate compared to the one obtained with **CaBaInd**, which is due to the less pronounced anisotropy of the yield surface.

5 Summary

The mechanical behaviour of pure magnesium was studied by use of a model for crystal plasticity. Simulations of plane-strain compression tests of single crystals and polycrystalline aggregates revealed that the experimentally observed phenomena for magnesium and its alloys, like a strong anisotropy and an important asymmetry of yielding in tension and compression, can be understood through the activation of only four deformation modes: basal, prismatic, and pyramidal slip as well as tensile twinning. The unlike yielding observed in textured material like rolled plates has been clearly identified to be controlled by twinning.

The approach which consists in studying single crystals and polycrystalline aggregates under plane-strain compression in order to identify the active deformation mechanisms, and extending afterwards to multiaxial test simulations in order to obtain a representation of the materials yield surface, seems to be successful. Model parameters for an anisotropic plastic potential have been calibrated referring to the computed material yield surface and the structural behaviour of rolled plates subjected to deep drawing could thus be simulated. The role of even small material anisotropy was shown to have an important impact on a cup in a deep drawing process. This underlines the importance of controlling material texture in components fabrication processes.

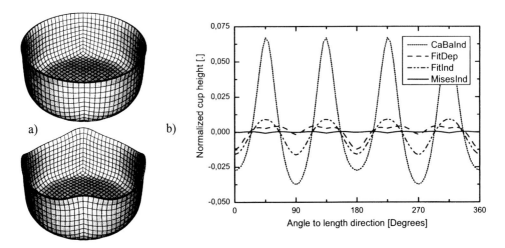

Figure 3: Deformed cup obtained through deep drawing with different yield behaviour (a); relative cup height quantifying the "earings" after deep drawing (b)

The work presented here contributes to a better understanding of the deformation behaviour of hcp metals on microscopic level. It can easily be extended to other hcp materials. It constitutes an attractive tool to be used in the improvement of fabrication processes and in the development of hcp alloys with specific desired mechanical properties.

P.S.: The input of W. Hosford for providing the original data of the tests is acknowledged.

7 References

[1] Kelley, E.W.; Hosford, W.F., Transactions of the metallurgical society of AIME 1968, 242, 5–13.
[2] Kelley, E.W.; Hosford, W.F., Transactions of the metallurgical society of AIME **1968**, *242*, 654–661.
[3] Cazacu, O.; Barlat, F., Int. J. Plasticity **2004**, *20*, 2027–2045
[4] Taylor, G. I., 1938. Plastic strain in metals. J. Inst. Metals 62, 307-324
[5] Hill, R., 1966. J. Mech. Phys. Solids 15, 95-102
[6] Rice, J. R., 1971. J. Mech. Phys. Solids 19, 433-455
[7] Hill, R., Rice, J. R., 1972. J. Mech. Phys. Solids 20, 401-413
[8] Peirce, D.; Asaro, R.J.; Needleman, A., Acta Met. **1982**, *30*, 1082–1119.
[9] Asaro, R.J., Adv. Appl. Mech. **1983**, *23*, 1.
[10] Asaro, R.J., J. Appl. Mech. **1983**, *50*, 921-934.
[11] Huang, Y., Report MECG-178, Div. Applied Science, Harvard University **1991**.
[12] Agnew, S.R.; Yoo, M.H.; Tomé C.N., Acta mater. **2001**, *49*, 4277-4289.
[13] Yi, S.-B.; Davies, C.H.J.; Brockmeier, H.-G.; Bolmaro, R.E.; Kainer, K.U.; Homeyer, J., Acta Materialia **2006**, *54*, 549-562.
[14] Staroselsky, A.; Anand L., Int. J. of PLasticity **2003**, *19*, 1843-1864.

Microstructure Formation in Magnesium Alloys Investigated by Phase-Field Simulation

J. Eiken, I. Steinbach, B. Böttger

1 Introduction

Magnesium has, compared to other cast metals, a very high potential for microstructure optimization [1]. The majority of today's commercial magnesium alloys contain aluminium, however they are still predominantly used in high-pressure die castings. An optimization of the microstructure adapted to lower temperatures rates would allow alloys such as AZ31 and AZ91 to be used more extensively in non high-pressure applications [2]. The prediction of microstructure parameters by means of numerical simulation, which can directly be related to engineering and mechanical properties, would present an indispensable tool in the design and optimization of magnesium alloy and processes for new critical applications [3]. Main focus is on the grain size, since a fine-grained microstructure is the most important prerequisite for good mechanical properties. Apart from the size, also the shape of the grains characterizes the microstructure. Depending on the process parameters, alloy composition and the nucleation density, the morphology of the grains may range from fine dendritic to almost globular structures. The latter are mostly preferred to increase the castability of the alloy and to reduce the tendency of hot tearing. In this study we apply the phase-field method to predict size and morphology of equiaxed grains dependent on the alloy composition and the process conditions. It is shown, how addition of calcium to the commercial alloy AZ31 influences both grain size and morphology.

2 Model Description

2.1 The Multiphase-Field Equations

The phase-field theory is an advanced computational approach which describes the evolution of so called „phase-fields" $\phi_\alpha(x,t)$, $\alpha = 1,\ldots,\nu$. In our model, these fields describe the distribution of either different phases or of grains with different orientations. If the respective phase or grain α locally exists $\phi_\alpha = 1$, else $\phi_\alpha = 0$. At the interfaces the values vary continuously from 1 to 0, thus the interfaces have a numerical thickness η, whose value can be adjusted for numerical convenience, but has to be small compared to the scale of characteristic microstructure lengths. The evolution in time is calculated by a set of phase-field equations, deduced by minimization of pairwise free energy functionals $F_{\alpha\beta}$ in double-obstacle formulation with respect to ϕ_α, see [4],[5],[6]. The resulting rate of the phase-field with time determines the local velocity of the interface between phase α and different adjacent phase(s) ß.

$$\dot{\phi}_\alpha = \sum_\beta \mu_{\alpha\beta}(\bar{n}) \left[\sigma^*_{\alpha\beta}(\bar{n}) K_{\alpha\beta} + w\Delta G_{\alpha\beta}(\bar{c}) \right]$$

with $K_{\alpha\beta} = \phi_\alpha \nabla^2 \phi_\beta - \phi_\beta \nabla^2 \phi_\alpha + \dfrac{\pi^2}{\eta^2}(\phi_\alpha + \phi_\beta)$ (1)

and $w = \dfrac{\pi}{\eta}\sqrt{\phi_\alpha \phi_\beta}$

In this equation $\mu_{\alpha\beta}$ is the mobility of the interface as a function of the interfacial orientation (here described by the interfacial normal vector). A hexagonal anisotropy function is used [7]. $\sigma_{\alpha\beta}$ is the effective anisotropic surface energy and $K_{\alpha\beta}$ is related to the local curvature of the interface. The interface is driven by the curvature contribution $\sigma_{\alpha\beta} K_{\alpha\beta}$ on the one hand and the thermodynamic driving force $\Delta G_{\alpha\beta}$ on the other hand. In most solidification processes the movement of the interface is controlled by a competition of these two contributions having opposite signs. The thermodynamic driving-force is a function of the local composition and couples the phase-field equations to the diffusion equations.

2.2 The Diffusion Equations

The concentration of the phase mixture at each location is given by the sum of the individual phase concentrations weighted by the respective phase-field parameter [6], [8]:

$$\bar{c}(\bar{x}) = \sum_{\alpha=1}^{v} \phi_\alpha(\bar{x}) \; c_\alpha(\bar{x})$$ (2)

To solve transport and redistribution a set of k coupled diffusions equations are deduced from the free energy functional [6],

$$\dot{\bar{c}}(\bar{x}) = \nabla \sum_{\alpha=1}^{v} \phi_\alpha \vec{\vec{D}}_\alpha \nabla \; \bar{c}_\alpha(\phi_\beta, \bar{c})$$ (3)

with $\vec{\vec{D}}_\alpha$ being the multicomponent diffusion coefficient matrix for phase α. The required phase concentrations $\bar{c}_\alpha$ can be determined from the phase-field parameters and the total concentrations using equation (2) and additionally the constraint of *quasi-equilibrium*. This constraint postulates that all phase diffusion potentials μ_α^i (defined as the chemical potentials of solute component i minus the chemical potential of the solvent component) are the same for coexisting phases, see [6],[9].

$$\bar{\mu}_\alpha(\bar{c}_\alpha(\bar{x})) = \bar{\mu}_\beta(\bar{c}_\beta(\bar{x}))$$ (4)

The quasi-equilibrium phase concentrations, which are required to solve the diffusion and the phase-field equations, are derived from thermodynamic calculations using approved databases. To speed up simulations, thermodynamic calculations are only run after certain intervals and the local data is stored and extrapolated. For the present applications, cross dependencies between the solute components are neglected to further reduce the computational effort.

2.3 Model for Heterogeneous Nucleation

To simulate heterogeneous nucleation, small non-resolved nucleant particles are distributed within the calculation area according to a given nucleant distribution function, specifying the number and radii of the particles. For the present simulation, exponential functions as can be seen in figure 1, are used. According to [10], [11] the required critical nucleation undercooling is calculated in dependence on the particle radius r:

$$\Delta T_{nuc} = \frac{2\sigma}{\Delta S r} \tag{5}$$

S is the entropy of fusion and the interfacial energy. When, at a particle position, the local undercooling (depending on the local composition) reaches the critical nucleation undercooling, a nucleus with critical radius is set. As long as the growing nucleus is too small to be accurately resolved by the numerical grid, the curvature is calculated analytically from the phase amount, later it is linearly transferred to the standard phase-field treatment. During its growth, the nucleus releases latent heat and redistributes solute, which hinders further nucleation. The interaction between heat extraction and growth-dependent latent heat release leads to a temperature drop to a certain maximal undercooling and a subsequent reheating. Only nucleation sites with smaller critical undercooling than the maximum reached undercooling are activated, while the others will simply be overgrown.

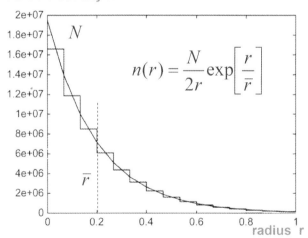

Figure 1: Example of a distribution function for nucleant particles

3 Simulations

3.1 Investigation of Basic Dependencies

The model, which is implemented within the phase-field software MICRESS [12] has been applied to ternary AZ31 (Mg-3%Al-1%Zn). The phase-field software is linked via the TQ interface of Thermo-Calc [13] to a newly assessed Calphad database of the Mg-Al-Zn-Mn-Ca system [14], [15], [16]. The melt is cooled down by a given heat extraction rate. Due to the high thermal conductivity, temperature gradients are low and neglected. Latent heat is averaged over the calculation area (1 mm × 1 mm). Simulations with varied input parameter have been performed. Figure 2a shows the resulting structures at 0.5 fraction of solid with the parameters given below. In a first variation the heat extraction rate has been increased. As a result the grain size is reduced and the morphology becomes more dendritic (figure 2b). This phenomenon is known from high-pressure die castings, where high cooling rates lead to very fine grain structures. In the second variation (figure 2c), a reduced grain size is achieved by a higher number of nucleant particles. In this case, the grain morphology becomes more globular. The search for an efficient grain refiner for aluminium bearing magnesium alloys is a major task for microstructure optimization [2], [17]. However, for effective refinement both nucleant particles and solute are required [18]. The effect of solute is exemplified here by a variation of the aluminium concentration. The enlarged constitutional undercooling restricts the growth of the earlier nucleated grains. As a result more grains can nucleate and the morphology becomes more dendritic. This effect is known for many solute elements, whose effect can be estimated by so called growth restriction factors [2], [18]. However, this estimation suffers from simplifying assumptions, e. g. the Scheil assumption, which is difficult to use. The complex interplay between anisotropic growth, solute redistribution and latent release during successive nucleation is much more complex and not yet fully understood. [1], [18].

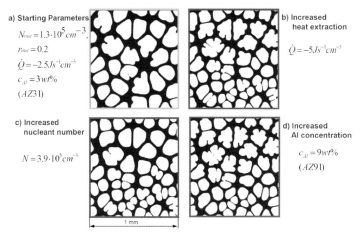

Figure 2: Basic effects of the input parameters on the grain structure

3.2 Investigation of the Effect of Calcium on the Grain Structure

In the following the effect of calcium on grain size and morphology is studied. Figure 3 shows simulated temperature curves for AZ31 with 0 %, 0.6 %, and 1.2 % Calcium. The maximal reached undercooling is slightly increasing with concentration, which his is commonly explained by increasing pile ups of calcium before the growth front which restricts the growth of the early nuclei, as discussed in [1], [2]. This is confirmed by the simulated calcium concentration profiles at the end of nucleation, plotted in figure 4 for $c_0^{Ca} = 0.6$ %, and $c_0^{Ca} = 1.2$ %. It shall be noted that the change in maximal reached undercooling was observed to be much smaller than the change in local constitutional undercooling at the interfaces. As a consequence of the higher maximum undercoolings, nucleant particles with smaller radii can become effective, which increases the number of nuclei and reduces grain size. Figure 5 gives the structures at 0.5 fraction of solid for the three initial calcium concentrations. The differences are small, but the tendency can clearly be seen by the final mean grain radii, which have been evaluated. The morphology of the grains becomes slightly more dendritic.

Further simulations have been performed for Calcium concentrations between 0% and 1.8% and for two different heat extraction rates. Maximum undercoolings and the final grain size are plotted in figure 6. The maximum undercooling increases in almost linear correlation with calcium concentration, while the decrease in grain size seems to slow down towards higher concentration values for the given conditions. A variation of the nucleant distribution function will probably affect these curves, which has not yet been studied. The shifted, but similar curves for both grain size and undercoolings for the two different heat extraction rates give evidence that the variation of calcium concentration and the variation of heat extraction rate do not interact significantly.

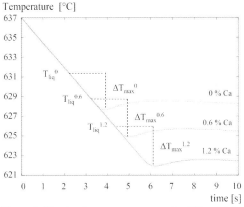

Figure 3: Simulated temperature curves for different initial calcium concentrations

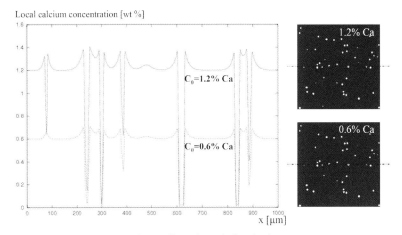

Figure 4: Calcium concentration profile at the end of nucleation

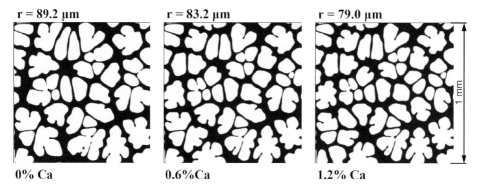

Figure 5: Structures resulting from simulations with varying calcium concentrations

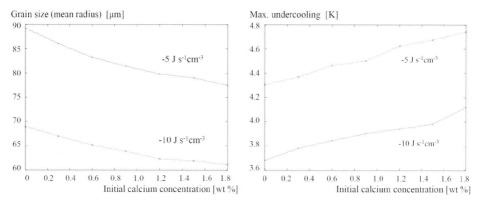

Figure 6: Grain size and maximum undercooling as a function of calcium concentration for two different heat extraction rates

4 Conclusions

A phase-field model with an integrated nucleation model and access to a Calphad database has been applied to the commercial Mg-alloy AZ31. The complex interaction of nucleation, anisotropic morphology evolution and multicomponent solute redistribution and diffusion has been simulated for equiaxed solidification. Basic factors controlling the microstructure have been studied. The effect of calcium concentration on the grain structure has been investigated.

The constitutional undercooling in front of the seeds just after nucleation was observed to be enlarged with concentrations and the maximum nucleation undercooling increases in almost linear correlation. For the given conditions, the grain size decreases continuously from 0–1.8 wt % calcium, but the maximum change in grain size is only 13 %. Towards increased calcium concentrations the change is slowing down. Simulations with a changed heat extraction rate revealed similar but shifted curves for the maximum undercoolings and the grain size.

5 Acknowledgement

We acknowledge funding by the German Research Foundation (DFG) in the Priority Programme SPP-1168.

6 References

[1] Y. C. Lee, A. K. Dahle, D. H. StJohn, Metallurgical.and Mat.Transactions A, November (2000), p. 2859
[2] D. H. StJohn et al., Metallurgical and Mat. Transactions A, July (2005), p. 1669
[3] H. Cao, M. Wessen, Metallurgical and Mat. Transactions A, January (2004), p. 309
[4] Steinbach et al. , Physica D 94 (1996), p. 135
[5] Steinbach, F. Pezzolla., Physica D 134 (1999), p. 385
[6] J. Eiken, B. Böttger, I. Steinbach, Physical Review E 73 (2006), p. 066122
[7] B. Böttger, J. Eiken, I. Steinbach, Acta Materialica 54 (2006), p. 2697
[8] J. Tiaden et al. , Physica D 115 (1998), p. 73
[9] S. G.Kim, W. T .Kim, T. Suzuki, Physical Review E 60 (1999), p. 7186
[10] T. E. Quested, A. L. Greer, Acta Materialica 52 (2004), p. 3859
[11] L. Greer, Advanced Engineering Materials (2003), p. 81
[12] www.micress.de
[13] www.thermocalc.se
[14] B. Böttger et al., Advanced Engineering Materials 8, (2006), p. 241
[15] R. Schmid-Fetzer, J. Groebner, Advanced Engineering Materials 3 (2001), p. 947
[16] R. Schmid-Fetzer et al., Advanced Engineering Materials 7 (2005) p. 1142
[17] K. U. Kainer, Stand der Technik und Entwicklungspotentiale für Magnesiumanwendungen, Wiley-VCH, (2000)
[18] M. Easton, D. H. StJohn et al., Metallurgical and Mat. Transactions A, July (2005), p. 1911

Effects of Calcium on Texture Evolution and Plastic Anisotropy of the Magnesium Alloy AZ31

T. Ebeling[1], Ch. Hartig[1], J. Bohlen[2], D. Letzig[2], R. Bormann[1,2]
[1] Hamburg University of Technology, Inst. of Materials Science and Technology, Hamburg, Germany
[2] GKSS Research Centre Geesthacht, Inst. for Materials Research, Geesthacht, Germany

1 Introduction

Magnesium wrought alloys have attained an increasing interest in research during the last years for saving weight of structural parts. However, the use of semi-finished components like sheets is still limited due to the insufficient formability at room-temperature, which can be traced back to the lack of adequate activation of slip modes [1]. As a consequence, a high plastic anisotropy appears during deformation of magnesium, and slight asymmetries in the texture of rolled sheets can have significant effects on their mechanical properties. For this reason, it is of fundamental interest for the processing of the magnesium alloys to gain more detailed insight into the deformation behavior of magnesium alloys. One basic approach can be a change in the alloy composition itself. In the present work we will use the well-known magnesium wrought alloy AZ31 and use a dilute addition of Calcium in order to study its effect on texture evolution during tensile and the resulting plastic anisotropy.

The predominant slip mode of magnesium is the basal slip mode which provides only two independent slip systems. Hence a fulfillment of the von-Mises criterion of five necessary independent slip systems requires an activation of the hard non basal slip modes [2]. Commonly, in addition to the basal mode, prismatic and <c+a> pyramidal slip modes as well as the $\{10\bar{1}2\}<10\bar{1}1>$ mechanical twinning mode are observed [3]. The prismatic slip offers two additional independent slip systems. The two commonly observed deformation modes that offer a deformation part in direction of the c-axis are the <c+a> pyramidal slip mode, which provides five independent slip systems, and the predominant $\{10\bar{1}2\}<10\bar{1}1>$ mechanical twinning mode. According to [4] the twinning mode occurs soon after the onset of plastic deformation and plays an important role in texture evolution of magnesium.

One basic approach to change the deformation behavior of magnesium alloys is a change in the alloy composition itself. In the present work, we will use the well-known magnesium wrought alloy AZ31 modified by a dilute addition of Ca in order to study its effect on texture evolution during tensile deformation and the resulting plastic anisotropy. The results will be compared with numerical simulations of the deformation behavior and texture development using a self consistent viscoplastic model (VPSC).

2 Experiment and Modeling

2.1 Experimental Procedures

AZ31 samples (nominal composition of 3 wt.% Al, 1 wt.% Zn, 0.3 wt.% Mn, balance Mg) with different additions of Ca (none, 0.7 wt.%) in solution treated (16h@350 °C) stress-relieved condition were hot-rolled in a two-high rolling mill (diameter 200 mm) at 400 °C. The rolling process consisted of 21 passes, each with a deformation degree of $\varphi = 0.1$. Between each two passes, the samples were re-heated for 3 minutes. After each five passes, a longer heat treatment of 30 minutes was carried out. Thus an initial thickness of the sheets of 10mm results in a final thickness of about 1mm. All sheets were tested in as rolled conditions.

For tensile-testing, specimens of the size $25 \times 5 \times 1$mm (gage length, width, thickness) were cut in rolling (RD, 0°) and transverse direction (TD, 90°). The tensile tests were carried out at room temperature using a universal ZWICK testing machine with a constant strain rate of $10^{-3} \cdot s^{-1}$. The longitudinal elongation was measured by means of an inductive extensometer. At least three samples of each material and direction were tested.

A metallography was carried out using a picric acid as etchant [5] to reveal grains in prepolished sections. The average grain size was determined from several micrographs using a computer-aided linear intercept measurement.

The texture samples were prepared from the middle part of each rolled sheet (diameter 25mm) and the middle part of the broken tensile specimens respectively (2 semicircles, diameter 9mm). The incomplete pole figures $\{10\bar{1}0\}$, $\{0002\}$, $\{10\bar{1}1\}$, $\{10\bar{1}2\}$, $\{11\bar{2}0\}$, $\{11\bar{2}2\}$ were measured up to a pole distance of $\alpha = 85°$ using a D8 X-ray diffractometer (BRUKER AXS Inc.) with reflection geometry and Cu-K_α radiation.

2.2 Modeling Procedures

A viscoplastic self consistent model was used to perform simulations. As boundary conditions, an elongation in load direction, zero shear strains and zero stresses perpendicular to the load direction were prescribed. The rate sensitivity for all slip and twinning modes was set to $n = 19$. The hardening of the involved individual slip and twinning modes was taken into account using the general voce hardening law (Formula 1) describing the critical resolved shear stress (CRSS) τ^s of the deformation systems as a function of the total shear strain Γ as follows:

$$\tau^s = \tau_0^s + \left(\tau_1^s + \theta_1^s \cdot \Gamma\right) \cdot \left[1 - \exp\left(\frac{\theta_0^s \cdot \Gamma}{\tau_1^s}\right)\right] \tag{1}$$

τ_0^s is the CRSS at the onset of plastic deformation, $\tau_0^s + \tau_1^s$ is the saturation shear stress and θ_0^s is the extrapolated work hardening for zero flow stress. θ_1^s is a hardening term which takes contributions form work hardening stage IV into account and was set to zero.

The simulations were done using two different combinations of deformation modes applying the systematic as described in Ref. [3]. While the first combination consists of the <a> basal, the <c+a> pyramidal and the twinning system, the second combination allows the <a> prismatic slip, too. To execute the simulations, the orientation distribution functions (ODF's) were converted to discrete orientations. For the simulations of tension in TD, the ODF's were rotated in tensile direction before.

Simulations with ODF's consisting of only 500 discrete orientations were executed first to get the voce-hardening parameters of the slip and twinning modes. The observed range was between 0.005 and 0.15 total true plastic shear strain. Parameter optimization was performed by calculating a theoretical stress-strain curve in the quoted range and minimizing the least squares differences between theoretical and experimental stress values. Thereby, suitable combinations of τ_0^s values were determined for each material in the first step. Subsequently, τ_1^s and θ_0^s were varied for all slip and twinning modes until the best fit of the simulated stress-strain-curve to the experimental curves was achieved.

With the obtained hardening parameters the simulation was repeated using the ODF's with 6000 discrete orientations to achieve detailed information of texture evolution. A verification of the deformation mode combination was then done by a comparison of the experimental texture evolution with the simulated one.

3   Results and Discussion

The microstructures of the two materials in as rolled conditions are shown in Figure 1. Both microstructures almost consist of almost recrystallised grains. In addition, mechanical twins are still visible. In both cases an average grain size of 10 μm is revealed. Visible stripes in the right figure consist of fractured Ca containing intermetallic precipitates.

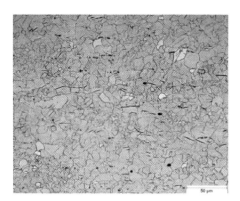
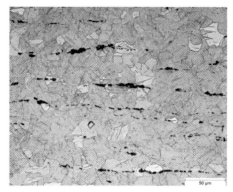

Figure 1: As rolled microstructure of AZ31 (left) and AZ31+0.7%Ca (right)

Mechanical properties were obtained from stress-strain curves and are collected in Table 1. It can be seen that the yield anisotropy is reversed by the addition of Ca. This effect already appears by an addition of 0.3% Ca (results not shown here). While AZ31 shows higher values in TD, the values in RD became higher by Ca-addition. The fracture elongation decreased with increasing Ca content.

The simulated stress-strain curves of AZ31 with an active <a> prismatic slip mode in RD and TD are compared with the experimental ones in Figure 2. The combination of deformation modes with an active <a> prismatic slip mode resulted in best fitting. In Table 2 the numerical results of the fitting are represented

Table 1: Strength values of the used materials in RD and TD

Sample material	Direction	$R_{p0.2}$ [MPa]	A [%]
AZ31	RD	226 ± 3	17.8 ± 0.7
	TD	261 ± 3	13.9 ± 4
AZ31+0.7% Ca	RD	236 ± 3	10.0 ± 2
	TD	218 ± 2	6.3 ± 4

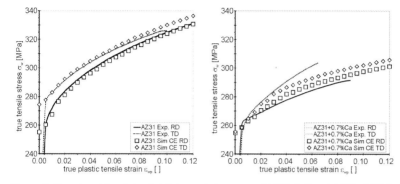

Figure 2: Experimental and simulated stress-strain curves of AZ31 (left) and AZ31+0.7%Ca (right)

Table 2: CRSS´s and hardening parameters according to Voce-Equation (Formula 1) of AZ31 (AZ31+0.7%Ca)

Deformation mode and symbol	Slip plane / Slip direction	τ_0^s [MPa]	τ_1^s [MPa]	θ_0^s [MPa]
Basal <a> (b)	$\{0001\} < 2\bar{1}\bar{1}0 >$	63 (60)	14 (10)	60 (20)
Prismatic <a> (pr)	$\{10\bar{1}0\} < 2\bar{1}\bar{1}0 >$	168 (160)	55 (20)	70 (20)
Pyramidal <c+a> (py)	$\{\bar{1}2\bar{1}2\} < \bar{1}2\bar{1}3 >$	176 (185)	100 (120)	300 (4000)
Twinning (tw)	$\{10\bar{1}2\} < 10\bar{1}1 >$	28 (40)	0 (0)	19 (19)

The only significant change caused by a Ca addition in the voce hardening parameters is the τ_0^s value of the mechanical twinning mode. Under ceteris paribus condition a higher CRSS results in less activity in both directions RD and TD. However, taking into account the different as-rolled textures, the activity of the twinning mode decreases drastically in RD while it increases in TD in case of the Ca-modified alloy.

Based on experimentally determined textures, the maximum intensity fiber in the ODF of the AZ31 alloy can be characterized as $\varphi 1$, Φ, $\varphi_2 = \pm 90°$; $< 20°$; $\varphi 2$ (Figure 3, left). In contrast the Ca-modified sheet shows a cold rolling like fiber φ_1, Φ, $\varphi_2 = \pm 90°$; $10°$; φ_2 with the typical splitting of the basal pole in RD (Figure 5, left). Furthermore a decrease in the texture intensity and a transverse expansion of the basal poles is found. The textures of the tensile samples in TD are shown in Figure 4 and Figure 6.

During the tensile test, the prismatic $\{10\bar{1}0\}$ planes rotate into a position perpendicular to RD resulting in a $\{0001\} < 10\bar{1}0 >$ main texture component (Figure 3-Figure 5, middle). Thus four prismatic $\{10\bar{1}0\}$ planes were favorably reoriented for prismatic slip during tensile testing. This reorientation appears sharper in the TD tensile samples of both materials. The Ca addition results in a decrease of this effect.

The simulated tensile texture evolution of AZ31 with an active prismatic slip mode is in good accordance with the experimental texture (Figure 3, right).

It is important to note that the simulations of the texture evolution of the AZ31 sample during tension in RD with an inactive prismatic slip mode show a φ_1, Φ, $\varphi_2 = \pm 90°$; $20°$; φ_2 texture. Consequently, they do not show the characteristic evolution of the $\{0001\} < 10\bar{1}0 >$ main texture component. The $\{0001\} < 10\bar{1}0 >$ main component develops only with an active prismatic slip mode, hence it is a result of the prismatic slip. These results demonstrate that already at room temperature, prismatic slip activity is necessary to explain the measured textures.

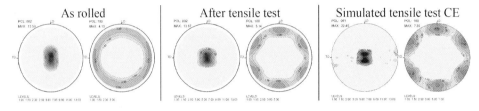

Figure 3: Experimental and simulated basal and prismatic pole figures of AZ31, tensile test in RD

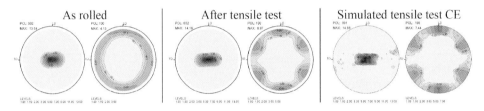

Figure 4: Experimental and simulated basal and prismatic pole figures of AZ31, tensile test in TD

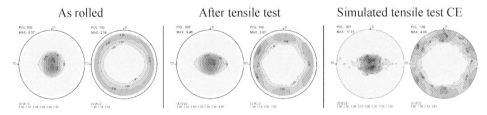

Figure 5: Experimental and simulated basal and prismatic pole figures of AZ31+0.7%Ca, tensile test in RD

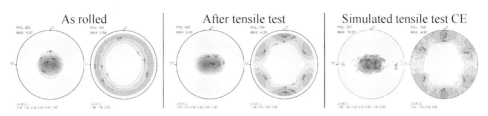

Figure 6: Experimental and simulated basal and prismatic pole figures of AZ31+0.7%Ca, tensile test in TD

The reversion of anisotropy in the tensile strength of the sheet materials can only be traced back to the different initial rolling textures. Therefore, simulations with the ODF's of AZ31 using the hardening parameters of AZ31+0.7%Ca and vice versa were carried out. They show clearly that the influence of the differences of voce hardening parameters is comparatively marginal while the initial texture is the main influence to the anisotropic behavior.

4 Conclusions and Summary

In the tensile test of the sheet samples it could be observed that an addition of Ca results in a reversed plastic anisotropy. Based on simulations of stress-strain curves this effect can be traced back to the different as-rolled textures, as the slightly different voce hardening parameters have only a marginal influence on to the anisotropic behavior. However, the influence of the Ca addition to the different initial rolling textures is still not clear yet. As a reason a different recrystalization behavior can be presumed. Further simulations and experiments have to be done.

The results of texture simulation based on a self consistent viscoplastic model agree with the experimentally observed texture development during tensile deformation, however only, if the activity of the <a> basal, <a> prismatic and <c+a> pyramidal slip modes as well as the twinning mode is taken into account. The experimental and the simulated tensile textures using that combination of deformation modes show a reorientation resulting in a $\{0001\}<10\bar{1}0>$ main component. Disabling the activity of the prismatic slip mode leads to a high misfit of the stress-strain simulation data to the experiment. Furthermore, the measured texture evolution could not be reproduced in the texture simulations hence the main component is a result of prismatic slip activity. Therefore, already at room temperature, the prismatic slip activity is necessary to develop the measured textures already at room temperature.

Further, the addition of Ca to the AZ31 leads to a decreased texture intensity in both the basal as well as the prismatic plane. The evolution of the $\{0001\}<10\bar{1}0>$ main component during tensile test appears less intense.

5 Acknowledgements

This work is part of the Virtual Institute „Key materials for light weight construction" of the Helmholtz Association of German Research Centres. In this respect, the authors are grateful to the "Helmholtz-Gemeinschaft" for the financial support. The authors also appreciate software support (VPSC6) by C.N. Tomé and the preparation of the cast slaps by G. Klaus of the RWTH Aachen.

6 Literature

[1] S.B. Yi, H.G. Brokmeier, K.U. Kainer, T. Lippmann in 6th Int. Conference on Magnesium Alloys and Their Applications, K.U. Kainer (editor), Wiley-VCH, **2003**, p.220–225.
[2] S.R. Agnew, M.H. Yoo, C.N. Tome, Acta Materialia 49 (**2001**), p.4277–4289.
[3] A. Styczynski, C. Hartig, R. Bormann, F. Kaiser, J. Bohlen, D. Letzig in 6th Int. Conference on Magnesium Alloys and their Applications, K.U. Kainer (editor), Wiley-VCH, **2003**, p.214–219.
[4] S.R. Agnew, C.N. Tome, D.W. Brown, T.M. Holden, S.C. Vogel, Scripta Materialia 48 (**2003**), p.1003–1008.
[5] V. Kree, J. Bohlen, D. Letzig, K.U. Kainer, Praktische Metallographie-Practical Metallography 41 (**2004**), p.233–246.

Modeling Microstructure and Mechanical Properties of Thin-walled Magnesium Castings

K.Weiss[1*], A. Wendt[1], Ch. Honsel[1], G. Arruebarrena[2], I. Hurtado[2], J. Väinölä[3], S. Mahmood[4], J. Townsend[4], A. Ben-Dov[5], A. Schröder[6], A. Pinkernelle[6]

[1] RWP GmbH, Roetgen, Germany
[2] Mondragon Goi Eskola Politeknikoa, Mondragon, Spain
[3] VTT Industrial Systems, Espoo, Finland
[4] Stone Foundries Ltd., London, United Kingdom
[5] Israel Aircraft Industries Ltd., Tel-Aviv, Israel
[6] INFERTA GmbH, Magdeburg, Germany
* Corresponding author

1 Introduction

In order to decrease air and noise pollution aerospace industry intensively searches for ways to lower the weight of airplanes. Successful integration of Mg alloys into the aerospace industry would reduce the airplane's weight, improve the noise damping and reduce the fuel consumption.

The use of structural materials in the aircraft industry is dictated by a number of factors including mechanical and physical properties (particularly density) as well as availability and cost. Magnesium alloys being the lightest of all of the commonly used structural metals may be considered as a very attractive alternative to the aerospace industry. In addition, magnesium alloys have other desirable features such as excellent machinability and damping capacity as well as good castability and higher strength-to-density ratio than other alloys. On the other hand, magnesium alloys exhibit a number of negative features including inferior strength – ductility relationship, in addition to a less than desirable fatigue strength, fracture toughness, yield and tensile strength [1].

The models and applications described in this paper were and are still being developed as part of an EU-supported research project named IDEA. The final aim of the research is to develop new magnesium alloys for aircraft industry, which fulfill the mechanical properties requirements especially in the heavily loaded areas of the parts. This paper describes the mechanical property model for sand casting, the validation procedure for the commercial alloy AZ91E and a first application.

The mechanical properties requirements of aircraft industry for secondary order structural Mg-castings are listed in Table 1 and compared to the properties of the alloy AZ91E and a new alloy developed in IDEA, the MRI207S. All data were measured in the IDEA project.

Table 1: Properties of Mg-alloys and requirements of aircraft industry

	UTS (MPa)	*YS* (MPa)	*E* (%)
Required Limiting Values	290	220	3.0
AZ91E – sand cast – T6	287	138	6.3
MRI 207S – sand cast – T6	304	220	4.0

2   Model Description and Implementation

It is generally accepted that mechanical properties of castings are determined by the microstructure. In the case of magnesium castings the microstructure is usually characterized by the grain size (GS) and the secondary dendrite arm spacing (SDAS). Relationships between tensile properties and these two parameters have already been determined since the forties of last century [2,3]. The effect of secondary phases, eutectic structures and interface properties between phases may be regarded as not significant compared to the influence of microstructure [4]. According to [4] the dominant influence of GS and SDAS on tensile properties increases with increasing Al content. The dominating influence of grain size on the mechanical properties of magnesium alloys is reflected by the high value of the Hall-Petch coefficient, which is approximately four times higher than that of aluminum alloys [5]. This assumption, however, does not absolve from investigating the effect of grain size to mechanical properties thoroughly for all alloys in question.

It is assumed that the grain size of magnesium alloys depends on the local solidification time according to the relation

$$GS = \alpha \, t_s^{1/\beta} \tag{1}$$

where t_s is the local solidification time and α and β are material constants. Suppose, *TYS* and *UTS* are the yield strength and the ultimate tensile strength of a material without presence of porosity. If *P* is the volume share of spherical pores *TYS* and *UTS* are reduced by

$$TYS_p = TYS \cdot (1 - P) \text{ and } UTS_p = UTS \cdot (1 - P) \tag{2}$$

as for example described in [6]. The more complex surface of non-spherical pores causes disruptions of the matrix and can be regarded as equivalent to a larger amount of porosity by a factor *s*. Thus, we obtain for the ultimate tensile strength

$$UTS_p = UTS \cdot (1 - s \cdot P) \tag{3}$$

The coefficient s (>1) is called the shape factor of the pore and describes the deviation of the pores shape from a spherical shape. The definition of shape factors can, for example be found in [6] or [7]. Some authors use the form factor *f* instead of the shape factor. The form factor is defined by

$$f = 1 - s \tag{4}$$

The well-known Hall-Petch relationship relates yield strength *TYS* with the average grain size *GS* through the equation

$$TYS = \sigma_{ys} + k_{ys}/(GS^{1/2}) \tag{5}$$

where σ_{ys} is the contribution from other strengthening mechanisms like for example inclusions. The proportionality factor k_{ys} is the so-called Hall-Petch coefficient. Also the ultimate tensile strength may be related to the grain size by an analogous relation with coefficients σ_{ts} and k_{ts}:

$$UTS = \sigma_{ts} + k_{ts}/(GS^{1/2}) \tag{6}$$

The elongation, E, depends on Young's modulus Y, the hardening exponent n, TYS and UTS as follows:

$$E = \{UTS/(Y^n * TYS^{1-n})\}^{1/n}. \tag{7}$$

Equations (1) to (7) allow simulating the local distributions of yield strength, ultimate tensile strength and elongation of castings.

In order to adjust the model to the commercial magnesium alloy AZ91E the parameters σ_{ys}, σ_{ts}, k_{ys} and k_{ts} had to be determined. For the Hall-Petch coefficient k_{ys}, literature gives values ranging from 210 to 600 MPa (µm)$^{1/2}$ and for σ_{ys} the values range from 53 to 84 MPa for the AZ91 cast alloy [8]. Typical values of the hardening exponent n are between 0.1 and 0.3.

A series of tensile specimens has been cast, and their mechanical properties as well as the microstructure were determined. Typical measured values are: Grain size: 176 µm, UTS: 150 MPa, TYS: 111 MPa, elongation E: 2.2 %. All values were measured in as-cast condition. Heat treatment leads to significant improvement of the properties but is not relevant for model verification. Casting was simulated with the FE-software WinCast®. The actually used values of the above mentioned coefficients were obtained by inverse simulation of the mechanical properties and lead to satisfying results, Figure 1. Before application to industrial castings the model and the obtained parameters are verified on a test casting.

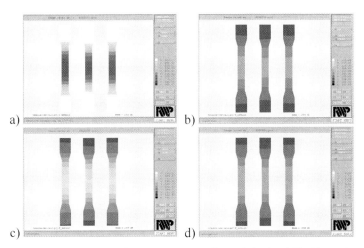

Figure 1: Three tensile specimens were cast at a time and simulated. The Figure shows grain size (a) with values between 170 and 180 µm, TYS (b), values between 110 and 112 MPa, UTS (c) between 148 and 152 MPa, and elongation (d), values at 1.8 %.

3 Model Validation

For verification a test plate, see Figure 2, was used. This plate consists of thick- and thin walled sections as well as of massive parts and a number of transitions between different wall thicknesses. The plate is mostly used for evaluation of different alloys and for comparing strengths and weaknesses of different casting technologies and alloys, but also for verification of numerical models. A series of plates has been cast and temperatures were measured in the positions marked in Figure 2a. Samples for investigating microstructure and porosity were excised from the positions shown in Figure 2b. Finally, tensile specimens were taken from the positions marked in Figure 2c.

Simulation of mold filling and solidification was performed using the software packages ConiferCast® and WinCast® and showed good agreement of simulated and measured temperatures, Figure 3. Simulation results were furthermore compared to microstructure, and mechanical properties in as cast condition. The simulation results are shown in Figure 3 and Figure 4. The grain size model predicts grain sizes rather well for local solidification times up to 100 s. For longer solidification times the actual rain sizes are overestimated. Since the mechanical property model depends on the grain size it also gives reliable results for local solidification times below 100 seconds. For the considered applications this confidence region is in general sufficient.

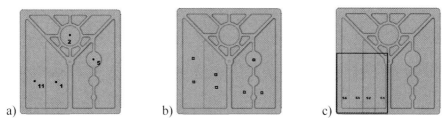

Figure 2: The AZ91E test plate a) positions of thermocouples, b) positions for microstructure samples, c) positions for excising tensile specimens

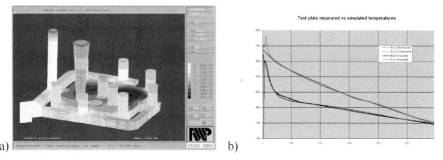

Figure 3: Mold filling and simulation of the sand-cast test plate were simulated. The comparison of measured and simulated results shows good agreement

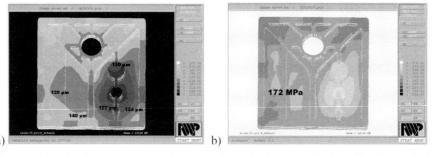

Figure 4: a) Simulated grain size (gravity die-casting). b) mechanical properties: UTS. The simulated values are given by the color scale. The numbers show the values measured in the corresponding positions. In all cases good agreement was achieved. Simulated values in the concerned area lie around 165 MPa.

As an example the simulated distribution of ultimate tensile strength is shown in Figure 4b. In the simulation the porosity distribution has been considered with a form factor of 0.6. The simulation slightly underestimates the measured value of 172 MPa.

4   Application to a Thin-walled Magnesium Sand Casting

A semi-structural housing was selected as a prototype casting, see Figure 5. The part is used as pressure bulkhead in the G150 business jet manufactured by Israel Aircraft Industries Ltd. Currently, the part is manufactured as an aluminum investment casting. In the IDEA project the housing is first manufactured as sand casting in the commercial AZ91E alloy in order to establish a first gating and feeding system. After successful casting it shall also be produced in the new developed magnesium alloy MRI207S.

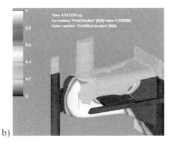

Figure 5: Prototype Casting „Housing": a) CAD-model and as AZ91E-casting, b) mold filling simulation at an intermediate step of gating design, simulated with ConiferCast®, c) optimized casting

Great challenges are the demands to a minimum wall thickness in the range of two to three millimeters. Several designs for gating and feeding have been studied. With the current design a minimum wall thickness of 2 mm was achieved. The simulations showed that the box could be successfully cast. The development of the gating and feeding design was widely supported by numerical modeling. Figure 5b shows the result of a filling simulation during the development of the gating system. The simulation indicated appearance of a cold shut and led to the choice of a different gating design.

Figure 6 shows predicted defect areas, expected grain size and mechanical properties. Grain size varies from 55 to 140 µm, and the mechanical properties reach values of approximately 160 MPa for UTS, 110 to 120 MPa for TYS and 2.2 % for elongation. The simulated porosity distribution has been taken into account in the prediction of mechanical properties. An average form factor of 0.8 of the pores has been assumed when calculating the ultimate tensile strength. Predicted microstructure and mechanical properties will be verified on excised samples. Next step after verification of the results will be the simulation and manufacture of the housing with the new casting alloy MRI207S.

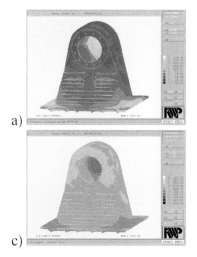
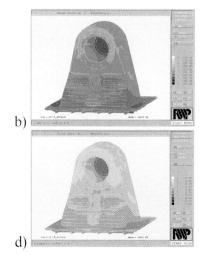

Figure 6: Simulation of the housing: a) Grain size distribution (55–140µm), b) TYS distribution over the casting (around 120 MPa), c) UTS distribution (around 160 MPa), c) Elongation (around 2.2 %). The simulations are performed with the FE software package WinCast®.

5   Model Extension and Prospects

The applicability of the described model for simulation of mechanical properties has been shown for the commercial Magnesium alloy AZ91E. The model has been verified on test parts and applied to an aircraft casting. This work has been carried out in the still ongoing EU research project IDEA. Current developments concern the extension of the model to the new developed alloy MRI207S. Following, the model will be used to adapt the design of gating and feeding for the housing shown in Figures 5 and 6 to the new alloy. Also in this case the model will be verified on excised specimens, but in addition full scale tests will be carried out by the aircraft manufacturer IAI Ltd.

6   Acknowledgement

This study was supported by the EC under contract number FP6-503826

7   References

[1] F.H. Froes et al, in Proceedings of the Annual World Magnesium Conference – IMA 2000, 2000, Vancouver, Canada, 56–63.
[2] E. F. Emley, Principles of Magnesium Technology, Pergamon Press, Oxford, London, Edinburgh, New York, Paris, Frankfurt, 1966, 365–377.
[3] C.H. Caceres and D.M. Rovera, J. Light Met., 2001,1, 151–156.

[4] H. Cao and M. Wessén, Metallurgical and Materials Transactions A, 2004, 35A, 309–319.
[5] E. Aghion and B.Bronfin, in *Magnesium alloys – Science, Technology and Applications*, (ed.: A.Aghion and D.Eliezer), ISRAELI Consortium for the Development of Magnesium Technologies) Haifa, Israel, 2004, 1–44.
[6] M. Todte, Ph.D. Thesis, Otto-von-Guericke- University, Magdeburg, Germany, 2002, 45–49.
[7] P. Simmons, Ph.D. Thesis, St. Anne's College, Oxford, UK, 2004, p. 52.
[8] C.H. Caceres et al., Materials Science and Engineering, 2002, A325, 344–355.

Melting, Casting and Recycling

Advancements in High Pressure Die Casting of Magnesium

H. Gjestland, H. Westengen
Hydro Magnesium Competence Centre, Porsgrunn, Norway

1 Abstract

The high pressure die casting (HPDC) process is characterized by rapid die filling and a subsequent rapid cooling of the molten metal. These characteristics are favourable for magnesium die casting alloys. Due to the high cooling rate the microstructure formed reveals a fine dendrite and grain structure, which in turn leads to substantial hardening and improved ductility.

The local cooling rate of the metal is highly dependent on the geometry of the casting, and the process parameters. The HPDC process allows a large flexibility in design of complex geometrical shapes. The latter factor means, however, that the cooling rate will not be uniform throughout the casting. The varying cooling rate will lead to local variations in the mechanical properties in the die cast component. These variations are due to an inherent property of the material, in contrast to casting defects like microporosity, non-metallic inclusions, filling defects, and formation of hot cracks. The mechanical properties of the casting are also affected by the pre-solidification of metal in the shot sleeve.

The porosity and some of the filling defects are frequently associated with gas entrapment in the die and a counter pressure built up in the die during injection of the metal. In aluminium die casting, evacuation of the die is a commonly applied technology to reduce these defects. For magnesium die casting, vacuum assisted die filling is not common. However, lately it is reported that vacuum is applied for critical applications with success.

Hydro Competence Centre for Magnesium has studied the HPDC process in a 420 tons die casting machine dedicated for research work. In the present paper the correlation between the thermal conditions through the process and the resulting microstructure and the mechanical properties in the casting is discussed. The effect of vacuum assisted die filling will also be discussed.

2 Introduction

High-pressure die casting is a versatile process for forming metal shapes from molten material, most typically nonferrous metals such as aluminium, magnesium, and zinc. The process is characterized by the use of steel moulds operated at low temperatures relative to the temperature of the metal being cast, and the application of high pressures (up to 1000 bars) to fill the moulds and feed the structure during solidification. The moulds can be designed to produce complex shapes with a high degree of accuracy and repeatability. The die castings are typically cast to near the final required shape, with a minimum of subsequent machining required. Maintaining a low die temperature promotes rapid cooling of the castings and results in a fine grain structure, providing a relatively high ratio of strength to weight. The application of high pressure during the filling of the die allows for effective filling of extremely thin sections relative to other casting processes. These positive features make die casting the mass production process of choice when the parts to be produced are required in a high volume. Some examples of magnesium die castings are shown in Figure 1.

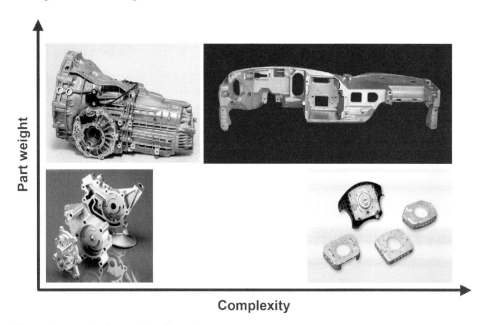

Figure 1: Examples of magnesium die castings

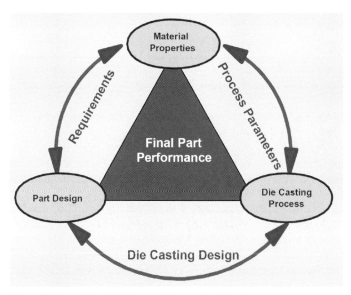

Figure 2: Interrelationship between design, material and process

The market for applications of magnesium die castings has increased during the last 10 years with average growth of 10 % pr year. This growth is made possible by the technological improvements in the high pressure die casting process and the mastering of the understanding of the interrelationship between the process, the material, and the design. This interrelationship is illustrated in Figure 2. By taking into consideration all the elements in the figure, the performance of the final part can be optimized.

In this paper the die casting process will be discussed with a focus on advancements in the understanding of important factors in the process and their impact on the material properties of the metal. The investigations are carried out at Hydro Magnesium Competence Centre.

3 The High Pressure Die Casting Process

Some important elements in high pressure die casting are:
- Melt handling with control of the thermal history of the metal from the furnace to the die cavity
- Cooling / solidification of the metal in the shot sleeve during dosing and in the slow portion (seconds) of the injection shot
- Injection (milliseconds) of molten / semisolid metal to the uncoated steel die under high pressure
- Cooling / solidification of the metal in the cavity during and after die filling and the metal pressure held at levels up to 1000 bar after die filling; feeding and segregation occurring under external / internal pressure following injection and solidification

The microstructural elements that are influenced include grain size, dendrite arm spacing, type and distribution of intermetallic particles, and especially the segregation arising from the composition of alloying additions. These elements are affected both by inherent factors like geometry, as well by process related parameters that can be either systematic / controllable or stochastic / uncontrollable.

4 Factors that Influence the Mechanical Properties

The grain size and the secondary dendrite arm spacing are both strongly dependent upon the solidification rate as shown in Figure 3 for the alloy AZ91D [1].
Grain sizes typically found in high pressure die cast AZ91D magnesium are 5–100 micrometers, and from Fig. 3 this corresponds to solidification rates in the wide range of 1000–10 °C/s. The heat content in the metal is extracted through the steel die. Locations close to the surface of the component will thus experience a high solidification rate, while the inner locations in thicker sections will solidify with a lower solidification rate, giving coarser grains. This structure variation is inevitable, but can to some extent be controlled by component and die design.

A more uniform structure is important to assure more homogeneity in the yield strength and the elongation throughout the part because of the strong relationship between these parameters.

In hexagonal structures such as magnesium, the tensile yield strength follows a Hall-Petch relationship as shown in Figure 4, and the hardening from the decreasing grain size is the most significant hardening mechanism for high pressure die cast magnesium alloys.

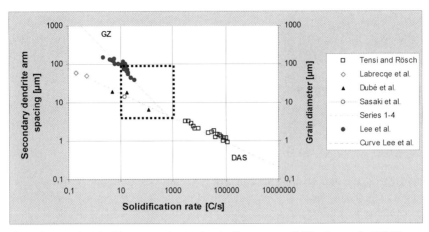

Figure 3: Secondary dendrite arm spacings and grain diameter vs. solidification rate in AZ91D

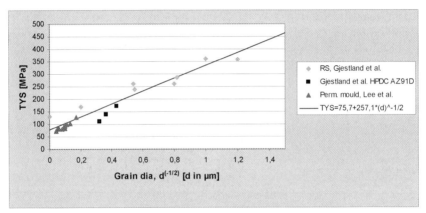

Figure 4: Hall-Petch relation for tensile yield strength of AZ91D [2,3]

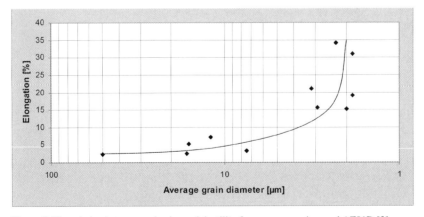

Figure 5: The relation between grain size and ductility for pure magnesium and AZ91D [3]

As a result of the hexagonal crystal structure, magnesium alloys have a limited number of slip systems at ambient temperatures. This causes inhomogeneous deformation, and fracture may occur at relatively low macroscopic deformations due to slip localization and the build up of locally high stresses at the grain boundaries. The plastic deformation becomes more homogeneous with decreasing grain size, as more grains will have a favourable orientation relative to the deformation direction. This will lead to larger macroscopic strains before fracture, as illustrated for magnesium alloy AZ91D in Figure 5.

5   Pre-solidification in the Shot Sleeve and Cavity Filling

When hot metal is poured into the relatively cold shot sleeve in a die casting machine, the metal will experience a significant heat loss. Locally at the sleeve wall, all superheat in the metal dissipates prior to the injection stage, thus initiating solidification. Shot sleeve solidification is comprised of the formation of a solid skin at the shot sleeve wall and dendritic crystals growing from the wall into the melt, forming a mushy zone.

Figure 6 illustrates a computer model of this solidification. As metal flows from the pour hole against the die, heat is extracted to the sleeve wall, and the volume fraction of pre-solidified metal increases at the die-end of the sleeve. The case used in the simulation shown in Figure 7 is from a Bühler 42D cold chamber machine. The magnesium alloy AM60B, is dosed by means of a gas displacement pump where the metal is supplied at a temperature of 680 °C into the pour hole of the H13 tool steel shot sleeve with a diameter ⌀60 mm and a length of 345 mm. The fill time to 50 % fill fraction in the sleeve is 1.25 s. The dwell time for the metal from the end of the dosing to start of the cavity filling (2nd phase of shot) is 1.15 s.

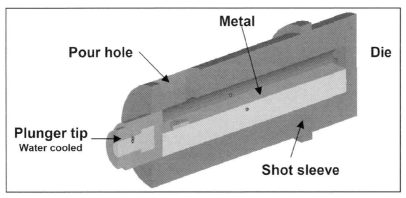

Figure 6: Computer modelling of pre-solidification in the shot sleeve

As seen from Figure 7a the metal has solidified on the cold walls of the shot sleeve within this short period of time, and there is a temperature gradient in the metal both radially and axially, Figure 7b. In the front of the shot sleeve, close to the start of the runner system, the temperature in the hottest area has dropped to about 630 °C, which is close to the liquidus temperature of the alloy (619) °C. The average volume fraction of solidified metal in the sleeve prior to injection of the metal into the cavity is approximately 14 %.

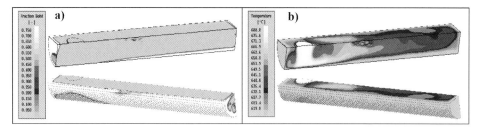

Figure 7: (a) Fraction solid distribution after filling of the shot sleeve and (b) the corresponding temperature distribution. The pouring hole is to the left in the pictures, while the right end is connected to the start of the runner system [9]

The plunger movement during injection will break off the primary dendritic crystal arms, in the following termed floating crystals, and the crystals will follow the residual melt flow into the cavity. It is reasonable to expect that the metal that was originally close to the die end of the shot cylinder will enter the cavity first [4, 5, 6]. If this metal is still above the liquidus temperature, it will be free from pre-solidified crystals. It is assumed that this metal ends up at the surface of the casting as seen in Figure 8b, and in positions far away from the gate locations. The microstructure at the surface will then be given by the high cooling rate at the die cavity wall. This is illustrated in case II in Figure 8a.

As filling proceeds, more coarse pre-solidified crystals will follow the metal flow from the shot sleeve into the die cavity. At the end of the filling stage, the metal with the longest dwell time in the shot sleeve will enter the die cavity. This metal will consequently have the highest fraction of pre-solidified coarse crystals and will end up close to the gate as in case I in Figure 8. This mechanism will give a decreasing gradient of coarse crystals from the gate towards the end of the flow path as shown in the box casting in Figure 9.

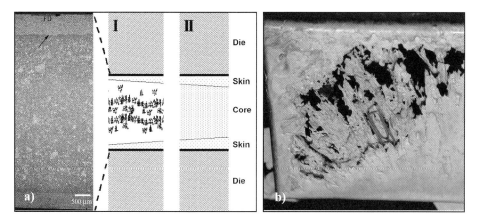

Figure 8: Model for skin formation and distribution of pre-solidified crystals. a) The micrograph to the left shows a cross section of a tensile bar normal to the flow direction (FD) with a solidified layer at the die wall, b) Solidification on the die wall during filling of a box casting [10] is illustrated in an interrupted casting shot. A paper clip is put in between the surface skins in a fill test casting to demonstrate the solid skin at the surface.

The thickness of the outer skin will be given by the balance between heat removed through the die surface and the heat flowing through the core of the cross section. Close to the gates, where the metal flux normally is high, a very thin, or vanishing, skin is expected to form. Further away from the gate, and in areas with less metal flux, the surface skin is expected to become thicker. This is shown in Figure 9. The highest volume fractions are typically located a) throughout the whole cross section close to the gates, and b) in the inner core of the casting wall away from the gates [7]. Far away from the gate, location D, almost no pre-solidified crystals are seen.

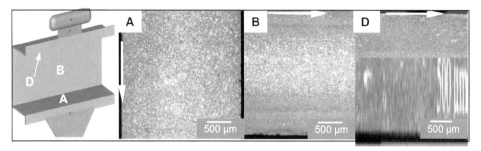

Figure 9: Floating crystals in AM60B (grains with bright contrast) in a box casting. A) Wall close to the gate, B) in the bottom and D) wall far from the gate. Arrows indicate flow direction [7]

4.1 Influence of the Pre-solidified crystals on the Properties of a Casting

Figure 10 shows a longitudinal cross section in a HPDC tensile bar after tensile testing. The floating crystals, pre-solidified in the shot sleeve, are located in the centre zone of the specimens gauge length. The outer zone of the bar shows a microstructure of small grains almost free from these crystals. Between the two zones, pore bands are formed.

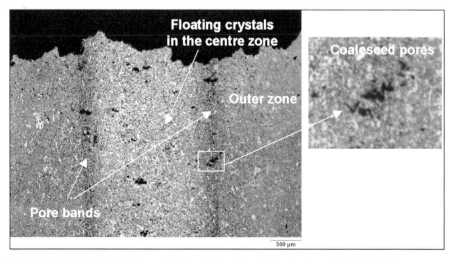

Figure 10: Cracks initiated in the pore bands of a tensile bar during tension, alloy AM60B [8]

The floating crystals and the porosity bands impact the elongation [8]. During straining, pores close to each other can coalesce and cause fracture when the local stress exceeds the yield strength of the material. No cracks seem to be initiated in the fine grained surface zones. To assure reliability in die cast components, the presence of the floating crystals should therefore be minimized.

6 Efforts to Avoid Pre-solidification in the Shot Sleeve

Pre-solidification in the shot sleeve is affected by the heat transfer between the metal and the sleeve wall, the heat conductivity in the sleeve wall, the temperature of the sleeve and the contact time between the metal and the sleeve. Computer simulations have been carried out to estimate the volume fraction of the pre-solidified material formed under different thermal conditions in the shot sleeve, using the case from the Bühler die casting machine mentioned above. The results are shown in Table 1.

Table 1: Computer calculation of fraction pre-solidified AM60B as a function of thermal conditions in the shot sleeve

No	Sleeve condition	Volume fraction of pre-solidified metal
1	Heat transfer coefficient $h(T) = 5000$ W/m² Heat conductivity in the sleeve $\lambda = 35.4$ W/m °C Temperature of the sleeve $T = 200$ °C	14 %
2	Heat transfer coefficient $h(T) = 600$ W/m² Heat conductivity in the sleeve $\lambda = 35.4$ W/m °C Temperature of the sleeve $T = 200$ °C	0 %
3	Heat transfer coefficient $h(T) = 5000$ W/m² Heat conductivity in the sleeve $\lambda = 7.4$ W/m °C Temperature of the sleeve $T = 200$ °C	5 %
4	Heat transfer coefficient $h(T) = 5000$ W/m² Heat conductivity in the sleeve $\lambda = 35.4$ W/m °C Temperature of the sleeve $T = 440$ °C	2.5 %

The reference is condition 1 which is a commonly used H13 tool steel sleeve where as much as 14 % of the metal is likely to solidify at the wall prior to injection. The three other conditions represent different ways of changing the thermal properties of the sleeve in order to reduce the pre-solidification.

The shot sleeve conditions presented in Table 1 were tested in practice in a 420 metric ton Bühler Evolution die casting machine.

- Condition no 1: The shot sleeve was made from H13 tool steel.
- Condition no 2: Boron nitride was sprayed inside the H13 tool steel sleeve.

- Condition no 3: A H13 tool steel shot sleeve was equipped with a metal matrix composite (MMC) liner of Ti-SiC.
- Condition no 4: A H13 tool steel shot sleeve was equipped with heating cartridges in the sleeve wall.

Micrographs of cast tensile bars are shown in Figure 11. The H13 sleeve, Figure 11a, gives pre-solidified crystals distributed over the cross section, while there is almost no pre-solidification visible in the micrograph of the bar cast with boron nitride coating in the sleeve, Figure 11b. Only a small volume fraction of the pre-solidified crystals can be seen in the bars cast with the MMC-lined and the heated sleeve, Figures 11c-d. These results reveal a potential for controlling the pre-solidification in the shot sleeve.

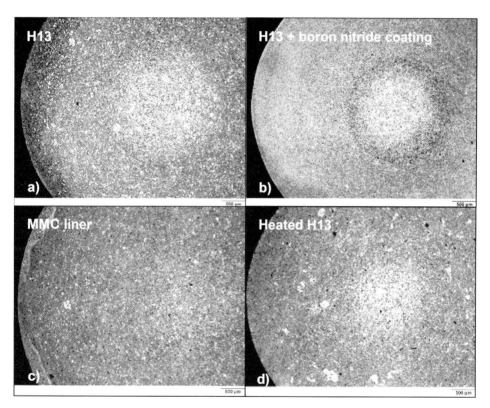

Figure 11: Microstructure in tensile bars cast with different thermal conditions in the shot sleeve, alloy AM60B. The 'white particles' are the pre-solidified crystals

Figure 12 shows the mechanical properties of the four series of tensile test bars representing the different shot sleeve conditions. The bar graphs reveal that the average values are improved when the volume fraction of pre-solidified material is reduced.

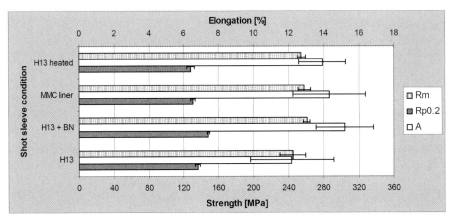

Figure 12: Bar graphs showing the average mechanical properties and one standard deviation in tensile bars cast with different thermal conditions in the shot sleeve, alloy AM60B

A large series of tensile bars in the alloy AM60B were cast and tensile tested using the H13 sleeve, the MMC-lined sleeve and the heated H13 sleeve. Due to the difficulty in uniformly applying the boron nitride coating, appreciable scatter was anticipated in the results using this sleeve condition.

The results were processed with statistical tools to evaluate the scatter in the results. Figures 13a-c shows frequency bar graphs for the elongation values in the three series.

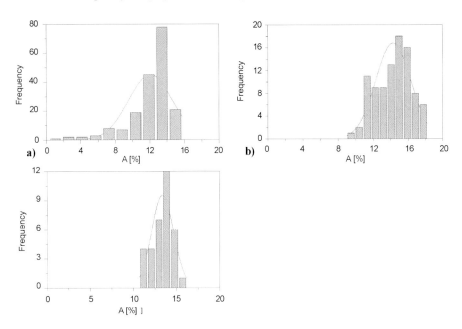

Figure 13: Frequency bar graphs for the elongation in tensile specimens cast **a)** with a H13 shot sleeve (200 specimens of AM60B), **b)** with a MMC-lined shot sleeve (93 specimens of AM60B) **c:** Frequency bar graphs for the elongation in tensile specimens cast with a heated H13 shot sleeves (34 specimens of AM60B)

The H13 series, Figure 13a, shows a left skewed distribution with very low minimum elongation values. In the two other series, the 'tail' with the small values is removed, leaving a distribution with much less scatter than in the first series, Figures 13b-c.

The most important impact on the elongation value is the reduction in the scatter when pre-solidification is reduced.

7   Vacuum Assisted Die Casting

The porosity and portion of the filling defects sometimes observed in high pressure die cast components are frequently associated with gas entrapment in the die and the counter pressure building up in the die during injection of the metal. In die casting of aluminium, vacuum assisted die filling is a commonly applied technology to reduce these defects. For magnesium die casting, vacuum technology is less common, but is sometimes used for critical applications. Vacuum technology can never substitute for good die casting design practice in the engineering of the die cavity, runners, gates and overflows, as well as optimum machine operation.

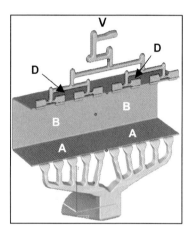

Figure 14: U-shape profile with vacuum channels. A, B and D are locations where properties and structures have been evaluated, V is the location for the vacuum valve

Several systems for moderate level vacuum, 100–200 mbar, for die casting are currently commercially available. The Hydro Magnesium Competence Centre has initiated an investigation to study the effect of vacuum on high pressure die casting of magnesium alloys. A Fondarex™ vacuum system has been implemented on the Bühler Evolution 42D die casting machine. A newly designed U-shaped die is equipped with vacuum channels and a vacuum valve as shown in Figure 14.

7.1   Challenges for Vacuum Assisted Die Casting of Magnesium

Trials with a standard shot profile for this U-shape casting resulted in a relatively poor vacuum level (~620mbar) in the die at the onset of 2nd phase injection, due to the short pumping time. No improvements were observed in the mechanical properties. To achieve a better vacuum level in the die, it is necessary to extend the evacuation time. But as discussed above, the dwell time for the metal in the shot sleeve is an important factor to reduce the pre-solidification in the shot sleeve. The challenge is therefore to find the best compromise between these two requirements.

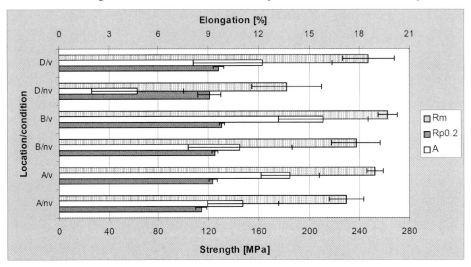

Figure 15: Mechanical properties of AM60B cast bars. A, B and D refer to locations in the U-profile. v = vacuum, nv = no vacuum

The plunger velocity in the slow phase of the shot (1st phase) was then reduced. The resulting extended evacuation time allows a vacuum level of about 375 mbar. However, as a standard H13 tool steel shot sleeve was used, the prolonged dwell time for the metal in the shot sleeve resulted in more pre-solidification in the shot sleeve and a large scatter in the mechanical properties as discussed above.

The reduced heat loss due to the lower heat conductivity of the MMC-liner made it possible to employ the reduced plunger velocity giving a vacuum level of ~375 mbar without any serious pre-solidification in the shot sleeve.

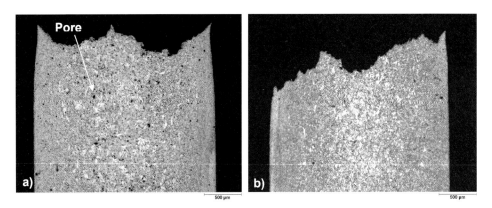

Figure 16: Microstructure of castings processed with and without vacuum assistance, a) no vacuum, location A in Figure 14, A_m = 9,76 %, b) vacuum, location A in Figure 14, A_m = 15.5 %

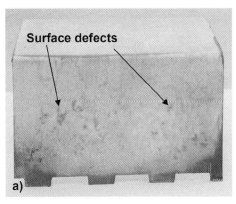 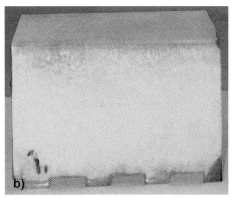

Figure 17: Surface appearance of castings processed a) without vacuum and b) with vacuum assistance

Figure 15 shows the improvement in the mechanical properties of the alloy AM60B, especially in the elongation when vacuum is applied, and Figures 16a-b reveal the microstructure with the reduced porosity in some locations. These trials indicate that even a moderate vacuum level (~375 mbar) has an effect on the mechanical properties.

Another interesting observation is the surface quality improvement when vacuum is applied. This is shown in Figures 17a-b where dye penetrant has been used to more clearly develop the surface defects like cold flows. A possible explanation for this improvement could be that the chance for entrapped air pockets between the metal and the die wall is reduced or eliminated when vacuum is applied.

8   Summary

- The mechanical properties of magnesium alloys are strongly affected by the grain size. Small grains will both improve the ductility and the strength of castings. The rapid cooling rate in high pressure die casting is highly efficient in securing excellent mechanical properties.
- Use of conventional shot sleeves manufactured from of H13 tool steel results in a significant volume fraction of pre-solidified material in the sleeve due to heat loss in the metal prior to injection. These pre-solidified crystals, which are α-magnesium dendrite fragments, may lead to reduction and a large scatter in the mechanical properties in the cast components.
- Pre-solidification in the shot sleeve can be decreased by a) reducing the heat transfer from the metal to the sleeve wall by insulating the inner sleeve wall, b) reducing the heat conductivity in the sleeve wall, c) reducing the heat transfer by maintaining a higher temperature in the sleeve wall. These efforts in turn will lead to improved mechanical properties in the casting.
- Vacuum assisted die filling has been demonstrated to give improved mechanical properties and a better surface quality of the casting.
- Conventional high pressure die casting is a production process that has proven to be well suited for magnesium. To utilize the full potential of the process, a good understanding of the interrelationship between the material, the process and the component design are important. In critical applications, thermal control in the shot sleeve and vacuum assisted die filling can be of valuable help to meet the requirements of the component.

9 References

[1] Dubé, D., Couture, A., Carbonneau, Y., Fiset, M., Angers, R., Tremblay, R., Secondary dendrite arm spacings in magnesium alloy AZ91D: from plaster moulding to laser remelting, Int. J. Cast Metals Res., **1998**, 11, pp. 139–144

[2] Choong Do Lee et al., Effects of Solidification Rate on Microstructure and Mechanical Properties of AZ91D Magnesium Alloy, J. Kor. Inst. Met. & Mater., Vol 38, No. 9, pp. 1225–1232, **2000**

[3] Nussbaum, G., Sainfort, P., Regazzoni, G., Gjestland, H., Strengthening Mechanisms in the Rapidly Solidified AZ91 Magnesium Alloy, Scripta Metallurgica, Vol 23, pp. 1079-1084, **1989**

[4] Garber, L. W., Theoretical analysis and experimental observation of air entrapment during cold chamber filling, Die Casting Engineer, May-June **1982**, pp 14–22.

[5] Laukli, H. I. et al., The effect of solidification of metal prior to injection in HPDC on the grain size distribution in a complex die casting, *NADCA 2002*

[6] Wang L. et al, Simulation of Flow Pattern and Temperature Profile in the Shot Sleeve of a High Pressure die Casting Process, *NADCA 2001*

[7] Laukli, H. I., High Pressure Die Casting of Aluminium and Magnesium Alloys – Grain Structure and Segregation Characteristics, Doctoral Theses, Norwegian University of Science and Technology, **2004**

[8] Gjestland, H., Crack initiation and crack propagation in HPDC tensile bars, Hydro Magnesium Competence Centre, Internal report, August **2003**

[9] Sannes, S., Gjestland, H., Westengen, H., Laukli, H. I., Lohne, O., Die Casting of Magnesium Alloys – The importance of Controlling Die Filling and Solidification, *paper no. 2003-01-0183, SAE* **2003**, *Detroit, USA*

[10] Sannes, S., Gjestland, H., Westengen, H., Laukli, H. I., Lohne, O., Magnesium Die Casting for High Performance, *6th Int. Conf on Magnesium Alloys and Their Applications, vol. 6, pp 725–731*, ed. K.U. Kainer, **2003**, Frankfurt, Germany

A New Analytical Methodology for the Assessment of Grain Refinement in Magnesium Alloys

David H. StJohn [1], Peng Cao [2], Ma Qian [3] and Mark A. Easton [4]

CAST Cooperative Research Centre
[1] School of Engineering, The University of Queensland, Brisbane, 4072, Australia
[2] Department of Engineering, The University of Waikato, Private Bag 3105, Hamilton, New Zealand
[3] BCAST (Brunel Centre for Advanced Solidification Technology), Brunel University, West London, UK
[4] Department of Materials Engineering, Monash University, Melbourne, 3800, Australia

1 Abstract

Although zirconium is an excellent grain refiner for magnesium alloys that do not contain aluminium, it is expensive and research has focused on reducing the amount of zirconium required for good refinement. On the other hand, for those magnesium alloys that contain aluminium it has been much more difficult to develop a commercially viable and reliable grain refiner.

This paper presents a new methodology for investigating the grain refinement performance of alloys and master alloys and then uses this methodology to shed new light on the mechanisms occurring during superheating and native refinement, and the effect of iron and manganese on grain refinement. With this new methodology, and the knowledge it generates, the potential for the discovery of new or improved refiners is significantly enhanced.

2 Introduction

Solidification is one of the most important microstructure formation processes for metallic alloys. The initiation of solidification, namely nucleation, almost always occurs heterogeneously in large masses of liquid metals. The ability to control such a process lies at the heart of the development of a novel technology or microstructure [1]. This is because nucleation controls to a large extent the initial structure type, size scale, and spatial distribution of the product phases [2]. In the metal casting industry, it is common practice to control the process through the introduction of carefully selected nucleating agents to a liquid metal, referred to as inoculation, for the formation of fine, uniform and equiaxed non-dendritic grain structures. Achievement of such a grain structure is desirable for the many benefits it brings in the solidification process itself and the final properties of the cast products. For magnesium alloys grain refinement has further importance, in that a fine grain size can fundamentally change the formability of a magnesium material, which has been a primary barrier that restricts the wider application of magnesium alloys in wrought forms. For instance, grain refinement below a grain size of about 115 μm for pure magnesium can significantly enhance the ductility compared to coarse grained (i.e. 1000 μm) pure magnesium materials [3]. Samples of Mg-9Al-1Zn with a grain size of about 1μm can exhibit elongations of more than 1,000 % at room temperature [4]. Further to these established benefits of grain refinement, recent work [5] has shown that fine and spherical primary magnesium crystals are directly obtainable from the liquid state simply by grain refinement, without liquid agitation. This offers a simple alternative to semisolid processing of magnesium alloys. Another recent development is that well-controlled grain refinement can directly lead to spherical grain structures in magnesium alloys such as AM-SC1 and ZE41 cast under different cooling rates [5]. This provides further opportunities to improve the structural uniformity of a magnesium alloy. In all probability, grain refinement presents itself as one of the most important factors for the further development of magnesium alloys.

Central to the successful inoculation or grain refinement of a liquid metal is the identification of a potent nucleating substrate (i.e., those characterised by a small contact angle θ [6]). Because of the lack of data on θ, crystallographic similarity, in most cases, has been the only measurable criterion that may be used to assist the selection process. The criterion in general explains most identified effective nucleating substrates as good crystallographic matching favours the formation of a low energy coherent interface. However, experimental observations have well established that the effectiveness of a nucleating substrate is not just determined by crystallographic matching. Many other factors such as the substrate size and geometry, the interactions between the substrate phase and the alloy chemistry including various impurity/minor elements, the undercooling experienced by the melt during solidification, the settling speed of the substrate in the melt, etc., could all come into play. It is desirable and necessary then to develop an alternative methodology for both the assessment and development of new grain refiners under complex metallurgical circumstances encountered by a commercial alloy.

This paper presents a new methodology for investigating the grain refinement performance of alloys and master alloys and then uses this methodology to shed new light on the mechanisms occurring during superheating and native refinement, and the effect of iron and manganese on grain refinement. With this new methodology, and the knowledge it generates, the potential for the discovery of new or improved refiners is significantly enhanced.

The new methodology for the assessment of grain refinement was developed predominantly from studies of the grain refinement of aluminium alloys [7] and recently applied to studies of grain refinement in magnesium alloys [8]. It is based on using a simple graphical representation of the relationship between grain size and the growth restriction factor Q to reveal information about the grain refining effectiveness of alloy systems and master alloys. The following section describes the principles underlying this simple representation. Section 3 then applies this methodology to analyzing the results from experiments on Mg-Al alloys to reveal new knowledge about the grain refinement of these alloys. As a result of applying this analytical method a number of important factors need to be considered if it is going to be possible to reliably compare studies by different researchers on a range of alloy systems and master alloys. These factors are considered in Section 4.

3 A New Analytical Methodology for the Study of Grain Refinement

Due to the importance of grain refinement to a broad range of aluminium alloys, considerable work has been carried out for over half a century to determine the mechanisms by which grain refinement occurs [9–13]. It is now generally accepted that both the potency of the nucleant particles (defined here as the undercooling required for nucleation, ΔT_n) and the segregating power of the solute (defined as the growth restriction factor, Q) are critical in determining the final grain size [13–17].

It has been found [7] that the grain size can be related to the factors defined above by an equation of the form

$$d = \frac{1}{\sqrt[3]{\rho f}} + \frac{b_1 \Delta T_n}{Q} \tag{1}$$

where ρ is the density of nucleant particles and f is the fraction of these particles that are activated, across a wide range of aluminium alloys and for the existing data on magnesium alloys. Hence, the first term of Equation 1 is related to the availability of nucleant substrates and the second term is related to the potency of the nucleant particles and the development of constitutional supercooling which restricts grain growth and facilitates further nucleation. A theoretical framework for the second term was developed by Easton and StJohn [16].

The model was derived by assuming that in most melts constitutional supercooling is the dominant undercooling available once the thermal undercooling generated by the mould walls on pouring, has quickly dissipated. To simplify the model it was also assumed that the actual temperature gradient in the melt is negligible or very low. This is reasonable due to the high thermal conductivity of the melt. Thus the maximum constitutional supercooling can be approximately estimated as

$$\Delta T_{cs} = m_l c_0 \left(1 - \frac{1}{(1-f_s)^p} \right) \tag{2}$$

where m_l is the slope of the liquidus line, c_0 is the composition of the alloy, f_s is the solid fraction solidified and $p = 1 - k$ where k is the partition coefficient. By taking the first derivative of Equation 2 at $f_s = 0$ it is found that the initial rate of development of constitutional supercooling is equal to the growth restriction factor Q, i.e.

$$\frac{d\Delta T_{cs}}{df_s} = m_l c_0 (k-1) = \text{the Growth Restriction Factor, } Q \tag{3}$$

This relationship is true for either equilibrium or non-equilibrium Gulliver-Scheil solidification [12].

Further, if it is assumed that a nucleation event occurs when ΔT_{cs} reaches ΔT_n then f_s must correlate with the grain size as f_s represents the amount of grain growth that occurs before the next nucleation event occurs. Taking the limiting version given by Equation 3, this suggests that the amount of growth, which can be related to f_s, is proportional to $\Delta T_n/Q$ and hence the second term $b_1 \Delta T_n / Q$ in Equation 1.

One of the most powerful uses of Equation 1 is to analyze grain size data to determine grain refinement mechanisms [7]. If the same density of a particular nucleant particle is added to alloys with a range of Q values and the data is plotted against $1/Q$ a relationship of the form

$$d = a + b/Q \tag{4}$$

is found. The gradient, b, is proportional to the nucleant potency and the intercept, a, is related to the number of active particles present. Hence a more powerful nucleant added with the same number of active particles will have the same intercept but a lower gradient (Fig 1a). However an increase in the nucleant particle density will decrease the intercept, a, but the gradient, b, will remain constant (Fig 1b).

Undertaking experiments to gather grain size data for a range of Q values allows the grain size data to be plotted against the inverse of Q. This new methodology for the study of grain refinement provides information on the potency of the nucleant particles whether we know what type of particles they are or not. It also provides information on the effectiveness and efficiency of the particles that are deliberately added or naturally present in the melt. This information is beneficial in comparing different grain refining systems and in elucidating the mechanisms of grain refinement responsible for the observed grain refiner performance.

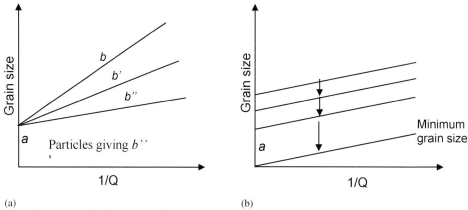

Figure 1: (a) shows the effect of changing the nucleant particle potency on the gradient b while keeping the number of active particles constant and (b) shows the effect of adding more of the same type of particles on the value of *a* while the potency, *b*, remains constant. After [7]

4 Application of the New Methodology

This section examines the results of grain refinement experiments using the new methodology as described by Equation 4 to determine whether new light can be shed on the mechanisms responsible for grain refinement. It will also be shown that this methodology has a number of limitations which will be discussed in the final section of this paper.

4.1 Mg-Al Alloys

4.1.1 Effect of Alloy Purity on Grain Size

Recent work [18] concludes that the finer grain size (native grain refinement) achieved in high purity Mg-Al alloys compared to commercial purity alloys (Figure 2), is due to Al_4C_3 particles naturally present in the melt. However, in commercial purity Mg-Al alloys it is proposed that impurity elements such as Fe and Mn combine with Al to form an intermetallic layer on the Al_4C_3 particles which has a much lower potency leading to the larger grain sizes observed in commercial purity alloys. Figure 2 also shows that the intercept, *a*, for high purity alloys is larger than, or in a similar range to, the value of *a* for commercial purity alloys. A larger value of *a* implies there are less nucleant particles in the high purity alloys which is a reasonable expectation. However, the potency is significantly better for the high purity alloys (much lower value of *b*) resulting in a much finer grain size for the alloys investigated.

For this study high purity aluminium (Fe: 0.001 %, Si: 0.001 %, Cu: 0.0015–0.003 %, Zn: 0.001 % and Ti: 0.001 %) and high purity magnesium (99.98 %) ingots were used to prepare high purity Mg-Al alloy melts. To avoid the uptake of iron and carbon that often accompanies the use of a mild steel crucible, melting of raw materials was conducted in aluminium titanite (Al_2TiO_5) crucibles at 730 °C. Grain size measurement was done on the central region of conical samples taken from the top of the melt using a BN-coated cone ladle with the dimensions of ϕ20 mm × ϕ30 mm × f 25 mm.

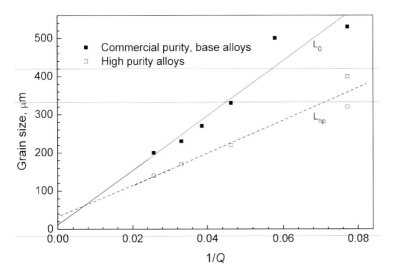

Figure 2: Grain size versus $1/Q$ for a range of commercial purity and high purity Mg-Al alloys. After Ref. [18].

4.1.2   Grain Refinement by Superheating

Commercial purity (99.7 %) magnesium ingots (supplied by Norsk Hydro) and aluminium ingots were used as raw materials to prepare Mg-Al alloy melts (Mg-3Al, Mg-6Al and Mg-9Al). Melting was conducted in an electrical resistance furnace under SF_6 cover gas. Superheating was carried out by raising the temperature of the melt from 730 °C to 900 °C, holding it for 30 minutes and then withdrawing the crucible from the furnace for rapid cooling (the cooling rate is about 20 °C/min) to the sampling temperature (730 °C). Fig. 3 presents the grain size data obtained from this experiment [19].

Using the new analytical method the grain size data for the commercial purity Mg-Al alloys was plotted against $1/Q$ before and after superheating. It is clear that superheating caused significant grain refinement in each of these three alloys. For example, the grain size of Mg-9Al decreased from 180 μm to 100 μm after the superheating treatment. The gradient of the line of best fit for the superheated alloys is nearly half that for the base alloys (Table 1), which suggests that after superheating the potency of the nucleant particles has increased significantly.

The explanation for native refinement (Section 3.1.1) suggests that superheating of commercial purity alloys causes grain refinement by removing the contaminated coating and exposing fresh Al_4C_3 surfaces to the melt and hence improving the potency of these nucleant particles. If cooling is too slow then the impurity layer reforms removing the effect of superheating. By examining Figure 3 it can be concluded that superheating produces a finer grain size because the particles have become more potent without increasing the number of potent particles (i.e. *a* increases slightly rather than decreases (Figure 3), and b decreases in value). Thus the mechanism of impurity layer removal is consistent with the measurements.

The values of a and b for high purity Mg-Al alloys are very similar to the values for the superheated commercial purity alloys (Table 1). This indicates that the nucleant particles present in the melt are likely to be the same as in the superheated melt, possibly Al_4C_3. However, all we can say for certain is that the potency is the same as the nucleant particles may be different phases that have a similar contact angle for the nucleation of the magnesium phase. As potency is defined as $b_1 \Delta T_n$ in Equation 1 it is probable that this term is related to the disregistry between the nulceant particle and the magnesium phase and/or the contact angle.

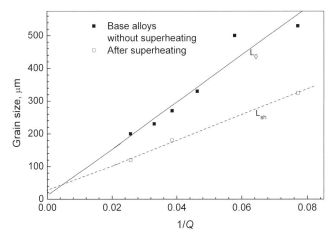

Figure 3: Grain size versus $1/Q$ for a range of commercial purity Mg-Al alloys before and after superheating. Data are from Ref.[19].

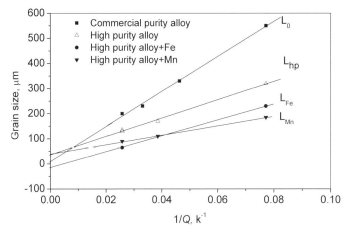

Figure 4: Grain size versus $1/Q$ for a range of Mg-Al alloys where L_0 is for commercial purity alloys, L_{hp} for high purity alloys, L_{Fe} for high purity alloys with the addition of Fe and L_{Mn} for high purity alloys with the addition of Mn. Data are from Refs.[20,21].

4.1.3 Effect of Fe and Mn Additions on the Grain Size of High Purity Alloys

In order to understand the role of Fe and Mn in commercial purity alloys it is useful to study the independent effects of these two elements when added to high purity Mg-Al alloys. The results are presented in Figure 4. The surprising result is that the individual addition of these elements resulted in grain refinement rather than grain coarsening. It was proposed [20,21] that new types of nucleants were introduced or formed in the melt – an Fe-Al compound when Fe was added and an ε-AlMn phase when a splatter Al-Mn master alloy was added. It can be implied from Figure 4 that the Fe-Al nucleant particles have a very similar potency to the native nucleant particle (Al_4C_3), while ε-AlMn is more potent than Al_4C_3. However, there are only two data points used to plot the Fe addition line and therefore any conclusions drawn regarding the potency or type of the particles responsible for the observed grain refinement when Fe is added, are unreliable.

Table 1: The parameters a and b for lines of best fit in Figures 2 to 4.

Line	a	b	R^2
L_0	8.1	7217	0.924
L_{sh}	22.9	3933	0.998
L_{hp}	26.5	4307	0.931
L_{Mn}	40.2	1869	0.998
L_{Fe}	−17.5	3208	1

4.2 Grain Refinement by Zr

The addition of Zr to Mg alloys that do not contain Al, results in very good grain refinement. This is because Zr dissolves such that very high values of Q are generated by only a fraction of one percent of soluble Zr while leaving high potency particles of Zr undissolved which have an excellent orientation relationship with the magnesium matrix [8]. Figure 5 shows the grain size versus $1/Q$ plot for a range of total Zr contents.

In the case of Zr, plotting the grain size data against $1/Q$ is problematic because the Q values are not accurate mainly due to the difficulties in the measurement of the dissolved Zr content [22] and the number of particles increase while the amount of soluble Zr is increased. Thus to determine the slope b for a constant number of Zr particles, the data in Figure 5 needs to be corrected as shown. Although the soluble Zr content can be analyzed, determination of the number of Zr particles requires detailed metallography to count the number of Zr particles visible in a cross section.

This example shows that the interpretation of grain refining mechanisms from grain size data can be misleading and work remains to be done to obtain an accurate and reliable b value for Zr so that we can rank alternative refining particles against the effectiveness of Zr refiners. Despite this limitation the data from a range of experiments was analyzed in a recent paper [8] providing information on a number of factors affecting the performance of Zr as a grain refiner.

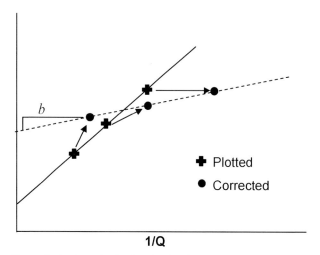

Figure 5: A schematic plot showing that the measured grain size data for the Mg-Zr system where Q has been calculated from the total Zr content, is skewed compared to the data after correction of Q for undissolved Zr and the increased number of Zr particles after each further addition of Zr. After [8].

4.3 Grain Refinement by the Addition of SiC Particles

The new methodology was also recently applied to SiC particles [23] and it was found that SiC particles were moderately good nuclei for commercial purity Mg-Al alloys (Figure 6). The SiC particles were more effective than the native nuclei in the alloys and were particularly efficient at grain refining Mg-1Al where the native grain size was very large. Interestingly, despite the grain size results that indicated stable nucleation behaviour, microstructural analysis showed that the SiC particles decomposed in the melt. This observation is supported by other work [24] which proposes that the SiC particles convert to Al_4C_3. Further support for this proposition could have been provided if the grain size measurements were taken from samples cast under exactly the same conditions as the experiments described in sections 3.1.1 and 3.1.2. Because the casting conditions were different the a and b values cannot be compared.

5 Concluding Remarks

The above examples show that by plotting the grain size data in the form of $a + b/Q$ we can gain information about the relative potency of particles and the number of these particles that can act to nucleate grains. This methodology also shows how sensitive the grain size is to changes in the constitution of the melt as defined by Q. Additionally, it can reveal the relative effect of changing the value of b (ie particles of different potency) or increasing the number of particles on the grain size achieved.

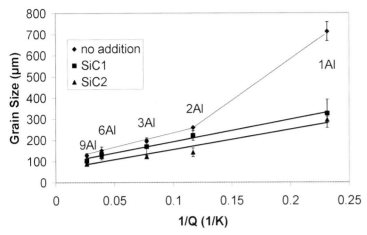

Figure 6: The grain sizes plotted against $1/Q$ for the alloys. SiC1 refers to a 0.15wt% addition of SiC and SiC2 refers to a 0.3wt% addition of SiC. After [23].

However, this work shows that to be able to compare alloys, alloy systems, master alloys, types of particle additions and the amount of particles added we need to use a constant set of casting conditions. Our early work did not do this so we cannot directly compare the results from aluminium alloys with those of the magnesium alloys or between work by different research groups. After considering this dilemma, the casting conditions that would be suitable as a standard test, are based on the set up used for thermal analysis developed by Bäckerud [25]. This set of casting conditions and experimental arrangement also prevents the formation of chill crystals which are the result of thermal undercooling. However, this casting arrangement uses a graphite crucible which would not be suitable for magnesium alloys. A possible modification would be to use a preheated steel crucible that delivers a cooling rate of 1 °C/s. An additional advantage of this casting method is that it can be used with little modification to determine the effect of cooling rate, mould temperature, etc. on grain size and whether or not chill crystals are making a contribution to the measured grain size.

To be able to have confidence in the values of a and b, at least five data points are needed over a sufficient range of Q values. However, as seen from this work useful information can be obtained from three data points over a broad range of Q values but any less is unreliable. Sometimes it can be difficult to have a sufficiently broad range of Q values when the solubility of alloying elements in the melt is low.

There are potential problems that need to be considered when examining the results which relate to situations were the particle numbers are not known or they are not stable. For example, the added particles may dissolve to some extent (eg Zr, SiC) and thus the value of Q may change and also particle numbers may be affected. Also, the addition of particles may add impurities that can poison the potency of particles as is suspected to be the factor reducing the potency of particles added to commercial purity Mg-Al alloys. In these cases the analysis can still provide useful insights into the mechanisms occurring even if the actual values of a and b are not accurate.

If the above points are taken into consideration then the data obtained from experiment can be compared and conclusions drawn on the nature of the potency of the nucleant particles and the effectiveness and efficiency of master alloys. Also, comparisons can be made between different alloy systems to determine their potential for grain refinement. For example, aluminium alloys are generally considered to be alloys that are very responsive to the addition of potent master alloys. However, it can be noted by comparing the grain size results of aluminium and magnesium alloys that magnesium alloys can be refined to much smaller grain sizes (eg 40 μm compared with 100 μm). This is supported by the low values of a indicating that Mg-Al alloys have a very large number of nucleant particles naturally present in the melt.

As concluded above, a constant set of casting conditions is necessary for valid comparisons. However, changes in cooling rate, mould material and casting design are likely to have a significant effect on the contribution of constitutional supercooling and thermal undercooling to the formation of grains in the bulk of the melt as well as the formation of chill crystals. These aspects will be investigated in future work.

Despite improvements to the conceptual framework and analytical skills for examining grain refinement there are still many questions remaining unanswered. It is likely that the study of grain refinement of magnesium alloys will continue for some time as has occurred in the study of grain refinement of aluminium alloys. The methodology presented in this paper provides a unique tool to maximize the information arising from further studies.

6 Acknowledgements

The CAST Cooperative Research Centre was established under and is supported in part by the Australian Government's Cooperative Research Centre's Programme. MQ acknowledges the support from EPSRC, UK.

7 References

[1] Kelton, K.F., Greer, A.L., Herlach, D.M., Holland-Moritz D. MRS Bulletin **2004**; 30: 940–944.
[2] Perepezko, J.H. in: Metals Handbook, 9th ed., vol. 15, Casting, Metals Park: ASM **1988**, p.101–108.
[3] Emley, E.F., Principles of Magnesium Technology. Pergamon, Oxford **1966**, p. 483–493.
[4] Lahaie, D, Embury, J.D., Chadwick, M.M., Gray, G.T. Scripta Metall. Mater. **1992**; 27: 139–142.
[5] Qian, M. Acta Mater. **2006**; 54: 2241–2252.
[6] Turnbull, D. Acta Metall **1953**; 1: 8–14.
[7] Easton, M.A., StJohn, D.H. Metall. Mater. Trans. A **2005**; 36: 1911–1920.
[8] StJohn, D.H., Qian, M., Easton, M.A., Cao, P., Hildebrand Z. Metall. Mater. Trans. A **2005**; 36: 1669–1679.
[9] Cibula, A. Journal of the Institute of Metals **1949**; 76: 321–360.
[10] Crossley, F.A., Mondolfo, L.F. Journal of Metals **1951**; 191: 1143–1151.
[11] Maxwell, I., Hellawell, A. Acta Metall. **1975**; 23: 229–237.
[12] Easton, M.A., StJohn, D.H.: Metall. Mater. Trans A **1999**; 30: 1613–1623.
[13] McCartney, D.G. International Materials Reviews **1989**; 34: 247–260.
[14] Johnsson, M. Zeitschrift fur Metallkunde **1994**; 85: 781–785.

[15] Greer, A.L., Bunn, A.M., Tronche, A., Evans, P.V., Bristow, D.J. Acta Mater. **2000**; 48: 2823–2835.
[16] Easton, M.A., StJohn, D.H. Acta Mater. **2001**; 49: 1867–1878.
[17] Desnian, P, Fautrelle, Y., Meyer, J.-L., Riquet, J.P., Durand, F. Acta Metall. Mater. **1990** ; 41: 1513–1523.
[18] Cao, P., Qian, M., StJohn, D.H. Scripta Mater. **2005**; 53: 841–844.
[19] Cao, P. PhD thesis, the University of Queensland, Australia, 2005
[20] Cao, P. Qian, M., StJohn, D.H. Scripta Mater. **2004**; 51:125–129.
[21] Cao, P., Qian, M., StJohn, D.H. Scripta Mater. **2006**; 54: 1853–1858.
[22] Qian, M., StJohn, D.H., Frost, M.T. Mater. Sci. Forum **2003**; 419–422: 593–598.
[23] Easton, M.A. Schiffl, A., Yao, J.-Y., Kaufmann, H. Scripta Mater. **2006**; 55:379–382.
[24] Lu, L., Dahle, A.K., StJohn, D.H. Scripta Mater. **2006**; 54: 2197–2201.
[25] Bäckerud, L., Krol, E., Tamminen, J. Solidification Characteristics of Aluminium Alloys vol. 1, Wrought alloys, Skanaluminium, Oslo, Norway, **1986**. p. 63.

Creation of Fine and Spherical Magnesium Crystals via Control of Nucleation and Growth Instabilities

Ma Qian
BCAST (Brunel Centre for Advanced Solidification Technology), West London, UK

1 Introduction

Microstructures are at the centre of materials science and engineering. The creation of a new microstructure implies the creation of a new material and therefore potential new applications. Solidification is one of the most economic and versatile microstructure formation processes for metallic materials. One of the central tasks of solidification control is to achieve fine, uniform and equiaxed non-dendritic grain structures. This is often attained through effective inoculation or grain refinement in commercial production. This work discusses the creation of spherical grains or crystals through control of nucleation and growth instabilities.

The creation of spherical crystals is of particular interest to semisolid metal processing. Through more than three decades of development, semisolid metal (SSM) processing has been established as a commercially useful manufacturing route for the production of near-net shape components of high integrity and performance. The key to the process is to obtain a semisolid slurry free of dendrites, with the solid being present as non-agglomerated, fine and spherical particles, and with minimum entrapped liquid in the solid [1, 2]. Such semisolid slurries are obtainable by a number of approaches [3], most notably by direct or indirect liquid agitation, or by partial remelting of a feedstock material that possesses a fine equiaxed grain structure, which can be obtained by recrystallisation, grain refinement, spray casting or low superheat casting in some cases. It is therefore desirable to be able to produce spherical primary crystals directly from the liquid state without liquid agitation. In addition, the creation of an as-cast spherical grain structure will provide further opportunities to improve the structural uniformity compared to an equiaxed grain structure at the same grain size.

The purpose of this work is to demonstrate, both theoretically and experimentally, that fine and spherical magnesium grains are directly obtainable from the liquid state, without liquid agitation, simply by grain refinement, either when the melt is cast from a semisolid or superheated state. Historically, chemical grain refinement has been tried for semisolid structure formation [4]. However, these efforts have been limited to attaining a fine, equiaxed non-dendritic grain structure in the billet, followed by reheating the billet to a semisolid state to obtain thixo-formable slurries [5]. These are not direct liquid processing routes, in which grain refinement only plays an intermediate and supplementary role.

2 Theoretical basis

2.1 The M-S stability criterion and its applicability to the growth of magnesium crystals

The rate of growth of a solid phase during solidification is essentially determined by the rate of heat conducted away from the solid-liquid interface. Solidification proceeds as long as the interface temperature is below the melting point. Mullins and Sekerka (M and S) [6] considered the morphological stability of a solid sphere growing in a uniformly supercooled melt by introducing a perturbation in the spherical shape. The principal approximations used were: (i) isotropy of bulk and interface properties; (ii) description of the thermal fields by Laplace's equation; (iii) local equilibrium at the interface. To satisfy approximation (ii), the system under consideration must fulfil [6]

$$|C_v(T_m - T_\infty)/L_v| \ll 1 \qquad (1)$$

where C_v is the specific heat per unit volume of the liquid, L_v the latent heat of freezing per unit volume of the solid, T_m the melting point, and T_∞ the temperature of the melt.

The stability problem was then examined by introducing a spherical harmonic $Y_{lm}(\theta,\varphi)$ of infinitesimal amplitude δ in the original sphere of radius R ($\delta/R \ll 1$), in the spherical coordinates (θ,φ)

$$r = R + \delta Y_{lm}(\theta,\varphi) \qquad (2)$$

where $Y_{lm}(\theta,\varphi)$ is the angular part of a solution to Laplace's equation [7], l is a positive integer defined by $\lambda \cong 2\pi R/l$, in which λ is the distance between nodes of the spherical harmonic, and $-l \leq m \leq l$ [7]. By solving the thermal fields outside and within the distorted sphere to obtain the velocity v and then equating the coefficient of $Y_{lm}(\theta,\varphi)$ in the expression of v and that in the time derivative of Eq.(2), i.e. $v \approx dR/dt + (d\delta/dt)Y_{lm}(\theta,\varphi)$, the rate of growth of the amplitude $d\delta/dt$ was defined. Mullins and Sekerka [6] found that the spherical shape was unstable (i.e. $d\delta/dt > 0$) when at least the second harmonic ($l = 2$) grew or when R exceeded a critical size R_c:

$$R_c = \frac{2\Gamma_T(7 + 4k_s/k_L)}{[(T_m - T_\infty)/T_m]} = (7 + 4k_s/k_L)r^* \qquad (3)$$

where $\Gamma_T = \gamma_{SL}/L_v$ and $r^* = 2\gamma_{SL}T_m/(L_v\Delta T)$, in which γ_{SL} is the crystal-melt interfacial free energy, L_v the latent heat of freezing per unit volume of the solid, k_s the thermal conductivity of the solid, k_L the similar quantity for the liquid, T_m the bulk melting temperature, ΔT the undercooling and r^* the critical nucleus radius defined by the classical theory of nucleation. Equation (3) is the M-S stability criterion for spherical growth in a uniformly supercooled melt. It predicts that all solid spheres undergoing thermally controlled growth are morphologically unstable if their radii are larger than $(7 + 4k_s/k_L)r^*$.

A detailed examination of the three assumption conditions used for deriving the M-S criterion shows that each assumption can be met for the growth of a magnesium crystal in molten magnesium [8], particularly in the presence of an effective grain refiner. This means that Eq. (3) can be used to control the growth of magnesium crystals in molten magnesium.

2.2 Implications of Eq. (3) for the production of spherical magnesium crystals

Substituting the values of k_L, k_s, γ_{SL}, L_v and T_m for magneisum [8] into Eq. (3) gives

$$R_c = 13.67 r^* = \frac{4.84}{\Delta T} \ (\mu m) \qquad (4)$$

The dependency of R_c on ΔT has been plotted in Fig. 1. As can be seen, R_c is highly sensitive to ΔT. For example, when $\Delta T = 1$ K, all spherical magnesium crystals larger than 9.6 μm in diameter will be unstable. By contrast, when $\Delta T = 0.1$ K, all particles smaller than 96 μm in diamter will be spherically stable. It is thus essential to ensure a small undercooling (e.g. ≤ 0.5 K) for spherical growth during solidification.

For a given cooling rate, the magnitude of ΔT for heterogeneous nucleation generally reflects the effectiveness of the substrates existing in the melt. Zirconium and magnesium have the same type of crystal structure and nearly identical lattice constants. In addition, zirconium has a reasonable solubility limit in molten magnesium. This allows solutal zirconium atoms to also play a strong growth restriction role in the grain refining process. These factors make Zr a nearly perfect grain refiner for magnesium [8]. It has been found that optimum grain refinement occurs when the melt contains both a high level of dissolved Zr and a sufficient number of suitably sized, undissolved Zr particles [9, 10]. In fact, under such conditions, the grain refining effect of Zr for magnesium is so powerful that the undercooling is often too small to be readily detectable. Assuming the undercooling varies in the range 0.1-0.2K, Eq. (3) predicts that $d_c \approx 48 - 96$ μm. This implies that grain refinement by zirconium should be able to lead to spherical grains with an average diameter close to 100 μm under normal cooling conditions. This forms the theoretical basis of the present work.

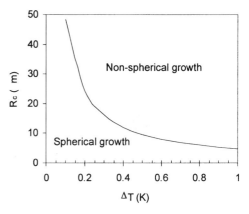

Figure 1 Critical radius for spherical growth of a magnesium crystal as a function of ΔT.

3 Experimental procedure

The experimental alloy selected had a nominal composition of Mg-3.8%Zn-2.2%Ca, which was based on ZK40 with an addition of 2.2%Ca. The Zn content was purposely controlled below 4% and the Ca content above 2%, as elaborated previously [11]. Melting was conducted in a coated steel crucible, protected by 0.4%SF_6 in N_2. A 6kg batch of the alloy was made by alloying pure zinc at 680°C and pure calcium at 780°C to molten pure magnesium. In actual experiments, the weight of each heat of melt prepared was ~1,250g by remelting a slice of the ingot in a small stainless steel crucible.

Zirconium was introduced to the melt at 730°C in the form of a Zirmax master alloy (Mg-33.3%Zr). After alloying with Zr, a steel ladle, which was preheated at 730°C and capable of containing ~250g molten magnesium, was used to scoop the melt from the crucible. The melt in the ladle was subsequently allowed to cool in the air and the temperature was monitored using a thermocouple. The melt was cast into a conical steel mould (cavity dimensions: φ36mm×φ22mm×45mm; wall thickness: 2.2mm) at different semisolid casting temperatures.

Samples for metallographic examination were all cut from the middle height of each cone. They were polished and etched with an acetic-picral solution (10ml acetic acid, 4.2g picric acid, 10ml H_2O and 140ml ethanol). The dissolved Zr in each sample was determined using an wet chemical approach developed previously [9].

4 Results and discussion

As can be expected from the dendritic growth pattern of magneisum crystals, conical samples of the base alloy semisolid-cast from 626°C (liquidus: ~632°C) without grain refinement showed large primary magnesium dendrites, Fig. 2 (a,b). Also observed is the fine, dendritic matrix formed in the conical mould following casting. Grain refinement with zirconium resulted in completely different structures for both the primary phase and the matrix phase, when cast under similar conditions. These are demonstrated in Fig. 3 (a,b), which were captured from the edge and the central region of the cone from a transverse section. Figure 4 (a, b) shows similar views captured from a longitudinal section that was cut through the axis of the cone. Both sections demonstrate that spherical primary magnesium particles were obtained throughout the sample after grain refinement.

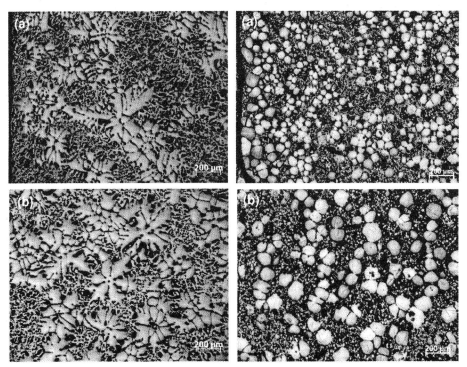

Figure 2 Primary magnesium dendrites observed in a conical sample of the Mg-3.8%Zn-2.2%Ca alloy cast without grain refinement from 626°C (liquidus: ~ 632°C): (a) the edge and (b) the central region.

Figure 3 Spherical primary magnesium particles observed in a conical sample of the Mg-3.8%Zn-2.2%Ca alloy cast at 626°C after grain refinement with 2.5%Zr (dissolved Zr: 1.09%): (a) the edge and (b) the central region.

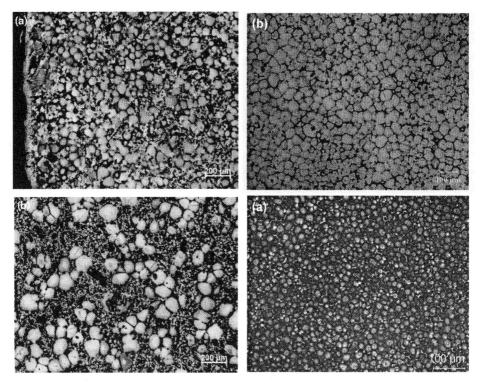

Figure 4 Spherical primary magnesium particles observed on a longitudinal section that was cut through the axis of the same conical sample shown in Fig. 3: (a) the edge and (b) the central region. The positions of (a) and (b) were purposely selected to correspond approximately to those of Fig. 3 (a,b), respectively.

Figure 5 Spherical grain structures obtained in: (a) a chill bar sample (ϕ28 mm × 200 mm) of a Mg-RE based alloy (AM-SC1) and (b) a sand-cast plate sample (180mm × 140mm × 15 mm) of a Mg-4.0%Zn-1.5%RE (ZE41). In both cases, the melt was cast at 730°C after grain refinement with 1%Zr in the form of a Mg-25%Zr master alloy [12].

Apart from casting from a semisolid state, the concept has also been tested for permanent mould casting and sand casting, where the inoculated melt was directly cast into the mould from a superheated state (730°C). Figure 5 (a) shows a view of the grain structure achieved in a chill bar sample of a creep resistant magnesium alloy (AM-SC1), cast at 730°C, with a 1%Zr addition in the form of a Mg-25Zr master alloy [12]. Spherical magnesium grains with an average diameter of 25 µm was obtained throughout the sample. Figure 5 (b) shows a view of the grain structure obtained in a sand-cast plate sample of a Mg-RE-Zn alloy (ZE41) after refinement with zirconium, where a nearly spherical grain structure with an average diameter of 50 µm was similalrly observed.

Experimental work [8] has revealed that the formation of spherical primary magnesium grains is due to the combined effects of both dissolved and undissolved zirconium, which give rise to a high nucleation rate and a small undercooling. A desired level of soluble zirconium is about 0.6% at 730°C and that o the undisslved Zr content should reach about 1%, which ensures a sufficient number of suitably sized, undissolved Zr particles in the melt prior to solidification.

Inspection of the large spherical primary magnesium crystals in the sample shown in Fig. 3 at high magnificaitons revealed that many of the spherical crystals exhibited a perturbed periphery [8]. This provids direct evidence for the M-S perturbation concept for spherical growth. The existance of readily noticeable perturbations suggest that such magnesium crystals have started losing their spherical morphologies in the late stages of solidification. Further development of the perturbed rims will turn each particle into a rosette-like or dendritic morphology.

5 Summary

Following the Mullins-Sekerka instability theory for spherical growth, it has been shwon that fine and spherical magneisum crystals are obtainable directly from the liquid state, without liquid agitation, through effective control of nucleation and growth by grain refinement with zirconium. The inoculated melt can be cast either from a semisolid state or a superheated liquid state. This provides a simple alternative to the preparation of semisolid slurry for semisolid processing. The spherical grain structures obtained under sand casting and permanent mould casting conditons provide further opportunites to improve the uniformity of a magnesium casting.

The formation of spherical magnesium grains is attributed to the combined effects of both dissolved and undissolved zirconium, which give rise to a high nucleation rate and a small undercooling. In addition, it is likely that the growth anisotropy of magnesium crystals has been suppressed by the dissolved Zr atoms in the melt. As a quantitative model, when all conditions required for the Mullins-Sekerka stability analysis are satisfied, the model provides a reasonable indication of the spherical particle dimensions that may be produced from a magnesium melt by grain refinement with zirconium.

Acknoweledgements – This work was supported by EPSRC and the Royal Society.

6 References

[1] Flemings M. C., Metall. Trans. A **1991**, *22A*, 957-981.
[2] Fan Z., Inter. Mater. Rev. **2002**, *47*, 49-85.
[3] Atkinson H. V., Progress in Materials Science, **2005**, *50,* 341-412.
[4] Kirkwood D. H., Inter. Mater. Rev. **1994**, *39*, 173-189.
[5] Loué W. R.; Suéry M., Mater. Sci. Eng. A **1995**, *203*, 1-13.
[6] Mullins W. W., Sekerka R. F., J. Appl. Phys. **1963**, *34*, 323-329.
[7] Arfken G. B.; Weber H. J., Mathematical Methods for Physicists (4th ed.). San Diego: Academic Press; **1995**, pp.737-739.
[8] Qian M., Acta Mater. **2006**, *54*, 2241-2252.
[9] Qian M.; StJohn D. H.; Frost M. T., Mater. Sci. Forum **2003**, *419-422*, 593-598.
[10] StJohn D. H.; Qian M.; Easton M. A.; Cao P.; Hildebrand Z., Metall. Mater. Trans. A **2005**, *36*, 1669-1679.

[11] Qian M.; Das A., Scripta Mater. **2006**, *54*, 881-886.
[12] Qian M.; StJohn D. H.; Frost M. T.; Magnesium alloys and their applications (Ed. K. U. Kainer), Weinheim: WILEY-VCH; **2003**, pp.706-712.

The Role of Fluorine Solubility in the Protection of Molten Magnesium

M. Syvertsen*, K. Aarstad**, G.Tranell*, and T.A. Engh**
*SINTEF Materials and Chemistry, Trondheim, Norway
**The Norwegian University of Science and Technology (NTNU), Trondheim, Norway
((Abbildungen aus Autoren-PDF einbauen))

1   Abstract

During experiments where molten magnesium is protected with fluorine containing gases, a residual protection effect has been observed when the supply of gas is cut off. That is, the molten metal does not oxidise the moment the protective gas flow is cut off. It is possible that this observed residual effect can be attributed to a limited solubility of fluorine in the molten magnesium, which may supply the surface oxide layer with necessary fluorine to remain protective for some time.

The solubility of fluorine in pure magnesium and in the alloys AM50 and RZ5 has hence been determined by equilibrating the respective alloy in magnesium fluoride crucibles under Argon atmosphere, with subsequent analysis of the fluorine concentration in the metal. The solubility measured in both pure and alloyed magnesium in the temperature range 650 °C to 950 °C ranged between approximately 10 to 100 ppm by weight.

2   Introduction

Molten magnesium is rapidly oxidised in an oxygen-containing atmosphere – a strongly exothermic reaction which may cause burning if the liquid surface is left unprotected. In present industrial handling of molten magnesium, SF6 – which will inhibit the oxidation by facilitating the formation of a protective surface film - is thus commonly added to the atmosphere under which the liquid metal is kept.

The use of fluorine containing compounds for protection of molten magnesium was suggested and patented by Reimers already in the early 1930's [1]. It was, however, not until the 1970's, following the work of Fruehling [2] and subsequently Couling [3,4], that SF_6 was introduced as an active agent in cover gases for magnesium melt protection. Prior to the introduction of SF_6, magnesium was protected with alkali metal halide fluxes, SO_2 or even elemental sulphur. Although SF_6 has been used for three decades, it has only been recently that the mechanisms – through which certain fluorine containing gases protect liquid magnesium from uncontrolled oxidation – were thoroughly investigated and partially revealed [5–7].

In the last decade, global awareness and legislation regarding green house gas emissions have made continued use of SF_6 in the magnesium industry unsustainable [8]. Alternatives for the magnesium operators include reverting to (or in some countries continuing) the use of SO_2 or finding new protection methods. Given the toxicity of SO_2, it is not commonly regarded as an optimal protective agent and hence, alternatives to SO_2/SF_6, with regard to both technical and environmental performance as well as economic considerations, have been / are being sought. In order to find suitable replacements to SF_6, an understanding of the fundamental protection mechanisms is vital: For any given metal / metal oxide system, the „coverage" of the metal oxide on the corresponding metal surface – the ratio between the oxide and metal molar volumes – may be described by the Pilling-Bedworth Ratio (PBR) [9], as defined by equation (1):

$$PBR = \frac{m_{oxide}/\rho_{oxide}}{n\, m_{metal}/\rho_{metal}} \quad (1)$$

By definition, m_{oxide} is molar mass of oxide, ρ_{oxide} is the density of oxide m_{metal} is molar mass of the metal, ρ_{metal} is the density of the metal, and n is the number of metal atoms needed to form an oxide molecule. With *PBR* values <1, the oxide layer is porous, providing poor metal surface coverage. For systems with $1 < PBR < 2$, the oxide layer is generally dense, giving good coverage (as with for example Al_2O_3). For a $PBR > 2$; the large difference between the molar volume of oxide and that of the metal creates tensions which may give cracks in the oxide layer (e.g. Fe_2O_3). For MgO, the *PBR* equals 0.81, i.e. the oxide provides poor coverage of the metal surface. In combination with the strongly exothermic nature of the Mg oxidation reaction, uncontrolled oxidation and subsequent fire will take place if Mg metal is left unprotected in oxygen containing atmospheres at temperatures above its melting point. Previous work has shown that fluorine is an active element in the protection of the melt, by helping to build a stable surface film [5,6,10]. Based on these observations it is concluded that the main mechanisms and reactions involved in magnesium melt protection by SF_6 are:

- Parallel reactions between oxygen and SF_6 (and potentially also its thermal decomposition products) on the one hand and Mg on the other hand facilitating the formation of a dense, protective film with a Pilling Bedworth ratio >1.
- The main, initial reactions between Oxygen / SF_6 and magnesium vapour take place in the gas phase at the gas/liquid metal interface. The homogenous reactions in the gas phase are kinetically more favourable than heterogeneous reactions involving the Mg liquid phase. After extended exposure to and/or high concentrations of SF_6, reactions between fluorine – in or diffusing through the film – and liquid magnesium take place.
- The reaction products from the homogenous and heterogeneous reactions are: – mainly crystalline MgO in continuous film – amorphous or very small (nanosized) crystalline fluorine in a continuous film – Fluorine present to a small extent in the MgO crystal lattice – A Fluorine-rich phase in between magnesium oxide grains – Magnesium fluoride particles at the interface between the film and the bulk metal In the context of trying to find alternatives to SF_6, an exploratory investigation into alternative methods to **gaseous** magnesium melt protection has been carried out. The investigation involved measuring the solubility of fluorine in liquid magnesium and subsequently calculate and test whether the dissolved fluorine would supply the molten magnesium surface with enough fluorine to form a protective film.

3   Experimental Set-up

To study the solubility of F in Mg, the experimental approach was to equilibrate pure liquid magnesium in a solid magnesium fluoride crucible under Argon atmosphere. The temperature was varied from 675 to 950 °C.

3.1 The Crucible

MgF2 from Merck, quality 5836, was mixed with polyvinyl alcohol. The polyvinyl alcohol acts as an adhesive, and 2 % by weight was dissolved in 250 ml water and heated to 80 °C. The solution was poured into a bottle together with 250 g magnesium fluoride. The bottle was placed on a ball grinder for two hours to make sure that the solution was mixed completely. The solution was then left in a vacuum furnace at 80 °C over night to evaporate the water. What remained was crushed and pressed in a rubber mould at 2000 bar. It was necessary to sinter the crucible to give strength. Therefore, the crucible was held at 1050–1100 °C for about an hour. The size of the crucible after sintering was about 50 mm in outer diameter and a height of 80 mm.

3.2 The Furnace

Figure 1 shows a schematic of the furnace set-up. The furnace is a resistance element tube furnace. The crucible was charged with 60–80 g 99.99 % pure magnesium, AM50 (4.5–5.3 % Al, 0.28–0.5 % Mn), or RZ5 (3.5–5 % Zn, 0.8–1.7 % RE and 0.4–1.0 % Zr) alloy and placed in the middle of the furnace where the temperature gradients are negligible. The melt depth was approximately 50 mm.

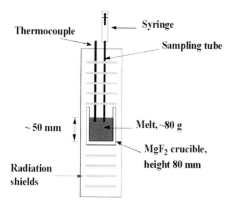

Figure 1. The inside of the furnace used. Also shown is the syringe and sampling tube

Before the experiments were started, the furnace was evacuated to 1 mbar. The chamber was filled with argon (purity 99.99 %), evacuated once more, filled with argon again so that the pressure was a little higher than atmospheric pressure. The off-gas valve was subsequently opened and argon allowed to flow through the furnace. The gas flow was controlled by a Bronkhorst flowmeter and set to approximately 70 Ncm^3/min. The temperature in the melt was measured with a thermocouple placed inside an alumina tube lowered into the melt.

3.3 Sampling

The flow of argon was increased before the lid of the sampling hole was removed and the sampling tube was lowered into the furnace. This was done to create an excess pressure inside the furnace so that air is not drawn into the furnace during sampling. Samples were taken by placing a syringe at the end of an alumina tube with inner and outer diameters 3 and 5 mm, respectively. The tube was carefully lowered into the melt, and liquid metal drawn into the tube. Samples were taken at least 2 cm above the bottom of the crucible, to avoid fluoride particles near the bottom, except in special cases with high temperature where the evaporation of magnesium was so high that there was not much metal left. Further details on the experimental procedure and first rounds of results with pure magnesium, has previously been reported [10].

3.4 Analysis

The samples were partly analysed with GDMS (Glow Discharge Mass Spectrometer) by SHIVA Technologies Inc. in Syracuse, USA, and partly with SIMS (Secondary Ion Mass Spectrometry) by Chalmers Technical University in Gothenburg, Sweden.

4 Experimental Results and Discussion

During the investigation, there were serious issues in finding a reliable/accurate method of analysing the metal samples for fluorine, necessary to determine its solubility in liquid magnesium. Measurements with the SINTALYZER (an electro chemical method) as reported in [16] were not possible to repeat and thus, new experiments were carried out with all three alloys and samples analysed by GDMS. In a third attempt to verify the absolute concentrations of fluorine in the pure magnesium samples, the SIMS was applied. Both the SIMS and the GDMS methods gave reproducible, but different in magnitude results. Figure 2 shows the solubility of fluorine in pure magnesium in equilibrium with MgF_2, as measured with SIMS and GDMS respectively.

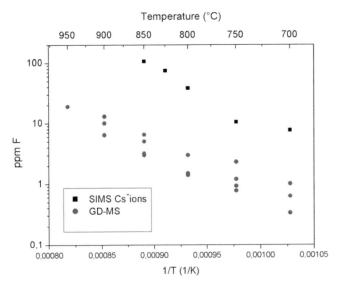

Figure 2: Solubility of fluorine in industrially pure magnesium in equilibrium with magnesium fluoride as a function of temperature, analysed with two different analytical methods

As can be seen from the figure, the results vary considerably. However, both show an increasing trend for the solubility of fluorine with increasing temperature. The results from the GDMS analysis seem to lie approximately a factor ten lower than the SIMS results. The GDMS laboratory did not have accurate standards [10] to trust the absolute values of the F-concentration. The SIMS laboratory used an implantation of ^{19}F ions into one of the samples sent to them as standard. F-measurements before and after implantation enabled the determination of fluorine in the original sample. This procedure are believed to give reliable measurements in the absolute fluorine levels in the equilibrated samples.

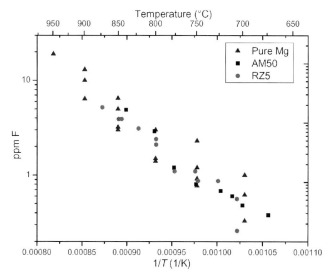

Figure 3: Solubility of fluorine in pure magnesium and magnesium alloys in equilibrium with magnesium fluoride as a function of temperature, as analysed by GDMS

5 Applications-Protection through Fluorine Introduced in the Liquid Magnesium

Although it has previously been concluded that the initial, fast, reactions between magnesium, Oxygen and Fluorinated species mainly take place in the gas phase when using gaseous magnesium protection, it was also observed that reactions between liquid magnesium and fluorine do take place to form MgF_2. It has been speculated that it may be possible to protect liquid magnesium from excessive oxidation by saturating the metal with fluorine, which may help form the MgO/MgF_2 film on the metal surface.

5.1 Is the Measured Fluorine Solubility High enough to give Melt Protection?

It is not easy to give an exact answer to this question. Since tabulated values for physical properties of molten magnesium is sparse, and our measured solubilities are quite scattered, the best one can hope for is to calculate some length or time scales and see if these are within reasonable order of magnitudes. In order to do this one has to look at the transport of fluorine from the bulk melt to the surface.

5.1.1 Thickness of Diffusion Boundary Layer

In the melt there are always some temperature gradients. These gradients give convection and hence mixing of dissolved elements. However, close to a surface there is a laminar boundary layer where the flow is parallel to the surface. Across this boundary layer the transport is governed by Brownian diffusion, driven by concentration differences. The flux (number of moles per area and time) of a solute is given by Fick's first law:

$$\dot{n} = -D \frac{dc}{dz} \tag{2}$$

where D is the diffusion coefficient for the solute in the solvent and dc/dz is the concentration gradient in the z-direction (perpendicular to the surface). Since there has been a lack of solubility data of fluorine in liquid magnesium, also the value of the diffusion coefficient is missing. It is possible to estimate the diffusion coefficient of fluorine in liquid magnesium from kinetic theory [13].

$$D_F = \frac{k_B T}{2\pi \mu d} \left(\frac{m_{Mg} + m_F}{2 m_F} \right)^{0.585} \tag{3}$$

In the equation, k_B is the Bolzmann constant, T the melt temperature, μ the melt viscosity, d the diameter of a fluorine atom, and m_i the molar mass of substance i. With tabulated physical data for viscosity of liquid magnesium, ionic radius of fluorine, and molar masses [14], and a melt temperature of 983 K, the diffusion coefficient becomes

$$D_F = 8.1 \cdot 10^{-9} \text{ m}^2/\text{s} \tag{4}$$

The concentration gradient can be approximated to be equal to the concentration difference over the boundary layer divided by the thickness of the boundary layer, δ.

$$\dot{n} = D \frac{c_{bulk} - c_{interface}}{\delta} \approx D \frac{0.7 \cdot c_{bulk}}{\delta} \tag{5}$$

where, in the rightmost term, the interface-concentration of fluorine is assumed to be 30 % of the bulk concentration. The justification for this assumption arises from calculation of the Pilling-Bedworth Ratio (*PBR*). As previously described, a dense oxide layer requires 1 *PBR* 2. And, in order to get an average PBR, of a mixture of MgO (*PBR* = 0.81) and MgF$_2$ (*PBR* = 1.45), equal to unity, the amount of MgF$_2$ in the oxide layer must be 30 % (by volume). Therefore, the equilibrium fluorine concentration at the interface will be 30 % of the bulk concentration (= concentration in the melt near the pure MgF$_2$-surface). This is shown schematically in Figure 4.

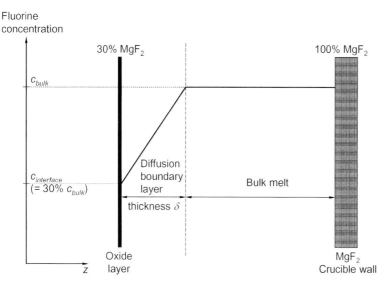

Figure 4: Fluorine concentration profile and diffusion boundary layer thickness in the melt. The oxide layer is made up of 30% MgF_2 and 70 % MgO by volume (26 % MgF_2 by mass).

Then, if the calculated thickness of the diffusion boundary layer is extremely thin (~sub-microns), the measured solubility, and fluorine flux, is not enough to maintain a protective oxide layer. Previously it has been reported that the average oxide growth rate during the first minute is [10]

$R = 2.5$ nm/s (6)

The flux of fluorine needed to give the measured rate is

$\dot{n} = 7.2 \cdot 10^{-5}$ mol/m^2s (7)

Then with the use of a measured solubility (SIMS-data) of

$c = 10$ ppm $= 0.84$ mol/m²s (8)

The thickness of boundary layer where diffusion must occur should not exceed

$d_{max} \approx 66$ μm (9)

According to Guthrie [12], the diffusion boundary layer in molten metals are often in the size order of 10^{-6}–10^{-5} m. And therefore, the measured fluorine concentration should be able to give a high enough flux of fluorine to the surface to give enough protection against rapid oxidation. For a lower oxygen activity, as in a CO_2 atmosphere, the reaction rate between magnesium and oxygen is suppressed. It is also commonly known that magnesium alloys, such as AM50 or RZ5 do oxidize less violently than pure magnesium. Therefore it should be possible to protect liquid fluorine saturated magnesium alloys from rapid oxidation by using CO_2 gas as the only melt protection.

6 Conclusions

The solubility of fluorine in liquid pure magnesium and the alloys AM50 and RZ5 has been investigated through equilibration experiments. At common magnesium processing temperatures around 700 °C, the fluorine solubility in pure magnesium is measured to approximately 10 ppm. From a fluorine concentration point of view, calculations show that liquid magnesium can theoretically be protected against uncontrolled oxidation by saturation of fluorine.

7 Acknowledgements

The authors wish to gratefully acknowledge the financial support of the International Magnesium Association, the US and Canadian EPA's, as well as the Norwegian Research Council.

8 References

[1] H.A. Reimers, „Method for Inhibiting the Oxidation of Readily Oxidizable Metals", US Patent 1,972,317, 1934.
[2] J.W. Fruehling, „Protective Atmospheres for Molten Magnesium", PhD Thesis, The University of Michigan, 1970.
[3] S.L. Couling, F.C. Bennett and T.E. Leontis, „Melting Magnesium under Air/SF6 Protective Atmosphere", *Light Metal Age*, 1977, pp. 12–21.
[4] S.L. Couling, „Use of Air/CO2/SF6 Gas Mixtures for Improved Protection of Molten Magnesium", *Proc. of the 36th Annual World Conference on Magnesium*, Oslo, Norway, 1979, pp. 54–57.
[5] S.P. Cashion, „The Use of Sulphur Hexafluoride (SF6) for Protecting Molten Magnesium", PhD Thesis, Dept. of Mining, Minerals and Materials Engineering, University of Queensland, Australia, 1998.
[6] G. Pettersen, E. Øvrelid, G. Tranell, J. Fenstad and H. Gjestland, „Characterisation of the Surface Films Formed on Molten Magnesium in Different Protective Atmospheres", *Materials Science and Engineering*, 2002, vol. A332, pp. 285–294
[7] K. Aarstad, G. Tranell, G. Pettersen and T.A. Engh, „Various Techniques to Study the Surface of Magnesium Protected by SF6", *Magnesium Technology 2003*, ed. H. Kaplan, The Minerals, Metals and Materials Society, 2003, pp. 5–10.
[8] Kyoto Protocol, http://unfccc.int/resource/docs/convkp/kpeng.html.
[9] N.B. Pilling and R.E. Bedworth, *Journal of the Institute of Metals*, 1923, vol. 29, pp. 529–591.
[10] K. Aarstad, „Protective Films on Molten Magnesium", PhD Thesis, The Norwegian University of Science and Technology, Trondheim, Norway, 2004
[11] K. Aarstad, M. Syvertsen and T.A. Engh, „Solubility of Fluorine in Molten Magnesium", *Magnesium Technology 2002*, ed. H. Kaplan, The Minerals, Metals and Materials Society, vol. 3, pp. 39–42.
[12] R.I.L Guthrie, *Engineering in Process Metallurgy*, Oxford University Press, Oxford England, 1989, pp: 286–290.
[13] T.A. Engh, Principles of Metal Refining *Oxford University Press,* Oxford England, 1992.
[14] CRC Handbook of Physics and Chemistry

Reduction and Avoidance of Ecologically Harmful Cover Gases for Handling Molten Magnesium

Fr.-W. Bach*, M. Schaper*, M. Hepke*, A. Karger*, J. Werner**
*Institute of Materials Science, Leibniz Universität Hannover, Germany
**FGK, Höhr-Grenzhausen, Germany

1 Abstract

The aim of a joint research with the partners Linde AG, Audi AG and Laukötter Gusstechnik GmbH was the development and implementation of an economic and environmental-friendly protection system for molten magnesium. The main emphasis of these melt protection systems is the reduction or avoidance of the greenhouse cover gases used at present. Due to the specifications in the Kyoto protocol this development is necessary.

Based on examinations of alternative cover gases and melt protection systems a new protective method, called CO_2-snow-technique, was developed. The CO_2-snow precipitates at the molten magnesium bath, lowering the surface temperature required for melt protection. Due to the sublimation of the CO_2-snow when hitting the molten magnesium surface the gas expansion displaces all oxygen from the bath surface. Moreover, the temperature reduction reduces the magnesium's proneness to evaporation.

Another main emphasis was the reduction of the required cover gases quantity for molten magnesium protection. To this point metallic and ceramic floatation devices for reducing the bath surface were examined.

The combination of these two protecting methods and the fact that CO_2 has a GWP (Global Warming Potential) of 1 makes this innovative melt protection systems interesting for industrial applications. For this reason examinations of the industrial handling of this technique were carried out.

2 Introduction

Due to the shortage of fossil resources the constructive material magnesium as an economical and ecological effective material obtains particularly in the automobile industry more and more importance. By using lightweight components in automobiles the fuel consumption can be clearly reduced through weight saving. But the application possibilities of this material are not already deployed in the technology. One of the reasons is the high chemical reactivity especially with oxygen. It requires special protective measures against ignition and combustion during the melting and casting process. The inevitable use of ecologically harmful substances for the melting protection reduces the weight specific use of the constructive material magnesium concerning the ecobalance by this means.

3 Melting Protection

Basically for the covering of magnesium meltings the methods are possible which can be applied individually but also in combination [1]:

1. Covering with molten covering salts
2. Doping of the melting with coating forming, passivated elements
3. Melting handling under reactive gas atmosphere, formation of thin protective films
4. Melting handling under inert gas atmosphere

The covering of the magnesium melting with salts and the doping of the melting with coating forming elements includes especially quality problems so that an application of these protective mechanisms only in particular cases is possible. To guarantee the quality and the purity of industrial castings, basically protective gases are used. Normally these protective gases are reactive gases which form a thin protective layer on the molten bath surface and therefore prevent a surface reaction and an exhausting. At present SO_2 und SF_6 are the established and industrial used gases.

3.1 Alternative Melting Protection Concepts

Through the greenhouse potential and therefore the constituted legal restriction (Kyoto journal) for the use of SF_6 (sulphur hexafluoride), this for over 30 years established protective gas will be no more available for the industrial processing of magnesium meltings from 2008-01-01 [2]. An industrial applicable alternative is at present sulphur dioxide (SO_2). Sulphur dioxide is a poisonous gas and due to this additional burden and the danger for the employees it is not attractive.

The development of alternatives for SO_2 and SF_6 containing protective gas mixtures is focused at present especially on other fluorinated materials with a lower potential of the greenhouse effect. In the last years the fluorinated greenhouse gas HFKW-134a and a perfluor ketone ($C_6F_{12}O$) were examined as alternatives in different installations [3, 4, 5].

Both protective gases are applied like SF_6 togehter with carrier gases (dry air or CO_2). But these substitute materials are not a final solution. The greenhouse potential of HFKW-134a must be evaluated as very high although the GWP (Global Warming Potzential) value is about the factor 20 lower in contrast to SF_6. Furthermore at high temperatures poisonous decomposition products can be originated during the application of fluorinated ketones as well as HFKW-134a [5, 6, 7].

Another possible alternative of melting protection systems for an emission reduction, is the coating of the magnesium melting with CO_2 snow [8]. By depositing the CO_2 snow on the melting bath surface, the exhausting tendency can be reduced compared with the CO_2 gas used as protective gas. Through the sublimation on the surface of the magnesium melting, it will be cooled down which prevents a rupture of the protective layer at temperatures over 580° C [9]. Furthermore through the gas expansion on the melting bath surface, the oxygen will be displaced whereby a melting loss of the surface will be avoided [7]. Presently the high consumption of CO_2 impedes an effective application of this innovative protective system.

4 Experiments

4.1 Testing Aim

A total prevention of protective gases is nearly not achievable with the cast processing of magnesium. Therefore the aim is to minimize the used portion of environmental harmful gases or to provide a substitute for these protective gases. The aim of the investigations of the Institute of Materials Science is the minimization of the open melting surface through floating bodies, to reduce the reaction area of the magnesium melting. Also a reduction of the exhausting behaviour is connected with the minimization of the melting surface. The application of protective mediums could be decreased like this or arranged more effective. One aspect of the investigations is also the use of CO_2 snow as protective medium. As floating bodies different materials and geometries can be applied.

4.2 Protective Capacity Through Floating Hollow Balls and Inert Gas

The protective capacity of pure inert gas as a substitute for environmental harmful protection mediums is not sufficient due to the high exhausting of the magnesium melting. The main focus of these investigations is a combination of the protective effects. The fact that floating hollow balls on the melting impede the exhausting of the magnesium melting, will be connected with the protective effect of an inert gas. With regard to a more simple handling, hollow balls of unalloyed steel will be applied as floating bodies.

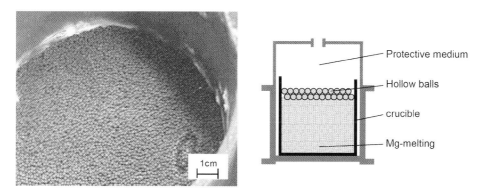

Figure 1: with steel hollow balls covered bath surface

In the following described tests the magnesium melting (figure 1) will be covered with the three steel hollow ball charges under argon atmosphere, like it is mentioned in table 1. The exhausting will be determined by the alteration of the weight.

Table 1: Applied steel ball charges

Charge	hollow ball density	bulk density	medium diameter
1	0.61 g/cm³	0.39 g/cm³	3.9 mm
2	0.62 g/cm³	0.40 g/cm³	2.7 mm
3	0.88 g/cm³	0.56 g/cm³	1.3 mm

In figure 2 is shown that the metal losses through exhausting due to the loading of steel hollow balls onto the melting bath surface, will be extremely reduced. It can be determined that the smallest hollow balls (Ø 1.3 mm) provide the most effective protection against the exhausting of the magnesium melting into the protective furnace gas. The smaller the size of the hollow balls, the more dense is the ball layer which will be formed on the melting bath surface. The bigger hollow spaces in the ball layer, like with balls of 3.9 mm, result in more higher metal losses whereby a reduction of the exhausting of up to 97 % is achievable in comparison to the gassing with pure argon.

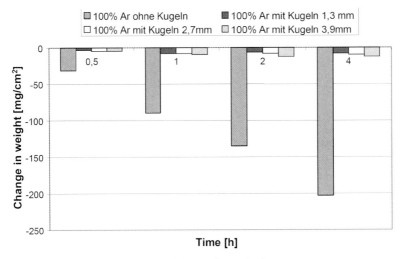

Figure 2: Change in weight with and without surface reduction

Through the coating of the melting bath surface with steel hollow balls the exhausting of the liquid metal will be nearly completely prevented. By introducing the argon the oxygen of the air will be superseded so that no oxidation of the melting can take place. The surface remains solid and the formation of dross on the surface will not take place.

4.3 Protective Capacity through Floating Hollow Balls and Active Gas

Another investigation aspect is the minimization of the free melting bath surface in connection with reactive gases. If the surface which is in contact with the oxygen will be reduced, less protective gas can be used to guarantee a sufficient melting protection.

In tests it could be proved via visual analysis that for a sufficient protective capacity the concentration of SF_6 in the carrier gas (in this case N_2) can be reduced about approximately 60 %.

4.4 Ceramic Hollow Balls

The application of steel hollow balls on the melting surface contains the risk of a contamination of the melting through a possible reaction between the hollow ball material and the liquid magnesium. Therefore ceramic hollow balls are tested as alternative to the steel balls.

Table 2: Properties of test relevant ceramic materials in comparison to unalloyed steel

	Density	conductivity	thermal spalling resistance	resistance in AZ91
Steel	7.8–7.9 g/cm³	50–70 Wm^{-1}K^{-1}	+	+
MgO	3.2–.4 g/cm³	7–11 Wm^{-1}K^{-1}	+	+
Al$_2$O$_3$	3.4–3.9 g/cm³	14–30 Wm^{-1}K^{-1}	+	++

To prevent the contamination of the melting with external elements, ceramics are used which elements are present in the melting. The ceramics which are listed in table 2 have not shown any considerable decomposition reactions in liquid AZ91. Especially the aluminium oxide has proved as solid opposite the attack through the magnesium melting. The very low wetting by the magnesium melting must also be emphasized. Furthermore the thermal conductivity as well as the temperature change solidity have good values for the application as protective barrier in connection with the CO_2 snow procedure.

5 Conclusion and Prospect

By reducing the free melting surface a saving of active protective gases and through a nearly total prevention of the exhausting, an application of inert gases for the cast processing of magnesium is possible. Further potential for emission avoidance can also be found in a combination with the CO_2 snow procedure and the application of ceramic hollow balls. Through the reduction of the reactive surface and the low thermal conductivity of the ceramic hollow balls which are floating on the melting, a reduction of the CO_2 consumption of this alternative melting protection method can be expected.

6 Acknowledgement

We would like to say thank you to the German Federal Foundation Environment (Deutsche Bundesstiftung Umwelt) which promotes this project financially.

7 References

[1] Kammer, C., Magnesium Taschenbuch, Aluminium-Verlag Düsseldorf, 2000, p. 461–467
[2] Rat der Europäischen Union, Verordnung über bestimmte fluorierte Treibhausgase, Dossier 2003/0189A, Brüssel, 2005

[3] Andrews, G., Synthetic Gas Use in Non-Montreal Protocol Industries, Australian Greenhouse Office, 2000
[4] Ricketts, N. J., Cashion, S. P., Hydrofluorcarbons as a Replacement for Sulphur Hexafluoride in Magnesium Processing, TMS Conference, 2001, p. 31–36
[5] Milbrath, D. S., Development of 3M Novec 612 Magnesium Protection Fluid as a Substitute for SF_6 over Molten Magnesium, Conference SF_6 and the Environment, San Diego, 2002
[6] EG-Sicherheitsdatenblatt (Entwicklungsprodukt L15566), 3M Deutschland GmbH
[7] Schwaab, K., Plehn, W., Fluorierte Treibhausgase in Produktion und Verfahren, Umweltbundesamt Berlin, 2004, p. 201–205
[8] Biedenkopf, P., Laukötter, M., Karger, A., Schneider, W., Protecting Liquid Mg by Solid CO_2: New Ways to Avoid SF_6 and SO_2, TMS Conference, 2005, p. 39–42
[9] VDG, Integrierter Umweltschutz in Gießereien, VDG, IN-GUSS 2003, p. 26, 27

Ceramic Foam Filtration of Magnesium Alloy Melt

Bong Sun You[1], G.H.Wu[2] and Chang Dong Yim[1]
[1] Korea Institute of Machinery and Materials, Changwon, Gyeongnam 641-010, Korea
[2] State Key Laboratory of Metal Matrix Composites, Shanghai Jiao Tong University, Shanghai 200030, P.R.China

1 Introduction

High chemical activity of magnesium makes it oxidize rapidly and burn on air to bring a lot of inclusions into the Mg melt. In order to improve corrosion resistance and mechanical properties, impurities which were included in the metal during manufacturing and metal handling should be controlled effectively [1]. The purification of magnesium alloys can broadly be classified into flux and fluxless methods [2, 3]. While the researches on the Mg melt purification concentrate on flux purification processes in the past, flux-free purification process of magnesium alloys became an interesting topic in the past few years because of several advantages, i.e. (1) Cleans the metal of oxides as well as eliminates any potential for flux contamination, (2) Removal of dissolved gases from the melt, (3) Decreased melt losses as compared with flux refining, (4) Elimination of chlorides which may tie up alloying elements.

At present, filtration technique is one of the most important processes among fluxless methods. One of the conventional techniques to remove the inclusions in the casting production is to install a filter at a proper site of the gate system. Porous iron disc, steel wool and multi-layer steel mesh are often used as filtering medium [4, 5]. The major defect of these mediums is that the absorption of the medium surface with inclusions is very poor. Foam ceramic filter is a kind of high efficient filtering medium for alloy melt. These ceramic mediums have been extensively used to purify the liquid aluminum alloys, copper alloys, cast iron, cast steel and high temperature alloys. However, there are very few research reports about the application of ceramic foam filtration on Mg alloy melt. Although Bakke [6, 7] has studied the Mg melt filtration behavior with Al_2O_3 ceramic foam filters, his research concentrates on the filtration efficiency of the inclusions. At present, no study on the effects of ceramic foam filter purification on the microstructure and mechanical properties of AZ91 are reported. Moreover, there is lack of scientific information on the reactions involved between the ceramic foam filter and Mg melt.

The present investigation was devised to research the effects of ceramic foam filtration incorporating gas bubbling on the microstructure and mechanical properties of AZ91 and AZ31 alloys.

2 Experimental Procedure

The AZ91 and AZ31 magnesium alloys used in the experiment were commercial alloys made by DSM in Israel. Tools, raw stuff used in the experiment were heated to 200 °C in the stove before experiment in order to eliminate the water remaining in them. 6 Kg Mg stuff was melted in a 2.5 KW crucible electric resistance furnace at a time. The liquid metal surface was protected by 0.5 % SF_6 in CO_2 gases. During melting, a rotary impeller was inserted into the melt and an inert gas, Ar, was introduced through the impeller under the metal temperature of about about

700–730 °C. The hole size of the rotary impeller head was 0.2 mm in diameter. The rotating speed was 120 rpm. The sparging times were 30 minutes and flow rates were 0–2 liters/min.

After gas bubbling, the melt was poured into 5 metallic moulds at pouring temperatures of 730±3 °C. The permanent molds were coated with boron nitride and preheated to 250 °C. The mold was removed from the oven once preparations were complete, and pouring started. The ceramic foam filters were placed over the sprue of the permanent molds. The filters used in these experiments include MgO, Al_2O_3, ZrO_2 and SiC ceramic foam filters, as shown in Fig. 1. Filter manufacturers commonly characterize the pore size of the filters by the number of pores per linear inch, ppi. Filters with pore sizes 20 ppi were used. Approximate pore size was thus 1.25 mm.

The castings were cut to obtain the tensile samples as shown in the reference [8]. The tensile specimens were tested using an Instron Series ¢ù automated testing machine according to ASTM B557M specification. The gage length, width and thickness of the samples are 25 mm, 6mm and 3mm, respectively. All the tests were carried out at a strain rate of 1.5 mm min^{-1}. Three specimens under the same condition were used to be tested and the final values of test were obtained in term of their average value. The microstructural analysis was carried out using an optical and a scanning electron microscope equipped with an energy dispersive X-ray spectrometer (EDS).

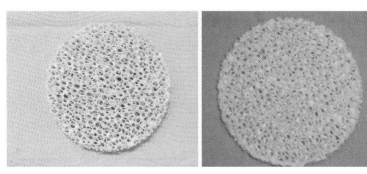

a) Al_2O_3 filter   b) MgO filter
Figure 1: Different kinds of filters

3   Results and Discussion

Fig. 2 and 3 show the effects of the filter porosity and thickness on the mechanical properties of AZ91 alloy, respectively. Fig. 4, in which the filter is 20 mm in thickness, indicates that with the increase of ppi, the σ_b and δ increased, especially, the δ increased significantly. The use of 20 ppi filter increased the σ_b and δ from 175.3 MPa and 2.74 % to 189.1 MPa and 4.02 % by 7.9 % and 46.7 %, respectively. That is, the δ can be improved greatly by the filtration. On the other hand, it can be seen in Fig. 2 that the filter did not have as dramatic of an effect on σ_s. Housh and Petrovich[9] reported that use of the filter significantly reduced the amount of large inclusions remaining the melt. And the obvious solution is to reduce the filter hole size to trap smaller inclusions. From Fig. 2, it can be concluded that the σ_b and δ can be improved with the decrease of inclusion in the metal. But, the inclusion content maybe has little effect on the σ_s.

Fig. 3, in which the filter is 20 ppi, shows the effects of filter thickness on the mechanical properties. Fig. 3 indicates that the use of 10 mm thickness filter increased obviously the σ_b and δ from 175.3 MPa and 2.74 % to 185.6 MPa and 3.72 % by 5.9 % and 35.8 %, respectively. When the filter thickness increased from 10 mm to 20 mm, the σ_b and δ increased from 185.6 MPa and 3.72 % to 189.3 MPa and 4.02 % by 2 % and 8 %, respectively. On the other hand, it can be seen in Fig. 3 that the filter thickness did not have as dramatic of an effect on the σ_s.

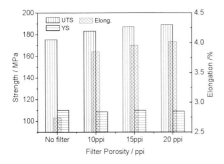
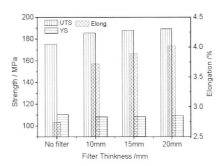

Figure 2: Effects of filter ppi on the mechanical properties

Figure 3: Effects of filter thickness on the mechanical properties

Figs.4-5 shows the effects of gas bubbling incorporating filtration on the mechanical properties of AZ91 alloy. Fig. 4 shows the effect of filter ppi on the mechanical properties, in which argon sparging at 2 l/min for 30 min under the temperature of 730 °C was performed before filtration with 20mm thickness filter. From this figure, it is clear that with the increase of ppi, the σ_b and δ increased. Compared with no filtration, with 20 ppi, the σ_b and δ improved from 180.6 MPa and 3.53 % to 192.1 MPa and 4.22 % by 6.4 % and 19. 5%, respectively. However, the σ_s has not obvious difference.

Fig. 5 describes the effects of filter thickness on the mechanical properties with the gas bubbling incorporating filtration process, in which argon sparging at 2 l/min for 30 min under temperature of 730 °C was performed before filtration with 20 ppi ceramic foam filter. It can be seen that with the increase of filter thickness, the σ_b and δ increase, especially, δ increase obviously. Compared with no filter, with the 10mm filter, the σ_b and δ increase from 180.6 MPa and 3.53 % to 189.7 MPa and 3.99 % by 5.0 % and 13.0 %, respectively. However, when filter thickness increased from 10mm to 20mm, the σ_b and δ only increase from 189.7 MPa and 3.99 % to 192.1 MPa and 4.22 % by 1.3 % and 5.7 %, respectively. It concluded that filter thickness only have small effect on the mechanical properties.

The microstructures of the unrefined AZ91 magnesium alloys are showed in the Fig. 6. The white matrix phase is α-Mg, the gray grains are β phase ($Mg_{17}Al_{12}$), and black particles are nonmetallic inclusions. This figure shows that there exist more inclusions in the microstructure of unrefined magnesium alloys. Figs.10a–c show the inclusion clusters, dispersoid, film morphologies, respectively. Significantly larger inclusion clusters and dispersoids are seen in the melts. Also, lesser amounts of inclusion films are noted. It should be pointed out that these photomicrographs provide information regarding the inclusion type, size, and morphology in unrefined AZ91 alloy. Individual inclusions such as MgO particles are in the order of 10–30 μm in size; clusters of these inclusions range in size from 50–500 μm, and film oxides such as MgO extend to 200–400 μm in length. This entails that substantial efforts should be paid to remove these inclusions. The microstructure of the specimen purified by the combination of ceramic filter and gas bubbling is shown in Fig. 7. It is clear that the inclusions are reduced substantially and the cleanl melt can be achieved.

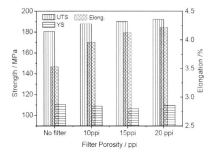
Figure 4: Effects of filter ppi on the mechanical properties, incorporating gas bubbling

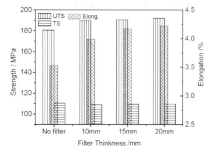
Figure 5: Effects of filter thickness on the mechanical properties, incorporating gas bubbling

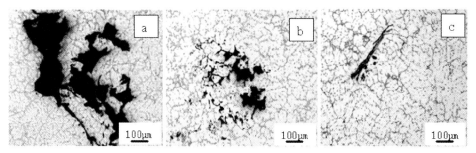
Figure 6: Microstructures of the unrefined AZ91 alloys, a) cluster-like inclusion, b) dispersoid-like inclusion, c) film-like inclusion

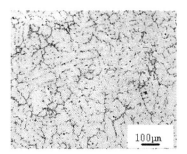
Figure 7: Microstructure purified by the combination of ceramic filter and gas bubbling

The fracture surfaces of the unrefined sample after tensile tests are shown in Fig. 8. The images show a more brittle fracture surface with large oxide clusters in the fracture surface in as-melted samples. The fracture path follows the inclusions and clearly exposes them. The fracture surface of the sample after gas bubbling and filtration is shown in Fig. 9. No inclusions are found on the fracture surface. The fracture is quasi-cleavage crack. The micro-morphology feature of the fracture seems belong to river pattern of cleavage crack, but it cannot be accepted as the genuine cleavage crack. Accurately, it belongs to transcrystalline fracture, which is mixed with small cleavage plane, steps or tear arises and rivers. Compared with Fig. 8 and 9, it can be concluded that filtration process have a certain effect on the fracture pattern of AZ91 alloys.

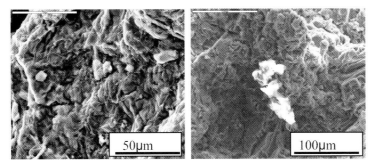

Figure 8: The fracture surfaces of the unrefined sample

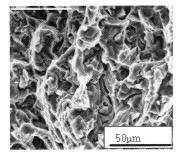

Figure 9: The fracture surface of the sample after gas bubbling and filtration

Fig. 10 shows the effects of gas bubbling incorporating ceramic foam filtration on the ultimate tensile strength σ_b and elongation δ of AZ31, in which argon sparging at 2 l/min for 30min under the temperature of 730 °C was performed. The results show that the purification process of ceramic foam filters incorporating gas bubbling process can improve obviously the ultimate tensile strength σ_b and elongation d of AZ31 alloy. The results show that ceramic foam filtration is more effective than gas bubbling process to improve σ_b and d of AZ31. It is clear that with increase of filter ppi, the σ_b and d increase. The filtration effective sequence of different filtrating materials is as follows: MgO > Al_2O_3. With 20 ppi MgO filter incorporating gas bubbling treatment under Ar flow rate of 2 l/min and temperature of 730 °C, the σ_b and δ of AZ31 can be improved from 144.6 MPa and 5.21 % to 180.8 MPa and 9.06 % by 25 % and 73.9 %, respectively.

Fig. 11 shows the fracture morphologies of AZ31 alloys with or without ceramic foam filtration, respectively. It can be seen that the fracture surfaces of the unrefined AZ31 sample after tensile tests show a brittle fracture pattern with large oxide clusters in the fracture surface. The fracture path follows the inclusions and clearly exposes them. But, no inclusions are found on the fracture surface of the sample filtrated by MgO ceramic foam filter. The fracture is quasi-cleavage crack. The results show that filtration process has somewhat effect on the fracture pattern of AZ31 alloys.

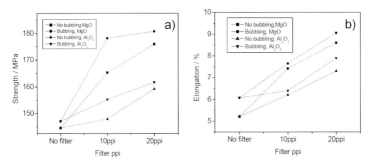

Figure 10: Effects of gas bubbling incorporating filtration on the mechanical properties of AZ31; a) ultimate tensile strength, b) elongation

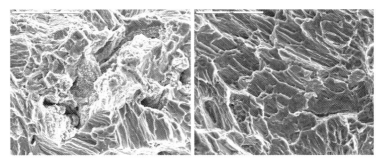

No filtration   MgO filtration

Figure 11: Fracture morphology of AZ31

4   Summary

1. Ceramic foam filter is a kind of high efficient filtering medium for Mg alloy melt. The ceramic foam filter can effectively improve the σ_b and δ of AZ91 alloy, especially δ. The results show that filtration is more effective for AZ91 scrap than that of AZ91 fresh alloy. However, the filters don't have as dramatic of an effect on σ_s. The research results indicated that the σ_b and δ increase with the increase of ppi and thickness. But, the experiments found that the filter of more than 20 ppi tend to choke the flow of metal to lead to low filter efficiency. Hence, considering the success of filtration process, the filters of no more than 20 ppi are suitable for filtrating the Mg melt. And the study suggested that 15–20 ppi are the optimal porosity sizes.
2. Both filtration and inert gas sparging techniques must be utilized together to effectively clean magnesium melt of inclusions and dissolved hydrogen gas so that the mechanical properties can be improved significantly.
3. Gas bubbling and filtration haven't changed the microstructure of alloys. But the fracture surfaces of the unrefined sample show a more brittle fracture characteristic with large oxide clusters in the fracture surface. The fracture path follows the inclusions and clearly exposes them. The fracture surface of the sample after gas bubbling and filtration show that no

inclusions are found on the fracture surface. The fracture is quasi-cleavage crack. It can be concluded that filtration process have a certain effect on the fracture pattern of AZ91 alloys.
4. With 20ppi MgO filter incorporating gas bubbling treatment under Ar flow rate of 2 l/min and temperature of 730 °C, the σ_b and δ of AZ31 can be improved from 144.6 MPa and 5.21 % to 180.8 MPa and 9.06 % by 25 % and 73.9 %, respectively.

5   Acknowledgement

This work was financially supported by the research program in the Korean Institute of Machinery and Materials (NRL).

6   References

[5] A. G. Haerle, R. W. Murray, W. E. Mercer, B. A. Mikucki, The effect of Non-Metallic inclusions on the properties of die cast magnesium, SAE technical paper 970331, Society of Automotive Engineers, Inc. international congress and exposition, Detroit, (1997) 75–84

[6] H. S. Tathgar, P. Bakke, T. A. Engh, Impurities in magnesium and magnesium based alloys and their removal, in:K.U.Kainer(Eds.), Magnesium alloys and their applications, WGV Verlagsdienstleistungen GmbH Inc., Weinheim, 2000, pp. 767–779

[7] H. T. Gao, G. H. Wu, W. J. Ding, Y. P. Zhu, Effects of Purification Fluxes Containing Boride on AZ91 Alloy, Materials Science Forum, 488-489(2005) 25–30

[8] S.E.Housh, V.Petrovich, Magnesium refining: a fluxless alternative. SAE technical paper 920071, Society of Automotive Engineers, Inc, Detroit, (1990) 1–7

[9] D. Emadi, J.P.Thomson, K.Sadayappan, M.Sahoo, Effects of grain refinement and melt filtration of the mechanical properties of sand and permanent mold cast magnesium AZ91 alloy, in: N.R.Neelameggham, H.I.Kaplan, B.R.Powell(Eds.), Magnesium technology 2005,T.M.S (the minerals, metals and materials society), Pennsylvania, (2005) 335–339

[10] P.Bakke, A.Nordmark, Filtration of magnesium by ceramic foam filters. in: Euel R.Cutshall (Eds.) Light Metals 1992, The minerals, metals and materials society, Warrendale, 1991, pp. 923–935

[11] P.Bakke, T.A.Engh, Magnesium filtration using ceramic foam filter, and subsequent quantitative microscopy of the filters, Materials and Manufacturing Processes. 9(1)(1994) 111–139

[12] Y. S. Wang, Study on Hot Tearing Behavior of Magnesium Alloys [D]. Shanghai: Shanghai Jiao Tong University, 2002

[13] S.E.Housh, V.Petrovich. SAE technical paper 920071, Society of Automotive Engineers, Detroit, MI 1990

Development of an Investment-Casting Process of Mg-Alloys for Aerospace Applications

G. Arruebarrena[1], I. Hurtado[1], J. Väinölä[2], S. Dévényi[3], J. Townsend[4], S. Mahmood[4], A. Wendt[5], K. Weiss[5], A. Ben-Dov[6], A. Schröder[7], A. Pinkernelle[7]

[1] MGEP, University of Mondragon, Mondragon, Spain
[2] VTT Industrial Systems, Espoo, Finland
[3] SpecialValimo Oy, Espoo, Finland
[4] Stone Foundries Ltd., London, United Kingdom
[5] RWP GmbH, Roetgen, Germany
[6] Israel Aircraft Industries Ltd., Tel-Aviv, Israel
[7] INFERTA GmbH, Magdeburg, Germany

1   Introduction

Investment casting offers high-quality with good dimensional accuracy, being therefore the most suitable for aircraft applications. Due to the facts that commercial ceramic shells react with Mg melts and that appropriate inhibitors are not available, magnesium alloys are not widely used in industrial investment-casting. In order to overcome this situation and enable the aircraft industry to take advantage of the light weight properties of magnesium, the process is analyzed within the European FP6 IDEA project, which aims at developing new high-strength magnesium casting alloys for aeronautic applications. The aspects of the project considered in this paper are:

- Study of metal mould reactions with the purpose of developing a new shell mould for Mg-investment casting.
- Replacement of SF_6 as protection gas by other inhibitors.
- Development of investment castings with high mechanical properties.

The reactions with the conventional shells containing colloidal silica (SiO_2) make the use of these refractories impossible in the casting of magnesium [1]. With the aim of finding alternative techniques to reduce the degree of the metal-mould reactions, different slurry materials and inhibitors were studied with the commercial magnesium alloy AZ91E. The less reactive mould materials were found to be magnesia (MgO) and yttria. The cost of yttria is very high and therefore, the industrial use is very limited. MgO was found to be the best alternative for industrial level, which is comparable with commonly used zirconia's. More detailed information about this research has already been published [2].

Investment casting process is better suited for thin walled complex castings. Development work, however, is expensive and time consuming due to the fact that wax pattern dies are expensive and time consuming to manufacture and that the investment casting process itself is quite lengthy. Anyway, by employing methods such as Rapid Prototype Modelling and casting simulation softwares, the lead-time could be considerably shortened. Effective use of such tools can often lead to good quality castings in the very first trials.

In the present work, once the slurry was prepared, various tests were performed using alternative inhibitor gases. Tensile test specimens and a test piece were used for the characterization of the material and the optimization of the process parameters. The characterization of the material was fulfilled with tensile tests and microstructural analysis.

All the experimental activity was supported with simulation, using ConiferCast-Flow3D® simulation package and WinCast® finite element simulation software for determining the process parameters and predicting filling and solidification patterns, as well as final mechanical properties.

2 Experimental Procedure

The commercial AZ91E alloy was used for the slurry selection and the inhibitor investigation. In this last task, different concentrations of inhibitors were used and the surface oxidation was analyzed by visual inspection. With the aim of improving the effectiveness of the inhibitor and reduce the SF_6 usage, the inhibitor gas was introduced through the mould wall. The mould was placed into a special steel chamber that was filled either with CO_2 or argon to avoid the effect of external oxygen inside the mould. The mould temperature was 420–450 °C and the metal temperature 740–750 °C.

AZ91E and a new alloy developed by MRI for aeronautic applications, the MRI207S, were cast for the optimization of process parameters and characterization of the properties.

A large number of test castings were produced to find optimum casting parameters for investment casting process. Tensile bars and a test-part (Figure 1) were used for a better evaluation of the process. For the mechanical and microstructural characterization, both tensile bars and tensile specimens machined out from the test-parts were used. The dimensions were defined by ASTM-E8M for samples of 6 mm diameter and the testing was made with a strain rate of $1.5 \cdot 10^{-4}$ s^{-1} and an extensometer with a gauge length of 25 mm.

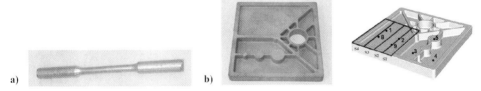

Figure 1: Test parts; a) tensile bars; b) Test-part and the area of machining of tensile specimens

Nine control points were defined for the grain size and porosity distribution in the test-part. Microstructural analysis was made in polished as-cast samples, and etched with nital for grain-size measurements and without etching for porosity measurements. Automatic image analysis was used for both of them.

Test castings were simulated and simulation results were compared with experimental results. By means of simulation and a few casting experiments, optimum casting parameters were found for investment casting process in AZ91E and MRI207 alloys.

3 Results and Discussion

3.1 Inhibitors

The first objective was to find the minimum SF_6 concentration for avoiding the oxidation in the casting or even the burning of the shell. Table 1 shows the results of these tests with different SF6 concentrations and different chamber inhibitors.

Table 1: Results of the tests for minimizing the usage of SF_6.

Mould Inhibitor	Chamber Inhibitor	Observations in the Casting	Results
1 % SF_6 / CO_2	Argon	Rapid burning	Completely oxidized
3 % SF_6 / CO_2	Argon	Rapid burning	Completely oxidized
6 % SF_6 / CO_2	Argon	Delayed burning	High level of surface oxide
9 % SF_6 / CO_2	Argon	No burning noted	Some oxidation noted
1 % SF_6 / CO_2	CO_2	No burning noted	High level of surface oxide
3 % SF_6 / CO_2	CO_2	No burning noted	Low level of surface oxide
6 % SF_6 / CO_2	CO_2	No burning noted	Minimal oxide
9 % SF_6 / CO_2	CO_2	No burning noted	Oxide free

Best results were obtained by immersing the mould in a CO_2 atmosphere instead of an argon atmosphere when casting and using 9 % SF_6 concentration in CO_2 as inhibitor. Special care has to be taken when designing the venting of the mould to allow all oxygen to be evacuated from the mould cavity, since air pockets remain easily inside the cavity even if the mould is flushed with protective gas. Strong oxidation or even burning of the melt can start from these air pockets.

The next step was to find alternatives to the use of SF_6, and the effectiveness of different inhibitors was analyzed. More information about this investigation can be found in [2]. BF_3 is formed inside the cavity and also possible gas holes inside the mould are filled with protective gas. No additional protection with SF_6 or any other gas was used in the mould in these tests.

First trials showed no signs of surface oxidation and the surface color was clean. Results were confirmed by non-destructive testing by penetrant flaw detection and X-rays. No oxides were revealed on X-ray and only one casting exhibited hot tears.

After future process optimization, BF_3 can be regarded as a non ozone depleting replacement for SF_6 as an investment casting magnesium inhibitor

3.2 Mechanical Properties

All the results were compared to the requirements of the aeronautic industry, in this case defined by Israel Aircraft Industries (IAI) for the applicability of magnesium alloys in aircrafts. The mechanical properties of tensile bars are shown in Table 2. From several cast alloys, the MRI207 was selected for being the most suitable alloy due to its higher tensile yield strength and melt stability. The MRI207 was superior in properties to the commercial alloy AZ91E. All the alloys were tested in as-cast and T6 condition and results are shown in Table 2.

Table 2. Tensile results in tensile bars

	UTS (MPa)	YS (MPa)	E (%)
Defined Limit Values	290	220	3
AZ91E – T6	235 ± 4	127 ± 4	4 ± 1
AZ91E – F	137 ± 4	94 ± 3	1.25 ± 0.2
MRI207S – T6	285 ± 3	215 ± 3	3 ± 0
MRI207S – F	205 ± 12	132 ± 3	6.3 ± 1.7

The grain size and porosity measurements in the test part were used for verification of the simulation results. Figure 2 shows the distribution of grain size and porosity in different control points for the alloy AZ91E.

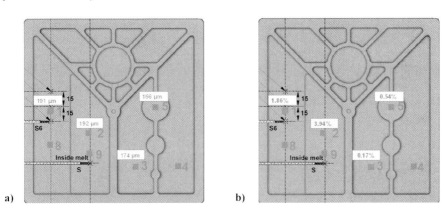

Figure 2: The read numbers denote points used for temperature measurement, microstructure and defect analyses. a) Grain size and b) porosity distribution in the investment-cast AZ91E test-part.

The mechanical properties of the test part were poor due to oxides found inside the part. Even if there was not any surface oxidation, some oxides appeared inside the part, mostly due to a non-adequate filling system or the melt oxidation prior to casting. Since the inferior mechanical properties, tensile testing values are not reliable and only the grain size and porosity distribution can be used for the evaluation of the test-parts.

With the aim of completing the information of investment-casting, further analysis are planned to be performed in prototypes. Nevertheless, with its better mechanical properties, the new MRI207 alloy looks promising for aircraft applications.

3.3 Simulation Tools

Filling and solidification simulations were performed to optimize process parameters. The simulations were compared with the results of solidification temperature measurements, mechanical tests and grain size and porosity measurements. The obtained agreement for the commercial AZ91E alloy was good as can be seen in the prediction of porosity and grain size measurements in Figure 3. More information about the performed simulation work can be found in [3] and [4]

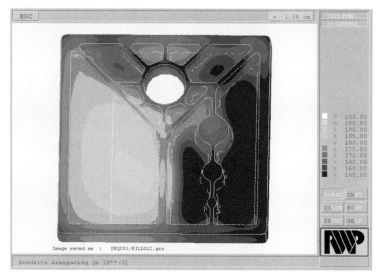

Figure 3: Simulation of the grain size distribution in AZ91E alloy using WinCast®; Simulated values are between 160-200 μm

Defect zones were also detected where they were predicted by simulation as shown in Figure 4.

Figure 4: Prediction of defects with ConiferCast-Flow3D®; a) Simulation for prediction of defects, colors showing temperature, and velocity vectors showing the direction of flow. Areas, which are prone to problems, are marked in blue, which are the points where the fluid fronts meet and possible cold shuts, filling problems or even cracks can be seen; b) Visible defect area, with misruns.

4 Conclusions and Outlook

- Alternative slurry materials and inhibitors are being analyzed for investment casting process, with promising first results. Obtained cast surfaces are visually cleaner and brighter than when SF_6 is used as inhibitor.
- New magnesium alloys showed promising properties for their use in aeronautic industry.
- Simulation tools are a powerful tool for prediction of defects and mechanical properties.

- More tests will be performed for the acceptance of the new alloys and the investment casting process. For this purpose, prototypes are being cast and tested.

5 Acknowledgment

This work is being carried out in the frame of a STREP project (FP6-503826) G. A. and I. H. thank the Diputación Foral de Gipuzkoa for the financial support. G. A. also acknowledges the Basque Government for providing her with a predoctoral fellowship.

6 References

[1] M. Rosefort, S. Korte, A. Bührig-Polaczek, Proc. of the 6th Int. Conf. in Mg Alloys and Their Applications, WILEY-VCH, Weinheim, p. 752–757, 2004.
[2] C. Cingi, J. Väinölä, J. Orkas, 45th Foundry Conf. 2005, Portoroz, Slovenia, Sept. 2005.
[3] A. Wendt, Ch. Honsel, K. Weiss, G. Arruebarrena, A. Schröder, J. Townsend, S. Devenyi, MCWASP 2006, Opio, France.
[4] K. Weiss, A. Wendt, Chr. Honsel, G. Arruebarrena, I. Hurtado, J. Väinölä , S. Mahmood, J. Townsend , A. Ben-Dov, A. Schröder, A. Pinkernelle, to be published in 7th Int. Conf. Mg Alloys and their Applications, Dresden, Germany, 2006 Nov. 6–9.

Evaluation of the Hot Tearing Susceptibility of Selected Magnesium Casting Alloys in Permanent Molds

J.P. Thomson, M. Sadayappan and M. Sahoo
Natural Resources Canada, CANMET-Materials Technology Laboratory, Ottawa, Ontario, Canada
D.J. Weiss
Eck Industries, Inc., Manitowoc, Wisconsin, USA

1 Abstract

The hot tearing tendency of several new creep resistant magnesium casting alloys was evaluated at CANMET-Materials Technology Laboratory (MTL) using permanent molds, focusing on the effects of alloy type and mold temperature. The alloys studied include three AE alloys (AE44, AE63, AE35), AJ62, AZ91 and three MRI alloys (MRI 201S, MRI 202S, MRI 206S). For comparison purposes, similar studies were performed on aluminum alloys 535, 206 and 319.

The results suggest that hot tearing of magnesium alloys can be reduced by casting in molds having a temperature 350 °C or higher. AZ91, an alloy with very poor creep resistance but good resistance to hot tearing, chosen as a good bench mark alloy, performed the best in this study. Of the creep resistant alloys tested, the AE alloys had very poor resistance to hot tearing with AE35 being the worst. The MRI alloys and AJ62 were less susceptible to hot tearing than the AE alloys. MRI 206S was the alloy that was the most robust at low mold temperatures. No hot tearing was observed when casting MRI 206S in molds having a low temperature of 340 °C.

2 Introduction

The project team involved in the USAMP cradle casting project, which uses the low pressure permanent mold (LPPM) process, indicated that some exploratory research should be carried out to measure the hot tearing behaviour of some selected magnesium alloys and compare with that of aluminum alloy 319. This will help the team to select the most suitable alloy for the second phase of the cradle casting project. The team indicated that they may select some new magnesium alloys that are being developed for high temperature applications, especially die casting alloys. CANMET-MTL was contracted to carry out the hot tearing tendency evaluation of several magnesium alloys. Their hot tearing tendencies would be compared to hot tearing data determined for three aluminum alloys (319, 535 and 206). The magnesium alloys evaluated were: AE35, AE44, AE63, MRI 201S, MRI 202S, MRI 206S, AZ91 and AJ62.

3 Experimental

The cast iron permanent molds used for this hot tear evaluation are shown in Figure 1. Both molds have a 2.7 cm diameter vertical riser. The 3-bar mold riser feeds horizontal rods having lengths of 24, 19 and 14 cm. The 4-bar mold riser feeds horizontal rods having lengths of 14, 10, 6 and 2.5 cm. All rods have a diameter of 1 cm. To facilitate restraint, there is a ball having a diameter of 2 cm at the end of each rod. These molds were designed as a very severe test of an alloys susceptibility to hot tear or hot crack during solidification.

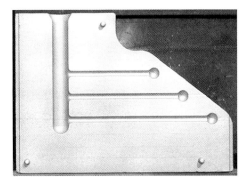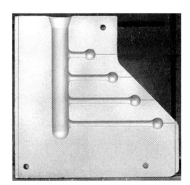

Figure 1: Photographs of the molds used in this study.

The molds were heated to 250–300 °C to apply coatings. The base coat used was insulating Dycote 6 from Foseco and the top coat was HGLS 07290402, from Hill and Griffith Company, which contains a fluorspar addition. An additional 1% CaF_2 was added to the top coat for these trials. The molds were then heated to 350–430 °C for casting. The same coating was used for all alloys. Table 1 shows the melting procedure and molten metal processing for each alloy. The molds were flushed with $CO_2 + 0.5\%SF_6$ before each casting. A heated hand held spoon was used for pouring. Once the spoon had a significant build-up of MgO, it was replaced, usually after making 4 or 5 castings. The metal superheat for each trial ranged between 125 and 160 °C.

Table 1: Melting Practice for alloys tested

Alloy	Melting Procedure
319	Melted and held in a resistance furnace
	Grain refined using Al-5%Ti-1%B
535	Melted and held in an induction furnace
206	Both alloys were pre-grain refined
	206 received additional grain refinement using Al-5%Ti-1%B
AE44	Melted and held in a resistance furnace using $CO_2 + 0.5\%SF_6$ as a cover gas
AE35	Grain refined using C_2Cl_6
AE63	
AZ91	
AJ62	
MRI 201S	Melted and held in a resistance furnace using $CO_2 + 0.5\%SF_6$ as a cover gas
MRI 202S	All alloys grain refined using 6.5% Mg-15%Zr
MRI 206S	MRI 201S receives additional refinement using 1.3% Mg-22%Y-1.3%Nd
	Grain refinement additions were carried out according to instructions by the ingot supplier

4 Results and Discussion

The stress generated during solidification is highest in the longest rod and least in the smallest rod. Therefore, if an alloy is prone to hot tearing or hot cracking, more rods will break compared to an alloy that is not prone to hot tearing or hot cracking. The location of breaks was also noted. The theory followed in this work is that if the casting breaks at the end of the rods where the restraining ball is located or in the middle of the rod, the alloy is susceptible to hot cracking. Hot cracking occurs when there is stress / strain build-up due to the thermal contraction and this build-up cannot be resolved by feeding alone. If breaks are seen near where the rod is attached to the riser, the alloy is prone to hot tearing. Breaking at this location can be avoided if the feeding from the riser is adequate. The data for the number of hot tears is reported in this work. The severity of breaking was not considered in this work.

The data is represented by the number of hot tears and the number of hot cracks per casting. For this work, a casting is considered as both the 3 and 4 bar mold combined. So it is possible to have a maximum of 7 hot tears per casting. If a ranking is close to 7, then the alloy is very prone to hot tearing while alloys that have a lower number are less prone. The hot tearing data is represented in Figure 2 for the the magnesium alloys. A casting exhibiting hot tearing is shown in Figure 3.

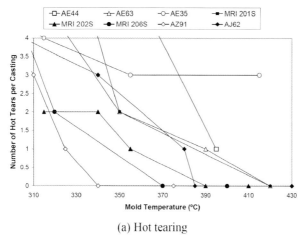

(a) Hot tearing

Figure 2: Plots of Hot Tearing as a function of mold temperature

Two of the aluminum alloys tested showed no hot tearing or hot cracking during these tests. 319 and 535 did not crack when cast in molds having a temperature of 350 °C and 220 °C respectively. The 206, which is known to have poor resistance to hot tearing, broke when cast in molds having a temperature of 290 °C. However, hot tear testing conducted on this alloy for another project has shown that hot tearing can be eliminated when the mold temperature exceeded 400 °C.

The AE magnesium-base alloys, which contain rare earths, tested were AE44, AE35 and AE63. Alloy AE44 exhibited hot tearing when cast in molds having temperatures less than 395 °C. The hot tearing tendency of alloy AE35 was much higher and this alloy did not produce a good casting even when the mold temperature was as high as 415 °C. Of the three AE alloys, AE63 performed relatively better even though the lowest temperature that can be tolerated was 395 °C. The AE family of alloys likely contain trace amounts of beryllium (Be) which is added to ingots to hinder oxidation during melting and casting operations. It should be mentioned that beryllium is also known to hinder the effectiveness of grain refinement.

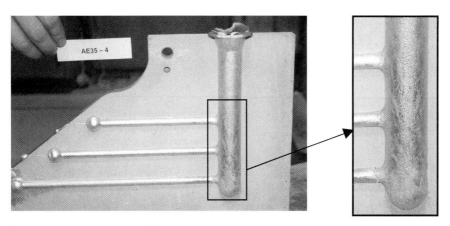

Figure 3: Hot tearing in alloy AE35

AZ91, even though it is not a creep resistant alloy, was tested in this investigation and found to be the benchmark for magnesium alloys. Castings free of cracks were observed even when the mold temperature was as low as 335 °C.

AJ62, another aluminum bearing creep resistant alloy containing strontium was found to be marginally better than AE63. No hot tearing was observed when the mold temperature was 380 °C.

The MRI magnesium-base alloys, containing rare earths, were also tested. Of the three alloys, MRI 201S, 202S and 206S tested, MRI 206S was the best alloy and its performance was close to AZ91 which is the bench mark in this study. The other two alloys performed much similar to the other creep resistant magnesium alloys. The minimum mold temperature to have a crack free casting for MRI 202 alloy was 390 °C. A crack free casting was produced in MRI 201S only when the mold temperature was 420 °C.

Table 2 summarizes the minimum mold temperature needed to have no hot tearing in the magnesium alloys tested. From all the creep resistant magnesium alloys tested, MRI 206S can be cast at the lowest mold temperature (340 °C) to observe no hot tearing. All other creep resistant magnesium alloys need a mold temperature of 390 °C or higher to be free of hot tearing using these molds.

Some typical microstrucures are shown in Figure 4 to depict the nature of crack propagation which is found to be intergranular / interdendritic.

Table 2: Hot Tearing Data

Alloy	Mold Temperature for No Hot Tearing (°C)
AZ91	=335
MRI 206S	=340
AJ62	=380
MRI 202S	=390
AE63	=395
AE44	=395 (1 crack)
MRI 201S	=420
AE35	=415 (3 cracks)

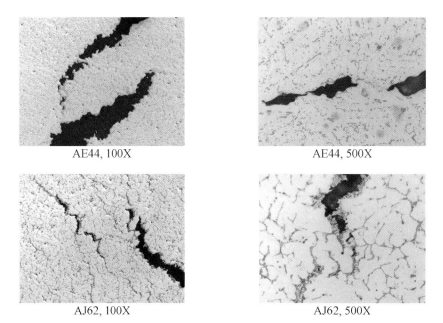

AE44, 100X AE44, 500X
AJ62, 100X AJ62, 500X

Figure 4: Typical microstructures of selected magnesium alloys showing hot tear cracks and the mode of crack propagation

5 Conclusions

From this work it can be concluded that the occurrence of hot tearing is reduced when the mold temperature is 350 °C or greater. Molds having a temperature of 340 °C or higher eliminated the occurrence of hot tearing in alloy MRI 206S. These preliminary results show that MRI 206S could be the creep resistant alloy of choice to use for complicated castings since it requires the lowest mold temperature before hot tearing occurs.

6   Acknowledgements

This material is based upon work supported by the Department of Energy National Energy Technology Laboratory under Award Number DE-FC05-02OR22910. This report was prepared as an account of work sponsored by an agency of the United States Government. Neither the United States Government nor any agency thereof, nor any of their employees, makes any warranty, express or implied, or assumes any legal liability or responsibility for the accuracy, completeness, or usefulness of any information, apparatus, product, or process disclosed, or represents that its use would not infringe privately owned rights. Reference herein to any specific commercial product, process, or service by trade name, trademark, manufacturer, or otherwise does not necessarily constitute or imply its endorsement, recommendation, or favoring by the United States Government or any agency thereof. The views and opinions of authors expressed herein do not necessarily state or reflect those of the United States Government or any agency thereof.

The authors wish to thank the following ingot suppliers for contributing to this project: Dead Sea Magnesium for the MRI alloys, Magnesium Elektron for AJ62 and Norsk Hydro for the AE alloys. Nick Fantetti's assistance with the foundry trials for the MRI alloys is greatly appreciated. Some experiments were performed with the help of Daryoush Emadi. The dedication of Peter Newcombe and Luc Millette from the Experimental Casting Laboratory from CANMET-MTL is also appreciated. Special thanks to Renata Zavadil and Lee-Ann Sullivan for metalography.

Temperature Dependence of the Viscosity of Magnesium Alloy AZ91D with Additions of Calcium in Liquid State and Effect of Thermal Treatment of these Melts on the Structure that Forms after their Crystallization

I. S. Abaturov[1], P. S. Popel[1], I. G. Brodova[2], V. V. Astafiev[2], Li Peijie[3]
[1] Ural State Pedagogical University, Ekaterinburg
[2] Institute of Metal Physics, Ural Branch of the Russian Academy of Sciences, Ekaterinburg
[3] Tsinghua University, Beijing

1. Introduction

Alloys of the Mg-Al-Zn system, the most popular of which are AZ91 and its analogs, are widely used in present-day technology, mainly owing to their high castability, good mechanical properties and corrosion resistance [1]. Their properties have been intensively studied lately, first of all in connection with their use in automotive industry as alloys for die casting [2,3]. The structure and the properties of these alloys can be improved by introducing small amounts of various elements, including calcium.

It is well known that calcium reduces the tendency of Mg-Al-Zn alloys to oxidation in the process of melting, improves their tightness, thermal-treatment susceptibility, heat resistance, but at the same time the presence of more than 0.2% Ca in these alloys (here and in the subsequent discussion the concentration is given in mass percentage) results in their embrittlement [4]. In introducing 0.05-0.40% Ca into pure magnesium and Mg-Mn and Mg-Zn alloys one could note the grinding of grains [5,6]. The essential modifying effect of calcium introduced in amounts of 0.2% and the ensuing increase of ductility has also been noted for Mg-Zn-Si and Mg-Al-Zn-Si alloys [7,8].

The effect of small additions of Ca (0.1-1.0%) on the grain size, the microstructure, hot-shortness and mechanical properties of the AZ91D alloy is investigated in reference [9]. The authors have found that the introduction of 0.1-0.4% calcium is accompanied by a considerable grinding of grains and a decrease in the size of the dendrite cell, but with a further increase of its concentration the structure dispersivity does not practically increase. The alloy hot-shortness increases with increasing concentration of calcium x_{Ca}, and particularly fast up to the same 0.4%. The limit of strength and the elongation gradually decrease in the whole investigated interval of x_{Ca}, and conversely, the limit of fluidity increases.

The mechanism of the effect of calcium additions on the structure and properties of magnesium alloys is for the time being poorly understood. The authors [9] connect it with the limitation of the growth of the β-phase of $Mg_{17}Al_{12}$ under the action of the first portions of calcium and with the appearance of a new phase of Al_2Ca at a further increase of x_{Ca}. However, the first causes of these phenomena remained non-established. It seemed probable to us that additional information on the peculiarities of effect of different amounts of calcium on the structure of the AZ91D alloy might be obtained from the results of investigating the properties of the initial melts, in particular, their viscosity.

Therefore, the present paper investigates temperature dependences of viscosity for the alloy AZ91D without additions of calcium and with its concentrations of 0.3, 0.4, 0.6, 0.8 and 1.0% in the temperature range from 600 to 840°C. On the basis of the obtained results regimes of Thermal treatment of melts have been suggested, and the effect of this treatment on the structure of cast samples that form after their crystallization has been studied.

2. Experimental Procedure

The paper investigates the same samples as in [9]. They were prepared on the basis of the commercial alloy AZ91D containing 90.01% Mg, 9.00% Al, 0.73% Zn, 0.19% Mn, 0.058% Si, 0.0037% Cl, <0.0003% Be, 0.007% Fe, 0.0036% Cu and 0.0019% Ni, into which the required quantity of the master alloy Mg-30% Ca was introduced at 730°C. the melting was realized in an electric resistance furnace in a protective atmosphere of $CO_2 - 0.5\%$ SF_6. After stirring for 2 min at the indicated temperature the melt was allowed to stand for 10 min and then poured into steel cylindrical moulds 30 mm in diameter and 65 mm in height heated previously up to 280-300°C.

The measurement of properties of liquid magnesium alloys is complicated by the intense evaporation of magnesium in the course of the experiments. The attempt to overcome this difficulty by performing experiments under inert gas excess pressure (about 1 atm) was not successful. In the subsequent series of experiments the authors used crucibles of stainless steel with a spiral cover sealed with a copper ring. This solved the problem of magnesium evaporation. However, a chemical analysis of the samples before and after measurements pointed to the appearance of steel components in them, nickel in particular, which was indicative of the interaction between the melt and the crucible. Besides, it turned out that liquid AZ91D wetted the walls of a crucible. It resulted in a considerable distortion of the sample shape, which complicated the calculation of viscosity. A felicitous solution of the problem proved to be the introduction of an additional crucible of graphite 15 mm in diameter and 43 mm in height into a cylindrical container of stainless steel 13X30 (Russian Standard) with a spiral cover sealed with a copper ring. The hermetic sealing of the container was realized under a jet of pure helium. Experiments have shown that in this case no evidence of interaction between the melt and the refractory and visible contamination of the sample surface is observed. In most cases after measurements samples were freely withdrawn from the crucible, which made it possible to use crucibles repeatedly.

The viscosity was measured by the method of damped torsion oscillations of a crucible with a melt on the setup whose scheme is presented in [10]. The procedure of measuring the period and the decrement of the oscillation damping was considerably modernized. For the registration of periods use was made of an electronic stopwatch with a photo sensor. The oscillations of a beam of light reflected from the mirror of a suspended system on a transparent ruler were recorded by a video camera on a digital medium. Then the video image was treated in the still-picture regime for every oscillation for determining its amplitude. Thus the amplitudes of 25 successive oscillations were recorded, whereupon the exponential dependence of the amplitude on the vibration number was constructed with the help of a computer by the least squares method. The logarithmic decrement was determined by this exponent of a power. The values of the kinematic viscosity v for a melt were calculated by the procedure of Shvidkovsky for low-viscosity liquids [10]. As a result of modernization of the viscometer it was possible to decrease the error of determination of v from 3-5 to 1-2%.

Before every measurement the melt was allowed to stand at a given temperature for not less than 12 minutes.

Metallographic investigations of the structure of samples of the base alloy and those containing 0.4 and 0.8% Ca before and after thermal treatment in the liquid state were made on an optical microscope "Neophot-32" with a computer attachment for a quantitative phase analysis and photographing of objects. For revealing the morphology of the intermetallic compounds that were present in the structure of alloys a study was made of metal microsections without chemical picking. The internal microstructure of grains and their size were determined after chemical picking in a 10% solution of hydrofluoric acid.

The matrix microhardness was measured with a microhardness gauge ПМТ-3 (Russian Standard) at a load of 20 g and calculated as the average value of ten measurements.

3. Experimental Results

In Figs 1-2 one can see the structures of the initial samples. They consist of grains of a solid solution on the basis of magnesium and intermetallic phases located inside and at the boundaries of the matrix grains (Fig.1a). These compounds contain aluminum ($Mg_{17}Al_{12}$, Mg_4Al_3 phases), silicon (Mg_2Si phase) and calcium (Mg_2Ca phase). The largest volume of them is occupied by crystals of $Mg_{17}Al_{12}$, near which there is a lamellar eutectics ($Mg+Mg_2Si$) of liquation origin. In samples with a high content of calcium there additionally appear dark crystals of Mg_2Ca of rounded form (Fig.1b).

An increase in the content of Ca in alloys results in a twofold comminuting of grains (see Table 1), which is in agreement with the results from [9]. Besides, one can note a certain tendency to the dispersion of intermetallic phases. The matrix microhardness does not practically change and is equal to 820-850 MPa.

Temperature dependences of the kinematic viscosity ν of the base alloy AZ91D and samples containing 0.3, 0.4, 0.6, 0.8 and 1.0% Ca obtained in the course of heating from 600 to 840°C and a subsequent cooling are presented in Fig.2.

Within the limits of the declared measurement error the curves $\nu(T)$ of the base alloy and that with 0.3% Ca obtained in heating and cooling coincide (the non-coincidence of the first point of the heating curve and the last point of the cooling curve is evidently connected with the two-phase structure of a sample at temperatures lower than the liquidus). In a sample with 0.4% Ca the ramification of the heating and the cooling curve manifests conspicuously in the whole investigated temperature range (viscosity hysteresis).

With a further increase in the calcium concentration the ramification of the dependences $\nu(T)$ becomes less pronounced, at high temperatures there appears their coinciding segment, and ramification point decreases to 700°C at 0.6% Ca and disappears altogether in a sample with 0.8% Ca. However, with calcium content of 1% hysteresis of viscosity with a ramification point of about 775°C is again registered distinctly.

It should also be mentioned that at calcium concentrations of 0.4 and 1.0% on heating curves a well-defined maximum is registered in the vicinity of 720-730°C, and in the last case it manifests on the cooling curve too.

Previously the ramification of temperature dependences of the viscosity and some other properties of melts was noted repeatedly in studying liquid steels, cast irons, aluminum and nickel alloys [11]. B.Baum et al. showed that after superheating a liquid metal above the ramification point one could observe considerable changes in the structure of the ingot that formed after its crystallization. As far as we know, previously there were no similar investigations of magnesium alloys.

Therefore the present paper studies the effect of superheating the alloys AZ91D and AZ91D with additions of 0.4 and 0.8% Ca above the ramification point of temperature dependences of the viscosity on the structure of a cast metal.

The initial samples were previously re-melted, heated in the liquid state to 650°C and crystallized in cooling condition similar to those observed in the final part of viscometric experiments. After the indicated re-melting owing to a relatively low cooling rate during crystallization one could observe the roughening of the structure of both the base alloy (the grain size increased with respect to the initial state to 200 μm) and the samples with calcium. The phase composition, however, remained unchanged (Fig.3).

After superheating the melts to 850°C, i.e. above the ramification point of temperature dependences of the viscosity of samples containing calcium one could note a further growth of grains of the base alloy (to 290 μm) and partial dissolution of the compound $Mg_{17}Al_{12}$. the last result is confirmed by the increase of microhardness (see Table 1) evidently caused by solid-solution hardening at the expense of additional alloying of magnesium by aluminum.

In samples that contain calcium a superheat in the liquid state above the ramification point of the curves $v(T)$ is also accompanied by the roughening of the structure (Table 1). Besides, one can note the dispersion of intermetallic phases at the boundaries of grains and a decrease in the volume fraction of eutectics. The modifying effect of calcium is also retained after a superheat, but, as in the case of samples that have not been heat treated in the liquid state, an increase in the content of Ca above 0.4% is not accompanied by an amplification of this effect.

4. Discussion of the Results

The authors believe that the most original result of the presented work is ramification of temperature dependences of the viscosity of calcium-containing melts, which manifests itself more or less distinctly depending on the concentration of Ca. A chemical analysis of samples before and after a viscometric experiment has shown that in the process of heating to the maximum experimental temperature no considerable changes in their composition are observed. There were also no signs of interaction between the melts and the crucible or any noticeable evaporation of magnesium. Besides, as has already been mentioned, in the base melt, which did not contain calcium, no viscosity hysteresis was observed.

All these circumstances lead to the conclusion that in melts AZ91D containing Ca in heating one can observe an irreversible rearrangement of structure. The temperature of this rearrangement depends on the concentration of calcium.

On the qualitative level this process may be interpreted in the framework of notions of the metastable microheterogeneity of liquid metal solutions formulated in [ ]. According to these notions, after the melting of a multiphase crystalline sample in the forming melt for a many hours there may exist disperse particles close in composition to the corresponding phases and weighted in a dispersion medium enriched with the main component. These particles are in metastable equilibrium with the dispersion medium. When the temperature is raised to the values corresponding to the ramification points of temperature dependences of the properties, this equilibrium is upset, and one can observe irreversible dissolution of disperse particles of a definite type.

In our case in the initial crystalline structure of samples containing up to 0.4% Ca there are 3 above mentioned compounds, which makes it possible to predict appearance in a melt of at least three types of disperse particles of different composition. The absence of ramification of temperature dependence of the base melt viscosity allows us to assume that a superheat to

840°C is not sufficient for irreversible dissolution of these particles. Evidently, the first portions of calcium introduced into the system dissolve in the matrix and owing to their interfacial activity are concentrated in the surface layers of disperse particles reducing the interfacial tension at their boundaries, which determines the thermal stability of these particles. At a calcium concentration of 0.4% in proximity to the maximum temperature of the melt heating the interfacial tension at the boundaries of disperse particles of one of the types decreases to such low values that there begins spontaneous dispersion of these particles, which usually precedes their dissolution. This is just the reason for the pronounced discrepancy between the curves $v(T)$ obtained in heating and cooling. With a further increase in the content of calcium in the system the dissolution of the particles under discussion has time to be completed at the approach of the ramification temperature, as evidenced by appearance of their coinciding segment at high temperatures. The absence of ramification of these curves in a sample with 0.8% Ca may be treated as the dissolution of particles of the indicated type immediately after the melting of the initial sample. To a content of Ca of about 1% there evidently corresponds the limit of its solubility in a melt. Beginning with this concentration, in a microheterogeneous system there appear disperse particles of a composition close to Mg_2Ca, which are inherited from corresponding phase of the initial ingots and are irreversibly destroyed when heated to 775°C as demonstrated again by a clearly defined ramification of temperature dependences of the viscosity.

It goes without saying that this interpretation of the results of viscometric measurements is of hypothetical character and to a considerable agree is based on data obtained previously in studying some other commercial melts and melts synthesized in a laboratory. To confirm or to refuse this, it is necessary to make additional investigations of the binary and ternary systems which are the basis for the melt AZ91D.

The above-mentioned dispersion of intermetallic inclusions at the boundaries of grains after superheating melts above the temperature of destruction of disperse particles may be connected on the qualitative level with this effect. Indeed, after their dissolution the process of formation of the corresponding phase during crystallization is kinetically hampered, it begins at a higher undercooling with respect to the liquidus and proceeds at an increased rate, which eventually results in formation of a more disperse system.

5. Conclusion

Thus, in the process of doing the present work the authors managed to solve some serious methodical problems connected with the high volatility of magnesium and its alloys in the liquid state. As a result, well-reproduced temperature dependences of the viscosity of melts AZ91D and AZ91D with additions of calcium have been obtained. Indications of the irreversible structural changes that are observed in these melts at temperatures depending on the content of Ca have been discovered for the first time. It is shown that the thermal treatment of samples in the liquid state, which includes their heating to temperatures exceeding that of irreversible rearrangements, is accompanied by considerable changes in the cast structure that forms after the crystallization of the samples.

The authors are grateful to the Russian Foundation for Basic Research for the financial support of these investigations in the framework of Grant No. 05-03-32653.

References

1. A.N.Khovanov, V.G.Davydov, F.M.Elkin, Tech. Light Alloys. 2003, 2-3, 9.
2. H.Friedrich, S.Schumann, J. Mater. Process. Technol., 2001, 117, 276.
3. B.Tang, P.Li, D.Zeng, Tech. Light Alloys, 2003, 2-3, 17.
4. M.B.Altman, M.E.Dritz, M.A.Timonova, M.V.Chuhrov, Magnesium alloys 1, Metals Science of Magnesium and its Alloys, USSR, Moscow, 1978.
5. I.F.Kolobnev, V.V.Krymov, A.P.Poliansky, Foudryman's Reference Book, USSR, Moscow, 1975.
6. Y.C.Lee, A.K.Dahle, D.H.Stjohn, Metall. Mater. Trans., 2000, A 31, 2895.
7. J.J.Kim, D.H.Kim, K.S.Shin, N.J.Kim, Scr. Mater., 1999, 41/3, 333.
8. G.Y.Yuan, M.Liu, W.Ding, I.Akihisa, Mater. Sci. Eng., 2003, A 357, 314.
9. P.Li, B.Tang, E.G.Kandalova, Mat. Lett., 2005, 59, 671-675.
10. I.G.Brodova, P.S.Popel, G.I.Eskin, Liquid Metal Processing: Applications to Aluminum Alloy Production, Series Part: Advances in Metallic Alloys, Vol.1, Taylor&Francis, London and New York, 2002, p.269.
11. B.A.Baum, G.A.Hasin, G.V.Tjagunov et al., Zhidkaja stal' (Liquid Steel), Moscow: Metallurgija, 1984, p.228.

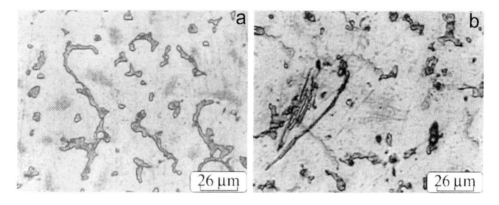

Figure 1: Structure of initial sample of AZ91D alloy: a) content of Ca 0.4 %, b) content of Ca 0.8 %

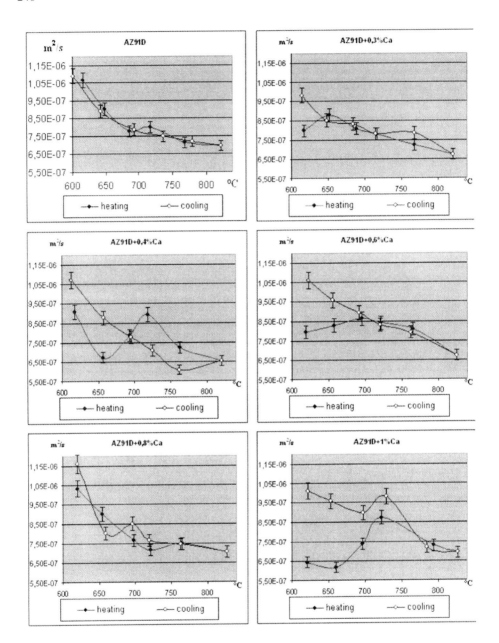

Figure 2: Temperature dependence of the kinematic viscosity of the base alloys AZ91D and samples containing 0.3, 0.4, 0.6, 0.8 and 1.0 % Ca obtained in heating from 600 °C to 840 °C and a subsequent cooling.

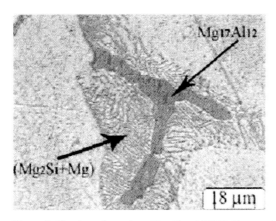

Figure 3: Structure of samples of the alloy AZ91D after remelting in a viscosimeter.

Table 1. The microhardness and the grain size of an alloy versus the composition and the means of thermal treatment in the liquid state

Composition and heat treatment	Microhardness, [MPa]	Grain size, [μm]
AZ91D (initial)	800	120
AZ91D re-melted at 650°C	700	200
AZ91D re-melted with superheating to 840°C	750	280
AZ91D+0.4% Ca (initial)	820	75
AZ91D+0.4% Ca re-melted at 650°C	765	180
AZ91D+0.4% Ca re-melted with superheating to 840°C	710	230
AZ91D+0.8% Ca re-melted at 650°C	840	180
AZ91D+0.8% Ca re-melted with superheating to 840°C	650	250

Advances in Magnesium Injection Molding (Thixomolding®)

M. Scharrer, A. Lohmüller, R. M. Hilbinger, H. Eibisch, R. Jenning, M. Hartmann, R. F. Singer
Neue Materialien Fürth GmbH, Fürth

1 Introduction

Magnesium Injection Molding is a relatively new and environmental-friendly technique for production of high quality magnesium parts that are net or near net shape. Due to a low casting temperature near the liquidus or even in the semi-solid regime, less porosity, lower distortion and closer tolerances in comparison to high pressure die casting can be achieved [1].

The present paper will introduce further advantages of the Thixomolding technology. As discussed below, the closed process in combination with intensive mixing in the barrel is beneficial for casting alloys containing elements that are known to be prone to oxidation, segregation or evaporation in conventional die casting.

Table 1: Overview on the most important alloying elements in magnesium alloys and their tendencies to segregation, oxidation and evaporation

Element	Al	Zn	Mn	Ca	Si	Sr	RE
Density (g/cm^3)	2,7	7,1	7,4	1,6	2,3	2,5	~6,5
Segregation	X	X	X	X	X		
Oxidation	X	X	X				
Evaporation	X						

Table 1 gives an overview of the most important alloying elements in magnesium alloys. During melting the large difference in atomic weight leads to segregation of the heavy elements towards the bottom of the crucible. Additionally the formation of intermetallic phases with high melting points, especially Mg_2Si, Mn_5Si_3 and Al_XMn_Y, can be very strong. As a result, a sump is developed in the crucible, which has to be removed from time to time. The elements Ca, Sr and Al tend to oxidation and form a slag on the melting bath surface. Magnesium itself has a relatively high vapor pressure of about 10 mbars at 700 °C. So the magnesium tends to evaporate in air. Zink has an even higher vapor pressure of about 100 mbars at 700 °C. As a consequence there is always a steady loss of zink due to evaporation. All resulting deviations from the nominal alloy composition have to be observed and adjusted by the foundry man.

The effects described above cause a lot of problems in foundry practice. In magnesium injection molding, all these problems don't occur because of a small melt volume, intensive mixing during transport in the screw and a closed system. In addition the thixomolding process is characterized by a low temperature level of the melt that can reduce die sticking, hot cracking and improve die life. Beyond this the processing technology is similar to polymer injection molding, which allows the development of hot runner systems that gives thixomolding another unique selling point in comparison to die casting. Scrap can be greatly reduced, heat energy and raw material is saved and flow length during filling can be reduced dramatically.

In this study processing of a variety of commercial magnesium alloys (AZ-, AJ-, AS-series) using thixomolding has been investigated. For AZ91 the influence of several processing parameters on the mechanical properties is discussed in detail (e. g. die temperature, processing temperature and solid phase content). It is shown that excellent strength and elongation can be achieved over a wide range of solid phase contents. For all alloys tensile properties that are typical for thixomolding are presented. The trials have been performed at processing temperatures near the liquidus temperature. A comparison with data from the literature for die casting is introduced. Furthermore first results for a prototype hot runner system developed at NMF are presented.

2   Experimental

The casting trials were carried out on a 220 t thixomolding machine from Japan Steel Works (JSW) type JLM220-MG. Figure 1 shows a schematic drawing of a thixomolding machine.

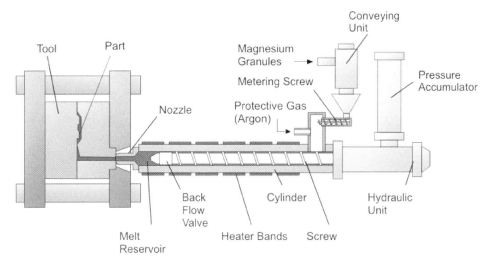

Figure 1: Schematic drawing of a thixomolding machine

The chipped raw material, delivered by ECKA granules, is measured into the cylinder using a metering screw. In the cylinder the magnesium granules are heated up and sheared. The heating energy is supplied by electric heater bands. To prevent the magnesium from oxidation argon gas is applied. By the rotation of the screw of about 100–200 rpm the magnesium is transported to a melt reservoir in front of the back flow valve. After closing the mold, the screw is moved forward with velocities of about 1–4 m/s and the molten metal is shooting into the cavity. After filling the mold a hold pressure up to 1000 bars can be achieved during solidification.

Casting trials were done using a mold for separately cast test bars according to ASTM B557-02a. With one shot two tensile bars were cast. The bars had a gage length of 50,8 mm and an overall length of 228 mm. Mechanical testing was done using a Zwick/Roell Z100 tensile testing machine. Table 2 shows the alloy compositions of the raw material measured with a glow discharge optical emission spectroscope (GDOES) GDProfiler from Horiba Jobin Yvon at

NMF. Additionally the barrel temperatures used for the comparison with die casting are mentioned. The concentrations of the elements Fe, Ni, Cu were below the ASTM B93-2004 limits.

Table 2: Alloy compositions of the raw material and barrel temperatures during casting

Element	Mg	Al	Zn	Ca	Mn	Sr	Sn	Si	T_{barrel}(°C)
AZ91D	bal.	9,3	0,7		0,2				600
AZ70	bal.	6,6	0,5		0,2				610
AJ52	bal.	5,0	0,1		0,3	1,8			625
AJ62	bal.	6,5	0,1		0,2	2,1			620
AS41low	bal.	3,7	0,2		0,2			0,7	635
AS41high	bal.	4,9	0,2		0,2			0,6	630

3 Results

3.1 Magnesium-aluminum-alloys (AZ-alloys)

An important characteristic of the thixomolding technique is the possibility to cast the material in the semi-solid state. The solid phase content is controlled by the processing temperature. Former studies have already shown that high volume fractions of solid phase lead to lower porosity, especially in thick walled sections of the castings [9]. Figure 2 shows the mechanical properties of AZ91D as a function of casting temperature. All casting trials were done with a die temperature of 150 °C. Casting parameters like injection speed and hold pressure were kept constant (full symbols). With lower casting temperatures, thus higher fraction of solid, the tensile strength and the elongation seem to descend. One of the reasons for this is the reduced heat content of the melt, which results in casting defects like cold flow. This effect has already been discussed in detail [4]. By adjusting the process parameters, the mechanical properties can be significantly raised, essentially to the values for fully liquid material (open symbols). All data are within the scatter band for die casting [3]. Especially tensile strength and elongation are near or above the upper limit.

Another important issue concerning the mechanical properties of magnesium alloys is their strong dependency on the grain size. In cast magnesium parts the grain size can be controlled by the solidification rate. With lower die temperatures, smaller grain sizes can be achieved [2].
Figure 3 shows the mechanical properties of AZ91D as a function of die temperature. The casting temperature was 590 °C, resulting in a volume fraction of solid of about 12 %. Yield strength rises from 145 MPa at 275 °C die temperature to 175 MPa at 50 °C. Metallographic studies showed that the grain sizes vary from about 10 µm for the lowest temperature to 20 µm at the highest die temperature. Elongation and tensile strength rise with lower die temperatures too. At 100 °C the highest values with 270 MPa and 8 % were achieved.

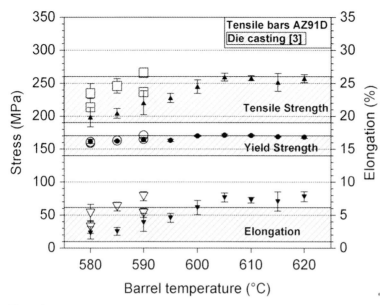

Figure 2: Mechanical properties of AZ91D as a function of casting temperature. The shaded areas represent results from the literature for die casting [3]

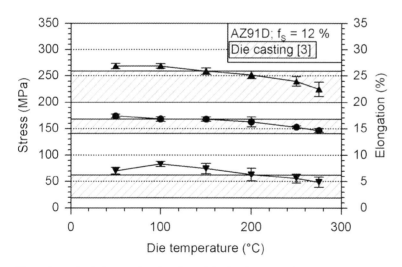

Figure 3: Mechanical properties of AZ91D with a fraction solid of about 12 % as a function of die temperature

In Figure 4 the mechanical properties of the alloys AZ70 and AZ91D, cast directly above the liquidus temperature, are shown in comparison to literature data for die casting [5].

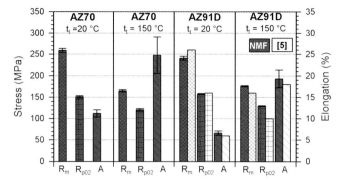

Figure 4: Mechanical properties of the alloys AZ70 and AZ91D at room temperature and 150 °C

While the properties at room temperature for AZ91D are within the same range for both processes, the thixomolded specimens are quite a bit better at 150 °C (R_{p02}: 130 MPa, A: 19 %).

An interesting alternative to AZ91D is the alloy AZ70. At room temperature yield and tensile strength are comparable with AZ91D, while elongation is significantly higher ($A_{20°C}$: 11 %, $A_{150°C}$: 25 %).

3.2 Magnesium-aluminum-strontium-alloys (AJ-alloys)

For components that are used at higher temperatures, Mg-Al-Sr-alloys have gained importance within the last years. By adding strontium the formation of the phase $Mg_{17}Al_{12}$, which is believed to be detrimental with respect to creep, is suppressed. Instead other phases like Al_4Sr are formed. In this context the new BMW six-cylinder-aluminum-magnesium-crankcase received much attention [11]. In spite of their other advantages, processing of the new AJ-alloys does not seem to be straightforward (high casting temperatures, etc.) [5].

Thixomolding trials have been carried out with no problems whatsoever for AJ52 and AJ62. Figure 5 shows the mechanical properties of these alloys at room temperature and 150 °C in comparison to conventional die cast material [6]. At room temperature and 150 °C, all mechanical properties of the thixomolded specimens are significantly better than the reference. In particular thixomolded AJ52 shows higher elongations up to 20 % at 150 °C.

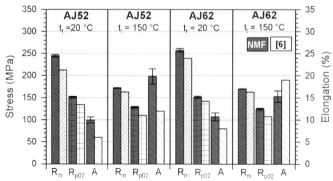

Figure 5: Mechanical properties of the alloys AJ52 and AJ62 at room temperature and 150 °C

3.3 Magnesium-aluminum-silicon-alloys (AS-alloys)

Another way to obtain higher creep resistance than AZ91D is by lowering the aluminum content and adding silicon. Because of the lower aluminum content less $Mg_{17}Al_{12}$ is formed. The silicon addition leads to formation of the thermally stable Mg_2Si phase, which has a melting point of about 1085 °C. Silicon-containing alloys have been used extensively in the VW Beetle. Presently the Daimler Chrysler 7G-TRONIC transmission case is produced from AS31 [10].

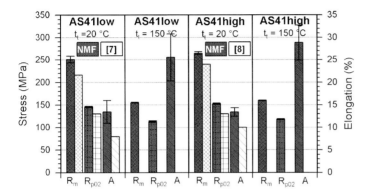

Figure 6: Mechanical properties of the alloys AS41low and AS41high at room temperature and 150 °C

In this study two AS41-derivates with different aluminum contents were thixomolded. The composition of the AS41low is within the range for AS41 and AS31. Figure 6 shows their mechanical properties. At room temperature the yield strength of AS41low is 130 MPa and elongation 13 %. AS41high has 151 MPa and 13 % elongation. For both alloys these values are much better than the reference data [7], [8]. For 150 °C no reference data were available.

3.4 Hot Runner Technology

The above mentioned die for test bars was modified to integrate a first attempt at a hot runner system. The runner system is heated up to the same temperature as the barrel. Compared to die casting the lower processing temperatures are beneficial when it comes to selection of the materials. Liquid metal corrosion of the die materials is reduced while strength is increased. Special attention has to be paid to the thermal separation of the heated and cooled parts of the runner. Sealing of the melt in the barrel and shot chamber is achieved by a cold plug, which is pushed out before every shot.

Figure 7 shows the first parts cast with the new system in comparison to the conventional system. The weight of the normal part is 198 g. Thixomolded tensile bars with the hot runner system have a weight of 139 g. As such there is a weight reduction of 30 % for this part. The cycle time was reduced too, because the thickest section in this part dominating the solidification time is the runner section.

Figure 8 shows the mechanical properties of test bars cast with the conventional system and with the hot runner system. They are both on the same level due to the fact that the conventional gating system already avoids cold flow and leads to excellent properties. Bigger advantages in the mechanical properties are expected for more complex parts with higher flow length.

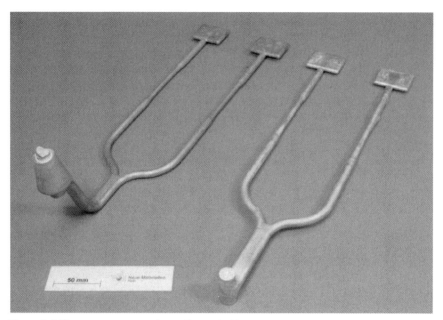

Figure 7: Different runner systems for tensile bars in thixomolding: conventional runner (left side) and hot runner (right side)

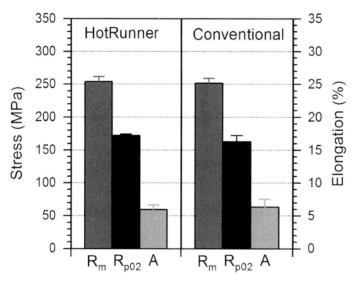

Figure 8: Mechanical properties of AZ91D cast with the conventional runner system and the hot runner system

4   Conclusions

Casting experiments in a thixomolder were carried out using a variety of commercial magnesium alloys (AZ-, AJ-, AS-series). Mechanical testing was conducted at room temperature and 150 °C. Casting runs could be realized without any problems with regard to segregation, evaporation and oxidation. Mechanical properties of AZ91D have a strong dependence of die temperature because of the cooling rate. Mechanical properties of thixomolded material are found to be equal to or even better than for high pressure die cast material.

A first hot runner system for a test bar die was developed. Mechanical properties are on the same level than with cold runner system. 30 % weight reduction was realized and cycle time was also reduced. As such, the hot runner system leads to high quality and low cost magnesium parts. In the near future a mold with a multi point hot runner system will be developed, based on the performance of the one point hot runner system.

5   References

[1]   Carnahan, R.D., in ASM Specialty Handbook Magnesium and Mg Alloys, **1999**, p. 90–97
[2]   Cacares, C.H.; Davidson, C.J.; Griffiths, J.R.; Newton, C.L., Material Science and Engineering A325 (**2002**), p. 344–355
[3]   Kompetenz in Magnesium Druckguss, Andreas Stihl AG & Co. KG, **2002**
[4]   Lohmüller, A.; Scharrer, M.; Hilbinger, M.; Jenning, R.; Hartmann, M.; Singer, R.F., Giessereiforschung 57 (**2005**) Nr. 1, p. 2–9
[5]   Aghion, E.; Bronfin, F.; von Buch, R.; Schumann, S.; Friedrich, H., JOM, November **2003**, p. 30–33
[6]   Baril, E.; Labelle, P.; Pekguleryuz, M.O., JOM, November **2003**, p. 34–39
[7]   Eigenschaften von Magnesium Druckgusslegierungen, Hydro Magnesium, **2005**
[8]   Datenblatt Magnesium Druckguss Legierungen, Hydro Magnesium, **2001**
[9]   Lohmüller, A.; Scharrer, M.; Jenning, R.; Hilbinger, M.; Hartmann, M.; Singer, R.F., Magnesium, Proceedings of the 6th international conference magnesium alloys and applications, **2003**, p. 738–743
[10] Barth, A., lecture at the 62nd annual world magnesium conference, Berlin, **2005**
[11] Landerl, C.; Fischersworring-Bunk, A.; Wolf, J.; Fent, A, VDI-Berichte Nr. 1830, **2005**, p. 69–91

Microstructure and Mechanical Properties of Rheo-diecast Mg-alloys

S. Ji. G. Liu, Y. Wang, Z. Fan
BCAST (Brunel Centre for Advanced Solidifaction Technology) Brunel University, Uxbridge, Middlesex, UK

1 Introduction

The majority of magnesium components are currently produced by high pressure diecasting (HPDC). Although HPDC can offer many advantages such as low cost, high efficiency and suitability for high volume production, it also encounters several critical problems such as high porosity levels, inadequate mechanical properties and sensitive to alloy composition [1]. Therefore, further growth in magnesium application will largely depend on the successful development of new processing technologies, which are capable of producing high quality components, whilst maintaining the advantages in HPDC [2].

Rheo-diecasting (RDC) is a newly developed technology based on the semisolid metal (SSM) processing principles for production of high integrity components. It adopts a twin-screw shearing mechanism to produce semisolid slurry, containing fine and spherical particles of the primary phase, and shapes the semisolid slurry into a high integrity component with conventional HPDC machine [3,4]. In this paper, we report the microstructure and mechanical properties of a variety of Mg-alloys, produced by the RDC process.

2 The RDC process

Figure 1 schematically illustrates the RDC equipment. The RDC process is a one-step SSM processing technique for manufacturing near-net shape components directly from liquid Mg-alloys. The equipment consists of two basic functional units, a twin-screw slurry maker and a standard cold chamber HPDC machine. The twin-screw slurry maker has a pair of screws, which are co-rotating, fully intermeshing and self-wiping. The basic function of the slurry maker is to convert the liquid metal into high quality semisolid slurry through solidification under intensive forced convection. In combination of the viscosity of semisolid slurry, the high rotation speed of screws, and the complexity of the screw geometry, the fluid flow inside the twin-screw slurry maker is characterised by high shear rate and high intensity of turbulence [5]. As a consequence of such fluid flow characteristics, the solidification of the liquid alloy takes place inside the slurry maker under following conditions: (1) uniform

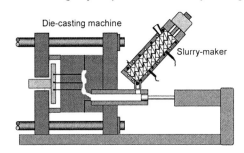

Figure 1: A schematic illustration of the rheo-diecasting equipment.

temperature, (2) uniform composition, (3) well dispersed nucleation agents, (4) rapid heat extraction.

The twin-screw slurry maker works in a batch manner, typically providing slurry every 30secs. The semisolid slurry is immediately transferred into a HPDC machine for final component shaping. No modification is required to the standard HPDC machine. To avoid oxidation of magnesium melt, a protective gas mixture of N_2 containing 0.4 vol.% SF_6 has been used in the melting furnace and the twin-screw slurry maker.

3 Solidification behaviour and resultant microstructure

In the RDC process, solidification takes place in two distinct stages [6], as illustrated schematically in Fig. 2. The primary solidification occurs inside the twin-screw slurry maker under intensive shearing to produce semisolid slurry, while the secondary solidification of the remaining liquid in semisolid slurry includes the solidification occurring during the slurry transfer to the shot sleeve, mould filling and solidification in the die cavity.

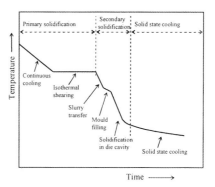

Figure 2: Schematic illustration of the solidification process in the RDC process.

The primary solidification in the RDC process can also be divided into two substages, continuous cooling and isothermal shearing. During continuous cooling, which lasts about 15 s, heterogeneous nucleation occurs continuously and uniformly throughout the entire volume of the melt, with a small undercooling. Due to the extremely uniform temperature and composition in the melt, every single nucleus survives, achieving 100% nuclei survival rate. The growth rate at this stage is extremely fast, with the maximum particle size being capped by the melt temperature. As a result of intensive forced convection, the growth of the nuclei takes place in a spherical manner. The fast and spherical growth leads to the formation of fine and spherical particles with a narrow size distribution. The isothermal shearing stage is basically a coarsening process. This is achieved through dissolution of the smaller particles. At this stage, coarsening due to mass transport from high curvature region to the low curvature region in the same particle is not significant, because of the large shape factor achieved by spherical growth at the continuous cooling stage. Figure 3 shows the typical microstructure of Mg-alloys produced by the RDC process. The relatively large and spherical particles are primary α-Mg particles ($α_1$) produced during primary solidification.

The secondary solidification starts when the semisolid slurry is delivered to the shot sleeve. Due to the low heat capacity and relatively low temperature of the shot sleeve, heterogeneous nucleation occurs in the intensively sheared liquid, and nearly all the nuclei survive. However, the nuclei grow dendritically in the absence of shearing. The resultant dendrites are fragmented when they pass through the narrow gate during mould filling, thus the dendrite fragments are observed in the final microstructure ($α_2$). The remaining liquid in the semisolid slurry eventually solidifies in the die cavity, where the liquid still has largely uniform temperature and composition fields due to the intensive shearing in the slurry maker and through the gate. Under the large cooling rate (about 10^3 K/s) provided by the metallic die block, nucleation is expected to take place throughout the entire remaining liquid, with a high nucleation rate. Numerous nuclei compete to grow, and the solidification finishes before the instability occurs, producing fine α_3 particles. The α_3 is smaller in size than both α_1-Mg and α_2-Mg. In all processed Mg-alloys the size of the α_3-Mg phase varies in a narrow range from 6.5 to 8.5 μm, irrespective of the volume fraction of the α_1-Mg phase and the type of alloys.

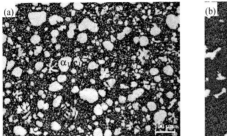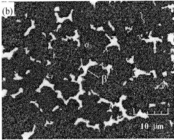

Figure 3: Typical microstructures of primary Mg-phase formed in solidification of Mg-alloys.

The variation of solid fraction formed during primary solidification can be achieved by changing the barrel temperature between liquidus and solidus of the alloy. For AZ91D alloy, when the processing temperature decreases from 597 to 585°C the solid fraction increases from 0.091 to 0.314. The processing windows for the different alloys vary according to their liquidus and solidus temperature. However, the particle size and morphology of primary phase remains be fairly constant. Figure 4 shows the typical microstructures of AM50 and AZ91D alloys processed at different temperatures. The primary particle (α_1-Mg) sizes are around 40 μm and the shape factor is about 0.8, both factors are independent of processing temperatures and alloy compositions [7,8].

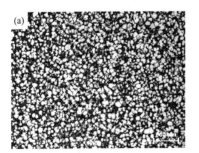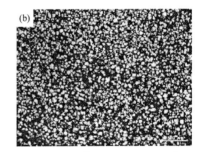

Figure 4: Typical microstructures of rheo-diecast Mg-alloys: (a) AM50 at 613°C; (b) AZ91D at 585°C.

Figure 5: A montage of micrographs showing the microstructure of AM50 alloy on the cross section of a 6mm bar produced by the RDC process.

One of the most important features of SSM processing is the ability to achieve uniform microstructure across the casting section. Figure 5 presents a montage of micrographs showing the as-cast microstructure through the entire cross-section of a 6 mm tensile test sample. It is clear that the primary particles are distributed uniformly throughout the entire section, and no apparent segregation is observed under the optimised processing conditions. The sample surface is sharp and tidy, indicating a very good surface finish produced by the RDC process [7]. Large pores due to gas entrapment are eliminated and the often-observed hot cracks in HPDC samples of Mg-alloys are significantly reduced in the RDC samples. Only isolated hot cracks with a few microns in size can be observed occasionally. A direct comparison of porosity level in AZ61 castings produced by HPDC process and RDC process is shown in Fig. 6. The results clearly show a significant reduction of porosity level in the RDC castings [9].

The RDC process is flexible to Mg-alloys with varying compositions, which can be either cast or wrought alloys.

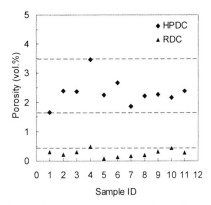

Figure 6: A comparison of porosity levels in tensile samples of AZ61 alloy produced by HPDC process and RDC process.

4 The mechanical properties of the RDC Mg-alloys

The mechanical properties of both AZ and AM series Mg-alloys produced by the RDC process are summarised in Table 1, in comparison of the results obtained from HPDC for cast AZ91D and AM50 alloys. Alloys processed by RDC show an improved ultimate tensile strength (UTS) to those obtained from the HPDC process. More importantly, the RDC process offers a substantial increase in tensile elongation for all experimental alloys.

For both AM and AZ series alloys, an increase in Al content increases yield strength and decreases ductility, while the UTS is fairly constant. The increase in strength and the decrease in ductility with the increase in Al content are fairly consistent for all experimental alloys. In AZ series alloys, the elongation of AZ91D samples can reach 7.4%. Therefore, RDC process can be used to process AZ series Mg-alloys with lower Al content to improve ductility. In AM series alloys, the elongation of AM90 is still at 10.2%. In comparison with AM50, AM90 exhibits an improved processability. Therefore, RDC process can be used to process AM series alloy with higher Al content for improved processability.

Table 1. Mechanical properties of varied Mg-alloys

Alloy	Process	UTS (MPa)	YS (MPa)	Elongation (%)
AZ61	RDC	253	126	12.3
AZ81	RDC	236	143	9.2
AZ91	RDC	248	145	7.4
	HPDC	212	146	3.3
AM50	RDC	249	122	20.5
	HPDC	195	129	6.7
AM70	RDC	251	133	13.3
AM90	RDC	253	150	10.2

5 Heat treatment of the RDC Mg-alloys

Due to the low level of porosity in the RDC samples, heat treatment can be used to further enhance the mechanical properties, as shown in Table 2. Compared with the as-cast condition (RDC), T4 heat treatment improves ductility substantially but decreases strength. Both T6 and T5 slightly improve strength but with some sacrifice of ductility. It appears that the conventional heat treatment of AZ91D alloy only results in limited benefit to mechanical properties [10].

Tx is a new heat treatment procedure developed recently [11], which aims to break up the $Mg_{17}Al_{12}$ β-phase network. Figure 7 shows the microstructure of a RDC AZ91D sample heat treated at 390 °C for 1 hr. The $Mg_{17}Al_{12}$ β-phase network has been broken up into essentially isolated particles. Therefore, Tx offers the samples with strength better than T6 and ductility close to T4.

Table 2. Mechanical properties of the RDC AZ91D under different heat treatment conditions

Properties	YS (MPa)	UTS (MPa)	Elongation (%)
RDC	145	248	7.4
RDC+ T4	91	230	10.7
RDC+ T5	133	236	6.2
RDC+ T6	134	255	6.4
RDC+ Tx	125	259	9

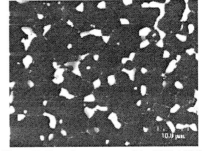

Figure 7: The microstructure of the RDC AZ91D alloy after heat treatment at 390°C for 1 h.

Figure 8: AZ91D component produced by RDC process and the microstructure at different locations of component, all the micrographs have the same magnification.

6 Component trials

A production trial has been carried out using an industrial component die [12]. The section thickness of the component varies between 2 and 6 mm, the runner thickness is 10mm and the biscuit is 60mm in diameter. Figure 8 presents a photograph of the casting, and a series of micrographs showing the microstructures at various positions in the casting. The trial results indicate that the components produced by RDC have a very good surface finish, close to zero porosity and a very fine and uniform microstructure throughout the entire casting, independent of the section thickness. Under support of the DTI (Dept. of Trade & Industry), further industrial production trials are in progress. The RDC process is being used to produce seat frame component from magnesium alloy.

7 Summary

(a) Rheo-diecasting has been developed for the production of Mg-alloys components with high integrity. Rheo-diecasting can be easily achieved by adding a twin-screw slurry maker to the existing cold chamber diecasting machine.

(b) Solidification in the rheo-diecasting process takes place in two distinctive stages; one is primary solidification in the twin-screw slurry maker under intensive forced convection, and the other one is secondary solidification in the die cavity with high cooling rate.

(c) The fluid flow inside the twin-screw slurry maker is characterised by high shear rate and high intensity of turbulence, which in turn promote effective nucleation and spherical growth during primary solidification, resulting in a fine and uniform microstructure.

(d) Secondary solidification takes place uniformly throughout the entire die cavity, producing an extremely fine and uniform microstructure.

(e) Under optimised conditions the rheo-diecast samples have very low levels of porosity, and a fine and uniform microstructure throughout the entire section of castings.

(f) The rheo-diecasting process produces castings with improved strength and ductility in the as-cast condition, compared with high pressure diecasting techniques. Heat treatment can be used to further enhance the mechanical properties of rheo-diecast samples.

References

[1] Fredrich H.; Schumann S., *Proceedings of IMA Magnesium Conference*, International Magnesium Association, Brussels, Belgium, **2001**, p. 8.
[2] Fan Z., Inter. Mater. Rev. **2002**, 47, 49–85.
[3] Ji S.; Fan Z., Metall. Mater. Trans. **2002**, 33A, 3511–3520.
[4] Fan Z.; Ji S.; Liu G., Materials Science Forum, **2005**, 488-489, 405-416.
[5] Ji S.; Fan Z.; Bevis M. J., Mater Sci Eng A, **2001**, A299, 210-217.
[6] Fan Z.; Liu G., Acta Mater. **2005**, 53, 4345–4357.
[7] Ji S.; Zhen Z.; Fan Z., Mater. Sci. Tech. **2005**, 21, 119-1024.
[8] Ji S.; Ma Qian; Fan Z., Metall. Mater. Trans. **2006**, 37A, 779-787.
[9] Zhen Z.; Ma Qian; Ji S.; Fan Z, Scripta Mater. **2006**, 54. 207-211.
[10] Wang Y.; Liu G.; Fan Z., Acta Mater. **2006**, 54, 689-699.
[11] Wang Y.; Liu G.; Fan Z., Scripta Mater. **2006**, 54, 903-908.
[12] Fan Z., Sci Eng A, **2005**, A413-414, 72-78.

Development of an AM50 Based Magnesium Recycling Alloy

D. Fechner, C. Blawert, P. Maier, N. Hort, K. U. Kainer
Center for Magnesium Technology, Institute for Materials Research, GKSS-Research Center Geesthacht, Geesthacht, Germany

1 Introduction

Magnesium scrap can be used for metallurgy such as steel desulphurisation and aluminum alloying as well as for new magnesium products. So far metallurgy has been the main use for secondary magnesium and recycling with the aim of alloy production has not been of great economical importance. One reason for that is the low quantity of primary magnesium produced worldwide for the last 50 years compared to steel and aluminum. Sufficient amounts of clean and sorted post consumer scrap were not available and recycling was mainly done by die-casters who were able to remelt their sprues and runners [1, 2]. Things will change as the EU directive on end-of life vehicles requires a recycling rate of 80 % per vehicle and year with the beginning of the year 2006 which will increase to 85 % in 2015 [3]. So far only clean scrap of a single alloy – also known as class-1 scrap - has been used for the production of secondary magnesium [4–11]. At present a market for secondary alloys like for steel and aluminum does not exist for magnesium. The main reason is the low corrosion resistance of magnesium in presence of small amounts of Cu, Ni, and Fe. Primary magnesium alloys therefore have to meet very low thresholds for these three elements. However, it is quite likely that a great amount of the predictable automotive post consumer scrap can not be considered as class 1. In Al containing alloys Fe can be removed with Manganese. High levels of other impurities can be adjusted by dilution. Scharf et. al. [12] showed that additions of Cu, Fe, Ni, and Si similar to the thresholds of AZ91B do not have a significant effect on the corrosion properties compared to AZ91D. There are possible applications which do not require high corrosion resistance. Vehicle interior parts are not exposed to high humidity. Housings for tools, computers and cameras are generally covered by a protective layer. A further argument for recycling is the price of several alloying elements that would be wasted otherwise. The importance to develop secondary alloys has already been claimed some time ago, but only minor research activities have been carried out on this field [13, 12].

2 Experimental

AM50 was used as the base of four modified alloys. Following additions were added to the base alloy: Sr, Ca, Si and further Ca + Si. The unmodified AM50 alloy was cast as a reference material. The base material AM50 was heated up to 730 °C in a mild steel crucible. For all casting operations a mixture of argon and 0,2 % SF_6 was used to prevent melt oxidation. Alloying elements were added in their pure form and the melt was subsequently cast into a permanent mould that had been preheated to 200 °C. The mould had the geometry of a rectangle plate, with 350 mm height, 450 mm width, and 30 mm thickness. The chemical composition of the alloys was investigated via OES using a SPECTROLAB Spark Analyzer, see Table 1.

Table 1: Chemical composition of the alloys in wt. %, Be in ppm, Mg – balance

Alloy	Al	Zn	Mn	Sr	Ca	Si	Fe	Cu	Ni	Be
AM50+Sr	4.9	0.01	0.31	0.32	–	0.01	0.002	0.001	0.001	12
AM50+Si	4.95	0.01	0.298	–	–	0.0971	0.0019	0.0009	0.0006	11
AM50+Ca	5.045	0.0099	0.2905	–	0.127	0.0145	0.0021	0.001	0.0005	11
AM50+Si+Ca	4.99	0.0102	0.311	–	0.127	0.177	0.0022	0.0009	0.0006	12
AM50	4.9	0.01	0.31	–	–	0.01	0.002	0.001	0.001	12

The as-cast material was solution heat treated for 24 hours at 400 °C and aged for 6 hours at 200 °C. As there is no standard heat treatment procedure for the AM50 it was derived from the AM100's according to [14]. The microstructure was investigated using optical microscopy, EDX and XRD. The tensile tests were performed on a ZWICK Z050 tensile testing machine according to [15, 16]. A set of three samples was used for each average value. In the as-cast and heat treated condition tests have been performed at 20 °C, and 150 °C. For the high temperature tests, a convection furnace was adjusted to the tensile testing machine. After mounting, the samples were hold for 7 to 10 minutes to ensure an overall homogeneous temperature. The metallographic preparation for the material in the as-cast and heat treated condition was performed following the standard procedure described in [17].

To evaluate the corrosion properties, salt spray tests were performed. All corrosion specimens were grinded with silicon carbide paper (1200 grit) and cleaned in alcohol prior to corrosion testing. Salt spray testing for 24 hours with 5 % NaCl solution (pH 6.5) according to DIN 50021 was used as the standard test.

3 Results

The grain size of the as-cast material was around 150–300 μm. There was no significant influence on the grain size by the heat treatment. Due to small volume fractions the identification of second phase precipitates was not always possible. In each as-cast alloy homogeneously distributed white precipitates were observed by optical microscopy. During solution heat treatment most of this phase dissolved in the matrix. Therefore, it is considered to be $Mg_{17}Al_{12}$ or β-phase. Figure 1 and 2 indicate the dissolution of $Mg_{17}Al_{12}$ by disappearing of the white precipitates, present in Figure 1 only. This observation was confirmed via XRD and EDX. In the silicon modified alloy Mg_2Si particles were found in the as-cast and heat treated condition. In strontium and calcium modified alloys EDX analysis pointed amongst others at the presence of Al_4Sr and Al_2Ca. Due to their low volume fraction these phases were not found in XRD. Second phase precipitates were distributed nearly homogeneously in contrast to die-cast material where the second phases are usually found on grain boundaries.

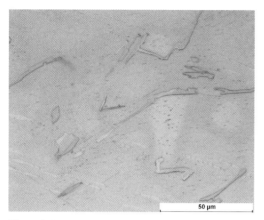
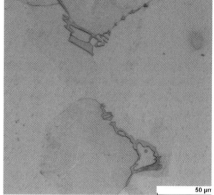

Figure 1: Sr modified AM50 alloy in the as-cast condition

Figure 2: Sr modified AM50 alloy after solution heat treatment

Figure 3 shows the tensile properties in the as-cast and heat treated condition at room temperature. Figure 4 shows the tensile properties of the same alloys at 150°C. Thereby YS is yield strength, UTS is ultimate tensile strength, A is the fracture elongation, ac the as-cast condition, T4 stands for solution heat treated and RT for room temperature.

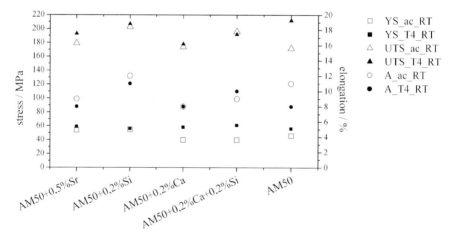

Figure 3: Tensile properties at room temperature

At room temperature in the as-cast condition alloying always improved the UTS but not necessarily the YS. Alloying with 0.5 % Sr and 0.2 % Si clearly improved the mechanical properties compared to the base alloy. calcium additions of 0.2 % Ca alone and together with 0.2 % Si seemed to have a detrimental effect as the yield strength decreased compared to pure AM50. Solution heat treatment raised the yield strength and ultimate tensile strength of nearly every alloy. The silicon modified alloy only showed a slight improvement where YS of the two calcium containing alloys increased for about 20 MPa. Judged by the YS it seems as if differences between the alloys have been balanced through annealing

The same development can be observed at elevated temperatures. Solution heat treatment improved the strength and thus the heat resistance of nearly all alloys. No difference is visible for the silicon modified alloy whereas YS of the calcium modified alloys is approximately 15 MPa higher in the T4 condition. Again T4 treatment seems to have balanced differences in YS between the alloys.

Aging following the solution heat treatment did not improve the mechanical properties significantly.

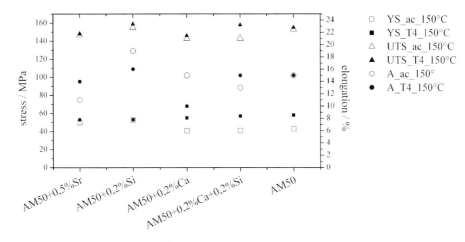

Figure 4: Tensile properties at 150 °C

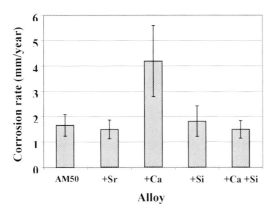

Figure 5: Corrosion rates after 24 hours of exposure to salt fog (DIN 50021)

4   Discussion

The corrosion results show large variations within the same alloy indicating a not uniform distribution of alloying elements, precipitates. The corrosion rate can vary by a factor of two for the same alloy, making predictions of the alloying influence on the corrosion behaviour of the

AM50 alloy difficult. The salt spray test results are depicted in Figure 5. The results indicate that additions of Si and Sr are not critical for the corrosion resistance. No major differences compared to AM50 were observed and the average corrosion rates were about 1.5 mm/year. Ca additions led to an increase of the corrosion rate to 4 mm/year. It is interesting to note that the combined addition of Si and Ca kept the corrosion rate at the low level of 1.5 mm/year. Similar results were found for Cu and Si in AZ91 [12].

Sr, Si and Ca are common alloying elements for improving high temperature properties in magnesium alloys. It has been shown that the phases Al_4Sr [18], Mg_2Si [19] and Al_2Ca [20], are responsible for good heat resistance. Main reasons for that are their high melting points and, in case of Sr and Ca, the suppression of the $Mg_{17}Al_{12}$, or β-phase [21, 20]. Since the strength of the latter decreases at high temperatures it causes usually only good room temperature properties. For heat resistant alloys the suppression of the $Mg_{17}Al_{12}$ phase is related to the ratio of Al to Ca and Sr, respectively. In the literature $Mg_{17}Al_{12}$ has been reported for alloys AJ62x and AJ62Lx, but not for alloys AJ52x and AJ51x [22]. Because of the relative low concentration of Sr and Ca in alloys prepared for this work, the β-phase was likely to occur in all the alloys at least in the as-cast condition. Therefore the slight improvement of the alloy's heat resistance in the as cast condition might be due to the relative low content of strengthening alloy additions. Usually, around 2 wt. % of Sr, 2 wt. % of Ca and 1 wt. % of Si are used in common heat resistant alloys.

The improvement of mechanical properties after solution heat treatment is most probable due to dissolving of the â-phase. This is confirmed by EDX measurements in the matrix of two alloys in the as-cast and heat T4 condition. The unmodified AM50 and the alloy with strontium additions show an elevated Al content in the matrix after heat treatment.

5 Summary

From the corrosion point of view alloying elements such as Si and Sr are not critical or even slightly beneficial for the corrosion performance of AM50. In contrast Ca has a negative effect, but this effect might be reduced if further alloying elements, e.g. Si, are added together with Ca.

The tensile tests have shown that even small amounts of Sr, Ca and Si in particular can result in a strengthening effect of the tensile properties. However, for improved performance at elevated temperatures, higher amounts of Sr, Ca, and Si are required.

The experiment's results also promote another way of magnesium recycling. For applications which do not require high purity quality, mixed alloying elements can provide sufficient mechanical properties. Alternatively primary magnesium could be used to reduce the amount of impurities contained in post consumer scrap and to save expensive alloying elements at the same time.

6 References

[1] Ditze, A., ALUMINIUM, 75, **1999**, p. 157–160
[2] Hanko, G.; Lochbichler C.; Riederer, W.; Macher, G., Magnesium, proceedings of the 6th International Conference Magnesium Alloys and Their Applications (Ed.: K.U. Kainer), **2004**, p. 980–987
[3] European parliament: Official Journal of the European Communities, October, **2000**, p. 34–42

[4] Hong Tao Gao; Guo Hua Wu; Wen Jiang Ding; Yan Ping Zhu, Journal of Materials Science, Vol. 39 , **2004**, p. 6449–6456
[5] Hanko, G.; Macher, G., Magnesium Technology 2003 (Ed.: H.I. Kaplan), TMS, **2003**, p. 29–32
[6] Koichi Kimura; Kouta Nishii; Motonobu Kwarada, Materials Transactions, Vol. 43, No. 10, **2002**, p. 2516–2522
[7] Chino, Y.; Yamamoto, A.; Iwasaki, H.; Mabuchi, M.; Tsubakino, H., Materials Science Forum, Vol. 419–422., **2003**, p. 671–676
[8] Makoto Inoue; Masao Iwai; Shigeharu Kamado; Yo Kojima; Tadao Itoh; Mitsuo Sugama, Transactions of the Materials Research Society Japan, 24 [3], **1999**, p. 349–352
[9] King, J.F.; Hopkins A.; Thistlethwaite, S., Proceedings of the 3rd International Magnesium Conference (Ed.: G. W. Lorimer), **1996**, p. 51–61
[10] Shalev, G.; Moscovitch, N.; Bronfin, B.; Rubinovich, Z.; Aghion, E., 12th Magnesium Automotive and End User Seminar, 13-14th September **2004**
[11] Argo, D.; Forakis, P.; Lefebvre, M., Magnesium Technology 2003 (Ed.: H.I. Kaplan), TMS, **2003**, p. 33–37
[12] Scharf, C.; Blawert, C.; Ditze, A., poceedings of the 6th International Conference Magnesium Alloys and Their Applications (Ed K.U. Kainer), **2004**, p. 988–994
[13] Ditze, A.; Schwertfeger, K.; Scharf C.; Mutale, C.-T., Materialwissenschaft u. Werkstofftechnik, Vol. 32, **2001**, p. 31–35
[14] ASTM, Designation B 661–93
[15] DIN50125, **1991**
[16] EN 10002, **1991**
[17] Kree, V.; Bohlen, J.; Letzig, D.; Kainer, K.U., Practical Metallography, 41 (5), **2004**, p. 233–246
[18] Fischersworring-Bunk, A.; Landerl, C.; Fent, A.; Wolf, J., Proceedings of 62nd IMA, **2005**, p. 51–60
[19] Dargusch, M.S.; Bowles, A.L.; Pettersen, K.; Bakke, P.; Dunlop, G.L., Metallurgical and Materials Transactions A., Vol. 35A, June **2004**, p. 1905–1909
[20] Keun Yong Sohn; Wayne Jones J.; Allison, John E., Magnesium Technology 2000 (Ed.: H.I. Kaplan; Hryn, J. and Clow, B.), TMS, 2001, p. 271–278
[21] Pekguleryuz, M.O.; Baril, E., Magnesium Technology 2001 (Ed.: J. Hryn), TMS, **2001**, p. 119–125
[22] Baril, E.; Labelle, P.; Pekguleryuz, M.O., Journal of Metals, November **2003**, p. 34–39

The Properties of Magnesium Alloy Strip Produced by Melt Drag Process

Shinichi Nishida, Mitsugu Motomura
Waseda University, Shinjuku-ku Okubo, Japan

1 Introduction

Magnesium alloy is the lightest structural material and the applications are increasing in Japan. Mainstream of magnesium alloy applications are castings such as die cast because of their poor formability at room temperature, so the wrought magnesium alloys should be used more to expansion the market. But the magnesium alloy material for plastic forming such as sheet is too expensive because it is hard to roll. So that the new techniques to produce the sheet of magnesium alloy for reducing the cost of that is strongly required. [1]

The strip casting techniques is well suited to produce the sheet of poor formability material like a magnesium alloy. These techniques are able to the required thickness sheet directly from molten metal and the near net shape will be achieved. Since H. Bessemer invented the twin roll strip casting process at 1865, a lot of studies of strip casting process about stainless steel and aluminum alloy have been studied. [2]

And the melt drag process is one of the single roll rapidly solidified strip casting process invented by Battele Development Corp. in USA at 1986. In generally the twin roll strip casting process are better known as the magnesium alloy strip casting process, however we focused on the melt drag process from the perspective that are simple process, easy to control the solidification, easy to increase the operation speed and etc. comparing to twin roll casting process. [3,4]

We have succeeded in producing the magnesium alloy strip directly from the molten AZ31 magnesium alloy by the melt drag process. The strip has been produced continuously at roll speed from 1 m/min to 90 m/min without oxidization. And we have improved the strip thickness distribution by semi-solid forming using forming roll in the melt drag process. [5]

In this paper, we report about the properties of the strip produced by this process. It is very important to research the properties of the as-cast sheet of magnesium alloy in detail, because magnesium alloy is affected by the crystal orientation. The investigated properties are such as the relationship between the strip thickness and the experimental conditions, the strip surface conditions, the crystal grain size distribution at each position (near roll side surface, center and near free solidified surface), the relationship between the average strip thickness and the average crystal grain size, the Vickers hardness, the mechanical properties of as-cast strip, the element mapping by EPMA, XRD pattern and the strip thickness distribution using the forming roll.

This research was investigated at Waseda university kagami memorial laboratory for materials science and technology in Tokyo Japan.

2 Melt Drag Process

2.1 Experimental device

The photograph of experimental device is shown in Figure 1. This device is typically entirely covered with aluminum plates. Since the molten magnesium is very active, the device can be sealed up around the molten magnesium to fill the protective gases. The protective gases used inside the device are CO_2 and SF_6 to prevent the burning of molten magnesium. The oxygen density is kept under 7 %.

  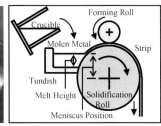

(a) Experimental device   (b) Around the rolls   (c) Schematic illustration
Figure 1: The experimental device

The schematic illustration of melt drag process with forming roll is shown in Figure 1(c). The material used for the solidification roll is Cu and the diameter is 300 mm. The solidification roll is water-cooled and the any lubricant or release agent isn't used. The material used for the forming roll is SUS304 in terms of stiffness, and the diameter is 130 mm. The forming roll isn't water-cooled because the object of the forming roll isn't to cool the strip.

The forming force is adjusted by the springs for supporting the forming roll and the initial gaps between the rolls. The angle between the solidification roll and the forming roll is 90°. The each roll speed is same. The material used for the crucible is low carbon steel that is hot dip aluminized and removed the remained aluminum. Meld drag process needs the tundish to pool the molten metal. In this study, the material used for the tundish is insulator made of silica calcium, and is coated by boron nitride. The melt height is adjusted a objective value by hand. The exit width of the tundish is almost same the width of the produced strip.

2.2 Experimental Procedure and Conditions

The magnesium alloy in the crucible is melted in the electric furnace. The surrounding atmosphere in the furnace is CO_2. After melting the crucible is transferred in the experimental device and the melt drag process is run. The pouring temperature, it is defined as the molten metal temperature just before pouring in the experimental device, is measured by the K-type thermocouple. The strip produced by the melt drag process has two surfaces. We call the surface; roll side surface and the free solidified surface, respectively.

The experimental conditions are shown in Table 1. Each parameter was shown in Figure 1(c).

Table 1: Experimental conditions

Experimental material	AZ31
Roll speed [m/min]	1~90
Pouring temperature [°C]	660~780
Meniscus position [mm]	145
Melt height [mm]	25
Protective gas	CO_2 and SF_6
CO_2 gas flow rate [l/min]	25
SF_6 gas flow rate [l/min]	0.1
The tundish exit width [mm]	60
Forming force [kgf]	35~350

Strip thickness was measured at 11 points for the cross of the strip with 5 mm intervals. For all experimental condition, 3 samples were chosen and all the data were averaged as the average strip thickness. The microstructure of the strip was observed by optical microscope. The compositions of the etchant were picric acid 4.2 g, glacial acetic acid 10 ml, water 10 ml and ethanol 70 ml. The etching time was 15 sec. The crystal grain size was measured by image analysis software; Image-Pro Plus. The tensile test was according to JIS Z2241. The model numbers of the SEM used for the investigation were JSM-6500F and JED-2300F, which were made by JEOL. Accelerating voltage was 15 kV, and prove current was 7.94E-10 A. The XRD pattern was measured with RINT TTR made by Rigaku.

3 Results and Discussions

3.1 Strips and Strip Thickness

Figure 2 shows the produced strips by the melt drag process. It is possible to produce the continuous strips at the large range roll speed by the melt drag process. The strips are not oxidized, and the metallic luster is observed.

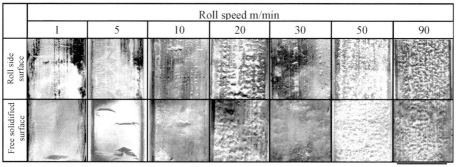

Figure 2: The strips produced by the melt drag process (pouring temperature: 780 °C)

Some defects are observed at roll side surface such as voids, ripple mark and peeling. These defects will make the free solidified surface the bumpy surface. In particular, the voids are very often observed at low roll speed. Trapping the air between the roll and the strip may cause the voids. At low roll speed the strip thickness is very thin and the amount of the trapped air is large, so the trapped air can't go out before the solidification of the strip because of the low pressure of itself. These defects may be improved by the detailed research on the contacting conditions between the roll and strip and the controlling the solidification.

Figure 3 shows the relationship between the strip thickness and the experimental conditions. The range of strip thickness in this study is 1.8 mm ~ 8.8 mm. In general the roll speed is faster and the melt temperature is higher, the strip thickness is thinner. The strip thickness is determined by the contacting time between the roll and molten metal in the tundish, so that the roll speed mainly affect the strip thickness.

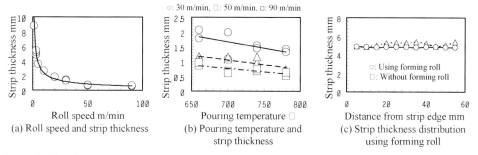

Figure 3: The relationship between the strip thickness and the experimental conditions.

Figure 3(c) shows the strip thickness distribution using the forming roll. The purpose of the forming roll is the semi solid forming of the strip surface in the melt drag process. The forming roll can make the strip surface smooth. At the present study it is revealed that the very low forming force can roll the semi solid strip surface without the defect such as cracks. The free solidified side surface of the strip produced by the melt drag process is coarse, so the research of the effective process to produce the flat strip is strongly desired.

3.2 Microstructure of the Strip

Figure 4 shows the microstructure and the crystal grain size distribution of the produced strip by melt drag process. The microstructure is equiaxed structure. Average crystal grain size is 53 μm. The strip solidify from the roll side surface to the free solidified surface, so that the crystal grain size near the free solidified surface may be coarser than that neat the roll side surface. In addition the micro porosities are not observed.

Figure 5 shows the relationship between the average crystal grain size and the strip thickness. The crystal grain size is affected by the cooling rate of the molten metal. The roll side surface of the thick strip produced at lower roll speed is smooth. So the contacting condition between the roll and the strip may be good and the cooling rate of the strip may increase. The thin strip produced at higher roll speed has a small heat capacity, so the cooling rate of the strip may also increases.

Figure 6 shows the results of X-ray diffraction analysis and the EPMA element mapping of aluminum. According X-ray diffraction analysis, the β-phase is not detected. And the crystal orientation is randomized and the prismatic planes strongly lie on the surface in the strip. This strip is suggested that the formability is better than the rolled sheet. [6]

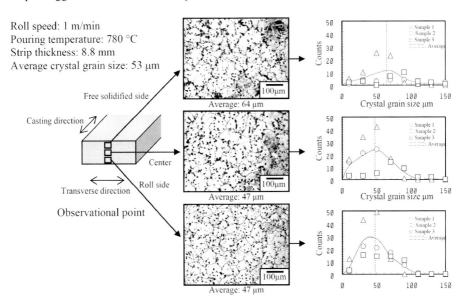

Figure 4: The microstructure and the crystal grain size distribution

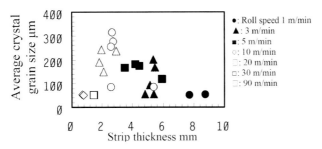

Figure 5: The relationship between the average crystal grain size and the strip thickness

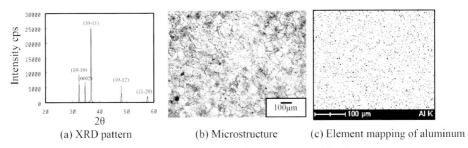

(a) XRD pattern   (b) Microstructure   (c) Element mapping of aluminum

Figure 6: The X-ray analysis. (roll speed: 10 m/min, Pouring temeprature: 660 °C)

3.3 Mechanical Properties and Vickers Hardness

Figure 7 shows the result of tensile test and Vickers hardness test. The tensile strength and the elongation of the as-cast AZ31 magnesium alloy strip by melt drag process are very low comparing to the rolled sheet because the crystal orientation of the as-cast strip is randomized and the basal plane is not in a direction parallel to the surface of the strip. The value of Vickers hardness shows almost constant. The melt drag process is the rapid cooling solidification process and the segregation is not observed (Figure 6(c)). So the Vickers hardness may be constant.

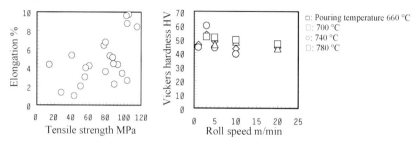

Figure 7: Mechanical properties and Vickers hardness

4 Conclusions

The properties of AZ31 magnesium alloy strip produced by the melt drag process were investigated. The results obtained are as follows:
- Producing the continuous and no oxidized magnesium alloy strip by the melt drag process is possible.
- Range of strip thickness in this study is 1.8 mm ~ 8.8 mm.
- Forming roll makes the strip surface smooth by semi solid forming.
- Crystal grain size is affected by the contacting condition between the roll and strip.
- Crystal orientation of the strip is randomized.
- Mechanical properties of the as-cast strip were revealed.

We express our grateful appreciation for the assistance of SANKYO TATEYAMA ALUMINUM Co. Ltd. in taking of magnesium material.

5 References

[1] Handbook of Advanced Magnesium Technology (Ed.: Y. Kojima), The Japan Magnesium Association, Tokyo, Japan, **2000**, Chapter 8.
[2] H. Bessemer, U. S. Patent 49053, **1865**.
[3] Maringer R. E., Materials Science and Engineering **1988**, 98, p. 13–20.
[4] N. Zapuskalov, ISIJ International **2003**, 43, 1115–1127
[5] S. Nishida, M. Motomura, the Journal of Japan Institute of Light Metals **2005**, 55, p. 315–320.
[6] K. Iwanaga, H. Tashiro, H. Okamoto, K. Shimizu, Journal of Materials Processing Technology **2004**, 155-156, p. 1313–1316.

Experimental and Numerical Investigation of Texture Development During Hot Rolling of Magnesium Alloys

Rudolf Kawalla[1], Christian Schmidt[1], Hermann Riedel[2], Aruna Prakash[2]
[1] Freiberg University of Mining and Technology, Freiberg; [2] Fraunhofer Institute for Mechanics of Materials (IWM), Freiburg

1 Introduction

Due to the deformation mechanisms of magnesium and the typical basal texture rolled magnesium sheets show a significant asymmetry of flow stress in tension and compression which is disadvantageous for subsequent processing and leads to anisotropic properties. In order to avoid this undesired behavior it is necessary to achieve non-basal texture during rolling, or at least, to reduce the intensity of the basal texture component. The reduction of the anisotropy caused by the basal texture is very important for subsequent forming processes. The knowledge of microstructure and texture development during forming opens up possibilities to influence those processes by thermo mechanical treatment and by utilizing recrystallization phenomena and promises significant progress in terms of production and processibility.

This work aims at optimizing the hot rolling process with special consideration of texture effects by means of numerical simulation. The development of the model is carried out in close cooperation with the experimental work. Therefore, the experiments described in the following shall be seen as a contribution to the clarification of the forming mechanisms and the accompanying specifics and effects of texture development. At the same time those results are used for the development and verification of a simulation model, which shall enable the user in the future to predict and influence the texture development during forming of magnesium alloys.

2 Experimental Procedure

The experiments comprised compression tests with conventionally produced feedstock as well as rolling tests with cast-rolled material (TRC–twin roll casting).

The samples (Ø = 10mm, length = 18mm) for the compression tests were made from a multi-stage caliber rolled bar of continuous casting AZ31. Subsequent to a 10 min heat treatment at 520 °C the compression test was carried out on a servohydraulic press in rolling direction at seven different temperatures between 280 °C und 500 °C with a logarithmic deformation degree of $\varphi = 1$ and a strain rate of 10 s^{-1}. In a second test series samples of the same kind were compressed with small deformation degree of $\varphi = 0.025$ to $\varphi = 0.15$ in order to investigate the distinctive behavior of the flow curve in its slope.

The strips (thickness 6mm) used as feedstock for the rolling tests were manufactured with the cast-rolling technology (TRC) by the MgF Magnesium Flachprodukte GmbH Freiberg [1]. Their chemical composition is in accordance with AZ31 with slightly reduced Al content. The results for the development of microstructure and texture dwelled on in 3.3 were obtained in rolling trials which were carried out on the finishing line of the semi-continuous hot rolling mill at the Institute of Metal Forming at the Freiberg University of Mining and Technology. The rolling trails consisted of tests in which two different initial rolling speeds (IRS: 1 m/s, 5 m/s) at three different initial rolling temperatures (IRT: 300 °C, 400 °C, and 500 °C – all strips preheated to 500 °C) were applied on a rolling process with three passes and an overall deformation degree of $\varphi = 1.3$. Rolling was performed in cast-rolling direction.

All samples were quenched in water right after deformation to conserve the microstructure after deformation which is not only of interest for the model development. The texture and microstructure investigations were carried out in two different metallographic sections, at 1/2 and at 4/5 of the specimen height, with the objective of detecting an existing gradient of properties.

3 Results

3.1 Compression Tests

After compression tests the specimens' shape was more oval than round which is an indication for anisotropic properties already after caliber rolling. The texture related cause of the oval shape of the specimen assumed after macroscopic examination is confirmed by the examination of the initial texture [2]. It showed that after caliber-rolling the majority of crystals were oriented with its (0002)-crystallographic direction perpendicular to the rolling direction and towards the direction of the preferred widening. In this case, the phenomena of twinning which clearly acts as an additional forming mechanism can be seen as cause of that behavior. The polfigures of the deformed specimens show a texture with a basal character (Figure 1). Deformation accomplished only by main slip systems would lead to a strong basal texture. Asserting a different result it is assumed that secondary slip systems must have been involved in the forming process [3]. The reduction of the basal texture component at higher temperatures supports that supposition. However, the basal component of the texture remains predominantly and decreases only slightly at higher temperatures. Hence, the texture development can not sufficiently be affected only by a variation in forming temperature.

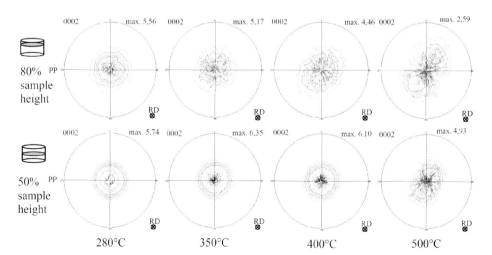

Figure 1: Texture development subject to temperature

Analyzing the flow curves of those tests an anomaly appearing at all temperatures could be found in the range of $\varphi = 0...0.2$. That anomaly is reflected in several inflection points in the slope of the flow curve at small deformation degrees [2]. So far it oftentimes was suspected that with increasing temperature and the thereby activation of additional slip systems the addiction to mechanical twinning would decrease in favor of the new activated slip systems and not occur at high temperatures anymore. In order to ascertain whether mechanical twinning or recrystallization phenomena are responsible for that behavior, additional compression tests at 280 °C and 400 °C with small deformation degrees were conducted thereupon (Fig. 2).

Those additional experiments showed a good compliance between the development of twins in the microstructure and the change in flow stress in the flow curve. Therefore, it could be asserted that mechanical twinning appears even at high temperatures and it was found responsible for the flow curve behavior [2]. It is therefore of relevance when evaluating deformation processes in a wide range of temperatures. This, however, holds only true if the initial orientation of grains is pleasant for twinning.

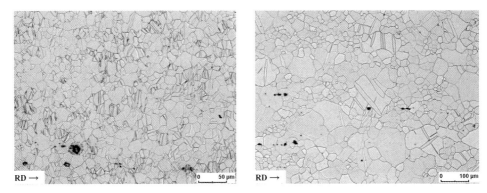

Figure 2: left: 280 °C, $\varphi = 0.05$, right: 400 °C, $\varphi = 0.025$

3.2 Simulation

In the first approach a recrystallization model has been combined with a visco-plastic self-consisted (VPSC) model [4–6]. The entire model [7] calculates the texture development as a result of deformation and recrystallization, and the resulting anisotropic plastic properties. Deformation is produced by crystallographic slip on various slip systems as well as by mechanical twinning. The influence of twinning on texture development is described by a predominant twin reorientation (PTR) model [8]. The recrystallization model is based on the fact that no new texture components are generated by recrystallization. Only the distribution of the individual components originating form the deformation is changing.

The above mentioned model was adapted in a first step to the flow curves of the compression tests where two different combinations of deformation mechanisms were tested [2]. The first combination comprised slip on basal planes and second order pyramidal planes as well as mechanical twinning. In the second combination slip on prismatic systems was additionally permitted but did not improve the adaptation. The consideration of recrystallization, however, seems to be essential. Even though the calculated flow curves do not match exactly with the experimental ones yet, they nevertheless describe the run already quiet well.

In the next step the polfigures of the compression tests were calculated and compared with the experimental texture (Fig. 3). The comparison shows that the experimental texture is reached in the simulation already after 30 % of strain. It intensifies significantly as the deformation simulation continues [2]. For this reason further experiments and improvements in terms of adaptation will be necessary.

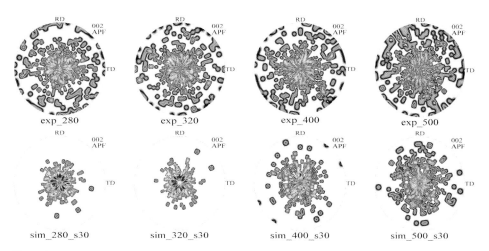

Figure 3: Experimental polfigures (first row) after compression tests at 280 °C, 320 °C, 400 °C, 500 °C; beneath: simulated polfigures at 30 % of deformation

3.3   Rolling Tests with Cast-rolled Feedstock

The typical structure of a cast-rolled Mg strips is a result of the superposed solidification and transportation process during cast-rolling. The melt starts to solidify at the surface of the rollers and the solidification continuous toward the middle of the strip, which results in a structure with about 45° tilted segregation sections between equiangular dendrites (Fig. 4). The solidification is finished before the strip reaches the center line of the roller whereby the strip experiences a partly rolling. That preformed, crushed casting structure possesses a remarkable formability, that rolling with 75 % reduction per pass can be possible [1].

Starting with that structure, which did not significantly change during preheating up to 500 °C and cooling down to the initial rolling temperatures (IRT), the microstructure during the first pass at 1m/s IRS at 300 °C IRT changed into a microstructure with deformed grains superposed by dynamic recrystallization. The same holds true for the second pass whereas in the third pass due to the lower forming temperature not only no recrystallized structure could be observed anymore but also twinning was now clearly apparent. With rising initial temperature the share of recrystallized grains increased especially after the first and second pass. After the third pass at 400 °C (IRT), however, a deformation structure without twins or recrystallization was found, whereas at 500 °C (IRT) dynamic recrystallization had already started to alter the microstructure slightly.

Figure 4: Cast-rolled strip, macro etching

In contrast to that behavior the samples rolled with 5m/s IRS possessed already after the first pass at 300 °C IRT a strongly dynamically recrystallized microstructure with a mean grain size of about 4 µm. The share of deformation structure was significantly lower compared to 1m/s IRS. After the second pass the recrystallized share of microstructure and its mean grain size increased until only recrystallized structure was found after the third pass. A similar behavior was found for the higher IRT, however, the mean grain size of the recrystallized grains increased with rising temperatures [9]. Twinning could only be observed insignificantly after the first pass at 300 °C and 400 °C IRT.

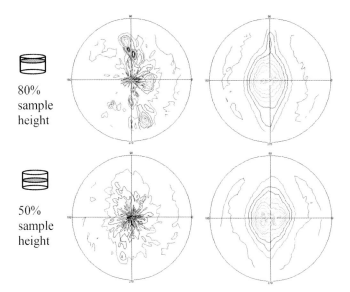

Figure 5: left: cast-rolled Mg strip, right: after second pass at 400 °C IRT (rolling temperature of pass 355 °C), 60 % overall reduction

The previously assumed texture related reason for the good forming ability of cast-rolled magnesium is confirmed by the polfigures of the cast-rolled condition (Fig. 5 left). In that connection the special structure (Fig. 4) originating from the cast-rolling process plays an important roll. Even thought the mid layer of the strip possess already a quiet homogeneous texture with only some small peeks, the surrounding layer, however, is fare more responsible for the good formability. Its texture shows that most of the crystals in that layer are orientated with its basal planes tilted by about 45° to the sheet normal (Fig. 5 top left). Especially those crystals are extremely advantageously orientated for dislocation movement in a first forming process as their preferentially activated basal planes have the highest Schmid factor in that condition.

This is also indicated by the resulting texture after the first rolling pass. Even though a basal character develops, its intensity however, is weaker than after compression tests. Additionally, the distribution of intensities scatters considerably in and against rolling direction. In the surrounding layer the basal texture is clearly less developed. As rolling continuous the basal texture component increases in the mid layer but only slightly in the surrounding layer. The scatter around rolling direction remains as Fig. 5 (right hand side) exemplifies. Furthermore, unlike the compression tests showed, a temperature dependency of the basal texture development could be observed in those experiments.

4 Summary

The present research work has been concentrated on the clarification of selected problems of deformation mechanisms for the aspired deformation-recrystallization model. The model at the current state of development is already able to describe many experimental results at least quantitatively. Among other things the experimental results showed that twinning occurs even at high temperatures and is responsible for the anomaly in the slope of the flow curve in the range of small deformation degrees. During compression tests which were mainly required for the adaptation of the model a temperature dependency of the basal texture development could barely be observed. This, however, does not apply to the rolling trails with cast-rolled feedstock. In addition, the reasons for the remarkable forming ability of cast-rolled magnesium strips could be revealed by examining the distinctive initial texture and tracing its development during rolling.

Subject of further work will be the detection and utilization of recrystallization phenomena. The cognitions of the experiments will be taken into account for the model development in order to reproduce all experimentally found mechanisms.

5 Acknowledgements

The financial support of the project by the German Research Foundation (Deutsche Forschungsgemeinschaft DFG, KA 1591/8-1 and RI 329/27-1) in the framework of the priority program SPP 1168 is gratefully acknowledged.

6 References

[1] C. Schmidt, R. Kawalla, B. Engl, N.D. Cuong: An innovative method to produce magnesium strip. Magnesium – broad horizons, Moscow, **2005**, proceedings
[2] C. Schmidt, R. Kawalla, T. Walde, H. Riedel, A. Prakash: Experimental and numerical investigation of texture development during hot rolling of magnesium alloy AZ31. Thermec **2006**, Vancouver, proceedings

[3] S.R. Agnew, M.H. Yoo, C.N. Tomé: Acta mater. 49 (**2001**), p. 4277– 4289
[4] R.A. Lebensohn and C.N. Tomé: Acta Metall. Mater. 41 (**1993**), p. 2611–2624.
[5] T. Walde: Dissertation at Karlsruhe University, **2004**, (carried out at Fraunhofer-IWM)
[6] [6]T. Walde, H. Riedel: Materials Science Forum 426-432 (**2003**) p. 3679–3684
[7] T. Walde, H. Riedel: Solid State Phenomena 105 (**2005**), p. 285–290.
[8] C.N. Tomé, R.A. Lebensohn, U.F. Kocks: Acta metall. mater. 39 (**1991**), p. 2667–2680
[9] R. Kawalla, M. Ullmann, M. Oswald, C. Schmidt: Properties of strips and sheets of magnesium alloy produced by cast-rolling technology. 7th International Conference on Magnesium Alloys and their Applications, Dresden, **2006**, proceedings

Basic Research in the Contacting Condition between the Molten Magnesium Alloy and the Roll on the Melt Drag Process

Shinichi Nishida, Mitsugu Motomura, Takeshi Yamazaki, Yoshitaka Ando, Yuki Takano
Waseda University, Shinjuku-ku Okubo, Japan

1 Introduction

Magnesium alloy is the lightest structural material and the applications are increasing in Japan. And the melt drag process is one of the single roll rapidly solidified strip casting process. We have succeeded produced the magnesium alloy strip directly from the molten magnesium by the melt drag process. But there were some defects on the roll side surface of the strip, such as voids, ripple mark and peeling. These defects were likely to be caused by the contacting conditions between the molten magnesium and the roll. And it is supposed that the contacting conditions are affected by the roll material and the roll surface conditions. The effective factors of the roll material on the strip surface conditions may be the heat conductivity and the chemical attraction including the surface treatment. And the effective factors of the roll surface conditions may be the surface roughness and the mechanical structure to exhaust trapped air. The optimum roll material and roll surface condition will be determined by investigating the effect of these factors. [1–11]

However it is not easy to change these factors of the roll in melt drag process. And so we have designed and manufactured the model device that we can examine easily the effect of process factors and process conditions in melt drag process. The substrate substituted for the roll. And the experimental design was applied. The factors such as the melt temperature, the melt height, the substrate material and the surface treatment, the substrate surface roughness and the substrate speed. The significance wasn't clearly found, but the tendency was obtained. The smallest substrate side surface roughness of strip was obtained when the substrate roughness was about Rzjis 10 μm. The shot blasted substrate surface roughness prevented the strip surface defects.

This research was investigated at Waseda university kagami memorial laboratory for materials science and technology in Tokyo Japan.

2 Strips Produced by Melt Drag Process

The strip produced by the melt drag process has two surfaces. We call the surface; roll side surface and the free solidified surface, respectively. Figure 1 shows the roll side surface of the strips produced by the melt drag process. Some defects on the roll side surface are often observed; such as such as voids, ripple mark and peeling. The voids tend to be caused at over roll speed 20 m/min. The size of voids become small and the number of voids become increase as the roll speed increases. The ripple mark tends to be caused at low roll speed because of the discontinuous solidification. The peeling is observed at low melt temperature. These defects frequently appear at a time and are likely to be caused by the contacting conditions between the molten magnesium and the roll. And it is supposed that the contacting conditions are affected by the roll material and the roll surface conditions. [7]

(a) Voids (b) Ripple mark (c) Peeling (circled area)

Figure 1: The defects of the roll side surface of the strips produced by the melt drag process

3 Experiment

The invented model device, experimental procedure, conditions and experimental design are as follows.

3.1 Model Experimental Device of Melt Drag Process

Figure 2 shows the schematic illustrations of the model experimental device. The substrate substituted for the roll. So it is easy to change the factors of the substrate that is the roll in melt drag process, such as the material, roughness and surface finishing.

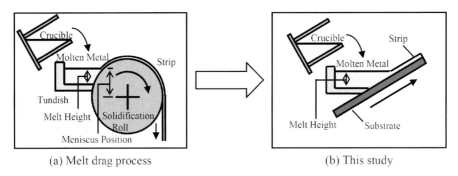

(a) Melt drag process (b) This study

Figure 2: The schematic illustrations

Figure 3 shows the photographs of model device. This device is entirely covered. Since the molten magnesium is very active, the device can be sealed up around the molten magnesium to fill the protective gases. The protective gases used inside the device are CO_2 and SF_6 to prevent the burning of molten magnesium. The oxygen density is kept under 7 %.

(a) Model device　　　　　　　　(b) Around the tundish

Figure 3: The photographs of model device

3.2　Experimental Procedure and Conditions

The magnesium alloy in the crucible is melted in the electric furnace. The surrounding atmosphere in the furnace is CO_2. After melting the crucible is transferred in the experimental device and the melt drag process is run. The pouring temperature, it is defined as the molten metal temperature just before pouring in the experimental device, is measured by the K-type thermocouple. The experimental conditions are shown in Table 1. Each parameter was shown in Figure 2(b). The surface treatment was chosen among major ones. The Rzjis 3 and 10 μm substrate roughness was applied by sanding. And the 30 Rzjis roughness was applied by shot blasting.

Table 1: Experimental conditions

Level number	1	2	3	4	5	6	7
Experimental material	AZ31						
Pouring temperature: Tm [°C]	760	800	840				
Melt height: h [mm]	15	25	35				
Substrate material: Ma	Cu	A1050	A5052	SUS430	SUS-TiN	SUS-Al	SUS-Zn
Heat conductivity [W/mK]	394	231	137	26.2	21.7	238	133
Substrate surface roughness: R [Rzjis]	3	10	30				
Substrate speed: V [m/min]	3	10	30				
Angle of dip: θ [deg]	20	30					
Tundish gap: t [mm]	0.5	1.0					

The experimental design was applied to investigate the effect of the process factor. Table 2 shows the factors and levels in this study. A total of 27 times experiment was run with the orthogonal array L27 (313) used dummy levels.

The roll side surface roughness was measured with laser displacement measuring equipment made by KEYENCE. Moving average method eliminated the waving of the produced strip. Strip width is about 30 mm that nearly equals the exit width of the tundish. Strip thickness was measured at 5 points for the cross of the strip with 3 mm intervals. [6]

Table 2: Orthogonal array

Experiment number	T_m [°C]	h [mm]	V [m/min]	R [μm]	Ma	θ [°]	t [mm]
1	760	15	3	3	Cu	20	0.5
2	800	25	10	3	A1050	30	0.5
3	840	35	30	3	A5052	30	0.5
4	760	25	3	3	SUS430	30	1.0
5	800	35	10	3	SUS-TiN	30	1.0
6	840	15	30	3	SUS-Al	20	1.0
7	760	35	3	3	SUS-Zn	30	0.5
8	800	15	10	3	Cu	20	0.5
9	840	25	30	3	A5052	30	0.5
10	840	25	10	30	Cu	30	0.5
11	760	35	30	30	A1050	20	0.5
12	800	15	3	30	A5052	30	0.5
13	840	35	10	30	SUS430	20	0.5
14	760	15	30	30	SUS-TiN	30	0.5
15	800	25	3	30	SUS-Al	30	0.5
16	840	15	10	30	SUS-Zn	30	1.0
17	760	25	30	30	Cu	30	1.0
18	800	35	3	30	A5052	20	1.0
19	800	35	30	10	Cu	30	1.0
20	840	15	10	10	A1050	30	1.0
21	760	25	3	10	A5052	20	1.0
22	800	15	30	10	SUS430	30	0.5
23	840	25	10	10	SUS-TiN	20	0.5
24	760	35	3	10	SUS-Al	30	0.5
25	800	25	30	10	SUS-Zn	20	0.5
26	840	35	10	10	Cu	30	0.5
27	760	15	3	10	A5052	30	0.5

4 Results and Discussions

Figure 4 shows the produced strips by the model device. It is possible to produce the flat strips by this process. The strips are not oxidized, and the metallic luster is observed.

Some defects are often observed at substrate side surface such as voids, ripple mark, peeling, shrinkage crack and granular defect. The defect free strip was produced with shot blasted substrate.

Figure 5 shows the trends about strip thickness. The pouring temperature and the substrate speed become higher, the strip thickness becomes thinner. These are because of the reducing the solidification amount per unit of the time. The substrate roughness becomes smoother, the strip thickness slightly becomes thinner. This is the reason of reducing the anchor effect. The strip produced by this process only gets up the substrate, so the frictional force is necessary to produce the continuous strip. The appropriate surface roughness may be effective the controlling strip thickness.

Experiment number	4	5	6	16	17	20
Free solidified surface						
Substrate side surface						
Kind of defect	Granulating	Peeling	Voids	Voids Ripple mark		Ripple mark

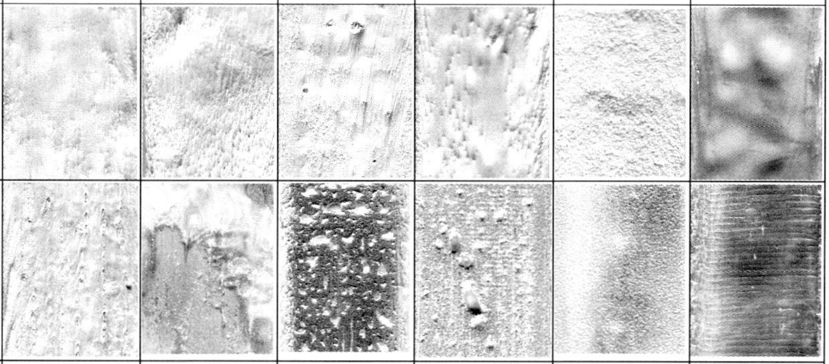

10mm

Figure 4: The photographs of produced strips

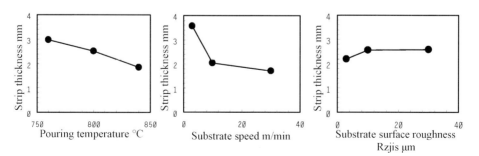

Figure 5: The trends about strip thickness

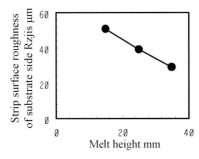
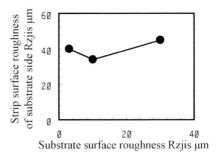

Figure 6: The trends about strip surface roughness of substrate side

Figure 6 shows the trends about the strip surface roughness of substrate side. The melt height becomes higher, the strip roughness becomes smoother. It means that the strip surface defects of the substrate side decreased. The high melt height, that is the high pressure to the substrate, is effective to prevent the surface defects. The smallest substrate side surface roughness of strip was obtained when the substrate roughness was Rzjis 10 μm. The strip surface defects were caused at the condition of the substrate surface roughness Rzjis 3 μm. When it is 30 μm, the strip surface is affected by the substrate roughness. However the produced strip at the substrate roughness Rzjis 30 μm is defect free. The Rzjis 30 μm substrate was made by shot blasting. The appropriate substrate surface roughness and the pattern may effectively act as the air discharger between the substrate and the molten metal. It is interesting that the smooth surface mold is not necessary to produce the defect free strip in the strip casting process. Most strip surface defects are supposed to result from the trapping of the air between the mold and the molten metal. More investigation of the air discharging mechanism should be needed. [9]

5 Conclusions

The model device of melt drag process was developed and the trends affected by the process factors were investigated. The results obtained are as follows:

- Producing the no oxidized magnesium alloy strip with the invented device is possible.
- Strip thickness is affected by the factors such as the melt temperature, the substrate speed and the substrate surface roughness.
- Strip surface roughness of the substrate side become smooth at the condition of high melt height and the appropriate substrate surface roughness.
- Preventing the strip surface defects of the substrate side is possible by making the air discharger on the substrate surface.

We express our grateful appreciation for the assistance of SANKYO TATEYAMA ALUMINUM Co. Ltd. in taking of magnesium material.

6 References

[1] Handbook of Advanced Magnesium Technology (Ed.: Y. Kojima), The Japan Magnesium Association, Tokyo, Japan, **2000**, Chapter 1.
[2] H. Bessemer, U. S. Patent 49053, **1865**.
[3] Maringer R. E., Materials Science and Engineering **1988**, 98, p. 13–20.
[4] H. Nakae, kinzoku **2002**, 72, p. 57–66.
[5] H. Nakae, kinzoku **2002**, 72, p. 147–153.
[6] H. Nakae, Shin H. C., S. Matsuo, Journal of Japan Foundry Engineering Society **2002**, 74, p. 644–649.
[7] I. Ohnaka, H. Yasuda, M. Takahashi, *in Proceedings 8th Japan Germany Seminar*, Sendai, Japan, **1993**, p. 166–172.
[8] N. Zapuskalov, ISIJ International **2003**, 43 1115–1127.
[9] L. Strezov, J. Herbertson, G. R. Belton, Metallurgical and materials transactions B **2000**, 31B, p. 1023–1030.
[10] D. K. Choo, H. K. Moon, T. Kang, S. Lee, Metallurgical and materials transactions A **2001**, 32A, p. 2249–2258.
[11] S. Nishida, M. Motomura, the Journal of Japan Institute of Light Metals **2005**, 55, p. 315–320.

Method for Horizontal Continuous Casting of Magnesium Alloy Slab

K. Tada
Sanyu Seiki Co., Ltd., Yoshii-town

1 Introduction

A conventional method for producing a laminate plate is generally a method, in which a slab manufactured through casting (slab manufactured by a continuous casting method) is subjected to rolling. In order to produce a laminate plate from a slab having crystal grain boundaries without the occurrence of flaking and edge cracks of a surface, a large number of (in general, over 25) processing steps are required, including grinding of a surface layer of a slab and cutting of side edges thereof, and further, hot rolling, cold rolling, and heat treatment. Accordingly, this conventional method has a problem in that the cost of a rolling plate member is high. In addition, in the case of a slab with poor workability which is made of magnesium (hereinafter, referred to as Mg) or magnesium alloy (hereinafter, referred to as Mg alloy), it is extremely difficult with the above method to plastically work the slab into a laminate plate without the occurrence of flaking or edge cracks of a surface thereof. This is one of the causes of high cost. Against the background described above, there has been an increased market demand for expanded products formed through press working molding of Mg laminate plates or Mg alloy laminate plates with high productivity and highly reliable product quality. However, a safe and stable casting method of an Mg slab or Mg alloy slab, which is a raw material for a laminate plate, has not been sufficiently established. Thus, at the present, domestic mass production of slabs has not been realized.

The present research relates to a method for horizontally drawing-out type (referred to as horizontal) continuous casting for horizontally drown out and continuously casting a Mg slab or a Mg alloy slab, and relates to an apparatus for horizontal continuous casting which is used for the method. [1,2] The aim of this research is to establish a basis for the fabrication of Mg slabs, by studying experimental variables in the horizontal continuous casting method. The equipment for the casting of Mg alloy slabs was constructed by authors. The optimum casting conditions were determined using AZ31B magnesium, with particular reference to the effects of casting variables such as casting speed, mold temperature and cooling position on the surface. The structure of the solidified slabs was examined.

2 Experimetal Procedure

2.1 Casting Apparatus

In the apparatus for horizontal continuous casting of a Mg alloy slab which is shown in Figure 1, a sheltered board is provided to cover an area extending from a portion above a pool in a tundish to a portion above an outlet of a mold, thereby shielding the area from the outside. A cooling jetting tool, which jets a cooling water onto a Mg alloy slab, drawn out from the outlet to thereby cool the slab, is provided on the outer side with respect to the outlet of the mold, and an gas jetting tool is arranged on the inner side with respect to the outlet of the mold, as shown in Figure 1. Further, a schematic illustration of an outlet point of the mold is shown in Figure 2.

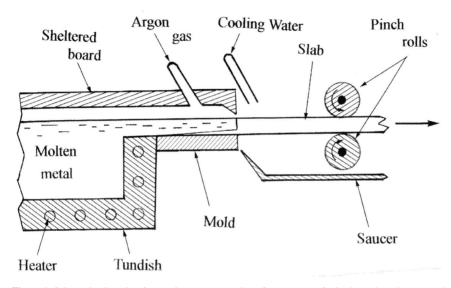

Figure 1: Schematic view showing a primary cross section of an apparatus for horizontal continuous casting of an Mg slab or Mg alloy slab according to the present research

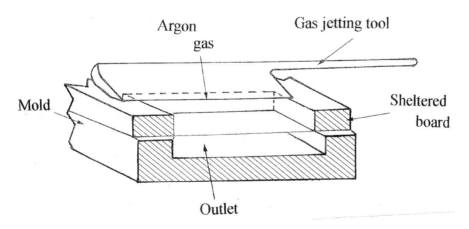

Figure 2: Schematic view of an outlet end of the mold of the apparatus of Figure 1

Heaters are built in a bottom wall and a peripheral side-wall of the tundish. The outlet side of the tundish and the mold each have a gutter shape (upwardly facing concave shape) having an upper opening. The sheltered board covers the upper opening portions, thereby covering the upper surfaces from the outside. Thus, a molten magnesium (hereinafter, referred to as molten

Mg) in the tundish and the Mg alloy slab in the mold don't contact with the air. Table 1 is a chemical composition table of an expanded material, AZ31B magnesium which is used in this study

Table 1: Chemical composition table of an expanded material, AZ31B magnesium

	Chemical composition [wt%]								
	Al	Zn	Mn	Si	Fe	Cu	Ni	Be	Mg
AZ31B	3.17	0.92	0.28	0.02	0.002	0.002	0.0004	0.0005	Bal

In the gas jetting tool having a wide width, a supply opening is located above a free surface of the molten Mg in the mold. A sealing gas such as an Argon gas is supplied from the supply opening toward the free surface side to form a gas curtain for sealing a gap between the free surface and a part of the sheltered board, which surrounds the upper surface of the mold. The gas seals a portion above the entire width of the free surface of the molten Mg alloy in the mold to prevent oxygen from flowing into the mold. As a result, the gap is sealed, and the cooling water, droplets, spray, and the like are prevented from running back to the mold side. Further, a slit-shaped nozzle, round pipe nozzle, or the like can be used for a jetting outlet of the cooling jetting tool having a wide width. The jetting outlet is arranged downward such that the cooling water is jetted onto the Mg alloy slab drawn out from the outlet of the mold. The width of the jetting outlet is substantially the same as or slightly narrower than the width of the Mg alloy slab such that the water is jetted all over the Mg alloy slab. Furthermore, Stainless steel (SUS 430 etc.) is suitably used for the mold, and it is preferable that a mold be used in which an inner surface thereof has been subjected to molten aluminum plating. The mold may have a built-in heating device such as the heater. The mold of 100 mm long and of inside width 120 mm is heated by electric heating coils wound from nichrome wire of 1.2mm diameter. A chromel-alumel thermo-couple is placed in between the mold and heaters to monitor the mold temperature. The operation of the heater is made switchable between on and off.

On the other hand, the apparatus is arranged with a second gas jetting tool between the outlet of the mold and the cooling water jetting tool. The gas jetting tool is provided at an acute angle. A second sealing gas such as air, argon gas, nitrogen gas, incombustible gas is jetted from a jetting outlet toward the Mg alloy slab which is drawn out from the outlet of the mold to thereby form an air curtain or gas curtain over the Mg alloy slab. The gas curtain seals the portion above the entire width of the Mg alloy slab. The seal also can prevent the cooling water, droplets, fine splashes, and the like from running back to the mold side (in a two-staged manner). A water receiving portion (referred to as saucer) having an upper opening is provided on the lower downstream side with respect to the outlet of the mold. A drawing device, which draws out the Mg alloy slab from the outlet, is provided on the outer side of the outlet of the mold. Two or more pairs of pinch rolls for withdrawal of the slab, which sandwiched the Mg alloy slab from both the upper and lower sides to draw out the slab, were appropriately used as the drawing device.

2.2   Casting Method

About 45 kg of AZ31B magnesium was melted in a crucible, whose temperature was maintained at 953 K. or more. Molten Mg was ejected into the heated tundish by slowly lowering the level control rod into the crucible. The molten Mg (at 973 ± 20 K.) in the tundish in Figure 1 was supplied to the mold which was heated by the heater, and a surface temperature of the mold is set at a temperature exceeding or not exceeding a solidification temperature (hereinafter, referred to as s. t.) of the AZ31B magnesium. The Mg alloy slab under solidification in the mold was drawn out from the outlet of the mold to the outside. At the time of the start of drawing-out, the tip point the molten Mg under solidification, was coupled with the dummy member, which was set at the outlet position of the mold. The dummy member was drawn out, whereby the Mg alloy slab under solidification is drawn out. Thereafter, a cooling water was jetted onto the Mg alloy slab. The cooling water jetting position should be regulated on the Mg alloy slab on the outside of the mold outlet and in the vicinity of the outlet. A sealing gas was jetted toward the outlet side of the mold over the entire width of a free surface of the molten Mg on the inner side of the mold outlet and in the vicinity of the outlet. Further, between an outlet of the mold and the cooling water jetting part, a second sealing gas was jetted toward the cooling water jetting side over the entire width of the Mg alloy slab. Moreover, the sealing gas was jetted toward the outlet side of the mold along the free surface of the molten Mg, and the second sealing gas was jetted toward the cooling water jetting side along an upper surface of the Mg alloy slab. The cooled Mg alloy slab was continuously drawn out by means of a drawing device. The thickness of the slab was controlled between 5 mm and 15 mm by changing the supply of molten Mg. The water flow rate used for this casting operation was set at 5.83×10^4 mm^3/s, while the water temperature ranged averaging 283 K. To study the influence of casting variables, a slab with 100 mm to 120 mm width, was produced in each casting operation. The casting speed ranged from 1 mm/s to 4 mm/s. Properties such as the surface condition and the solidified structure of the slab were examined.

3 Results and Discussion

3.1 Description of the Related Art

As a method of manufacturing a strip of a nonferrous metal such as aluminum alloy or tin alloy or the like, a horizontal continuous casting method has been developed in the prior art. [3] This method is commonly called an OSC (Ohno strip casting) method. In the OSC method, a horizontal, heated mold is used instead of a conventional cooling mold. Further, some conventional strip manufacturing apparatus [4] is an application of the principle of OSC method. The OSC method is superior to conventional manufacturing methods.

However, in the case where the mold is a metal such as stainless steel, when Mg slab or Mg alloy slab is cast, it was necessary to heat the mold in advance to a higher temperature than needed for casting, while taking into account a cooling amount, since the cooling water is applied onto the strip on the mold to cool the strip, not only the strip but also the mold is cooled. The mold needed to be heated to a temperature higher than a melting point of a pure Mg or the s.t. of the Mg alloy by approximately 573 K. to 623 K. Such a high temperature leads to the shorter life of the mold and the heater element, and the large trans-formation of the mold due to thermal expansion. In the case of a marked transformation, the flow of the molten Mg which is supplied to the mold, is separated to both the side surface sides the mold from the upper curved portion of the Mg or Mg alloy slab since the mold is greatly deformed.

As a result, it is impossible to produce the Mg slab or Mg alloy slab with a wide width (for example, a width of 100 mm or more). Further, under the condition of such a high temperature, the leak of a heater temperature control current occurs. Furthermore, in the case of Mg or Mg alloy which is easy to be oxidized, a surface of a Mg alloy slab burns at the same time the slab is drawn out from the mold, and the surface becomes black due to the generation of the large gray and black oxide. Besides, when oxygen flows into the mold, a molten Mg burns through reaction with oxygen in the air. When the combustion contacts with the cooling water, the water is decomposed to generate hydrogen and oxygen, possibly leading to an explosion. As a result, in manufacturing of an AZ31B slab with the prior method, a surface thereof is uneven.

3.2 Field of this research

In the present method, the key factors for successful production of Mg alloy slabs are the gas sealing conditions and the mold temperature. Further, the position and shape of the solid-liquid interface are important factors in the easy control and stable operation of the continuous casting for long time on the mass production scale. As described above, the argon gas is jetted into the gap between the free surface of the molten Mg and the sheltered board surrounding the upper

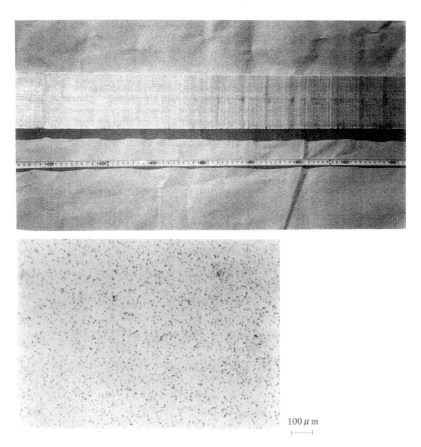

Figure 3: Surface appearance of the AZ31B magnesium slab and microstructure of the section taken in the width direction of the AZ31B magnesium slab

surface of the mold, thereby forming the gas curtain in the gap. Therefore, it is desirable that the sealing gas be jetted so as to lick the free surface. Further, the air or gas jetting tool is provided between the outlet of the mold and the cooling medium jetting tool. That is, such prevention is preferably achieved by the two curtains, the above curtain and the gas curtain formed by the sealing gas jetted from the gas jetting tool. An ordinary compressor may be used as a jetting device. Thus, the usage amount of expensive argon gas or the like can be vastly saved. When the mold is heated by the heater, solidification of the molten Mg, which is supplied from the pool in the tundish into the mold, progressed more slowly. Thus, at the time of the start of drawing-out, a tip point of the molten Mg can be coupled with a dummy member relatively slowly with a margin of time. Also, the thickness of a solidified shell, which is generated along an inner wall surface of the mold, can be reduced. Thus, the Mg alloy slab, which has been drawn out from the outlet of the mold, is cooled immediately after being drawn out from the outlet of the mold. Accordingly, it doesn't only occur that the slab surface is oxidized to be black until the slab is cooled after being drawn out from the outlet, but also the continuous casting control is very easy due to the position and the shape of the solid-liquid interface can be clearly observed by the naked eye at any time during the casting. As a result, the AZ31B slab, which was manufactured with the manufacturing method and the apparatus of the present research, widely differed from a conventional slab and had a smooth outer circumferential surface as shown in Figure 3.

The figures have revealed that the AZ31B slab, has a solidification structure of which section has a mirror surface and in which eutectic crystals of dissolved elements, aluminum and zinc, are uniformly and minutely distributed in a form of fine particles in Mg primary crystals serving as basis metal. The slab has no casting defects, in its inner side, such as holes, gas porosities, segregation of the dissolved elements or impurity elements, and inclusions, and the slab is a high-quality slab having excellent mechanical property and excellent characteristics such as rolling and pressing workability.

4 Conclusions

The Authors succeeded in stable casting of AZ31B slab which has no crystal grain boundary, the crystal grain boundary being formed on a surface or a side edge portion toward an interior portion of the slab and being a main cause of occurrence of flaking and edge cracks of the surface in hot or cold rolling working, which doesn't have inner holes, gas porosities, inclusions, and the like, and which isn't oxidized to become black, without the danger of explosions. The slab has a smooth surface condition and high quality without a gas porosity, a segregation or the like. Since the molten metal is rapidly cooled and solidified at a relatively high cooling speed (10 to 10^2 K/sec.), the Mg alloy slab can be produced which has a microstructure, in which the dissolved elements (Al, Zn) and an intermetallic compound (Mg-Mn) are minutely and uniformly distributed, and which has high corrosion resistance. Further, the slab has a size of a width of 100 to 120 mm x a thickness of 5 to 20 mm x a span of 10 m or more.

5 References

[1] K. Tada and K. Miyazaki, „Method for Horizontal Continuous Casting of Magnesium Slab or Magnesium Alloy Slab and Apparatus Therefore", Japanese Patent, No. JP3668245, 15 April **2005**.

[2] K. Tada and K. Miyazaki, „Method for Horizontal Continuous Casting of Magnesium Slab or Magnesium Alloy Slab and Apparatus Therefore", U.S. Patent Application Publication, No.2005/0224146 A1, 13 October **2005**.
[3] K. Tada and A. Ohno, „Production of Aluminum Strip using an Open Type, Horizontal, Heated Mould", *Aluminium, Vol. 69, December 1993*, Germany, **1993**, p.1092–1093.
[4] A. Ohno, „Apparatus for Horizontal Continuous Casting of Metal Strip", Japanese Utility Model Application Laid-open, No. Hei 4-125046, 13 November **1992**.

Wrought Alloys

Development and Application of Wrought Magnesium Alloys in China

Fusheng Pan[1,a], Aitao Tang[1,a], Siyuan Long[1], Mingbo Yang[2]

[1] Chongqing Engineering Research Center for Magnesium Alloys, Chongqing University, Chongqing 400045, P. R. China
[a] fspan@cqu.edu.cn, [b] tat63@sina.com.cn
[2] College of Materials Science and Engineering, Chongqing Institute of Technology, Chongqing 400050, P.R.China

1 Abstract

The current status of development and application of wrought magnesium alloys in China have been summarized. The development of new types of wrought magnesium alloys in China is focused in AZ alloys and ZK alloys, with more attention to development of high formability alloys. The additions of rare earth elements are often used to improve the properties of wrought magnesium alloys. Strontium is found to be another very beneficial element to modification of microstructure of wrought magnesium alloys, which is also very rich in China. The investigation on second phases in wrought magnesium alloys has become a hot research field, and many new results about compounds in these alloys have been obtained. A lot of effort has been done to enlarge the application fields of wrought magnesium in China, and more attention has been paid to increasing their applications in transportation vehicles. The recent progresses in the processing technologies for wrought magnesium alloys are also discussed. The twin roll casting of some wrought magnesium alloys is being investigated in some Chinese universities.

2 Introduction

It is well known that rolled sheets or extruded profiles allow a broader application by using post-processing shape giving technologies, and semi-finished components made of wrought magnesium alloys exhibit a better quality and homogeneity of the microstructure than their cast counterparts. Together with the generally improved mechanical properties they are obviously an important extension for an increasing usage of magnesium in industrial applications. As a result of both potential high performance of wrought magnesium alloys and rich magnesium resource, the development and application of wrought magnesium alloys have an increasing interest in the past five years in China. A lot of work has been carried out mainly in Chongqing University, Shanghai Jiaotong University, the Research Institute of Metals of Chinese Academy of Science and Northeastern University. The research funds for wrought magnesium alloys from central government have increased apparently in the past five years, which are mainly from the Ministry of Science and Technology of China and the Natural Science Foundation of China. Some local governments, such as Chongqing government, also have given much support to wrought magnesium alloys. The biggest project on wrought magnesium alloys supported by the Ministry of Science and Technology of China is „Development of high performance wrought magnesium alloys and their application", which were undertaken by Chongqing University and the Research Institute of Metals of Chinese Academy of Science and led by Fusheng Pan in Chongqing University.

3 Effects of Alloying Elements on the Microstructure and Mechanical Properties and Development of New Alloys

The research results in China showed that the mechanical properties of as-extruded Mg-Zn-Zr alloys were improved significantly by Nd and Y additions, especially the elevated temperature strength, which was above 150 MPa in ultimate tensile strength at 250 ºC. AZ31-RE alloy was also found to exhibit better heat resistance than AZ31 alloy. Nd addition was found to bring about two kinds of precipitation phases. One is AlNd phase, the other is AlNdMn phase, which were identified as $Al_{11}Nd_3$ and Al_8NdMn_4 by X-ray diffraction and TEM. The Al-10.5%Sr master alloy was found to produce obvious modification on the as-cast microstructure of the AZ31 alloy. In the as-cast ZK60 alloys, the eutectic was prone to have fish-bone-like shape with the increase of Zn content. When the eutectic grew up to contact with each other, then the network eutectic formed. Zr addition was found to fine not only the grain size but also the eutectic intensively. However, when Zr content is more than 0.6wt-%, a compound containing Zr or pure Zr phase would appear, which reduced the Zr content in the solution and consequently coarsened the crystal grain. The $β-Mg_{17}Al_{12}$ phase particles were found to enhance the grain refinement of Mg-Al-Zn magnesium alloys during ECAE process. As a given deformation, the grain size in AZ91 Mg alloy was finer than that in AZ61 Mg alloy after ECAE due to more $β-Mg_{17}Al_{12}$ phase particles in AZ91 Mg. The addition of Ca could refine both the grain and $Mg_{17}Al_{12}$ phase of AZ91 magnesium alloy. Small amount Ca addition to AZ91 alloy could make $Mg_{17}Al_{12}$ phase become lamellar structure from network distributed at grain boundaries, which had stronger effect on preventing the movement of grain boundaries to some extent at elevated temperatures effectively. The fine I phase particles and $Mg_3Zn_4Er_2$ precipitation phase in Mg-6Zn-0.5Zr alloy were found to suppress the migration of DRX grain boundaries, which could increase the deformation activation energy and refine dynamic recrystallization structure.

New types of Mg-Y-Zn alloys have been developed in China. The strength of the extruded alloys was increased significantly to 400MPa at ambient temperature, which was considerable high strength for magnesium alloy. With the increase of temperature, the strength of alloy decreased and the elongation increased progressively. In this change pattern, there was an abrupt fall of strength from 200 °C to 250 °C. Even so, some new alloys still have UTS of about 290 MP at 250 °C. By comparison of Mg97Y2Zn1 alloy and Mg96Y3Zn1 alloy, it was found that the increase of only yttrium content resulted in increase of strength and decrease of elongation at elevated temperature (Fig.1). However, by comparison of Mg97Y2Zn1 alloy and Mg95.5Y3Zn1.5alloy, it was obvious that increasing both yttrium and zinc contents could result in an increase of both strength and elongation in mechanical properties. In the extruded Mg-Y-Zn alloys, the LPS structure of 14H structure was observed, which was found to modify both strength and elongation. The excellent mechanical properties of new alloys were thought due to multifactor effects including the grain refinement, solid solution strengthening, and formation of LPS structure and fine $Mg_{24}Y_5$ phase. Nd addition has been used to improve the microstructure and properties of AZ80 alloy. The morphologies of β-precipitates and effects of aging time on the mechanical properties of AZ80 alloy containing Nd element are given in Fig. 2 and 3.

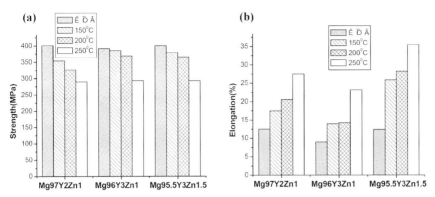

Figure 1: The mechanical properties of extruded Mg–Y–Zn alloys at both ambient temperature and elevated temperature. (a) the ultimate tensile strength (b) elongation

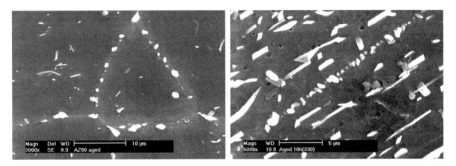

Figure 2: Precipitation of β phase in boundaries and inside grains in AZ80 alloy

With the help of phase diagram calculations, a new ultra-light magnesium alloy with a nominal composition of Mg-12Li-3Al-1Zn-x alloys was developed by casting, forging and rolling procedures in laboratory. Different combinations of strength and ductility can be obtained, depending on processing and heat-treatment route. The alloy could be cold rolled into sheets with thickness ranging from 1.5 to 0.1 mm, which can be easily formed at room temperature through deep punching after a suitable aging treatment. Typical values of tensile properties of the test alloy measured at room temperature are listed in Table 1. When aged at 473 K for 30 minutes after cold rolling, a maximum value in ductility was obtained at an acceptable strength level, which would give rise to an obviously improved formability. Prototype mobile telephone covers had been successfully punched at room temperature by using the roll-aged sheet of 0.6 mm thickness.

Table 1: Typical tensile properties of Mg-12Li-3Al-1Zn-x alloys

Process Condition (Thickness)	$\sigma_{0.2}$, MPa	σ_b, MPa	δ, %
Forged plate (10mm)	155	175	32
Worm rolled plate (6mm)	150	170	35
Cold rolled Sheet (1.5mm)	190	230	13
Roll-aged sheet (1.5mm)	145	175	39

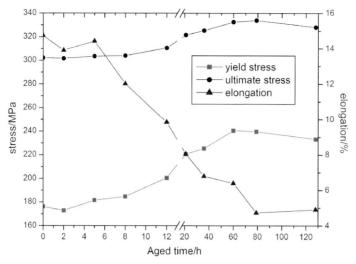

Figure 3: Effects of aging time on the mechanical properties of AZ80 alloy containing rare earth element

Texture evolution in Mg-13wt%Li-X alloy cold-rolled from 1.35 mm to 0.34 mm thickness had been investigated by using pole figures and orientation distribution functions (ODFs). Punching tests were conducted to reveal the effect of texture nature on formability. It was found that:

1. the textures of the as-received sheet were characterized by α fiber texture, a γ fiber texture and a cubic texture in both cold-rolled and annealed sheets;
2. with the decrease of thickness though rolling, the intensity of the γ fiber texture continuously increased.
3. the good punching behavior of the cold-rolled sheet was found to correspond to the appearance of a well-developed γ fiber texture.

In order to develop new types of alloys containing manganese, Al-Mn phases had been investigated. The powders of Al-Mn phases were obtained from the as-cast, as-rolled and as-extruded AZ magnesium (AZ31, AZ61) by using the extraction technology. X-ray diffraction (XRD), scanning electron microscopy (SEM) and energy dispersion X-ray detector (EDX) were used to investigate the structure and morphology of these compound powders. The results showed that manganese existed in both the way of Al_8Mn_5 phases and dissociated manganese particles in the AZ magnesium alloys, and rolling or extrusion processing did not change the type of the Al_8Mn_5 phases. SEM observation showed that Al_8Mn_5 phases were formed mainly in dendrite and leaf shape with size of 15–100 μm in the as-cast AZ magnesium alloys, and were broken into fine spherical fragments with size of about 4–5 μm after rolling or extrusion.

4   Deformation of Wrought Magnesium Alloys

A lot of work has been carried out on the changes of microstructure and mechanical properties of alloy AZ31 after ECAE process in China. The results showed that after four or five extrusions at between 498–523 K, the ductility increased due to grain refinement, but the strength decreased. With increasing processing temperature, the strength continued to decrease. Therefore, a new ECAE processing route, a two-step ECAE, was developed to control the microstructure and mechanical properties. Ultra-fine grains (0.5 μm diameter) could be produced, and both the ductility and yield stress were significantly improved as a result of this processing. Although ECAE improves microstructure and mechanical properties, there are still many questions which need to be answered about cost, tooling design, extrusion rates and ratios for possible applications. Little work was done on the ECAE of magnesium matrix composites in China.

AZ31 magnesium alloy was found to have good ductility when it was rolled at higher than 300 °C. However, if the rolling temperature was below 300 °C, the ductility of the sheets decreased dramatically and serious edge cracks, sometimes middle part cracks, appeared on the surface of sheets during the rolling. After warm rolling, there were low density twins in AZ31 sheets. However, many twins could been found if the warm rolled sheets were imposed cold-rolling with small reduction (5 %) by using many passes.

The superplastic forming limit diagram (FLD) at 573K and $3.3 \cdot 10^{-4}$ s^{-1} was established for AZ31B alloy. The forming limit diagram offers the theoretical guide for the design of forming mould and forming technology of AZ31B magnesium alloy sheet. The load p at 723 K and $1 \cdot 10^{-3}$ s^{-1} as a function of equivalent true strain with strain rate sensitivity m 0.36 was established by using Gauss-Newton nonlinear regression.

$$\frac{p_1}{p_{1\max}} = 1 - \ln(1+0.36)[1.37043(\varepsilon - 0.29) + 1.84183(\varepsilon - 0.29)^2]$$

There were four stages in the straining path for superplastic forming of AZ31B sheet, which were the increasing stage with the increase of loading, homogeneous forming stage after peak load, strain drafting stage and planar strain stage. Either in tensile-compressive deformation or in bi-axis tensile deformation, the judgment criterion for local necking of superplastic deformation was found to be $d\varepsilon_2 = 0$.

The effect of temperature on the microstructure and texture evolution during compression of AZ31 was found to be obvious. In the TD (transverse direction) samples of AZ31 alloy, the dominant deformation mode changed from twinning to slip at temperatures above 250 °C. In the ND (normal direction) samples, deformation took place only by slip, with the amount of low angle misorientations inside each grain increasing with increasing deformation temperature. The crack propagation primarily was found to occurs at the α-Mg/β-Mg$_{17}$Al$_{12}$ interfaces and grain boundaries in the Mg-Al-Zn magnesium alloy during the tensile deformation. CEC (cyclic extrusion compression) process could be used to refine the grains of AZ61 Mg alloy effectively. The microstructure could not only be refined from 50 μm to 0.8 μm after 15 passes at 573 K but also be homogenized and equiaxed.

Grain growth, slip dislocation band, deformation twin, grain boundary migration and precipitates were obviously observed in the microstructure of AZ61 alloy after fatigue at elevated temperature. FCP rate increased with increasing temperature. Because of the growth of grain size at 120, a bend occurred in the da/dN verse ΔK curve, which corresponded to a fracture transition from a mixed intergranular and transgranular fracuture to a transgranular fracture. FCP rate improved significantly with increasing relative humidity.

A new method, unidirectional bending repeatedly (UBR), was applied to improve the texture and properties of magnesium alloy sheets in Chongqing University. The samples were deformed repeatedly by the unidirectional bending at room temperature and some of samples were annealed. The formability, texture and microstructure of the samples before and after UBR were characterized and compared. It was found that, by UBR, the basal texture of magnesium alloy sheet was reduced dramatically and the formability of AZ31B magnesium alloy sheets at the room temperature was improved considerably.

Based on the stress-strain data obtained from isothermal compression experiments, a warm deformation mathematical model for AZ31 alloy was developed in China. The warm extrusion of a magnesium alloy radiator was simulated using a rigidity-plasticity finite element method so that the flow law of the alloy and extrusion forces during isothermal extrusion could be obtained. Furthermore, the metal flow of tubular-shaped products of magnesium alloys with five different sized dies during extrusion was simulated. The results showed that the metal flow of the tube products with inner diameter of 200 mm was more uniform with the dissipation.

5 Thin Strip Casting of Wrought Magnesium Alloys

AZ31 strips with good microstructure and 1-3mm thickness has been produced by vertical twin roll casting in China. With the increase of the casting speed, the grains became finer but more non-uniform. With the decrease of the casting temperature, the as-cast microstructure also became finer, and changed from equiaxed grains to sphere-like grains and further to flat grains. It was found that when the casting temperature was too high, many dendritic crystals were observed in microstructure. However, when the casting temperature was too low, a globular structure appeared. When the casting temperature was slight higher than liquidus temperature, the microstructure was mainly composed of fine spherical grains. AZ31B strips having dimensions of 6mm thickness and 600–800mm width had been successfully produced in Chinese companies. The suggested casting process parameters were 640 × 12m/min for AZ31 alloy.

A general mathematical model which connects macroscopic heat and mass transfer with microscopic nucleation and grain growth during the twin-roll thin strip solidification process has been developed. In this model, the latent heat release was treated by an enthalpy method, and network and node were divided by streamline boundaries. The heterogeneous nucleation model and the competition growth mechanism between columnar and equiaxed dendrites were introduced, together with a revision of the dynamic KGT model, which was originally established under the pure diffusion condition to simulate dendrite tip growth during rapid solidification. On the basis of their mathematical model, the model was found to be able to simulate the effects of pouring temperature, casting velocity and pool height on the proportion of the columnar grain zone, dendrite spacing and the columnar grain growth angle The theoretical basis for the optimization of the process parameters and the control of the solidification structure and quality was provided. Fig. 4 is a growth schematic view of columnar dendrite under relative flow condition

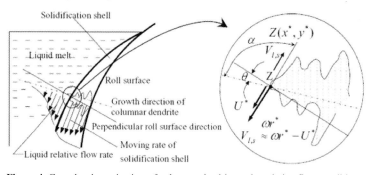

Figure 4: Growth schematic view of columnar dendrite under relative flow condition

6 Database Prototype of Magnesium Alloys

A database prototype of magnesium alloys based on the Internet has been developed in China, which is useful tool for scientific research and engineering. This database system can provide much available information about Mg-alloys such as mg-alloys chemical composition description, mg-alloys phase structure, mg-alloys standard, mg-alloys process, and mg-alloys Properties, etc. The combination of internet technology, database and materials science combined together is our research emphasis to deal with the problem of interconnection, communication and interoperation of Mg-alloys database in distributed Internet environment. In the future works we shall focus our attention on the Mg-alloys phase structure module including Mg-alloys phase data, magnesium alloys diagram and so on. At the same time, further research would be made to construct models on the correlation between composition and properties by using semi-empirical methods based on the magnesium alloys database system. Fig. 5 is the calculated results for effects of Y additions on the mechanical properties of Mg-Zn-Y alloys.

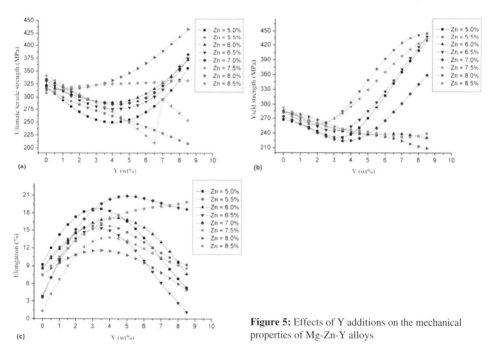

Figure 5: Effects of Y additions on the mechanical properties of Mg-Zn-Y alloys

7 Products of Wrought Magnesium Alloys

Although the application of wrought magnesium alloys in China is still very limited, a lot of new products made of wrought magnesium alloys have been produced and manufactured in laboratories or in factories. Fig. 6 is some products made of wrought magnesium alloys, which were developed recently. Among them, the production of a hollow thin-wall section parts with big size is very difficult. A lot of work is being carried in China to enlarge application of wrought magnesium alloys in industry and to seek higher production speed and lower cost. The most potential application fields are still in transportation tools and 3C products in China.

Figure 6: Part of wrought magnesium alloy products in China

8 Summary

The R&D activities on wrought magnesium alloys have increased obviously in the past five years in China in order to increase understanding of wrought magnesium alloys and related processing technologies and to enlarge the application fields in industry. The central government and local governments gave an increasing support to these activities in China's 10th five year plan. More attention for development of new types of wrought magnesium alloys is paid to additions of rare earth elements. Some new processing technologies, such as two-step ECAE process, unidirectional bending process and warm deformation technology, have been developed. The thin strip casting technology is near to industrial production. The database prototype of magnesium alloys developed has been used to predict the effects of process parameters and alloy compositions on the mechanical properties of magnesium alloys. A lot of work is being carried out for higher production speed, lower production cost and wider application of wrought magnesium alloy products. However, little work has been done on forging of magnesium alloys in China, which is another potential process method for components.

9 Acknowledgements

The present work was supported by both the High-tech Research and Development Program of China (863) (No.2001AA331050) and Chongqing Science and Technology Commission.

Wrought Mg Alloy for Civil Aircraft Application – A Process Chain Approach

M. Kettner[1], U. Noster[1], H. Kilian[1], R. Gradinger[1], W. Kühlein[2], A. Drevenstedt[3], F. Stadler[4], E. Ladstaetter[5], A. Lutz[5]

[1]ARC Leichtmetallkompetenzzentrum Ranshofen GmbH (LKR), Ranshofen, [2]AMAG extrusion, Ranshofen, [3]SAG Euromotive GmbH, Ranshofen, [4]AHC-Oberflächentechnik, St. Pantaleon, [5]Fischer Advanced Composite Components, Ried.

1 Introduction

1.1 Overview

In the automotive sector public concern with respect to the environment and pollution has motivated agreements between governments and industry, which aim to reduce the output of environmentally-harmful substances [1, 2, 3].

Original equipment manufacturers hope to achieve this objective via the following strategies [4]:

- Increase engine efficiency.
- Reduce friction in the drive train.
- Reduce weight.
- Improve aerodynamic body design.

Efforts to reach these target need enhancement because end users require more safety and comfort in cars, inevitably generating weight increases [5].

Probably the best indication of the value of weight reduction is the amount of money a certain industry sector is willing to spend to achieve it. In 2005, a weight gain of one kilogram cost about US$6 in the automotive industry, and about US$600 in the civil aircraft industry. In military and space applications it reportedly cost about US$6,000 and $60,000, respectively [6].

Because of this financial incentive strategies have been developed aiming at light-weight design [7]:

- Increase the load capacity of structures by improving the distribution of forces and the shape at the lowest possible material need.
- Substitute one material with another.
- Combine different materials to reach a needed stiffness and load capacity at reduced weight.
- Systematically select different components with optimized adoption to the system.

Comparing only the densities of prominent structural materials, Mg (1.8 g/cm³) seems to allow a considerable weight reduction compared to Al (2.7 g/cm³) or steel (7.8 g/cm³). However, density is not the only parameter relevant for the design of structural parts. The Young´s modulus, i.e. where higher values indicate higher stiffness of the according material, is 45 GPa for Mg, 69 GPa for Al and about 200 GPa for Fe. At first glance this comparison may not seem advantageous for Mg. However, slight modifications in geometry can improve a section modulus.

Therefore, as shown below, in comparison with Al a lower weight can be achieved at the same stiffness.

1.2 Mg Cast Products

Mg casting alloys convinced the automotive industry. This is reflected in an increase in demand for die cast Mg ingot from about 60,000 t in 1995 to 152,000 t in 2004 [8]. It must be noted that in 2004 about 85 %, i. e. 127,500 t, of the die castings went into automotive applications [4].

In the aircraft industry Mg castings can be found predominantly in the military sector. Among only a few civil applications, they are used in helicopters [9]. Up to now Boeing and Airbus have not accepted structural Mg parts in their planes at all due to a high corrosion tendency and a flammability risk. Civil aircrafts do not seem to offer opportunities for the implementation of Mg castings. For example, a Boeing 757 contains only 1 % castings, but 28 % extrusions, 20 % sheets, 18 % forgings, 16 % plates and 13% bars. It should also be noted that the most prominently applied material in the Boeing 757 is Al, with 78 %, as compared to steel with 12 %, Ti with 6 %, composites with 3 % and other materials with 1 % [10].

1.3 Mg Wrought Products

Considering the overall share of wrought alloys in the Boeing 757, the situation for Mg wrought products seems to be advantageous when compared to Mg cast alloys. The starting situation is further improved by the large amount of Al, a metal which, at least for certain applications, would be a candidate for a substitution by Mg.

However, Mg wrought alloys only account for about 1 % of total annual Mg production and are not applied in the automotive or aircraft areas in relevant quantities [11, 12]. Improving this situation is a challenge, as the technologies for the production of Mg wrought alloys are considered extremely underdeveloped. Requirements to be met are the development of [13]

- new alloys with improved corrosion and mechanical properties
- new production technologies
- reliable protective systems

Prejudices in the aircraft industry with respect to Mg also have to be overcome.

1.4 Considerations with Respect to the Alloy

The corrosion behaviour and flammability of Mg are prominent arguments against its use in aircraft applications [10]. According to [14] corrosion protection has to be taken into consideration for all Mg applications.

Another severe handicap of Mg alloys is the increased anisotropy of yielding behaviour due to the strong texture generated by the fabrication process (extrusion or rolling): values for tensile yield stress can differ markedly from those of compressive yield stress [15, 16]. This deformation anisotropy may be reduced by retaining a fine grain size in the extruded material, which also improves the mechanical properties [17, 18].

Extrusion at elevated temperatures is accompanied by dynamic re-crystallization and grain growth. To allow the necessary process temperature and avoid large grains, grain growth must be retarded by the presence of second-phase particles [19].

It is well known that the addition of zirconium (Zr) results in retarded grain growth [20, 21]. Zr additions to Mg alloys provide good nucleant substrates and enhance constitutional undercooling, which restricts dendrite growth and facilitates further nucleation [22].

In certain industries additional factors need to be considered. For the Zr-containing alloy ZK60 the extrusion speed is too low to allow economical production, at least for automotive applications. Even AZ31, the most widely-used Mg wrought alloy, with good extrudability, achieves at the maximum only half of the extrusion speed of the Al alloy 6063. Therefore industry is striving to develop alloys which provide a good balance of strength, ductility, extrudability and corrosion resistance [23].

1.5 Current Production Technologies

In the following only DC casting of billets and the extrusion process are addressed.

1.5.1 Direct Chill Casting

The input material for the extrusion process, the billet, is cast in the direct chill (DC) casting process. There molten metal is poured into a ring-shaped mould open at top and bottom. The melt is cast from the top into the mould, solidifies partially and leaves the mould at the bottom. The heat is removed in two stages, the first being via contact with the water-cooled mould, the second via exiting of the water from the mould onto the billet [24].

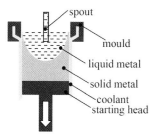

Figure 1: Principle of vertical direct chill casting

Two topics are of major importance in this process. Liquid Mg displays a high reactivity with the atmosphere which, in comparison to Al, necessitates more elaborate melt protection [25]. A mixture of liquid Mg and water carries a high probability of a strong exothermic reaction, posing a considerable safety hazard [26], which has to be considered in the cooling process.

Machines for direct chill (DC) casting of Mg can be multi-strand [25], though no publication is known which states a precise number of strands being cast simultaneously. Considering the demand for Mg wrought alloy, and the number of locations interested in casting DC products and their safety concerns, it may be assumed that no more than four strands are cast at any one time. Comparing this to the e.g. 48 strands that are cast in Al on one casting table alone [26], it may be appreciated that the conversion cost of one Mg billet is considerably higher than that of one Al billet.

1.5.2 Extrusion

Three technologies are available for the extrusion of Mg. In direct extrusion (Figure 2a) a plunger forces the heated billet through the mould. Indirect extrusion (Figure 2b), where the mould is driven over the billet, has the advantage over direct extrusion in that there is no relative movement between billet and container, thus reducing shear force and homogenizing material flow. However, at a given billet diameter, the achievable cross-section is smaller in indirect extrusion [26]. In hydrostatic extrusion (Figure 2c) the billet is placed in a container surrounded by a hydrostatic fluid, on which pressure is applied. Then the billet is forced through the mould without direct contact between billet and container, which again reduces friction and additionally lubricates. Therefore higher extrusion speeds are possible than in direct and indirect extrusion. However, for complex sections such as multi hollows hydrostatic extrusion is not a feasible alternative to conventional extrusion [28].

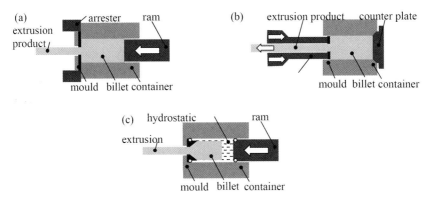

Figure 2: Principle of direct (a), indirect [26] (b) and hydrostatic [28] (c) extrusion

2 InnMag

2.1 Overview

The above introduction indicates that, from a material standpoint, Mg wrought alloys may provide a promising solution to weight reduction demands in the civil aircraft industry. From an economic perspective, too, Mg application seems feasible. The following issues, among others, still have to be addressed, however:

- Alloy: high strength, low anisotropy, retarding grain growth.
- Corrosion: to be prevented.
- Flammability: to be prevented.

This is being done within the framework of an Austrian-funded multi-firm project called „InnMag", where the process chain from alloying to assembly of parts for civil aircraft is demonstrated. The overall target of the project is to qualify two current Mg interior parts and demonstrate the feasibility of their manufacture. Table 1 lists the project partners and their respective areas of interest.

Table 1: InnMag project partners

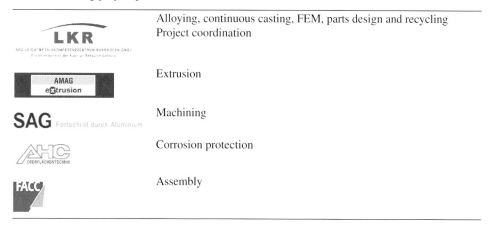

LKR	Alloying, continuous casting, FEM, parts design and recycling Project coordination
AMAG extrusion	Extrusion
SAG	Machining
AHC	Corrosion protection
FACC	Assembly

Three parts were identified by FACC together with a European aircraft manufacturer where a substitution of the current material, the Al alloy 6061, with a Mg alloy would be of interest. The parts are components of an overhead compartment (Figure 3). Two of them were chosen with the objective of increasing experience in extrusion (Figure 4a), and the other with the same aim in machining (Figure 4b) [30].

Figure 3: Overhead compartment

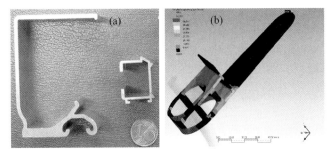

Figure 4: Selected parts: C-profile and PSU rail (a), yz-bracket (b)

2.2 Dimensioning/Design

Due to boundary conditions defined by adjacent parts, the possibilities for design change are rather limited. However, the following example shows a feasibility study with the aim of reaching the same stiffness for a magnesium part as for the already-used aluminium part, and simultaneously of gaining weight savings of 15 %.

In a first step the dimensioning of the C-profile was carried out analytically. The limits of the extrusion process were taken into account via simulation. As a final design check structural FEM simulation was performed. Selected results are shown in Figure 5.

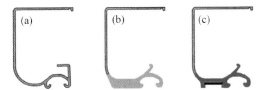

Figure 5: Mechanical dimensioning of C-profile: (a) Pure substitution, (b) 15 % reduction in mass, (c) hybrid construction

For the C-profile a pure substitution of the Al alloy with the Mg alloy resulted in a weight reduction of 35 % but also reduced the stiffness to the same degree. In order to achieve the same stiffness as for the current Al part, a hybrid construction was computed which resulted in a weight reduction of 20 % at a slightly increased stiffness in comparison to the Al solution.

For the second part, the yz-bracket, stress distribution according to the load situation was computed by FEM analysis. After a first run, where the Al alloy was substituted with the Mg alloy without any further changes, the expected density-related weight reduction of 35 % was achieved. Stresses and strength of the material at certain positions were analyzed and it was found that further weight reductions could be achieved by changing the design. Elongated holes were introduced and a gauge partially reduced (Figure 6). This resulted in a further weight reduction of 10 %. A second FEM analysis pointed out that stresses at load are still in the elastic area. Therefore substitution with another material and design optimization resulted in a part which is 41 % lighter overall than the original part.

Figure 6: Yz-bracket in Mg alloy with equivalent stresses at positions where material was reduced

2.3 Alloy Decision

AZ31 (Table 2) is an extensively-researched alloy that can be readily cast and extruded. Therefore it was decided to use it as reference material. The primary alloy, however, had to be chosen with respect to improvements in anisotropy and grain growth. As mentioned before, Zr-containing alloys are said to meet this requirement. According to [31], a Zn-based alloy containing Zr and RE elements (ZK10, Table 2) should display extrudability similar to that of AZ31, but better mechanical properties (Table 3). It was also assumed that the financial constraints in the automotive industry (see above) do not apply to the same extent to the aircraft industry.

Table 2: Composition of ZK10 [31] and AZ31 [33]; [32]

		Al [%]	Mn [%]	RE [%]	Zn [%]	Zr [%]
ZK10	min.			0.2	1.35	0.5
	max.		0.15	0.25	1.45	0.6
AZ31	min.	2.5	0.20		0.6	
	max.	3.5			1.4	

As Mg has not yet been applied to civil aircraft, no specific qualification criteria are available. Therefore the properties of the material to be replaced were taken as the benchmark. For mechanical properties a theoretical comparison revealed promising results (Table 3): at room temperature the mechanical properties of relevant Al and Mg alloys are comparable.

Table 3: Composition and mechanical properties of the alloys in question: 6061 and AZ31 according to [33], ZK10 according to [31]

	E [MPa]	r [g/cm³]	R_m [MPa]	$R_{p0.2}$ [MPa]	A [%]
6061-T4	69,000	2.7	228	131	22
6061-T6	69,000	2.7	290	255	12
AZ31B-F	45,000	1.77	260	200	15
ZK10-F	45,000	1.77	309	288	25.5

2.4 Melting

Zr displays certain benefits. However, when Zr is added to Mg melts the Zr particles dissolve to different extents depending on the initial Zr particle size and processing conditions [34]. The corresponding considerations and laboratory scale experiments are explained elsewhere [35]. Results from the laboratory scale experiments with respect to the severe grain refining effect of Zr are displayed in Figure 7. Prior to addition the grain size was ~240 μm. 15 minutes after addition the grain size was 75 μm, and ~80 μm after 135 minutes.

a) b)

Figure 7: Optical micrograph showing the microstructure of MgZnREMn alloy (a) before and (b) 15 minutes after addition of Zr master alloy in a laboratory scale experiment

2.5 DC Casting

The DC casting was carried out at the facilities of LKR. A detailed description of setup and process can be found in [35]. An analysis of the cast billets delivered an extremely homogenous distribution of grain size over casting length as well as over the cross-section of the billet for ZK10. The smallest grain size was 37 μm and the largest was 40 μm. For AZ31 a pronounced influence of the cooling condition could be detected: the grain size near the surface was 150 μm, whereas in the centre of the billet a grain size of 200 μm was identified.

a) b)

Figure 8: Optical micrographs showing the grain size near the cast surface for alloy ZK10 (a) and AZ31 (b) ((a) and (b) at the same magnification)

2.6 Homogenizing

As result of earlier experiments it was decided to apply no homogenizing treatment to AZ31 billets. For ZK10 a computational analysis was performed using Thermocalc to determine relevant temperatures. For the actual composition it was found that Mg-Zn phases, Mn, a Mg12RE phase and Zr would dissolve at 150, 260, 430 and 470 °C, respectively.

For ZK10 homogenizing trials were performed at 330 °C and 420 °C for 6, 12 and 24 hours respectively. No significant changes could be seen via the light microscope. Therefore the ZK10 billets, too, were not homogenized prior to extrusion.

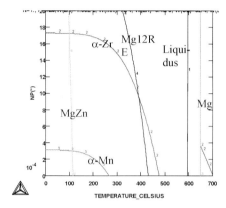

Figure 9: Thermocalc® computation according to Scheil

2.7 Extrusion

AZ31 and ZK10 billets were extruded to the C-profile at the facilities of AMAG extrusion. A direct 16 MN press and multi-strand tools were used (Figure 10). With a 6- and an 8-strand tool extrusion ratios of 1:55 and 1:41 were achieved. However, in the first industrial tests the surface quality was not equal for all strands. Therefore it was decided to commence with a two-strand tool and an extrusion ratio of 1:47.

To allow a better comparison of the mechanical properties with those of alloys used in other projects at LKR, ZK10 and AZ31 were also extruded to rods with a diameter of 30 mm. These tests were conducted at the Extrusion Research and Development Center, TU Berlin, at an extrusion ratio of 1:47. The mechanical properties achieved are shown in Table 4. Generally, for both alloys values for yield and ultimate tensile strength are higher and values for elongation are lower than mentioned in literature (compare with Table 3). However, it must be considered that mechanical properties are affected by the shape of the product and by the process conditions. The process parameters will be optimized in ongoing work. Therefore a direct comparison to other given values does not seem appropriate.

Figure 10: Extrusion of C-profile

Table 4: Mechanical properties of solid cylindrical samples; yield test according to EN 10002-1; compression test close to ASTM E9-89a

		R_m [MPa]	$R_{p0,2}$ [MPa]	A [%]
AZ31-F	tensile	318	270	14
ZK10-F	tensile	328	300	20

2.8 Surface Protection and Flammability of Mg

As mentioned earlier, the corrosion resistance and flammability of Mg are concerns of the aircraft industry. However, in the current application the material must also demonstrate a certain amount of wear resistance. To address the issues of corrosion and wear resistance a MAGOXID-COAT® [36] was applied to the extruded product (Figure 11a). The adequately coated parts passed a 96-hour salt spray test according to DIN 50021 SS. As demonstrated on other Mg parts, the coating also shows good performance in the Taber abrazer test.

With respect to the flammability concern, a test was carried out on a PSU rail manufactured from AZ31. Tests on an uncoated part and on a part coated with MAGOXID were performed. These were carried out closely to Federal Aviation Regulations (FARs) (Figure 11b). A burner was moved below the part and held at the position for a specified amount of time. Neither the coated nor the uncoated part ignited, and therefore they passed the test. To gain an understanding of the flammability limits a stronger and not required second test was performed. A thermocouple was placed in an uncoated PSU rail which then was heated with welding torches. Up to a temperature of 500 °C no ignition occurred.

These results meet the expectations of the project partners and indicate that flammability does not hinder Mg application in aircraft.

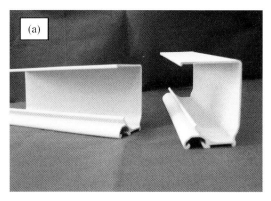 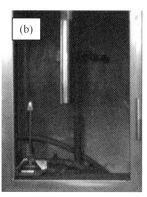

Figure 11: a) Surface and wear protection, b) flammability test

3 Summary

In the project InnMag partners from industry and research are working together to show the possibility of using Mg extrusions in the interiors of civil aircraft. Partners in the funded project are located at the process chain, from alloy development to interior parts assembly, and have experience in the same line of work for Al extrusions. Planned result of the project is the qualification of two Mg interior parts for civil aircraft application.

The feasibility of processing one Mg part was demonstrated by all project partners. In this contribution, the results of casting and extrusion in particular were shown:

- In comparison to AZ31 alloy ZK10 displayed an extremely fine grain size in the as-cast condition
- AZ31 and ZK10 could be readily extruded by the InnMag consortium.
- The yield point for alloy ZK10 was higher than expected from literature, and about 10 % higher than that of alloy AZ31 processed via the same route as alloy ZK10.
- Ultimate and tensile strength for both AZ31 and ZK10 in an as-extruded condition are higher than for 6061-T4, which is the material currently used for the application.
- Mechanical properties for ZK10-F are better than for 6061-T6.
- The substitution of an Al alloy with a Mg alloy allows reduction of weight at a stiffness comparable to the Mg alloy.
- Corrosion of Mg can be disregarded for the application in question if an appropriate surface coating is applied.
- Ignition of Mg application in question does not occur at temperatures up to 500 °C.

4 Acknowledgements

The authors would like to thank the program "Take Off" of the Austrian bmvit and FFG for funding.

5 References

[1] Communication from the Commission to the Council and the European Parliament, COM(95) 689 final, Brussels, 20.12.1995.
[2] Commission Recommendation of 5 February 1999 on the reduction of CO_2 emissions from passenger cars (1999/125/EC).
[3] Assembly Bill No. 1493, Approved by Governor July 22, 2002.
[4] J. Willekens, Presentation at Seminar Magnesium, Anwendungen, Potentiale, May 2005, Geesthacht.
[5] K.-H. von Zengen, MP Materialprüfung, 2005, 47, 11–12.
[6] N. Moskovitch, in Proceedings IMA Annual Conference Berlin, May 2005.
[7] H.-G. Haldenwanger, U. Walther, in Proceedings „Verbundwerkstoffe und Werkstoffverbunde", Chemnitz, Sep. 2001.
[8] D. Webb, in Proceedings IMA Annual Conference Berlin, May 2005
[9] www.magnesium-elektron.com.
[10] Wendt, K. Weiss, A. Ben-Dov, M. Bamberger, B. Bronfin, in Proceedings Magne-sium Technology 2005, TMS 2005, 269–273.
[11] D. Wieser, in Seminar Fortschritte durch neue Werkstoffanwendungen, March 2003, Munich.
[12] S. R. Agnew, JOM, May 2004, 20–21.
[13] Aghion. B. Bronfin, Materials Science Forum, 2000, 350–351, 19–28.
[14] MIL-HDBK-5H, Dec. 1998
[15] U. Noster, B. Scholtes, HTM 58 (2003), 322–327.
[16] U. Noster, B. Scholtes, Zeitschrift für Metallkunde 94 (2003), 559–563.
[17] D. Letzig, J. Swiostek, J. Bohlen, K. U. Kainer, in Proceedings Magnesium Technology 2005, TMS 2005, 55–59.
[18] C. Davies, M. Barnett, JOM, May 2004, 22–24.
[19] C. J. Bettles, M. A. Gibson, JOM, May 2005, 46–49.
[20] M. R. Barnett, D. Atwell, C. Davies, R. Schmidt, in Proceedings 2nd International Light Metals Conference 2005, St. Wolfgang, 161–166.
[21] E. F. Emley, Principles of Magnesium Technology, Pergamon Press, London, 1966.
[22] M. Easton, F. Pravdic, R. Schmidt, in Proceedings 2nd International Light Metals Conference 2005, St. Wolfgang, 69–74.
[23] A. Luo, A. K. Sachdev, in Proceedings Magnesium Technology 2006, TMS 2006, 333–339.
[24] F. Pravdic, D. Leitlmeier, C. Wögerer, A. Sigmund, in Proceedings 2. Ranshofener LMT 2002, 150–159.
[25] Magnesium Direct Chill Casting: A Comparison with Aluminium, P. W. Baker, P. T. McGlade, in Proceedings Light Metals 2001, TMS 2001, 855-862.
[26] Magnesium Taschenbuch (Ed.: C. Kammer), Aluminium-Verlag, Düsseldorf, 2000.
[27] www.wagstaff.com.

[28] J. Bohlen, J. Swiostek, W. H. Sillekens, P.-J. Vet, D. Letzig, K. U. Kainer, in Proceedings Magnesium Technology 2005, TMS 2005.
[29] J. Bohlen, D. Letzig, K. U. Kainer, W. H. Sillekens, P. J. Vet, in Proceedings 12th Magnesium Automotive and End User Seminar, Sep. 2004.
[30] W. Billinger, G. Haider, E. Ladstätter, R. Gradinger, in Proceedings 3. Ranshofener Leichtmetalltage, 2004, 117–129
[31] W. Sebastian, K. U. Kainer, H. Haferkamp, P. Juchmann, German Patent 199 15 276 A1, 1999.
[32] All compositions in this work are expressed in wt.%.
[33] www.matweb.com.
[34] D. StJohn, M. Easton, in Proceedings 2nd International Light Metals Conference 2005, St. Wolfgang, 63–68.
[35] M. Kettner, F. Pravdic, W. Fragner, K. U. Kainer, in Proceedings Magnesium Technology 2006, TMS 2006, 133–138.
[36] www.ahc-oberflaechentechnik.de.

The Dynamic Recrystallization of Commercially Pure Mg and Three Mg-Y Alloys

R. Cottam*, J. Robson*, G. Lorimer* and B. Davis[+]
*School of Materials, University of Manchester, Manchester, U.K.
[+] Magnesium Elektron Ltd., Manchester, U.K.

1   Introduction

The texture of magnesium and magnesium alloys plays a significant role in formability due to the plastic anisotropy of the unit cell. The classical basal texture produced as a result of hot rolling has been shown to exhibit poor forming characteristics [1]. Equal channel angular extrusion performed at elevated temperatures has shown to produce excellent tensile ductility as a result of a combination of the texture and fine grain size [2, 3] (the industrial application of this process is still in its infancy). For both of these thermo-mechanical treatments dynamic recrystallization occurs during processing. The textures that form as a result of the recrystallization of Mg alloys and other hexagonal crystal systems has received attention in the literature [4–8]. These papers have shown that oriented nucleation of recrystallization is responsible for the formation of the recrystallized textures and that the texture forms as a result of the operative deformation mechanisms. It was also shown that initial texture relative to the application of the applied stresses and strain plays a role in texture formation for highly textured magnesium alloys. It is well known that solute elements play a role in how dynamic recrystallization occurs [9, 10] particularly by effecting boundary mobility, but their role on texture formation is not well understood. Solute elements also effect the activity of different slip systems in magnesium alloys [11, 12] which may also influence the resulting texture. Understanding the relationship between the mechanisms of deformation, the role that the initial texture plays, the influence of solute additions and the subsequent texture that forms as a result of dynamic recrystallization, for magnesium alloys may allow new textures to be designed.

2   Experimental Procedure

Three binary Mg-Y alloys and commercially pure Mg were prepared by Magnesium Elektron at the Manchester site with the following compositions: Mg: 99.9 wt%Mg; Mg-0.23 wt%Y; Mg-0.84 wt%Y; Mg-2.71 wt%Y. The alloys were produced as billets 7.5 cm diameter, solution treated at 525 °C for 8 h and extruded into 3 cm diameter bar at 390 °C. The initial textures of the extruded billets were measured by EBSD using a Philips XL30 FEG SEM with a spot size of 4, working distance of 20 mm and an accelerating voltage of 20 kV. The extruded grain size was determined using the linear intercept method.

Channel die samples 10 mm × 11 mm × 12 mm were cut from the bar with the compression direction parallel to the extrusion direction. The samples were coated in powdered graphite which acted as a lubricant during the deformation. An initial strain rate of $1 \cdot 10^{-3}\,\mathrm{s}^{-1}$ was employed to deform the samples. The addition of yttrium increased the temperature at which dynamic recrystallization occurred and different test temperatures were used for each alloy, Table 1. For each test temperature and alloy, samples were compressed to true strains of 0.1, 0.3, 0.6 and 0.9 to evaluate the development of microstructure.

Table 1: Test temperatures used during channel die compression of pure magnesium and the three Mg-Y alloys

AlAlloy	Temperature (°C)		
Mg	20200	25250	30300
W Mg0.23wt%Y		25250	30300
W Mg0.84wt%Y			40400
Mg2.71wt%Y			45450

The deformed samples were cut ground and polished. The samples were etched using a solution of 5 g picric acid, 100 mL ethanol, 10 mL water and 5 mL acetic acid. For samples compressed to a true strain of 0.9 the macro-texture and micro-texture was measured by EBSD using the same parameters as the characterization of the extruded texture.

3 Results and Discussion

3.1 As-Extruded Alloys

After extrusion Mg had a grain size of 57 μm, Mg 0.23 wt%Y 51 μm, Mg 0.84 wt%Y 24 μm and Mg 2.71 wt%Y 21μm. In all three alloys the grain structure was fully dynamically recrystallized during extrusion. All alloys and the pure magnesium show a typical extruded magnesium texture [13] where the basal poles are radially distributed perpendicular to the extrusion axis. Increasing the yttrium concentration from 0.23to 2.71 wt%Y not only refines the grain structure of the material during dynamic recrystallization (DRX) that occurred during extrusion but also weakened the extruded texture.

3.2 Mechanical Testing

The true stress – true strain response during channel die deformation is shown in Figure 1. All curves initially show hardening followed by a region of steady state deformation and then some secondary hardening up to the highest true strain used in the test (0.9). It is notable that in both the pure magnesium and alloy specimens there is little evidence of the strain softening that often accompanies dynamic recrystallization [14], Figure 1. However, this softening is also in competition with hardening associated with dislocation accumulation, twin formation and texture changes. For the initial texture and deformation conditions employed here, this appears to offset the softening due to DRX, and gives the flat response observed in the stress strain curve. Once DRX has refined the grain structure and twinning has reoriented the material into a new orientation, work and texture hardening begin again to dominate the stress strain response and the strain hardening exponent increases (secondary hardening).

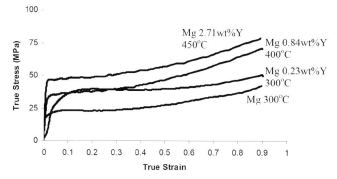

Figure 1: True stress - true strain response for the three Mg-Y alloys and pure magnesium

3.3 Optical Microscopy

The microstructures were observed at various levels of plastic strain. After 0.1 strain and at all temperatures the pure magnesium and Mg 0.23 wt%Y showed extensive deformation twinning (Figure 2), a result consistent with the texture and the direction of the applied stress [13]. Mg 0.84 wt%Y showed significantly less twinning and Mg 2.71 wt%Y showed no twinning even after a true strain of 0.3, Figure 2. After a strain of 0.3 all alloys showed dynamic recrystallization was well advanced (Figure 2), which corresponded with the steady state region of the deformation (Figure 1). From a true strain of 0.3 to 0.9 the fraction of dynamically recrystallized grains increased until full recrystallization had been achieved.

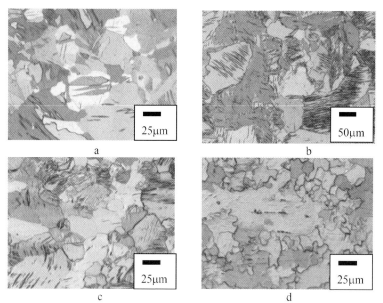

Figure 2: Optical micrographs after channel die compression; a – pure magnesium compressed to a true strain of 0.1 at 250 °C; b – Mg 0.23 wt%Y compressed to a true strain of 0.1 at 250 °C; c – Mg 0.84 wt%Y compressed to a true strain of 0.1 at 400 °C; d – Mg 2.71 wt%Y compressed to a true strain of 0.3 at 450 °C

3.4 Macroscopic Texture

Figure 3 shows the (0002) pole figures as a result of channel die deformation of pure Mg and the Mg-Y alloys. A range of recrystallization textures have been produced as a result of channel die deformation. Mg deformed at 300 °C and Mg 0.84 wt%Y exhibit the classical basal texture, Figure 3. The formation of this texture has been attributed to formation of the $\{10\bar{1}2\}$ tensile twins reorienting the basal poles of the parent grains from perpendicular to the compression direction to approximately parallel to the compression direction followed by dynamic recrystallization at the twins [5]. This is consistent with the observed microstructural development.

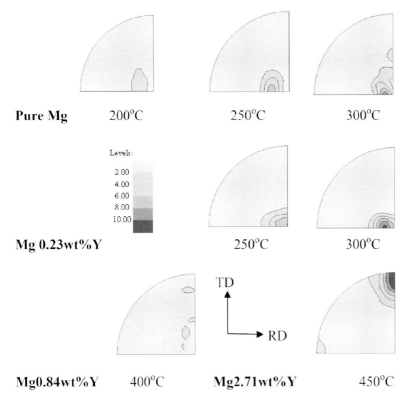

Figure 3: Pole figures after channel die deformation for the alloys and pure Mg at a range of temperatures

Pure magnesium deformed at 200, 250 °C and 300 °C show a split of the basal texture in the rolling direction. Splitting of the basal texture in the „rolling direction" (the unconstrained direction in the channel die test) is also observed. This type of split in the texture is commonly observed in rolled magnesium alloys. Texture modelling shows that this split may be attributed to the operation of the <c+a> slip mode in magnesium alloys [11]. This deformation mode is expected to become significantly active in pure magnesium at around 200 °C [15], the lowest temperature used in this work, and is expected to be enhanced by the addition of Y [11].

The highest yttrium alloy (2.71 wt%Y) shows a completely different texture, with strong alignment of the basal poles in the transverse direction. In this alloy twining was not observed and there is no reoreientation of basal poles by twinning along the compression axis. Since the initial texture is highly unfavourable for basal slip during channel die compression, and no twinning occurs, other deformation mode must be operative. Yttrium is known to enhance the activity of both the prismatic and pyramidal (<c+a>) slip modes [11]. The observed texture in the high yttrium alloy can be explained by operation of these slip modes as the primary deformation mechanisms, rather than twinning. The lack of twining means that, unlike in the pure Mg and 0.23 wt%Y alloys, the basal poles are not strongly reoriented towards the compression axis but remain concentrated around the TD-RD plane (rotated towards TD). This texture has been shown to be stable in other hcp metals such as zinc[7]. The slow strain rate of testing will decrease the likelihood of deformation twinning. Under rolling conditions, where twinning is likely to occur, this texture is unlikely to be observed.

After channel die deformation Mg 0.84 wt%Y exhibits a range of orientations from the basal texture to the texture produced by deformation of Mg 2.71 wt%Y, Figure 3. Micrographs of the evolution of microstructure with strain, reveal that initially the amount to deformation twinning is lower than for pure Mg and Mg 0.23 wt%Y. Therefore the twinned regions that dynamically recrystallize can explain the basal orientations. The other orientations may be a result of a shift in the ratio of the critical resolved shear stresses for basal, prismatic and pyramidal slip [4] as a result of the solute addition. An addition of 0.84 wt%Y appears to be close to the critical addition required for the marked change in deformation behaviour.

3.5 Microscopic Texture

Figure 4 shows the Euler maps with the unit cell direction of each grain superimposed on teach grain for pure magnesium and two of the Mg-Y alloys. The Euler map for Mg 0.84 wt%Y deformed at 400 °C, Figure 4b, shows that when dynamic recrystallization occurs at a twin it results in the recrystallized grains forming in an orientation significantly different to the parent grain.

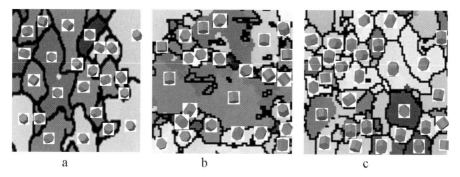

Figure 4: Euler maps after channel die deformation to a true strain of 0.9; a – pure Mg deformed at 200 °C; b – Mg 0.84 wt%Y deformed at 400 °C; c – Mg 2.71 wt%Y deformed at 450 °C

The three Euler maps show that neighbouring dynamically recrystallized grains do not necessarily share an orientation that is similar to the macroscopic texture. Further studies are required to clarify whether oriented nucleation is responsible for the texture or the contribution of the fine grains to the macroscopic deformation, as identified by Jonas [16]. This is part of our ongoing work in this subject.

4 Conclusions

Three new deformation textures have been produced as a result of channel die deformation of pure Mg and the three Mg-Y alloys. In pure magnesium and alloys with low Y content (0.23wt%Y) a basal texture with a split towards the rolling direction is observed, resulting from a combination of twinning and <c+a> slip, consistent with previous observations in the highest Y alloy. For Mg2.71wt%Y, no twining occurred and a completely different texture is therefore formed, with the basal poles aligned in the TD direction. The rotation of initial to this orientation was attributed to <c+a> slip and prismatic slip. Mg0.84wt%Y exhibited a texture that appears to be a transition between Mg0.23wt%Y and Mg2.71wt%Y. This suggests that in this alloy, the critical resolved shear stress for non-basal slip modes and twinning are similar.

5 References

[1] Agnew, S.R., Duygulu, O., International Journal of Plasticity, 2005, 21, 1161–1193.
[2] Koike, J., Kobayashi, T., Mukai, T., Watanabe, H., Suzuki, M., Maruyama, K.,Higashi, K., Acta Materialia, 2003, 51, 2055–2065.
[3] Agnew, S.R., Horton, J.A., Lillo, T.M.,Brown, D.W., Scripta Materialia, 2004, 50, 377–381.
[4] Agnew, S.R., Mehrotra, P., Lillo, T.M., Stoica, G.M.,Liaw, P.K., Acta Materialia, 2005, 53, 3135–3146.
[5] Xiong, F.,J., Davies, C.H., in Magnesium Technology 2005, TMS 2005. 217–222
[6] Barnett, M.R., Materials Science Forum, 2003, 419–422, 503–508.
[7] Solas, D.E., Tome, C.N., Engler, O.,Wenk, H.R.., Acta Materialia, 2001, 49, 3791–3801.
[8] Yi, S.B., Zaefferer, S., Brokmeier, H.-G., Materials Science and Engineering A, 2006, 424, 275–281.
[9] Suehiro, M., Liu, Z., Agren, J., Metallurgical and Materials Transactions A, 1998, 29, 1029–1034.
[10] Le Gall, R., Acta Materialia, 1999, 47, 4365–4374.
[11] Agnew, S.R., Yoo, M.H., Tome, C.N., Acta Materialia, 2001, 49, 4277–4289.
[12] Akhtar, A.,Teghtsoonian, E., Acta Metallurgica, 1969, 17, 1339–1349.
[13] Barnett, M.R., Keshavarz, Z., Beer, A.G.,Atwell, D., Acta Materialia, 2004, 52, 5093–5103.
[14] Barnett, M.R., Atwell, D.,Beer, A.G., Materials Science Forum, 2004, 467–470, 435-440.
[15] Mathis, K., Nyilas, K., Axt, A., Dragomir-Cernatescu, I., Ungar, T.,Lukac, P., Acta Materialia, 2004, 52, 2889–2894.
[16] Jonas, J.J., Toth, L.S., Scripta Metallurgica et Materialia, 1992, 27, 1575–1580.

Data Correction for Constitutive Analysis of Wrought Magnesium Alloys Based on Compression Tests

F.A. Slooff, J.P. Boomsma, J. Zhou, J. Duszczyk, L. Katgerman
Delft University of Technology, Delft, The Netherlands

1 Introduction

Numerical modelling is nowadays a powerful tool and can be used to characterize material behavior under specified deformation conditions efficiently. The trustworthiness of the output from modelling depends to a large degree on the reliability of input data. The material data necessary for modelling can be obtained from basic material testing (compression being the most common one) over a wide range of temperatures and strain rates for a consistent representation of the material behavior. As far as industrial deformation processes such as forming and extrusion is concerned, very limited data have been reported on strain rate behavior of wrought alloys. Especially problems occur for deformation data of high strain rate > 10/s typical for industrial presses, since not every testing device is suitable for achieving these high strain rates. Fortunately, the state of the art Gleeble 3800 system used for the current research is capable of applying strain rates up to 100/s. Not only the wide range of strain rates that can be used for experiments makes this machine unique; also the accurate temperature working range does.

Previous studies have shown the usefulness of compression tests for modelling [1, 2] and others use similar tests to characterize the material [3, 4, 5]. Often the data gathered from testing cannot be used straightforward for further analysis, since usually isothermal data are required. Especially at higher deformation rates, (adiabatic) deformation heating becomes an issue and needs to be accounted for. However, only rarely some limited information is provided to explain what kind of data correction has been performed. Therefore, in the current work the data correction method after state of the art uniaxial compression tests is explained, since for a proper constitutive analysis, correct input data is of vital importance. In the present research, uniaxial hot compression tests were performed on two magnesium wrought alloys over a wide range of temperatures and strain rates aiming at characterizing its deformation behavior, even at high strain rates. The two alloys are AZ61 and AZ80, which mainly differ in aluminum content, resulting in a different β-phase content. By incorporating two different alloys in the research, the corrections applied are not specifically alloy based.

2 Experimental Details

The as-received material was a magnesium alloy containing approximately 4.5 wt% Al, 1 wt% Zn and 0.3 wt% Mn in the as-cast condition. In order to have this material in the pre-extruded state for further investigation, it was extruded at 350 °C at a constant ram speed of 2 mm/s using a die with a round opening, giving an extrusion ratio of 9.4. The extruded rod with a diameter of 15.8 mm was machined to cylindrical specimens of 12 mm in length and 10 mm in diameter. These specimens were used in the Gleeble 3800 system for uniaxial hot compression tests.

The Gleeble 3800 system is capable of performing uniaxial hot compression tests at various strain rates from 0.01 to 100/s over a wide temperature range, from room temperature to the melting temperature of the magnesium alloy. To reduce the effect of friction, both anti-seizure nickel paste and graphite foil were applied. Thermocouples were welded on the specimens for temperature control and heating of the specimen was performed via resistance heating. For these experiments, true strain rates of 0.01, 0.1, 1, 10 and 100/s were chosen at temperatures of 250, 300, 350, 400, 450 and 500 °C. The specimens were deformed until a true strain of one was achieved, after which air quench was applied.

3 Results

The data gathered from the compression tests were analysed and corrected for time shift, load cell ringing and deformation heating, to be suitable for constitutive analysis. The corrections were required, since the data could not be directly used for immediate constitutive analysis since it would greatly affect the outcome. In the next section the corrections are explained.

3.1 Time Shift

First of all, it was apparent that some time shift occurred during the tests; different channels were not synchronised and therefore correction was required. For example, the registered force increased before any displacement occurred by the machine, as shown in Figure 1. Since the jaw is the moving part of the experimental setup and the force is a result of this motion, the force should not change before the actual deformation was initiated. The effect of the time shift was noticeable especially at high a strain rate, where the flow stress displayed a clear drop before the test ended. This was a clear result of the time shift between the force and displacement of the jaw.

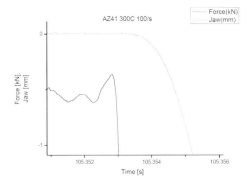

Figure 1: Example of time shift as recorded during experiments

In order to correct for this phenomenon, the force and jaw are plotted as shown in the above Figure 1 and the force is shifted in time to correspond to the beginning of the deformation. The shifted force is than used to recalculate the true flow stress. By correcting for the synchronization, the flow stress displayed normal behavior and the drop at the end of the test in the original data is removed.

3.2 Load Cell Ringing

Another item required attention was the observed oscillation in the flow stress. This oscillation was more prone in the higher strain rate tests and was first thought to be caused by microstructural changes in the material. An example of this oscillation is given in Figure 3 by the thin black line. However, close examination of the different flow stress curves showed a specific oscillation frequency and could be related to the experimental setup. The manufacturer of the equipment confirmed [6] the occurrence of load cell ringing at higher strain rates (~100/s), since some significant mass needs to be accelerated and stopped within a fraction of a second. This kind of deformation cause the system shaft to vibrate, excited by the impact of the test. Since the load cell is attached to the shaft, the load cell signal displays an oscillation pattern. Smoothening can be obtained by making use of equation (1):

$$f_0 = \frac{1}{n\Delta t} \qquad (1)$$

In this equation, f_0 is the resonance frequency of the system, n the number of data points and Δt is the time interval between two adjacent data points. With this Fast Fourier Transformation (FFT) filtering technique, the load cell ringing imposed on the load cell signal can be removed.

It appeared that using the right data processing software is adequate to reduce the effect of load cell ringing. The Fast Fourier Transformation (FFT) filter is embedded in the Origin software and is recommended to use for smoothening, since it is a quick and convenient way of applying the desired correction.

3.3 Deformation Heating

With increasing deformation rate, more energy is put into the material in a short amount of time, causing an increase of temperature of the specimen. Since isothermal stress-strain data is required for constitutive analysis of the two alloys, the stresses need to be corrected for the difference in temperature throughout the test. Due to the shortening of the test at higher strain rates, thermocouples become unreliable since their response time influences the correct registration of the temperature in regard of amount of deformation. Therefore, other means are needed to obtain the temperature of the specimen during testing.

It appeared that for tests with strain rates of 10/s and up, the deformation heating becomes adiabatic [7]. By introducing equation (2) it is possible to calculate the temperature rise during testing with the aid of the measured flow stress.

$$\Delta T = \frac{\left(\eta(0.9 \sim 0.95)\int \sigma \mathrm{d}\varepsilon\right)}{\rho C_p} \qquad (2)$$

In this equation, η is the adiabatic factor and can be adjusted at strain rates lower than 1/s. Since the temperature calculation is required for higher strain rates, the adiabatic factor is 1; the deformation is completely adiabatic. Furthermore, an efficiency factor can be adjusted between 0.9 and 0.95. The other input parameters are the area of the stress-strain curve, the density ρ and the specific heat C_p.

The measured temperature at strain rates below 1/s displayed some variation during the experiments, and therefore temperature correction for low strain rate was also required. Using all the measured temperatures of the experiments at strain rates of 1/s and below and the calculated temperatures at strain rates of 10/s and above, the stress can be plotted against the reciprocal temperature. For low strain rates the plots of $\ln\sigma$ versus $1000/T$ (measured temperature) were linearly fitted in order to interpolate the stresses at the preset temperatures. For high strain rates, the plots of $\beta\sigma$ versus $1000/T$ (measured temperature) were used to interpolate the stresses. Since the data consist of time-stress-strain-temperature information, each plot needs to be made at similar strain and therefore, multiple plots are necessary. Figures 2a and 2b show two graphs with the temperature correction at low and high deformation rates.

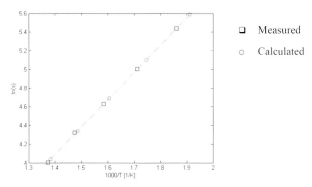

Figure 2a: Linear fit of $\ln(\sigma)$ vs $1000/T$, used at strain rates of 1 s^{-1} and lower to correct flow stresses for heat fluctuations during deformation at a strain of 0.205

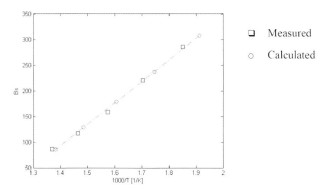

Figure 2b: Linear fit of $\beta\sigma$ vs $1000/T$, used at strain rates of 10 s^{-1} and greater to correct flow stresses for adiabatic heat effects during deformation at a strain of 0.205

With the isothermal stresses, new true stress-true strain graphs can be made for each experiment. Figure 3 shows the impact of correcting for the time shift, oscillations and temperature on the stress-strain behaviour. These data can now be used for further constitutive analysis to obtain the Zener-Hollomon parameter.

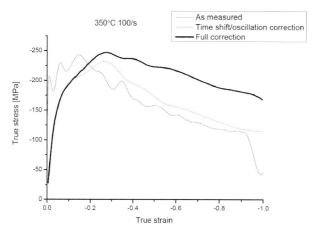

Figure 3: Stress versus strain at 350 °C and 100 s^{-1} with and without corrections

Constitutive equations are used in hot working to calculate the stresses for deformation at set strain rates. In the 1960s an equation was introduced that includes both equations in the limits. This equation (4) uses the Zener-Hollomon parameter Z, which combines two control variables $\dot{\varepsilon}$ and T in the hot working process.

$$A[\sinh(\alpha\sigma)]^n = \dot{\varepsilon}\exp\left(\frac{Q}{RT}\right) = Z \qquad (3)$$

In this equation, α is the stress multiplier, an additional adjustable constant which brings $\alpha\sigma$ into the correct range to make constant T curves in $\ln[\dot{\varepsilon}]$ versus $\ln[\sinh(\alpha\sigma)]$ plots linear and parallel. A and n are constants, R is the universal gas constant and Q the activation energy. With the correction of the stress due to temperature changes during deformation, it is possible to obtain constant T lines for a determination. Plotting $\ln[\sinh(\alpha\sigma)]$ versus $\ln[\dot{\varepsilon}]$ for the available temperatures, provided that the spacing between the measured points is homogeneous. Plotting at fixed strain provides the best results for a when the standard deviation of the combined direction coefficient of the $\ln[\sinh(\alpha\sigma)]$ versus $\ln[\dot{\varepsilon}]$ plots at different temperatures is as small as possible.

4 Conclusions

- All the corrections made can be applied to both AZ61 and AZ80 and the calculated flow stress can be used for further constitutive analysis. Therefore, the method does not need adjustment for each different alloy.
- Time shift between applied deformation and recorded force resulted in oddly shaped stress-strain curves for AZ61 and AZ80 alloys. Shifting the force in respect to the displacement in the time dimension resulted in proper stress-strain curves for higher strain rates at different temperatures.

- Load cell ringing caused by system shaft excitation at high strain rate deformations of AZ61 and AZ80 can successfully be corrected by applying an FFT filter, such is embedded in data processing software.
- Adiabatic deformation heating can be calculated from measured stress-strain data and can be used to calculate isothermal stress-strain data for both AZ61 and AZ80.
- Small temperature fluctuations during deformations at strain rates of 0.0–1/s can be accounted for to obtain isothermal stress-strain data for both AZ61 and AZ80 similar to that of the adiabatic deformation heating.

5 Acknowledgment

The authors gratefully acknowledge the *Innovatieve Onderzoeksprogramma* (IOP) for funding this research under project code IOT 3003.

6 References

[1] [1]Z.Q. Sheng, R. Shivpuri, Modeling Flow Stress of Magnesium Alloys at Elevated Temperature, Mat Sci and Eng A 419, 2006, pp 202–208.
[2] [2]L. Li, J. Zhou, J. Duszczyk, Determination of a constitutive relationship of AZ31B mag-nesium alloy and validation through comparison between simulated and real extrusion, J. of Mat. Process. and Techn., 172, 2006, pp. 372–380.
[3] [3]A.G. Beer, M.R. Barnett, Influence of Initial Microstructure on the Hot Working Flow Stress of Mg-3Al-1Zn, Mat Sci and Eng A 423, 2006, pp. 292–299.
[4] [4]M. Güden et al, Effect of Strain Rate on the Compressive Mechanical Behavior of A Con-tinuous Alumina Fiber Reinforced ZE41A Magnesium Alloy Based Composite, Mat Sci and Eng A 425, 2006, pp. 145–155.
[5] [5]S.-B. Yi et al, Deformation and Texture Evolution in AZ31 Magnesium Alloy During Uniaxial Loading, Acta Mater 54, 2006, pp. 549–562.
[6] [6]Dynamic Systems Inc., Elimination of Load Cell Ringing during High Speed Deforma-tion by Mathematical Treatment, Gleeble Systems Application Note 019, 2001, pp. 1–4.
[7] [7]R.L. Goetz, S.L. Semiatin, The Adiabatic Correction Factor for Deformation Heating During Uniaxial Compression Test, J of Materials Engineering and Performance 10, 2001, pp. 710–717.

Change in Formability of AZ31 and AZ61 from Room to Elevated Temperatures

H. Takuda, H. Fujimoto, T. Hama, Y. Sakurada
Kyoto University, Kyoto

1 Introduction

The use of magnesium and its alloys has been limited to a narrow range because of the poor formability. However, they have adequate ductility at elevated temperatures. AZ31 and AZ61 are the magnesium-based alloys that are commercially produced and most commonly used. It is reported that AZ31 and AZ61 have good formability at elevated temperatures between 423 and 573 K [1–4]. Therefore, the warm forming of AZ31 and AZ61 at the temperature range is being investigated from a practical point of view [5–11]. However, we are lacking in the fundamental knowledge of the mechanical properties and the formability of the magnesium-based alloys from room to elevated temperatures, yet.

In this study, the change in the formability of the magnesium-based alloys AZ31 and AZ61 from room to elevated temperatures is examined. The tension tests are carried out for various temperatures and strain rates. The measured data are analysed and the formulae for the limit strain and the flow stress are proposed. Furthermore, the change in the microstructure during deformation is observed.

2 Experimental Procedure

2.1 Materials

The materials used in this study are two kinds of commercial magnesium-based alloy round bars of AZ31(Mg-3%Al-1%Zn) and AZ61(Mg-6%Al-1%Zn) with a diameter of 16 mm. The tensile specimens were cut from the bars in such a way as to be parallel to the extruding direction. Figure 1 shows the dimensions of the tensile specimens. Figure 2 shows the microstructures of the materials at the cross section vertical to the extruding direction. The grains of the AZ31 and AZ61 are equiaxial, and the average grain diameters are 30 and 20 µm, respectively.

2.2 Uniaxial Tension Tests

For the specimens the uniaxial tension tests were carried out by means of a thermo-mechanical simulator, THERMECMASTOR-Z. The specimens were heated by the coil located at the center of specimens with high-frequency induction. The specimens reached the target temperatures within a minute, were kept at the temperatures for a minute, and then the tests were started. During the tests the temperatures of the specimens were controlled to within ± 5 K from the target temperatures of 298, 348, 373, 398, 423, 448, 473, 498 and 523 K. The specimens were elongated with constant crosshead velocities of 20, 2 and 0.2 mm s^{-1}.

The deformation of specimens concentrates at the center part, i.e. the heated part. Therefore, the strain cannot be evaluated from the displacement of the crosshead. Only the fracture strain, ε_f, was evaluated from the measurement of the reduction in area, as

$$\varepsilon_f = \ln(S_0/S_f) \tag{1}$$

where S_0 is the initial area of the specimen and S_f is the area at the fracture site. The strain rate, $\dot{\varepsilon}$, was given with the average value to the fracture strain.

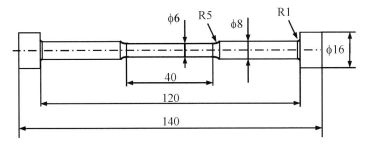

Figure 1: Dimensions of tensile specimens

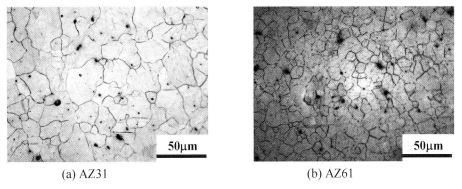

(a) AZ31 (b) AZ61

Figure 2: Microstructures of the materials before tension tests

3 Results and Discussion

3.1 Fracture Strain

Figure 3 shows the relationships between the fracture strain and the strain rate for various temperatures in semi-log scale. For both AZ31 and AZ61, the fracture strain and the effect of strain rate on the fracture strain increase with temperature. The fracture strain increases with decrease in the strain rate, while almost no influence of strain rate exists at room temperature. The linear relationship is observed between the fracture strain and the strain rate for each temperature, and the following approximation may be given,

$$\varepsilon_f = a \ln \dot{\varepsilon} + b \tag{2}$$

where the parameters a and b correspond to the inclinations of the straight lines in Figure 3 and the values of ε_f at $\dot{\varepsilon} = 1$ s^{-1}, respectively. However, the parameters cannot but become the functions of temperature.

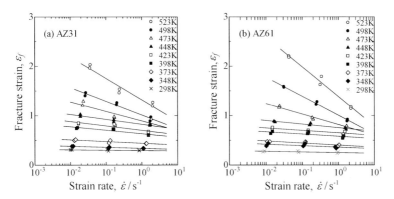

Figure 3: Relationships between fracture strain and strain rate for (a) AZ31 and (b) AZ61

In order to avoid the complication of the formula, the temperature-compensated strain rate, i.e. the Zener-Hollomon parameter, Z, is used here.

$$Z = \dot{\varepsilon} \exp(Q/RT) \tag{3}$$

where T is the temperature, R is the universal gas constant (= 8.31 J mol^{-1} K^{-1}), and Q is an apparent activation energy for the deformation. In this analysis, for the sake of convenience, the activation energy for magnesium self diffusion, 136 kJ mol^{-1} [12], is uniformly used as the value of Q for AZ31 and AZ61. Figure 4 shows the relationships between the fracture strain and the Zener-Hollomon parameter for AZ31 and AZ61 in semi-log scale. The fracture strain increases with decrease in the Zener-Hollomon parameter, notably in the range below 10^{14}. The relationships can be approximated by the following equation,

$$\varepsilon_f = \exp(A/Z + B). \tag{4}$$

The material constants A and B are evaluated to be 48.2 s^{-1} and –3.34 for AZ31 and 57.8 s^{-1} and –4.01 for AZ61. The lines in the figure indicate the calculated values by this equation.

There is almost no difference between AZ31 and AZ61 in the fracture strain. However, the fracture strain of AZ61 is slightly larger than that of AZ31 in the range of Z below 10^{14}, and is slightly smaller in the range beyond 10^{14}. The boundary value corresponds to the temperature of 448 K in the present experimental conditions.

3.2 Flow Stress

The measured data show that the yield stress decreases with increase in temperature and with decrease in strain rate, and that the relationship between the yield stress, σ_y, and the strain rate can be expressed for each temperature by the equation,

$$\sigma_y = K\dot{\varepsilon}^m .\tag{5}$$

Figure 5 shows the relationships between the strain rate sensitivity exponent, m, and the temperature for AZ31 and AZ61. The m-value increases with temperature. The m-value of AZ61 is slightly larger than that of AZ31 in the temperature range beyond 448 K. It is commonly accepted that the m-value is closely related with the elongation, i.e. the fracture strain. The tendency shown in this figure corresponds well to that of the fracture strain shown in Figure 4.

Figures 6 and 7 show the relationships of the Zener-Hollomon parameter with the yield stress and the tensile strength, respectively. These figures reveal that also the yield stress and the tensile strength of AZ31 and AZ61 can be given by simple equations of the Zener-Hollomon parameter.

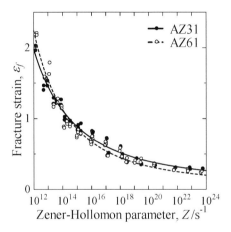

Figure 4: Relationships between fracture strain and Zener-Hollomon parameter

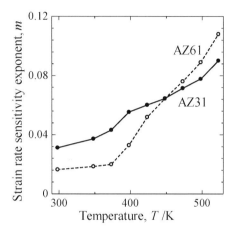

Figure 5: Relationships between strain rate sensitivity exponent and temperature

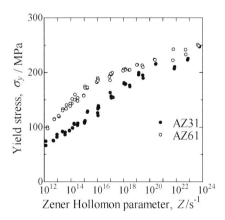

Figure 6: Relationships between yield stress and Zener-Hollomon parameter

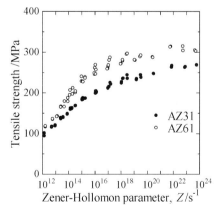

Figure 7: Relationships between tensile strength and Zener-Hollomon parameter

Figure 8: Microstructures of AZ31 after tension tests for various temperatures

3.3 Microstructure

The microstructural changes of AZ31 and AZ61 were observed to be qualitatively the same. Therefore, they are explained by means of the microstructures of AZ31 at the section parallel to the tensile direction, as follows. Figures 8(a)–(d) show the microstructures of AZ31 after the tension tests for 373, 423, 473 and 523 K, respectively. The vertical direction of the figures corresponds to the tensile one. Below 373 K, there is no change in the grain size during deformation, while many twin crystals appear (a). The grains are somewhat elongated to the tensile direction and the twin crystals decrease at 423 K (b). At 473 K fine grains appear only near the grain boundaries (c). This phenomenon is seen also at 448 K. The grains become notably small in the whole area at 523 K (d). The fine grains are considered to be generated by the dynamic recrystallization. The partial recrystallization begins to occur at 448 K. The microstructural change is considered to be closely related with the fracture strain and the flow stress mentioned before.

4 Conclusions

In this study, the change in the formability of the magnesium-based alloys AZ31 and AZ61 from room to elevated temperatures was examined. The tension tests were carried out for various temperatures and strain rates, and the mechanical properties were analysed. It turned out that the fracture strain and the flow stress could be expressed by simple equations of the Zener-Hollomon parameter. Furthermore, the microstructural change during deformation was observed, and the temperature ranges for twin crystal, grain elongation and dynamic recrystallization were clarified.

5 References

[1] J. Kaneko, M. Sugamata, M. Numa, Y. Nishikawa, H. Takada, J. Japan Inst. Metals 2000, 64, 141–147.
[2] S. Aida, H. Tanabe, H. Sugai, I. Takano, H. Ohnuki, M. Kobayashi, J. Japan Inst. Light Metals 2000, 50, 456–461.
[3] H. Somekawa, M. Kohzu, S. Tanabe, H. Higashi, Mater. Sci. Forum 2000, 350–351, 177–182.
[4] M. Kohzu, F. Yoshida, H. Somekawa, M. Yoshikawa, S. Tanabe, K. Higashi, Mater. Trans. 2001, 42, 1273–1276.
[5] S. Lee, Y.-H. Chen, J.-Y. Wang, J. Mater. Process. Technol. 2002, 124, 19–24.
[6] F. Kaiser, D. Letzig, J. Bohlen, A. Styczynski, Ch. Hartig, K.U. Kainer, Mater. Sci. Forum 2003, 419–422, 315–320.
[7] M. Kohzu, F. Yoshida, K. Higashi, Mater. Sci. Forum 2003, 419–422, 321–326.
[8] M.S. Yong, B.H. Hu, C.M. Choy, A.V. Kreij, Mater. Sci. Forum 2003, 437–438, 435–438.
[9] F.-K. Chen, T.-B. Huang, J. Mater. Process. Technol. 2003, 142, 643–647.
[10] F.-K. Chen, T.-B. Huang, C.-K. Chang, Int. J. Mach. Tools Manuf. Des. Appl. 2003, 43, 1553–1559.
[11] H. Palaniswamy, G. Ngaile, T. Altan, J. Mater. Process. Technol. 2004, 146, 52–60.
[12] P.G. Shewmon, J. Met. 1956, 8, 918–922.

Examination of the Forming Properties of Magnesium Wrought Alloys MRI 301F and AZ80A

A. Löffler[1], S. Schumann[1], K.-U. Kainer[2], B. Bronfin[3]
[1] Volkswagen AG, Wolfsburg, Germany
[2] GKSS Forschungszentrum GmbH, Geesthacht, Germany
[3] Dead Sea Magnesium Ltd., Beer Sheva, Israel

1 Introduction

Magnesium alloys are known to be the lightest structural materials and can therefore also be used in automotive applications to reduce weight, which in turn results in a reduction of fuel consumption and CO_2 emissions. Currently, HPDC magnesium alloys are predominantly used in the transportation industry. However, casting alloys do not allow fully exploit the weight benefits offered by magnesium due to unavoidable pores, segregations and other structural defects. It is evident that just using magnesium wrought alloys the maximum benefits in terms of lightweight material construction can be achieved. However, the current application of magnesium wrought alloys in the manufacture of cars is very limited because of its poor formability. It is forecast that the magnesium proportion will increase to 25 kg per vehicle [1] in the medium term. Compared with aluminium, the level of knowledge regarding the forming behaviour of magnesium is very limited. This is particularly true for the newly developed creep-resistant alloys, where the introduction of temperature-stable secondary phases is an obstacle to high ductility. However, in order to expand the operation limits of magnesium, particularly in terms of higher application temperatures, such creep-resistant alloys are essential.

2 Experimental Examinations

This paper compares the commercial wrought alloy AZ80A with the new, creep-resistant wrought alloy MRI 301F which has been developed in cooperation between Dead Sea Magnesium and Volkswagen AG [2]. MRI 301F is a heat-resistant and creep-resistant magnesium alloy containing Zn, Zr, Y, Nd and Ca and has mechanical properties which are comparable with the WE43 and WE54 alloys, however at significantly lower cost. The need to develop alternatives to using the AZ80 alloy is related particularly to its insufficient high temperature properties and susceptibility to corrosion cracking. However, these two properties are very important for forged wheels (temperatures during driving of up to approx. 200 °C [3]). A comparison of alloys will be carried out by means of forming experiments, metallographical examinations, texture measurements and mechanical testing.

2.1 Model Wheel Manufacturing Process Chain

One-piece magnesium forged wheels are manufactured by the process chain which includes casting, extrusion, upsetting, forging and metal spinning. In this work the entire process chain was analysed with regard to the microstructure evolution (Fig.1).

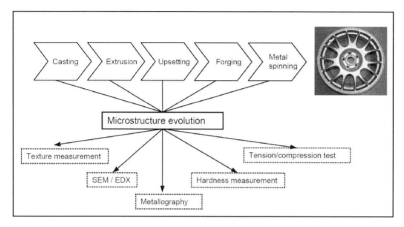

Figure 1: Schematic diagram of the analysis of the microstructure evolution through the model wheel manufacturing process chain

The alloy MRI 301F was received in the form of gravity die cast billet while AZ80A alloy was received in the form of DC cast billet. All of the stages shown in Fig. 1 were carried out using model wheels with a diameter of 120 mm (Fig. 2). A hydraulic forging press was used for this objective. The model wheels were forged in the closed die without flash. The forging temperature was 450 °C for MRI 301F and 350 °C for AZ80A.

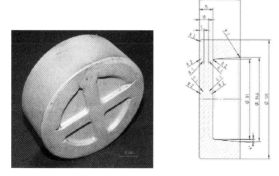

Figure 2: Forged model wheel and model wheel geometry

The metal spinning of the rim well of the model wheels was carried out on a 2-roller facility, in which the model wheel, clamped on a mandrel, is drawn through the laterally fed metal spinning facility (Fig. 3). The material is therefore displaced by the rollers against the feed in the longitudinal direction (cf. Fig. 3). By varying the roll gap, the degree of forming ($\varphi = \ln h1/h0$)

can be adjusted within precise limits. In the case of higher degrees of forming, the entire forming process is distributed over several forming stages. Depending on the alloy system, the temperatures were varied in the region of 300–450 °C.

Figure 3: Metal spinning facility for the model wheels (1: Feed, 2: Roller advance)

2.2 Microstructure Examinations

The effect of the manufacturing process and the process parameters on the microstructure and the mechanical properties was analysed. The microstructure was analysed for each processing condition in the wheel manufacturing process chain (Fig. 1), by using both light-microscopy and SEM equipped with EDS spectrometer. The specimen preparations were made in accordance with the procedure given in [4]. In addition, texture measurements were conducted, in order to obtain indications regarding the orientation of the crystallites in the samples. The texture measurements were carried out with a TEX-2 neutron diffractometer. Due to the high penetration depth and the very low dispersion of the neutron beam, it was possible to record the entire pole figures, i.e. the complete angle range. The mechanical properties of the examined alloys depending on the processing condition were tested by means of universal hardness measurements. The surfaces of the samples were ground and polished to 3 µm beforehand, in order to avoid the effects of surface roughness. The measurements were carried out with a load of 100 N in accordance with the DIN EN ISO 14577 standard.

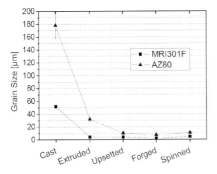

Figure 4: Grain size of MRI301F and AZ80 along the process chain

3 Results and Discussion

3.1 Microstructure Evolution

Micrographs of various conditions through the process chain are shown in Figs. 5 and 6. It is clear from the micrograph that in as-cast condition MRI 301F contains some porosity as a result of the gravity die casting (Fig. 5). Thus, that this material must be extruded, in order to be suitable for subsequent wheel manufacturing. The continuously cast alloy AZ80 did not show any porosity in the as-cast condition. The typical pronounced lamellar nature of the $Mg_{17}Al_{12}$ phase is becoming clear (Fig. 6a). MRI 301F had in as cast condition fine grains with an average diameter around 50 µm, which can be related to the effect of the zirconium added. On the other hand, AZ80A had a very coarse grain with an average grain size of 178 µm (Fig. 6). The pores of MRI 301F in the as-cast condition were eliminated during the subsequent extrusion. It was possible to obtain very fine and relatively homogeneously distributed grains with an average diameter of 4 µm in the MRI 301F alloy. Compared to MRI 301F, AZ80A alloy exhibited relatively coarse grains with average diameter around 30 µm in the extruded condition. In this alloy only after subsequent upsetting of the extrusion billets were obtained grains with an average diameter of 6 µm that can be considered as comparable with the grain structure of the extruded alloy MRI 301F. As can be seen from Fig. 4, the following forming processes do not result in additional grain refinement of AZ80A. The inhomogeneous grain size distribution in AZ80A was observed throughout the entire process chain. On the other hand, this effect was not observed in the case of the MRI 301F alloy. The micrographs show a very homogeneous, fine microstructure.

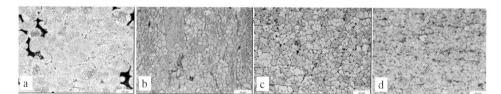

Figure 5: Micrographs of the MRI301F alloy depending on the processing condition (Expl.: a: Casting, b: Extruded, c: Forged, d: Rolled out)

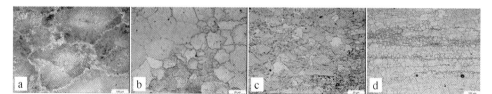

Figure 6: Micrographs of the AZ80 alloy depending on the processing condition (Expl.: a: Casting, b: Extruded, c: Forged, d: Rolled out)

3.2 Texture Development

The pole figures of the basal planes (0002) for MRI 301F depending on the processing condition are shown in Fig. 7. It is evident that in the as-cast condition the texture is random, i.e. it is not possible to demonstrate any pronounced orientation of the crystallites (Fig. 7a). The basal and prismatic planes cannot be clearly separated from one another. This type of pole figure is typical for the as-cast condition, which is determined by the random solidification of the crystallites from the liquid state by nucleation and subsequent nucleus growth. The intensities at 0.9 to 1.1 are within the range of the grey structure (corresponding to intensity 1). The minor deviations can be considered as statistical errors. In contrast, as expected, the extrusion produces a pronounced texturing. The basal planes are oriented in the extrusion direction (Fig. 7b). The maximum intensity of the orientation distribution is displaced somewhat in the 180° direction, which can be accounted for by the slightly off-centre sampling. This effect could be intensified by means of an inhomogeneous flowing during extrusion over the cross section. In addition, it is shown in [5] that the length of the extrusion billet also has an effect on the developing texture. The cause of this is the increasing temperature rise inside the block as a result of the forming heat occurring and the associated tendency towards flow type c. Due to the hexagonal structure, the 1st order prismatic planes (1010) are oriented at right angles to the extrusion direction, which explains the pronounced tension/compression anisotropy of extruded magnesium alloys.

Pre-upsetting of the extruded billets leads to a concentration of the intensities of the basal plane in the 90° direction as a result of the 90° rotation of the extrusion billets during simple upsetting (Fig. 7 c). The prismatic planes are accordingly oriented at a 90° angle thereto at 0° and 180°. As the basal planes are oriented in the extrusion direction following extrusion, they are therefore parallel to the upsetting direction during upsetting, because the extrusion billets were upset in the longitudinal direction. This results in preventing the sliding of the basal planes as a basic mechanism of the plastic deformation of magnesium. It can therefore be assumed that intensified deformation occurred through twinning during forming. In the case of the model wheel examined, the textures from the upsetting are intensified both in the centre and at the edge. It can be seen from the pole figures from the edge area that the material in this area had to flow back around the header punch, in order to shape the area of the rim well of the model wheels. This results in the slight twisting of the texture diagrams of the sample from the centre and the edge of the model wheel. During the subsequent metal spinning of the rim well, though, a classic rolling texture appears. This means that the basal planes are very strongly oriented in the rolling direction. The prismatic planes, on the other hand, are located at a 90° angle relative thereto.

In order to compare the alloys, pole figures for MRI 301F and AZ80 are shown in the rolled out condition in Fig. 8. The orientation distributions, i.e. the position of the basal and also of the prism planes, do not differ. Here again, the basal planes are strongly oriented in the rolling direction and the prismatic planes are oriented transversely thereto. It was not possible to demonstrate an alloying effect on the plastic deformation. However, it was not possible to find any effects of the processing parameters during metal spinning on the ensuing textures. This means that a change in the forming temperature, forming speed or number of rolling stages does not have any effects on the orientation of the crystallites. Only the deformation direction plays a decisive role in determining the texturing of the samples measured. What is striking, however, and what can be judged to be a clear alloying effect are the intensities of varying strengths with which the orientations of the basal planes occur. The intensity is 14.6 mri (multiple random intensity) for MRI 301F and 5.9 mri for AZ80. It can be concluded from this that the basal planes of AZ80 are less strongly oriented in the rolling direction. This could result in a less strongly pronounced tension/compression anisotropy.

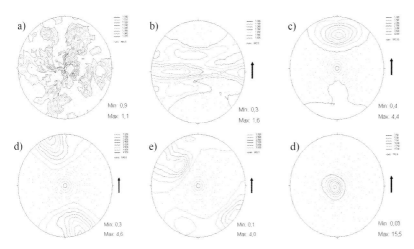

Figure 7: Pole figures of the basal plane (0002) of the MRI301F alloy depending on the processing condition (a: Cast, b: Extruded, c: Upset, d: Forged model wheel centre, e: Forged model wheel edge, f: Rolled out)

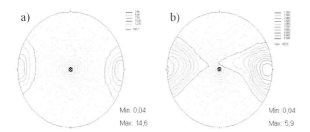

Figure 8: Pole figures of the basal plane (0002) of the rolled out MRI 301F (a) and AZ80 (b) alloys

3.3 Mechanical Properties

The mechanical properties have been analysed by means of Martens hardness measurements. The Martens hardness for the respective processing condition along the process chain is given in Table 1. The result obtained clearly show the relationship between the grain size and the hardness. As expected, AZ80 had the lowest hardness because of the very coarse grain in the as-cast condition. The Martens hardness of MRI 301F is comparable with that of AZ80 with the same grain size in each case (cf. MRI 301F – Ups. with AZ80 - MW). It becomes clear that the Martens hardness increases as the grain refining increases. This Hall-Petch grain size relationship is shown graphically in Fig. 9. The ascent of the straight line for AZ80 is steeper than that for MRI 301F. This indicates that, as a strength-increasing effect, the grain refining has a greater effect in the case of AZ80 than for the MRI alloy.

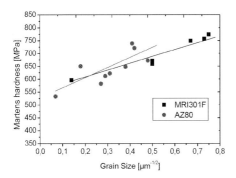

Figure 9: Hall-Petch dependence along the process chain

Table 1: Martens hardness along the process chain

Condition	MRI301F		AZ80	
	GS [µm]	MH [N/mm^2]	GS [µm]	MH [N/mm^2]
As cast	52	596	178	533
Ext.	4	659	32	650
Ups.	4	671	6	739
MW	2,2	749	4,4	672
Roll	2,9	660	11,8	602

Expl: GS: Grain size, MH: Martens hardness
Ext: extruded, Ups: upset,
MW: Model wheel: Roll: rolled out

4 Summary

The comparative study of newly developed MRI 301F alloy with the standard wrought alloy AZ80 have shown that both alloys exhibit good formability and are suitable for manufacturing of forged road wheels. The texture measurements demonstrated that although the two alloys belong to completely different alloy systems, they do not differ in their deformation mechanisms. However it was found that the different intensity of the orientation of the basal planes should be related to the alloying effect.

5 Acknowledgement

The authors thank Prof. Heinz-Günter Brokmeier from the GKSS Research Center, Geesthacht for the accomplished texture measurements.

6 References

[1] Stauber R., Cecco C.: Moderne Werkstoffe im Automobilbau, Werkstoffe im Automobilbau, Special edition of the ATZ and MTZ magazines, 2005
[2] Bronfin B. et al.: European Patent EP 1,329,530, Aug. 2004
[3] Duning R.: Aluminiumräder für die Automobilindustrie, Euroforum Conference „Stahl, Aluminium und Magnesium im Wettbewerb", Bonn, 1998
[4] Kree V. et al: Metallographical examination of magnesium alloys, Practical Metallography 41/5, 2004
[5] Schwarzer R.: Texture in hot extruded, hot rolled and laser welded magnesium base alloys, Proc. of the 2nd Int. Conf. on Texture and Anisotropy of Polycrystals, Metz 2004

Die Forging of Commercial and Modified Magnesium Alloys

J. Swiostek*, J. Bober**, C. Blawert*, D. Letzig*, W. Hintze**, K.U. Kainer*
* GKSS Research Center Geesthacht GmbH, Magnesium Innovation Center – MagIC, Max-Planck-Strasse 1, D-21502 Geesthacht, Germany
** Technical University TU-Hamburg-Harburg, Institute Production Management and Technology, Denickestraße 17, D-21073 Hamburg, Germany

1 Introduction

Magnesium wrought alloys are of special interest for structural parts due to the possibility to obtain improved and more homogenous microstructure and better mechanical properties compared to cast components. The share of magnesium forgings is still relatively small and they are used dominantly for special applications because of high cost und difficult procurement of the feedstock [1,2,3,4]. In this study commercially established wrought magnesium alloys from different alloy series were taken to determine their processing behaviour by die forging process. For these experiments well-known commercial alloys such as zinc containing ZK30 and ZK60, rare earth containing WE43 and aluminium containing AZ80 were used [5,6]. Furthermore modified ZK60 alloy with small amount of cerium rich misch metal (RE) and modified AZ80 alloy with additions of cerium rich misch metal (RE) or calcium were tested. A small addition of RE to the zinc containing alloys could improve the mechanical properties due to the formation of intermetallics Mg-Zn-RE [7]. On the other hand additions of rare earth and calcium elements to aluminum based alloys could cause an improvement of the mechanical properties, high temperature strength and corrosion resistance due to generation of intermetallics Al-RE and Al-Ca, too [8,9,10]. In this study the influence of alloying additions on the process and the obtained properties of the forged components is determined and compared.

2 Experimental

All forging experiments were carried out using pre-extruded magnesium alloys. Commercial DC-cast feedstock with a diameter of 95 mm from the alloys ZK30, ZK60, ZK60RE (ZK60 modified with 2 % RE), WE43, AZ80, AZ80Ca (AZ80 modified with 2 % Ca), AZ80RE (AZ80 modified with 2 % RE), were machined to billets with a diameter of 90 mm and heat-treated for 12h at 400 °C. All billets were air–cooled afterwards. After that the material was hydrostatically extruded at 350 °C with an extrusion ratio of about 8 and an extrusion rate of 3 m/min. The obtained rods were machined to a feedstock with 30 mm diameter and 40 mm length. All forging experiments were executed with a mechanical press applying a nominal force of 100 kN (**Fig. 1**). Round forgings with an easy shape were produced using pre-heated tools (**Fig. 2**).

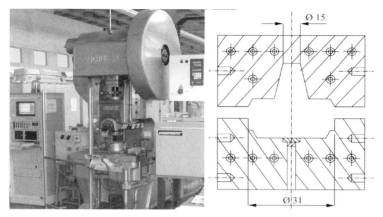

Figure 1: For forging trials used Schueler-1-MN-press

Figure 2: Schematic diagram of the die device used for forging experiments. The die was electrical preheated

The chemical composition of the used alloys and the process parameters of the forging trials are given in **Table 1.** All experiments with ZK30, ZK60, ZK60RE, WE43, AZ80, AZ80Ca, AZ80RE were carried out at in a temperature range from 450 °C down to 200 °C in steps of 50 °C. These experiments were performed to determine the best forging temperature for the following die forging trials presented in this paper. In order to avoid materials failure during the process the die was pre-heated at 220 °C. A graphite and water suspension was used as lubricant. Forging forces were electronically measured during each forging trial.

Table 1: Alloy composition and parameters of the die forging trials.

Alloy	Chemical composition [wt.%]	Forging temperature [°C]	Die temperature [°C]
ZK30	Mg - 3.0Zn - 0.58Zr	450, 400, 350, 300, 250, 200	220
ZK60	Mg - 5.5Zn - 0.58Zr	450, 400, 350, 300, 250, 200	220
ZK60RE	Mg - 5.5Zn - 0.58Zr - 2.1RE	450, 400, 350, 300, 250, 200	220
WE43	Mg - 0.7Zr - 4Y - 3.4RE	450, 400, 350, 300	220
AZ80	Mg - 8.5Al - 0.51Zn - 0.31Mn	450, 400, 350, 300, 250, 200	220
AZ80Ca	Mg - 8.5Al - 0.9Zn - 0.31Mn - 2.1Ca	450, 400, 350, 300, 250, 200	220
AZ80RE	Mg - 8.5Al - 0.9Zn - 0.30Mn - 1.9RE	450, 400, 350, 300, 250, 200	220

During first forging trials carried out at 450 °C warm cracks could be observed on the surface of the forgings in case of alloys ZK60, AZ80 and AZ80Ca (**Fig. 3**). Afterwards the forging temperature was reduced to 400 °C and 350 °C. At this temperatures the surface of all tested materials was smooth without any cracks (**Fig. 5**). Next forging trials were carried out at 300 °C. In this case only the surface of the alloy WE43 was affected by cold cracks. For this reason next forging experiments at lower temperatures were performed without WE43. At 250 °C cold cracks could be observed in the case of ZK60, ZK60RE, AZ80Ca and AZ80RE on the edge of the samples (**Fig. 4**) and at 200 °C all forgings showed cold cracking. In **Fig. 6** is the forging force dependent on the forging temperature for each alloy shown. Only for alloy WE43 higher forging forces were needed. In case of another alloys the forging forces were similar.

Figure 3: Hot cracks after forging experiment at 450 °C

Figure 4: Samples forged at 250 °C with cold cracks at the edge

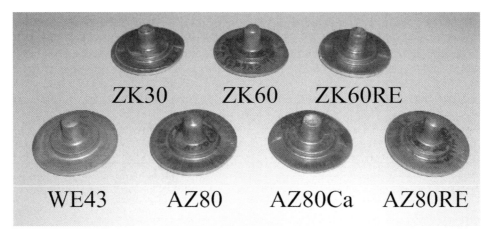

Figure 5: Received parts received at 350°C. The surface of the material is affected without any cracks

The trials were accompanied by microstructural analyses which were carried out on polished and etched central sections of the parts using optical microscopy. For sample preparation an etching solution based on picrid acid was applied [11]. Tensile tests were performed at room temperature following DIN 50125 instructions. All samples were prepared in the transversal orientation of the forging direction. These experiments were done by a Zwick^{TM} Z050 testing machine with a maximum applied force of 50 kN whereas a strain rate of 10^{-3} s^{-1} was adjusted in all cases. Hardness tests were estimated with Vickers HV5. The corrosion behaviour was tested at discs ($\varnothing$ 30 mm, height 5 mm) which have been tooled from the forged specimens. Prior to salt fog testing, the surfaces were cleaned in an ultrasonic bath using acetone. The salt spray testing were performed with a 5 % NaCl solution at a pH of 6.8 for 48 hours according to

DIN 50021. The chamber was opened once after 24 hours for an inspection of the specimens. In contrast to DIN 50021 the specimens were placed under a 90° angle towards the horizontal. The corrosions rates were determined from weight measurement before and after the salt fog test (cleaned for 30 min in chromic acid).

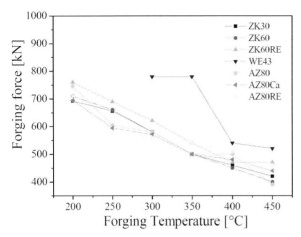

Figure 6: Comparison of the forging force for tested alloys

3   Results and Discussion

3.1   Microstructure

After the forging trials at 350 °C micrographs were cut from the centre of the sample. They are illustrated in **Figs. 7–13** where the longitudinal direction of all micrographs represents the forging direction. The average grain size was determined using several micrographs from each alloy. Only partially recrystallisation was observed for the ZK30, ZK60 and ZK60RE alloy (**Fig. 7–9**). In all samples from ZK-alloys, an inhomogeneous microstructure is found including grains with a wide variation of sizes. Very small grains with 2–4 µm were observed in combination with large not-recrystallisated grains. For alloys WE43 and AZ80 a fine grained microstructure was obtained with an average grain size of about 5–9 µm diameter (**Fig. 10–11**). Compared to the other alloys after the forging process relatively coarse grains with a diameter about 15–20 µm were observed for the calcium and RE modified AZ80Ca and AZ80RE alloy (**Fig. 12–13**). In this case a fully recrystallised microstructure was obtained. Additionally the formation of intermetallic phases in case of the modified ZK60RE, AZ80Ca und AZ80RE alloys were observed.

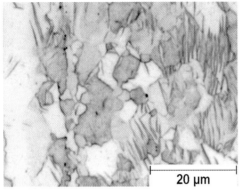
Figure 7: Microstructure of ZK30 forged at 350 °C (forging dir. ↑)

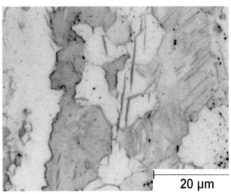
Figure 8: Microstructure of ZK60 forged at 350 °C (forging dir. ↑)

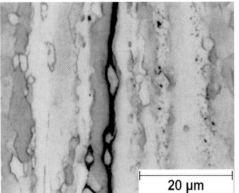
Figure 9: Microstructure of ZK60RE forged at 350 °C (forging dir. ↑)

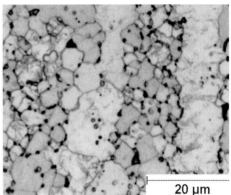
Figure 10: Microstructure of WE43 forged at 350 °C (forging dir. ↑)

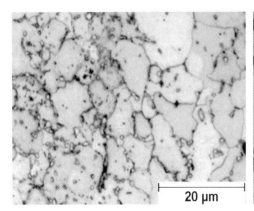
Figure 11: Microstructure of AZ80 forged at 350 °C (forging dir. ↑)

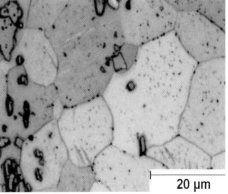
Figure 12: Microstructure of AZ80Ca forged at 350 °C (forging dir. ↑)

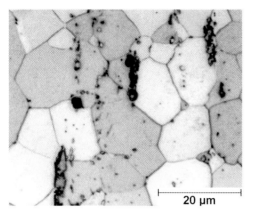

Figure 13: Microstructure of AZ80RE forged at 350 °C (forging dir. ↑).

3.2 Mechanical properties

Figs. 14-15 show the results of mechanical properties of the forged components (forged at 350 °C) have been received from tensile tests performed at room temperature and with a constant strain rate $\dot{\varepsilon}$ 10^{-3} s^{-1}. In Fig. 14 the ultimate tensile strength (UTS) of the materials is presented for all magnesium alloys used in this study. UTS of the alloy WE43 and AZ80 are slightly higher (10–15 %) compared to the data of another tested alloys. This behaviour could be explained with a fine grained and homogenous microstructure of the WE43 and AZ80 alloy. The tensile yield strength (TYS) of the alloys ZK30, ZK60, AZ80, AZ80Ca and AZ80RE is similar and varied between 180-200 MPa. In case of the modified ZK60RE alloy and WE43 alloy the TYS was at about 230 MPa and 290 MPa.

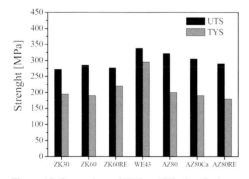
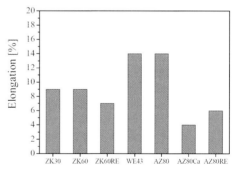

Figure 14: Comparison of UTS and YS after die forging at 350 °C

Figure 15: Comparison of elongation after die forging at 350 °C

In Fig. 15 the elongation to fracture is presented for each investigated material. The highest elongation to fracture was measured for the commercial alloys WE43 and AZ80. In the case of modified aluminium based alloys the elongation was about 4–6 %. This behaviour can be explained with the presence of brittle additional Al-Ca and Mg-Al-RE intermetallic phase in case of modified alloys [7,9,10,12]. For commercial ZK30 and ZK60 alloys was the elongation about 9 % and slightly higher than for modified ZK60RE alloy. This behaviour was similar like for modified aluminium alloys where elongation was lower for modified alloys. Further investigations are necessary to verify the influence of RE and calcium phases on the microstructure and resulting mechanical properties of the modified aluminium and zinc based alloys. The highest hardness about 75–80 HV5 was obtained for commercial AZ80 and WE43 alloys and could be correlated with the highest ultimate tensile strength in case of this alloys.

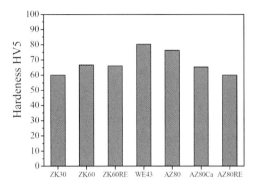

Figure 16: Hardness after die forging at 350 °C

3.3 Corrosion Behaviour

Fig. 17 shows the results of salt spray testing of various magnesium alloys forged at 350°C. For comparison the weight loss data was converted into corrosion rate ignoring the fact that not all specimen reveal uniform corrosion. The results suggest that the increasing zinc content in the ZK series is detrimental for a good corrosion resistance. With increasing zinc content a stronger tendency to enhanced localised corrosion (pitting) was observed. Much better corrosion resistance was found for the AZ80 and WE43 alloys. Latter showed the best performance, without any visible pitting or filiform corrosion. The AZ80 alloy, also showing the lower corrosion rate, suffered from minor filiform corrosion. However alloying with rare earth elements is suitable to further improve the corrosion resistance of AZ and ZK based alloys. In contrast, the addition of calcium to the AZ80 alloy has a slightly negative effect.

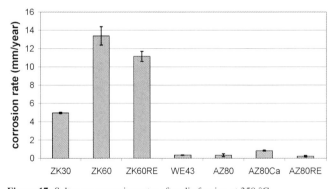

Figure 17: Salt spray corrosion rates after die forging at 350 °C

4 Conclusions

The behavior of magnesium alloys during die forging was investigated. It was demonstrated that the forging process offers the possibility to produce parts with simple geometry. Especially at a the forging temperature of 400 °C and 350 °C all received parts were not affected by hot or cold cracks and this temperatures could be recommended for ideal processing. In the case of commercial and modified AZ-alloys and WE43 alloy a homogenous and recrystallised microstructure was observed. For zinc based alloys only partially recrystallisation was observed leading an inhomogenous grain size. The highest ultimate tensile strength and hardness was obtained for commercial alloy WE43 and AZ80 followed by ZK30, ZK60 and modified alloys. The results of corrosion tests suggest that the aluminium containing AZ-series and WE43 alloy have a significantly better corrosion resistance than zinc based ZK-series.

5 References

[1] Steins, R.; Birkenstock, A.; Lindner K.H. , *in Proceeding 3. Ranshofer Leichtmetalltage 2004*, LKR Verlag **2004**, Ranshofen, Österreich
[2] Fischer, G.; Becker, J.; Stich, A. , Materialwissenschaft und Werkstofftechnik **2000**, 31, p. 993-999
[3] Viehweger, B. , *in Proceeding „MIA - Magnesium im Automobilbau", Drittes WING-Seminar* , November **2004**, Jülich, Germany
[4] Wenfang, S.; Wensheng, L. , *in Proceeding IMA Conference*, Berlin **2005**
[5] ASM Specialty Handbook: Magnesium and Magnesium Alloys (Eds. M.M. Avedesian, H. Baker), ASM International, **1999**
[6] Emley, E.F. , Principles of magnesium technology, Pergamon Press, Oxford, **1966**
[7] Ma, C.; Liu, M.; Wu, G.; Ding, W.; Zhu, Y. , Materials Science and Engineering A **2003**, Volume 349, Issues 1-2, 25 May, Pages 207–212
[8] Bakke, P., Westengen, H. , *in Proceeding of the 2nd International Light Metal Technology Conference 2005*, LKR Verlag **2005**, Ranshofen, Österreich, p. 57–62
[9] Lü, Y.; Wang, Q.; Zhu, Y. , Material Science and Engineering A **2000**, 278, p. 66–76
[10] Fan, Y.; Wu, G.; Gao, H.; Zhai C. , Material Science Forum **2005**, 488–489, p. 869–872
[11] Kree V.; Bohlen J.; Letzig D.; Kainer K.U. , Practical Metallography **2004**, 5, p. 233–246
[12] Gröbner, J.; Kevorkov D.; Chumak, I.; Schmidt-Fetzer R. , Z. Metallkunde **2003**, 94, p. 976–982

Die-Forged Disks for Automobile Wheels in Magnesium Alloys

Boris V. Ovsyannikov
Kamensk Uralsky Metallurgical Works J.S.Co.

Kamensk Uralsky Metallurgical Works (KUMZ in Russian transcription) is one of the major suppliers of forged and die-forged semi-finished products in aluminum and magnesium alloys for aerospace complex, ship-building, car-building and other sectors of Russian industry, and also is a reliable partner for many leading foreign companies of a related profile.

Development of the technical and technological base, unique experience and knowledge of KUMZ specialists guarantee production of material meeting all requirements of the modern market and specific customers.

At KUMZ manufacturing semi-finished products in magnesium wrought alloys provides for a complete technological cycle including alloy preparation, ingot casting, and further fabrication of versatile products by metal forming methods: extruding, forging, die-forging.

It is possible to single out several distinctive groups per types of semi-finished products manufactured in magnesium alloys:

1. Die-forgings of aircraft wheel hubs and flanges practically for all planes made in Russia. Their production has begun in 1960. In total, more than 20 semi-finished items have been manufactured in alloy MA14 with weight from 9 to 230 kg.
2. Die-forgings and sections of construction elements for airframes (levers, beams, arms) in alloys MA14, MA2-1, MA2-1pch. Such die-forgings are made with high rate of non-machined surface with machining only along docking interfaces. Totally, more than 100 semi-finished items have been mastered with weight from 0.5 to 20 kg.
3. Large die-forgings like shells, end plates, conical barrels with maximum diameter of 2,300 mm and maximum weight of 900 kg in superlight magnesium-lithium alloy MA 21 (IMV2) for manufacturing rocket airframes.
4. Ring forgings and die-forgings for bulkheads of spacecrafts in alloy Ma2-1pch. Diameter of ring blanks is from 600 to 1,850 mm, weight range is 70–800 kg.
5. Die-forgings of automobile and motorcycle wheels for export supplies in alloys MA2-1 and MA14, MA20.

When producing die-forged wheels in magnesium alloys KUMZ utilize a so-called „fractional" deformation due to constrained plasticity of magnesium alloys with high deformation rates.

For manufacturing die-forged wheels the mill has approbated magnesium alloys of systems Mg-Zn-Zr (MA14, MA15, MA19, MA20, ZK60) and Mg-Al-Zn-Mn (MA2-1, AZ80A). These magnesium alloys possess sufficiently high ductility in a range of optimal temperatures and rates of deformation. Temperature interval of optimal ductility by extruding is determined based on ductility diagrams of a specific alloy.

The first operation in a technological chain is obtaining extruded billets of Ø 200-300 mm from ingots of Ø 600-650 mm. It is a fundamental difference of production process of magnesium die-forgings from aluminum ones. Since cast ingot ductility is insufficient for stiff schemes of deformation, as billet upsetting and forging, extruding ensures good workability of cast structure due to uniform compression and, in result, ductility enhancement of the billet to be forged.

Then an extruded bar is cut to sized blanks, which size is determined by dimensions of a finished part.

When extruding billets for a purpose of utilization for further forming (upsetting, hot drop forging) susceptibility of magnesium alloys (due to hexagonality of crystalline structure) to appearance of surface tears as a result of even slight tensile stresses is taken into account. Therefore, ingot structure is controlled before extruding, and extruded blanks are turned before further upsetting in order to exclude any surface defects. This operation together with extrusion process materially complicates production technological cycle, reduces yield, and, in result, increases expenses by manufacturing die-forgings.

Above all, 3 factors influence on magnesium wrought alloy die-forging structure formation: temperature, extent and rate of deformation. However, extent of stressed condition, chemical and phase compositions also have a certain impact.

Completely re-crystallized structure with equiaxial grain size of 10–30 microns is inherent to Zr-free extruded rods (alloys MA2-1, MA5), for Zr-contained alloys it is typical to have partially re-crystallized structure and pronounced lineage structure along extruding direction.

Irrespective of re-crystallization level, it is typical for extruded blanks to have formation of axial texture due to sliding plane orientation along deformation axis, which determines anisotropy of properties, especially of yield strength. Areas of non-recrystallized structure differ from re-crystallized ones by good anisotropy. Anisotropy of properties and structure is regarded as a drawback of extruded semi-finished products in magnesium alloys.

For a purpose of changing grain orientation in the structure of a pre-extruded blank as a result of the metal flow direction change, upsetting of turned and cut-to-size cylindrical blanks is applied to obtain a disc of certain height and diameter.

After upsetting, as a rule, structure discontinuity is still left in hot deformation condition (metallographically it is determined on presence of twins in non-recrystallized areas and non-uniform grain shape).

Technology of hot drop forging is used for further structure refinement, forming and obtaining required properties. Forging is performed in two steps: preliminary and final, at that, blank heating temperature and time, soaking by preset temperature and temperature ranges of die-forgings are strictly observed.

For magnesium wrought alloys it is necessary to especially strictly maintain temperature range of die pre-heating (up to 350–300 °C), as insufficient tooling heating will impede deformation and result in crack formation.

Since alloy MA2-1 used for wheel forging is not hardened by heat-treatment, a level of strength properties will be mainly determined by the extent of structure re-crystallization, therefore final forging is completed at decreased temperatures, that allows obtaining fine-grained structure and satisfactory mechanical properties.

Such deformation conditions to produce die-forged wheels are also applicable to alloys MA14 and AZ80A, as under alloy heat-treatment hardening effect is insignificant.

Utilized technology of manufacturing die-forgings for magnesium wheels allows obtaining parts with satisfactory mechanical properties, notwithstanding the fact that deformation scheme does not completely eliminate a problem of property anisotropy along cross-section of die-forgings.

Thus, for wheel die-forgings in alloy Ma2-1 strength properties in ST direction are high σ_B = 290–296 MPa, $\sigma_{0,2}$ = 200–215 MPa; in L and LT directions resistance to deformation is slightly lower: σ_B < 280 MPa, $\sigma_{0,2}$ = 150–170 MPa.

Influence of alloying magnesium wrought alloys by neodymium on properties of die-forgings in alloys Ma19 and MA20 has been studied. Neodymium alloying has not given improvement of properties, strength properties are σ_B = 220–248 MPa, $\sigma_{0,2}$ = 117–200 MPa. Low strength and plastic properties have been obtained also with die-forgings in alloy AZ80A. Plastic properties of all utilized alloys are good: elongation values are within 13–20 %.

Information on strength properties of die-forgings in various magnesium wrought alloys is given in Table 1.

One of the main wheel quality indicators is bending testing with rotation. Bench tests for wheels in magnesium alloys Ma2-1, Ma20 have demonstrated excellent results in determination of fatigue resistance to cyclic loads and superiority of magnesium wheels over aluminum ones more than two times as much. After bending testing with rotation (wheels sustained doubled number of cycles under full load, in comparison with standard requirements) wheels have been exposed to impact loading along a wheel rim. After the impact the magnesium wheel rim has got a deep dent, which has been managed to repair by an ordinary bench hammer: the rim has got a regular shape, which is impossible in case of aluminum die-forged wheels.

Bench tests according to GOST R 50511 have shown superiority in fatigue resistance to cyclic loads of magnesium wheels over aluminum ones more than two times as much, which is characteristic for excellent performance attributes of forged magnesium discs for automobile wheels. Similar testing results have been obtained by OZ, Italy, on magnesium automobile wheels with tire setting diameter of 13" and profile height of 12–14", which are supplied by KUMZ for utilization on Formula 1 sports racing cars of.

Corrosion resistance being less than with aluminum alloys relates to drawbacks of magnesium alloys: they are subject to general and galvanic corrosion. Due to this reason, stronger requirements are demanded to surface protection of such wheels. Parts are oxidized, lacquer coating is applied to oxidized surface, special reducing bushings to exclude magnesium contact with fixing bolts are inserted into attaching points by pressing. At present this problem is soluble, since new materials for coating are created, and corrosion protection methods are developed.

Besides, utilized technologies and observance of precautions by manufacturing die-forgings of magnesium and their machining eliminate fire risk.

All these drawbacks of magnesium wheels, including also their costs (or market price) are covered by advantages.

Advantages of wheels in magnesium alloys in comparison with aluminum ones are evident:

1. They are lighter by 35 %, it gives reduction of car unsprung mass, improvement of acceleration dynamics.
2. Magnesium has high heat conductivity, it gives temperature decrease of brake mechanisms and hubs.
3. Magnesium alloys have high damping ability (several tens of times as much, than with aluminum alloys).
4. Fatigue strength of magnesium discs is two times as much, than of discs in aluminum alloys, hence, operation properties are higher.

Furthermore, wheel strength depends on construction, alloy quality, and execution of technological requirements at each stage; wheel weight is also determined by its design.

KUMZ specialists have been steadily performing analysis of properties obtained by testing wheels in magnesium alloys, continue work under improvement of technological process and design, applying Deform software, which allows modeling die-forged wheels with regulated structural and mechanical properties.

Nowadays magnesium wheels are an exclusive product designed for expensive and sports cars and motorcycles. Still the latest trends at the metal market may make magnesium wheels competitive also for serial cars.

Table 1: Mechanical properties of die-forged automobile wheels

Alloy grade	Die-forging code	Sampling direction	Mechanical properties			
			σ_B, N/mm^2	$\sigma_{0.2}$, N/mm^2	δ, %	Hardness, HB
MA2-1	AMK-112M	ST	287–296 / 291	196–216 / 205	15–17 / 16	60.5
		LT	264–277 / 268	142–153 / 147	11.5–16.5 / 13.3	
		L	264–274 / 267	148–167 / 155	9.2–12.0 / 10.4	
MA2-1	AMK-70M prototype	ST	291–301 / 296	209–222 / 215	16.0–19.0 / 17	54
		LT	272–284 / 277	153–199 / 171	16.0–19.0 / 17.3	
MA20	AMK-9M4	ST	224–242 / 237	117–168 / 142	16.5–28.0 / 21.4	95
		L	240–241 / 241	167–169 / 168	17–20.5 / 18.8	
		LT	218–228 / 225	112–121 / 118	14.0–23 / 19.3	
MA20	AMK-13M2	ST	218–236 / 229	121–154 / 133	13.5–25.5 / 17.1	85
		L	209–214 / 212	76–87 / 83	10–15.5 / 12.3	
		LT	226–228 / 207	135–149 / 141	11–24 / 19	
MA20	AMK-22M5	ST	208–243 / 226	135–170 / 154	11–28 / 21	95
		L	182–187 / 184	54–62 / 59	19–22 / 21	
		LT	208–225 / 215	109–123 / 117	14–21 / 17	

Alloy grade	Die-forging code	Sampling direction	Mechanical properties			
			σ_B, N/mm^2	$\sigma_{0.2}$, N/mm^2	δ, %	Hardness, HB
AZ 80A	AMK-12M	ST	196–311 259	110–188 153	4.8–10.0 7.1	64.6
		L	316–331 323	223–243 233	11.0	
		LT	239–242 240	140–154 147	4.8–6.4 5.6	
AZ 80A	AMK-111M	ST	173–286 223	118–197 147	2.8–6.0 4.03	68.2
		L	308–317 313	231–235 233	9.2–11.0 9.6	
		LT	184–189 186	120–128 124	3.2–5.6 4.2	
AZ 80A	AMK111M	ST	210–268 240	126–157 142	3.2–7.2 5.3	59.4
		L	293–300 297	207–214 210	6.0–7.6 6.5	
		LT	182–196 190	114–121 118	3.2–4.8 4.2	
MA2-1	AMK-183	ST	262–293 278	165–227 192	10.0–19.0 15	69
		L	281–283 282	182–184 183	17.0–19.0 17.5	

A Study on the Microstructure of Strip Cast Magnesium

H. Palkowski, P.M. Rucki, H.-G. Brokmeier, S.-B. Yi
TU Clausthal, Clausthal-Zellerfeld

1 Introduction

Technology developments of the 21st century are characterized by a trend to reduce the use of non-renewable energy resources. This trend is due to high oil prices and a changing attitude towards our environment. One way to decrease the use of non-renewable energy resources is the use of light materials, such as Magnesium, which meets the demands of designers and engineers with a low mass and high strength. Especially magnesium sheet can be an alterna-tive for steel and aluminum car-bodies in the future. High production costs and poor corrosion resistance of magnesium alloys made it difficult to introduce in the automobile industry. One way to decrease production costs is to use a cost-effective technology for magnesium sheet production. Direct Strip Casting (DSC) which is presently being developed at the University of Technology in Clausthal is this kind of cost-effective technology because it allows casting small thicknesses, thereby reducing the need for rolling passes.

This publication presents the solidification and cooling conditions of the strip for DSC and evaluates the microstructure of the strip in dependence of the process parameters. The solidi-fication and cooling conditions are influenced by mould design, casting procedure, casting temperature, as well as solidification and cooling velocity of the strip. The analyzed micro-structure parameters are distribution of grains, intermetallic participates and the texture across the thickness of the strip.

2 Direct Strip Casting DSC

The caster used at the Clausthal University of Technology is a laboratory device with a cast-ing width of approximately 160 mm. Molten magnesium is fed from the supply system, with a constant melt flow onto a moving belt which is cooled by water from below. Strip thickness is adjusted by melt flow in combination with speed of the cooling belt. During solidification the strip is transported to the roller table by the moving belt. The belt is driven and fixed by two rolls which are rotating synchronous. The schematic sketch of the laboratory caster is shown in Figure 1.
The length of the strip can reach approx. 5 m, and a thickness of approx. 10 mm. A view of the produced strip is shown in Figure 2. Determining factor for the microstructure is the solidification process of the magnesium and later the cooling of the strip. The molten magnesium solidifies on the belt which is cooled from below by water emerg-ing from nozzles located in the cooling chamber. The belt is guided and stabilized by small sec-tioned rolls. The whole cooling chamber is under low air pressure, which seals the cooling chamber and stabilizes the moving belt. The cooling section is shown in Figure 3. The cooling-area is 1.5 m long, the belt has a thickness of 0.8 mm and is made of low carbon steel. The belt is coated with boron nitride coat-

ing, which on one hand prevents sticking to the belt but on the other hand declines the heat transfer coefficient be-tween the belt surface and magnesium melt and strip.

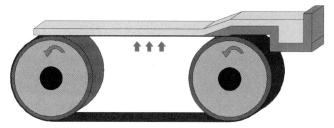

Figure 1: Single Belt Caster

Figure 2: Cast AZ31 magnesium strip

a)   b)   c)
Figure 3: Cooling chamber of DSC, a) nozzles and sectioned rolls, b) working water spray, c) cooling belt

3 Experiment

The microstructure of the strip was determined for AZ31 for various casting temperatures and cooling conditions. Three different casting temperatures were adjusted, namely 680 °C, 700 °C and 730 °C. Thermocouples were fixed 1mm above the cooling belt and the cooling profile of the coasted material was registered. The thermal profiles for three cast strips with different thicknesses are shown in diagram „a" in Figure 4. Casting temperature was 700 °C and strip thickness was varied between 10, 15 and 20 mm. The first registered temperature is almost 100 degrees lower than the melt temperature of 700 °C. This suggests that solid-liquid inter-face between the cooling belt and the melt causes a dramatic drop in the temperature and the thermocouple is not able to register this fast decline. Subsequently the melt solidifies on the lower zone of the strip and a solid-solid interface between the cooling belt and the strip develops. The heat transfer coefficient changes to a much lower value, the dramatic drop of the temperature is suppressed and the rest of the melt solidifies with the velocity described by the diagrams in Fig. 4.

The profile in diagram „a" in Figure 4 shows the melt solidifying and the strip cooling off on the belt and later suspended in air. The difference between cooling on the belt and in air is apparent for every cooling curve which was registered. The higher cooling rate on the curve (up to 40 s) shows the strip cooling off on the belt. It is apparent that the cooling rate depends on the thickness of the strip and is much lower for thinner strips especially while cooling on the belt.

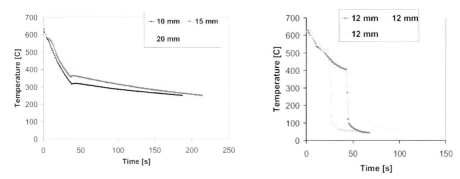

Figure 4: Thermal profile of solidification and cooling of strip a) typical cooling, b) water spray cooling

In subsequent studies the microstructure of the strips which are cooled by water was examined in order to find out how the cooling time affects the development of the microstructure. Magnesium strips with different thickness and cast at different melt temperatures were cooled in cold water for different time intervals. The cooling curves for different trials are illustrated in diagram „b" in Figure 4.

4 Results

Grain size, distribution of precipitations and texture of cast material are among the parameters that influence rolling and hence sheet quality. The fine microstructure of AZ31 improves the formability and thus allows using the higher strains while processing this alloy [5]. The texture influences the deformation mechanism in the rolled strip by activating or inactivating slip and twinnig systems. Thus the rolling parameters must consider texture development [6]. The precipitations in AZ31 have different effects on the properties of the cast material. The $Mg_{17}Al_{12}$ precipitations are a brittle phase, which is harmful for the formability. One should always try to dissolve this phase with an annealing process at about 350 °C [4]. The Al-Mn precipitations are also brittle and harmful for the formability of AZ31, but they are useful in particle stimulated nucleation while dynamic recrystallisation. They don't dissolve into the matrix during dynamic recrystallisation up to at least 400 °C and remain effective in produc-ing a new fine microstructure. The Al-Mn intermetallics can also suppress the grain growth via the pinning or drag mechanisms during annealing treatment [1]. In addition, the homoge-nous distribution and small size of Al-Mn particles make it a strengthening phase [3].

4.1 Grains

Every strip irrespective of casting parameters characterizes inhomogeneous grain size and grain morphology over the thickness. The strip has a typical cast–microstructure; columnar grains in the vicinity of the cooling belt and equiaxed grains away from it. Especially strips cast at higher temperatures, e.g. 730 °C have expanded columnar grains. The length of these dendrites reaches 800 μm and they spread approx. 300 μm. This can be explained by the fact that with low melt temperatures under cooling can be rapidly established on the solidification front and the new crystals are nucleated in a way that the growth of the dendrites is hindered and even stopped. For strips cast at the same melt temperature but cooled under different conditions (4 to 10 K/s cooling rate and presence of water spray cooling as illustrated in Figure 4) no substantial differences in grain size and morphology were observed and no correlation between cooling conditions and microstructure formation were found. Polished micrograph sections over the thickness of two samples cast with different melt temperatures are shown in Figure 5.

The average size of the grains in most of the samples is about 200 μm but it is worth mentioning that some samples show a finer microstructure with an average grain size about 100 μm. This phenomenon however can't be correlated with the cooling and solidification conditions but rather related to segregation or grain refining effect in this alloy which can occur under same circumstances.

4.2 Texture

Three samples with $50 \times 50 mm^2$ were prepared from the different positions of the strip for the texture measurement using X-ray diffraction technique. For both sides of the strip, the bot-tom-side and the upper-side, pole figure measurements were carried out at three different po-sitions for each sample; i.e. 9 measuring points for each side of the strip. Figure 6 shows the {10-10} and (0002) pole figures from the bottom- and upper-sides of the strip after addition of 9 measured pole figures. Because of the quite coarse grain structures and the accompanied low statistical reliability of the measurements, though the pole figures are represented after addition of 9 measured ones, the texture evolution will be discussed only in terms of the qualitative analysis in the following section.

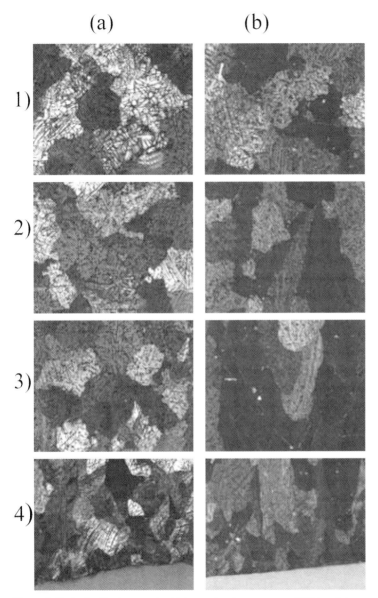

Figure 5: Grain distribution over the thickness a) melt temperature 680 b) melt temperature 730 °C; 1: 12 mm from the bottom, 2: 6 mm, 3: 3 mm, 4: bottom

The upper-side of the strip shows the randomly distributed intensity in {10-10} as well as in (0002) pole figures, Figure 6 (a), which can be typically found in randomly (or very weak) textured materials. In contrast, the bottom-side shows clearly the development of the texture component in which most grains have their **c**-axis parallel to the cooling direction (strip thickness direction). Moreover, the slight shift of this texture component into the strip direction is also shown, see {10-10} pole figure from the bottom side. However, further studies are necessary for understanding the quantitative texture evolution as well as for getting statistically more reliable texture information.

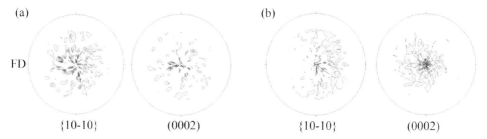

Figure 6: Measured pole figures from the upper-side (a) and from the bottom-side (b) of the strip. Pole figure center is corresponding to the strip thickness direction, i.e. cooling direction. FD: flowing direction (longitudinal direction).

4.3 Precipitations

In AZ31 cast-material there are generally two kinds of precipitations $Mg_{17}Al_{12}$ and Al-Mn. In our strips both kinds of precipitations are present and they are predominantly within the grains, although the big intermetallics are also present at the grain boundaries. $Mg_{17}Al_{12}$ is spherical and the Al-Mn precipitations are longitudinal. The same phenomenon found in case of the grain size is observed in all cases. There are inhomogeneous distributions of precipitations over the thickness of the samples as illustrated in the Figure 7.

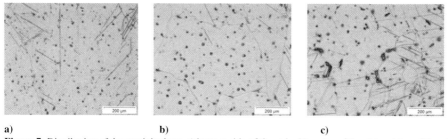

Figure 7: Distribution of the precipitations, a) bottom side of the strip, b) center of the strip, c) top of the strip

In order to better understand the inhomogeneous distribution of intermetallics Figure 8 shows the precipitations categorized according to their area on the polished micrograph sec-tion and the position within the strip. On both diagrams it is apparent that the number of precipitations with an area from 3,7 to 24 µm is highest in the strip. The highest concentration of the precipitations is close to the bottom side of the strip. The largest intermetallics category 6, 7 and 8 appear in the center and at the top and not at all at the bottom.

Comparing the two diagrams in Figure 7 it is noticeable that the distributions are very similar except for the fact that there are hardly any intermetallics at the bottom of the strip which was cast at a higher melt temperature. The area of the highest content of the precipitations seems to be moved inside the strip. The reasons for this could be growing columnar dendrites in this area which do not allow the different phases to precipitate.

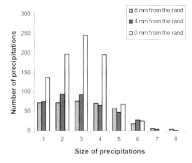 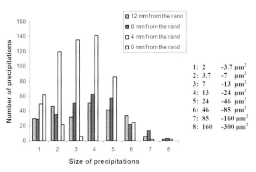

Figure 8: Distribution of the precipitation, size of the analyzed micrograph section 428 × 684 µm, a) melt temperature 680 °C, thickness of the strip 10 mm b) melt temperature 730 °C, thickness of the strip 15 mm

In experiments no substantial differences were recognized in the distribution of precipitation between strips which were cooled under different conditions.

5 Conclusions

1. There is anisotropy of grain size, texture and precipitations for the microstructure across the thickness of the strip.
2. The influence of cooling conditions after the solidification of the strip on the microstructure is hardly noticeable and there is only a bad correlation of the cooling conditions with the growth of the microstructure.
3. The melt temperature affects the grain size. Low melt temperature hinders the growth of the columnar dendrites and a correlation between the melt temperature and the distribution of the precipitations in the cast strips was found.

6 References

[4] Laukli, H.I.; Lohne, O.; Arnberg, L.; Gjestland, H.; Sannes. S., Magnesium Proceeding of the 6th International Conference, **2005**, 183–189.
[5] Hort, N.; Huang, Y.; U., Kainer, Advanced Engineering Materials, **2006**, 8, No.4, 235–240.
[6] Jiang, L.; Huang, G.; Godet, S.; Jonas, J.J.; Luo, A.A., Materials Science Forum, **2005**, 488–489, pp. 261–264.
[7] Hang-liang, Z.; Shao-kang, G.; Fei-yan, Z.; Qing-kui, Li.; Li-guo, W., Trans. Nonferrous Met. Soc. China, **2005**, 15, pp. 144–148.
[8] Yang, X.; Sanada, M.; Miura, H.; Saki, T., Material Science Forum, **2005**, 488–489, pp. 223–226.
[9] Gottstein, G.; Al Samman, T., Materials Science Forum, **2005**, 495–497, pp. 623–632
[10] Palkowski, H.; Wondraczek, L., Z. Metallkd. **2004**, 95, pp. 1080–1086.

Properties of Strips and Sheets of Magnesium Alloy Produced by Casting-Rolling Technology

Rudolf Kawalla, Madlen Ullmann, Matthias Oswald, Christian Schmidt
Institute of Metal Forming, Freiberg University of Mining and Technology, Germany

1 Introduction

Because of the hexagonal structure of magnesium only small deformation degrees are practicable at room temperature. Therefore, the production of magnesium alloy sheets is carried out at elevated temperatures

The rolling of continuously casted slabs to sheets or strips requires numerous rolling passes with several annealings in between. By using near-net shape profiles the following rolling process can be minimized. The production of near-net-shape strip in the continuously working casting-rolling process offers a very economic production of magnesium sheets on strip-rolling mills for an appropriate coil mass. [1, 2, 7, 8]

In the following, some results of the recent investigations at the Institute of Metal Forming at the Freiberg University of Mining and Technology for the development of a strip rolling technology for the production of sheets in deep drawing quality made of cast-rolled material are introduced. The main aim of these investigations is the realisation of a rolling technology with optimal sheet properties. Thereby, both, the rolling process on a continuous rolling mill and the reversing strip rolling, e. g. on hot Steckel mills, are considered. [3, 4, 5]

In principle the question of suitable temperature control as well as an optimal rolling speed and pass reduction has to be clarified. Especially the rolling process at high rolling speeds in combination with high pass reductions allows rolling at low initial temperatures in the range of warm forming.

2 Experimental procedures

The production of cast-rolled strips with the casting-rolling process (TRC-process)[1] was carried out with the pilot plant at the MgF Magnesium Flachprodukte GmbH in Freiberg [1, 2]. For the rolling trials many laboratory mills with suitable furnaces were available at the Institute of Metal Forming at the Freiberg University of Mining and Technology. The following results for the development of mechanical properties and microstructure were obtained in rolling trials with cast-rolled strips, which were carried out solely on the finishing line of the semi-continuous hot rolling mill seen in **Figure 1**.

Before rolling all the cast rolled samples were annealed in the same way. During rolling the influence of different initial temperatures, rolling speeds, and pass numbers on the strip edge and surface quality and the mechanical properties was investigated systematically. After rolling the strips were immediately quenched in water.

The chemical compositions of the investigated AZ alloys are summarized in **Table 1**.

---

1. TRC...Twin Roll Casting

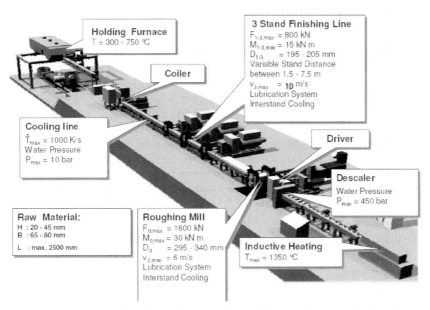

Figure 1: Semi-continuous hot rolling mill at the Institute of Metal Forming at the Freiberg University of Mining and Technology

Table 1: Average chemical composition of the investigated cast-rolled strips

material	Al	Zn	Mn	Zr	Cu	Si	Fe	Ni	Sn	Be	Pb	Mg
A	2,19	0,683	0,317	<0,001	<0,001	0,021	0,0034	<0,001	<0,005	0,0014	<0,005	96,78
B	2,28	0,74	0,373	<0,001	0,0025	0,0252	0,0035	<0,001	<0,005	0,0012	0,005	96,57

3 Results

3.1 Edge and Surface Quality

The formation of strip edges during rolling is considerably defined by the edge preparation and the forming conditions in the rolling line, in particular by temperature control. Trimming of cast-rolled strips with a milling cutter allows an accurate rolling with high reductions (over 50 % per pass) and rolling speeds (~10 m/s) at relatively low initial temperatures of 200 °C. In comparison to that, edges prepared with shears show clearly cracks, applying the same rolling conditions (see **Figure 2**).

Especially cold work rolls lead to a formation of magnesium coating on the work rolls due to detaching process of small particles off the strip surface. By the use of suitable lubricants the rolling force and the rolling torque can be reduced considerably as own investigations have shown (see **Figure 3**). At the same time the surface of the strip is influenced positively. The formation of a coating on the work rolls is avoided by an additional dividing effect of some lubricants.

 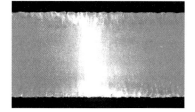

a) milled edges   b) cut edges

Figure 2: Influence of the edge preparation on the formation of cracks after rolling (initial temperature: 200 °C, rolling speed: 10 m/s, 2 passes (50 %, 50 %), rolled without lubrication)

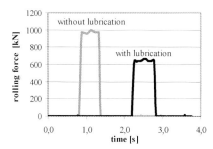

Figure 3: Influence of roll gap lubrication on rolling force (reduction: 30 %, average rolling speed: 1m/s, initial temperature: 410 °C)

3.2   Influence of the Rolling Conditions on the Mechanical Properties of the Rolled Sheets

The mechanical properties of the rolled strips are strongly influenced by the rolling conditions (see **Figure 5**). Low initial temperatures (~150 °C) lead to the formation of an inhomogeneous microstructure with many twins and deformed grains. This affects the mechanical properties. Relatively low elongations at high strengths are reached. However, not only the initial temperature is of importance, but also the rolling speed (see **Figure 5b**), because of its impact on the temperature gradient. The lower the rolling speed the lower the finishing temperature (see **Figure 4**).

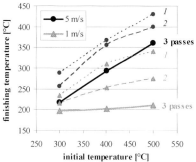

Figure 4: Influence of the rolling speed (1 m/s ($\dot{\varphi}_V = 62\ \text{s}^{-1}$), 5 m/s ($\dot{\varphi}_V = 315\ \text{s}^{-1}$)) on the finishing temperature depending on the initial temperature

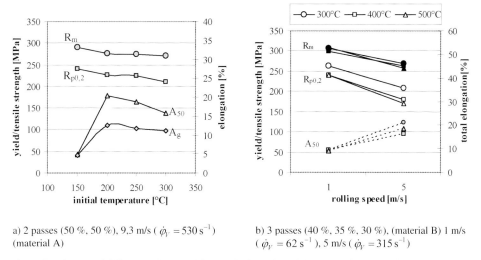

a) 2 passes (50 %, 50 %), 9,3 m/s ($\dot{\varphi}_V = 530\,s^{-1}$) (material A)

b) 3 passes (40 %, 35 %, 30 %), (material B) 1 m/s ($\dot{\varphi}_V = 62\,s^{-1}$), 5 m/s ($\dot{\varphi}_V = 315\,s^{-1}$)

Figure 5: Influence of different rolling conditions to the formation of the mechanical properties.

An increasing rolling speed leads to lower strength values and rising elongations, whereas an initial temperature of 500 °C results in inferior results compared to an initial temperature of 300 °C. This is explained by the microstructure (see chapter 3.3, **Figure 7**). High rolling speeds combined with low initial temperatures cause optimal results. By means of **Figure 5a** it can be seen that at 200 °C initial temperature and a rolling speed of 9,3 m/s for example, rolling in only 2 passes leads to most favourable properties. The mechanical properties are already good enough that it can be done without subsequent annealing.

Additionally, the number of passes and the reduction per pass have an influence on the mechanical properties (see **Figure 6**). With increasing reduction the strength values rise while elongations decrease or remain constant.

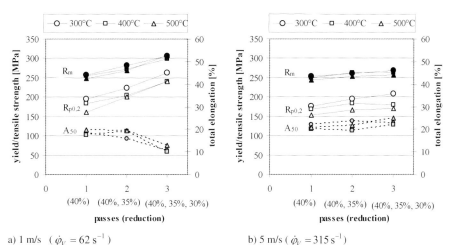

a) 1 m/s ($\dot{\varphi}_V = 62\,s^{-1}$)

b) 5 m/s ($\dot{\varphi}_V = 315\,s^{-1}$)

Figure 6: Mechanical properties depending on the number of passes (reduction per pass), the initial temperature, and the rolling speed (material A)

3.3 Microstructure of Rolled Sheets

Among the investigated sheets and strips the finest and most uniform microstructures are reached by rolling in 2 passes at 9,3 m/s and an initial temperature of 200 °C as well as rolling in 3 passes at 5 m/s and an initial temperature of 300 °C. The mean grain size $\bar{d}$ in both cases is ~ 4,5 µm. Because of grain growth the size of dynamically recrystallized grains increases with increasing rolling temperature (see Figure 7).

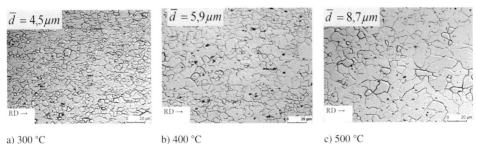

a) 300 °C   b) 400 °C   c) 500 °C

Figure 7: Influence of the initial temperature on the mean grain size $\bar{d}$ for rolling in 3 passes (5 m/s) (material A), RD…rolling direction

During rolling below or at 150 °C dynamic recrystallization does not take place completely. In fact, there is an inhomogeneous microstructure with many deformed grains and shear bands as well as non-uniform grains observable. An example for such a microstructure shows **Figure 8a**, whereas even at 300 °C no completely and homogeneously recrystallized microstructure developed, but inhomogeneous microstructures with many twins, deformed grains, and some partially recrystallized grains would be found for low rolling speeds (see **Figure 8b**).

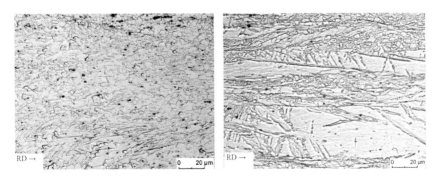

a) 150 °C, 2 passes, 9,3 m/s (material B)   b) 300 °C, 1 m/s, 3 passes (material A)

Figure 8: Microstructure at low rolling temperatures (a) or at low rolling speeds (b)

4 Summary

The results of the recent investigations at the Institute of Metal Forming at the Freiberg University of Mining and Technology for the development of a strip rolling technology for the production of sheets in deep drawing quality made of cast-rolled material show the following:

1. The rolling process at high rolling speeds in combination with high pass reductions allows rolling at low initial temperatures in the range of warm forming.
2. Among the investigated sheets and strips the finest and most uniform microstructure ($\bar{d}$ = 4,5µm) is reached by rolling in 2 passes at 9,3 m/s and an initial temperature of 200°C as well as rolling in 3 passes at 5 m/s and an initial temperature of 300°C.
3. With the choice of a suitable lubricant the rolling force and rolling torque can be reduced considerably. At the same time the surface of the strip is influenced positively. The formation of a coating on the work rolls is avoided by an additional dividing effect of some lubricants.
4. The formation of strip edges during rolling is basically defined by the edge preparation and forming conditions in the rolling line, in particular by temperature control. Trimming of cast-rolled strips with a milling cutter allows an accurate rolling with high reductions and rolling speeds at relatively low initial temperatures. Edges prepared with shears, however, show clearly cracks at the same rolling conditions.

5   Acknowledgement

The authors thank the Saxony State Ministry for Economic Affairs and Labor (SMWA) - technology promotion (Development Bank of Saxony, SAB) for the financial support.

6   References

[1] B. Engl: Potential uses and new production technology for magnesium sheet. Steel Grips 1 (2003), p. 413–418
[2] B. Engl, R. Kawalla: Entwicklung eines neuartigen kostengünstigen Verfahrens zur Herstellung von Magnesiumband; Freiberg MEFORM 2006, Tagungsband, p. 96–117
[3] M. Oswald, C. Schmidt, S. Waengler, N.D. Cuong: Einfluss der Umformbedingungen beim Walzen von Magnesiumgießwalzband aus der Gießhitze auf die Feinblech- und Bandqualität. Freiberg MEFORM 2006, Tagungsband, p.128–143
[4] M. Ullmann, M. Oswald, N. D. Cuong: Werkstoff- und technologische Kennwerte für Feinbleche am Bsp. von Magnesium, Freiberg MEFORM 2006, Tagungsband, p. 65–80
[5] R. Kawalla, M. Oswald, C. Schmidt, N.D. Cuong.: 12. Sächsische Fachtagung Umformtechnik SFU, Leichtbau durch Umformtechnik. Tagungsband, Dresden, 2005,

[6] p. 231–243
[7] H. Utsunomiya, T. Sakai, S. Minamiguchi, H. Koh: High-speed heavy rolling of magnesium alloy sheets, Magnesium Technology 2006, TMS, p. 201–204
[8] L. Löchte, H. Westengen, J. Rodseth: An efficient route to magnesium alloy sheet: twin roll casting and hot rolling, Magnesium Technology 2005, TMS, p. 247–252
[9] Sung S. Park; Jung G. Lee; Hak C. Lee; Nack J. Kim: Development of wrought Mg alloys via strip casting, Magnesium Technology 2004, TMS, p. 107–112

Single Pass Large Draught Rolling of Some Magnesium Alloys below 473K by High Speed Rolling

Tetsuo Sakai, Hiroshi Utsunomiya, Satoshi Minamiguchi, Hiroaki Koh
Osaka University, Suita, Japan

1　Introduction

Magnesium alloy sheets had to be rolled at elevated temperatures to avoid edge cracking. The need for processing at elevated temperatures requires additional cost of magnesium alloy sheet products and is a significant impediment to their introduction to automotive industries. The poor workability of magnesium alloy is ascribed to its hcp crystallography and insufficient activation of independent slip systems.

They are mostly shaped by casting or thixo-molding. Wrought magnesium products are still limited, though increase rapidly. Thin sheets and strips are expected to be used in quantity for electronic applications and automotive parts in near future.

Magnesium sheets are usually produced by rolling from cast slabs. In cold rolling, applicable reduction in thickness is less than 10 % due to the poor ductility at room temperature. Therefore the thickness is mostly reduced in hot or warm rolling. In the process, multi-pass operation with small reduction accompanied by intermediate annealing is employed to suppress edge cracks or fracture of the material and to maintain the workability [1,2]. Rolls are often heated to minimize the temperature drop. Because of all these procedures, fabrication of magnesium alloy sheets is less productive and magnesium alloy sheets are more expensive than cast magnesium products as well as other metal sheets.

If magnesium alloy sheets can be rolled to a large reduction by single pass rolling, not only the productivity is improved but also the grain refinement by recrystallization is possible due to large amount of stored energy.

The authors have proposed high speed rolling for fabrication of AZ31B magnesium alloy sheets [3-5]. During rolling of a sheet, heat is generated by plastic deformation and by friction between the material and rolls. Meanwhile the heat transfers from the material to cold rolls. With increasing the roll speed, the duration where the material and the rolls are in contact becomes shorter. This results in effective temperature rise of the sheet during rolling. So fracture or cracks due to the low-temperature brittleness can be suppressed considerably by high speed rolling. Furthermore, higher strain rate deformation at lower temperature, i.e., high Zener-Hollomon parameter, is advantageous for grain refinement by dynamic recrystallization.

In the present study, typical commercial magnesium alloys AZ31B, ZK60A and AZ91D sheets were processed at the rolling speed of 1000 m/min and 2000 m/min. The significant improvement in the workability of conventional magnesium alloys with high speed rolling is discovered.

2 Experimental

A two-high laboratory high speed rolling mill with 530 mm rolls was used. The roll peripheral speed can be varied from 200 m/min to 2600 m/min. Detailed specification can be found elsewhere [6-8], where the authors studied microstructure and texture evolution of various steels and other metals systematically.

Commercial 2.5mm thick AZ31B (Mg-3%Al-1%Zn-0.4%Mn) annealed sheets were received. Commercial AZ91D (Mg-9.1%Al-0.8%Zn-04%Mn) and commercial ZK60A (Mg-5.6%Zn-0.5%Zr) sheets with the thickness of 2.5mm sliced from an extruded bar were also received. Specimens with 30mm in width and 300mm in length were cut from the sheets and subjected to the rolling experiment. The AZ91D sheets were solution treated at 412 °C for 86.4 ks and the ZK60A sheets were solution treated at 500 °C for 7.2 ks followed by water quenching.

Prior to the rolling, a specimen was held for 900 s at 100, 150, 200 or 350 °C in an electrical tube furnace, then supplied to the mill through the pinch roller. Cold rolling of AZ31B sheets was also performed without the prior heating. The rolling experiment was conducted in single-pass operation with reduction in thickness from 30 % to 60 %. The peripheral speed of the rolls was 1000 and 2000 m/min for AZ31B and 1000 m/min for AZ91D and ZK60A. At the rolling speed of 2000 m/min, the estimated strain rate during 50 % rolling is 1.5×10^3 s^{-1}. The specimen was quenched immediately (precisely 7 ms that was estimated from the roll peripheral speed and the distance between the roll exit and the quenching apparatus) after rolling by the water spray closely attached to the exit of the mill. The rolls were neither heated nor lubricated. Roll surface was degreased with ethanol before rolling.

3 Results

3.1 Rolling of AZ31B

The overall view of AZ31B sheets rolled by single pass at 100 and 150 °C to the various reductions at the rolling speed of 1000 m/min is shown in Fig.1. Reduction in thickness is indicated at the bottom end of each specimen. These photographs show that the reduction in thickness of 60 % could be attained by single pass at high speed rolling even at 100 °C. Small edge cracks are observed at 100 °C rolling. However, these cracks may not be serious because much wider sheet will be rolled at industry and edges will be trimmed after rolling. At 150 °C, cracks become very small.

Effects of rolling temperature and reduction on the appearance of AZ31B sheets rolled at 2000 m/min by single pass operation are shown in Fig.2. These photographs clearly show that 60 % reduction is applicable by single pass rolling at 2000 m/min even at room temperature. Although regularly and periodically arranged longer edge cracks due to shear bands diagonally inclined to the rolling plane are observed, a sound rolled sheet is obtained in the center region of width. Above 100 °C, edge cracks become smaller. At 100 °C with smaller reduction of 34 %, cracks formed at edges propagates to longitudinal fracture at the center are observed. This type of defect called „scissors cracks" is often observed in heavy rolling of ductile metals. It is found that AZ31B sheets exhibits ductile behavior in high speed rolling.

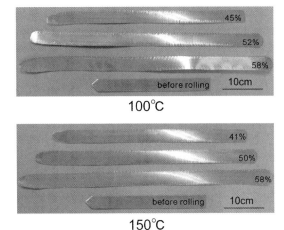

Figure 1: Appearance of AZ31B sheets rolled by single pass at 1000 m/min

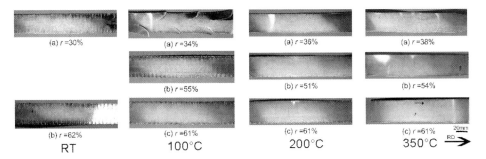

Figure 2: Effect of rolling temperature and reduction on appearance of AZ31B sheets in single pass rolling at 2000 m/min

Occurrence of defects at 2000 m/min and 1000 m/min are summarized in Fig. 3. No transverse fracture occurs under all the rolling conditions performed. Above 150 °C, sufficiently sound sheets are obtained at the rolling speed of 2000 m/min. Fig.3(b) shows the mapping of defects observed on AZ31B sheets rolled at 1000 m/min. Similar to the rolling at 2000 m/min (Fig.3(a)), sufficiently sound sheets can be obtained after single pass large draught rolling. It is found that the high speed rolling faster than 1000 m/min significantly improves the workability of magnesium alloy sheet and that the applicable reduction in single pass rolling increases to 60 %. This does not mean that the limiting reduction is about 60 %. The limit is not clear in the present study because further reduction was not applicable due to the elastic deformation of the rolls.

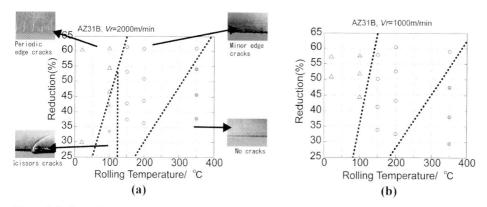

Figure 3: Defects of AZ31B sheets rolled at (a) 2000 m/min and (b) 1000 m/min

3.2 Rolling of ZK60A

Among magnesium alloys, AZ31 is known as the most ductile alloy. It remains uncertain whether the above-mentioned high limiting reduction by high speed rolling can be attained for other magnesium alloys that are less ductile but stronger than AZ31B. Appearance of ZK60A sheets rolled at 1000 m/min by single pass operation is shown in Fig.4. A large reduction of about 60 % can be applied to ZK60A alloy sheets even at 100 °C by high speed rolling, while cracks at both edges are larger than that of AZ31B sheets. Scissors cracks are observed on the sheet rolled 43 % at 100 °C, which suggests that the ZK60A alloy behaves as ductile metals in high speed rolling. Figure 5 shows the mapping of defects observed on ZK60A magnesium alloy sheets rolled at 1000 m/min. ZK60A alloy, which is normally aged after extrusion, shows the highest room temperature strength of the commonly used wrought magnesium alloys. A ZK60A sheet is commercially manufactured by extrusion and understood that it is more difficult to be rolled. However, as is shown in Fig.4, ZK60A alloy sheet is successfully rolled to high reduction by high speed rolling. The tendency of the occurrence of edge cracks in ZK60A sheets is almost similar to that of AZ31B.

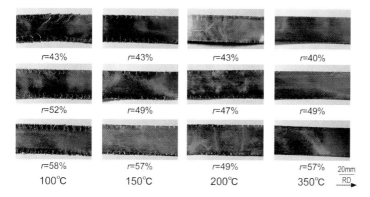

Figure 4: Effect of rolling temperature and reduction on appearance of ZK60A sheets in single pass rolling at 1000 m/min

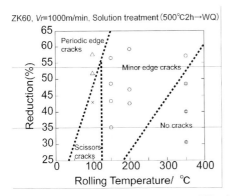

Figure 5: Defects of ZK60A sheets rolled at 1000 m/min

3.3 Rolling of AZ91D

AZ91D alloy shows extremely poor ductility at lower temperatures. It is usually formed by die casting or thixo-molding even in manufacturing thin walled parts. If thin sheets of AZ91D alloy can be manufactured by rolling, the production cost is reduced, the range of application extends and consumption of magnesium alloy increases. Figure 6 shows the appearance of AZ91D sheets rolled at 200 °C at 1000 m/min. Reduction in thickness greater than 40 % could not be applied in high speed rolling, while 35% reduction by single pass rolling was possible. On the sheet rolled 40 %, transverse fracture commonly observed in rolling of magnesium alloys does not occur, while large scissors cracks are observed. This suggests that an AZ91D sheet deforms in a ductile manner in high speed rolling. The single pass reduction of 35 % could not be applied to AZ91D sheets at 200 °C by other rolling methods than high speed rolling. The limiting reduction or workability may be further improved by optimizing rolling condition.

It is found that the high speed rolling is effective for single pass large draught rolling of the wide range of magnesium alloys.

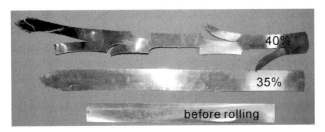

Figure 6: Appearance of AZ91D sheets rolled at 200 °C at 1000 m/min

4 Discussion

The limiting reduction in thickness of magnesium alloy sheets achieved by single pass rolling was surprisingly increased by high speed rolling. Even at room temperature, 60 % reduction was applicable at the rolling speed higher than 1000m/min. The optical micrograph of the AZ31B sheet quenched immediately after rolling to 55 % reduction at 100 °C is shown in Fig.7. This microstructure is observed at the plane perpendicular to the transverse direction. The remarkable feature observed is a band of fine recrystallized grains diagonally inclined to the rolling direction. These fine recrystallized grains may have dynamically nucleated along a shear band where extremely large shear strain accumulated and temperature rise due to almost adiabatic heating occurred. If the dynamic recrystallization did not occur at shear bands, cracks might initiate and fragmentation of rolled sheet might be caused. The softening due to dynamic recrystallization at shear bands caused excellent ductility of magnesium alloy sheets during high speed rolling at lower temperature.

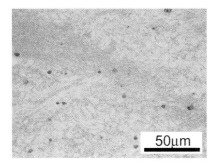

Figure 7: Microstructure of the AZ31B sheet quenched immediately after rolling to 55 % reduction at 100 °C at the rolling speed of 2000 m/min

5   Conclusion

The limiting reduction in thickness of magnesium alloy sheets achieved by single pass rolling is surprisingly increased by high speed rolling faster than 1000 m/min. Reduction larger than 60 % can be applied to AZ31B and ZK60A alloy sheets by single pass even at room temperature by high speed rolling. AZ91D can be rolled to 35 % reduction at 200 °C. The improvement of workability can be explained by dynamic recrystallization initiates at shear bands during high speed rolling. The high speed rolling is proved to be a promising method for mass production of various magnesium alloy sheets at a reduced cost.

6   References

[1]   C.S. Roberts, Magnesium and Its Alloys, John Wiley and Sons, Inc., NJ, USA, 1960, 171–177.
[2]   E.F. Emley, Principles of Magnesium Technology, Pergamon Press Ltd., Oxford, UK, 1966, 550–556.
[3]   H. Utsunomiya, T. Sakai, S. Minamiguchi and H. Koh, Magnesium Technology 2006, (Ed., 4) A. A. Luo et al.), TMS, 201 (2006)
[4]   T. Sakai, H. Utsunomiya, H. Koh and S. Minamiguchi, Proc. Thermec 2006,(Ed., K. Tsuzaki), 2006, in press.

[5] T. Sakai, H. Utsunomiya, S. Minamiguchi and H. Koh, Proc. Int. Symp. on Magnesium Technology in the Global Age, (Ed. O. Pekguleryuz), CIM, 2006, in press.
[6] K. Kato, Y. Saito and T. Sakai, Transactions ISIJ, 1984, 24, 1050–1054.
[7] T. Sakai, Y. Saito and K. Kato, Transactions ISIJ, 1987, 27, 520–525.
[8] T. Sakai, Y. Saito, K. Hirano and K. Kato, Transactions ISIJ, 1988, 28, 1028–1035.

Evaluation of Forming Behaviour and Tribological Properties of Magnesium Sheet-metal Directly Rolled From Semi-continuously Casted Feedstock

B. Viehweger[1], M. Düring[1], A. Karabet[1], H. Hartmann[2], U. Richter[2], G. Pieper[3], F. Newiak[3]
[1]Brandenburg University of Technology Cottbus, Chair of Design and Manufacturing, Cottbus, Germany
[2]Research and Quality Center Brandenburg GmbH, Eisenhüttenstadt, Germany
[3]GP innovation GmbH, Lübbenau, Germany

1   Introduction

Within the last years the reduction of fuel consumption became one of the most important engineering objectives for car design. It is assumed that weight savings of 100 kg can lead to a reduction of fuel consumption of about 0.5 l/km [1]. Opposite to this objective, growing customer demands for more comfortable car equipment cause an increase of overall weight of about 10 kg/year [2]. Actually, this can not be equalised by application of more economic and energy efficient engines and power transmission components. Therefore, weight saving potentials, considering the entire car design, have to be detected and activated. About 20 percent of the weight of a car is a result of the car body [1]. Hence, more sophisticated design and production techniques, such as tailored blanking and hydro-forming, have successfully entered the car body design and contribute to the principal objective of weight reduction. Furthermore, materials substitution becomes more important and is actually realised in form of design approaches applying advanced high strength steels or light metal alloys, such as aluminium alloys. The lightest metallic materials for engineering purposes, especially Mg-wrought-alloys, also have been focus of various examinations concerning the determination of potentials for their future use in car body design.

The general physical and mechanical properties of Mg-wrought-alloy AZ31 exemplarily show that a weight reduction in automotive components design becomes basically possible by means of materials substitution. Considering only the physical density of the Mg-wrought-alloy AZ31, weight savings can reach values of about 35 percent in comparison to aluminium- or up to 70 percent in comparison to steel-alloys. With respect to the density-specific young's modulus and yield strength of commonly used steel materials, the Mg-wrought-alloy AZ31 offers comparable, respectively superior properties for engineering purposes (fig. 1). In particular the Mg-wrought-alloy AZ31, in form of sheet metal, seem to offer chances for a future application in automotive design [3, 4, 5]. Because of its most promising portfolio of material properties, production costs and availability, which could result in medium-term material substitution in several automotive applications, this Mg-wrought-alloy has been the principal focus of research interests concerning sheet-metal application within the BMBF-funded research project "MIA – Magnesium im Automobilbau" [6]. AZ31-sheet-metal is widely known for poor forming behaviour at room temperature. Achievable drawing ratios for commercially available batches of AZ31-sheet-metal are limited to values of 1.4 [7]. The formability of Mg-alloys in general, is mainly determined by the hcp-crystal-structure. Therefore, under cold forming conditions slip movements are supported by available basal plane slip directions. Supplementary twinning in

2nd order pyramidal planes is observed. An increase of forming temperatures of values above 225 °C leads to an enhancement of formability. This is explained by the activation of additional slip directions in pyramidal and prismatic planes [8].

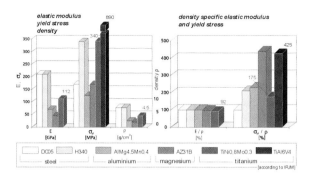

Figure 1: Mechanical properties of Mg-alloy AZ31 compared to several engineering materials

2 Evaluation of Forming Behavior of AZ31-sheet-metal

2.1 Rolling of AZ31-sheet-metal at the BUT

Within research project „MIA – Magnesium im Automobilbau" sheet-metal from Mg-wrought-alloy AZ31 has been rolled at the Brandenburg University of Technology using semi-continuously casted feedstock, manufactured at Otto Fuchs KG. The use of commonly applied intermediate stage of extrusion was thus no more necessary. This approach can offer potentials for cost-savings in production of Mg-sheet-metal. The thickness of supplied AZ31-sheet-metal was 1.0 mm. Commercially available AZ31-sheet-metal exhibits average grain sizes of about 8–10 µm. Despite of typical initial grain size values for semi-continuously casted AZ31-material of about 250-350 µm the rolled AZ31-sheet-metal shows a fine-grained microstructure characterised by values of about 10–15 µm. Experimentally manufactured sheet-metal was fully recrystallised. A nearly perfect basal-plane orientation of elementary cells has been confirmed by means of complementary performed texture measurements. Within the rolling plane an isotropic allocation of mechanical properties is existent.

2.2 Evaluation of Formability of Own Sheet-metal

The formability of AZ31-material, directly rolled from semi-continuously casted material has been evaluated by means of technological forming tests. Forming limit diagrams (FLD's) have been acquired and cup drawing tests have been carried out. The experimental tool consists of die, blankholder and punch, which can be heated separately form each other. The temperature control was automated by means of electronic control system.

The tests have been carried out using hemispherical punch with radius of 50 mm and punch velocity of 20 mm/s. The obtained forming limit diagrams are displayed in fig. 2. As expected, forming results achieved at forming temperatures of at least 175 °C are of practical importance. It is also apparent, that elevated forming temperatures, exceeding values of 250 °C, will not lead to an enhanced formability anymore. Commercially available AZ31-sheet-metal products of Russian and American origin, rolled from extruded stock, have been applied as reference material (fig. 3). The experiments demonstrated obviously, that own sheets, directly rolled from semi-continuously casted material, exhibit the forming limits values almost equal to that of commercially available sheet products.

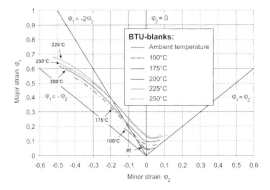

Figure 2: Forming limit diagram for AZ31-sheet-metal, experimentally manufactured at the Brandenburg University of Technology

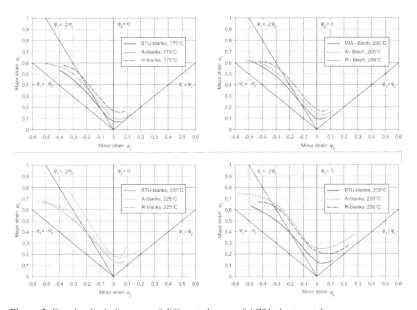

Figure 3: Forming limit diagrams of different charges of AZ31-sheet-metal

In order to evaluate the suitability of own AZ31-material for deep-drawing multiple conventional cup drawing tests have been carried out. Maximum drawing ratios as well as draw-depths have been determined. The punch diameter was 100 mm and punch velocity 20 mm/s. The results have shown that a suitable temperature range for conventional deep drawing has to exceed values of 225 °C. The achievable draw depth was 80 mm for a forming temperature of 225 °C, respectively 90 mm for 240 °C. Corresponding values for the maximum drawing ratio are 2.1 and 2.2 (fig. 4). Moreover, even better results concerning the achievable draw depth and maximum drawing ratio were realized by additional cooling of the punch. This results in an increased strength in skirt areas of the cup. The transmission of drawing forces, supplied by the punch and conducted to the bottom of the cup, into the flange area is eased. The estimated values of draw depths at room temperature were comparatively low and are located in the range of about 15 mm. The achieved maximum drawing ratios and draw depths of the BUT-sheet-metal at increased temperatures meet the formability requirements of widespread variety of workpiece-geometries.

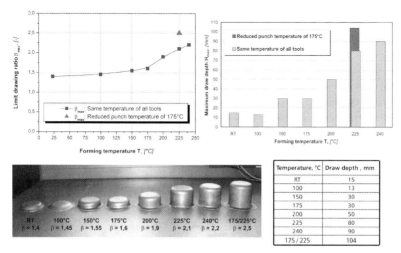

Figure 4: Maximum drawing ratio and draw depths for own AZ31-sheet-metal

2.3 Microstructural Changes During Hot Forming

In order to analyse the microstructural changes during hot forming several samples have been cut out from typical areas of deep-drawn cups. The results of optical light microscopy have shown that nucleation of new grains occurs at grain boundaries and the recrystallisation process is already initiated at forming temperatures above 175 °C and logarithmic plastic strains of about 0.35. As a result of unequal plastic strain distribution across the cup cross section an inhomogeneous microstructure is formed (fig. 5).

Particularly in the areas of high deformation degrees such as flange and cup skirt intensive dynamic recrystallisation takes place. Thereby complete recrystallised structure was not observed. Further investigations regarding the influence of microstructural changes during hot forming on the mechanical properties of formed parts have to be done.

Figure 5: Inhomogeneous microstructure due to recrystallisation (cup cross section)

2.4 Evaluation of Flow Stress Curves

For an exact numerical simulation of forming processes, the material characteristics and values need to be known precisely. The flow stress curves measured by means of tactile strain gauge are usually obtained only till uniform elongation and then extrapolated mathematically by means of different laws to higher strains. These mathematical approaches are based upon assumtions and ideal material laws and therefore describe the reality only insufficient. Moreover, they describe the actual forming behaviour of magnesium alloys very inaccurate due to their softening yield behaviour at elevated temperatures and diffuse necking already at low plastic strains. After reaching the uniform strain the determination of specimen's cross section by means of conventionel tactile strain gauge becomes incorrect. The local strains and their development can not be measured with standardized tensile test.

Hence optical deformation analysis system ARAMIS has been applied for determination of flow stress curves in temperature range of 175–250 °C and constant strain rates of 0,0008–0,08 s^{-1}. Using this system the local deformations in surface direction and, based on the consistent volume, thickness reduction and thus the actual cross section development of tensile specimens could be continuously measured during the entire test. Applied measuring method enabled determination of flow stress curves beyond uniform elongation and thus provided extended input data for numerical simulation. The measuring system is based upon the principle of identification of stochastic patterns on the speciment's surface. Synchronized images are recorded at different load stages using two CCD cameras. Subsequently 3D coordinates and displacements of single points of stochastic pattern are calculated using photogrammetric evaluation procedures. Climatic chamber with two heatable coplanar glass panels and interior lighting for optical accessibility has been installed on the testing machine. From calculated deformations in gauge and thickness direction as well as force values recorded during the test, the true stresses have been calculated with respect to the true strains. Flow stresses have been determined as the ratio of force values and specimen's cross sections (equation 1).

$$kf = \frac{F}{b \cdot s} = \frac{F}{b_0 \cdot e^{\varphi_2} \cdot s_0 \cdot e^{\varphi_3}} \tag{1}$$

The analytical expression for crosshead velocity control (equation 2) has been applied in order to ensure the constant strain rate during the entire tensile test. KF is a correction factor, which has to be determined experimentally and is different depending on forming temperature and nominal strain rate value.

$$V = \dot{\varphi} \cdot e^{-\dot{\varphi} \cdot KF \cdot t} \tag{2}$$

The flow stress curves, measured and calculated by means of optical measurement system, are exemplarily illustrated in fig. 6. They could be evaluated till plastic strains of about 0.7–0.9. For numerical simulation the flow stress curves have been interpolated till strains of 1.0.

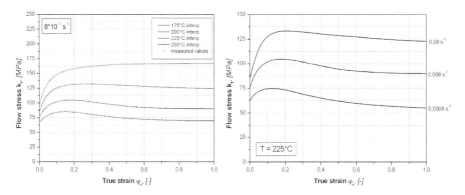

Figure 6: Optically measured flow stress curves, AZ31-sheet-metal

3   Tribological Investigations

In order to determine the influence of surface quality on friction coefficients of AZ31-sheet-metal the application of vacuum-suction-blasting technology of GP innovation GmbH has been chosen as a model technique. This technology enables an eased design of different surface textures in an experimental scale by varying suitable blasting agents of different sizes as well as blasting parameter. Controlled creation of surface textures was desired in order to influence the friction behaviour of AZ31-sheet-metal during forming process.

2D and 3D roughness values of different blasted specimens have been evaluated. Besides conventional values the Abbott-Firestone curves have been determined. Tempered strip drawing tests without bending have been carried out at the Research and Quality Center Brandenburg GmbH in order to determine friction coefficients of different specimens. The tests have been carried out using the contact temperature of 220 °C, drawing speed of 50 mm/s, drawing distance of about 400 mm and lubricant CLW WA 03. Surface pressure has been continuously increased during the test. The objective was the evaluation of friction coefficients at concrete test parameters. Low friction coefficients as well as their stable development during forming process have been desired. Commercial sheet-metal of Russian origin has been applied as reference material. The measured Abbott-Firestone curves are illustrated in fig. 7.

The surface textures of untreated and blasted specimens are very different. Respectively different friction coefficients have been measured in strip drawing tests (fig. 8). Own sheet-metal has relative flat surface with pronounced roughness valleys. Contrary to this the commercially available sheet-metal shows flat surface with many peaks. The untreated own sheets perform much better as sheet-metal of Russian origin due to the fact, that they have much more lubrication pockets (lubrication reservoirs) isolated from each other and thus can enclose and hold more lubricant in the roughness valleys (high S_{vk} value). Moreover, less peaks (low S_{pk} value) must be leveled by own sheet-metal. Commercially available sheet products have much lower number of lubrication pockets and exhibit due to high peak number the occurrence of cold welding („powdering") in tool during forming process and as result higher friction coefficients.

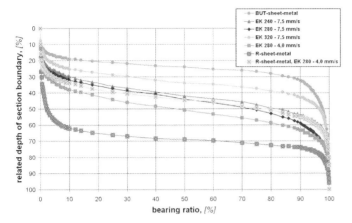

Figure 7: Measured Abbott-Firestone curves of different sheet-metal charges

The blasted specimens show lower friction coefficients as the untreated sheet-metal. The reason for reduced friction is more homogeneous roughness structure (homogeneous peak and valley distribution) as well as favorable profile shape and as a result lower contact area magnitude (fig. 7). All blasted surfaces show similar friction behavior. The blast speed had in the researched range of blasting parameter only a little influence on the friction coefficient. The variation of blasting agents of different grain sizes had more pronounced influence on friction behavior. The investigations demonstrated obviously that cconsiderable improvement of tribological properties of Mg-sheet-metal could be achieved by specimens treated under certain blasting parameters using vacuum-suction-blasting technology.

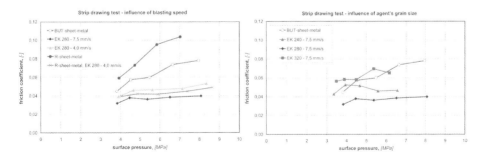

Figure 8: Results of tempered strip drawing tests

4 Conclusions

In order to evaluate the formability of AZ31-sheet-metal, directly rolled from semi-continuously casted feedstock, forming limit diagrams as well as limiting drawing ratios and maximum draw depths have been determined. The experiments demonstrated obviously that own sheets exhibit formability values almost equal to that of commercially available products. The technological tests have been carried out from room temperature up to 250 °C. According to obtained forming limit diagrams, the best forming results can be achieved at 225 °C. The maximum drawing ratio of 2.4 and draw depth of 90 mm could be achieved at 240 °C using conventional cup drawing test. The obtained material properties and experimental data build an important fundamental for layout of forming processes and workpieces. The analysis of microstructure had shown that dynamic recrystallisation begins at forming temperatures above 175 °C and logarithmic strains of 0.35. The flow stress curves have been obtained using optical measurement deformation system in temperature range of 175–250 °C and at constant strain rates of 0,0008–0,08 s^{-1}. Local deformations and thus the actual cross section development of tensile specimens could be continuously measured during entire test. Applied measuring method enabled more precisely determination of flow curves beyond uniform elongation and thus provided extended data for numerical simulation of forming processes. The influence of mechanical treatment of AZ31-sheet-metal using vacuum-suction-blasting technology on friction behavior has been investigated. Considerable improvement of tribological properties of Mg-sheet-metal could be observed by specimens, treated under certain blasting parameters. The blasted specimens showed in general lower friction coefficients as untreated sheet-metal.

5 Acknowledgements

The authors are very grateful to the German Federal Ministry of Education and Research for funding this research work (BMBF project 03N3106K).

6 References

[1] Friedrich, H.; Schuhmann, S.: Forschungsstrategien für ein zweites „Magnesium-Zeitalter" im Fahrzeugbau; in: Materialwissenschaft u. Werkstofftechnik; vol. 32 **2001**; p. 6–12.
[2] Lösch, S.; Straube, O.: Entwicklungsstand zur Verarbeitung von Magnesiumblechen im Karosseriebau; in: conference-proceedings – Leichtbauwerkstoffe in der Verkehrstechnik; Cottbus **2005**; p. 125–136.
[3] Doege, E.; Droeder, K.; Elend, L.-E.: Tiefziehen und Clinchen von Magnesium-Feinblechen in: conference-proceedings - Fortschritte mit Magnesium im Automobil – Leichtbau; Bad Nauheim **2000**; p. 123–141.
[4] Wagener, H.-W.; Hosse-Hartmann, J.: Zum Tiefziehen von Magnesium-Blech; in: UTF science; vol. 2 **2001**; p. 28–34
[5] Viehweger, B. et al: Hydromechanisches Tiefziehen und Hochdruckblechumformung als Verfahren zur Herstellung komplexer Bauteile aus Magnesiumfeinblechen des Typs AZ31B-O; in: Materialwissenschaft und Werkstofftechnik; vol. 35 **2004**; p. 440–446.

[6] Viehweger, B.; Leyens, C.; Becker, J.; Bogon, P.; Gers, H.; Pieper, G.; Kiefer, A.; Roll, K.; Straube, O.: Aktuelle Entwicklungen bei Magnesium-Knetlegierungen; Automotive Materials; vol. 2 **2005**; No. 2; p. 34–37.

[7] Wagener, H.-W.; Hosse-Hartmann, J.: Tiefziehen von Magnesiumblech bei Raumtemperatur und mit partieller Erwärmung; in: conference proceedings - EFB T21 Prozessoptimierung in der Blechverarbeitung; Hannover **2001**.

[8] Doege, E.; Jannsen, S., Elend, L. E.: Grundlagen zum Umformverhalten von Magnesium; in: Kammer, C.; editor: Magnesium Taschenbuch; Aluminium Verlag; Düsseldorf **2000**; p. 357–359.

Investigations on Formability and Tribology of Magnesium Sheets with Regard to Warm Forming Processes

J. Hecht, M. Merklein, M. Geiger
Chair of Manufacturing Technology, Friedrich-Alexander University Erlangen-Nuremberg, Germany
A. Stich
AUDI AG, Ingolstadt, Germany

1 Introduction

The use of lightweight magnesium sheets in the automotive field offers a great potential to realize a weight reduction without concessions to the mechanical properties of the vehicle's structure. At the present time the application of magnesium alloys is mostly limited to parts, manufactured by casting processes. Due to advantageous material properties and a fine-grained structure without porosity, formed sheets of magnesium promise to offer better mechanical properties than castings. For the evaluation of forming processes by means of Finite Element Analysis, fundamental knowledge of the beginning of yielding, the forming limit curve as well as of the friction coefficient to be applied, is necessary. Since these quantities are affected by the process temperature, they also have to be determined at the respective temperature. Therefore commonly used testing techniques have to be adapted for the processing at elevated temperature.

2 Determination of Forming Limits

For assessing the feasibility of forming processes within the Finite Element Analysis, a comparison of the predicted strain distribution is commonly performed with the characteristic forming limit curve. Since the latter one is specific for each material, it has to be determined experimentally. Thereby the general aim is to determine the crucial strain condition which marks the onset of the necking process.

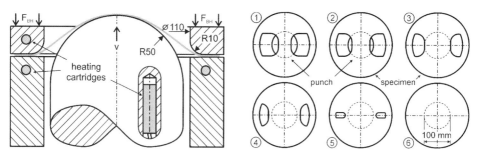

Figure 1: Hemispherical punch (left) for stretching the specimen with varying shapes (right)

With respect to comparability, it is necessary to stretch the sheet in a proper way to guarantee a constant relationship between the major and minor strain. The concept of Keeler [1] and Backofen of stretching sheets over rigid punches allows the determination of strain conditions in the field of positive and negative minor strain values. Different geometries of the involved punch are in use. For the investigation of the magnesium sheets a hemispherical punch was chosen, as Figure 1 (left) shows. To control the temperature of the set-up, heating cartridges are inserted into the punch, into the blank holder and into the drawing die. A furnace next to the set-up enables a short transfer of the heated blank to the heated tooling.

In order to control the minor strain, the shape of the specimen is varied, as depicted in Figure 1 (right). Being stretched by the punch, i. e. the specimen with number 6 represents an equibiaxial strain condition. The other specimens show symmetric cuts of different shapes and sizes. Thus parallel bands of different width remain in the centre of the specimens. Thereby the minor strain decreases with the rising size of the cuts. While the punch is being displaced at constant velocity, the resulting deformation of the specimen is recorded by two CCD cameras. The images are analyzed by an optical strain measurement system and the values of major and minor logarithmic strain become visible.

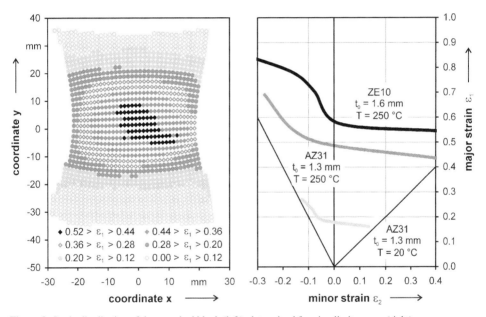

Figure 2: Strain distribution of the stretched blank (left); determined forming limit curves (right)

In [2] several authors propose to restrict the determination of the maximum forming limit condition to the area close to the pole of the specimen. Since the radial tensile stresses, caused by the stretching of the sheet, concentrate towards the center of the specimen, the failure ought to occur right at the pole. Within the temperature range of 200 °C to 350 °C, PTFE-based foils as well as graphite-based lubricants show an acceptable suitability to minimize friction between the punch and the specimen. The graduation of the illustrated points in Figure 2 (left), provided by the optical strain measuring system, shows the concentric pattern of the distribution of the major strain with the maximum at the center. The information on the major and minor strain condition at the maximum is taken as a specific point within the forming limit diagram. The chain of all characteristic points, referable to a specific specimen, is being fitted by a continuous curve, using standard mathematic approaches. Following the described procedure, the curves shown in Figure 2 (right) were obtained. The formability of the well known AZ31 alloy at room temperature is limited by the characteristic hexagonal closed packed crystal structure of magnesium. However, due to the activation of pyramidal slip planes at temperatures exceeding 200 °C, a strong increase of formability can be achieved. This effect is further increased with rising temperature and enables the very high forming limit curve which is shown for the temperature of 250 °C. An even higher formability is shown by the alloy ZE10. The observed improvement in formability is referable to the continuous development of magnesium sheets within the past years. While the presented AZ31 sheets were manufactured in 2003, the ZE10 alloy represents the state-of-the-art of the year 2005.

3   Investigations on Tribology

In sheet forming processes the draw-in of material from the flange into the die cavity is strongly influenced by the clamping force and the friction coefficient between sheet and tool. Besides the commonly used strip drawing test, the cup deep drawing test offers the possibility to determine friction coefficients too [3]. With special regard to the flange areas, where the sheet is stretched in the direction towards the draw-in and compressed in perpendicular direction, the cup drawing test appears to be more adequate to simulate the special conditions during the deep drawing process. For this purpose a heated cup drawing tool was designed (see figure 3 left). In order to be able to vary the contact pressure between the blankholder and the sheet within a widespread area, it was provided with a pneumatic and a hydraulic cushion system, offering blank holder forces within the range of 0.5 kN to 50 kN.

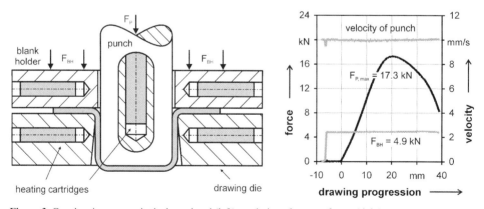

Figure 3: Cup drawing process in the heated tool (left); evolution of process forces (right)

For the determination of friction coefficients the analytical approach of [4] was chosen, which provides the following equation for the maximum drawing force during the cup drawing:
The maximum drawing force $F_{p,max}$ depends on the blank diameter d_p, the mean flow stresses k_{fm1} and k_{fm2} in the flange and at the drawing ring, the initial sheet thickness t_0, the blank holder force F_{BH}, the radius of the drawing die r_R and the friction coefficient μ. As shown in Figure 3

(right), the cup drawing tool, equipped with load cells for the continuous recording of the punch force and the blank holder force, enables the determination of the necessary quantities for solving the upper equation. The diameter of the punch is 50 mm, the diameter of the drawing die is 54 mm. The thickness of the investigated magnesium sheets is 1.3 mm. The corner edge radii of the punch and the die are 10 mm.

In favor of the comparability of the experiments, a uniform blank diameter of 110 mm (drawing ratio: ~2.2) was predefined. Following the procedures for the interpretation of the tests, described in detail in [5], several lubricants in the field of oils, greases, PTFE-based foils and sprays, as well as dry lubricants were investigated. With regard to the specific thermal stability, tests were performed at 200 °C, 250 °C and 300 °C. The determined friction coefficients for an extract of the lubricants are shown in figure 4 (left).

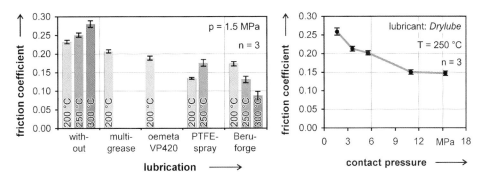

Fig. 4: Comparison of different lubricants and impact of contact pressure

The contact pressure right at the maximum of the punch force was 1.5 MPa. The unlubricated contact between the magnesium sheet AZ31 and the tool can be considered as worst case scenario and as reference for the measure of quality of a lubricant. The lubricants *multigrease*® and the synthetic oil *oemeta*® *VP420* show a limited thermal stability at temperatures higher than 200 °C, whereas PTFE-based lubricants and dry lubricants (i.e. *Beruforge*) also can be used at temperatures of 250 °C and 300 °C respectively. The oils and greases were applied with the help of a rotating sponge to guarantee a constant amount of applicated lubricant on the upper and lower surface of the sheet. The scatter of the determined friction coefficients increases with rising temperature. In addition to the temperature, the influence of the contact pressure was investigated. The result can be seen for the dry lubricant *Drylube*® at 250 °C in figure 4 (right). The friction coefficient obviously decreases with rising contact pressure. As reported by [6], this behavior can be referred to hydrostatic and hydrodynamic effects due to the lubricant pockets enclosed between the surface asperities of the material.

4 Implementation of a Double-Sheet Hydroforming Process

Dedicated for the design of warm forming processes, the gained knowledge in the field of formability and tribology was applied to the manufacturing of a demonstrator part of complex shape. Following the request to make use of the characteristic advantages of the double sheet hydroforming technique, a clam-shell similar geometry was designed, as pointed out in figure 5. The shape was designed in the style of a nodular element and was provided with different geometrical details offering challenges for the forming process. These were performed with respect to the required comparability of the results to other geometries. The technique of forming both sheets of a part within one tool is illustrated in figure 5 (right).

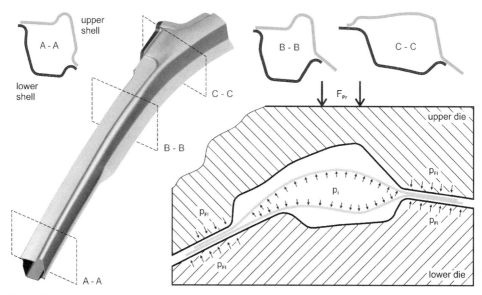

Fig. 5: Shape of the projected part (left); equilibrium of internal pressure and contact pressure in the flange

In contrast to regular deep drawing processes, where a single sheet is drawn by the progression of the plunger of a press, double sheet hydroforming implies that the upper and the lower sheet of a part are affected by a uniform hydrostatic fluid pressure which makes them to adapt the geometry of the forming tool. As a consequence the latter one comprises only two dies which fulfill the purpose of forming the respective sheet and serving as blankholder to the opposite sheet at the same time. Thus the integration within one forming tool leads to reduced tooling costs, which is of special importance when limited numbers of a part have to be produced.

The forming process is based on the stretching of the blank as well as a material draw-in from the flanges into the die cavity. For the ability to apply a pressure p_i on the sheets, the flanges of the sheets have to be subjected to a certain contact pressure p_{FL} in order to guarantee a metallic sealing between the flanges and the tool as well as between the mutual flanges. In favor of a robust process the contact pressure has necessarily to be at the very least of the magnitude of the interior pressure. According to figure 5, during the process a press of appropriate size has to apply the force F_{Pr} on the tool that has to provide the above mentioned contact pressure in the flanges as well as an equivalent force to the opening force, rising up by the pressure affected surface of the sheets. The characteristic evolution of the pressure and the force during a cycle can be observed in figure 6.

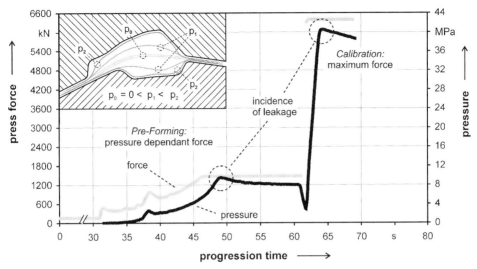

Fig. 6: Evolution of the characteristic process quantities fluid pressure and press force

After the closing of the tool, the press remains in idle motion for approximate 30 seconds which proved to be sufficient to heat up the blanks just by contact to the heated tool. Then the so-called *Pre-Forming* step launches and fluid is pumped by the pressure intensifier between the sheets. The force F_{Pr} is increased as a function of the fluid pressure which guarantees, that the draw-in from the flanges into the die cavity is not suppressed. When the predefined force for the *Pre-Forming* is reached and kept constant, the fluid pressure shortly continuous to further increase until leakage occurs due to insufficient contact pressure in the flanges. In comparison to the fluid pressure at this occasion, the forming of the sheets already achieves a remarkable high degree, similar to the condition related to fluid pressure p_2, shown upper left in figure 6. For the *Calibration* stage the force F_{Pr} is increased to the maximum to gain a sharp pressure increase up to a maximum value, restricted by a final leakage again.

For the forming process magnesium sheets (AZ31) with a thickness of 3.0 mm were dispatched. Providing low friction coefficients the dry lubricant Beruforge served to optimize the material draw-in. However at a temperature of 200 °C the results were not yet satisfying. First, the limited force of 6.300 kN, provided by the given hydraulic press, lead to a maximum sealable fluid pressure of approximate 40 MPa, which was not sufficient for the forming to the full extend. Second, the bending induced strains, referable to the deliberately chosen sharp radii at the die inlet corners sometimes caused a premature failure. At a temperature of 250 °C a reproducible forming process became possible. The thermally activation of additional sliding planes in the hexagonal closed packed crystal structure made the material to cope with the bending strain. The parallel decrease of the flow stress caused a supplementary material softening which improved the forming results to the level depicted in figure 7.

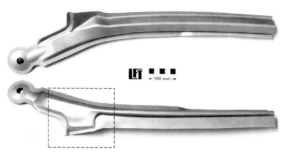

Fig. 7: The manufactured magnesium part

5 Concluding Remarks

In association with innovative forming processes at elevated temperature, the outstanding formability of magnesium sheets pave the way to manufacture sheet metal parts of low weight, high complexity and advantageous mechanical behavior. For the minimization of friction between the sheet and the die, dry lubricants proved to be a useful solution since they do not tend to evaporate at elevated process temperatures and can be easily removed due to their water-based nature. As presented in the case of the clam-shell similar part, the temperature induced softening of the material offers the crucial contribution to form the part with a press of given size to a higher degree. So, the increased expenses, connected to the processing at elevated temperature, can be partly compensated by a reduced press size.

6 References

[1] Keeler, S.P.: Plastic instability and fracture in sheet stretched over rigid punches. ASM Trans. 56, **1964**, p. 25–48.
[2] Liebertz, H.; Duwel, A.; Illig, R.; Hotz, W.; Keller, S.; Koehler, A.; Kroeff, A.; Merklein, M.; Rauer, J.; Staubwasser, L.; Steinbeck, G.; Vegter, H.: Guideline for the determination of forming limit curves. *Proceedings of IDDRG 2004*, Germany, p. 216–224.
[3] Witthüser, K.-P.: Untersuchung von Prüfverfahren zur Beurteilung der Reibungsverhältnisse beim Tiefziehen, PhD, University of Hanover, Germany, **1980**.
[4] Siebel, E., Beisswänger, H.: Tiefziehen, München-Hanser, Germany, **1955**.
[5] Tolazzi, M.; Vahl, M.; Geiger, M.: Determination of friction coefficients for the finite element analysis of double sheet hydroforming with a modified cup test; *Proceedings of ESAFORM 2003*, Salerno, Italy, **2003**, p. 479–482.
[6] Azushima, A.: Direct Observation of Contact Behavior to Interpret the Pressure Dependence of the Coefficients of Friction in Sheet Metal Forming, *Annals of CIRP*, Vol. 44/1, **1995**, p. 209–212.

Superplastic Forming of Magnesium Alloys: Composition and Microstructure Effects

R. Boissière, J. J. Blandin

Institute National Polytechnique de Grenoble (INPG), Génie Physique et Mécanique des Matériaux (GPM2), UMR CNS 5010, ENSPG, BP46, 38402 Saint-Martin d'Hères, France

Abstract

The possibility to perform superplastic forming (SPF) of magnesium alloys has received increasing attention in the recent past. To do that, fine grained and stable microstructures are required. This work investigates some effects of composition and microstructure on the superplastic properties of magnesium alloys. Thanks to appropriate thermomechanical treatments (including equal channel angular extrusion), various microstructures have been produced, resulting in important changes of the experimental domain (temperature, strain rate) in which superplastic properties can be obtained. Deformation mechanisms in the superplastic regime are discussed and a particular attention is also given to the damage resistance of the alloys during superplastic deformation. It is shown, in particular thanks to the use of X-ray microtomography for imaging the cavities, that depending on both the alloy composition and the microstructure, this resistance may vary significantly resulting in a variation of the fracture strain.

1. Introduction

Due to its hexagonal close packed structure, magnesium alloys exhibit generally poor formability at room temperature. Indeed, it is considered that only basal slip is activated at room temperature, due to a critical resolved shear stress (CRSS) for non basal slips much larger than for basal one. It is also a reason why in addition to dislocation slip, twinning often occurs. Moreover, the necking resistance during forming of Mg sheets can be also limited by relatively low strain hardening exponents [1].

A way to promote the formability of Mg alloys is to form at a high enough temperature since for temperatures of about 200°C or higher, the difference between CRSS for pyramidal or prismatic slips and the basal one is noticeably reduced. Moreover, at high temperature, for appropriate microstructures and experimental conditions (temperature, strain rate), magnesium alloys can exhibit superplastic properties. In SuperPlastic Forming (SPF) conditions, strains typically larger than one can be reached. Such forming capacities are useful when significant strains are required, even locally, in industrial components. The capacity to carry out superplastic forming at an industrial scale generally needs to satisfy several conditions: the ability to produce fine grains, the superplastic properties in an experimental interval remaining compatible with industrial requirements, a damage sensitivity which does not lead to the use of superimposed pressure during forming.

The aim of this paper is to give pieces of information about these points in the case of AZ (Mg-Al) alloys with a particular attention given to the effects of composition (i.e. Al content) and microstructure (i.e. grain size).

2. Capacity to produce fine grains

In the last years, grain refinement by thermomechanical processing (extrusion, rolling) of magnesium alloys has received attention [2-4]. By extrusion of rolling of Mg-Al alloys, grain sizes typically from few microns to about 15 µm are frequently produced after complete recrystallization. Fig. 1a shows for instance the microstructure of an AZ91 alloy after extrusion carried out at 370°C with an extrusion ratio of about 15. A mean grain size less than 5 µm is measured. It has been widely shown in the case of aluminium alloys that a way to get ultra fine grains is to accumulate strain during thermomechanical processing. To do that, various techniques (so-called severe plastic deformation -SPD- techniques) have been investigated, like friction stir processing, accumulative roll bonding or equal channel angular extrusion (ECAE). Most of these techniques have been recently applied with success to magnesium alloys [5-7]. In particular, the refinement efficiency of ECAE (despite the associated difficulties for industrial scaling up of this process) of Mg alloys has received sustained attention. Fig. 1b shows a TEM observation of the microstructure of an AZ91 alloy after ECAE processing, displaying a structure size close to 0.5 µm.

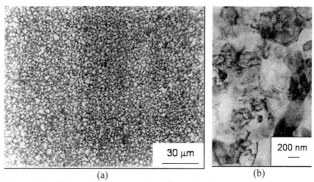

Fig. 1: Examples of fine grained magnesium alloys after thermomechanical processing

In both conventional and severe plastic deformation techniques, the fine grains are produced by recrystallization, even if the mechanisms of deformation may differ. The production of fine grains requires to accumulate large enough strains to get a strong driving force for recrystallization. However, as already mentioned, for Mg alloys, such critical strains are difficult to obtain at room temperature. This means that thermomechanical processing of Mg alloys must include steps in a temperature interval for which non basal slips are activated. A consequence of this requirement is that, depending on the composition of the alloy (particularly in terms of solute content), recovery processes can play a significant role by limiting the efficiency of recrystallization to produce fine grains. Consequently, for instance that in the case of SPD techniques applied to AZ alloys, the resulting structure size depends not only on processing conditions like the total accumulated strain, the temperature or the route but is also closely related to the composition of the alloy. For given processing conditions, the resulting grain sizes decrease when the aluminium amount increases, this solute affecting both the content in second phase particles and the recovering ability of the alloy. For relatively low aluminium amounts, like in the AZ31 alloy which is of particular interest in terms of industrial applications, it is thus relatively difficult to produce ultra fine grains. By conventional processing, like rolling, grain size of about 10 µm are generally obtained and even by ECAE, structure sizes larger than those produced in the AZ91 alloy are obtained.

3. Superplastic properties

Superplastic properties are closely related to the ability for given experimental conditions to activate grain boundary sliding (GBS) rather than dislocation creep. It is well known that dislocation creep is not sensitive to grain size effect whereas GBS is strongly favored by a reduction in grain size. This means that for given experimental conditions (temperature, strain rate), there will be a critical grain size under which GBS is expected to be the dominant mechanism of deformation. This will correspond to the "superplastic region".

In the case of magnesium alloys with microstructures displaying grain sizes between 5 and 15 μm, superplastic properties are generally obtained in a quite large temperature interval, typically between 300°C and 450°C and for strain rates in the range $10^{-4} - 10^{-3}$ s^{-1}. Compared to aluminum alloys with quite similar grain sizes, the superplastic domain for Mg alloys is roughly in the same strain rate interval but for lower temperatures. In terms of elongation to fracture in tension, values between 400 and 700 % can be achieved, which is also more or less in the same range to what is reached for superplastic aluminum alloys.

In the case of very fine microstructures produced by SPD techniques, the temperature for superplastic forming can still be reduced. For instance Fig. 2 displays a view of an ECAE processed AZ91 sample after superplastic deformation in tension at 250°C, exhibiting an elongation to fracture e_f of about 1000 %. Values of e_f of about 2000 % were even reported very recently in the case of a ZK60 alloy after ECAE [8].

The superplastic behavior of magnesium alloys can be rationalized according to:

$$\frac{\dot{\varepsilon}}{\dot{\varepsilon}_0} = \frac{A}{kT} \left[D_L + \alpha \frac{\pi \delta}{d} D_{gb} \right] \left[\frac{b}{d} \right]^p \left[\frac{\sigma - \sigma_{th}}{G} \right]^n \quad (1)$$

with $\dot{\varepsilon}$ the strain rate, σ the flow stress, σ_{th} a threshold stress, b the Burgers vector of dislocations, G the shear modulus, T the temperature, $\dot{\varepsilon}_0$, A, k constants, p the grain size exponent, n the stress exponent, D_L the lattice diffusion coefficient, D_{GB} the grain boundary diffusion coefficient, δ the grain boundary thickness and α a constant. This constant would be equal to 1 for pure diffusion creep but in the case of a deformation controlled by GBS, it has been previously shown that a value close to 0.017 was more appropriate to fit the experimental behaviours of superplastic magnesium alloys [9]. For ultra fine grains, a threshold stress must sometimes be taken into account but in the case of the ECAE processed AZ 91 alloy shown in Fig. 1b, σ_{th} was lower than 1 MPa at 250°C and 10^{-3} s^{-1}.

A value of the stress exponent close to 2 is frequently measured for superplastic deformation of magnesium alloys whatever the composition or the grain size [7,9]. This value suggests that GBS is preferentially accommodated by dislocation climb in Mg alloys. However, depending on the conditions of deformation, lattice or grain boundary diffusion can control the climb. In the case of magnesium alloys, due to high grain boundary diffusivity, one can show that in most superplastic conditions, grain boundary diffusion is expected to play a key role. This is particularly true for the ultra fine grained microstructures produced by severe plastic deformation but even for conventional microstructures, the contribution of GB diffusion remains probably significant. This result differs to what can be estimated in the case of superplastic aluminium alloys for which lattice diffusion is frequently dominating. This particular role of grain boundary diffusion in superplastic Mg alloys has some practical consequences. Indeed, when grain boundary diffusion dominates, the grain size exponent is expected to be equal to 3, resulting in an important sensitivity of the flow stress for forming upon the grain size. Fig.3 displays the variation of $(\dot\varepsilon kT/D_{gb}Gb)(d/b)^3$ with $(\sigma-\sigma_{th})/G$ for ECAE processed AZ91 microstructures, confirming the value of a grain size exponent equal to 3. A. This result supports the idea that grain refinement can noticeably increase the superplasticity domain of Mg alloys as long as the size of the grains can be preserved during superplastic deformation.

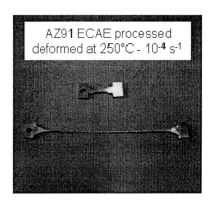
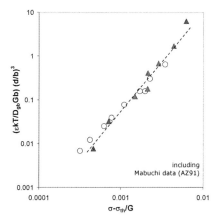

Fig. 2: Tensile sample of an AZ91 alloy processed by ECAE and deformed at 250°C and 10^{-4} s^{-1}

Fig. 3: Variation of $(\dot\varepsilon kT/D_{gb}Gb)(d/b)^3$ with $(\sigma-\sigma_{th})/G$ for ECAE processed microstructures

4. Damage resistance

Two main features control the maximum ductility which can be reached during superplastic deformation: the plastic stability, which is related to the capacity for the alloy to resist to necking by strain rate hardening (i.e. in this case the key parameter is the stress exponent n, the lower the value of n, the higher the necking resistance) and the strain induced damage resulting from cavity nucleation, growth and coalescence. In the case of aluminium alloys, it has been widely shown that superplastic deformation frequently induces strain induced cavitation leading to fracture. For magnesium alloys, recent studies [11,12] have suggested that damage can be induced during superplastic deformation but that the resistance to cavitation can vary drastically depending on the alloy (i.e. composition, microstructure...) but also upon the conditions of deformation (temperature, strain rate) for a given alloy.

Various techniques can be used to characterise strain induced cavitation like the variation of relative density of the alloy or quantitative metallography from polished sections. However, as a result of the important role played by GBS, irregular shaped voids sometimes develop during superplastic deformation. In such a framework, it is particularly delicate to get reliable experimental data concerning the damage characteristics from quantitative metallography. A way to overcome this difficulty is to use X-ray micro-tomography technique, which provides 3D images of the bulk of materials. It has been shown that this technique is particularly well adapted for damage characterisation during superplastic deformation, providing 3D images with a resolution close to 1 μm of the population of cavities [13,14].

Fig. 4 shows such a three dimensional view of the cavitation induced by superplastic deformation of an AZ31 alloy superplastically deformed at 400°C after a strain of about 1.3. This image confirms that magnesium alloys can be sensitive to damage during superplastic deformation and that quite large cavities with complex shapes can be obtained for strains close to fracture strain. Such cavity shapes confirm also the predominant role of GBS as the main mechanism of deformation. Indeed, when the alloy deforms by dislocation creep, elongated cavities parallel to the stress axis are preferentially formed.

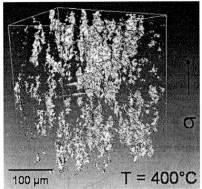
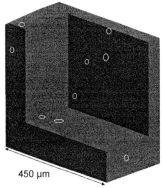

Fig. 4: 3D view of damage after superplastic deformation of an AZ31 alloy (400°C, ε ≈ 1.3)

Fig. 5: 3D view of damage in an ECAE processed AZ91 after superplastic deformation (250°C, ε ≈ 1.4)

Fig. 5 displays a 3D view of the microstructure of the ECAE processed AZ91 alloy deformed at 250°C, after a superplastic strain of about 1.4. Very few cavities are observed (white circles in fig. 5). Nevertheless, thanks to the X-ray micro-tomography analysis and despite very low cavitation levels, a continuous increase with strain of the cavity volume fraction was measured, confirming the possibility to detect low cavity volume fraction by this technique. This damage increase was attributed to cavity nucleation rather than cavity growth since the mean cavity diameter remained roughly constant whereas the number of cavities per mm^3 increased with strain [7]. Moreover, round shapes of the cavities were maintained up to large strains, supporting the idea that cavity growth was controlled by diffusion process when superplastic properties are obtained for such low temperatures (250°C) as a result of the ultra fine grains produced by severe plastic deformation techniques. One can also remark that this good resistance to cavitation of the AZ91 alloy was unexpected since conversely to alloys with lower aluminium content (like the AZ31 alloy), which can be roughly considered as single phase alloys in the superplastic regime, about 10 vol. % of $Mg_{17}Al_{12}$ particles can be present in AZ91 alloy during superplastic deformation at 250°C. It is well known that cavitation is frequently promoted by second phase particles but in the case of Mg-Al alloys, due to their low melting point, the $Mg_{17}Al_{12}$ particles may admit some ductility at the temperature of superplastic deformation, which probably reduces their efficiency to act as nucleation sites for cavities [15].

5. Conclusions

Due to the difficulty to form magnesium alloys at room temperature, the optimisation of high temperature forming processes may contribute to increase the use of magnesium alloys in industry. In this framework, SuperPlastic Forming can appear as an attractive solution for some specific applications. Fine grained Mg alloys can be produced thanks to various techniques leading to large superplastic elongations to fracture for experimental conditions compatible with industrial requirements. During superplastic forming of Mg alloys, grain boundary diffusion plays an important role. Magnesium alloys can be sensitive to strain induced cavitation but this sensitivity appears to depend on both the alloy composition and the grain size affecting the experimental conditions of deformation.

6. References

[1]   S.R. Agnew, O. Duygulu, Int. J. Plast. 21 (2005) 1161.
[2]   Y. Chino, R. Kishihara, K. Shimojima, H. Hosokawa, Y. Yamada, C. Wen, H. Iwasaki, M. Mabuchi, Mater. Trans. 43 (2002) 2437.
[3]   N.V. Ravi Kumar, J.J. Blandin, C. Desrayaud, F. Montheillet, M. Suéry, Mater. Sc. Eng., A359 (2003) 150.
[4]   J.A. del Valle, M.T. Perez-Prado, O.A. Ruano, Mater. Sci. Eng., A355 (2003) 68.
[5]   T. Mukai, M. Yamanoi, H. Watanabe, K. Higashi, Scripta Mater., 45 (2001) 89.
[6]   L. Yuanyuan, Z. Datong, C. Weiping, L. Ying, G. Guowen, J. Mater. Sc., 39 (2004) 3759.
[7]   A. Mussi, J.J. Blandin, L. Salvo, E.F. Rauch, Acta Mater., (2006) in press.
[8]   R. Lapovok, R. Cottam, P.F. Thomson, Y. Estrin, J. Mater. Res., 20 (2005) 1375.
[9]   H. Watanabe, T. Mukai, M. Kohzu, S. Tanabe, K. Higashi, Acta Mater., 47 (1999) 3763.
[10] J.V. Aguirre, H. Hosokawa, K. Higashi, Mater. Sc. Forum, 419-422 (2003) 545.
[11] C.J. Lee, J.C. Huang, Acta Mater., 52 (2004) 3111.
[12] C.F. Martin, C. Josserond, L. Salvo, J.J. Blandin, P. Cloetens, E. Boller, Scripta Mater., 42 (2000) 375.
[13] L. Dupuy, J.J. Blandin, Acta Mater., 50 (2002) 3253.
[14] J.J. Blandin, Mater. Sc. Forum, 426-432 (2003) 551.

On the Superplastic Forming of the AZ31 Magnesium Alloy

Fadi Abu-Farha & Marwan Khraisheh
Center for Manufacturing & Mechanical Engineering Department, University of Kentucky, Lexington, KY 40506, USA

1 Introduction

Environmental and economical issues have been increasingly demanding reduced fuel-consumption and exhaust-emission vehicles. Despite the different means to satisfy these demands, reduction of mass is the most influential and least costly one, if large reductions of 20-40% are to be achieved [1]. Leading car manufacturers have shown that more than 50% of fuel consumption is mass dependent [2–4]. And since magnesium is the lightest constructional metal on earth, it is naturally a prime target for many investigators, promising significant weight reduction if it could successfully replace steel and aluminum. A number of magnesium auto parts have been evolving recently, yet they mainly fall into the cast-components category [5]. Unless magnesium usage is expanded to cover other areas, mainly sheet metal forming, feasible weight reduction will be limited. The metal's inferior ductility at room temperature, however, is a key factor in limiting such an expansion.

AZ31 Mg alloy is commercially available in sheet form, and possesses good mechanical properties. High strength-to-weight ratio in particular provoked the interest in this alloy for structural components. Warm forming has been carried out to enhance the formability of the alloy, as demonstrated by many investigators who successfully formed various components, some for automotive applications [6–8]. Yet, a more attractive attribute of this alloy is its superplastic behavior at higher temperatures [9–11]. Superplasticity stretches the limits of formability of magnesium alloys beyond conventional, offering more opportunities for magnesium usage in the automotive sector.

Recently, a large number of studies investigating superplastic deformation aspects in the AZ31 Mg alloy have evolved. The various mechanical aspects of deformation and microstructural changes have been fairly covered by these various studies [11–16]. Yet, no available study covers and combines both the mechanical and microstructural aspects of its deformation over a wide range of temperatures and strain rates. Our goal is to develop a multi-scale constitutive model that can describe the superplastic deformation of the material under various conditions, taking anisotropy and microstructural evolution into account. The framework of the constitutive model has been already developed by the investigators based on the continuum theory of viscoplasticity [17–18]. Calibration of such a model requires various tests covering a wide-range of operating conditions, mainly temperature and strain rate.

This study presents a comprehensive investigation of the superplastic behavior in the AZ31-H24 magnesium alloy. Uniaxial tensile tests are carried out at constant strain rates, varying between $2 \cdot 10^{-5}$ and 10^{-2} s^{-1}. Each band of strain rates is covered at temperatures between 325 and 450 °C, in a 25 °C increment. In addition, 225 °C is also covered for comparison, as a typical warm forming temperature. Tensile tests are carried out in order to correlate flow stress to plastic strain under the various forming conditions. Strained specimens will be examined for the microstructural changes in the material, quantified in terms of grain growth and cavitation.

Some of the preliminary results on cavitation are presented here. On the other hand, strain rate jump tests are conducted over similar ranges of temperature and strain rate. Four jumps are carried out, and strain rate sensitivity index m is evaluated at each jump. The outcome of these tests is used to investigate the effects of plastic strain, strain rate and forming temperature on the strain rate sensitivity index of the alloy.

For better understanding of the material behavior during actual forming practices, the analysis has to be expanded to multi-axial loading cases, biaxial stretching in particular. Therefore, a pneumatic bulge forming setup has been built, and currently being used to blow-form AZ31 Mg sheets at various temperatures and strain rates. The setup is presented here.

2 Experiments

High temperature uniaxial tensile tests were carried out using a 5582 INSTRON load frame, equipped with electrical resistance heating chamber (*furnace*) that provides a maximum temperature of 610 °C. 5 KN load cell was used for load measurement. Strain measurement was based on the direct displacement of the cross-head beam, which is fairly satisfactory in superplastic studies due to the high plastic strains involved. 3.22 mm thick AZ31B-H24 commercial magnesium alloy sheets were used to prepare the testing specimens. 19×6.35 mm gauge section specimens were machined at 0° with respect to the rolling direction.

Several problems were encountered when conventional sliding-wedge type grips are used for high temperature tensile testing; slippage, material flow and inaccurate strain measurements are the most significant. To overcome these problems, a new set of grips was built specifically for these tests, and proved to be effective and accurate. The new grips and the test specimen are shown in Figure 1. More details about these grips and gripping/testing issues during high temperature tensile testing can be found in previous publications [10, 19].

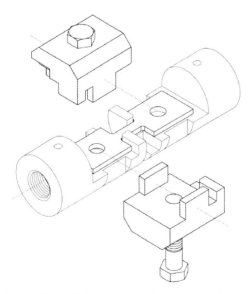

Figure 1: New high temperature grips and test specimen

3 Uniaxial Tensile Tests and Mechanical Behavior

Constant strain rates uniaxial tensile tests were conducted at temperatures varying between 325 and 450 °C, with 25 °C increment. Though it is a warm forming temperature, 225 °C is also considered for evaluating ductility enhancement achieved by superplasticity. For each temperature, strain rates between $2 \cdot 10^{-5}$ and 10^{-2} s^{-1} were covered, extracting a band of stress/strain curves, like the ones shown in Figure 2-a. The corresponding deformed specimens are shown in Figure 2-b, which clearly indicates the enhancement in ductility and deformation uniformity as strain rate decreases.

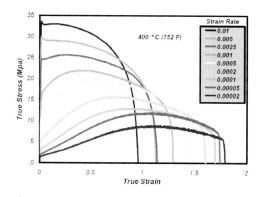
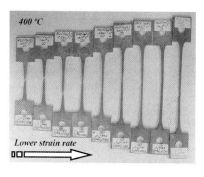

a) b)
Figure 2: Stress/strain curves and the corresponding deformed specimens at 400 °C

Figure 3-a presents all the flow stress/strain rate curves obtained at the various forming temperatures. The well-known sigmoidal-shaped behavior is observed. At each specific temperature, flow stress sensitivity to strain rate is strongly depicted. In other words; the higher the strain rate, the higher the flow stress of the material. A quick assessment of this sensitivity can be made by considering the slope of each curve at any specific strain rate. It is observed that for all the temperatures but 225 °C, significant increase in the curve's slope takes place as strain rates decreases below 10^{-3} s^{-1}. Yet, a more accurate assessment in this regard requires a quantitative evaluation of the strain rate sensitivity index m by means of strain rate jump tests, which is covered later in this work. On the other hand, the maximum attainable elongation is inversely proportional to strain rate, as demonstrated by Figure 3-a. And since 200 % elongation is usually considered indicative of superplasticity, it can be inferred that 10^{-3} s^{-1} is roughly the threshold for superplastic behavior in the alloy at all the considered high temperatures (> 325 °C). Clearly, superplasticity cannot be achieved by warm forming at 225 °C, except for extremely low strain rates.

Data points presented in Figure 3 are re-plotted against temperature for the various strain rates, as shown in Figure 4. Two observations can be extracted from the two sets of curves. First; for a fixed temperature, varying strain rate has more effect on fracture-strain, compared to the case when strain rate is held fixed and the temperature is varied. Second; for any strain rate, there is a limiting temperature beyond which no further ductility-enhancement can be achieved. Figure 4-b in particular demonstrates the conclusion that forming the AZ31 Mg alloy at temperatures higher than 425 °C brings no benefit when ductility is concerned.

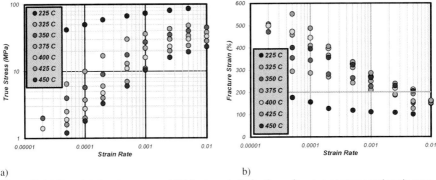

a) b)
Figure 3: (a) Stress/strain rate curves and (b) fracture strain for the various temperatures and strain rates

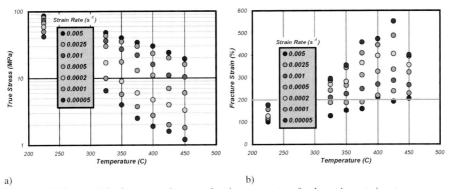

a) b)
Figure 4: (a) Stress and (b) fracture strain versus forming temperature for the various strain rates

3.1 Cavitation

Most superplastic alloys develop internal cavitation during deformation, which not only lead to premature failure, but also adversely affect the mechanical properties of the formed part. Figure 5 for instance, shows the clear evidence of cavitation in specimens deformed to different strains at 400 °C. Quantitatively, cavitation is expressed in terms of the area fraction of voids, denoted f_a. These and other preliminary observations induced the need to further investigate the subject. The effect of different forming parameters on cavitation is being currently carried out.

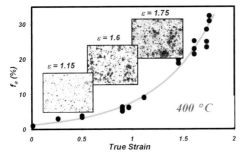

Figure 5: Escalation of cavitation with plastic strain at 400 °C

4 Strain Rate Jump Tests and Sensitivity Index

High strain rate sensitivity of flow stress to strain rate is one of the most intrinsic aspects of superplastic deformation, which accentuates the need for accurate determination of the strain rate sensitivity index m in superplastic studies. In a previous work, it was illustrated how inaccurate and misleading the value of the sensitivity index can be, when evaluated by taking the slope of the stress/strain rate curve at any specific point [19]. Therefore, strain rate jump tests were conducted, covering strain rates between $1 \cdot 10^{-5}$ and $2.5 \cdot 10^{-2}$ s^{-1}. The whole band of strain rates was covered at temperatures between 325 and 450 °C, just similar to the uniaxial tensile tests. Also, 225 °C was considered for comparison, as a warm forming temperature. To enhance the accuracy of our evaluation, the jump between every two successive strain rates was carried out at four plastic strain values; upward jumps at 0.2 & 0.4, and downward jumps at 0.3 & 0.5. The average of the four m values was set corresponding to the mean of the two strain rates between which the jumps were performed. A summary of the results is shown in Figure 6, where m is plotted against both strain rate and temperature.

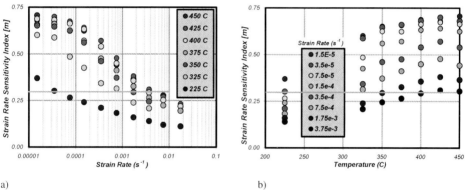

a) b)

Figure 6: Strain rate sensitivity index m versus (a) strain rate and (b) forming temperature

For a fixed temperature, m is in general inversely dependent on strain rate, up to a certain point within the low strain rate region, as shown in Figure 6-a. By considering the horizontal line corresponding to 0.3, it is roughly concluded that 225 °C does not provide enough strain rate sensitivity for superplastic behavior to take place, except for extremely low strain rates. Additionally, any strain rate lower than 10^{-3} s^{-1} guarantees superplastic behavior at all the high temperatures. This highly agrees with the previous discussion in which the boundaries of the superplastic region were set based on fracture-strain values.

Though temperature's increase seems to improve m, a clearer look on its effect may be gained by considering Figure 6-b. Raising the temperature beyond 425 °C adversely affects m, which coincides with the conclusion regarding the effect of temperature on the maximum attainable fracture-strain, shown in Figure 4-b. Another interesting conclusion supporting the previous discussion; varying strain rate significantly affects the m value at a fixed temperature, while holding a fixed strain rate value yields less effect on m when temperature is varied.

5 Bulge Forming

The uniaxial loading case can be scarcely considered as sufficient in describing the actual material behavior during sheet metal forming practices. Load multi-axiality and anisotropy are topics to be considered, if a more accurate analysis is intended. Therefore, a pneumatic bulge forming setup has been designed and built to study the biaxial stretching of sheet metals in superplastic forming processes. A section in the bulge forming die setup is shown schematically in Figure 7-a; the open die is shown here. Multiple dies of different geometries can be simply interchanged to form sheets into various shapes, as shown in Figure 7-b. AZ31 Mg sheets are formed at elevated temperatures onto the die using argon gas at controlled pressures, by means of electronically controlled regulator. After clamping the sheet onto the die, the whole setup is heated to the desired temperature, before bulge forming is started.

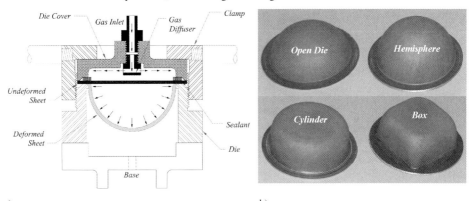

Figure 7: (a) Schematic of the open bulge forming die setup (b) various geometries formed using the setup

6 Acknowledgment

The support of the U.S. National Science Foundation, CAREER Award # DMI-0238712, is acknowledged.

7 References

[1] G. Cole, Proceedings of the 56th Annual Meeting of the International Magnesium Association, Rome, Italy, June 6-9, 1999, 21–30.
[2] D. Engelhart, C. Moedel, Technologien um das 3l-Auto, Brunswick, Germany, November 16-18, 1999, 11–21.
[3] A. Jambor, M. Beyer, Materials & Design, 1997, 18, 203–209.
[4] S. Schumann, F. Friedrich, Mg Alloys and their Applications, Wolfsburg, Germany, April 28–30, 1998.
[5] H. Friedrich, S. Schumann, The Second Israeli International Conference on Mg Science and Technology, Sdom, Israel, February 2000, 9–18.
[6] E. Doege, K. Dröder, Journal of Materials Processing Technology, 2001, 115, 14–19.

[7] E. Doege, L-E. Elend, F. Meiners, 33rd ISATA: Automotive & Transportation Technology, Dublin, Ireland, September 25–27, 2000, 87–94.

[8] E. Doege, W. Sebastian, K. Dröder, G. Kurz, The Second Global Symposium on Innovations in processing and Manufacturing of Sheet Materials, TMS 2001, 53–60.

[9] L. Tsao, C. Wu, T. Chuang, Materials Research and Advanced Techniques, 2001, 92, 572–577.

[10] F. Abu-Farha, M. Khraisheh, Proceedings of the 8th ESAFORM Conference on Material Forming, Cluj-Napoca, Romania, April 27-29, 2005, 627–630.

[11] B. Lee, K. Shin, C. Lee, Materials Science Forum, 2005, 475–479, 2927–2930.

[12] H. Watanabe, A. Takara, H. Somekawa, T. Mukai, K. Higashi, Scripta Materialia, 2005, 52, 449–454.

[13] D. Yin, K. Zhang, G. Wang, W. Han, Materials Letters, 2005, 59, 1714–1718.

[14] Y. Chino, H. Iwasaki, Journal of Materials Research, 2004, 19, 3382–3388.

[15] A. Bussiba, A. Ben Artzy, A. Shtechman, S. Ifergan, M. Kupiec, Materials Science & Engineering, 2001, A302, 56–62.

[16] X. Wu, Yi Liu, Scripta Materialia, 2002, 46, 269–274.

[17] M. Khraisheh, H. Zbib, C. Hamilton, A. Bayoumi, International Journal of Plasticity, 1997, 13, 143–164.

[18] F. Abu-Farha, M. Khraisheh, Materials Science Forum, 2004, 447-448, 165–170.

[19] F. Abu-Farha, M. Khraisheh, Journal of Materials Engineering & Performance, in press.

Extrusion of Different AZ Magnesium Alloys

S. Mueller, K. Mueller, T. Huichang, P. Wolter, W. Reimers

Technical University Berlin, Institute for Materials Science and Technologies and Extrusion Research and Development Center, Berlin

1 Introduction

AZ31 profiles often do not yet meet the requirements of the automobile or aircraft industry for the mechanical properties. Because an increasing percentage of aluminum leads to an increase in the ultimate tensile strength of the magnesium alloy [1] the first part of this paper deals with the effect of an increasing aluminum fraction on the indirect extrusion process and the mechanical properties of the extrudates.

The combination of texture and twinning in AZ31 causes a strength differential effect (SDE) where the yield stress under compression is well below that under tension [2]. Since the texture after the extrusion has a significant influence on the SDE another option to improve the mechanical properties of extruded Mg alloys is the optimization of the texture through a variation of the extrusion process. In the second part of this paper the extrusion of AZ31 into a counter pressure and the influences on the texture and therefore on the mechanical properties will be presented.

2 Indirect Extrusion

2.1 Extrusion Trials

The indirect extrusion trials were performed on the 8.32MN extrusion press of the Extrusion Research and Development Center of the Technical University Berlin. The extrusion press is equipped with load cells so that the total as well as the die force is recorded through out the whole extrusion process. For these trials a special die with a thermocouple in the bearing length was used so that the product temperature can be measured during the deformation. Continuous cast billets a AZ31, AZ61 and AZ80 were used. Three extrusion speed and temperature parameter sets were chosen to analyze to influence of speed and temperature on the process and the product. In the following only the results for the parameter set 250 °C and 0.9 m/min product speed are reported since the extrusion with a counter pressure was performed under a comparable parameter set. The results for all three parameter sets are reported in [3].

In the beginning of the extrusion AZ magnesium alloys exhibit a strong peak in the extrusion force before the extrusion force reaches a steady state condition. The aluminum content does not have a significant influence on the extrusion forces in the steady state condition, whereas the peak force as well as the maximum product temperature and the product temperature after 100 mm ram displacement increases with increasing aluminum content due to the increased deformation resistance. (Fig. 1a and b).

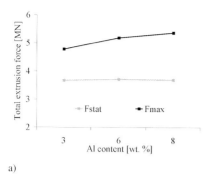 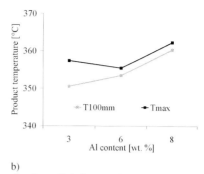

Figure 1: a) Influence of the aluminum content on the extrusion force, **b)** Influence of the aluminum content on the product temperature

2.2 Microstructure

Fig. 2 shows the dynamically recrystallized grain structure of the three alloys that were extruded. The AZ31 alloy exhibits a microstructure with distinctive grain boundaries and only some precipitates in the grains. The microstructure is not very homogeneous and has an average grain size of 11.5 µm. With increasing aluminum content the microstructure becomes more homogeneous and finer so that the AZ61 alloys has an average grain size of 11µm. This is caused by the increasing amount of γ-phase along the grain boundaries In the case of AZ80 these discrete precipitates are interconnected to form a network. This network of $Mg_{17}Al_{12}$ precipitates reduces the grain growth rate so that the AZ80 alloy exhibits an average grain size of 9 µm.

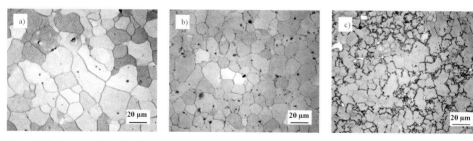

Figure 2: Influence of the aluminum content on the extrusion force and product temperature
a) AZ31, b) AZ61, c) AZ80

2.3 Texture

The texture analyses were performed by X-ray diffraction on the cross-sections of the specimens so that the inverse pole figures refer to the extrusion direction. The φ and ψ-scans were carried out for the $(10\bar{1}0)$, $(10\bar{2}0)$, $(10\bar{1}0)$, $(10\bar{1}2)$ and (0002) planes. All samples regardless of the aluminum content showed almost the same extrusion texture. This is a double fiber texture with the $<10\bar{1}0>$ and the $<10\bar{2}0>$ as the fibers, perpendicular to the extrusion axis. The <0001> basal planes are oriented radially to the extrusion axis. The only difference that can be observed is a slight decrease of the texture intensity with increasing aluminum content. In Fig. 3 the inverse pole figures for the different alloys are presented.

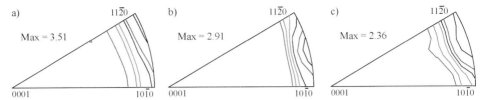

Figure 3: Texture after the extrusion; inverse pole figures perpendicular to the extrusion direction
a) AZ31, b) AZ61, c) AZ80

2.4 Mechanical Properties

The strengthening effect of the aluminum content on the mechanical properties of the extrudates can be observed under tensile and under compressive loading. With increasing aluminum the yield stress as well as the ultimate strength under tension and compression increases as well whereas the fracture strain under compression and tension decreases with increasing aluminum due to the larger amount of γ-phase along the grain boundaries. Since the increase of the yield stress under compression is larger than under tension the strength differential effect (SDE) which is calculated by [4]

$$\text{SDE} = 2\frac{\left|\sigma_{comp}\right| - \left|\sigma_{ten}\right|}{\left|\sigma_{comp}\right| + \left|\sigma_{ten}\right|} \tag{1}$$

can be reduced from –0.3 to –0.15.

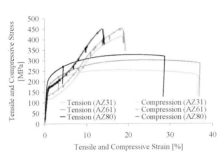

Figure 4: Stress-strain curves of the extruded alloys

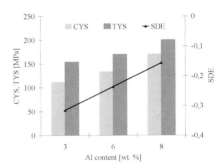

Figure 5: Influence of the aluminum content on the CYS, TYS and SDE

3 Indirect Extrusion with a Counter Pressure

3.1 Extrusion Trials

Besides the increase of strengthening alloying elements in Mg alloys another possibility to reduce the SDE is to increase the compressive yield stress through a specific change in texture. The primary twinning system for Mg alloys $\{10\bar{1}2\}<10\bar{1}\bar{1}>$ is a tensile twinning system due to a c/a ratio < √3. These twins are active under compression along the extrusion direction and inactive under tension along this direction. Since the activated twinning system under compression leads to a reorientation of the planes within in the twin by almost 90° [5] the activation of this twinning system already during the extrusion process leads to a favorable orientation of the hexagonal crystals in the extrudate for a compressive load.

The activation of the twinning system during the extrusion has been implemented through a hydrostatic counter pressure in which the extrudate is being extruded. Fig.6 shows schematically the extrusion process with a counter pressure. The additional extrusion force hat is needed because of the counter pressure does not exceed the peak force in the beginning of the indirect extrusion, therefore it is possible to extrude Mg alloys with a counter pressure on all extrusion presses that are capable to indirect extrude Mg alloys (Fig. 7).

The extrusion trials with a counter pressure were performed with continuous cast AZ31 billets, a product speed of 0.9 m/min, billet temperatures of 250 °C and 300 °C and a counter pressure between 0 and 700 bar. In the following the results for the trials at 250 °C with a counter pressure of 700 bar are presented.

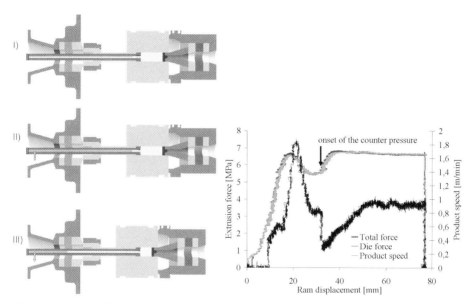

Figure 6: Extrusion with a counter pressure (schematic drawing)

Figure 7: Extrusion diagram of the extrusion of AZ31 with a counter pressure of 700 bar

3.2 Texture

The texture analysis was carried out under the same parameter as described above. The comparison of the inverse pole figure of an indirect extruded sample and an indirect with counter pressure extruded sample exhibits an additional texture pole for the basal plane perpendicular to the extrusion direction (Fig. 8). This texture component is caused by the activation of the $\{10\bar{1}2\}<10\bar{1}\bar{1}>$ twinning system because of the applied counter pressure.

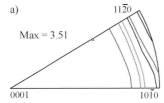 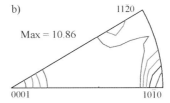

Figure 8: Texture after the extrusion; inverse pole figures perpendicular to the extrusion direction a) indirect extrusion, b) indirect extrusion with counter pressure

3.3 Microstructure

The extrusion into a counter pressure does not only influence the texture of the extrudates it also has a positive influence on the microstructure. Fig. 9b) shows the micrograph of the cross-section of an extrudate that was extruded into a counter pressure. In comparison to Fig. 9a) that shows the cross-section of an indirect extruded sample without a counter pressure a smaller grain size of 9µm can be observed although the microstructure is still inhomogeneous.

As it is already known extruded AZ31 products exhibit elongated grains along the extrusion direction [6, 7]. These elongated grains can also be found in the products of the extrusion into a counter pressure. But in contrast to the indirect extrusion without a counter pressure the elongated grains in the products of the extrusion with a counter pressure are inclined to the extrusion axis (Fig. 9c)).

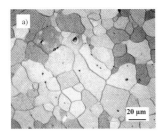 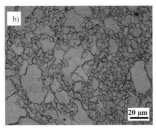 

Figure 9: Microstructure after the indirect extrusion, a) without a counter pressure cross-section, b) with a counter pressure cross-section, c) with a counter pressure longitudinal section

3.4 Mechanical Properties

The effect of the counter pressure on the texture and the microstructure results in a positive influence on the mechanical properties of the extrudates. Under a tensile load the extrusion into a counter pressure positively influences foremost the ultimate tensile strength and the fracture strain compared to the mechanical properties of extrudates without a counter pressure. Under a compressive load the compressive yield stress can be improved by 50 % compared to just indirect extruded samples whereas the fracture strain under compression decreases (Fig. 10).

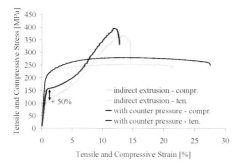

Figure 10: Mechanical properties of indirect extruded AZ31 with and without a counter pressure

4 Conclusions

The mechanical properties of indirect extruded AZ magnesium alloy products exhibit a strength differential effect which is characterized by a lower yield stress under compression than under tension. This SDE is caused by the combination of twinning and the texture after the extrusion. Two possible methods have been described in this paper. One is the increase of strengthening alloying elements. In the case of the AZ alloy family the aluminum content has been increased from 3 % to 6 % to 8 %. Whereas the aluminum content did not have a significant influence on the texture the increased $Mg_{17}Al_{12}$ precipitates along the grain boundaries resulted in a finer and more homogenous microstructure as well as an increase in the yield stress under tension and compression whereas the stronger increase of the yield stress under compression lead to a reduction of the SDE from –0.3 for AZ31 to –0.15 for AZ80.

The second method is the extrusion into a counter pressure. Because of the reorientation of almost 90° after the twinning which is caused by compression the texture could influenced by the activation of the twinning through the counter pressure. The counter pressure caused a new texture pole in the extrudates where the basal planes are oriented perpendicular to the extrusion axis. Due to the c/a ratio of AZ magnesium alloys the $\{10\bar{1}2\}<10\bar{1}\bar{1}>$ twinning system can only be activated under compression along the extrusion axis if the basal planes are oriented parallel to the extrusion direction. Therefore the extrusion into a counter pressure could increase the yield stress under compression by 50 %.

5 Acknowledgements

The first part of this paper has been supported in the framework of the Virtual Institute „Key materials for light weight construction". The financial contribution of the Hermann von Helmholtz Association is gratefully acknowledged. Furthermore the authors thank Dr. Jan Bohlen of the Center for Magnesium Technology of the GKSS for the fruitful discussions as wells as the Otto Fuchs KG and the GKSS for providing the cast magnesium billets.

6   References

[1] C. Kammer, Magnesium Taschenbuch, Aluminum Verlag, Düsseldorf, 2000, 1. edition
[2] S. Mueller, K. Mueller, M. Rosumek, W. Reimers, Aluminium International Journal, 82, 5, 2006, pp. 438–442.
[3] S. Mueller, K. Müller, H. Tao, W. Reimers, International Journal of Materials Research, accepted for publication 13.06.2006
[4] J.L. Raphanel, J.-H. Schmitt, P. van Houtte, Materials Science And Engineering, A108, 1989, pp. 227–232.
[5] J. P. Simon, Journal of physics F, Metal physics, 10, 1980, pp. 337–345.
[6] S. Mueller, K. Mueller, M. Rosumek, W. Reimers, Aluminium International Journal, 82, 4, 2006, pp. 327–330.
[7] J. Swiostek, J. Bohlen, D. Letzig, K.U. Kainer, Magnesium, Proceedings of the 6th International Conference Magnesium Alloys and Their Applications, 2003, pp. 278–284.

Effect of Billet Temperature and Ram Speed on the Behavior of AZ31 During Extrusion

M.A. Leeflang, J. Zhou, J. Duszczyk
Department of Materials Science and Engineering, Delft University of Technology, Delft

1 Introduction

Magnesium is the lightest of all the structural metals with its gravity only about 75 % of iron and 35 % of aluminum. When properly alloyed, magnesium exhibits not only good machinability, recycleability and thermal conductivity but also high specific strength. As such, it has a great potential to find itself in a wide variety of applications, especially in the automotive industry, which has recently been forced to reduce the emissions of cars and improve their fuel economy [1-3]. At present, wrought magnesium products are underdeveloped, if these are compared to magnesium castings that claim over 95 % of the total magnesium products [4].

At room and moderately elevated temperatures, h.c.p. structured magnesium has a limited number of slip systems for deformation. However, under hot deformation conditions as during rolling, forging and extrusion, prismatic and pyramidal slip systems are activated and it becomes deformable within certain limits [5]. In addition, during hot deformation, dynamic recrystallization takes place, which plays a significant role in raising ductility and reducing flow stress, thus facilitating deformation to proceed without requiring enhanced external forces for large deformation. It has been observed that during the hot deformation of a number of wrought magnesium alloys recrystallized grains nucleate at near grain boundaries and twin boundaries at an early stage. Under large strains extensive dynamic recrystallization may result in the development of an equiaxed grain structure [6–9].

It is known that during extrusion of AZ31, billet temperature and extrusion speed have large influences on the microstructure of the extrudates [10]. It is however unclear which parameter is more dominant. Moreover, the vast majority of the research on magnesium extrusion carried out so far concerns the process to produce a round bar or rod, which has fundamental importance to characterize the process but suffers from a lack of practical significance. The presented study attempted to determine the variations of extrusion pressure and the as-extruded microstructure with billet temperature and ram speed in the case of extruding AZ31 to produce a cross-shaped profile.

2 Experimental Details

A commercially produced magnesium alloy AZ31 was supplied in the form of a pre-extruded billet with a diameter of 51 mm. Its chemical composition was 3.45 % aluminum, 1.18 % zinc, 0.51 % manganese (by weight) and balance magnesium. After being machined, the billet with a diameter of 49.5 mm and a length of 200 mm was preheated for an hour and then extruded into a cross-shaped profile (Figure 1) at a reduction ratio of 8.8 using a 250 MT extrusion press with a container having a diameter of 50 mm. Billet temperatures selected for the experiments were

350, 400 and 450 °C with the tooling temperature (container, die and follower pad) 50 °C lower. At each billet temperature, extrusion was carried out at the ram speeds of 2, 4, 6 and 8 mm/s. During extrusion, measurements were taken, including container and die temperatures by thermocouples placed close to the liner and die bearing, respectively, profile temperature by a multi-wavelength pyrometer placed 200 mm after the die exit, extrusion pressure by a pressure sensor, ram displacement and speed by rulers.

From the extrudates, samples were cut perpendicularly to the extrusion direction near the end of ±1600 mm long profiles. After being cold mounted in an epoxy resin, the specimens at the core and the periphery of the transverse section (resp. A and B, Figure 1) were polished automatically, etched with an acetic picric etchant and examined with an optical microscope.

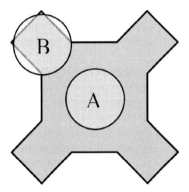

Figure 1: Cross-section of the profile and examined regions

3 Results and Discussion

The peak pressures measured during extrusion under different conditions are given in Table 1.

Table 1: Peak extrusion pressures measured during actual extrusion experiments

Extrusion Run no.	Billet temperature [°C]	Extrusion tooling temperatures [°C]			Ram speed [mm/s]	Peak pressure [MPa]
		Ram	Container	Die		
1	350	300	302	290	2.1	1078
2	400	350	352	336	2.2	902
3	400	350	348	335	4.2	990
4	400	350	362	341	6.1	934
5	450	400	410	388	2.1	703
6	450	400	405	385	4.2	745
7	450	400	397	379	6.3	747
8	450	400	412	383	8.5	786

At the billet temperature of 350 °C, the applicable ram speed was restricted to 2 mm/s. At this ram speed, the peak extrusion pressure was close to 90 % of the pressure capacity of the extrusion press used. An increase in ram speed caused a breakthrough pressure to exceed the pressure limit of the press and the extrusion run could not be completed. With an increase in billet temperature to 400 °C, the peak pressure did not approach the pressure limit of the press until a ram speed of 6 mm/s. At an even higher billet temperature of 450 °C, extrusion could be performed up to a ram speed of 8 mm/s without causing any surface defects.

The variation of the peak extrusion pressure with ram speed is shown in Figure 2. It is clear that there is no unique relationship between the peak pressure and ram speed at the billet temperature of 400 °C. The higher container and die temperatures of Extrusion Run 4 may have caused the breakthrough pressure to decrease slightly, but this could not account for the entire drop in the peak pressure (compare Extrusion Runs 3 and 4). Likely at 400 °C, when ram speed was high enough, an additional deformation mechanism or softening mechanism became operative, thus lowering the peak pressure.

The extrusion runs at a billet temperature of 450 °C displayed a similar trend to those at 400 °C. The increase in ram speed from 4 to 6 mm/s caused no increase in peak pressure, even though the container and die temperatures during Extrusion Run 7 were lower than those during Extrusion Run 6. This seems to support the hypothesis of an additional operative deformation mechanism or softening mechanism at the ram speed of 6 mm/s, which disturbs the expected increments of the peak pressure with rising ram speed.

The relationship between the peak pressure and billet temperature, as shown in Figure 3, is clear; the peak pressure at a ram speed of 2.1 mm/s and billet temperature of 450 °C is only about 65 % of the extrusion run at a billet temperature of 350 °C. At a billet temperature of 450 °C and ram speed of 8.5 mm/s, the peak pressure is still significantly lower that that of the extrusion run at a billet temperature of 350 °C and ram speed of 2.1 mm/s. The strong pressure dependence on temperature must be related to the energy required to initiate thermally activated softening processes. This dependence can be utilized in optimizing the extrusion process conditions to achieve a maximum throughput.

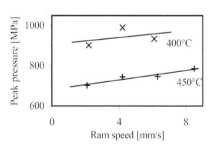 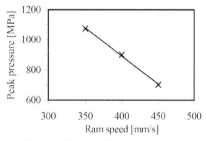

Figure 2: Variation of the peak pressure with ram speed at 400 and 450 °C

Figure 3: Relationship between the peak pressure and billet temperature at a preset ram speed of 2 mm/s

In extruding soft aluminum alloys, a moderate pressure decline after a blunt pressure peak is considered to be caused by the reducing friction between the liner and the shortening billet. In extruding dispersion-strengthened aluminum alloys, the pressure experiences a sharp peak followed by a steep decrease and finally reaches the steady state as strain hardening and dynamic recovery reach a dynamic balance. In the case of extruding magnesium, however, the shape of the extrusion pressure curve is quite different. A typical pressure / ram displacement diagram

(Run 1 at 350 °C / 2 mm/s) is shown in Figure 4. It can be seen that pressure decreases gradually after the peak without showing typical steady-state deformation. Because magnesium is not adhesive to tool steel, the friction between liner and billet must be much smaller than that during aluminum extrusion. Therefore, the predominant contributor to the decrease in extrusion pressure must be progressive strain softening during extrusion.

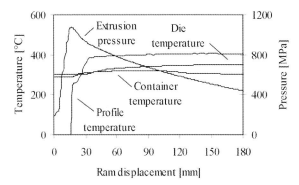

Figure 4: Extrusion diagram of Extrusion Run 1

The microstructures of the extrudates (Figures 5, 6 and 7) suggest strain softening through dynamic recrystallization. Fine, recrystallized grains were found at the grain boundaries of the original coarser grains at the core of the extrudate, at low ram speeds and all the billet temperatures. The present observations support the description of dynamic recrystallization of magnesium; with deformation proceeding, grain boundaries become serrated and new grains are formed in the vicinity of serrated grain boundaries. In the outer regions, however, the profile portrayed an overall larger volume fraction of recrystallized grains than at the core, which might be related to higher strains at the periphery of the extrudate. In this region, higher local temperatures in combination with higher local strains and strain rates could have promoted dynamic recrystallization and led to a near equiaxed grain structure [8,10].

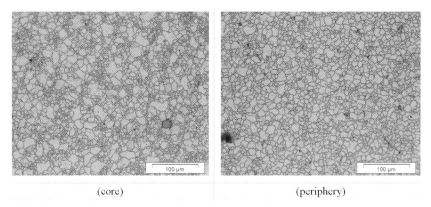

Figure 5: Microstructures of AZ31 extruded at a billet temperature of 350 °C and a preset ram speed of 2 mm/s

It was interesting to observe twins within coarse grains at the edge of the extrudate (see Figure 7). Previous studies on the microstructural evaluation of magnesium alloys during hot compression tests showed that above a temperature of 350 °C, twinning during deformation was of little importance [11]. Another study on the as-extruded AZ31 microstructure, however, showed twins at the thin section of a profile [12]. It seems that the local conditions of stress, strain and strain rate at the profile periphery may be favorable for twining [13] during extrusion. It was observed in the present investigation that at higher ram speeds, grains were near equiaxed, suggesting that recrystallization might have started at the intersection points of two twins and at the intersection of twins with grain boundaries [7,9]. Thus, another recrystallization mechanism, twinning recrystallization, was operative in the edge region of the extruded profile.

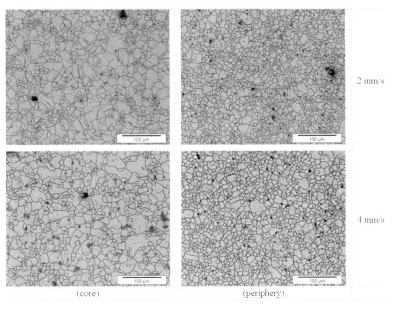

Figure 6: Microstructures of AZ31 extruded at a billet temperature of 400 °C and different ram speeds

Increasing ram speed led to the growth of fine grains at all the temperatures. It however did not affect their volume fraction significantly. At a given ram speed, for example, at 2 mm/s, the recrystallized grains appeared to grow with increasing billet temperature in all regions. The volume fraction of recrystallized grains also increased with billet temperature. This indicated that both ram speed and billet temperature had substantial influences on the dynamic recrystallization of the material [10], but the influence of billet temperature was notably stronger.

4   Conclusions

1. The peak extrusion pressure depends strongly on billet temperature. This provides an opportunity to increase extrusion throughput by maximizing ram speed at the highest possible billet temperature before the incipient melting point is reached during extrusion.
2. Strain softening after the extrusion pressure peak is significant, as a result of dynamic recrystallization. A larger volume fraction of fine recrystallized grains are achieved when billet temperature and ram speed are both low.

3. A combination of a higher billet temperature and a higher ram speed leads to a more equiaxed microstructure with slightly larger grain sizes.

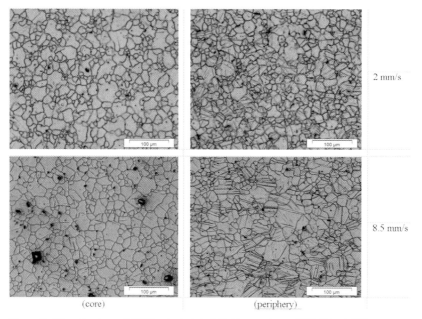

Figure 7: Microstructures of AZ31 extruded at a billet temperature of 450 °C and different ram speeds

5 References

[1] D. Eliezer, E. Aghion, F. H. Froes, Adv. Perfor. Mater. 1989, 5, 201–212.
[2] B. L. Mordike, T. Ebert, Mater. Sci, Eng. A 2001, 302, 37–45.
[3] L. L. Viguier, M. H. Babiker, J. M. Reilly, Energ. Pol. 2003, 31, 459–481.
[4] C. Jaschik, H. Haferkamp, M. Niemeyer, Magnesium Alloys and their Applications (Ed. K. U. Kainer), Wiley – VCH, Weinheim, 2000, pp. 41–46.
[5] E. F. Emley, Principles of Magnesium Technology, 1st ed., Pergamon Press, Oxford, 1966, pp. 483–487.
[6] H. T. Zhou, X. Q. Zeng, L. L. Liu, J. Dong, Q. D. Wang, W. J. Ding, Y. P. Zhu, Mater. Sci. Techn. 2004, Vol. 20, 1397–1402
[7] O. Sitdikov, R. Kaibyshev, T. Sakai, Mater. Sci. Forum, 2003, 419–422, 521–526.
[8] X. Yang, H. Miura, T. Sakai, Mater. Sci. Forum, 2003, 419–422, 515–520.
[9] M. M. Myshlyaev, H. J. McQueen, A. Mwembela, E. Konopleva, Mater. Sci, Eng. A. 2002, 337, 121–153.
[10] T. Murai, S. Matsuoka, S. Miyamoto, Y. Oki. Mater. Process. Techn. 1003, 141, 207–212.
[11] U. Noster, B. Schotes, Z. Metallkd, 2003, 94, 559–563.
[12] Q. Wang, B.C. Li, Z.X. Feng, Mater. Sci. Forum, 2005, 488-489, 519–522.
[13] R. E. Reed – Hill, R. Abbaschian, Physical Metallurgy Principles, 3rd ed., PWS, Boston, 1994, 548–549.

Influence of Extrusion Conditions on the Superplastic Properties of a Mg-8%Li Alloy Processed at Room Temperature by Equal-Channel Angular Pressing

Mitsuaki Furui[1], Hiroki Kitamura[1], Hiroshi Anada[1] and Terence G. Langdon[2]
[1]Department of Material Systems Engineering and Life Science, Faculty of Engineering
University of Toyama, Toyama, Japan
[2]Departments of Aerospace & Mechanical Engineering and Materials Science
University of Southern California, Los Angeles, CA, U.S.A.

1 Abstract

Experiments were conducted on a two-phase Mg-8% Li alloy to evaluate the high temperature mechanical properties under three different processing conditions corresponding to the cast condition, the cast condition followed by extrusion and the cast condition followed by extrusion and then processing by equal-channel angular pressing (ECAP). The processing by ECAP was conducted at room temperature for a total of 4 passes using a die having an internal channel angle of 35°, where this is equivalent to a total imposed strain of ~2. The results show extrusion leads to a refinement of the grains and there is also additional grain refinement when processing by ECAP. High superplastic elongations are attained after processing by extrusion and ECAP. The results demonstrate the tensile ductilities increase with decreasing extrusion temperature but they are essentially independent of the extrusion speed.

2 Introduction

Magnesium alloys are used in a wide variety of structural and functional applications due to their attractive properties such as low density and high specific strength and elastic modulus [1,2]. However, the use of Mg alloys in more complex applications is limited because of problems associated with ductility, corrosion and creep resistance. It has been demonstrated that ductility enhancement may be achieved in Mg alloys by refining the grain structure using conventional thermo-mechanical treatments [3,4], powder metallurgy [5,6] or severe plastic deformation [7,8]. Among the severe plastic deformation methods, it is well-known that equal-channel angular pressing (ECAP) is capable of producing uncontaminated fully-dense and relatively large bulk materials with submicrometer or even nanometer grain structures [9,10]. It has been found also that the addition of lithium to magnesium significantly improves the plastic workability of the magnesium alloys. The limit of reduction in thickness for the cold rolling of a Mg-8% Li alloy exceeds 80 % [11]. Therefore, it is anticipated that the Mg-8% Li alloy will be an excellent candidate material for processing by severe plastic deformation at lower temperatures.

Earlier experiments showed the potential for achieving superplastic ductilities in magnesium alloys such as Mg-0.6% Zr by using a two-step processing route, designated EX-ECAP, in which the alloy is initially extruded and then subjected to grain refinement using ECAP [12,13].

A recent report demonstrated that excellent superplastic properties were achieved in the Mg-8% Li alloy, including a maximum elongation of ~970 % at 473 K when using an initial strain rate of $1.0 \cdot 10^{-4}$ s^{-1}, by extrusion and subsequent ECAP through 2 passes with a pressing temperature of 473 K [14]. However, no information is available regarding the influence of the extrusion conditions on the superplastic properties of the Mg-8% Li alloy.

The results available to date demonstrate that a preliminary extrusion step prior to ECAP provides a valuable tool for increasing the effectiveness of the grain refinement. Nevertheless, there have been no attempts to identify the preferred extrusion conditions leading to the greatest grain refinement and the optimum superplastic ductilities. The present investigation was initiated to address this deficiency by examining the influence of the extrusion speed and the extrusion temperature on the microstructures and tensile properties of the Mg-8% Li alloy.

3 Experimental Material and Procedures

The experiments were conducted on the Mg-8mass% Li alloy containing two phases: an α-phase consisting of a solid solution of Li in hcp Mg and a β-phase consisting of a solid solution of Mg in bcc Li. Commercial high purity Mg and Li were used in the preparation of the Mg-8% Li alloy. The melting of the alloy was conducted in a vacuum high-frequency induction furnace in an argon atmosphere without any flux. The molten alloy was cast into a steel mould with a diameter of 140 mm and a height of 180 mm. The cast material was homogenized at 573 K for 86.4 ks in an argon atmosphere. A small portion of the alloy was cut from one of the ingots for testing in the as-homogenized condition. For the other ingots, the ends were removed to give lengths of 100 mm and average diameters of 50 mm and they were then extruded at 373, 473 or 573 K at speeds of 1, 5 and 10 mm/s to give rods with diameters of 10 mm corresponding to a reduction ratio of 25:1. Rods were cut for ECAP processing with diameters of 10 mm and lengths of 60 mm. The sequential processing of Mg alloys by extrusion and ECAP corresponds to the EX-ECAP processing route which is the ideal two-step procedure for achieving a superplastic forming capability in Mg alloys [12].

The ECAP processing was conducted using a solid die having an internal channel with a diameter of 10 mm bent through an angle of $\Phi = 135°$. There was an additional angle of $\Psi \approx 20°$ representing the outer arc of curvature where the two parts of the channel intersect. It can be shown from first principles that these values of Φ and Ψ lead to an imposed strain of ~0.5 on each separate passage through the die [15]. The rods were pressed through the die at a rate of ~7.5 mm/s. Samples were pressed for 4 passes at room temperature, equivalent to a strain of ~2, using route Bc where the rods are rotated by 90° between passes [16].

Tensile samples were machined from billets having three different conditions: for the unpressed material either in the cast condition or in the cast+extrude condition and for the pressed material in the cast+extrude+ECAP condition. All tensile samples had gauge lengths of 4 mm, cross-sectional areas of 3×2 mm^2 and with the gauge lengths lying parallel to the extrusion or pressing directions. All samples were pulled to failure at a temperature of 473 K using a testing machine operating at a constant rate of cross-head displacement and initial strain rates in the range from $1.5 \cdot 10^{-3}$ to $1.5 \cdot 10^{-1}$ s^{-1}.

The microstructures of the specimens were observed on planes cut parallel to the working directions using optical microscopy. These specimens were prepared by sectioning, mechanically polishing, and then etching using 10 % HNO_3 solution in ethanol to reveal the interphase boundaries. The widths of the α and β phases were measured perpendicular to the working directions using the linear intercept method.

4   Experimental Results

Representative microstructures for the three different processing conditions are shown in Figure 1 where (a) is the Cast condition, (b) is the Cast + Extrude condition where the sample was extruded at 373 K at a speed of 1 mm/s and (c) is the Cast + Extrude + ECAP condition where the sample was extruded under the same conditions as in (b) and then pressed for 4 passes at room temperature. In Figure 1 (a), the α-phase is the lighter discontinuous phase, the β-phase is the darker continuous phase and thus in the as-cast condition the α-phase is elongated and dispersed within a matrix of the β-phase. The volume fractions of the α and β phases were measured as ~55% and ~45% in this condition, respectively. By contrast, in Figures 1 (b) and (c) there is significant microstructural refinement and the α and β phases form banded structures with the bands lying reasonably parallel to the extrusion and pressing directions. There is also a clear difference between the Cast + Extrude and the Cast + Extrude + ECAP conditions because Figure 1 (c) shows the α and β phases are finer and generally more uniform after ECAP.

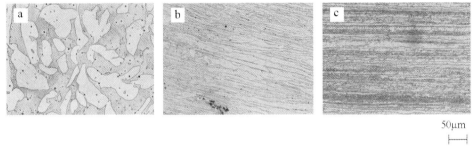

50μm

Figure 1: Optical micrographs of the Mg-8% Li alloy (a) in the Cast condition, (b) in the Cast +Extrude condition and (c) in the Cast +Extrude+ECAP condition: the extrusion and ECAP directions are indicated

To obtain more detailed information on the nature of the microstructural refinement associated with the extrusion and ECAP processes, additional microstructures are shown in Figure 2 for the Cast + Extrude + ECAP condition. All of these microstructures were obtained after ECAP through 4 passes at room temperature. Both the α and β phases coarsen with increasing extrusion temperature but there is no apparent change with increasing extrusion speed by comparison with the extrusion temperature.

In order to obtain a more precise characterization of the microstructures for each condition, the width distributions of the α and β phases were recorded and plotted in the form of histograms as shown in Figures 3 and 4 for the Cast + Extrude and the Cast + Extrude + ECAP conditions, respectively; the extrusion speed is 1 mm/s for all of these plots and the average widths of the histograms are denoted by $\overline{W}_\alpha$ for the a phase and $\overline{W}_\beta$ for the b phase. There is an increase in the average widths of the two phases when the extrusion temperature is increased from 373 to 573 K where this increase is by a factor of <3. Finally, processing by ECAP after the extrusion step gives a further, but relatively minor, reduction in the average width of the phases by a factor of <2.

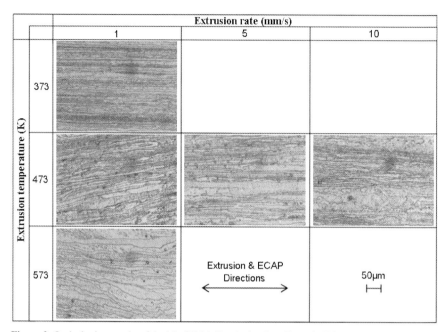

Figure 2: Optical micrographs of the Mg-8% Li alloy in the Cast+Extrude+ECAP condition after processing using various extrusion conditions: the extrusion and ECAP directions are indicated

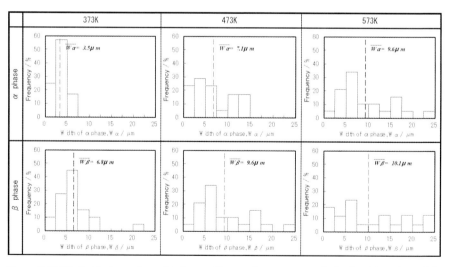

Figure 3: Width distributions of a and b phases with extrusion temperature in the Cast+Extrude condition: the average widths of both phases, $\overline{W}_\alpha$ and $\overline{W}_\beta$, are indicated

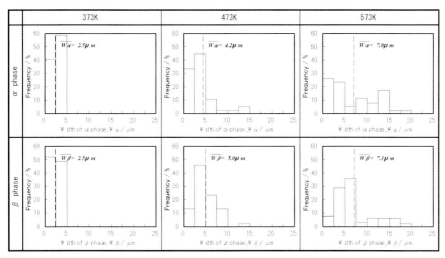

Figure 4: Width distributions of α and b phases with extrusion temperature in the Cast+Extrude+ECAP condition: the average widths of both phases, $\overline{W}_\alpha$ and $\overline{W}_\beta$, are indicated

Another striking feature of these histograms is the consistent increase in the widths of the distributions with increasing extrusion temperature in both the Cast + Extrude and the Cast + Extrude + ECAP conditions. Furthermore, at the lowest extrusion temperature of 373 K, the distribution is especially narrow for the Cast + Extrude + ECAP condition, thereby suggesting the presence of a reasonably uniform microstructure under these conditions.

Figure 5 shows the influence on the measured elongations to failure of (a) the extrusion temperature and (b) the extrusion speed, where all samples were pulled to failure at testing temperature of 473 K using an initial strain rate of $1.5 \cdot 10^{-3}$ s^{-1}. The elongations to failure are always higher, by a factor of 2 or more, in the specimens processed to incorporate the additional step of ECAP. Furthermore, this is consistent with the microstructural observations in Figures 3 and 4 where the phase widths are smaller in the Cast + Extrude + ECAP samples. The ductilities also increase with decreasing extrusion temperature and thus with decreasing phase width. By contrast, Figure 5 (b) shows that the extrusion speed has an insignificant effect on the ductilities achieved after ECAP, at least within the limited range from 1 to 10 mm/s.

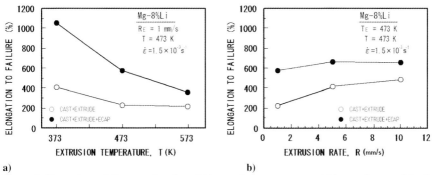

Figure 5: Elongation to failure as a function of (a) extrusion temperature and (b) extrusion speed for the Mg-8% Li alloy tested at 473 K using a strain rate of $1.5 \cdot 10^{-3}$ s^{-1}

5 Summary and Conclusions

1. Microstructural observations show there is a very significant reduction in the average widths of the phases due to extrusion and there is an additional microstructural refinement due to ECAP. An exceptionally fine and uniform microstructure was achieved after ECAP when the extrusion was conducted at the lowest temperature of 373 K.
2. The elongations to failure in tensile testing increase with decreasing extrusion temperature but they are reasonably independent of the extrusion speed within the range from 1 to 10 mm/s.

6 References

[1] D. M. Lee, B. G. Suh, B. G. Kim, J. S. Lee, C. H. Lee, Mater. Sci. Technol. **1997**, 13, 590–595.
[2] Y. Kojima, Mater. Sci. Forum **2000**, 350-351, 3–17.
[3] T. Imai, S. W. Lim, D. Jiang, Y. Nishida, Scripta Mater. **1997**, 36, 611–615.
[4] W. J. Kim, S. W. Chung, C. S. Chung, D. Kum, Acta Mater. **2001**, 49, 3337–3345.
[5] J. K. Solberg, J. T. Torlkep, O. Bauger, H. Gjestland, Mater. Sci. Eng. **1991**, A134, 1201–1203.
[6] H. Watanabe, T. Mukai, K. Ishikawa, M. Mabuchi, K. Higashi, Mater. Sci. Eng. **2001**, A307, 119–128.
[7] M. Mabuchi, H. Iwasaki, K. Yanase, K. Higashi, Scripta Mater. **1997**, 36, 681–686.
[8] A. Yamashita, Z. Horita and T. G. Langdon, Mater. Sci. Eng. **2001**, A300, 142–147.
[9] R. Z. Valiev, N. A. Krasilnikov, N. K. Tsenev, Mater. Sci. Eng. **1991**, A137, 35–40.
[10] R. Z. Valiev, T.G. Langdon, Prog. Mater. Sci. **2006**, 51, 881–981.
[11] K. Matsuzawa, T. Koshihara, Y. Kojima, J. Japan. Inst. Light Met. **1989**, 39, 45–51.
[12] Z. Horita, K. Matsubara, K. Makii, T. G. Langdon, Scripta Mater. **2002**, 47, 255–260.
[13] K. Matsubara, Y. Miyahara, Z. Horita, T. G. Langdon, Metall. Mater. Trans. A **2004**, 35A, 1735–1744.
[14] M. Furui, C. Xu, T. Aida, M. Inoue, H. Anada, T. G. Langdon, Mater. Sci. Eng. **2005**, A410–411, 439-442.
[15] Y. Iwahashi, J. Wang, Z. Horita, M. Nemoto, T. G. Langdon, Scripta Mater. **1996**, 35, 143–146.
[16] M. Furukawa, Y. Iwahashi, Z. Horita, M. Nemoto, T. G. Langdon, Mater. Sci. Eng. **1998**, A257, 328–332.

Severe Plastic Deformation of AZ31

S. Mueller, K. Mueller, W. Reimers
Technical University Berlin, Institute for Materials Science and Technologies and Extrusion Research and Development Center, Berlin

1 Introduction

Despite the fact that magnesium is the lightest construction metal it is still not widely-used because of its unfavorable mechanical properties under compression. This strength differential effect (SDE) is characterized by a lower yield stress under compression than under tension [1]. The SDE is caused by mechanical twinning which again is influenced by the texture and the grain size.

One way to influence the SDE is changing the extrusion method from direct to indirect to hydrostatic [2]. This will improve the SDE but there is still a significant asymmetry between the yield points under tensile and under compressive load. Equal Channel Angular Extrusion (ECAE) trials with aluminum alloys show a finer grained and more homogeneous microstructure after the extrusion [3]. So in this paper it is demonstrated that the ECAE can be used to improve the mechanical properties of Mg alloys under compressive load. By using ECAE the extruded specimen is under severe shearing deformation without any changes of its geometrical dimensions.

2 Linear ECAE

In order to be able to conduct the ECAE on a commercial extrusion press the ECAE process had to be modified. In the first step the conventional ECAE process was changed to a linear multiple shearing process. The tool set that was developed is shown in Fig. 1.

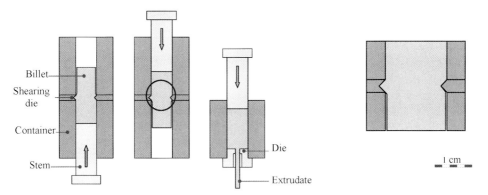

Figure 1: Tool set for the linear multiple shearing process

The ECAE tool used for this study has three shearing planes with angles of 135°, 90 and 135°. The total strain intensity for the conventional ECAE is given by [4]

$$\varepsilon_n = 1.15 \cdot N \cdot \cot\frac{\phi}{2} \qquad (1)$$

where N is the number of passes and ϕ the angle between the channels. Since the die used in this study provides three shearing planes during one pass the equivalent strain for the linear multiple shearing process can be calculated by

$$\varepsilon_n = 1.15 \cdot N \cdot (\cot\frac{\phi_1}{2} + \cot\frac{\phi_2}{2} + \cot\frac{\phi_3}{2}) \qquad (2)$$

where ϕ_1 and ϕ_3 are the 135° angles at the beginning and the end of the die and ϕ_2 is the 90° angle of the die. The total strain intensity after N cycles is presented in Table 1.

Table 1: Total strain intensity after N cycles for the multiple linear shearing process

Number of cycles N	1	2	3	4	5	6	7	8	9	10
Total strain intensity ε_n	2,1	4,2	6,3	8,4	10,5	12,6	14,7	16,8	18,9	21

2.1 Microstructure

The microstructure evolution of the samples which were deformed by linear ECAE depends on the process temperature as well as on the number of cycles. The influence of the number of repetitions is shown in Fig. 2. Whereas the microstructure of the as-cast billet exhibits a very coarse grain size with a grain size of 250 µm (Fig. 2 a) the grain size decreases after just 4 cycles of linear ECAE to 3µm. The further deformation up to 10 cycles linear ECAE does not result in a much finer grain size. The grain size after 10 cycles is 2.8 µm. An increasing number of cycles rather leads to a more homogeneous microstructure. As it can be observed from Fig. 2b) and c) the microstructure of the sample that was deformed by 4 cycles linear ECAE still exhibits some large grains with a grain size of 35 µm. After 10 cycles linear ECAE the number as well as the size of large grain has declined to a grain size of 12 µm.

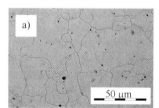 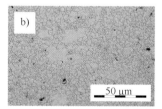 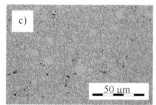

Figure 2: Microstructure before and after the linear ECAE a) as-cast, b) 4 cycles linear ECAE at 250 °C, c) 10 cycles linear ECAE at 250 °C

2.2 Texture

As it has already been reported by Kim et al. [5] for the AZ61 alloy and Agnew et al. [6] and Yoshida et al. [7] for the AZ31 alloy the texture after the conventional ECAE process is characterized by an inclination of the basal planes to the extrusion direction at 45° as well as a misorientation of 30° around the c-axis. Almost the same texture exists after the linear ECAE process (Fig. 3). As Fig. 3c) shows the misorientations in the texture lead to an orientation of the $11\bar{2}3$ plane perpendicular to the extrusion direction.

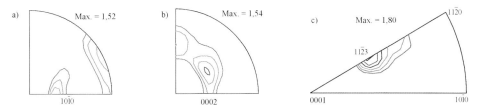

Figure 3: Texture after 10 cycles linear ECAE; measured on the cross-section perpendicular to the extrusion direction a) $10\bar{1}0$ pole figure, b) 0002 pole figure, c) inverse pole figure in extrusion direction

2.3 Mechanical Properties

The influence of the linear ECAE on the grain size as well as on the texture of the AZ31 alloy also leads to a significant influence on the mechanical properties. Compared to the only indirect extruded samples the predeformation by 4 cycles linear ECAE leads to an increase of the CYS of more than 70 % while the TYS decreases around 10 % (Fig. 4). The influence of the process temperature is shown in Fig. 5. By decreasing the process and billet temperature of the linear ECAE process from 300 °C to 200 °C the SDE can be reduced to nearly 0.

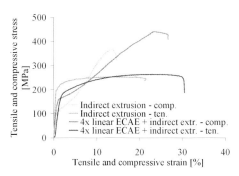
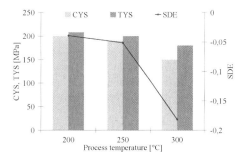

Figure 4: Compression and tension curves for indirect extrusion and linear ECAE plus indirect extrusion at 250 °C

Figure 5: CYS, TYS and SDE in dependence of the linear ECAE process temperature

3 Reciprocating Extrusion

As promising as the results of the linear ECAE trials have been there are still some disadvantages. As it can be observed from Fig. 6 a dead metal zone developed at the internal 90° angle of the linear ECAE tool. Furthermore, because of the construction of the linear ECAE tool it was not possible to perform the ECAE and the extrusion out of the same container. Therefore the predeformed billet had to be reheated to the extrusion temperature which resulted in a grain coarsening (Fig. 7).

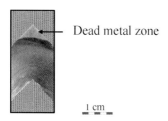

Figure 6: Dead metal zone in the linear ECAE tool

Figure 7: Microstructure after 4 cycles linear ECAE and reheating to 300 °C for 1 h

The predeformation of the billets had thus been altered to the reciprocating extrusion with two moving stems. During this process the billet is extruded through a kneading die and afterwards upset by the opposing stem. After a designated number of cycles one stem can be replaced by an extrusion die so that the predeformation and extrusion can be performed with out any discontinuity out of the same container (Fig. 8).

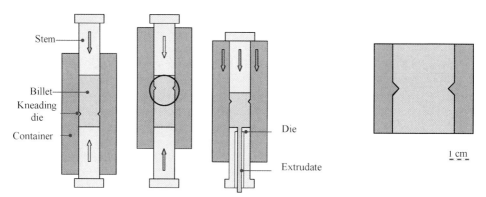

Figure 8: Tool set for the reciprocating extrusion process

The total strain intensity for the reciprocating extrusion is calculated by

$$\varepsilon_n = 4 \cdot \ln \frac{D}{d} \tag{3}$$

With a container diameter of 30 mm and a diameter of the kneading die of 24 mm the total strain after N cycles is shown in Table 2.

Table 2: Total strain intensity after N cycles for the reciprocating extrusion

Number of cycles N	1	2	3	4	5	6	7	8	9	10
Total strain intensity ε_n	0,89	1,79	2,68	3,57	4,46	5,35	6,25	7,14	8,03	8,93

3.1 Microstructure

The reciprocating extrusion leads after 4 cycles to a fine grained microstructure with a grain size around 3 µm (Fig. 9). An increase in the number of cycles up to 10 does not result in a finer or more homogeneous microstructure. The grain size after 10 cycles is around 4 µm. Additionally in the micrograph after 10 cycles shearbands are observed (Fig. 10).

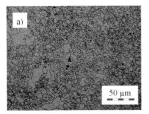

Figure 9: Microstructure after 4 cycles reciprocating extrusion and indirect extrusion at 300 °C

Figure 10: Microstructure after 10 cycles reciprocating plus indirect extrusion at 300 °C

3.2 Texture

The texture of the extrudates after indirect extrusion either with or without a predeformation by reciprocating extrusion is mainly the same. Both show a double fiber texture with the $10\bar{1}0$ and $10\bar{2}0$ planes perpendicular to the extrusion direction while the basal planes are oriented parallel to the extrusion direction (Fig. 11 and 12). Because of the finer grain size of the predeformed billet material the texture intensity of the extrudate after 4 cycles of reciprocating and indirect extrusion is about 90 % higher than the intensity of the just indirect extruded product.

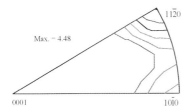
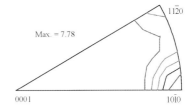

Figure 11, 12: Inverse pole figures; measured on the cross-section perpendicular to the extrusion direction; a) after indirect extrusion at 250 °C, b) after 4 cycles reciprocating plus indirect extrusion at 250 °C

3.3 Mechanical Properties

As Figure 13 shows 4 cycles of reciprocating extrusion before the actual indirect extrusion cause an assimilation of the CYS and TYS; compared to the mechanical properties after indirect extrusion only the CYS increases more than 100 % while the TYS decreases only around 5 %. By lowering of the process temperature from 300 °C to 250 °C a CYS which is higher than the TYS and therefore an SDE above 0 can be achieved (Fig. 14).

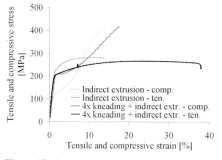
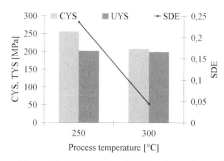

Figure 13: Compression and tension curves for indirect extrusion and reciprocating plus indirect extrusion at 300 °C

Figure 14: CYS, TYS and SDE in dependence of the reciprocating extrusion process temperature

4 Conclusions

Both described methods for a predeformation of the extrusion billets, linear ECAE and reciprocating extrusion, cause a significant finer grain size as well as a more homogeneous microstructure. Due to the microstructural changes the twinning under compression can be obstructed and therefore the SDE of the extrudates after linear ECAE or reciprocating extrusion plus indirect extrusion is improved compared to just extruded products. Whereas an SDE >0 could already be achieved after 4 cycles of reciprocating extrusion 10 cycles of linear ECAE were necessary to reach an SDE close to 0 (Fig. 15).

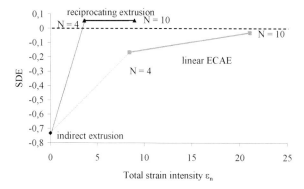

Figure 15: SDE dependence from the predeformation method and the total strain intensity

5   Acknowledgements

This work has been supported by the Otto Fuchs KG, Honsel GmbH, WEFA Singen and Stiftung Industrieforschung Cologne, Project: S591.

6   References

[1] J.L. Raphanel, J.-H. Schmitt, P. van Houtte, Materials Science And Engineering, A108, 1989, pp. 227–232.
[2] S. Mueller, K. Mueller, M. Rosumek, W. Reimers, Aluminium International Journal, 82, 4, 2006, pp. 327–330.
[3] Richert J., Richert M., Aluminium International Journal, 62, 8, 1986, pp. 604–607.
[4] V.M. Segal, Materials Science & Engineering, A197, Issue 2, 1995, pp. 157–164.
[5] W.J. Kim, C.W. An, Y.S. Kim, S.I. Hong, Scripta Materialia, 47, 2002, pp. 39–44.
[6] S.R. Agnew, J.A. Horton, T.M. Lillo, D.W. Brown, Scripta Materialia, 50, 2004, 377–381.
[7] Y. Yoshida, L. Cisar, S.Kamado, Y. Kojima, Materials Transactions, 44, 4, 2003, 468–475.

Production and Properties of Small Tubes Made from MgCa0,8 for Application as Stent in Biomedical Science

Th. Hassel, Fr.-W. Bach
Institute of Materials Science, Leibniz University of Hanover (Germany)
A. N. Golovko
Metal Forming Department, National Metallurgical Academy of Ukraine

1 Introduction

The development of a biodegradable, cardiovascular implant (stent) made from an absorbable magnesium calcium alloy demands an accurately defined production process. The fabrication of magnesium stents requires special concepts to produce thin walled tubes which consists of several deformation processes. Tube extrusion and the further processing by drawing is a very promising concept to develop absorbable magnesium stents. Hot extruded tubes with a diameter of 6,5–6,6 mm and a wall thickness of 500–1000 μm were reduced by a drawing process without and with use of a mandrel. Investigations on the influence of the drawing process sequence and some heat treatments between the drawing steps on the mechanical properties were carried out. The drawing ratio, the temperature of the tube and the drawing die and their effects on the tensile strength, the elongation and the microstructure of tube samples were investigated. Production details for the optimal deformation and temperature parameters of the drawing process of magnesium alloys with a calcium contents of 0,8 mass% are given. The results show that MgCa0,8 nevertheless of its low alloyed character are up to the mark which is necessary for the application as a stent. Investigations on the degradation properties also demonstrate that this alloy will be dissolved slower and with a moderate rate of hydrogen formation than pure magnesium. The results show that an absorbable Mg alloy is available which can be easily produced and exclusively contains essential mineral nutrients with a none toxic character.

2 Materials and Methods

For this investigations a MgCa0,8 alloy was cast in the light metal foundry of the Institute of Materials science. The alloy is melted from pure Magnesium with additions of a MgCa30 pre alloy and the Calcium contents amounts 0,8 wt%. The cylindrical bars are turned and a centric hole is been drilled into the centre of the cylinder.

2.1 Extrusion of Small Tubes

The extrusion of tubes on vertical extrusion (VE) presses provides a smaller wall thickness difference in the tube cross section. The vertical hydraulic press of the Institute of Material Science at the University of Hanover was used for the experimental part of this work [1, 2]. The press works with a nominal force of 800 kN. The container has an inside diameter of 29 mm and its length amounts 120 mm. The maximum extrusion velocity (constant) is 1,25 mm/s. Due to the small diameter of the container the averages container pressure upon a dummy block reaches 1200 N/mm² and is nearly equivalent of industrial extrusion presses. Extrusion was performed with a long mandrel through a double-cone die, instead of a porthole type die, for decreasing the force of extrusion and to prevent formation of longitudinal welding seams in the tube. The billets were pre extruded rods (diameter 20 mm) which were upsetted to cylinders (diameter 29 mm) with a conic head in the container of the VE-press. Such an operation increases the tensile stress and plasticity of extruded semi finished products of magnesium alloys [3]. To lead the mandrel into the billet a centric hole was drilled trough the billet with the diameter of the mandrel. The turning of the exterior was not necessary. As lubrication was molybdenum disulfide spray was used and applied on the working surfaces of mandrel and die.

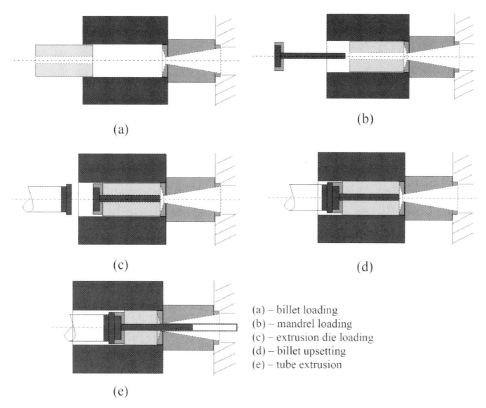

(a) – billet loading
(b) – mandrel loading
(c) – extrusion die loading
(d) – billet upsetting
(e) – tube extrusion

Figure 1: Hot Tube Extrusion

2.2 Thinning of the Tubes by a Drawing Process Chain

The extruded tubes (diameter $D = 6,5$ mm, wall thickness $d = 500$ μm) are processed by hot tubular slide drawing without mandrel and a following hot tubular slide drawing with mandrel to very thin tube (diameter $D = 1,9$ mm, wall thickness $d = 150$ μm). The Drawing process is composed of several steps of drawing. A possible skeleton schema of this process chain is showed in Figure 2. Thereby the next 5 drawing steps are implemented by tubular slide drawing without mandrel to reduce the diameter of the tube. The increment of the diameter reduction may amounts 0,7–0,5mm.

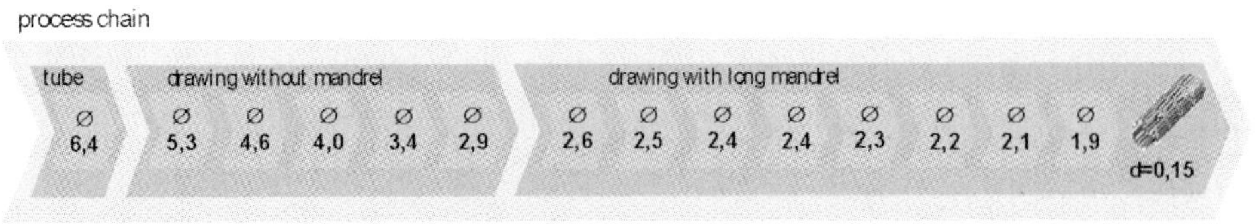

process chain

tube	drawing without mandrel					drawing with long mandrel							
Ø 6,4	Ø 5,3	Ø 4,6	Ø 4,0	Ø 3,4	Ø 2,9	Ø 2,6	Ø 2,5	Ø 2,4	Ø 2,4	Ø 2,3	Ø 2,2	Ø 2,1	Ø 1,9

d=0,15

Figure 2: Process chain of the thin tube production

The drawing die was heated to a temperature of 250–300 °C. Higher temperatures leads very often to the interruption of the process by breaking or upsetting the tube. The drawing speed amounts 10–50 mm/s and the diameter is reduced to 2,9 mm. The reduction of the wall thickness demands smaller deformation increments because the material has to flow between the drawing die and the mandrel. As mandrel polished steel wire with different diameter was used to draw the MgCa tube. Temperature, speed and lubrication were the same like the drawing steps 1–5. The magnitude of the deformation has to be much smaller and the drawing die diameter should be reduced to 0,05–0,1 mm per drawing step. To remove the mandrel from the tube after the last drawing step it is drawn through a template. This complex process step demands an accurate handling of the equipment and can be improved by a heat treatment of the tube [4, 5]. The heat treatment increase the ductility of the material.

3   Results and Discussion

During the processing of the MgCa0,8 alloy its microstructure changes significantly. The grain size of the cast material amounts 80–150 µm and decrease by the tube extrusion to 30–35 µm.

The first step of drawing also reduce the grain size to 10–15 µm. The further machining by drawing do not more increase the grain size and its stays nearly constant at 10-15µm. Figure 3 shows this microstructure modification. The at the extruded material observed difference between the grain shape in extrusion direction and across to it not occurs after the drawing process. The drawn products show in extrusion direction and across the same grain size. This indicates that the structure is reorganized during the process by recrystallisation processes [6]. This also cause the constant grain size during the drawing process, because of the destroying the old grains by the deformation and the new building structure by dynamic and static recrystallisation into the deformation zone and after the exit of the hot tube form the die. Figure 4 show the development of the grain size during the process chain.

The mechanical properties of the tubes strongly depends from the hat treatment of the material. After the extrusion the tubes show tensile strength of 1040–160MPa and an elongation of 13–15%. These values changes dramatically without any heat treatment to an increased strength up to 280–300MPa and decreased elongation less than 1 %. This properties are not adequate to apply this material as vascular implants. The implementation of heat treatments between the process steps of drawing enhance the mechanical properties significantly. Figure 5 show the dependence of the heat treatment of a tube with 3,9 mm in diameter before and after several heat treatments. The tubes are heat treated at 300 °C for one hour without any protecting atmosphere in the furnace. The elongation increases from 1.5 % to 14 % after the first treatment and 19 % after the second treatment the elongation returns to the magnitude of the extruded tube (19 %). With the increasing elongation the tensile strength decreases. Figure 6 show the development of the strength during the heat treatments. The decrease of the strength must be tolerated and balanced by the implant construction.

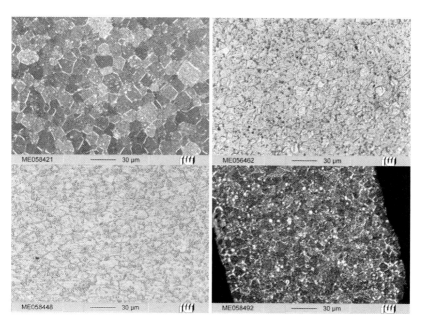

Figure 3: Micrograph of the extruded tube structure (top left), of the structure after the first drawing step (top right); of the structure after the last drawing step without mandrel (bottom left) and the structure of the finished tube material (alloy: MgCa0,8)

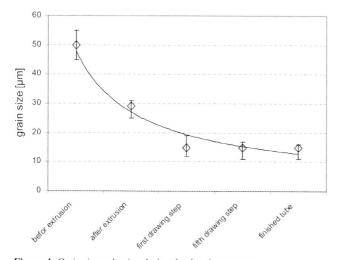

Figure 4: Grain size reduction during the drawing process

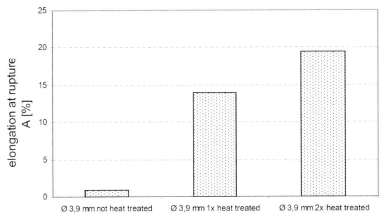

Figure 5: Influence of the heat treatment on the elongation of drawn tubes made from MgCa0,8

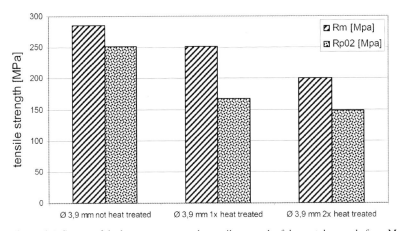

Figure 6: Influence of the heat treatment on the tensile strength of drawn tubes made from MgCa0,8

4 Summary

The investigations show that the development of the combined extrusion and drawing processes enables the production of small and thin tubes made from MgCa0,8 alloy. The changing of the microstructure leads to a very fine and unidirectional grain structure at the tubes. Small grain sites of 10–15 µm are observed. The mechanical properties get worse during the production process. This can be prevented by heat treatment steps between the drawing processes. It is shown that the mechanical properties nearly achieve the values of the extruded material by the heating in furnace. The developed process chain enables the production of small tubes made from MgCa0,8 with adequate mechanical properties. The diameters of the tube is variable by choosing a special mandrel diameter and the wall thickness can be effectively reduced to 150–200 µm. This is equivalent to the required strut thickness of a vascular stent.

5 References

[1] Fr.-W. Bach, Th. Hassel, A. Golovko, Ch. Hackenbroich, A. Meyer-Lindenberg: Resorbierbare Implantate aus Magnesium durch Mikrolegieren mit Calcium, deren Verarbeitung und Eigenschaften. Biomaterialien, 6. Jahrgang, Heft 3, Okt. 2005, S. 163.

[2] Fr.-W. Bach, A.N. Golovko, Th. Hassel: Characteristic Features of Mg-Ca Alloy Tubes Extrusion by Vertical Hydraulic Press. Metalurgical and Mining Industry, Nr. 1, 2005, S. 48–51.

[3] Luo A.L., Sachdev A.K.: Mechanical properties and microstructure of AZ31 magnesium alloy tubes. Magnesium Technology, 2004

[4] Emley E.F.: Principles of magnesium technology, Oxford, London, Edinburgh, New York, Paris, Frankfurt: Pergamon Press, 1966.

[5] Beck A.: Magnesium und seine Legierungen, 2. Auflage, Berlin: Springer, 2001.

[6] Bauser M., Sauer G., Siegert K.: Strangpressen. 2. Auflage. Düsseldorf: Aluminium-Verlag, 2001.

Microstructural Evolution

The Effect of Zinc and Gadolinium on the Precipitation Sequence and Quench Sensitivity of Four Mg-Nd-Gd Alloys

L. R. Gill[*#], G. W. Lorimer[*], P. Lyon[+]

[*] Manchester Materials Science Centre, The University of Manchester, Manchester, U.K.
[+] Magnesium Elektron, Manchester, U.K.
[#] Now at Department of Chemical and Materials Engineering, University of Alberta, Edmonton, Canada

1 Introduction

Magnesium rare earth alloys show excellent age hardening properties, due to the decreasing solid solubility of rare earth elements with decreasing temperature; a critical requirement for age-hardening. It is known that the poor casting quality of binary Mg-Zn alloys can be improved with the addition of rare earth elements [1]. Additions of gadolinium to yttrium and neodymium-based systems have been shown to improve the mechanical properties of the alloy [2].

A study has been made of the effects of zinc and gadolinium concentration on the precipitation sequence and quench sensitivity of four Mg-Nd-Gd-Zn alloys The research presented contributed to an alloy development project of Elektron 21 undertaken by Magnesium Elektron. The alloy was developed for use for specialist applications such as helicopter/jet engine gearboxes.

2 Experimental

Alloy production and solution heat treatment were carried out at Magnesium Elektron Ltd. The alloys were cast into sand moulds to form plates 20cmx20cmx3cm. The alloy compositions are listed below:
1. Mg-2.7wt%Nd-0.007wt%Zn-0.35wt%Gd-0.48wt%Zr
2. Mg-2.8wt%Nd-0.6wt%Zn-0.29wt%Gd-0.46wt%Zr
3. Mg-2.8wt%Nd-1.3wt%Zn-0.28wt%Gd-0.47wt%Zr
4. Mg-3.0wt%Nd-0.64wt%Zn-1.98wt%Gd-0.44wt%Zr

The plates that were used to determine the ageing response were solution treated for eight hours at 520 °C followed by a hot water quench. The alloys were then aged in a salt bath at 200 °C. The plates used for the Jominy end quench tests were machined to BS 4437:1987 specimen size. Thermocouples of 0.5 mm diameter were placed into 0.9 mm diameter holes, drilled along the length of the test piece to the centerline of the sample. The thermocouples were connected to a data logger and the computer package Labview Data Acquisition recorded the cooling rates along the Jominy bar.

Furnace cooling was carried out by solution treating as-cast 1cm3 samples for four hours at 520 °C in a fan furnace, the samples were cooled by turning the furnace off and leaving the door closed, the cooling rate to 150 °C was 0.2 °C s^{-1}. Vickers hardness testing was carried out on flats ground to 1200 grit along each side of the Jominy bars. Samples for TEM were electropolished using a solution of 1000ml methanol, 200 ml 2-butoxyethanol, 22.4 g magnesium perchlorate and 10.6 g lithium chloride at. –40 °C and a voltage of 30 V. TEM examination was carried out using a Philips CM200 or an FEI Tecnai F30 FEGTEM.

3 Results

3.1 Ageing Response at 200 °C

Figure 1 shows the ageing response of the four alloys after ageing at 200 °C. The four horizontal lines in the left-hand corner correspond to the hardness of each of the alloys after solution treatment. Alloy 1 had the lowest hardness and alloy 4 had the highest. Alloy four showed the highest peak hardness where as alloys 1,2 and 3 had similar values. Alloy 3 retained its hardness at longer ageing times better than alloys 1 and 2.

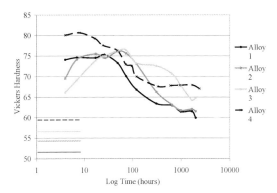

Figure 1: Ageing response of the four alloys at 200 °C

Increasing the zinc and/or gadolinium content of the alloys increased the hardness of the alloys after solution treatment and increasing the gadolinium content increased the hardness of the alloy after solution treatment to a greater extent than increasing the zinc content. This was mostly due to the solid solution strengthening effect of the higher concentration of zinc and gadolinium.

The peak hardness of the alloys is increased by increasing the amount of gadolinium. Increasing the zinc content does not increase the peak hardness value but it does increase the alloys hardness at extended ageing times.

3.2 Precipitates Formed During Ageing at 200 °C

Figures 2 and 3 show a TEM micrograph and the associated diffraction pattern from the precipitates that formed in alloys, 1, 2 and 4 during ageing between 4 and 24 hours. The precipitates were ellipse-shaped plates with their long axis parallel to the [0001]Mg direction; the habit planes of the precipitates were the primary and secondary prism planes of the magnesium matrix. The presence of maxima at intervals halfway between the matrix spots in the diffraction pattern (figure 3) is evidence of a DO19 structure. Precipitate with a DO19 in magnesium-neodymium alloys have been reported by other researchers [3–5]. The precipitates were designated β".

Figures 4 and 5 show a TEM micrograph and the associated diffraction pattern from the precipitates formed in alloys 1, 2, 3 and 4 during ageing between 24 and 1680 hours. The precipitates formed as plates on the planes and had a FCC structure with orientation relationship [011]pte // [0001]Mg, $(\bar{1}1\bar{1})_{pte}//(11\bar{2}0)_{Mg}$. The lattice parameters of the precipitates were 0.72+/-0.04nm, the same as that reported by previous authors [5–11] in WE, Mg-Nd and Mg-Nd-Zn alloys. The precipitates were designated β'.

Figures 6 and 7 show the precipitates that formed in alloy 3 between 4 and 1680 hours and between 120 and 650 hours in alloy 2. The precipitates had an HCP structure with lattice parameters, $\mathbf{a} = 0.55$ nm and $\mathbf{c} = \mathbf{c}_{Mg}$ and orientation relationship, $\{1\bar{1}00\}_{pte}//\{11\bar{2}0\}_{Mg}$, $[0001]_{pte}//[0001]_{Mg}$. The precipitates were found to be plate-shaped which is in agreement with work by previous researchers [6,12,13]. The precipitates were designated 1.

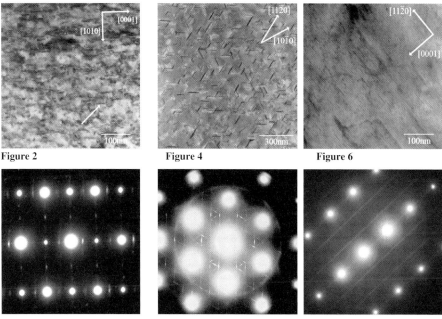

Figure 2

Figure 4

Figure 6

Figure 3: Figures 2 and 3 show alloy 1 after ageing for 26 hours at 200 °C. The electron beam is parallel to $[1\bar{2}10]_{Mg}$

Figure 5: Figures 4 and 5 show alloy 1 after ageing for 120 hours at 200 °C. The electron beam is parallel to $[0001]_{Mg}$

Figure 7: Figures 6 and 7 show alloy 3 after 4 hours at 200 °C. The electron beam is parallel to $[1\bar{1}00]_{Mg}$

3.3 Effect of Quenching Rate on Ageing Response

Figures 8 to 11 show the ageing curves from the alloys at different positions along the Jominy bar. The cooling rates are the average at that position for all the quench tests to 150 °C. At peak hardness it can be seen that the fastest quench rate gives the highest hardness, and the peak hardness value decreases with decreasing quench rate for alloy 3 and to a lesser extent alloys 2 and 4. Alloy 1 does not show this trend. Alloy 3 appears to be more quench sensitive than alloy

1, as are alloys 2 and 4 albeit to a lesser extent. This suggests that increasing the zinc content increased the quench sensitivity of the alloys.

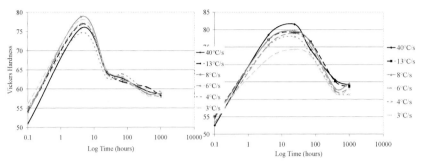

Figure 8: Alloy 1, ageing response at 200 °C following different quench rates

Figure 9: Alloy 2, ageing response at 200 °C following different quench rates

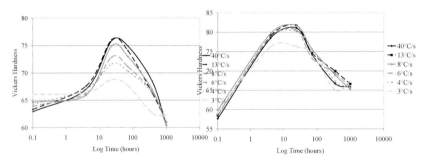

Figure 10: Alloy 3, ageing response at 200 °C following different quench rates

Figure 11: Alloy 4, ageing response at 200 °C following different quench rates

Examination of the microstructures of each of the alloys subjected to a cooling rate of 40 °C s^{-1} did not reveal any precipitation. Figures 12 and 13 show evidence of β' precipitates that formed during a quench rate of 3° C s^{-1}, these were observed in alloys 1, 2 and 4. Upon subsequent ageing for 24 hours at 200 °C the precipitation that occurred in the zinc-free alloy 1, at a quench rate of 3° C s^{-1} was the same as that which occurred at 40 °C s^{-1}. Figures 14 and 15 show evidence of 1 and β" in the microstructure of alloy 4 (white spots and grey spots in figure 15, respectively) after ageing and a quench rate of 3 °C s^{-1}, these were also observed in alloy 2. The two alloys appeared to be more quench sensitive than the zinc-free alloy 1. Figures 16 and 17 show alloy 3 after quenching at a rate of 3 °C s^{-1}. Large 1 precipitates had formed and these were associated with the alloy's high quench sensitivity.

These results show that the precipitates that form after ageing in alloys 2 and 4 are altered if the alloys are subjected to a slower quench rate; 1 appears to form at the expense of β". It has been shown [14] that precipitates with a prism habit plane are more effective at strengthening an HCP alloy than precipitates that form on the basal planes. Therefore, β" precipitates strengthen more effectively than 1 precipitates due to their prism habit planes as opposed to the basal planes occupied by the 1 precipitates. Alloy 1 does not appear to form the 1 precipitate therefore it is possible that this precipitate will only form in zinc-containing alloys.

Figure 18 shows alloy 1 after a furnace cool. The larger precipitates were identified as β' and the smaller ones were β". Figure 19 shows β' precipitates that formed in alloy 4 after a furnace cool and figure 20 shows 1 precipitates, both sets precipitates were also observed in alloy 2.

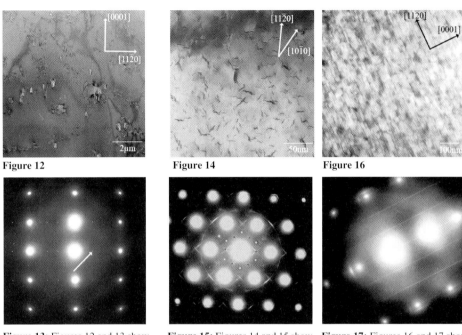

Figure 12

Figure 14

Figure 16

Figure 13: Figures 12 and 13 show a dark field image and its associated diffraction pattern from alloy 2 after solution treatment, and cooling at a quench rate of 3 °C s^{-1}. The electron beam was parallel to the direction. The arrow indicates the spot figure 12 was taken from.

Figure 15: Figures 14 and 15 show alloy 4 after quenching at 3 °C s^{-1} and ageing for 24 hours at 200 °C. The electron beam was parallel to the [0001]Mg direction. Two sets of precipitate spots are shown, one indicated by white spots and the other by dark grey spots.

Figure 17: Figures 16 and 17 show alloy 3, after solution treatment, and cooling at a quench rate of 3 °C s^{-1}. The electron beam is parallel to the [1$\bar{1}$00]$_{Mg}$ direction.

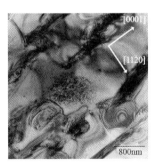

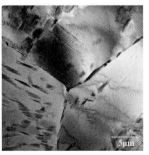

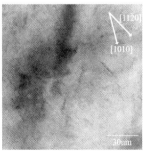

Figure 18: Alloy 1 solution treated for four hours at 520 °C followed by a furnace-cool. The electron beam is parallel to the [1$\bar{1}$00]$_{Mg}$ direction.

Figure 19: Alloy 4 solution treated for four hours at 520 °C followed by a furnace-cool

Figure 20: Alloy 4 solution treated for four hours at 520 °C followed by a furnace-cool. The electron beam is parallel to [0001]Mg

4 Conclusions

The precipitation sequences in each of the alloys during ageing at 200 °C were:

1. $Mg_{ssss} \rightarrow \beta''_{(DO_{19})} \rightarrow \beta'_{(FCC)}$
2. $Mg_{ssss} \rightarrow \beta''_{(DO_{19})} \rightarrow \beta^1_{(HCP)} \rightarrow \beta'_{(FCC)} \rightarrow \beta$
3. $Mg_{ssss} \rightarrow \beta^1_{(HCP)} \rightarrow \beta'_{(FCC)}$
4. $Mg_{ssss} \rightarrow \beta''_{(DO_{19})} \rightarrow \beta'_{(FCC)}$

Increasing the gadolinium content of the alloy improved the ageing response alloy through solid-solution strengthening.

Increasing the zinc content of the alloy increased the ageing response in the over-aged condition due to the enhanced strengthening offered by the coexistence of 1 and β' precipitates. In the under-aged condition, the ageing response was reduced as 1 offered inferior strengthening compared to β" which was probably due to its habit plane.

Alloys 1, 2 and 4 formed β' precipitates during a slow quench rate. Upon subsequent ageing the precipitation that occurred in the zinc-free alloy 1 was the same as that which occurred after a faster cooling rate. After ageing, the precipitation occurring in alloys 2 and 4 was changed, with 1 forming at the expense of β", therefore the two alloys appeared to be more quench sensitive than the zinc-free alloy. Large 1 precipitates formed in the high-zinc alloy 3 during a slow quench and these were the contributing factor to the alloys apparent high quench sensitivity. A furnace-cool changed the precipitation occurring during cooling in alloys 1, 2 and 4. Alloy 1 formed β" and β' and alloys 2 and 4 formed 1 and β'.

5 References

[1] E. F. Emley, Principles of Magnesium Technology, 1st edition , Pergamon Press, 1966.
[2] P. J.Apps, H. Karimzadeh, J. F. King, G. W. Lorimer, Scripta Materialia, 2003, 48, 1023–1028.
[3] T. J. Pike, B. Noble, Journal of Less Common Metals, 1973, 30, 63–74.
[4] M. Hisa, PhD Thesis, 1995, The University of Queensland.
[5] H. Karimzadeh, PhD Thesis, 1985, University of Manchester.
[6] R. Wilson, C. J. Bettles, B. C. Muddle, J. F. Nie, in Magnesium Alloys 2003 (Ed.: Y. Kojima, T. Aizawa, K. Higashi, S. Kamado), , Trans Tech Publications, Switzerland, 2003, 267–272.
[7] P.J. Apps, PhD Thesis, 2001, University of Manchester.
[8] J. F. Nie, B. C. Muddle, Scripita Materialia, 1999, 40, 1089–1094.
[9] J. F. Nie, B. C. Muddle, Acta Materialia, 2000, 48, 1691–1703.
[10] P. A. Nuttall, T. J. Pike and B. Noble, Metallography, 1980, 13, 3–20.
[11] K. J. Gradwell, MSc Thesis, 1970, The University of Manchester.
[12] D. H. Ping, K. Hono, J. F. Nie, Scripita Materialia, 2003, 48, 1017–1022.
[13] J. F. Nie, X. Gao, S.M. Zhu, Scripta Materialia, 2005, 53, 1049–1053.
[14] J. F. Nie, Scripta Materialia, 2003, 48, 1009–1015.

Secondary Precipitation in the Magnesium Alloy WE54

Yuchang Su†, Brian M. Gable, Barry C. Muddle, Jian-Feng Nie

ARC Centre of Excellence for Design in Light Metals, Department of Materials Engineering, Monash University, Clayton, Victoria, Australia
†School of Materials Science and Engineering, Central South University, Changsha, Hunan, P.R. China

1 Introduction

Precipitation hardened magnesium-rare earth (RE) alloys, particularly those based on the Mg-Y-Nd system, offer attractive properties for aerospace and automotive applications [1]. Unfortunately, there are reports of embrittlement after long-term exposure to moderate temperatures [2]. As for all age-hardenable alloys the type, size and distribution of the secondary precipitate phases dictate the behavior of Mg-RE alloys. Many researchers [2-8] have studied the precipitation sequence for Mg-Y-Nd alloys, which is commonly reported:

$$\text{Mg}_{\text{supersaturated solid solution}} \rightarrow \text{GP zone} \rightarrow \beta'' \, (DO_{19}) \rightarrow \beta' \rightarrow \beta_1 \rightarrow \beta \ .$$

However, the decomposition process from a supersaturated solid solution (sss) in Mg-RE alloys is not fully understood and uncertainties remain concerning the formation and kinetics of metastable precipitates, particularly those formed at low to moderate aging temperatures.

The object of this study is to characterize the effect of duplex aging on the hardening response of the WE54 alloy on the precipitate type, distribution, number density and size in order to identify the origin of embrittlement. Concurrently, this study offers more attractive thermal treatments for improvements in room temperature strength and introduces a physical basis for the origin of secondary precipitation.

2 Experimental Methods

The alloy WE54 investigated has a nominal composition of Mg-(5.0-5.5)Y-(1.5-2.0)Nd-(1.5-2.0)RE-0.4Zr (wt.%), with the results from chemical analysis given in Table 1. The as-received billet was obtained from the MEL Company, England. Samples were machined from the billet and were solid solution treated at 525 ºC for 8h followed by cold water quenching. Samples were then aged in salt or oil baths for various times, with Vickers hardness testing monitoring the general age-hardening response using a 5 kg load. Initial investigations concerned the isothermal aging response at temperatures between 160–250 ºC, with subsequent duplex treatments taking place over this same temperature range.

After aging, 3mm diameter discs were punched from the heat treated specimens and metallographically ground to a thickness of ~150 µm. Subsequent electropolishing was performed on a twin-jet unit in a solution of 5.3 g lithium chloride, 11.6 g magnesium perchlorate, 500 ml methanol and 100 ml 2-butoxy-ethanol, cooled to –50 ºC, and polished with an applied current of 0.1 A. Characterization of precipitate phases focused on the type, distribution and size via a

200 kV transmission electron microscope (TEM). All images were recorded with the incident electron beam parallel to the $[0001]_{Mg}$ direction.

Table 1: Alloy composition determined via chemical analysis of the as-received billet (wt%)

Y	Nd	Zr	Dy	Er	Gd	Yb	La	Ce	Fe	Cu	Ni
5,11	1,72	0,29	0,26	0,18	0,13	0,13	0,05	0,04	<0.003	<0.002	<0.001

3 Experimental Results

3.1 Isothermal Aging Treatments

A series of isothermal age-hardening curves are demonstrated in Figure 1 for a temperature range from 160–250 °C. These curves follow the general trend of accelerated aging kinetics, but lower potential age-hardening as the aging temperature is increased. Most importantly the 250 °C aging curve demonstrates that there is no measurable change in hardness after reaching peak hardness at ~18h, which presumably is a signature of microstructural stability that is desirable for creep resistance. Both the 160 °C and 200 °C aging curves indicate that these microstructures continue to evolve over the time intervals investigated, with the 160 °C treatment continually hardening during more than over 5000 h of exposure.

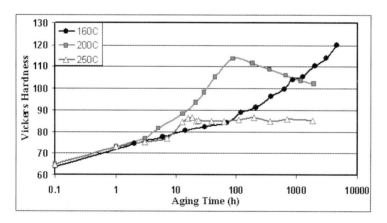

Figure 1: Isothermal age-hardening curves for WE54 aged between 160–250 °C

The microstructural evolution for WE54 isothermally aged at 250 °C has been carefully monitored in previous work and is summarized in [6]. For isothermal aging conditions at 250 °C precipitation involves formation of three separate metastable phases prior to formation of the equilibrium phase. These metastable precipitate phases include $\{11\bar{2}0\}_{Mg}$ platelets of an unidentified structure, b' and b1. The $\{11\bar{2}0\}_{Mg}$ platelets are very fine in size and often form in isolation or in arrays associated with coarse, spherical particles of the b' phase. They are gradually replaced by b1 phase during prolonged aging. The β_1 phase forms as coarse plates on $\{10\bar{1}0\}_{Mg}$ planes of the magnesium matrix and its formation is always in association with particles of b'. The b1 phase eventually transforms in situ to the equilibrium phase b.

Precipitation in WE43 (Mg-4Y-3.3RE-0.5Zr wt%) treated at 150 °C has been studied via TEM and SAXS, as reported in [8]. Isothermal aging at relatively low temperatures, *e.g.* 150 °C, led to an extremely high number density of fine plate-like precipitates. Previous work [8] has concluded that these precipitates are initially DO_{19} structures that upon further aging transform into the plate-like and β" and globular β' precipitates. These initial structures have been described as mono-planar with diameters of only 5–20 nanometers after prolonged aging at 150 °C.

3.2 Multi-Step Aging Treatments

To investigate the phenomenon of lower temperature embrittlement, samples initially aged at 250 °C for various times up to the peak-aged condition of 18h were exposed to the secondary aging temperature of 160 °C. The resulting hardening response is summarized on Figure 2.

As seen on Figure 2 hardening was continuous for each specimen when aged at 160 °C, regardless of the extent of exposure at 250 °C. Not surprisingly there is a general trend of decreased potential for secondary hardening related to longer aging times at 250 °C. The overall hardening through 5 kh of aging at 160 °C resulted in ~100% gain in hardness for the sample isothermally aged at 160 °C, while the specimen initially peak-aged at 250 °C for 18 h exhibited moderate gains in hardness of ~25 %. These gains in hardness imply that even after peak-aging at 250 °C there is still solute available to further precipitation.

An investigation into the microstructural evolution of the two-stage aging specimens reveals that there is a distinct non-uniform distribution of precipitation that results, as seen in Figure 3. The coarse β' and β phases from the initial 250 °C treatment are still present and qualitatively have not changed in size or number density with the additional 160 °C treatment. The non-uniformity of the microstructure includes denuded zones adjacent to the coarse β precipitates as well as very fine precipitation occurring in those areas isolated between the coarse β precipitates. The form and distribution of this fine secondary precipitation is consistent with a phase having the DO_{19} structure identified in a previous investigation [8], although the identity of the present phase has yet to be confirmed.

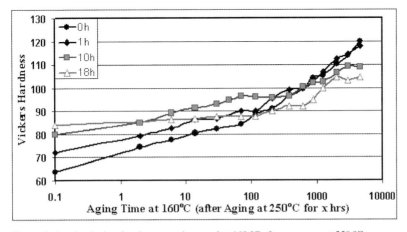

Figure 2: Age-hardening development when aged at 160 °C after exposure at 250 °C

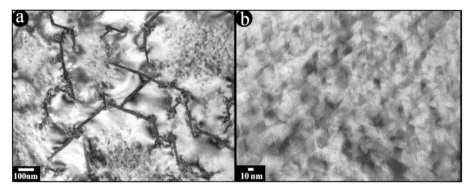

Figure 3: Representative microstructure after duplex aging at 250 °C 18h and 160 °C 2000 h a) illustrating the non-uniform distribution of β precipitates, and b) the fine precipitates found in isolated regions that formed during the secondary aging step

Microanalysis of several areas as illustrated in Figure 4(a) using energy dispersive X-ray spectrometry (EDXS) indicates that concentration of Nd and Y varies from the distance away from those precipitates formed at 250 °C, Figure 4(b). As expected, the primary β precipitate has higher concentration of Nd and Y. There is no detectable Nd in the denuded precipitate-free zones, however, there is some Y present, although to a lesser extent than in the precipitates themselves. The EDXS analysis of the fine precipitates suggests that they are comprised of both Nd and Y, in similar fashion to the β phase. Due to precipitate length-scale considerations relative to the TEM foil dimensions it is not possible to quantitatively compare the chemistry of the coarse β phase with that of the fine structures.

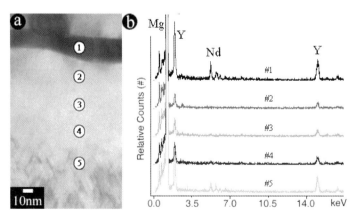

Figure 4: EDXS analysis profile from a β precipitate formed during the initial aging treatment at 250 °C through the denuded zone into the secondary aged region of fine precipitation

4 Discussion

The age hardening data and TEM images clearly indicate that the microstructure of the WE54 alloy is evolving during exposure at 160 °C, regardless of pre-aging condition. These observations provide strong evidence for the source of embrittlement of these alloys during low to moderate temperature exposure. The enhanced hardness, coupled with non-uniform precipitate distribution and particle-free zones adjacent to coarse precipitates will each contribute to a reduction of ductility and toughness versus that of the original isothermally peak-aged material.

The microstructural and chemical analysis also gives an indication to the source of the precipitate stability and thus the attractive creep resistance of WE54 when aged at 250 °C. As evidenced through the formation of these denuded zones in the multi-step aged specimens the solute, in particularly the Nd, is exhausted from the matrix neighboring the β-type precipitates. Due to soft impingement of the various β-type phases some solute remains isolated at distances too great for diffusion to compensate on the time scale investigated in this study, as illustrated in Figure 5. This limits the future coarsening of the strengthening phases, rendering them relatively stable due to their inability to consume more solute, limiting them to self-coarsening and/ or phase transitions [6]. However, when the driving force for nucleation is enhanced through a low/moderate temperature exposure these solute-rich regions of isolated matrix are then sites for the secondary precipitation, furthering the non-uniformity of the structure and possibly acting as the source of the reported embrittlement. This secondary aging phenomenon due to isolated solute parallels that which is now well known and exploited in several age-hardenable Al alloy systems [9–14].

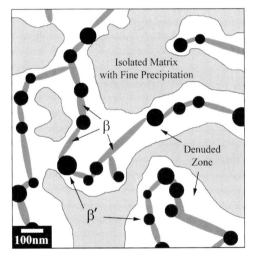

Figure 5: Sketch summarizing the origin for the isolated matrix regions. This is a simplified 2-D reconstruction of an actual TEM image captured as a part of this investigation after duplex aging of 250 °C for 18h and 160 °C for 2000 h.

5   Conclusions

This investigation has revealed that secondary precipitation is prevalent in WE54 when aged at 160 °C after initial aging at 250 °C, even for the isothermal peak-aged condition. TEM results suggest that this secondary aging is an artifact of solute that has been isolated through the coarse formation of β-type precipitates that is available for further precipitation when the driving force for nucleation is high. It is believed that the formation of these fine phases and non-uniformity of the precipitate structure are contributing factors to the low/moderate temperature embrittlement observed for this class of alloy.

6 Acknowledgements

The authors would like to acknowledge funding through the Australian Research Council. YCS gratefully acknowledges Monash University and Central South University for the travel grant and stipend that made this research collaboration possible.

7 References

[1] J.F. King, Advanced Materials Technology International, Vol. 12, 1990.
[2] M. Ahmed, G.W. Lorimer, P. Lyon and R. Pilkington, Proc. Magnesium Alloys and Their Applications, ed. B.L. Mordike and F. Hehmann, DGM Informationsgesellschaft Verlag, Germany, 1992, 301–308.
[3] T.J. Pike and B. Noble, J. Less Common Metals. 1973, 30, 63–74.
[4] P. Vostry, I. Stulikova, B. Smola, M. Cieslar and B.L. Mordike, Z. Metallkde 1988, 79, 340–344.
[5] T. Hilditch, J.F. Nie and B.C. Muddle, Proc. Magnesium Alloys and Their Applications ed. B.L. Mordike and K.U. Kainer, Werkstoff-Informationsgesellschaft, Frankfurt, Germany, 1998, 339–344.
[6] J.F. Nie and B.C. Muddle, Scripta Mater. 1999, 40, 1089–1094.
[7] J.F. Nie and B.C. Muddle, Acta Mater. 2000, 48, 1691–1703.
[8] C. Antion, P. Donnadieu, F. Perrard, A. Deschamps, C. Tassin and A. Pisch, Acta Mater. 2003, 51, 5335–5348.
[9] X. Gao, J.F. Nie, B.C. Muddle, in Proc. Materials 98, ed. M. Ferry, Institute of Materials Engineering Australasia, Australia, 1998, 573–578.
[10] Polmear I.J., Light Alloys - Metallurgy of the Light Metals, 3rd ed., 1995, 96–206.
[11] .N. Lumley, A.J. Morton and I.J. Polmear, Acta Mater. 2002, 50, 3597–3608.
[12] R.N. Lumley, I.J. Polmear and A.J. Morton: Mater. Sci. Tech. 2003, 19, 1483–1490.
[13] C.R. Hutchinson, P. Cornall and M. Goune, Mater. Sci. Forum 2006, 519–521, 1029–1034.
[14] K.M. Nairn, B.M. Gable, R. Stark, N. Ciccosillo, A.J. Hill, B.C. Muddle and T.J. Bastow, Mater. Sci. Forum 2006, 519–521, 591–596.

Influence of Production Technology and Heat Treatment on the Precipitation Processes in Mg-Y-Nd-Zn-Zr Alloy

J. Pelcová, B. Smola, I. Stulíková
Faculty of Mathematics and Physics, Charles University, Prague, Czech Republic

1 Abstract

The investigation of microstructure development and phase transformations in the course of isochronally increasing temperature was performed by means of electrical resistometry. Two magnesium alloys containing various amounts of Zn, Nd, Y and Zr were produced by different production technology (spray forming, powder metallurgy or squeeze casting) and thermal treatment. The results of electrical resistometry were correlated with microhardness measurement HV0.1 performed in the same mode as resistometry measurements. The microstructure analysis and precipitate identification were done by the transmission electron microscopy in the selected states of materials.

The preparation technology of the alloy changed markedly temperature ranges of precipitation processes as well as the type and structure of phases formed. The main and only minimum of relative resistivity changes corresponding to the main precipitation process was observed between 330 °C–360 °C for squeeze cast alloys contrary to the more complicated resistivity response to the same annealing of spray formed alloy or material prepared by powder metallurgy. The microhardness response to the isochronal annealing confirms the precipitation origin of the main resistivity changes.

2 Introduction

Zinc is often used in commercial Mg based alloys. During decomposition of supersaturated solid solution of binary Mg-Zn alloys, many metastable phases are formed (Guinier-Preston zones, $MgZn$, $MgZn_2$, Mg_2Zn, etc.). Zirconium is added to refine the grain size and to participate in the development of phases leading to a higher ratio of proof stress to tensile strength and increase the creep resistance.

Most Mg-Zn alloys contain rare earth elements, which form eutectic systems with magnesium and improve the castability. Grain boundary networks of relatively low melting point eutectics are formed in these alloys. A strengthening effect of RE on wrought Mg-Zn-Zr-RE due to the formation of RE-containing particles was observed [1], which suppresses dynamic recrystallisation during extrusion. The addition of 3 wt.% Nd to the Mg-Zn-Zr alloy can effectively improve the yield strength and ultimate tensile strength of the alloy at higher temperatures as a result of grain refinement and the formation of the $Mg_{12}Nd$ phase [2].

The use of low-cost Mg-base alloys is limited owing to their moderate mechanical and creep properties at elevated temperatures. These could be improved not only by the composition modification but also by the use of modern processing technologies (composites, rapidly solidified alloys, nano-particle reinforced alloys, etc.).

3 Experimental Details

Two magnesium alloys containing various amount of Zn, Zr and rare earths (Nd +Y) were studied. The alloy Mg3Zn1Nd0.5Zr was prepared by squeeze casting or the spray forming technique with subsequent extrusion. The temperature of the melt in spray forming was 740°C; the process gas was Ar +1 vol.% O_2. Squeeze casting was done in a protective atmosphere of Ar + 1 % SF_6. Extrusion was carried out at 350 °C with a reduction of 50:1 after a one-hour preheating at 300 °C. The composition is in Table 1.

Table 1: Composition of the alloy Mg3Zn1Nd0.5Zr

Alloy	Zn [wt%]	RE* [wt%]	Zr [wt%]
Squeeze cast + extruded	4.19	0.98	0.58
Spray formed + extruded	3.08	0.91	0.32

* Nd + small amount of Y

The alloy Mg0.5Zn3Nd2Y0.5Zr was prepared by squeeze casting or by powder metallurgy technique (PM) and compacted by subsequent extrusion. Squeeze casting was done again in a protective atmosphere of Ar + 1 % SF_6. The temperature of the melt before atomizing was 740 °C. The atomizing was carried out in Ar + 1 % O_2 atmosphere. Extrusion used for compaction of powder was carried out at 350 °C with a reduction of 50:1 after a one-hour preheating at 300 °C. The composition is listed in Table 2.

Table 2: Composition of the alloy Mg0.5Zn3Nd2Y0.5Zr

Alloy	Zn [wt%]	Nd [wt%]	Y [wt%]	Zr [wt%]
Squeeze cast	0.35	3.21	1.75	0.42
Powder + extruded	0.42	2.84	1.65	0.40

The relative changes in resistivity during isochronally increasing temperature in the range 20 °C to 510 °C were determined in steps of 30 °C / 30 min. Each annealing step was followed by quenching into liquid nitrogen, for annealing temperatures up to 240 °C, and into water of room temperature for higher annealing temperatures. Heat treatment was carried out in a stirred oil bath up to 240 °C and in a furnace with an argon protective atmosphere at higher temperatures. The four contact specimens in the shape of a letter H were used in the resistivity measurements carried out in liquid nitrogen after each heating step. Relative resistivity changes $\Delta\rho/\rho_{0?}$ were obtained to an accuracy of 10^{-4} using the dc four-point method with a dummy specimen in series. The effect of parasitic thermo-electromotive forces was suppressed by current reversal.

Changes in the micro-hardness HV0.1 were measured following the same thermal treatment as resistivity measurements to reveal the thermal stability of the mechanical properties related to the micro-structural development.

Transmission electron microscopy, electron diffraction (ED) and X-ray microanalysis (EDX) were used to determine the structure and morphological characteristics of the phases precipitated (using JEOL JEM 2000FX electron microscope and Link AN 10000 micro-analyzer). The specimens for TEM were prepared also by the same isochronal annealing procedure as those for electrical resistivity and hardness measurements.

4 Results and Discussion

4.1 Mg3Zn1Nd0.5Zr Alloy

Relative resistivity changes $\Delta\rho/\rho_{02}$ and microhardness response HV0.1 to isochronal annealing were investigated in spray formed and extruded and squeeze cast and extruded Mg3Zn1Nd0.5Zr alloy. The results are compared in Fig. 1.

Both materials contain particles the c-base centered orthorhombic phase containing Zn and Nd with parameters a = 0.99 nm, b = 1.15 nm, c = 0.98 nm in the initial state. This phase formed during the extrusion process at 300 °C and is known as a pseudo-ternary T-phase in Mg-Zn-RE alloys [3, 4]. Individual oval shaped particles (Fig. 2a.) of this T-phase were observed in squeeze cast alloy contrary to conglomerates of fine rectangular particles in the spray formed material. Precipitates containing Y and Zn (size up to 30 nm) and a relatively high dislocation density were observed in some grains of as extruded spray formed material.

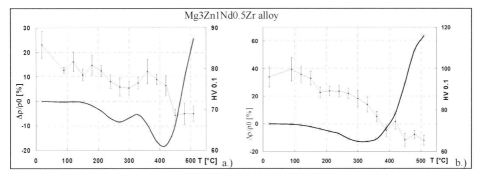

Figure 1: Relative resistivity and hardness responses to isochronal annealing of Mg3Zn1Nd0.5Zr alloy
a) spray formed and extruded alloy, b) squeeze cast and extruded alloy

A coarsening of the T-phase and recovery of mentioned dislocation substructure was observed after annealing up to 270 °C in the spray formed and extruded material. This process leads to a slight decrease of resistivity and slight microhardness degradation. The development of fine, dense dispersed precipitates of Zn-, Y- and Nd- containing phase was detected after annealing up to 420 °C, which resulted in the resistivity decrease in the range from 330 °C to 420 °C. Slight microhardness increase (5 %) was associated with this process. Annealing of the spray formed and extruded specimen to higher temperatures leads to the increase of resistivity over the initial value (more than 25 %). It is, most probably, caused by the dissolution of precipitates containing Nd and simultaneous precipitation of the phase containing Zn-Zr (needles and ellipsoids) observed by TEM in specimens after annealing up to 480 °C . Though the contribution of 1 at. %Nd in Mg matrix to the residual resistivity is relatively high [5–7], it can not compensate the decrease caused by the depletion of Zn and Zr solute in the matrix due to precipitation of Zr-Zn phase. The grain size (1–2 µm) did not change during isochronal annealing.

The response of relative resistivity changes to isochronal annealing of the squeeze cast and extruded material can be described by continuous slight decrease of resistivity values up to 330 °C (minimum –13%). The coarsening of the T-phase and development of fine Zn-Y-Nd particles in grain boundaries was observed in specimens annealed up to 300 °C (Fig. 2b). The

Zn-Y-Nd precipitates were also observed in the grain interiors (Fig. 2c) after annealing up to 420 °C together with Zn-Zr needle shaped particles and stable $Mg_3Y_2Zn_3$ phase (fcc, a = 0.698 nm). For higher annealing temperatures, a very pronounced resistivity increase was observed (more than 60 % above initial value) similar to that in the spray formed and extruded material. It indicates increasing concentration of solutes in the matrix. Contrary to spray formed material, the grain grew from ~2 μm at initial state to about 4 μm after annealing up to 480 °C and the microhardness continuously decreased during the whole annealing process in the squeeze cast and extruded material. This result indicates a better thermal stability of mechanical properties and fixed microstructure of spray formed and extruded material.

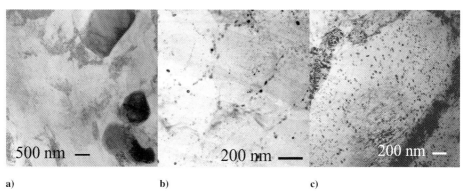

a)   b)   c)

Figure 2: Microstructure of Mg3Zn1Nd0.5Zr squeeze cast alloy
a) particles of T-phase in as prepared state, b) fine Zn-Y-Nd particles in grain boundaries after annealing up to 300 °C, c) fine Zn-Y-Nd particles in the grain interior after annealing up to 420 °C

4.2   Mg0.5Zn3Nd2Y0.5Zr Alloy

Relative resistivity changes $\Delta\rho/\rho_0$ and microhardness response HV0.1 to isochronal annealing were investigated in squeeze cast alloy and in extruded material prepared by powder metallurgy. The results are compared in Fig. 3.

The presence of the eutectic (α-Mg matrix and β_1 phase - fcc, a = 0.74 nm) and the triangular arrangement of β_1 prismatic plates in grain interiors was observed in the initial state of the squeeze cast alloy (Fig 4a). The triangular arrangement of prismatic plates is well known from Mg-RE system [8–10]. The resistivity responds to the isochronal annealing up to 210 °C by a slight decrease in squeeze cast alloy. This decrease is most probably caused by further precipitation of metastable β_1 phase in the triangular arrangement and leads to microhardness improvement. The main minimum of resistivity (–14 %) is reached after annealing up to 300 °C and is connected with the transformation of metastable β_1 phase to the stable $Mg_{41}Nd_5$ phase of the size up to 0.5 μm. A microhardness does not changed during this process. Isochronal annealing to higher temperatures (over 300 °C) leads to the continuous increase of resistivity values that markedly increases over the initial value (increase about 30 %). This resistivity increase was connected with the rapid microhardness decrease (–40 %) and was caused, most probably, by dissolution of precipitates.

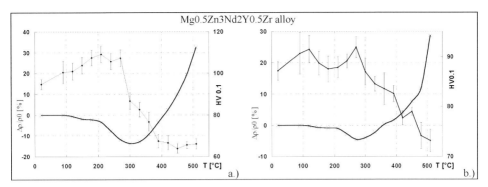

Figure 3: Relative resistivity and microhardness response to isochronal annealing of Mg0.5Zn3Nd2Y0.5Zr alloy a) squeeze cast alloy, b) powder metallurgy + extruded alloy

The extruded material prepared from powders contained the particles of equilibrium $Mg_{41}Nd_5$ phase with Y, Zn content (tetragonal structure, a = b = 1.476 nm, c = 1.039 nm), that developed during the extrusion process. A slight resistivity decrease after annealing up to 210°C similar to squeeze cast alloy was observed. The microhardness values were almost stagnating up to 270 °C, where the main minimum of resistivity was reached. This decrease of resistivity (–5 %) is most probably caused by precipitation fine rectangular and oval shaped particles containing mainly Y in the grains and by precipitation and coarsening of $Mg_{41}Nd_5$ stable phase particles, which was persisting up to 480 °C. Both types of precipitates (Fig. 4b) were present after annealing up to 420 °C. After isochronal annealing to higher temperatures (over 420 °C) the needle shaped particles containing Zn-Zr were formed similar to those observed in spray formed Mg3Zn1Nd0.5Zr alloy mentioned above. The grain boundaries were decorated by the eutectic consisting of the stable $Mg_{41}Nd_5$ phase and clusters of particles with high amount of Y after annealing up to 510 °C (Fig. 4c).

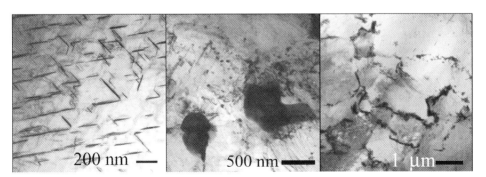

Figure 4: Microstructure of isochronally annealed Mg0.5Zn3Nd2Y0.5Zr alloy
a) triangular arrangement of β_1 phase in as prepared squeeze cast alloy, b) coarse particles of stable $Mg_{41}Nd_5$ phase and fine Y containing phase in powder prepared alloy annealed up to 390°C, c) eutectic consisting of the stable $Mg_{41}Nd_5$ phase and particles with high amount of Y in powder prepared material after annealing up to 510 °C

5 Conclusions

The processing technology and thermal treatment significantly influence the microstructure of as prepared material and result in changes of phase transformations during isochronal annealing.

Varied morphological formations of the T-phase were observed in the spray formed and extruded and squeeze cast and extruded Mg3Zn1Nd0.5Zr alloy. The main relative resistivity decrease was ascribed to the precipitation of fine, dense dispersed Zn and Y containing phase. The grain size was stable in the spray formed alloy contrary to the grain growth observed in the squeeze cast and extruded material. Better thermal stability of microhardness in spray formed and extruded alloy was observed.

The squeeze cast Mg0.5Zn3Nd2Y0.5Zr alloy contains the eutectic formed by α-Mg and β_1 phase and the triangular arrangement of β_1 prismatic plates in the as prepared state. Powder prepared alloy contained fine and individual coarse oval shaped particles of stable $Mg_{41}Nd_5$ phase containing also Nd and Zn formed during extrusion. The main resistivity decrease is produced by β_1 to $Mg_{41}Nd_5$ transformation and $Mg_{41}Nd_5$ coarsening.

6 Acknowledgements

Financial support by the Czech Science Foundation (GACR project 106/03/D110) and by Ministry of Education of Czech Republic (the research program MSM 0021620834) is gratefully acknowledged.

7 References

[1] Luo, Z.P., Song, D.Y., Zhang, S.Q., J. Alloys Comp. **1995**, 230, 109–114.
[2] Wu, W., Wang, Y., Zeng, X., Chen, L., Liu, Z., J. Mater. Sci. Lett **2003**, 22, 445– 447.
[3] Wei, L.Y., Dunlop, G.L., Westengen, H., Metall. Mater. Trans. A **1995**, *26A*, 1947–1954.
[4] Wei, L.Y., Dunlop, J. Mater. Sci. Lett. **1996**, 15, 4–7
[5] Vostrý, P., Stulíková, I. , Smola, B. , Kiehn, J. , Buch, F. von, Z. Metallkde **1999**, 11, 888–891.
[6] Bijvoet, J., de Hon, B., Dekker, J. A., Solid State Comm. **1963**, 1, 273–240.
[7] Geritsen, A.N., Phys. Rev. B **1981**, 23, 2531–2535.
[8] Stulíková, I., Vostrý, P., Mordike, B.L., in Proceedings of the 5th Heat Treatment of Materials congress, Volume III, Budapest **1986**,1993–2000
[9] Ahmed, M., Pilkington, R., Lyon, P., Lorimer, G.W., in Magnesium alloys and their application, ed. Mordike, B.L., Hehman, F., **1992**, 251–257
[10] Smola, B., Stulíková, I, Kovove Mater. **2004,** 42, 5, 301–315

Microstructure of Elektron 21 Magnesium Alloy after Heat Treatment

A. Kielbus
Silesian University of Technology, Katowice, Poland

1 Introduction

Magnesium alloys belong to the lightest structural alloys. They are characterised by low density and good mechanical properties. Mainly for these reasons, magnesium alloys have a widespread application in the motor vehicle and aircraft industries. The disadvantages are poor properties in elevated temperatures and high reactivity [1, 2]. Magnesium alloys with addition of yttrium characterize high strength in elevated. However alloys that contain yttrium have high associated cost due to the difficulties in founding. Therefore there is a need for an alternative alloy which has similar properties to Mg-Y alloys, but with foundry handling and associated costs like non-yttrium containing alloys [3, 4]. Magnesium alloys containing neodymium and gadolinium are interesting as light structural materials with high mechanical properties at room and elevated temperatures [5].

Elektron 21 is new magnesium based casting alloy containing neodymium, gadolinium and zinc for used to at 200 °C. It is a Mg-RE-Zn-Zr alloy designed for aerospace and speciality applications. This alloy has high strength, good corrosion resistance and excellent castability. Elektron 21 is fully heat treatable (T6). This alloy generates a useful range of mechanical properties which can best be summarized by comparison with other Mg alloy. Neodymium has a positive effect on tensile strength at elevated temperatures and reduces porosity of casts and susceptibility to cracking during welding. Gadolinium, like neodymium shows a decreasing solid solubility as temperature falls, indicating potential for precipitation strengthening. Addition neodymium to Mg-Gd alloys reduces the solid solubility of Gadolinium. It improves precipitation hardening response, at lower levels of gadolinium than the binary system offers. Zinc is added to magnesium alloys in sufficient quantities to achieve precipitation strengthening. It improves strength without reducing ductility [4]. Zirconium, which does not form any phases with magnesium or alloying elements, contributes to the obtaining of a fine grain structure and improves the mechanical properties at an ambient temperature and castability and corrosion performance [6]. Elektron 21 is being used in both civil and military aircraft and also in automobile (motorsport) industry [4].

The strength of this magnesium alloys with RE is achieved essentially via precipitation strengthening. These alloys precipitate from the solid solution according to the sequence of phases: α-Mg$\rightarrow\beta''\rightarrow\beta'\rightarrow\beta$. Sometimes between β' and β the β_1 phase has been observed [3].

The β'' phase is metastable and fully coherent with the matrix. It has a DO_{19} crystal structure (hexagonal, a = 0,642 nm and c = 0,521 nm). The intermediate β' phase is also metastable and semi coherent with the matrix. It has an orthorhombic crystal structure (a = 0,640 nm, b = 2,223 nm and c = 0,521 nm). The β_1 phase has a face-centred cubic structure with lattice parameter a = 0,74 nm. The equilibrium β phase is face-centred cubic (a = 2,223 nm) [3, 7].

2 Experimental Procedure

The material for the research was a casting Elektron 21 magnesium alloy. The alloy was purchased from Magnesium Elektron, Manchester, UK. The chemical composition of this alloy is provided in Table 1.

Table 1: Chemical composition of the Elektron 21 magnesium alloy in wt.-%

Gd	Nd	Zr	Zn	Mn	Fe	Ag	TRE	Mg
1,2	2,7	0,49	0,4	0,001	0,003	0,01	4,2	balance

Solution treatment was performed at 520 °C/8h in an electrically heated furnace with water cooling. Ageing treatments were performed at 200 °C/16h and 300 °C/48h and then quenched in air. The specimens were prepared by the standard technique of grinding and polishing, followed by etching at 4.2 ml picrate acid, 70 ml ethanol and 10 ml water. For the microstructure observation, an OLYMPUS GX71 metallographic microscope and a HITACHI S-3400N scanning electron microscope with a Thermo Noran EDS spectrometer equipped with SYSTEM SIX were used. To measure the stereological parameters, a program for image analysis „MET-ILO" was used. Samples for TEM investigations were given in Gatan PIPS ion beam. Examinations were carried out in JEM 2010 ARP microscope.

3 Results and Discussion

3.1 Microstructure of Elektron 21 Alloy in As-cast Condition

The Elektron 21 alloy in as-cast condition is characterized by a solid solution structure α with precipitates of $Mg_{12}(Nd_x,Gd_{1-x})$ intermetallic phase on grain boundaries (Fig.1).

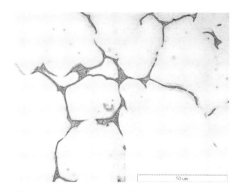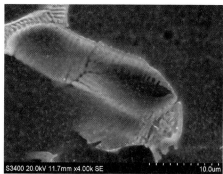

Figure 1: Microstructure of Elektron 21 alloy in as cast state

The $Mg_{12}(Nd_x,Gd_{1-x})$ phase is a modification of $Mg_{12}Nd$ (Fig.2) phase with neodymium substituted by gadolinium without destroying the crystal structure, due to reasonably small difference in the atomic radii of gadolinium $r_{Gd} = 0,1802$ nm and neodymium $r_{Nd} = 0,1821$ nm. The results of SEM EDX quantitative analysis confirm that whole amount of neodymium and gadolinium enters the $Mg_{12}(Nd_x,Gd_{1-x})$ phase containing apart from Mg, 8,17 at% Nd and 2,51 %at Gd. A quantitative evaluation of the Elektron 21 alloy microstructure has shown that the mean area of the solid solution a grain equals $\overline{A}$ = 649 μm², and the mean surface fraction of intermetallic phase is A_A = 6.55 %.

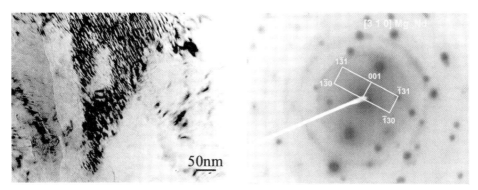

Figure 2: The TEM image and corresponding diffraction pattern of $Mg_{12}Nd$ phase

The mapping of Mg, Nd and Gd in the $Mg_{12}(Nd_x,Gd_{1-x})$ phase visible in the SE image can be seen in Fig.3.

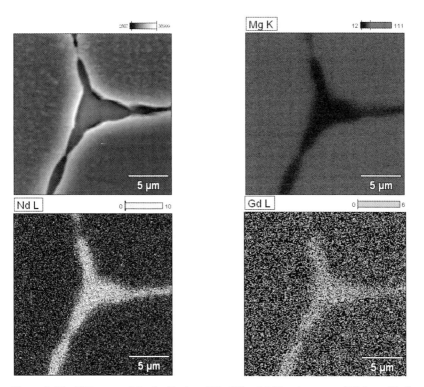

Figure 3: The SE image and the distribution of Mg, Nd and Gd in microareas of Elektron 21 alloy

3.2 Microstructure of Elektron 21 Alloy after Solution Treatment

After solution treatment at a temperature of 520 °C/8h the $Mg_{12}(Nd_x,Gd_{1-x})$ intermetallic phase dissolved in the matrix. The volume fraction of the $Mg_{12}(Nd_x,Gd_{1-x})$ phase decrease to $A_A = 0.65\ \%$. The mean area of the a-Mg grain equals $\overline{A} = 3091\ \mu m^2$ and is higher (~5-times) compared to the as cast state (Fig. 4).

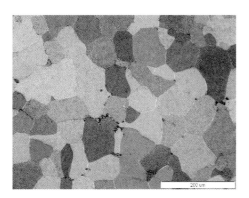 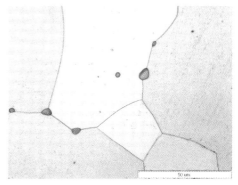

Figure 4: Microstructure of Elektron 21 alloy after solution treatment 520 °C/8h/water

3.3 Microstructure of Elektron 21 Alloy after Ageing Treatment

3.3.1 Ageing 200 °C/16 h/air

The Elektron 21 alloy after aged 16 h at 200 °C is characterized by the precipitates of zinc zirconides with fine precipitates into matrix. TEM examinations of zinc zirconides reveal a distribution of rod-shape and equiaxed precipitates. The EDX analysis of this phase has shown, that Zn/Zr ratio is 2:1, what means that it is probably Zn_2Zr phase (Fig.5).

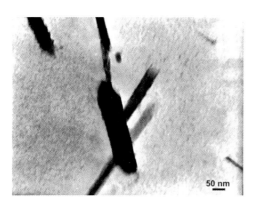 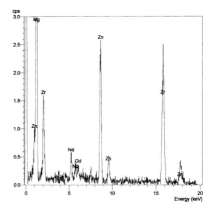

Figure 5: The TEM image and EDS analysis of zinc zirconide

A fine dispersion of β' precipitates in matrix are present as needles. The size of the precipitates was very small, that it was difficult to obtain electron diffraction patterns from individual particles to identify their structure. It is possible that analyzed phase was β' (Fig.6). Due to the small precipitate size accurate EDX analysis was not possible.

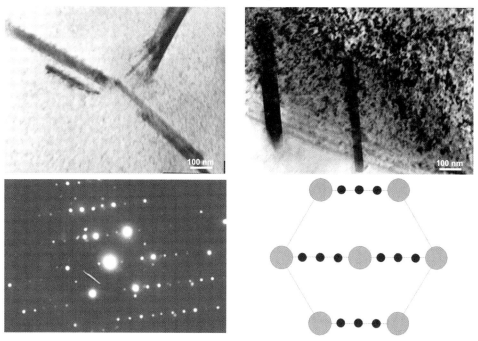

Figure 6: a,b) TEM bright-field images of microstructures taken from the alloy aged at 200 °C for 16h, **c)** corresponding (SAD) pattern, **d)** simulated pattern

3.3.2   Ageing 300 °C/48h/air

The Elektron 21 alloy after aged 48 h at 300 °C is characterized by the β and probably $β_1$ precipitates of in the solid solution matrix. A bright field micrograph typical for the microstructure after aged at 300 °C/48 h is show in Fig.7. Precipitates inside grains are identified as β (isomorphous to Mg_5Gd) (Fig.8). The EDX analysis of β phase has shown, that this phase containing apart Mg, 6,99 %$_{at}$ Nd, 1,08 %$_{at}$ Gd and 1,16 %$_{at}$ Zn, what can correspond the stoichiometry $(Mg,Zn)_9(Nd_x,Gd_{1-x})$. Also the precipitates containing smaller content of Nd, Gd and Zn have been observed. These precipitates can correspond to the $β_1$ phase.

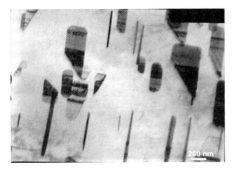 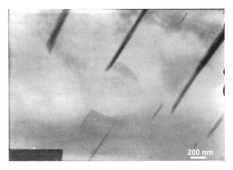

Figure 7: TEM bright-field images of microstructures taken from the alloy aged at 300 °C for 48 h

Figure 8: The corresponding (SAD) patterns taken from the alloy aged at 300 °C for 48 h

4 Conclusions

Based on the experimental results obtained, the following conclusions can be drawn:
1. The Elektron 21 alloy in as-cast condition is characterized by a solid solution structure α with precipitates of $Mg_{12}(Nd_x,Gd_{1-x})$ intermetallic phase on grain boundaries.
2. After solution treatment at a temperature of 520 °C /8 h the $Mg_{12}(Nd_x,Gd_{1-x})$ intermetallic phase dissolved in the matrix.
3. After ageing treatment at 200 °C/16 h the Elektron 21 alloy is characterized by the precipitates of Zn_2Zr phase and a fine dispersion of β' precipitates in matrix.
4. After ageing treatment at 300 °C/48 h the Elektron 21 alloy is characterized by the by the β and probably β_1 precipitates of in the solid solution matrix.

5 Acknowledgements

This work was supported by the Polish Ministry of Education and Science under the research project No. 3 T08C 060 28.

6 References

[1] Mordike B., Journal of Material Processing Technology **2001**, *117*, 391–394.
[2] Smola B., Stulikova I., Pelcova J., Mordike B., in: *Proceedings of the 6th International Conference Magnesium Alloys and their Applications*, Wiley-Vch Verlag **2003**, 43–46.
[3] Lorimer G., Apps P., Karimzadech H., King J., Materials Science Forum **2003**, 419–422, 279–284.
[4] Lyon P., Syed I., Wilks T. , Magnesium Technology **2005**, 303–308.
[5] Rokhlin L., Nikitina N., Dobatkina T., Journal of Alloys and Compounds **1996**, 239, 209–213.
[6] Avedesian M., Baker H., Magnesium and Magnesium Alloys, ASM Speciality Handbook, **1994**.
[7] Nie J., Muddle B., Scripta Materialia **1999**, *40*, No. 10, 1089–1094.

Structural Evolution on Thermal Treatments of EV31 Alloy

M. Massazza*, G. Riontino*, D. Lussana*, A. Iozzia*, P. Mengucci§, G. Barucca§,
A. Di Cristoforo§, R. Ferragut+, R. Doglione&
*Università di Torino, § Università Politecnica delle Marche, +Politecnico di Milano, &Politecnico di Torino

1 Introduction

The recent introduction in the transport market of the EV31 alloy has reduced the problems connected to the corrosion resistance and to the production costs [1]. In the aerospatial sector, the EV31 offers an effective alternative to the more commonly used WE43, exhibiting similar mechanical properties [2]. Recent studies performed on alloys having compositions similar to the EV31 have evidenced [3] the importance of small additions of Zn in the age hardening response, while direct observations by 3D atom probe [4] have allowed to characterize the precipitation sequence in an alloy with the same solute atoms than the EV31. In spite of the relatively large literature on Mg-based alloys, few papers dealt with the possibility of planning thermal treatments able to ensure optimal and reproducible mechanical properties. In the past, some Authors of the present paper tried to extend the use of Differential Scanning Calorimetry (DSC) to describe the structure evolution of Mg-Rare Earth (RE)-based alloys, aiming to design programmed thermal treatments able to increase the hardness [5,6].

The present paper reports some results of the structure evolution on annealing an EV31 alloy. This has been made initially through calorimetric measurements and microhardness tests in order to follow the phase transformations and their effects on mechanical properties. TEM observations performed on thermal states selected by DSC provide an essential contribution to identify crystallography and morphology of the phases. Positron annihilation spectroscopy, with the variants of lifetime (LS) and coincidence Doppler broadening (CDB) measurements, associates the formation of different types of matrix-precipitate interfaces to the positron trapping around the lattice vacancies, whose interaction with solute atoms mainly affects their diffusion.

2 Experimental

Extruded bars of nominal composition Mg+2.8%Nd+1.4%Gd+0.3%Zn+0.5%Zr(wt%) have been supplied by Teksid Aluminum, in the as-cast (AC) and T6 conditions (16 hours at 200 °C after 8 hours at 520 °C and air calm cooling) according to the manufacturer specifications [2]. From slices of as-received material (about 1.5 mm thick), small discs of about 5 mm in diameter and about 50 mg in weight have been punched. Samples have been solutioned for 8 hours at 525 °C in a vertical furnace in air, then quenched in water at room temperature. Thermal treatments of the water-quenched (WQ) samples have been made in an air-ventilated oven. After removing with emery papers a thin layer of surface oxide, microhardness Vickers indentations with a load of 3 N have been made before calorimetric measurements. A TA DSC/Q100 instrument has been used, with a reference of pure Al in Al sample holders, at a suitable scanning rate with a protective atmosphere of pure argon. TEM observations have been performed with a Philips CM200 microscope at 200 kV, after grinding and final thinning with an ion polishing system. XRD patterns have been collected with a Philips PW3020 diffractometer with an anti-cathode of Cu.

Positron annihilation measurements were performed with a pair of identical samples, mounted in the standard sandwich geometry with a ^{22}Na positron source sealed between Kapton foils. The measurements were carried out at liquid nitrogen temperature (LNT). The momentum resolution (FWHM) of the CDB set-up was $3.6 \cdot 10^{-3}$ m_0c (other details in Ref.7). The data were analyzed according to the linear combination procedure of reference samples that is discussed in Ref.8. The weight associated to the moment distribution so determined are reported in Table 1: each w_i^* is the weight of the distribution attributed to positron trapped at vacancy-like defects in the i-th component of the alloy, and wMg is relative to bulk Mg. The w_i^* coefficients can be taken as the average atomic ratio i at the annihilation site. The separation between w_{Mg} and w_{Mg}^*, as well as that between w_{Nd}^* and w_{Gd}^*, is not well defined, so the total contribution is reported. PALS measurements were performed at LNT with a standard fast-fast coincidence system (time resolution - FWHM - 250 ps). The mean positron lifetime was obtained by fitting the spectra with a single exponential after subtraction of the source component; the results are reported in Table 1.

3 Results and Discussion

An *ab ovo* description of the material under investigation has been necessary due to the lack of literature data. Wide-angle X-ray diffraction analyses (Fig.1) of the solution treated sample WQ show the magnesium solid solution with only traces of equilibrium phases, so indicating an almost complete solutioning. On the contrary, in the AC and T6 samples equilibrium and metastable phases are present.

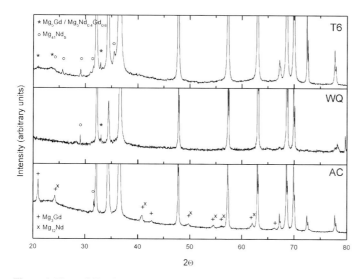

Figure 1: X-ray diffraction patterns of the EV31 alloy in the T6, WQ and AC conditions

The CDB coefficients of Table 1 obtained for the WQ sample (Fig. 2) together with the average positron lifetime suggest a weak presence of vacancies in a solute rich environment. Probably simple solute-vacancy pairs survive after quenching.

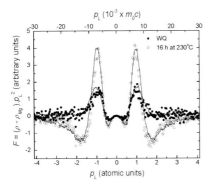

Figure 2: CDB spectra for water-quenched samples and successively aged 16h at 230 °C (relative difference to bulk Mg weighting by a factor p_L^2). The solid line represent a linear combination fit (see Ref.8).

Table 1: Weights (errors) of the linear combination fitting of CDB data and average positron lifetimes

Thermal treatment	$w_{Mg} + w_{Mg}^*$	$w_{Nd}^* + w_{Gd}^*$	$\dfrac{w_{Nd}^* + w_{Gd}^*}{w_{Mg} + w_{Mg}^*}$	Lifetime (ps)
WQ	0.88 (0.02)	0.12 (0.02)	0.14 (0.02)	222 (1)
(T6) 16 h at 200°C	0.69 (0.02)	0.31 (0.02)	0.45 (0.07)	227 (1)
16 h at 230°C	0.74 (0.02)	0.26 (0.02)	0.35 (0.05)	247 (1)
7 days at 150°C	0.75 (0.02)	0.25 (0.02)	0.33 (0.04)	221 (1)

Preliminary calorimetric scannings at different rates allowed to choose 10 K/min as the most suitable scanning rate to evidence the phase transformations occurring on heating the alloy. Furthermore, a comparison between the starting as-cast condition (AC), the WQ state and the normally adopted artificially aged T6 state can be driven from Fig. 3.

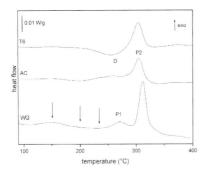

Figure 3: DSC traces of the alloy in the T6, AC and WQ conditions. Scanning rate: 10 K/min

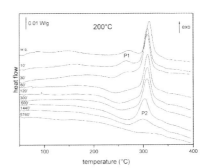

Figure 4: Calorimetric traces of WQ samples annealed at 200 °C for the labelled times

In this figure, three main features can be evidenced: 1) a large exothermal signal P2 at around 300 °C for all the states, that can be assigned to the main precipitation on heating; 2) a large dissolution (endothermal) signal D centred at around 250 °C for the T6 sample, and 3) a small exothermal hump P1 at around 260 °C, more evident in the as-quenched sample. While the thermal evolution for the T6 sample can be easily described as an initial dissolution of phases formed during heating at 200 °C, and the subsequent formation of the main precipitates, the traces of WQ have suggested two further thermal treatments, besides the classic artificial ageing at 200 °C. Annealings at 150 and 230 °C has been performed on as-quenched samples in order to isolate the exothermal effects at the corresponding temperatures on the DSC traces. The results are reported in Figs. 4, 5 and 6. In order to correlate the structure evolution on annealing with the hardness variations, in Fig. 7 the microhardness values for the three annealing temperatures are reported as a function of time.

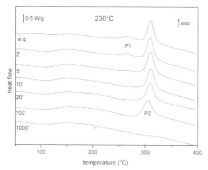

Figure 5: Calorimetric traces of WQ samples aged at 230 °C for the labelled times

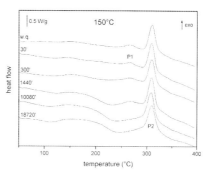

Figure 6: Calorimetric traces of WQ samples aged at 150 °C for the labelled times

3.1 Ageing at 200 °C

This is to follow the phase transformations leading to the commercial T6 state. On the basis of the calorimetric scans, the choice of this temperature can be justified by: a) the completion of a possible small and extended exothermic signal centred at about 150 °C in the WQ and, at a less extent, in the AC samples and b) the beginning in the commercial T6 state (Fig. 3) of a large endothermic signal that could foreshadow the dissolution of phases formed on annealing. The hardness reaches a maximum after nearly 2 hours (Fig. 7), when the signal P1 is almost completely disappeared (Fig. 4) and so the underlying hardening phase is almost completely formed. At higher annealing times, the hardness decreases together with the signal P2: the phase giving rise to this signal progressively increases on annealing, and it gives a negative contribution to the hardness. TEM observations made on the sample aged 16 hours (T6 state) show only the presence of the β" phase coherent with the Mg matrix ($a = 2a_{Mg}$, $c = c_{Mg}$) with dimensions ranging from 2 to 25 nm in length and a width between 1 and 4 nm (Fig.8). The CDB coefficients of Table 1 show that the CDB signal of rare earths and the positron lifetime (~227 ps) obtained after the T6 thermal treatment are increased with respect to WQ. These results indicate increased positron trapping in solute-rich environment, as expected in consequence of the formation of new positron traps. In this case, the ratio (0.45 ± 0.07) is something over, within the experimental uncertainty, the average atomic ratios of the β" phase, whose composition is expected to be Mg3X (X: Rare Earth). The positron lifetime value could indicate the formation of new positron traps in the misfit interface of an intermediate phase, most probably β'.

Less can be said about the phase giving rise to the signal P1, but it is probable a partial evolution of β" into β' with the progressive disappearance of the signal P1 on annealing. As the annealing proceeds, the two phases evolve into β1/β to which the peak P2 can be associated.

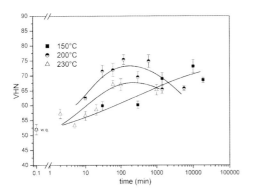 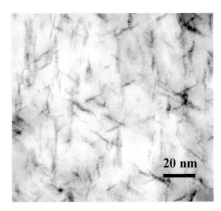

Figure 7: Vickers Hardness Numbers as a function of time for different ageing temperatures. Load: 3 N

Figure 8: TEM bright field image of the sample aged 16 h at 200°C (T6). Zone axis $[0001]_{Mg}$

3.2 Ageing at 230 °C

At this temperature, taking also into account the thermal drag due to the scanning rate, the hump P1(see Fig. 3) has not yet manifested in the WQ and AC samples, while in the T6 temper the large dissolution signal superimposes to its formation. After 20 minutes of annealing, P1 is nearly completely disappeared (Fig. 5). TEM observations show again the formation of β" after 20 minutes. After 16 hours of annealing, very large aggregates are observable (Fig. 9). Microdiffraction investigations (not reported here) performed on these aggregates allow to identify them as the â1 phase. The structure evolution at this temperature is not dissimilar from the one at 200 °C, as can be seen from the calorimetric traces. The microhardness trend is also the same, within the experimental errors, with a more rapid initial increase on annealing at the highest temperature. The peak ageing value is slightly lower than at 200 °C: the more rapid formation of $β_1/β$ at 230 °C, revealed by an initial decreasing of the signal P2 after 100 minutes of annealing and by its complete disappearance after 1000 minutes, reduces the hardness increase.

The value of the positron lifetime (~247 ps), which is above the value of Mg bulk (220 ps) and approximately the same obtained for defects of Mg in a saturation regime (~245 ps), suggests the presence of an incoherent matrix-precipitate interface like that of the equilibrium β phase. Figure 2 shows the CDB spectra; the continuous line is the result of the fitting (see Ref. 8). The coefficients (Table 1) give information of a high presence of defects in a solute rich environment.

3.3 Ageing at 150 °C

The choice of this ageing temperature is in part due to the possible small exothermic signal in the WQ sample (Fig. 3), and in part to considerations already made for the WE43 alloy [6] on possible separation of the contribution of two initially undifferentiated phases. From the traces in Fig.6 it is not possible to separate these contributions by the successive precipitations or dissolutions on scanning, although at the highest annealing times the endothermic signal D becomes more extended than initially.

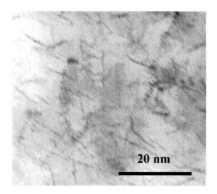

Figure 9: TEM bright field image of the sample aged 16 h at 230 °C. Zone axis $[0001]_{Mg}$

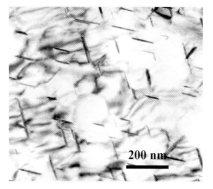

Figure 10: TEM bright field image of the sample aged 7 days at 150 °C. Zone axis $[0001]_{Mg}$

The microhardness starts to increase slowly (Fig.7), and then tends to saturate at a level not distant from the other two annealings: no overageing occurs at the highest annealing times, as indirectly shown by the persistence at these times of the signal P2 (no formation of hardness limiting phases). TEM observations confirm the presence of β", whose dimensions are very small after 1 day of annealing. After 7 days at 150 °C the β" phase shows higher dimensions with a length between 2 and 15 nm and a width ranging from 1 to 2 nm (Fig.10).

On the other hand, the CDB results indicate increased positron trapping in solute-rich environment, as expected in consequence of the formation of new positron traps compared to WQ. In this case, the ratio $\frac{w^*_{Nd} + w^*_{Gd}}{w^*_{Mg} + w^*_{Mg}}$ (0.33 ± 0.04) is consistent with the average atomic ratios of the β" phase. The positron lifetime value indicates the presence of coherent particles.

4 Conclusions

- The phase transformations on ageing an EV31 alloy after solutioning and quenching follow the sequence: solid solution → β" → β' → β$_1$ → β.
- The peak hardness value is reached on ageing at 200 °C for times much lower than the conventional T6 treatment.

5 References

[1] Lyon P.; in „Investment Casting Inst. 53rd Techn. Conf. & Expo", Dearborn (USA), **2005**
[2] sheet MEL-Magnesium Elektron n.455
[3] Nie J. F., X. Gao X., Zhu S. M.; Scripta Materialia **2005**, 23, 1049–1053
[4] Ping D.-H., Hono K., Nie J.F.; Scripta Materialia **2003**, 48, 1017–1022
[5] Riontino G., Lussana D., Massazza M.; J.of Ther. Anal. and Calorim. **2006**, 83, 643–647
[6] Riontino G., Lussana D., Massazza M., Zanada A.; J. Mater. Sci. **2006**, 41, 3167–3169
[7] Calloni A., Dupasquier A., Ferragut R., Folegati P., Iglesias M. M., Makkonen I., Puska M. J.; Phys. Rev. B **2005**, **72**, 054112
[8] Folegati P., Dupasquier A., Ferragut R., Iglesias M.M., Makkonen I., Puska M. J.; Phys. Stat. Sol. C (in press)

Microstructural Stability of Magnesium Alloys for Aerospace Applications

S. Avraham[1*], B. Bronfin[2], G. Arruebarena[3], M. Bamberger[1]
[1]Materials Engineering Department, Technion, Haifa, Israel.
[2]Magnesium Research Institute, DSM, Beer Sheva, Israel.
[3]Mondragon Goi Eskola Politeknikoa, Loromeni 4, 20500 Mondragon, Spain.
* Contact author : shaula@tx.technion.ac.il

1   Introduction

Magnesium alloys are very attractive for engineering applications due to their low density and consequently high specific mechanical properties. Integration of magnesium components in the aerospace industry may result in reduced structure weight, improved noise damping, reduced fuel consumption and reduced air pollution. A saving of one pound in weight can result with an economical benefit that ranges between 300 and 3000 USD, depending on aircraft type [1]. Although magnesium alloys have been used in the past in the aircraft industry their current usage is very limited [2].

   This is the reason stimulating the „Integrated Design and Product Development for the Eco-efficient Production of Low-weight Aeroplane Equipment" (IDEA) program, which were launched on January 2004 in the EC 6th framework program (FP6), to focus on demonstrating the applicability of magnesium castings in the aviation industry [3]. The IDEA project, in which magnesium producers, foundries, aircraft manufacturers, research organizations and universities take part aims at developing new magnesium alloys and technologies that would expand magnesium applications in the aircraft industry.

   The requirements for the magnesium components were determined by the end user, aircraft manufacturer. Tables 1 and 2 present the requirements to tensile and axial fatigue properties the Mg castings, respectively [3]. Semi-structural components like housing and pedal were selected as demonstrators by Israel Aircraft Industries Ltd (IAI).

   The present paper addresses the microstructure stability and mechanical properties of two new alloys that were developed in the framework of the IDEA project.

2   Experimental Methods

Two new alloys designated as MRI 219 and MRI 207 were investigated. MRI 219 (Mg-Al-Sn alloy) is designed for high-pressure die casting (HPDC) applications while MRI 207 (Mg-Y-Gd-Nd alloy) is dedicated for gravity casting processes (sand casting, permanent mould casting and investment casting). MRI 219 is a high strength HPDC alloy for room temperature applications. MRI 207 is designated both for ambient and elevated temperature service. Table 3 present conditions in which both alloys were studied. solid solution treatment of alloy MRI 207-T4 was subjected to solid solution treatment of at 522 °C for 7 hours.

   The structural stability of alloy MRI 219-F was tested by the immersion of the sample in a oil bath at 125 °C. Thermal exposure of MRI 207-F and MRI 207-T4 samples cast state con-

Table 1: Tensile properties requirements to magnesium alloy castings [3]

Alloy	TYS [MPa]	UTS [MPa]	E [%]
A357, gravity die cast	215	255	3
AZ91D, high-pressure die cast	160	240	4
Target for Mg alloys to be developed	220	290	3

Table 2: Axial fatigue properties requirements to magnesium alloy castings [3]

Number of cycles	10^4	10^5	10^6
UTS [%], R=0.2	88	68	50
UTS [%], R=-1	61	44	28

Table 3: Summary of the analyzed samples

Alloy designation	Sample state
MRI 219-F	High-pressure die cast
MRI 207-F	Permanent mold cast
MRI 207-T4	Permanent mold cast and solid solution treated

ducted by immersion in a salt bath at 200 °C. The samples were immersed for different time periods (1, 2, 4, 8, 12, 16, 20, 24, 28 and 32 days). After the predefined immersion period the treatment was terminated by quenching the samples in water (RT ≅ 20 °C). The prediction of the phase formation during solidification was performed by computational thermodynamics (CT) method [4], the simulations were carried out using the Thermo-Calc software (Q version). The simulation results are based on thermodynamic considerations and a specially developed database [5]. DSC measurements were carried out with a Setaram Labsys™ DSC system.

XRD analysis of alloys was conducted by a PW-3020, Philips X-ray automatic powder diffractometer using Cu Kα, operated at 40 mA and 40 kV. Scanning electron microscopy (SEM) analysis was conducted using a FEI Quanta 200 microscope (20 kV, WD 10.5 mm) equipped with a Link Isis EDS system (6506, Oxford Instruments, UK, Z>4).

Tensile tests were performed according to ASTM-E8M. The diameter of the gravity die-cast bars and the HPDC bars were 9 and 6 mm, respectively. Tests were performed at strain rate of $1.5 \cdot 10^{-4}$ s^{-1} using a INSTRON 4206 testing machine. The Vickers microhardness measurements were carried out as per ASTM E92-82. The measurements were conducted using a DMH-2 microhardness tester (MATSUZAWA SEIKS Co. LTD Japan). The load level was set at 50 g and the load time was 15 sec.

3 Results & Discussion

3.1 Microstructural Stability of MRI 219-F Alloy

Figure 1 presents the results of the CT simulation for equilibrium phase evolution during cooling of MRI 219 alloy. The α-Mg phase starts to solidify during cooling of the melt at the liquidus temperature (585 °C). The Al$_8$Mn$_5$ intermetallic compound starts to precipitate at 660 °C but its amount is negligible. The solidification terminates at 452 °C. It is accompanied with the precipitation of Mg$_2$Sn from the α-Mg matrix. The precipitation of the γ-Mg$_{17}$Al$_{12}$ intermetallic starts around 385 °C. Figure 2 presents the results of the CT simulation for non-equilibrium phase evolution during solidification of alloy MRI 219. α-Mg starts to solidify at the liquidus temperature (585 °C). Al$_8$Mn$_5$ starts to precipitate at 662 °C but its amount is negligible. The γ-Mg$_{17}$Al$_{12}$ intermetallic phase is formed at 434 °C in a eutectic reaction, while another intermetallic phase Mg$_2$Sn forms above this temperature (446 °C). According to DSC results the liquidus temperature is 591 °C, which is in good agreement with the simulated results.

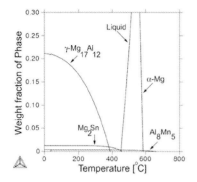

Figure 1: Equilibrium phase evolution during cooling, MRI 219-F

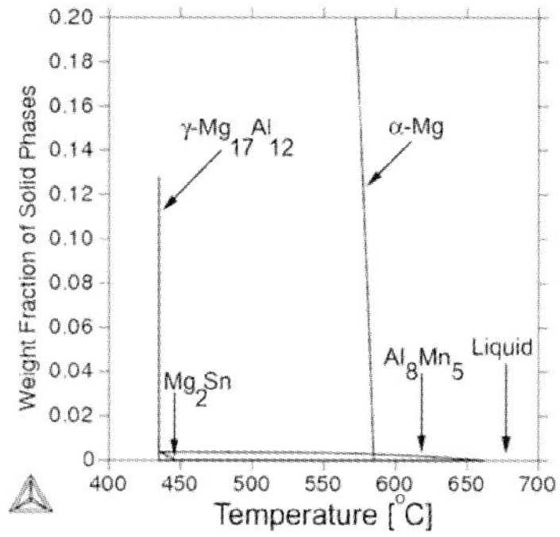

Figure 2: Non-Equilibrium phase evolution during solidification, MRI 219-F

Figure 3 presents the XRD pattern recorded for MRI 219-F alloy The major phases are α-Mg and γ-Mg$_{17}$Al$_{12}$. As a result of the Sn presence in the alloy the formation of Mg$_2$Sn is apparent. Figure 4 presents the XRD pattern of a sample exposed at 125 °C for 32 days. An increase in the amount of Mg$_2$Sn is discernible. This coincides with the CT simulation (Figure 1) in which solid state precipitation of Mg$_2$Sn is expected.

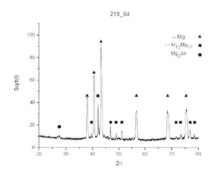
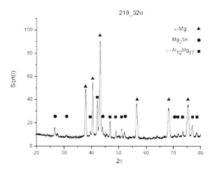

Figure 3: XRD pattern of MRI 219-F

Figure 4: XRD pattern of aged MRI 219-F

Figure 5 presents a back scattered electron (BSE) SEM micrograph of MRI 219-F. The EDS microanalysis determined the composition of the present phases. The microstructure of alloy MRI 219-F consists of α-Mg matrix and the γ-Mg$_{17}$Al$_{12}$ intermetallics located at grain boundaries. The bright precipitates that are seen in several locations are Mg$_2$Sn. At the α-Mg grain boundary the presence of a thin bright layer is observed. EDS results showed that this layer is enriched with Sn (coring effect). Figure 6 presents a BSE SEM micrograph of MRI 219-F after thermal exposure at 125 °C for 32 days. It is evident that the relative amount of the grain boundary γ-$Mg_{17}Al_{12}$ phase increases. This is accompanied with fine discontinues precipitation of Mg_2Sn intermetallics at the vicinity of the grain boundaries. The results of XRD analyses confirmed that thermal exposure is followed by precipitation of γ-$Mg_{17}Al_{12}$ and Mg_2Sn particles from the super saturated regions in the α-Mg matrix.

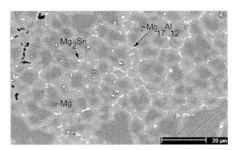

Figure 5: BSE micrograph of MRI 219-F alloy

Figure 6: BSE micrograph of MRI 219-F alloy after thermal exposure at 125 °C for 32 days

As can be seen from SEM micrographs (Figures 5 & 6) thermal exposure at 125 °C for 32 days resulted in the coarsening of the γ-$Mg_{17}Al_{12}$ phase at grain boundaries. The results of XRD and SEM examinations confirmed the CT simulations, which predicts solid state precipitation of Mg_2Sn during exposure at 125 °C and the formation of the γ-$Mg_{17}Al_{12}$ as a eutectic phase.

3.2 Microstructural Stability of MRI 207 Alloy

Table 4 presents a summary of the XRD results for the exposed and aged MRI 207 samples. In the microstructure of MRI 207-F alloy, in addition to the α-Mg solid solution the Mg_3Gd phase is also presented. Thermal exposure at 200 °C for 32 days results in an increased amount of Mg_3Gd precipitate. Solid solution treatment T4 results in dissolving the Mg_3Gd phase. However, subsequent aging of MRI 207-T4 sample led to the formation of Mg_3Gd precipitates.

Table 4: Phases detected by XRD after different thermal treatments (α-Mg excluded)

Alloy	Initial state	Thermally exposed/ aged at 200 °C for 32 days
MRI 207-F	Mg_3Gd	Mg_3Gd
MRI 207-T4	---	Mg_3Gd

Figure 7 presents a BSE SEM micrograph of MRI 207-F alloy. The microstructure consists of a α-Mg matrix and a eutectic phase. The bright layer at the GB is associated with the coring effect during solidification. Enrichment with Gd and Zr were detected at the center of the α-Mg matrix. At the grain boundary regions the Gd content increased while the Zr concentration diminished. On the other hand, Y and Nd enrichments were detected at the GB. The composition of the eutectic phase well as fine discontinues precipitates at the GB were identified as Mg_3(Gd,

Nd). Figure 8 illustrates a BSE SEM micrograph of MRI201-F alloy after thermal exposure at 200 °C for 32 days. The presence of Zr and Gd enrichment was detected at the center of the α-Mg matrix. Fine precipitation of Mg_3(Gd,Nd) intermetallics are formed in the matrix and as a continued layer at the GB.

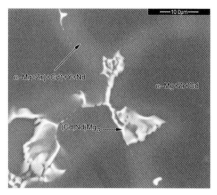
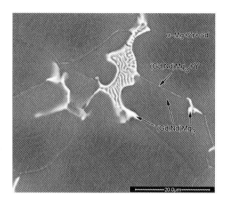

Figure 7: BSE micrograph of alloy MRI207-F

Figure 8: BSE micrograph of MRI207-F alloy after thermal exposure at 200 °C for 32 days

3.3 Mechanical Properties

Exposure of MRI 219-F at 125 °C for 32 days leads to increase of microhardness values from 78 [Hv] to 87 [Hv] (Figure 9). MRI 219-F alloy appears to be insensitive to elevated temperature of 125 °C. Figure 10 presents the microhardness stability results for MRI 207 alloy during thermal exposure and aging at 200 °C up to 32 days. The microhardness value of MRI 207-F appears to be constant in respect to time. The T4 treatment increases the microhardness of the alloy during intermediate aging time from 24 to 288h by 25 %, followed by a decrease in microhardness at longer aging times. The increase in hardness is associated with precipitation of Mg_3(Gd,Nd) intermetallics from the supersaturated magnesium matrix The subsequent coarsening after more than 288 h of exposure results in decrease in microhardness.

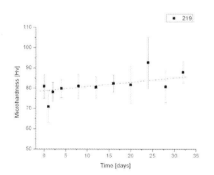
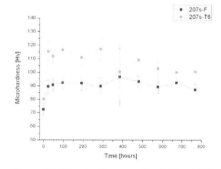

Figure 9: Microhardness results for alloy MRI 219-F during thermal exposure at 125 °C for 32 days

Figure 10: Microhardness results of alloys MRI 207-F and MRI 207-T4 during thermal exposure and aging at 200 °C for 32 days

Figure 11 compares tensile properties of MRI 219-F and MRI207 alloys with commercial AZ91E and AZ91D alloys. It is evident that MRI 219 alloy outperforms AZ91D alloy in strength properties. Based on microhardness results, MRI 219-F is expected to be more stable than AZ91D. The tensile properties of MRI 207 alloy are superior to that of AZ91E alloy. The properties of MRI 207-T6 conform to the tensile properties requirements set by the end user. In addition, MRI 207 alloy exhibits enhanced structural stability.

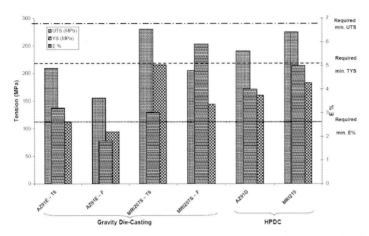

Figure 11: Tensile properties of MRI 207 and MRI 219 in comparison with AZ91E & AZ91D alloys

4 Summary & Conclusions

The newly developed alloys MRI 207 and MRI 219 exhibit improved mechanical properties and microstructural stability compared to existing commercial alloys.

Based on computational thermodynamics method it was predicted that in the MRI 219 takes place the formation of Mg_2Sn intermetallics, which result in precipitation hardening effect. This was also confirmed by the XRD and SEM experimental observations. It was shown that in the MRI 207 alloy the precipitation of Mg_3 (Gd, Nd) intermetallics occurs in the form of fine precipitates within the α-Mg matrix and as a thin continues phase at the grain boundaries.

5 Acknowledgement

This study was supported by the EC under contract number FP6-503826 and partially by the fund for the promotion of research at the Technion.

6 References

[1] F.H. Froes, D. Eliezer and E. Aghion, in Magnesium 2000 Proceedings of the Second Israeli International Conference on Mg Science and Technology, **2000**, Dead Sea, Israel, p.43–49.

[2] Wendt, K. Weiss, A. Ben-Dov, M. Bamberger and B. Bronfin. Mg Technology 2005, N. Neelameggham, H. I. Kaplan and B. R. Powell Editors, TMS 2005 annual meeting, San Francisco, USA, pp. 269–273, 13-17 February 2005
[3] IDEA – magnesium castings for the aerospace industry, http://idea-fp6.net/
[4] R. Schmid Fetzer and J. Grobner, Adv. Eng. Mater., 3[12]: 947–61, 2001.
[5] M. Bamberger, J. Mater. Sci., 41[10]: 2821–2829, 2006.
[6] A.A. Luo, Int. Mater. Rev., 49[1]: 13–30, 2004.

The Refinement of Precipitate Distributions in an Age Hardenable Mg-Sn Alloy Through Microalloying

C. L. Mendis[1], C. J. Bettles[1,2], M. A. Gibson[2], S. Gorsse[3] and C. R. Hutchinson[1]

[1]ARC Centre of Excellence for Design in Light Metals, Department of Materials Engineering, Monash University, Clayton, 3800, Victoria, Australia
[2]CAST CRC, CSIRO Manufacturing and Infrastructure Technology, Private Bag 33, Clayton South MDC, Clayton, 3169, Victoria, Australia
[3]Institut de Chimie de la Matière Condensée de Bordeaux et Ecole Nationale Supérieure de Chimie et de Physique de Bordeaux, 33608 Pessac, Cedex, France

1 Introduction

The Mg-Sn system has the potential to age harden through the formation of ~10 vol. % of Mg_2Sn. The Mg_2Sn phase has a high melting temperature and alloys based on this system may show promise in applications requiring elevated temperature creep resistance [1]. However, studies of precipitation in this system are very limited. Van der Planken considered the age hardening response of a Mg-1.65Sn (at. %) alloy at 200, 250 and 300°C using microhardness tests and despite the relatively large fraction of Mg_2Sn particles formed, the hardening response was very poor [2]. Derge *et al.* [3] examined the orientation relationship (OR) between the Mg matrix and the Mg_2Sn equilibrium phase and reported that at 200°C the Mg_2Sn particles form with a plate-shaped morphology on the $(0001)_{Mg}$ basal planes of the matrix.

In precipitation hardenable alloys where Orowan looping is active, the critical resolved shear stress for dislocation motion scales inversely with λ, where λ is the mean particle spacing in the slip plane [4]. Since λ scales as $R/\sqrt{f_v}$, where R is the effective particle size and f_v is the particle volume fraction, halving the particle size at constant volume fraction corresponds to an approximate doubling of the hardening increment due to precipitate/dislocation interaction. Efforts to refine particle size as a means of increasing the material strength have therefore received much interest. Of the approaches so far developed, attempts based on microalloying additions are particularly attractive since they can be integrated into the overall bulk alloy chemistry and additional processing steps are not usually needed. Detailed electron microscopy and atom probe field ion microscopy have now shown that the large effects of microalloying additions are related to the formation of heterogeneities (clusters or precipitates) on the matrix lattice at which the nucleation of hardening precipitates may be enhanced, therefore giving rise to a refined distribution.

This work considers the enhancement of the age hardening response of Mg-Sn alloys through refinement of the precipitate distribution by using microalloying additions.

2 Choice of Microalloying Elements

A qualitative thermo-kinetic criteria for choosing microalloying additions for age hardenable alloys has recently been proposed by Mendis *et al.* [5]. The approach distinguishes between the processes of heterogeneity formation on the matrix lattice (an energetically downhill process) and the process of heterogeneous nucleation of the hardening precipitates at these heterogeneities (an energetically uphill process). In selecting microalloying additions emphasis is placed on the downhill process of heterogeneity formation and two cases of microalloying additions are considered for the Mg-Sn system: Single (Mg-Sn-(X)) and double additions (Mg-Sn-(X-Y)). Selection is based on the following criteria that are considered necessary for the formation of heterogeneities effective at catalyzing the nucleation of Mg$_2$Sn precipitates.

1. A driving force for heterogeneity formation must exist (*i.e.* a clustering or co-clustering tendency of the microalloying elements upon cooling from the solution treatment temperature).
2. There should not be any repulsive tendencies between the solute forming the strengthening precipitates (Sn) and the microalloying additions forming the cluster or precipitate at which nucleation is desired.
3. The natural length (λ^*) and time (t^*) scales for cluster or precipitate formation must be finer and faster than the length (λ) and time (t) scales associated with precipitation of the strengthening precipitate in the equivalent alloy without microalloying additions. Otherwise, the distribution of hardening precipitates would not be refined.

In disordered solid solutions, the atomic configuration is never perfectly random. There is a tendency to cluster or order and this tendency increases with decreasing temperature. An indication of the tendency can be inferred from the topology of phase diagrams [6,7] and we have used the topology of binary Mg-X, Sn-X and X-Y phase diagrams to infer the clustering or ordering tendencies of potential microalloying additions. The special topological feature indicating clustering tendencies of X atoms on the Mg lattice is a large miscibility gap in the Mg-X system (Column 1, Table 1). For X-Y co-clusters, it is the presence of a highly stable ordered X-Y compound in the X-Y system (Column 1, Table 2). Repulsive tendencies between Sn and X are indicated by large miscibility gaps in the Sn-X systems (Column 2, Tables 1 and 2). Condition 3 (the kinetic condition) requires knowledge of the respective kinetics of clustering and precipitation which are not well understood. However, the rate of clustering will probably scale with diffusivity and the potential microalloying elements have been listed in order of estimated diffusivity [8] in Column 4 of Tables 1 and 2. This simple qualitative analysis leads to the choice of Na and In-Li as potential microalloying additions for the Mg-Sn system.

Table 1: Methodology for selecting microalloying additions for Mg-Sn-(X) system

Phase separating Mg-X systems	Phase separating Sn-X systems	Possible choices for Mg-Sn-(X)	Prioritized choices for Mg-Sn-(X)	Final Choice for Mg-Sn-(X)
Hf, Mo, Na, Nb, Rb, Ti, V	Al, Be, Cr, Ga, Ge, Rb, Zn	Hf, Mo, Na, Nb, Ti, V	Na, Hf, Ti, Nb, Mo, V	Na

Table 2: Methodology for selecting microalloying additions for Mg-Sn-(X+Y) system

Ordering X-Y Systems	Phase separating Sn-X systems	Possible choices for Mg-Sn-(X-Y)	Prioritized choices for Mg-Sn-(X-Y)	Final Choice for Mg-Sn-(X+Y)
Al-Co, Al-Ni, Al-Pd, Au-Mn, Au-Zn, Be-Co, Be-Ni, Cd-Li, Ga-Li, Hg-Li, In-Li, Ir-Ti, Mn-Pd, Mn-Pt, Rh-Ti	Al, Be, Cr, Ga, Ge, Rb, Zn	Au-Mn, Cd-Li, Hg-Li, In-Li, Ir-Ti, Mn-Pd, Mn-Pt, Rh-Ti	In-Li, Cd-Li, Hg-Li, Au-Mn, Rh-Ti, Mn-Pd, Mn-Pt, Ir-Ti	In-Li

3 Experimental Procedure

A total of four alloys were prepared from high purity Mg (99.9 %), Sn (99.9 %), In (99.9 %) and Sn-11.8Na (at. %) or Mg-19.8Li (at. %) master alloys, by induction melting in a mild steel crucible under an argon atmosphere and casting into a carbon coated mild steel mould held at approximately 300 °C. The compositions and heat treatments are listed in Table 3.

Table 3: Alloy compositions (at. %) and Heat Treatments

Base Alloys	Microalloyed Compositions	Heat Treatment
Mg-1.3Sn	Mg-1.3Sn-0.13Na	345 °C for 2h, ramp to 500 °C at 1.3 °C/min, hold 6 h, CWQ.
Mg-1.9Sn	Mg-2.0Sn-0.48Li-0.53In	Homogenised at 520 °C for 144 h, solution treated for 1h at 500 °C and CWQ

Isothermal aging was carried out at 200 °C and the age hardening response was monitored using Vickers hardness measurements (5 or 10 kg). Thin foil specimens for transmission electron microscopy (TEM) were prepared by electropolishing with a solution containing LiCl, $MgClO_3$, 2-buytoxy ethanol and methanol at -45 °C, 0.125 A and 100 V.

4 Results

4.1 Base Alloys: Mg-1.3Sn and Mg-1.9Sn (at. %)

The hardening behaviours of the two binary alloys as a function of aging time at 200 °C are shown in Fig. 1. The responses are typical of age hardenable alloys; an incubation period before a rise to peak hardness, followed by a decrease in hardness during overaging. The time to peak hardness was shorter in the alloy containing 1.9Sn as expected (240 h compared to 1000 h). In both cases the maximum increment in hardness over and above the as-quenched value was ~10VHN.

The microstructures corresponding to peak hardness were examined and found to be very similar in both alloys. Bright field (BF) TEM micrographs of the Mg-1.9Sn alloy with the electron beam oriented approximately parallel to the $<11\bar{2}0>_{Mg}$ and the $[0001]_{Mg}$ directions are shown in Figs. 2a and d, respectively. Microbeam electron diffraction patterns (MBED) were obtained from the precipitates and all could be indexed as the equilibrium FCC $_{Mg_2}$Sn phase lying on the basal plane of the matrix with an orientation relationship with the matrix consistent with those previously reported [2]. However, the morphology is lath-like and particle elongation is in $<11\bar{2}0>_{Mg}$ directions. The mean particle lengths, widths and thicknesses are summarized in Table 4. The precipitate distributions were relatively coarse and the number densities of particles were found to be ~$6 \cdot 10^{17}$ and ~$8 \cdot 10^{17}$ m^{-3} for the alloys containing 1.3Sn and 1.9Sn respectively.

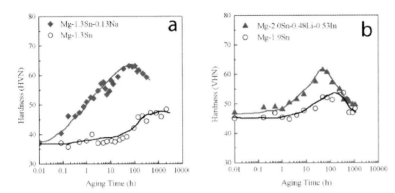

Figure 1: The room temperature hardness as a function of isothermal aging time at 200 °C for a) Mg-1.3Sn and Mg-1.3Sn-0.13Na and b) Mg-1.9Sn and Mg-2.0Sn-0.48Li-0.53In alloys

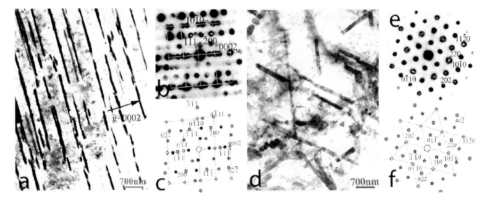

Figure 2: (a and d) BF TEM micrographs showing the solid-state precipitates in the Mg-1.9Sn (at. %) alloy aged at 200 °C for 240h, (b and e) MBED patterns typical of those recorded from Mg$_2$Sn precipitates and (c and f) schematic representation of the indexed MBED patterns in (b and e). (a-c) are parallel to $<11\bar{2}0>_{Mg}$ and (d-f) parallel to $[0001]_{Mg}$

4.2 Microalloyed Compositions: Mg-1.3Sn-0.13Na and Mg-2.0Sn-0.48Li-0.53In (at. %)

The hardening responses of the two microalloyed compositions, as a function of aging time at 200 °C, are shown in Fig. 1. The as-quenched hardnesses of the alloys containing microalloying additions are both only slightly higher than their corresponding binary alloys, however in both cases the response to aging is greatly enhanced and accelerated. The maximum increment in hardness observed in the Na-containing alloy was ~27VHN. This is ~2.7 times the increment observed in the corresponding binary alloy. This hardness is reached after ~58 h compared with ~1000 h in the binary alloy (Fig. 1a). The time to peak hardness in the In+Li containing alloy was ~48 h, compared with 240 h in the corresponding binary alloy (Fig. 1b), and the maximum increment in hardness observed in the In+Li containing alloy was ~14VHN. This is ~1.5 times that observed in the corresponding binary alloy.

BF TEM micrographs from the peak aged condition are shown in Fig. 3. The microalloying additions did not have any effect on the identity or orientation relationship of the precipitates formed, however an obvious refinement of the precipitate distribution is observed.

The particle number density measured in the Na-containing alloy is ~5.6 · 10^{19} m^{-3}; this is a hundredfold increase over that observed in the corresponding binary alloy. The number density measured in the In+Li-containing alloy was ~2.6 · 10^{19} m^{-3}; a thirtyfold increase over that observed in the corresponding binary alloy. In the Na-containing alloy, the mean particle length is ~230 nm whereas in the corresponding binary alloy it was ~1500 nm (Table 4). Similarly in the In+Li containing alloy the mean particle length was found to be ~375 nm compared with ~800 nm in the corresponding binary alloy (Table 4).

The large increase in the number density of particles, and the concomitant decrease in particle size, is thought to be due to the increased nucleation rate of Mg_2Sn particle at heterogeneities on the matrix lattice formed by clusters or small precipitates of the microalloying elements, Na and In+Li. The exact nature of these heterogeneities will be investigated in future work.

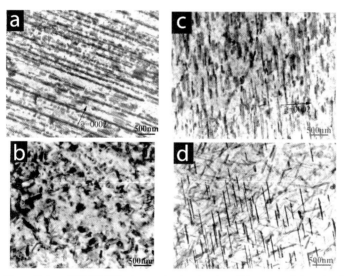

Figure 3: (a and b) BF TEM micrographs showing the solid-state precipitates in a Mg-1.3Sn-0.13Na (at. %) alloy aged for 58h at 200 °C, (c and d) BF TEM micrographs showing the solid-state precipitates in Mg-1.9Sn-0.48Li-0.53In (at. %) alloy aged for 48h at 200 °C. (a and c) parallel to $<11\bar{2}0>_{Mg}$. (b and d) parallel to $[0001]_{Mg}$

Table 4: Quantitative microstructural measurements of the alloys investigated at maximum hardness at 200 °C

Alloy Composition (at.%)	$N_v \times 10^{18}$ (m^{-3})	$\bar{l}$ (nm)	$\bar{w}$ (nm)	$\bar{t}$ (nm)	$\bar{l}/\bar{w}$	$\bar{l}/\bar{t}$
Mg-1.3Sn	0.60	1500±400	500±80	54±11	3	28
Mg-1.3Sn-0.15Na	56.4	230±50	90±22	25±8.5	2.55	9.2
Mg-1.9Sn	0.85	800±230	170±80	55±31	4.7	14.5
Mg-2Sn-0.48Li-0.53In	25.5	375±42	77±28	26±4.5	4.9	14.4

5 Conclusions

- Relatively simple thermo-kinetic criteria for choosing microalloying elements for precipitation hardenable alloys are summarized and applied to a model Mg-Sn alloy.
- Na additions were shown to increase the number density of Mg$_2$Sn precipitates formed at 200 °C by two orders of magnitude resulting in an increase of ~270 % in the hardening increment.
- In+Li additions were shown to increase the number density of Mg$_2$Sn precipitates formed at 200 °C by approximately an order of magnitude resulting in hardening increment increases of ~150 %.

6 Acknowledgements

This work is supported by the Australian Research Council (ARC) under the Discovery Projects scheme (CRH; Project ID: DP0557517) and as part of the ARC Centre of Excellence for Design in Light Metals. CJB and MAG acknowledge the support of CAST CRC for part of this work. CAST CRC was established under and is supported by the Australian Government's Cooperative Research Centres Scheme.

7 References

[1] Bowles, A.L., Dieringa, H., Blawert, C., Hort, N., Kainer, K.U. Mat. Sci. Forum. 488–489 135 (2005).
[2] van der Planken, J. J. Mat. Sci. Letts. 4 927 (1969).
[3] Derge, G., Kommell, A.R., Mehl, R.F. Trans. AIME. 124 367 (1937).
[4] Brown, L.M., Ham, R.K. *Strengthening Methods in Crystals*. London, Applied Science Publishers, 1971.
[5] Mendis, C. L., Bettles, C. J., Gibson, M. A., Gorsse, S. and Hutchinson, C. R. Philosophical Magazine Letters (*in press*, 2006).
[6] DeHoff, R.T. in *Thermodynamics in Materials Science,* New York, McGraw-Hill, 1993, p. 196.
[7] Lupis, C.H.P. in *Chemical Thermodynamics of Materials,* New York, Nth-Holland, 1983, p. 437.
[8] Brown, A.M., Ashby, M.F. Acta metall. 28 1085 (1980).

Microstructural Evolution and Phase Formation in Novel Mg-Zn Based Alloys

B. Rashkova[1,2], J. Keckés[2], G. Levi[3], A. Gorny[3], M. Bamberger[3], and G. Dehm[1,2]

[1] Max-Planck-Institut für Metallforschung Stuttgart, Germany
[2] Erich Schmid Institut für Materialwissenschaft, Österreichische Akademie der Wissenschaften and Department Materialphysik, Montanuniversität Leoben, Austria
[3] Department of Materials Engineering, Technion, Haifa, Israel

1 Introduction

Based on the requirements for novel Mg alloys, Al-free Mg-Zn alloys with Sn, Sb, and Y, as alternative alloying elements, were investigated in order to obtain Mg-Zn based alloys with improved thermal stability. The motivation for adding Sn is the expected formation of a Mg_2Sn phase which has a melting temperature of 770 °C, and thus should improve the creep properties [1]. Alloying with Sb is expected to further improve the precipitation hardening of the Mg-based alloy by the formation of fine particles, such as Mg_3Sb_2 [2]. Y is added to improve the strength of the alloy at higher temperatures by forming precipitates of different symmetry, additionally Y may delay the onset of overaging [3, 4]. In this study results concerning the phase formation and the microstructural evolution during thermal treatments are reported for a Mg-Zn-Sn alloy containing additions of Y and Sb.

2 Experimental

Mg-alloys containing 4.5 wt% Zn and 3.8 wt% Sn were prepared by melting high purity Mg, Zn and Sn, additionally alloyed with 1.1 wt% Sb and 0.6 wt% Y, respectively. Solution treatment and aging of the alloys has been described elsewhere [5, 6]. The microstructure of the alloys was investigated by X-ray diffraction (XRD), scanning electron microscopy (SEM) equipped with an energy-dispersive X-ray spectrometer (EDS), and transmission electron microscopy (TEM) techniques including high-resolution TEM [7]. The samples were prepared by a similar route to that given in [8]. Vickers hardness measurements (50g load) were performed on various microstructures of as-cast and aged samples.

3 Results and Discussion

3.1 Mg-Zn-Sn alloy

3.1.1 *Microstructure after Solution Treatment and Age Hardening at 175 °C*

The precipitation hardening sequence was studied by annealing solution treated specimens in the range 175–250 °C for up to 96 h. Hardness studies of the Mg-Zn-Sn alloy [5] revealed that the hardness increased from 55 to 88 HV after aging for 2h at 175 °C, followed by a rapid decrease to 73 HV and then a second peak of 87 HV after aging for 16 h. Finally, the hardness decreased to 70 HV after 48h, and remained constant at this level up to 9 6h. At 225 °C overaging occurs very rapidly, thus the alloy is not suitable for precipitation hardening and service at such high temperatures. Based on SEM-EDS and XRD measurements it was concluded that the occurrence of the first peak can be attributed to the precipitation of $MgZn/MgZn_2$ and second peak to the formation of Mg_2Sn. From the observed 2Θ values of the XRD peaks discrimination between MgZn and $MgZn_2$ is difficult due to several overlapping peaks.

Figure 1 shows bright-field TEM micrographs of the microstructure after 2h and 16h of aging at 175°C. The first stages of precipitation reveal fine precipitates with a rod-like shape. The precipitates are similarly oriented within the Mg grain indicating that a well-defined orientation relationship exists between the precipitates and the matrix. The information about the chemical composition was obtained by EDS in the TEM. Element specific peaks reveal that the precipitates contain Mg and Zn.

HRTEM studies in combination with Fourier transformation of the lattice images [9] reveal that the rod-like precipitates are $MgZn_2$. The $MgZn_2$ / α-Mg interface appears to be coherent. The orientation relationship between the $MgZn_2$ precipitates and the Mg matrix is:

$$(11\bar{2}0)<10\bar{1}0>_{MgZn_2} |(0002)<11\bar{2}0>_{\alpha-Mg} \tag{1}$$

Identification of the precipitates based on their chemical composition was done by taking the elemental maps using electron energy loss spectroscopy and by EDS measurements in the TEM. After 24h spherical shaped particles containing Sn were identified by TEM. The chemical composition, selected area diffraction (SAD) and XRD measurements indicate that this phase is Mg_2Sn. The microstructure of the samples aged for 48h remains similar, except that in addition to fine $MgZn_2$ particles large MgZn particles exist. HRTEM studies of these samples reveal that the fine $MgZn_2$ particles maintain the same topotaxial orientation relationship with the Mg matrix as noted in Eq.1. Upon growth the $MgZn_2$ phase transforms into the thermodynamically more favourable MgZn phase. The reason why $MgZn_2$ precipitates first is the smaller activation energy barrier for nucleation compared to MgZn. This is most likely a result of the small interfacial energy for $MgZn_2/Mg$, estimated to be $\sigma = 25$ mJ/m^2 [10]. Calculation of the particle-matrix misfit of $MgZn_2$ in Mg matrix with the orientation relationship denoted in Eq. 1 yields $\delta = 0.04$ along the $<10\bar{1}0>$ zone axis of MgZn2 and in the perpendicular direction $<1\bar{2}10>$ of MgZn2 $\delta = -0.06$. Obviously, the orientation relationship with its relatively low misfit results in a low strain energy for $MgZn_2$.

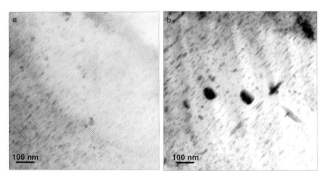

Figure 1: TEM bright-field micrographs of the Mg-Zn-Sn alloy aged at 175 °C for (a) 2h and (b) 16 h

Therefore, the phase nucleating first may not necessarily be the equilibrium phase, MgZn, with the lowest free energy but that with the lowest free energy barrier, i.e. the coherent $MgZn_2$ phase with a low value of total interfacial energy.

3.1.2 As-cast Microstructure

The as-cast microstructure was found to consist of $MgZn_2$/MgZn and Mg_2Sn precipitates within the α-Mg grains and eutectic of Mg+MgZn. The TEM micrographs presented in Fig. 2 show an example of the precipitates in the α-Mg grain. Elemental maps of Mg, Zn and Sn were acquired to reveal the local distribution of the chemical constituents. In the elemental map shown in Fig. 2b the Mg_2Sn particles are marked in blue and the Zn-rich precipitates in red. We assume that Mg-Zn precipitates with a rod-like shape are $MgZn_2$ and those with an ellipsoid shape are Mg-Zn, as determined from the results obtained from the aged samples. It can be clearly seen that the microstructure of the as-cast alloy is inhomogeneous with a wide range in particle size distribution compared to solutionized and aged samples.

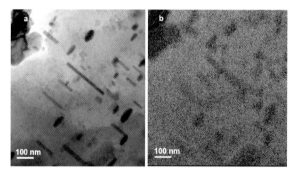

Figure 2: (a) TEM bright-field image of the as-cast Mg-Zn-Sn microstructure, and (b) the corresponding elemental map: the Mg matrix is shown in green, Zn in red, and Sn in blue

3.2 Mg-Zn-Sn-Sb and Mg-Zn-Sn-Y alloys

3.2.1 As-cast Microstructures

In order to improve the microstructural stability of the Mg-Zn-Sn base-alloy, which is susceptible to rapid overaging, Sb and Y were added. SEM and optical microscopy reveal a dendritic structure of both as-cast alloys consisting of Mg + MgZn eutectic and Mg_2Sn particles similar to the as-cast microstructure of the Mg-Zn-Sn base-alloy. The formation of the binary phase Mg_3Sb_2 in the Mg-Zn-Sn-Sb alloy is clearly identified by SEM-EDS and XRD. In the Mg-Zn-Sn-Y alloy an unknown Mg-Sn-Y phase was detected by local TEM-EDS and by XRD [6].

TEM micrographs of the as-cast alloys are shown in Figures 3, 4. The results confirm the presence of $MgZn_2$, MgZn and Mg_2Sn precipitates (Fig. 3a) in agreement with the microstructure of the Mg-Zn-Sn base-alloy. TEM-EDS analyses reveal that the rod-like and ellipsoid precipitates are Zn-rich phases. The Mg_2Sn phase usually possesses a plate-like morphology. Spherical Mg_2Sn particles were also found. The Mg_3Sb_2 phase (Fig. 3b) was identified by TEM-EDS and SAD.

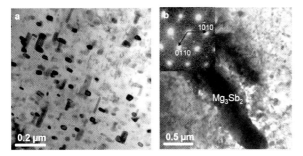

Figure 3: TEM bright-field micrographs of the as-cast Mg-Zn-Sn-Sb. (a) Microstructure of Mg-Zn-Sn-Sb alloy with precipitates embedded in a Mg grain, (b) Mg_3Sb_2 phase and corresponding SAD pattern

TEM studies of the Mg-Zn-Sn-Y samples (Fig. 4) reveal a coarse microstructure in the as-cast alloy. A broad size distribution of the particles is observed. Based on TEM-EDS analysis the very fine precipitates are the Mg-Zn phase, which we assume is again $MgZn_2$. TEM-EDS analysis and SAD patterns obtained from individual round-like particles reveal that these precipitates are Mg_2Sn. TEM-EDS analysis of the elongated particle (Fig. 4a) yields a Mg-Sn-Y with an atomic ratio of 1:1 for Sn to Y. Some of those elongated particles agglomerate in composition bunch-like shapes (Fig. 4b). The diffraction data of the Mg-Sn-Y phase does not fit to any published PDF data, and may be a new and/or metastable phase. According to XRD and SAD measurements and using computer simulations [11], the Mg-Sn-Y crystallographic structure was identified as primitive orthorhombic with lattice parameters a = 6.97 Å, b = 10.33 Å, and c = 13.27 Å. A SAD pattern obtained from the Mg-Sn-Y particle tilted in a $[11\bar{6}3]$ zone axis is shown in Fig. 4a.

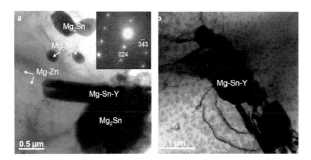

Figure 4: TEM bright-field micrographs of the as-cast Mg-Zn-Sn-Y alloy. (a) Microstructure of the Mg-Zn-Sn-Y alloy with fine Mg-Zn precipitates, Mg_2Sn particles and the ternary Mg-Sn-Y phase with the corresponding SAD pattern in a $[11\bar{6}3]$ zone axis, (b) agglomerates of the Mg-Sn-Y phase.

3.2.2   Microhardness Evolution

XRD studies of the Mg-Zn-Sn-Sb and Mg-Zn-Sn-Y alloys after solution treatment reveal the complete dissolution of the phases, except the Mg_3Sb_2 and Mg-Sn-Y phases. Microhardness of the solution treated alloys show a solid solution hardening of 25 HV for the Sb-containing alloy and 19 HV for the Y-containing alloy, compared to pure Mg.

Aging of the alloys at 175 and 225 °C up to 16 days shows 2 pronounced peaks which can be attributed to the precipitation of $MgZn_2$ and Mg_2Sn particles based on XRD measurements and previous studies [5]. Aging at 175 °C reveals a maximum increase in hardness from 52 to 72 HV and from 50 to 67 HV for Sb-, and Y-containing alloys, respectively. The Mg-Zn-Sn-Sb alloy aged at 225 °C shows the same evolution of hardness with annealing time as at 175 °C, with lower peak values. The hardness of the Y-containing alloy after aging for 16 days at 225 °C (64 HV) is higher than that measured for the Mg-Zn-Sn (55 HV) and Mg-Zn-Sn-Sb (57 HV) alloys which already start to overage after 2 days at the same temperature [6].

3.2.3 Aged Microstructures

TEM studies of the microstructure of the Sb-, and Y-containing alloys after solutionizing and subsequent aging at 225°C are shown in Figures 5a–d. After aging for 30 minutes, both alloys exhibit inhomogeneously distributed Mg-Zn precipitates. Based on EDS analyses and the rod-like morphology we assumed that these precipitates are $MgZn_2$. As in the Mg-Zn-Sn base-alloy, all precipitates embedded in a Mg matrix are similarly oriented indicating a well-defined orientation relationship. However, compared to the Mg-Zn-Sn alloy, the microstructures of the Sb-, and Y-containing alloys yield a broad size distribution and a low density of the precipitates after short aging time (compare Figs. 1a, 5a, and 5c). This result explains the low microhardness values measured for the Mg-Zn-Sn-Sb and Mg-Zn-Sn-Y alloys. Figure 5b shows the microstructure of the Mg-Zn-Sn-Sb sample after 12 days of aging. A coarsening of the Mg-Zn and Mg_2Sn particles can be clearly seen. After solutinizing and aging, TEM-EDS and SAD studies reveal the same 2–3 µm large Mg_3Sb_2 particles as in the as-cast material. The Mg-Sn-Y phase, which was not dissolved during the solution treatmen,t remained unchanged after aging for 12 days (Fig. 5d). Several µm-size agglomerates of Y-containing particles, like in the as-cast state, were also found in the aged samples. The present results indicate that Y addition to Mg base-alloy inhibit the aging process.

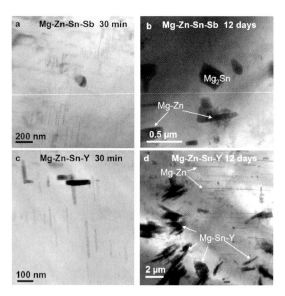

Figure 5: TEM bright-field images of the microstructure of (a, b) Mg-Zn-Sn-Sb and (c, d) Mg-Zn-Sn-Y alloys after aging at 225 °C for 30 min, and for 12 days, respectively

4 Summary

The microstructure of the as-cast and thermally exposed Mg-Zn-Sn alloy is heterogeneous, with $MgZn_2$/MgZn and Mg_2Sn precipitates inhomogeneously distributed within the Mg grains. Intergranular Mg_2Sn and an eutectic consisting of α-Mg and MgZn were observed.

In the samples aged at 175 °C for 1 h to 96 h $MgZn_2$ particles with a rod-like shape form initially, followed by plate-like $MgSn_2$ precipitates. The $MgZn_2$ phase shows a well-defined orientation relationship and a coherent interface with the Mg matrix. The formation of two types of precipitates is responsible for the occurrence of two pronounced hardness maxima. In contrast to the as-cast microstructure the aged alloy shows potential for application.

The addition of Sb has no effect on the hardness and microstructure of the current alloys. The addition of Y to the Mg-Zn-Sn base-alloy inhibits the aging process.

5 Acknowledgement

The study was partially supported by the German Israeli foundation for scientific research and development (GIF) under contract number I-704-43.10/2001 and B. and N. Ginsburg Research Fund. The funds are acknowledged for the support.

6 References

[1] A. Nayeb-Hashemi and J. Clark in Binary Alloy Phase Diagrams, (Eds.: T. Massalski, H. Okamoto, P. Subramanian, and L. Kacprzak), ASM International, 1992, 2571.
[2] Y. Guangyin, S. Yangshan and D. Wenjiang, Materials Science and Engineering A 308, 2001, 38.
[3] A. Singh and A. Tsai, Scripta Materialia 49, 2003, 143.
[4] A. Singh, M. Nakamura, M. Watanabe, A. Kato and A. Tsai, Scripta Materialia 49, 2003, 417.
[5] S. Cohen, G. Goren-Muginstein, S. Avraham, B. Rashkova, G. Dehm, and M. Bamberger, Z. Metallkd. 96 (2005) 9.
[6] A. Gorny, G. R. Goren-Muginstein, A. Katsman, and M. Bamberger, in Mg Technology 2006, (Ed.: A. Lou), TMS annual meeting, 2006, 387.
[7] G. Dehm, F. Ernst, J. Mayer, G. Möbus. H. Müllejans, F. Phillipp, C. Scheu and M. Rühle, Z. Metallkd., 1996, 87, 11.
[8] A. Strecker, U. Bäder, U. Salzberger, M. Sycha, and M. Rühle, in Dreiländertagung für Elektronenmikroskopie, Insbruck, Austria, S.D2.P, 2001, 144.
[9] B. Rashkova, S. Cohen, G. Goren-Muginstein, M. Bamberger, and G. Dehm, in 7th Multinational Congress on Microscopy, Portoroz, Slovenia, (2005) 183.
[10] A. Katzman, S. Cohen, and M. Bamberger, 2006 *unpublished*.
[11] R. Shirley, The CRYSFIRE System for automatic Powder Indexing: User's Manual, Lattice Press, Guildford, England 2000.

Zirconium-rich Coring Structure in WE54 Alloy after Heat Treatment

A. Kiełbus
Silesian University of Technology, Katowice, Poland

1   Introduction

Mg-Y-Nd magnesium alloys are characterised by high-strength and good creep-resistant alloys for automotive and aerospace applications [1]. The Mg-Y-Nd system provides a combination of mechanical and corrosion properties, which allows the development of commercial alloys: WE43 and WE54. The WE54 alloy contains 5 %$_{wt.}$Y, 1.7 %$_{wt.}$Nd 3.1 %$_{wt.}$HRE and 0.55 %$_{wt.}$ Zr. The rare earth elements have beneficial effect of on the creep properties and thermal stability of structure and mechanical properties of magnesium alloys. Zirconium, which does not form any phases with magnesium, contributes to the obtaining of a fine grain structure and improves the mechanical properties at an ambient temperature [2]. The solubility of zirconium is 3,8 %$_{wt.}$ at the melting temperature and decrease to 0,2 %$_{wt.}$ at room temperature (Fig.1) [3].

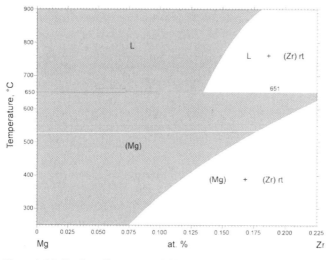

Figure 1: Mg-Zr phase diagram up to 0,22 %$_{at.}$ Zr [3]

Magnesium alloys with zirconium characterize of microstructure with zirconium-rich core areas inside grains [4,5]. Since both Mg and Zr are hexagonal and have nearly identical lattice parameters, that the zirconium particles are nucleants for magnesium grains. In alloys saturated with Zr in the liquid phase, during cooling precipitation of α-Zr particles is observed. At ~657 °C (peritectic temperature) zirconium particles form nuclei for the peritectic reaction between Zr particles and Zr-saturated Mg liquid to produce solid solution α-Mg with high Zr content [3].

2 Experimental Procedure

The material for the research was a casting WE54 magnesium alloy. The alloy was purchased from Magnesium Elektron, Manchester, UK. The chemical composition of this alloy is provided in Table 1.

Table 1: Chemical composition of the WE54 magnesium alloy in $\%_{wt}$

Y	HRE	Nd	Zr	Zn	Si	Fe	Mn	Cu	Mg
5	3,1	1,7	0,55	<0,01	<0,01	0,002	<0,01	<0,01	balance

Solution heat treatment was performed at 525 °C/8 h, quenched in water. Ageing treatments were performed at 150, 300 and 400°C. The duration of treatment varied between 16 and 1000 hours. The microsections for structural examination were subjected to grinding with sandpaper of 250 to 1200 granulation. Next, they were polished. The surface of the specimen was etched in a reagent containing 4.2 ml picrate acid, 70 ml ethanol and 10 ml water. For the microstructure observation, a OLYMPUS GX71 metallographic microscope and a HITACHI S-3400N scanning electron microscope were used.

3 Results and Discussion

3.1 Microstructure of WE54 Alloy in As-cast Condition

The WE54 alloy in as-cast condition was characterized by a solid solution structure α with some fine-dispersion precipitates of MgY, $Mg_{24}Y_5$ and probably $Mg_{14}Y_2Nd$ (equilibrium β phase) intermetallic phases inside and on grain boundaries. Also the zirconium-rich core areas have been observed. Zirconium-rich core areas are ellipsoidal or nearly circular (Fig. 2).

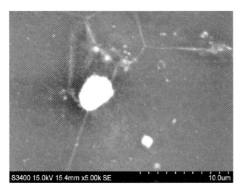
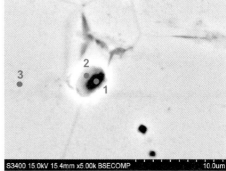

Figure 2: Zirconium rich areas in as cast WE54 alloy

The EDS analysis of the chemical composition has shown a change of the zirconium content from 70.85 $\%_{at.}$ in the core (point 1, Fig. 2), through 32,54 $\%_{at.}$ in an area located near solid solution grain (point 2, Fig. 2), to 0 $\%_{at.}$ in the solid solution grain (point 3, Fig. 2). This means, that zirconium was located only in zirconium-rich core areas.

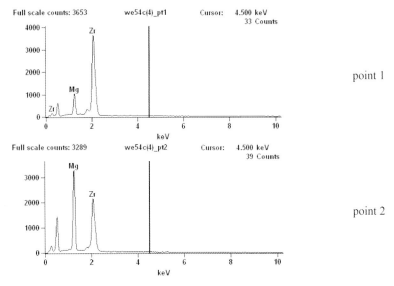

Figure 3: EDS spectra of zirconium-rich core areas from Fig.2

Table 2: Results of EDS chemical composition from Fig. 2

point	Mg		Zr		Y		Nd	
	%wt.	%at.	%wt.	%at.	%wt.	%at.	%wt.	%at.
1	9.88	29.15	90.12	70.85	-	-	-	-
2	35.58	67.46	64.42	32.54	-	-	-	-
3	90.75	97.70	-	-	5.54	1.63	3.72	0.67

3.2 Microstructure of WE54 Alloy after Solution Treatment

After solution treatment performed at 525 °C/8 h by water-cooling, the intermetallic phases and zirconium were dissolved in the matrix (Fig.4).

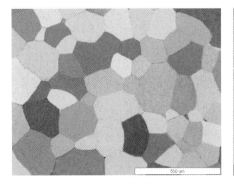 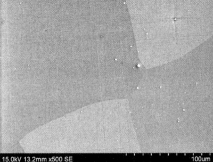

Figure 4: Microstructure of WE54 alloy after solution streatment

The EDS analysis confirm that whole amount of zirconium enter the solid solution α–Mg containing apart from magnesium, 1,43 %$_{at.}$ Y, 0,31 %$_{at.}$ Nd and 0,18 %$_{at.}$ Zr in the core. According to the binary Mg-Zr diagram the solubility of zirconium in magnesium is ~0,2 %$_{at.}$ at the 525 °C temperature (Fig.1). This means, that zirconium dissolved in the solid solution α-Mg.

3.3 Microstructure of WE54 Alloy after Ageing Treatment

3.3.1 Ageing 150 °C/1000 h/air

The ageing treatment at 150 °C/1000 h with air-cooling applied after solution treatment caused precipitation of the strengthening phases and Zr-rich core areas. The solubility of zirconium in magnesium decrease to ~0,05 %$_{at.}$ at the 150 °C. Therefore, precipitation of zirconium-rich core areas will occur at the areas with zirconium concentrations more than 0,05 %$_{at.}$ and will continue till the concentrations is reduced. Zr-rich core were observed inside solid solution grains or at the grain boundaries. Many grains contain more then one zirconium-rich core (Fig.5).

Figure 5: Microstructure of the WE54 alloy after ageing at 150 °C/1000 h/air

The mapping of Zr and Mg in the zirconium-rich core areas visible in the SE image can be seen in Fig.6. Figure 7 contains a linear analysis across Zr-core. It can be seen that distribution of zirconium follows that of magnesium.

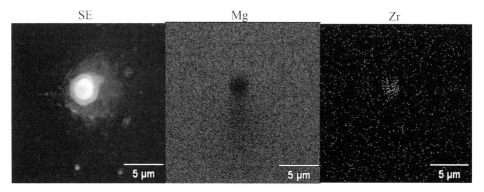

Figure 6: The SE image and the distribution of Mg and Zr in microareas of WE54 alloy

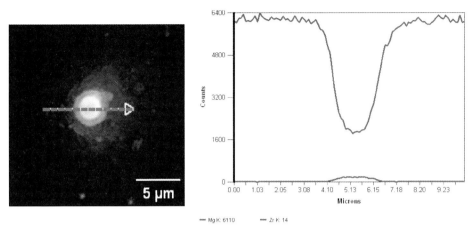

Figure 7: The linear analysis of Mg and Zr along the marked line in the SE image. a) BSE image; b) linear analysis

3.3.2 Ageing 300 °C/48 h/air

The ageing treatment at 300 °C/48 with air-cooling caused precipitation of the equilibrium β phase. Also the zirconium-rich core areas have been observed. However, due to higher solubility of zirconium in magnesium, the quantity of zirconium-rich core was smaller (Fig.8).

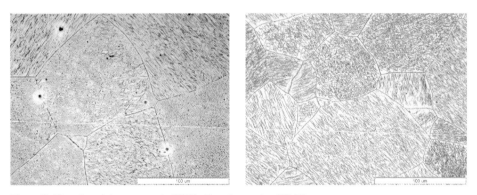

Figure 8: Zirconium rich core areas in WE54 alloy after ageing 300 °C

3.3.3. Ageing 400 °C/22 h/air

The ageing treatment at 400 °C/22 h/air caused precipitation of globular phases, containing neodymium and yttrium. The solubility of zirconium in magnesium increase to about 0,12 %$_{at.}$ at the 400 °C. Therefore, zirconium was dissolved in solid solution α-Mg and zirconium-rich core areas has not been observed. (Fig.9).

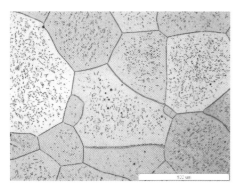 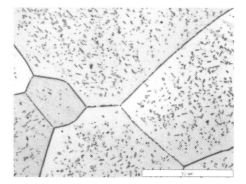

Figure 9: Microstructure of the WE54 alloy after ageing at 400°C/22h/air, without Zr-rich cores

4 Conclusions

Based on the experimental results obtained, the following conclusions can be drawn:
1. The WE54 magnesium alloy structure in as cast condition was characterized by a solid solution structure α with some fine-dispersion precipitates of MgY, $Mg_{24}Y_5$ and $Mg_{14}Y_2Nd$ intermetallic phases with ellipsoidal zirconium-rich core areas.
2. The solubility of zirconium in magnesium is ~0,2 %$_{at.}$ at the 525 °C temperature. This caused, that zirconium dissolved in the solid solution α-Mg after solution treatment.
3. After ageing treatment at 150 °C and 300 °C, due to reduced solubility of zirconium in magnesium, zirconium-rich core areas were formed. They were observed inside solid solution grains or at the grain boundaries.
4. The solubility of zirconium in magnesium increase to about 0,12 %$_{at.}$ at the 400 °C. Therefore, zirconium was dissolved in solid solution α-Mg and zirconium-rich core areas has not been formed.

5 Acknowledgements

This work was supported by the Polish Ministry of Education and Science under the research project No. 3 T08C 060 28.

6 References

[1] Li Z., Zhang H., Liu L., Xu Y. , Materials Letters **2004**, *58*, 3021–3024.
[2] Avedesian M., Baker H., Magnesium and Magnesium Alloys, ASM Speciality Handbook, **1994**.
[3] Friedrich H.E., Mordike B.L., Magnesium Technology, Springer-Verlag Berlin Heidelberg, **2006**.
[4] Qian M., StJohn D.H., Frost M.T., Scripta Materialia **2004**, 50, 1115–1119.
[5] Qian M., StJohn D.H., Frost M.T., Scripta Materialia **2002**, 40, 649–654.

Microstructural Characterization of Die Cast Mg-Al-Sr (AJ) Alloy

M. Kunst[1], A. Fischersworring-Bunk[1], C. Liebscher[2], U. Glatzel[2], G. L'Esperance[3],
P. Plamondon[3], E. Baril[4], P. Labelle[5]
[1]BMW Group, Munich, Germany
[2]Lehrstuhl Metallische Werkstoffe, Universität Bayreuth, Bayreuth, Germany
[3]Ecole Polytechnique de Montreal, Montreal (Quebec), Canada
[4]National Research Council, Boucherville (Quebec), Canada
[5]Composites Development Centre of Quebec, St-Jerome (Quebec), Canada

1 Abstract

The microstructure of a six cylinder Mg/Al composite crankcase cast with three alloy formulations of the Mg-Al-Sr alloy AJ62 has been investigated.

Microstructural observations were carried out using light optical microscopy and SEM in order to determine the impact of alloy composition and cooling conditions on the microstructure. The microstructure stability during long-term heat treatment was examined using TEM. Both, precipitates and α-Mg matrix were analyzed by TEM-EDS and X-ray diffraction.

Depending on the chemical composition, different precipitates were observed. In addition to the lamellar eutectic morphology of the Al_4Sr phase the precipitation of a massive type phase was observed, but with different chemical composition to the previously reported ternary compound $Al_3Mg_{13}Sr$. Analyzing the alpha-Mg matrix by TEM-EDS revealed a relatively high Al supersaturation of the matrix. The effects of alloy composition and cooling conditions on the microstructure are discussed in terms of particle morphology, phase volume fraction and stability.

2 Introduction

Motivated by weight saving and reduced fuel consumption, alloys with improved elevated temperature properties have attracted considerable interest over the last decade and resulting in new alloy systems.

One of these new alloys is the Mg-Al-Sr system which emerged as potential heat resistant alloy. The challenge in composing a creep resistant die casting alloy is to keep enough hardening elements to give adequate solid solution strengthening, maximize the volume fraction of precipitated phases, optimize the spatial distribution of dispersions without introducing brittleness and providing metallurgical stability under elevated temperature exposure.

A recent launched innovative solution already in series production since 2004 is BMW's in-line six cylinder engine with its hybrid magnesium-aluminum crankcase [1] using the AJ62x alloy. It is a new alloy castable in a high pressure die cast process attractive for the high volume application and providing a very good balance in mechanical properties, i.e. very good creep resistance, static and fatigue properties for the engine crankcase design temperature target of 150 °C. Details on the mechanical properties have previously been reported [2, 3, 4].

To assess the impact of variability in local chemical composition and microstructure to be expected in a large and complex component as the engine crankcase cast in the high pressure die cast process, an in depth investigation of the alloy performance around the compositional target was performed. Another question to be addressed is related to the microstructural stability of the alloy under exposure to elevated temperature for a significant period of time.

3 Materials

In this study the microstructure of magnesium samples machined out from a high pressure die cast composite Mg/Al crankcase was investigated. Components with different chemical compositions were cast by BMW Group in Landshut, Germany, conducted on a 4500 ton Bühler cold chamber machine [4].

Three different alloy compositions were formulated based on the AJ62 ASTM alloy specification [5]. Compared to AJ62 the Al content was increased for the formulation AJ62+Al while decreasing the Sr content, and vice versa for AJ62+Sr, adjusting different Al/Sr ratios for all three alloys. The nominal compositions and Al/Sr ratios are listed in table 1.

Table 1: Chemical compositions used for the high pressure die casting

Alloy	Al content [wt.%]	Sr content [wt.%]	Al/Sr ratio
AJ62+Al	6.1	2.2	2.8
AJ62	6.0	2.3	2.6
AJ62+Sr	5.5	2.8	2.0

4 Experimental Procedure

4.1 Heat Treatment

Heat treatments were performed for 1500 hours at 175 °C to study the long term stability concerning morphology and chemical composition of the microstructure. The machined samples were heat treated under ambient atmosphere.

4.2 Microstructure Characterization

Scanning electron microscopy (SEM) and transmission electron microscopy (TEM) were used to characterize the microstructure. The TEM analyses were performed on a 200 kV field emission gun (FEG) JEOL JEM-2100F transmission electron microscope equipped with an energy dispersive X-ray spectrometer (EDS). The EDS was used to obtain the chemical composition of the different phases.

The X-ray analyses, performed to the chemical composition of the different phases, were quantified using the Cliff-Lorimer relation [6]

$$\frac{C_A}{C_B} = K_{AB} \times \frac{I_A}{I_B}$$

where C_A and C_B are the concentrations of elements A and B, K_{AB} is the Cliff-Lorimer factor specific to these two elements and I_A and I_B are the net intensities of elements A and B measured from the X-ray spectrum. The Cliff-Lorimer factors were obtained by acquiring 30 spectrums on standards of $MgAl_2O_4$, $SrTiO_3$ and Al_2O_3. The value of the K_{MgAl} factor was determined to 1.04, while the one for K_{AlSr} was 0.40. The ratio of the intensities was plotted on a graph as a function of the sum of the intensities in order to extract the ratio for a thickness of 0 (sum of the intensities = 0) with zero absorption. The analyses were corrected for absorption using a technique described by Goldstein et al. [6].

Samples for SEM were ground and then polished using diamond suspensions with a particle size of 6, 3, and 1 µm in ethanol and a suspension of SiO_2 powder with size of 0.25 µm in ethanol.

TEM thin foils were prepared using a Hitachi FB 2000A focused ion beam (FIB). To prepare a thin foil, a small sample of the size 5 × 8 × 1 mm was inserted in FIB system. After protecting the area of interest by a tungsten (W) coating, the material was milled using a gallium (Ga) ion beam accelerated to 30 keV. The milling of the sample has to be performed in several steps until the thin foil with a thickness of about 2 µm can be transferred to the TEM grid using a micromanipulator. After welding the thin foil to the TEM grid the final milling is performed, which brings the sample to a thickness of about 100–200 nm.

X-ray diffraction patterns were used to confirm the phase identification of the samples and to characterize the influence of the heat treatment. These investigations were conducted with a Bruker AXS D5005, equipped with a Co-anode.

5 Results and Discussion

5.1 Microstructure in As-cast Condition

The microstructure of the high pressure die cast crankcase samples is characterized by a primary α-Mg solid solution with globular morphology and one or more secondary precipitated phases in between these regions. This cellular structure results from the rapid cooling during die filling and solidification. While the melt is flowing rapidly into the die, the first α-Mg nuclei are formed. Meanwhile the residual melt is enriched with high melting point elements resulting in second phases [9].

The morphology of the second phases of both AJ62 and AJ62+Al samples is lamellar eutectic, which consists of Al_4Sr and secondary α-Mg lamellae, shown in figure 1. This morphology of the AJ62 microstructure and the presence of the lamellar eutectic have been reported previously [3]. In contrast, the second phase morphology of the AJ62+Sr samples is different, although the chemical composition of the AJ62+Sr alloy lies within the AJ62 specification. The precipitates of AJ62+Sr are divorced eutectic and massive type, as shown in figure 1. A detailed description of the observed phase types and of their morphology is given in figure 2.

XRD analyses of the crankcase samples confirmed the appearance of an additional phase in the AJ62+Sr alloy. For AJ62 and AJ62+Al samples two different phases, Al_4Sr and α-Mg, have been positively identified in the XRD pattern, shown in figure 3. The peaks of an additional phase were observed in the AJ62+Sr XRD pattern, which could not be identified. Baril et al. [3] associated these peaks with the massive type ternary phase, which was tentatively named $Mg_{13}Al_3Sr$.

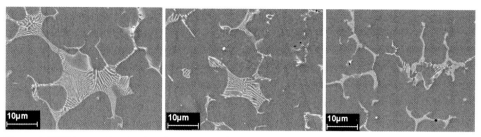

Figure 1: SEM images showing the microstructure of AJ62 (left), AJ62+Al (middle) and AJ62+Sr (right) samples

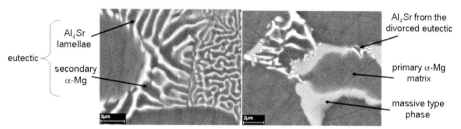

Figure 2: Morphology types of the AJ alloy microstructure

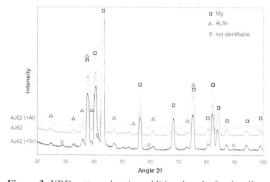

Figure 3: XRD pattern showing additional peaks for the alloy AJ62+Sr

5.1.1 Chemical Composition of the Mg Solid Solution

Chemical compositions of all microstructure constituents, primary α-Mg solid solution, Al_4Sr and massive type phase, were analyzed using TEM-EDS.

The average Al content measured in the primary α-Mg solid solution of the AJ62 and AJ62+Sr samples is (4.0 ± 0.3) wt.%. The composition is constant within one grain, but varies slightly from grain to grain. In contrast, a higher average Al content, (5.3 ± 0.5) wt.%, was measured for the α-Mg solid solution of the AJ62+Al samples, which again appears constant within the grains. Sr was not detected in the Mg solid solution in any of the crankcase samples using TEM-EDS.

Compared to previous publications reporting an Al content of about 1–2 wt.% in the α-Mg solid solution of separately cast samples [3, 7], the Al content in the α-Mg solid solution of the crankcase samples is remarkably higher. This high supersaturation is most probably related to the specific filling and cooling conditions during high pressure die casting of the crankcase [9]. The maximum solubility for Al in Mg at room temperature is below 1 wt.% Chartrand, P.; Pelton, A., Journal of Phase Equilibria, 1994, 15, 591–605., hence a high supersaturation is supposed to be detrimental to creep strength due to the discontinuous precipitation of $Mg_{17}Al_{12}$. Nevertheless, precipitates of $Mg_{17}Al_{12}$ were not detected in the „as cast" crankcase samples.

5.1.2 Chemical Composition of the Al-Sr Precipitates

TEM-EDS measurements in the Al-Sr precipitates of the samples AJ62, AJ62+Al and AJ62+Sr revealed an average Al content of (52.1 ± 2.4) wt.%. The average Sr content of this phase is (40.6 ± 1.2) wt.%. This composition translated to atomic concentrations results in an Al/Sr ratio relatively close to 4, which corresponds to Al_4Sr.

The Mg content of the Al_4Sr precipitates is stable at (11.2 ± 2.8) at.%. The maximum solubility of Mg in Al_4Sr is 25 at.% [10]. The Mg present in the precipitates does not seem to change the Al/Sr ratio. This would mean that the Mg is in solid solution in the precipitates.

5.1.3 Chemical Composition of the Massive-type Phase

The chemical composition of the massive type phase present in the AJ62+Sr samples is relatively constant with an average of (57 ± 4) wt.% Mg, (21 ± 2) wt.% Al and (22 ± 2) wt.% Sr. This average concentration translates into a stoichiometry very near Mg_9Al_3Sr [9]. Baril et al. [3] reported the appearance of a bulky constituent only for AJ52, but with a stoichiometric composition of $Mg_{13}Al_3Sr$. This result indicates a lower Mg concentration in the ternary compound of AJ62+Sr than for AJ52. Probably this decrease can be related to the higher Al and Sr content in AJ62+Sr and the different Al/Sr ratio (2.6 for AJ52 [3] and 2.0 for AJ62+Sr) [9]. Our hypothesis is that the ternary compound can precipitate in a wide chemical range depending on the alloy composition and the specific solidification conditions.

The observations show that the specific Al/Sr ratio causing precipitation of the massive type phase changes with the absolute content of alloying elements. Comparing results of the present study and literature data, a borderline for the Al/Sr ratio causing precipitation of the massive type phase can be proposed depending on the absolute Al content (figure 4).

Despite of the difference in chemical composition compared to the massive-type phase of the AJ52, the additional peaks in the XRD pattern related to this phase appear at the same Bragg angles 2θ as those reported by Baril et al. [3].

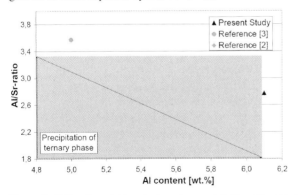

Figure 4: Al/Sr ratio and Al content causing precipitation of ternary phase

5.2 Microstructure after Heat Treatment

5.2.1 Change in Morphology

SEM observations after a 1500 hours heat treatment at 175 °C exhibit only a slight coarsening of the eutectic Al_4Sr lamellae, whereas the size and the morphology of the α-Mg matrix remain stable. SEM images before and after heat treatment are shown in figure 5.

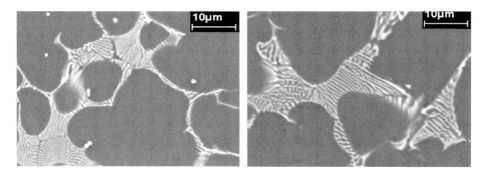

Figure 5: Microstructure of AJ62 sample before and after heat treatment (1500 hours at 175 °C)

This pronounced stability is also confirmed by XRD pattern, performed before and after the 1500 h heat treatment at 175 °C, shown in figure 6. A change in the relative volume fraction of α-Mg matrix and Al_4Sr phase was not detected by quantifying the XRD peaks. The slight variation in intensity of some peaks after heat treatment could be related to the change in chemical composition of the Al_4Sr lamellae and the α-Mg matrix.

5.2.2 Change in Chemical Composition

The Al content of the AJ62 α-Mg solid solution after the 1500 hours heat treatment at 175 °C remains stable at about 5 wt.%. However, a slight decrease of the Al concentration to approximately 3 wt.% was detected near the Al-Sr precipitates. Still dissolved Sr was not detected in the α-Mg matrix by EDS analysis. Finally, the negligible solubility for strontium in magnesium reported by Chartrand and Pelton [8] was confirmed by TEM investigations even after long term heat treatment. Thus dissolution of the Al_4Sr precipitates in the α-Mg matrix during heat treatment does not occur.

Concerning the chemical composition of the Al-Sr precipitates a change during heat treatment can be stated. The Al content in these precipitates increased to (60.2 ± 3.6) wt.%, whereas the Sr content decreased to (35.2 ± 1.7) wt.%. This change affects the ratio of the atomic concentrations of the precipitates, which increases to (5.6 ± 0.6) while the ratio for Al_4Sr should be 4.

Furthermore the Mg content of the Al-Sr precipitates in the AJ62 heat treated samples is lower than before heat treatment. It appears that the heat treatment led to a change in the chemical composition of the precipitates. Since Sr does not diffuse into the matrix, some Al would then need to diffuse from the matrix into the precipitates to increase the Al/Sr ratio. The fact that the Al content of the matrix is lower near the precipitates is an indication that some Al diffused from the matrix into the precipitates. This could also increase the size of the precipitates.

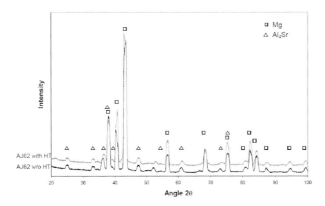

Figure 6: XRD pattern for AJ62 before and after heat treatment

6 Conclusion

The AJ alloy formulations analyzed in this study reveal an average Al content of the Mg matrix near 4 wt.%, except for the sample AJ62+Al where it is 5.3 wt.%. The Al concentration appears stable within one grain. The negligible solubility of Sr in the Mg matrix was confirmed by TEM-EDS analyses. $Mg_{17}Al_{12}$ precipitates were not observed in all the samples, even after long-term heat treatment.

In the as cast condition the atomic Al/Sr ratio of the Al_4Sr precipitates is near 4; the Mg content of these precipitates is about 8 to 10 wt.%. During heat treatment the Al/Sr ratio of the precipitates increases up to 5-6 due to diffusion of Al from the matrix into the precipitates. Despite, the XRD analysis did not reveal volume growth of the Al-Sr phases.

The ternary non-lamellar constituent present in the AJ62+Sr formulation exhibits an average composition that translates into a stoichiometry near Mg_9Al_3Sr.

Summarizing, the Al_4Sr precipitates can be regarded as the major hardening phase in AJ62. In conventional precipitation hardened alloys the particle strengthening effect decreases due to Ostwald-Ripening. With increasing temperature dissolution of the precipitates occurs and causes softening of the material. In contrast, the AJ62 alloys investigated in this study exhibit a very stable microstructure.

7 Acknowledgements

The authors would like to acknowledge BMW Group for sponsoring this project. The authors thank all the engineers and technologists from BMW Group who were involved in the die casting campaign. One of the authors (MK) is grateful for financial and organizational support by the Bavarian State Chancellery for his stay at $(CM)^2$, École Polytechnique de Montréal, Canada.

8 References

[1] Fischersworring-Bunk, A.; Landerl, C.; Fent, A.; Wolf, J., 62nd Annual World Magnesium Conference, **2005**, 49–58.
[2] Bakke, P.; Fischersworring-Bunk, A.; de Lima, I.; Lilholt, H.; Bertilsson, I.; Abdulwahab, F, Labelle, P., SAE Technical Paper, **2006**, 2006-01-70.
[3] Baril, E.; Labelle, P.; Pekguleryuz, M., JOM, **2003**, *55*, 34–39.
[4] Labelle, P.; Baril, E.; Fischersworring-Bunk, A.; SAE Technical Paper, **2004**, 2004-01-0659.
[5] ASTM B93 SB93/B93M-05, *Standard Specification for Magnesium Alloys in Ingot Form for Sand Castings, Permanent Mold Castings, and Die Castings.*
[6] Goldstein, J.I.; Williams, D.B.; Cliff, G., in: *Principles of Analytical Electron Microscopy*, Plenum Press, **1986**.
[7] Pekguleryuz, M.; Baril, E., The Minerals, Metals & Materials Society, **2001**, 119–125.
[8] Chartrand, P.; Pelton, A., Journal of Phase Equilibria, **1994**, *15*, 591–605.
[9] Kunst, M.; Fischersworring-Bunk, A.; Glatzel, U.; L'Espérance, G.; Plamondon, P.; Baril, E.; Labelle, P., *Proceedings of the International Symposium on Magnesium Technology, October 1–4, 2006,* Montréal, Québec, Canada, COM **2006**.
[10] Makhmudov, M.M.; Bodak, O.I.; Vakhobov, A.V.; Dzhurayev, T.D., Izv. Akad. Nauk SSSR, Met., 1981, 6, 216–220.

Procedure of Quantitative Description of Mg-Al Alloy Structure After Heat Treatment

Janusz Adamiec[*], Andrzej Kiełbus[*], Jan Cwajna[*], Janusz Paśko[**]
[*]Silesian University Of Technology, Department Of Materials Science, Katowice, Poland
[**]ZM WSK Rzeszów SA, Rzeszów, Poland

1 Abstract

The objective of the research conducted was to develop a comprehensive procedure of a quantitative evaluation of the microstructure of magnesium casting alloy (AZ91D) after heat treatment. The procedure encompasses: a methodology of metallographic specimens preparation, methods of microstructure detection, selection of image acquisition methods, methodology of image analysis and adjustment of morphological and stereological parameters to the quantitative description of volume fraction of phases and structural components of the AZ91 alloy after solution heat treatment and ageing.

2 Introduction

Mg-Al alloys apart from the two basic elements (magnesium and aluminium) also contain manganese, zinc and silicon. In the microstructure of the casting alloy AZ91D in an as-cast state, one can observe: solid solution α with phase β ($Mg_{17}Al_{12}$) precipitations on grain boundaries and regions of lamellar mixture $\alpha+\beta$. In addition, some regions of the so called divorced eutectic, precipitations of $MnAl_4$, and Al_8Mn_5 phases as well as Laves' Mg_2Si phases are to be found (Fig. 1a).

However, after solution heat treatment at a temperature above the solubility limit temperature of Al in a solid state, a solid solution of aluminium is present in magnesium (α) and possibly, some regions of undissolved massive phase β ($Mg_{17}Al_{12}$) (Fig. 1b).

After ageing, two types of precipitations occur in the AZ91D alloy: massive and lamellar. In most cases, the precipitations occur simultaneously. The continuous precipitation is a result of nucleation and growth of individual $Mg_{17}Al_{12}$ phase particles, which leads to changes in the matrix composition. Whereas lamellar precipitations nucleate on the boundaries of the solid solution grains and when growing, they take the form resembling nodules [1]. Mg-Al alloys containing 5–10 at.-% of Al, are dominated by the massive precipitations of $Mg_{17}Al_{12}$ phase (Fig. 1c). However, it has been found that the morphology of precipitations of the $Mg_{17}Al_{12}$ phase in Mg-Al alloys depends on the chemical composition (Al content) and temperature (Fig. 1d). It has been shown that when (Fig. 1d):

- at $T < T_{c1}$ temperature – only massive precipitations of phase $Mg_{17}Al_{12}$ occur in the alloy;
- in the temperature range of $T_{c1} < T < T_{d1}$ – both massive and lamellar precipitations of the $Mg_{17}Al_{12}$ phase occur in the alloy, where as the temperature rises, the number of lamellar precipitations increases;
- in the temperature range of $T_{d1} < T < T_{d2}$ – only lamellar precipitations of phase $Mg_{17}Al_{12}$ occur in the alloy;

- in the temperature range of $T_{d2} < T < T_{c2}$ – again, both massive and lamellar precipitations of the $Mg_{17}Al_{12}$ phase occur in the alloy – however at the same time, along with the rise in temperature, the number of lamellar precipitations increases;
- in the temperature range of $T_{c2} < T < T_s$ (solubility limit temperature – solvus) – only massive precipitations of phase $Mg_{17}Al_{12}$ occur in the alloy.

Critical temperature T_{c1} occurs mainly in alloys containing 18,8 at.-% of Al. Other temperatures occur in all commercial alloys [2]. In Mg-Al alloys, continuous precipitation is prevailing at a high temperature (close to solvus line) and at a low temperature, whereas in the range of temperatures in-between, discontinuous precipitation prevails [2].

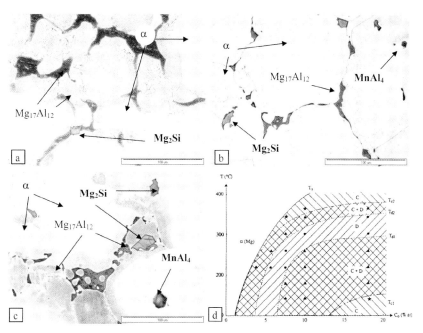

Figure 1: GA8 alloy microstructure: a) input specimen – after casting, b) specimen N2 after solution heat treatment (table 1, entry 1), c) specimen N3 after solution heat treatment and ageing (table 1, entry 2), d) Influence of Al content and temperature on the morphology of $Mg_{17}Al_{12}$ phase: C – continuous precipitation region, D – discontinuous precipitation region [1]

Nevertheless, there is no explicit data regarding the impact of technological parameters on the microstructure and the effect of microstructure on technological, mechanical and physicochemical properties of the AZ1D alloy. Progress in this area may be achieved as a result of improvement of the alloy microstructure description. A quantitative description of the microstructure is necessary to determine the unambiguous dependencies in the cause-and-effect chain: chemical composition – production technology – structure – functional properties.

3   Research Material

The research was conducted with the use of fragments of a cast made of the AZ91D alloy (7.5–9 % Al., 0.2–0.8 % Zn, 0.5–0.5 % Mn) which was subjected to heat treatment in accordance with the parameters shown in Table 1. In order to work out the methodology, the fragments were cut out from the cast's side surface after solution heat treatment (Table 1, entry 1) and after both solution heat treatment and ageing (Table 1, entry 2). The region was considered representative and sufficient to develop the procedure for a qualitative and quantitative evaluation of the AZ91D alloy structure after heat treatment.

Table 1: Heat treatment parameters for the GA8 alloy

Entry	Specimen design.	Solution heat treatment				Ageing			
		Heating	Temp.	Time	Cooling	Heating	Temp.	Time	Cooling
1	N2	25°C/h	360°C	3h	air	–	–	–	–
		25°C/h	415°C	24h	air				
2	N3	25°C/h	360°C	3h	air	50°C/h	170°C	8h	air
		25°C/h	415°C	24h	air				

The specimens' rough cut was made with Phoenix cut-off machine. The precise cutting was performed with a Beuhler Isomet 5000 cut-off machine. The specimens were first ground and then polished according to the recommendations of Buehler company [3].

Attempts to reveal the AZ91D microstructure after heat treatment were made on the surface of the metallographic specimen. The best images of the alloy microstructure after solution heat treatment and after both solution heat treatment and ageing are presented in Fig. 1b,c.
It has been found that:
- the best etching reagent for the alloy after solution heat treatment is reagent: 10 ml HF + 96 ml H_2O. Etching with this reagent allows for a clear identification of phases in the structure and for its qualitative description (Fig. 1b);
- in order to evaluate the grain size in the alloy after solution heat treatment, the etching reagent to be applied is: 20ml acetic acid + 80 ml H_2O + 5g $NaNO_2$;
- the best reagent for the alloy after solution heat treatment and ageing to be used for a qualitative and quantitative evaluation is: 5–20 ml acetic acid, 80–95 ml H_2O; this reagent also allows the determination of the grain size in the structure. A good etching reagent is also: 10ml HF+96 ml H_2O, which allows individual phases in the structure to be detected (Fig. 1c);
- it is recommended that in all cases, observations on a light microscope in bright field at a 500× magnification should be carried out for a qualitative evaluation, whereas the 100× to 200× magnification should be used for a quantitative evaluation.

4 Quantitative Evaluation of the GA8 Alloy Structure after Heat Treatment

The quantitative evaluation of phases detected in the GA8 alloy after heat treatment was performed using a quantitative image analysis facility. The facility consists of a light microscope, OLYMPUS GX-71, equipped with an automatic table for image stitching in XYZ axes and AnalySIS Pro® software as well as MetIlo® software.

Image acquisition was conducted in accordance with the methodology presented in Chapter 2, using the bright field technique. For the evaluation of phases' surface fraction, the

AnalySIS Pro® program was used. It was assumed that, based on the metallographic investigations, the surface fractions of the following phases would be analysed: phase $Mg_{17}Al_{12}$ of massive morphology, lamellar $Mg_{17}Al_{12}$ phase and regions enriched with Al, phase Mg_2Si and the area occupied by solid solution α. A static analysis of results repeatability and of the minimum number of measuring fields has shown that measurements should be conducted at the surface of at least five measuring fields. The procedure and conditions of image detection were elaborated on the example of specimen N3, i.e. the alloy after ageing and solution heat treatment. The detection of phases and the measurements are shown in Fig. 2a, whilst Fig. 2b presents graphically the change in fraction of particular phases in the alloy after heat treatment. The results of the tests are juxtaposed in Table 2.

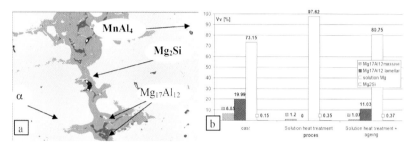

Figure 2: Assessment of phases' surface fractions in the AZ91D alloy structure after heat treatment: a) detected image, b) set of results

Table 2: Assessment of phases' surface fractions in the AZ91D alloy microstructure after heat treatment

Solution heat treatment				
Phase	Surface fraction of A_A [%]			Variability index [%]
specimen N2	Minimum	Maximum	Mean	
$Mg_{17}Al_{12}$ massive	0,79	1,89	1,20	29
Solution α	96,60	98,19	97,62	0,5
Mg_2Si	0,29	0,39	0,35	10
specimen N3	Solution heat treatment + ageing			
massive $Mg_{17}Al_{12}$	0,49	2,32	1,07	43
lamellar $Mg_{17}Al_{12}$ + regions enriched with Al	4,28	16,27	11,03	27
Solution α	74,56	90,02	80,75	5
Mg_2Si	0,25	0,72	0,37	32

variability index = standard empirical deviation / arithmetic mean · 100 %

Complementary to the quantitative description of the AZ91D alloy structure after heat treatment is the measurement of grain size in the specimens after solution heat treatment.

Specimen N2 was used to develop the methodology. The metallographic specimen was prepared in compliance with the procedure recommended in Chapter 2 and next, etched with a reagent containing 20 ml of acetic acid and 80 ml of H_2O, and 5g of $NaNO_2$. Image acquisition was performed using a light microscope, Olympus GX 71, in a bright field, at a magnification of 200×.

For grain size measurement, an automated image analysis facility based on the MetIlo® software was used. A set of grey image transformations until obtaining a binary image is shown in Fig. 3.

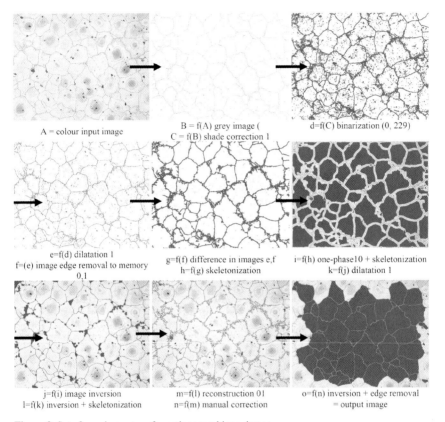

Figure 3: Set of grey image transformations to a binary image

For the description of grain, the following morphological parameters were used:
- mean area of the grain plane section $\bar{A}$ [mm^2]
- mean grain circumference P [µm],
- mean maximum diameter of the grain plane section $\bar{d}_{max}$
- mean minimum diameter of the grain plane section $\bar{d}_{min}$
- dimensionless grain shape coefficient $\bar{\xi} = 4\pi A / P^2$ (P: grain circumference in [mm]),
- grain elongation ratio $\bar{\delta} = d_{max} / d_{min}$

The quantitative evaluation results are juxtaposed in Table 3. Also, static distributions of those parameters were determined and the variability index was calculated as a quotient of standard deviation from the mean value times 100 %, for parameters ($\bar{A}$), $\bar{\xi}$ and $\bar{\delta}$, which allows evaluation the grain size inhomogeneity and the grain shape.

Table 3: Results of the quantitative description of the size and shape of the GA8 alloy grain after solution heat treatment.

Phase specimen N2	Roztwór sta³y Al w Mg (a?)			Variability index [%]
	Minimum	Maximum	Mean	
$\bar{A}$ [mm²]	172	10284	3080	95
P [µm]	49,5	555	207	59
$\bar{d}_{max}$	19,3	156	71,1	55
$\bar{d}_{min}$	13,1	108	50,9	59,3
$\bar{\xi} = 4\pi A / P^2$	1,13	2,22	1,46	16,5
$\bar{\delta} = d_{max} / d_{min}$	0,397	0,884	0,697	14,4

variability index = standard empirical deviation / arithmetic mean · 100 %

Based on a statistical analysis, it should be assumed that in order to obtain correct and repeatable results, which enable a description of the alloy structure after solution heat treatment, not less than 300 grains should be measured.

5 Final Comment

The procedure developed in a new, original tool in studying the relations in the cause-and-effect chain: chemical composition – technology – structure and properties of the AZ91D alloy. It allows a quantitative evaluation of Mg-Al alloys and its results can be applied to investigate the influence of structure on functional properties of those alloys as well as to qualitatively evaluate the process of alloys manufacture and to qualify the products made of them.

6 Acknowledgements

This work was supported by the Polish Ministry of Education and Science under the research project No. 6 ZR7 2005 C/06609.

7 References

[1] Zhang M., Kelly P.M.: „Crystallography of $Mg_{17}Al_{12}$ precipitates in AZ91D alloy.", Scripta Materialia 48 (2003) 647-652
[2] Duly D., Simon J.P., Brechet Y.: "On the competition between continuous and discontinuous precipitations in binary Mg-Al alloys." Acta Meta. Mater. Vol. 43, No. 1,m pp. 101-106, 1995.
[3] Buehler SUM-MET „The science behind Materials preparation", Buehler LTD, 2004

The Evolution of Dislocation Structure as a Function of Deformation Temperature in UFG Magnesium Determined by X-ray Diffraction

T. Fabián, Z. Trojanová, R. Kužel
Charles University, Prague (Czech Republic)

1 Abstract

Ultrafine-grained pure magnesium (UFG-Mg) was deformed in compression at various temperatures at a constant initial strain rate. The deformed samples were investigated by the high resolution X-ray diffraction peak profile analysis. The diffraction peaks were fitted by theoretical profile functions where the strain profile is scaled for strain anisotropy by the dislocation contrast factors. The contrast factor parameters are evaluated in terms of the fundamental Burgers vector types in hexagonal crystals. In the paper the evolution of the dislocation density and the Burgers vector types with the temperature of deformation are discussed.

2 Introduction

Magnesium, the lightest structural metallic element, and its alloys possess a great potential to be developed as structural materials provided that significant improvements in properties are achieved. Coarse grained Mg materials have some limitations such as low strength, low ductility and poor corrosion resistance. According to the Hall–Petch relationship [1,2], it can be shown that grain size refinement is the most effective method for improving the strength of materials. It is important to mention here that with grain size refinement, ductility has also been found to increase [3]. Several papers [3–7] have reported that high ductility and even low-temperature superplasticity may be obtained by grain size refinement of magnesium or magnesium alloys. The role of <c+a> slip in plastic deformation of hexagonal close-packed metals (hcp) is important for fundamental understanding of dislocation mechanisms involved in a generalized plasticity of polycrystals. The significance of a pyramidal slip system, {hkil} $\langle 11\bar{2}3 \rangle$, in hcp crystals is emphasized by the fact that it can provide five independent deformation modes [8,9]. Therefore, any hcp crystal exhibiting pyramidal <c+a> slip fulfils the von Mises criterion for a general homogeneous plastic deformation, regardless of which other systems are present. There are two main ways by which non-basal <c+a> slip dislocations can be generated (a) at/near surfaces and interfaces such as grain boundaries and incoherent twin boundaries, and (b) at heterogeneous sites in the grain interior such as the junction associated with <a> and <c> dislocations [10,11]. Compressive deformation of nanocrystalline Mg (with the grain size of 42–45 nm) was studied by Hwang and co-workers [12]. They found that the main deformation mechanism operating during straining at room temperature is grain boundary sliding with the threshold stress.

In this paper, density of <a> dislocations, fraction of <c+a> dislocations of the basal dislocation density estimated for UFG-Mg are reported for samples deformed at elevated temperatures.

3 Experimental Procedure

Samples used in this study were prepared by ball milling procedure in an inert atmosphere and subsequent consolidation and hot extrusion. Uniaxial compression tests were carried out at temperatures between room temperature and 300 °C using an INSTRON testing machine. Cylindrical specimens with the diameter of 10 mm and length of 15 mm were deformed with an initial strain rate of $2.8 \cdot 10^{-4}$ s^{-1}. The temperature in the furnace was kept with an accuracy of ± 1 °C.

The diffraction profiles are evaluated by assuming that peak broadening is caused by smallness of crystallite size and strain caused by dislocations. The Williamson–Hall plots (integral breadth β vs. $\sin\vartheta$) were constructed. The measured physical profiles are fitted by theoretical profiles calculated on the basis of well-established profile functions of size and strain. Both, strain and strain anisotropy are accounted for by the dislocation model of lattice distortions [13,14]. In a crystal containing dislocations the mean square strain is done [13]

$$\langle \varepsilon_{g,L}^2 \rangle \cong (\rho C b^2 / 4\pi) f(\eta), \tag{1}$$

where ρ is density of dislocations and b their Burgers vector, C is the dislocation contrast factor, $f(\eta)$ is the Wilkens function, where $\eta = L/R_e$ (R_e is the effective outer cut-off radius of dislocations), L is the Fourier length defined as $L = n a_3$ ($a_3 = \lambda/2(\sin\vartheta_2 - \sin\vartheta_1)$), n are integers starting from zero, λ is the wavelength of X-rays and $(\vartheta_2 - \vartheta_1)$ is the angular range of the measured diffraction profile. In the case of hexagonal polycrystals, where dislocations with various Burgers vectors take place in deformation, the average contrast factor may be obtained by the weighted linear combination of the individual contrast factors for the active sub slip systems:

$$\langle C b^2 \rangle = \sum_{i=1}^{N} f_i \langle C^i \rangle b_i^2, \tag{2}$$

where N is the number of the different activated sub slip systems, $<C^i>$ is the average dislocation contrast factor corresponding to the ith sub slip system and f_i are the fractions of the particular sub slip systems by which they contribute to the broadening of a specific reflection. The average contrast factors for a single sub slip system in hexagonal crystals are [15]:

$$\langle C_{hk,\ell} \rangle = \langle C_{hk.0} \rangle \left[1 + q_1 x + q_2 x^2 \right] \tag{3}$$

where $x = (2/3)(1/ga)^2$, q_1 and q_2 are parameters depending on the elastic properties of the material, $C_{hk.0}$ is the average contrast factor corresponding to the hk.0 type reflections and a is the lattice constant in the basal plane. The q_1 and q_2 parameters and the values of $C_{hk.0}$ have been calculated hexagonal crystals and compounds in [15]. Two Burgers vectors types were taken into account: $<\mathbf{a}> = \frac{1}{3}\langle \bar{2}110 \rangle$ and $<\mathbf{c}+\mathbf{a}> = \frac{1}{3}\langle \bar{2}113 \rangle$. Using the scheme described by Kuzel and Klimanek [16] N$<\mathbf{a}>$ = 4 and N$<\mathbf{c}+\mathbf{a}>$ = 5. Once the Burgers vector types are determined the value of $<C_{hk.0} b^2>$ and the dislocations density r can also be calculated, for further details see [15]. The experimental values of q_1 and q_2 denoted as can be estimated by the whole profile fitting procedure (details see [17]).

4 Results and Discussion

The Williamson–Hall plots in terms of integral breadth b vs. magnitude of the reciprocal lattice vector **G** (**G** represents $\sin\vartheta$) were are introduced in Figs. 1–4 for samples deformed at various temperatures. Squares correspond to the values calculated from the experimental data [17]. Circles correspond to the values calculated on the basis of the model conception [15]. The fitting procedure allow to estimate density of basal <**a**> dislocations, density of <**c+a**> dislocations and the grain size depending on deformation temperature. Results are introduced in Table 1.

From Table 1 it follows that the total dislocation density decreases with increasing deformation temperature, while the fraction of <**c+a**> dislocations is constant. Grain size growth has been also observed with increasing deformation temperature. Máthis et. al . [18] used the similar method to study the evolution of the dislocation density and the Burgers vector types with the temperature of deformation on the coarse grained Mg.

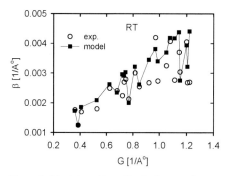

Figure 1: The integral breadths for the sample deformed at room temperature

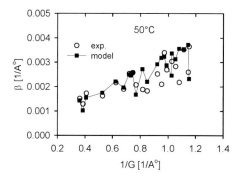

Figure 2: The integral breadths for the sample deformed at 50 °C

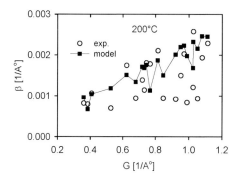

Figure 3: The integral breadths for the sample deformed at 200 °C

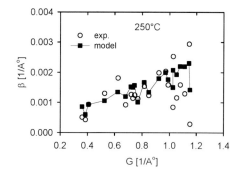

Figure 4: The integral breadths for the sample deformed at 250 °C

Authors of [18] found that at room temperature and at 100 °C the dominant dislocation type is <**a**> or mainly basal. At higher temperatures the fraction of <**a**> decreases, whereas the fraction of <**c+a**> increases. The fraction of <**c**> dislocations remains practically unchanged. The average total dislocation density increased considerably upon deformation as compared to the value in the as-cast state.

Table 1: Basal dislocation density $\rho_{<a>}$, density of **<c+a>** dislocations $\rho_{<c+a>}$ and grain size estimated for various deformation temperatures T_D

Deformation temperature [°C]	23	50	200	250
Density of basal dislocations $\rho_{<a>}$ 10^{14} [m^{-2}]	4.04	3.0	1.5	1.2
$\rho_{<c+a>}/\rho_{<a>}$	0.1	0.1	0.1	0.1
Grain size [nm]	150	200	350	400

However, with increasing deformation temperature the increment of dislocation density decreased strongly in accordance with dynamic recovery. The concomitant increase of the fraction of **<c+a>** dislocations and the decrease of the average dislocation density with deformation temperature was explained by the dynamic recovery mechanism/-s. Results obtained in our study for UFG-Mg showed similar decrease of the basal dislocation density. The fraction of the **<c+a>** dislocations remains in the UFG-Mg constant while in the coarse grained material it increases. It is very probably caused by the different deformation mechanism in the UFG-material.

Deformation experiments showed that the stress-strain curves at temperatures higher than 150 °C exhibit a steady state character [19]. Hardening and softening processes are in a dynamic equilibrium. A steady-state density of dislocations is determined by a dynamic balance between dislocation generation during plastic deformation and annihilation due to recovery process. Acoustic emission measurements detected no signal from the deformed sample at room temperature. It indicates that no dislocation pile-ups have been formed. In order to produce a pile-up of dislocations the grains need to contain Frank-Read source operating in a cyclic way. Dislocations generated from very close spaced sources (grain boundaries) are close to sinks (such as grain boundaries). Therefore, the dislocation annihilation rates are high. A high importance of the grain boundaries in the deformation mechanism/-s was also showed by measurements of the internal friction spectrum [20]. Decreasing basal dislocation density and density of the **<c+a>** dislocations with increasing test temperature indicate significant activity of the grain boundary sliding.

5   Conclusions

The basal dislocation density and density of the **<c+a>** dislocations have been determined in in UFG-Mg deformed at different temperatures between room temperature and 250 °C. High resolution X-ray diffraction peak profile measurements have been carried out. A numerical procedure has been developed for evaluating experimental values of the dislocation contrast factors in terms of the different, well established dislocation types, **<a>** and **<c+a>**. It is found that the rate of **<c+a>** and **<a>** dislocation remains constant to be of 0.1, in the deformation temperature range studied (room temperature –250 °C). Density of both types of dislocations decreases with the increasing deformation temperature. It is very probably due to increasing intensity of the grain boundary sliding. Results obtained for UFG-Mg are different from ones estimated for coarse grained Mg because of different deformation mechanism(-s). The grain size determined from the model increases with increasing deformation temperature.

6 Acknowledgements

This work received a support from the Research Project 1M 2560471601 „Eco-centre for Applied Research of Non-ferrous Metals" that is financed by the Ministry of Education, Youth and Sports of the Czech Republic.

7 References

[1] E.O. Hall, Proc. Phys. Soc. B64 (1951) 747.
[2] N.J. Petch, Iron Steel Inst. 174 (1953) 25.
[3] M. Mabuchi, K. Ameyama, H. Iwasaki, K. Higashi, Acta Mater. 46 (1999) 2024.
[4] L. Lu, L. Froyen, Scr. Mater. 40 (1999) 1117.
[5] L. Lu, M.O. Lai, Y.H. Toh, L. Froyen, Mater. Sci. Eng. A 334A (2001) 163.
[6] J.M. Wu, Z.Z. Li, J. Alloys Compd. 2999 (2000) 9.
[7] S. Hwang, C. Nishimura, P.G. McCormick, Scr. Mater. 44 (2001) 2457.
[8] G.W. Groves, A. Kelly, Phil. Mag. 8 (1963) 877.
[9] U.F. Kocks, D.G. Westlake, Trans. AIME 239 (1967) 1107.
[10] M.H. Yoo, Metall. Trans. A 12 (1981) 409.
[11] M.H. Yoo, S.R. Agnew, J.R. Morris, K.M. Ho, Mater. Sci. Engn. A319–321 (2001) 87–92.
[12] S. Hwang and C. Nishimura, P.G. McCormick: Scripta mater. 44 (2001) 1507–1511.
[13] M. Wilkens, Phys Stat Sol (a) 1970; 2: 359.
[14] T. Ungar, A. Borbely, Appl Phys Lett 1996; 69: 3173.
[15] I.C. Dragomir and T.J. Ungár, J. Appl. Cryst. 35 (2002) 556–564.
[16] R. Kuzel, P. Klimanek, J Appl Cryst 1989; 22: 299.
[17] R. Kuel, DIFPATAN—Program for Powder Pattern Analysis. World wide web:
[18] http://www.xray.cz/ecm-cd/soft/xray/indexdifp.html
[19] K. Máthis, K. Nyilas, A. Axt, I. Dragomir-Cernatescu, T. Ungár, P. Lukáč, Acta Mater. 52 (2004) 2889–2894.
[20] Z. Trojanová, P. Lukáč, Z. Száraz: Rev. Adv. Mater. Sci. 10, 2005, 34–40.
[21] Z. Trojanova, J. Alloys Compd., in press

Microstructure and Thermal Stability of Ultra Fine Grained Mg-4Tb-2Nd Alloy Prepared by High Pressure Torsion

J. Cizek, I. Prochazka, B. Smola, I. Stulikova, M. Vlach
Faculty of Mathematics and Physics, Charles University, Prague, Czech Republic
R. K. Islamgaliev, O. Kulyasova
Institute of Physics of Advanced Materials, Ufa State Aviation Technical University, Ufa, Russia

1   Introduction

Low density Mg-based alloys allow for a significant weight reduction which rises the effectiveness in a broad range of industrial applications. Unfortunately, the use of most of Mg alloys is limited to low temperature applications due to degradation of their mechanical properties at temperatures above 200 °C. There are several approaches how to overcome this problem. The particularly promising way is the use of non-traditional rare earth alloying elements [1]. The Mg-Tb-Nd ternary alloy represents a novel age hardenable Mg-based alloy with enhanced strength and favorable creep properties even at elevated temperatures [2]. The supersaturated solid solution of Tb in Mg decomposes in the following sequence of consequently precipitating phases: $\alpha'(hcp) \rightarrow \beta''(D0_{19}) \rightarrow \beta_1(fcc) \rightarrow \beta(cubic)$ [2]. Despite the favorable strength and thermal stability, a disadvantage of this alloy consists in a poor ductility insufficient for most of the potential industrial applications. Grain refinement is a well-known method how to improve ductility of metallic materials. An extreme grain size reduction is achieved by severe plastic deformation (SPD), see [3] for review. The methods based on SPD produce bulk materials with ultra fine grain (UFG) structure. The largest grain refinement is achieved by high pressure torsion (HPT) [3]. The HPT technique has been already successfully employed for preparation of UFG Mg10%Gd alloy with grain size ≈ 100 nm [4]. Small grain size, which lies in the nanocrystalline range, leads to a significant volume fraction of grain boundaries which represent obstacles for movement of dislocations. It causes a significant hardening of UFG metals in addition to the age hardening effect caused by fine precipitates. As a consequence, the UFG metals exhibit a favorable combination of very high strength and a reasonable ductility.

The aim of this work is microstructure characterization of Mg-Tb-Nd alloy prepared by HPT and its comparison with the microstructure of corresponding coarse grained material. Subsequently, we compared the precipitation sequence in the UFG sample and the coarse grained alloy. A typical feature of UFG structure is very high number of lattice defects introduced by SPD. Obviously, defects introduced by SPD play crucial role in the UFG structure. Detailed characterization of these defects represents, therefore, an important task in microstructure investigations of the UFG materials. For this reason we employed positron lifetime (PL) spectroscopy in the present work. PL spectroscopy represents a well established non-destructive technique with very high sensitivity to open-volume defects like vacancies, dislocations, etc. [5]. Thus, PL spectroscopy is an ideal tool for defects studies of UFG materials. PL spectroscopy was combined with transmission electron microscopy (TEM), electrical resisitivity and microhardness investigations.

2 Experimental

Specimens of technically pure Mg (99.9 %) and Mg-4%Tb-2%Nd (Mg4Tb2Nd) alloy were investigated. The Mg4Tb2Nd alloy was prepared by squeeze casting using the technically pure Mg. The as-cast material was subjected to a solution annealing at 525 °C for 6 hours. This treatment is sufficient to dissolve the alloying elements completely [6]. The solution annealing is finished by quenching into water of room temperature. To fabricate the UFG structure, the as-received Mg and the solution treated Mg4Tb2Nd alloy were deformed by HPT at room temperature using 5 rotations under a high pressure of 6 GPa. After detailed characterization of the as deformed microstructure, the specimens were subjected to step-by-step isochronal annealing (20°C/20 min). Each annealing step was finished by quenching into water of room temperature and subsequent investigations performed at room temperature.

A fast-fast PL spectrometer similar to that described in [7] with time resolution of 160 ps was used in this work. The TEM observations were carried out on a JEOL 2000 FX electron microscope operating at 200 kV. The Vickers microhardness, HV, was measured at a load of 100 g applied for 10 s using a LECO M-400-A hardness tester. Electrical resistivity was measured at 77 K by means of the dc four-point method with a dummy specimen in series. Relative electrical resistivity changes $\Delta\rho/\rho$ were obtained to within an accuracy of 10^{-4}. The effect of a parasitic thermo-electromotive force was suppressed by a change in polarity.

3 Results and Discussion

The positron lifetimes τ_i and the relative intensities I_i of the components resolved in the PL spectra of studied specimens are listed in Table 1. The well annealed Mg reference specimen exhibits a single component PL spectrum with lifetime $\tau_B = 225$ ps which agrees well with the calculated Mg bulk lifetime [8]. Thus, defect density in the reference Mg specimen is negligible and virtually all positrons annihilate from the free state. The as cast Mg4Tb2Nd alloy exhibits two component PL spectrum. The shorter component with lifetime τ_1 comes from free positrons, while the component with a longer lifetime τ_2 represents a contribution of positrons trapped at defects. The lifetime τ_2 agrees well with the lifetime of positrons trapped at dislocations in Mg [8]. Thus, it can be concluded that the as-cast alloy contains dislocations introduced in the course of casting and shaping of the specimens. Dislocation density ρ_D in the specimen can be calculated using the two state trapping model [5]

$$\rho_D = \frac{1}{\nu_D} \frac{I_2}{I_1} \left(\frac{1}{\tau_B} - \frac{1}{\tau_2} \right), \qquad (1)$$

where ν_D stands for the positron trapping coefficient for dislocations in Mg. In the present work we use $\nu_D = 1 \cdot 10^{-4}$ m^2 s^{-1} [9]. The Eq. (1) then yields $\rho_D = 8 \cdot 10^{12}$ m^{-2} for the as cast Mg4Tb2Nd.

The solution treated alloy exhibits coarse grains with mean diameter of over 500 μm. One can see from Table 1 that the most of positrons in the solution treated Mg4Tb2Nd annihilate from the free state and contribute to the shorter component. However, there is also a weak component with lifetime $\tau_2 \approx 280$ ps, which is close to the lifetime of positrons trapped at Mg vacancy [8]. It indicates that a small fraction of positrons is trapped in quenched-in vacancies. The free vacancies in Mg are not stable at room temperature, therefore, the observed defects are vacancies bound to Tb or Nd atoms. Similar type of defects, i.e. the quenched-in vacancies bound to Gd atoms, were found also in the solution treated Mg-Gd alloy [8]. No contribution of positrons trapped at dislocations was found in the PL spectrum of the solution treated Mg4Tb2Nd. It testifies that dislocations were annealed out under detectable limit (below 10^{-12} m^{-2}). This conclusion is supported also by TEM observations. It should be also mentioned that no precipitates were observed by TEM in the solution treated alloy.

Table 1: Positron lifetimes τ_i and relative intensities I_i of the components resolved in the PL spectra

Sample	τ_1 (ps)	I_1 (%)	τ_2 (ps)	I_2 (%)
well annealed Mg (280°C/30min)	224.8 ± 0.5	100	–	–
Mg4Tb2Nd – as cast	193 ± 2	44 ± 2	255 ± 2	66 ± 1
Mg4Tb2Nd – solution treated	220 ± 1	91 ± 1	280 ± 20	9 ± 1
Mg4Tb2Nd – HPT deformed	180 ± 2	14.9 ± 0.4	256 ± 2	85.1 ± 0.4

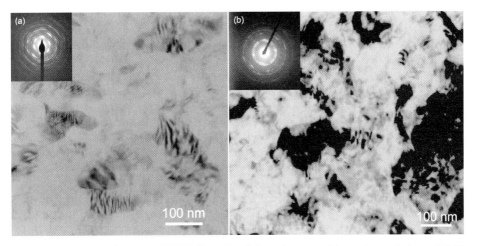

Figure 1: TEM images of UFG Mg4Tb2Nd alloy: (a) as deformed structure, (b) sample annealed up to 140 °C

A representative TEM image of HPT-deformed Mg4Tb2Nd alloy is shown in Fig. 1a. The specimen shows a homogeneous UFG structure with grain size around 100 nm. The dominant component in the PL spectrum exhibits lifetime $\tau_2 \approx 256$ ps which corresponds to positrons trapped at dislocations in Mg. Thus, majority of positrons are trapped at dislocations introduced by SPD. The dislocation density $\rho_D = 3 \cdot 10^{13}$ m^{-2} was calculated using Eq. (1). A high density of dislocations can be see also on the TEM image. It is too high to resolve individual dislocation lines. The dislocations are homogeneously distributed throughout whole grains. The electron diffraction pattern testifies long angle miss-orientation of neighboring grains. The specimen exhibits (00.1) type texture. No precipitates were found in the as-deformed alloy by TEM. The strong grain refinement and a high number of dislocations leads to a substantial hardening which can be seen in a rise of microhardness: the HPT-deformed alloy exhibits about 140 % higher microhardness compared to the solution treated specimen. Temperature dependence of the intensity of trapped positrons in the solution treated and the HPT-deformed alloy subjected to isochronal annealing are shown in Figs. 2a and 2b, respectively. Temperature dependence of microhardness HV for both the specimens is shown in Figs. 3a,b.

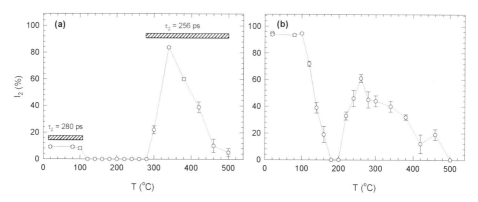

Figure 2: Temperature dependence of the intensity I_2 of positrons trapped at defects: (a) solution treated Mg4Tb2Nd alloy, (b) HPT deformed Mg4Tb2Nd alloy

Let us start to discuss the precipitation effects in the solution treated alloy. The quenched-in vacancies bound to the alloying elements are annealed out by annealing at 120 °C. It is seen in Fig. 2a as disappearance of the long lived component with lifetime $\tau_2 = 280$ ps and intensity I_2. The sample exhibits a single component PL spectrum in the temperature range (120–280) °C, i.e. there are no active positron traps in this temperature interval. From Fig. 3a, it becomes clear that the precipitation of the β" phase particles starts around 80 °C and causes a remarkable hardening. The β" phase particles are fully coherent with the Mg lattice and, thereby, no open volume defects are introduced by precipitation of the β" phase. The positron affinity of the β" phase is not known so far, but our results show clearly that it is most probably not favorable for positrons. As a consequence, there is no positron trapping in the β" precipitates and PL spectroscopy is not sensitive to the precipitation of the β" phase. TEM investigations revealed out that fine spherical β" phase precipitates transform into fine plates in the temperature interval (180–240) °C. One can see in Fig. 3b that it has a strong hardening effect with the peak hardness at 210 °C. Fine β"phase plates precipitates in a triangular configuration parallel with the prismatic planes {11.0}. Further annealing up to 270 °C leads to growth of the plate shaped precipitates (diameter 20–30 nm) reflected by a decrease of microhardness and resistivity. At higher temperatures the β" phase is transformed into $β_1$ phase with fcc structure. Plates (200–500 nm in diameter) of the $β_1$ phase were observed by TEM in the alloy annealed up to 330 °C. The formation of the $β_1$ phase is accompanied by appearance of a defect component with lifetime $\tau_2 = 256$ ps in PL spectra. Intensity of this component steeply increases with temperature up to a maximum at 340 °C, see Fig. 2a. Thus, new positron traps are created by formation of the $β_1$ phase precipitates. Positrons are most probably trapped at misfit defect at the precipitate-matrix interfaces. However positron trapping inside the precipitates can not be excluded as well. An increase of resistivity and a decrease of I_2 above 340 °C is due to dissolution of the $β_1$ phase particles. This behavior is postponed in the temperature range (390–450) °C by formation of the β phase (plate-shaped particles 2–3 μm in diameter). The precipitation of the β phase causes a slight hardening seen in Fig. 3a. Above 450 °C the β phase precipitates dissolve and the solid solution is restored.

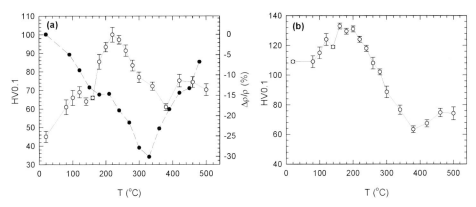

Figure 3: (a) solution treated Mg4Tb2Nd alloy – temperature dependence of microhardness (open circles) and the relative change of electrical resistivity (full circles); (b) HPT deformed Mg4Tb2Nd alloy – temperature dependence of microhardness

The microstructure development of HPT-deformed Mg4Tb2Nd includes not only the precipitation effects, but also the recovery of the defects introduced by SPD. The intensity I_2 of positrons trapped at dislocations exhibits an abrupt decrease in the temperature range (100–180) °C, see Fig. 2b. It gives a clear evidence for a recovery of dislocations which takes place in this temperature interval and was confirmed also by TEM (Fig. 1b). A single component PL spectrum above 180 °C indicates that the dislocation density dropped well below ≈ $10^{12} m^{-2}$. The precipitation of the β" phase, which takes place at similar temperatures as in the coarse grained alloy, causes a remarkable hardening, see Fig. 3b. As has been already explained, PL spectroscopy is insensitive to the precipitation of the coherent β" phase. One can see in Fig. 2b that the intensity I_2 starts to increase again in the sample annealed up to 220 °C and exhibits maximum at 260 °C. The recovery of dislocations was completed already at 180 °C. Hence, this increase of I_2 is not connected with dislocations, but occurs due to positron trapping at defects introduced by precipitation of the $β_1$ phase particles. The lifetime of the defect component which appeared above 220 °C lies again around 256 ps. It supports the picture that this component comes from positrons trapped at the misfit defects at the $β_1$ phase incoherent interfaces. Precipitation of the $β_1$ phase is reflected also by an increase of microhardness. After annealing above 260 °C, the behavior of I_2 is reversed and it gradually decreases in similar manner as in the coarse grained alloy. The difference between the HPT-deformed and the coarse grained alloy consists in the fact that precipitation of the $β_1$ phase starts not at about 270 °C but already at 220 °C. As a consequence, the maximum of I_2 and the peak hardness in the HPT-deformed alloy are shifted to about of 80°C lower temperatures compared to the coarse grained alloy. Thus, the precipitation of the $β_1$ phase and most probably also the shape transformation of the β" phase start at significantly lower temperature in the HPT-deformed alloy. It has two reasons: (i) The extremely small grain size leads to a significant volume fraction of grain boundaries. The defects in grain boundaries serve as centers for nucleation of the second phase particles. (ii) Diffusivity of the Tb and Nd atoms is enhanced by a possibility to diffuse along grain boundaries. Both the factors facilitate the precipitation effects in the HPT-deformed alloy and shift the precipitation of the $β_1$ phase to lower temperatures.

4 Conclusions

The capability of HPT to achieve an extreme grain size refinement of was demonstrated in the present work on Mg4Tb2Nd alloy. The HPT-deformed alloy exhibits a grain size around 100 nm and a high density of homogeneously distributed dislocations. The UFG structure leads to a significant rise of hardness of the HPT-deformed alloy. Temperature development of microstructure of the HPT-deformed alloy was studied and compared with the coarse grained sample. Full recovery of dislocations in the HPT-deformed alloy takes place in relatively narrow temperature interval (100–180) °C. The precipitation sequence in the alloy with UFG structure differs from that in the coarse grained alloy. Namely the precipitation of the β_1 phase starts and shape transformation of the β'' phase take place at remarkably lower temperatures.

5 Acknowledgements

This work was financially supported by the Czech Science Foundation (contract 106/05/0073), The Ministry of Education, Youth and Sports of Czech Republics (project MS 0021620834), and INTAS.

6 References

[1] Mordike, B.L., Mat. Sci. Eng. A **2002**, *324*, 103–112.
[2] Neubert, V., Stulikova I., Smola, B., Mordike B.L., Vlach, M., Bakkar, A., Pelcova J., in: Proc. of ISPMA 10, Prague, September 2005, to be published in Mater.Sci. Eng. A.
[3] Valiev, R.Z., Islamgaliev, R.K., Alexandrov, I.V., Prog. Mat. Sci. **2000**, *45*, 103.
[4] Cizek, J., Prochazka, I., Stulikova, I., Smola, B., Kuzel, R., Cherkaska, V., Islamgaliev, R.K., Kulyasova, O., in: *Magnesium, Proceedings of the 6th International Conference Magnesium Alloys and Their Applications*, K.U. Kainer (Ed.), Wiley-VCH Weinheim **2003**, p. 202–207.
[5] Hautojärvi, P., Corbel, C., in: *Proceedings of the International School of Physics „Enrico Fermi"*, A. Dupasquier, A.P. Mills (Eds.), Course CXXV, IOS Press, Varena, **1995**, p. 491–562.
[6] Stulikova, I., Smola, B., Zaludova, N., Vlach, M., Pelcova, J., Kovove Mater. **2005**, *43*, 272–279.
[7] Becvar, F., Cizek, J., Lestak, L., Novotny, I., Prochazka, I., Sebesta, F., Nucl. Instr. Meth. A **2000**, *443*, 557–577.
[8] Cizek, J., Prochazka, I., Becvar, F., Stulikova, I., Smola, B., Kužel, R., Cherkaska, V., Islamgaliev, R.K, Kulyasova, O., phys. stat. sol. (a) **2006**, *203*, 466–477.
[9] Abdelrahman, M., Badawi, P., Jpn. J. Appl. Phys. **1996**, *35*, 4728–4729.

Amorphous Precipitates Sandwiched by Long Period Stacking Structures in the Melt-quenched $Mg_{98}Cu_1R_1$ (R = Y and Gd)

K. Konno, M. Matsuura, M. Yoshida
Miyagi National College of Technology, Natori, Miyagi, Japan
M. Nishijima and K. Hiraga
Institute for Materials Research, Tohoku University, Katahira, Aoba-ku, Sendai, Japan

1   Introduction

The Mg-Cu-Y ternary alloy system is known to exhibit a stable super-cooled liquid state, i.e. $Mg_{65}Cu_{25}Y_{10}$ bulk metallic glass [1,2]. On the other hand, the Mg-Zn-Y alloys system with high mechanical properties have a long period stacking (LPS) structure, i.e. rapidly solidified powder microstructure (RS P/M) $Mg_{97}Zn_1Y_2$ alloy [3,4]. Matsuura et al. reported that a melt-quenched $Mg_{98}Cu_1Y_1$ alloy have the peculiar morphology in Mg-matrix grains [5]. The morphology is aligned parallel to the c-plane of the Mg-matrix crystal, and have the uniformly distribution consisting of a disk-shaped amorphous core sandwiched between 14H-typed LPS structure. Furthermore, the finding of LPS structure in the Mg-Cu-Y system is the first time with the exception of the Mg-Zn-Y system.

The investigations of Matsuura et al. for the microstructure and the morphology development process of the melt-quenched $Mg_{98}Cu_1Y_1$ alloy interest in the development condition, the thermal stability and the mechanical properties of the peculiar morphology. In the present work, we will report the microstructure and the mechanical properties for the changes cooling rate of melt-quenched $Mg_{98}Cu_1Y_1$ alloy, and for the $Mg_{98}Cu_1Y_1$ and $Mg_{98}Cu_1Gd_1$ alloys after thermal annealing.

2   Experimental Procedures

$Mg_{98}Cu_1Y_1$ and $Mg_{98}Cu_1Gd_1$ were prepared by a single-roll melt-spinning technique. The purity of the elements was 99.95 % for Mg, 99.99 % for Cu, 99.5 % for Y and Gd. The elements were melted together by induction heating in a graphite-crucible under the 0.09MPa Ar gas, and then injected through a nozzle with the diameter of 0.5mm and quenched to a rotating Cu roll. In order to know effect of quenching rate from the melt on hte structures and mechanical properties, surface velocity of rotating Cu roll was adjusted from 10 to 51m/s for $Mg_{98}Cu_1Y_1$ while fixed velocity of 41m/s for $Mg_{98}Cu_1Gd_1$. The melt quenched samples were annealed at 523K for 30min in the quarts tube under the 0.1MPa Ar gas, and then rapidly cooled down to the room temperature from 523K in the water. The results of X-ray diffraction using Cu Kα radiation show that very weak peaks of Mg_2Cu are observed for the $Mg_{98}Cu_1Y_1$ alloys, but not for the $Mg_{98}Cu_1Gd_1$ alloy. The peaks of the Mg_2Cu alloy show a decreasing tendency as cooling rate increases. The microstructures of ribbon samples were observed using scanning electron microscopy (SEM), transmission electron microscopy (TEM) with JEM-2000EX operating at 200kV and high resolution transmission electron microscopy (HRTEM) with JEM-4000EX operating at 400kV having a resolution of 0.17nm. The ribbon samples were thinned by grinding and ion milling for TEM observations. HAADF-STEM images were taken by a 300 kV electron microscope (JEM-3000F) equipped with a field emission gun in the scanning trasmission electron microscope mode. In HAADF-STEM observations, beam probe with a half width of about 0.2 nm was scanned on samples. Energy dispersive X-ray spectroscopy (EDS) was performed with the JEM-3000F microscope operated at 300 kV. Tensile tests for the melt quenched ribbon samples were carried out using a tensile tests machine of the maximum load 5.8 N at room temperature with an initial strain rate of $2 \cdot 10^{-5}$ s^{-1}.

3 Experimental Results

Figure 1 (a) shows HAADF-STEM image of $Mg_{98}Cu_1Y_1$ alloy melt-quenched with the surface velocity of 41m/s. Bright areas in Fig. 1 (a) indicate Cu- and/or Y-rich regions. In this figure of the $Mg_{98}Cu_1Y_1$ alloy many fine precipitates are dispersed with aligned in same direction which is parallel to the c-plane of the matrix grain. Structure details of the precipitates inside grains in the $Mg_{98}Cu_1Y_1$ alloy have been studied by HRTEM and results are shown in Figs. 1 (b), (c) and (d). Figure 1 (b) shows that the precipitates have the peculiar morphology; it has a disk-like shape core sandwithed between two crystalline phase with lattice fringes. The schematic drawing of this precipitates is shown in this figure. The disk planes of all precipitates are aligned along the horizontal direction, which is parallel to the c-plane of the hcp Mg crystal. The fine structure of the precipitate can be confirmed in Fig. 1 (c). Figure 1 (d) is a enlarged image of the A region in Fig. 1 (c), which shows that the core part of the precipitate is an amorphous structure showing a typical granular-like pattern. The image and electron diffraction pattern of Fig. 2 indicate that the B area in Fig. 1 (c) is a 14H-typed LPS structure with an ABACBCBCB-CABAB stacking sequence of close-packed planes. Above results indicate that the core part of a disk-like shape consists of amorphous phase and 14H-typed LPS crystals sandwich it. We called it „LAL-particle" because it consists of LPS/Amorphous/LPS. Compositions of the amorphous and LPS areas were evaluated by EDS as 67 at% Mg, 28 at% Cu and 5 at% Y ($Mg_{67}Cu_{28}Y_5$) and 80 at% Mg, 15 at% Cu and 5 at% Y ($Mg_{80}Cu_{15}Y_5$), respectively. The reported composition of the 18R-typed LPS in an $Mg_{97}Zn_1Y_2$ alloy is $Mg_{87}Zn_7Y_6$ [6], i.e. the composition of Zn of the 18R-typed LPS is about half of Cu for the present $Mg_{98}Cu_1Y_1$ alloy.

Figure 3 shows a TEM image of the precipitates inside the grain region in the melt-quenched $Mg_{98}Cu_1Gd_1$ alloy with the surface velocity of 41m/s. The shape of the precipitates for $Mg_{98}Cu_1Gd_1$ is similar to that for $Mg_{98}Cu_1Y_1$, which is parallel to the c-plane of the hcp Mg crystal. The electron diffraction shows that the LPS phase have a 14H-typed structure. EDS results for $Mg_{98}Cu_1Gd_1$ alloy show that the amorphous and LPS region has compositions of 87.4 at% Mg, 9.1 at% Cu, 3.5 at% Gd ($Mg_{87.4}Cu_{9.1}Gd_{3.5}$) and 91.5 at% Mg, 5.4 at% Cu, 3.1 at%Gd ($Mg_{91.5}Cu_{5.4}Gd_{3.1}$), respectively. The reported compositions of the 14H-LPS structures in the warm-extruded $Mg_{96.5}Zn_1Gd_{2.5}$ alloy estimated by EDS spectra are 81 at%Mg, 11 at% Zn, 8 at% Gd ($Mg_{81}Zn_{11}Gd_8$) [7]. The composition of Zn and Gd of the 14H-typed LPS for the warm-extruded $Mg_{98.5}Zn_1Gd_{2.5}$ are about twice of Cu and Gd for the present $Mg_{98}Cu_1Gd_1$ alloy. Additionally, the composition of the amorphous core is far from that for the reported bulk $Mg_{65}Cu_{25}Gd_{10}$ glassy alloy [8].

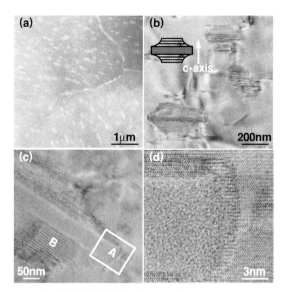

Figure 1: (a) HAADF-STEM images of melt-quenched $Mg_{98}Cu_1Y_1$ alloy. (b) HRTEM image of precipitates in a Mg-matrix grain in the $Mg_{98}Cu_1Y_1$ alloy. The c-axis of the Mg-matrix grain is indicated. A schematic drawing showing the morphology of the precipitate is inserted. (c) Enlarged HRTEM image of a precipitate. (d) Enlarged HRTEM image of the part A in (c), showing an amorphous structure of the disk-shaped core.

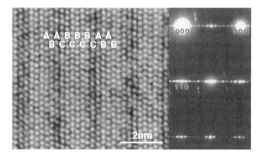

Figure 2: Enlargd HRTEM images of the part B in Fig. 1 (c) and electron diffraction pattern taken from the part B, showing a 14-typed LPS structure with an ABACBCBCBCABAB stacking sequence

Effect of cooling rate from melt on the fromation of the LAL-particles was studied by changing the surface velocity of a Cu roll for the melt-quenched $Mg_{98}Cu_1Y_1$ alloys. Figure 4 shows TEM images of the surface velocity of 10m/s and 51m/s for $Mg_{98}Cu_1Y_1$ alloys. The LAL particles were also found in other surface velocity between them, and configuration of the LAL perticles does not depend on the cooling rate. The TEM results indicate that LAL perticles for the $Mg_{98}Cu_1Y_1$ alloy can be formed in the wide cooling rate of 10~51m/s in surface velocity. DSC observation indicates that the amorphous core of the LAL-particle crystallizes at around 475K. Thermal stability of the LAL-particles is studied by annealing $Mg_{98}Cu_1Y_1$ and $Mg_{98}Cu_1Gd_1$ ribbons at 523K for 30min. Figures. 5 and 6 show the TEM images for the melt-quenched (a) and annealed (b) alloys of the $Mg_{98}Cu_1Y_1$ and $Mg_{98}Cu_1Gd_1$, respectively. The TEM results showed that the amorphous core region disappears and consequently LPS region expands.

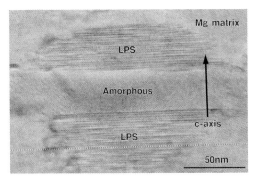

Figure 3: TEM image of LAL particle in Mg-matrix grain in the $Mg_{98}Cu_1Gd_1$ alloys for the surface velocity of 41 m/s. The c-axis of the Mg-matrix grain is indicated

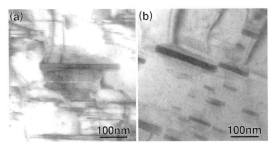

Figure 4: TEM images of the $Mg_{98}Cu_1Y_1$ alloys for the surface velocity of 10 m/s (a) and 51m/s (b)

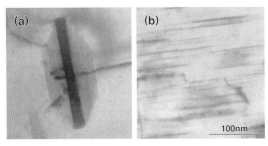

Figure 5: TEM image of the melt-quenched alloy (a) and the annealed alloy at 523 K for 30 min (b) of $Mg_{98}Cu_1Y_1$ for surface velocity of 41 m/s

The changes of the grain size and the 0.2 % proof stress ($\sigma_{0.2}$) with cooling rate from the melt for both as-quenched and annealed ones for $Mg_{98}Cu_1Y_1$ alloys are shown in Fig. 7. The averaged grain size of the as-quenched and annealed alloy gradually decrease with increasing the cooling rate. While, the $\sigma_{0.2}$ increases with the increase in cooling rate. The grain size increases by annealing and $\sigma_{0.2}$ also increases after annealing. The same tendency is seen for the $Mg_{98}Cu_1Gd_1$ alloy for surface velocity of 41m/s. The $\sigma_{0.2}$ for the annealed $Mg_{98}Cu_1Gd_1$ alloy was 147MPa that it is slightly larger than annealed $Mg_{98}Cu_1Y_1$ alloy.

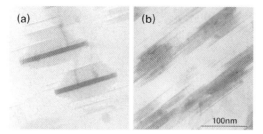

Figure 6: TEM image of the melt-quenched alloy (a) and the annealed alloy at 523 K for 30 min (b) of $Mg_{98}Cu_1Gd_1$ for surface velocity of 41 m/s

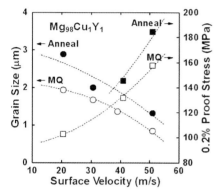

Figure 7: Changes in the averaged grain size and the 0.2 % proof stress of the melt-quenched (MQ) and the annealed $Mg_{98}Cu_1Y_1$ alloys with the surface velocity of Cu roll

4 Discussion

The LAL-particles found in the present work show following structural features; 1) the LAL-particles are aligned along the c-plane of the hcp Mg-crystal, 2) the amorphous core with a disk-like shape are sandwithed by two LPS crystals, and 3) the boundary between the amorphous and LPS region have a sharp edge. The solubility limit of Cu in the pure Mg is very low (maximum value 0.013 at%), while that of Y is about 3 at% at 840 K and decreases with decreasing temperature. Therefore, it is thought that the concentrated Cu and Y liquids segregate at inside Mg grains and the grain boundaries as Cu- and Y-enriched amorphous phases. The amorphous core of the LAL-particles have composition of $Mg_{67}Cu_{28}Y_5$, which is close to the $Mg_{65}Cu_{25}Y_{10}$ of the bulk metallic glass. The LPS region adjacent to the amorphous phase is probably formed by the transfer of Cu atoms ejected from the primary Cu-rich amorphous phase, consequently the LPS has concentration of $Mg_{80}Cu_{15}Y_5$. Therefore Cu concentration in amorphous phase is enriched more and amorphous phase is stabilized and left as the core of the precipitates. Additonally, strong bonding between Cu and Y is also responsible for the precipitation of the LAL-particles. Since the single addition of Cu into Mg produces spherical precipitants of Mg_2Cu with no LAL-particles [5]. The EDS spectra for the amorphous and LPS regions for the $Mg_{98}Cu_1Gd_1$ alloy showed compositions of $Mg_{87.4}Cu_{9.1}Gd_{3.5}$ and $Mg_{91.5}Cu_{5.4}Gd_{3.1}$, respectively. At the present study, honener, we cannot understand the reason why the compositions of the amorphous and LPS regions are far from the stable phases.

The LAL-particle of the melt-quenched $Mg_{98}Cu_1Y_1$ alloys was proved to be formed in the wide cooling rate with the surface velocity of Cu roll at 10~51 m/s. Therefore the increase of the 0.2 % proof stress with increasing cooling rate can be attributed to the grain refining rather than the LAL-particles. By the annealing at 523 K for 30 min, the amorphous region in the LAL-particles of the $Mg_{98}Cu_1Y_1$ and $Mg_{98}Cu_1Gd_1$ alloy disappear and consequently LPS phase expand. The concentrated Cu, Y and Gd in the amorphous phase in the LAL-particle seem to transform into the LPS phase after the annealing. The annealing of alloys with the LAL-particle increase in the grain size nevertheless 0.2 % proof stress increases. This fact suggests that the LPS phase rather than the LAL-particles contributes to improves the mechanical properties.

5   Conclusion

We have studied the microstructure and the mechanical properties for the cooling rate and the heat treatment of the melt-quenched $Mg_{98}Cu_1Y_1$ and $Mg_{98}Cu_1Gd_1$ alloy. The LAL-perticles of the melt-quenched $Mg_{98}Cu_1Y_1$ alloy are developed in wide range of the cooling rate, i.e. the surface velocity of 10~51 m/s. The maicrostructure of LAL-perticles does not show any change for the cooling rate, but the influence of the cooling rate appear in the mechanical properties. The 0.2 % proof stress incresing as the cooling rate increases seems to be attributed to the effect of grain refining. The LAL-perticles with 14H-typed LPS structure is also formed in the melt-quenched $Mg_{98}Cu_1Gd_1$ alloy. The LAL-perticles of the $Mg_{98}Cu_1Y_1$ and $Mg_{98}Cu_1Gd_1$ alloys have an arrangement along the c-plane of the hcp Mg crystal and a homogeneously distribution inside the grain region. Compositions of the amorphous and LPS areas of $Mg_{98}Cu_1Gd_1$ alloy were evaluated by EDS as $Mg_{87.4}Cu_{9.1}Gd_{3.5}$ and $Mg_{91.5}Cu_{5.4}Gd_{3.1}$, respectively. But these results are much different from the $Mg_{98}Cu_1Y_1$ that the amorphous and LPS region show the composition of $Mg_{67}Cu_{28}Y_5$ and $Mg_{80}Cu_{15}Y_5$, respectively. By the annealing at 523 K for 30 min, the amorphous core region of the $Mg_{98}Cu_1Y_1$ and $Mg_{98}Cu_1Gd_1$ disappears and the LPS region expands. The annealing induces the increase of the 0.2 % proof stress but also increases the grain size. It is thought that the expanded LPS region by the annealing contributes to an improvement of the mechanical properties.

6   References

[1] Inoue, A.; Kato, A.; Zhang, T.; T, Kim S.G.; Masumoto, T., Materials Transactions JIM, **1991**, *32*, p. 609–616.
[2] Inoue, A.; Nakamura, K.; Nishiyama, N.; Masumoto, T., Materials Transactions JIM, **1992**, *33*, p. 937–945.
[3] Kawamura, Y.; Hayashi, A.; Inoue, A.; Masumoto, T., Materials Transactions, **2001**, *42*, p. 1172-1176.
[4] Luo, Z. P.; Zhang, S. Q., J. Mater. Sci. Lett., **2000**, *19*, p. 813–815.
[5] Matsuura, M.; Konno, K.; Yoshida, M.; Nishijima, M.; Hiraga, K., Materials Transactions, **2006**, *47*, p. 1264–1267.
[6] Itoi, T.; Semiya, T.;Kawamura, Y.; Inoue, A., Scripta Materialia, **2002**, *50*, p. 3845–3855.
[7] Yamasaki, M.; Anan, T.; Yashimoto, S.; Kawamura, Y., Scripta Materialia **2005**, *53*, p. 799–803.
[8] Men, H.; Kim, D.H., J. Mater. Res. **2003**, *13*, p. 1502.

Texture Development in Different Routes ECAP Processed Mg-Si Alloys by Neutron Diffraction

Weimin Gan[1, 2], Mingyi Zheng[1], Xiaoguang Qiao[1], Shiwei Xu[1], Xiaoshi Hu[1], Kun Wu[1], Heinz-Günter Brokmeier[2, 3], Bernd Schwebke[2, 3], Andreas Schreyer[2], Karl Ulrich Kainer[2]

[1]School of Materials Science and Engineering, Harbin Institute of Technology, P. R. China
[2]GKSS-Research Centre, Geesthacht, Germany
[3]Institute of Materials Science and Engineering, Clausthal University of Technology, Germany

1 Introduction

Magnesium alloys are attractive for light-weight usages because of their low density of about 1.73 g/cm^3. However, their applications are often restricted due to their inherent deficiencies like low ductility, low stiffness, poor high temperature strength and creep resistance, etc. Addition of elements as Si, Y to Mg can improve its strength and ductility, especially at elevated temperatures [1, 2]. But the coarse and inhomogeneous distribution of the formed second phases greatly decreases the mechanical properties of these alloys.

Due to the large amounts of strain that can be introduced into a metal or alloy billet, equal channel angular pressing (ECAP) is a promising severe plastic deformation (SPD) technique for producing fine-grained bulk materials including Mg alloys [3–5]. Researches have shown that ECAP was effective for refining various Mg alloys with improved strength, ductility, hardness, and superplasticity [3–6]. Besides the refinement of grains size of matrix alloys and second phases, complicated texture will develop during ECAP processing due to the accumulative strain and strain imhomogeneity achieved after multi-passes deformation, and it is recognized that ECAP can produce a unique deformation textures [5, 7]. Four rotation routes have been widely used to study the effect of ECAP processing on microstructure and properties of various materials [3, 8]: route A, the billet is not rotated; route B_C, the billet is rotated 90° clockwise or counter-clockwise; route B_A, the billet is rotated 90° clockwise and counter-clockwise alternatively; route C, the billet is rotated 180°. Processing route significantly affects the refinement of grain size and its shape and consequently has the effect on the mechanical and physical properties [7–9].

Due to the high transmission of neutron diffraction for most materials comparing to normal x-rays diffraction, it has been proved that neutron diffraction is a powerful non-destructive analysis method for texture of different types of bulk materials [10, 11]. The aim of the present work is to investigate the effect of different ECAP processing routes on the texture evolution of Mg-Si alloys; and to investigate the relationship between texture and deformation mechanism.

2 Experimental Procedures

The as-cast Mg-Si alloy with a composition of 3.3 wt% Si was obtained by normal die-casting method. The alloy was melted in an electrical furnace under a protection of dynamic SF_6 and CO_2 mixed gas atmosphere, then cast into an air-cooled metal mould. The as-extruded Mg-Si bar was obtained by hot rectangular extrusion at 350 °C with an extrusion ratio of 6:1.

A rectangular billet with a dimension of 12 × 12 × 60 mm³ was machined from the cast ingot and parallel to the extrusion direction in the as-extruded bar for ECAP processing, respectively. ECAP die with an angle of 90° was used. Mo_2S was used for lubrication. The die was preheated and stabled at the testing temperature for 30 min before the ECAP specimen was inserted into the channel. Then it was held for 30 min before pressing; and at 350 °C for as-cast alloys and 320 °C for as-extruded alloys separately. The pressing speed is 20 mm/min. Deformation route A, route Bc, and route C were performed for both kinds of materials from 1 pass to 8 passes.

Optical microscopy (OM) was used to observe the microstructures of the deformed materials. The polished surface was etched by a solution of 10 ml HNO_3, 30 ml acetic acid, 40 ml water, and 120 ml ethanol. Neutron diffraction at TEX-2(GKSS Research Center, Germany) was used to measure the texture [10]. A bulk sample with a gauge of 6 × 12 × 12 mm³ was cut from the center part of the ECAPed materials for texture measurement. Four pole figures (00.2), (10.0), (10.1), (11.0) were measured; and neutron beam diffraction direction is parallel to the extrusion direction, as shown in Figure 1, the definition of the coordinate system. Four measured pole figures were used to recalculate pole figures and orientation distribution function (ODF) using Bunge's system definition of Euler angles by series expansion method with an expansion degree of $l_{max} = 22$ [10].

3   Results and Discussion

OM microstructures of the initial as-cast and as-extruded Mg-Si alloys are shown in Figure 2. As can be found that coarse non-granular primary Mg_2Si was formed in gains, shown by arrow in Figure 2(a); and Chinese script eutectic Mg_2Si was regularly dispersed in the large matrix grains. While grain size was greatly decreased after hot extrusion, primary and eutectic Mg_2Si were greatly broken and distributed along extrusion direction (Figure 2(b)). Twins were also observed.

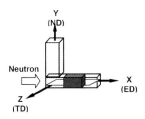
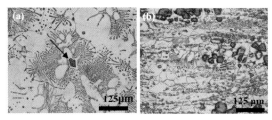

Figure 1: Definition of the coordinate system for texture measurement

Figure 2: OM initial Mg-Si microstructures (a) as- cast and (b) as-extruded

The recalculated (00.2) and (10.0) pole figures of the ECAPed cast Mg-Si alloys by three routes are shown in Figure 3. Grains in the initial cast alloy have a random distribution, and also the grain size is relatively large. Strong texture was formed just after 1 pass ECAP processing, as shown in Figure 3; the basal pole density distribution located with an angle about 20° to the extrusion and transition plane (Y plane in Figure 1), which indicating that basal planes in majority grains in samples were orientated at about 20° to the extrusion direction. With further 4 passes and 8 passes route A processing; the basal pole position has not changed a lot, while the pole intensity was increased from the as-cast of 5.29 m.r.d. (multiples of the random distribution) to 7.37 m.r.d and 8.38 m.r.d. for the 4P and 8P, respectively. Similar tendency of the results were obtained by other investigators [6, 7]. This is due to the successive distortions on consecutive passes [6, 7, 12], as shown in Figure 4. Route B_C processing is most attractive because it is recognized effective for grain refinement [2, 3, 13]. Textures developed after 4P and 8P by route B_C are shown together. It can be found that the texture was different comparing to route A, basal planes in majority grains tilted about 10° from the ND to Y plane (shown in Figure 1); and the intensity went down from 1P to 4P, then tended to be stable for the next pass.

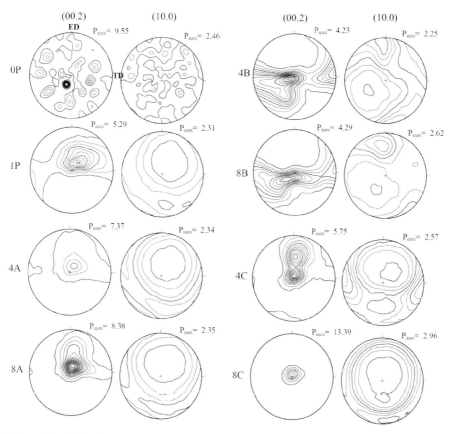

Figure 3: Recalculated (00.2) and (10.0) pole figures of the ECAPed cast Mg-Si alloys by different routes. The extrusion axis is to the top.

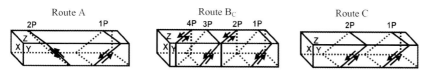

Figure 4: Schematic shown the shearing during successive pass with different routes

By route C processing, which leads to repetitive shearing on the same plane, as illustrated in Figure 4, two strong intensity points occurred in pole figure; while after 8P processing, there exited nearly a typical rolling texture in the samples. And the intensity was increased gradually from 5.75 m.r.d. by 4P processing to 13.39 m.r.d. by 8P processing, which meant that texture become stronger gradually. This is due to the successive shear deformation along one direction for route C.

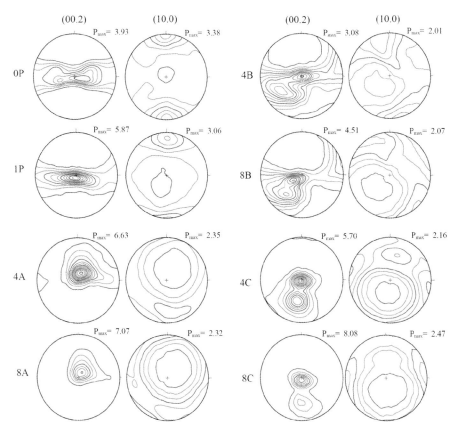

Figure 5: Recalculated (00.2) and (10.0) pole figures of the ECAPed as-extruded Mg-Si alloys by different routes. The extrusion axis is to the top.

Figure 5 shows the recalculated (00.2) and (10.0) pole figures of the three different routes ECAPed as-extruded Mg-Si alloys. In the rectangular extruded samples, basal planes in most of grains are parallel to the extrusion direction; while the intensity in (10.0) pole figure is a little high, which maybe due to the easy activities of the twinning (as can be found in Figure 2 (c)) and non-basal planes during high temperature extrusion [14]. After 1P ECAP processing, the basal planes tilted at about 8° parallel to the Y plane, which was similar as conventional rolling texture of Mg alloys. However, after 4P and 8P route A processing, the basal planes were changed to have a degree about 22° and 15° oft to the extrusion direction, respectively. This texture is similar as that obtained for the as-cast alloys, but the intensity is lower. Different textures were obtained by route B_C processing, there were two strong positions in pole figures, one was near to the center by 4P which indicated that most of the grains had a basal orientation parallel to the extrusion direction; but there exited an angle about 27° for 8P; another strong position occurred in pole figures, as shown in Figure 5. By route C deformation, similar texture was obtained as the cast alloys which the basal planes in most grains were almost parallel to the Y planes.

As for these three routes, texture intensity by route B_C was relatively lower than using the other two processing routes, and the texture tended to be stable after 4P with B_C. It is interesting to find that for both the as-cast and the as-extruded alloys after these three routes processing, almost similar textures were obtained by each route; but higher intensity obtained for the as-cast alloys. However, texture formed in the first few passes is greatly affected by the initial exited texture. For comparison, Figure 6 shows the ODFs of the 1P ECAPed as-cast and as-extruded samples, which indicates that the texture components in the ECAPed as-extruded alloys tend to be intensified to rolling texture; while for the ECAPed as-cast alloys, a new component is formed.

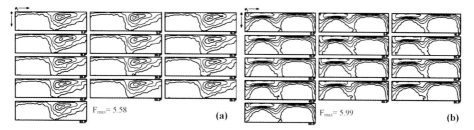

Figure 6: Constant φ_2 sections of ODF measured with neutron diffraction. (a) 1P ECAPed as-cast Mg-Si; (b) 1P ECAPed as-extruded Mg-Si.

4 Conclusions

For both the as-cast and as-extruded alloys, results showed that route Bc processing has the lowest pole density in each pole figures. The pre-exited texture has a great effect on the texture evolution for the fist few ECAP passes; except that the higher intensity for as-cast alloys, texture tended to be similar as the as-extruded alloys after 4P processing. As for the cast alloys, basal planes tilted about 20° to extrusion direction after 1P processing; though the texture type has not changed, the intensity increased after 4P; 8P processing made the basal planes nearly parallel to the extrusion-transition plane though the pole intensity have not changed. Especially, similar as the Mg alloys rolling texture, which basal planes parallel to the extrusion direction, was formed after 8P route C processing. As for the extruded Mg-Si alloys, two intensity peaks occurred after route Bc processing, which attributed to the 45° back-and-forth shear deformation during each successive 90° rotation. Briefly, ECAP processing of these two kinds of alloys tends to produce a certain type of basal texture, but just the orientation and strength is different for each deformation routes and alloys.

5 Acknowledgement

The author is very appreciated for the finical support of the DAAD-Helmholtz scholarship. We are grateful to the permission of the beam time at TEX-2 FRM-I, GKSS-Research center, Geesthacht, Germany.

6 References

[1] Mabuchi, M.; Higashi, K., Acta Mater. **1996**, *44*, 4611–4618.
[2] Zheng, M. Y., Qiao, X. G.; Xu, S. W.; Wu, K.; Kamado, S.; Kojima, Y., **Mater. Sci. Forum** 2005, *V. 488–489*, 589–592.
[3] Segal, V. M., Mater. Sci. & Eng. **1999**, *A271*, 322–333.
[4] Bing,Q.; Langdon, T. G., Mater. Sci. & Eng. **2005**, *A410–411*,435–438.
[5] Agnew, S. R.; Horton J. A.; Lillo, T. M.; Brown,D. W., Script Mater. **2004**, *50(3)*, 377–381.
[6] Kim, W. J.; Hong, S. I.; Kim, Y. S.; Min, S. H.; Jeong, H. T.; Lee, J. D., Acta Mater. **2003** *51(11)*, 3293–3307.
[7] Agnew, S. R.; Mehrotra, P.; Lillo, T. M; Stoica, G. M.; Liaw, P. K., Acta Mater. **2005**, *53(11)*, 3135–3146.
[8] Liu, T.; Wang, Y.D.; Wu, S.D.; Peng, R. Lin; Huang, C.X.; Jiang, C. B.; Li, S.X., **2004**, *51(11)*, 11057–1061.
[9] Valiev, R. Z.; Lowe, T. C.; Mukherjee, JOM **2000**, *April*, 37–40
[10] Brokmeier, H.-G., Physica B **1997**, *234–236*, 1144–1145.
[11] Yi, S. B.; Brokmeier, H.–G.; Bohlen, J.; Letzig, D.; Kainer, K. U., Physica B **2004**, *350*, 507–509.
[12] Furukawa, M.; Horita, Z.; Langdon, T. G., Mater. Sci. & Eng. **2002**, *A322*, 97–1092.
[13] Li, S. Y.; Beyerlin, I. J., Modeling Simul. Mater. Sci. Eng. **2005**, *13*, 509–530.
[14] Bohlen, J.; Yi, S. B.; Swiostek, J.; Letzig, D.; Brokmeier, H.-G.; Kainer, K. U., Script Mater, **2005**, *53*, 259–264.

Effect of Ca addition on the Microstructure Evolution of AZ31 Alloy During Thermomechanical Processing

E. Essadiqi[1], Jian Li[1], C. Galvani[1], P. Liu[1], F. Zarandi[2], S. Yue[2]
[1] Materials Technology Laboratory-CANMET, Ottawa, Ontario, K1A 0G1, Canada
[2] Depatrment of Metals and Material Engineering, McGill University, 3610 University, Montreal, Quebec, H3A 2B2, Canada

1 Introduction

Magnesium alloys are very attractive to the automobile industry due to their light weight, which has the potential to increase fuel efficiency. However, due to their hexagonal closed packed crystal structure, magnesium alloys possess poor formability. Grain refinement is probably the best method to increase both ductility and strength of magnesium alloys. Obtaining a homogeneous fine-grained microstructure is a challenging task. One of the difficulties arises from the presence of unstable grain boundaries that facilitate grain growth during heat treatment.

Previous attempts of grain refinement of magnesium alloys include equal channel angular extrusion (ECAE), accumulative roll bonding (ARB) and high-pressure torsion (HPT), etc. [e.g. 1-4]. However, these techniques are not suitable for magnesium sheet production, which is of great interest for automotive applications. Among the hot deformation processes suitable for large-scale sheet production, rolling combined with appropriate heat treatment is the most practical method. Proper control of rolling parameters can lead to a refined grain structure by continuous dynamic recrystallization [5,6]. However, during subsequent heat treatment or sheet forming at elevated temperatures, significant grain growth may occur which can lessen or eliminate the mechanical property gains associated with the initial fine grain size. The addition of a small amount of alloying elements such as Ca, Zr and Si to pure magnesium has been shown to result in efficient grain refinement by their growth restriction effects [7].

The addition of Ca has a marked effect on the microstructure and properties of Magnesium alloys. Previous studies of grain refinement of Mg alloys (pure magnesium, Mg-Zn and Mg-Mn) via Ca addition were mostly focused on the alloy casting processes [7-9]. Lee [7] suggested that the refinement was a result of grain growth restriction due to relatively slow solute atom diffusion at the solidification front, which reduced the rate of crystal growth. So far, the studies on the effect of Ca on the microstructure and the properties of both cast and wrought Mg alloys have either focused on the particle size refinement [e.g. 10, 11] or on magnesium alloys for high temperature applications [12-14]. The aim of the present study is to investigate the effect of the addition of a small amount of Ca on the microstructural evolution of AZ31 alloy during extrusion and hot rolling processing.

2 Experimental Details

Two Mg AZ31 alloys, one with 0.36 wt.% Ca and the other without Ca, were cast as ingots. These cast ingots were machined into cylindrical specimens 64 mm in diameter and 76 mm in height. Chemical compositions of the cast ingots are listed in Table 1. The specimens were homogenized at 350°C for 4 hours and extruded at 350°C into strips 4.7 mm in thickness and 31 mm in width. The extruded strips were hot rolled to a thickness of 2 mm at 400°C in 4 passes, then annealed at 400°C for 1 hour.

Table 1. Chemical compositions of cast alloy ingots (wt.%)

	Al	Mn	Zn	Ca	Si	Cu	Fe	Ni
AZ31	2.85	0.19	0.92	0.0	0.02	0.0	0.008	0.001
AZ31+Ca	3.14	0.19	0.95	0.36	0.02	0.0	0.006	0.0

Specimens from the as-cast, as-extruded, hot rolled and annealed states were cut out and cold mounted in low shrinkage epoxy resin. Examinations of the microstructure were performed in the longitudinal plane of the specimens. The as-polished specimens were used to assess the nature, size and distribution of the precipitates at different stages of processing. The samples were then etched using an acetic picral solution to reveal the grain structures. Optical images were acquired, followed by grain size and distribution measurements using a Clemax Vision image analysis system. Samples were also examined using a Philips XL-30 scanning electron microscope (SEM) equipped with an thin-window energy dispersive x-ray detector (EDS). Specimens for transmission electron microscopy (TEM) were prepared using a focused ion beam (FIB) system, and a FEG TEM was used to examine the fine precipitates.

3 Results and Discussion

The as-cast microstructures appear to be similar in both alloys with casting cell sizes of about 200 µm. In this paper, we focus on the microstructure at the center of the sheets. A bi-modal grain size distribution in the extruded and rolled states were observed. Also, the extruded samples contain some elongated grains along the extrusion direction. In general, the sample containing Ca appears to have a larger area fraction of small grains compared to that of the sample without any added Ca.

Image analysis was performed in order to quantitatively assess the effect of Ca on grain refinement. Average grain sizes of the small (less than 10 µm) and the large (larger than 10 µm) equi-axed grains were calculated in each case, and the area fractions of the three types of grains were also calculated. Elongated grains were counted separately). Up to ten images were analyzed in each case to ensure statistical significance. Table 2 summarizes the image analysis results.

Table 2. Grain size distribution from image analysis, in as-extruded samples

	Area% of small grains	Area% of large grains	Area% of elongated grains	Mean intercepts of small grains	Mean intercepts of large grains
AZ31-extruded	15.3%	78.7%	6.0%	5.7 μm (0.4)	13.1 μm (0.3)
AZ31+Ca-extruded	43.1%	27.6%	29.4%	3.9 μm (0.2)	10.3 μm (0.9)
AZ31 hot rolled	46.0%	54.0%	0%	5.1 μm (0.1)	10.1 μm (1.1)
AZ31+Ca hot rolled	66.2%	33.8%	0%	3.5 μm (0.1)	10.0 μm (0.9)
AZ31 hot rolled and Annealed 1.0 hr.*	100%		0%	10.0 μm (0.6)	
AZ31+Ca hot rolled and Annealed 1.0 hr.*	100%		0%	9.8 μm (0.4)	

* grains were not separated into small and large size for image analysis
** all hot rolling experiments are carried out on extruded sheets

The as-extruded specimens contain three types of grains: the small grains, the larger grains, and elongated grains. These elongated grains could have originated from the cast grains, which were flattened by the extrusion process and did not undergo recrystallization. Such a mixed microstructure has been previously reported [3]. The Ca-modified AZ31 contains a much larger area fraction of small grains after extrusion. This indicates that the addition of Ca has stabilized the smaller grains. The elongated grains do not exist in the hot rolled samples, which were dynamically recrystallized. Again, the Ca-modified hot rolled specimen shows a higher area fraction of small grains. However, the difference in average grain size between the extruded and rolled microstructures, both in the small grain and large grain zones, is negligible.

After a 1 hour anneal at 400°C, grain growth was observed. The grain size appears to have a normal distribution rather than the bimodal distribution in the hot-rolled state. There is virtually no difference in the grain size of the two alloys suggesting that the effect of Ca diminishes under this annealing condition.

Calcium has limited solubility in both magnesium and aluminum even at relatively high temperatures [15]. Therefore, the formation of calcium-rich precipitates may be responsible for the grain refinement in the Ca-containing alloy. The type and distribution of the precipitates were characterized using the as-polished samples in SEM. Chemical analysis using EDS indicated that the AZ31 alloy contained Al_xMn_y precipitates. With the Ca addition, the Al-Ca phase precipitated out in addition to the Al_xMn_y phase. This is in good agreement with the recent report by Suzuki et al [16], in which, they claim that the precipitation of Al_2Ca phase, rather than Mg_2Ca phase, occurs if the Ca/Al ratio is less than 0.8. The distribution of the precipitates tends to be banded along the extrusion/rolling direction. Much smaller precipitates, only observable by TEM, also exist in both alloys, as will be shown further on. Precipitates could act to inhibit recrystallization and pin grain boundaries through the Zener drag process [17] which reduces grain size.

Figure 1 shows higher magnification SEM images of the extruded sample with Ca addition. Various sizes of precipitates exist in the microstructure. Larger precipitates (brighter) in Figure 1(a) are Al_xMn_y and smaller precipitates, which are also shown in Figure 1(b) are Al-Ca precipitates. The chemical composition of very small precipitates can't be positively identified due to the limitation of the spatial resolution of the EDS system. These particles were further analyzed using a TEM.

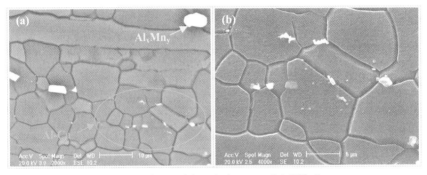

Figure 1. A SEM image showing various precipitates in the as-extruded AZ31+Ca.

Figure 2 shows the dynamically recrystallized microstructure of the extruded and hot rolled alloy with Ca. A large amount of precipitates exist as shown in figure 2(a). The majority of these precipitates are identified as Al-Ca intermetallic particles with a small number of Al_xMn_y particles present. Figure 2(b) shows that the area with a large amount of Al-Ca precipitates appear to have a much finer grain size compared to the areas that have relatively few precipitates. Thus, these precipitates could be acting as barriers during grain boundary migration. The effectiveness of the Al_xMn_y precipitates for grain refinement is unknown at this stage.

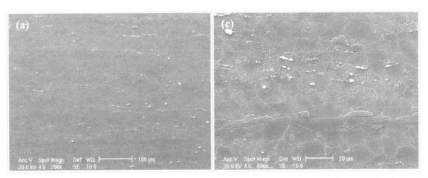

Figure 2. SEM images of dynamic recrystallized microstructure of (a) extruded and (b) extruded and hot rolled specimen with Ca addition

TEM specimens from the as-extruded samples, with and withour Ca, were investigated. Figure 3(a) shows a higher magnification view of the fine precipitates in the specimen without ca, on which EDS analyses were performed. Most dark spots are identified as Al_xMn_y precipitates. Some of the very small dark spots could be ion beam damage introduced during final ion beam milling. In addition, no precipitates are apparent on the grain boundaries.

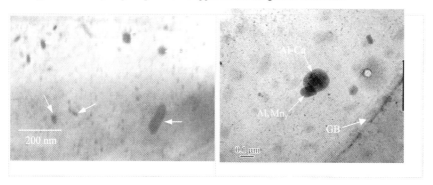

Figure 3. TEM analysis on samples a) without Ca addition in which some of the identified Al_xMn_y precipitates are labeled with arrow and b) with ca addition showing the presence of Al-Ca in contact with Al_xMn_y precipitate

Figure 3b shows high magnification TEM images of the AZ31 alloy with Ca addition. The large particle is both rich in Al and Ca and is in direct contact with an Al_xMn_y precipitate. Other small precipitates in this field are mostly Al_xMn_y intermetallic particles. Again, no precipitates were found on the grain boundary.

4 Summary

The addition of a small amount of Ca (0.36 wt.%) produced significant additional grain refinement in AZ31 alloy processed by extrusion and hot rolling. Most of the Ca precipitated out as fine Al-Ca phase (most likely Al_2Ca). The grain refinement could either be attributed to the Zener drag effect by the Al-Ca precipitates, which reduced grain boundary mobility during thermomechanical processing, or the Ca solute drag on the sweeping grain boundary during recrystallization and grain growth.

5 Acknowledgements

The authors would like to thank casting and forming team at CANMET, Mr. M.W. Phaneuf for access to FIB microscope in the FIBICS Inc. TEM work by Dr. V.Y. Gertsman of CANMET-MTL is highly appreciated.

6 References:

[1]. Furukawa M., Horita Z., Nemoto M., Langdon T.G., Mater. Sci. Eng. **2002**, 234, 82.
[2]. Prangnell, P.B. Bowen, J.R., Apps P.J., Mater Sci. Eng. **2004**, 375-377, 178.
[3]. Perez-Prado, M.T., Del Valle, J.A. and Ruano, O.A., Scripta Materialia, **2004**, 51, 1093-1097.
[4]. Chang, T.C., Wang, J.Y., Ming, C. and Lee, S., Journal of Materials Processing Technology, **2003**, 140 (1-3), 588-591.
[5]. Tan, J.C. and Tan, M.J., Materials Science and Engieering A, **2003**, 339 (1-2), 124-132.
[6]. Li, P., Tang, B. and Kandalova, E.G., materials Letters, **2005**, 59, 671-675.
[7]. Lee Y.C., Dahle, A.K. and Stjohn, D.H., Metall. Mater. Trans., **2000**, A31, 2895.
[8]. Chuhrov, M.V., Grain Resinement of Magnesium Alloys, RSSR, Moscow, **1972**.
[9]. Vetrano, J.S., Bruemmer, S.M., Pawlowski, L.M. and Robertson, I.M., Materials Science and Engineering A, **1997**, 238, 101-107.
[10]. Yuan, G.Y., Liu, Z.L., Wang, Q.D. and Ding, W.J., Materials Letters, **2002**, 56, 53-58.
[11]. Suzuki, A., Saddock, N.D., Jones, J.W. and Pollock, T.M. Acta Materialia, **2005**, 53, 2823-2834.
[12]. Zhang, Z., Tremblay, R. and Dube, D., Mater. Sci. Technolo., **2002**, 18, 433.
[13]. Pekguleryuz, M.O. and Baril, E., Mater. Trans. **2001**, 42(7), 1258.
[14]. Luo, A.A. Balogh, M.P. and Powell, B.F., Met. Mater. Trans., A. **2002**, 33, 567.
[15]. Ninomiya, R., Ojiro, T. and Kubota, K., Acta Metall., mater. **1995**, 43, (2), 669-674.
[16]. Suzuki, A., Saddock, N.D., Jones, J.W. and Pollock, T.M., Scripta Materialia, **2004**, 51, 1005-1010.
[17]. Furu, T., Marthinsen, K., and Nes, E., Mater. Sci. Forum, **1993**, 41, 113-115.

Flow Instabilities of Mg-Al4-Zn1 Alloy during High Strain-rate Deformation

J. Dzwonczyk, F. Slooff, J. Zhou, J. Duszczyk
Delft University of Technology, Delft

1 Introduction

Hot working is an efficient way to produce various products of high quality in terms of microstructure, mechanical properties and surface finish. However, while the quality of a product is no doubt of great importance, material-processing economics is equally important.

In general, the ability of a material to be shaped under a certain deformation condition without triggering any kind of defects should be specified before the final production parameters are set. It is usually problematic, because the workability of the material depends on many factors such as the local conditions of stress, strain, strain rate and temperature in combination with the self-condition of the material (its history and properties). These factors involve different deformation mechanisms, which cause different microstructural changes in the material. The mechanisms of hot working are rather complex and vary considerably from one alloy to another.

The material used in the present research is a new magnesium alloy AZ41, designed on the basis of the commercial AZ31 alloy suitable for hot bulk forming including extrusion, forging and rolling. Mg alloys with low density and high specific strength have in recent years gained growing interest from the transportation industry obligated to reduce the emissions of green house gases [1]. Nevertheless, much research and development work is still needed before various magnesium alloys can be used reliably, efficiently and cost-effectively.

Since AZ41 has not yet been introduced into the market, there are no literature data available on its deformability. The goal of present research was to reveal the changes in the microstructure of this alloy during hot compression tests performed at high strain rates and over a wide range of temperatures. The choice of these deformation conditions was based on the expectation that the applied strain rates would lead to undesirable structural changes occurring in the form of different flow instability manifestations, which would lead to different structures and properties and could be a site of failure in service. In addition, the deformation modes, mainly responsible for the resulting microstructure of the material, could vary with temperature applied. Therefore, the study was also aimed to determine which deformation mechanism would be responsible for the structural instability at a certain temperature.

2 Experimental Details

AZ41 ingot had a chemical composition of 4.76 wt.% Al, 0.85 wt.% Zn, 0.32 wt.% Mn, 0.012 wt.% Si, 0.008 wt.% Fe and balance Mg. It was re-melted and cast into extrusion billet using a permanent mould. Then, the billet was subjected to direct extrusion under the following conditions: ram speed 1.5 mm/s, container/die temperature 375 °C, billet temperature 425 °C and reduction ratio 9.2 (Ø 16 mm). After extrusion, the material was cooled in air.

Specimens with a height of 12 mm and a diameter of 10 mm were machined from the extruded rods and subjected to hot compression tests.

Isothermal compression was performed using Gleeble 3500 over a wide range of temperatures: 200 °C–500 °C with an interval of 50 °C and at strain rates: 10, 100 and 150 s^{-1}. Before testing, a nickel-based paste was applied on the top and bottom sides of specimens in order to minimize the friction between the deformed material and anvils and to avoid barreling at the initial stage of compression. Temperature rise during testing was monitored through a fine thermocouple spot-welded at the mid-height of each specimen surface. Testing temperature control was within ±2 °C. Under a given condition of strain rate and temperature, specimen was compressed up to a true strain of 1 in a closed chamber filled with nitrogen and then rapidly cooled with pressurized air so as to retain its as-deformed microstructure. The flow stress was recorded as a function of strain.

Samples for metallographic examination were cold-mounted in Demotec-30, fine ground up to 1200 mesh and polished in accordance with the polishing procedure for Mg alloys [2]. In order to reveal microstructural details, the samples were etched in a picric-based solution for 3 to 5 seconds and cleaned with 99 % ethanol. The microstructural analysis was performed on the sections parallel with and perpendicular to extrusion / compression axis.

3 Analysis of the Results

The observations reported are *sensu stricto* the microstructures representing part of a comprehensive investigation on the AZ41 alloy.

3.1 Structural Evolution

The deformation at low temperatures (200 °C, 250 °C) caused significant morphological inhomogeneities between sample cross-sections, indicating a high structural anisotropy. These inhomogeneities tended to reduce with rising temperature and strain rate. Moreover, the low temperature deformation did not remove the parent grains. The as-deformed grains inherited the original morphology in large, elongated forms with corrugated boundaries and distorted interiors.

The original grains appeared to be a convenient source for twinning, which mutually intersected them into smaller fractions. At higher temperatures, advanced recrystallization took place, removing the parent grains from the structure. It was also found that a large number of the original grains occupied the regions of strain localization, as a direct result of non-uniform strain distribution when strain rate applied was high. Regarding the influence of strain rate on the parent grains, it was observed that at 10 s^{-1}, the parent grains remained even after high temperature deformation and complete removal of the primary microstructure occurred not before 500 °C/10 s^{-1}, while higher strain rate deformation caused complete recrystallization of the material at 400 °C.

3.2 Twinning, Kink Bands and Dynamic Recrystallization

The specimens used in the present research had the basal plane parallel to the extrusion/compression direction and as a result the basal slip, which was expected to be a predominant slip mode, would not be favored. The activities of the prismatic slip system and the pyramidal slip system would also be restricted, since the critical resolved shear stresses required for their activation are much higher than for other gliding systems. The work hardening observed in the flow stress-strain curves (Figure 1a) can therefore be attributed to $\{10\text{-}12\}\langle 10\text{-}10\rangle$ twinning. Favored by the initial texture, mechanical twinning appeared to play a significant role in structural rearrangement during high strain rate deformation.

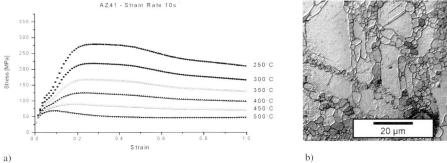

Figure 1: (a) Flow curves obtained from compression at 10s^{-1} and (b) recrystallization at 300 °C/100 s^{-1}

Primarily, needle-like (or lamellar) twins were observed in the grain interior at ±45° with respect to the compression axis (as {10–12} twins) in combination with kink bands and within the areas of extensive strain localization. Twins may stretch across a grain boundary into another grain, towards the next grain boundary, leading to deflections and also crossing mutually within the grain.

At low temperatures, such as in the condition of 200 °C/150 s^{-1}, twin boundaries appeared to be a major source for grain nucleation and fine cells were covering twins boundaries. It would be unstable, considering the higher boundary energy of twins. On the other hand, it has been reported [3] that small grains may act as natural obstacles to twins, since they possess a larger area of grain boundaries and therefore a larger driving force would be necessary to overcome these barriers. Indeed, the present research shows that with an increase in temperature the number of nuclei increases, covering a larger matrix area and dumping at the same time twinning appearance (Figure 1b). With an increase in temperature, twins became more lenticular. As dynamic recrystallization became more pronounced, the recrystallized grains replaced the twins. Exceptionally, after deformation at 500 °C/10 s^{-1} and 100 s^{-1}, twinning occurred within the areas of intercrystalline cracking. The summary made by Brown et al. [4] with regard to twin dependence on temperature and strain rate applied to hcp material deformation was in agreement with the present observations: at a given temperature, twinning was much more active at a higher strain rate, while at a given strain rate, deformation twinning was more active at a lower deformation temperature.

Furthermore, under deformation at a low temperature and a high strain rate, the kink bands (defined as in [5]), which led to distortions within single grains, were observed. Yang and others [6–8] investigated the function of kink bands in the initiation of the dynamic recrystallization process. The kink bands in AZ41 were present more extensively at higher strain rates and lower temperatures. Their occurrence led to grain interior distortions and after multiplied deformation these might result in evolution of new grains.

Recrystallization, which took place at most of given conditions, was influenced by two major factors: (i) the deformation condition in terms of temperature and strain rate and (ii) the morphology of the initial structure and texture:
- At low temperatures recrystallization was restricted. It was initiated mainly on the twin boundaries within the interior of the large parent grains, grain boundaries and in the most deteriorated regions, i.e. of large strain concentrations (Figure 2a).
- With temperature increased, recrystallization progressed and more regions of the deformed samples were covered with recrystallized grains. It, however, did not occur uniformly, having the most intensive occurrence at the former twin boundaries, leading at the same time to the growth of the grains with boundaries orientated unfavorably for nucleation (Figure 2b).
- With temperature further increased, complete recrystallization produced uniform, equiaxed grains as a result of twofold processes, i.e. grain nucleation and growth, leading to grain rearrangement within the whole sample, so that the parent grains and twins were not visible anymore and the structures exhibited a high uniformity (Figure 2c).

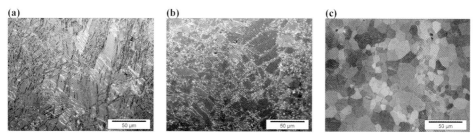

Figure 2: AZ41 micrographs after compression at (a) 200 °C/150 s^{-1}, (b) 300 °C/10 s^{-1} and (c) 500 °C/150 s^{-1}

With strain rate increased, recrystallization became far more complex. The microstructures revealed suggested that a certain combination of recrystallization processes might have occurred: (a) twin recrystallization accompanied by rotation recrystallization and (b) continuous recrystallization followed by grain growth.

The first one would be active at low temperatures, while the second one – consequently – over a higher temperature range starting at approximately 400 °C that is in agreement with the recrystallization temperature for magnesium.

Basically, twin recrystallization is composed of the twin boundary formation and their transformation into random boundaries, which may lead to local migration. Twin intersections involve grain nucleation followed by recrystallization. This kind of recrystallization was analyzed by Sitdikov [9]. Moreover, Ishikawa [10] reported similar observations on the AZ91 magnesium alloy. However, a high strain rate in combination with a moderate temperature seems to lead to more complex recrystallization. Therefore, additional recrystallization mode, i.e. rotation recrystallization may be taken into account as a supportive mechanism. Ion [11] discussed the rotation recrystallization in detail. In the AZ41 structures, the evidence of the rotation recrystallization would be the presence of kink bands, grain interior distortions and different orientations of grains in strain localization regions, which were revealed under a polarized light microscope and marked by different colors. Regarding recrystallization at high temperatures, it was found that at 400 °C and above a uniform grain growth took place, which could be attributed to the continuous recrystallization [12]. An average grain size was calculated and the results are shown in Table 1 (ASB – adiabatic shear bands, FL – flow localization). The largest grains were found in the sample deformed at 500 °C/10 s^{-1}, attributed to abnormal grain growth, as the time for deformation was the longest.

Table 1: The calculated average grain sizes for AZ41 at different deformation conditions

Deformation Condition	150 s^{-1}	100 s^{-1}	10 s^{-1}
200 °C	ASB	ASB	ASB
250 °C	ASB	FL	FL
300 °C	3.13 µm ± 0.53 FL	FL	FL
350 °C	4.84 µm ± 0.70 FL	3.98 µm ± 0.83 FL	FL
400 °C	6.09 µm ± 0.64 FL	5.78 µm ± 0.69 FL	FL
450 °C	10.32 µm ± 0.28	7.02 µm ± 0.3 FL	8.80 µm ± 0.50 FL
500 °C	10.60 µm ± 1.08	10.59 µm ± 1.40	14.95 µm ± 1.32

The recrystallization of the AZ41 alloy appeared to be strongly dependent on the deformation conditions and the initial microstructure. At a high strain rate, an additional factor - favorable texture, which by the activation of different gliding systems could support the progress of recrystallization, became important.

3.3 Flow Instabilities

The flow instabilities observed varied with deformation conditions (Figure 3a-c). At low temperatures, intensive adiabatic shear bands led to significant structure distortions and even fracture. The interior of the parent grains that had a high volume fraction in the specimens deformed at low temperatures, was distorted and gnarled. Additionally, high strain rate deformation caused kink bands. Temperature increase resulted in flow localization bands, which could be well distinguished from the adjacent, less deformed regions. Within the flow localization bands the imposed strain led to non-uniform recrystallization. Flow localization bands were found at most of the samples, indicating their stability with the temperature change, although, with temperature rising the flow localization bands tended to widen till they disappeared completely from the structure.

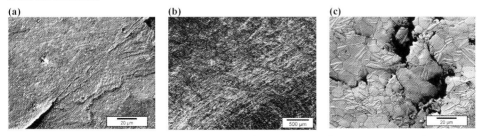

Figure 3: Flow instability manifestation at (a) 250°C/150s^{-1}, (b) 300 °C/10 s^{-1} and (c) 500 °C/100 s^{-1}

At a high temperature and a high strain rate, intercrystalline cracking was observed in the outer region of the samples and it was intensified as strain rate was lowered. Moreover, the intercrystalline cracking was accompanied by deformation twinning, which in this case was also an instability manifestation. It can be concluded that the shape, kind and degree of strain localization are strongly dependent on the strain rate applied, since the microstructural changes with temperature increasing were slower and did not vary considerably.

4 Conclusions

On the basis of the microstructural observations of the AZ41 alloy compression-tested at high strain rates of 10, 100 and 150 s^{-1} and at different temperatures, the following conclusions can be drawn:

- Deformation at low temperatures caused significant structural inhomogeneities. Due to limited slip systems operating, twinning became an important deformation mechanism and on the twin boundaries recrystallization was initiated. Moreover, kink bands were present in the parent grains inherited from the original structure and also influenced the extent of recrystallization. Flow instabilities in terms of adiabatic shear bands were pronounced and at a high strain rate these led to detrimental surface fracture.
- An increase in temperature caused structural evolutions in terms of more advanced recrystallization. Deformation instabilities occurred in the form of flow localization bands, which tended to disappear with increasing temperature and strain rate. Flow instabilities seemed to be more dependent on strain rate applied than on temperature.
- High temperature deformation led to complete recrystallization, resulting in homogeneous, equiaxed grains so that the parent grains were not visible anymore. At the same time, intercrystalline cracking associated with mechanical twinning manifested the flow instabilities. Cracking appeared to be more pronounced at a lower strain rate.
- Different deformation mechanisms can be related to strain rate and temperature. Strain rate seems to have a more pronounced influence on the resulting structure.

5 References

[1] R. R. Braeutigam, R. G. Noll, Report of Economic Statistics **1984**, *66(1)*, 80.
[2] V. Kree, J. Bohlen, D. Letzig, K. U. Kainer, Practical Metallography **2004**, *41(5)*, 233.
[3] J. Dzwonczyk, Influence of different extrusion methods on microstructure and mechanical properties of AZ31 alloy, PhD dissertation **2004**, 69.
[4] D. W. Brown, et al., Materials Science Forum **2005**, *vols. 495–497*, 1037–1042.
[5] F. J. Humphreys, M. Hatherly, Recrystallization and related annealing phenomena, Elsevier Science Ltd. **1996**, 21.
[6] X. Yang, H. Miura, T. Sakai, Materials Science Forum **2003**, *vols. 419–422*, 515–520.
[7] X. Yang, H. Miura, T. Sakai, Materials Science Forum **2004**, *vols. 467–470*, 531–536.
[8] X. Yang et al., Materials Science Forum **2006**, *vols. 503-504*, 521–526.
[9] O. Sitdikov, R. Kaibyshev, T. Sakai, Materials Science Forum **2003**, *vols. 419–422*, 521–526.
[10] K. Ishikawa, H. Watanabe, T. Mukai, Materials Letters **2005**, *59*, 1511–1515.
[11] S. E. Ion, F. J. Humphreys, S.H. White, Acta Metallurgica **1982**, *30*, 1909–1919.
[12] G. Huang et al., Materials Science Forum **2005**, *vols. 488–489*, 215–218.

Microstructure and Mechanical Properties of Rolled Magnesium Sheets for Deep-drawing Applications

Fr.-W. Bach, M. Hepke, M. Rodman, A. Rossberg
Institute of Materials Science, Leibniz Universität Hannover, Germany

1 Introduction

The achievable advantages by using magnesium sheets in the automobile industry are generally known but at present it is not possible to realise them in series production. Due to this reason in different research works and small-lot productions the lightweight potential of components of magnesium wrought alloys was studied in the last years. In the centre of attention was mainly the production of semifinished products of magnesium wrought alloys for deep-drawing applications. The main obstacles for the industrial application of deep-drawn components are the varying quality concerning the surface condition and the anisotropy of the mechanical properties as well as the high material costs. This article will show that through an optimisation of the rolling which is appropriate for the material involved, an improved quality of the semifinished products could be realised by reducing the costs. The central point of the effected tests is the reduction of the anisotropy of the mechanical properties through an appropriate alloy selection, adapted temperature control and change of the rolling direction.

2 Experimental Procedure

The most important criterion for the application of magnesium sheets in the deep-drawing process concerning the final product are the keeping of geometrical tolerances, a high surface quality and a high level of the mechanical properties. To fulfil these demands the applied magnesium sheets must show special characteristical properties. Hereby high importance has the fine grain structure, the homogeneity, the low anisotropy of the mechanical properties and the minimum texture formation. The properties of the magnesium sheet are affected by different factors of the whole process chain which are mutually reacting.

2.1 Alloying Influence

An important factor is the alloying influence. As a standard alloy for forming applications basically the magnesium wrought alloy AZ31 is used. But for more complex deep-drawing applications this alloy comes up against its limiting capacities. At high strains AZ31 tends to the formation of an inhomogeneous structure and to a texture formation. An improvement of the forming ability can basically be achieved through grain refining, stabilization of the dispersion phases and homogenizing. To avoid high material costs and a high portion of low melting dispersion phases which can have negative consequences on the ductility, only alloys with a low portion of alloying elements were included into the tests. As alloying elements zircon (Zr) as grain refiner and rare-earth metals (SE) for the stabilization of the dispersion phases were applied [1, 2].

Table 1: Composition of the used alloys

	Al [wt%]	Mn [wt%]	Zn [wt%]	SE [wt%]	Zr [wt%]
AZ31	2.9–3.1	0.3–0.5	0.9–1.1	-	–
AZE310	2.9–3.1	0.3–0.5	0.9–1.1	0.3–0.4	–
AE21	1.7–1.8	0.3–0.6	–	0.8–1.0	–
ZEK100	–	–	1.3–1.4	0.15–0.25	0.1–0.2

2.2 Initial Structure

Another factor is the initial state of the semifinished products before the rolling process. Basically this will be determined by the corresponding production process. As starting material cast semifinished products were examined (mould casting, continuous chill casting and strip casting material). The central points of the examinations, illustrated in this article, concerning the level of influence of the initial state on the rolling suitability are basically the initial grain size and the homogeneity of the structure and its alteration through the rolling. But also the geometry and the surface quality of the initial semifinished products as well as the portion of pores and sinkholes are of decisive significance. The number of reduction stages can be clearly reduced by measurements which are close to final contours and by reducing the surface defects and casting crust, the cutting finishing of the initital semifinished product before the rolling process can be minimized [3].

The final properties of the magnesium sheet will be achieved by including the mentioned factors, at least through the regulation of the parameters of the rolling process, the regulation process and the following thermal treatment.

3 Results and Discussion

3.1 Microstructure of the Rolled Semifinished Products

The structure of the cast semifinished products of the alloy AZ31 and ZEK 100 can be seen in figure 1. The cast semifinished products were produced with the three mentioned casting methods at the Institute of Materials Science in a laboratory scale.

The measurements of the semifinished products are 190 × 160 × 18 mm for the mould casting, Ø 90 × 500 mm for the continuous chill casting and 180 × 60 × 8.5 mm for the continuous strip casting. For the production of the continuous chill casting material and the strip casting material, a casting process, described in [4] according to the principle of Hazelett® [5] and an Upward Direct Chill Casting process described in [6], was used.

By means of the illustrated macrophotographs it can be seen that with all three used casting methods, the alloy ZEK 100 which is grain refined with zircon shows a clearly reduced medium grain size compared with the standard alloy AZ31. It also can be seen that the medium grain size of both alloys is reduced from the mould casting to the continuous chill casting till the continuous strip casting. The reason of the grain size reduction is the increasing rate of cooling of the cast material which is with the continuous strip casting the highest [3].

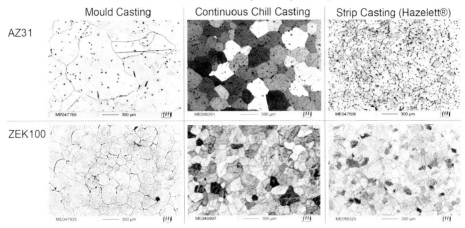

Figure 1: Structure of mould casting, continuous casting and continuous strip casting semifinished products of the alloys AZ31 and ZEK100

A further aspect of research was to determine the extent of grain refinement by alloying rare-earths [2]. This effect could not be proved in one of the used casting processes of the alloys AE21 and AZE310 in relation to AZ31.

3.2 Microstructural Development through Rolling Transformation

The basic material for the tests of the microstructural development were mould cast and at 450 °C and 8 hours homogenized plates with final dimensions of $160 \times 60 \times 6$ mm.

Investigations of the grain size development during the rolling process with different strains per pass, shown in figure 2, exhibit that in all cases the grain refining first takes place more intensively at the surface than in the middle of the sample. Only at a summed strain per pass of 90 %, a regular structure is given. With high strains per pass the grain size reduction is nearly linear. In contrast to this small strains per pass result with the alloy AE21 in a rapid grain size reduction at a summed strain per pass of approximately 60 %. This effect can be put down to an incomplete dynamical and statical recrystallisation [7]. The inserted energy is not sufficient to constitute new grain borders out of the low number of accumulated dislocations. This restriction of the recrystallization can be especially observed with thin sheets of aluminium containing alloys in the form of shear bands.

In figure 3 the shear band formation can be seen by means of micrographs of rolling samples of AZ31 and ZEK100. The orientation of the shear bands corresponds approximately to the direction of the main shearing stress. Detailed tests have shown that the grains in the area of the dark presented lines are considerable smaller and the microhardness of these areas is essentially higher. This demonstrates a concentration of the formation along the lines. This inhomogeneity which is a result of the rolling transformation also cannot be removed by a thermal treatment.

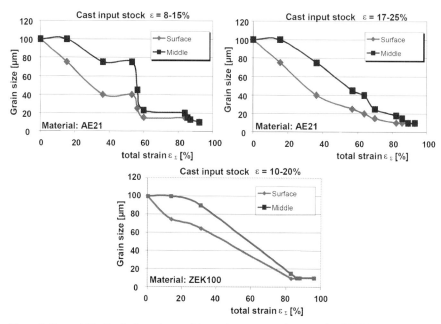

Figure 2: Recrystallisation in dependence of the strain per pass at the example of AE21 and ZEK100

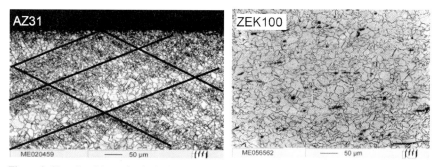

Figure 3: Shear band formation in the heat-treated AZ31 in comparison to ZEK100

With the same formation this effect cannot be determined with a sample of ZEK100. The constant forming behaviour of ZEK100 bases on the higher fine-grained structure and the more solid crystals through the zircon addition. In figure 4 the thermographs of the contact zone in the roll gap during the rolling process are compared. The alloy AZ31 shows clearly a temperature increase of 15–20 °C in the areas which are more formed which is caused by a concentrated sliding of the dislocations into narrow crystalline areas. With the alloy ZEK100 the forming heat distributes regularly in the contact zone.

Due to the concentrated forming, in the area of the shear bands only a limited residual ductility exists. Through a further loading in rolling direction a breakdown takes place. This pre-damage causes the opposed anisotropy of aluminium containing and aluminium free magnesium alloys. In figure 5 the tensile strengths of exemplary alloys at temperatures from room temperature to 225 °C are compared.

Figure 4: Thermograph during the rolling of AZ31 and ZEK100 at 100 °C

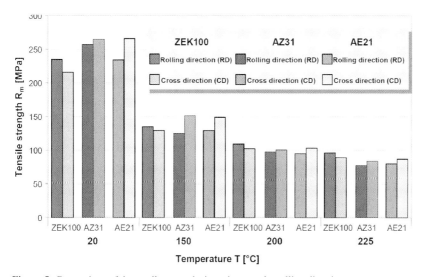

Figure 5: Comparison of the tensile strengths in and across the rolling direction

Texture analysis confirm that the formation of aluminium containing alloys basically takes place through the basis sliding and the portion of the pyramidal and prismatic sliding with alloys free of aluminium is higher whereby the texture sensitivity of the alloy ZEK100 will be reduced.

3.3 Influence on the Properties through the Directional Process

Actual investigations deal with the influence of the roller straightening process on the mechanical properties of the magnesium sheet. The analysis of resulting stress-strain diagrams shows a non common influence of other metals of a multiple cyclical bending load on the yield strength which can be seen in figure 6. Current investigations show that through an adapted rolling regime in the last reduction stage, adequate directional parameters and a thermal treatment this effects can be reduced.

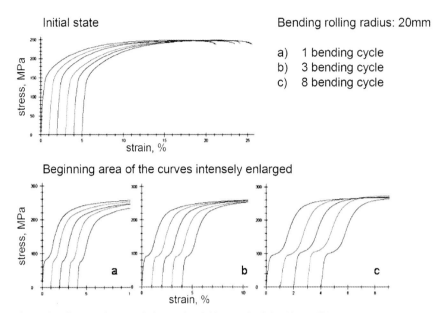

Figure 6: Influence of the regulation on the yield strength of the alloy AZ31

4 Conclusion

The application of continuous strip casting input stock enables a clearly abbreviated rolling process. But with aluminium containing alloys the shear band formation must be avoided through an adapted roll pass plan to reduce the anisotropy of the mechanical properties in contrast to aluminium free and with the grain refiner zircon provided alloys. The influence of the directional process must be considered but requires more extensive investigations.

5 References

[1] Friedrich, H.; Mordike, V., Magnesium technology: metallurgy, design data, applications, Springer Verlag Berlin, 2005
[2] Kammer, C., Magnesium Taschenbuch, Aluminium-Verlag Düsseldorf, 2000, p. 155–167
[3] Kawalla, R.; Engl, B., Magnesiumblechherstellung – Stand und Perspektiven, MEFORM 2006, Freiberg, p. 81-95
[4] Bach, Fr.-W.;Hepke, M.;Rossberg, A., New Strip Casting Prozess for Magnesium Alloys, DGM Continuous Casting, Neu Ulm, 2005, p. 81–86
[5] Hazelett, C. W., US-Patent No. 2 640 235 – Metal manufacturing apparatus, 1953
[6] Bach, Fr.-W.; Rossberg, A., Kontinuierliche Gießtechniken für Magnesiumhalbzeuge aus Knetlegierungen: Optimierte Dünnband- und Stranggussverfahren, 13. Magnesium Abnehmerseminar, Aalen, 2005
[7] Altman, M. B.;Belov, A. F.; Dobatkin, V.I.; Dritz, M.E.; Kvasov, F.I.; Tumanov, A. T., Magnesiumlegierungen Vol. 2, Moskau, „Metallurgie", 1978, p. 209

The Influence of Strain Path on Texture Evolution in Magnesium Alloy AZ31

T. Al-Samman, G. Gottstein
Institut für Metallkunde und Metallphysik, RWTH-Aachen, Aachen, Germany

1 Abstract

To investigate the deformation and recrystallization behaviour of the commercial magnesium alloy AZ31 during hot forming, thermo-mechanical processing by hot cross rolling, channel-die deformation and uniaxial compression was applied. Besides investigating the influence of different deformation temperatures (200 °C–400 °C) and strain rates ($10^{-2}\,s^{-1}$–$10^{-4}\,s^{-1}$) on deformation behaviour, the effect of the starting texture was also investigated and showed a strong impact on texture evolution. The deformation mechanisms and thus, texture formation for each deformation process were found to be distinctly different. Low temperature dynamic recrystallization was observed at 200 °C resulting in ~ 1 μm equiaxed grains. At higher temperatures (400 °C) deformation proceeded mainly by **<c+a>**-pyramidal slip and the final microstructure revealed a homogeneous grain size distribution.

2 Introduction

The absence of magnesium sheet in mainstream production at present is mostly due to the lack of insufficient formability. A major problem encountered in wrought alloy components is the strong directionality of properties, i.e. anisotropy, which is caused by the deformation characteristics of the hexagonal crystal structure. The limited number of active slip systems in hcp metals results in the formation of a strong crystallographic texture after mechanical processing. It is known that the change of deformation path introduces important changes in both the crystallographic texture and microstructure [1]. The aim of this study was to investigate the deformation behaviour of a commercial wrought magnesium alloy AZ31 during different modes of deformation in order to optimize microstructure and crystallographic texture. For this purpose hot working was applied for two reasons. One is to activate new slip systems by reducing their critical resolved shear stress (thermal activation) and hence increase the number of independent slip systems to accommodate the necessary shape change during deformation. The other is to introduce dynamic recrystallization (DRX) in order to investigate its influence on texture development.

3 Evolution of Texture in TRC-AZ31 during Hot Rolling

The material used for rolling experiments was a twin roll cast (TRC) magnesium alloy AZ31 provided by Hydro Aluminium R&D Bonn [2]. The chemical composition (wt. %) is the following: 3.2 Al, 0.86 Zn, 0.38 Mn, 0.016 Si, 0.002 Fe, 0.002 Cu, 0.001 Ni, Mg (balance). The

TRC material had a basal texture with a split maximum intensity of 6.8 times random, as well as a preferred <10-10>-orientation parallel to RD (Fig. 1a). Before rolling, the TRC-material was annealed for 2h at 400°C in order to homogenize the microstructure, dissolve precipitates and weaken the texture by means of static recovery and recrystallization. Fig. 1b shows the texture after annealing. As evident, the intensity of the basal poles decreased by nearly 60% and the {10-10}-planes were no longer preferred in RD. Unidirectional and cross rolling were conducted at 400°C up to 90% final thickness reduction. After each rolling pass the samples were reheated for 5 min to regain the rolling temperature. For cross rolling, the rolling direction was changed after each pass by rotating the sample by 90° around ND.

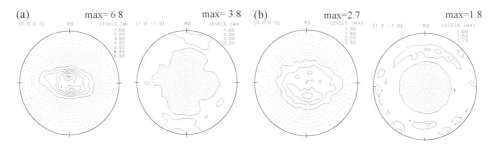

Figure 1: Initial textures of TRC-AZ31 (a) as-received and (b) after heat treatment, represented by the basal (left side) and prism (right side) pole figures. Textures were measured in the mid-thickness of the rolling plane

The influence of thickness reduction per rolling pass on texture and microstructure evolution was investigated and compared for both unidirectional and cross rolling. The results are shown in Fig. 2 and Fig. 3. The shown pole figures were recalculated from the ODFs under the assumption of triclinic sample symmetry.

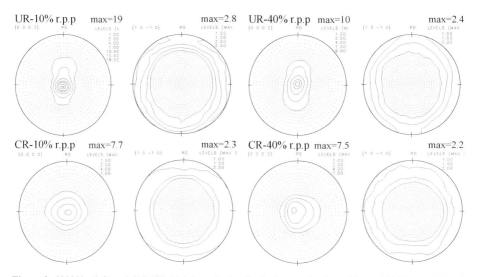

Figure 2: {0002}- (left) and {10-10}-(right) recalculated pole figures after hot rolling at 400 °C up to 90 % final thickness reduction, using 10 % (maximum number of rolling passes till final thickness) and 40 % thickness reduction per pass (minimum number of rolling passes till final thickness). Top row: unidirectional rolling; bottom row: cross rolling; RD=RD1, TD=RD2. Rolling velocity v=22 m/min ($\dot{\varepsilon} \sim 42\ \mathrm{s}^{-1}$)

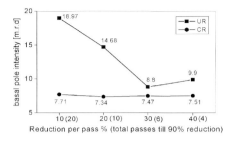 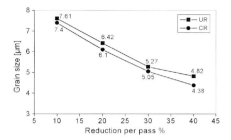

Figure 3: (a) Basal pole intensity at 90 % for unidirectional and cross rolling at different reductions per rolling pass; (b) Influence of reduction per rolling pass on the final grain size for unidirectional and cross rolling

4 Non-basal Slip, Twinning and DRX During Plane Strain Compression

The combined effect of non-basal slip and dynamic recrystallization on the texture development in a commercial magnesium alloy AZ31 was investigated. Plane strain compression (PSC) was chosen as the deformation mode using a channel-die device since in contrast to rolling it allows to conduct experiments at defined deformation conditions, i. e. constant temperature and strain rate and the samples can be quenched immediately after the test. The starting material had been produced by rolling and had a pronounced texture with the basal planes aligned perpendicular to the compression plane and the **c**-axis parallel to the rolling direction (Fig. 4).

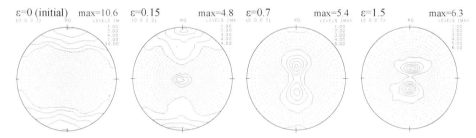

Figure 4: {0002}-recalculated pole figures showing initial hot rolling texture and texture development at 400 °C, 10^{-4} s^{-1} during PSC at selected strains

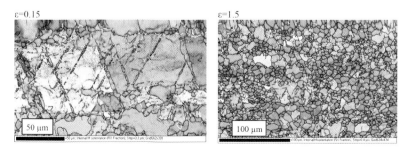

Figure 5: OIM maps showing microstructure development upon PSC at 400 °C, 10^{-4} s^{-1} at early and advanced stages of deformation

PSC was conducted at 400 °C at a constant strain rate of 10^{-4} s^{-1}, a condition where, a combined influence of <c+a>-pyramidal slip and DRX occurs. Tests were terminated at selected strains, and the sample was immediately quenched to freeze texture and microstructure for subsequent investigation of texture and microstructure at various stages of deformation. The results are presented in Fig. 4 and Fig. 5 in terms of basal pole figures and OIM maps, respectively.

5   Deformation Behaviour of Initially-textured AZ31 During Uniaxial Compression

The material used in uniaxial compression was an extruded commercial magnesium alloy AZ31B supplied by Otto Fuchs KG. The chemical composition (wt. %) was the following: 2.92 Al, 0.84 Zn, 0.33 Mn, 0.02 Si, 0.004 Fe, 0.001 Cu, 0.001 Ni, Mg (balance). For compression tests cylinders of size Ø 15 mm × 25 mm were machined from the extruded material in three different orientations with the compression axis parallel, perpendicular and 45° aligned to the extrusion direction, respectively. The initial orientations are presented in Fig. 6 in terms of recalculated basal pole figures. Compression tests were performed in a temperature range of 200 °C–400 °C at different constant strain rates from 10^{-2} s^{-1} to 10^{-4} s^{-1} up to a final true strain of 1.2. The influence of texture on deformation behaviour and deformation mode was investigated. The results are shown in Fig. 7.

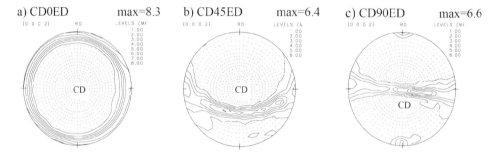

Figure 6: Starting textures for uniaxial compression; a) compression axis CD parallel to extrusion direction ED, b) CD aligned 45° to ED and c) CD perpendicular to ED

6   Discussion

6.1   Unidirectional and Cross Rolling

The rolling textures after hot rolling at 400 °C up to 90 % total reduction are given in Fig. 2. The rolled plates exhibit mainly an asymmetric basal texture. The shape (c-axis shifted toward RD or TD) and location of the main texture components as well as their intensities depend on the mode of rolling and on the thickness reduction (strain) per rolling pass which determines the total number of rolling passes till final thickness. Experiments were conducted using reductions per pass ranging from 10 % to 40 % of the actual thickness. Reductions of less than 10 % per rolling pass require a very high number of rolling passes till final thickness, whereas reductions of more than 40 % per rolling pass can cause a premature fracture of the sample. For unidirectional rolling, the texture intensity decreased rapidly with increasing reduction per pass, i. e. with decreasing total number of rolling passes (Fig. 3a). In the case of cross rolling this tendency was not observed. In contrast to unidirectional rolling, cross rolled samples exhibited weaker texture intensities, and the strain per rolling pass showed no influence on texture formation. It worth noting that alternating the strain path during cross rolling results in a rotation of the main texture component in the basal pole figure, always in the previous rolling direction, either RD1 or RD2. Keeping the strain path unchanged in unidirectional rolling increasingly accumulates the basal orientation perpendicular to ND (Fig. 2) owing to a c-axis rotation towards ND. Apparently, this is the reason for the strong texture intensity upon unidirectional rolling. Singh and Schwarzer [3] observed similar results and suggested that cross rolling induces higher deformation than unidirectional rolling due to the change in strain path which would result in early recovery or recrystallization and weaken the texture. Microstructure development (onset of deformation twinning and grain size) showed similarities between both rolling modes regarding its dependency on the reduction per rolling pass. Higher strain per rolling pass caused a conspicuous grain refinement. The finest average grain size was ~4 µm upon cross rolling with 40 % reduction per pass (Fig 3b). Furthermore, cross rolled samples exhibited a much better surface quality than samples rolled unidirectionally. This was particularly observed at high reductions per pass, i.e. 30 % and 40 %.

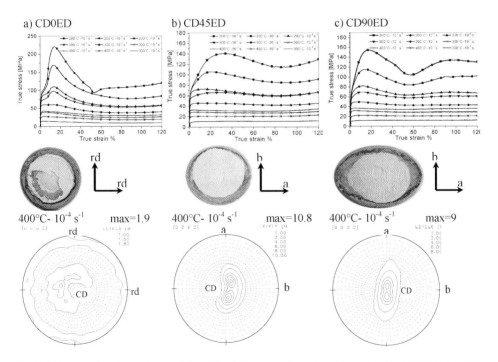

Fig. 7: Influence of starting texture on deformation behaviour during uniaxial compression (a) compression axis CD parallel to extrusion direction ED, (b) CD aligned 45° to ED and (c) CD perpendicular to ED; top row: flow curves at various conditions (temperature and strain rate), centre row: top view of compression specimens after completion of test ($\varepsilon = 1.2$), bottom row: final texture upon compression at 400 °C, 10^{-4} s^{-1} (rd= radial direction)

6.2 Plane Strain Compression

Fig. 4 reveals the evolution of texture in AZ31 upon high temperature channel-die deformation at early ($\varepsilon = 0.15$), moderate ($\varepsilon = 0.7$) and advanced ($\varepsilon = 1.5$) stages of deformation. The starting texture shown in Fig. 4 is designed to suppress basal slip under compression since the basal planes are aligned parallel to the compression axis. Due to the geometry of the channel-die device, prismatic slip is also suppressed because it would cause sample broadening. Under these conditions the grains are favourably oriented for twinning and <c+a>-pyramidal slip, both important mechanisms for the accommodation of strain along the c-axis. As evident from Fig. 4, at $\varepsilon = 0.15$, the intensity increase of basal poles near the compression axis is attributed to a characteristic 86° rotation from the initial orientation towards CD caused by {10-12}-tensile twinning. As this new orientation develops during ongoing deformation, it disfavours deformation twinning and the volume of grains favourably oriented for twinning will decrease. Since the deformation was carried out at high temperature and very low strain rate, ($T = 400°C$ and $\dot{\varepsilon} = 10^{-4}\ s^{-1}$) the CRSS of pyramidal slip is very much lower than at low temperatures. Hence after twinning is exhausted, <c+a>-pyramidal slip with a sufficient number of independent slip systems becomes the main deformation mechanism for straining along the c-axis. This accounts for the splitting of the basal pole intensity as observed at moderate and high strains in Fig. 4. The role of <c+a>-pyramidal slip was frequently reported earlier [4,5]. Fig. 5 reveals the microstructure at low and high strains. At the beginning of deformation it is characterized by huge grains in the form of bands aligned parallel to the extension direction, containing deformation twins and shear bands. Recrystallized grains can be recognised at the grain boundaries. At high strains the microstructure is dominated mainly by dynamically recrystallized grains with an average grain size of ~16 µm.

6.3 Uniaxial Compression

The starting textures, shown in Fig. 6 had conspicuously different orientations and caused different deformation behaviour. The orientations differed in such a way that the activation of crystallographic slip on basal and non-basal planes as well as the activation of twinning was easier for some orientations than for others. This gave rise to the different plastic deformation behaviour as evident from the stress-strain curves. In orientations with the basal planes parallel to the compression direction (Fig. 6a, CD0ED), there is practically zero resolved shear stress on the basal planes and thus, basal slip is suppressed. Depending on testing conditions, i.e. temperature and strain rate, either mechanical twinning or non-basal slip (both favourably oriented) will have more influence in the beginning of deformation, reorienting grains into more favourable orientations. By contrast, the starting orientation CD45ED (Fig. 6c) is strongly favoured for basal slip and hence exhibits a low work hardening rate at low strains than the other two starting orientations. Sample CD45ED (Fig. 6b) develops a <10-10> ED fiber texture. The two main orientations (basal planes parallel and perpendicular to CD, respectively) suppress basal slip. However, in comparison to Fig. 6a (CD0ED) twinning is favoured only in one orientation with the basal planes parallel to CD. This will introduce less twin boundaries in the microstructure and leads to less dislocation hardening. Also, the fact that part of the grains in the fibre around ED can deform easily by basal slip adds to the same conclusion. The above mentioned competing mechanisms of slip and twinning had a substantial influence on the macroscopic shape change of the compressed specimens (Fig. 7a-c). Samples with an initial orientation

CD0ED possessed a perfectly round shape after completion of the tests, whereas a strong anisotropy was observed in the final shape of the CD90ED-samples. At 200 °C and 300 °C all three orientations gave rise to characteristic basal fibre textures. At 400 °C and a strain rate of 10^{-4} s^{-1} (most favourable conditions for <c+a>-pyramidal slip associated with DRX) CD0ED-samples showed a very weak texture. The CD90ED-orientation developed a basal texture with the basal poles shifted towards the **a**-direction (**a**: main axis of the elliptical cross section of the CD90ED-compressed sample). An asymmetric texture with strong basal pole splitting was found in CD45ED-samples.

7 Summary

Plastic deformation behaviour and the evolution of texture in a commercial magnesium alloy AZ31 during thermo-mechanical processing were investigated. The obtained results are summarized as follows:
1. Unidirectional hot rolling of twin roll cast AZ31 results in asymmetric basal textures with strong intensities. An increasing strain per rolling pass and thus a reduced the total number of rolling passes weakens texture strength. This does not play a role for cross rolling. Compared to unidirectional rolling, cross rolling showed better surface quality regarding crack formation and exhibited weaker orientation densities. Hot rolling of TRC-AZ31 at 400 °C with 30 % thickness reduction per rolling pass offers a good combination between surface quality, texture intensity, grain size and production time.
2. When at incipient deformation an orientation is strongly disfavoured for easy slip, such as basal slip, mechanical twinning and <c+a>-pyramidal slip become important. Both mechanisms assist straining of grains along their c-axis and affect the strain hardening rates. Reorientation of grains into more favourable orientations for basal slip results in a softening behaviour in the stress-strain curves.
3. The different extrusion textures used for uniaxial compression showed strong compression asymmetry and caused substantially different deformation behaviour.

8 Acknowledgements

The financial support of the Deutsche Forschungsgemeinschaft DFG (GO 335/27) is gratefully acknowledged. The Authors would also like to thank Hydro-Aluminium R&D and Otto-Fuchs AG for the kind donation of the AZ31 material.

9 References

[1] S. Suwas, A. Singh, K. Rao, T. Singh, Z. Metallkd. 93 (2002) pp. 928–935
[2] L. Löchte, H. Westengen, J. Rødseth, in: N. Neelameggham, H. Kaplan, B. Powell(Eds.), Magnesium Technology 2005, TMS, USA (2005) pp. 247–252
[3] AK. Singh, RA. Schwarzer, Z. Metallkd. 96 (2005) pp. 345–351
[4] S. Agnew, C. Tomé, D. Brown, T. Holden, S. Vogel, Scr. Mat. 48 (2003) pp. 1003–1008
[5] M. Yoo, S. Agnew, J. Morris, K. Ho, Mat. Sci. Eng. A319-321 (2001) pp. 87–92

Microstructural Evolution by Hot-Working of Extruded and Continuously Casted Mg-Wrought-Alloys AZ31 and AZ80

B. Viehweger[1], L. Schaeffer[2] and M. Düring[1]
[1]Brandenburg University of Technology, Chair of Design and Manufacturing, Cottbus, Germany
[2]Universidade Federal do Rio Grande do Sul, Centro de Technologia, Porto Alegre - RS, Brazil

1 Introduction

Due to the low specific weight Mg-alloys offer potentials for weight-savings in a widespread variety of automotive applications [1, 2]. In particular high quality forgings made of Mg-alloys exhibit superior mechanical properties, but up to now the predominant contingent of automotive parts is produced in form of castings. Only a minor part is in use as high-quality forging for body, powertrain and supporting assemblies of cars. A lack of knowledge concerning processing technologies for Mg-alloys is considered as the prime reason for this situation.

Available Mg-material of alloy type AZ31 and AZ80 for forging is mainly supplied by commercial producers in form of semi-continuously casted and subsequently extruded rods. The extrusion process ensures a fine-grained Mg-material of enhanced forming properties and causes a pronounced fibre texture [3]. While plastic forming occurs slip is preferred on (0001)-basal planes. Between room temperature and 225 °C this is the primary forming mechanism additionally supported by deformation twinning. At forming temperatures above 225 °C slip can act on additionally activated pyramidal planes $\{10\bar{1}1\}$. The forming of Mg-alloys generally leads to a basal–plane lattice orientation in flow direction. In consequence the activation of different deformation modes is caused in dependence on employed temperatures and loading directions for subsequent forming processes [4]. Furthermore, anisotropic mechanical properties are resulted. At room temperature compressive loads, acting perpendicular to the extrusion direction of rods, enable eased slip in numerous pre-orientated basal planes. Nevertheless, published values in literature exhibit by far higher values for obtained tensile yield strength than for compressive yield strength for loads acting parallel to the extrusion direction. These differences may be explained by deformation twinning in $\{10\bar{1}2\}$-second order pyramidal planes. The comparison of tensile and compressive yield strength values in an orientation perpendicular to the extrusion direction of rods does not show such significant differences [5]. Finally, the microstructural texture in form of elongated grains and precipitation bands, mainly caused by the extrusion process, influence the forming behaviour of AZ31 and AZ80 material in different forming directions.

2 Characteristics of Investigated Mg-Material

Static mechanical properties of examined Mg-material have been acquired by means of cylindrical upsetting test at room temperature. In order to determine the impact of heat treatment on achievable strength values various tempers - F, T4, T6 - have been taken into consideration. The solution treatment has been carried out at 390°C for the AZ31 and 410 °C for the AZ80 material. A dwell time of 12 hours has been applied. For the subsequent aging of solution treated Mg-material a temperature of 150 °C has been set for 12 hours for AZ31 respectively 170 °C for 12 hours for AZ80. Obtained values for compressive yield strength, ultimate compressive strength and elongation to fracture are displayed in tab. 1.

Table 1: Mechanical properties of extruded Mg-material

Compression tests	$\sigma_{0,2C}$ [MPa]		UCS [MPa]		El [%]	
AZ31 – extruded	l	r	l	r	l	r
as received (F)	94,7	56,3	397,6	278,0	11,8	15,8
solution treated (T4)	73,7	52,6	376,8	287,0	11,5	16,6
artificial aged (T6)	75,1	48,5	393,8	263,3	12,6	14,2
AZ80 – extruded	l	r	l	r	l	r
as received (F)	159,3	116,8	431,3	352,9	11,4	12,8
solution treated (T4)	104,3	82,0	420,1	366,2	12,5	16,5
artificial aged (T6)	159,2	126,7	457,6	373,6	10,5	11,7

The examined material is characterised by a fine-grained microstructure of average grain sizes of 16 µm for the AZ31- respectively 20 µm for the AZ80-alloy. Certain areas of investigated Mg-stock exhibit larger, in extrusion direction elongated grains. With regard to the grain size distribution, the extrusion process for the alloy AZ80 is less critical than for AZ31 [6]. The most apparent difference between the two alloys is the existence of γ-phase-precipitations ($Mg_{17}Al_{12}$) in AZ80-stock due to 8 at% aluminium alloying addition in comparison to 3 at% in case of the AZ31-alloy (fig. 1).

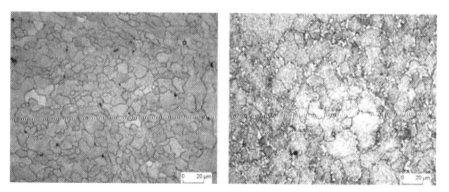

Figure 1: Micrographs of examined extruded Mg-material AZ31 and AZ80 (500:1)

In contrast the static mechanical properties of casted Mg-material are assumed to be nearly isotropic. For the AZ80-material, extruded as well as semi-continuously casted, it has been pointed out that static mechanical properties strongly depend on applied temper. Compressive yield strength values after solution treatment have been noticeably increased by artificial aging. For the AZ31-alloy same effects could not be confirmed. Only decreased compressive yield strength values, obtained on extruded AZ31-material, have demonstrated the impact of softening on their mechanical properties as a result of static crystal recovery and recrystallisation processes. Displayed values for the elongation to fracture correspond to this fact. Displayed mechanical properties for the continuously casted Mg-material clearly indicate that their direct use for engineering purposes related to enhanced strength is nearly impossible. Primary characteristic values for compressive yield strength do not comply with most of the technical requirements of engineering solutions and therefore do not contribute to weight-saving purposes.

Table 2: Mechanical properties of semi-continuously casted Mg-material

Compression tests	$\sigma_{0,2C}$ [MPa]	UCS [MPa]	El [%]
AZ31 – semi cont. casted			
as received (F)	38,4	263,5	21,8
solution treated (T4)	34,8	267,2	21,6
artificial aged (T6)	34,3	273,0	21,6
AZ80 –semi cont. casted			
as received (F)	89,6	268,8	10,6
solution treated (T4)	55,6	261,5	13,8
artificial aged (T6)	82,5	284,0	12,2

The micrographs taken from received semi-continuously casted Mg-material exhibit average grain sizes of 250 µm for both alloys. For the as-received AZ80-material the presence of γ-phase-precipitations has been confirmed as a result of segregation processes during cooling of homogenised bars (fig. 2).

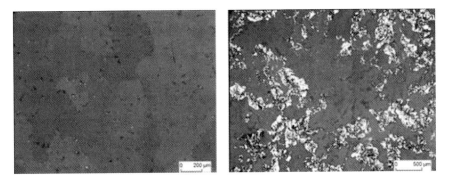

Figure 2: Micrographs of semi-cont. casted Mg-material AZ31 and AZ80 (500:1)

3 Flow Stress Curves of Investigated Mg-Materials

In order to obtain necessary data for the prediction of forces for bulk forming processes of Mg-alloys AZ31 and AZ80 flow stress curves have been acquired by means of cylindrical upsetting tests at forming temperatures of 300, 350, 400 and 450 °C as well as strain rates of 10^{-1} and $10\ \text{s}^{-1}$. Cylindrical specimens in dimensions of Ø12.5 mm × 20mm were cut from extruded rods in orientations longitudinal and perpendicular to the extrusion direction. Considering the extruded AZ31-material it was observed that plastic forming up to a true strain limit of 1 is nearly possible for the whole examined parameter range. Cracks on the surface of specimens, formed at 300 °C, have been detected at true strain values near 0.4 for a strain rate of 1 s^{-1} respectively 0.25 for a strain rate of 10 s^{-1}. Concerning attained strains to fracture the AZ80 must be considered as to be less critical for warm forming purposes. Achievable true strain values for a forming temperature of 300 °C and a strain rate of 10 s^{-1} usually exceed true strain values of 0.4. Exemplarily flow stress curves, obtained for both alloys in an upsetting direction parallel to the extrusion direction, are displayed in fig. 3 for a strain rate of 1 s^{-1}.

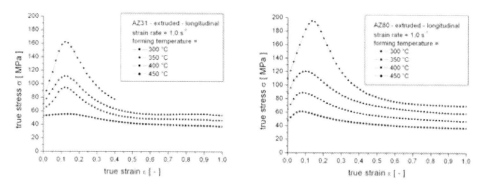

Figure 3: Flow stress curves for extruded AZ31 and AZ80 material (longitudinal)

Due to low stacking fault energy values of Mg-alloys their flow stress curves are characterised by a recrystallisation peak at low strain values. Both Mg-alloys initially exhibit pronounced strain hardening effects. The strain hardening effect is accompanied by dynamic softening that becomes more important as plastic forming proceeds. The critical point of initial recrystallisation, around true strain values of 0.15, is characterised by softening as a result of nucleation processes [7, 8]. This finally results in decreased flow stress values, as much as softening overlays the hardening effect. Achievable compressive strains at higher temperatures are minor for AZ31 than for AZ80. Furthermore, obtained flow stress values are higher for extruded AZ80- than for AZ31-material.

The dependence of the activation of varying deformation modes of extruded Mg-material under warm forming conditions on the loading direction has been demonstrated by means of cylindrical upsetting tests, applying cylindrical specimens taken from extruded rods in an orientation perpendicular to the extrusion direction. Obtained results are displayed in fig. 4. The flow stress maxima clearly differ from values obtained in longitudinal specimen direction. The maximum flow stress at a forming temperature of 300 °C does not exceed values of 120 MPa for the AZ31- respectively 140 MPa for the AZ80-material. This may be also explained as a result of deformation twinning at higher forming temperatures.

In order to acquire reference values cylindrical upsetting tests have been performed applying semi-continuously casted Mg-material. As shown in fig. 5 the softening peak is less pronounced as in case of extruded material. The nucleation sites are mainly distributed in form of a "necklace structure" at the grain boundaries. Furthermore excessive shear movements in grain boundary areas seem to cause gradients in dislocation density. While observed near grain boundary areas enter in excessive recrystallisation activites, due to existing high quantity of created nucleation sites, the core areas of mother-grains augment their dislocation density but do not recrystallise. For the semi-continuously casted AZ31-material the application of a forming temperature of 350 °C and a strain rate of 10 s^{-1} allows achievable true strain values of 0.9. The application of a reduced forming temperature of 300°C finally decreases achievable true strain values up to 0.9 for a strain rate of 1 s^{-1} respectively 0.5 for 10 s^{-1}. Although the application of an augmented forming temperature of 450 °C is hardly applicable - the melting temperature of μ-phase precipitation is 437 °C - the semi-continuously casted AZ80-material must be considered as to be less critical for warm forming purposes. Due to better nucleation conditions the flow stress peak is more pronounced what indicates enhanced softening caused by recrystallisation.

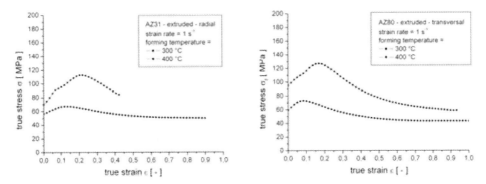

Figure 4: Flow stress curves for extruded AZ31 and AZ80 material (transversal)

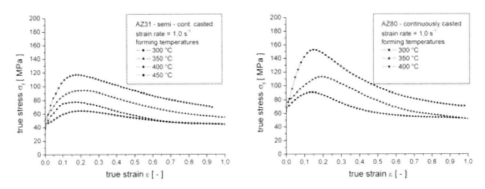

Figure 5: Flow stress curves for semi-cont. casted AZ31- and AZ80-material

4 Microstructure after Hot-Forming

In order to determine the dependence of microstructural evolution, caused by hot forming on applied forming parameters and examined alloy, a quantitative analysis of microstructure has been carried out. Therefore applied specimens of already mentioned dimensions have been cut in an orientation perpendicular to the extrusion direction of received AZ31- and AZ80-rods. The upsetting of the specimens has been carried out comparatively at temperatures of 300 and 400 °C and at a strain rate of 1 s^{-1}. After forming the specimens were directly quenched in water. The time between upsetting and quenching has been kept as short as possible. The average delay between these two processes was about 10 seconds. Main objective of this procedure was the separation of occurring metadynamical softening processes, in particular metadynamical recrystallisation. Set up maximum true strain values of 0.1, 0.2 and 0.5 were of particular interest for further examinations. The evaluation of microstructural evolution at corresponding points of the flow stress curves should be enabled.

The results, obtained for the extruded AZ31-feedstock, exemplarily show that at a forming temperature of 400 °C and a plastic strain of 0.1 twinning occurs. As plastic forming proceeds deformation twins become more obvious as well as the recrystallisation process is initiated due to the creation of nucleation at grain boundary areas. At a logarithmic plastic strain value of 0.5 the corresponding micrographs do not show any deformation twins as a result of proceeding recrystallisation. In comparison to this, it is remarkable that at a plastic strain of 0.1 and a forming temperature of 300 °C deformation twins appear in smaller dimensions as already explained for a forming temperature of 400 °C [9]. Recrystallised grain sizes are around 30 µm for 400 °C and less than 15 µm for a forming temperature of 300 °C.

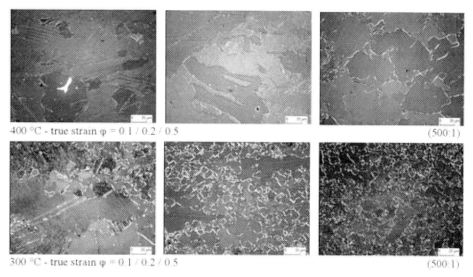

Figure 6: Micrographs of AZ31-extruded / strain rate = 1 s^{-1}

The recrystallisation behaviour of specimens, machined out of received AZ80-material, has been examined in the same way. At forming temperatures of 400 °C formed specimens do not exhibit any twinning. In fact primary recrystallisation seems to dominate the material's softening behaviour directly after the forming process starts. Flow stress curves, determined from specimens that have been cut in a orientation radial to the extrusion direction of received rods, confirm that during plastic forming, even at smaller true strain values, the equilibrium state of hardening- and softening mechanisms is achieved (fig. 7). The softening behaviour of AZ80-alloys at forming temperatures of 300 °C is characterised by twinning. Primary recrystallisation becomes obvious when plastic forming exceeds true strain values of 0.5.

The analysis of AZ31-specimen formed up to the true strain value of 1.0 by the application of forming temperatures of 300 and 400 °C demonstrates that their microstructure is characterised by fine grains of an average grain size of 10 and 15 µm. For the AZ80-alloy values of 10 respectively 20 µm were detected.

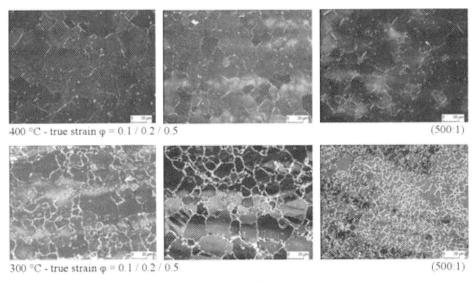

Figure 7: Micrographs of AZ80-extruded / strain rate = 1 s^{-1}

For the evaluation of the forming behaviour of semi-continuously casted Mg-material temperatures of 300, 350 and 400 °C have been taken into account. Specimens were hot worked up to the true strain value of 1.0. The results of metallografical analysis show that in particular the initial grain size as well as lacking prior work hardening influences their forming behaviour. A remarkable characteristic is that in relation to the grain volume smaller percentage of grain boundary areas results in decreased nucleation activity. The nucleation in coarse grained material is less randomly distributed and occurs mainly in form of a typical "necklace"-structure. For the AZ31 alloy this can be confirmed according to fig. 8 for all forming temperatures. In recrystallised grain boundary areas shear movements are detectable what contributes to the softening behaviour as plastic forming proceeds. In addition, these shear areas may be pointed out as a reason for further nucleation activities. Commonly published arguments for the description of influences of the initial grain size on the forming behaviour also include the fact that stored energy tends to increase with a decrease of initial grain size [8].

The results, obtained using the semi-continuously casted AZ31-feedstock could not point out remarkable differences in recrystallisation behaviour depending on varying forming temperatures, although the percentage of recrystallised grains tends to increase with increasing temperatures. Opposite to this, the forming of extruded AZ-alloys under given parameters is characterised by a lack of recrystallisation at forming temperatures of 300 °C even at a true strain value of 1. The stress relief that displays the forming behaviour of Mg-materials whilst plastic forming is obviously provided by shear movements between the grains. Near the grain boundary areas no further nucleation is apparent. In addition to this it is assumed that only dynamically initiated recovery processes can contribute to the material's softening. However, the forming at comparatively elevated temperatures of 350 and 400 °C is characterised by a fine-grained microstructure, whereas already the explained dependence of recrystallized grain size on the forming temperature could be confirmed as well.

Figure 8: Micrographs of AZ31 /AZ80 semi cont. casted-true strain $\varphi = 1$, strain rate = 10^{-1} s^{-1}

5 Conclusion

In order to optimise forging processes for manufacturing high-quality forgings made of Mg-alloys AZ31 and AZ80 various facts have to be taken into consideration. Due to the hexagonal shape of elementary cells forming at room temperature is difficult. Therefore a warm forming process is needed. The investigation concerning the formability of extruded Mg-alloys between 300 and 450 °C have shown that local strain rate values up to 10 s^{-1} are not critical for open die forging. Only at 300 °C cracks occurred at higher strain values und strain rates. It has been pointed out that the extruded AZ80-alloy is less critical for warm forming purposes at higher strain rates than the extruded AZ31-alloy. The reduction of necessary forces for the forming processes by applying temperatures above 400 °C is not a suitable way due to resulting decreased critical strains to recrystallisation and increased recrystallised grain sizes at higher strain values. Therefore an optimised microstructure can be achieved only by applying forming temperatures below 350 °C for the last press stage. At this temperature the formability of both Mg-alloys allows the set up of higher strain rates for economically streamlined forging processes and best achievable strength caused by finely-recrystallised grains. Additionally, remaining work hardening values can be controlled by subsequent quenching and heat treatment processes. As a result of the examinations it became clear that the recrystallisation of extruded AZ80-material occurs much faster in comparison to investigated AZ31-material.

The direct forging of semi-continuously casted Mg-alloy AZ31 is more complicated. Even applying a forming temperature of 350 or 400 °C dynamic recrystallisation does not lead to a fully recrystallised microstructure for the AZ31 alloy and therefore remains inhomogeneous. In comparison the semi-continuously casted AZ80-material does no contribute to softening by recrystallisation at a forming temperature of 300 °C up to a true strain value of 1. Nevertheless, excessive recrystallisation activities have been observed at forming temperatures of 350 and 400°C due to better nucleation conditions. A forming temperature above is not suitable for warm forming of the semi-continuously casted AZ80-material.

6 Acknowledgements

The authors are very grateful to the German Federal Ministry of Education and Research for funding this work.

7 References

[1] Friedrich, H.; Schuhmann, S.: Forschungsstrategien für ein zweites "Magnesium-Zeitalter" im Fahrzeugbau, in: Materialwissenschaft u. Werkstofftechnik, Vol. 32 (2001), 6–12.
[2] Fischer, G., Becker, J., Stich, A., Gesenkschmieden hochfester Magnesium-knetlegierungen für Bauteile der Automobil- und Luftfahrtindustrie, Materialwissenschaft u. Werkstofftechnik, Vol. 31 (2000), 993–999.
[3] Doege, E; Janssen, S.; Elend, L.-E.: Schmieden, in: Magnesium-Taschenbuch, Aluminium-Verlag, Düsseldorf (2000), 359–382.
[4] Barnett, M.R.: Influence of deformation conditions and texture on the high temperature flow stress of magnesium AZ31, Journal of Light Metals, Vol. 1 (2001), 167–177.
[5] Hilpert, M. et al, Influence of thermomechanical processing on microstructure, texture and fatigue performance of the high-strength magnesium alloy AZ 80, in: Mordike, B. L., Kainer, K. U., editors, Magnesium Alloys and their Applications, Werkstoff-Informationsgesellschaft, Frankfurt (1998), 319–324.
[6] Swiostek, J. et.al.: Comparison of Microstructure and Mechanical Properties of Indirect and Hydrostatic Extruded Magnesium Alloys, in: Magnesium Alloys and Their Applications, Wolfsburg (2004), 278–284.
[7] Barnett, M.R.: Hot Working Microstructure Map for Magnesium AZ31, Material Science Forum, Vol. 426–432 (2003), 515–520.
[8] El-Magd, E.; Abouridouane, M.: Einfluss der Umformgeschwindigkeit und -temperatur auf das Fließverhalten der Magnesiumlegierung AZ80, Zeitschrift für Metallkunde, Vol. 92 (2001), 1231–1235.
[9] Humphreys, F. J.; Hatherly, M.: Recrystallisation and Related Annealing Phenomena, Pergamon, Oxford (2000), 2002.

Mechanical Properties

Ductile Failure of Magnesium AZ31: Twinning when the c-axis is Compressed

M. R. Barnett
Centre for Material and Fibre Innovation, Deakin University, Geelong, Australia

1 Introduction

The extent to which magnesium alloys can be subjected to room temperature deformation is limited. The main reason given for this in the literature is that there is an insufficient availability of deformation systems. However, the material does display *some* plasticity, which testifies to the fact that it does, at least initially, have enough deformation systems for generalized plastic deformation. There is something about these systems, modes or some other feature of the material, that leads to rapid ductile failure.

Certainly, the deformation of this material is dominated by easy basal slip [1], which supplies only two independent systems. The most difficult deformation is thus that in which tension or contraction occurs along the c-axis, perpendicular to the basal plane. In single crystals, these types of deformations have been observed to be accompanied by significant amounts of deformation twinning. When the c-axis is extended, $\{10\bar{1}2\}$ twinning occurs [1]. When the c-axis is compressed, a number of twinning modes have been reported, one of which is the $\{10\bar{1}1\}$ mode [1]. The former is frequently observed in polycrystals and the latter is increasingly also being reported to occur in rolled and extruded products [2].

As with the single crystal studies, examples of $\{10\bar{1}1\}$ twinning in polycrystals are frequently associated with the occurrence of secondary twinning on the $\{10\bar{1}2\}$ mode in the twinned interiors [3]. This second twinning step frequently places the basal plane in an orientation favourable for basal slip. Thus, this mode can lead to flow localization and rapid failure of single crystals. The present work examines the possibility that a similar occurrence might be important to understanding the failure of polycrystals.

2 Experiments

Commercial samples of extruded and rolled AZ31 were received for use in the present work. A series of tensile tests were carried out in the extrusion direction of the extruded bar, the mean linear intercept grain size of which was ~9 µm. Selected samples were prepared for optical microscopy and electron backscattering diffraction (EBSD). The as-received AZ31 sheet was subjected to 1% rolling reduction followed by annealing to generate a structure comprised of coarse grains with a mean linear intercept of ~80 µm. These coarse grains made it a simpler matter to observe twinning. A series of interrupted tensile tests were performed in the rolling direction. After an initial elongation of 5 %, the sample edge was polished, etched and subjected to impression by micro-hardness testing. These indents provided fiducial markers for subsequent strain measurement. The sample was also subjected to EBSD analysis, which enabled the twin type to be estimated.

3 Results / Discussion

The microstructure revealed by optical metallography near to the region of failure in the extruded sample tested at room temperature is shown in Figure 1. It is clear that the fraction of twinning increases from left to right in this image and that near to the fracture face long regions of what appear to be bands of shear can be seen. These appear to be formed, at least initially, by the co-operative operation of twinning in adjacent grains. Such a scenario is consistent with the idea that these twins comprise regions of favourable basal plane alignment which makes them softer than the matrix and likely to undergo greater amounts of deformation. When this local deformation attains a high value, and when adjacent regions become coupled, a shear band is formed.

Figure 1: Twinning and shear bands evident in a sample of AZ31 bar tested to failure along the extrusion direction (horizontal) at room temperature at a strain rate of 0.01 s^{-1}

In the case shown in Figure 1, no void formation is obvious behind the fracture front, and this is common in materials undergoing failure by shear localization. However, in a number of other instances, isolated failures within twinned regions could be seen. An example of this is shown in Figure 2. The formation of voids preferentially in twinned regions is also consistent with the idea that the twins create regions where the deformation localizes.

To determine the nature of the twins forming in the extruded bar, EBSD examination was performed on the deformed samples. It turned out that the twin interiors were frequently difficult to index; that is, they did not permit clear Kikuchi patterns to be obtained. This too is what one would expect if the twins were regions of concentrated deformation. However, in a number of cases, fragments of the twin interiors were able to be indexed and an example is given in Figure 3.

The twins in Figure 3 are twins that form first on the $\{10\bar{1}1\}$ plane followed by secondary twinning on the $\{10\bar{1}2\}$ plane. This was ascertained by the habit of the twin and by the nature of the boundary misorientation relationship. The reorientation produced by the two twins in succession has been shown to give a 38° rotation around a $<1\bar{2}10>$ axis [4]. Boundaries corresponding to this reorientation are given in Figure 3. The basal pole figures are also shown and it can be seen that, for a (horizontal) tensile stress, the basal planes are more favourably aligned after twinning than they were before. (Favourable alignment can be inferred from the proximity of the basal plane to the planes of maximum shear stress, which lie at 45° to the tensile direction).

Figure 2: Example of a void forming in a twin in AZ31 bar tested to failure at rom temperature

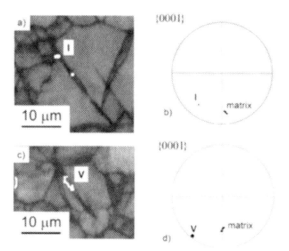

Figure 3: EBSD analysis of $\{10\bar{1}1\}$ double twins. a) and c) EBSD maps showing double twin boundaries (38° $<1\bar{2}10>+/-7°$) in white and shaded according to the Kikuchi band contrast. b) and c) Basal pole figures showing that the basal plane is more favourably aligned for slip in the twin interior. The tensile direction is horizontal and the pole figure reference frame is the same as for the images.

In an attempt to measure the strain localizing in $\{10\bar{1}1\}$ double twins, a grid of hardness indents was impressed on a coarse grained sample of rolled AZ31 sheet following 5% deformation. The sample was then strained to failure. In most instances the grids were intersected by twins. The example shown in Figure 4a reveals regions of intense flow in which shear strains of the order of 3 can be observed. (The hardness indents were originally aligned along lines more or less parallel to the edges of the image.)

The pole figures corresponding to the grain shown in Figure 4a are given in Figure 4b. The reference frame for these is the same as for Figure 4a. The dashed lines mark the expected trace of the intersection of the plane with the sample surface. The three dashed lines share a similar orientation to the key microstructure features in Figure 4a. The feint diagonal lines in Figure 4a lie close to the expected trace of the basal plane and are thus most likely to be basal slip lines. The two families of sharp diagonal features are most probably twins that have formed on the $\{10\bar{1}1\}$ plane. The ideal line orientations are all rotated slightly clockwise to the observed lines. This is likely to be due to slight differences in sample alignment. The inset shows feint basal slip lines within the twin interior. These are more favourably aligned for slip, which can explain the high degree of localization.

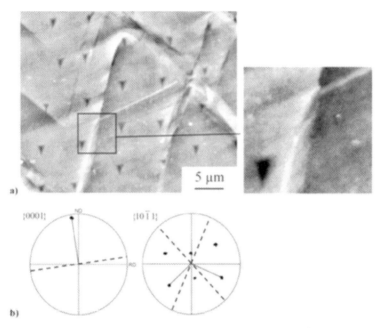

Figure 4: a) Regions of intense local shear (shear strain >3) in regions of more favourable basal plane alignment (see inset). Rolled AZ31 sheet, tensile test direction horizontal. Diagonal feint lines are basal slip lines. b) Pole figures showing the inclinations (dashed) expected for the intersections of basal slip and $\{10\bar{1}1\}$ twins with the sample surface.

4 Conclusion

The present work illustrates the occurrence of $\{10\bar{1}1\}$ twinning followed by $\{10\bar{1}2\}$ secondary twinning in commercial grade polycrystalline wrought AZ31. It is shown that these double twins can lead to favourable alignment of the basal planes, which leads to high degrees of local flow. Such local flow is expected to contribute to the ductile failure of these materials.

5 Acknowledgements

The help of Zohreh Keshavarz, Brian Gerard, Simon Jacob, Rob Pow and John Vella is gratefully acknowledged. This work was supported by an ARC Discovery Project grant.

6 References

[1] E. W. Kelley, W. F. Hosford, Trans. Metall. Soc. AIME 1968, 242, 5–13.
[2] M. R. Barnett, M. D. Nave, Z. Keshavarz, Metall. Trans. A 2005, 36A, 1697–1704.
[3] M. D. Nave, M.R. Barnett, Scripta Mater. 2004, 51, 881–885.
[4] B. C. Wonsiewicz, W. A. Backofen, Trans. Metall. Soc. AIME 1967, 239, 1422–1431.

Dynamic Strain Ageing in Magnesium Alloys

P. Lukác, Z. Trojanová
Charles University, Praha

1 Introduction

Plastic instabilities of different types are observed in many alloys when an alloy is deformed over a certain temperature range at a certain strain rate. Many phenomena as the Portevin-Le Châtelier effect, a negative strain rate sensitivity, a positive dependence of the flow stress on temperature and post relaxation effect are considered as plastic instabilities. They are a consequence of dynamic strain ageing (DSA). Solute atoms may diffuse during deformation to moving dislocations and the segregation of solute atoms at dislocation occurs. The slip resistance to the dislocation movement increases due to the pinning of dislocations.

In comparison to fcc alloys, only little attention has been paid to dynamic strain ageing phenomena in Mg alloys [1–5]. Gärtnerová et al. [2] observed the Portevin-Le Châtelier (PLC) in Mg-0.7 wt.% Nd alloy at temperatures between 200 and 300 °C. They also observed a negative strain rate dependence of the yield stress for the alloy deformed at 250 °C. Zhu and Nie {3} observed serrations in the flow stress curves, (PLC), in WE54 magnesium alloy in a temperature range 150–225 °C, which is slightly lower than that reported by Gärnterová et al. [2]. Zhu and Nie [3] carried out the strain rate change test at 200 °C. The specimen exhibits negative sensitivity of the flow stress to strain rate for all the downward strain rate changes whereas a negative strain rate sensitivity for the upward strain rate changes was only observed in regimes of serrated flow [3]. Trojanová et al. [5] investigated dynamic strain ageing during stress relaxation in selected magnesium alloys containing rare earth elements. To the authors' knowledge no systematic studies have been preformed to investigate DSA in Mg alloys.

The aim of the present paper is to report the observation of plastic deformation in commercial magnesium alloys ZE41 and QE22 and to analyze the obtained information in relation to the DSA effect.

2 Experimental

The materials used in this study were magnesium alloys ZE41 with a nominal composition of Mg-4Zn-1Nd (wt.%) and QE22 with a nominal composition of Mg-(2.3–2.7)Ag-2.4Nd (wt.%). Test specimens were machined from the castings and they had a rectangular cross section of 6x6 mm^2 and a gauge length of 12 mm. Compression tests were performed in a screw-driven Instron testing machine over the temperature range from room temperature to 300 °C at various but constant crosshead speeds giving an initial strain rate between $2.7 \cdot 10^{-6}$ and $1.3 \cdot 10^{-2}$ s^{-1}. The temperature was controlled to within ±1 K. Test data were collected by computer data-acquisition system.

Stress relaxation tests were performed from room temperature to 300 °C at different strains. During deformation, the testing machine was stopped and the specimen was allowed to relax for 300 or 600 s. The flow stress after stress relaxation was higher or lower than that at the beginning of the relaxation. Strain rate change tests were also performed. The strain rate was changed alternatively by a factor of 10 and 0.1 (in some cases by a factor 5 and 0.5).

3   Experimental Results

The true stress- true strain curves of both ZE41 and QE22 alloys at temperatures higher than 250 °C are flat. The flow stress σ is practically constant, which indicates a dynamic balance between hardening and softening. The difference between the yield stress and maximum stress is close to zero. The yield stress does not decrease smoothly with temperature but shows a slight inflection (a local maximum and a local minimum). The ZE41 alloy exhibited serration in the stress-strain curves in a temperature range of 150–200 °C and at a strain rate of $1.3 \cdot 10^{-3}$ s^{-1}. The serrated flow in the QE22 alloy was observed in a temperature range 175–225 °C [5,6]. Outside these temperature ranges, the flow curves were smooth. Please type in the manuscript text here. Please type in the manuscript text here. Please type in the manuscript text here. Please type in the manuscript text here. Please type in the manuscript text here. [5,6]

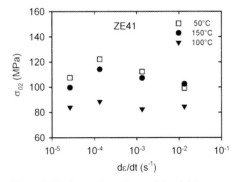 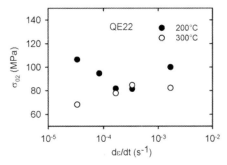

Figure 1: Strain rate dependence of the yield stress measured for ZE41 alloy and various temperatures

Figure 2: Strain rate dependence of the yield stress measured for QE22 alloy and two temperatures

Figure 1 shows the strain rate dependence of the yield stress of ZE41 alloy deformed at three temperatures 50, 100 and 150 °C. The values of the yield stress s_{02} were determined as the flow stress at 0.2 % offset strain. It is obvious that the yield stress does not increase with increasing strain rate, as it is usually the case, but at the strain rates higher than $1.3 \cdot 10^{-4}$ s^{-1}, the yield stress decreases at 50 and 150 °C [5]. At 100 °C the yield stress is practically independent of the strain rate. It is also obvious that the values of the yield stress at 100 °C are lower than those at 150 °C at all imposed strain rates. Figure 2 shows the strain rate dependence of the yield stress for QE22 alloy at 200 and 300 °C. The yield stress at 200 °C decreases with increasing strain rates between $3.3 \cdot 10^{-5}$ and $3.3 \cdot 10^{-4}$ s^{-1}. Similar strain dependence of the yield stress was observed for Mg-0.7Nd alloy deformed at 250 and 300 °C [5]. The results indicate dynamic strain ageing.

The stress relaxation tests were performed at different strains during compression of the specimen. Figure 3 shows the true stress-true strain curve for QE22 specimen deformed at 200 °C at $1.7 \cdot 10^{-4}$ s^{-1}. It can be seen serrations. The flow stress after the stress relaxation, σ_1, was higher or lower than the flow stress at the beginning of the relaxation, σ_0. In some cases, yield points were observed after reloading after a relaxation. The value of $\sigma = \sigma_1 - \sigma_0$ is a function of the flow stress (strain) at which the stress relaxation started, duration of the stress relaxation test and temperature. Figure 4 shows the plot of $\Delta\sigma$ against stress for ZE41 alloy for two temperatures of 150 and 200 °C [6]. It can be seen that the values of $\Delta\sigma$ at the beginning of deformation are positive, then they decrease with stress and if the relaxation occurs at stresses higher than a certain stress, the values of $\Delta\sigma$ are negative. Variations of $\Delta\sigma$ for QE22 alloy as a function of strain at selected temperatures are presented in Fig. 5. It is obvious that the values of $\Delta\sigma$ depend on strain and temperature for a constant time of relaxation (in both case relaxation time was 300 s).

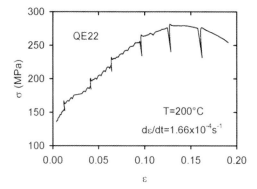

Figure 3: Serrations at the stress strain curve obtained at 200 °C

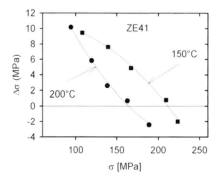 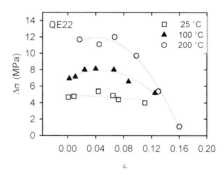

Figure 4: Post relaxation effect estimated for ZE41   **Figure 5:** Post relaxation effect estimated for QE22

The true stress-true strain curves for QE22 alloy illustrating the effect of changing strain rate by a factor of 10 (and 0.1) at 50 and 200 °C are shown in Fig. 6. The specimen deformation started at $3.3 \cdot 10^{-4}$ s^{-1} and the strain rate was changed to $3.3 \cdot 10^{-5}$ s^{-1} after a certain strain. Then, the strain rate was changed alternatively. It can be seen that the upward (downward) strain rate changes at 50 °C cause an increase (a decrease) in the flow stress. On the other side, a decrease (an increment) in the flow stress is observed after the upward (downward) strain rate changes at 200 °C i.e. negative strain rate sensitivity is observed. It should be mention that the serrated flow is observed at 200 °C. The negative strain rate sensitivity was also observed when the specimens were deformed in the strain rate change tests (at 200 °C) and the strain rate was changed between $1.7 \cdot 10^{-4}$ and $1.7 \cdot 10^{-3}$ s^{-1}. A yield point occurs after strain rate changes. Zhu and Nie [3] have also reported the occurrence of a yield point after each change in the strain rate. Figure 7 shows the strain rate change test results obtained at various flow stresses (between 120 and 240 MPa) when the strain rate was changed by a factor 0.1 as a function of the temperature at which the specimens were deformed.

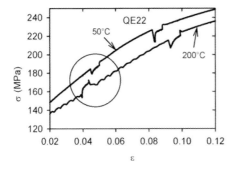
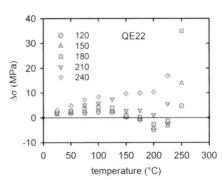

Figure 6: Strain rate changes at two temperatures. Negative strain rate sensitivity at the lower strain

Figure 7: Strain rate changes at various stresses (values given in MPa)

4   Discussion

The results of this work show that dynamic strain ageing influences the deformation behavior of ZE41 and QE22 magnesium alloys. We assume that ageing occurs while dislocations are temporarily held at obstacles waiting for thermal activation to overcome them. Solute atoms diffuse to dislocations arrested at obstacles.

The mean waiting time t_w is connected with the strain rate by the Orowan equation

$$\Delta \varepsilon / dt = b \rho_m \Lambda / t_w = \Omega / t_w \tag{1}$$

where b is the magnitude of the Burgers vector, ρ_m is the density of mobile dislocations and Λ is the mean free path of dislocations. If it is assumed that the forest dislocations are the rate controlling obstacles then $\Lambda = \rho^{-1/2}$, where ρ is the forest dislocation density. The elementary plastic strain $\Omega = b \rho_m \rho^{-1/2}$ is a strain dependent quantity; it may have a local maximum at a certain strain [7,8]. The local concentration of solute atoms c at dislocation lines as a function of the waiting time can be expressed as

$$c = c_M [1 - \exp(-t_w / t_0) p] \tag{2}$$

where c_M is the limiting value of solute atoms on the dislocation and the relaxation time t_0 depends on the diffusion coefficient ($t_0 \sim D^{-1}$), on solute concentration and on the binding energy between a solute atom and a dislocations. The exponent p is 2/3 and 1/3 for bulk and pipe diffusion, respectively [8,9].

The plastic strain rate may be given by an Arrhenius law

$$\Delta \varepsilon / dt = \Omega v_0 \exp(-G/kT) \tag{3}$$

where v_0 is the attack frequency, k is the Boltzmann constant and T is the absolute temperature. The obstacle strength increases owing to DSA and hence the activation free enthalpy for overcoming of an aged dislocation segment $G = G0 + \Delta G$. $G0$ is the free activation enthalpy for the thermally activated process in the absence of DSA and ΔG is a function of the stress acting on the dislocation segments and the solute concentration at the dislocation lines [10]. Thus, the free activation enthalpy depends on the stress and on the ageing process.

The evolution with strain of the mobile and forest dislocation densities may be described by a system of coupled differential equations [7,8]. By considering the strain dependence of the elementary strain, it is possible explain the observed features observed in the present work: (a) the critical strains for the onset (and the disappearance of serrations, (b) the decrease of the yield stress with increasing strain rate in a certain strain rate range and at certain temperatures, and (c) the negative values of the strain rate sensitivity of the flow stress (SRS) at certain strains (stresses) and at certain temperatures. The SRS is defined as $S = [\sigma/\ln(d\varepsilon/dt)]$. It should be mentioned that the critical strains are defined from the condition of vanishing of the SRS [7].

The flow stress for the dislocation motion in an alloy may be considered to be a sum of two components, i.e.

$$\sigma = \sigma_d + \sigma_f \quad (4)$$

where σ_d is the dislocation component due to strong obstacles (e.g. forest dislocation, grain boundaries, precipitates) and σ_f is the friction stress due to the interaction between moving dislocations and solute atoms. We consider that solute atoms influence both components of the flow stress in the dynamic strain ageing regime [10]. In this case, the flow stress may be decomposed on a non-aged part sna and a dynamic strain ageing part σa. The stress increment due to DSA is given by the following equation

$$\sigma = (f_1 + f_2)c(t_w). \quad (5)$$

The first term is related to the dislocation-dislocation interaction influence by DSA (ageing of dislocation interaction influenced by DSA) and it is proportional to the pinning force. Solute atoms locking dislocations during the stress relaxation test cause a stress increase after stress relaxation. An increase in the flow stress is needed to move the pinned dislocations after stress relaxation. It is reasonable to assume that $\Delta\sigma$ is proportional to solute atoms at dislocation lines, which depends on temperature, on strain and on stress relaxation time. At a given test temperature, the number of solute atoms at dislocations lines is constant. But the density of dislocations increases with strain. Thus, the stress increase after stress relaxation should depend on strain in the dynamic strain regime, which is experimentally observed. At higher strains and higher temperatures, motion of **(c+a)** dislocations in the second-order pyramidal slip systems occurs [11–13]. The moving **(c+a)** dislocations may cross slip and dislocation annihilation may follow. Different dislocation reaction between **a** and **(c+a)** dislocations may also occur. Some dislocation reactions can contribute to softening. Dynamic recovery causes a decrease in the internal stress. Thus, the flow stress a the beginning deformation after stress relaxation may be lower than the flow stress at the beginning of the stress relaxation, which is also experimentally observed.

5 Acknowledgements

The authors are grateful for the financial support offered by the Grant Agency of the Academy of Sciences of the Czech Republic under grant A201120603. A part of this work was financed by the Ministry of Education of the Czech Republic under the Research Project 1M2560471601 "Eco-centre for Applied Research of Non-ferrous Metals"

6 References

[1] M. C. Chatuverdi, D. J. Lloyd, Phil. Mag. 1974, 30, 1199.
[2] V. Gärtnerová, Z. Trojanová, A. Jäger, P. Palcek, J. Alloys Compd. 2004, 378, 180.
[3] S. M. Zhu, J. F. Nie, Scripta Mater. 2004, 50, 51.
[4] C. Corby, C. H. Cáceres, P. Lukác, Mater. Sci. Eng. 2004, A387, 22.
[5] Z. Trojanová, P. Lukác, K. U. Kainer, V. Gärtnerová, Adv. Eng. Mater. 2005, 7. 1027.
[6] Z. Trojanová, P. Lukác, Kovove Mater. 2005, 43, 73.
[7] P. L. Kubin, Y. Estrin, Acta Metall. Mater. 1990. 38, 697.
[8] J. Balík, P. Lukác, Acta Metall. Mater. 1993, 41, 1447.
[9] J. Friedel, Dislocations, Pergamon Press, Oxford, 1964.
[10] J. Balík, P. Lukác, Kovove Mater. 1995, 36, 3.
[11] P. Lukác, Czech. J. Phys. B 1985, 36, 275.
[12] S. R. Agnew, M. H. Yoo, C. N. Tome, Acta Mater. 2001, 49, 4077.
[13] K. Máthis, K. Nyilas, A. Axt, I. Dragomir-Cernatescu, T. Ungár, P. Lukác, Acta Mater. 2004, 52, 2889.

Behaviors of High Temperature Deformation and Texture Development of AZ31 and AZ61 under Uniaxial Compression

L. Helis, K. Okayasu, H. Fukutomi
Yokohama National University

1   Introduction

Light-weight high-strength magnesium alloys have been a subject of extensive research in recent years. The major problem for the practical application of the alloys is the poor workability at low temperatures. Addition of alloying element contributes to improve the ductility of the alloys to some extent, the improvement, however, is not sufficient at present. Development of the proper texture is a promising solution for the further improvement in workability. Texture development of AZ31 under various deformation conditions was experimentally examined in the previous study [1]. It was found that texture varies depending on deformation conditions although the textures are not sharp in general, indicating that texture control is possible for the alloy.

Since the improvement in workability might be achieved by the combination of alloying and texture effects, it is important to investigate the effect of alloying on the texture development systematically. In the present study, the behaviors of deformation and texture development are studied on AZ31 and AZ61. Special attention was paid on the relationship between deformation mechanism and the texture characteristics by taking the effect of aluminum amount into consideration.

2   Experimental Procedures

The materials used in the present study are commercial magnesium alloys AZ31 and AZ61. The alloys were obtained in the form of hot extruded bars with diameter of 16 mm (AZ31) and 15 mm (AZ61). Cylindrical specimens with diameter of 12 mm and length of 18 mm were machined from the extruded rods for uniaxial compression tests (hereafter referred to as A-type specimens). It was found that these rods have sharp textures having the basal plane parallel to the rod axis. Therefore, another set of cylindrical specimens, with 9 mm in height and 6 mm in diameter, was machined from extruded bars in a way, where basal planes frequently aligns about 45 degree against the compression direction (hereafter referred to as B-type specimens); orientation favorable for the activation of basal slip system.

All cylindrical specimens were annealed at 723 K for 2 h and furnace cooled. The texture remained after the annealing. The mean grain size after the annealing was 42 µm in the case of AZ31 and 50 µm in the case of AZ61. Specimens for microstructure examination were prepared by the same technique as was used in the previous study [1].

Uniaxial compression tests on A-type of cylindrical specimens were performed at final strain rates ranging from $5.0 \cdot 10^{-5}$ to $1.0 \cdot 10^{-3}$ s^{-1} up to true strains ranging from –0.2 to –1.5. The tests were conducted in the temperature range of 573–773 K in the case of AZ31, and 523-

673 K for AZ61, slightly lower than those for AZ31, as ternary phase diagram shows a melting point around 730 K for the almost identical alloy [2]. B-type specimens were compressed at 623 K, under the constant crosshead speed and strains ranging from –0.7 to –1.3. All specimens were quenched in oil immediately after the compression, in order to prevent the microstructure from changing after deformation.

Texture measurements on mid-plane sections, were carried out by the Schulz reflection method using nickel filtered Cu K_α radiation. Five pole figures were constructed using diffracted X-ray intensities measured on $10\bar{1}0$, 0002, $10\bar{1}1$, $10\bar{1}2$ and $11\bar{2}0$ reflections, and based on them, ODF was calculated by the Dahms-Bunge method [3]. The main components and texture sharpness were evaluated by the examination of the inverse pole figures derived from the ODF.

3 Results

3.1 Stress-strain Curves

Figure 1(a) shows stress-strain curves for the deformation of A-type specimens at temperature of 623 K, up to a strain of $\varepsilon = -1.0$ at various strain rates. Figure 1(b) shows two stress-strain curves for the B-type specimens deformed up to a strain of $\varepsilon = -0.7$. For both figures, solid lines represent the results for AZ61 and broken lines present the results for AZ31. In the case of A-type specimens, no large difference from AZ31 was found in the level of flow stresses. Flow stress increases with decreasing temperature and increasing strain rate and the peak in stress at initial stage of deformation is followed by work softening. AZ61 showed obvious high temperature yielding.

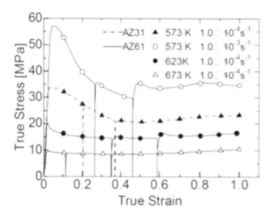

Figure 1: Stress-strain curves for AZ31 and AZ61 deformed at 623 K. (a) A-type specimens (b) B-type specimens

Stress-strain curves for the deformation of A-type specimens unloaded twice during the compression are shown in Fig. 2. Specimens were reloaded immediately after the first unloading. Before the second reloading, the specimens were kept in the furnace for 1 hour. No high temperature yielding is observed for AZ31 at the second reloading, while it can be clearly seen on AZ61 both at higher and lower stresses than that for AZ31.

583

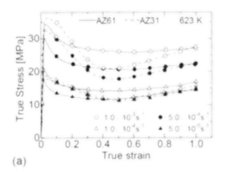 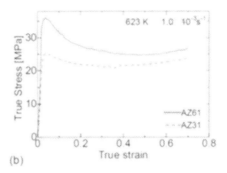

Figure 2: Stress-strain curves for A-type specimens unloaded twice during the compression

3.2   Microstructure Observation

According to the ternary phase diagram [2], precipitation is not expected at the temperatures higher than 573 K in AZ31, the lowest temperature used for compression tests on this alloy. On the other hand, presence of precipitates in the microstructure has to be taken into account for deformation of AZ61 at temperatures below approximately 600 K. Figure 3 shows the microstructure formed by the heating at 573 K for 1 hour after the annealing at 723 K for 2 hours. The specimen was quenched in oil after the heating; this gives the microstructure, right before deformation at 573 K. Precipitates segregated on the grain boundaries can be seen. No precipitates were observed on the specimens quenched after heating at 623 K and 673 K.

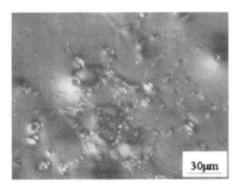

Figure 3: Microstructure of AZ61 after heating at 573K for 1 hour and quenched in oil

Microstructure observation of AZ61 specimens after deformation above 623K showed similar results as for AZ31, described in detail in the previous work [1]. Mean grain size on the compression plane of all deformed specimens is different from the one of as annealed specimen, and was decreased with increasing peak stress, suggesting the occurrence of dynamic recrystallization (hereafter abbreviated as DRX). Figure 4 shows the peak stress vs. grain size relationship examined on AZ31 and AZ61. It is seen that the mean grain size of the two alloys is almost the same when stresses are high, while the mean grain size of AZ61 becomes larger with a decrease in peak stress.

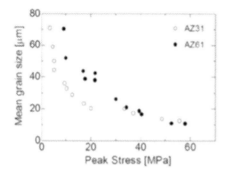

Figure 4: Grain size after deformation as a function of peak stress

3.3 Texture Masurements

As annealed specimens of both alloys revealed sharp fiber texture, with main component rotated 90 degrees away from the basal pole, typical texture for extruded bars. Results on AZ31 texture formation, described in detail in previous study [1], can be briefly summarized as follows. The main component of the texture, measured on a compression plane, appeared about 33–38 degrees away from the basal pole after deformation characterized by low values of Zenner-Hollomon parameter. The tilting angle of the main component decreased down to about 20 degrees with an increase in Zenner-Hollomon parameter and finally construction of basal texture was detected after deformations at the lowest temperature and high strain rates.

In the case of AZ61, the main component of the texture exists at similar positions to the case of AZ31 after deformation at temperatures of 523 K and 573 K. Namely, the main component of the fiber texture exists about 20–30 degrees away from basal pole. In some cases the position of main component of the texture remained unchanged from as annealed state, even after deformation up to a final strain of –1.0.

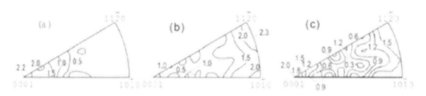

Figure 5: Inverse pole figures for (a) AZ31 deformed at 623 K, strain rate $1.0 \cdot 10^{-3}$ s^{-1} and strain –1.0; (b) AZ61 deformed at 623 K, strain rate $1.0 \cdot 10^{-3}$ s^{-1} and strain –1.0; (c) AZ61 deformed at 623 K, strain rate $1.0 \cdot 10^{-3}$ s^{-1} and strain –1.1

Figure 5 shows inverse pole figures after deformation of AZ31 (a) and AZ61 (b, c) at 623 K. Densities of the poles for the compression planes are given. Mean pole density is used as a unit. Deformation conditions are: (a) $1.0 \cdot 10^{-3}$ s^{-1} up to a strain of –1.0; (b) $1.0 \cdot 10^{-3}$ s^{-1} up to strain of –1.0; (c) $1.0 \cdot 10^{-3}$ s^{-1} up to a strain of –1.1. Basal texture was formed in AZ31, while main texture component remained rotated 90 degrees away from basal pole after identical deformation for AZ61 (Fig. 5(b)). But it can be said, that texture was randomized in comparison to as annealed state, and basal texture appeared after deformation to final strain of –1.1(Fig. 5(c)). In the case of AZ61 deformed at 673 K, main component was observed in a range of 29–34 degrees away from basal pole, independently of deformation condition, even after deformation up to strain of –1.5. Also maximum pole density was increased compared with lower temperatures, but was still lower than the one of as annealed specimen.

4 Discussion

4.1 Effect of Solute Content on the Microstructure Formation

As shown in Fig. 2, the effect of unloading followed by reloading on the stress-strain curve is different between AZ31 and AZ61. In the case of AZ31, the level of stress at the second reloading is almost the same as that before the unloading. While in the case of AZ61, clear high temperature yielding is seen. This suggests that the effect of solute drag in AZ61 is higher than that in AZ31. The same trend is seen in the stress-strain curves for type B specimens. Comparison of Figs.1 (a) and (b) shows that large difference in the peak stress exists in AZ31, but almost no difference is observed in AZ61. This suggests that the slip deformation predominantly proceeds by basal slip system in AZ31, while no big difference exists in CRSS between basal and other slip systems in AZ61. Although the addition of zinc as well as the increase in deformation temperature contributes to the reduction in the difference in CRSS, the results shown in Fig. 2 might suggest that the equivalent activity of various kinds of slip systems can be attributed to the effect of solute atmosphere.

According to Fig. 4, the size of grains of AZ61 is larger than that of AZ31 when the peak stress is below 30 MPa. It is generally considered that the size of grains is controlled by the rates of nucleation and the growth of new grains. When the effect of solute drag is strong, it is reported that dislocations are distributed homogeneously and formation of subgrains becomes difficult [4]. In this case, the driving force for the migration of grain boundaries becomes higher, because relaxation of strain field around dislocations by the rearrangement of dislocations does not proceed. If the nucleus of new grains is formed among subgrains, the nucleation becomes harder with an increase in the effect of solute atmosphere. Thus the size of grains in AZ61 becomes larger than that of AZ31 in the lower stress region in Fig. 1.

4.2 Effect of Deformation Condition on the Texture Formation

In the previous paper on AZ31, it was pointed out that the change in the main component of the texture corresponds to the change in the so-called n value. The same tendency as the case of AZ31 is found in AZ61. Namely, when the n value is close to 5 where the values of Zener-Hollomon parameter (Z parameter) are high, the main component of the texture appears about 20 degrees away from the basal pole, while the main component of the texture is found about 33 degrees far from the basal pole when the deformation is conducted at the higher temperature region such as 673 K; the deformation under low Z condition. The orientation about 33 degrees away from the basal pole gives almost the same values of Schmid factor for the basal and pyramidal slip systems, indicating that the orientation might be stable for the compression deformation, although the texture sharpness is reduced by DRX. Accumulation of pole density at this position is seen even when the main component of the texture exists at (0001).

4.3 Formation of Basal Texture

As for the workability of sheet material, formation of basal texture should be avoided. From this point of view, it is important to clarify the formation mechanism of basal texture. In the present study, the formation of basal texture is seen on AZ61 after the compression beyond a strain of 1.0 at 623 K under a strain rate of $1.0 \cdot 10^{-3} s^{-1}$. Comparison of Figs. 5(b) and (c) suggests that there seems to be an orientation change from the initial position about 90 degrees away from the basal pole towards the inner area of the stereographic triangle. In order to examined the possibility that the formation of the basal texture might be attributed to the continuous lattice rotation due to crystal deformation, pole density distribution along the line connecting the position of initial main component of the fiber texture and the basal pole. The result is given in Fig. 6. The distribution of pole density is examined on the strains 0.7, 1.0, 1.1, 1.25 and 1.5. As shown in Fig. 6, the pole density does not change monotonously. It decreases with the change in the angle from the basal pole from 90 to about 40 degrees, and then increases with further decrease in the angle towards 0 degrees. It seems that the decrease in the pole density at 90 degrees corresponds to the increase in the pole density at 0 degrees. This may suggest, that formation of the basal orientation should be attributed to the discontinuous process such as twinning.

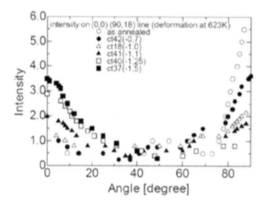

Figure 6: Distribution of pole density in dependence on angle from the basal pole for AZ61, deformed at 623 K up to various final strains

5 Conclusions

The effects of solute contents and deformation conditions on the deformation behavior and texture development were experimentally studied on AZ31 and AZ61. Grain size of AZ61 after dynamic recrystallization is larger than that of AZ31, and the texture of AZ61 changes slower than AZ31. Compression tests consisting of unloading-reloading processes suggest that the big difference exists in the effects of solute atmosphere on the motion and the microstructure of dislocations between the two alloys. It was concluded that the differences in the behaviors described above are attributable to the differences in the activities of slip systems and the expected dislocation arrangement. Although the formation mechanism is still not clear, it is experimentally confirmed that the basal texture is produced by a discontinuous process and not by the continuous process such as lattice rotation due to slip deformation.

6 Acknowledgment

The research was conducted under the support by the Japan Light Metals Foundation. The authors greatly appreciate the foundation.

7 References

[1] L. Helis, K. Okayasu, H. Fukutomi, Mater. Sci. Eng. A, **2006**, 430, 98–103.
[2] W. Köster, W. Dullenkopf, Z. Metallk., **1936**, 28, 363–367.
[3] M. Dahms, H.J. Bunge, J. Appl. Crys., **1989**, 22, 439–445.
[4] R. Horiuchi, M. Otsuka, Trans. JIM, **1972**, 13, 284–293.

Slip during Tension Testing of Mg-3Al-1Zn Sheet

Z. Keshavarz, M.R. Barnett
Centre for Material and Fibre Innovation, Geelong Technology Precinct, Deakin University, Pigdons Rd., Geelong, VIC 3217, Australia

1 Introduction

The poor formability of magnesium sheet at room temperature is due to the lack of enough independent deformation modes such as slip. The following planes are important in deformation by slip: (0001) basal, $\{1\bar{1}00\}$ prismatic and $\{11\bar{2}2\}$ second order pyramidal <c+a>) [1]. Single crystal studies have shown [2, 3] that the critical resolved shear stresses (CRSS) for the non-basal modes are considerably higher than for basal slip. However, small amounts of non-basal slip have been seen after room temperature deformation of pure magnesium [4].

Agnew et al. [5], Barnett [6] and Staroselsky et al. [7] mention the necessity of considering prismatic and/or second order pyramidal slip in crystal plasticity models for the simulation of the mechanical response of Mg-3Al-1Zn (AZ31) at room temperature. In recent work [8], a semi-analytical Sachs model was developed for the flow stress of AZ31. The minimum values obtained for CRSS/m (m is the Schmid factor) for the four modes (basal, prismatic, pyramidal slip and twinning) for uniaxial tension was calculated and these are presented in Figure 1 [8]. The CRSS ratios employed to generate the plots are those found to be optimal in our previous study and are basal slip: twinning: prismatic slip: pyramidal <c+a> slip = 1:0.7:2:15. The predictions are given as a function of Φ (the inclination of the c-axis from the stress axis z(3)). The locus of the lowest values is shown in bold in Figure 1 and this permits the active deformation mode to be identified. It can be seen that this analysis predicts prismatic slip for orientations with a c-axis within 15 deg of perpendicular to the tensile direction.

The activation of non-basal slip at room temperature in AZ31 has also some support in TEM observations [9, 10]. Studies to date have employed optical and TEM microscopy and crystal plasticity modelling. Not much work appears to have been done electron back scattering diffraction (EBSD). The present work aims to use this technique in conjunction with line trace analysis to determine deformation mode activity.

2 Experimental Methods

Flat tensile specimens with a cross section of 10 mm^2 and a gauge length of 25 mm were machined parallel to the rolling direction of a commercial sample of AZ31 (3% Al, 1% Zn) with an initial average linear intercept grain size of 7 µm (Figure 2a). Before machining, the specimens were annealed for 30 min at 350 °C and cooled in air to obtain a stress-free microstructure.

The texture of the as-received rolled AZ31 was measured by means of EBSD in a field emission gun scanning electron microscope (W-filament SEM Leo 1530) (Figure 2b). The texture of the sheet is typical, which means the c-axis of most of the grains is preferentially aligned close to the normal direction to the sheet.

Samples for EBSD mapping and texture measurements were ground with 1200 grit SiC paper. Polishing was done with diamond polishing through 15, 6, 3 and 1 µm. Then the specimens were mechanically polished with a Colloidal Silica Slurry and etched with a solution of 10 ml HNO_3, 30 ml acetic acid, 40 ml H_2O and 120 ml ethanol for about 10 sec. Tensile testing was carried out using Kammrath-Weiss in situ tensile stage in a SEM. The stage is small enough to be tilted 70 deg, which allows EBSD patterns to be recorded. The deformation was performed at $5 \cdot 10^{-4}$ mm/s.

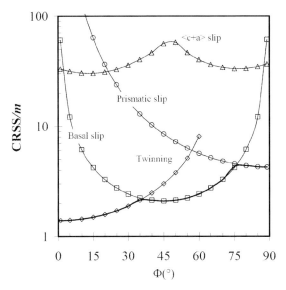

Figure 1: Predictions of the values for CRSS/m as a function of the inclination of the c-axis to the stress direction for tension. The calculations were made assuming CRSS ratios for basal slip, twinning, prismatic slip and <c+a> slip of 1:0.7:2:15. [8]

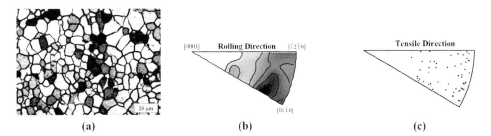

Figure 2: (a) Initial microstructure and (b) texture (Rolling Direction inverse pole figure intensity levels 0.5, 2 and 2.5) of AZ31 sheet. Orientations investigated for slip are given in (c).

In order to investigate the active slip systems at room temperature, observation by means of SEM and EBSD of the surface of the tensile samples was performed following plastic strains of 0.04, 0.08, 0.10, 0.14 and 0.19. A true stress-true strain curve showing the corresponding stresses is given in Figure 3. To permit identification of active slip systems, the crystallographic orientation was determined by EBSD mapping and the slip trace inclination to the rolling direction was measured from SEM micrographs. After deformation, the surface was not smooth and this meant that the EBSD indexing success rate was quite low (~40 %). The orientation of the grains in which slip trace analysis was performed, are shown on an inverse pole figure in Figure 2c. In all, 54 sets of slip traces were examined.

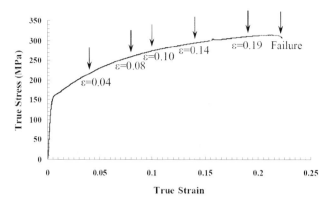

Figure 3: Stress-strain curve of rolled AZ31 (velocity of $5 \cdot 10^{-4}$ mm/s), showing strains at which microscopic analysis was performed.

3 Results

Examples of the different slip modes observed in the present study are presented in Figure 4. Expected traces of the intersection of the basal, prismatic and 2nd order pyramidal planes with the sample surface were inspected for each orientation. The trace closest to the observed trace is shown on the pole figures.

The orientations in which basal, prismatic and <c+a> slip systems were considered to be active are presented in Figure 5 by way of inverse pole figures. In all bar a few cases, slip mode activity is observed when the Schmid factor is in excess of ~0.4.

The deviation of the measured slip trace from the nearest computed mode fell within 6 deg. The angular deviations of the measured traces from the second nearest computed traces were determined. The deviation from the next nearest computed slip line was in the range 2–35 deg. For the cases where the deviation of the second nearest slip line was in the range of 2–6 deg, the decision of the most active mode was made according to the mode with the highest Schmid factor. (This only occurred in a few instances.)

According to Figure 1, it is clear that prismatic slip is expected to occur in tension for values of Φ greater than 75 deg. This is consistent with the present results in all bar two cases (Figure 6). In the present tensile tests, ~63 pct of the grains displayed prismatic slip, 33 pct showed basal slip and ~4 pct displayed <c+a> slip.

4 Discussion

Detection of slip lines and indexing nearby material were difficult in the present study for three reasons. One, the surface became considerably roughened with increasing strain, which interfered with the indexing of EBSD patterns. Two, the use of the in-situ stage necessitated a high working distance which reduced the ability to resolve slip lines. Three, presence of intermetallic particles masked slip lines. The slip identified in the present study was therefore confined to regions where EBSD indexing was possible and where the slip lines were most distinct.

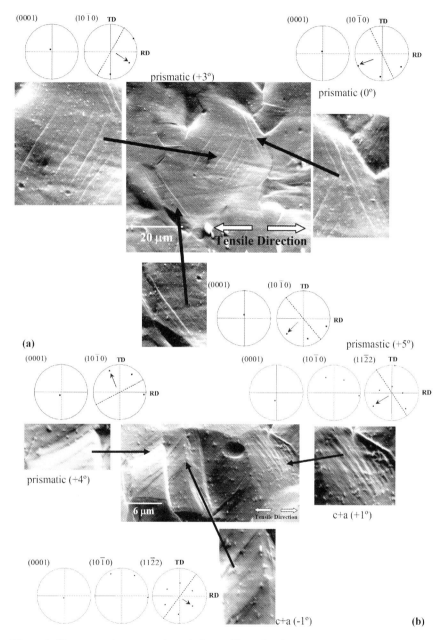

Figure 4: Slip traces and accompanying pole figures illustrating the method used to identify the most likely active deformation mode after strains of (a) 0.08 and (b) 0.19. The line on the pole figures corresponds to the trace of the slip plane (arrowed pole) closest to the observed slip trace. The deviation between the two is given in parenthesis. (RD–Rolling Direction, TD–Transverse Direction).

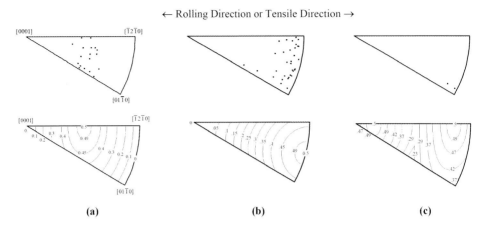

Figure 5: Inverse pole figures of the tensile direction along with iso-curves of the maximum Schmid factor for orientations in which (a) basal, (b) prismatic and (c) c+aslip were observed.

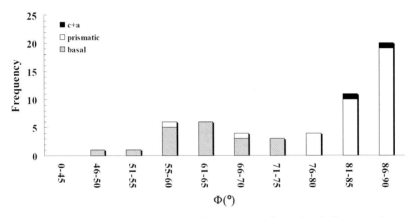

Figure 6: The classification of grains showing slip traces according to their inclination to the stress axis.

Given the well-known ease of basal slip [11], the fraction of basal slip seen in the present study was surprisingly low. This is probably due, in part, to the inability of the current technique to resolve slip on planes that are near or parallel to the observation plane.

The current findings show the importance of prismatic slip in room temperature deformation of AZ31. As mentioned above, this is in agreement with what has been reported in recent modelling work. Koike et al. [10] emphasized the activity of non-basal slip near grain boundaries but in the current study prismatic slip was observed well within the grain interior. This suggests that its activation may not be dependent on proximity grain boundaries for average grain size of the order of those employed in the present study (7 μm).

Agnew et al. [12] emphasized the importance of <c+a> slip in promoting c-axis contraction and reported that tensile twinning appears not to be a major contributor to the macroscopic strain of in-plane AZ31 tension samples. The observation of slip lines consistent with <c+a> slip in the present work lends some support to their arguments. However, the uncertainty associated with the present technique leaves open the possibility that the few <c+a> slip lines seen here are actually another mode.

5 Conclusions

The present work provides evidence for the operation of significant levels of prismatic slip during tensile straining of AZ31 sheet at room temperature.

6 Acknowledgments

This work was supported by the provision of research scholarship by Deakin University and a grant from ARC Discovery Project.

7 References

[1] C. S. Roberts, Magnesium and Its Alloys, John Wiley & Sons, Inc., New York, 1960, 180.
[2] B. C. Wonsiewicz, W. A. Backofen, Trans. Metall. Soc. AIME 1967, 239, 1422–1431.
[3] R. E. Reed-Hill, W. D. Robertson, Acta Metall. 1957, 5, 728–737.
[4] F. E. Hauser, P. R. Landon, J. E. Dorn, Trans. Am. Soc. Metals 1958, 50, 856–883.
[5] S. R. Agnew, in Magnesium Technology 2002, TMS Annual Meeting (Ed.: H. I. Kaplan), Seattle, USA, TMS, 2002, 169–174.
[6] M. R. Barnett, Metall. Mater. Trans. A 2003, 34A, 1799–1806.
[7] A. Staroselsky, L. Anand, Int. J. Plast. 2003, 19, 1843–1864.
[8] M. R. Barnett, Z. Keshavarz, X. Ma, Metall. Mater. Trans. A 2006, 37, 2283–2293.
[9] S. R. Agnew, O. Duygulu, in Magnesium Technology 2004 (Ed.: A. A. Luo), Charlotte, North Carolina, U.S.A, TMS, 2004, p. 61–65.
[10] J. Koike, T. Kobayashi, T. Mukai, H. Watanabe, M. Suzuki, K. Maruyama, K. Higashi, Acta Mater. 2003, 51, 2055–2065.
[11] F. E. Hauser, P. R. Landon, J. E. Dorn, Trans. ASM 1956, 48, 986–1002.
[12] S. R. Agnew, C. N. Tomé, D. W. Brown, T. M. Holden, S. C. Vogel, Scripta Mater. 2003, 48, 1003–1008.

Microstructure and Deformation Behaviour of Mg-4Li Alloy

Zdeněk Drozd, Zuzanka Trojanová
Charles University, Faculty of Mathematics and Physics, Prague

1 Introduction

Magnesium-lithium alloys are considered to be the lightest metallic structural materials. Lithium is a suitable alloying element that decreases density and increases ductility of hexagonal magnesium alloys [1]. The solubility of Li in Mg is about 5.5 wt.%. The alloys which are studied in this work contain 4 wt.% Li and have hcp structure. By an addition of Li to Mg critical resolved shear stress (CRSS) for basal slip increases with Li concentration by solid solution hardening [2]. The CRSS for Mg-4wt.% is almost independent on the temperature above room temperature while the CRSS for non-basal slip decreases at temperature below 200 °C as shown in [2]. Li also decreases both *a* and *c* lattice parameters of Mg-Li solid solution [3].

Very effective tool for studying of thermally activated processes is the stress relaxation (SR) test. The specimen is deformed at a constant strain rate up to predetermined stress or strain. Then the machine is stopped and the stress is allowed to relax. The stress decrease is measured as a function of time. The relaxation of the stress occurs when the accumulated elastic deformation, both in specimen and deformation system is transformed into plastic deformation of the sample. The rate of the change of the stress $\dot{\sigma} = d\sigma/dt$ is related to the plastic strain rate $\dot{\varepsilon} = d\varepsilon/dt$

$$\dot{\varepsilon} = \dot{\sigma}/M, \tag{1}$$

where M is a function of the elastic modulus of the machine, elastic modulus of the sample and geometrical factors of the sample. Plastic deformation during the SR test is realised by dislocation motion. The motion of dislocations through a crystal is generally discontinuous process. The dislocation meets many kinds of obstacles located in the slip plane which have be to surmounted possibly with the help of thermal activation. Then, the applied stress necessary for deformation of the crystal may be divided into two components

$$\sigma = \sigma_i + \sigma^*, \tag{2}$$

where σ_i is an athermal component often called internal stress:

$$\sigma_i = \alpha G b \rho_t^{1/2}, \tag{3}$$

where G is the shear modulus, α describes the dependence of the effective interaction strength on thermal activation, b is the Burgers vector and ρ_t is the total dislocation density. The effective stress σ^* acts on dislocations during their thermally activated motion. The effective stress (also called thermal stress) is sensitively dependent on the temperature and strain rate. Assuming that a thermally activated process controls the dislocation motion, the strain rate can be expressed by the Arrhenius equation

$$\dot{\varepsilon} = \dot{\varepsilon}_0 \exp\left[-\frac{\Delta G(\sigma^*)}{kT}\right], \qquad (4)$$

where $\dot{\varepsilon}_0$ is a pre-exponential factor, $\Delta G(\sigma^*)$ is the change in the Gibbs free enthalpy depending on the effective stress σ^*, k Boltzmann constant and T is the absolute temperature. The stress dependence of the free enthalpy may be expressed by a simple relation

$$\Delta G(\sigma^*) = \Delta G_0 - V\sigma^*, \qquad (5)$$

where ΔG_0 is the Gibbs free energy of the dislocation activation in the absence of the stress and V is the activation volume.

The SR test can be analysed using the following theoretical relationship between stress relaxation rate $\dot{\sigma}$ and stress σ [5]

$$\ln(-\dot{\sigma}) = C + n\ln\sigma, \qquad (6)$$

where C is a constant and n is the stress sensitivity parameter defined as

$$n = \left(\frac{d\ln\dot{\varepsilon}}{d\ln\sigma}\right)_T. \qquad (7)$$

The stress sensitivity parameter n has a connection with the activation volume V:

$$nkT = V\sigma. \qquad (8)$$

The SR curves are usually analysed assuming that the mobile dislocation density ρ_m and internal stress σ_i are constant during the SR test, which are main conditions for the correct determination of the activation volume and consequently for clearing up the microscopic mechanism controlling the dislocation mobility in a material.

The aim of this paper is to investigate deformation behaviour of superlight Mg-Li alloys and study thermally activated processes in these materials using the stress relaxation tests.

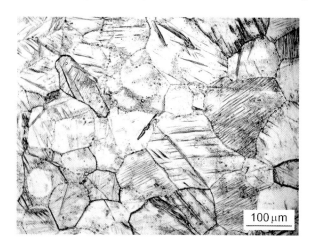

Figure 1: Microstructure of the thermally treated sample

2 Experimental Procedure

The alloy Mg-4wt.%Li has been investigated in this study. Samples were thermally treated at 350 °C for 1h with subsequent quenching into water of ambient temperature. The mean grain size was estimated by the light microscopy to be 100 μm. Rectangular prismatic samples with the base of 5 × 5 mm and the height of 10 mm were deformed in compression in an Instron testing apparatus with the constant machine speed giving the initial strain rate of $1.7 \cdot 10^{-4} \text{s}^{-1}$. Experiments were performed in the temperature range from room temperature up to 200 °C. The protective argon atmosphere has been used in the furnace at temperatures higher than 100 °C. The temperature was kept with the accurateness better then ±1 °C using the Eurotherm 2208 regulator. Successive stress relaxation tests (duration of 300 s) at different stress levels were performed in order to determine the thermal activation parameters – the parameter of stress sensitivity n and the activation volume.

3 Experimental Results

Microstructure of the thermally treated sample is introduced in Figure 1. The temperature dependence of the yield stress σ_{02} and the maximum stress σ_{max} are introduced in Figure 2. Both the yield stress and the maximum stress decrease with increasing testing temperature. A small local maximum in the temperature dependence of the yield stress has been observed at other Mg alloys [5,6] and it indicates strain ageing phenomena.

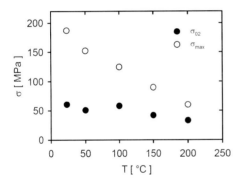

Figure 2: Temperature dependence of the yield stress and the maximum stress

An example of the stress-strain curve obtained during successive stress relaxation (SR) tests (the SR curves were taken out) is given in Figure 3. Examples of the SR curves in a $\ln(-\dot{\sigma})$ vs $\ln(\sigma/\sigma_0)$ representation are introduced in Figure 4 for room temperature. The values of SR starting stresses (in MPa) for SR curves are introduced in Figure 4.
Values of the apparent activation volume V_{ap} has been calculated according to equation (8) considering the value n obtained from stress relaxation curve (plotted in logarithmical representation). This value V_{ap} estimated for polycrystals is proportional to the true activation volume V_d characterizing thermally activated dislocation motion by the Taylor factor ψ i.e. $V_d = \psi \cdot V_{ap}$. The values the apparent activation volume V_{ap} obtained from stress relaxation at various temperatures are introduced in Figure 5 for various temperatures in b^3 units (b is the Burgers vector

of dislocations) has been used. It can be seen that the activation volume decreases with increasing starting stress of the SR test. The decrease of the activation volume with the flow stress and values of the activation volume in the order of 10^{-27} m^3 are typical value for metallic polycrystals.

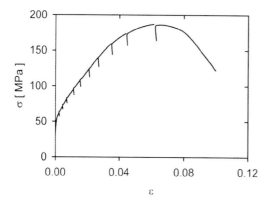

Figure 3: Stress strain curve with the cut SR

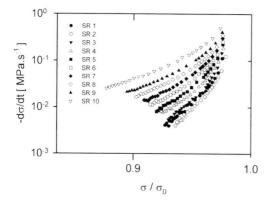

Figure 4: SR curves, obtained at room temperature

Table 1: Starting stresses of the relaxation curves introduced in Figure 4

SR 1	SR 2	SR 3	SR 4	SR 5	SR 6	SR 7	SR 8	SR 9	SR 10
61.7	71.2	81.2	95.5	107.5	123.2	139.5	159.4	173.8	187.0

Temperature dependence of the activation volume V_a is introduced in Figure 6. It has been calculated for the first SR (in the vicinity of the yield stress) and for the maximum stress. While the activation volume corresponding to the onset of the straining decreases with increasing temperature, the activation volume estimated at the maximum stress is nearly constant in the temperature interval studied.

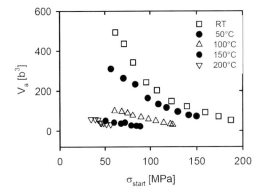

Figure 5: Stress dependence of the activation volume estimated for various temperatures

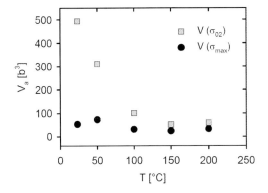

Figure 6: Temperature dependence of the activation volume estimated in the vicinity of yield and maximum stress

4 Discussion

The most probable thermally activated mechanisms in magnesium single crystals deformed at low temperatures are considered to be the intersection of dislocations or non-conservative motion of jogs in screw dislocations. Conrad et al. [6] have proposed the intersection of dislocations as the rate controlling mechanism in Mg single crystals. On the other hand, Sharp and Christian [7] assumed that small clusters of point defects produced by non-conservative motion of jogs in screw dislocations are the dominant obstacles. In both cases the values of the activation volume were of the order of 1000 b^3. In both cases Mg single crystals were of high purity. Feltham [8, 9] determined the activation volume in polycrystalline Mg (99.999 purity) from the stress relaxation experiments. The activation volumes lie between 130 b^3 and 240 b^3. Lukáč [10] have reported a strong dependence of the activation volume of Mg single crystals on the concentration of Cd as solute. The values of the activation volume are of the order of 100 b^3 and at the beginning of deformation the activation volume is decreasing linearly with $c^{-2/3}$ where c is the concentration of Cd. It can be shown that the activation volume of Mg-Zn single crystals estimated by Akhtar and Teghsoonian [11] exhibits the same concentration dependence as in the case of Mg-Cd. The Mg-4Li alloy is solid solution. Relatively high values of the activation volume at lower temperatures and beginning of plastic deformation indicate that the single solute atoms are probably not the main thermally activated obstacle for dislocation motion. The activation volume of alloys studied in papers [10] and [11] is only proportional to $c^{-2/3}$. It means that small clusters of the solute atom are active as dislocation obstacles. The geometrical interpretation of the activation volume considers

$$V = bdL, \tag{9}$$

where b is the magnitude of the Burgers vector and $d \approx 2b$ is the obstacle width and L effective distance between obstacles along the dislocations. According to [12] the activation volume depends on the grain size. It increases with the increasing grain size as $V_a \sim d^{1/2}$. Relatively high grain size in the alloy used is very probably reason for higher values of the activation volume at the beginning of the plastic deformation. Small clusters of solute atom are probably the main obstacles at the beginning of the plastic deformation. With increasing strain, the obstacles density due to dislocation-dislocation interaction increases. Small values of the activation volume estimated at the vicinity of the maximum stress are nearly equal for all temperatures. It indicates, that the dislocation obstacles generated during the straining.

5 Conclusions

Mg-4wt.%Li alloy were thermally treated at 350 °C for 1h. The alloys were deformed in compression in the temperature range of 20 to 200 °C at the constant strain rate. Stress relaxation tests were performed with the aim to study thermally activated process(es) controlling the deformation behaviour. The stress and temperature dependences of the activation volume were estimated from the stress relaxation curves. The main obstacles for the thermally activated dislocation motion are at the beginning of deformation small clusters of the solute atoms. Obstacles formed due to dislocation-dislocation interactions predominate at higher strains.

6 Acknowledgements

This work received a support from the Research Project 1M 2560471601 „Eco-centre for Applied Research of Non-ferrous Metals" that is financed by the Ministry of Education, Youth and Sports of the Czech Republic.

7 References

[1] F. W. Bach, M. Schaper, C. Jaschik, Mater. Sci. Forum 2003, 1037, 419–422.
[2] H. Yoshinaga, R. Horiuchi, Trans. JIM 1963, 4, 134–141.
[3] D. Hardie, N. Parkins, Philos. Mag. 1959, 4, 815–825.
[4] H. Yoshinaga, R. Horiuchi, Trans. JIM 1963, 4, 134–141.

[5] V. I. Dotsenko, Phys. Stat. Sol. (b) 1979, 93, 11–43.
[6] H. Conrad, R. W. Armstrong, H. Wiedersich, G. Schoeck, Philos. Mag. 1961, 177–188.
[7] J. V. Sharp, J. W. Christian: Phys. Stat. Sol. (b) 1965, 11, 831–843.
[8] P. Feltham, Phys. Stat. Sol. (b) 1963, 3, 1340–1346.
[9] P. Feltham, Philos. Mag. 1963, 8, 989–996.
[10] P. Lukáč, Phys. Stat. Sol. (a) 1992, 131, 377–390.
[11] A. Akhtar, E. Teghtsoonian, Acta Metall. 1969, 17, 1339–1349.
[12] Z. Trojanová, Z. Drozd, P. Lukáč, K. Máthis, H. Ferkel, W. Riehemann: Scripta Mater. 2000, 42, 1095–1100.

Method or Determination of Specific Damping Capacity Based on Relaxation Oscillations of Cantilever Beam in Magnesium Alloys

B. Landkof, A. Kotler, H. Abramovich
Faculty of Aerospace Engineering, Technion-Israel Institute of Technology, Haifa

1 The Importance of Material Intrinsic Damping Capability on Suppression of Aerospace Structural Vibrations

Mechanical vibrations can be developed by many different sources. In airplanes, missiles and satellites the vibrations are due to jet and rocket engines and to aerodynamic buffeting.

Most of vibration-control methods isolate the vibration-making components or attempt to dissipate the vibration produced. While both methods are effective, they usually require valuable space or add unwanted weight to the aerospace system. A third approach-producing structural and moving component from high damping metals avoids such problems and opens up new a possibility of vibration control. In order to use this approach, a reliable experimental method for measuring material intrinsic damping properties should be established. This method has to emphasize isolation of these properties from *structural damping*. One of the measured properties is "specific damping capacity" of material, which is given by the following expression:

$$\psi = SDC = \frac{W_n - W_{n+1}}{W_n} \times 100\%, \tag{1}$$

where: W_n - amplitude of the nth cycle; W_{n+1} - amplitude of the $n + 1$st cycle, as shown in Fig. 1.

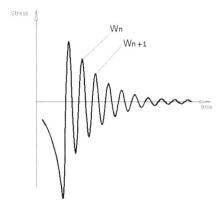

Figure 1: The relaxation oscillations

Most metals and composites found in aerospace structures do not have high damping capacity of the stress amplitudes associated with vibrations. The specific damping capacities of steels, aluminum alloys and other high-strength alloys are usually less than 1 %. One of the many metals known as the most energy absorbing is gray cast iron, which has SDC in the range 5 % to 10 %. In many existing publications, magnesium wrought alloys show very high SDC, reaching in some cases 50 %.

In order to verify this data a common experimental method for comparison of SDC for different aerospace structural materials must be use. The current investigation aims to establish such a method.

The main and easily observed effect of damping response is that associated with free vibrations. If, for illustration, we deform a cantilever beam and then release it from the rest, it will begin to oscillate regularly, and the amplitude of each successive oscillation will be smaller than the one before, when the ratio of successive amplitudes is regarded as a measure of the amount of damping.

Such damping brings the system to rest in a finite time. Clearly, the vibrations of damped structures will normally die away with time at a rate that may be use as a measure of the amount of damping. In fact, the well-known measure of damping is the logarithmic decrement δ, which is related to the ratio of the nth cycle and to the $n + N$th cycle amplitudes by:

$$\delta = \frac{1}{N} \ln \frac{W_n}{W_{n+N}}, \tag{2}$$

where W_n is the amplitude of the nth cycle and W_{n+N} is the amplitude of the $n + N$th cycle. It can be seen at once that only if the envelope of the decaying oscillation is an exponential curve of the form, as illustrated in Fig.1.1

$$W(t) = W_0 e^{-dt} \tag{3}$$

the modal damping d will relate to logarithmic decrement δ by

$$d = f_n \delta. \tag{4}$$

This is known as the case of viscous damping (at least as an approximation).

Essentially, the approach is to introduce into the system a "device" such as a damping force, which is proportional to relative velocity, like in the single degree of freedom system. The mass spring-dashpot (viscous damper) system is actually used and probably has been for many demonstrations. The advantage of the concept is its physical and mathematical simplicity, which allows one to obtain solutions using elementary mathematical methods without any of the apparent paradoxes or mathematical difficulties that other models of damping can lead too. Its disadvantage is that this model, only very rarely represents the real world. The equation of motion of the model shown in Fig. 1.2 is

$$m\frac{d^2w}{dt^2} + C\frac{dw}{dt} + kw = F(t), \tag{5}$$

which has the advantage of being linear, having constant coefficients with respect to time, and being simple.

After all external sources of damping have been accounted for there are still a very large number of mechanisms remaining where vibration energy can be dissipated within the volume of a material element, as it is cyclically deformed. We shall not, in any way endeavor, to review all of these mechanisms, only a few of which are dominant at any one time. All these mechanisms are associated with internal reconstructions of the micro and/or macro structure, ranging from crystal lattice to molecular scale effects. Included are magnetic effects (magnetoelastic and magnetomechanical hysteresis), thermal effects (thermoelastic phenomena, thermal conduction, thermal diffusion, and thermal flow), and atomic reconstruction (dislocation, concentrated defects of crystal lattices, phonoelectronic effects, stress relaxation at grain boundaries, phase processes in solid solutions, blocks in polycrystalline materials, etc.).

Regardless of the precise physical mechanisms involved, all real materials dissipate some energy, no matter how little, during cyclic deformation. Such effects are often highly nonlinear, so detailed analysis of response with such damping mechanisms is usually very difficult. However, experimental measurements of behavior of samples of specific materials can be, qualitatively and sometimes quantitatively assessed in terms of the measured specific damping capacities ψ,

$$\psi = \frac{2d}{f_n} 100\% .\tag{6}$$

Specific damping capacities ψ are very small for most conventional structural materials and somewhat higher for certain very unique high damping alloys. For these alloys, one damping mechanism or another is enhanced. The highest damping capacity exhibit polymeric rubber like materials, which are usually not used as basic constructional elements [4].

2 Test Setup

2.1 Measurement System and Sensors

For scanning of relaxation oscillations phenomenon standard laboratory equipment was used:
- Measurement system 6000 "Vishay",
- Strain gages "Kyowa KFR-2-120-C1-23".

2.2 Requirements and Design of Specimens

The basic condition for continuation of this investigation is the development of such test specimen, which would allow making measurements with sufficient precision and reliability of results using available equipment and within natural, existing environmental noises. First of all the sample was partitioned into two functionally different parts, as it has been illustrated in Fig.2:

- Base part for clamping;
- Cantilever beam part for measurement of vibrations.

The base part provides conditions for clamping without deformation of the main cantilever beam. Base part does not require high precision in manufacturing, but some other conditions, such as sufficient relative stiffness and gradual transition with minimal radius between one part and another. Cantilever beam which is needed for measurement of vibrations, require relative more high precision in manufacturing. The geometrical dimensions of the cantilever beam were derived from the first mode of resonance condition of the oscillations. In order to decrease the influence of the effect of coupling of various frequencies of natural oscillations, the ratio of width and thickness has been chosen as 3:1.

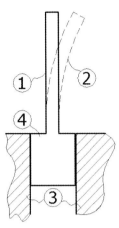

Figure 2: Specimen and clamping. 1 - Beam prior to bending, 2 - Beam during bending, 3 - Clamp, 4 - Base

The resonance frequency for mode n is given by:

$$f_n = \frac{HC_n}{l^2}\sqrt{\frac{E}{12\rho}}, \qquad (7)$$

where: C_n = coefficient for mode n, of clamped-free beam, n = mode number: 1, 2, 3…,

Thickness has been chosen based on the manufacturing limitations. The selected range of frequencies 300 ± 50 Hz was satisfied by the ratio conditions of dimensions chosen, gave sufficient stability to external disturbances and to the duration of oscillatory process.

2.3 Test Procedures

Actuation of oscillations was executed by a none normalized manual bending of a cantilever beam at it tip. Maximal values of a transient response were higher then the required of measured stress. The main test included several runs with 3–5 specimens of each alloy. Each specimen was tested, and three similar measurements were taking into account.

For alloys with high damping capacity, values of modal damping for each peak of maximum and minimum were determined, as illustrated in Fig. 1. For alloys with low damping capacity, values of modal damping were determined as an average value for 5 or 10 peaks of maximum and minimum.

3 Presentation of Results

As illustrated in a sample Fig. 3, this method allows determination of specific damping capacity values with a fluctuation of 10–15 %. With the increase in stress, fluctuation of results is growing; therefore, more peaks of high stress are needed, which in turn, result in more tests. For presented region of stress, behavior of specific damping capacity versus stress may be successfully and linearly approximated.

For convenience, a comparison of the received results with results of the other authors, of Specific Damping Capacity values corresponding to discrete values of stresses is shown in Figure 4.

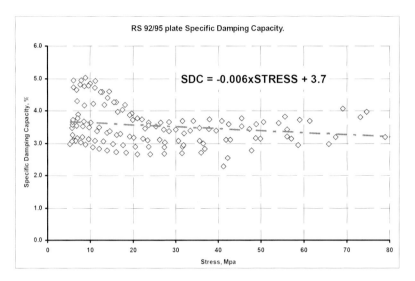

Figure 3: RS 92/95 alloy (plate) specific damping capacity vs. stress

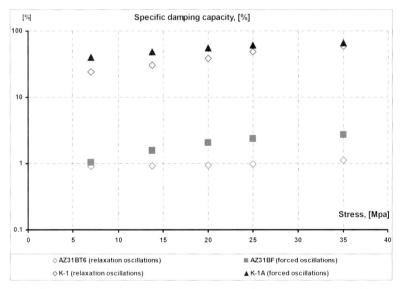

Figure 4: Graphical results comparison of selected magnesium alloys stressed by cantilever-beam bending, received from relaxation oscillations and forced oscillations [2]

4   Conclusions and Recommendations

1. This investigation offers a new approach to the damping properties measurement of aerospace materials.
2. The proposed method decouples the measurements of the cantilever free vibration decay from the clamping boundary conditions. This is done by the special design of the test specimen.
3. The results were linearly approximated and the table of "specific damping capacity" of various magnesium alloys versus discrete stress values was created.
4. Further experimental work will be done to cover additional magnesium alloys according to the requirements of "Aeromag" program.
5. From the magnesium alloys tested, AZ31BT6 shows the poorest damping properties, very similar to Al6061T6, while AZ91T6 the highest. Further investigation is needed in order to show that the growing content of Al in the alloy improves the damping properties.
6. The rapid solidified alloys RS66, 92, 93, show good damping capabilities.
7. The K1 alloy shows excellent "Specific Damping Capacity", but it has a low tensile strength, therefore it is not applicable to aerospace structures.
8. Further investigation is needed, if someone wants to prove the connection of microstructure to damping properties.

5   References

[1] "Standard Method for Measuring Vibration Damping Properties of Materials", ASTM Standard E 756, ASTM, 1983.
[2] Avedesian, M.M. and Barker, H., "Magnesium and Magnesium Alloys", ASM International Materials Park, OH 44073-0002, 1999.
[3] Snowdon, J.C., "Vibration and Shock in Damped Mechanical Systems", John Wiley & Sons, Inc, 1968.
[4] Nashif, A.D., Jones, D.I.G., Henderson, J.P., "Vibration Damping", John Wiley & Sons, Inc, 1985.
[5] Wei, C.Y., Kukureka, S.N., "Evaluation of Damping and Elastic Properties of Composites and Composite Structures by the Resonance Technique", Kluwer Academic Publishers, 2000.
[6] Korontzis, D.Th., Vellios, L., Kostopoulos, V., "On the Viscoelastic Response of Composite Laminates", Kluwer Academic Publishers, 2001.

Acoustic Emission Study of the Mechanical Anisotropy of AZ31 Magnesium Alloy Sheet

P. Dobroň[1*], J. Bohlen[2], F. Chmelík[1], P. Lukáč[1], D. Letzig[2], K.U. Kainer[2]
[1] Department of Physics of Materials (former Dept. Metal Physics), Charles University, Prague 2
[2] GKSS-Forschungszentrum GmbH, Geesthacht

1 Abstract

The mechanical anisotropy of hot rolled AZ31 magnesium alloy has been investigated by measurements of the acoustic emission (AE). Specimens of different orientations between tensile axis and rolling direction were deformed at room temperature, at an initial strain rate of 10^{-3} s^{-1}, in order to study the microstructure changes during plastic deformation. The activity of different slip systems (especially of basal slip) and deformation twinning has been found to have a significant effect on the mechanical anisotropy. The AE parameters are correlated with the stress-strain curves and discussed in terms of possible deformation processes.

2 Introduction

Magnesium and its alloys belong to the lightest construction materials and therefore they are prospective for a use in many technical applications, especially in the automotive industry, where a weight reduction leads to improving fuel efficiency and car performance. However, the limited number of active dislocation slip systems at room temperature (RT) in hexagonal close-packed (hcp) Mg alloys is responsible for poor strength and formability. Presently, the wrought magnesium alloys offer better mechanical properties in comparison to magnesium cast alloys; nevertheless they also show an anisotropic mechanical behaviour which is based on the texture of this material and an initially strong orientation dependence of mechanical properties in the hcp-cell of magnesium itself. This is due to a lack in initially active deformation modes based in distinctive differences in the critical resolved shear stresses (CRSS) for different dislocation and twinning systems as e. g. reported in [1-2]. The results demonstrate that the CRSS in the basal plane is much lower than that in the other systems. The texture affects the activity of basal slip and also the activation of hard deformation modes such as non-basal slip and also deformation twinning. Reorientation of basal planes due to twinning in hcp metals affords opportunity for dislocation slip in these planes [3].

The acoustic emission (AE) is defined as transient elastic waves generated by sudden release of energy due to local dynamical changes in the material structure such as dislocation slip and twinning [4]. A direct correlation of AE parameters with the stress-strain curve yields information on the dynamic processes involved in plastic deformation of magnesium and its alloys.

The present paper will make use of this method for a study of the mechanical anisotropy of hot rolled AZ31 magnesium sheet.

3 Experimental Procedure

A commercial AZ31B rolled sheet (Mg + 3 wt. % Al + 1 wt. % Zn) in a strain hardened and stress-relieved condition (H24 in terms of ASTM descriptions) with a thickness of 1.0 mm was used for this study.

The specimens (125 mm × 20mm, thickness 1 mm) were deformed in a universal testing machine Zwick® Z50 in tension at RT and at an initial strain rate of $10^{-3}\,s^{-1}$. Tensile tests were performed in various orientations υ of the specimen axis with respect to rolling direction (RD = 0°).

The computer controlled DAKEL-XEDO-3 AE system was used to monitor AE on the basis of two-threshold-level detection, which yields a comprehensive set of AE parameters involving count rates N_{C1} and N_{C2} (count number per second [5]) at two threshold levels (giving total AE count and burst AE count by proper setting – see below). The burst AE occurs mainly as a consequence of an instable fashion of plastic deformation (e.g. twinning) or degradation of materials. A miniaturized MST8S piezoelectric transducer (diameter 3 mm, almost point AE detection, a flat response in a frequency band from 100 to 600 kHz, sensitivity 55 dB ref. 1 V_{ef}) was attached on the specimen surface with the help of silicon grease and a spring. The total gain was 90 dB. The AE signal sampling rate was 4 MHz, the threshold voltages for the total AE count N_{C1} and for the burst AE count N_{C2} were 730 and 1450 mV, respectively. The full scale of the A/D converter was ± 2.4 V.

4 Experimental Results

The microstructure and texture of the tested alloy is reported in [6]. The average grain size is 15 μm.

The texture is of a strong basal-type with the majority of grains oriented such that the basal planes are close to the sheet plane. The angular distribution of basal planes is higher if looking to the RD rather than to the TD. Furthermore, a slight off-basal double-peak towards rolling direction indicates a deformation type of texture. The true stress – true strain curves of the sheets deformed in tension with an angle of 0°, 30°, 45°, 60° and 90° (transversal direction, TD) with respect to RD are depicted in Figure 1. Deformation curves are smooth and show distinctive orientation dependence, while conventional tensile yield strength (YS) and ultimate tensile strength (UTS) are lowest in RD and increasing with the angle towards TD. A deformation plateau on YS occurred for specimens with orientation angle between υ = 30° and 90°.

Figure 2 shows the dependence of YS and UTS on the orientation angle υ between tensile axis and RD. It can be seen that with the increasing angle υ both YS and UTS increase. Furthermore, for υ higher than 30°, YS increases faster than UTS.

Work hardening rate vs. true strain curves for some angles υ between RD and TD are plotted in Figure 3. Between υ = 30° and TD, the curves exhibit a local minimum as a consequence of the occurrence of the deformation plateau; the more pronounced plateau the more distinctive minimum. Note, that the work hardening rate decreases with increasing angle υ.

The Figures 4-6 show a correlation between the engineering stress-strain curves and the AE count rate curves for various orientations of the specimen axis to RD. Measurements of the AE activity at two threshold levels (N_{C1}, N_{C2}) are helpful to recognize AE signals having a burst character with large amplitude. The AE activity decreases with increasing orientation angle υ. For various directions between RD and TD the AE count rate exhibits two maxima closed to the macroscopic yield point, which are followed by a decrease in the count rate. The AE decreases after the second maximum correspond to the plateau on the deformation curves; the more pronounced plateau the lower AE activity.

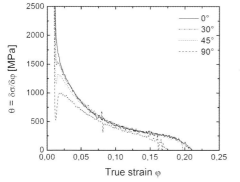

Figure 1: The true stress *vs.* true strain for hot rolled AZ 31 sheet in the H24 condition deformed in 0°, 30, 45°, 60° and 90° to rolling direction.

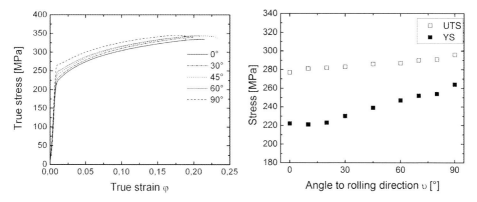

Figure 2: Mechanical properties as a function of an angle υ to rolling direction.

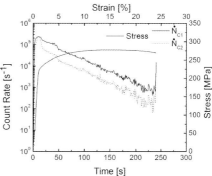

Figure 3: Work hardening rates as a function of true strain for various orientations between 0° and 90° to rolling direction.

Figure 4: The correlation between the engineering stress-strain and the AE count rate *vs.* time curves in rolling direction (0°).

The ratio $\dot{N}_{C2} / \dot{N}_{C1}$ *vs.* time depicted in Figure 7 indicates the relative amount of AE burst signals as part of the total AE activity during plastic deformation. These dependences also give hint on two local peaks, where the first one is more pronounced at a higher orientation angle υ and the second one, related to the macroscopic yield point, decreases with increasing orientation angle. The following decay in the ratio $\dot{N}_{C2} / \dot{N}_{C1}$ also depends on the orientation angle and from an angle $\upsilon = 30°$ exhibits a local minimum. Afterwards, the curves do not show any significant differences.

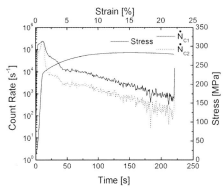 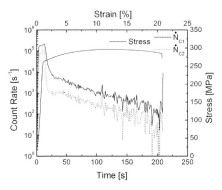

Figure 5: The correlation between the engineering stress-strain and the AE count rate vs. time curves in orientation angle of 30° to rolling direction.

Figure 6: The correlation between the engineering stress-strain and the AE count rate vs. time curves in transverse direction (90°).

5   Discussion

Tensile tests of the AZ31 alloy in H24 condition performed at RT exhibit the lowest YS in RD and with an increase in orientation angle υ, the values of YS increase, which can be explained by a different activity of deformation mechanisms owing to the rolling texture of the sheet. Some simulations on AZ31 in H24 condition [2, 7] confirm, that beside of basal slip, prismatic slip as well as pyramidal <$c+a$> slip are activated. Furthermore, deformation twinning, primarily in (1012), may occur. The compatibility of plastic deformation in hcp metals and its activation depends on the grain size, the orientation of grains and also on the deformation mode.

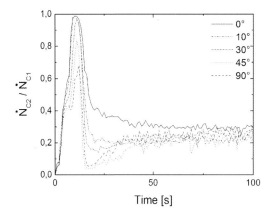

Figure 7: Relative occurrence of burst emission signals (N_{C2}) as part of the total AE activity (N_{C1}) vs. time for same orientations between 0° and 90° to rolling direction.

More favorite orientation of basal planes to RD indicates a more important role of basal slip during deformation, which results in lower value of YS in this direction than in other directions

(Figures 1–2). More rapid strain hardening observed from an angle $\upsilon = 30°$ up to TD (Figure 2) can be explained by a more intense activity of non-basal slip systems such as prismatic and pyramidal slip during deformation. The plateau on the deformation curves is likely occur due to the formation of the Lüders bands which are formed when a number of dislocations move one after another in a chain reaction known as the cataclysmic release of dislocations and it is evident that their formation is associated with a slip. This also indicates a partly inhomogeneous plastic deformation of AZ31.

Hardening caused by dislocation storage and softening due to recovery processes can also be seen in Figure 3, for υ higher than 30° the work hardening rates exhibit local peaks followed by a drop, which can be caused by a higher activity of non-basal slip systems.

The AE peak related to the macroscopic yield point is explained by a massive dislocation multiplication and twinning, which are excellent sources of AE [8]. The following decrease in the AE count rates can be assumed as a consequence of strain hardening. Increasing density of forest dislocations due to interactions between basal <a> and pyramidal <c+a> dislocations, which reduces the flight distance. A decrease in the free path of moving dislocations leads to a reduction of the AE activity. The slowest decay in the AE count rate is found in RD and it becomes faster with increasing υ. This can be understood by taking into account a more distinctive role of twinning in RD than in other directions.

The first maximum depicted in Figure 7 can indicate an activation of primary $\{10\bar{1}2\}$ $<10\bar{1}\bar{1}>$ twins (also observed in [9]) because this twinning mode has the lowest shear and involves simple shuffles in the shear plane [10]. Consecutive glide in the reoriented grains and an activation of differently oriented twins to the primary twins (such as in $\{10\bar{1}2\}$ $<10\bar{1}\bar{1}>$ system, observed in [9]) contributes to strain hardening and results in lower amount of high amplitude AE signals. Keeping in mind that twinning unlike continuous dislocation glide will generate AE mainly of a burst type character [8] it is evident that twinning should play an important role in RD generating the highest ratio of burst emissions to the total emission. This is demonstrated by the orientation dependence of the second peak. This peak becomes less pronounced with increasing angle υ and therefore hints a lower activity of twins at the beginning of the plastic deformation. Following decrease of the AE activity can also be caused by the interaction between slip dislocations and deformation twins, which is in accordance with work hardening rates shown in Figure 3.

6 Conclusions

The hot rolled AZ31 sheets in the H24 condition with various orientation of the specimen axis to rolling direction were tested in tension at room temperature with concurrent acoustic emission (AE) measurements. The AE technique was used to evaluate the processes responsible for the observed deformation behaviour. The AE count rates showed two peaks close to the macroscopic yield point followed by a subsequent decrease in the AE activity. The results can be explained by the orientation-dependent activation of non-basal slip systems and deformation twinning during plastic deformation. The observed effect of the specimen axis orientation to the rolling direction demonstrates the importance of the activation of primary $\{10\bar{1}2\}$ and $<10\bar{1}\bar{1}>$ consequently differently oriented twins at beginning of the deformation resulting in the mechanical anisotropy.

7 Acknowledgements

This work received a support from the Research Project 1M 2560471601 "Eco-centre for Applied Research of Non-ferrous Metals" that is financed by the Ministry of Education, Youth and Sports of the Czech Republic. Additional support was also provided by the Grant Agency of the Czech Republic under Grant No. 103/06/0708.

8 References

[1] S. Ando, H. Tonda, Mater Sci Forum 2000, 43, 350–361.
[2] A. Staroselsky, L. Anand, Int. J. Plast. 2003, 19, 1843–1864.
[3] P. Zhang, B. Watzinger, P. Q. Kong, W. Blum, Key Eng. Mater. 2000, 171-174, 609–616.
[4] R. Heiple, S. H. Carpenter, J. Acoustic Emission 1987, 6, 177–204.
[5] Standard Practice for Acoustic Emission Examination of Fiberglass Reinforced Plastic Resin, ASTM E 1067-85. Tank/Vessels, May 31, 1985.
[6] J. Bohlen, F. Chmelík, P. Dobron, D. Letzig, P. Lukác, K. U. Kainer, J. Alloys Compd. 2004, 378, 214–219.
[7] S. R. Agnew, in: H.I. Kaplan (Ed.), Magnesium Technology 2002, Proceedings of the Minerals, Metals and Materials Society TMS, Seattle, 2002, p. 169.
[8] M. Friesel, S. H. Carpenter, J. Acoustic Emission 1984, 3, 11–18.
[9] K. Máthis, F. Chmelík, M. Janeček, B. Hadzima, Z. Trojanová, P. Lukáč, Acta mater., accepted for publication.
[10] S. Lay, G. Nouet, Philos. Mag. A, 1994, 70, 261–270.

# Surface Rolling on the High-Strength Wrought Magnesium Alloy AZ80

P. Zhang[1], J. Lindemann[2], C. Leyens[2]

[1]National Engineering Research Center of Light Alloy Net Forming, School of Materials Science and Engineering, Shanghai Jiaotong University, Shanghai 200030, China
[2]Lehrstuhl Metallkunde und Werkstofftechnik, BTU-Cottbus, Postfach 101344, D-03013, Cottbus, Germany

1 Introduction

Due to its high strength, excellent workability and relatively low cost, AZ80 has become to the most commonly used wrought magnesium alloy in the industry. Recently, AZ80 is considered as a potential candidate for the application as suspension parts in future automobiles [1]. For this application, good high cycle fatigue (HCF) performance must be achieved.

It is well known that the fatigue performance of structural metallic materials could be improved through mechanical surface treatments, such as shot peening and surface rolling (roller burnishing and deep rolling) [2–4]. Several studies have demonstrated that mechanical surface treatments can improve the HCF strength of magnesium alloys [5-8]. However, in order to apply these techniques into practical magnesium applications, further indepth investigations are needed.

In the present work, the influence of surface rolling (including roller burnishing and deep rolling) on fatigue performance of the high-strength wrought magnesium alloy AZ80 has been investigated. The rolling force has widely been varied in order to obtain optimum conditions of surface rolling with regard to fatigue performance.

2 Experimental

The high-strength wrought magnesium alloy AZ80 (nominal composition in wt.%: 8Al, 0.5Zn, 0.2Mn, balance: Mg) used in the present work was produced by Otto Fuchs Metallwerke, Meinerzhagen, Germany. The alloy was forged to a rectangular bar with a cross section of 60 mm $\times$ 11 mm. The alloy has a single α-phase structure with an average grain size of 30 μm. Texture measurement results show that the basal planes are oriented predominantly parallel to the longitudinal–transversal (L–T) plane.

Specimens were machined with the load axis parallel to longitudinal direction of the rectangular bars. Tensile tests were performed on threaded cylindrical specimens having gage lengths of 20 mm at initial strain rates of $8.3 \cdot 10^{-4}$ s^{-1}. Tensile test results are shown in Table 1. For fatigue testing, both smooth and notched specimens were used. The smooth specimens were round hour-glass shape with the gage diameter of 6 mm (notch factor of k_t = 1). The notched specimens had a circumferential 60° V-notch, the notch radius was either 0.43 or 0.3 mm. After machining, a layer with a thickness of about 50 μm was removed from the surface of the specimens by electrolytical polishing (EP) in order to avoid any influence of machining on the fatigue results.

Table 1: Tensile results on the wrought magnesium alloy AZ80

Material	Testing direction	$\sigma_{0.2}$ (MPa)	UTS (MPa)	EL (%)	RA (%)
AZ80	L	226	337	18.2	25.2

Roller burnishing (RB) was performed on smooth specimens using a one-roll hydraulic system with 6 mm hard-metal ball operating on a conventional lathe (Fig. 1a). Deep rolling was conducted on notched specimens using a three-roll hydraulic system (Fig.1b). The spindle speed was 36 min^{-1}; rolling forces were varied in the range of 50–400 N. In addition, the specimens used for DR had a notch root radius of 0.43. The rolls used for deep rolling had a 55° cross section and a tip radius of 0.3 mm. Thus, during deep rolling, the notch root radius of the specimens was reduced from 0.43 to roughly 0.3 mm, resulting in a geometrical notch factor of $k_t = 2.7$.

The surface properties of specimens after surface rolling were determined by roughness measurements through profilometry, measurements of the microhardness–depth profiles and residual stress measurements by means of the incremental hole drilling method [9]. Fatigue tests were performed under rotating beam loading (R = –1) at a frequency of about 100 Hz in air.

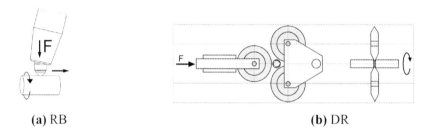

(a) RB   (b) DR

Figure 1: Schematic sketch of surface rolling

3 Results and Discussion

3.1 Fatigue Behaviour at EP Condition

The S-N curves of AZ80 at the reference condition (EP) are shown in Fig.2. It can be seen that fatigue strength is considerably reduced by geometrical stress concentrations (Fig. 2a), i.e. the fatigue strength decreases from 100 MPa to 65 MPa as the notch factor increases from 1.0 to 2.7. In order to determine the influence of the geometrical notch factor on the fatigue strength, the data in Fig.2a are re-plotted in terms of maximum notch root stresses (σk_t). The fatigue strength for the notched specimens is about 165 MPa, which is significantly higher than that of smooth specimens (Fig. 2b). From Fig. 2b, the notch sensitivity factor q [10] can be derived to be 31 %, which is much smaller than the value of 100% obtained by Kuester et. al [7] at the same test condition for AZ80, indicating that the improvement in ductility of AZ80 results in the reduction of the notch sensitivity.

Fracture surfaces of the reference specimens (EP) are illustrated in Fig. 3. It is found that for both smooth and notched specimens, the fatigue cracks nucleate on the surface. The smooth specimen has a single crack nucleation site at low stress amplitudes, while multi-nucleation is observed at higher stress amplitudes. On the contrary, the notched specimens have a significant higher number of crack nucleation sites on the surface, independent of the stress amplitude.

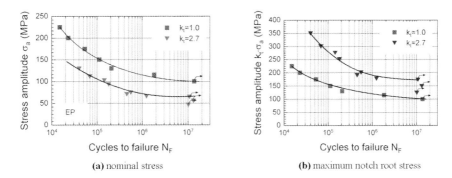

(a) nominal stress (b) maximum notch root stress

Figure 2: S-N curves of AZ80 with different notch factors

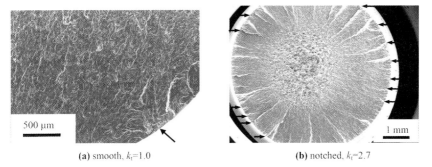

(a) smooth, $k_t=1.0$ (b) notched, $k_t=2.7$

Figure 3: Fracture surfaces of the specimens at EP condition (crack nucleation site is indicated by an arrow)

3.2 Roller Burnishing

The influence of RB on the resulting surface roughness values is plotted in Fig. 4. Compared to shot peening [8], it can be seen that RB results in a relatively smoother surface. Microhardness-depth profiles after RB are shown in Fig. 5. Due to work hardening induced by RB, the near-surface microhardness increases significantly. From the microhardness profiles in Fig. 5, the thickness of the deformation layer can be estimated to be about 700 µm.

The residual stress-depth profiles in AZ80 after RB are demonstrated in Fig. 6. One can see that that RB induced compressive residual stresses. The maximum compressive residual stress of 340 MPa is located at the surface when the specimen is roller burnished with a roller force of 200 N. As the rolling force increases from 200 to 300 N, the location of the maximum compressive stress shifts to subsurface, and its value decreases to 220 MPa.

The effect of rolling force on the fatigue life is illustrated in Fig. 7. The best fatigue performance after rolling burnishing was found at intermediate rolling forces of 200–300 N, indicating that at lower rolling forces the work-hardened surface layer is possibly too thin to improve the fatigue properties and at higher rolling forces the surface defects such as overlaps and microcracks probably deteriorate the fatigue life. The improvement in fatigue life is more than two orders of magnitude at the optimum rolling force of 200 N.

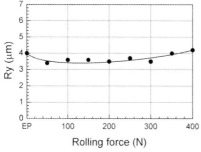

Figure 4: Surface roughness profile after RB profiles after RB

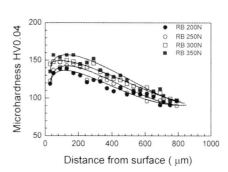

Figure 5: Microhardness-depth

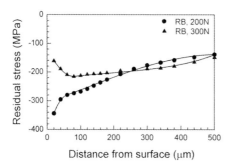

Figure 6: Residual stress–depth profiles after RB rolling force of RB.

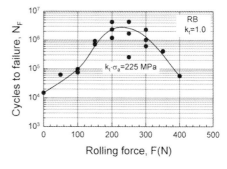

Figure 7: Fatigue life vs. rolling force of RB.

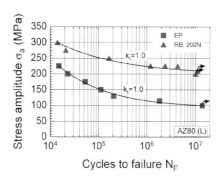

Figure 8: S-N curves of AZ80 after RB at the optimum condition. (rotating beam loading in air).

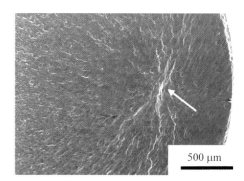

Figure 9: Fracture surfaces of the smooth specimens after RB (crack nucleation site is indicated by arrow).

The S-N curves at the optimum condition for RB are shown in Fig. 8. The fatigue strength increases from 100 to 210 MPa after RB, the improvement is about 110 %. Compared to shot peening [8], RB is more effective in improving the fatigue performance of AZ80.

Fracture surface of roller burnished condition (RB) is shown in Fig. 9. Subsurface fatigue crack nucleation is observed. The depth of the crack nucleation site is about 700 μm below the surface, in agreement with the depth of plastic deformation estimated by the microhardness measurements.

3.3 Deep Rolling

Fig. 10 shows microhardness–depth profiles after DR. A pronounced increase in microhardness in the near-surface region of the specimens after DR is noticed. Increasing the rolling force from 50 to 150 N leads to the increase in microhardness. Compared to RB, the increment of microhardness produced by DR is in the same level. However, the thickness of the deformation layer after DR is smaller, about 400 μm.

The fatigue life at $k_t \sigma_a$ = 350 MPa of notched specimens with various rolling forces is illustrated in Fig. 11. In general, the response of fatigue life to rolling force is similar in both RB and DR. The run-outs (10^7 cycles) are found already at intermediate rolling forces of 100–250 N. Therefore, the rolling force of 150 N is taken as the optimum condition for DR.

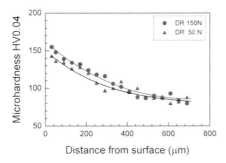

Figure 10: Microhardness-depth profiles after DR force of DR.

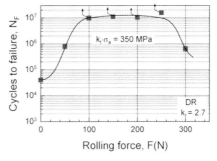

Figure 11: Fatigue life vs. rolling force of DR.

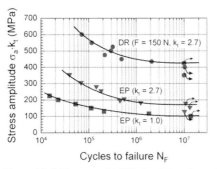

Figure 12: S-N curves of AZ80 after DR at the optimum condition (rotating beam loading in air).

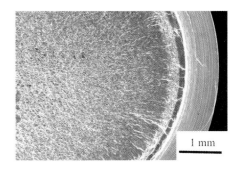

Figure 13: Fracture surface after DR.

The S-N curves of notched specimens are shown in Fig. 12 comparing deep rolled with electrolytically polished conditions. DR dramatically increases the notch HCF strength from 160 to 425 MPa, the corresponding improvement is about 160 %.

Kuester et al. [7] have reported that the fatigue strength of wrought magnesium alloy AZ80 increases from 100 MPa (EP) to 350 MPa (DR) after DR. Obviously, the present fatigue strengths at both EP and DR conditions are much higher than those obtained by Kuester [7]. This is possibly associated with the ductility of two materials. The improved ductility may reduce the notch sensitivity and causes a greater degree of work-hardening during DR, resulting in higher fatigue strengths.

The typical fatigue fracture surface of DR condition is shown in Fig. 13. As often observed on fatigue fracture surfaces of deep rolled specimens, the circumferential fatigue cracks can be clearly seen. The circumferential fatigue crack is not fully understood yet. It has been suggested that the width of the circumferential cracks possibly corresponds to the critical depth for crack propagation in notched specimens [3,11].

4 Conclusions

The influence of surface rolling (roller burnishing and deep rolling) on the fatigue behaviour of the high-strength wrought magnesium alloy AZ80 has been investigated. It is found that the HCF performance of AZ80 can be significantly improved by surface rolling. For smooth specimens, the improvement in fatigue strength by RB is about 110 % at the optimum condition. While for notched specimens, the HCF strength increases from 165 to 425 MPa after DR with the optimum rolling force, the corresponding improvement is about 160 %.

5 Acknowledgements

Thanks are due to BMBF for financial support. The authors would like to thank Otto Fuchs Metallwerke, Meinerzhagen, Germany, for providing the magnesium alloy.

6 References

[1] Viehweger, B., Leyens, C., Becker, J., Bogon, P., Gers, H., Pieper, G., Kiefer, A., Roll, K., Straube, O.ATZ – Automobiltechnische Zeitschrift, 2005(10), Jg. 107, 922.
[2] Gregory J.K., Wagner L., In: Wagner L, Editor. Shot Peening, Weinheim: WILEY-VCH, 2003, p.349.
[3] Lindemann J, Grossmann K, Wagner L., Z. Metallkd, 2003, 94, 711.
[4] Lang K.H., Schulze V., Vöhringer O., In: Wagner L, Editor. Shot Peening, Weinheim: WILEY-VCH, 2003, p.281.
[5] Zhang P., Lindemann J., Scripta Mater. 2005, 52, 1011.
[6] Hilpert M., Wagner L., In: Kainer K.U., Editor. Magnesium Alloys and their Application, WILEY-VCH, 2000, p.463.
[7] Kuester B., Hilpert M., Kiefer A., Wagner L., In: Wagner L, Editor. Shot Peening, Weinheim: WILEY-VCH, 2003, p.281.

[8] Zhang P., Lindemann J., Scripta Mater. 2005, 52, 485.
[9] Schwarz T., Kockelmann H., VDI Report 940, 1992, p.99
[10] Forrest P.G., Fatigue of Metals, Pergamon Press, Oxford, 1962.
[11] Drechsler A., Doerr T., Wagner L., Mater. Sci. Eng. 1998, A234, 217.

Rolling of AM50 Magnesium Cast Alloy

J. Göken[a], I. Stulíková[b], B. Smola[b], K. Steinhoff[a], N. Hort[c], V. Očenášek[d]
[a]Metal Forming Technology, University of Kassel, 34125 Kassel, Germany
[b]Faculty of Mathematics and Physics, Charles University, 12116 Prague 2, Czech Republic
[c]GKSS Research Center, 21502 Geesthacht, Germany
[d]Research Institute for Metals, 250 70 Odolena Voda, Czech Republic

1   Introduction

Magnesium as the lightest known metallic engineering material is offering an enormous potential for light weight constructions. Especially in automotive industries there is a demand for weight savings to fulfill requirements from the legislative to reduce CO_2 emissions [1], to save fuel and to provide materials that can be recycled easily. [2] In the past magnesium cast alloys have been successfully developed and are fairly good introduced to vehicles even in power train applications with service temperatures up to 150 °C and more. It can be expected that the use of magnesium alloys will increase and other materials will be replaced by magnesium components. This has already influenced the modification of known alloys as well as the development of new cast alloys with improved high temperature stability and/or ductility. [3]

But there is still a lack of information regarding the use of magnesium alloys as wrought materials especially in sheet applications. This concerns a limited knowledge on the processability as well as a limited number of suitable wrought alloys. While the use of cast components is fairly well introduced in a wide number of applications which are ranging from steering wheels, instrumental panels and even to hybrid engine blocks the use of magnesium wrought materials for production of sheet, forgings or extrusions is still limited. [4] Moreover, the difference between magnesium cast and wrought alloys is not yet clearly defined. Therefore the deformation behavior of alloys which are by origin designated as cast alloys as well as the behavior of typical wrought alloys need to be investigated with respect to their heat treatment response and their ability of being deformed by several subsequent steps as performed by rolling. [5]

In the present study the commercially available cast magnesium alloy AM50 has been used to produce sheet by hot rolling. Preheated material has been deformed in several passes until the sheet reached a final thickness of about 0.05 mm. A special concern was given to the development of the deformation behavior. The microstructural evolution during the deformation process was observed and accompanied by texture measurements, TEM investigations and tests of the mechanical properties.

2   Experimental

AM50 ingots (Hydro[TM] Magnesium, 4.9 wt.% Al, 0.01 wt.% Zn, 0.31 wt.% Mn, 0.01 wt.% Si, 0.001 wt.% Cu, 0.001 wt.% Ni, rest Mg) have been used to manufacture the feedstock materials for the production of sheet. Thus bars with 50 mm x 20.5 mm x 600 mm have been machined. These bars were solutionized at 500 °C for 2 h prior to rolling with cooling in air afterwards. Before rolling performed with a Wezel[TM] BW 300 laboratory mill at the Chair of Metal Forming at the University Kassel using a rolling speed of 110 mm/s per pass the bars were heated again up to 500 °C for 30 min before each pass. A lubricant based on zinc oxide mixed with montmorrillonite was used in the rolling experiments. In total the material underwent 30 passes in reverse rolling to reach a final thickness of about 0.05 mm. The forming ratio was varied during the forming process which is illustrated in Fig. 1a where the change of forming ratio between two sequent rolling passes is plotted versus the number of pass. A total forming ratio of about 6.02 was obtained in this study. The resulting reduction of the sheet thickness is shown in Fig. 1b including a classification in three sheet categories. It is worth mentioning that our chosen process route with the cast magnesium alloy AM50 made sheets of foil dimensions possible.

In order to get information about the microstructural evolution during forming various analyses (tensile tests, hardness measurements, optical micrographs, TEM and texture analysis) were performed. Tensile tests (Zwick™ 1446 tensile machine) were carried out at room temperature and with a tensile testing velocity of 1 mm/min. The metallographic preparation for the material in the as cast/solutionized and rolled condition was performed at the GKSS Research Center following a standard procedure as described in [6]. For texture analysis a X-ray diffractometer (Siemens™ D 5000) was used.

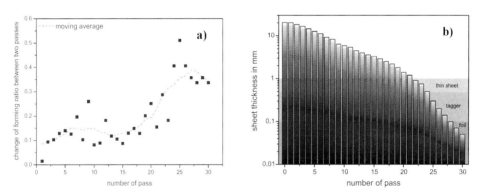

Figure 1: a) Forming ratio during the rolling process; b) reduction of the sheet thickness during the rolling process

3 Results and Discussion

From the micrographs of the material in the as cast (Fig. 2a) and solutionized condition (Fig. 2b) it is clearly visible that the heat treatment is able to remove precipitates within the grains and from the grain boundaries. The precipitates consist of $Mg_{17}Al_{12}$ which can be dissolved by the heat treatment and more stable Al-Mn precipitates that can withstand the heat treatment regime. Beside dissolution of precipitates only a slight growth of grains was observed. The grain size before heat treatment was 1.22 ± 0.85 mm while after heat treatment the grain size changed to 0.87 ± 0.52 mm. Therefore, the heat treatment at temperatures close to 500 °C appears to be a suitable solutionizing treatment prior to the deformation processes.

In Fig. 3a the microstructure of AM50 after rolling pass 7 is displayed from which the grain size is determined as 35.80 ± 29.44 µm. The strong scattering of the mean grain size can be attributed to recrystallization which is obviously seen. Isolated strong twins are observable. In contrast to that after thirteenth pass (Fig. 3b) practically no recrystallized areas occur. The mean grain size is about 43.53 ± 15.86 µm and the micrograph is characterised by larger grains and a more homogeneous microstructure which is due to grain growth supposed to be the main mechanism for microstructural evolution at this state. Finer and more homogeneously distributed twins can be found.

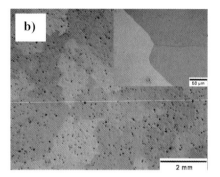

Figure 2: Microstructure of a) the as cast material and b) after solutionizing heat treatment (2 h at 500 °C)

Figure 3: Microstructure of AM50 after a) 7th rolling pass and b) after 13th rolling pass

Figure 4: Microstructure of AM50 after a) 16th rolling pass and b) after 22nd rolling pass

A higher resolution of the optical micrograph was taken for microstructural analysis of AM50 after pass 16 (Fig. 4a). The grain size of 40.32 ± 15.33 µm was near the corresponding value after rolling pass 13. This means that using almost the same forming ratio (Fig. 1a) no significant change of the grain size after forming occurred. The influence of the chosen intermediate heat treatment and the influence of forming are expected to compensate each other.

In Fig. 4b the microstructure of AM50 after pass 22 is shown. With regard to Fig. 1a it can be found that nearly a twofold forming ratio was applied. This leads to a massive presence of twins and a reduction of the grain size to 29.91 ± 10.68 µm. Considering the same heat treatment procedure between two sequent rolling passes, a further reduction of the grain size must be caused by dynamic recrystallization due to intensified forming.

As know from Hall-Petch-relationship [e.g. 7] the change of grain size has an influence on the tensile strength which is illustrated in Fig. 5a where the tensile strength is plotted versus the number of rolling pass. As determined from Figs. 3a and 3b the grain size is hardly affected by the solution treatment. The following rolling passes lead to an increase in tensile strength. The recovery of microstructural defects due to intermediate heating after forming seems to be small. After that a plateau between pass 7 and pass 17 occurs. This is in accordance with Figs. 3a and 4a showing only a small variation of the grain size. With changing forming ratio (starting at about pass 18) an increase of the tensile strength was again observed. With rising number of rolling passes a maximum of the tensile strength was measured. The development of the tensile strength during rolling process is significantly dependent on the applied forming ratio (Fig. 1a).

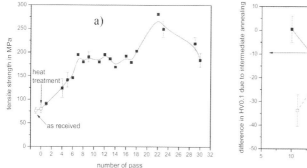
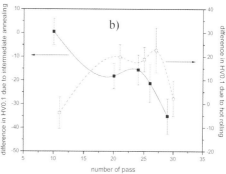

Figure 5: a) Development of tensile strength during rolling process; b) influence of intermediate annealing and rolling on the hardness of AM50

The interplay between intermediate heat treatment and subsequent forming and the individual impact on the microstructure in terms of the hardness has to be investigated more detailed. In Fig. 5b the left ordinate describes the difference in hardness due to subsequent intermediate annealing. On the right ordinate the difference in hardness caused by subsequent hot rolling is given. Both y-values are plotted versus the number of pass. As expected, the intermediate heat treatment reduces the hardness which can be attributed to recovery. This recovery is more effective at lower passes up to about pass 20. When the forming ratio increases the hardness reduction by annealing does not change as strong as before. At highest number of rolling passes the heat treatment becomes more dominantly. The hardness data after hot rolling show an increase from pass 10 to pass 20, after pass 27 a reduction of the hardness occurs as also observed for the tensile strength (Fig. 5a). The drastic decrease of the hardness change can not be ascribed only to the drop of the forming ratio which is still about twice higher than the value at pass 20 where an increase of the difference in HV0.1 was connected with.

In Fig. 6 the quadratic difference in HV0.1 due to intermediate annealing is plotted versus the quadratic hardness after rolling. The abscissa characterises the dislocation density which is approximately proportional to quadratic hardness [8]. From this plot it can be concluded that the intermediate annealing is more effective at increased dislocation density, which occurs at higher number of passes.

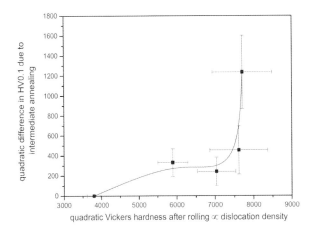

Figure 6: Influence of hardness change due to intermediate annealing with increasing dislocation density

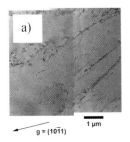 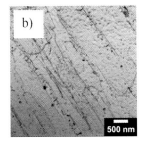 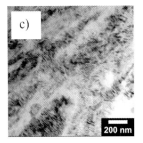

Figure 7: a) Dislocation structure in as cast AM50; b) dislocation structure of AM50 after solutionizing heat treatment (2 h at 500 °C) – near [11–20] pole; c) after pass 27: BF image in (1–100) reflection – near [0001] pole

Two deformation mechanisms are significant during hot rolling of the AM50 alloy, namely dislocation glide and twinning. TEM has shown only a low dislocation density both in as cast condition (Fig. 7a) and also after solution treatment (Fig. 7b). Isolated twins characteristic for Mg, namely {10-10}<10-1-1>, were observed in the solutionized alloy. Dislocation density increases considerably during rolling (Fig. 7c) and twinning plays an important role, especially in rolling steps over 11th one (Fig. 4).

Texture analysis evidences the formation of a distinct arrangement of grains parallel to the rolling plane (Fig. 8). After solution treatment and in cast condition (not shown here) a random texture is observed. During rolling the well known egg-shaped pole figure in (0002) reflection develops [9]. This type of texture is considered to develop only if both basal and non-basal slip systems are active together with twinning. Dislocations with c-component Burgers vector were observed after the 30th rolling step. Further investigations concerning the weighting of the individual deformation mechanisms are necessary and will be subject of a following paper.

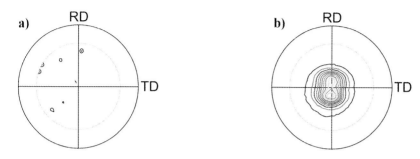

Figure 8: Texture measurement: a) after solution treatment and b) after 30th rolling pass; RD: rolling direction, TD: transverse direction

4 Conclusions

Stepwise rolling of the cast magnesium alloy AM50 alloy led to the production of a very thin sheet (thickness ~ 0.05 mm). It could be shown that dislocation glide in basal and non-basal slip systems together with twinning were the active deformation mechanisms. A typical hot rolling texture developed during the stepwise rolling. Static recovery taking place during the intermediate annealing before each rolling step is highly significant for achieving such a high rolling reduction.

5 Acknowledgements

I. Stulíková and B. Smola gratefully acknowledge financial support by Czech Science Foundation (project 106/06/0252) and Ministry of Education of Czech Republic (Research Program MSM 0021620834). The authors also want to thank Prof. B. Scholtes from the University of Kassel who enabled the performance of texture measurements. They are also wish to acknowledge the contribution of Mr. Volker Kree from the GKSS Research Center who was responsible for the metallographic sample preparation.

6 References

[1] Communications from the Commission to the Council and the European Parliament: Implementing the Community Strategy to Reduce CO_2 Emissions from Cars: Fifth annual Communication on the effectiveness of the strategy, COM **2005** 269.
[2] Villalba, G.; Segarra, M.; Chimenos, J.M.; Espiell, F., Ecol. Econ. **2004**, *50*, 195-200.

[3] Mordike, B.L.; Ebert, T., Mater. Sci. Eng. A **2001**, *302*, 37–45.
[4] Swiostek, J.; Göken, J.; Letzig, D.; Kainer, K.U., Mater. Sci. Eng. A **2006**, *424*, 223–229.
[5] Göken, J.; Hort, N.; Blawert, C.; Steinhoff, K.; Kainer, K.U., in: *Magnesium Technology 2006*, TMS Annual Meeting, San Antonio, USA, **2006**, p. 205–210.
[6] Kree, V.; Bohlen, J.; Letzig, D.; Kainer, K.U., Praktische Metallographie **2004**, *41*, 233–246.
[7] Ono, N.; Nowak, R.; Miura, S., Mater. Lett. **2004**, *58*, 39–43.
[8] Shan, F.L.; Gao, Z.M.; Wang, Y.M., Thin Solid Films **1998**, *324*, 162–164.
[9] Pérez-Prado, M.T.; del Valle, J.A.; Contreras; J.M., Ruano, O.A., Scripta Mater. **2004**, *50*, 661–665.

Static and Dynamic Dent Resistance of AZ31 Magnesium Alloy Sheet

J. Kaneko, M. Sugamata, E. Yukutake
College of Industrial Technology, Nihon University, Narashino, Chiba 275-8575, Japan

1 Introduction

Formation of dents must be avoided for the sheet panel products in any case. Dents are formed on the sheet panels due to plastic yielding under static or dynamic external forces. The dent resistance of the sheet panel can be evaluated by the critical force for the formation of the dents, the dent depth under a constant applied force or energy, or the depth of dents formed under a constant indentation stroke. However, the standard methods for measurements of the dent resistance of the sheet panels have not been established [1].

The dent resistance of the steel sheet panels has been studied extensively [1-4] because of their practical importance. Although magnesium sheets have been known to show superior dent resistance compared to the other metal sheets, reported experimental data for the dent resistance of magnesium sheet panels are insufficient and limited to the dynamic dent resistance [5].

In this work, both static and dynamic dent resistance of AZ31 magnesium alloy sheets has been studied at room temperature and compared that of aluminum alloys, brass and mild steel. The dent resistance was tested for the flat circular sheet blank because of its simple geometry. The static dent resistance of the sheet panel has been examined by slowly moving a punch against a circular sheet panel clamped at its circumference thereby obtaining a punch load-stroke curve. The dynamic dent resistance has been tested by measuring the dent depth formed by dropping a metal or ceramic ball onto the circular sheet panels with varied impact energies.

2 Experimental Procedures

AZ31B magnesium alloy sheets in both O and H24 temper were used as the test material. The H24 rolled sheet of 0.8mm in thickness and 1200 mm in width produced by SCI in USA was originally supplied. For O-temper, the H24 sheet was annealed at 573K for 1800s Deformation twins were observed within each grain of H24 sheets, whereas they were completely disappeared in O-tempered sheet. Equiaxed grains of the average size of 17μm were observed on the sheet plane of the O-sheet. The experimental results of formability tests on the same sheet materials have been reported in our previous paper [6] together with metallographic observation.

For the reference materials, the following sheets of 0.8mm in thickness were used: aluminum alloy sheets of 1100-O, 5052-O and 5052-H34, cold rolled mild steel sheet of the

Table 1. Tensile properties of test sheets

Test material	Tensile strength (MPa)	0.2% proof stress (MPa)	Young's modulus (GPa)	Elongation (%)
AZ31-O	256	156	44.8	24
AZ31-H24	291	200	44.8	17
5052-O	211	106	70	22
5052-H34	260	200	70	9
A1100-O	106	60	70	31
SPCE	328	180	210	43
C2801-O	341	110	108	62

Table 2. Balls used for dynamic dent tests

Ball material	Diameter (mm)	Mass (g)	Drop Height (mm)
Carbon steel	20	28.2	
	30	112	200, 250, 300
	40	262	
Al$_2$O$_3$	20	22.5	
Cr steel	20	32.5	200, 300, 400
Tungsten carbide	20	59.4	

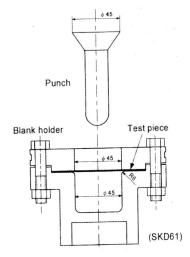

Fig.1. Tool geometry for static dent tests

drawing quality (JIS SPCE) in mill-annealed temper, and brass sheet of 60Cu-40Zn (C28000) in O-temper. Tensile properties of all the tested sheets are listed in **Table 1**.

For both static and dynamic dent tests, a flat circular blank of 90 mm in diameter was clamped at its circumference as shown in **Fig.1**. In case of static dent test, a punch was moved at a constant rate of 1 mm/min against the clamped blank and the punch stroke and punch load were measured during the test. A hemispherical punch with head radius of 10 mm was used. Two types of static dent tests were carried out; constant punch stroke and constant punch load tests. In constant punch stroke tests, a constant punch stroke of 1 or 2 mm was given against the blank sheet and the resultant dent depth was measured. The magnitude of spring back was determined by subtracting the dent depth from the punch stroke. In constant punch load tests, the punch was moved against the blank sheet up to a constant load of 0.5 or 1.0 kN and the resultant dent depth was measured.

In dynamic dent tests, steel balls of different diameter and balls of the same diameter but of different density was dropped onto the flat circular sheet blanks from the different drop height and the resultant dent depth was measured. Balls used in the dynamic dent tests are listed in **Table 2** along with tested drop height. Resultant dent depth was discussed with respect to the impact energy. In both static and dynamic tests, dent depth was measured at its deepest position by using a dial gage. Triplicate tests were performed for each condition and the average values were obtained as the test results.

3 Results and Discussion

3.1 Static Dent Resistance

Examples of punch load-stroke curves obtained are shown in **Fig.2** for various test sheets. Steel sheet showed a clear deflection at a punch load of about 0.14kN thereby initiation of the dent formation is indicated. However, no such deflection was observed in the punch load-stroke curves of magnesium, aluminum and copper alloy sheets. Since the critical load for the dent initiation could not be determined in these sheets, dent resistance was evaluated by the dent depth under a constant punch stroke or punch load in static dent tests.

The dent depth at punch strokes of 1 and 2mm after contact is shown in **Fig.3** for all the tested sheets. AZ31 magnesium sheets in both O and H24 temper showed dents less deep than the other metal sheets, although AZ31-O showed slightly deeper dents than AZ31-H24. At a constant punch stroke, both A1100-O and SPCE sheets showed the lowest dent resistance among the tested sheets.

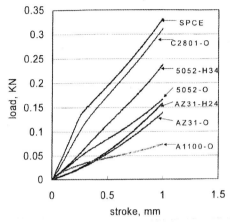

Fig.2. Punch load-stroke curves in static tests

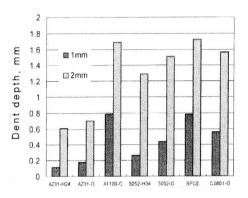

Fig.3. Dent depth at 1 and 2 mm punch stroke

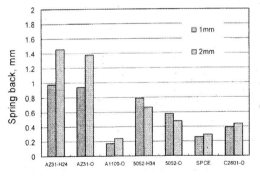

Fig.4. Spring back in static dent tests

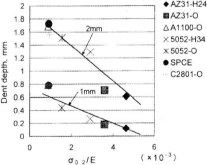

Fig.5. Dent depth vs. σ_y/E

The difference between the punch stroke and dent depth corresponds to the spring back of the sheet panel. Obtained values of spring back are shown in **Fig.4**. AZ31 sheets showed the largest spring back. In bending of metal sheets, the magnitude of spring back is proportional to σ_y/E, where σ_y is yield stress or 0.2% proof stress and E is Young's modulus [7]. The dent depth at punch strokes of 1 and 2 mm is plotted against the σ_y/E value in **Fig.5** for all tested sheets. It is clear that the dent depth decreases with increasing σ_y/E. Thus, dents in AZ31 sheets is less deep because of their larger spring back due to lower σ_y/E values at a constant punch stroke in static dent tests.

Dent depth at punch load of 0.5 and 1.0 kN is shown in **Fig.6** for all the tested sheets. Again, the dent depth of AZ31 is least, whereas the 1100 aluminum sheet showed the largest dent depth. In this case, the dent resistance is related to deformation resistance to form dents. Although strength of AZ31 in uniaxial tensile deformation is comparable to that of 5052-H34, steel and brass sheets as shown in Table 1, deformation resistance of AZ31 to form dents, that is, deformation resistance in biaxial tension is considered to be high due to its high r-value [6].

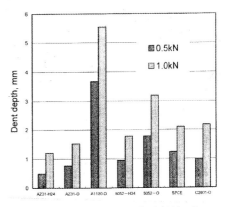

Fig.6. Dent depth at 0.5 and 1.0 kN load

3.2 Dynamic Dent Resistance

The depth of dents formed by dropping steel balls of different sizes listed in Table 2 is shown in **Fig.7** for drop height of 300 mm. Dent depth of each test sheet increased with increasing weight of the balls. Dents in AZ31 sheets are least deep except for the smallest ball of 20mm in diameter. Steel sheet showed the lowest value of the dent depth in case of the smallest ball. The 1100 aluminum sheet showed the largest dent depth. Both AZ31-O and –H24 showed the dent depth about 1/3 and 1/2 of 5052-O and –H34 aluminum sheets, respectively.

In case of steel balls of different sizes, the effect of ball weight on dent formation is overlapped by that of the ball radius. Hence, dynamic dent tests were performed by dropping the balls of different weight but the same diameter of 20 mm as listed in Table 2, and obtained experimental results are shown in **Fig.8** for drop height of 300 mm. No dent was formed for steel sheet in case of an alumina ball of 22.5g in weight. Dents are least

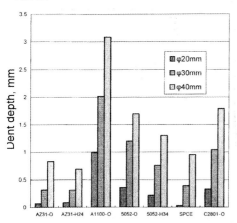

Fig.7. Dynamic dent depth by balls of various sizes (drop height: 300mm)

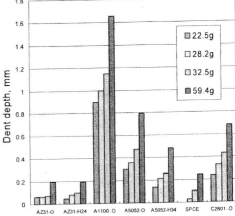

Fig.8. Dynamic dent depth by balls of various weights (drop height: 300mm, ball diameter: 20mm)

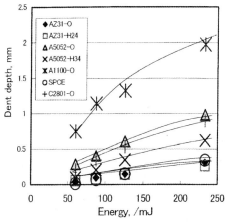
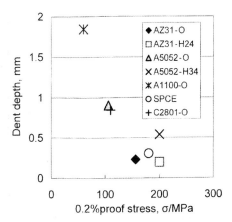

Fig.9. Dynamic dent depth vs. impact energy   **Fig.10.** Dynamic dent depth vs. 0.2% proof stress

deep for the steel sheet in case of a steel ball of 28.2g in weight, but those of AZ31 sheets are less deep for heavier balls. The differences in dynamic dent depth between AZ31-O and H24 are much smaller than those between 5052-O and –H34.

Dent depth of each test sheet is plotted against impact energy of dynamic dent tests in **Fig.9** for the results shown in Fig.8. The dent depth of each test sheet shows a parabolic increase with increasing the impact energy. In the present experimental range, the dynamic dent resistance of AZ31 sheets is almost the same as the steel sheets but appreciably higher than aluminum alloys and brass sheets when compared for the same sheet thickness.

Dent depth formed by steel ball of 30mm in diameter is shown in **Fig.10** against the 0.2% proof stress for drop height of 200mm. For the most part, dent depth decreases with increasing proof stress. However, dent depth of AZ31-H24 is about 1/3 of that of 5052-H34 although both sheets show almost the same proof stress. Impact energy is consumed by both elastic and plastic deformation in dynamic dent tests. Since elastic energy consumed is proportional to σ_y^2/E, a portion consumed by elastic energy is larger for magnesium of a lower E value than for aluminum. This explains why AZ31-H24 showed higher dent resistance than 5052-H34.

4 Conclusions

(1) Dent resistance of AZ31 magnesium sheets was higher than that of aluminum alloys, brass and mild steel sheets under a constant punch stroke in static dent tests, due to larger spring back.
(2) Dent resistance of AZ31 magnesium sheets was also higher than that of aluminum alloys, brass and mild steel sheets under a constant punch load in static dent tests due to higher deformation resistance in biaxial tension.
(3) Dent resistance of AZ31 magnesium sheets was higher than that of aluminum alloys, brass and mild steel sheets in dynamic dent tests under various impact energies.
(4) AZ31 magnesium alloy sheets showed higher dent resistance than aluminum alloys, brass and mild steel sheets in both static and dynamic conditions when compared for the

same sheet thickness. Thus, magnesium alloy sheets are suitable for light-weight sheet panels.

5 References

[1] Handbook on Press Forming, Nikkan Kougyo/Japan, 1987, 296-308.
[2] Yutori, Y.; Nomura, S.; Kokubo, I; Ishigaki, H., J. Japan Soc. Tech. Plasticity, **1980**, *21*, 168-174.
[3] Gunnarsson, L.; Schedin, E., J. Materials Processing Technology, **2001**, *114*, 168-173.
[4] Holmberg, S.; Thilderkvist, P., Materials and Design, **2002**, *23*, 681-691.
[5] Busk, R.S., Magnesium Products Design, Mercel Dekker/U.S.A., **1987**, 252-253.
[6] Yukutake, E.; Kaneko, J.; Sugamata, M., J. Japan Soc. Tech. Plasticity, **2003**, *44*, 276-280.
[7] Hosford, W.F.; Caddell, R.M., Metal Forming; Mechanics and Metallurgy, Prentice-Hall/U.S.A., **1983**, 251-254.

Acknowledgements: Authors are grateful to M. Kubota, H. Ouguchi and S. Saito for their experimental assistance.

Elaboration and Mechanical Behaviour of a Mg-based Bulk Metallic Glass

S. Puech[1], J. J. Blandin[1], J. L. Soubeyroux[2]

[1] Institut National Polytechnique de Grenoble (IPNG), Génie Physique et Mécanique des Matériaux (GPM2), UMR CNRS 5010, BP 46, 38402 Saint-Martin d'Hères, France

[2] Laboratoire de Cristallographie / CRETA, CNRS, Grenoble, 25 avenue des Martyrs, BP1166, 38402 Grenoble, Cedex 9, France

1. Introduction

Grain size refinement is a well known technique to improve mechanical properties of metallic alloys. Fine grained materials can be produced by recrystallization after severe plastic deformation or thanks to rapid solidification techniques. At high enough cooling rates from liquid state, amorphous alloys can even be produced. In this case, the "liquid" structure is preserved down a critical temperature for which it is frozen. Up to 80's, to avoid primary crystallization, very high cooling rates were required, leading to metallic glasses only under the form of ribbons. Today, thanks to appropriate compositions, bulk metallic glasses (BMG) can be produced. These materials exhibit attractive mechanical properties, in particular high strengths and large elastic domains at room temperature. Fig. 1 compares the tensile strengths and the Young's modulus of conventional metallic alloys with the corresponding bulk metallic glasses. One can see that for all the studied alloys, the tensile strength is sharply increased for the amorphous structure whereas the Young's modulus is not significantly affected.

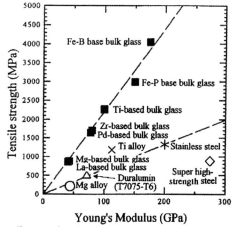

Figure 1: Comparison of tensile strength and Young's modulus between conventional metallic alloys and the corresponding bulk metallic glasses [1].

In this context, Mg-based BMG appears of special interest since they can provide new light alloys for structural applications with net shape fabrication possibilities. A large number of studies have been carried out to find good glass former compositions for Mg alloys. Frequently, Mg-Cu-Y or Mg-Cu-Gd compositions have been selected. For instance, a $Mg_{65}Cu_{25}Y_{10}$ glassy alloy was firstly reported by Inoue [2]. In the present work, a $Mg_{65}Cu_{25}Gd_{10}$ composition initially suggested by Men [3] has been investigated, under the form of a 6 mm diameter amorphous rod [4]. The resulting mechanical properties at room temperature were studied. Moreover, the possibility to form the alloy in the glass transition interval was also investigated.

2. Elaboration and characterization

Elements with purity better than 99.9% were used as starting materials. Copper and gadolinium were first melted together in order to obtain a binary eutectic easier to melt with magnesium. The master alloy was re-melted several times to assure homogeneity. Amorphous rods of 6 and 4 mm diameter were obtained by copper mould casting in argon atmosphere, as illustrated by Fig. 2 [4].

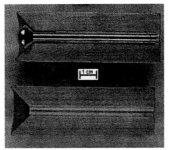

Figure 2: Typical rods produced in the copper mould.

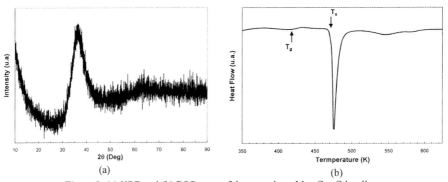

(a)   (b)

Figure 3: (a) XRD and (b) DSC scans of the amorphous $Mg_{65}Cu_{25}Gd_{10}$ alloy.

Fig. 3-a displays XRD patterns taken from the cross-section of the rod confirming the amorphous structure of the as-cast alloys (i.e. no diffraction peaks) and Fig. 3-b shows the corresponding DSC curves in continuous heating conditions (heating rate : 10 K/min.). A glass transition temperature T_g = 418 K was measured, followed by a main crystallization peak with an onset crystallization temperature T_x equal to 472 K. These values are in good agreement with those reported for the same glassy alloy and confirm that Gd helps to improve $\Delta T = T_x - T_g$, (one frequently used criterion for glass forming ability), in comparison to the $Mg_{65}Cu_{25}Y_{10}$ and $Mg_{60}Cu_{30}Y_{10}$ alloys [2-5].

3. Mechanical properties at room temperature

At room temperature, BMG exhibit generally brittle behaviours in both compressive and tensile conditions. In the case of the $Mg_{65}Cu_{25}Gd_{10}$ alloy tested in compression, a maximum stress of about 650 MPa was measured, which is much larger to what is generally obtained for conventional Mg alloys: for Mg-Y-RE (i.e. WE43) alloys, yield strength less than 300 MPa are obtained in peak aged conditions [6] and even in the particular case of ultra fine microstructures produced by severe plastic deformation techniques, such strength levels are not reached [7]. One can note that the high mechanical resistance of the glass was also confirmed by micro-harness measurements for which values of 265 HV was obtained. The Young's modulus was measured by ultra sonic techniques and a value close to 45 GPa was obtained, confirming that the amorphous structure does not modify significantly this mechanical property. One can again underline that these values of maximum strength and Young's modulus lead to a particularly large elastic domain since assuming a Hooke's law, a value of elastic strain of about 0.015 can be estimated.

The mechanisms of fracture of BMGs are still under debate but typical features are often observed. The most frequently reported observation is related to the development of a vein pattern, as it can be seen in the SEM observation shown in Fig. 4-a of the fractured Mg-based BMG. This morphology suggests a rise of temperature in shear bands leading to a decrease in viscosity and a confined ductility of the glass [8]. However, in the case of magnesium based BMG, the sample does not fracture along a unique shear plane oriented roughly at 45° in reference to the stress axis (as it can be observed for instance in the case of Zr based BMG) but in a large number of pieces. The veins are generally observed on the larger fractured parts whereas rivers are preferentially observed on the small parts, as illustrated by Fig. 4-b. Such rivers can be also observed in the case of Zr based BMG but only after partial crystallization of the glass for witch brittleness is enhanced in comparison with the amorphous state [9].

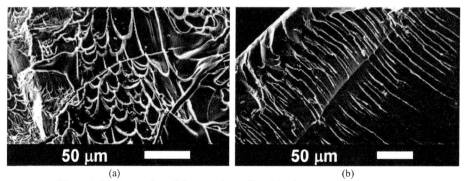

(a)   (b)
Figure 4: Fracture surface of the amorphous alloy: (a) vein pattern, (b) rivers pattern.

4. Forming at temperatures near T_g

Due to their brittleness, Mg based BMGs cannot be formed at room temperature. Casting of BMG is one way to get components but is not always straightforward since it requires to optimize cooling rates and castability, keeping in mind that these alloys exhibit particularly large viscosities in the liquid state. In consequence, forming in solid state can appear as a particularly interesting way. Moreover, as it can be obtained in the case of oxide glasses or polymers, BMG exhibit a very large capacity of forming at temperatures higher than the glass transition temperature. This great forming ability results from a large plastic stability associated to the Newtonian rheology (i.e. viscosity independent upon strain rate) which can quite easily obtained in this temperature domain.

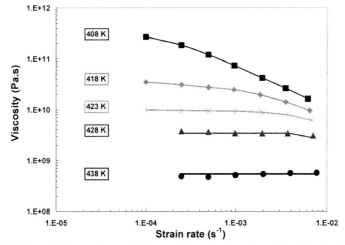

Figure 5: Variation with temperature and strain rate of the viscosity of the studied $Mg_{65}Cu_{25}Gd_{10}$ glass.

Fig. 5 shows the variation with temperature and strain rate of the viscosity of the $Mg_{65}Cu_{25}Gd_{10}$ glass. Compression tests were performed on cylinders of 4 mm diameter and 6 mm length and before tests, an holding time of 300 s was applied to ensure temperature stabilization. Strain rate between 10^{-4} and 10^{-2} s^{-1} were carried out for temperatures in the interval [T_g -10 K, T_g + 20 K]. Newtonian rheologies are obtained in a large part of the investigated domain and a transition from Newtonian to Non-Newtonian (i.e. viscosity decreasing with increasing strain rate) behaviour is observed only when temperature decreases or when strain rate increases. Three remarks can be done concerning these results: *i.* the extrapolated Newtonian viscosity at T_g is close to 5.10^{10} Pa.s, which is significantly lower than the value of 10^{12} Pa.s frequently reported at T_g for other BMG [10]; *ii.* in the Newtonian domain, the viscosities depend strongly on temperature since viscosity is decreased by approximately a factor 10^3 between 408 K and 438 K; *iii.* At 438 K, viscosities less than 10^9 Pa.s are measured which means that forming can be carried out under very low stresses if moderate strain rates are acceptable. It also suggests that such Mg based BMG are not adapted for high temperature conditions of use.

The choice of the best experimental forming conditions in the case of BMG is not straightforward since glasses are fundamentally out of equilibrium materials. In particular, at $T > T_g$, crystallization can occur after a given incubation time. If such crystallization occurs during the forming process, it can modify strongly the viscosity. An example of this situation for the studied Mg based BMG is shown in Fig. 6. A constant strain rate (5×10^{-4} s^{-1}) compression test is performed at $T_g + 20K$: the viscosity remains constant up to a strain of about 0.8, corresponding to a total holding time of 30 minutes, and then increases strongly. One can however note that the alloy remains ductile in the studied experimental domain.

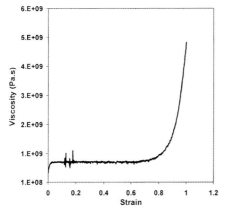
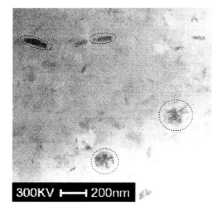

Figure 6: Strain stress curve at 438 K for a 5×10^{-4} s^{-1}.

Figure 7: TEM Bright field image of a 40 minute transformed sample.

Such an increase of viscosity can be correlated to crystallisation and has already been reported for others BMG's [10]. Generally, but not systematically, when crystallisation becomes too high, the glass can no longer be deformed even at high temperature and the sample breaks down. In consequence, it is very important for each glass not only to get information about its thermal stability but also to investigate the impact of such a crystallisation on the mechanical properties at room temperature. For the $Mg_{65}Cu_{25}Gd_{10}$ glass, isothermal DSC analysis performed at $T_g + 20$ K reveals that crystallisation started after a 20 minutes incubation time and ended for roughly 50 minutes. These data were confirmed by TEM observations carried out after partial crystallization treatments. For instance, Fig. 7 displays a TEM observation of the alloy after a treatment of 40 min. at 438 K. Two kinds of crystallites can be observed: rod shaped crystallites with characteristic size of 50 by 100 nm and spherical crystallites of roughly 200 nm in diameter. Other observations suggested that such sizes were quickly obtained, the increase of volume fraction of crystallites resulting mainly from an increase of their number (rather than of their size). The natures of these two populations of crystallites were identified by XRD, leading to Cu_2Gd and Mg_2Cu respectively.

The effect on properties in use of such partial crystallization was also investigated since in some cases, the presence of nanocrystals can still increase the fracture stress [9]. However, for the studied $Mg_{65}Cu_{25}Gd_{10}$ glass, even for limited quantities of crystals, the fracture stress continuously decreases and the initial value of fracture stress equal to 650 MPa for the amorphous alloy was reduced to about 300 MPa after complete transformation.

5. Conclusion and perspectives

A $Mg_{65}Cu_{25}Gd_{10}$ bulk metallic glass was successfully elaborated by copper mould casting. Very high strengths (> 600 MPa) were obtained in compression at room temperature associated with a large macroscopically elastic strain (≈ 0.015) but also to a brittle behaviour. Forming capacity at $T > T_g$ has been investigated. For such temperatures, thanks to Newtonian rheologies, the glass exhibits a particularly good forming ability with low viscosities. However, in some cases, crystallization can occur during forming and increase sharply the viscosity. For the studied Mg-based BMG, the nature of the crystals has been identified.

Work is under progress to reduce the brittleness of such bulk metallic glasses at room temperature. In the case of Zr based BMG, some results have been obtained by addition of appropriate elements, which in some cases, lead to a composite containing both ductile crystalline and brittle amorphous phases. In the case of Mg-based BMG, some attempts were carried out adding iron to the nominal composition. In this case, macroscopic plasticity could be obtained at room temperature, maintaining high strength levels [11].

6. Acknowledgements

The authors thank Dr. P. Donnadieu from LTPCM / INPG for her help in the TEM observations.

7. References

[1] A. Inoue, Acta Mater., 48, (2000) 279
[2] A. Inoue, A. Kato, T. Zhang, S. G. Kim and T. Masumoto, Mater.Trans., JIM, 32 (1991) 609.
[3] H. Men and D.H. Kim, J. Mater. Res., 18 (2003) 1502.
[4] J.L. Soubeyroux, S. Puech, J.J. Blandin, P. Donnadieu, J. All. Comp., (2006), in press.
[5] U. Wolff, N. Pryds, E. Johnson and J.A. Wert, Acta Mater., 52 (2004) 1989.
[6] M. Mabuchi, Y. Chino and H. Iwasaki, Mater. Trans., 43 (2002) 2063.
[7] J.J. Blandin, Mater. Sc. Forum, 426-432 (2003) 551.
[8] Z.F. Zhang, G. He, J. Eckert and L. Schultz, Phys. Rev. Lett., 91 (2003) 45505.
[9] L.Q. Xing, C. Bertrand, J.-P. Dallas and M. Cornet, Mater. Sc. Eng., A241 (1998) 216.
[10] M. Blétry, P. Guyot, Y. Brechet, J.J. Blandin and J.L. Soubeyroux, Mater. Sc. Eng., A387-389 (2004) 1005.
[11] H. Ma, J. Xu and E. Ma, Appl. Phys. Lett., 83 (2003) 2793.

Fracture Toughness Behavior of Pressure Die Cast Magnesium Alloys Under Static and Impact Loading

G. Dietze, D. Regener
Otto-von-Guericke-University, Magdeburg

1 Introduction

Restricted natural resources involving an intentional use of energy demand an economical application of materials and energy supports. However, safety aspects and modern consumer comfort in automobile constructions cause a rising weight of cars. Therefore the development of lightweight components is one of the main aims in automotive industries. Magnesium alloys as the lightest structural materials are very suitable in regard to the reduction of vehicle weight and a consequent fuel economy.

In priority the most of commercial magnesium alloys are used because of their high specific strength or low density. It is necessary to have a comprehensive knowlegde about the damage tolerance of these alloys for an expanded use in the field of safety relevant components. However, only a few publications are occupied with investigations on wrought and sand cast magnesium alloys [1-3]. The determination of fracture toughness values of pressure die cast magnesium alloys is more difficult due to the inhomogeneity of the microstructure. It depends on the wall thickness and the existence of numerous defects like gas pores and microshrinkage.

This paper shows results of research on the field of fracture toughness investigations under static as well as impact loading to help closing the deficiency of material data.

2 Experimental

All specimen investigated were manufactured from plate-like components with dimensions of 200 x 75 x 10 mm, cast on a cold chamber die-casting machine GDK 200. The hp alloys AZ91, AM50 and AE42 were placed at disposal. Table 1 shows the chemical composition. Compared to producer data [4], only the values of Mn in AZ91 and AM50 and of Al in AM50 are not in accordance.

Table 1: Chemical Composition of alloys investigated

Alloy	%Al	%Mn	%Zn	%Si	%Cu	%Ni	%Fe	%RE
AZ91	9.3	0.1	0.8	0.015	0.0007	0.0006	0.0005	
AM50	6.1	0.3	0.02	0.01	0.0007	0.0007	0.0015	
AE42	4.2	0.2	0.01	0.002	0.002	0.001	0.002	2.5

Tensile tests were carried out on flat tensile specimens (gauge length 55 mm, thickness 10 mm) for the assessment of the strength and deformation behavior.

In comparison with tensile data determined on separately cast specimens the values of mechanical properties are considerably lower [5]. This is attributed to the strong influence of size and distribution of gas pores and microshrinkage.

Table 2: Tensile properties at room temperature of specimen taken from plates (The values of separately cast specimens [5] in brackets)

Alloy	$R_{p0.2}$ [N/mm²]	R_m [N/mm²]	A [%]	E [GPa]
AZ91	128 (150)	175 (207)	1.3 (2)	42 (48)
AM50	102 (124)	161 (210)	2.7 (7)	50 (48)
AE42	119 (130)	141 (194)	1.0 (5)	46 (45)

The determination of the crack resistance behavior under static loading occurred with help of the multispecimen test method. Fatigue pre-cracked SENB- (10 x 10 x 55 mm) and CT-specimen (W=20 mm, H=24 mm and B=10 mm) were used. The stable crack extension was measured on the fracture surface after heat tinting (AZ91 350°C/2h, AM50 and AE42 430°C/2h). The data were evaluated according to [6, 7].

Specimen, tested under impact loading, had the Charpy dimensions 10 x 10 x 55 mm. The notch bend tests occured on un-notched, notched and pre-cracked specimens in a pendulum impact machine (maximum of impact energy = 300 Joule) with a reduced impact energy from 9.69 Joule (drop angle 20°) and 46.68 Joule (drop angle 45°).

The preparation-steps for microstructure investigations consisted of wet grinding until 4000 mesh silicon carbide paper, polishing with finally 1 μm diamond suspension and etching with 3 % alcoholic nital solution. The microstructure and the fracture appearances were assessed on the basis of SEM-observations.

3 Results and Discussion

3.1 Microstructure

The microstructural investigations of the hpdc alloys AZ91 and AM50 show the typical network of primary α-Mg solid solution, surrounded by a divorced eutectic consisted of α-Mg and β-phase $Mg_{17}Al_{12}$ (Fig. 1a and b). EDS analysis certifies, that the amount of Al in the eutectic

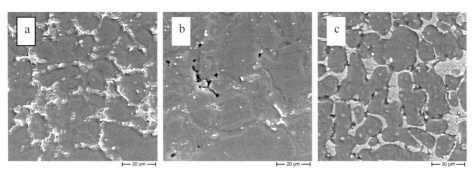

Figure 1: Microstructure of the investigated alloys AZ91 (a), AM50 (b) and AE42 (c)

α-Mg is higher (9-11 wt %) than in the primary α-Mg (4-6 wt %). The alloy AM50 has a higher ductility because of a reduced content of Aluminium connected with a lower amount of the brittle β-phase $Mg_{17}Al_{12}$ compared with AZ91. In dependence on the amount of Mangan precipitations of the type Al_8Mn_5 were also observed [8].

The alloy AE42 has a globular α-matrix surrounded from fine lamellar eutectic with different precipitations of the typ MgRE and AlRE (Fig. 1c).

The development of microshrinkage was often confined to the grain boundaries, especially observed in the alloy AE42. This is attributed to the solution conditions and the difficulties with die casting of alloys, which contain rare earth elements.

3.2 Fracture Toughness Behavior Under Static Loading

The evaluation of the crack resistance behavior under static loading based on load-displacement-curves as well as the measurement of the crack extension of several specimens. As a result the crack-resistance curve J-Δa is obtained, using the energy integral J as loading parameter (Fig. 2 – SENB-specimens, Fig. 3 – CT-specimens). It is obvious, that the alloy AM50 has the best energy absorption capability with the highest energy-integral-values. Because of the comparatively higher amount of the brittle β-phase $Mg_{17}Al_{12}$ the crack resistance of the alloy AZ91 is strongly reduced. The alloy AE42 takes in its energy capability an intermediate position. However, the comparison of crack-resistance-curves of both shapes of specimens shows, that the crack resistance behavior under tension load of the CT-specimens is quite better than under bending load for SENB-specimens.

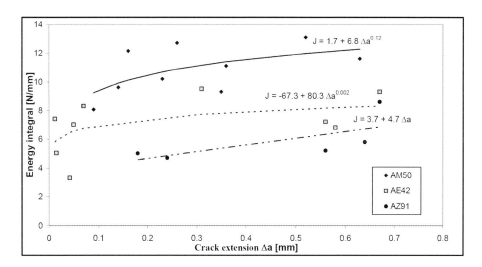

Figure 2: Crack resistance curves obtained from SENB-specimens

The materials sensitivity to crack initiation was characterized by the $J_{0.2}$-value, the fracture resistance J at 0.2 mm stable crack extension. The determination of J_i-values is not possible because of the difficulty to detect the blunting of the crack. Table 3 represents the results of investigations of the crack resistance. The lowest $J_{0.2}$-value of the alloy AZ91 points out the lowest damage tolerance. In contrast the alloy AM50 reveals the most favorable $J_{0.2}$-value. Reasons for this distinctive material behavior are differences in the microstructure including various concentration and distribution of intermetallic phases, internal defects and cracks. Especially the crack resistance of the alloy AE42 is extreme reduced due to a large number of internal defects.

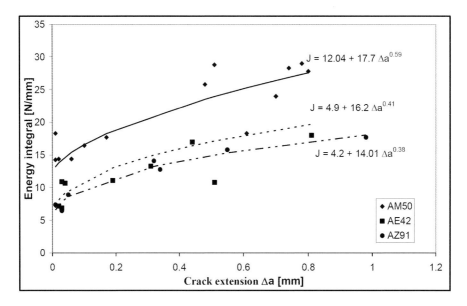

Figure 3: Crack resistance curves obtained from CT-specimens

Table 3: Crack initiation values $J_{0.2}$

Alloy	$J_{0.2}$ [N/mm]		$\Delta J/\Delta a$	
	SENB	CT	SENB	CT
AZ91	4.7	11.8	4.4	7.3
AE42	7.4	13.5	3.3	10.3
AM50	10.2	20.1	4.5	12.3

3.3 Behavior Under Impact Loading

The determination of dynamic fracture toughness values was not possible due to the high impact energy of the available impact testing equipment. All pre-cracked samples were broken during testing even when the lowest dynamic load under a drop angle of 20° was applied. Therefore, the calculation of the impact value K_{max} was carried out by considering the maximal absorbed impact energy. Figure 4 shows load-displacement-curves of pre-cracked samples of the magnesium alloys investigated which were obtained during testing with a reduced impact energy of 9.69 Joule (drop angle 20°).

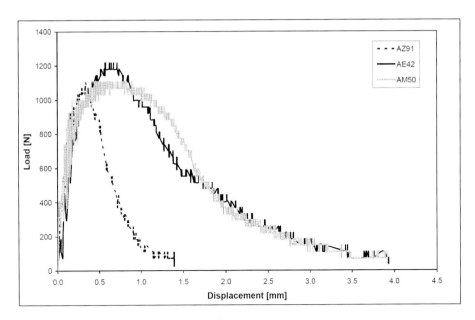

Figure 4: Load-displacement-curves obtained under impact loading

As can be deduced from Figure 4 all alloys show an elastic-plastic material behavior. It is obvious, that the energy absorption capability of the alloys AZ91 and AM50 is comparable with that obtained under static loading. However, the alloy AE42 reveals a different material behavior. Whereas the energy absorption capability of AE42 under static loading is only average, its behavior under dynamic loading is very similar to that of alloy AM50.

In table 4 the impact values K_{max}, which were determined by integration of the area under the load-displacement-curves, are summarized, confirming the statement above. Note that the K_{max} values are mean averages of five individual tests.

Table 4: Impact energies of the alloys investigated

Alloy	Drop angle 20°			Drop angle 45°		
	K_{max} [J] un-notched test bar	K_{max} [J] notched test bar	K_{max} [J] pre-cracked test bar	K_{max} [J] un-notched test bar	K_{max} [J] notched test bar	K_{max} [J] pre-cracked test bar
AZ91	4.03	0.7	0.7	4.8	1.1	1.4
AM50	not broken	2.9	1.8	25.3	2.9	2.7
AE42	not broken	2.7	1.8	15.9	2.3	3.2

Major differences were observed for un-notched samples, tested with reduced impact energy of 46.68 Joule (drop angle 45°). The alloy AM50 required the highest impact energy, followed by the alloy AE42 and alloy AZ91, as expected. The impact energies of notched and pre-cracked test samples didn't show considerable differences (see table 4). With almost 3 Joule the alloys AM50 and AE42 exhibit a better energy absorption capability as the alloy AZ91 (1

Joule). This can be attributed to differences in the microstructure, especially to a varying amount of the brittle β-phase $Mg_{17}Al_{12}$.

It is generally evident, that values for the impact energy of notched and pre-cracked test samples of AM50 and AE42 are almost the same. That means, that the influence of the higher porosity in notched samples of the alloy AE42 is only marginal in case of dynamic loading.

4 Conclusions

The alloys AZ91, AM50 and AE42, investigated in this study, exhibit a dendritic microstructure with differences in the concentration and distribution of intermetallic phases according to their chemical composition. In addition, numerous cast induced defects were observed in all alloys. These defects distinctly affected the crack resistance behavior. The lowest resistance to crack initiation and propagation under static loading was observed for AZ91 leading to a strongly reduced fracture toughness, which could be attributed to a higher amount of the brittle intermetallic β-phase $Mg_{17}Al_{12}$. The significantly lower concentration of the β-phase in AM50 led to a high crack resistance of the alloy. The crack resistance behavior of the alloy AE42, however, remains behind the expectations. The reason is the presence of a large number of internal defects, which extremely impairs the good intrinsic properties of this alloy.

The described crack resistance behavior is confirmed by investigations under impact loading. However, the negative effect of cast induced defects is less pronounced than under static loading.

5 References

[1] [1] C. J. Padfield, T. V. Padfield in Light Metals for Automotive Industry, SAE SP-1683, Warrendale, PA, USA, **2002**, 27-33
[2] [2] ASM-Handbook Volume 19: Fatigue and Fracture, ASM International, Materials Park;OH, 877-878
[3] [3] S. Barbagello, E. Cerri in Engineering Failure Analysis 11 (**2004**) 127-140
[4] [4] Hydro Magnesium: Datenblatt Magnesium-Druckgusslegierungen 06/00
[5] [5] D. Regener, G. Dietze, H. Heyse in *11. Sommerkurs Werkstofftechnik Werkstoffentwicklung – Werkstoffanwendung, 14./15.06.2002*, p. 131-140
[6] [6] H. Blumenauer in Werkstoffprüfung, Dt. Verl. f. Grundstoffindustrie, Leipzig, **1994**
[7] [7] ISO 12135: Metallic materials- unified method of test fort he determination of quasistatic fracture toughness, **2002**
[8] [8] G. Dietze, D. Regener in *11. Sommerkurs Werkstofftechnik Werkstoffentwicklung – Werkstoffanwendung, 14./15.06.2002*, p. 121-130

Influence of the Microstructure and Local Mechanical Properties of Magnesium Die Cast Components Made of MgAl6Mn on the Quality of Numerical Simulations

Elke Lieven, Ralf Koch
Takata-Petri AG, Aschaffenburg
Wolfgang Böhme, Simone Schwarz
Fraunhofer-Institut Werkstoffmechanik (IWM), Freiburg

1 Introduction

Numerical simulation is an important tool to describe the mechanical behavior of components during different phases of developing and production [1]. Therefore, it is important that the material properties of the component are well characterized. The structural simulation needs on one hand the exact design geometry and on the other hand the mechanical properties of the material of the component for the relevant loading conditions. The quality of the results is strongly dependent on the quality of the input data.

The structural simulation of components made of magnesium die cast alloy is mainly based on material data that is determined by separately cast tensile test specimens. However, the mechanical properties of die cast components are locally often different from the specimens because of different casting conditions within a complex geometry. This creates different microstructures and pore distributions, and mechanical properties vary with the thickness of the component, the distance from the gate system, and the distance between cast skin and inner area.

This paper describes the material characteristics in different areas of a component made of magnesium die cast alloy MgAl6Mn. The Fraunhofer Institute for Mechanics of Materials (IWM) uses mini-tensile specimens prepared out of the final component to investigate local mechanical properties for static and dynamic loads. To characterize and evaluate the different mechanical properties of different parts of the component, micro structure analysis and fracture surface analysis is used. The influence of different microstructures on the stress-strain behavior is shown. Based on this data, the correlation between simulation and real component behavior is demonstrated.

2 Determination of Material Properties of Magnesium Components

It is very common to determine the material properties of die cast components of Magnesium alloys by separately cast specimens [2, 3]. But, the local material properties of a structural component may vary due to different local cooling conditions during die casting, e.g., depending on local thickness and distance from the die cast system. In order to determine local material properties, the specimens have to be extracted locally from the components, which sometimes leads to very small specimens. The determination of local true stress strain curves by small round bar tensile specimens is described here. The material properties were determined for quasistatic and for crash relevant high rate loads, also microstructure and fracture analyses were performed.

2.1 Material and Specimens

The mini round bar tensile specimens with gauge diameters of only 1mm and gage length of 10mm (see Figure 1) were extracted mainly at three positions (see Figure 2) from five steering wheels (called Part 1-5) made of die cast magnesium alloy MgAl6Mn and finally prepared by spark erosion.

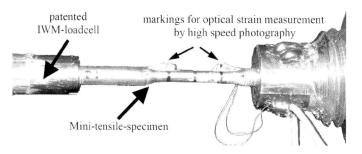

Figure 1: Instrumented Mini-tensile-specimen for high rate testing.

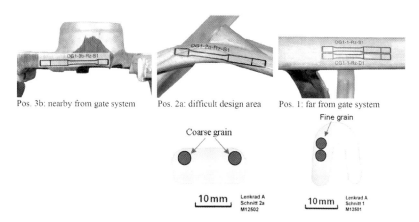

Figure 2: Specimens extracted from three different positions.

2.2 Determination of Static and Dynamic Stress Strain Curves

The quasistatic tensile tests at room temperature were performed under displacement control with a 40 kN testing machine at a piston velocity of 0,01 mm/s. Force and displacement were determined by a conventional load cell and a clip gauge with a gauge length of $L_o = 10$ mm.

The tests at high strain rates were performed with a high rate tensile testing machine (load capacity up to 100 kN, velocities up to 20 m/s). With piston velocities of $v_o = 2,5$ m/s, nominal strain rates $\dot{\varepsilon}_{nom} = 200/s = v_o/L_c$ were achieved (parallel gauge length $L_c = 12,5$ mm). The resulting average strain rates, as can be calculated from the strain at fracture A_5 divided by the time-to-fracture t_f by $\dot{\varepsilon}_{av} = A_5/t_f$, were somewhat lower and about 150/s. The average strain rates of the quasistatic tests were about 0,0006/s.

Since no necking was observed during these tests, the evaluation is based on the measurement of the elongation of the gauge length L_o. The elongation during the high rate tests was measured by a special 24 spark high speed camera with an extremely good optical resolution as well as a high time resolution of up to 1 μs [4]. The force was measured "quasi-locally" by a patented IWM-load-cell with reduced oscillations [5].

The technical and true stress strain curves were calculated from the measured force-displacement-curves. Figure 3 shows the determined true stress-strain curves for three different positions according to (Figure 2). It should be mentioned, that these curves are lower than those usually obtained by separately cast specimens (see Figure 3). Further, there are significant differences visible:
- The highest strength and largest strain at fracture are observed for Pos. 2a and nearly as good for Pos. 3b. At Pos. 1, far way from the gate system the lowest strength and strain at fracture were measured.
- The stress-strain-curves are remarkably higher at "dynamic" high rates of loading compared to static ones and the strain at fracture is practically the same.
- Pos. 2a Part 3 shows significant higher stress-strain curves for static and dynamic load (Figure 3).

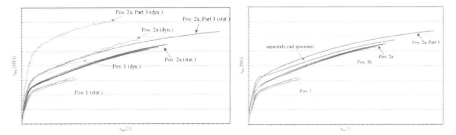

Figure 3: Determined true stress-strain curves on the right static results for Pos. 1, 2a, 3b and separately cast specimen on the left comparison static/dynamic for two Pos. 1, 2a.

2.3 Microstructure and Fracture Analyses

The microstructure analysis of steering wheel A in Table 1, Pos. 1 shows a coarse grain rate of 7 to 14%. The rate of fine grain is around 67%. In Pos. 2a the rate of coarse grain is significantly higher. It is between 24% and 28%. The rate of fine grain is only 40% to 50%.

The variation of coarse grain of Pos 3b is high, with rates from 17% to 28%. The rate of fine grain shows a high variation as well (50% - 70%). Pores and sinkholes are in all positions lower than 1%.

The results of fracture surface analysis and micro structure analysis are shown in Table 2. Pos. 2a Part 3 shows a higher rate of eutectic compare to Pos. 2a Part 4. This sample show further the highest rate of instoichiometric eutectic.

3 Influence of the Numerical Simulation

Figure 5 shows the non-linear simulation results of section force displacement with the different quasistatically evaluated material properties compared to the test results.

Table 1: Microstructure analysis of Steering Wheel A.

	Pore/ sink-holes	Coarse grain	Fine grain	Eutectic	Instoichiometric eutectic
Pos. 1 stat	0.46	13.88	67.16	3.4	15.1
Pos. 1 dyn	0.27	6.76	67.37	3.7	21.9
Pos. 2a stat	0.18	24.59	48.13	5.6	21.5
Pos. 2a dyn	0.33	27.68	41.69	4.5	25.8
Pos. 3b stat	0.55	28.65	52	4.5	14.3

Table 2: Fracture surface analysis and micro structure analysis.

	Pore / sinkhole		Microstructure			
	Crack surface	Micro-graph	Coarse grain	Fine grain	Eutectic	Instoichiometric eutectic
Pos. 1 stat. Part 1	39.78	1.72	15.28	71.96	2.58	8.46
Pos. 2a stat. Part 3	31.44	0.26	29.61	46.75	*3.96*	19.42
Pos. 2a stat. Part 4	25.23	1.13	23.68	57.09	2.9	15.2
Pos. 3b stat. Part 3	22.32	0.27	9.69	75.88	2.72	11.44

The best correlation is realized with the separately cast specimens, followed by the material properties at Pos. 2a Part 3.

The section force mirrors the stress-strain behavior of the material input database. The influence of the Young's modulus and Yield Point is high [2], because of the prior plastic performance behavior of the MgAl6Mn (Figure 4).

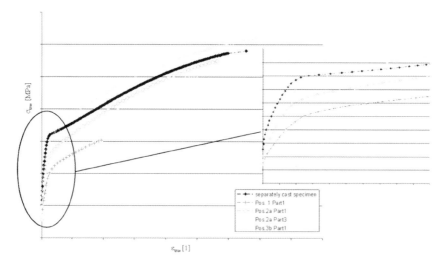

Figure 4: Material data for Pamcrash 2005 on the left and on the right focus the area from elastic to plastic.

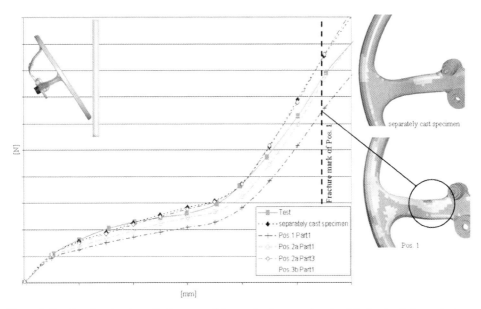

Figure 5: Results of section force displacement of the quasistatic test and on the right the simulation equivalent plastic strain results for the maximum specimen material data.

The comparison of the damage results by the quasistatic test shows that the steering wheel has no critical area in the simulation up to the marked displacement, when using separately cast specimen data. However you use the data with the lowest stress-strain from Pos. 1 the components become fractured (Figure 5). The safety factor here is very high because the tested components do not fracture through the entire displacement range. The critical area of the material characteristics of Pos. 1 is not the fracture location of the simulation. The consequence of the input of material data from Pos. 1 is an over dimensioning of a component for one test condition and an accumulation of material which has an influence on the die cast for Pos. 1. The results show that a numerical simulation works with separately cast specimen. The local material properties of a component give the developer the experience to assess the results of the simulation.

4 Summary

It is shown that the influence of different parameters like the test conditions, loadcase (static or high speed) or the sample's location of the gate system or the design have an influence on the result of the stress-strain characteristics of a component.

Local material properties of a component do not show the same level as separately cast specimens. Specimens taken far from gate system (Pos. 1) show lower and shorter stress-strain curves and elongation than near the gate system (Pos. 3b). Test with a high feed (dynamic) show higher a flow curve than static tests. Results of metallographic and fractographic analysis: Specimen Pos. 1 (far from gate system) shows 7 and 15 % lower coarse grain rate. Fine grain rates are highest, between 65 and between 70% . Specimen Pos. 2a (difficult design area) shows the highest rate of coarse grain. The rate of 24-28% was found for all specimens taken from several steering wheels. The fine grain rate was 40-50%. The micro structure of specimen

Pos. 2a of Part 3 show a significantly higher rate of eutectic compared to all other specimens. Additionally the microstructure shows a higher rate of the stoichiometric mixed phase (instoichiometric eutectic). Therefore, the mechanical properties show higher values than separately cast specimens and the other position.

The numerical simulation results show that it is correct to use the separately cast specimens for Phase 1 (virtual development) [1]. The local material properties of a component give the developer the experience to assess the results of the simulation and the known the influence of design areas which have an influence on the die cast results. The known of difference in the stress-strain characteristics are helpful in identifying problems in Phase 3 (testing) [1] as well.

5 References

[1] Lieven, E ; Wuth, M.: Product: Product Development Process Optimization by Linking Design numerical Simulation and Process Development, Magnesium 2003, November **2003**, Wolfsburg

[2] Wuth, M.; Lieven, E.; Böhme, W.: Determination of Material Properties and Numerical Simulations to Predict the Mechanical Performance of Die Casted Components, Magnesium 2000, September **2000**, München

[3] Böhme, W.; Hug, M.: Bestimmung von technischen Kennwerten, E-Modul und Querkontraktionszahl einer Aluminium- und einer Magnesium-Gußlegierung, IWM-Bericht V27/2000, Freiburg, Februar **2000**

[4] Böhme, W.; Sun, D.-Z.; Schmitt, W.; Hönig, A., Application of Micromechanical Material Models to the Evaluation of Charpy Tests, ASME-Symposium: Advances in Local Fracture/Damage Models for the Analysis of Engineering Fracture Problems, Scottsdale, Arizona, April 28 - May 1, **1992**, Eds.: J. H. Giovanola and A. J. Rosakis, AMD-Vol. 137, Book No. H00741, pp. 203-216, **1992**

[5] Böhme, W.; Hug, M.: Vorrichtung zur schwingungsarmen Kraftmessung bei schnellen, dynamischen Zugversuchen an Werkstoffproben, Deutsches Patent: DE 10 201 861, Anmeldetag: 18.01.2002; Europäisches Patent: EP 1 466 157 B1, Veröffentlichungstag **03.08.2005**

Effect of Anodized Layer on Fatigue Behavior of AM60 Magnesium Alloy

Y. Miyashita*, Sabrina A. Khan**, Y. Mutoh** and T. Koike***
*Nagaoka National College of Technology, **Nagaoka University of Technology, ***Yamaha Motor Co. Ltd.

1　Introduction

According to the previous study on fatigue behavior of a magnesium alloy, fatigue strength is reduced in high humidity condition [1], where corrosion pit nucleates and grows to an initial crack. The nucleation and growth of corrosion pit can occur under the interactive effect of corrosion environment and cyclic deformation. Surface treatment may be one of the solutions for improving corrosion-fatigue resistance of materials. Anodizing has been widely used as a surface treatment of magnesium alloys. It is effective for protection from the environment and is also used as a pre surface treatment for painting. However, anodized materials have often shown lower fatigue strength [2-4]. The detailed fatigue mechanism of anodized materials has not yet been clear. In the present study, fatigue tests of AM60 die-cast magnesium alloy specimens with two different thicknesses of anodized layer were carried out. The interrupted fatigue test was also carried out to discuss fatigue crack nucleation and growth mechanisms of the anodized material.

2　Experimental Procedure

AM60 die-cast magnesium alloy was used as a base metal. Thickness of the as-cast plate was 3 mm. Microstructure near the surface region is shown in Fig. 1. Chemical composition and mechanical properties for the base material is shown in Tables 1 and 2, respectively. Bending specimens with dimensions of 3×10×33mm are cut from the as-cast plate. A specimen with as-cast surface was used for the fatigue test to obtain base-line fatigue property of the base metal. In case of anodized specimens, the as-cast surface was polished by an emery paper with grit No. of #1500 and then the surface was cleaned by shot blasting. After anodizing, the surface was painted. Cross sectional observation of the anodized specimen is shown in Fig. 2.

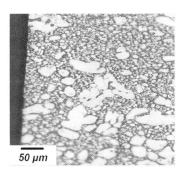

Figure 1: Microstructure of the AM60 magnesium alloy used.

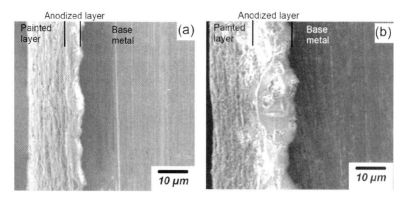

Figure 2: Cross sectional observations of anodized specimens with thickness of anodized layer of (a) 5μm and (b) 15μm.

Three layers, that is base metal / anodized layer / painting layer, can be observed from the figure. A target thicknesses of anodized layer in the anodizing process were 15 μm[2] and 5μm. Figure 3 shows the results of EDX analysis near the surface region on the cross section of anodized specimen with thickness of 15 μm. The anodized layer can be determined based on Mg and O distributions. Carbon was mainly contained in the painting layer.

Fatigue tests were conducted with a hydraulic fatigue machine (Shimadzu, EHF-1) under four point bending. Inner and outer span lengths were 10 and 30 mm, respectively, as shown in Fig. 4. Fatigue tests were conducted in the controlled air with a temperature of 20°C and a relative humidity of 55%RH, using a sinusoidal wave form with a frequency of 20Hz and a stress ratio R=0.1.

Table 1: Chemical composition of the base metal AM60.

Al	Mn	Si	Cu	Zn	Ni	Mg
5.5-6.5	0.13	0.5	0.35	0.22	0.03	Bal.

Table 2: Mechanical properties for the base metal AM60.

Tensile strength, MPa	0.2% proof stress, MPa	Elongation, %	Young's modulus, GPa
224	103	9.0	43

3 Results and Discussion

Figure 5 shows S-N curves for as-received and anodized specimens. The fatigue limit obtained for as-received specimen was 73 MPa. The fatigue limit slightly decreased with an increase in thickness of anodized layer. Fatigue limits obtained for anodized specimens with thicknesses of 5 μm and 15 μm were 58 and 53 MPa, respectively.

SEM observations of fracture surface for the as-received specimen is shown in Fig. 6. The arrow in the figure shows the crack nucleation region. As can be seen from the figure, a fatigue crack nucleated at the surface for the as-received specimen. A casting defect or an inclusion was not observed at the crack initiation region. In case of anodized specimen, a fatigue crack nucleated from the anodized layer or the interface between anodized layer and base metal as shown in Figs. 7 and 8. Rough surface of the base metal and pores inside the anodized layer were observed at the fracture origin for an anodized specimen with a thickness of 15 μm. Rough surface of the base metal was also observed for an anodized specimen with a thickness of 5 μm, but pores inside the layer were not clearly observed in crack nucleation region.

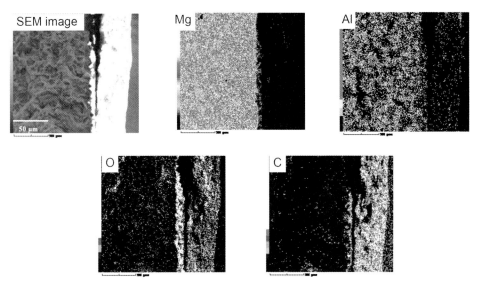

Figure 3: Results of EDX analysis for anodized material with thickness of 15 μm.

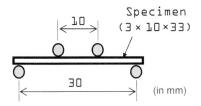

Figure 4: Schematic illustration of the four point bending test.

It is considered that a total fatigue life is divided into two periods, that is crack nucleation life and crack propagation life [5]. Crack propagation life can be calculated based on Paris's power law as,

$$N_f = \int_{a_0}^{a_f} \frac{da}{A(\Delta K)^m}, \tag{1}$$

where a_0 is the initial crack length, a_f is the final crack length for fatigue fracture, ΔK is the stress intensity factor range, A and m are constants which are obtained from the fatigue crack growth curve. When it is assumed that the geometry factor in the calculation of stress intensity factor can be unity, fatigue life for fatigue crack propagation, N_f is given as,

$$N_f = \frac{1}{A\left(\Delta\sigma\sqrt{\pi}\right)^m}\int_{a_0}^{a_f} a^{-\frac{m}{2}}\mathrm{d}a, \qquad (2)$$

where $\Delta\sigma$ is the range of applied stress. The value of m is assumed to be 2 according to the previous result of fatigue crack growth test for AZ61 [6]. Fatigue crack nucleated in the anodized layer or at the interface as mentioned above. Initial crack length could not be measured in the present work. However, if the initial crack length is assumed to be the same order of the thickness of a brittle anodized layer, initial crack length for the specimen with 15 μm thick anodized layer would be three times larger than that for the specimen with 5 μm thick anodized layer. It can be also assumed that crack growth rate in the base metal after crack nucleation is the same for both the specimens regardless of the thickness of anodized layer. Then the ratio between fatigue crack propagation life of the specimens with 5 μm thick and 15 μm thick anodized layers, N_{f5}/N_{f15} under the same applied stress range is given as,

$$\frac{N_{f5}}{N_{f15}} = \frac{a_f - a_0}{a_f - 3a_0}. \qquad (3)$$

If the final crack length, a_f is assumed to be 1.5 mm, which is the half of the thickness of the specimen, $N_{f5}/N_{f15} \approx 1$. Namely, difference in fatigue crack propagation life between two specimens with 5 μm and 15 μm thick anodized layers will be very small.

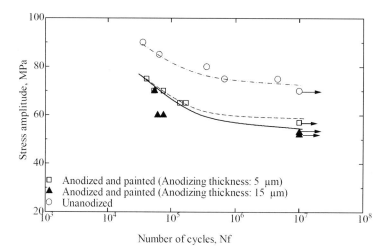

Figure 5: S-N curves for as-received and anodized AM60 magnesium alloy.

In order to discuss fatigue mechanism of anodized specimen, interrupted fatigue tests at an applied stress amplitude of 70 MPa were conducted. In the interrupted test, three point bending with a span length of 30 mm was used to limit the crack nucleation region to be smaller than in case of four point bending. Other test conditions were the same as the fatigue strength tests, as shown in above. In the interrupted test, fatigue loading was stopped at 30%, 40%, 50% and 60%

of the total fatigue life shown in Fig. 5. After the test was interrupted, the specimen was cut in the longitudinal direction into three pieces and the cross sections were observed by an SEM.

Figure 6: Observation of fracture surface of as-received specimen. (σ_a= 75 MPa, N_f = 674, 150 cycles). The arrows are indicating the crack nucleating sites.

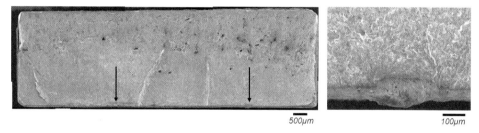

Figure 7: Overall fracture surface of an anodized specimen. (Anodizing thickness: 15μm, σ_a= 60 MPa, N_f = 61,326). The arrows are indicating the crack nucleating sites.

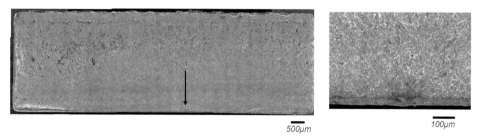

Figure 8: Overall fracture surface of an anodized specimen. (Anodizing thickness: 5μm, σ_a= 70 MPa, N_f = 53,710). The arrow is indicating the crack nucleating site.

Figure 9 shows examples of cross sectional observations for the specimens interrupted at 50% of the total fatigue life. Fatigue crack path is clearly marked with a line in the figure. A fatigue crack propagated was longer in the specimen with 15 μm thick anodized layer. Relationship between number of cycles and crack length is shown in Fig. 10. A fatigue crack nucleates earlier for the specimen with thicker anodized layer. Crack initiation life was estimated as about 40% and 30% of the total fatigue life for the specimens with 5 μm and 15 μm thick anodized layers, respectively. This tendency coincides with fatigue lives for the anodized specimens obtained in higher stress region as shown in Fig.5. Fatigue life for the specimen with the thinner anodized layer is slightly longer than that for the specimen with the thicker anodized layer.

Assuming the same threshold stress intensity factor range, ΔK_{th} the fatigue limits, σ_w can be estimated based on the fracture mechanics relation,

$$\sigma_w = \frac{\Delta K_{th}}{2\sqrt{\pi a_0}} \qquad (4)$$

by using the initial crack length of 5 µm and 15µm as discussed above. If the threshold stress intensity factor range is assumed to be 0.83 [6], the estimated fatigue limits are 105 MPa and 60 MPa for 5 µm and 15 µm thick anodized layers, respectivery. The estimated fatigue limit for the specimen with 15 µm thick anodized layer is reasonable compared to the experimental result as shown in Fig.5. In case of the specimen with a thicker anodized layer, a crack nucleated at the anodized layer can be considered as an initial crack. On the other hand, the estimated fatigue limit for the specimen with 5 µm anodized layer is much higher than the experimental result, even compared to that for the base metal. This may imply that a crack nucleated in the anodized layer for the specimen with thinner anodized layer does not dominate the fatigue limit. However, another factor due to the anodizing process, e. g. rough surface of the base metal can reduce the fatigue limit.

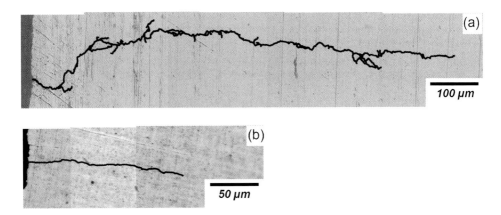

Figure 9: Cross-sectional observation of anodized specimen after applied fatigue load of 50% of its total fatigue life. (a) 15 µm anodized layer, (b) 5 µm anodized layer.

4   Summary

Anodized materials showed lower fatigue strength compared to the base metal. Fatigue crack nucleated at the anodized layer or the interface between anodized layer and base metal. An anodized specimen with thinner anodized layer showed slightly longer fatigue life compared to that with thicker anodized layer. The fatigue crack nucleation lives were about 40 % and 30% of total fatigue life for the specimens with anodized layer thicknesses of 5µm and 15 µm, respectively. The crack nucleated in the anodized layer can be assumed to be an initial crack and reduces fatigue limit for the specimen with thicker anodized layer. However, in case of the specimen with thinner anodized layer, another factors, such as interface roughness, due to the anodizing process may influence on the fatigue limit.

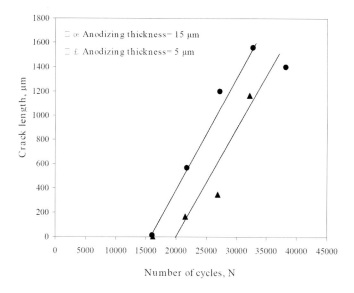

Figure 10: Relationship between number of cycles and crack length in the interrupted fatigue test.

5 References

[1] Zainuddin bin Sajuri, Yukio Miyashita Yoshiharu Mutoh Journal of Japan Institute of Light Metal, **2002**, *52-4*, 161-166.
[2] Sabrina Alam Khan, Yukio Miyashita, Yoshiharu Mutoh and Toshikatsu Koike, Proceedings of 54th Japan Society of Materials Engineering annual meeting, **2005**,397-398.
[3] Shuji Nagai, Kazuo Matsuzawa, Tsuneo Saga, Masahiro Akimoto, Proceedings of 96th Japan Society of Light Metal annual meeting, **1999**, 83-84.
[4] J. Eifert, J. P. Thomas and R. G. Raterick, Jr., Scripta Materia, **1999**, *40-8*, 929-935.
[5] Zainuddin Bin Sajuri, Takashi Umehara, Yukio Miyashita and Yoshiharu Mutoh, Advanced Engineering Materials, **2003**, *5-12*, 910 – 916.
[6] Z. B. Sajuri, Y. Miyashita and Y. Mutoh, Materials Science Forum, **2003**, *419-422*, 81-86.

Structural Durability of MRI 153M Die-Cast Components

A. Esderts, P. David
Institute for Plant Engineering and Fatigue Analysis, Clausthal University of Technologies, Clausthal-Zellerfeld, Germany
C. Berger, M. Gugau, T. Troßmann, J. Grimm
Institute for Materials Technology, Darmstadt University of Technology, Darmstadt, Germany

1 Introduction

As limiting factors for a further growth in the application of magnesium alloy, very often the poor corrosion resistance (under certain conditions), the strength and creep-resistance of "standard-alloys" at elevated temperatures as well as the higher costs are attributed to. For instance, the most commonly used magnesium alloy AZ 91 shows a distinct decrease of fatigue strength at elevated temperatures, predominantly when exceeding 120 °C [1]. Other developed Mg-alloys, which counterbalance these disadvantages, are so far not widely used in mass production, some because of economical reasons. So there is still a need in order to develop Mg-alloys, which accomplish the requirement of high temperature strength, creep resistance and corrosion resistance by comparable price to competing materials.

With the development of MRI 153M the aspect of higher strength values at higher temperatures in a price range comparably to the "standard-alloy" AZ91 should be faced. Thus, an increase in the use of magnesium alloys in automotive components – especially in power train applications – could be expected.

2 Experimental Procedure

2.1 Material and Specimen Geometry

In order to determine the fatigue behavior of the Mg-alloy MRI 153M, component-like rib specimens were cast in a hot-chamber die-cast process. The die was the same one that was already used in [2], when the same sample geometry was cast in AZ91 and AM60, respectively. So the geometry can be stated as identical.

Rib structures are typical elements which can often be observed in fabricated die-cast components in order to enhance the stiffness of a structure. The ribs of these specimen were situated between a right-angled arrangement of two plates with each a width of 40 mm, a depth of 60 mm and a thickness of 4 mm, whereas the ribs had a thickness of 2.5 mm and 4.0 mm, respectively, Fig. 1: Casted component and single specimen with hardening rib. One cast object contained 6 of these angles with different notch radii (r = 3 mm and r = 12 mm), which were cut to individual rib testing specimen. In the course of the investigation, the tests were only conducted at rib samples with the thickness of 2.5 mm.

The examinations comprise tests in the as-cast condition as well as with removed casting skin. Therefore, the rib samples with notch radius r = 12 mm were machined to a notch radius of r=3 mm to reach the identical shape and geometry of the unmachined (and casting skin containing) samples. Subsequently the machined surface as well as the remaining ridge was smoothened by stepwise grinding with SiC paper from grain size 800 up to 4000.

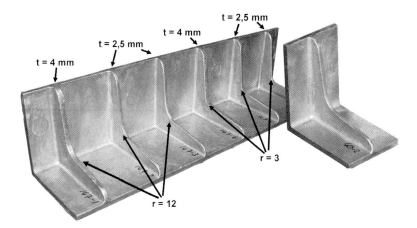

Fig. 1: Casted component and single specimen with hardening rib

2.2 Environment

Since in practical application, automotive components are not only subjected to ambient conditions, but are usually exposed to an interaction between mechanical loading, thermal strain as well as corrosive fluids, tests were conducted under various conditions superimposing all these factors. For that reason, the fatigue behavior was tested besides ambient air also under elevated temperatures (120°C and 150 °C) as well as using solutions of sodium chloride (5 wt%, 0,05 wt%). It is worth to mention, that all solutions were prepared with analytical-reagent-quality reagents in deionized (DI) water [3].

2.3 Test Procedure

The tests at ambient air and elevated temperatures were performed at the Institute for Plant Engineering and Fatigue Analysis, Clausthal University of Technologies. All investigations concerning corrosion fatigue (by the presence of NaCl-solutions of different concentration) were conducted at the Institute for Materials Technology, Darmstadt University of Technology.

In case of the test rig for the investigation without corrosive medium, the resulting bending load, the specimen was subjected to, was transmitted by a pneumatic cylinder by means of a plate spring which works as a nearly frictionless joint, Figure 2: Correlation between force and strain under bending loading. The lower end was fixed upon the testing machine. All tests were stress-controlled. The displacement was measured by a laser sensor. The test abortion was calibrated so the final crack length was between 3 and 5 mm.

For the experiments concerning corrosion fatigue with a superimposed corrosive solution, a servohydraulic testing system was used. The cycling frequency was set to 10 Hz. By sustaining 10^6 cycles without occurring of a crack initiation, the frequency was increased to 15 Hz in order to reduce testing times. The test rig for clamping the component-like rib-specimen was identical to the setup described above. Additionally, a corrosion chamber was developed, which enabled the continuous sprinkling of the sample with NaCl-solution.

3 Test Results

The examinations comprise various tests varying parameters like stress ratio, temperature, surface finish (with and without casting skin), notch radius and corrosive medium with different concentration (5 wt% NaCl; 0,05 wt% NaCl).

This paper only deals with tests under constant amplitude loading in order to record S-N-curves. In this context, it is focused on the documentation of the results which were derived by a stress ratio of R=0. The final results including lifetime test, strain controlled tests and the proposal of a calculation method for lifetime estimation will be published in June 2007.

3.1 Load and Local Stresses

The correlation of the bending load and the local stresses was examined by strain gages. They were fixed parallel to the hardening rib. Their exact position and the resulting force-stress hys-

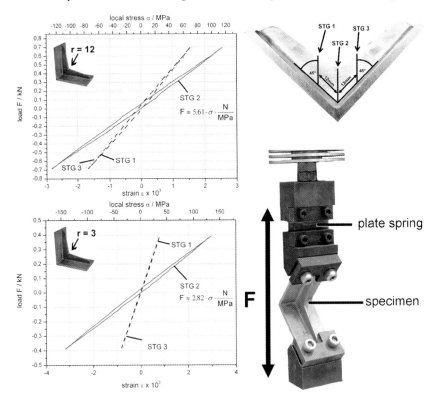

Figure 2: Correlation between force and strain under bending loading

teresis is show in Figure 2: Correlation between force and strain under bending loading. For comparison between the r = 3 mm and r = 12 mm specimen, the transmitted force is converted into stress via linear transformation.

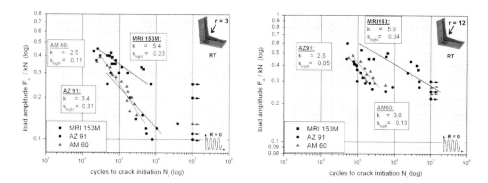

Figure 3: S-N-curves for AZ 91D, AM 60 and MRI 153M at room temperature, R = 0, ambient air

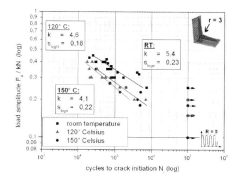

Figure 4: The influence of higher temperature, r = 3, R = 0, ambient air

3.2 Woehler Tests at Ambient Atmosphere

The S-N-curves of MRI 153M at room temperature and ambient air can directly be compared to AZ 91D and AM 60, Fig. 3. These reference results were published during a former project using the same specimen geometry, [2]. In conclusion MRI 153M shows little longer lifetime than the reference alloys. The knee point is not statistically firm. The regression line of the r=12 mm specimen of MRI 153M shows a greater number of cycles than the limiting. For better clearness it was drawn to the greater value anyway.

The tests at elevated temperatures were perfomed at 120 ° C and 150 ° C, which is the application limit for MRI 153M (manufacturer information). Expectedly the influence of higher temperature decreases the fatigue strength of the material, Fig. 4. The reduction in fatigue strength is about one scatter range, which can be considered as very moderate. Between 120 °C and 150 °C no significant difference can be observed.

3.3 Influence of Superimposed Corrosive Load on the S-N-curves at Room Temperature (Corrosion Fatigue)

Despite the fatigue tests in ambient environment and at elevated temperatures, which were already illustrated and discussed above, further studies have been conducted concerning corrosion fatigue of the magnesium alloy MRI 153M. Since magnesium alloys are generally regarded as susceptible to chloride-anions, the experimental setup was modified to examine the influence of continuously sprinkling sodium chloride solution on the fatigue properties. The solutions were prepared in deionized water varying the concentration of the NaCl-content from 5 wt% (with respect to the salt water spray test, which is often practiced in automotive application) to 0,05 wt%. The results are displayed in Figure 5: Influence of continuously sprinkling sodium chloride solution of different concentration on the Woehler-curve of rib specimen with casting skin (left) and with removed casting skin (right).

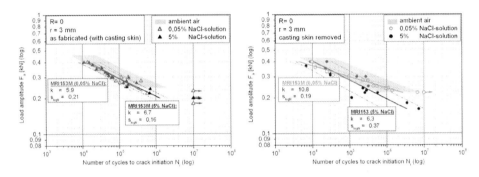

Figure 5: Influence of continuously sprinkling sodium chloride solution of different concentration on the Woehler-curve of rib specimen with casting skin (left) and with removed casting skin (right)

As reference for better estimation, the scatter band of the results at ambient air are additionally presented in the diagrams. One can recognize, that in the case of the samples with casting skin (Fig. 5, left), no substantial influence of the corrosive fluids on the fatigue lifetime was found in comparison to the results at ambient air, neither using 0,05wt% NaCl-solution nor 5wt% NaCl-solution. No failure occurred beyond 10^6 cycles. What might be surprising at first glance can be explained inspecting the fracture surface and condition of the casting skin within the notch radius of the rib, Fig. 6. The casting skin is presented by a significant scarred and scaled surface, respectively. These scars can be regarded as notches, which facilitate crack propagation emanating from the surface of the casting skin. So the geometrical condition of the surface of the casting skin - as a result of fabrication - is dominating over, or at least superimposing with, the time dependent local corrosion attacks caused by the sodium chloride solution.

The machined specimen with removed casting skin show more susceptibility to corrosive attacks, Figure 5: Influence of continuously sprinkling sodium chloride solution of different concentration on the Woehler-curve of rib specimen with casting skin (left) and with removed casting skin (right), right. While the slope of the S-N curve using 5wt% NaCl-solution is almost equivalent to the results obtained with casting skin, the curve is not as steep for 0,05wt% NaCl-solution. In both cases, crack initiation still occurred exceeding 10^6 cycles. It is apparent, that with higher numbers of cycles (and therefore longer exposure to the corrosive fluids) the influence of the different concentrations of the sodium chloride solution can be verified.

An inspection of the fracture surfaces and front of the ribs, respectively, indicates that the different concentration of the corrosive agents results in different failure mechanisms, Figure 6: Fracture surface and condition of the casting skin of ribs in the as-fabricated condition. Even for longer exposure to 0,05 wt% NaCl-solution, no significant macroscopic forms of corrosion of certain depth, like filiform corrosion or notches caused by corrosive attacks, could be observed, Figure 7: Fracture surface and condition of the rib front with removed casting skin, left. The surface was rather characterized by uniform corrosion resulting in dulling, staining or discoloration of the metal surface. On the contrary, the use of 5wt% NaCl-solution induces localized notches or grooves of distinct depth predominantly emanating from the edges of the rib, which initiate the failure of the rib structure, Figure 7: Fracture surface and condition of the rib front with removed casting skin, right.

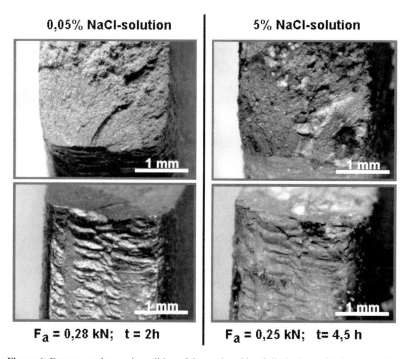

Figure 6: Fracture surface and condition of the casting skin of ribs in the as-fabricated condition

4 Conclusion

In order to characterize the fatigue properties of the Mg-alloy MRI 153M, S-N-curves were generated under constant amplitude loading using component-like rib specimen. The examinations comprised various tests varying parameters like stress ratio, temperature, surface finish (with and without casting skin), notch radius and corrosive medium with different concentration (5 wt% NaCl; 0,05 wt% NaCl).
- Comparing the Woehler-curve of MRI 153M with the equivalent die-castings of the materials AZ91 and AM60 in their individual casting conditions (microstructure) in ambient air, MRI 153M shows superior fatigue life.

- Conducting the tests under elevated temperatures (T = 120 °C and T = 150 °C, respectively), the material MRI 153M exhibits a slight decrease of the S-N-curve in comparison to room temperature. The difference between the results of T = 120 °C and T = 150 °C can be regarded as negligible.
- Simulating a corrosive environment, the experimental results indicate that the condition of the specimen surface (e.g. casting skin, casting seam) as well as the microstructure can be even more dominating the fatigue behavior of a component than the presence of a corrosive agent.
- Using NaCl-solution in the concentration of 5 wt% during the experimental procedure, localized corrosion in form of pitting and filiform corrosion with distinct depth is promoted. Reducing the content of sodium chloride to 0,05wt%, even certain Mg-alloys are able to form passive films based on Mg-hydroxides – also in chloride containing fluids. So a uniform corrosion attack, which can be regarded as less detrimental concerning a reduction of corrosion fatigue strength, can be promoted depending on the conditions in the experiment as well as during service [3].

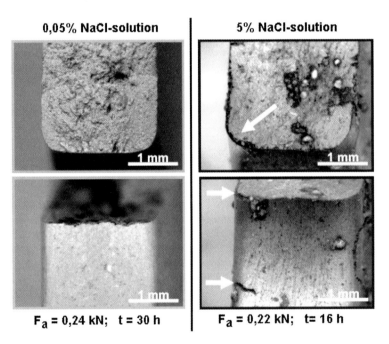

Figure 7: Fracture surface and condition of the rib front with removed casting skin

5 Acknowledgement

The investigations described have been conducted within the scope of the research project 14158, financed by the FKM (Forschungskuratiorium Maschinenbau) and AiF (Arbeitsgemeinschaft industrieller Forschungsvereinigungen). The material was donated by Volkswagen AG, Wolfsburg, Germany. The test bars were cast at Andreas Stihl AG, Waiblingen, Germany. The authors wish to express their thanks for this support.

6   References

[1] Polmear, I.J.; Magnesium Alloys and Applications. Mater. Sci. Technol. 10, 1994
[2] Renner, F.; Zenner, H., Fatigue and Fracture of Engineering Materials and Structures; 2002, 25, p. 1157-1168
[3] Troßmann, T.; Dissertation, TU Darmstadt, 2005
[4] Mordike, B. L., v. Buch, F.; Stand der Technik und Entwicklungspotenziale von Magnesium und seinen Legierungen. DGM-Seminar Magnesium, 1997

Fatigue Crack Propagation Behavior of Mg-Al-Zn Alloys

R. C. ZENG [1,2,3], E. H. HAN[1], W. KE[1], W. Dietzel[3], K. U. Kainer[3]
1 Environmental Corrosion Center, Institute of Metals Research, Shenyang, China
2 School of Material Science and Engineering, Chongqing Institute of technology, Chongqing, China
3 GKSS-Forschungszentrum Geesthacht GmbH, Geesthacht, Germany

1 Abstract

An investigation of the fatigue crack propagation (FCP) behavior of as-extruded magnesium alloys AZ80, AZ61 was made by using single-edge notched plate specimens. This paper describes the influence of loading frequency, load ratio, aluminum content, heat treatment, temperature and relative humidity on FCP rate. It is demonstrated that the FCP rate increases to a great extent with a decrease in loading frequency from 10 Hz to 1 Hz. An increase in load ratio results in acceleration in FCP rate. Depending on not only the aluminum content but also on the microstructure and thickness of the oxide film, the FCP rate of AZ80 is higher than that of AZ61 at 1 Hz but lower at 10 Hz. Ageing treatment facilitates the FCP rate. Increasing the ambient temperature from room temperature to 60 °C and 120 °C accelerates the FCP rate, particularly at 120. A bend occurred in the curves of FCP rate versus cyclic stress intensity factor at 120. At first, the FCP rate increased sharply, and then went up slowly. Hydrogen embrittlement may be responsible for the acceleration of the FCP rate of Mg-Al magnesium alloys in higher relative humidity.

2 Introduction

Magnesium alloys are getting a wider application in transportation industry due to their high specific strength. Although nowadays mostly used for static parts such as cases, housings, brackets, panels etc., these materials have a potential to be used as load-bearing components, e.g. wheels, which will be subjected to fatigue loading [1-4]. The materials, however, usually have voids and other defects to some degree. Therefore, it is of primary importance to understand the FCP behavior of magnesium alloys in various situations. This study aims to investigate the effect of test parameters (e.g. loading frequency and load ratio), material condition (such as chemical composition and heat treatment), and environmental factors (temperature and relative humidity) on the FCP behavior of AZ80 and AZ61.

3 Experimental

Single-edge notched plate specimens of 32×8 mm^2 cross-section and 150 mm length (notch depth 3.5 mm) were machined from continuously extruded sheets of AZ80 (Mg-9.0 % Al-0.7%Zn) and AZ61 (Mg-6.0% Al-0.7%Zn). An artificial aging (-T5) heat treatment was performed for 16 hours at 177 °C in a vacuum heat furnace. The mechanical properties of the Mg-Al alloys are shown in **Table 1**.

Table 1: Mechanical properties of Mg-Al alloys

Material	σ_b [MPa]	σ_y [MPa]	E [GPa]	Elongation [%]	Area reduction [%]
AZ80-F[a]	333.0	235.0	45.0	7.0	10.0
AZ80-T5[b]	357.0	271.0	45.0	3.2	6.5
AZ61-F	274.0	187.0	45.0	21.2	13.4

(a)-F stands for as received. (b)-T5 stands for artificial aging

Constant load amplitude static fatigue tests were performed using an EHF-EB10-20L servo-hydraulic fatigue machine in ambient air (room temperature (RT): 16 °C–20 °C, relative humidity (RH): 40 %–70 %). Elevated temperatures (between 60 °C and 120 °C) were adjusted in an insulated chamber by heating an electricity resistance wire. A frequency range of 0-10Hz, maximum load of 7KN, and a load ratio of zero under a sinusoidal waveform were applied. The value of the crack length, a, was in-situ measured by a traveling microscope. The stress intensity factor range ΔK was calculated using following equation [5]:

$$\Delta K = \frac{\Delta p}{BW}\sqrt{a}\, f\!\left(\frac{a}{w}\right), \tag{1}$$

where (a/w) is the crack length divided by the specimen width and

$$f(a/w) = 1.99 - 0.41(a/w) + 18.7(a/w)^2 - 38.48(a/w)^3 + 53.85(a/w)^4. \tag{2}$$

Δp is the load range, B is the specimen thickness.

4 Results and Discussion

4.1 Effect of Frequency

Fatigue crack propagation rates are typically plotted as log-log graphs of da/dN versus ΔK. The da/dN versus ΔK curves of AZ80 are shown in **Figure 1**. It shows that the curves follow the Paris equation and that the frequency has a significant impact on the FCP rate. The lower the frequency is, the higher is the FCP rate. This situation is similar to fatigue of metal materials in corrosive solutions. Thus, air can be regarded as a corrosive environment for magnesium alloys. The test results showing the fatigue crack growth rates at an ambient temperature in air may be represented by:

$$da/dN = D(t)(\Delta K)^m, \tag{3}$$

where $D(t)$ depends on the sinusoidal cyclic stress-intensity frequencies. The values of the constants m and C for AZ80 at different frequencies are given in **Table 2**.

According to **Equation 3** and **Table 2**

$$da/dN = (3 \cdot 10^{-10} + 9.7 \cdot 10^{-9}/f)(\Delta K)^{1.9}. \tag{4}$$

Oxide films play an important role in FCP. In fact, MgO can readily transform into $Mg(OH)_2$ in moist air. The lower the frequency is, the thicker the oxide film will grow. The thick oxide film is so brittle that the fatigue crack can easily initiate and grow. The FCP rate obviously depends on the frequency as well as on crack closure effects [6].

Table 2: Values of the constants m and C for AZ80 at different frequencies

No.	R	f [Hz]	m	C
	0	10	1.9	$1.3 \cdot 10^{-9}$
	0	1	1.8	$1.0 \cdot 10^{-8}$

3.2 Effect of Load Ratio

Figure 2 shows the effect of load ratio on the FCP rate of AZ80. The FCP rate increases with increasing load ratio. This result may be correlated with the crack closure effect. On one hand, crack closure could be easily induced by corrosion products at the lower R ratio. On the other hand, mismatch between extrusions and protrusions on the fracture surface could result in crack closure due to permanent plastic deformation and irreversible slip occurring during peak stress. Therefore, the crack closure induced by alloy roughness, combined with oxide films is proposed to impede the FCP, thus decreasing the FCP rate. The curves in **Figure 2** may be expressed as:

$$\frac{da}{dN} = C \left\{ \frac{(\Delta K)^m}{(1-R)^{cl}} \right\}. \tag{5}$$

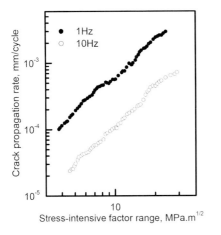
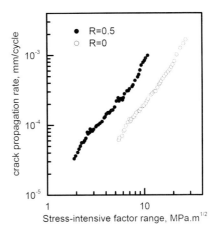

Figure 1: Effect of frequency on the FCP rate of AZ80 **Figure 2:** Effect of load ratio on the FCP rate of AZ80.

Table 3 gives the values of the constants m and C for AZ80 under various load ratios.

$$\frac{da}{dN} = 2.5 \cdot 10^{-9} \left\{ \frac{(\Delta K)^{1.9}}{(1-R)^{2.3}} \right\}. \tag{6}$$

Table 3: Values of the constants m and C for AZ80

No.	R	f [Hz]	m	C
	0	5	1.9	$2.5 \cdot 10^{-9}$
	0.5	5	1.8	$1.3 \cdot 10^{-8}$

3.3 Effect of Aluminum Content

Figure 3 displays the effect of the aluminum content at different loading frequencies. It is demonstrated that the FCP rate of AZ80 is higher than that of AZ61 at a frequency of 1 Hz, whereas it is lower than that of AZ61 at a frequency of 10 Hz. The FCP rate is controlled by three factors: plastic zone size, oxide films, and microstructure. Firstly, the size of the plastic zone, r_y, can be estimated from the stress-field equation under plane strain condition as [7]:

$$r_y = \frac{1}{6\pi}\left(\frac{\Delta K}{\sigma_y}\right)^2. \tag{7}$$

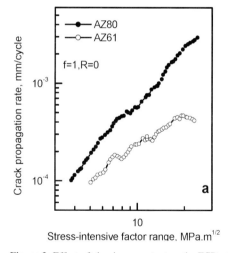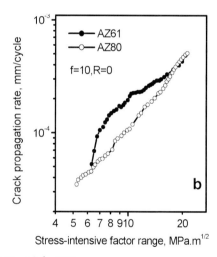

Figure 3: Effect of aluminum content on the FCP rate: (a) f = 1 Hz, (b) f = 10 Hz

According to Table 1:

$$r_{yAZ80} / r_{yAZ61} = (\sigma_{yAZ61}/\sigma_{yAZ80})^2 = (187/235)^2 = 0.63. \tag{8}$$

Equation 8 indicates that the size of the plastic zone at the crack tip in AZ80 is smaller than that of AZ61 because the yield strength of AZ80 is higher than that of AZ61. Therefore, the plasticity induced crack closure effect is weaker for AZ80 than for AZ61. Secondly, the crack propagation velocity depends on the oxide films the thickness of which significantly depends on the aluminum content. The lower the aluminum content of the alloy, the thinner the oxide film becomes. That means that the oxide film of AZ80 is thinner than that of AZ61. Thirdly, the inhomogeneous microstructure of the alloy leads to crack deviation and branching observed during the AZ61 tests. Crack closure, induced by deviation and branching, significantly impedes the FCP at higher ΔK as confirmed by the curves of AZ61 in Figure 3b.

In summary, the crack closure effect caused by combination of the above three variables more remarkably impedes the FCP of AZ61 than that of AZ80 at lower frequencies. Thus, the FCP rate of AZ61 is lower than AZ80 at 1 Hz (see Fig.3a), and vice versa.

3.4 Effect of Heat Treatment

The effect of the heat treatment on the FCP rate is demonstrated in Figure 4. The artificial aging can accelerate the FCP velocity. There is an argument on whether the heat treatment accelerates the FCP rate. Based on this, the aged microstructure is composed of a higher volume fraction of the lamellar precipitates β-phase ($Mg_{17}Al_{12}$) in the α-matrix. Kobayashi et al [6] proposed that the precipitates had a negative effect on the FCP rate. This is, however, not always the case. Bag et al [8] suggested that ageing treatment reduces the FCP rate of as-cast alloys, mostly due to the larger deviation and branching of the crack from the plane of maximum stress caused by the inhomogeneous microstructure. In fact, the fatigue crack propagation rate depends on both secondary phases and the uniformity of the microstructure of materials. An ageing heat treatment usually increases the FCP rate if the microstructure is relatively homogenous and the crack propagation path become smoother. Otherwise, an ageing heat treatment may decrease the FCP rate if microstructure is inhomogeneous. And crack closure, produced by the crack biased and branching, impedes the crack propagation.

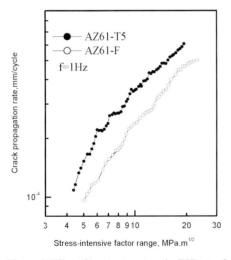

Figure 4: Effect of heat treatment on the FCP rate of AZ61

3.5 Effect of Temperature and RH

Figure 5 exhibits the effect of temperature on the FCP of AZ61. The FCP rates for AZ61 increases slightly at 60°C, and largely increases at 120°C. A bend occurred in the curves of the

FCP rate versus stress intensity factor at 120 °C. At first stage, the FCP velocity rises steeply; and then, in a second stage, it increases more slowly. The effect of RH on the FCP rate is shown in Figure 6. It shows that the FCP rate apparently grows with the increasing RH. Hydrogen embrittlement may be primarily responsible for this acceleration in humid air.

The values of the constants m and C for AZ61 are given in Table 4.

Table 4: The values of the constants m and C for AZ61 ($f = 1Hz$, $R = 0$)

No.	m	C	Remark
	1.1	$1.8 \cdot 10^{-8}$	RT.
	2.4	$1.6 \cdot 10^{-9}$	wet air
	1.0	$3.2 \cdot 10^{-8}$	120 °C
	0.9	$4.8 \cdot 10^{-8}$	120 °C
	1.1	$1.7 \cdot 10^{-8}$	60°C
	1.0	$3.2 \cdot 10^{-8}$	T5

4 Conclusion

The FCP rates increase when lowering the loading frequency, increasing the load ratio, when applying an ageing heat treatment, or if temperature and/or relative humidity are increased. The effect of the Al content on the FCP rates depends on the combined effect of loading frequency, inhomogeneous microstructure and precipitates.

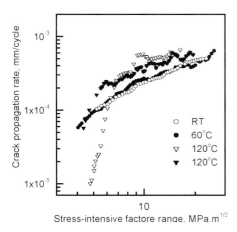

Figure 5: Effect of temperature on the FCP rate of AZ61

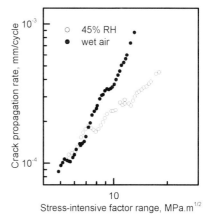

Figure 6: Effect of relative humidity on the FCP rate of AZ61

5 Acknowledgements

This work was supported by the National Hi-Tech Research and Development Program under grant No.2001AA331050. Thanks go to Guangling Magnesium Industry Science and Technology Co. Ltd. for providing the tested materials.

6 References

[1] C. Potzies, K. U. Kainer, Adv. Eng. Mater. 2004, 6(5), 281–289.
[2] R.C Zeng, E.H. Han, Y.B. Xu, W. Ke, Chinese Journal of Materials Research. 2003, 17(3), 241–245. (In Chinese)
[3] R.C Zeng, E.H. Han, W. Ke, Mater. Sci. Forum, 2005, 488–489, 721–724.
[4] R.C. Zeng, J. Zhang, W. J. Huang, W. Dietzel, K. U. Kainer, C. Blaeert, W. Ke, Trans. Nonferrous Met. Soc. China, 2006, 16, s763–s771.
[5] S. Suresh, Fatigue of materials (Translated Z. G. WANG), Defense Industry Press, Beijing, 1999, p434.
[6] Y. Kobayashi, T. Shibusawa, K. Ishikawa, Mater. Sci. Eng, 1997, A234–236, 220–222.
[7] J.M.Barsom, S.T.Rolfe, Fracture and Fatigue Control in Structures, Prentice-Hall, Inc., New Jersey, 1987, p72.
[8] A. Bag, W. Zhou. J. Mater. Sci. letters, 2001, 20, 457.

Microstructure-based Characterization of the Cyclic Deformation Behavior of the Magnesium Die-cast Alloys MRI 153M and MRI 230D

B. Ebel-Wolf, F. Walther, D. Eifler

Institute of Materials Science and Engineering, University of Kaiserslautern, Kaiserslautern, Germany

1 Introduction

The automotive industry makes big efforts to use magnesium alloys in mass production cars [1], for example in the engine, where the technical needs can not be satisfied by well known magnesium alloys like AZ91D. Thus, new magnesium alloys like MRI 153M and MRI 230D were developed. The majority of fatigue research at magnesium alloys was performed lifetime-orientated, partially regarding the influence of environmental conditions, mean stresses, notches and multiple step loading, e.g. [2, 3]. There are only very few systematic investigations dealing with the cyclic deformation behavior of magnesium alloys, e.g. [4].

In this paper, the cyclic deformation behavior of the new magnesium alloys MRI 153M and MRI 230D as well as AZ91D as reference material was characterized in continuous load increase tests (LIT) and in single step tests (SST) by means of mechanical stress-strain hysteresis, temperature and electrical resistance measurements. On the basis of generalized Morrow, Manson-Coffin and Basquin equations a physically based lifetime calculation method „PHYBAL" was developed. With the cyclic stress-strain curve taken from one LIT and the plastic strain amplitude at $N = 10^4$ cycles of two additional SST, S-N curves can be precisely calculated.

2 Materials and Experimental Methods

The investigated Mg alloys MRI 153M, MRI 230D and AZ91D were supplied by Dead Sea Magnesium Ltd., Israel, as die-casted specimens. This ensures a fine-grained microstructure and a low pore fraction of the specimens, which were investigated with the cast skin. At ambient temperature the yield strength $R_{p0.2}$ (tensile strength R_m) of the magnesium alloys is 147 MPa (262 MPa) for AZ91D, 169 MPa (235 MPa) for MRI 153M and 176 MPa (220 MPa) for MRI 230D.

The microstructure consists of an interdentritic magnesium solid solution (α phase) with different precipitations. Partially, the precipitations are surrounded by an Al-supersaturated magnesium solid solution, as can be seen in the SEM micrographs in Figure 1. For AZ91D, the interdentritic β phase mainly consists of the brittle face-centered cubic $Mg_{17}Al_{12}$ phase. The exact chemical composition of the MRI alloys is undisclosed. Their precipitations belong to the Mg-Al-Zn-Ca system and are expected to be predominantly $Al_2(Ca, Zn)$. The precipations of the MRI alloys are larger but more filigree than the ones of AZ91D. With 11.3 % the mean fraction of precipitations is comparable for the three investigated Mg alloys.

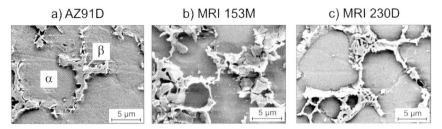

Figure 1: Microstructure of the Mg alloys AZ91D (a), MRI 153M (b) and MRI 230D (c)

The hardness of the magnesium solid solution is 66 HV0.1 and the hardness of the precipitations reaches values of about 154 HV0.01 for AZ91D, 147 HV0.01 for MRI 153M and 142 HV0.01 for MRI 230D.

Stress-controlled continuous load increase (LIT) and single step tests (SST) with a load ratio of $R = -1$ and a frequency of $f = 5$ Hz were performed at ambient temperature on servohydraulic testing systems. In LIT the stress amplitude σ_a was continuously increased with a rate of $d\sigma_a/dt = 1.2 \cdot 10^{-3}$ MPa/s until specimen failure. SST were performed until specimen failure or reaching a maximum number of cycles of $N_{max} = 2 \cdot 10^6$. Besides the plastic strain amplitude determined in mechanical stress-strain hysteresis measurements, the change of the specimen temperature T and the change of the electrical resistance ΔR were measured to ensure a detailed microstructure-based characterization of the cyclic deformation behavior of the die-casted magnesium alloys AZ91D, MRI 153M and MRI 230D.

3 Results

3.1 Load Increase Tests (LIT)

As shown in former work for AZ91D in the sand-casted condition [5], load increase tests (LIT) allow a reliable estimation of the endurance limit.

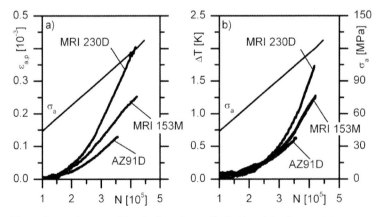

Figure 2: Development of the plastic strain amplitude (a) and the change of the temperature (b) in load increase tests for AZ91D, MRI 153M and MRI 230D

Figure 2 shows the development of the plastic strain amplitude $\varepsilon_{a,p}$ (a) and the change of the temperature ΔT (b) versus the number of cycles N in LIT for the investigated Mg alloys. After an initial linear increase, the $\varepsilon_{a,p}$ and ΔT values increase exponentially, whereby the values of the MRI alloys differ more and more from AZ91D. Despite higher $\varepsilon_{a,p}$ and ΔT values, the MRI alloys reach higher lifetimes. The endurance limit estimated in these LIT on the basis of the transition from a linear to an exponential increase of the $\varepsilon_{a,p}$ and ΔT values yield 53 MPa for AZ91D as well as 56 MPa for MRI 153M and 59 MPa for MRI 230D.

3.2 Single Step Tests (SST)

The cyclic deformation behavior in single step tests (SST) is characterized for MRI 153M, MRI 230D and AZ91D by cyclic softening for the first $1-2 \cdot 10^3$ cycles followed by cyclic hardening for the greatest part of the lifetime. Figure 3 exemplarily shows the development of the plastic strain amplitude $\varepsilon_{a,p}$ (a) and the change of the temperature ΔT (b) of MRI 153M for stress amplitudes a between 70 and 100 MPa. Maximum $\varepsilon_{a,p}$ (ΔT) values of $0.8 \cdot 10^{-3}$ (3 K) are reached for MRI 153M in the state of maximum cyclic softening.

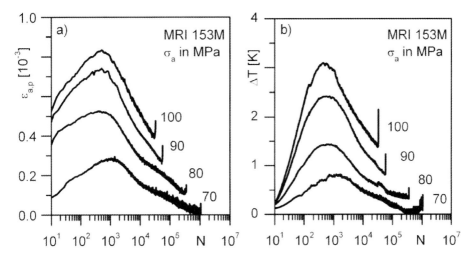

Figure 3: Cyclic deformation curves (a) and cyclic temperature curves (b) of MRI 153M

In Figure 4a, the plastic strain amplitude $\varepsilon_{a,p}$, the change of the temperature ΔT as well as the change of the electrical resistance ΔR are plotted versus the number of cycles for a SST with MRI 230D at $\sigma_a = 105$ MPa. The development of $\varepsilon_{a,p}$ and ΔT values indicate initial cyclic softening followed by cyclic hardening until macroscopic crack growth. In contrast, after an initial increase of the ΔR values followed by a saturation state, the ΔR-N curve is characterized by exponentially increasing values for $N > 7 \cdot 10^3$ cycles. This exponential increase of ΔR indicates failure in a very early fatigue state, much earlier than $\varepsilon_{a,p}$ and ΔT.

Despite higher maximum $\varepsilon_{a,p}$, ΔT and ΔR values of the MRI alloys compared to AZ91D in LIT (Figure 2) and SST, higher lifetimes are consistently observed for the MRI alloys (0b). This can be explained by more pronounced cyclic hardening of the MRI alloys [7].

The test with MRI 230D and $\sigma_a = 60$ MPa leads to $N = 2 \cdot 10^6$ cycles without failure. This stress amplitude correlates very well with the endurance limit estimated in the LIT (59 MPa), see Figure 2.

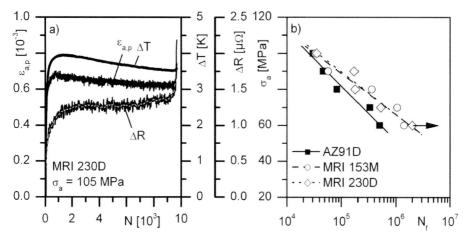

Figure 4: Cyclic deformation, temperature and electrical resistance curve for MRI 230D (a) and S-N curves of AZ91D, MRI 153M and MRI 230D (b)

3.3 Physically Based Lifetime Calculation „PHYBAL"

The <u>phy</u>sically <u>ba</u>sed <u>l</u>ifetime calculation „PHYBAL" is based on deformation-induced changes of the physical quantities $\varepsilon_{a,p}$, ΔT and ΔR, which are measured at only 10^4 cycles [6].
The power law according to Morrow is generally used to describe cyclic stress-strain (CSS) curves. In a generalized form it can be written as follows, whereby M(N) may be $\varepsilon_{a,p}$ as well as ΔT and ΔR.

$$\sigma_a = a(M) \cdot (M(N))^{b(M)} \qquad (1)$$

The development of the $\varepsilon_{a,p}$, ΔT and ΔR values measured in a LIT accumulate the influence of the preceded cyclic load. In this investigation, the Morrow curve determined for discrete stress-strain or stress-temperature data of a LIT was calculated for single step loading using the σ_a-$\varepsilon_{a,p}$ or σ_a-ΔT relation of two SST. One stress amplitude of the two SST is slightly higher than the endurance limit estimated in the LIT and the other one is slightly lower than the stress amplitude leading to failure in the LIT. As exemplarily shown in 0a for MRI 153M, the CSS curve of the LIT is plotted on the basis of $\varepsilon_{a,p}$ values for discrete stress amplitudes σ_a besides the σ_a-$\varepsilon_{a,p}$ data at $N = 10^4$ cycles of the two SST. This approach is necessary to receive the ratio $Q(\varepsilon_{a,p})$ between the measured $\varepsilon_{a,p}$ values in the LIT and the two SST. The ratio $Q(\Delta T)$ based on ΔT measurements in the LIT and the two SST was similarly determined:

$$Q(\varepsilon_{a,p}) = \frac{\varepsilon_{a,p,SST}}{\varepsilon_{a,p,LIT}}, \quad Q(\Delta T) = \frac{\Delta T_{SST}}{\Delta T_{LIT}} \qquad (2)$$

Assuming a linear decrease of the ratio $Q(\varepsilon_{a,p})$ for an increasing stress amplitude (Figure 5a), the CSS curve for single step loading can be calculated by multiplying the $\varepsilon_{a,p}$ values of the LIT with the corresponding ratio $Q(\varepsilon_{a,p})$ (Figure 5b).

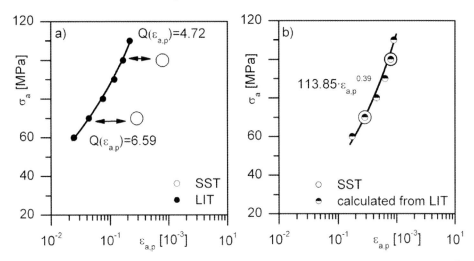

Figure 5: Cyclic stress-strain curve for a load increase test and σ_a-$\varepsilon_{a,p}$ data of two single step tests at $N = 10^4$ cycles (a) and cyclic stress-strain curve calculated from the load increase test for single step loading (b) of MRI 153M

The CSS curve calculated for single step loading can be described by Eq. 1 leading to the parameters $a(M)$ and $b(M)$. The fatigue strength exponent $g(M)$ in Eq. 3 can be calculated according to Morrow with the cyclic hardening exponent $b(M)$:

$$g(M) = \frac{-b(M)}{5 \cdot b(M) + 1} \tag{3}$$

With $g(M)$ from Eq. 3 and with the σ_a-N_f relation of one SST, $f(M)$ in the Basquin equation (Eq. 4) can be determined (Eq. 5):

$$\sigma_a = f(M) \cdot (2N_f)^{g(M)}. \tag{4}$$

$$f(M) = \frac{\sigma_a}{(2N_f)^{g(M)}} \tag{5}$$

With the parameters $f(M)$ and $g(M)$ it is possible to calculate the S-N curve according to Eq. 4 with only one single LIT and two SST on the basis of the plastic strain amplitude or the change of the temperature.

In Figure 6 lifetimes calculated with „PHYBAL" on the basis of $\varepsilon_{a,p}$ or T from one LIT and two SST are compared to experimentally determined data. As can be seen clearly, the calculated and experimental S-N curves agree excellently. It is evident that the application of this new lifetime calculation method is combined with an enormous saving of experimental time and costs.

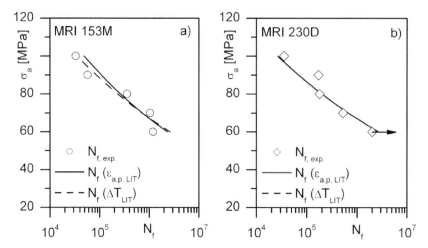

Figure 6: Calculated and experimental S-N curves according to Eq. 6 for MRI 153M (a) and MRI 230D (b)

4 Summary

In load increase tests the endurance limit of the Mg alloys AZ91D, MRI 153M and MRI 230D can be reliably estimated with one single specimen. Despite higher plastic strain amplitudes and higher changes of the temperature, the MRI alloys allow higher stress amplitudes and higher numbers of cycles to failure compared to AZ91D. The cyclic deformation behavior under single step loading is characterized by cyclic softening followed by pronounced cyclic hardening over the greatest part of the lifetime.

The comparison of experimentally determined lifetimes and S-N curves calculated on the basis of the plastic strain amplitude and the change of the temperature from one load increase and two single step tests match excellently. The new lifetime calculation concept „PHYBAL" leads to an enormous saving of experimental time and costs.

5 Acknowledgments

The financial support of this work by the German Research Foundation (Deutsche Forschungsgemeinschaft) is gratefully acknowledged. Special thanks to Dr B. Bronfin, Dead Sea Magnesium Ldt., for providing the magnesium alloy specimens investigated.

6 References

[1] V. Kaese, L. Greve, S. Jüttner, M. Goede, S. Schumann, H. Friedrich, W. Holl, W. Ritter, in: Magnesium - Proc. of the 6th Int. Conf. Magnesium Alloys and Their Applications (2003) 949–954.
[2] H. Mayer, M. Papakyriacou, B. Zettl, S. Vacic: Int. J. Fatigue 27 (2005) 1076–1088.
[3] C. M. Sonsino, K. Dieterich: Int. J. Fatigue 28 (2006) 183–193.

[4] G. Eisenmeier, B. Holzwarth, H. W. Höppel, H. Mughrabi: Mat. Sci. Eng. A 319–321 (2001) 578-582.
[5] B. Ebel-Wolf, F. Walther, D. Eifler: Materialprüfung 47 (2005) 462–467.
[6] P. Starke, F. Walther, D, Eifler: Int. J. Fatigue 28 (2006) 1028–1036.
[7] B. Ebel-Wolf, F. Walther, D. Eifler: Cyclic deformation behaviour and lifetime calculation of the magnesium die-cast alloys AZ91D, MRI 153M and MRI 230D, Int. J. Mater. Research (2006), submitted.

Tensile-Compressive Creep Asymmetry of Die Cast Magnesium Alloys AM50, AE44 and AJ62A

S. Xu, M.A. Gharghouri*, and M. Sahoo

CANMET-Materials Technology Laboratory, Natural Resources Canada, 568 Booth Street, Ottawa, Ontario, Canada K1A 0G1
* Canadian Neutron Beam Centre, National Research Council Canada, Chalk River Laboratories, Chalk River, Ontario, Canada K0J 1J0

1 Introduction

In Mg alloy structural components that are exposed to elevated temperatures, the critical stress may be tensile in some regions and compressive in others (such as in bolt joints). The creep behaviour of cast Mg alloys may be significantly different under tension and compression [1–4]. Agnew et al. found that the compressive creep resistance of die cast AM60B alloy was much better than its tensile creep resistance at 150 °C [1,2]. The same trend was also reported in high pressure die cast (HPDC) AM50 at 150 °C [3]. The earlier assumption of the effect of cast microvoids [2] to explain the creep asymmetry was abandoned later [1]. Instead, it was proposed that tension would enhance and compression would suppress dynamic precipitation of β ($Mg_{17}Al_{12}$) phase, and the precipitation of this phase would produce a strain, which adds to tensile strain and reduces compressive strain [1]. It was noted that the bolt-load retention (BLR) behaviour is more consistent with the compressive creep results compared to the tensile creep results [3]. It was reported [4] that Sr, an element added to Mg alloys for improving creep resistance, has much more effect on the compression stress retention results than on tension creep results. If the dynamic precipitation of β ($Mg_{17}Al_{12}$) was at the root of tensile-compressive creep asymmetry, Mg alloys in an over-aged condition would exhibit lower creep asymmetry. It is unknown if other mechanisms contribute to the observed creep asymmetry. In a non-Al containing HPDC alloy, MEZ (Mg-1.92wt.%RE-0.33Zn-0.26Mn), tensile and compressive creep results were reported [5]; in the minimum creep rate vs stress plot, there was a transition in slope in tension which was absent in compression and this behaviour is similar to that observed in AM60B alloy [1]. For the MEZ alloy, no significant differences in creep behaviour between tension and compression were noted below the transition stress [5].

This work has been undertaken to determine the creep behaviour of die cast Mg alloys for modeling bolt-load retention behaviour, and to investigate the creep asymmetry mechanisms. This paper summarizes creep data of several die cast Mg alloys of interest to car engine cradle applications under tension and compression. Preliminary results from neutron diffraction are presented and the creep deformation asymmetry mechanisms are discussed.

2 Materials and Experimental Procedures

Dog-bone samples of AM50, AE44 and AJ62A were separately die cast. The nominal composition of AM50 Mg alloy (wt%) is 94.7 Mg, 4.9 Al and 0.4 Mn. AE44 and AJ62A are creep-resistant alloys. The specification of AE44 alloy (wt%) is 3.5–4.5 Al, 0.15–0.50 Mn, ≤0.20 Zn,

≤0.01 Cu, ≤0.005 Fe, ≤0.10 Si, ≤0.001 Ni, and 3.5–4.5 Re (bal. Mg). The nominal composition (wt%) of AJ62A is 91.7 Mg, 6 Al, 2 Sr and 0.3 Mn. Compressive creep samples were cut from gage sections with a diameter to length ratio of 1:3. The details of sample production, microstructure, mechanical properties, experimental procedures and some preliminary results are reported elsewhere [6,7]. Creep tests were carried out on the as-cast samples and on samples given an aging treatment. The aging treatment was carried out at 180 °C for 250 hours, which is a similar aging treatment used for HPDC AZ91D for an over-aged stable microstructure [8]. A few HPDC samples showed non-parallel surfaces after compressive creep tests. This was thought to be due to a shifting during loading. Surprisingly, these samples did not show a significant difference in creep strain compared with other samples tested under the same conditions (i.e. stress and temperature). To determine whether the results were affected, an improved technique was employed at Westmoreland Mechanical Testing and Research Inc. (WMTR). A Teflon specimen holder was used to ensure that the compressive creep sample was firmly seated and centered in the test fixture while being placed into the furnace and loaded. The creep results under 60 MPa at 175 °C for the AE44 samples showed the same creep asymmetry as observed in this study. Therefore, it is considered that the observed creep asymmetry is not significantly affected by sample alignment. It should be pointed out that buckling may have an effect on compressive creep samples involving large creep strains, but the buckling limits for the samples were not determined.

Neutron diffraction can be used to measure texture and residual stress. Thanks to their high penetrating power, neutrons provide crystallographic texture data averaged over the entire volume of a bulk specimen non-destructively, and with little sample preparation. Neutron diffraction is also a well established method for the measurement of residual stresses in crystalline materials; it has been applied successfully to metals, ceramics, and composites [9].

3 Results and Discussion

3.1 Creep of As-Cast Samples

Figure 1 shows creep curves for the HPDC AM50 samples at 125 °C and 150 °C. The creep strains under tension are taken as positive and the strains under compression are taken as negative. Smaller creep strains are observed under tension than under compression, i.e. better tensile creep strength than compressive creep strength. Creep tests of the low pressure die cast (LPDC) AM50 samples were also performed. The same tensile-compressive creep asymmetry trend is observed for a high creep stress of 84 MPa but for a low creep stress of 35 MPa, the creep strains under tension and compression are similar after 250 hours. Creep resistance of the LPDC AM50 is better than that of the HPDC AM50 due to differences in grain size and porosity level [6].

AE44 and AJ62A are creep resistant Mg alloys and their creep strains are much smaller than those observed in the AM50 samples. They show the same tensile-compressive creep asymmetry trend as the AM50 samples (Fig. 2). A comparison of tensile and compressive creep data of as-cast samples is given in Table 1. The creep stress to yield strength ratios are also included in Table 1. Most of the creep tests were performed under 84 MPa at 125 °C. All tests were interrupted before fracture. Creep properties are reported either in time to 1 % creep strain (h) or in strain at specified creep test durations. The creep properties reported may still correspond to the transient creep stage for some test conditions. All creep data show better tensile creep strength than compressive creep strength except for the low pressure die cast AM50 samples at a low creep stress of 35 MPa at 125 °C as described above. The compressive creep strains are 2.5 to 7 larger than tensile creep strains. The compressive creep resistance of the AM50 samples would be higher than that of tensile creep if the ratios of stress to yield were the same.

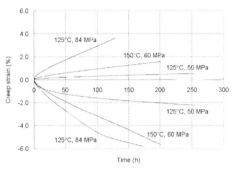
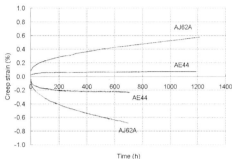

Figure 1: Creep curves of HPDC AM50 samples under tension and compression

Figure 2: Creep curves of AE44 and AJ62A under tension and compression at 125 °C and 84 MPa

The tensile-compressive creep asymmetry observed in this study for the three Mg alloys is the opposite to that previously reported for AM50 [3], AM60 and AZ91 [1,2]. The reason for this discrepancy is not known and may require an examination of details of sample and testing. The dynamic precipitation mechanism proposed by Agnew et al. [1] cannot be directly applied to the creep asymmetry observed in this study, which is opposite to the mechanism of dilation caused by dynamic precipitation.

3.2 Creep of Aged Samples

The purpose of creep tests using aged AM50 samples was to minimize the effects of dynamic precipitation on creep behaviour. Creep data of the aged AM50 samples are given in Table 2. For the aged LPDC AM50 samples tested at 84 MPa and 125 °C, the short-term compressive creep behaviour seems to be better than the tensile creep behaviour, in contrast to the results from the as-cast samples. For the aged HPDC AM50 samples at 125 °C, the short-term creep data show little creep asymmetry at 84 MPa though there was considerable scatter; of three samples tested in compression at 50 MPa, one shows better tensile than compressive creep behaviour, while the other two show the opposite trend. Therefore, the trend of tensile-compressive creep asymmetry of the aged samples is not clear for the short-term creep tests under high creep stresses, and certainly is much smaller than and different from the creep asymmetry observed in as-cast samples (the ratio differences being less than 1.7 for the LPDC AM50). These results show a difference in creep behaviour between as-cast and aged samples, and could be considered to support the theory that in as-cast samples the dynamic precipitation during creep plays an important role in the observed tensile-compressive creep asymmetry.

Table 1: Comparison of tensile and compressive creep data of as-cast samples

Alloy/ Process	Creep Loading Mode	s (MPa)	T (°C)	σ/σ_{ys}[1]	Creep Properties
					Time to 1% creep strain (h)
AM50/ LPDC	Tensile	84	125	1.40	88.4 (103.4, 73.3)[2]
	Compressive	84	125	1.69	36.1 (19.1, 53.1)
		69	125	1.40	235.7 (273.2, 198.1)
					Creep strain at 250 hours (%)
AM50/ HPDC	Tensile	35	125	0.58	0.035
	Compressive	35	125	0.71	0.037 (0.043, 0.030)
	Tensile	84	125	0.90	43.1 (85.0, 22.5, 21.8)
	Compressive	84	125	1.07	14.2 (12.9, 23.6, 6.0)
		71	125	0.90	28.9 (59.1, 17.9, 9.7)
					Creep strain at 250 hours (%)
	Tensile	50	125	0.54	0.410 (0.291, 0.529)
	Compressive	50	125	0.64	2.89 (2.21, 5.16, 1.28)
					Creep strain at 200 hours (%)
	Tensile	60	150	–	1.29 (1.55, 1.03)
	Compressive	60	150	–	5.45 (5.26, 5.63)
					Creep strain at 700 hours (%)
AE44/ HPDC	Tensile	84	125	0.75	0.070 (0.070, 0.072, 0.067)
	Compressive	84	125	0.93	0.175 (0.235, 0.137, 0.153)
					Creep strain at 400 hours (%)
	Tensile	60	175	–	0.063
	Compressive	60	175	–	0.118
					Creep strain at 300 hours (%)
AJ62A/ HPDC	Tensile	84	125	0.75	0.164 (0.010, 0.317)
	Compressive	84	125	0.93	0.443 (0.401, 0.450, 0.478)

[1] Yield strength data of LPDC AM50 were measured at 120 °C
[2] Figures in brackets are individual test values

3.3 Mechanism of the Tensile-Compressive Creep Asymmetry

The mechanisms contributing to creep asymmetry in cast Mg alloys are not yet clear and some apparently contrary results have been reported as briefly discussed above. Any texture in the Mg samples may influence deformation under tension and compression. Residual stress may also be present in rapidly solidified samples and it might play a role in the creep asymmetry. Texture and residual stress have been investigated in this study. (0002) and (1010) pole figures for the grip sections of as-cast HPDC AE 44, AJ62A and LPDC AM50 were measured and examples are shown in Fig. 3. The centre of the pole figures corresponds to the stress axis of the specimens. The texture is effectively random, with no preferred orientation. However, the pole figures are „spotty", with distinct regions of higher intensity. Such features are typical of materials exhibiting a relatively large grain size. Preliminary results from the residual stress measurements suggest that there is no variation along the gauge length. Strain measurements along the diameter are yet to be done.

Table 2: Comparison of tensile and compressive creep data of aged AM50 samples at 125 °C

Process	Creep Test	Stress (MPa)	σ/σ_{ys}	Creep Properties
LPDC				**Time to 1% creep strain (h)**
	Tensile	84	1.06	55.3, 71.9
	Compressive	84	1.22	111.7, 77.0
HPDC				**Time to 1% creep strain (h)**
	Tensile	84	1.17	203, 17.4, 15.5
	Compressive	84	1.28	10.4, 11.4, 185.0, 26.8
				Creep strain at 250 hours (%)
	Tensile	50	0.51	0.51, 0.48, 0.50
	Compressive	50	0.47	0.59, 0.51, 1.49

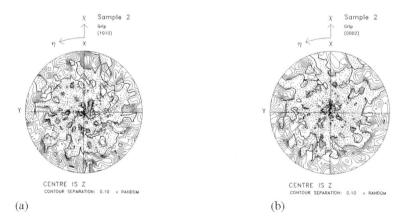

(a) (b)

Figure 3: Pole figures for the as-cast HPDC AE44 specimen

Dynamic precipitation may also play an important role. Precipitation in Mg alloys may cause dilation and, therefore, tensile loading may be expected to induce more precipitates than compressive loading. If precipitation hardening occurs [e.g. 10], then this could explain why the materials exhibit better creep properties in tension than in compression. Dynamic precipitation depends on both temperature and stress. For as-cast LPDC AM50 at 125 °C and 35 MPa, no creep asymmetry was observed, probably because dynamic precipitation was not sufficient to cause a measurable difference. Quantitative phase analysis can also be performed using neutron diffraction, and remains to be explored.

4 Summary

Better tensile than compressive creep strengths were observed for die cast Mg alloys AM50, AE44 and AJ62A, except for the low pressure die cast AM50 samples at a low creep stress of 35 MPa at 125 °C which revealed no difference in tensile and compressive creep strengths. The creep data of the aged samples showed a large scatter and the difference between tensile and compressive creep was smaller than that in as-cast samples; they also did not show the tensile-compressive creep asymmetry revealed in the as-cast sample. Neutron diffraction showed random texture in the die-cast Mg samples.

5 Acknowledgements

This work is part of the bolt-load retention work for the USAMP-SCMD project under the leadership of Mr. Dick Osborne, General Motors. This research was also partially funded the Canadian Lightweight Materials Research Initiative (CLiMRI). Mr. Mike Evans and Gibbs Die Casting Corporation are gratefully acknowledged for providing the HPDC samples. We are thankful to Mr. Doug Bruce and Westmoreland Mechanical Testing and Research Inc. for performing mechanical tests. Norsk Hydro Co. and Noranda Inc. are gratefully acknowledged for providing AE44 and AJ62A alloys. We also would like to gratefully acknowledge Mr. J.P. Thomson for producing LPDC AM50 samples, and Drs. V.Y. Gertsman and J. Li for microstructural examinations during the course of the project.

6 References

[1] Agnew, S.R.; Viswanathan, S.; Payzant, E.A.; Han, Q.; Liu, K.C.; Kenik, E.A., in *Proc. Conf. on Magnesium Alloys and their Applications*, Ing. K.U. Kainer, ed., Munich, Germany, Sept. 26-28, **2000**, pp. 687–692.
[2] Agnew, S.R.; Liu, K.C.; Kenik, E.A.; Viswanathan, S., in *Proc. Conf. Magnesium Technology 2000*, H.I. Kaplan, J. Hryn and B. Clow, eds., TMS, Warrendale, PA, **2000**, pp. 285–290.
[3] Sohn, K.Y.; Jones, J.W.; Allison, J.E., in *Proc. Conf. Magnesium Technology 2000*, H.I. Kaplan, J. Hryn and B. Clow, eds., TMS, Warrendale, PA, **2000**, pp. 271–278.
[4] Powell, B.R.; Luo, A.A.; Rezhets, V.; Bommarito, J.J.; Tiwari, B.L., SAE Technical Paper Series 2001-01-0422, **2001**.
[5] Moreno, I.P.; Nandy, T.K.; Jones, J.W.; Allison, J.E.; Pollock, T.M., Scripta Mater., Vol. 48, **2003**, pp. 1029–34.

[6] Xu, S.; Li, J.; Gertsman, V.Y.; Thomson, J.P.; Sahoo, M., in *Proc. Inter. Symp. on Light Metals and Metal Matrix Composites*, D. Gallienne and R. Ghomashchi, eds., as held at 43rd annual Conference of Metallurgists of CIM (COM 2004), the Metallurgical Society of CIM, CIM, Hamilton, ON, Aug. 22-25, **2004**, pp. 41–55.
[7] Xu, S.; Gertsman, V.Y.; Li, J.; Thomson, J.P.; Sahoo, M., Canadian Metallurgical Quartterly, Vol. 44, No. 2, **2005**, pp. 155–165.
[8] Regev, M.; Botstein, O.; Bamberger, M.; Rosen, A., Materials Science and Engineering, Vol. A302, **2001**, pp. 51–55.
[9] Hutchings, M.T.; Withers, P.J.; Holden, T.M.; Lorentzen, T., Introduction to the Characterization of Residual Stress by Neutron Diffraction, CRC Press, Taylor & Francis Group, Boca Raton, **2005**.
[10] Regev, M.; Rosen, A.; Bamberger, M., Metallurgical and Materials Transactions, Vol. 32A, **2001**, pp. 1335–1345.

Mechanical Strength and Creep Properties of Heat Treated AZ Alloys

S. Mueller, B. Camin, W. Reimers
Technical University Berlin, Institute for Materials Science and Technologies, Berlin

1 Introduction

An improvement of the mechanical properties of extruded AZ magnesium alloys can be achieved by optimizing the extrusion parameters, product speed, billet temperature and by the choice of the extrusion method [1]. The mechanical properties of the extrudates are furthermore influenced by the aluminum content of the AZ alloy [2]. The strengthening effect of increasing aluminum content can be intensified by a heat treatment after the extrusion. Therefore, a heat treatment roadmap was developed in order to improve the mechanical strength as well as the creep behavior of hot extruded AZ61 and AZ80 alloys.

2 Heat Treatment Roadmap

The annealing temperature and time are the critical parameters for the precipitation hardening. Since the annealing temperature is supposed to be close below the eutectic temperature which is 436 °C for AZ80 the annealing temperature was set to 400 °C. In order to determine the optimum annealing time extruded AZ80 samples were annealed at 400 °C for times between 8 and 68 min and afterwards quenched in water. After an annealing of 8 min. the intermetallic γ-phase $Mg_{17}Al_{12}$ which formed a network along the grain boundaries after the extrusion is starting to dissolve (Fig. 1a) and b)). After an annealing time of 68 min. the γ-phase as well as the finely disperse distributed AlMn precipitates are almost totally dissolved in the matrix (Fig. 1b)). According to [3] a complete homogenization at this annealing temperature will be reached after an annealing time of 20 h; but a 20 h annealing is not practical because of a cost increase as well as a probable grain growth. Since a moderate extension of the annealing time is reasonable to achieve better mechanical properties the optimum annealing time was set to 100 min.

 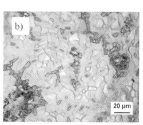 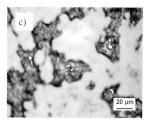

Figure 1: Microstructure of AZ80; a) after indirect extrusion, b) after 8 min annealing, c) after 68 min annealing

The precipitating in Mg alloys is carried out through diffusion. Since the diffusibility of the alloying elements in Mg alloys is below that of Al alloys cold age-hardening is not possible for

AZ61 and AZ80. The age-hardening temperature was set to 150 ± 3°C. The effect of the age-hardening time is shown in Fig. 2. An increase in the microhardness for AZ80 could already be observed after 0.4 h ageing, whereas the microhardness for AZ61 does not increase until 4 h ageing at 150 °C. There is also a significant gap between the maximum hardness that could be reached, 77 HV for AZ61 and 93 HV for AZ80.

The complete heat-treatment roadmap that was used for these studies is shown in Fig. 3.

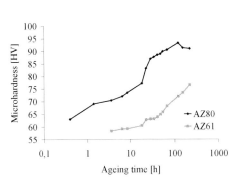
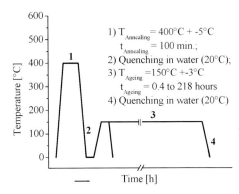

Figure 2: Age-hardening curves for AZ61 and AZ80   **Figure 3:** Heat-treatment roadmap

3   TEM Analysis of Heat Treated AZ80

The TEM analysis was performed on AZ80 samples after annealing and ageing for 8 h and 218 h. In the beginning of the age-hardening the precipitates are discoidal and probably parallel to the (0001) basal plane [4] since the E modulus in C_{44} = 16.4 MPa is the lowest [5, 6] (Fig. 4a)). With increasing ageing time the shape of the precipitates changes to a more spherical form. After 218 h age-hardening two different sized precipitates can be identified. The small precipitates have a size of 0.1 µm and are mainly Mg-Al precipitates (Fig. 4b)), whereas the larger precipitates with a size of 2 µm are Al-Mn precipitates (Fig.4c)).

4   Mechanical Properties

After the heat-treatment selected AZ80 samples were tested for their mechanical properties. The tests included tension and compression tests at room temperature as well as creep tests at 190 °C and 270 °C.

4.1   Tension and Compression Tests

Figure 5 and 6 show the results of the tension and compression test. It can be observed that for both loadings an increase of the ultimate strength after 48 h age-hardening of ~ 18 % and of the yield stress around 40% can be achieved. Whereas only a 9 % decrease of the fracture elongation under compression comes along with the increase of the yield stress and ultimate strength the decrease of the fracture elongation under tension is 40 % after 48 h and more than 70 % after

68 h age-hardening. Other than the fracture elongation under tension the extension of the age-hardening time to 68 h does not have a significant influence on the yield stress and ultimate strength.

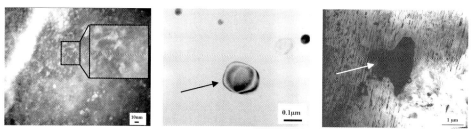

Figure 4: Precipitates after heat-treatment, AZ80; a) after 8 h, b) after 218 h; small precipitates, c) after 218 h large precipitates

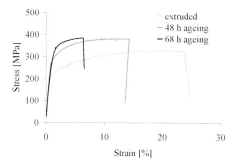
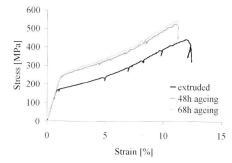

Figure 5: Stress-strain curves under tension after extrusion, 48 h and 68 h ageing

Figure 6: Stress-strain curves under compression after extrusion, 48 h and 68 h ageing

4.2 Creep Tests

Creep measurements until fracture were carried out in a novel constant tensile creep device which was designed for in-situ tomography investigations under synchrotron conditions [7]. For tomography small size creep specimens are needed. Therefore specimens with 1 mm in diameter, 26 mm in whole length were machined from hot extruded magnesium alloy AZ80, heat treated (ht) and non heat treated (nht). The samples were creep tested at 190 °C with a constant load of 70 MPa and 90 MPa as well as at 270 °C with constant loads of 25 and 40 MPa. It has to be noted that the creep data measured on the miniature specimen differs from the creep data obtained from specimen with a standard geometry. However, the data can be used for qualitative comparisons. The obtained creep curves are shown in Figure7 a) and b).

Figure 8a and 8b show the creep rates as a function of the normalized time ($t_{creep}/t_{rupture}$). Whereas the heat treated samples show similar fracture strains between 4.3 and 5.6% at all creep conditions tested here, the non heat treated sample exhibits a strain at fracture of 16.3 %.

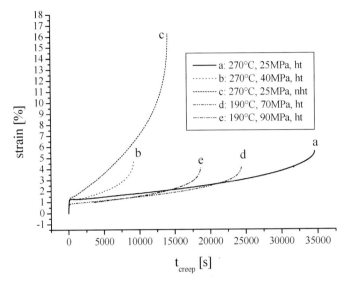

Figure 7a: Creep curves (strain as function of time)

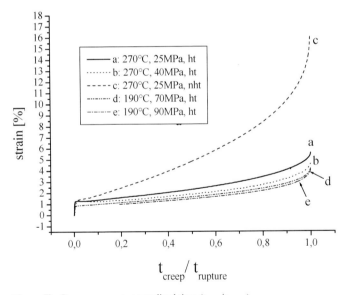

Figure 7b: Creep curves vs. normalised time ($t_{creep}/t_{rupture}$)

The primary creep stage of all samples does not extend 10 % of the lifetimes whereby the strain of 1–1.5 % is nearly the same for all samples and test conditions investigated here. The extension of the secondary as well as of the tertiary creep stage differs depending on load, temperature and material heat treatment. Again the ht samples all show significant differences to the nht sample. For the ht samples especially the extension of the secondary creep stage is increased and therefore the lifetime is enhanced (Table 1).

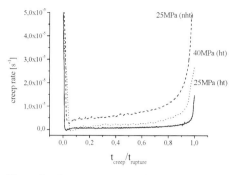 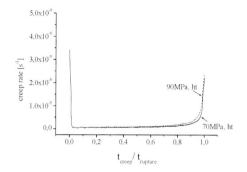

Figure 8a: Creep rate vs. normalized time (at 270 °C)   **Figure 8b:** Creep rate vs. normalized time (at 190 °C)

Table 1: Results of the creep tests

load	temp.	creep time	elongation at rupture	creep rate	extension of creep stage [% of lifetime]		
[MPa]	[°C]	[s]	[%]	[s⁻¹]	primary	secondary	tertiary
25, nht	270	14080	16.3	$5.5 \cdot 10^{-6}$	10	25	65
25, ht	270	34610	4.7	$7 \cdot 10^{-7}$	10	75	15
40, ht	270	9180	5.6	$1.5 \cdot 10^{-6}$	10	45	45
70, ht	190	24400	4.3	$5 \cdot 10^{-7}$	10	65	25
90, ht	190	18620	4.1	$7 \cdot 10^{-7}$	10	65	25

SEM observations of the fracture surfaces show intergranular (more brittle) mode of failure of the ht samples independent of the load and temperature. This suggests that grain boundary sliding is responsible for the failure. It has to be noticed that these samples yield comparatively small fracture strain values which is in agreement with [9], where it is stated that grain boundary sliding does not significantly contribute to the elongation at fracture. In contrary the nht-sample failed in mixed fracture intergranular and dominating cleavage-type transgranular mode [8] (Fig. 9).

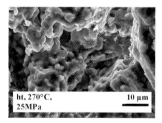 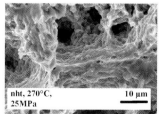 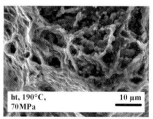

Figure 9: Fracture surfaces

This finding advises that the creep deformation is enabled by the deformation of the grain volume. This deformation mechanism delivers higher creep rates and higher creep strains than the grain boundary sliding which is in agreement with the shorter lifetime and the larger elongation of the nht samples.

5 Conclusions

By the heat-treatment of the Magnesium alloys AZ61 and AZ80 better mechanical properties under tension, compression and creep can be achieved.

Whereas the increase of ultimate strength and yield stress under tension and compression is about the same the loss of fracture elongation under tension is almost double of that under tension.

Creep experiments at 270 °C and 190 °C with heat treated and non heat treated AZ80 show nearly no differences in the primary creep stage where the strain of 1–1.5 % and a extension of ~10% of the sample lifetimes were observed. This suggests that the strain hardening mechanisms in heat treated and non heat treated AZ80 are the same.

In the secondary creep stage the creep rate of the non heat treated AZ80 is one order of magnitude higher than the ht AZ80 at 25MPa, 270 °C. The SEM fractographs show an intergranular fracture mode of the AZ80 ht samples and a combined intergranular and dominating transgranular fracture mode for the non heat treated samples.

6 References

[1] S. Mueller, K. Mueller, M. Rosumek, W. Reimers, Aluminium International Journal, 82, 4, 2006, pp. 327–330.
[2] S. Müller, K. Müller, H. Tao, W. Reimers, International Journal of Materials Research, accepted for publication 13.06.2006.
[3] Schumann, H,; Metallographie, Deutscher Verlag für Grungstoffindustrie Leipzig, 13th ed., 1991.
[4] P.G. Partridge, Metallurg. Reviews 12, 1967, p. 169.
[5] J.B. Clark, Acta Met. 16, 1968, p. 141.
[6] K.H. Matucha, Structure and Properties of nonferrous Alloys. VCH Weinheim, 1996, pp. 115–212.
[7] A. Pyzalla, B. Camin, T. Buslaps, M. di Michiel, H. Kaminski, A. Kottar, A. Pernack, W. Reimers, Science, 2005, Vol 308, pp. 92–95.
[8] Y. H. Wei et al., Materials Science and Engineering A360 (2003), pp. 107–115.
[9] http://www.vieweg.de/freebook/3-8348-0078-3_1.pdf (01.08.2006).

Creep Behavior of Squeeze Cast and T5 Heat Treated Mg4Y2Nd1Sc1Mn

F. Hnilica[a)], V. Očenášek[b)], B. Smola[c)], I. Stulíková[c)]

a) Faculty of Mechanical Engineering, Czech Technical University in Prague, Czech Republic, b) VÚK s.r.o, Panenské Bøe_any., Czech Republic, c) Faculty of Mathematics and Physics, Charles University, Prague, Czech Republic

1 Introduction

Magnesium alloys are lightweight, relatively strong materials especially promising for automotive applications. Special attention is given to the development of alloys suitable for high temperature applications. Good creep resistance and long-term stability of mechanical properties up to high temperatures are required. Such properties can be provided by microstructures containing a population of thermally stable precipitates that can hinder the motion of dislocation during long-term loading in the corresponding high-temperature range. Therefore, besides their thermal stability and resistance to coarsening, other precipitate population characteristics such as volume fraction, fine dispersion, spatial distribution, orientation relationships, and aspect ratio, are also of great importance [1, 2].

The commercial alloys WE43-T6 and WE54-T6 exhibit very good creep properties. These alloys can be used at temperature up to 250 °C. It has been demonstrated [3–5] that high creep resistance up to 350 °C can be achieved by additional alloying with Sc and Mn. The results reported in the present paper resume our previous creep studies of these alloys [2–6]. The knowledge about alloy Mg4Y2Nd1Sc1Mn is complemented by new results on microstructure and the stress relations of fracture and deformation parameters. The mechanisms of creep deformation and damage of the alloy are studied. Creep behavior of the studied alloy is compared with WE43-T6 alloy at same conditions, i.e. temperature and stress range.

2 Experimental

The alloy Mg4Y2Nd1Sc1Mn containing 3.0 wt. % Y, 1.96 wt. % Nd, 0.35 wt. % Sc, 1.40 annealing at 200 °C for 40 hours was used to produce a T5 temper [6]. Cylindrical creep test wt. % Mn was squeeze cast under a protective gas atmosphere (Ar + 1 % SF_6). Specimens of 6 mm in diameter and gauge length of 30 mm were used. The creep tests were carried out at constant load ranging from 30 to 80 MPa at 300 °C. The tests continued up to final fracture. Fractographic analysis of the fractured pieces was performed using a JEOL JSM 5410 scanning electron microscope. The deformation and damage modes were evaluated also by examinations of the surface of creep specimens. Samples cut at cross sections perpendicular to the fracture surface and parallel to test piece axis were prepared for metallographic analysis by light and scanning electron microscopy. The microstructure and creep deformation mechanisms were studied also by transmission electron microscopy (TEM) using a JEOL JEM 2000FX microscope equipped with LINK AN 1000 microanalyser.

3 Results and Discussion

Figure 1 shows the typical creep curves for loading at 300 °C and different stresses. The curve of the commercial squeeze-cast WE43-T6 alloy is added for the sake of comparison.

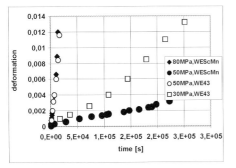
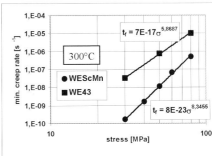

Figure 1: Typical creep curves for exposures at 300 °C and various stresses

Figure 2: Stress dependence of minimum creep rate for exposures at 300 °C

The minimum creep rate $\dot{\varepsilon}$ is plotted against the applied stress σ on a logarithmic scale in Fig. 2. The stress exponent n of the creep rate ($n = (\partial \ln \dot{\varepsilon})/(\partial \ln \sigma)$) is equal to 8.4 for Mg4Y2Nd1Sc1Mn alloy (labeled WEScMn in Figures).

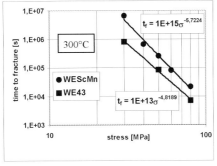
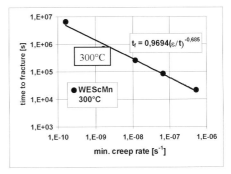

Figure 3: Stress dependence on time to fracture for exposures at 300 °C

Figure 4: Time to fracture vs. minimum creep rate at 300 °C

The results also show that a linear relations exists between the life-time t_f and the stress when plotted in log-log scale, the stress exponent $m = -5.7$ for WEScMn alloy (Fig. 3). The experimentally measured time to fracture t_f is plotted as a function of the minimum creep rate $\dot{\varepsilon}$ in Fig. 4. The relation between these two parameters indicates the validity of a generalized form of Monkman – Grant equation $\dot{\varepsilon}^{0.615} t_f = 0.97$.

The microstructure of all alloy specimens crept up to the fracture at 300 °C exhibits the same phase composition. Grain boundary (GB) eutectic consisting of the α-Mg matrix and the equilibrium β phase isomorphic with Mg_5Gd phase (fcc structure, a = 2.23 nm), formed during squeeze casting, persist long exposure to 300 °C – see Fig. 5. In the vicinity of GBs parallel to $\{10\bar{1}0\}$ all equivalent prismatic planes of the α-Mg matrix (diameter about 1.5 μm, thickness ~ 30 nm), see Figs 5 and 6.

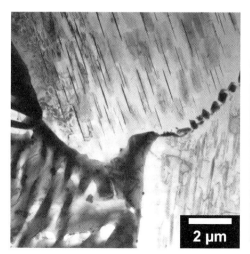 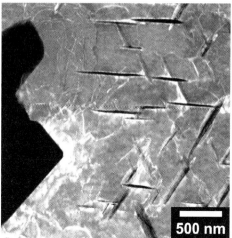

Figure 5: GB eutectic and stable β phase prismatic plates in the GB vicinity. Specimen crept to the fracture at load 60 MPa

Figure 6: Triangular arrangement of stable β phase prismatic plates and basal dislocations passing through. Specimen crept to the fracture at load 80 MPa

Tiny discs of the Mn_2Sc phase parallel to the basal planes of the α-Mg matrix (diameter ~ 15–25 nm, thickness ~ 3 nm) precipitated and persisted without a significant coarsening during creep exposure up to fracture, regardless of the load (and consequently of the life time) –see Fig. 7. Basal dislocations pass through the triangular arrangement of the prismatic plates of equilibrium phase during creep deformation – cf. Fig. 6. They leave geometrically necessary dislocations in the interfaces of the α-Mg matrix and prismatic plates – Fig. 8. In the some grains dislocations with the c component of the Burgers vector were observed.

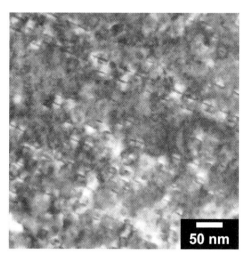 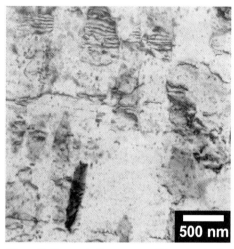

Figure 7: Fine basal discs of the Mn_2Sc phase in the specimen crept to the fracture at 60 MPa

Figure 8: Geometrically necessary dislocations in the interface of the α-Mg matrix and â phase prismatic plates accommodating plastic creep deformation. Specimen crept to the fracture at load 30 MPa

The fractographic examination indicated that the mechanisms of damage of all samples were qualitatively the same. For this reason the description of the damage mechanism will be common for all examined fracture surfaces. Fig. 9 shows a global view of the fracture surface.

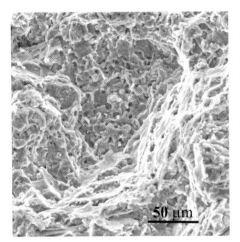 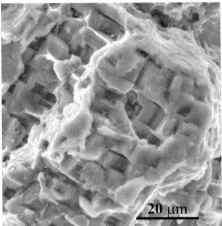

Figure 9: General appearance of the fracture surface of crept specimen

Figure 10: Example of various fracture facet morphologies formed on the boundaries of the eutectic phase

The damage propagated (progressed) through the dendrite cell boundaries by cavity initiation, growth and coalescence. Examples of cavity coalescence are in Figures 10 and 11. These figures are an example of the miscellaneous morphology of fracture facets formed due to the various microstructure of the eutectic phase on dendrite cell boundaries. The mechanism of damage initiation and propagation described above is illustrated also by the microstructure of the cross-section at proximity of the fractured surface (Fig. 12).

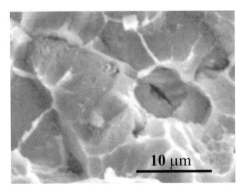 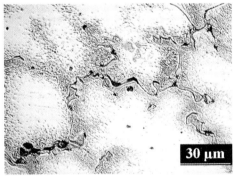

Figure 11: Example of the coalescence of creep cavities on fracture surface

Figure 12: Large creep cavities and micro-crack occurring along the eutectic phase, metallographic sample of the cross-section

4   Discussion

A comparison of creep curves of the WE43 alloy after the T6 treatment with those of the modified T5 treated alloy containing Sc and Mn – Fig. 1 shows the pronounced decrease of creep rate in the MgYNdScMn alloy. Minimum creep rates of both alloys obey power-law stress dependences at 300°C and 30–80 MPa with stress independent exponents (n = 5.9 for WE43-T6 and n = 8.3 for MgYNdScMn-T5 alloy). Minimum creep rate of the MgYNdScMn alloy is of about one order of magnitude lower at 80 MPa and 300°C and this difference is even more pronounced at lower stresses due to the higher stress exponent – Fig. 2. Power-laws were experimentally confirmed also for time to fracture-stress dependences with stress exponents m not depending on stress (m = –4.8 in WE43-T6 alloy and m = –5.7 in MgYNdScMn-T5 alloy). Time to fracture in the investigated stress range is almost one order of magnitude larger in the modified WE alloy with Sc and Mn than in the WE43 alloy (Fig. 3).

The value of the stress exponent in the minimum creep rate vs. stress dependence for the modified WE alloy (n = 8.3) indicates the dislocation mechanism of creep. The direct dislocation structure observation by TEM clearly supports this mechanism.

Higher creep resistance of MgYNdScMn alloy than that of WE43 alloy is caused by the presence of high density of tiny basal discs of the Mn_2Sc phase in the former alloy. The initial precipitation structures of the alloys differ:

- triangular arrangement of the prismatic plates of the transient $β_1$ phase (fcc, a = 0.74 nm) in WE43 T6 treated alloy [7];
- triangular arrangement of relatively coarse prismatic plates of equilibrium β phase and Mn2Sc phase in the form of fine basal discs in the MgYNdScMn alloy [6].

In the course of the heating on 300°C and during creep exposure $β_1$ phase plates transform to the β phase ones in the WE43 alloy and the only structural difference remaining is the persistence of Mn_2Sc basal discs in the MgYNdScMn alloy (cf. Fig 7). These basal discs can be very effective in inhibiting cross slip of basal dislocations and are strong obstacles to the non-basal slip [2]. Substantially lower Sc content in the alloy investigated here than in the alloy studied previously [6] (0.35 wt. % vs. 1.28 wt. %) does not significantly influence volume fraction of Mn_2Sc basal discs, as this depends mainly on the Mn content.

The triangular arrangement of the β phase prismatic plates is effective obstacle to the slip of basal dislocations as proved by the geometrically necessary dislocations left in the interface of α-Mg matrix and β phase prismatic plates (cf. Fig 8). The observation of the dislocations with c-component of the Burgers vector indicates activity of non-basal slip systems in this alloy at 300°C. Non basal slip also enhances the life to fracture in creep tests at high temperatures.

The more frequent occurrence of GB eutectic in T5 treated MgYNdScMn alloy, together with different slip orientations in neighboring dendritic cells causes predominant cavity initiation on β phase particles, their growth and coalescence. As a result fracture proceeds by intercellular cracking.

5 Conclusions

1. The creep tests results show a linear relations between the minimum creep rate and life time vs. stress in log – log scale with stress exponent n = 8.4 and m = - 5.7, respectively. The generalized Monkman – Grant relation, $\dot{\varepsilon}^{0.615} t_\mathrm{f} = 0.97$, is valid.
2. The value of the exponent n = 8.4 and the results of TEM examinations of the deformed microstructure give evidence of the dominant role of the dislocation creep in the control of creep deformation.
3. The excellent high temperature creep resistance of the WE type alloy modified by the small additions of Sc and Mn is due to a dense dispersion of fine basal discs of the Mn_2Sc phase, which is very effective inhibitor of cross slip of basal dislocations and non-basal slip.
4. Creep failure occurs by nucleation, growth and coalescence of cavities situated predominantly on the boundaries of dendrite cells decorated by GB eutectic.

6 Acknowledgement

All authors gratefully acknowledge financial support by Czech Science Foundation (project 106/06/0252), I. Stulíková and B. Smola also support of Ministry of Education of Czech Republic (the research program MSM 0021620834).

7 Reference

[1] Stulíková, I, Smola, B., Pelcová, J., Mordike, B. L., in: Proc. of 6 th Int. Conf. Magnesium Alloys and Their Applications (Ed.: K.U.Kainer), 2003, Wiley-VCH Verlag GmbH&Co.KGaA, Weinheim
[2] Nie, J. F., Scripta Mater. 2003, 48, 1009–1015
[3] Smola, B., Stulíkové, I, Pelcová, J., Mordike, B. L., Y. Metallkde., 2003, 94, 553–558
[4] Stulíkové, I, Smola, B., von Buch, F., Mordike, B. L., Mat.-wiss. U. Werkstofftech. 34, 102-108, 2003
[5] Smola, B., Stulíkové, I, Pelcová, J., von Buch, F., Mordike, B. L., Y. phys. stat. sol. (a) 191, No. 1, 305–316, 2002
[6] Stulíková, I., Smola, B., Pelcová, J., Mordike, B.L., Z. Metallkd. 2005, *96*, 821
[7] Antion, C., Donnadieu, P., Perrard, F., Deschamps, A., Tassin, C., Pisch, A., Acta Mater. 2003, *51*, 5335

Influence of Specimen Orientation on Creep of Mg–10 Vol. % Ti PM - Composite

F. Dobeš[1], P. Perez[2], K. Milika[1], G. Garces[2], P. Adeva[2]
[1]) Institute of Physics of Materials, Brno
[2]) National Center of Metallurgical Investigations, Madrid

1 Introduction

Microstructure of many materials – either intentionally or owing to production history – is not isotropic. Consequently, mechanical properties are not isotropic, too. A detailed knowledge of dependence of these properties on orientation within material may be important for an exact design of construction parts. An investigation of orientation dependence may also contribute to an identification of mechanisms that control the respective property. As far as it is known to the present authors, published investigations of this topic are relatively scarce: e.g., the influence of fiber orientation on creep properties in metal matrix composites was studied by Yawny et al. [1] and by the present authors [2]. In our contribution, we would like to give the results of measurement of creep properties of MgTi alloy prepared by powder metallurgical processing. We will try to interpret the results in terms of strengthening by needle-like particles of secondary phase.

2 Experimental

A magnesium matrix composite reinforced with 10 vol.% titanium particles was prepared through a powder metallurgical route. The magnesium and titanium particle sizes were less than 45 microns and 25 microns, respectively. The powders were mixed for 3 h at 100 rpm in a planetary mill. Then, the mixture was cold-pressed at 310 MPa. The compacts were hot-extruded into rods at 673 K using an extrusion ratio of 18:1. Results of the subsequent characterization of the composite material by optical microscopy, scanning electron microscopy, X-ray diffraction and tensile tests are given elsewhere [3, 4].

Cylindrical specimens of diameter 5 mm and height 9 mm were machined from the extruded bars. The specimens were cut in such a way that their longitudinal axis (i.e. the direction of compressive creep stress) and the axis of extruded bar contained a predestined angle θ from 0° to 90° in steps of 15°. Constant load compressive creep tests of the alloy were performed at temperatures 573 and 623 K, respectively. A stepwise loading was used: in each step, the load was changed to a new value after stationary creep rate had been established. The terminal values of the true stress and the true strain rate were evaluated for the respective step. Protective atmosphere of dried and purified argon was used. During the test, temperature was kept constant within ±1K. Creep curves were PC recorded by means of special software. The sensitivity of elongation measurements was better than 10^5.

3 Results

Examples of measured dependences of the creep rate $\dot{\varepsilon}$ on the applied stress σ for three different orientations of specimens are given in Fig. 1. The dependences can be described by the power function in the form

$$\dot{\varepsilon} = A\sigma^n, \qquad (1)$$

where A is a temperature dependent factor and n is exponent. The values of exponent n evaluated by means of the least square method are summarized in Fig. 2. Relatively high values of n are typical for creep in metallic materials strengthened by dispersion of secondary phase. A clear tendency to the decrease of n with the increasing disorientation angle between extrusion axis and specimen axis can be noted.

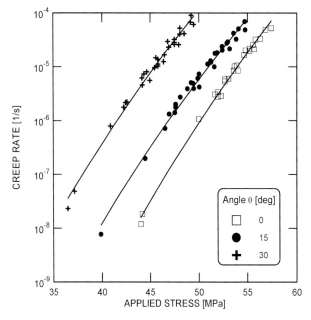

Figure 1: Dependence of creep rate on applied stress for three different orientations at 573 K

The equation (1) was used also for an evaluation of the influence of orientation on creep rate. The creep rates corresponding to the applied stress 40 MPa were calculated by means of optimized values of A and n for all orientations and both testing temperatures. The results are given in Fig. 3. The dependence of the creep rate is very sensitive to the orientation especially at low inclinations from extrusion axis. The greatest creep resistance is observed in specimens with stress axis parallel to the extrusion axis.

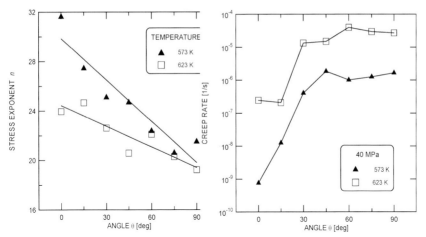

Figure 2: Dependence of stress exponent n on disorientation angle

Figure 3: Dependence of creep rate on disorientation angle

4 Discussion

Microscopic observations revealed three distinct anisotropic features of the experimental alloy [3, 4]: (i) elongated grains, (ii) crystallographic texture and (iii) elongated titanium particles.

Ad (i) Grains are elongated in the extrusion direction, with an aspect ratio of about 2. It is generally accepted, that the grain size and shape influences the rate of diffusional creep but not the rate of dislocation creep [5]. The diffusional creep can be excluded as possible rate-controlling mechanism due to the observed high values of stress exponent.

Ad (ii) The alloy exhibited a fiber texture with the basal plane parallel to the extrusion direction. For such a type of texture, the deformation resistance in the extrusion direction should be the lowest. This is in contrast with the obtained experimental results.

Ad (iii) Titanium particles are very often highly deformed; they are elongated in the extrusion direction to such an extent that they can be considered as long fibers. Their existence seems to be the most plausible reason for an explanation of the observed creep behavior.

Theory published by Bao et al. [6, 7] suggests to include the effect of fibres directly into the applied stress function as a strengthening coefficient λ

$$\dot{\varepsilon} = A_{CC} \left(\frac{\sigma}{\lambda} \right)^n, \tag{2}$$

where A_{CC} is a temperature dependent factor. An estimate of λ is

$$\lambda = 1 + 2(2 + S) V_f^{3/2}, \tag{3}$$

where S is the aspect ratio (the ratio of length to diameter of fibers) and V_f the volume fraction of fibers, respectively.

The mechanical behavior of discontinuous fiber composites can alternatively be described by means of the shear-lag models [8, 9]. One version of these models leads to eqn. (2) with the strengthening coefficient [9]

$$\lambda = \left(\frac{2}{3}\right)^{\frac{1}{n}}\left(\frac{n}{2n+1}\right)\left[\left(\frac{2\sqrt{3}V_f}{\pi}\right)^{-\frac{1}{2}}-1\right]^{-\frac{1}{n}} V_f(S)^{\frac{n+1}{n}}+1-V_f. \qquad (4)$$

Both eqns. 3 and 4 are valid for fibres parallel with the stress direction. For other orientations, the aspect ratio S should be substituted by its effective value S_{eff}. Hong and Chung [10] assumed that the load transfer efficiency of fibers with aspect ratio S and the disorientation angle θ is dependent on

$$S_{\text{eff}} = S\cos\theta . \qquad (5)$$

Few years later, Ryu and Hong [11] suggested a refined equation

$$S_{\text{eff}} = S\cos^2\theta + \left(\frac{3\pi-4}{3\pi}\right)\left(1+\frac{1}{S}\right)\sin^2\theta . \qquad (6)$$

We can try to compare – at least qualitatively – the present experimental data with the above theories. The calculations were done with the stress exponent n found by linear regression of experimental values (cf. Fig. 2). The volume fraction of particles $V_f = 0.1$ was used. Examples of results of calculations are given in Fig. 4 in terms of reciprocal value λ raised to power n (this is proportional to creep rate at given applied stress). For these calculations the stress exponent n for temperature 573 K and the particle aspect ratio $S = 5$ was applied. It can be seen that both eqns. (3) and (4) can explain the increase of the creep rate with the increasing disorientation angle of particles very well. Note that the refined expression for the effective aspect ratio (eq. 6) is in substantially better agreement with the experimental data for great disorientation angles.

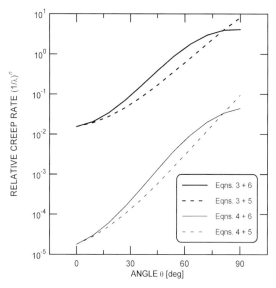

Figure 4: Dependence of calculated relative creep rate on disorientation angle for theories of Bao et al. [6, 7], Kelly et al. [9] and Hong et al. [10, 11]

5 Acknowledgements

The paper was prepared within the joint research program of the Spanish National Research Council CSIC and the Academy of Sciences of the Czech Republic. The partial financial support of the Grant Agency of the Czech Republic within the project 106/06/1354 is gratefully acknowledged.

6 References

[1] Yawny, A.; Kausträter, G.; Skrotzki, B.; Eggeler, G., Scripta Mater. **2002**, *46*, 837–842.
[2] Milika, K.; Dobeš, F., Journal of Alloys and Compounds **2004**, *378*, 167–171.
[3] Perez, P.; Garces, G.; Adeva, P., Composites Science and Technology **2004**, *64*, 145–151.
    Perez, P.; Garces, G.; Adeva, P., to be published.
[4] Poirier, J.P., Plasticité à Haute Température des Solides Cristallins, Editions Eyrolles, Paris, **1976**.
[5] Bao, G.; Hutchinson, J.W; McMeeking, R.M., Acta Metall. Mater. **1991**, *39*, 1871–1882.
[6] Rösler, J.; Bao, G.; Evans, A.G., Acta Metall. Mater. **1991**, *39*, 2733–2738.
[7] Mileiko, S.T., J. Mater. Sci. **1970**, *5*, 254
[8] Kelly, A.; Street, K.N., Proc. Roy. Soc. **1972**, *A328*, 267.
[9] Hong, S.H.; Chung, K.H., Key Engineering Materials **1995**, *104–107*, 757–764.
[10] Ryu, H.J.; Hong, S.H., in *Creep Behavior of Advanced Materials for 21st Century*. (Eds.: Mishra R.S., Mukherjee, A.K. and Murty, K.L.), TMS, Warrendale, Pa., USA, **1999**, 159-170.

Effect of Finely Dispersed Particles on Mechanical Properties of Magnesium Alloys

Mizuno, S., Tamura, Y., Tamehiro, H., Funami, K. and Takaya, M.
Chiba Institute of Technology, Japan

1   Introduction

Not only magnesium (Mg) is the most lightweight (specific gravity 1.74) metal among industrial metal materials but also it has high specific strength. In addition, magnesium attracts much attention as a circulation type super lightweight materials in the next generation because it possesses good recyclability peculiar to metal materials and magnesium resources occur abundantly in the sea water, etc. As a result, applications of magnesium alloys are rapidly increasing in the field of automobile and portable electronic devices, and furthermore the extensive use as various structural materials is very much being expected. It is, however, difficult to achieve remarkable improvement in mechanical properties or to give a new function for magnesium alloys with a conventional melting and casting method.

In this study, in order to overcome the above-mentioned difficulties in magnesium alloys, the effect of grain-refinement and fine dispersion of various chemical compounds on the mechanical properties (hardness) and microstructure are examined by using powder metallurgy (mechanical alloying (MA) and spark plasma sintering (SPS) method) [1].

2   Experimental Procedure

Pure magnesium powder (purity: 99wt%, size: less than 150m) and magnesium boride (MgB_2) powder were used as raw materials. Their powder mixture of total 8g was charged in the container of agate (capacity: 80ml), together with 20 agate balls (diameter: 10mm) and the inside of the container was substituted with purified argon gas after deaeration. The amount of MgB_2 was made to change from 0 to 3.0, 5.0 and 10.0wt%. Mechanical alloying (milling) was carried out at the table turn 200rpm for 200h by using planetary ball mill (Fritsch GmBH). After mechanical alloying, the powder was sintered into a disk of 20mm diameter x approximately 5mm high in a graphite die by using spark plasma sintering (SPS) system (SPS-1030, Sumitomo Coal Mining Co., Ltd). The sintering was performed at the heating temperature 673K for 15min with the pressure force of 15.4kN.

In a SPS specimen, microstructure was observed after etching with 3% acetic acid aqueous solution and Vickers hardness (load: 98N) was measured. Distribution of elements was analyzed by EPMA and CMA (computer aided micro-analyzer). In addition, a dynamic coefficient of friction was measured by a ball on disk type tribometer (FPR-2000, Rhesca Co., Ltd).

3 Experimental Results

No difference can be recognized in mechanically alloyed powders with different MgB_2 content by visual observation. Pure magnesium powder and MgB_2 added powders were consolidated by the SPS condition of 673K x 15min to the extent that there was no defect microscopically, and all the SPS specimens show metallic lustrous surface after polishing similar to a melting specimen.

Figure.1 shows the relationship between MgB_2 content and hardness in SPS specimens. As the MgB_2 content increases, the hardness linearly increases. The Mg-10wt%MgB_2 specimen became higher in hardness about 30 HV than that of pure magnesium specimen. The optical microstructure tends to become finer with increasing MgB_2 content, as shown in Figure.2.

Figure.3 shows the relationship between MgB_2 content and dynamic coefficient of friction in SPS specimens. As the MgB_2 content increases, the coefficient of friction linearly decreases. The coefficient of friction remarkably decreases from 0.5 of pure magnesium specimen to 0.2 of Mg-10wt%MgB_2 specimen.

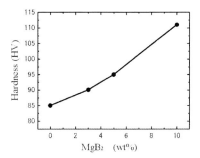

Figure 1: Relationship between MgB_2 content and hardness in SPS specimens

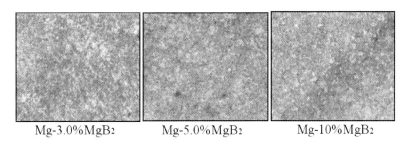

Figure 2: Optical microstructure of SPS specimens

4 Conclusions

The following results were obtained by this experiment.

1. The mechanically alloyed powder of pure magnesium and MgB_2 added magnesium was consolidated by the SPS method to the extent that there was no defect microscopically.

2. The grain-refinement and fine dispersion of MgB_2 compounds in the magnesium matrix increase hardness and simultaneously reduce the dynamic coefficient of friction.

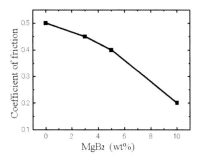

Figure 3: Relationship between MgB_2 content and dynamic coefficient of friction in SPS specimens

5 References

[1] Tamura,Y., Tamehiro,H., Funami, K. and Takaya, M.: Autumn Symposium, 2005, No.103, P.207 (Institute of Japan Light Metal).

Corrosion and Surface Treatment

Advances in the Active Environment Impact on the Viscoelastic Behavior of Magnesium Alloys

Ya. B. Unigovski, E. M. Gutman
Ben-Gurion University of the Negev, Beer-Sheva
Z. Koren
Israeli Institute of Metals, Technion, Haifa

1 Introduction

The study of viscoelastic behavior of metals under a constant load (creep) or constant strain (strain relaxation) assisted by anodic dissolution is very important for increasing the lifetime of metals, since in real applications mechanical and corrosion processes occur simultaneously, accelerating each other in a synergistic manner. For instance, it is especially important for Mg alloys due to their low creep and corrosion resistance, which may lead to a high sensitivity to stress corrosion. The phenomenon called earlier [1] *corrosion creep* (CC) was investigated in certain die-cast Mg alloys in 0.1N $Na_2B_4O_7$ buffer solution (pH = 9.3). In contrast to the data in the air demonstrating only the first stage of the creep process at room temperature, in corrosive solutions, secondary and tertiary creep due to the plasticization effect of the solution was observed. In corrosive solutions, the creep life and elongation-to-fracture of Mg alloys decreases with increasing aluminum content in alloys [1, 2]. However, the mechanisms of additional plasticity of metals in a corrosive solution under static loading (plasticization) in comparison to that in air are not clear.

Different mechanisms of the creep behavior of pure copper and Cu-based alloys in corrosive solutions were proposed [3–5], e.g., the generation of divacancies by the dissolution process during the anodic polarization of pure Cu, and of a corresponding reduction in the surface energy by the cathodic polarization [3]; growth of the creep rate of brass in 3.5% NaCl with the increase in the anodic current density as a result of dislocation climb induced by vacancy supersaturation during the anodic polarization [4]. However, Van Der Wekken [5] concludes that the accelerated creep rates of copper single crystal specimens during the anodic dissolution in acetate buffer solutions under a constant load are essentially due to an increasing stress. Under the conditions of a constant stress achieved on the specimen, the effect largely disappears in single crystals, as well as in polycrystalline wire specimens.

There are only a few studies aimed at the study of stress relaxation behavior of Mg in corrosive solutions [6, 7]. For example, constant strain testing of a Mg-Al alloy in solutions containing various concentrations of sodium chloride and potassium chromate shows that a more extensive pitting associated with more aggressive solutions means that the initiation of stress corrosion cracks becomes more difficult [6]. Earlier we have shown that Mg and die-cast Mg alloys have a relatively poor stress relaxation resistance in corrosive solutions, especially at elevated temperatures [7].

Thus, the literature data on the environment-assisted creep and stress relaxation are very contradictory. In the present work the effect of corrosive medium on the corrosion resistance, creep and stress relaxation behavior, pitting and cracking of pure magnesium and Mg alloys has been investigated.

2 Experimental

Round samples (5.9-6.0 mm in diameter, gauge length of 32 and 75 mm) of conventional die-cast Mg alloys AZ91D (8.4% Al, 0.85% Zn, 0.17% Mn, Mg-balance); AM50 (Mg - 5.1% Al, 0.15% Zn, 0.57% Mn) and AS21 (Mg - 2.3% Al, 0.23% Mn, 1.10% Si) were produced from ingots (Dead Sea Magnesium, Israel) on a cold-chamber machine with the locking force of 2,000 kN (IIM – Israel Institute of Metals, Technion). The rheological properties of AZ91 were obtained on a 400 kN cold-chamber machine of IIM by means of a twin-axes stirrer rotating in a spinning crucible located at the center of an electrically heated furnace [8]. A part of samples of AZ91D and AM50 alloys were coated by non-chromate anodic coating ALGAN-2 (Algat Co., Israel). Typical mechanical properties of the studied materials – ultimate tensile stress (*UTS*), tensile yield stress (*TYS*) and elongation-to-fracture *e* are given in Table 1.

Table 1: Mechanical properties of materials

Material	Conditions	UTS, MPa	TYS, MPa	e, %
Pure Mg	Machined	62	37	5.7
Conventional AZ91	As-cast	225	170	2.8
Rheoformed AZ91	Machined	220	150	2.0
Conventional AM50	As-cast	229	136	11.7
Conventional AS21	As-cast	221	134	9.7

Specimens were studied after rinsing in distilled water and alcohol and wiping with acetone. As active aqueous corrosive environments, $0.1N$ $Na_2B_4O_7$ buffer solution saturated with magnesium hydroxide $Mg(OH)_2$ (pH = 9.3); the same solution with the addition of 20 % KCl; 3 % Na_2SO_4 and NaCl-based solutions were used. All solutions were prepared from analytical grade chemicals and distilled water. The cell for CC tests was made from a transparent plastic (PMMA) or lightly powdered natural rubber latex with a hermetical sealing of the cell during the tests by rubber bands [1, 2].

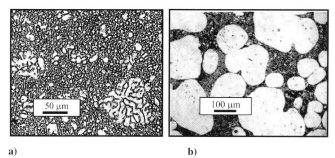

a) b)
Figure 1: Optical micrographs of conventional (a) and rheoformed (b) AZ91 alloy

Creep tests were carried out on Model 3 SATEC creep tester (Satec, Inc., USA) at 25 ± 2 °C in air and in solutions at different stress values. Stress relaxation tests with the duration of 50 and 100 hours were performed on the universal testing machine Zwick-1445 (Zwick GmbH & Co, Germany) in the tension mode at the temperatures of 25 ± 2 °C and 70 ± 1 °C and a constant strain during the entire test. The initial stress σ_o in stress relaxation tests varied from 0.67 to 1.15 of tensile yield strength of the alloys. Regular AZ91D die-castings produced at 630 °C have a typical dendritic microstructure showing Mg(α)-phase grains with a typical diameter varying from 5 to 20 microns (Fig. 1a). However, rheoformed alloy produced at 580 °C shows large spheroid grains of the primary Mg(α)-phase with a typical diameter of 50–250 micron surrounded by a fine structure of Mg(α)+Mg$_{17}$Al$_{12}$(β) phases (Fig. 1b).

3 Results and Discussion

3.1 Corrosion Creep Tests

In air at stresses below TYS, both pure Mg and die-cast AZ91D alloy show only the primary stage typical of low-temperature creep (Fig. 2), which occurs at temperatures below about a half of the melting point, while high-temperature creep is predominantly a steady-state creep. However, in corrosive solutions, especially containing Na$_2$B$_4$O$_7$, fracture (marked in Fig. 2 by cross points) occurred even under stresses less than 70 % of TYS (Fig. 2, Table 2). In typical corrosion creep tests carried out on pure Mg at room temperature under the stress of 28 MPa, the lifetime of Mg varied from 12 hours in 3.5% NaCl to ~ 250 hours in the buffer solution (Fig. 2a).

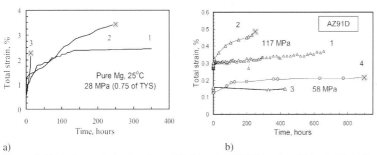

Figure 2: Typical creep behavior of pure Mg (a) and AZ91D Mg alloy (b) in air (1), buffer solution (2), 3.5% NaCl (3) and in buffer solution + 20%KCl (4, 5). Cross points show the fracture; (1-3, 5) – conventional and (4) rheoformed AZ91; 1-3 – as-cast and 4, 5 – machined samples

Creep tests of conventional and rheoformed alloys both in air and in solutions (0.1N Na$_2$B$_4$O$_7$ + 20% KCl and 3% Na$_2$SO$_4$) under the stress of 58 MPa, showed a minor strain during long-term tests (Fig. 2b). In sodium sulfate solution, stressed Mg alloys experienced fast stress corrosion cracking. Therefore, in this solution, time-to-rupture of both types of alloys is less significant than in borate-based solutions (Table 2).

Anodic coating delays metal dissolution facilitating plastic deformation of the stressed metal, and for this reason alone increases the corrosion creep life of metals during long-term creep tests in corrosive media (Fig. 3). For example, in the buffer solution under the stress of 0.89 of tensile yield strength, time-to-fracture of anodized alloys in comparison with uncoated specimens increases from 55 to 160 hours and from 340 to 637 hours for AZ91D and AM50 alloys, respectively (Fig. 3). An increase in the plasticity of coated AM50 in air as compared to a conventional alloy (see Fig. 3b) is connected, probably, with a decrease in the alloy surface energy during the anodizing process in basic electrolytes. A removal of a part of the dense die-cast surface layer for the depth of 10-20 microns can also result in a certain increase in the creep strain.

Table 2: The environment effect on time-to-rupture of Mg alloys in the corrosion creep tests

	Stress, Mpa / % of TYS	Solution	Surface conditions	Time-to-rupture, h
Rheoformed AZ91	58/75	3% Na_2SO_4	Machined	5
Rheoformed AZ91	58/75	$0.1N\ Na_2B_4O_7$ + 20% KCl	Machined	898
Conventional AZ91D	58/34	$0.1N\ Na_2B_4O_7$ + 20% KCl	Machined	431*[)]
Conventional AZ91D	117/69	3% Na_2SO_4	Machined	60
Conventional AZ91D	117/69	$0.1N\ Na_2B_4O_7$	As-cast	240

*[)] Fracture was occurred at an increase in the stress up to 140 Mpa

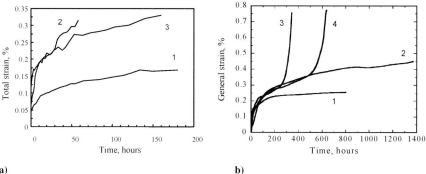

Figure 3: Typical creep behavior of die-cast and anodized AZ91D (a) and AM50 (b) alloys in air and in the buffer solution in as-cast (2) and anodized (3) states. (a): 1 – air, 2, 3 – die-cast and anodized AZ91D, respectively, in the solution, 25 °C, 159 MPa; (b): 1, 2 – die-cast and anodized AM50, respectively, in the air, 3, 4 – die-cast and anodized AM50, respectively, in the solution, 25 °C, 121 MPa;

One can observe microstructure, pits and cracks (filled with epoxy), e.g., in anodized AZ91D after corrosion creep tests in buffer solution, in Fig. 4. A crack originates from a pit, extends beyond the coating and propagates into the surface layer of metal. Creep-rupture of AZ91D and AM50 in the buffer solution originates, more likely, in an intercrystalline manner.

3.2   Corrosion Stress Relaxation

Earlier it was reported that a significant corrosion stress relaxation in Mg alloys was observed only at elevated temperatures [7]. The sensitivity to corrosive solutions may be characterized by the difference $\Delta\sigma_{rem}$ between the remaining stresses corresponding to that in air and in a corrosive solution. The highest $\Delta\sigma_{rem}$ value was observed in AS21 alloy, and the lowest one – in AM50 alloy. For instance, at an initial stress of 0.8 of TYS $\Delta\sigma_{rem}$ in air in comparison to 3.5% NaCl was equal to 20; 9.7 and 2.9% for AS21, AZ91D and AM50 alloys, respectively (Table 2, Fig. 5).

In air, a higher relaxation resistance of AS21 alloy (a higher remaining stress) in comparison with AM50 alloy is connected with the presence of needle-shaped intermetallics Mg_2Si improving viscoelastic properties. On the contrary, in an aggressive environment (3.5% NaCl) under stress, needle-shaped silicides significantly decrease relaxation resistance of this alloy as compared with AM50 (Table 2). Such intermetallics promote strain hardening and, thus, increase chemical potential of metal atoms and their mechanochemical dissolution [9].

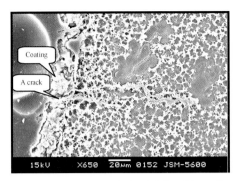

Figure 4: Microstructure in the vicinity of the surface and cracks in coated AZ91D after a CC-test in the buffer solution (159 MPa)

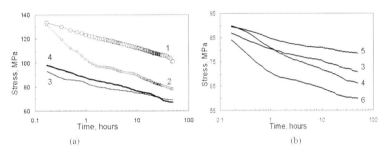

Figure 5: Stress relaxation in AZ91D (1, 2), AM50 (3, 4) and AS21 (5, 6) alloys in air (1, 3, 5) and in 3.5% NaCl (2, 4, 6). σ_0/TYS: 1.0 (a) and 0.8 (b); 70 °C

Table 2: Effect of the environment on the remaining stresses in Mg alloys after 50 hours, 70 °C

	σ_0/TYS	Remaining stress, %		$\Delta\sigma_{rem}$, %
AS21	0.8	86.4	66.4	20.0
AZ91D	0.8	74.1	64.4	9.7
AM50	0.8	78.4	75.5	2.9

SEM micrographs presented in Fig. 6 show the microstructure and cracking of AZ91D alloy after corrosion relaxation tests in 0.1N Na$_2$B$_4$O$_7$ solution saturated with Mg(OH)$_2$. In spite of the presence of relatively long cracks both on the cast surface and in the center (Fig. 6b) of a specimen, the latter was not broken after 50 hours of testing in the buffer solution. In contrast to the above mentioned intercrystalline cracking in a CC-test, the crack presented in Fig. 6b shows that cracking of AZ91D alloy originates in a transcrystalline manner. One can observe β-phase (white) on the grain boundaries of α-Mg in AZ91D, and, in addition, a certain fragmentation of grains after stress relaxation tests (Fig. 6a).

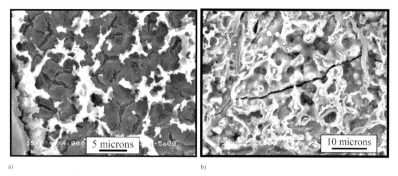

Figure 6: SEM microstructure in the vicinity of cast surface (a) and transcrystalline crack in the center (b) in AZ91D alloy after 50 hours in the buffer solution (25 °C, $_o$ = 195 MPa)

4   Conclusions

1. In air, only the primary stage typical of low-temperature creep of the studied metals was observed. However, in corrosive solutions, marked elongation and fracture occurred even under stresses less than 70% of *TYS*. It was found that very small pits were nucleated during CC-tests at the primary stage of creep.
2. It is shown that the stress relaxation method manifests a high sensitivity to structural homogeneity of alloys, especially in corrosive solutions. At elevated temperature, all studied die-cast Mg alloys shows a strong stress relaxation in a 3.5 % NaCl corrosive solution. The highest sensitivity to a corrosive environment under a constant strain was observed in AS21 and AZ91D alloys, and the lowest one – in AM50 alloy.

5   References

[3]   E. M. Gutman, Ya. Unigovski, A. Eliezer, and E. Abramov, J. Mat. Synthesis and Processing, 2000, 8, Nos.3/4, 133-138.
[4]   Ya. Unigovski, Z. Keren, A. Eliezer and E.M. Gutman, Mater. Sci. Eng. 2005, A398, 188-197
[5]   R. W. Revie and H. H. Uhlig, Acta Metallurgica, 1974, 22, Issue 5, 619-627.
[6]   B. Gu, W. Y. Chu, W. Chu, L.J. Qiao and C. M. Hsiao, Corr. Science, 1994, 36, Issue 8, 1437-1445.
[7]   C. J. van der Wekken, Acta Metallurgica, 1977, 25, Issue 10, 1201-1207.
[8]   K. Ebtehaj, D. Hardie and R. N. Parkins, Corrosion Science, 1988, 28, Issue 8, 811-821.
[9]   Ya. Unigovski, L. Riber, A. Eliezer, E.M. Gutman, in Proceedings Int. Conference Magnesium Alloys and Their Applications, 19-23 November, 2003, Wolsburg, Germany, WILEY-VCH, **2003**
[10] Z. Koren, H. Rosenson, E. M. Gutman, Ya. B. Unigovski, A. Eliezer, J. of Light Metals, 2002, 2, 81-87.
[11] E. M. Gutman, Mechnochemistry of Solid Surfaces, World Scientific, Singapore - New Jersey – London, 1994, 322.

Evaluation of Mg SCC Using LIST and SSRT

N. Winzer[1], G. Song[1], A. Atrens[1], W. Dietzel[2], C. Blawert[2] and K.-U. Kainer[2]
[1]Materials Engineering, The University of Queensland, Brisbane, Australia
[2]GKSS-Forschungszentrum Geesthacht GmbH, Germany

1   Introduction

Our recent critical review [1] described two commonly used methods for characterizing SCC: the Linearly Increasing Stress Test (LIST) and the Slow Strain Rate Test (SSRT). LIST apparatus are passive systems that steadily increase the nominal load. SSRT apparatus incorporate an open-loop control system to increase the actual load by increasing the displacement. Neither method accurately simulates operating conditions; however, they allow more rapid characterization of SCC susceptibility compared to constant load tests and permit direct measurement of critical parameters such as threshold stress (σ_{SCC}) and crack velocity (V_C) when coupled with equipment to measure crack extension.

SCC of Mg alloys is inherently transgranular yet the predominant mechanism is still highly equivocal. Hitherto proposed mechanisms include H Enhanced Decohesion (HEDE), H Enhanced Localized Plasticity (HELP), Adsorption Induced Dislocation Emission (AIDE) and Delayed Hydride Cracking (DHC). These mechanisms have been summarized by Lynch [2]. DHC is the most commonly proposed [3–5]. It involves repeated stages of stress-assisted diffusion of H toward the crack tip leading to hydride formation and fracture. This has been refuted by statements that the H-diffusion rate is too low to support measured crack velocities [6]. Considering the wide range of crack velocities reported by previous workers it is possible that more than one mechanism is prevalent and that the predominant mechanism is related to crack velocity. The primary commonality between these mechanisms is that they involve cathodically produced H entering the metal substrate. The loading conditions applied by the SCC test apparatus may influence the propensity for mechanically or chemically induced film rupture to occur. Consequently, the choice of test method may affect the tendency for H to enter the substrate and cause crack initiation.

2   Experimental Method

All tests were performed using high purity AZ91 alloy (>90.11 wt%Mg, 8.99 wt%Al, 0.78 wt%Zn, 0.21 wt%Mn). Cylindrical tensile specimens with 5 mm diameter waisted gauge were machined from as-cast ingots. The gauge surface was polished with 600 grit emery paper and the specimens were degreased in ethanol before use. The test environments were 5 g/L NaCl solution, double-distilled water and laboratory air. All solutions were made from reagent grade chemicals and double-distilled water. All tests were conducted at open-circuit conditions.

The LIST apparatus is illustrated in Figure 1. The specimen is attached to one end of a lever arm. To the opposite end of the arm a known mass is attached such that the tensile load applied to the specimen increases linearly as the distance between the fulcrum and the mass is increased by means of a screw thread and synchronous motor. The SSRT apparatus (Figure 2) maintains a constant strain rate by means of an open-loop control system; the average strain is measured by two high-resolution LVDTs in parallel with the specimen whilst a geared synchronous motor increases the strain and thus the load accordingly.

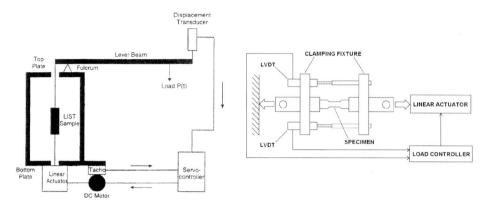

Figure 1: Schematic illustration of the LIST apparatus **Figure 2:** Schematic illustration of the SSRT apparatus

SCC susceptibility was characterised according to threshold stress, which was determined using the DC potential drop (DCPD) method as per Atrens et al [7–11]. The DCPD method involves applying a low constant current to the specimen and measuring the change in specimen resistance as the cross-section is reduced, primarily by crack propagation, according to the relationship $\Delta R = \rho l / \Delta A$, where ΔA is the change in cross-sectional area, ρ is the resistivity and l is the length of the specimen. LIST tests were performed with a basic DCPD system as described in Winzer et al [12]. The resistance of the specimen was measured with insufficient resolution to accurately determine the SCC parameters. SSRT tests were performed using a more advanced DCPD system as described in Dietzel et al [13]. A pulsed and reversing current was applied to the specimen to minimise thermoelectric offset voltages due to coupling of dissimilar metals in a variable-temperature environment. Variations in the resistivity of the specimen due to temperature were accounted for by using a reference specimen. These features enabled the SCC parameters to be characterised with considerably higher resolution and greater repeatability.

3   Results

Figures 3 and 4 show the stress-strain behaviour and DCPD measurements for the SSRT tests in distilled water and in air. There is a considerable reduction in the UTS relative to air for all of the samples exposed to distilled water. Moreover, for the specimens exposed to distilled water the threshold stress and UTS decreased with decreasing strain rate. The threshold stress was interpreted as being the point where the DCPD curve becomes non-linear. The threshold stresses at $3 \cdot 10^{-8}$ s^{-1}, 10^{-7} s^{-1} and $3 \cdot 10^{-7}$ s^{-1} were approximately 55 MPa, 65 MPa and 75 MPa respectively. A similar trend occurred for 5 g/L NaCl; however, for these strain rates the UTS and threshold stresses for distilled water were significantly less than for 5 g/L NaCl.

Figure 5 shows a comparison between two specimens under LIST and SSRT conditions at $7.3 \cdot 10^{-4}$ MPa/s and 10^{-7} s^{-1}, respectively, under identical environmental conditions. LIST tests generally failed soon after crack initiation, whereas cracks under SSRT conditions propagated for a long period of time. SSRT tests were typically 2 to 3 times longer in duration and were sometimes stopped after a significant reduction in nominal stress had occurred.

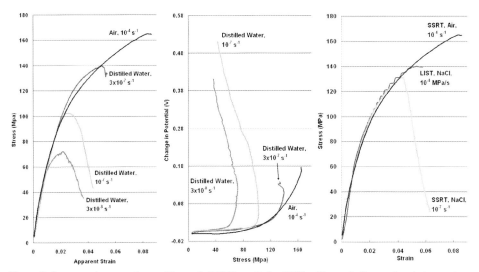

Figure 3: Stress-strain curves for AZ91 in distilled water and air

Figure 4: DCPD results for AZ91 in distilled water and air

Figure 5: Comparison between stress-strain results for LIST and SSRT

Fracture surfaces for SSRT specimens in distilled water at $3 \cdot 10^{-8}$ s^{-1}, 10^{-7} s^{-1} and $3 \cdot 10^{-7}$ s^{-1} were partly comprised of relatively flat regions containing coarse stepped markings typically 5 µm apart as shown in Figure 6A. The parallel edges feature considerable amount of jogging. Secondary cracks were also evident. The stepped regions penetrated the specimen up to 1 mm from the surface with the depth of penetration and the total area of the regions increasing with decreasing strain rate. The remaining fracture surface is generally comprised of broken, jagged regions as shown in Figure 6B. The specimen fractured at $3 \cdot 10^{-7}$ s^{-1} also contained some randomly interspersed dimpled regions consistent with ductile tearing (Figure 6C).

Figures 6D-F show the fracture surface for 5 g/L NaCl at 10^{-7} s^{-1} under SSRT conditions. There is extensive pitting with some pits penetrating approximately 400 µm from the surface (see Figure 6D). The pitting is generally adjacent to relatively flat fracture surface regions containing coarse parallel markings (similar to those observed for distilled water) and elongated dimples (see Figure 6E). These regions comprise a high proportion of the total fracture surface area. The parallel markings are unidirectional within a discrete area; however, they may differ from other regions. There are also highly-localised regions containing fine parallel markings that are approximately 0.5 µm apart (see Figure 6F). The remaining fracture surface is comprised of the same broken, jagged features observed for distilled water and is undermined with some secondary cracks.

Fracture surfaces for AZ91 in 5 g/L NaCl solution under LIST conditions were macroscopically non-crystallographic and largely comprised of cleavage-like regions, typically <20 µm across, randomly interspersed between broken, jagged regions with some dimpled regions. There were also several localised areas at edges containing coarse parallel markings similar to those observed for SSRT specimens in distilled water. Fracture surfaces produced under LIST conditions were shown in Winzer et al [12].

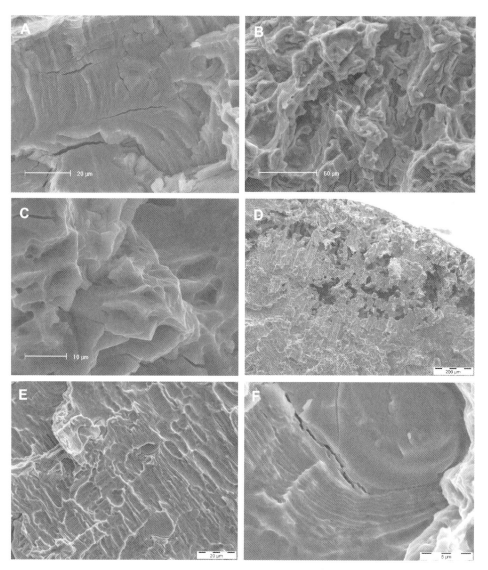

Figure 6: SEM micrographs of specimens fractured in distilled water at 10^{-7} s^{-1} (A and B), distilled water at $3 \cdot 10^{-7}$ s^{-1} (C) and 5g/L NaCl at 10^{-7} s^{-1} (E and F)

4 Discussion

The coarse, parallel markings observed for tests in distilled water and 5 g/L NaCl solution are consistent with those observed by Stampella et al [14] for pure Mg anodically polarized in

Na$_2$SO$_4$ solution, Meletis and Hochman [3] for pure Mg in NaCl-K$_2$CrO$_4$ solution and Chakrapani and Pugh [15] and Wearmouth et al [17] for Mg-7.5Al in NaCl-K$_2$CrO$_4$ solutions. That these regions were located invariably at or close to gauge surfaces suggests that they correspond to crack initiation sites. Meletis and Hochman proposed that the parallel facets and joining perpendicular steps were primarily due to cleavage on $\{2\bar{2}03\}$ planes parallel to the direction of crack propagation; since there are six $\{2\bar{2}03\}$ planes in the HCP crystal lattice cleavage may result in crystallographic 3-dimensional fracture surfaces. This may explain the observed jogging. The fine parallel markings observed are similar in size to markings on cleavage steps observed by Meletis and Hochman and Chakrapani and Pugh and attributed to micro-tearing at crack arrest sites.

Figures 3 and 4 show for AZ91 in distilled water at $3 \cdot 10^{-8}$ s^{-1} to $3 \cdot 10^{-7}$ s^{-1} there is a reduction in UTS relative to air that is indicative of SCC. Ebtehaj et al [16] and Wearmouth et al [17] proposed that for Mg-Al alloys in solutions containing approximately equal concentrations of chloride and chromate ions maximum susceptibility occurs at a discrete range of intermediate strain rates; at low strain rates repassivation retards H-ingress at mechanical film rupture sites whereas at high strain rates ductile tearing predominates. In this instance, there was no evidence of ductile tearing for the fracture surface corresponding to steady-state crack propagation between $3 \cdot 10^{-8}$ s^{-1} and $3 \cdot 10^{-7}$ s^{-1}. The fact that the UTS and threshold stress decrease with decreasing strain rate suggests that SCC susceptibility is dependent on the balance between mechanical film rupture and repassivation between $3 \cdot 10^{-8}$ s^{-1} and $3 \cdot 10^{-7}$ s^{-1}. It may be important to note that this study includes only a limited range of strain rates; a distinct minimum threshold stress or UTS may be identified if the range of strain rates were extended.

Specimens fractured in distilled water had lower UTSs than those in 5 g/L NaCl. No pitting was observed after exposure to distilled water whereas exposure to 5 g/L NaCl resulted in severe pitting. This suggests that H ingress in distilled water is dependent on mechanical film rupture. For specimens exposed to the 5 g/L NaCl solution, H ingress may have been partly inhibited by excessive dissolution as proposed by Ebtehaj et al [16] and Makar et al [18].

That LIST tests in SCC-causing environments result in small, localised regions of parallel markings suggests that when the applied stress reaches σ_{SCC}, H-embrittlement causes the crack to propagate until some critical crack length is reached a short distance from the surface. Here the crack propagation mechanism is overwhelmed by some non-H-assisted process such as cleavage or microvoid coalescence. In contrast, the SSRT tests result in considerably larger regions of parallel markings. This may be attributed to a tendency for stress relaxation coupled with crack tip opening displacement; as the crack propagates some distance stress may be reduced below the threshold or the local stress at the crack tip may be reduced such that the HE processes are retarded. These tendencies underpin the principal differences between the two methods; SSRT tests lend themselves to easier fractography and fracture surface analysis whereas LIST tests are more rapid owing to their tendency for fast fracture once the critical crack length is reached.

5 Summary

The described methods have been used in an ongoing program to develop a mechanistic understanding of the environmental and mechanical factors influencing SCC of Mg alloys. From these results it is apparent that the LIST method provides more rapid determination of the threshold stress, whereas the SSRT method provides a larger fracture surface area for analysis and increased crack propagation time for determining the overall steady-state crack velocity. This latter trait may be particularly important in determining the underlying mechanism for SCC in Mg alloys, since the predominant mechanism is somehow related to the crack velocity. SCC testing in distilled water and 5 g/L NaCl under SSRT conditions showed that maximum susceptibility occurred in distilled water at the slowest strain rate.

6   Acknowledgements

The authors wish to thank the GM Technical Centre at Warren MI, the Australian Research Council (ARC), the Australian Research Network for Advanced Materials (ARNAM) and GKSS-Forschungszentrum Geesthacht GmbH.

7   References

[1]   N. Winzer, A. Atrens, G. Song, E. Ghali, W. Dietzel, K.U. Kainer, N. Hort and C. Blawert, Adv. Eng. Mater., 2005, 7, No. 8, 659–693
[2]   S.P. Lynch, Hydrogen Effects on Material Behaviour and Corrosion Deformation Interactions (Ed.: N.R. Moody, A.W. Thompson, R.E. Ricker, G.W. Was, R.H. Jones), TMS, 2003, 449–466
[3]   E.I. Meletis, R.F. Hochman, Corrosion, 1984, 40, 39–45
[4]   D.G. Chakrapani, E.N. Pugh, Metall. Trans. A , 1976, 7A, 173–178
[5]   G.L. Makar, J. Kruger, K. Sieradzki, Corros. Sci., 1993, 34, No. 8, 1311–1342
[6]   S.P. Lynch, P. Trevena, Corrosion, 1988, 44, 113–124
[7]   A. Atrens, C.C. Brosnan, S. Ramamurthy, A. Oehlert, I.O. Smith, Meas. Sci. Technol., 1993, 4, 1281–1292
[8]   A. Oehlert, A. Atrens, J Mater. Sci., 1998, 33, 775–781
[9]   A. Oehlert, A. Atrens, J Mater. Sci., 1997, 32, 6519–6523
[10] A. Oehlert, A. Atrens, Corros. Sci., 1996, 38, 1159–1170
[11] A. Oehlert, A. Atrens, Acta Metall. Mater., 1994, 42, 1493–1508
[12] N. Winzer, G. Song, A. Atrens, W. Dietzel and C. Blawert, Corrosion and Prevention 2005, ACA, 37
[13] W. Dietzel, K.-H. Schwalbe, Z. Materialprüfung, 1986, 28, No 11, 368–372
[14] R.S. Stampella, R.P.M. Proctor, V. Ashworth, Corros. Sci., 1984, 24, No 4, 325–341
[15] D.G. Chakrapani, E.N. Pugh, Corrosion, 1975, 31, 247–252
[16] K. Ebtehaj, D. Hardie, R.N. Parkins, Corros. Sci., 1993, 28, 811–829
[17] W.R. Wearmouth, G.P. Dean, R.N. Parkins, Corrosion, 1979, 29, No 6, 251–258
[18] G.L. Makar, J. Kruger, K. Sieradzki, Corros. Sci., 1993, 34, 1311–1342

Corrosion Behaviour of Magnesium Alloys: Material Specific Corrosion Testing

S. Bender, E. Boese, A. Heyn, J. Goellner
Otto-von-Guericke-University, Magdeburg

1 Introduction

Corrosion problems are a major issue that often prevents magnesium alloys from being used. Therefore, understanding the mechanism and improving the corrosion resistance of magnesium alloys are important for their applications [1, 2]. No suitable methods are readily available at the present time for the assessment of the corrosion resistance of magnesium alloys. Methods such as the salt spray test, which originally were developed for materials with coatings, do not adequately take into account the specific character of the corrosion of magnesium and magnesium alloys. The objective of this work is to develop a material specific method for corrosion testing of magnesium and magnesium alloys, which is based on an existing electrochemical examination method. During application of a static or dynamic electrochemical load to a working electrode the influence of unstable hydroxide layers typically formed on magnesium and magnesium alloys may be suppressed with the help of a rotating disc electrode. The electrochemical noise measurements reflect the submicroscopic processes of the corrosion, although diffusion controlled process takes place as well [3]. Therefore, different magnesium alloys are investigated under open circuit conditions to detect very small changes in the corrosion system (alloy, medium, design) with high precision.

2 Experimental

For the investigations of the corrosion behaviour of magnesium alloys the following materials were available as shown in Table 1.

Table 1: Chemical composition of alloys investigated (manufactures data)

Magnesiumalloy type	Alloy	% Al	% Zn	% Mn	% Si	% Fe	% Cu	% Li
Diecast	AM20	2,37	0,12	0,365	0,006	0,005	0,022	
Diecast	AM20Li59	0,8	0,37	0,165	0,015	0,034	0,041	9,83
Diecast	AM20Li110	0,64	0,28	0,172	0,015	0,018	0,028	12,40
Diecast	AM20Li123	0,70	0,27	0,184	0,017	0,016	0,027	11,52
Wrought alloy	AZ31	2,36	0,89	0,210	0,024	0,004	0,005	
Diecast	AZ91 DC	9,3	0,79	0,120	0,020	0,004	0,007	
Pig casting	AZ91 PC	9,1	0,81	0,125	0,049	0,002	0,009	

It can be seen that the concentration of Fe (max. 0,004 %) and Cu (max. 0,025 %) in the alloys AM20 and AM20+Li is higher than the maximal permissible value. Specimens were manufactured from the materials being available.

2.1 Experimental Methods

2.1.1 Polarization Measurements

The polarization measurements were carried out in a three-electrode cell with NaCl solution (0.01 M, 24 ± 3°C) as electrolyte. The working electrode was a rotating disc (RDE); a saturated Ag/AgCl (E_H = 197 mV) electrode was used as reference electrode and a platinum electrode as counter electrode. The polarization curves were scanned with a rate of 60 mV/min from -2000 mV to about 0 mV. Once the current density reached the threshold of 50 A/m², the measurement was ended. The electrochemical cell was connected with a potentiostat-galvanostat. A grinding machine stood right next to the equipment, because it is important, that the ground working electrode is immediately dipped in the electrolyte to prevent the formation of oxide layers on the surface.

2.1.2 Electrochemical Noise Measurements

Electrochemical noise investigations were done to determine the potential, potential noise, the current and current noise. A specific experimental set-up was designed to carry out electrochemical noise measurements. These measurements were conducted in a three-electrode cell consisting of two working electrodes, which are macroscopically identical specimens, and a reference electrode (Ag/AgCl). The specimens were electrical connected with simple clamps to minimize the set-up time. Due to sensitivity of the measurements, a faraday cage was used to suppress external influences. The experiments were carried out using a combined zero-resistance ammeter (ZRA) that was combined with a high-impedance, potential-measuring device. The DC (direct current) part of the current and potential was filtered via a low-pass filter (cut-off frequency of 1 Hz). The AC (alternating current) part was filtered via a band-pass filter (cut-off frequencies of 0.1 Hz and 10 Hz), so that the frequency range effectively measured was 0.1–10 Hz. The sampling rate of the 16 Bit data acquisition was chosen to be 20 samples per second due to the cut-off frequency of the filter. All experiments were carried out in a 0.05 M Na_2SO_4 solution to avoid chloride ions at different pH values. One method for the quantification of noise data over a fixed period of time is the calculation of the standard deviation (S). The noise resistance and the charge were determined for the assessment of the results. The noise resistance (R_N) can be calculated from the standard deviation (S) of potential noise and current noise. The noise resistance was calculated over a time interval of 30 min. The charge (Q) can be calculated by integration over the time.

3 Results

3.1 Polarization Measurements

The commonly used polarization measurement method in combination with RDE had to be modified for the development of a magnesium specific corrosion testing method. In order to achieve this, different test parameters were varied to determine their influence on the corrosion behaviour of magnesium alloys. After preliminary investigations, the following test parameters for polarization measurements were dedicated as most effective: a rotation speed of 2000 rpm was chosen because at this speed a stable flow field develops, which removes the reaction products and therefore causes reproducible corrosion data. Aqueous electrolytes with pH 9 were chosen to slow the dissolution rate (slower than at pH 7) and prevent formation of dense layers on the surface of the magnesium alloys (which occurs at higher pH values). During the polarization measurements argon was bubbled into the electrolyte to avoid an adsorption of CO_2 from ambient air and, consequently, a decrease of pH value. Subsequently, the corrosion behaviour of different magnesium alloys was tested under optimized conditions.

3.2 Polarization Measurements on Mg-Al-Zn Alloys and Mg-Al-Mn Alloys

In Table 2 the influence of the composition of different Mg-Al-Zn alloys on the position of free corrosion potential E_{corr} and current density i_{corr} determined from Tafel plots is illustrated.

Table 2: Results of Mg-Al-Zn alloys investigated with optimized parameters

	AZ91 DC	AZ91 PC	AZ31
E_{corr} [mV$_{Ag/AgCl}$]	-1580 ± 2	-1570 ± 15	-1470 ± 15
i_{corr} [A/m²]	0,008 ± 0,002	0,009 ± 0,001	0,014 ± 0,001

AZ91 DC has the lowest free corrosion potential at –1580 mV$_{Ag/AgCl}$, followed by AZ91 PC at –1570 mV$_{Ag/AgCl}$, and AZ31 at –1470 mV$_{Ag/AgCl}$. It can be seen that the values of i_{corr} do not significantly differ for AZ91 DC (0.008 A/m²) and AZ91 PC (0.009 A/m²), but AZ31 shows the highest current density of approximately 0.014 A/m².

In Table 3 the dependence of Mg-Al-Mn alloy on the position of free corrosion potential E_{corr} and current density i_{corr} determined from Tafel plots is illustrated.

Table 3: Results of Mg-Al-Mn alloys investigated with optimized parameters

	AM20	AM20+Li
E_{corr} [mV$_{Ag/AgCl}$]	–1540 ± 20	–1620 ± 10
i_{corr} [A/m²]	0,010 ± 0,001	0,018 ± 0,002

The values of the current density indicate that AM20 has a better corrosion behaviour compared to AM20 containing Li. AM20 modified with Li shows a higher current density of approximately 0.018 A/m² as compared to AM20 with approximately 0.010 A/m².

3.3 Electrochemical Noise Measurements

The materials tested above were examined further by measurements of the electrochemical noise. The electrochemical current and potential noise were measured under open circuit conditions in 0.05 M Na_2SO_4 solution without chlorides.

3.3.1 Electrochemical Noise Measurements on Mg-Al-Zn Alloys

In Figure 1 the electrochemical potential noise (EPN) and the electrochemical current noise (ECN) of the three Mg-Al-Zn alloys are illustrated.

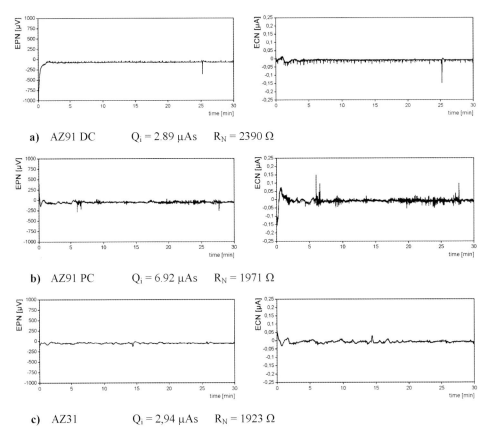

Figure 1: Electrochemical noise of three Mg-Al-Zn alloys under open circuit conditions over 30 minutes in Na_2SO_4 (0.05 M, pH 9); a) AZ91 DC, b) AZ91 PC, c) AZ31

The relationship between the resulting charge, the noise resistance, the microstructure, and the Al content of the alloys will now be discussed. In the first two plots of Figure 1: Electrochemical noise of three Mg-Al-Zn alloys under open circuit conditions over 30 minutes in Na2SO4 (0.05 M, pH 9); a) AZ91 DC, b) AZ91 PC, c) AZ31 **a**, the results for AZ91 DC are shown. AZ91 DC has the finest microstructure and the highest Al-concentration of the three tested Mg-Al-Zn alloys. Therefore, the current noise is less intensive and the respective charge Q_i has, with 2.89 µAs, the lowest value. AZ91 PC (Figure 1 **b**) has nominally the same elemental composition as AZ91 DC but its grain size is larger. As can be seen, the potential noise and current noise reach higher intensities and, therefore, the resulting charge increases to 6.92 µAs. AZ31 (Figure 1 **c**) is a wrought alloy and contains only 3 % Al. The current noise is slightly higher than that for AZ91 DC and lower than that for AZ91 PC. The resulting charge of AZ31 (2.94 µAs) is comparable to the charge of AZ91 DC (2,89 µAs).

3.3.2 Electrochemical Noise Measurements on Mg-Al-Mn Alloys

In Figure 2, the electrochemical potential noise (EPN) and the electrochemical current noise (ECN) of Mg-Al-Mn alloys alloyed with different Li concentrations are illustrated.

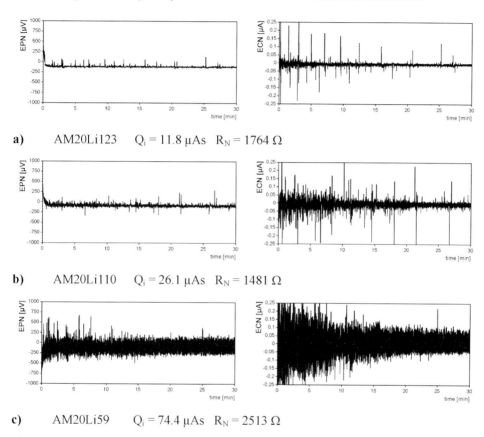

a)   AM20Li123 $Q_i = 11.8\ \mu As$ $R_N = 1764\ \Omega$

b)   AM20Li110 $Q_i = 26.1\ \mu As$ $R_N = 1481\ \Omega$

c)   AM20Li59 $Q_i = 74.4\ \mu As$ $R_N = 2513\ \Omega$

Figure 2: Electrochemical noise of three Mg-Al-Mn alloys under open circuit conditions in Na_2SO_4 (0.05 M, pH12); a) AM20Li123, b) AM20Li110, c) AM20Li59; the Fe-concentration at the AM20 and AM20+Li alloys increases from a) to c)

As can be seen in Figure 2 a-c, the potential noise and the current noise obviously increase with increasing Fe-concentration.

4 Summary

Although in the existing literature polarization measurements have been criticized, the present polarization measurements with the RDE show reproducible results,. It has been reported that magnesium has abnormal polarization behaviour. Particularly, under an anodic polarization condition, the broken areas in the surface film and other parameters could change with applied polarization potential or current density. However, the advantage of polarization measurements with the RDE as compared to the typically used salt spray test is the ease of obtaining an instantaneous corrosion rate and the removal of water soluble and not strongly adherent corrosion products from the exposed surface. Regarding the current density, polarization measurements with the RDE are suited to characterize different magnesium alloys. The corrosion behaviour of the available magnesium alloys in 0.01 M NaCl at pH9, with Ar and 2000 rpm, can be distinguished. The cast alloy AZ91 DC with a current density of 0.008 A/m² shows the best corrosion resistance followed by AZ91 PC with a current density of 0.009 A/m² and AZ31 with 0.012 A/m². The cast alloy AM20 shows a current density of approximately 0.010 A/m². The Mg-Al-Mn alloys alloyed with Li show higher values of the current density (approximately 0.018 A/m²) than the pure AM20. The too high Fe concentration of AM20+Li resulting from the fabrication process of these alloys is seen as a major influence factor because all tested Mg-Al-Mn-Li alloys are no hp alloys. This is the reason that no influence of Li concentration on the corrosion behaviour of Mg-Al-Mn-Li alloys could be determined. Regarding the different Fe concentration of Mg-Al-Mn-Li alloys, no dependence of the polarization curves and polarization data on the Fe-concentration could be found. However, the the anodic current density increases most for the alloy AM20 modified with Li, which has the lowest Al concentration (2 %) and the highest Fe concentration (<0.034 %). This shows clearly that by exceeding the maximal permissible value of Fe content the corrosion behaviour of the alloy is no longer dependent on the other alloying elements, but the Fe content determines the corrosion rate.

The electrochemical noise measurements without external polarization confirm the results of the polarization measurements with RDE. But the results are more convincing because of the high resolution of this method. Even minute amounts of dissolution on the metal surface can be recorded. The advantages of the electrochemical noise technique are obvious, since noise investigations indicate a tendency to and occurrence of corrosion before macroscopic damage develops. The criterion for determining the influence of the alloying elements of the available magnesium alloys, such as decreasing Al concentration and increasing Fe concentration, is based on a clear rise in the charge calculated from the noise. A dependence of the charge on different Fe concentrations in the Mg-Al-Mn alloys modified with Li could be found. The charge increased with increasing Fe concentration. Furthermore, a dependence of the charge on Al concentration and fabrication process is found. The charge increased with decreasing Al concentration. The alloys AZ91 DC and AZ91 PC have nominally the same composition, but the charge for AZ91 PC is higher due to the macro grained structure in comparison to AZ91 DC, which has a fine-grained structure.

5 References

[1] Ghali, E.; Uhlig's Corrosion Handbook, 2000, p. 793–830.
[2] Song, G. L.; Atrens, A.; Advanced Engineering Materials, 1999, 1, p. 11–33.
[3] Göllner, J.; Burkert, A.; Heyn, A.; Materials and Corrosion, 53, 2002, 9, p. 656–662.

Corrosion-Fatigue of AZ31 Wrought Magnesium Weldments

C.E. Cross[1], G. Ben-Hamu[2], D. Eliezer[2], and P. Xu[1]
[1] Federal Institute for Materials Research and Testing, Berlin, Germany
[2] Ben-Gurion University of the Negev, Beer-Sheva, Israel

1 Introduction

The use of wrought magnesium alloys, as compared to more conventional die-cast alloys, continues to show potential for extended weight reduction applications in the automobile industry due to improved mechanical properties [1]. In order to make full use of wrought plate and sheet in automotive applications will require the assembly and joining of formed components, including the possibility for wrought-to-die cast hybrid combinations [2]. Arc welding, including gas-metal arc and gas-tungsten arc processes, has been demonstrated to be a viable method for joining magnesium [3-6]. It follows that the corrosion and corrosion-fatigue behaviour of wrought alloys and their weldments will need to be fully characterized. However, such a knowledge base for wrought magnesium is limited to date, particularly with regard to welded joints.

Ellermeir et al. [7] have investigated the influence of different welding processes (laser, electron beam, plasma, and gas-metal arc) on the corrosion resistance of magnesium alloys exposed to salt water immersion. Tests revealed different location-dependent corrosion phenomena related to a) differences in aluminum concentration between filler and base materials, b) oxidation and evaporation processes taking place at the weld pool surface, and c) presence of a heat-affected zone (HAZ). Corrosion was found to initiate preferentially in the HAZ, and welds with a narrow HAZ exhibited higher corrosion resistance.

Thate et al. [8] report that the relative corrosion resistance of base materials AZ31 and AZ61 determines the corrosion behaviour of the corresponding weldment. The corrosion resistance of the base metal is reported to increase with increasing aluminum content [9], which was also observed when comparing AZ31 with AZ61 [7, 10]. The use of different welding processes (gas-metal arc, gas-tungsten arc and plasma arc) was not observed to influence corrosion test results [8].

Fatigue strength of magnesium alloy weldments in air, for arc welds made on rolled plate, reaches approximately 50–60 % that of the base material [11, 12]. Likewise, welded joints on extruded AZ31 have fatigue strength efficiencies around 80 %. The notch effect of the weld toe determines the fatigue strength of the weldment [11], where use of filler metal improves the weld bead profile and fatigue properties.

Investigations of corrosion-fatigue have been reported for wrought magnesium in salt water environments [13–17]. NaCl salt solutions are typically saturated with a $Mg(OH)_2$ buffer to stabilize the pH and produce a more uniform passivation layer. This has led to a marked increase in the corrosion-fatigue life of extruded AZ31, whereas for other alloys (e.g. AM50) no influence of $Mg(OH)_2$ has been found [15].

Reports regarding corrosion-fatigue behaviour of wrought magnesium alloy weldments are unknown to date. The main objective of this research was therefore to fill this gap in knowledge, making a comparative study of fatigue behaviour for AZ31 wrought magnesium weldments in both air and salt water environments. The information presented here represents the initial results from an on-going investigation.

2 Experimental

Test specimens for S-N (Wöhler) type corrosion-fatigue tests, shown in Figure 1, were made from 3 mm thick AZ31 wrought magnesium (continuous cast and hot rolled), designed according to ASTM E466-02. Welding was done using a variable polarity, gas-tungsten arc, cold wire feed process (VP-GTA-CWF) using AZ61 filler metal and a square, butt-joint configuration. The variable polarity process was selected because it provides exceptional control for oxide removal in the electrode-positive cycle, combined with good penetration in the electrode-negative cycle. Welds were made full-penetration in a single pass, using the welding parameters given in Table 1. Weld reinforcements were removed prior to testing, so that corrosion-fatigue behaviour could be investigated without notch effects imposed by the weld toes.

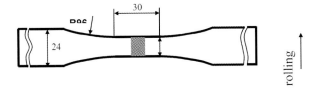

Figure 1: Specimen for S-N corrosion fatigue test per ASTM E466-02 (weld is shown shaded, dimensions are in mm)

Table 1: VP-GTA-CWF welding parameters

electrode negative, current and time	135 amps, 16 ms
electrode positive, current and time	80 amps, 4 ms
variable polarity frequency	50 Hz
torch travel speed	5 mm/s
shielding gas (torch, trailing, and back side)	Ar
AZ61 filler wire size and feed rate	1.2 mm dia., 42 mm/s

Uniaxial fatigue tests were performed at 30 Hz in tension under load-control (load ratio 0.1) using a hydraulic MTS machine. Tests were performed at room temperature in both air and a 3.5% NaCl solution saturated with $Mg(OH)_2$. When using the 3.5% NaCl solution without the $Mg(OH)_2$ buffer, the pH of the salt solution gradually increases over time from 5 to 9.8. With the buffer, however, the pH value of the solution remains constant at approximately 10.5. The corrosion cell was made from plexiglas, with silicone used to seal the connection with the test specimen. This cell was connected to a 10 liter tank, with the solution circulated at a rate of 0.3 L/min. The test specimen was immersed in solution over a length of 80 mm.

Chemical composition was determined from a weld cross-section using an SEM-EDX scanning microprobe analyser (see Table 2). It was observed that the weld metal picked-up significant amounts of aluminum, zinc, and silicon from dilution with the filler wire, with a slight loss in manganese. Optical metallography was performed on polished cross-sections using a acetic-picric acid etch. A microhardness traverse was made across the weldment at 0.5 mm intervals. Corrosion of the weldment was characterized using potentiodynamic and Kelvin probe analysis for each of the three zones (base metal, HAZ, and weld metal). Potentiodynamic curves were obtained using a 263 EG&G Potentiostat/Galvanostat, with a voltage scan of 10 mV/min.

Table 2: Measured composition of AZ31 weldment showing major alloying elements (wt.%, average)

	Al	Zn	Mn	Si
Base Metal	2.07	0.68	0.45	0.01
Weld Metal	4.37	1.16	0.31	0.06

3 Results and Discussion

3.1 Weld Microstructure

Welds were observed to have a uniform bead along the weld length, with no undercutting and only a few scattered pores (under 0.5 mm dia.) detected with radiography in a 200 mm length of weld. Figure 2 shows a weld cross-section comparing base metal, heat-affected zone, and weld metal microstructures. Significant grain growth occurred in the heat-affected zone (HAZ), increasing from an average grain diameter of 8 µm (±3 µm) in the base metal to 38 µm (± 16 µm) in the HAZ. The coarse second phase in the base metal and HAZ has been identified as Al_8Mn_5. The weld metal consists of coarse equiaxed grains (43 µm dia., ± 15 µm) with a coarse ß ($Al_{12}Mg_{17}$) eutectic phase constituent. This equilibrium ß phase exists primarily along grain boundaries, and it is very nearly continuous.

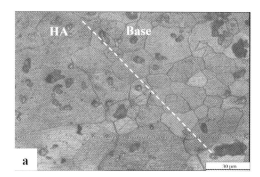 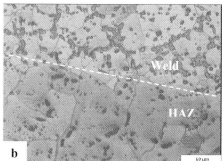

Figure 2: Metallographic cross-section of AZ31 weld showing a) HAZ-base metal boundary and b) HAZ-weld metal boundary

Traverse microhardness measurements are shown in Figure 3, showing the weld metal to be slightly lower in hardness than the base metal: 40 HV for the weld metal versus 50 HV for the base metal. Simple cross-weld tensile tests confirmed that fracture occurs in the weld metal, reaching 238 MPa tensile strength (84% of base metal tensile strength).

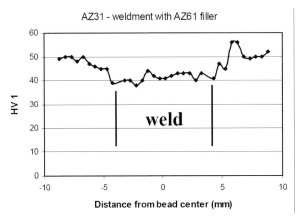

Figure 3: Microhardness traverse across weld (Vickers)

3.2 Corrosion Behavior

Polarization curves for the investigated weldments immersed in a 3.5% NaCl solution saturated with $Mg(OH)_2$ are shown in Figure 4. Corresponding corrosion and pitting potentials for each of the three weld zones, averaged over five tests, are compared in Table 3. Corrosion rate estimations, calculated from polarization curve results, are also given. Magnesium alloys normally start pitting at a potential slightly more negative than their corrosion potential in NaCl solutions, as indicated by visible evolution of hydrogen from the specimen during testing. The pitting potential is an important electrochemical parameter as it indicates the tendency for localized corrosion. A more negative pitting potential means that localized corrosion is more likely. From Table 3 it is observed that the base metal has the most negative pitting potential and, hence, should be more prone to pit. This was found to be the case in fatigue test specimens, where the most severe pitting attack occurred in the base metal.

The weld metal was found to have the most negative corrosion potential (Table 3). This was also observed in Kelvin probe measurements, taken across a freshly polished weldment in a humid air environment, as shown in Figure 5. However, this difference was found to diminish over time, indicating formation of a stable passive layer. The extensive formation of the ß phase in the weld metal, believed to be cathodic relative to the matrix, should promote localized galvanic attack in the initial period of corrosion [18]. Its over-all effect, however, is reported to improve corrosion resistance [19, 20]

Table 3: Comparison of corrosion parameters derived from polarization curves

	corrosion current (mA)	corrosion potential (V)	pitting potential (V)	corrosion rate (mpy)
base metal	33.6	−1.670	−1.650	118
heat-affected zone	47.3	−1.659	−1.635	165
weld metal	48.5	−1.675	−1.645	170

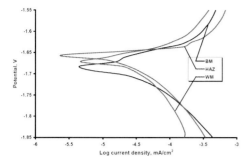 

Figure 4: Polarization curves for a VP-GTA-CWF AZ31 weldment in 3.5% NaCl saturated with $Mg(OH)_2$

Figure 5: Kelvin probe traverse across weldment showing weld metal (center) to have more negative corrosion potential in humidified air environment

3.3 Corrosion-Fatigue

Figure 6 compares S-N curves for AZ31 wrought plate, in welded and unwelded conditions, for both air and buffered NaCl environments. Each data point represents the result from one test. The fatigue data for the base metal in air is similar to what has already been reported in the literature for AZ31 [12].

It is observed that the fatigue life of AZ31 base metal in salt solution is significantly reduced compared to that in air at high loads. This is likely due to the presence of corrosion pits serving to initiate fatigue cracks [16, 21]. At low loads this difference becomes diminished, likely due to the interaction of corrosion at the crack tip. Weldments also have a reduced fatigue life compared to the base metal, with welds tested in salt solution displaying the lowest fatigue life. All welded specimens fractured through the weld metal, with the exception of two specimens (low load, high cycle) which failed in the base metal.

4 Conclusion

This investigation has compared the fatigue and corrosion-fatigue behaviour of 3 mm thick rolled AZ31 plate and its corresponding weldments, tested both in air and buffered salt solution. Use of the VP-GTA-CWF welding process has proven successful in generating welds of high quality using AZ61 filler wire. Fatigue life of AZ31 weldments in air is significantly lower than the base metal for all loads examined. Fracture of weldments occurred predominantly in the weld metal, which is lower in strength and has a nearly continuous network of grain boundary beta phase. Fatigue life of AZ31 base metal and weldments in buffered NaCl solution is lower than corresponding values for air.

However, this difference becomes increasing small in the low load-high cycle regime where corrosion is likely to play a role in crack propagation. The weld metal, with its higher aluminium content, shows a corrosion potential that is more negative than either the HAZ or base metal on a freshly polished surface. However, after passivating, the weld metal shows better corrosion resistance and the base metal is found to have the most severe pitting.

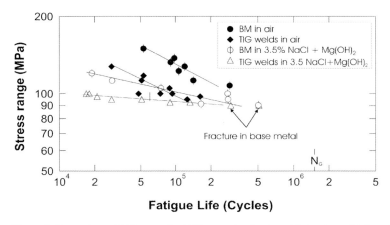

Figure 6: Comparison of S-N curves for AZ31 wrought plate in welded and unwelded condtion, and in air and buffered NaCl solution

5 Acknowledgements

The authors are grateful to Th. Michael (BAM) for weld process development and Th. Böllinghaus (BAM) for technical discussions, and particularly wish to thank B. Engl and Magnesium Flachprodukte for providing material.

6 References

[1] B. Engl, Light Metal Age 2005, no. 10, 14–19.
[2] A. Ben-Artzy, I. Makover, I. Dahan, M. Kupiec, Y. Salah, A. Heler, A. Shtechman, A. Bussiba, Y. Weinberg, High Performance Metallic Materials for Cost Sensitive Applications (eds. F.H. Froes, et al.), TMS, 2002, 219–226.
[3] M. Rethmeier, B. Kleinpeter, H. Wohlfahrt, Welding in the World 2004, 48 (3/4), 28–33.
[4] S. Lahtabai, K.J. Barton, D. Harris, P.G. Lloyd, D.M. Viano, A. McLean, Magnesium Technology 2003 (ed. J.I. Kaplan), TMS, 2003, 157–162.
[5] A. Munitz, C. Cotler, A. Stern, G. Kohn, Materials Science and Engineering-A 2001, 302, 68–73.
[6] A. Stern, A. Munitz, G. Kohn, Magnesium Technology 2003 (ed. J.I. Kaplan), TMS, 2003, 163–168.
[7] J. Ellermeier, K. Eppel, C. Berger, DVM-Tagungsband: Magnesium-Bauteilfestigkeit, Fulda, 2003, 153–173.
[8] W. Thate, J. Zschetzsche, Werkstofftechnik 2003, band 93, heft 10, 699–704.
[9] E. Aghion, B. Bronfin, 3rd Int. Magnesium Conf. London, Institute of Materials, 128–134.
[10] K.E. Wendler, Proc. Zukunftorientierter Einsatz von Magnesium im Automobilbau, 2002, Uni. Erlangen-Nürnberg, 1–14.
[11] H. Krohn, S. Singh, Schweißen von Magnesiumlegierung für dem Automobilbau, DVS-report 204, 197–201.

[12] T. Masato, H. Somekawa, K. Higashigi, K. Higashi, H. Iwasaki, T. Hasegawa, A. Mizuta, Fatigue Materials Transactions 2004, 45, 419–422.
[13] E. Aghion, D. Eliezer, Proc. Israeli Consortium for the Development of Magnesium Technology, Haifa, 2004.
[14] A. Eliezer, E.M. Gutman, E. Abramov, Y. Unigovski, E. Aghion, Proc. Magnesium Alloys and their Applications 2000, Wiley Pub., 499–507.
[15] E.M. Gutman, A. Eliezer, Y. Unigovski, E. Abramov, Materials Science Forum 2003, v.419–422, 115–120.
[16] Y. Unigovski, A. Eliezer, E. Abramov, Y. Snir, E.M. Gutman, Materials and Engineering-A 2003, 360, 132–139.
[17] A. Eliezer, J. Haddad, Y. Unigovski, E.M. Gutman, Static Materials and Manufacturing Processes 2005, 20, 75–88.
[18] G. Ben-Hamu, A. Eliezer, E.M. Gutman, Electrochimica Acta, in press (2006).
[19] G. Song, A. Atrens, X. Wu, T. Bao, B. Zhang, Corrosion Science and Engineering 1998, 40, 1769–1791.
[20] O. Lunder, J.E. Lein, T.K. Aune, K. Nisansioglu, Corrosion 1989, 45, 741–748.
[21] S.A. Michel, R. Kieselbach, M. Figliolino, Fatigue Fracture of Engineering Material Structures 2005, 28, 205–219.

Investigation of the Corrosion of Welded Magnesium Joints

C. Berger, K. Eppel, J. Ellermeier, T. Troßmann
Darmstadt University of Technology, Darmstadt, Germany
U. Dilthey, H. Masny, K. Woeste
RWTH - Aachen University, Aachen, Germany

1 Introduction

The request for weight saving in the production of moving components is, among other things, accomplished by the application of light-weight metals. The light metal "magnesium" is the metallic engineering material with the lowest density and it is, due to the high density-related strength parameters, predestined for material light weight structures. Apart from the large number of established cast material applications, magnesium alloys – if applied in modern joining processes – also offer a high potential for structures made of magnesium semiproducts. Joining methods are required which are applicable for magnesium cast parts and also for semiproducts made of magnesium wrought alloys. Hybrid joining methods (combination of mechanical, thermal joining and bonding) have a decisive part in the realisation of current and future lightweight concepts. If, in particular, magnesium materials are used, the application of mechanical joining methods is, however, limited. It is, therefore, assumed that, due to the high economic viability and further increase of applications, welding processes will be, also in the future, of vital importance.

The large number of welded light weight structures, especially in the automotive industry, show the importance of the systematic examination of thermal joints for magnesium applications. Apart from the economic viability and the process reliability of the joining methods, the required mechanical properties of the joint must be guaranteed over the total component life time. Special attention must be paid to the ageing and corrosion behaviour of the joint.

As far as the corrosion properties of magnesium alloys are concerned, the high electro-chemical reactivity and the connected low corrosion resistance under certain conditions must be observed. Under practice-oriented, neutral pH and/or slightly alkaline environmental conditions, for example outdoor exposure, the corrosion behaviour is decisively determined by the properties of protective layers for which, again, the chemical material composition, metallurgical and environmental influential parameters are of importance. A further, important influence on the corrosion behaviour stems from production induced contaminations which are located in the magnesium material and on its surface. These are frequently leading to the development of galvanic couples and to the obstruction of passivating processes with local activation [1–4].

Through the melting, solidification and incorporation or relocation of contaminations, the welding process exerts considerable influence on the material properties which are relevant for the corrosion behaviour. Therefore, and particularly in the case of magnesium alloys, a differentiated consideration of the process-specific influences and their effects on the corrosion properties of the welded joint must take place, especially if applied as design material for structural parts [5, 6, 7].

In the following, the results from corrosion tests on welded plates (AZ31, AZ61 (wrought alloys) and AZ91, AM50 (cast alloys)) are specified which, depending on the used welding method, were welded with different heat input. In section 2 it is referred to particularities of magnesium alloys with regard to the application of different **welding methods** and to important aspects of the corrosion behaviour of welded magnesium plates. Moreover, we go further into the question of which **microstructural changes** take place during cooling after the welding process and of their effect on the corrosion behaviour (section 3). Subsequently, the effects of the **surface condition**, influenced through the welding process, on the corrosion behaviour are specified and remedial measures are explained (section 4). Finally, the risk of **stress-corrosion cracking** caused through weld residual stresses is explained (section 5).

The results presented are mainly extracts from the AiF-sponsored research projects "Influence of energy-per-unit length, energy density and welding zone protection during welding on the corrosion resistance of magnesium alloys with and without dynamic loading (AiF No. 12786)" [8] and "Influence of pre- and post-treatment of magnesium semiproducts on the quality and the corrosion behaviour of welded magnesium materials (AiF No. 14451)".

2 Results from the Welding Tests

Within the scope of the AiF research project No. 12786 the following welding methods were used for the thermal joining of magnesium alloys:

- electron beam welding in vacuum (EBW)
- laser beam welding
- electron beam welding in atmosphere (NV-EBW)
- plasma welding and
- metal inert gas welding (MIG).

These welding methods differ from each other with regard to different process parameters; thus the separate consideration of the methods and their effects on the material, particularly on its corrosion properties is required, Figure 1.

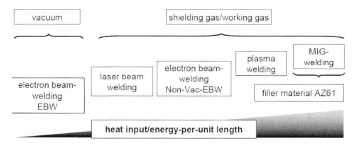

Figure 1: Welding methods, influential parameters

In the course of the welding tests which were carried out in the Welding and Joining Institute (ISF) of the RWTH Aachen it became apparent that it is possible to join all investigated magnesium alloys with the applied welding method and to yield a good weld quality. Compared with cast alloys, the magnesium wrought alloys entailed – this applies to all welding methods – better results particularly as far as the porosity of the welded specimens is concerned.

Electron beam welding in vacuum offers the process-specific advantages of high power density in the focussed beam and also optimal protection of the joining point from the air influences. The oxide layer is steadily vaporised. Due to the small beam cross-sections the developing seam width and the gap bridging ability are, process-typically, insignificant.

With the method **electron beam welding in atmosphere, (NV-EBW),** the oxide layer is also easily removable. The achievable high welding speeds which from the economic point of view are desirable, lead to the fact that the process reacts comparatively sensitive to slight parameter variations. The undercut formation which is caused by the beam characteristic must be avoided through the application of filler material.

In **laser beam welding** the pre-treatment of the parts is often necessary in order to have sufficiently good weld quality since thick oxidic layers and/or the casting skin/rolling scale are often not satisfactorily removed and/or penetrated by the laser beam. The energy of the Nd:YAG laser is, due to its wave length stronger absorbed by the magnesium base material and the coupling into the magnesium material is therefore easier, compared with the laser beam of the CO_2 laser, and it is therefore recommended for the joining of this material.

Joints made with the **plasma welding** method show a regular weld geometry with little weld sinkage and a smooth surface. In plasma positive pole welding the oxide layer is removed very reliably which is a main reason for the good weld quality.

The **metal inert gas welding process** reacted very sensitive towards slight external influences which resulted in a low reproducibility of the welds. The energy-per-unit length input, the drop detachment and the shielding gas coverage rate among these influences.

3 Influence of the Microstructure in the Welding Zone on the Corrosion Behaviour in the Welding Zone

By means of metallographical investigations of the aluminium and zinc-bearing magnesium alloys, which were carried out among the investigations concerning the corrosion behaviour in the Institute for Materials Technology of the Darmstadt University of Technology, it was documented that a fine-grained structure, similar to a cast microstructure, with a high proportion of secondarily solidified eutectic phase (which is observed as a grey band between the light α–solid solution) develops in the weld. This is particularly pronounced if a method with a low input of energy-per-unit length is applied since the fast cooling and solidification of the molten metal promotes the grain refinement; cognisable by comparing the laser beam weld (low energy-per-unit length) and the NV-electron beam weld (higher energy-per-unit length), Figure 2.

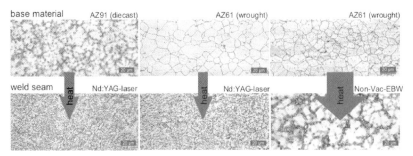

Figure 2: Etched cross-sections, AZ91, AZ61, laser beam and/or NV-EB welded

The used alloys and welding methods with a low energy-per-unit length were not characterised by a pronounced heat affected zone. Particularly with increasing heat input through the welding process the melting of grain boundaries and grain coarsening may occur in the heat affected zone.

The representative current density / potential curves of the plasma-welded specimens of the alloy AZ31 in a 0,05% NaCl solution which was saturated with magnesium hydroxide show, in a ground surface condition and in comparison with the base metal specimens only slight differences in their passivation properties, Figure 3.

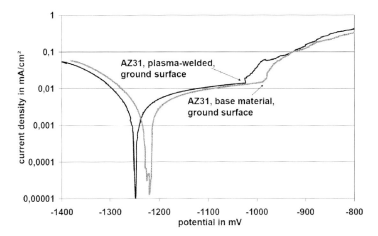

Figure 3: Current density/potential curves, AZ31, ground surface, comparison of specimens with and without (plasma) weld, 0,05 % NaCl saturated with magnesium hydroxide

The curves show comparable features which describe the passivation properties. Among those are the pronounced passivation range (range where the investigated specimen reacts to the increase of the potential only with a slight increase of the current density) and the transpassivation potential (breakthrough potential, jump of the current density after the passivation range). The range which has been tested electrochemically reaches over the entire welding zone (base material, heat affected zone and weld). A significant deterioration of the passivation properties, in comparison with the ground base material, was also not observed on the laser beam welded specimens with ground surface.

It may be deduced for the practical application of the examined welding methods and alloys that through the influence on the weld microstructure alone no significant, negative influence on the corrosion resistance occurs. These findings are confirmed through the steady corrosion attack in the range of the welding zones, in immersion tests and corrosion tests according to VDA 621-415 on specimens (laser-, electron beam welded) with a ground surface. A very high energy input during welding, however, may increase the influence on the adjacent base material (development of the heat affected zone). Through the dissolution of grain boundaries coarse grains may develop which would locally promote a faster progression of corrosion.

3 Influence of the Surface Properties on the Corrosion Behaviour of the Welded Zone

Despite the microstructure formation in the welding zone is uncritical for the corrosion resistance, untreated specimens (surface as-welded) are frequently subject to local corrosion attacks. Through local melting during the welding process and dependent on the applied method (energy-per-unit length, pre-treatment) changes of the material surface occur besides the influence on the microstructure. The immersion tests showed that these influences on the surface in the welding zone have the effect that the investigated plates (made of magnesium alloys) with untreated surface show local corrosion attacks which are similar to pitting. This applies particularly to the tested cast alloys which had been welded with a high heat input. The observed appearance of corrosion attacks is concentrated on the edge of the weld, Figure 4.

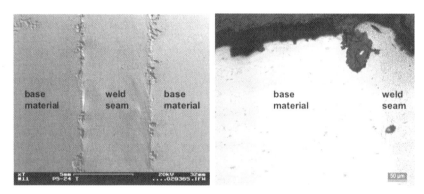

Figure 4: 1 h Immersion test: AZ91 plasma welded in 5 % NaCl-solution at 30 °C, surface as-welded, left: SEM-image, right: metallographic crosssection

This is, particularly in the case of cast alloys, to be ascribed to increased pore formation in the weld edge areas and to (welding-) process-induced locally increased concentrations of contaminations. Such contaminations on the base material surface (induced by production of the semiproducts) may, in parts, flake in the form of slag or remain in the molten pool where they are locally concentrated through pool motion during welding. Increased contamination in the corrosion attack zones with the elements carbon (C) and sulphur (S) were substantiated by EDX analyses. That means that methods with high energy-per-unit length which lead to the development of voluminous molten pools and thus promote the specified relocation and local concentrations of contaminations may result in increased corrosion at the transition zone between weld and base material. Particularly components made of magnesium alloys with a high content of near-surface contaminations are characterised by significantly deteriorated corrosion properties in the welding zone. In the immersion tests, the welded plates of the alloy AZ61 which were in untreated surface condition showed slightest and steadiest corrosion of all tested alloys. Glow discharge spectroscopy (GDOS) was used to demonstrate that the plates (alloy AZ61) showed the smallest quantities of near-surface contaminations (C, S) of the examined materials.

For practical application it is deduced that the use of semi-finished products with lowest possible near-surface contaminations – like here the alloy AZ 61 – is at least one of the most important prerequisites for a corrosion-resistant welding zone in the case of magnesium alloys. In addition and/or as an alternative the demands on a steady quality of the joints require an appropriate pre-/post treatment, e.g. grinding, blasting or pickling in order to ensure a homogeneous surface and to avoid corrosion-induced notches which are particularly critical when mechanical stresses are applied.

4 Measures to Improve Corrosion Behaviour in the Welding Zone

The influence of pre-treatment of the joining parts on the welding process and the corrosion resistance of welded magnesium alloys are examined within the scope of the AiF research project (AiF No. 14451). The results which have been received so far point to a fundamental improvement of corrosion properties through the ablation of potentially contaminated surface areas before welding by means of grinding or pickling.

Through a post-treatment of NV-EB welded plates of the alloy AZ 61 (pickling after welding) it was demonstrated that through the ancillary-process, also if a starting material with slight subsurface contaminations was used, an improvement of the passivation properties (increase of the passivation range, shift of the transpassivation (breakthough-) potential) in the welding zone and thus higher corrosion resistance may be accomplished, Figure 5.

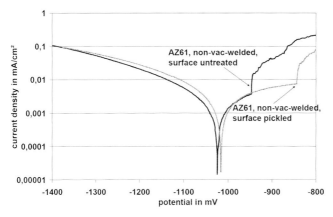

Figure 5: current density/potential curves, AZ61, NV-EB-weld, comparison of untreated specimens and of specimens which were pickled after welding, 0,05 % NaCl saturated with magnesium hydroxide

5 Influence of Residual Stresses on the Corrosion Behaviour in the Welding Zone

Under specific conditions, stress-corrosion cracks may develop in magnesium-aluminium alloys. Stress-corrosion cracking (SCC) is understood as crack development which may occur in materials with stress-corrosion cracking susceptibility under the influence of critical corrosive environmental conditions and simultaneous tensile stresses. Production-induced residual stresses may act as tensile stresses, for example, through the application of thermal joining methods. After a 50-week outdoor exposure (in accordance with VDA 621-414) non-vac-electron beam welded and laser beam welded magnesium plates with the alloys AZ91 and AZ61 showed cracks. Stress-corrosion cracking is causative for crack initiation. What happens is a local disturbance of the passive state on the surface through the tensile residual stresses caused by welding (passivation is a prerequisite for stress-corrosion cracking). They were demonstrated by means of radiography measurements on specimen plates which were joined with different welding methods. In the plasma-welded and in the non-vac-welded plates a tendency towards higher tensile residual stresses than in laser-beam welded plates was demonstrated.

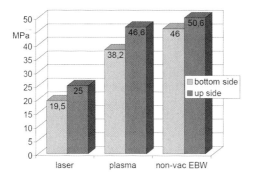

Figure 6: Radiography measurement of residual stresses, AM50, mean values of the longitudinal stresses from 5 series of measurements, weld centre

The failure pattern shows the inter- and transgranular crack path which is typical of the stress corrosion cracking in magnesium alloys, Figure 7.

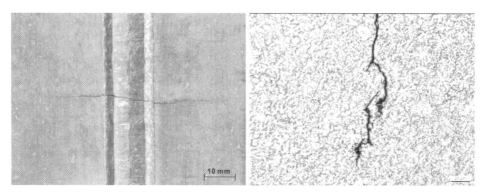

Figure 7: Stress corrosion crack AZ61, 50 weeks outdoor exposure to air according to VDA 621-414, left: Non-Vac-welded, right: laser beam welded

Within the scope of the investigations, the plates which were joined by the plasma welding method showed no stress-corrosion cracking, despite the likewise high residual tensile stresses. This is, among other things, ascribed to the more favourable weld geometry and the different cooling conditions, compared with the laser beam and non-vac-welded specimens without filler material.

The risk of stress corrosion cracking is a considerable impediment to the application of magnesium alloys under corrosive conditions. There is substantial need for research in the systematic investigation and the avoidance of stress corrosion cracking, particularly in welded plates.

6 Summary

Several methods are applicable for the welding of magnesium alloys and, particular with wrought alloys, they lead to good results as far as the weld quality is concerned. Especially laser beam welding and plasma positive pole welding have a large potential for the application in the field of magnesium lightweight construction.

If welding methods with a low heat input are applied, the reduction of the corrosion resistance of welded magnesium alloys is less induced by the microstructural changes in the welding zone but more through the influence on the surface properties. Near-surface contaminations in the semiproducts which are to be joined play a decisive role: Thus the demand for just slightly contaminated semi-finished and/or suitable cleaning pre-treatment of the parts which are to be joined is derived. The corrosion behaviour in the welding zone is also improved through follow-up treatment, for example pickling after welding.

The residual stresses in the weld which have been caused through welding may lead, particularly with unfavourable weld geometry, also without superimposed external loads, to stress-corrosion cracking and are, therefore, an obstacle for the application of welding of magnesium alloys. There is substantial need for research in the systematic investigation of the causes of the development of stress-corrosion cracking on welded magnesium alloys and its avoidance through suitable, fabrication-related or metallurgical measures.

7 Acknowledgement

The research projects AiF 12786 and 14451 were sponsored by the Bundesministerium für Wirtschaft und Technologie (BMWi) via the German Federation of Industrial Cooperative Research Associations "Otto von Guericke" e.V. (AiF) and the "Gesellschaft für Korrosionsschutz e.V. (GfKORR)". We wish to express our thanks to BMWi, AiF and to GfKORR for their promotion and financial support for this research project.

8 References

[1] Song, G.; Recent Progesses in Corrosion and Protection of Magnesium Alloys. Advanced Engineering Materials, No. 7, 2005.

[2] Troßmann, T.; Verhalten von Magnesium-Legierungen bei korrosiver und mechanisch-korrosiver Beanspruchung. Dissertation D17, TU-Darmstadt, 2005.

[3] T. Troßmann, K. Eppel, M. Gugau, C. Berger in Proccedings of the 7th Int. Conference on Magnesium Alloys and their Application, 2006

[4] Gugau, M., Troßmann, T., Berger, C.; Einsatzmöglichkeiten und Grenzen von Magnesiumlegierungen unter komplexen Beanspruchungen. DVM/FKM/VDI-Tagung, Fulda, 2003.

[5] Kopp, J.; Magnesium im Automobilbau – Korrosionserscheinungen an Karosseriebauteilen aus Magnesium verursacht durch legierungs- und verfahrensbedingte Fremdmetallverunreinigungen sowie lokal unterschiedliche Gefügezustände. Dissertation, Uni Erlangen, 1996.

[6] Ellermeier, J., Eppel, K., Berger, C.; Einfluss der Schweißverfahren auf die Korrosionsbeständigkeit von Magnesiumlegierungen. DVM/FKM/VDI-Tagung, Fulda, 2003.

[7] Eppel, K., Ellermeier, J., Woeste, K., Berger, C., Dilthey, U.; Korrosion von Magnesium-Schweißverbindungen. Schweißen und Schneiden (Nr.7, 2006), DVS-Verlag GmbH, Düsseldorf.

[8] Berger, C., Dilthey, U., Ellermeier, J., Eppel, K., Woeste, K.; Einfluss von Streckenenergie, Energiedichte und Schweißzonenschutz beim Schweißen auf die Korrosionsbeständigkeit ohne und mit schwingender Beanspruchung von Magnesiumlegierungen. Abschlussbericht, AiF 12786 N.

Effect of Salt Spray Corrosion on the Tensile Behavior of Wrought Magnesium Alloy AZ31

Sp.G. Pantelakis, N.D. Alexopoulos and A.N. Chamos

Laboratory of Technology & Strength of Materials, University of Patras, Greece

1 Introduction

Magnesium is the lightest structural engineering metal, and therefore, most attractive for structural applications where weight saving is of serious concern. Improvements in mechanical properties, corrosion resistance [1, 2] and the development of advanced manufacturing processes have led to increased interest in magnesium alloys and many efforts are in progress to produce magnesium alloys for aerospace and automotive applications [3-5]. Nevertheless, the corrosion properties require special attention in view of the fact that magnesium is at the active end of the galvanic series. In the past decade, significant improvements have been achieved on the issue of corrosion resistance by reducing the heavy metal (Fe, Cu and Ni) impurity level, and this has led to the development of the 'high-purity' alloys [6, 7].

The effect of both, the impurity level and the different corrosive environment on the material's corrosive behavior has been investigated in detail [7-10] for the case of commercial, cast magnesium alloys, e.g. alloy AZ91. It is worth noting that although the evaluation of the effect of prior corrosion to the mechanical performance of structural engineering alloys used in lightweight structures is of primary concern for the widely used wrought aluminum alloys, e.g. [11, 12], this issue is still only rarely faced in the scientific literature for the case of wrought magnesium alloys.

The present paper aims to contribute on assessing the corrosion resistance of the AZ31 magnesium alloy. The tensile mechanical performance of pre-corroded wrought magnesium alloy AZ31-O is evaluated. Salt spray fog environment has been used for the accelerated corrosion of the material. The corrosion damage was evaluated by means of pitting density and pit depth measurements for the various corrosion exposure times. Tensile specimens of the investigated alloy were cut in L and LT directions and were exposed for seven different corrosion times in a salt spray chamber. After corrosion exposure the specimens were subjected to tensile test. The dependency of the tensile properties on the severity of corrosion damage has been experimentally derived and fitted by means of appropriate expressions.

2 Experimental Investigation

The material investigated in this work was an AZ31 magnesium alloy, received in sheet form with nominal thickness of 2.0 mm. The chemical composition of the investigated alloy is consistent to the DIN-EN 10002 specification. The material was provided in the 'O' heat treatment condition, which was annealing at 300 °C for 30 min. It is worth noting that in practical applications magnesium alloys are not used as bare materials but are protected against corrosion by using coatings. Furthermore, the salt spray test is an accelerated laboratory corrosion test and, hence causes a corrosion attack which as compared to the expected corrosion in real service conditions is much more severe. The aim of the present investigation is to provide fundamental knowledge on the corrosion resistance of magnesium alloys in aggressive environments, as it is the salt spray exposure.

Metallography and tensile specimens were cut in both longitudinal (L) and longitudinal transverse (LT) directions from the sheets. The metallography specimen was a rectangular one having 100x50 mm dimensions. Tensile specimens were machined according to the ASTM E8M specification with 50 mm gauge length and 12.5 mm gauge width at the reduced cross section.

Prior to the corrosion exposure, the surface of the tensile specimens was chemically cleaned in order to remove any oily lubricant, left from the rolling process of the material. Two subsequent solutions were used. The first treatment was immersion of the specimens for 1 min to a solution contained 10 % HNO_3 and ethanol. Immediately after the first treatment, the specimens were immersed for another minute in a solution contained 10 % NaOH and distilled water. Afterwards the specimens were dried and immediately inserted to the corrosive environment.

For the corrosion tests, the accelerated salt spray fog environment has been used. The salt spray tests were conducted according to ASTM B117 specification. After corrosion exposure the specimens were cleaned according to ASTM G1 specification.

A light microscope Leica DM LM had been used to measure the average and maximum pit depth of the metallography specimens for the investigated exposure times. The surface of each plate was divided into sections of 100 mm^2 in order to measure the pitting density and to calculate the pitting factor. The pits depth and density were measured and classified according to ASTM G46.

A servo hydraulic MTS 250 KN machine was used for the tensile tests. The tests were carried out according to ASTM E8M with a constant elongation rate of 2 mm/min. A data logger was used to store the data in a digital file. To get representative average values of the residual tensile properties, for each exposure time, a minimum number of three tensile tests has been carried out. In total, 46 tensile tests were performed and evaluated. The evaluated properties were: yield strength R_p (0.2% proof stress), tensile strength R_m, elongation to fracture A_f, and strain energy density W (tensile toughness). Strain energy density has been calculated as the integral of the engineering tensile stress-strain curves up to elongation to fracture. Involving engineering stress-strain curves instead of true stress-strain curves for calculating W is justified as the observed tensile necking at the elongation to fracture has not been appreciable.

3 Material Investigation

3.1 Surface Corrosion Evaluation

In the early stages of corrosion the initiation of the pits is confined to very few sites associated with intermetallic particles. The pits initiate already at 0.5 hour exposure, and the number and depth of the pits increased with increasing exposure time to the corrosion environment. Typical types of pits are presented in Figure 1 for the case of 3 and 24 hours salt spray exposure. The evident mechanism is the wide and shallow pits for the early stages of corrosion exposure, e.g. for the case of 3 hours corrosion in Figures 1a, b. The pitting mechanism seems to change by increasing the corrosion exposure to times higher than 12 hours. For the 24 hours corrosion exposure time, some wide and shallow pits are also present (Figure 1c), but the majority are narrow and deep pits (Figure 1d). The measurements of the pitting density and pitting depth of the investigated corrosion exposure times can be seen in the graphs of Figure 2. The pitting density increases almost linear with the increasing corrosion salt spray exposure time. The average depth of corrosion attack (pit depth), as well as the maximum pit depth can be seen in Figure 2b as a function of the investigated corrosion exposure times. The pit depth seems to increase exponentially with increasing corrosion exposure time. It should be mentioned that for the case of 72 hours exposure time, through-thickness holes could be seen (nominal thickness of the material = 2.0 mm).

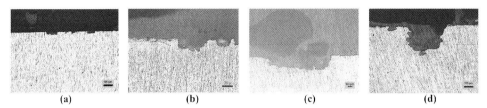

Figure 1: Metallography cross-sections of the pits for salt spray corrosion exposure for (a,b) 3 h and (c,d) 24 h

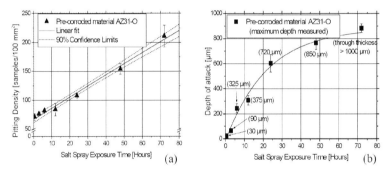

Figure 2: Pitting density and depth of attack measurements of the pre-corroded specimens of AZ31-O material.

3.2 Mechanical Performance Evaluation

Typical engineering stress-strain curves of the reference and pre-corroded specimens for various corrosion exposure times can be seen in Figure 3. The reference tensile properties of the investigated alloy can be also seen in Table 1. The values fit well to the range of values given in DIN-EN 10002 for the same alloy. The increasing corrosion exposure time decreases significantly the ductility of the material.

The decrease in tensile properties with increasing corrosion exposure times for the L direction can be seen in Figure 4. A continuous, almost linear decrease in the tensile strength can be seen with increasing corrosion exposure time; it reaches values 170 MPa for the case of exposure for 72 hours. This corresponds to almost 38 % reduction of the initial value. The corrosion exposure has a similar effect on the yield strength of the L direction. On the other side, a dramatic reduction in ductility is observed (Figure 4b), even at low corrosion exposure times, e.g. 6 hours. At higher corrosion exposure times, the decrease in both ductility properties, i.e. elongation to fracture and strain energy density becomes dramatic. The percentage decrease of the mechanical properties for the L direction with increasing exposure time is graphically depicted in Figure 5. The strength properties have a small degradation rate when compared to the ductility properties. As it can be seen in Figure 4, the reduction in elongation to fracture and strain energy density values was about 37% already for 6 hours exposure. For higher exposure times (i.e. 72 hours), the ductility decrease reaches 90 % of the initial value.

Table 1: Tensile properties for the reference material.

AZ31 - O	R_p [MPa]	R_m [MPa]	A_f [%]	W [MJ/m^3]
L- direction	165	263	19.45	48.00
LT-direction	193	264	20.00	50.80

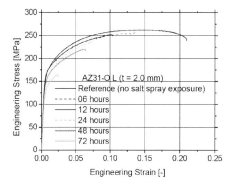

Figure 3: Engineering tensile stress-strain curves of the AZ31-O material for the reference and for the investigated pre-corroded specimens.

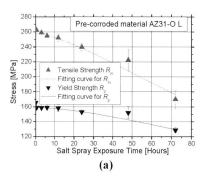
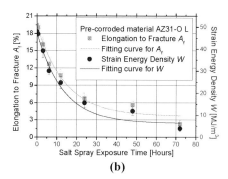

(a)　　　　　　　　　　　　　　　　(b)

Figure 4: (a) Yield and tensile strength and (b) elongation and strain energy density of AZ31-O material in L direction after exposure in salt spray environment.

Displayed in Figure 6 are the tensile mechanical properties as a function of the increasing corrosion exposure time for the LT direction. When compared to Figure 4, it is evident that both directions have the same qualitative degradation of the mechanical properties with increasing exposure time. As for the L direction, also for the case of LT direction, a small reduction in strength and a dramatic reduction in ductility is observed. Worth mentioning is the corrosion exposure time of 6 hours, where a 4 % decrease in the strength properties is observed for both directions in Figure 7. For the case of the ductility properties, the observed decrease is of the order of 38 %. For higher exposure times (72 hours) the strength properties decrease for almost 40 %, while a dramatic reduction of 90 % in the ductility properties is observed. It is noticeable that a dramatic ductility decrease has been also observed for a series of structural Al alloys like 2024, 6013, 7075 etc., after corrosion exposure. Extensive investigations have attributed the observed ductility decrease to the synergy of corrosion notches and corrosion induced hydrogen embrittlement. To identify the causes of the observed ductility reduction for the magnesium alloy under consideration is subject of further investigation.

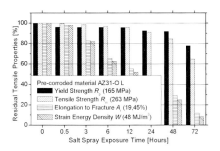

Figure 5: Residual tensile mechanical properties of the AZ31-O material in L direction after exposure in salt spray environment

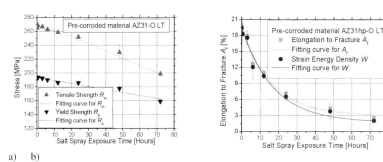

Figure 6: (a) Yield and tensile strength and (b) elongation and strain energy density of AZ31-O material in LT direction after exposure in salt spray environment

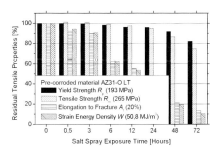

Figure 7: Residual tensile mechanical properties of the AZ31-O material in LT direction after exposure in salt spray environment.

4 Conclusions

- Corrosion exposure to salt fog leads to extensive pitting of the material. Pitting density seems to increases linearly with exposure time.
- The pitting morphology seems to change by increasing the corrosion exposure time to higher than 12 hours.
- Corrosion-induced degradation of mechanical properties occurs gradually. Yield and tensile strength decreased permanently at a decreasing rate. Tensile ductility decreases exponentially to extremely low final values.
- The lack of a quantitative correlation between accelerated laboratory corrosion tests and in-service corrosion attack or, atmospheric corrosion tests calls for additional investigation related to corrosion of Mg structures.
- The development of "high purity" magnesium alloys, including smaller contents of heavy metals Fe, Ni, Cu is needed to improve corrosion resistance.

5 References

[1] M. Avedesian, H. Baker, ASM Specialty Handbook, 2nd ed., Ohaio, USA, 1999.
[2] G. Song and A. Atrens, Adv. Eng. Mater. Rev. 1999, 1, 11–33.
[3] Anon. AEROnautical application of wrought MAGnesium (AEROMAG). EC/FP6 - Growth/Aeronautics, 2005–2008.
[4] Anon. Integrated Design and product development for eco-efficient production of low-weight Aeroplane equipment (IDEA). EC/FP6 – IP/Aeronautics, 2005–2008.
[5] Anon. Innovative Magnesiumverbundstrukturen im Karosseriebau. German National Funded Project, 1999–2002.
[6] T. Aune in Magnesium and Magnesium Alloys, Int. Mag. Assoc., Toronto, Ontario, Canada, 1983.
[7] K. Reitchek, K. Clark, J. Hillis, SAE Technical Paper No. 850417, 1985.
[8] O. Lunder, J. Lein, S. Hesjevik, T. Aune, K. Nisancioglu, Werkstoffe und Korrosion, 1994, 45, 331–340.
[9] R. Ambat, N.N. Aung, W. Zhou, Cor. Sci., 2000, 42, 1433–1455.
[10] G. Song, A. Atrens, M. Dargusch, Cor. Sci., 1999, 41, 249–273.
[11] Sp.G. Pantelakis, P.G. Daglaras, Ch.Alk. Apostolopoulos, Theor. And Appl. Frac. Mech., 2000, 33, 117–134.
[12] P.V. Petrogiannis, Al.Th. Kermanidis, P. Papanikos, Sp.G. Pantelakis, Theor. And Appl. Frac. Mech., 2004, 41, 173–183.

Characterization of Corrosion Interfaces in Extruded Mg-Al-Zn Alloys

Iulian Apachitei, Francesco Andreatta[1], Lidy E. Fratila-Apachitei, Joanna Dzwonczyk, Ahmed Berkani, Jurek Duszczyk

Delft University of Technology, Department of Materials Science and Engineering, Delft, The Netherlands
[1]present address: University of Udine, Department of Chemical Science and Technology, Udine, Italy

1 Introduction

Although magnesium alloys have received increased interest in the last decade for structural applications, the vast majority of the alloys used were cast and semi-solid formed components [1,2]. Despite the better mechanical properties of the wrought alloys [3], their production still poses technical challenges due to the poor formability of magnesium (high plastic anisotropy and limited room temperature elongation) imposing high deforming force and low deformation speed that make the processes not cost-effective. The applications for wrought alloys include automotive, aerospace and electronic industries that demand high ductility, high strength and high corrosion resistance at room and elevated temperature. The research is focused on the addition of new alloying elements (e.g. Zr, RE, Cu, Ag, Si) to the wrought alloy systems (Mg-Al, Mg-Zn, Mg-Th, Mg-Mn alloys) to improve their mechanical properties and efficiency of forming [4]. However, due to the high reactivity of magnesium alloys in chloride environments, their competitiveness as structural materials will depend on the corrosion resistance under (non-)stressed conditions.

Since the corrosion response is alloy dependent, investigations at different length scales and under different exposure conditions is required. These are enabled by the availability of local electrochemical techniques that allow examination of corrosion interfaces with a high spatial resolution. The scanning Kelvin probe force microscopy (SKPFM) and the microcell technique represent two of the most promising methods for investigation of alloys corrosion locally [5-12]. The first provides the Volta potential difference between structural features under dry conditions [5] whereas the second provides the corrosion response (corrosion potential, current, breakdown potential) of selective areas in contact with corrosive solutions [6]. These techniques have been used for aluminum alloys to identify the role of different second phase particles on initiation and propagation of corrosion [7-10]. For magnesium alloys, the research is relatively limited [5, 11-12] and deals mostly with AZ91 cast alloy. The results obtained with the cast alloys cannot be extrapolated to the extruded counterparts due to the significant differences in the two types of microstructures that can change the corrosion behavior [3,13].

This paper provides results on characterization of two different Mg-Al-Zn extruded alloys by SKPFM and microcell technique. The effects of second phase particles on alloy electrochemical response are discussed. In parallel, the macro-electrochemical response by potentiodynamic polarizations and the morphology of the corrosion attack following immersion tests are included.

2 Experimental

2.1 Magnesium Alloys

The chemical composition (wt%) of the two alloys used in the investigation was determined by X-ray fluorescence and is included in Table 1.

Table 1: Chemical composition of the magnesium alloys (wt%)

Alloy	Mg	Al	Zn	Mn	Ca	Si	Fe
Mg-4Al	Bal.	4.76	0.85	0.32	0.06	0.01	0.008
Mg-8Al	Bal.	7.81	0.44	0.23	0.04	0.03	0.009

The second alloy has the composition in the range of the AZ80 alloys whereas the first is an experimental composition between AZ31 and AZ60 alloys (i.e. AZ41), tested for extrusion formability. The alloys have been extruded under controlled conditions (T_{billet} = 400 °C for AZ80 and 450 °C for AZ41, ram speed 1 mm/s, reduction ratio 1:22.6) in rectangular profiles (30 mm × 3 mm).

2.2 Microstructural Characterization of Magnesium Alloys

The microstructure of the two alloys was examined by optical microscopy using an optical microscope type Olympus BX60M and an image analysis software to determine particles area fraction and particles size. Prior to examination, the samples were embedded in cold resin with the extruded surface exposed for analysis, ground to 600 SiC abrasive paper and polished with diamond suspensions of 9.0, 6.0, 3.0, 1.0 and 0.25 µm using an automatic Struers machine. The composition of the main particles in the alloys was previously determined by SEM/EDS [12].

2.3 Electrochemical Characterization

2.3.1 Macro-electrochemical Investigations

Open circuit potential (OCP), potentiodynamic polarisation and immersion tests were performed in 0.1M NaCl solution (pH = 6.7). For the OCP and polarisation experiments, the specimens cut to expose the extruded surface (10x5x3 mm^3) were electrically contacted with a screw welded wire and embedded in cold resin for polishing. The experiments were carried out using an EG&G PAR 273A potentiostat and a three-electrode cell with a standard calomel electrode (SCE) and a platinum counter electrode. The scans were performed at a rate of 0.0002 V/s and a step potential of 0.00015 V. For immersion tests, extruded specimens in different surface state, i.e. as-extruded, ground and polished were exposed to 0.1M NaCl solution, for different durations (e.g. 1.0, 3.0, 5.0, 8.0 hours). The morphology of the corrosion attack was examined by optical microscopy.

2.3.2 Micro-electrochemical Investigations

The SKPFM characterization was carried-out in air at room temperature using a Nanoscope III Multimode atomic force microscope equipped with an Extender TM Electronic Module. The scan height in the lift mode was 100 nm. The potential maps were sampled with a pixel density of 256×256 and with a scan frequency of 0.1 Hz. For all measurements, commercially available n⁺-silicon tips coated with $PtIr_5$ were used. At least two different polished samples for each alloy were examined and few tens of particles of each type were investigated by section analyses.

The experiments with the microcell technique were performed in 0.1M NaCl solution (pH = 6.7) using glass capillaries of 50 µm diameter, a standard calomel electrode as reference electrode and a Pt wire as counter electrode. The potentiodynamic polarisations were performed at a rate of 0.01 V/s. The equipment used was a Swiss microcell system. At least two different polished samples were examined for each alloy and tens of potentiodynamic polarisations were performed on different selected locations.

3 Results and Discussion

3.1 Alloys Microstructure

By optical microscopy (Fig. 1) it was revealed that both alloys contained particles distributed randomly on the surface with some of them having the tendency to follow the extrusion lines. In the AZ41 alloy (Fig. 1a) the particles were smaller than in the AZ80 alloy (Fig. 1b) with average sizes between 3-7 µm and 4-14 µm, respectively. The area fraction of the particles estimated by image data analysis using the equivalent circle diameter indicated a value of 0.47 ± 0.03 % for the AZ41 and 0.83 ± 0.2 % for the AZ80 alloy. One type of particles (dark gray particles pointed by arrows in Fig. 1) with mostly polygonal morphology, dominated the microstructure of the two alloys. They were identified as Al-Mn-Fe particles by SEM/EDS analysis [12]. In addition, the AZ80 alloy revealed the presence of $Mg_{17}Al_{12}$ intermetallic (β phase) with a broken, globular morphology distributed mostly along the extrusion lines (light gray encircled area in Fig. 1b). Other type of particles included Mg-Si based intermetallics but in significantly lower amount relative to the above mentioned particles.

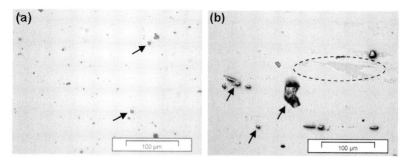

Figure 1: Microstructure of the AZ41 (a) and AZ80 (b) alloys

The examination of grains morphology revealed a dynamically recrystallized structure with equiaxed grains of about 21.2 ± 1.3 µm for the AZ80 alloy whereas the AZ41 alloy revealed a relatively non-uniform grain morphology with both equiaxed and elongated grains of about 23.32 ± 3.52 μm suggesting that the recrystallization was not complete and the extrusion parameters should be further optimized for this alloy.

3.2 Macro-electrochemical Characterization

3.2.1 Open Circuit Potential and Potentiodynamic Polarization Measurements

A difference of about 30 mV in the OCP after 24 hours exposure time was observed between the two alloys (Fig. 2a) with a shift to more positive values for the AZ41. The OCP of the alloys is primarily influenced by the composition of the solid solutions with cathodic alloying elements increasing it. The magnesium matrix is a solid solution of Mg, Al and Zn. The maximum solid solubility of zinc at room temperature is about 1.2 wt% [14] whereas of aluminum about 2.0 wt% [2]. However, in high aluminum containing alloys (e.g. AZ91) Zn can be found in higher concentration in the β phase than in the matrix [15]. Based on the composition of the two alloys (Table 1), the higher OCP for the AZ41 alloy may have been determined by the higher concentration of Zn in this alloy. The results are in line with previous data [15-16] showing that Zn can cause a more positive corrosion potential for Mg-Al series alloys whereas Al may decrease it.

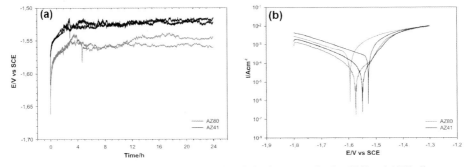

Figure 2: Open circuit potential and potentiodynamic polarization curves for the AZ41 and AZ80 alloys

The potentiodynamic polarization curves (Fig. 2b) revealed no difference in the breakdown potential and anodic current densities between the two alloys. The backward scan suggests a possible different repassivation kinetics for the two alloys.

3.2.2 Morphology of the Corrosion Attack

Typical surface morphology of the specimens following immersion tests in 0.1M NaCl solution is shown in Figure 3. Regardless of the surface preparation procedure, the corrosion attack was in the form of pitting followed by filiform growth in the case of AZ41 alloy (Fig. 3a-c) and a more uniform attack in the case of AZ80 alloy (Fig. 3d-f). The latter presented fewer but significantly larger pits on the surface after more than 3 hours of exposure (Fig. 3f). Further, for both alloys, the matrix around the Al-Mn-Fe particles was attacked (e.g. Fig. 3c). The filiforms grew mostly along the extrusion lines but also in radial direction (Fig. 3b) with no apparent relation with the presence of particles that were rather surrounded by the advancing filiforms (Fig. 3c). There were also areas on the surface with no filiforms. In the case of polished and embedded

AZ41 samples, the attack initiated mostly at the interface with the resin (Fig. 3a). Hydrogen evolution was associated with corrosion of both alloys.

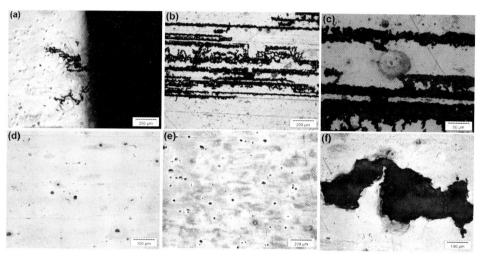

Figure 3: Optical micrographs of the specimens after immersion in 0.1M NaCl solution for different durations: (a) AZ41-1h (sample/resin interface); (b) AZ41-8h; (c) AZ41-8h (filiform arrest in the vicinity of an Al-Mn-Fe particle); (d) AZ80-3h; (e) AZ80-8h; (f) AZ80-8h (pit)

The non-uniform filiform corrosion attack on the AZ41 alloy may have been determined by the non-uniform grain morphology following extrusion. Similar corrosion morphology, usually encountered in the case of coated metals, was evidenced in the case of the AZ91-T4 cast alloy being caused by the crystallographic orientation following homogenization [17].

3.3 Micro-electrochemical Characterization

The SKPFM results are summarized in Figure 4 where an example of a Volta potential map (Fig. 4a), the line scans for intermetallic particles with the corresponding Volta potential difference (ΔV) relative to the adjacent matrix (Fig. 4b) and the ΔV for the particles in the two alloys (Fig. 4c) are included.

All the second phase particles detected in the alloys revealed a positive Volta potential difference relative to the matrix with the largest values for Al-Mn-Fe particles followed by Al-Mg (AZ80) and Mg-Si. The values for the Al-Mg particles are in line with those previously obtained for the AZ91 alloy [5]. The possible differences in the composition of the two solid solutions with respect to Zn did not change the Volta potential differences of similar micro-galvanic couples although the scattering of the data is high, especially in the case of Al-Mn-Fe particles. The largest values for the Al-Mn-Fe/matrix couple were detected in the AZ80 alloy that showed both a lower OCP value and a higher Fe/Mn ratio (0.039 vs 0.025) relative to the AZ41 alloy. The latter gives an indication of the Fe/Mn ratio in the Al-Fe-Mn particles [18]. This ratio may influence the Volta potential difference between the particles and the matrix, with larger values giving larger ΔV.

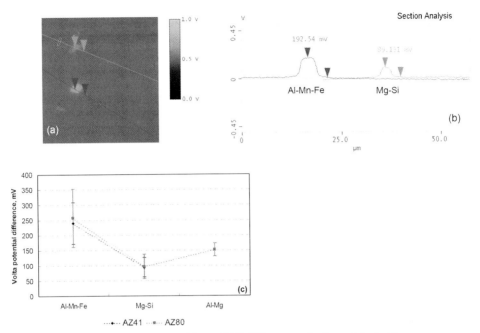

Figure 4: SKPFM results: (a) Volta potential map AZ41; (b) line scans for the particles in (a); (c) ΔV for the two alloys

Using the microcell technique, potentiodynamic polarisation curves for areas with no particles, areas with one type of particles, areas with more particles of the same type and areas with different type of particles were recorded. The main results for the AZ41 and AZ80 alloys are included in Figure 5.

For both alloys, the presence of the Al-Mn-Fe particles on the examined surface extended the passivity range and shifted the corrosion potential to more positive values when compared with areas containing no particles (i.e. matrix). The effect was more pronounced when large particles or more than one particle was present on the selected area (Fig. 5c). This finding suggests that, in chloride solutions at the used pH, the reactions mechanism in the presence of the Al-Mn-Fe particles, possibly associated with a change in the local pH and/or formation of passive corrosion products, may lead to an extended passivity range of the area locally. This finding requires further clarification by performing investigations under different pH conditions and with controlled concentrations of Mn in the alloys. Mn addition in the alloys renders the iron less active as local cathode by formation of Al-Mn-Fe particles [18]. Optimum concentrations of Mn for good appearance (no surface cracks) and mechanical properties were reported in the range 0.2–0.4 wt% for extruded AZ31 alloy [19]. The results obtained with the microcell on matrix areas support the macro-electrochemical results suggesting a more positive corrosion potential for the AZ41 matrix.

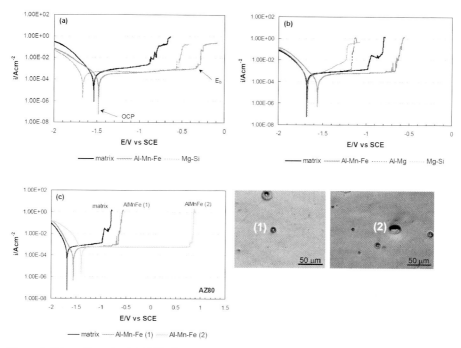

Figure 5: Microcell results: (a) AZ41 alloy; (b) AZ80 alloy; (c) effect of Al-Mn-Fe particle size on the micro-electrochemical response (AZ80). Attached the optical micrographs with the particles analysed in (c)

4 Conclusions

The corrosion response for two different extruded magnesium alloys (i.e. AZ41 and AZ80) was investigated using macro- and micro-electrochemical techniques coupled with optical microscopy.

The morphology of the corrosion attack at macro-scale was affected by the presence of Al-Mn-Fe particles around which the matrix was attacked. Further, the alloys revealed pits formation that evolved in filiforms in the case of the AZ41 alloy. The relatively fewer pits in the case of AZ80 grew in size with exposure time. The susceptibility of the AZ41 alloy to filiform corrosion may be related to the non-uniform grain morphology following the extrusion process that was not optimized for this alloy composition.

The micro-electrochemical investigations by SKPFM revealed positive Volta potential difference for all the intermetallic particles relative to the adjacent magnesium matrix, with values depending on particles composition. In NaCl solution at pH 6.7, the micro-electrochemical response of the areas bearing Al-Mn-Fe particles revealed a shift of both the corrosion potential and the breakdown potential towards more positive values extending the passivity range relative to areas containing no particles.

5   References

[1] I.J. Polmear, Mat. Sci. Technol. 1994, 10, 1–16.
[2] I.J. Polmear in Magnesium and Magnesium Alloys (Eds. M.M. Avedesian, H. Baker), ASM Specialty Handbook, USA 1999, p. 12–25.
[3] Ya. Unigovski, A. Eliezer, E. Abramov, Y. Snir, E.M. Gutman, Mat. Engg. 2003, A360, 132–139.
[4] Y.G. Na, D. Eliezer, K.S. Shin, Mat. Sci. Forum 2005, 488–489, 839–844.
[5] D. Brengtsson Blucher, J.-E. Svensson, L.-G. Johansson, M. Rohwerder, M. Stratmann, J. Electrochem. Soc. 2004, 151, B621–B626.
[6] T. Suter, H. Bohni, Electrochim. Acta 1998, 43, 2843–2849.
[7] T. Suter, R.C. Alkire, J. Electrochem. Soc. 2001, 148, B36–B42.
[8] V. Guillaumin, P. Schmutz, G.S. Frankel, J. Electrochem. Soc. 2001, 148, B163–B173.
[9] P. Schmutz, G.S. Frankel, J. Electrochem. Soc. 1998, 145, 2285–2295.
[10] N. Birbilis, R.G. Buchheit, J. Electrochem. Soc. 2005, 152, B140–B151.
[11] P. Schmutz, V. Guillaumin, R.S. Lillard, J.A. Lillard, G.S. Frankel, J. Electrochem. Soc. 2003, 150, B99–B110.
[12] F. Andreatta, I. Apachitei, A.A. Kodentsov, J. Dzwonczyk, J. Duszczyk, Electrochim. Acta 2006, 3551–3557.
[13] P.L. Bonora, A. Eliezer, F. Di-Gabriele, M. Andrei, E.M. Gutman, Mat. Sci. Technol. 2004, 20, 29–34.
[14] C.H. Caceres, A. Blake, Phys. Sta. Sol. (A) 2002, 194, 147–158.
[15] O. Lunder, J.E. Lein, T.Kr. Aune, K. Nisancioglu, Corrosion 1989, 45, 741–748.
[16] R. Ambat, N.N. Aung, W. Zhou, Corr. Sci. 2000, 42, 1433–1455.
[17] O. Lunder, J.E. Lein, S.M. Hesjevik, T.Kr. Aune, K. Nisancioglu, Werk. Korr. 1994, 45, 331–40.
[18] O. Lunder, T.Kr. Aune, K. Nisancioglu, Corrosion 1987, 43, 291–95.
[19] T. Murai, H. Oguri, S. Matsuoka, Mat. Sci. Forum 2005, 488–89, 515–518.

Influence of the Casting Method on Microstructure and Corrosion of AZ91 and AM50

Daniela Zander, Claudia Pieper, Uwe Köster
Dept. Biochem. & Chem. Eng., University of Dortmund, D-44221 Dortmund, Germany

1 Introduction

The corrosion of magnesium alloys is known to depend strongly on the casting technique used to prepare the material [1]. AM50 and AZ91 are standard magnesium alloys which can either be produced by common casting methods like die casting or by the newer thixoforming processes [2]. Due to the different temperature of the melt the two production methods thixomolding and die casting will lead to significant differences in the macrostructure (casting defects and porosity) as well as microstructure (different phases and their distribution).

In earlier papers by Nakatsugawa et al. [3], Mathieu et al. [4] and Pieper et al. [5] a higher corrosion resistance was found for thixoformed specimen as compared to die cast ones. Recent investigations showed that the corrosion behavior can be explained by the influence of four microstructural aspects: Porosity, volume fraction and distribution of the β-phase ($Mg_{17}Al_{13}$) and Al-content of the α-phase. An increased porosity is known to decrease the corrosion resistance. A high volume fraction of the β-phase, in particular when forming a continuous network in a fine-grained microstructure, can act as a barrier layer for further corrosive attacks. A higher Al-content in the α-phase will lead to a more noble potential as well as to the formation of a more protective oxide depending on the electrolyte. These controlling microstructural parameter are set by the casting technique used. Analyzing the corrosion behavior one has to take into account the strong microstructural gradient in the die-cast and in particular in the thixomolded material. The aim of this paper is to study the influence of thixomolding as compared to die casting on the microstructure and the corrosion behavior of AZ91 and AM50.

2 Experimental

The investigations were performed on die cast (dc) AZ91 (GKSS-Forschungszentrum, Geesthacht) in comparison to thixomolded (th) AZ91 (The Japan Steel Works, Ltd.) and die cast (dc) AM50 (Arge Metallguss). Die cast AZ91 was processed and investigated with thicknesses of 2 and 14 mm, more details are given in [6], whereas thixomolded AZ91 revealed a thickness of 0.5 and 1 mm.

The electrochemical tests using the DC (potentiodynamic polarisation) technique were carried out in deaerated solutions of pH = 11 without and with 0.01M NaCl as well as pH = 8. Since it was reported by Lunder et al. [7] that composition of the solution changes during corrosion due to the dissolved corrosion products the pH of the electrolyte has kept constant by formic acid as a titration solvent. For all measurements a three electrode electrochemical cell was used, with Hg/HgO or Ag/AgCl as reference electrodes and a platinum counter electrode. The potentiondynamic curves were obtained using a FAS1 potentiostat (Gamry Instruments) with a voltage scan rate of 0.2 mV/s (pH = 11) and 0.5 mV/s (pH = 8). All the potential values are reported with respect to the reference electrode ($E_{Hg/HgO \text{ or } Ag/AgCl}$ versus E_{SCE}). The Mg-based samples were embedded in an acrylic resin to provide electrical isolation of the sample surface, ground up to 1200 grid and polished up to 1 µm with an water free alcoholic solution in order to avoid deterioration by corrosion.

The microstructure was studied by X-ray diffraction (CuK$_\alpha$ radiation), optical microscopy and scanning electron microscopy (SEM+EDX, Hitachi S-4500) before and after corrosion. For optical microscopy and SEM studies, specimens were polished up to 1 µm and etched in 1% HNO$_3$. The fraction of the β-phase was estimated by relating the ratio of the main X-ray diffraction peaks of the α- and β-phase. The Al-content of the α-phase close to the surface was calculated from the shift of the magnesium lattice constants.

3 Results and Discussion

Microstructural investigations of die cast AM50 and AZ91 (figure 1a, b) revealed a similar grain size of α- and β-phase and porosity but differences in the Al-content at the grain boundaries of the α-phase and the phase distribution of the β-phase [8].

The most significant differences between die cast and thixomolded alloys are the coarse Mg crystals not dissolved during the thixomolding process and the almost uniform grain size of the newly crystallized Mg grains in thixomolded alloys (figure 1c); die cast alloys show a broad grain size distribution.

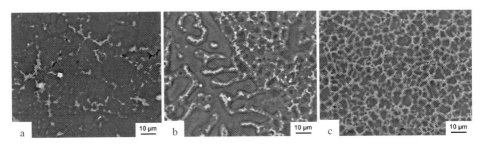

Figure 1: Microstructure (SEM) of die cast (a) AM50 and (b) AZ91 and thixomolded (c) AZ91 in the interior of a 2 mm thick plate

In thixomolded AZ91 the cell structure is finer in contrast to die cast AZ91 (figure 1b, c). The small Mg crystals are fully surrounded by a second phase just forming a continous net as already reported in literature [3, 9]. The volume fraction of the β-phase (Al$_{12}$Mg$_{17}$) is higher than in the die cast alloy. In thixomolded AZ91 typical flow patterns can be found exhibiting significant differences in their microstructure [8]. Microstructural investigations by X-ray revealed a similar fraction and continuous net of the β-phase but a decreasing gradient of the Al-content in the α-phase and the size of the α-grains from the gate towards the overflow side of thixomolded AZ91 with a thickness of 1 mm. The microstructural investigations of the center revealed a higher porosity than at the edges of the plate. Processing of thixomolded AZ91 with the smaller thickness of 0.5 mm could not reveal any gradient; only a high porosity in the center of the plate.

Potentiodynamic measurements of die cast AM50, AZ91 and thixomolded AZ91 in electrolytes of pH11 and pH8 reveal the corrosion parameters E_{corr}, i_{corr} in the active region and therefore the corrosion rate as well as the value of the average passive current i_p (figure 2, 4). Discussing the corrosion rate one has to take into account the strong cathodic hydrogen reaction as well as the different corrosion mechanism of magnesium and its alloys [10, 11].

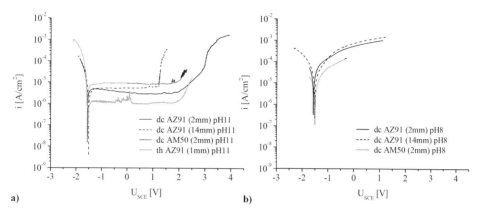

Figure 2: i-U-diagrams of die cast and thixomolded magnesium alloys in electrolytes with (a) pH11 and (b) pH8

An expanded passive region with low passive current densities in the range of 10^{-6} A/cm² was observed during potentiodynamic polarization of die cast AM50 and AZ91 in a strong alkaline electrolyte whereas only active dissolution was obtained in an electrolyte of pH8 (figure 2). In general die cast AZ91 revealed a better corrosion behavior regarding passivation and active dissolution in pH11 in contrast to the one in pH8 in comparison to AM50. However, figure 2a and b show small additional differences between AZ91 processed with different thicknesses of 1 and 14 mm. Die cast AZ91 (2 mm) shows the best passivation behavior but higher corrosion current densities and slightly more negative equilibrium potentials in both solutions.

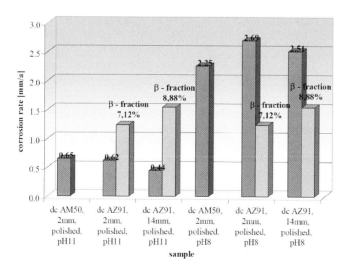

Figure 3: Corrosion rates of die cast magnesium alloys in electrolytes with (a) pH11 and (b) pH8

In the strong alkaline electrolyte the corrosion rate of die cast AM50 is slightly higher than of the corresponding AZ91 (14 mm) alloys but smaller in an electrolyte of pH8 (figure 3). Thus, indicates a different corrosion mechanism depending on the influence of the pH of the electrolyte and the microstructure. The corrosion rates increased significantly with decreasing pH for both magnesium alloys.

Since microstructural investigations showed similar porosity and distribution of the β-phase of die cast AM50 and AZ91 the main parameter influencing the corrosion are the volume fraction of the β-phase ($Mg_{17}Al_{13}$) and the Al-content of the α-phase: In general the β-phase is noble in comparison to the α-phase. The higher Al-content of the β-phase and at the grain boundaries of the α-phase leads to different corrosion mechanisms in pH11 and pH8. On one hand aluminum is known to dissolve in strong alkaline electrolytes, magnesium passivates under the same conditions whereas in neutral solutions aluminum passivates and magnesium dissolves. On the other hand aluminium dissolved in the α-phase changes the potential difference between the α-phase and the β-phase towards smaller values and improves the corrosion behavior [12].

Therefore, the significant increased corrosion rate of die cast AM50 and AZ91 with decreasing pH can be mainly related to the active dissolution of magnesium from the α-phase. The protecting passivation behavior of the β-phase in pH8 seems to be subordinate due to the formation of discontinuous distributed β-phase. In addition galvanic corrosion between the α-phase and the discontinuous distributed β-phase will accelerate the corrosion. It was reported [13] that the formation of a continuous net of the β-phase, e.g. by thixomolding, will improve galvanic corrosion between magnesium and the noble $Mg_{17}Al_{13}$ and lead to the subsequent formation of a closed and protective layer of the secondary phase in neutral electrolytes. However, the positive effect of the β-phase on the corrosion mechanism in alkaline electrolytes under consideration of the dissolution of aluminium might be due to the formation of a protective Mg-Al-oxide or hydroxide layer [14].

The thixomolded microstructure of AZ91 characterized by a finer cell structure, an increased volume fraction of the β-phase ($Mg_{17}Al_{13}$) and Al-rich α-phase forming a continuous net around the α-grains (figure 1c) leads to an improved corrosion behavior in strong alkaline electrolytes (figure 2a). A significant decrease of i_{corr} and the corrosion rate by one order of magnitude as well as an improved passivation behavior in comparison to die cast AZ91 in pH11 were obtained. It can be assumed that the dissolution of the Al-rich α-grains and the increased fraction of the secondary phase of thixomolded AZ91 play only a minor role for the corrosion in alkaline solutions than the formation of a net of thermodynamically stable $Mg_{17}Al_{13}$. In accordance to literature [3, 13] the magnesium corrodes first leading - in case of the fine microstructure of thixomolded alloys – to a new surface consisting of noble $Al_{12}Mg_{17}$. In addition the formation of $Mg(OH)_2$ as a protecting corrosion product is likely. In die cast alloys the discontinuous β-phase will fall out when undermined by corroded Mg grains.

Only small additions of NaCl in alkaline solutions lead to a change from passivation to active dissolution and pitting of thixomolded AZ91 (figure 4). In addition a strong influence of the sampling point (gate, center, overflow) and the thickness of plate on the corrosion current density and rate (figure 5) were observed.

It was observed that the corrosion rates decrease from the gate towards the overflow area not taking into account the center of thixomolded AZ91 with a thickness of 1 mm in pH11 with 0.01M NaCl (figure 5a).

The microstructural investigations indicate that the better corrosion behavior at the overflow side are due to the lower Al-content and smaller grain size of the α-phase. The extremely high corrosion rate at the center of the plate can be associated to the higher porosity. A similar Al-content which was observed over the whole thixomolded AZ91 plate with a thickness of 0.5 mm leads to similar corrosion rates from the gate towards the overflow side as shown in figure 5b except of areas with high porosity such as at the inner and outer center area. Since aluminium is known to dissolve in alkaline solutions it can be assumed that besides pitting one of the corrosion mechanisms is the increased solution of aluminium from the α-phase with increasing Al-content. High porosity leads to increased corrosion pitting. X-ray diffraction revealed the formation of $Mg(OH)_2$ and $Mg_6Al_2(OH)_{18} \cdot 4.5\, H_2O$ at the surface.

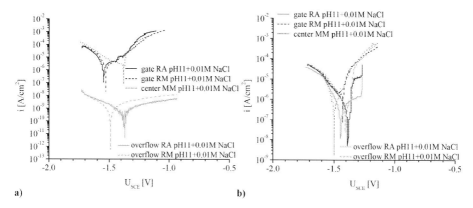

Figure 4: i-U-diagrams of thixomolded AZ91 of (a) 1 mm and (b) 0.5 mm thickness in pH11 with 0.01M NaCl

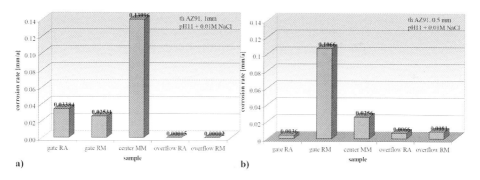

Figure 5: Corrosion rates of thixomolded AZ91 of (a) 1 mm and (b) 0.5 mm thickness in pH11 with 0.01M NaCl

4 Conclusion

Corrosion of thixomolded and die cast AZ91 and AM50 was investigated and discussed in regard to the microstructures involved. The investigations showed that the different behavior can be explained by the influence of four microstructural aspects: Porosity, volume fraction and distribution of the β-phase ($Mg_{17}Al_{13}$) and Al-content of the α-phase.

In general die cast AZ91 revealed a better corrosion behavior regarding passivation and active dissolution in pH11 than in pH8 in comparison to AM50. The corrosion rates increased significantly with decreasing pH for both magnesium alloys which can be mainly related to the active dissolution of magnesium from the α-phase. In addition galvanic corrosion between the α-phase and the discontinuous distributed β-phase accelerates the corrosion.

In alkaline solutions the corrosion rate of thixomolded AZ91 is one order of magnitude lower than of the die cast ones. The fine eutectic structure with the more noble continuous β-phase works as a barrier against further corrosion attack as well as the formation of $Mg(OH)_2$.

Only small additions of 0.01M NaCl lead to a change from passivation to active dissolution and pitting of thixomolded AZ91 in alkaline solutions. In addition a strong influence of the sampling point (gate, center, overflow), the processed thickness of tixomolded AZ91 and therefore of the Al-content on the corrosion current density and rate was observed.

5 Acknowledgement

The authors like to thank the Deutsche Forschungsgemeinschaft (German Research Foundation) for the financial support under the project Ko 668/25.

6 References

[1] H. Alves, U. Köster, E. Aghion, D. Eliezer, Materials Technology 2001, 16, 110-126
[2] R. Carnahan, R. Decker, D. Ghosh, C. VanShilt, P. Frederick, N. Bardley, Magnesium Alloys and Their Appl. 1992, 69-76
[3] I. Nakatsugawa, F. Yamada, H. Takayasu, T. Tsukeda, K. Saito, Proc. of the International Symposium on Environmental Degradation of Materials and Corrosion Control in Metals (Eds.: M. Elboujdaini, E. Ghali), Canadian Institute of Mining, Metallurgy and Petroleum, Montreal, Canada, 1999, 113-123
[4] S. Mathieu, C. Rapin, J. Hazan, P. Steinmetz, Corrosion Science 2002, 44, 2737-2756
[5] C. Pieper, U. Köster, H. Alves, I. Nakatsugawa, Proceedings of the 6th International Conference Magnesium Alloys and Their Applications, Wolfsburg 2003, (Eds. K.U. Kainer), Wiley-VCH Verlag, Weinheim, 2003, 586-591
[6] C. Blawert, V. Heitmann, E. Morales, W. Dietzel, S. Jin, E. Ghali, Canadian Metallurgical Quarterly, 2005, 44, 137-146
[7] O. Lunder, J.E. Lein, T. Kr. Aune, K. Nisancioglu, Corrosion 1989, 45, 741-748
[8] C. Pieper, Korrosion und Oxidation von Magnesiumlegierungen, Fortschritt-Berichte VDI Reihe 5 Nr. 711, VDI-Verlag, Düsseldorf, 2005
[9] E. Evangeista, S. Spiragelli, P. Cavaliere, Key Engineering Materials 2000, 188, 139-148
[10] G. Song, Advanced Engineering Materials 2005, 7, 563-586
[11] E. Ghali, W. Dietzel, K.-U. Kainer, Journal of Materials and Engineering and Performance 2004, 13, 517-529
[12] S. Mathieu, C. Rapin, J. Steinmetz, P. Steinmetz, Corrosion Science, 2003, 45, 2741-2755
[13] G. Song, A. Atrens, M. Dargusch, Corrosion Science 1999, 41, 249-273
[14] J. H. Nordlien, K. Nisancioglu, S. Ono, N. Masuko, Journal of the Electrochemical Society, 1996, 143, 2564-2572

Effect of Second Phase on Corrosion Behavior of AZ31-xCa Magnesium Alloys

Chang Dong Yim[1], Young Min Kim[1], Bong Sun You[1], Rang Su Jang[2], Su Geun Lim[3]
[1]Korea Institute of Machinery and Materials, Changwon, Korea
[2]Korea Iron & Steel Co., Ltd., Changwon, Korea
[3]Gyeongsang National University, Jinju, Korea

1 Introduction

Ca is very promising element because the addition of Ca significantly improves the high temperature strength and creep resistance of magnesium alloy [1-3] and it has low density and is low cost. Ca is also very effective for improving the ignition resistance of magnesium alloy melt [4, 5], which stimulated the development of non-combustible magnesium alloys. Magnesium and its alloys are very active in molten state and rapidly ignited or combusted when the clean surface of melt is in contact with oxygen in air. So, in general, protecting gas consisted of SF_6, CO_2 and air is blown to melt surface during melting and casting operations. But SF_6 is the strongest one of greenhouse gases which cause global warming. In near future, the usage of SF_6 will be prohibited strictly, which motivates the R & D for new melt protection methods. There are two major melt protection techniques to be developed. One is improvement of ignition resistance of magnesium alloy melt by addition of alloying elements [6-9] and the other is using other gases such as SO_2 and HFC-134a [10,11]. The method of addition of alloying elements has several advantages compared to using protection gas. There is no need for gas blowing equipment, which gives more freedom to designs of process and manufacturing equipment. The inherent ignition resistance of magnesium alloys increases, which makes a possibility of burning of various parts lower. As mentioned above, the ignition resistance of alloys containing Ca in molten state increases with increasing of Ca content. The addition of Ca changes not only the ignition resistance of melt but also the various properties such as tensile strength, elongation and corrosion resistance, etc. There have been many studies about the effects of Ca on ignition resistance and mechanical properties [4-13]. Although the corrosion resistance is one of important factors in aspects of application, there is little study about the corrosion behavior of magnesium alloys containing Ca. In this study, the effect of Ca on corrosion behavior of gravity cast AZ31-xCa alloys containing 0~5 wt.% Ca was investigated experimentally and analyzed in aspects of microstructural change by means of observation of microstructure and measurement of corrosion potential and average corrosion rate.

2 Experimental Procedure

AZ31B ingot of 3kg was inserted into a low carbon steel crucible and heated to 923K. Protecting gas consisted of SF_6:CO_2=1:10 was blown into the crucible at flow rate of 300 ml/min to prevent the ignition and/or combustion of melt. After the AZ31B ingot was melted completely, Ca pellet of 9~150g was added into the melt and heated up 993K. The melts was isothermally held at 993K for 30 minutes for Ca to be dissolved completely. A vacuum treatment was carried out to make the non-metallic inclusions floated to melt surface and non-metallic inclusions was removed by using a skimmer before casting. A gravity casting was produced by pouring the melt into a plate-type mold preheated to 573K. Table 1 shows the alloy compositions analyzed by inductively coupled plasma (ICP) spectrometer.

Table 1: Chemical compositions of gravity cast AZ31-xCa alloys (wt.%).

Alloys	Al	Zn	Mn	Ca	Mg
AZ31	3.20	0.92	0.13	-	Bal.
AZ31-0.3Ca	2.97	0.95	0.42	0.31	Bal.
AZ31-0.5Ca	2.91	0.92	0.42	0.49	Bal.
AZ31-0.7Ca	2.77	0.94	0.42	0.72	Bal.
AZ31-1.0Ca	2.89	0.75	0.44	1.03	Bal.
AZ31-2.0Ca	3.07	0.73	0.43	2.11	Bal.
AZ31-5.0Ca	2.90	0.68	0.42	5.10	Bal.

Specimens were cut into a proper size from gravity casting of AZ31-xCa alloys and mounted by using polymer resin. The specimens were mechanically polished by using emery papers of #100~2400 and pastes containing diamond particles of 1 μm diameter and then etched by using acetic picral solution (10ml acetic acid + 4.2g picric acid + 10ml distilled water + 70ml ethanol (95%)) in order to reveal the microstructure more distinctly. The microstructures of gravity cast AZ31-xCa alloys were observed by using scanning and transmission electron microscopes. For transmission electron microscopy (TEM) analysis, discs with 3mm in diameter were mechanically polished to 100 μm thickness, and followed by twinjet electro-polishing in a solution of ethanol with 1% perchloric acid at -30°C.

Electrochemical polarization tests were carried out using Princeton Applied Research Versa Stat. II to measure the corrosion potentials of gravity cast AZ31-xCa alloys. Working electrodes

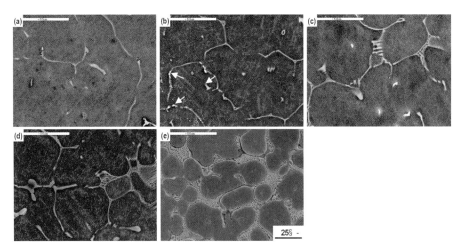

Figure 1: SEM photographs of gravity cast AZ31-xCa alloys with Ca content; (a) x=0.5 (b) x=0.7 (c) x=1.0 (d) x=2.0 (e) x=5.0.

were prepared by connecting a wire to one side of the sample and covering with cold resin. The opposite surface of the specimen was exposed to the solution. The exposed area was about 0.5 cm^2. The specimens were mechanically polished prior to each test, followed by washing with acetone. A polarization test was carried out in a corrosion cell containing 1000 ml of 3.5 wt.%NaCl solution using a standard three electrode configuration: Ag/AgCl as reference electrode with a carbon electrode as counter and the sample as the working electrode. Specimens were immersed in the test solution and a polarization scan was carried out toward more noble value at a rate of 1 mV s^{-1} for 1 hour.

A constant immersion test was carried out to evaluate the effect of Ca on the average corrosion rate of AZ31-xCa alloys. Sample preparation and immersion test were done in according to the procedure of ASTM standard G-I-72 [14]. The mechanically polished and preweighed specimens were exposed to a 5 wt.% NaCl solution for 120 hours. Final cleaning of the specimen at the end of the test was carried out by dipping it in a chromic acid solution (200g CrO$_3$ + 10g AgNO$_3$ + 1000 ml distilled water) at boiling condition followed by acetone washing. The weight loss was measured after each test and the average corrosion rate was calculated in mills per year.

3 Results and Discussion

Figure 1 shows the SEM photographs of AZ31-xCa alloys with Ca content. In the alloys containing below 0.7wt.%Ca, the eutectic regions were not observed distinctly and a second phase formed semi-continuously along grain boundaries of α-Mg phase. As the calcium content increased above 1wt.%, eutectic regions with fine lamellar structure were observed along grain boundaries. The volume fraction of eutectic region increased with increasing of Ca and in the alloy containing 5wt.%Ca, most of intergranular regions between α-Mg grains were consisted of eutectic structure.

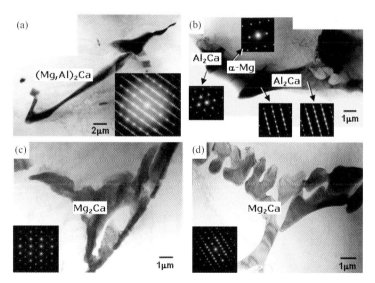

Figure 2: TEM photographs of gravity cast AZ31-xCa alloys with Ca content; (a) x=0.3 (b) x=0.7 (c) x=2.0 (d) x=5.0.

From the TEM analysis in Figure 2(a), the second phase observed in the alloy containing below 0.7wt.%Ca was defined as $(Mg,Al)_2Ca$ with dihexagonal (C36) structure as reported by Suzuki et al. [15]. In the alloy with 0.7wt.%Ca, it is found in Figure 2(b) that Al_2Ca phase with face centered cubic (C15) structure in addition to semi-continuous C36-$(Mg,Al)_2Ca$ phase formed along the grain boundaries of α-Mg phase as shown by arrows in Figure 1(b). As the calcium content increased above 1wt.%, second phase in eutectic regions along grain boundaries was defined as Mg_2Ca with hexagonal (C14) structure, which was also observed in the alloy containing 5 wt.%Ca (Fig. 2(d)). From TEM analyses, it can be concluded that the phase transition from $(Mg,Al)_2Ca$ with dihexagonal (C36) structure to Mg_2Ca with hexagonal (C14) structure in AZ31 alloy is occurred at around 0.7wt.%Ca.

Figure 3 shows the corrosion properties of gravity cast AZ31-xCa alloys as function of Ca content. The corrosion resistance of AZ31-xCa alloys increased with Ca content up to x=0.7, but decreased remarkably above 1wt.%Ca.

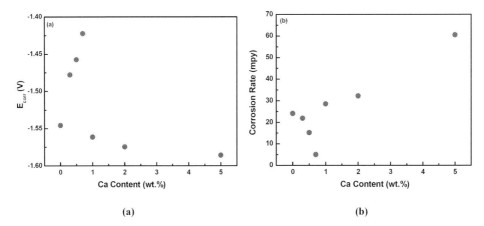

Figure 3: Corrosion properties of AZ31-xCa alloys; (a) corrosion potential (b) average corrosion rate.

Figure 4 shows the vertical section of AZ31-xCa alloy specimens after immersion test. In the alloys containing below 0.7wt.%Ca, the alloys were attacked homogeneously by corrosion. It seemed that the corrosion front was arrested by the second phase formed semi-continuously along grain boundaries of α-Mg phase. On the other hand, it seemed that the corrosion was initiated locally and typical pitting was observed in the alloys containing above 1wt.%Ca. As shown in figure 4(c), the corrosion propagated along grain boundary. The eutectic region between α-Mg grains was attacked preferentially by corrosion as shown in fig. 4(e).

This difference of corrosion resistance with Ca content was mainly resulted from change of morphology of second phase. When the second phase formed along grain boundary and the volume fraction of second phase is low as shown in figure 1(a), the corrosion is initiated at matrix α-Mg phase near second phase due to formation of internal micro-galvanic cell. The corrosion front does not penetrate the second phase because the second phase is invariably cathodic to the grain interior [16]. So the corrosion propagates through matrix α-Mg phase and the corrosion fronts from different grains meet together as the corrosion proceeds. At this time, the discontinuous second phase is undercut by connection of corrosion fronts and so the corrosion is not retarded effectively by second phase. When the second phase formed along grain boundary and the volume fraction of second phase is high enough to enclose the α-Mg grain as shown in figure 1(b), the corrosion is also initiated at matrix α-Mg phase near second phase and propagates through matrix α-Mg phase. But the corrosion fronts from the neighboring grains don't meet together by the second phase formed semi-continuously along grain boundaries and the corrosion is arrested at grain boundaries. In the alloys containing above 1wt.%Ca, the eutectic structure consisted of eutectic α-Mg and intermetallic second phases formed in the interdendritic region as shown in figure 1(c)~(e). The internal micro-galvanic cell is formed between eutectic α-Mg and second phase in interdendritic region. So the corrosion is initiated at eutectic α-Mg phase and propagates fast along eutectic α-Mg phase formed continuously along grain boundary. At this time, the discontinuous second phase in interdendritic region can not interrupt the propagation of corrosion effectively, which results in lower corrosion resistance.

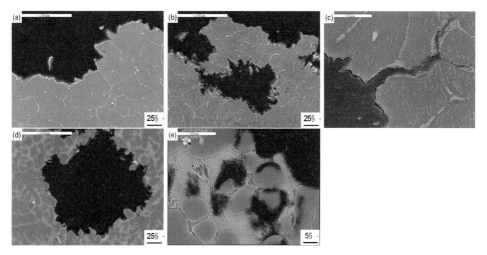

Figure 4: Vertical-sectional view of AZ31-xCa alloy specimens after immersion test: (a) x=0.7, (b) and (c) x=2.0, (d) and (e) x=5.0.

The addition of Ca into AZ31 magnesium alloy changes the species of second phase as well as the morphology of second phase as shown in figure 2. The corrosion potential of second phase could not be measured exactly because of the difficulty in separate production of second phase. In this study, it was assumed that the potential difference between $(Mg,Al)_2Ca$ and Mg_2Ca phases would be little due to similarity of crystal structures of both phases. So the change of second phase with Ca content from $(Mg,Al)_2Ca$ phase to Mg_2Ca phase would not affect the internal galvanic corrosion largely.

4 Summary

The corrosion behavior of gravity cast AZ31-xCa alloys was strongly dependent on the morphology of second phase. In the alloys containing below 0.7wt.%Ca, a micro-galvanic cell would form between the matrix α-Mg phase and the second phase formed semi-continuously at grain boundaries. The corrosion was initiated at α-Mg phase near second phase and propagated through α-Mg phase. At low volume fraction of second phase, the second phase did not envelop α-Mg phase completely. So the second phases were undercut by connection of corrosion fronts from neighboring grains and did not act as effective corrosion barrier. But in the case of high volume fraction of second phase, the second phase surrounded α-Mg phase nearly. So the corrosion was retarded when the corrosion front met the second phase at grain boundary, which resulted in higher corrosion resistance. In the alloys containing above 1wt.%Ca, a micro-galvanic cell would form between the eutectic α-Mg phase and discontinuous second phase in interdendritic region. The corrosion was initiated at the eutectic α-Mg phase and propagated fast through eutectic α-Mg phase. The discontinuous second phase in the eutectic region did not interrupt the propagation of corrosion effectively. With increasing of Ca content, the area fraction of eutectic region increased and the corrosion was propagated faster.

5 References

[1] Ninomiya, R.; Ojiro, T.; Kubota, K., Acta Metallurgica et Materialia **1995**, 43, 669-674.
[2] Terada, Y.; Sota, R.; Ishimatsu, N. ; Sato, T. ; Ohori, K., Metallurgical and Materials Transactions A **2004**, 35, 3029-3032.
[3] Hirai, K.; Somekawa, H.; Takigawa, Y.; Higashi, K., Materials Science and Engineering A **2005**, 403, 276-280.
[4] You, B.S.; Park, W.W.; Chung, I.S., Scripta Materialia **2000**, 42, 1089-1094.
[5] Choi, B.H.; You, B.S.; Park, I.M., Metals and Materials International **2006**, 12, 63-67.
[6] Ravi Kumar, N.V.; Blandin, J.J.; Suéry, M.; Grosjean, E., Scripta Materialia **2003**, 49, 225-230.
[7] Chang, S.Y.; Choi, J.C., Metals and Materials International **1998**, 4, 165-172.
[8] Kim, M.H.; Park, W.W.; You, B.S.; Huang, Y.B.; Kim, W.C., Materials Science Forum **2003**, 419-422, 575-580.
[9] Huang, Y.B.; Chung, I.S.; You, B.S.; Park, W.W.; Choi, B.H., Metals and Materials International **2004**, 10, 7-11.
[10] Cashion, S.; Ricketss, N., in: *Magnesium Technology 2000*, Nashville, U.S.A., **2000**, 77-81.
[11] Ricketts, N.; Cashion, S., in: *Magnesium Technology 2001*, New Orleans, U.S.A., **2001**, p. 31-36.
[12] Choi, B.H.; You, B.S.; Yim, C.D.; Park, W.W.; Park, I.M., Materials Science Forum, **2005**, 475-479, 2477-2480.
[13] Qudong, W.; Yizhen, L.; Xiaoqin, Z.; Wenjiang, D.; Yanping, L.; Qinghua, L.; Jie, L., Materials Science and Engineering A **1999**, 271, 109-115.
[14] *Annual book of ASTM Standards Part 3 and 4*, **1977**, 722.
[15] Suzuki, A.; Saddock, N.D.; Jones, J.W.; Pollock, T.M., Acta Materialia **2005**, 53, 2823-2834.
[16] Makar, G.L.; Kruger, J., International Materials Reviews **1993**, 38, 138-153.

Acknowledgement

This work was financially supported by the research program in the Korean Institute of Machinery and Materials (NRL).

The Role of Si and Ca in New Wrought Mg-Zn-Mn Based Alloys

G. Ben-Hamu, D. Eliezer and K. S. Shin[1]

Department of Materials Engineering, Ben-Gurion University of the Negev Beer-Sheva, Israel
[1]School of Materials Science and Engineering, Research Institute of Advanced Materials Seoul National University, Seoul, Korea

1   Introduction

The interest in lightweight materials in the structural applications has increased significantly due to their importance in the environmental and energy saving problems. Especially in the automotive industry, magnesium alloys, due to their low density and high specific strength, have become the key materials for enhancing the fuel efficiency. In fact, there has been a rapid growth in the structural applications of magnesium alloys during the past decade. However, the majority of this growth has been in the area of die-cast components and some semi-solid formed components [1]. In contrast, wrought magnesium alloys are used to approximately only 1% of total magnesium consumption. There are two major technical issues in order to expand the wrought magnesium market. First is a low production rate. A typical magnesium alloy must be extruded 5–10 times slower than a typical aluminum alloy. Second is a development of new wrought magnesium alloys with a combination of high strength, high ductility and high corrosion resistance. In recent years, the modification of alloy composition and/or heat treatment has been attempted for improved mechanical properties [2–4] and corrosion resistance [5-8] in the casting alloys. However, there is a growing need for high strength wrought Mg alloys in the automotive and aerospace industries. Four different alloy systems have mainly been utilized for the development of the wrought Mg alloys, i. e., Mg-Zn, Mg-Al, Mg-Th, and Mg-Mn alloys [9]. Among these, Mg-Zn alloys were found to have a large age hardening response, stemming from the precipitation of a transition phase (β'), and consequently offered a combination of good strength and ductility [10–15]. It has been reported, however, that grain refinement is difficult to achieve in Mg-Zn alloys [9]. Several alloying elements, including Zr, RE and Cu, have been added to Mg-Zn alloys to improve the mechanical properties; Zr for grain refining and strengthening [16,17], rare-earth (RE) for improved high temperature properties [17–19] and Cu for ductility improvement [20,21]. Mg-Zn-Si series is a new promising alloy system which is developed to meet the above requirements. The silicon addition to magnesium alloys causes an increased fluidity of the molten metal. The Mg_2Si formed by the addition of Si exhibits high melting point (1085 °C), high hardness (460 $HV_{0.3}$), low density (1.9 g cm^{-3}), high elastic modulus (120 GPa) and low thermal expansion coefficient (7.5 x 10^{-6} K^{-1}) [22]. This intermetallic phase is very stable and can impede grain boundary sliding at elevated temperatures. Yuan Guangyin et al.[23] investigate the micro structural features, tensile properties at both ambient and elevated temperature of 150 °C, impact toughness and creep resistance in order to get a better overall understanding of alloys in this system. They identify the most promising compositions Mg-Zn-Si alloyed with Ca. But up to now, limited research has been carried out on the corrosion behavior of this system. The objective of the present study is to develop new wrought Mg alloys with improved strength and ductility and investigate the corrosion resistance of the new alloys. The effects of Si and Ca addition of the new Mg-Zn-Mn-Si-Ca alloys were examined.

2 Experimental Procedures

Mg-Zn alloys were melted in a low carbon steel crucible and melt surface was protected with a gas mixture of CO_2+0.5% SF_6. The elemental Zn and Ca with 99.99% purity were added to the melt. Silicon was added to the melt in the form of Mg-10 wt.% Si mother alloy. The alloy designations used in the present study are as follows; ZSM6x1-yCa (ZSMX) for Mg-6 wt.% Zn-1 wt% Mn-x wt.% Si-y wt.% Ca. Maximum impurity levels: 0.004 wt% Fe 0.005 wt% Ni, 0.05 wt% Cu. The ingots were homogenized at 400 °C, water-cooled, and subsequently scalped to give 80mm diameter billets. After preheating, the billets were extruded with an extrusion ratio of 25:1 to give 16mm diameter cylindrical rods. Specimens for optical microscopy were mechanically polished followed by chemical etching with a 1% HNO_3 + 24% distilled water + 75% diethylene glycol solution. Thin foils for transmission electron microscopy (TEM) were prepared by chemical etching in a 5% HNO_3 + ethanol solution below 0 °C. The behavior of Mg-Zn-Mn-Si-Ca magnesium alloys was investigated by using potentiodynamic polarization (PD) measurements, linear polarization (L.P) measurements (DC polarization) and electrochemical impedance spectroscopy (EIS). All the electrochemical measurements were performed in 3.5% NaCl saturated with $Mg(OH)_2$ with a pH (=10.5) at which Mg can cover itself with more or less protective oxide or hydroxide which checks the dissolution reaction [24]. The electrochemical testing was employed to study the main features of the processes taking place at the alloy/solution interface. The effect of alloying elements on the Mg-Zn alloys corrosion resistance was studied. The corrosion resistance of Mg alloys was pointed out by EIS measurements performed during the free immersion time and under polarization and the effect of the different alloying elements was studied. The evolution of the electrode/electrolyte interface at different immersion times was also studied. Corrosion rates were derived from polarization data by the common method [25]. The electrochemical tests using both DC (potentiodynamic and

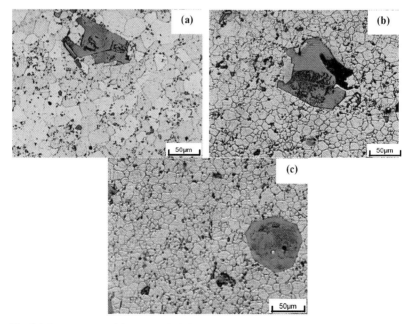

Fig. 1: Microstructures of the as-extruded alloys: (a) ZSM631+0.4Ca (b) ZSM651+0.8Ca (c) ZSM6101+0.4Ca

linear polarization) and AC techniques (EIS) were carried out in 3.5% NaCl saturated with Mg(OH)$_2$. For all measurements a three electrode electrochemical cell was used, with an SCE as reference electrode and a platinum counter electrode. The working electrode was prepared from the Mgalloy samples. Mg-alloy samples were embedded in an acrylic resin to provide electrical isolation of the sample surface. The samples were air dried at room temperature. The potentiodynamic and linear curves were obtained using a 273A EG&G Potentiostat/Galvanostat, with a voltage scan rate of 0.5 mV/s. The impedance measurements were carried out using a PARSTAT 2263 frequency response analyser coupled with the potentiostat. All the experiments were controlled by a PC, which was also used for the acquisition, storage and plotting of data. The scanned frequency ranged from 6 mHz to 100 kHz and the perturbation amplitude was of 5 mV (it was observed that a variation of the amplitude did not change the frequency response of the electrode/electrolyte interface). The impedance measurements were performed at open circuit potential (E$_{oc}$). A partial data fitting made with the Boukamp circuit equivalent software [26] for the charge transfer process produced the R$_P$ (polarization resistance) and C$_{dl}$ (double-layer capacitance) values.

3   Results and Discussion

Microstructure and Mechanical Properties of the Extrusions Alloys:

Figure 1 shows the microstructures of the ZSM631+0.4Ca (3 wt.% Si and 0.4 wt.% Ca), ZSM651+0.8Ca and ZSM6101+0.4Ca alloys in the as-extruded condition. From these microstructures, it can be seen that the addition of Si suppressed the grain growth during the extrusion process. However the suppressed of the grain growth influences by the addition of Ca more then the addition of Si to Mg-Zn-Mn alloy (Fig. 2)

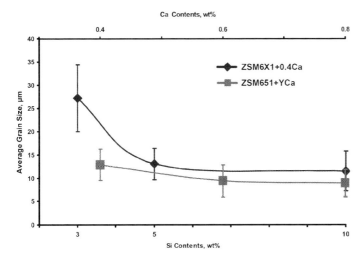

Fig. 2: Variation in Grain Size of ZSMX Mg alloy

Figures 3 show the TEM micrographs obtained from the ZSM651+0.4Ca alloy. It can be seen from the figure the precipitate of CaMgSi.

Corrosion Behavior of the Extrusions Alloys:

Potentiodynamic polarization (PD) measurements, linear polarization(L.P) (DC polarization) and electrochemical impedance spectroscopy (EIS) tests were used in this work in order to understand the effects of Si and Ca additions on the corrosion resistance of the Mg-Zn-Mn alloys. Figure 4 shows the effect of silicon on the corrosion rate of the Mg-6%Zn-1%Mn alloy. As shown in the figure, the addition of silicon decreased the corrosion rates of the *ZSM6x1*+0.4Ca alloys. It is known that silicon has little deleterious effect on the basic saltwater corrosion performance of pure magnesium because the electrochemical potential of the Mg_2Si precipitate (-1.65VSCE) is similar to that of magnesium (–1.66VSCE).[27] Figure 5 shows the effect of calcium on the corrosion rate of the Mg-6%Zn-1%Mn-5%Si alloy. As shown in the figure, the addition of calcium increases the corrosion rates of the ZSM651+YCa alloys. High content of Ca and Si at Mg-Zn-Mn alloy create more sits of CaMgSi phase in the grain. This sits active cathodic to the Mg matrix.[28]

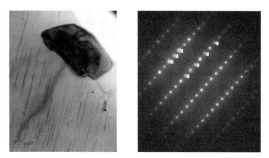

Fig. 3: TEM diffraction of CaMgSi precipitate from ZSM651+0.4Ca alloy.

Figure 6 shows the effect of extrusion process on the corrosion behavior of ZSMX alloys. The effect of Si and Ca was similar in the surface and the bulk of the extrusion pipe. Due to the different residual stress in the surface and the bulk of the extrusion pipe [29] the values of the corrosion rate not the same.

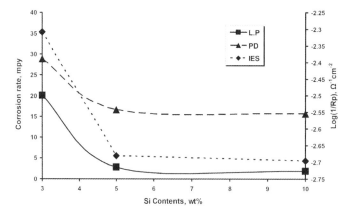

Fig. 4: Corrosion rate (~1/Rp) from EIS measurements and Corrosion rate from linear polarization and potentiodynamic methods vs. Si contents of ZSM6X1+0.4Ca Mg alloys in 3.5% NaCl saturated with $Mg(OH)_2$.

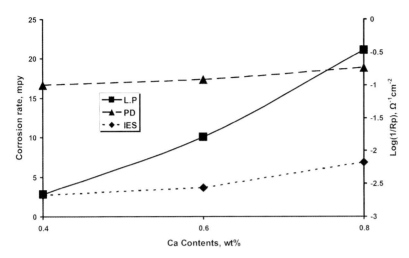

Fig. 5: Corrosion rate (~1/Rp) from EIS measurements and Corrosion rate from linear polarization and potentiodynamic methods vs. Ca contents of ZSM651+YCa Mg alloys in 3.5% NaCl saturated with $Mg(OH)_2$.

4 Summary

The grain size of the Mg-6%Zn-1%Mn alloy was significantly refined with the addition of silicon and calcium in the as-extruded condition. The addition of silicon to the Mg-6%Zn-1%Mn alloy increased the corrosion resistance. However, the addition of calcium decreased the corrosion resistance. The extrusion process influences on the corrosion behavior of ZSMX alloy due to the residual stress on the surface of the extrusion pipe.

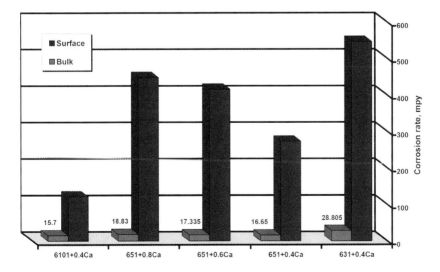

Fig. 6: The influence of the extrusion process on the corrosion behavior of ZSMX Mg alloy.

5 References

[1] C. D. Yim and K. S. Shin: Mater. Trans. Vol. 44(4) (2003), p. 558
[2] D. H. Kim, SJ. Song, H. Park and K.S. Shin: Met. Mater. Vol. 3 (1997), p. 51
[3] J.W. Kim, D.H. Kim, C. D. Yim and K. S. Shin: J. Kor. Inst. Met. Mater. Vol. 35 (1997), p. 1446
[4] J. J. Kim, D. H. Kim, K. S. Shin and N. J. Kim: Scripta mater. Vol. 41 (1999), p. 333
[5] C. D. Lee, C. S. Kang and K. S. Shin: Met. Mater. Vol. 6 (2000), p. 351
[6] C. D. Lee, C. S. Kang and K. S. Shin: Met. Mater. Vol. 6 (2000), p. 441
[7] C. D. Lee, C. S. Kang and K. S. Shin: Met. Mater.-Int. Vol. 7 (2001), p. 296
[8] Y. J. Ko, C. D. Yim, J. D. Lim and K. S. Shin: Mater. Sci. Forum Vol. 419-422 (2003), p. 851
[9] C. S. Roberts: *Magnesium and Its Alloys* (John Wiley & Sons, Inc., New York 1960)
[10] L. Sturkey and J.B. Clark: J. Inst. Met. Vol. 88 (1959-60), p. 177
[11] J. B. Clark: Acta. Metall. Vol. 13 (1965), p. 1281
[12] E. O. Hall: J. Inst. Met. Vol. 96 (1968), p. 21
[13] G. Mima and Y. Tanaka: Jpn. Inst. Met. Vol. 12 (1971), p. 71
[14] G. Mima and Y. Tanaka: Jpn. Inst. Met. Vol. 12 (1971), p. 76
[15] L. Y. Wei, G. L. Dunlop and H. Westengen: Metall. & Mater. Trans. A Vol. 26A (1995), p. 1705
[16] A. K. Bhambri and T. Z. Kattamis: Metall. Trans. Vol. 2 (1971), p. 1869
[17] T. E. Leontis: Trans. AIME, Vol. 180 (1949), p. 287
[18] L. Y. Wei, G. L. Dunlop and H. Westengen: Metall. & Mater. Trans. A Vol. 26A (1995), p. 1947
[19] V. Sustek, S. Spigarelli and J. Cadek: Scripta mater. Vol. 35 (1996), p. 449
[20] W. Unsworth and J. F. King: Magnesium Technology, The Inst. of Metal (1987), p. 25
[21] E. A. Ball and P. B. Prangnell: Scripta metall. et mater. Vol. 31 (1994), p. 111
[22] S. Beer, G. Frommeyer, E. Schmid, in Proceedings of Conference on 'Magnesium alloys and their applications', Oberursel, DGM, 1992, pp. 317-324.
[23] Yuan Guangyin, Liu Manping, Ding Wenjiang, Akihisa Inoue, Materials Science and Engineering A357 (2003) 314-320
[24] M. Pourbaix, Atlas of Electrochemical Equilibria in Aqueous Solutions, Cebelcor, Brussesls, 1974, p. 141.
[25] Boukamp, Equivalent Circuit Software, Users Manual, The Netherlands University of Twente, 1988, pp. 6-26.
[26] H. H. Uhlig, *Corrosion and Corrosion Control*, John Wiley & Sons Inc, New-York (1963).
[27] Y. G. Na, D. Eliezer and K. S. Shin: Mater. Sci. Forum Vol. 488-489 (2005), p. 839
[28] G. Ben-Hamu, D. Eliezer and K.S. Shin, Materials Science and Technology (2006), Article in Press
[29] G. Ben-Hamu, A. Eliezer, E. M. Gutman, Electrochimica Acta (2006), Article in Press

Corrosion Behavior of Die-Cast MRI 153M and AZ91D Alloys in Alkaline and Neutral Chloride Media

Sofiene Amira, Dominique Dubé, Réal Tremblay and Edward Ghali
Université Laval, Québec, Canada

1 Introduction

MRI 153M alloy, developed by the Magnesium Research Institute (MRI, Israel) is a new creep resistant magnesium die-casting alloy designed for automotive gearbox-housing components [1]. This alloy is a modified AZ91 with addition of Ca, Sr, and rare-earths, optimizing the properties of AZ91 (excellent castability, good balance of strength and ductility, good corrosion resistance), but improving its creep properties. Besides, the corrosion rate of the MRI 153M was measured by salt spray testing and was found to be slightly lower than that of AZ91 [1]. However further studies are needed to understand the corrosion behavior of this new alloy.

The purpose of this study is to compare the corrosion behavior of MRI 153M and AZ91D die-cast specimens in three media with different pH and chloride concentrations, using different experimental techniques.

2 Experimental

2.1 Materials

The magnesium alloy AZ91D contained 9,0% Al - 0,7% Zn - 0,25% Mn and the new MRI 153M alloy, considered as a modified AZ91 alloy, contained the following additional elements : <1,0% Ca, <0,2% Sr, and <1,0% RE (in wt%). For both alloys, experimental die-cast box-like parts were prepared from remelted billets gravity-cast previously in a steel mould (Figure 1). The main parameters for die-casting of AZ91D and MRI 153M box-like parts are given in Table 1.

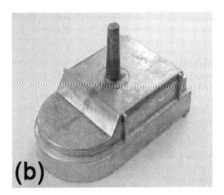

Figure 1: (a) Gravity-cast billet and (b) typical die-cast box-like part.

Table 1: Die-casting parameters for box-like parts.

Parameter	Alloy	
	MRI 153M	**AZ91D**
Preheating temperature [°C]	680	680
Die temperature [°C]	220	150
Shot sleeve speed [m.s^{-1}]	1,25	0,50
Injecting pressure [bar]	1200	900

2.2 Specimens Preparation

Specimens for electrochemical testing (10 mm x 10 mm) were cut from the 6,2 mm thick sidewall of die-cast box. Electrodes with a connecting wire to one side of the specimen were embedded with acrylic resin leaving 1,0 cm^2 of the surface in contact with the testing solution. The polished surface of specimens was obtained by using SiC paper to remove about 2,0 mm from the original surface. Testing specimens were afterward cleaned with distilled water and ethanol and then air-dried.

2.3 Corrosion Medium

Three solutions were used: (1) 0,05M NaCl + 0,1M NaOH + 0,1M H$_2$O$_2$ (30%) solution at pH 12,3 (the NaOH permits to adjust the pH of the solution and the peroxide is added as a strong oxidizing agent to induce a local pH increase which enables the precipitation of the protective magnesium hydroxide) ; (2) 0,05M NaCl solution at pH 6,8 ; and (3) 3,5% NaCl solution at pH 6,0. All experiments were carried out at 25°C without de-aeration or stirring.

2.4 Corrosion Testing

Potentiodynamic polarization experiments were performed with a potentiostat/galvanostat at a potential scanning rate of 0,166 mV.s^{-1}. A standard three-electrode configuration was used with

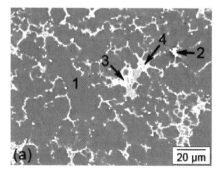 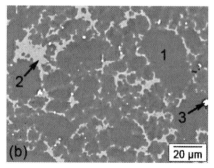

Figure 2: Backscattered electron images showing the microstructure of (a) MRI 153M-DC and (b) AZ91D-DC.

an Ag,AgCl/KCl$_{sat.}$ reference electrode, a platinum counter electrode and the specimen as a working electrode. For electrochemical noise analysis (ENA), the current noise was measured by a zero-resistance ammeter (ZRA) as the current flowing between two nominally identical working electrodes (WE) was kept at zero potential. The coupled potential or noise potential was measured by an Ag,AgCl/KCl$_{sat.}$ reference electrode. Scanning reference electrode technique (SRET) was used to detect cathodic and anodic sites at the surface of specimens immersed in the electrolyte.

2.5 Microstructure and Phases Characterization

The microstructure and distribution of elements in phases were estimated by an electron probe microanalyser (EPMA).

3 Results

3.1 Microstructures and Phases

The microstructures of die-cast MRI 153M and AZ91D specimens are shown in Figure 2. An EPMA analysis of phases is given in Table 2. Composition of very small secondary phase in die-cast specimens could not be determined accurately due to the interaction of the electron beam with the surrounding magnesium matrix. For both alloys, the fine microstructure is constituted with primary α-Mg grains surrounded by secondary phase particles. For the AZ91D specimens, the secondary phase particles are composed of an eutectic mixture of α-Mg and $Mg_{17}(Al,Zn)_{12}$ (β-phase) [2]. Mn-Al particles were also present. For the MRI 153M specimens, it was observed that Ca, Sr and RE elements were mainly found in secondary phases, forming a white lamellar structure. Although these elements seem to have a greater affinity for Al than for Mg, it can be expected that very small particles of new phases like Al-Ca, Al-Sr and Al-RE are

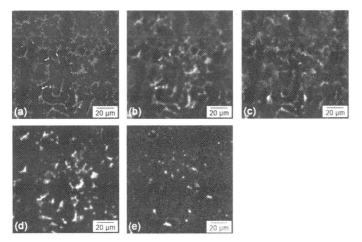

Figure 3: EPMA analyses of MRI 153M-DC specimen: (a) Backscattered electron image, and (b) Al, (c) Zn, (d) Ca, and (e) Sr elemental mappings.

formed at the expense of the β-phase [2]. EPMA maps presented in Figure 3, b and d, show that the Ca has combined with the Al to form Al-Ca particles (Al_2Ca) [1,3] or $(Mg,Al)_2Ca$ [4]. Al and Zn mappings (Figure 3, b and c) corroborate the combination of Al and Zn with Mg to form the $Mg_{17}(Al,Zn)_{12}$. Sr mapping (Figure 3e) confirms the presence of Sr-rich phase described as a hexagonal $(Mg,Al)_2(Sr,Ca)$ phase by Sato et al. [4].

Table 2: EPMA analysis of die-cast MRI 153M and AZ91D specimens.

Spot[1]	Element (at.%)							Comments
	Mg	Al	Zn	Mn	Ca	Sr	RE	
				MRI 153M-DC				
1	97,05	2,75	0,12	0,05	0,01	-	-	Primary α-Mg
2	29,82	49,82	0,18	15,58	3,33	0,02	1,22	Mixture of phases
3	66,84	27,74	2,14	0,02	3,08	0,16	-	Mixture of phases
4	59,19	30,72	1,38	3,50	3,34	1,69	0,15	Mixture of phases
				AZ91D-DC				
1	97,26	2,60	0,10	0,03	-	-	-	Primary α-Mg
2	70,69	27,60	1,64	0,05	-	-	-	α-Mg (eut.) + $Mg_{17}(Al,Zn)_{12}$
3	8,88	53,40	0,07	37,63	-	-	-	Al-Mn particle

[1] See Figure 2 for spot numbers.

3.2 Corrosion Behavior

3.2.1 In 0,05M NaCl + 0,1M NaOH + 0,1M H_2O_2 at pH 12,3 and 25°C

The potentiodynamic polarization curves for MRI 153M and AZ91D in the alkaline solution are shown in Figure 4. The MRI 153M specimen was passive in this solution while the AZ91D specimen showed no passivity. Figure 5 illustrates a typical pattern of the mean current noise of MRI 153M and AZ91D specimens for a 16 h immersion period. The current noise of MRI 153M shows less high amplitude fluctuations and tends more quickly to zero than that of AZ91D alloy. Typical SRET maps of MRI 153M and AZ91D over a 32 h period of immersion are presented in Figure 6. The MRI 153M specimens exhibit less active sites than AZ91D and these minor sites disappeared after 1 h of immersion. AZ91D specimens showed major cathodic sites adjacent to important anodic sites from the beginning of immersion and significant electrochemical activity disappeared only after 16 h.

3.2.2 In 0,05M NaCl Solution at pH 6,8 and 25°C

From the potentiodynamic polarization curves for the die-cast MRI 153M and AZ91D in the 0,05M chloride solution (Figure 7), the corrosion current density of the MRI 153M is lower than that of the AZ91D. Also, the linear parts of the cathodic curves for both specimens have a similar slope suggesting that the involved cathodic reactions are similar. A typical pattern of the mean current noise of MRI 153M and AZ91D specimens over a 24 h of immersion in the same solution is shown in Figure 8. The noise resistance (the ratio between the standard deviation of current and potential fluctuations) calculated for each alloy from noise data, was 2852 and 2180 $\Omega.cm^2$ for MRI 153M and AZ91D, respectively. For general corrosion and activation control, the noise resistance can be associated with the polarization resistance.

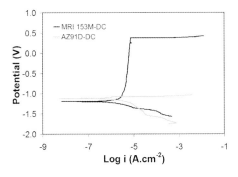
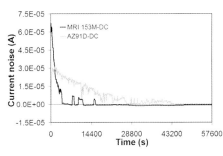

Figure 4: Polarization curves for MRI 153M-DC and AZ91D-DC specimens in the alkaline solution.

Figure 5: Current noise plots for MRI 153M-DC and AZ91D-DC specimens in the alkaline solution.

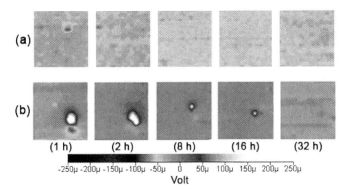

Figure 6: SRET maps for (a) MRI 153M-DC and (b) AZ91D-DC specimens in the alkaline solution.

3.2.3 In 3,5% NaCl Solution at pH 6,0 and 25°C

Figure 9 shows the potentiodynamic polarization curves for the die-cast MRI 153M and AZ91D specimens in 3,5% NaCl solution. The MRI 153M appeared to be slightly less active than the AZ91D with a corrosion potential of -1482 mV compared to -1516 mV, but its corrosion current density was found to be higher (27 $\mu A.cm^{-2}$ as compared to 22 $\mu A.cm^{-2}$). ENA showed that noise resistances for MRI 153M and AZ91D were 218 and 243 $ohm.cm^2$, respectively.

4 Discussion

This study has shown the beneficial effect of Ca, Sr and RE elements on the corrosion behavior of MRI 153M in the alkaline and the 0,05M NaCl solutions as compared to that of the AZ91D.

In the alkaline solution, the MRI 153M exhibited a good passivation behavior (E_p = 368 mV and i_p = 7,6 µA.cm^{-2}), while no passivation was displayed by the AZ91D. This behavior can be attributed mainly to the beneficial effect of Ca addition. Indeed, it has been reported that the addition of Ca (less than 1,0wt%) leads to the incorporation of Ca atoms in the corrosion product film, causing it to be more stable and protective [5]. Polarization experiments and ENA showed that the corrosion resistance of MRI 153M in the 0,05M NaCl was higher than that of AZ91D. In the case of AZ91D, it was reported that the β-phase acts as a barrier against corrosion since it forms a continuous network over the matrix [6]. In the MRI 153M, the Al-Ca phases formed along the grains boundaries acting as an additional barrier against corrosion with the β-phase [3], can explain its best corrosion resistance. Nevertheless, the corrosion rate of MRI 153M was found to be slightly higher than that of AZ91D in 3,5% NaCl solution and the same tendency was shown by ENA. Since the β-phase continues to act as a barrier against corrosion in such media [7], it can be expected that the Al-Ca phases do not act in a similar way. The reasons of such behavior have to be studied more deeply.

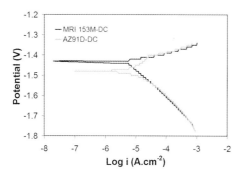

Figure 7: Polarization curves for MRI 153M-DC and AZ91D-DC specimens in 0,05M NaCl solution.

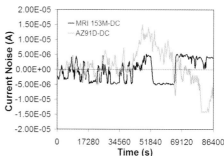

Figure 8: Current noise plots for MRI 153M-DC and AZ91D-DC specimens in 0,05M NaCl solution.

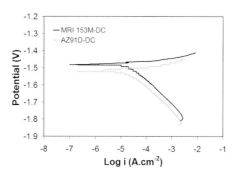

Figure 9: Polarization curves for MRI 153M-DC and AZ91D-DC specimens in 3,5% NaCl solution.

The influence of the other addition elements, i.e. Sr and RE elements, on the corrosion behavior of the die-cast MRI 153M alloy is hard to evaluate due to their low concentration and their presence in finely dispersed particles. However, some authors stated that the addition of RE elements with Ca in the AZ91 alloy improves their corrosion resistance [3], while others suggest that the presence of RE enhances a uniform enrichment of the passive layer in Al_2O_3 [5]. Other works showed that the addition of Sr to Mg-Al system improves their salt spray corrosion resistance [8].

5 Conclusion

- Analyses by EPMA showed that the die-cast MRI 153M is constituted of primary α-Mg grains surrounded by β-phase ($Mg_{17}(Al,Zn)_{12}$), Ca-rich and Sr-rich phases. The die-cast AZ91D is composed of primary α-Mg grains and secondary β-phase.
- The MRI 153M was passive in the alkaline solution, contrary to the AZ91D. This behavior can be attributed to the incorporation of Ca atoms in the corrosion product film, causing it to become more stable and protective.
- In the 0,05M NaCl solution, the MRI 153M was found to have a better corrosion resistance than that of the AZ91D, due to the barrier effect played by the Al-Ca phases.
- The corrosion rate of MRI 153M was slightly higher than that of AZ91D in the 3,5% NaCl solution.

6 Acknowledgements

The authors would like to thank the NSERC (National Sciences and Engineering Research Council of Canada) for financial support, Drs. N. Hort and K. Kainer (GKSS, Geesthacht, Germany), Dr. C.A. Loong (IMI-CNRC, Montreal, Canada), and Mr. M. Choquette (Laval University, Québec, Canada) for their contribution.

7 References

[1] F.V. Buch, S. Schumann, H. Friedrich, E. Aghion, B. Bronfin, B.L. Mordike, M. Bamberger, D. Eliezer, Magnesium Technology **2002** (Ed. H. Kaplan), TMS, 61-67.
[2] E. Aghion, B. Bronfin, M. Katriz, O. Bar-Yosef, E. Shmelkin, M. Lautzker, E. Lerer, D. Eliezer, N. Moscovitch, Magnesium Alloys : Science, Technology and Applications, (Eds. E. Aghion, D. Eliezer), The Israeli Consortium for the Development of Magnesium Technology, Israel, **2004**, 137-158.
[3] G. Wu, Y. Fan, H. Gao, C. Zhai, Y. P. Zhu, Mater. Sci. and Eng. A, **2005**, 255-263.
[4] T. Sato, B.L. Mordike, J.F. Nie, M.V. Kral, Magnesium Technology **2005**, TMS, 435-440.
[5] V. Neubert, A. Bakkar, Proc. 6th Inter. Conf. Magnesium Alloys and Their Applications, Germany (Ed. K.U. Kainer), **2003**, 638-645.
[6] G. Song, A. Atrens, M. Dargusch, Corr. Sci., **1998**, 41, 2, 249-273.
[7] R. Ambat, N.N. Aung, W. Zhou, Corr. Sci., **2000**, 42, 8, 1433-1455.
[8] M.O. Pekguleryuz, E. Baril, Magnesium Technology **2001** (Ed. J. Hryn), TMS, 119-125.

Study of the Corrosion Behaviour of New Mg Based Alloys

C. Juers, C.-E. Barchiche, E. Rocca, J. Hazan, J. Steinmetz
Université Henri Poincaré – Nancy I, Laboratoire de Chimie du Solide Minéral, Vandoeuvre Lès Nancy – FRANCE

1 Introduction

Magnesium and its alloys exhibit an interesting combination of very low density and high mechanical properties [1]. Consequently, they are excellent candidates for automotive and aeronautical applications where weight reduction is important in order to make fuel savings, especially if we compare them to aluminium and particularly to ferrous alloys mainly used today [2]. But magnesium alloys are highly susceptible to corrosion [3], particularly in chlorides or in salt-spray conditions. This has limited their use for a lot of applications where exposure to harsh service conditions is permanent. Among the magnesium alloys, the Mg-Al ones (such as AZ91 or AM50) are known to have the best mechanical properties/corrosion resistance balance. Their microstructure [4] shows a matrix of Mg containing a solid solution of Al (α phase) surrounded by the intermetallic compound $Mg_{17}Al_{12}$ (β phase).

Some new Mg alloys were developed for specific aeronautical applications in the framework of the IDEA European FP6 Project then compared to commercial AZ91E focusing in this study on electrochemical experiments and especially Electrochemical Impedance Spectroscopy.

2 Experimental

2.1 Materials

Two new Mg-Zn-Zr-Y-Gd-Re alloys that were developed in the framework of the IDEA project and designated as MRI204S and MRI207S were studied in comparison with AZ91E alloy that was taken as a reference. All the alloys were investment cast and then heat treated according to the optimal regime for each alloy.

The samples were tested for corrosion resistance without polishing and the only surface preparation was degreasing on the as received casting alloys which were heat treated. This surface preparation without mechanical polishing or etching was used so as to reproduce industrial conditions of use on pieces with complex shape where polishing is very difficult to apply. Consequently, the corrosion behaviour of the top surface of the alloys was studied (skin effect).

2.2 Electrochemical Tests

Electrochemical tests and EIS experiments were performed in a three-electrode electrochemical cell connected to a GAMRY PCI-4/300 potentiostat. This cell is a 70 ml one in which the working electrode (3 cm^2 of Mg alloy) is facing the counter electrode (a platinum grid) and a Hg/Hg_2Cl_2, KCl sat. electrode (E = +0.242 V/SHE) was taken as the reference electrode. All the working electrode potentials are given versus this reference.

The samples were subjected to 14 cycles of the following sequence of measurements:
- Open Circuit Potential for 2 hours.
- Electrochemical Impedance Spectroscopy. EIS experiments were carried out from 10^6 to 4.10^{-3} Hz with a 10 mV amplitude at the corrosion potential determined at the end of the previous open circuit.

After these cycles, that is to say after about 45 hours of immersion, anodic or cathodic potentiodynamic scans were recorded:
- The anodic from the corrosion potential E_{corr} up to E_{corr} + 1 V.
- The cathodic from the corrosion potential E_{corr} to E_{corr} - 0.5 V. All curves were recorded with a 1 mV/s scan rate.

All the electrochemical experiments were duplicated.

The corroding electrolyte used was ASTM D1384-87 standard water [5] with the composition of 148 mg.l^{-1} Na_2SO_4, 138 mg.l^{-1} $NaHCO_3$, 165 mg.l^{-1} NaCl (pH=8.3). This solution was chosen because of its moderate aggressiveness, and the results obtained can be correlated with long period exposure to atmospheric conditions.

3   Results and Discussion

3.1   Overall Electrochemical Behaviour

For the two new Mg alloys MRI204S and MRI207S in these conditions, the OCP increases during the first 12 hours of immersion before being stabilized at -1.57 for the former alloy and -1.62 V/ECS for the later (as can be seen on Figure 1: MRI204S in ASTM D1384-87 water (pH=8.3) – (a) OCP, (b) EIS, (c) i=f(E).a for MRI204S and Figure 2: MRI207S in ASTM D1384-87 water (pH=8.3) – (a) OCP, (b) EIS, (c) i=f(E).a for MRI207S).

For all the EIS data recorded, a high frequency (HF) and a low frequency (LF) capacitive loop can be evidenced on Nyquist plots (Figure 1: MRI204S in ASTM D1384-87 water (pH=8.3) – (a) OCP, (b) EIS, (c) i=f(E).b for MRI204S and Figure 2: MRI207S in ASTM D1384-87 water (pH=8.3) – (a) OCP, (b) EIS, (c) i=f(E).b for MRI207S). This can be attributed to an overall passive behaviour due to the oxide/hydroxide film formed on the surface of the alloys.

The current density – voltage curves recorded after 14 OCP/EIS cycles (equivalent to 45 hours immersion in the corroding electrolyte) on the new alloys are displayed in Figure 1: MRI204S in ASTM D1384-87 water (pH=8.3) – (a) OCP, (b) EIS, (c) i=f(E).c and Figure 2: MRI207S in ASTM D1384-87 water (pH=8.3) – (a) OCP, (b) EIS, (c) i=f(E).c. The anodic and cathodic branches were carried out separately. During the anodic polarization, a plateau can be observed. This plateau is about 250 mV for MRI204S while it is 300mV for MRI207S. The value of the passivation current recorded is 20 µA.cm^{-2} for the former alloy and 5 µA.cm^{-2} for the latter. From the anodic side, this plateau is limited by the pitting potential at about -1.2 V/ECS for the two alloys, which corresponds to the breakdown of the protective film formed on the samples. If we consider these characteristic data, we can conclude that MRI207S has more corrosion resistance but an overall passive behaviour can be clearly observed for the two new alloys tested.

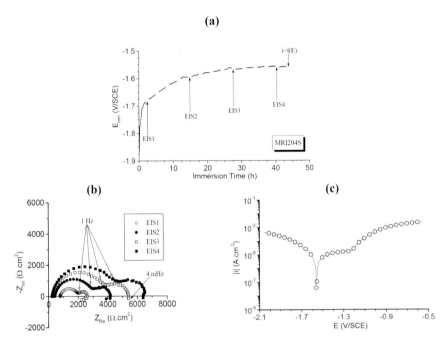

Figure 1: MRI204S in ASTM D1384-87 water (pH=8.3) – (a) OCP, (b) EIS, (c) i=f(E).

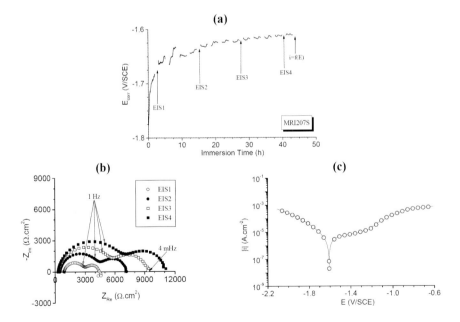

Figure 2: MRI207S in ASTM D1384-87 water (pH=8.3) – (a) OCP, (b) EIS, (c) i=f(E).

For the untreated AZ91E alloy in ASTM water, Figure 3: AZ91E in ASTM D1384-87 water (pH=8.3) – (a) OCP, (b) EIS, (c) i=f(E).a shows that the OCP is disrupted, near -1.4 V/ECS that is to say higher than for the two new alloys. All the EIS data recorded show one HF capacitive loop and one LF inductive loop on Nyquist plots (Figure 3: AZ91E in ASTM D1384-87 water (pH=8.3) – (a) OCP, (b) EIS, (c) i=f(E).b). Anymore, on the anodic polarization curve recorded after 14 OCP/EIS cycles (Figure 3: AZ91E in ASTM D1384-87 water (pH=8.3) – (a) OCP, (b) EIS, (c) i=f(E).c), no passivity plateau can be seen, which constitutes a different behaviour in comparison to that we have observed for the new alloys. The AZ91E provides low protective passivation properties and it can hardly resist to aggressive media such as the presence of chlorides as it is the case in the ASTM water. An accurate corrosion current density (I_{corr}) is uneasy to determine from these potentiodynamic curves due to the lack of linearity of the semi logarithmic plots.

These results indicate that pitting corrosion occurs for the AZ91E alloy and this is probably related to a higher Al content on the surface of the alloy in comparison to that of the bulk. The new alloys do not contain Al, consequently, this "skin effect" was not observed for both of them.

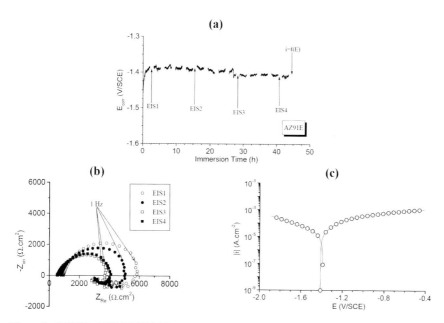

Figure 3: AZ91E in ASTM D1384-87 water (pH=8.3) – (a) OCP, (b) EIS, (c) i=f(E).

3.2 Electrochemical Impedance Spectroscopy Study

The Nyquist diagrams can be described by the equivalent electrical circuits shown in Figure 4: Equivalent circuits used for simulated results for the two new alloys (a) and for the commercial AZ91E one (b).a for the new alloys and Figure 4: Equivalent circuits used for simulated results for the two new alloys (a) and for the commercial AZ91E one (b).b for the commercial AZ91E one and a good correlation have been found between the simulated curves and the experimental results. The two loops indicate that two different processes occur on the electrodes for the three alloys tested.

For the new alloys, the HF capacitive loop can be attributed to the charge transfer reaction [2,6] between the metal and the oxidant, the water molecule, across the protective (or passive) film. This corrosion process, leading to the dissolution of metal, is characterized by a resistance noted here R_{HF} that can be assimilated to R_t, the charge transfer resistance, which can be determined from the intersection between the abscissa axis and the extrapolated first loop [7] and the higher it is, the lower the electron transfer is and thus the slower the corrosion rate is. It is the impedance parameter that was found to determine more accurately the corrosion rates for these corrosion systems. For the two new alloys, we can observe in Figure 5: Simulated results for the different alloys in ASTM D1384-87 water (pH=8.3) – (a) RHF = f(immersion time) – (b) CPE-QHF = f(immersion time).a that the R_{HF} values increase with immersion time and they are higher for the MRI207S alloy in comparison to the values observed for the MRI204S one, indicating better corrosion resistance for the former alloy in ASTM standard water.

The capacity values for the HF capacitive loop (C_{HF}) for the new alloys slightly decrease in a first time (indicating that even in ASTM water, some conditions of alkalinization and film formation are provided), then are stabilized around 15 µF.cm^{-2} and can be attributed to the double layer capacity (C_{dl}) where the relatively low value is due to the partially covered surface. C_{HF} reaches slightly lower values for MRI204S in comparison to MRI207S (see Figure 5: Simulated results for the different alloys in ASTM D1384-87 water (pH=8.3) – (a) RHF = f(immersion time) – (b) CPE-QHF = f(immersion time).b) and this is probably attributed to the higher thickness of the oxide layer formed on the former alloy.

The LF (low frequency) capacitive loop is generally thought to be due to the mass transfer of ions through the hydroxide/oxide layer [2].

Figure 4: Equivalent circuits used for simulated results for the two new alloys (a) and for the commercial AZ91E one (b).

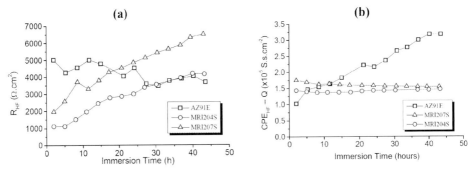

Figure 5: Simulated results for the different alloys in ASTM D1384-87 water (pH=8.3) – (a) R_{HF} = f(immersion time) – (b) CPE-Q_{HF} = f(immersion time).

A completely different behaviour can be observed for AZ91E alloy (Figure 5: Simulated results for the different alloys in ASTM D1384-87 water (pH=8.3) – (a) RHF = f(immersion time) – (b) CPE-QHF = f(immersion time).). The interpretation of the different phenomena that occur in this case is more complex due to the presence of the LF inductive loop which is related to the phenomenon of adsorption. Nethertheless, the HF capacitive loop is attributed in this case also to the charge transfer reaction between the metal and the oxidant (the water molecule) across the passive film. But in this case, if we consider the evolution of R_{HF} with immersion time on Figure 5: Simulated results for the different alloys in ASTM D1384-87 water (pH=8.3) – (a) RHF = f(immersion time) – (b) CPE-QHF = f(immersion time).a, we can observe that its value decreases with exposure and thus, the corrosion resistance is worsening with time. After 20 hours of immersion, R_{HF} reaches values that are lower in comparison to that found for MRI207S, inducing a slower corrosion rate for the later alloy. And after 30 hours of immersion, the MRI204S also exhibits better corrosion behaviour than AZ91E.

In parallel and in concordance to that, the evolution of the value of the HF capacity (C_{HF}), observed in Figure 5: Simulated results for the different alloys in ASTM D1384-87 water (pH=8.3) – (a) RHF = f(immersion time) – (b) CPE-QHF = f(immersion time).b for the AZ91E, increases with immersion time and becomes higher than the MRI204S already after 5 hours in the ASTM water and than the MRI207S after 10 hours.

4 Conclusions

The evaluation of the corrosion behaviour of the as received casting alloys was performed using electrochemical DC and AC methods in ASTM D1384-87standard water. EIS was used to determine more accurately the corrosion rates mainly from the analysis of the high frequency capacitive loop. Anodic polarization was applied in order to evaluate the pitting resistance of the alloys.

The EIS results show that the two new alloys MRI204S and MRI207S are slightly more corrosion resistant than the reference alloy AZ91E. This was observed first from the Nyquist plots which show two relatively well separated capacitive loops indicating a protection layer was formed, as well as from the passivation plateau which can be easily seen on the anodic polarization curves with passivation current density reaching values from 5 to 20 $\mu A.cm^{-2}$ for the two new alloys. On the other hand, the reference alloy AZ91E presents an inductive loop at the low frequency domain and no passivation state could be observed. Consequently, the new alloys are promising candidates for the development of new Mg alloys, especially the MRI207S which seems to be even better than the MRI204S

5 Acknowledgements

The authors would like to thank the European Union for funding this study in the framework of the IDEA project (FP6-503826).

The magnesium alloys were supplied thanks to the collaboration with MRI, Beer-Sheva (Israel), VTT, Helsinki (Finland) and Stone Foundries, London (UK).

6   References

[1] S. Mathieu, C. Rapin, J. Steinmetz, P. Steinmetz, Corrosion Science, 2003, 45, 2741-2755.
[2] S. Mathieu, C. Rapin, J. Hazan, P. Steinmetz, Corrosion Science, 2002, 44, 2737-2756.
[3] G.L. Song, A. Atrens, Advanced Engineering Materials, 1999, 1, 11-33.
[4] H. Umehara, M. Takaya, S. Terauchi, Materials Science Forum, 2003, 419-422, 883-888.
[5] Annual book of ASTM standards, Copyright ASTM, Pa 19103 (1986).
[6] G. Baril, N. Pébère, Corrosion Science, 2001, 43, 471-484.
[7] J. Hazan, E. Rocca, C.Juers, J. Steinmetz, Magnesium Technology 2005, edt Neal R. Neelameggham, Howard I. Kaplan and Bob R. Powell, San Francisco, California, USA, 457-461.

Influence of the Alloy Composition on the Mechanical and Electrochemical Properties of Binary Mg – Ca – Alloys and its Corrosion Behaviour in Solutions at Different Chloride Concentrations

T. Hassel (Sp), F.-W. Bach, C. Krause,
Institute of Materials Science, Leibniz University of Hanover (Germany)

1 Introduction

The application of magnesium alloys as a construction material is clearly complicated by its insufficient corrosion stability. But the process of corrosion can be an absolute requirement for a successful application of these materials. The development of temporary implants made from absorbable magnesium alloys is discussed [1] and researched [2] in the field of biomedical science for some years. Magnesium belongs to the group of trace elements and the human body contains approximately 35g per 70kg body weight. The daily demand averages approx. 375 mg / day. Toxic reactions caused by too much magnesium are generally unknown [3]. Magnesium offers beside its adequate mechanical properties compared e.g. to the absorbable polymers the great advantage that it will be completely absorbed. The aim is the development of a magnesium alloy with a minimized alloying element concentration. With as few as possible alloying elements a very gentle therapy should be possible. The investigated MgCa alloys show adequate mechanical properties to come into question of such an application. Because of the base character of magnesium and its extreme affinity to chemical solution in chloride media, magne-sium can be reabsorbed complete by the body. The mechanism of solution follows the reaction (eq. 1):

$$Mg + 2H_2O \rightarrow Mg(OH)_2 + H_2\uparrow, \tag{1}$$

whereas the metal is oxidized and hydrogen is produced. The electrolyte is constantly in contact with the base metal, because magnesium cannot form a close covering $Mg(OH)_2$ layer on the

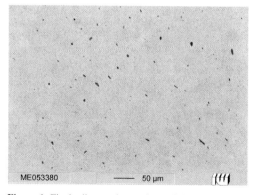

Figure 1: Finely dispersed eutectic precipitation of a MgCa0,4 alloy (etching with 2% nitric acid)

surface in aqueous media. This process leads to the complete dissolution of magnesium in Cl⁻ containing aqueous media. During the solution gaseous hydrogen is produced and laid off constantly (1g Mg » 1,081 H_2). The amount of hydrogen per time corresponds to the corrosion speed of the material. For this reason the suggestible magnitude has to be the corrosion speed. The implant has to reabsorb with a moderate rate of hydrogen formation because the formation of vapor locks must be avoided. This can only be influenced directly by alloying, because the dissolution of pure magnesium is too fast. Additionally the magnesium implant can be covered by a protecting surface layer (e.g. MgF_2) to delay the start of the bulk corrosion[4]. Normally all technical corrosion trials in the past were made in common salt solutions at concentratios above 2,5%. To compare the results of the in vitro corrosion tests with the in vivo degradation of this materials its is very important to know, which influence on the corrosion speed the chloride concentration will takes. Therefore the influence of the chloride concentration on the corrosion is been investigated by eudiometric and electrochemical investigations at variable electrolytes.

2 Materials and Methods

2.1 Binary System of Magnesium – Calcium

The investigations show the development of such an alloy based on the binary system magnesium–calcium. On the magnesium rich site of this system the maximum solubility of calcium in the magnesium lattice amounts at room temperature 0.8wt%. At a calcium concentration of 16.2wt% the alloy solidifies in an eutectic composition. Low alloyed MgCa – Systems consists of a α solid solution (α-phase) (magnesium with interstitial calcium) and an eutectic (α-phase + Mg_2Ca). From an amount of 0.2wt% calcium can be distinctly seen in micrographs[5]. The reason is the solidification imbalance at a height cooling rate, which occurs in degenerated eutectics. The micrograph (Figure 1) shows very fine dispersed eutectic on the surface.

The addition of a small amount of calcium firstly increases the corrosion stability and secondly minimize the grain growth. This refining process leads to smaller grains in casts. Disadvantageous is the rising sensitivity to constitute hot cracks during the deformation processes[5,6].

2.2 Production and Testing Methods

All investigated alloys are produced in our light metal foundry. Pure Magnesium (99,8%) and a MgCa30 pre alloy (30wt% Ca) are the primary materials to mix MgCa – alloys with different calcium concentrations. The mold is done by casting the melt from a steel pot at a temperature of 750 °C into a cylindrical steel mold which is pre heated up to 450 °C. The mold is protected by a Ar – CO_2 gas mix (82/18) and stirred for 45min. The Ca-content of the casted alloys amounts 0.4, 0.6, 0.8, 1.0, 1.2, 1.5, 1.7, 2.0, 2.5, 3.0 and 4.0wt%.

The cylindrical bars with a diameter of 140 mm and a length of 450 mm are worked to extrusion studs (D = 120 mm; L = 300 mm) by turning. In furnace the studs are homogenized at a temperature of 350 °C for 24 hours without any protecting gases. At a 10 MN extrusion press the studs are deformed to a rod of 16 mm in diameter. Fifteen tensile test specimens of each alloy composition are tested to statistically confirm the results. Electrochemical investigations on the same materials are made to measure the open circuit potential and the corrosion current density. In a three electrode cell and with an electrochemical workstation (IM6, Zahner, Germany) all electrochemical measurements are implemented. In epoxy resin embedded specimens are grounded, polished and attached into the specimen holder. All electrochemical investigations are made in electrolytes buffered with a TRIS solution to a constant pH value of 7.4 (physiologically pH value). To study the influence of the chloride concentration solutions of 0.05, 0.09, 0.9, 2.5 and 5 percent NaCl were used as electrolytes. Magnesium passivates at a pH value of 8,5–10,5 by the development of a dense $Mg(OH)_2$ surface layer. A platinum electrode is used as counter electrode. To measure the reference potential a silver/silver chloride electrode in saturated potassium chloride is used by installing it with a Hubber Luggin capillary. Deionized water is used for all corrosion measurements. The gas formation is investigated by eudiometric measurements at the same concentrations of the electrolyte and at the same pH value. Therefore cylindrical specimens (d = 16 mm; h = 5 mm) are immersed in 15 liter bowls which are filled with the different NaCl-solutions (buffered at pH = 7.4). Above the specimen a filled eudiometric container is arranged to catch the arising gas. To hold a constant electrolyte composition small pumps in the bowls mix the solution every time. The gas formation will be detected automatically by computer cameras every 15 minutes. These trials take place in air conditioned room at 21 °C with a constant lighting.

3 Results and Discussion

3.1 Mechanical Properties and Microstructure

Alloying of low calcium amounts, up to 4wt%, leads to an increase of the tensile strength up to approximately 210–240MPa compared to extruded pure magnesium (R_m <200MPa)[7]. The results show that for an increasing concentration of calcium the tensile strength also steadily increases. For low alloyed compositions the 0.2 elastic limit is about 80MPa lower than the tensile strength. This explains the relatively high plasticity (Figure 2). The difference of the tensile strength (R_m) and the elastic limit $R_{p0.2}$ decreases up to 40MPa for an increasing amount of calcium. From an amount of 2wt% of calcium no more significant increase of the yield strength can be observed. The workability also decreases for an increasing amount of calcium. At the direct extrusion of magnesium, super elevation of force is increasing significantly for an increasing amount of calcium. The processing of alloys containing an amount of calcium higher than 4wt% can only be processed by indirect extrusion because the container friction can be here disregarded as the limiting factor. For increased amounts of calcium an increased amount of eutectic with a melting temperature of 516.5 °C can be observed. This can lead to an increased amount of hot cracks if the temperature of deformation is the same as the melting temperature of the eutectic. Investigations on the ductility also allow statements about the plasticity. With increasing calcium content the elongation at rupture changes dramatically. At calcium concentrations below the solubility border (0.8wt% at room temperature) the elongation at rupture amounts 13 to 15%. The elongation decrease from 1.5wt% continuously. The lowest elongation of 5% shows the alloy with the highest calcium concentration (4.0wt%). The elongation at the tensile strength decreases continuously from 12% at 0.4wt% calcium to 4.5% at 4.0wt% calcium (Figure 2).

A calcium concentration above 1.5wt% embrittles the magnesium alloy and the plasticity get worse. Brittle intermetallic precipitations on the grain boundaries and into the grains are responsible for this deterioration of the elongation. The electrochemical properties are carried out by the measurement of the open circuit potential (OCP) and polarization curves. The OCP determination (time=5min) takes place directly before the polarization measurement starts. The OCP measurements show now singnifikant dependance to the Ca content, but with increasing NaCl content of the electrolyte it changes from –1,55V to –1,65V. We do not recognise this into the polarisation curves because there the zero potential changes more positive than the OCP (Figures 3 and 4).

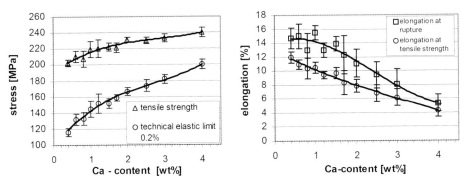

Figure 2: Tensile strength + 0.2 elastic limit and Elongation at rupture + elongation at the tensile strength of several MgCa alloys

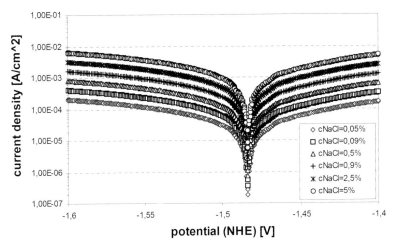

Figure 3: Current density vs. potential of a MgCa0,6 alloy at different electrolyte concentrations

Figures 3 and 4 show the polarisation curves of the MgCa0,6 and MgCa2,0 system at different salt concentrations. In Figure 5 the current densities of these alloys are displayed. The increasing of the NaCl concentration leads to a higher current density. The magnitude of the current density depends from the Ca content, because the higher the Ca content the higher the current density. This effect more articulates by increasing the salt concentration.

Final statements about the effect of calcium additions can be done only by corrosion tests at real conditions. The implemented eudiometric investigations yields that all investigated MgCa alloys corrode slower than the pure magnesium and depends strongly on the electrolyte concentration. The results of the long term trials are presented in Figure 6. The first diagramm shows the gas formation of the first 100 hours. The curves demonstrate the corrosion of a MgCa2,0 alloy and its dependance from the corrosion media competition. The increasing of the NaCl concentration leads to the increasing of the hydrogen formation to a concentration of 0,9%. With the further increasing of the NaCl content the gas formation stays nearly contant (0,9% and 2,5%) and decreases at the highest amount of NaCl (5%). This tendency show all investigated MgCa alloys. The gas formation during 700 hours also shows this tendency (Figure 6 below).

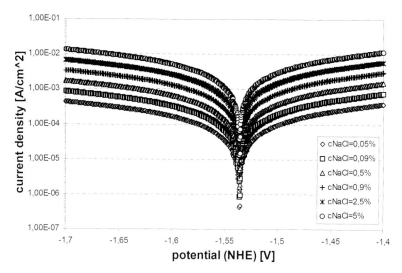

Figure 4: Current density vs. potential of a MgCa2,0 alloy at different electrolyte concentrations

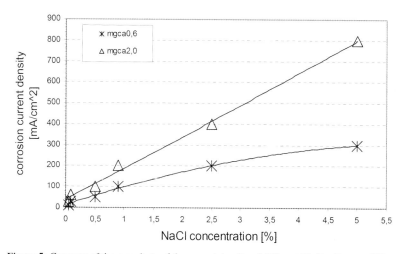

Figure 5: Compare of the magnitute of the current density of different MgCa alloys at different electrolyte concentrations

The decreasing of the gas formation can be caused on the covering of the surface by a precipitated chloride rich surface layer of corrosion products. This layer covers the free Mg surface and the attac of the electrolyte weaks. Thick and white corrosion products, precipitated on the specimen surface, was observed during the trials but not yet detailed investigated.

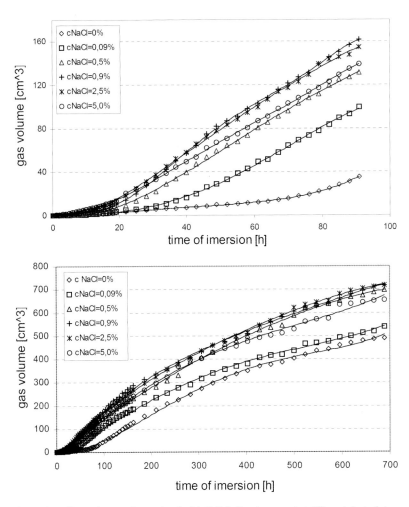

Figure 6: Eudiometric corrosion study of a MgCa2,0 alloy immersed at different electrolyt concentrations

4 Summary

The results of the investigations of the MgCa alloys show a good workability and adequate mechanical properties at a calcium concentration less than 1.5wt%. The increasing of the calcium content degenerates the plasticity but the extrusion quality is good. No hot crack occurs during the extrusion to an amount of 4wt% calcium and the rod surface is bright and plain. A negative shift of the OCP about –0,1V was found by increasing the electrolyte concentration for all investigated alloys. The polarization curves show the incraesing of the current with increasing the concentration. The results of the immersion tests show the same results at all investigated alloys, because the increasing of the NaCl concentration leads up to 0,9% to the increasing of the gas formation. At the higher concentration (2,5%) the hydrogen formation stays nearly constant and decrease with the increasing of the concentration up to 5%. This effect is not rearly clear, but the dense covering of the surface by chloride containing precipitations of corrosion products may be an idea to explain this effect. The detailed investigation of the corrosion products on the surface on specimen should clear up this question in the future investigations.

4   References

[1]   B. Heublein, R. Rohde, M. Niemeyer, V. Kaese, W. Hartung,, C. Röcken, G. Hausdorf and A. Haverich, „Degradation of Magnesium Alloys: A New Principle in Cardiovascular Implant Technology", Paper TCT-69, 11. Annual Symposium „Transcatheter Cardiovascular Therapeutics", The American Journal of Cardiology, Expcerpta Media Inc. New York, USA, 1999

[2]   H. Haferkamp, V. Kaese, M. Niemeyer, K. Phillip, T. Phan-Tan, B. Heublein and R. Rohde, „Exploration of Magnesium Alloys as New Material for Implantation", Mat.-wiss. u. Werkstofftech. 32, 116-120 (2001), Wiley-VCH Verlag GmbH, Weinheim, Germany

[3]   S. Golf, „Zur biologischen Vergleichbarkeit und biochemischen Wirksamkeit pharmazeutischer Magnesiumverbindungen", J. Miner. Stoffwechs. 4/1999, 11-21

[4]   T. Hassel, Fr.-W. Bach, C. Krause and P. Wilk, „Corrosion protection and repassivation after the deformation of magnesium alloys coated with a protective magnesium fluoride layer", Magnesium Technology 2005 Edited by TMS (The Minerals, Metals & Materials Society), 2005, 485-490

[5]   F.-W. Bach, A.N. Golovko and Th. Hassel, „Characteristic Features of Mg-Ca Alloy Tubes Extrusion by Vertical Hydraulic Press", Metalurical and Mining Industry, 2005, Nr.1, 48-51, ISBN: 0543-5749

[6]   F.-W. Bach, Th. Hassel and A.N. Golovko „The influence of the chemical composition and extrusion parameters on the mechanical properties of thin walled tubes made from magnesium-calcium alloys", Conference proceedings: suchasni problemy metalurgii, Naukovi visti, Tom 8, Plastychna deformacija metaliv, Dnipropetrovsk: "Systemni technilogii", 2005, 379-384

[7]   A. Beck, „Magnesium und seine Legierungen", Springer Verlag, Berlin, 1939

Study on Chloridic Corrosion of Two Magnesium Alloys WE43 and WE54

R.A. Khosroshahi, N. Parvini Ahmadi and A.Torabadi

Sahand University of Technology, Tabriz

1   Introduction

Because of their low density, high strength-to-weight ratio, good castability and damping capasity Mg alloys have become candidate materials for many applications in microelectronic, automobile and aerospace industries. The relative density of Mg is 1.74g/l and typical Mg alloys weigh ~35% lower than their Al counterparts at equal stiffness[1]. However they have relatively poor corrosion resistance, especially in chloride containing environments[2], since Mg is electrochemically the most active metal with the standard electrode potential of -2.375V[3]. In equeos solutions it shows a potential of -1.5V due to the formation of $Mg(OH)_2$ film on Mg [1] and dissolves rapidly by evolving hydrogen below pH 11, the equilibrium pH value for $Mg(OH)_2$[1]. In neutral solutions the presence of chloride ions lead to a local breakdown of the protective surface film and to pitting corrosion. The electrochemical reactions that occur during Mg corrosion in equeous solutions are[4]:

$$Mg \rightarrow Mg^{+2} + 2\overline{e} \quad \text{(anodic reaction)} \tag{1}$$

$$2H_2O + 2\overline{e} \rightarrow H_2 + 2O\overline{H} \quad \text{(cathodic reaction)} \tag{2}$$

$$Mg^{2+} + 2O\overline{H} \rightarrow Mg(OH)_2 \quad \text{(product formation)} \tag{3}$$

$$Mg + 2H_2O \rightarrow Mg(OH)_2 + H_2 \quad \text{(overal reaction)} \tag{4}$$

Many factors such as alloying elements or impurities influence the corrosion resistance of Mg alloys. Impurities of Fe, Cu or Ni lead to precipitations that are cathodic against the matrix and increase the corrosion rate. On the other hand Y and some rare earth elements such as Nd, and Ce can improve the corrosion resistance of the alloy due to their lower standard electrode potentials than Mg and formation of a protective film on the alloy[5]. The beneficial effect of Y and Nd on corrosion properties of Mg alloys were also reported by King [6]. The alloys WE43 and WE54, developed at MEL, have good corrosion properties because of their high purity and lack of corrosion promoting elements[7-8]. They exhibit excellent mechanical properties in the T6 condition[9]. The objective of this study was to determine the corrosion behavior of alloys WE43 and WE54 in chloride containing solutions to make a comparison between the alloys.

2   Experimental Procedure

2.1   Materials

Both WE43(Mg-4%Y-1.8%Nd-1.6%HRE-0.5%Zr) and WE54(Mg-4%Y-1.8%Nd-1.6%HRE-0.5%Zr) alloys, originally in the T6 conditions, were provided by MEL in the form of plates with approximate dimensions 200×200×25mm. Specimens with dimensions 10×10×10mm were prepared from the and their working sides (10×10mm) were treated with emery papers up to 2500 grits, cleaned ultrasonically in acetone and then dried in open air. A wire was connected to the opposite side of the samples and then covered with cold setting resin. The working sides of the specimens were polished to a mirror finish using diamond paste, fallowed by washing with distilled water and acetone prior to each experiment.

2.2 Corrosive Media

The effects of chloride ion concentrations were studied in solutions containing 0.01, 0.1, 1 and 3wt% NaCl at pH 9.5. The solutions were prepared using AR grade NaCl in distilled water. The pH of the solutions which had a volume of 500ml for potentiostatic and 250ml for potentiodynamic experiments, were adjusted with NaOH.

2.3 Electrochemical Corrosion Tests

Electrochemical polarization experiments were carried out using a standard three electrode configuration: saturated calomel was used as reference with a Pt as counter. Solutions were stirred by a magnetic stirrer during potentiostatic corrosion tests, while a potential of +150mV relative to the equilibrium potential had been applied to the alloy for 1 hour. Polarization resistance (Rp) was determined during potentiodynamic corrosion measurements using the Stern Geary method by scanning the potential at a rate of 1mV/s over the potential range of $E_{corr}\pm10mV$. The cathodic and anodic potentiodynamic scans were recorded from E_{corr}-200mV to E_{corr} and E_{corr} to E_{corr}+200mV, respectively. In potentiodynamic polarization measurements all the potentials were relative to saturated calomel reference electrode (SCE). When the samples were removed from the test solutions they were cleaned in a solution containing 200g/l CrO_3, 10g/l $AgNO_3$ and 20 g/l $Ba(NO_3)_2$ for 5 minutes[10]. After acid cleaning, they were rinsed with distilled water, cleaned ultrasonically in ethanol and dried in open air.

2.4 Surface Moorphology

The microstructure and surface morphology of the corroded samples were characterized using a Camscan 2300 scanning electron microscope equipped with an energy dispersive x-ray analysis.

3 Results

Potentiostatic test results for both alloys in different chloride containing solutions are shown in figure 1. A comparison of the figure 1(a) with 1(b) reveals that the anodic current of the alloy WE54-T6 is approximately 2 times more than that of WE43-T6 at high chloride concentrations. This is gradually decreased with decreasing chloride concentration.

Variations in corrosion currents as a function of imersion time in 0.01 and 3wt% NaCl solutions are also shown in figure 2.

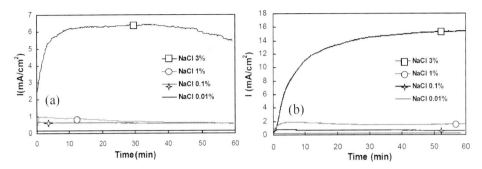

Figure 1: Variation in corrosion currents with time for a)WE43-T6 and b)WE54-T6 tested in various solutions

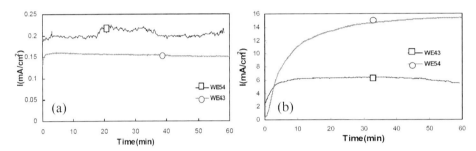

Figure 2: Variation in corrosion currents with time for the alloys tested in slutions containing a)0.01 and b)3(wt%) NaCl

The typical photomicrographs of the corroded surfaces in solutions containing 0.01 and 3wt% NaCl after removing the corrosion products are shown in figures 3 and 4, respsctively. According to these figures it seems that the alloy WE54-T6 was corroded more severely than WE43-T6 in both solutions. High magnification insert from the region shown boxed in 3(b) reveals pitting sites at the early stages of corrosion process.

The corrosion parameters derived from experimental data are reported in tables 1 and 2 for WE43-T6 and WE54-T6, respectively.

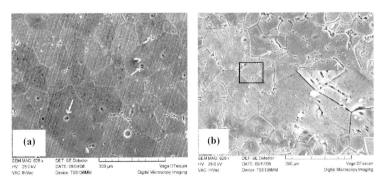

Figure 3: Photomicrographs of the corroded alloys a)WE43-T6 and b)WE54-T6 in solutions containing 0.01wt% NaCl after removing corrosion product

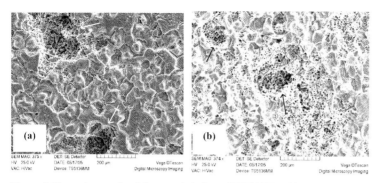

Figure 4: Photomicrographs of the corroded alloys a)WE43-T6 and b)WE54-T6 in solutions containing 3wt% NaCl after removing corrosion products

Table 1: Corrosion parameters derived from experimental data for the alloy WE43-T6

Chloride Containing Solutions (wt%) with pH=9.5	E_{OCP} (V)	R_P (Ωcm^2)	I_{corr} (μAcm^{-2})	b_a (V/dec)	b_c (V/dec)	r_{corr} (mpy)
3	-1.83	275.6	115.8	0.11	0.233	97.25
1	-1.792	946.2	40.79	0.136	0.256	39.07
0.1	-1.78	1116	24.5	0.126	0.126	20.84
0.01	-1.689	3018	7.35	0.102	0.102	6.25

Table 2: Corrosion parameters derived from experimental data for the alloy WE54-T6

Chloride Containing Solutions (wt%) with pH=9.5	E_{OCP} (V)	R_P (Ωcm^2)	I_{corr} (μAcm^{-2})	b_a (V/dec)	b_c (V/dec)	r_{corr} (mpy)
3	-1.728	385	164.2	0.107	0.247	139
1	-1.747	993	48.35	0.204	0.241	46.3
0.1	-1.699	852.3	43.84	0.161	0.184	42
0.01	-1.707	2712	11.51	0.137	0.151	11.02

According to tables, the corrosion rates of both alloys were increased with increasing NaCl. These were higher for WE54-T6 when exposed in various chloride containing solutions as compared to those of WE43-T6. The summarized corrosion data for the alloys tested in various NaCl containing solutions are shown in figure 6.

4 Discussion

The results of potentiostatic polarization measurements for the alloys WE43-T6 and WE54-T6 in different chloride containing solutions are shown in figure1. According to this figure anodic currents in solutions containing 3wt% NaCl were higher as compared with those of 0.01, 0.1 and 1wt% NaCl for both alloys. This implies higher susceptibility of the alloys to corrosion in 3wt% chloride containing solutions due to their higher elecerical conductivities [11]. Figure1 also shows that the corrosion currents of both alloys in solutions with 3wt% NaCl increase rapidly for the first 10 minutes. This can be related to an increase in exposed area due to local pitting. After 10 minutes the corrosion current of WE54-T6 increases contineously during immersion time, indicating cavity growth takes place in the alloy. While, it remains constant for another 30 minutes followed by gradual decrease for WE43-T6. In general, the alloy WE54-T6 shows higher anodic current densities in all solutions examined, indicating higher corrosion rates as compared to those of WE43-T6 alloy. Anodic currents of the alloys in 0.01and 3wt% NaCl solutions are separately shown in figures 2(a) and 2(b). Similar to high chloride containing solutions, the anodic current for the alloy WE54-T6 was also higher than that of WE43-T6 when exposed in 0.01% NaCl containing solution.. Typical post corroded surface features of both alloys exposed in 0.01 and 3wt% NaCl containing solutions are shown in figures 3 and 4, respectively. According to figure 3 both alloys have slightly been corroded and small corrosion pits formed on the surface of the alloys. The size and expension of of the micro corrosion pits in WE54 is higher as compared to those of WE43 indicating its higher susceptibility to corrosion. A high magnification insert from the region shown exposed in figure 3(b) shows pitting sites which resulted from micro galvanic corrosion at the early stages of exposure time. Figure 4 shows that the alloy WE54-T6 is corroded much more severly than WE43-T6 and more macro and micro pits which are not so deep are distributed on all over its surface than WE43-t6. The same corroded surface features has been reported for Zr cntaining Mg-RE alloy [1].

Microstructure of the alloys WE43 and WE54 in the T6 condition consisted of equiaxed grains with residual eutectic phase along some grain boundaries together with equilibrium β precipitates which were nucleated hetrogeneously on grain boundaries and dislocation lines during ageing treatment [10]. There was a larger volume fraction of residual second phase particles and β precipitates in WE54-T6 than WE43-T6 [12]. Addition of Y to Mg alloys improves their corrosion properties owing to the formation of a passive film on the alloy [5]. This may be prevented by precipitation of second phases due to increasing their sensitivity to galvanic corrosion [5]. The more susceptibility of WE54-T6 to corrosion can be related to the differences in either chemical composition or volume fraction of second phase particles present in the alloys which is higher in WE54 than WE43.

Chloride ion is one of the most surface activator which increases solubility of corrosion products and prevent their pile uping on the exposed surfaces. It is believed the microgalvanic cells will be formed on the exposed surfaces of the alloys due to the compositional differences exist between second phase particles and matrix of the alloys. This resulted in subsurface corrosion and high concentration of Mg^{+2} and Cl^{-1} ions on corroded sites. Similtaneously the concentration of H^+ ions increases due to hydrolizing of magnesium. The co.existance of H^+ and Cl^{-1} ions promotes the corrosion of the alloy according to the following reactions:

$$Mg \rightarrow Mg^{+2} + 2e^{-1} \tag{4}$$

$$Mg^{+2} + 2H_2O \rightarrow Mg(OH)_2 + 2H^+ \tag{5}$$

$$H^+ + Cl^- \rightarrow HCl \tag{6}$$

5 Conclusions

1. Corrosion rates of both alloys were increased with increasing NaCl and are always higher in WE54-T6 as compared with those of WE43-T6.
2. More susceptibility of WE54-T6 to corrosion than WE43-T6 is due to the difference in either chemical composition/volume fraction of second phase particles present in the alloy.
3. Pitting was found to be the dominant corrosion mode in both alloys due to mainly presence of chloride ions.

6 References

[1] G.L. Makar and J. Kruger, *Int. Mat. Rev.* 1993, 38(3), 138-153.
[2] H.H. Uhlig, R.W. Revie, Corrosion and Corrosion Control, third ed., Wiley, New York, 1985, 354-357.
[3] G.L. Makar and J. Kruger, J. of Electrochem. Soc., 1992, 8, 1403-1411.
[4] G.L. Song, A. Atrens, Adv. Engineer. Mat. **1(1)**, 1999, 1(1), 11-33.
[5] P.L. Miller and B.A. Shaw "Improving Corrosion Resistance of Magnesium by Nonequilibrium Alloying with Yttrium", Corrosion, 1993, 49, 947-950.
[6] King and Unssworth, Magnesium Technology in Institue of Metal, ASM, 1987, 27-28.
[7] R. Ambat, N.N. Aung, Evolution of Microstructural Effects on Corrosion Behavior of WE54 Magnesium Alloy, Corrosion Science, 2000, 42, 1433-1455.
[8] H. Karimzadeh, "Effect of Foundry Processing on the Corrosion Performance of High Purity Magnesium Sand Casting Alloys", Magnesium Alloys and their Applications, K.U. Kainer, Wiley-VCH, 2000, 451-456.
[9] M.Y. Ahmed, R. Pilkington, P. Lyon and G.W. Lorimer, Magnesium Alloys and their Applications, DGM, Informationsgesellschaft, Verlag, B.L. Mordike and F. Helmann (Eds.), 1992, 251-259
[10] ASTM Designation: G-190(reapproved 1994), Standard Practice for Preparing, Cleaning and Evaluating Corrosion Test Specimens, 12.
[11] P.M. Partovi, M.Sc. Thesis, Sahand University of Technology of Tabriz, 2003.
[12] R.A. Khosroshahi, Ph.D thesies, Manchester University, 1997.

Investigation of the Passivation Behavior of Magnesium Alloys by Means of Cyclic Current-Potential-Curves

T. Troßmann, K. Eppel, M. Gugau, C. Berger
Darmstadt University of Technology, Darmstadt, Germany

1 Introduction

Despite the fact that magnesium alloys have the lowest thermodynamic stability of all metallic construction materials the formation of passivating films can constrain the corrosion attack in a variety of solutions [1]. Even today comparatively few investigations have been performed concerning passivity and failure of passive films on magnesium alloys caused by specific anions.

A difficulty certainly consists in the fact that with the conventional approach of recording current-potential-curves only minor differences in the behavior of similar alloy groups (AZ, AM, AS, AE) can be observed. Usually this method is based on a continuously increasing polarization in the anodic range, where starting from the open circuit potential the states of active metal dissolution, passivation, passive areas (with low residual current densities) and transpassive dissolution or breakdown of the passivation are passed through. The difficulties in interpreting experiment (e.g. alloys may not be distinguishable) are owing to the strong rise of the current density in the anodic branch [2] and/or the immediate breakthrough of the fragile passive films at or near the open circuit potential (e.g. [3]) and therefore - despite passivating ambient conditions - passivity has a minor impact on the shape of the polarization curve.

Time-consuming potentiostatic polarization experiments, however, indicate that the region of passivity is sometimes extended in the anodic polarization direction [4]. This then identifiable phenomenon of passivity and local breakdown may be attributed to the ongoing formation of hydroxyl groups, film growth and alkalization of surface layers as well as the declining effect of harmful microstructural features, like anodic phases or impurities, during the long-term experiment.

It should be of some practical interest and use if the typical shape of a current-potential-curve could be more easily interpreted as an active metal dissolution or - interesting for alloy development - as a less passivating material condition or as a breakdown of passive films. Such conditions initiating the breakdown are of special interest in cases of superimposed mechanical loads, like static tensile stresses, initiating stress corrosion cracking (SCC), or fatigue loading, supporting corrosion fatigue (CF). In the case of CF the main criterion of a subdivision of the material behavior is derived by the damaging mechanism of the passive layer (CF in active/passive state under passivity preserving/supporting or breakdown stimulating conditions) [5]. For this reason an experiment executable on most conventional test hardware based on cyclic current-potential-curves is useful to acquire more information on the passivation behavior of magnesium alloys.

2 Experimental Procedures of Corrosion Tests

2.1 Materials

For the examinations die-cast and wrought high-purity magnesium alloys were used. The magnesium materials were delivered unmachined (with casting skin or rolling scale) in an as-fabricated condition. Apart from the examination of the specimens in the as-fabricated condition, the outer layer of the surface was removed by dry grinding with SiC paper. After grinding with a grain size of 1000 the specimens were cleaned with alcohol and tested in this condition. As a reference material for all comparative examinations in this paper the magnesium alloy AZ91 is used. More detailed investigations in [6] show that the represented method can be used for most cast and wrought magnesium alloys.

2.2 Solutions

All solutions used in the examinations, with the exception of tap water, were prepared with analytical-reagent-quality reagents in deionized (DI) water. The quantity of reagents required for a defined concentration was determined gravimetrically after drying (24 hrs in a drying oven at 80 °C). The concentration of the reagents is specified in ppm (parts per million) referred to the respective anion (e.g. chloride). Tests were conducted in a closed corrosion cell (600ml, area of tested specimen 1,4 cm^2) to control the atmosphere and the solution condition. The solution was saturated with gases (air, nitrogen, oxygen or carbon dioxide) for 1 hour before starting the test (meanwhile specimens were kept in the gas phase of the cell) and aerated with the respective gases during the whole test.

2.3 Test Procedure

For the recording of the changes in the material/electrolyte corrosion system a test method was designed based on current limited cyclic current-potential-curves. Following anodic polarization (scan rate 1000mV/h) beyond the range of the pitting potential, after passing a threshold current of +0,05 mA/cm^2 a reversal of the polarization direction in the direction of more base potentials is initiated to enable the repassivation of the material surface. The chosen current threshold prevents a massive damage of the surface on the one hand side but also ensures a breakthrough at the pitting potential.

After a another reversal of the polarization direction at a negative current density of -0,05 mA/cm^2 further complete polarization sweeps are conducted.

The observation of the tested surface during the experiment as well after the test normally reveals that the total amount of pitting sites does not increase significantly with rising number of polarization sweeps. That means pits initiated at the beginning tend to be deactivated during the cathodic polarization (until -0,05 mA/cm^2 are reached) and are reactivated after the reversal of the polarization direction. A cathodic threshold of only -0,01 mA/cm^2 is not sufficient to achieve this effect of repassivation.

3 Results of Cyclic Current-Potential-Curves

Examples of this modified testing method are displayed in figure 1 (left) for the alloy AZ91 with casting skin and in grinded state. With increasing number of completed sweeps the open circuit potential of the grinded specimen is shifted towards more noble values. Furthermore, in case of the grinded specimen it is possible to recognize a limited passivation of the magnesium alloy. After the breakdown potential is exceeded, the current density rises steeply at the cyclic pitting potential. After 2 sweeps a passive area up to -1100 mV$_H$ sets in. After further sweeps the pitting potential will even shift towards the range of -750 mV$_H$ with a decreasing of the current density in passive state, whereas the maximum pitting potential may not obligatory appear in the last sweeps. A cyclic pitting potential anodic respective to E_{OC} can even be recognized in solutions with much higher chloride contents, even in a neutral 5% NaCl solution used for the saltspray tests.

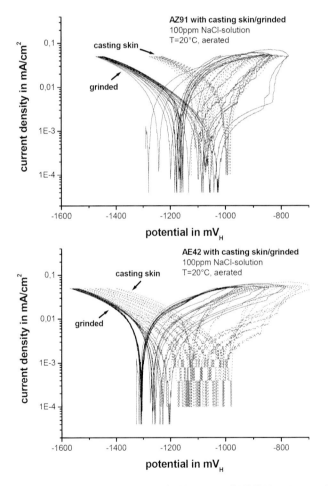

Figure 1: Cyclic polarization curves of AZ91 (top) and AE42 (down) as cast (dotted line) and grinded (solid line) in 100 ppm NaCl solution at 20 °C (left)

Figure 1 (top) also shows that, in contrast to the grinded specimen, no passivation is to be recognized on a surface with casting skin indicating rapid breakdown of passivity at or near the open circuit potential (E_{OC}). In this case merely a shift of the open circuit potential is taking place presumably induced by the alkalization of the boundary layer. In a $Mg(OH)_2$-saturated and therefore alkaline 100ppm NaCl solution passivity and breakthrough are visible in the first sweep in comparison to the "unbuffered" neutral solution after 2-3 sweeps. Under these beneficial conditions for passivation the cyclic pitting potential also shifts to more noble potentials. The same effect of an immediate occurrence of passivity takes place if the specimen is immersed in DI water for 1 hour before testing.

Further electrochemical investigations of the alloy AE42 with the proposed method showed, in contrary to the alloys AZ91, AS21, AZ31, ZK30 and AM50, that in case of a machined surface the span of passivity is decreased for the AE42 specimens, figure 1 (down). At the surface a uniform corrosion and no localized forms of corrosion, like pitting, appeared. Without the removal of the casting skin, the alloy AE42 (to be more precise: the specific samples used for this investigation) also showed a passive behavior with breakthrough and localized forms of corrosion developed.

Figure 2 opposes cyclic current-density-potential curves acquired for AZ91 in a 100 ppm NaCl solution and in Darmstadt tap water. The examination in tap water does not reveal any signs of a passivation and merely displays a slight shift of the E_{OC} in anodic direction. Both the results of the electrochemical examination as well as the occurrence of the corrosion attack suggest a dominating impact of the active metal dissolution as opposed to the chloride-containing solutions with a limited passivation and breakthrough. The latter curve illustrates an electrochemically induced failure of the passive layer owing to the chloride ions contained in the solution in contrary to the limited protection offered by carbonates formed in tap water (due to the dissolved carbon compounds, tap water reacts as a buffered solution) when chlorides or other breakdown stimulating reagents are present.

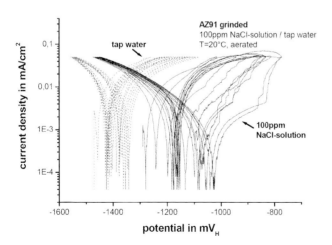

Figure 2: Cyclic polarization curves of grinded AZ91 in 100 ppm NaCl solution at 20 °C (solid line) and in tap water (dotted line) at 20°C

By means of cyclic current-potential-curves it is possible to demonstrate that merely by adding 10ppm calcium carbonate ($CaCO_3$, a typical constituent of tap water) to a 100ppm NaCl solution - similar to the behavior in tap water - a passivation of magnesium can be very much restricted or even suppressed [6]. Examinations of fatigue specimens tested in tap water revealed that a considerable active dissolution of the Mg solid solution takes place along the crack flanks and active slip planes (figure 3) in contrary to solutions without dissolved carbonate compounds [7]. For several alloys the attack is delayed at the grain boundaries, e.g. for AE42 at Rare Earth containing precipitations, but it is not stopped. If the surface films and later cracks are opened by mechanical loads (like CF) the cracks are widened by corrosion and the solution can progress to the crack tip and prevent the repassivation suggested by the results of the electrochemical tests.

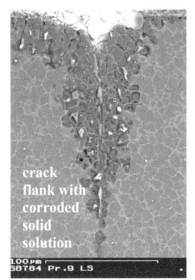 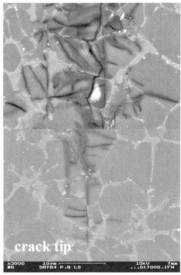

Figure 3: Films and cracks formed on a AE42 fatigue specimen tested in tap water (SEM; left: overview of a crack, V = 390:1; right: crack tip, V=4000:1)

Figure 4 illustrates the influence of a saturation of the solution with nitrogen on the shape of the polarization curve. Due to the constant gassing of the solution with nitrogen the amount of dissolved oxygen is minimized and the results therefore should be compared to the differing behavior in the same solution saturated with air (see figure 1, top). In contrast to the oxygen containing solution the passive area is limited in a neutral solution (pH7, figure 4, top). If the pH is increased to 11, the formation of passive films based on Mg-hydroxides is – as anticipated – supported and breakdown of passivity is clearly recognizable (figure 4, down).

Meanwhile passivation at a pH lower than 11 is influenced by further alloying elements, in this case aluminum, a similar behavior at a pH >11 can be recognized for all tested alloys. This means passivation of magnesium is solely based on films of dominantly $Mg(OH)_2$ and is not dominantly influenced by the oxygen content. Overall film formation including other oxides or hydroxides may absolutely depend on oxygen at a lower pH. If the contribution of the oxygen content on the corrosion behavior is determined this has to be kept in mind. Provided that (a fragile) passivity is an essential requirement for SCC and that passivation at a pH<11 is depending on oxygen and its reaction with alloying elements, these results may clarify the observations in [8], that SCC was not observed in N_2-saturated solutions.

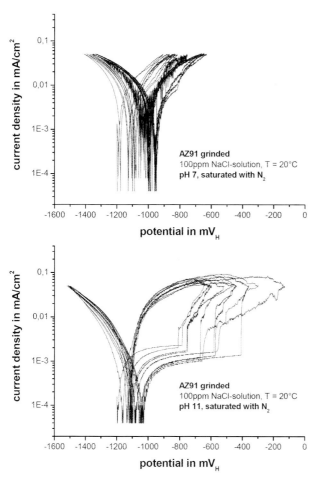

Figure 4: Cyclic polarization curves of AZ91 in a 100 ppm NaCl solution at 20 °C saturated with nitrogen at pH 7 (top) and pH 11 (down)

4 Concluding Remarks and Summery of Further Results

In order to enable a more detailed investigation of the changes in the corrosion system "material/electrolyte" - in particular the formation of passive layers as well as their local damaging and repassivation - cyclic current-potential-curves were recorded. During the sweeps changes in the passivating films, in the local electrolyte composition close to the surface boundary layer (in particular the formation of hydroxides and an increasing pH) as well as the change of the base material surface with development of local forms of corrosion, e.g. pitting, take place and are amplified by the chosen test conditions. The findings acquired with this modified test setup are in good agreement with results obtained using the (regarding the hardware) more expensive electrochemical impedance spectroscopy (EIS). Further investigations of several Mg-alloys with cyclic polarization curves in [6], [7] and [9] show

- a strong (for the investigated samples normally negative) influence of the casting skin or rolling scale on passivation which could be traced back by GDEOS (Glow Discharge Optical Emission Spectroscopy) to incorporated impurities,
- a decreasing current density in passive state and an increasing breakthrough potential with increasing aluminum content,
- a decisive influence of oxygen content on passivation of especially aluminum containing Mg-alloys at pH<11,
- that elevated temperatures above 40°C and solutions with dissolved carbon compounds constrain passivation,
- that carbonate films formed in solutions without other breakthrough stimulating substances, like chlorides, limit the than uniform corrosive attack, but only if no further mechanical loads are superimposed,
- that whenever passivation and breakthrough appeared the specimens showed localized forms of corrosion, like pitting or filiform corrosion, and that no repassivation under open circuit condition will occur,
- that localized electrochemical dissolution of magnesium leads to very often observed massive decline in corrosion fatigue strength and that passivity preserving or supporting conditions are beneficial to corrosion fatigue strength, but only in absence of breakthrough stimulating substances and
- that material or environmental conditions supporting a uniform attack (even if detrimental to overall corrosion or mass loss) can be beneficial in some cases of CF.

5 References

[1] G.L. Makar; J. Kruger in International Materials Reviews, 38, **1993**, 138–153
[2] G.L. Song, A. Atrens; D. Stjohn; J. Nairn; Y. Li in Corrosion Science, 39, **1997**, 855–875
[3] G.L. Song, A. Atrens in Advanced Engineering Materials, 12, **2003**, 837–858
[4] C. Berger; T. Troßmann; M. Gugau in Mat. Sc. a. Eng. Tech., 34, **2003**, 812–832
[5] H. Speckhardt; M. Gugau in Dubbel, Springer Verlag; 21st Ed; **2005**, Chapter E
[6] T. Troßmann, Dissertation, TU Darmstadt, **2005**
[7] S. Werdin, T. Troßmann, M. Gugau, K. L. Kotte in Mat. Sc. Eng. a. Tech., 36, **2005**, 659–668
[8] R.H. Jones et al in Stress-Corrosion Cracking, ASM International, **1992**, 251–164
[9] K. Eppel; J. Ellermeier; T. Troßmann; C. Berger; H. Masny; K. Woeste; U. Dilthey in Proccedings of the 7th Int. Conference on Magnesium Alloys and their Application, **2006**

Corrosion Resistance of an Electrodeposited Magnesium Layer from Grignard Reagents on Carbon Steel in Aqueous Chloride Media

S. Ben Hassen[1], L. Bousselmi[2], M, Razrazi[3], P. Berçot[3], E. Triki[1]
(1) Unité de Recherche Corrosion & protection des métalliques, ENIT, BP 37, Tunis-Belvédère. Tunisie
(2) Laboratoire Eau et Environnement, INRST, 43 av. Ch. Nicole, 1082 Cité Mahrajen, Tunis, Tunisie
(3) Laboratoire de Chimie Matériaux et interface, UFR Sciences et Techniques, Besançon. France

1 Introduction

The magnesium is one of most interesting metals for electroplating, compared with other metals used for coating, its first advantage is its lower density 1.74, another consideration is that magnesium is the third most abundant structural metal in the earth's crust. This metal cannot be plated from aqueous solutions because the hydrogen ions of the water are preferentially reduced over the active metal ions. The magnesium electrodeposition is possible from solutions based on Grignard reagent.

Reports on the electrolysis of Grignard solutions have focused on the fact that the conductivity of these solutions is moderately low and that in general, the deposit ranging in coulour from white to dark grey [1]. Studies made in „the National Institute of Standards and Technology" by Brenner et al [1] have indicated that a very good deposit of magnesium was obtained from a complex of alkyl halides and magnesium organoborate dissolved in ether solution, but in some cases there was codeposition of the bore with the magnesium. Other researchers [2–4] have worked equally with ether solutions containing organomagnesium halides, magnesium borohydride and Grignard reagents. Their best deposit was a magnesium alloy (90 % Mg, 10 % B) from a solution of magnesium bromide and lithium borohydride. Liebenow et al [5–7] have obtained a grey deposit of magnesium from three organomagnesium groups: organomagnsium halides, amidomagnesium halides and magnesium organoborates dissolved in terahydrofurane by imposing a current density of -1 mA/cm^2. The deposit regularity depends on the type of the substrate.

Basing on these results, solutions containing organomagnesium halides in ether solvents have proven most effective middles fore magnesium electrodeposition. However, their reactivity, their conductivity and the quality of the magnesium deposit vary considerably [2,8].

The aim of this work is to characterize the magnesium deposit on the carbon steel substrate and to investigate the corrosion behaviour of the coated substrate in sodium chloride medium 3%. The electrochemical impedance and the potentiodynamic polarization results show that the magnesium deposit provides a corrosion protection of the carbon steel substrate.

2 Experimental

The Grignard reagents methylmagnesium chloride, ethylmagnesium chloride and butylmagnesium chloride are obtained from Sigma-Aldrich as 2 M solutions in THF. All experiences undertaken with the organomagnesium electrolytes were manipulated under a controlled gas atmosphere of nitrogen in a glove compartment so as to avoid all reaction of the organomagnesium with the air. The surface of the carbon steel samples was polished with SiC paper, washed with anhydrous acetone and dried. The magnesium electrodeposition was realised by imposing a cathodic current of -1 mA cm^{-2} using a Mg reference electrode and platinum counter electrode [5–7]. For the corrosion test in aerated sodium chloride solution, the electrochemical cell was a saturated calomel electrode (ECS) as reference, platinum counter electrode and carbon steel as working electrode. The working area was 0,29 cm^2.

The polarization curves were realized by means of a potentiostat-galvanostat Radiometer Copenhagen PGZ 402 model, piloted by software Voltlab4.

The electrochemical impedance spectroscopy measurements were carried out using a Parstat 2273 model with a ZSimpWin version 3.20 software. The measuring frequency ranged from 10^2 Hz down to 10^{-2} Hz. The signal amplitude was 10 mV. All the experiments were performed at the corrosion potential. Scanning electron microscopy SEM was carried using a JEOL type 5600.

3 Results and Discussion

3.1 Coating Structure

The SEM observation of a steel sample on which a magnesium layer was applied (figure 1) shows the existence of an irregular thickness of the magnesium layer. The magnesium deposit has clearly irregular morphology. The EDS analyses present the existence of pronounced picks of oxygen that confirms the existence of MgO.

Figure 1: SEM micrographs of magnesium deposit on steel obtained from magnesium alkyl chloride inTHF (a) methylMgCl, (b) ethylMgCl and (c) butylMgCl

3.2 OCP and Polarisation Curves

Figure 2-(a) gives the open circuit potential curves in terms of time of magnesium carbon steel, and carbon steel with a magnesium deposit immersed in a 3 % sodium chloride solution. For the case of the magnesium, the potential reaches a value of -1634 mV/ECS and underwent fast corrosion with visible bubbling on the surface due to the reduction of protons H$^+$. As for the carbon steel sample, the potential accuses a fall to stabilize at the end of few minutes to an approximate value of -690 mV/ECS. The potentials in open circuit of the carbon steel covered by a magnesium layer obtained from the three Grignard solutions in THF, accuse a rising variation, corresponding to the decrease of the metal dissolution reaction, exceeding the carbon steel electrode potential. The potentials take stable values nobler than that of the steel in NaCl 3 %. These potentials took approximately 2 h to be achieved. The time for stabilization in this case is much longer, due to the need for the water to penetrate to the polymer and reach the magnesium particles making them active.

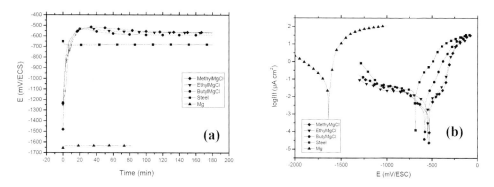

Figure 2: (a) Open circuit potential and (b) Polarization curves in NaCl 3 % of magnesium, uncoated and coated steel obtained from magnesium alkyl chloride in THF

The polarization curves relating to the three samples studied in the aggressive medium chloride sodium solution 3 % are recorded in figure 2-(b).The curve corresponding to the magnesium sample presents a symmetrical behaviour around the corrosion potential revealing that the metal is in its active state. The anodic branch corresponds to the oxidation of magnesium, whereas the cathodic branch reveals the reduction of water.

For the steel sample, the cathodic curve has a plateau corresponding to the diffusion limited reduction of dissolved oxygen followed by a logarithmic increase of the current probably due to the reduction of water. The anodic branch presents a plateau showing short passivation behaviour of the steel. The high slope that represents the end of the plot corresponds to the Fe oxidation. For the steel substrates coated with magnesium, the cathodic branches of the plots are practically parallel with a Tafel slope of 0,25 V/decade corresponding to the reduction of the oxygen. The anodic branch corresponds to the iron oxidation. The corrosion potentials are moved to more positive values compared to those of the steel.

Magnesium is classified at the bottom of the electrochemical series and in solution it leads to a high oxidation by protons. In NaCl 3 %, the pH is practically neutral (equal to 6.7) and in such pH Mg does not undergo passivation. Effectively, magnesium has a vast pH range over witch it remains active [9].The magnesium coating in this case does not behave as a physical barrier nor a sacrificial anode, if it was not the case, the potential values should be moved to intermediate values between those of magnesium and steel but the results showed that the potentials are nobler and the currents are lower and this behaviour can be explained as it follows:

In atmospheric standard conditions, the steel substrate forms a native layer of hydroxides and oxides. In contact with the organic solution, it leads to complicated reactions that take place with the species of the Grignard reagents in THF such as nucleophilic reactions between hydroxide, oxide and nucleophilic solvent molecules. Besides, the water traces which are present in all organic solutions (usually 10–100 ppm) are reduced to hydroxide, oxide and probably hydride within the surface by diffusing through the surface films to the active metal.

$$M(OH)2 + \begin{Bmatrix} ROR' \\ RCOOR' \\ RCOOR' \end{Bmatrix} \longrightarrow \begin{Bmatrix} (RO)_y M \\ (RCOO)_y M \\ (ROCO2)_y M \end{Bmatrix} + R'OH + \begin{Bmatrix} ROR' \end{Bmatrix} \qquad (1)$$

$$MO_x + RCOOR' \longrightarrow (R'O)_y M, (RCOO)_z M \qquad (2)$$

It has been shown in previous works [10,11] that during the magnesium electrodeposition from the Grignard reagents in ethereal solvents, organic species, that are generated in one hand from the Shlenk and the ionisation equilibrium and in an other hand from the magnesium and the dialkyl magnesium reductions. In this case species such as RMg^+, $RMgCl_{2-}$ and other complexes were adsorbed on the surface of the substrate and thus can be complexed within the magnesium deposit. This complicated layer of the magnesium deposit in addition of the adsorbed species and the reduced water molecules was then put in contact with the NaCl 3 % solution. However, other reduction reactions occurred in addition of the formation of the porous corrosion products. Indeed at the electrode interface the pH becomes alkaline and as a result it leads to the formation of the magnesium oxide and hydroxide layer providing a physical barrier protection of the steel substrate. This feature was verified by an indicator paper that was immediately put in contact with the surface. The pH value was about 11 [12].

3.4   Electrochemical Impedance Spectroscopy

Magnesium exhibited very low impedance with visible bubbling on the surface corresponding to a high corrosion rate. Nyquist diagram shows an inductive loop at lower frequencies, in agreement with other authors' observations [13–15]. This feature has been associated to the presence of an inductor in an electrical equivalent circuit, this inductor corresponds to adsorption and desorption phenomena occurring on the surface of the sample [12,13]. The electrical equivalent circuit that best fitted measured data and calculated values is shown in figure (3-a) [14].

Figure (3-b) shows Nyquist plot of steel sample, the feature of the curve was a slightly depressed semicircle with a diameter of about 1,6 k?. The high frequency part of the diagram presents an angle of 45°. The equivalent circuit contains the capacitance of the double layer C_{dl}, the charge transfer resistance and the diffusional component Z_D. When the magnesium deposit was applied on the steel substrate, the total impedance of the system increased significantly (figure 3-c) again in good agreement with the potentiodynamic observations. Each Nyquist plot was composed of a capacitive arc. Bode plots not shown in this paper, confirm that the electrode systems had two-time constants feature. Capacitive responses were fitted by constant phase element, Q, whose impedance is given by the following equation:

$$Z_Q = \frac{1}{Y_0(j2\pi f)^n},$$

in which Y_0 is the CPE constant, f is the frequency (Hz), $j = \sqrt{-1}$ and the exponent $n = \alpha/(\pi/2)$, α being the phase angle of the CPE (radians).

This behaviour can perhaps induce a diffusionnel component because of the existence of the 45° angle at the high frequency part of the diagrams. However, at this stage of investigation we can not confirm. The value of percent protection efficiency was also calculated with the following equation [16]:

$$\frac{R_p^{-1}(\text{uncoated}) - R_p^{-1}(\text{coated})}{R_p^{-1}(\text{uncoated})}.$$

The R_p and E % values (Table 1) show that the corrosion protection of the magnesium deposit obtained from the three organic solutions is classified in the following order methylMgCl > ethylMgCl > butylMgCl which corroborate with the electrochemical results.

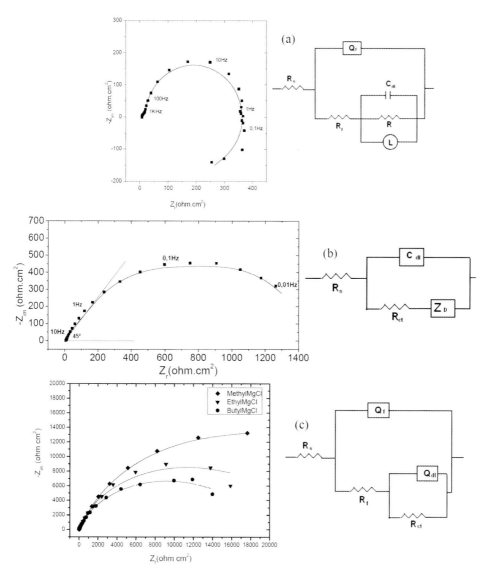

Figure 3: EIS spectra and equivalent circuits of magnesium (6-a), steel (6-b) and coated steel (6-c) after 3 h of immersion in 3% NaCl solution (Symbol: test, — fitting)

4 Conclusion

The magnesium deposit was obtained from three types of organic solutions methylMgCl ethylMgCl and butylMgCl in THF. The ocp, the polarization and the impedance curves have showed unexpected behaviour. This behaviour can be neither a sacrificial anode nor a physical barrier one. The formed layer has a protective role since we have increasing R_p and potential values and decreasing currents values. The best protection was offered by the deposit obtained from the methylMgCl solution. This protection is due to the formation of a complicated layer which contains different components. In a first stage, we have the native layer that reacts with the organic electrolyte. In a second stage the magnesium deposit with the permanently ad-layer of organic species which participate in the reduction reactions during the magnesium electrodeposition, in addition of MgO and $Mg(OH)_2$ and finally comes the corrosion products layer formed in contact with sodium chloride media may be reinforced by other reduced organic species.

Table 1: The R_p and E % values for uncoated and coated steel after immersion in NaCl 3 %

Electrode	$R_P \times 10^3$ (Ohm)	E (%)
Uncoated steel	1,62	–
Coated steel		
MethylMgCl	27,4	94,10
EthylMgCl	21,1	92,33
ButylMgCl	19,4	91,65

5 References

[1] Brenner, J.L. Sligh, Trans. Inst. Met. Finish **1971**, *49*, 71–78
[2] T. D. Gregory, R. J. Hoffman, R. C. Winterton, Journal of Electrochemical Society, **1990**, *137*, 775–780
[3] Mayer, Journal of Electrochemical Society, **1990**, *137*, 2806–2809
[4] J. H. Conner, W. E. Reid, G. B. Wood, Journal of Electrochemical Society, **1957**, *104*, 38–41
[5] C. Liebenow, Z. Yang, P.Lobitz, Electro. Com. **2000**, *2*, 641–645
[6] C. Liebenow, J. App. Electrochem, **1997**, *27*, 221–225
[7] C. Liebenow, Z. Yang, M.W. Wagner, P. Lobitz, Proc. Electrochem. Soc. **1997**, PV *97–18*, 597
[8] C. Chevrot, K. Kham, J. Perichon, Journal of Organometallic Chemistry. **1978**, *161*, 139–151
[9] M. Pourbaix, L'Atlas des Equilibres Chimiques à 25°C, Gauthier-Villard, Paris **1968**, 141
[10] Aurbach, R. Turgeman, O. Chusid, Y. Gofer, Journal of Electrochemistry Communications, **2001**, 3, 252–261
[11] Z. Lu, A. Schechter, M. Moshkovoch, D. Aurbach, J. Electrochem. Chem. **1999**, *466*, 203–217

[12] J. G. Kim, S. J. Koo, Corrosion 2000, 56 (4), 380–394
[13] T. Pajkossy, J. Electroanal. Chem. 1994, 364, 111–125
[14] G. Munoz, J. M. Flores, R. D. Romero, J. Genesca, Journal of Electrochimica Acta, **2006,** *51,* 1820–1830
[15] Baril, N. Pébère, Corrosion Science **2001**, *43,*471–484
[16] T. Tüken, B. Yazici, M. Erbil, Applied Surface Science **2006**, *252,* 2311–2318

Comparative Corrosion and Electrochemical Properties of Mg-Ni And Mg-Cu Alloy Systems After Cathodic Polarization

Shvets Vera A., Lavrenko Vladimir A., Talash Victor N., Khomko Tatiana V.

In this paper the comparative electrochemical and corrosion characteristics of Mg_2Ni, $MgNi_2$, Mg_2Cu and $MgCu_2$ intermetallics in 30 % KOH solution are presented. The mechanisms of cathodic and anodic processes have been determined using polarization curves, XRD analysis and Auger electron spectroscopy methods. It has been shown that in contrast to $MgNi_2$ and $MgCu_2$ the cathodic polarization curves of Mg_2Ni and Mg_2Cu samples are characterized by a hysteresis due to the formation on their surfaces the layer of hydride as a result of interaction with atomic hydrogen. In the case of Mg_2Ni, the sample outer layer consists of β-Mg_2NiH_4 and $Mg_2NiH_{3.85}$ phases; in the case of Mg_2Cu such layer incorporates the MgH_2 and $MgCu_2$ ones. Thus, it has been established that the Mg_2Ni intermetallic proved to be the least corrosion-resistant in 30 % KOH solution while the Mg_2Cu one has a lower rate of anodic oxidation after the cathodic polarization.

Intermetallics of Mg-Ni and Mg-Cu alloys systems are advanced materials for fabrication of hydrogen storage due to high hydrogen capacity, low density and low cost price [1-3]. Therefore, the study of corrosion and electrochemical properties of alloys above in alkali solutions under cathodic polarization and after it is the matter of particular interest. These investigations may also be useful for the study of possible application of these materials in other engineering fields. The latter is actual because of high enough corrosion resistance of hydrides, obtained on the surface of Mg-based intermetallics.

In this study four intermetallics were used: Mg_2Ni, $MgNi_2$, Mg_2Cu and $MgCu_2$. It should be noticed that Mg_2Ni and Mg_2Cu intermetallics are hydrogen absorbents while $MgNi_2$ and $MgCu_2$ are not. Such choice of samples allows to study the effect of cathodic polarization on corrosion of both types of intermetallics: forming and not forming the hydrides.

Electrochemical properties of the alloys above were studied using the samples obtained by sintering of fine-dispersion powders using potential-dynamic method at a potential sweep rate of 5 mV/s (P 5848 potential-static device). The electrolytic cell consisted of the auxiliary Pt electrode, the electrode for study and Ag/AgCl/KCl reference electrode. The surface state analysis before and after electrochemical treatment (both cathodic and anodic) was carried out using XPA and Auger electron spectroscopy (JAMP-10S). According to AES data, Mg_2Ni surface proved to be enriched with magnesium atoms while nickel content was 4-5 times lower the stoichiometry. The same study was carried out for $MgNi_2$ samples. Supposing that $MgNi_2$ doesn't absorb hydrogen because the MgO and NiO oxide mixture monolayer was present on the surface, then, according to AES data, 1.76 mol of Mg and 0.27 mol of Ni take 1.76 mol of Oxygen whereas 1.22 mol of Oxygen belongs to the chemisorbed layer on the sample surface. Suggested schemes of Mg_2Ni and $MgNi_2$ surface layer structure and oxygen chemisorption on them is represented in Fig 1.

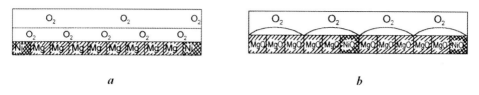

Figure 1: The chemisorption scheme of oxygen molecules on the surface of Mg_2Ni (*a*) and $MgNi_2$ (*b*)

It has been established that the thin film on $MgNi_2$ surface consists of MgO and NiO oxide mixture. The oxides cling to each other due to their isomorphous structures of NaCl type.

The direct (curves 1 and 2) and reciprocal (curves 1' and 2') cathodic polarization curves for Mg_2Ni and $MgNi_2$ are represented in Fig. 2.

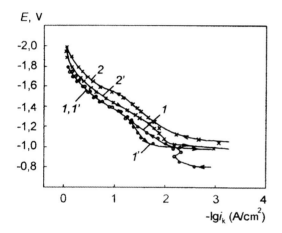

Figure 2: Cathodic polarization curves for $MgNi_2$ (1, 1') and Mg_2Ni (2, 2') in 30% KOH solution: 1', 2' - reciprocal behavior

The shape of these curves shows that the rate of cathode process being occurred on $MgNi_2$ surface, is by 2-5 times higher than on Mg_2Ni surface.

The cathode process taking place on $MgNi_2$ in the first stage in potential range from –0.8 V to –1.2 V is uneven because of adsorbed molecular oxygen (Fig. 1b). Below –1.2 V and down to –1.8 V this process is linear and corresponds to Tafel's equation along with hydrogen bubbles evolution. This stage is controlled by the charge transfer according to the equation (1)

$$H_2O + e = H_{ads.} + OH^-, \quad (1)$$

with the further surface recombination of adsorbed hydrogen atoms according to

$$H_{ads.} + H_{ads.} = H_2\uparrow. \quad (2)$$

In this case H_2 molecule formation comes about through the interaction between two "moving upon the surface" atoms according to Hinshelwood's mechanism [4]. Whereby the lower a binding energy between adsorbed atom or gas molecule and the surface the faster hydrogen formation occurs. At the same potential (Fig 2, curves 1 and 2) much higher reaction rate is observed for MgNi$_2$, than in the case of Mg$_2$Ni. It proves that adsorption energy of the gas on MgNi$_2$ surface is essentially lower than on the Mg$_2$Ni one. Long-term cathodic treatment during 2 h at the potential –1.8 V according to XPA results in no hydride formation on MgNi$_2$ surface. In the range of hydrogen formation both direct and reciprocal polarization curves totally coincide that proves, in turn, that no interaction between MgNi$_2$ surface and hydrogen during cathodic polarization occurs.

In contrast to MgNi$_2$, the cathodic process on Mg$_2$Ni is characterized by two regions on the direct polarization curve. The first one (Fig. 2, curve 2) in the potential range from –1.05 V to –1.5 V relates to retarded diffusion of oxygen and charge transmission according to (1). The second linear region of cathodic polarization curve from –1.5 V to –2.0 V responds to the process of very slow hydrogen bubbles evolution that is clearly observed visually. It's connected with the fact that essential part of hydrogen, forming on the electrode surface in this potential range according to reaction (1), reacts directly with Mg$_2$Ni surface free of oxides and chemisorbed oxygen (Fig. 1a):

$$Mg_2Ni + 4H_{ads.} = Mg_2NiH_4. \tag{3}$$

The β-Mg$_2$NiH$_4$ hydride formation is confirmed by XPA. High rate of this reaction is caused by low activation energy of hydrogen chemisorption on Mg$_2$Ni surface E_H = 19.6 kJ/mol [5] and high binding energy of hydrogen atom with surface of the intermetallic pointed out. XPA of Mg$_2$Ni subjected to cathodic polarization during 2h at the potential of –1.8 V in 30 % KOH showed the presence on the surface of β-Mg$_2$NiH$_4$ and Mg$_2$NiH$_{3.85}$.

The formation of hydrides mentioned above causes hysteresis of cathodic polarization curves – its reciprocal branch (Fig. 2, curve 2') absolutely diverges from the direct one.

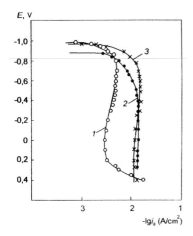

Figure 3: Anodic polarization curves for intermetallics MgNi$_2$ (1), Mg$_2$Ni (2) and Mg$_2$Ni after cathodic polarization (3)

Corrosion resistance of Mg$_2$Ni and MgNi$_2$ alloys before cathodic treatment and after it was studied using the method of anodic polarization curves. Anodic curves for initial MgNi$_2$ (1) and Mg$_2$Ni (2) samples are represented in Fig. 3 along with Mg$_2$Ni (3) sample after cathodic treatment. The curve 3 (Fig.3) shows that the Mg$_2$Ni intermetallic after cathodic polarization (Fig. 2, curve 1) has the largest range of active dissolution (from the corrosion potential of –1.01 V to –

0.75 V) in comparison with unhydrided Mg_2Ni sample (Fig. 3, curves 1, 2). Above –0.75 V the process of hydrided Mg_2Ni oxidation maintains its rate up to +0.43 V because of a surface passivation.

Anodic oxidation of Mg_2Ni after cathodic treatment may be represented with the following equations:

$$Mg_2NiH_4 + 4OH^- = MgO + Mg(OH)_2 + Ni^{2+} + H_2O + 4H^+ + 10e \qquad (4)$$

in the range of active dessolution, and

$$2MgO + 4OH^- = 2Mg(OH)_2 + O_2\uparrow + 4e \qquad (5)$$

in a passive range.

According to AES data, after oxidation the surface of unhydrided Mg_2Ni contains Ni-atoms by 17 times less than Mg. Elements molar ratio of oxygen, magnesium and nickel enables to conclude about formation of oxide film containing MgO and $Mg(OH)_2$.

As it may be seen from the curve 2 (Fig. 3), oxidation of initial Mg_2Ni takes place in two stages: the first one is the active dissolution in the potential range from the stationary potential to –0.50 V, the second one – passivation range from –0.50 V to +0.4 V. It should be noted that starting oxidation rate for initial Mg_2Ni sample is essentially lower than for one after cathodic polarization.

Thus, taking the AES data above into consideration, anodic oxidation of initial Mg_2Ni sample may be described by following equation:

$$Mg_2Ni + 4OH^- = 2MgO + Ni^{2+} + 2H_2O + 6e \qquad (6)$$

in the active dissolution range.

The lowest oxidation rate among all intermetallics of Mg-Ni system is characteristic for the unhydrided $MgNi_2$ one (Fig. 3, curve 1). On the base of AES data it may be concluded that the first stage of $MgNi_2$ oxidation (from the stationary potential to –0.9 V) should be describes by following reaction:

$$MgNi_2 + 2OH^- = MgO + 2Ni^{2+} + H_2O + 6e. \qquad (7)$$

Further up to +0.25 V the surface passivation takes place, being caused by formation of $Mg(OH)_2$ and NiO stable films:

$$MgNi_2 + 6OH^- = Mg(OH)_2 + 2NiO + 2H_2O + 6e. \qquad (8)$$

The third stage of anodic process for $MgNi_2$ at the potential above +0.25 V proved to be an oxygen evolution. The study of anodic behavior for the intermetallic pointed out after cathodic treatment showed that curves 2 and 3 (Fig. 3) virtually coincided. However, in the case of unhydrided Mg_2Ni, the rate of active dissolution is a little bit higher than for hydrided one.

The results of electrochemical investigation of anodic and cathodic processes for the intermetallics of Mg-Cu system are represented in Fig. 4. Herein has been shown that during cathodic polarization (curve 1) on the Mg_2Cu samples in the potential range from the stationary potential to –1.35 V the process is controlled by the stage of oxygen diffusion towards the surface according to reaction

$$2H_2O + 4e = 4OH^-. \qquad (9)$$

The following linear region down to –1.8 V corresponds to Tafel's equation while the process is controlled by the charge transfer according to (1). In spite of high cathodic currents at the potential –1.7 V only weak gas evolution is observed. This may be explained due to the interaction of the Mg_2Cu with adsorbed hydrogen ($H_{ads.}$). The hydrogenolysis reaction occurs resembling the analogical reaction in gas phase [6]:

$$2Mg_2Cu + 3H_2 = 3MgH_2 + MgCu_2. \quad (10)$$

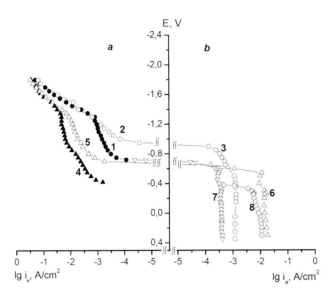

Figure 4: Cathodic (a) and anodic (b) polarization curves for intrmetallics Mg_2Cu (1, 2, 3, 7) and $MgCu_2$ (4, 5, 6, 8): 1,4 – forward behavior; 2,5 – reciprocal behavior; 3,6 – after cathodic polarization; 7,8 – initial samples

This was confirmed by XPA for Mg_2Cu samples polarized at the potential –1.8 V during 2 h in 30 % KOH. The $MgCu_2$ intermetallic layer was found on the Mg_2Cu surface that was a firm proof of the reaction (10) taking place.

The small hysteresis between direct and reciprocal (2) curves is observed due to the MgH_2 insulator film formation on the surface. However, this hysteresis is smaller than in the case of the Mg_2Ni due to the presence of $MgCu_2$ on the surface that is highly conductive. Meanwhile the thick film of the weakly conductive Mg_2NiH_4 is formed on the Mg_2Ni surface under cathodic polarization leading to obvious hysteresis (Fig. 2, curves 2,2').

The direct and reciprocal cathodic curves for the $MgCu_2$ (Fig. 4, curves 4 and 5) showed no hysteresis and, hence, no hydrides formation. In the range of active hydrogen evolution (from – 1.8 V to –1.4 V) direct and reciprocal curves coincided while, intensive hydrogen evolution was visually observed according to equation (1) and (2).

In this potential range the $MgCu_2$ surface acquired a light-silver colour, i.e. the surface reduction occurs. Above –1.4 V the electrode surface was dark, even black. The CuO black film formation in the cathodic potential range probably occurs according to the equation

$$MgCu_2 + H_2O + 2O_2 + 2e = MgO + 2CuO + 2OH^-. \quad (11)$$

Corrosion properties of the initial Mg_2Cu and $MgCu_2$ samples and ones after preliminary cathodic polarization are represented in Fig. 4 (curves 3, 6, 7 and 8). One can see on the curves 3 and 7 that the Mg_2Cu samples have by 15 times lower anodic dissolution rate than the preliminary hydrided $MgCu_2$ intermetallic (curve 6) and the initial one (curve 8).

It should be noticed that Mg_2Cu surface on both cases after anodic polarization had grainy etched structure. This allows to suppose that copper present on the sample surface goes into solution on the first stage of the active dissolution. Therefore, a surface became rich in magnesium. Further the magnesium oxidation occurs while magnesium oxide MgO is formed, the latter leading to the surface passivation. Taking into consideration the fact that anodic process rate of the initial Mg_2Cu sample (Fig. 4, curve 7) is twice lower than for preliminarily hydrided one (Fig. 4, curve 3), it may be concluded that preliminary cathodic treatment somehow prevents from a passive film formation on the Mg_2Cu surface.

In the case of the $MgCu_2$, the oxidation character of initial and hydrided samples differs only at the beginning of electrochemical process. Further, above +0.2 V potential, the oxidation rate for both samples – initial and hydrided – virtually coincide. The surface of both samples is covered with gray and yellow MgO film.

It has been established that Mg_2NiH_4, $Mg_2NiH_{3.8}$ and MgH_2 hydrides were formed as a result of cathodic polarization of Mg_2Ni and Mg_2Cu intermetallics in 30 % KOH solution. The hydrogenolysis process for the Mg_2Cu under these conditions was established. It has been shown that the preliminary cathodic polarization doesn't effect on corrosion rate of Mg_2Ni and $MgCu_2$ intermetallics, but enhances it a little bit for Mg_2Cu and $MgNi_2$ ones. It has been found out that the most corrosion resistant compound among intermetallic of Mg-Ni and Mg-Cu systems proved to be Mg_2Cu.

6 References

[1] Toshiro Kuji, Hiroaki Nakano, Tatsuhiko Aizawa. Hydrogen absorption and electrochemical properties of $Mg_{2-x}Ni$ (x = 0–0.5) alloys prepared by mechanical alloying // J. Alloys and Compounds – 2002. – Vol. 330–336. – P. 590–596.
[2] Kobotomori Toshiki, Ohnishi Keizo. The rise of hydrogen absorption capacity of alloys // Materia=Mater. Jap. [Nipon Rinozoky gakkai kaiho]. – 1995. – Vol. 34, No 3. – P. 276–278.
[3] Antonova M.M., Kiselev O.G., Khomko T.V. Hydriding of $Mg+Mg_2Ni+Ni$ mixtures // Powder Metallurgy. – 1993. – No. 4. – P. 58–64 (in Russian).
[4] Lavrenko V.A. Recombination of hydrogen atoms on solid surfaces. – Kiev: Naukova dumka, 1973. – 204 p. (in Russian).
[5] Han I.S., Lee J.Y. Study of hydriding kinetics of Mg2Ni under near isothermal conditions // Int. J. Hydrogen Energy. – 1987. – Vol. 12, No. 6. – P. 417–424.
[6] Semenenko K.N., Burnashova V.V., Verbetsky V.N. Interaction of hydrogen with intermetallic compounds. // Transactions of AS USSR. – 1983. – Vol. 270, No. 6. – P. 1404–1408 (in Russian).

The Role of Biological Environments on Magnesium Alloys as Biomaterials

A.Eliezer1 [a], F. Witte2 [b]

1Corrosion Research Center, Nano-Bio and Advanced Materials, Sami Shamoon College of Engineering, Bialik/Basel Sts, Beer Sheva 84100, Israel
2Laboratory for Biomechanics and Biomaterials, Department of Orthopaedic Surgery, Hannover Medical School, Anna-von-Borries-Str.1-7, 30625 Hannover, Germany
[a] amir@sce.ac.il, [b] F.witte@web.de

1   Introduction

Magnesium and magnesium alloys are of high interest as structural materials, due to their high specific strength they can also be applied as for implants. The main problem is that these materials are very reactive and may undergo environmentally induced degradation during service. Corrosion of magnesium alloys AZ91D and AM50 are investigated in vitro by Electrochemistry techniques into solutions imitating body's environment. Although actual corrosion resistance of a material can only be proven through long-term clinical trials in vivo, accelerated laboratory tests in vitro can be used to predict certain effects. Detailed investigations on the microstructure and in particular surface properties such as corrosion and surface biocompatibility in correspondence to structural biocompatibility will provide knowledge in order to propose directions for further development of advanced implant alloys and coatings. Previously [1–3] a deep research has focused on stress and unstressed magnesium alloys and in-situ corrosion properties of magnesium alloys. Microstructure analysis in regard to the observed corrosion behavior will provide conclusions for further trends for the design of advanced Mg-based structural alloys. Corrosion of magnesium alloys was introduced as a new degradation method for temporary implant materials in orthopaedic surgery [4]. The corrosion properties of metals depend on the surface reaction taking place in the respective environment. This in situ process protects the metal from further reaction. Since current standard laboratory tests for corrosion can not predict in-vivo corrosion [5], there is a demand for modified corrosion tests. This study considers parameters that might explain the disagreement of in-vitro and in-vivo corrosion [6] of magnesium alloys. In our research we have applied imitation of conditions of the internal environment of an organism with modeling the phenomena of a metabolism - homeostasis. Proteins are known to behave differently with various metals, since their role in a corrosive environment is governed by many factors such as surface chemistry of the metal, specific protein adsorption characteristics [7–10]. The research presented here uses the combination of multidisciplinary know how of electrochemical in-situ bio-reactor in order to study the environmental and proteins influence in order to achieve a correlation between in-vivo and vitro magnesium alloys implants.

2   Experimental Procedures

Die-casted magnesium alloys AZ91D and AM50 were investigated in vitro by Electrochemistry techniques in solutions imitating body's environments. The behaviour of magnesium alloys in

0.9% NaCl solution and Hanks balanced salt solution was studied using electrochemical methods by electrochemical impedance spectroscopy (EIS). The purpose of our research was to find the correlation between biological environments, corrosion exposure and magnesium alloys. During the experimental procedure we have used different physiological solutions, including fetal calf serum (FCS) at different temperatures as well as an external pH control. We have manufactured a unique system to form a bio-reactor in which key parameters of a solution were adjusted, and electrochemical cell, in which measurements were made.

It has allowed adjusting just as in an alive organism: temperature of a solution: 36.5–40 °C, pH: 7.35–7.45, volumetric speed of movement of a solution between a reactor (500 ml) and an electrochemical cell (500ml): 100 ml/min, speed of circulation a solution inside of an electrochemical cell: 300 ml / min. Measurements were taken using a potentiostat PARSTAT 2263 device (EG&G Princeton Applied Research) linked to a PC. The parameters were measured using three standard electrodes in an electrochemical cell with a calomel reference electrode and opposite a Pt electrode. In this study, the destructive method of cyclic polarization was used to form an accelerated pitting corrosion. The other two non-destructive techniques – Linear Polarization and Electrochemical impedance spectroscopy were applied before and after pitting was initiated by Cyclic Polarization. The general corrosion rate can then be calculated using Faraday's law.

3   Results and Discussion

After the adjustment the experimental unit and sample preparations the corrosion properties were researched and their results are shown in Table 1.

Table. 1: Results of Electrochemical measurements of magnesium alloys AM50 and AZ91 into solutions, $T = 37$ °C, pH = 7.4

Name	Co. Rate before CP [mpy]	Impedance before CP [Ohm]	Co. Rate after CP [mpy]	Impedance after CP [Ohm]	Solution
AM50	2.25	900	2.4	850	0.9% NaCl
AZ91	0.6	1800	0.9	1700	
AM50	1.1	1600	1	1100	0.9% NaCl+FCS
AZ91	0.4	2400	0.8	2200	
AM50	0.9	2600	2.13	1550	HANKS`
AZ91	0.5	3500	0.76	2300	
AM50	0.8	2800	1.1	2300	HANKS`+FCS
AZ91	0.55	3500	0.7	2700	

The following experiments shows, that only electrochemical measurements applied in this work may lead to achieve real demonstrations of in-situ in vivo-in-vitro relationship and correlation of corrosion processes. The corrosion rate increases on all samples and in all solutions after occurrence of irreversible pitting, arising owing to application of Cyclic Polarization.

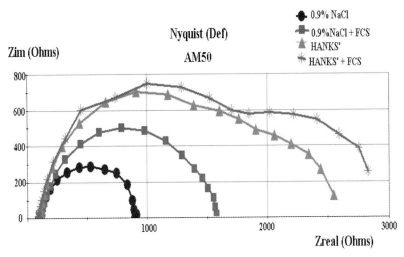

Figure 1: Impedance Spectroscope AM50 before Cyclic Polarization, T = 37 °C, pH = 7.4

Figure 1 describes the influence of the four selected environments on magnesium AM50 alloy before the exposure to pitting corrosion. It can be shown the impedance differs from one solution to another. The results in Figures 1-2 shows the contribution of the FCS on its phenomena on decreasing the corrosion rate as well as the biological environment and its interaction with the interface of the magnesium alloy. It can be seen that as the chlorine amounts increases the corrosion performance increases. An effect of the microstructure of different magnesium alloys on the interaction with FCS in different corrosive media was observed (Fig.2).

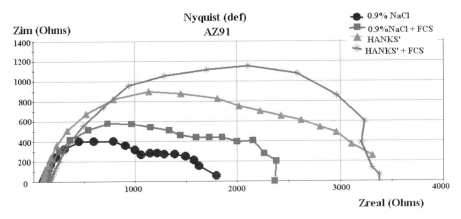

Figure 2: Impedance Spectroscope AZ91 before Cyclic Polarization, T = 37 °C, pH = 7.4

From these data we see of inhibition action of bioactive solution FCS and the best anticorrosive properties of alloy AZ91 even more evidently. It can also be seen that the Hanks acts as a stable buffer solution and is less reactive compared to saline 0.9% NaCl.

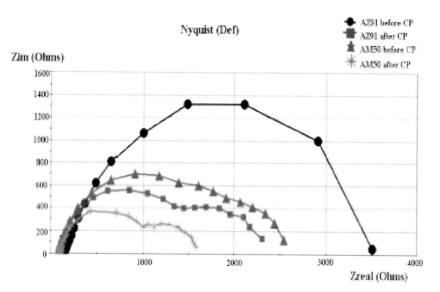

Figure 3: Impedance Spectroscope of AM50&AZ91 before and after Cyclic Polarization, HANKS' solution, $T = 37\,°C$, pH = 7.4

The protective properties of a passive film are characterized by the Impedance for both alloys better before pitting formation and acceleration, than after pitting (Fig. 1–Fig 3). The pitting irreversible damages the passive film. However, protective properties of a passive film both up to pitting, and after pitting in all solutions it is better on alloys AZ91, than on AM50. Corrosion aggression of Hank's-solution is less, than at solution 0.9 % NaCl, owing to presence in Hank's-solution of the some inhibition components. Additives of FCS solution in both of base solutions also reduce corrosion aggression of these solutions owing to inhibition absorption by surfaces of samples of biologically active components which are available within the solutions contents. The following SEM Figures 4–5 demonstrates the correlation between both magnesium alloys to the corrosion layer product formations. Due to the fact that AZ91 contains a vast amount of Mg17Al12 precipitations (β-phase) better corrosion performance is received. It can be also noticed in Figure 4 the formation of pitting corrosion appears more drastically for AM50 alloy. These results are for unstressed conditions and as mentioned before previous studies has testified that in stressed conditions AM50 performance increases compared to it's disabilities.

Figure 5 demonstrates a decrease of the corrosion product amount on the surface layer of AZ91 immersed in the presence of HANKS and FCS. This phenomena can be explained by the surface adsorption effect that forms a protective double layer on the surface layer. It can also be notice that a continuous Mg17Al12 phase assists the bonding and the interaction between the proteins and the surface. This leads also to the fact that there is a correlation between the corrosion analysis results and the SEM micrographs and the biodegradability properties of the alloys. Molecules of protein fiber from solution FCS render appreciable influence on corrosion properties to a surface magnesium alloys. Alloy AZ91 has the best corrosion properties under unstressed conditions due to a better passive layer and surface which contains a continuous β phase. Biodegradable materials should be designed based on the consideration of microstructure and environmental degradation processes.

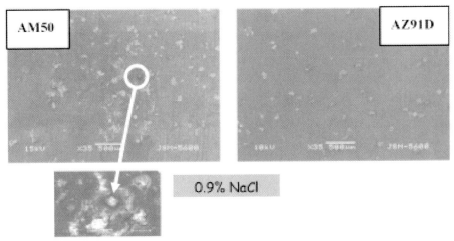

Figure 4: SEM micrographs of AM50 and AZ91 after electrochemical measurements in 0.9 % NaCl

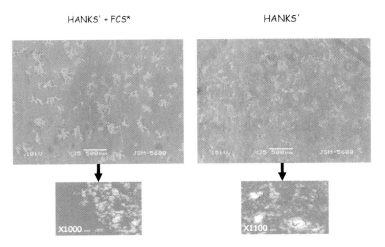

Figure 5: SEM micrographs of AZ91 after electrochemical measurements (*FCS percentage in solution stands on 10 % volume)

4 Conclusions

1. Corrosion properties of magnesium alloys in physiological solutions at presence of biologically active components are closely connected with absorption of protein on a surface of alloys.
2. Electrochemical methods of corrosion research „in vitro" in a combination to modeling a biological metabolism allow to predict rather precisely behavior of magnesium alloys in implants.

3. Electrochemical measurements methods of corrosion properties, including: Impedance, Linear and Cyclic Polarization allow us to investigate authentically corrosion properties of Mg alloys in various solutions simulating implants behavior in the human body environment.
4. The electrochemical measurements show that Mg alloys in 0.9 % NaCl are more exposed to pitting corrosion than in other solutions and AZ91 has a better corrosion resistance than AM50 which leads to the fact that the results confirm the bio-reactor system which was developed as an accurate device.
5. The Hanks balanced salt solution has succeeded to form a bio-stable film which allowed us to understand and to examine the effect of the present of the proteins as well as to determine a long term degradable environment compared to saline.
6. It was found that the presence of FCS in the tested solutions decreases the corrosion rate for both AZ91 and AM50 which can state that proteins has a major key role for producing future magnesium implants.
7. It may be assumed that this study will increase the understanding of corrosion modelling for other applications such as automotive and aerospace.

5 Acknowledgements

The authors would like to thank the BMBF for funding the project: ìDevelopment of an in-vitro biocorrosion simulator of in-vivo processes of biomedical implantsî 0313431 which partial results of this project are presented in this paper.

6 References

[1] A. Eliezer, E. M. Gutman, E. Abramov and E. Aghion: Corrosion Fatigue and Mechanochemical Behavior of Magnesium Alloys, *Corrosion Review,* 26, n.1-2 (1997) 1–26.
[2] P.L. Bonora, M. Andrei, A. Eliezer, E. M. Gutman: Mechanochemical Effect on Mg-Alloys by Impedance Measuements. *Journal of Materials Science Letters,* 20, no. 14 (2000) 1349–1351.
[3] R.M. Wang, A. Eliezer, E. M. Gutman: Microstructures and dislocations in the stressed AZ91D magnesium alloys. *Materials Science and Engineering A* 344(2002) 279–287
[4] Witte F, Kaese V, Haferkamp H, Switzer E, Meyer-Lindenberg A, Wirth CJ et al. In vivo corrosion of four magnesium alloys and the associated bone
[5] response. *Biomaterials* 2005; 26(17):3557–3563.
[6] Witte et al. Biomaterials 2006, 27:1013-8.
[7] Witte F, Goede F, Fischer J, Crostack HA, Nellesen J, Beckmann F et al.
[8] Degradable Magnesium Implants Enhance Early Cartilage Repair by Accelerated Subchondral Bone Regeneration. *ORS-Transactions* 2005; 30:1346.
[9] Sousa SR, Barbosa MA. Corrosion resistance of titanium CP in saline
[10] physiological solutions with calcium phosphate and proteins. *Clinical Materials* 1993; 14(4):287–294.
[11] Pourbaix M. Electrochemical corrosion of metallic biomaterials. *Biomaterials* 1984; 5(3):122–134.
[12] Williams DF. Review Tissue-biomaterial interactions. *J Mater Sci* 1987; 22:3421–3445.

Nanoparticle-based Inorganic Coatings for the Corrosion Protection of Magnesium Alloys

Wolfram Fürbeth, Florian Feil, Michael Schütze
DECHEMA e.V., Karl-Winnacker-Institut, Frankfurt am Main, Germany

1 Introduction

Magnesium alloys are increasingly used nowadays in a lot of applications. However, due to their high reactivity these alloys require a good protection against corrosion. Nowadays established coating methods, as e.g. chromating, may not be used anymore due to environmental regulations or are not sufficiently stable under mechanical or thermal influence (as e.g. organic coatings).

Besides a good corrosion protection dense inorganic coatings may offer also a high thermal and mechanical stability. However, the melting point of inorganic coating materials is often much higher than the thermal stability of most substrates. For example the application of inorganic coatings to magnesium alloys by heat treatment is limited to about 400 °C due to the low thermal stability of the magnesium alloys themselves and therefore is not possible by conventional methods.

The sintering activity usually increases with decreasing particle size [1]. Therefore nanoparticles are characterized by a very high sintering activity which allows to use them for the production of glass-like inorganic coatings by a thermal process at rather low temperatures even suitable for light metals [2]. Appropriate additives, forming a eutectic mixture with the silicon dioxide, like e.g. boron oxide or phosphorous oxide, may further lower the necessary sintering temperature and increase the chemical stability of the coatings.

In the work presented here environmentally compliant purely inorganic coatings have been applied to the widely used magnesium alloys AZ31 (wrought alloy) and AZ91 (cast alloy) by sol-gel technology. These coatings are based on commercially available SiO_2 nanopowders and dispersions with suitable additives as well as on mixed-oxidic nanoparticles made by the sol-gel process.

2 Results and Discussion

2.1 Coating Sols

By a base catalyzed sol-gel process (e.g. with NH_3) starting from alkoxides (e.g. tetraethoxysilane, triethylborate, triethylphosphate) a particulate sol of mostly monodisperse spherical oxide nanoparticles can be formed (figure 1). The particle size can be customized by the reaction conditions (concentration, pH, solvent, water content, ...) from some nanometres up to microns. The electrochemical double layer thereby avoids aggregation of the particles, however, if the solid content in the sol becomes too high agglomeration or early gel formation may occur. Particulate sols may be used directly for coating formation. Higher solid contents are possible by redispersion of centrifugally isolated nanoparticles together with suitable stabilizers (e.g. NH_3).

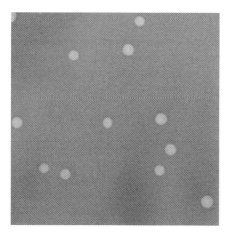

Figure 1: TEM picture of monodisperse non-agglomerated mixed-oxidic particles, ⌀ 200 nm (85 % SiO_2, 10 % B_2O_3, 5 % P_2O_5)

Alternatively the coating sol may also be produced based on commercially available silicon dioxide nanoparticles or water-based SiO_2 dispersions (e.g. *Ludox, Aerodisp, Levasil, ...*). Such particles are available nowadays with a size between some nanometres and microns and a narrow size distribution. However, the sintering activity of pure SiO_2 nanoparticles is not high enough to allow moderate coating treatment temperatures. Therefore additional sintering additives in the form of soluble salts (e.g. H_3BO_3, $Na_2B_4O_7$, NaH_2PO_4, KH_2PO_4, $Al(NO_3)_3$) have to be used.

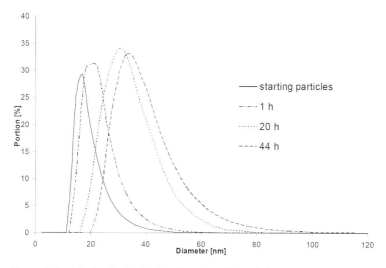

Figure 2: Particle size distribution in a dispersion of Levasil, borax, Na_2HPO_4 in H_2O as a function of time; composition: 9.6 % SiO_2, 0.75% B_2O_3, 0.2% P_2O_5, 0.6% Na_2O (dynamic light scattering, measurements in highly diluted dispersions)

In coating sols based on commercial SiO_2 dispersions particles react with the sintering additives and agglomerate (see. figure 2). Depending on composition such sols are forming a homogeneous inorganic gel in some minutes or days. Homogeneous gel formation is observed especially with sols containing B_2O_3, P_2O_5 or Al_2O_3. Formation of e.g. Si-O-B bonds could be proven by IR spectroscopy. Bulk material from gels with only 2 % Al_2O_3 could be sintered to clear glasses at about 700 °C, yet. By further adjusting the composition the sintering temperature can certainly be further decreased.

2.2   Dip- and Brush-coatings

The coating sol may be applied to the cleaned and activated substrate by dipping or brushing. For dipping the coating thickness is dependant on the viscosity of the sol as well as on the speed of withdrawing the sample from the sol. If pulled out slowly, coatings will be very thin. As magnesium alloys are not stable under acidic conditions sol stabilisation at low pH is not possible.

After drying of particulate sols homogeneous but highly porous gel layers are obtained. The porosity increases with the mean particle size [3–5]. In order to obtain a high density of the green layer the solid content of the sol should be as high as possible. However, with increasing solid content also the viscosity of the sol increases leading to a worse wetting behaviour as well as thicker coatings.

Since shrinkage of the coatings is limited to one dimension during drying and densification the maximum coating thickness is strongly limited. For pure inorganic SiO_2 films the critical thickness without cracking is mostly below 1 µm. Further complicating is the great difference of the thermal expansion coefficients between magnesium ($26 \cdot 10^{-6}$/K) and the silicate layer (quartz glasses: $55 \cdot 10^{-8}$/K, window glasses: $76 \cdot 10^{-7}$/K). This leads to stresses during sintering and therefore to cracking or delamination. Nanoparticles may, however, increase the critical coating thickness [6,7]. Suitable additives can be used to adjust the thermal expansion coefficient of the coating to the substrate and this way to avoid cracking. A further possibility is to decrease the stresses in the cross-linked coating by the use of network modifiers [8] to make the coatings more flexible. The whole coating process is schematically shown in figure 3.

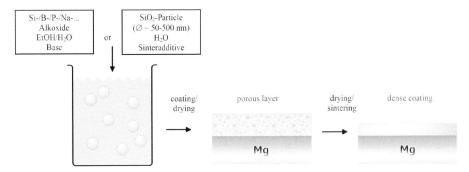

Figure 3: Schematic drawing of the coating process with particulate sols on magnesium alloys

A comparison of dip- and brush-coatings (figure 4) shows that in both ways thin transparent coatings may be applied whereas brush-coatings can become thinner (up to 300 nm). If the higher coating thickness is due to the long exposure of the substrate to the sol and reactions taking place in this time could not be investigated, yet. Dip-coatings of higher thickness (up to 3 µm) display some cracks due to the strong shrinkage during drying. Smaller particles with the same solid content of the sol lead to a lower coating thickness which, however, can more easily be kept crack-free.

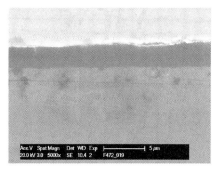 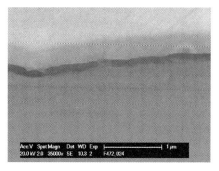

Figure 4: SEM pictures from cross-sections of coatings on AZ31 based on 20 nm SiO_2 particles + 6.7% B_2O_3 + 1.8% P_2O_5 + 5.4% Na_2O; sintered at 400 °C; left side: dip-coating, right side: brush-coating

2.3 Electrophoretic Deposition

An alternative route of coating application is the use of electrophoretic deposition (EPD). This method is based on the fact that in an electric field in liquids particles will move towards the oppositely charged electrode and will coagulate there. By this method coatings with high green density may be obtained also from dispersions with low solid content. However, only solid matter from the dispersion will be deposited. Therefore all sintering additives must be contained in the particles. The obtained coating thickness is dependent on concentration, time, voltage and applied current. A further advantage of EPD is that coatings may also be applied to surfaces with complex geometries.

Electrophoretic deposition has been performed with mixed-oxidic particles made by the sol-gel process. For this purpose particulate sols were directly used or isolated mixed-oxidic particles were redispersed. Because of the possible electrolysis of water EPD was done from alcoholic solutions, however about 10 % of water have been added to increase dispersion stability and conductivity. As an electrolyte and stabiliser ammonia has been used. At pH 9-10 SiO_2 particles are negatively charged. The magnesium sample to be coated was therefore the anode. At a solid content of only 2 % and a constant voltage of 5 V coating deposition was done in up to 30 minutes.

Especially at the beginning of the deposition process a strong decrease of the current can be observed. This shows that the anode surface is covered by isolating nanoparticles. The amount of particles deposited per time decreases with the current flowing and longer deposition times do not lead to a significantly higher coating thickness. In this way coatings of up to 6 µm could be obtained. However, despite the high green density obtained with this method thicker coatings also tend to crack formation so that after sintering crack-free coatings could only be obtained up to 1 µm so far.

The particle size is very decisive for the maximum coating thickness reached by EPD. Small particles (~20 nm) will completely cover the surface much faster so that at very low thicknesses

(<100 nm) the current strongly decreases and no further deposition takes place. Larger particles (>200 nm) will lead to thicker crack-free coatings of up to 1 µm composed of only some particle layers (figure 5 left side). Coatings from particles only containing SiO_2 and P_2O_5 keep their particulate structure even after 2 hours of sintering at 400 °C (figure 5 right side). However, in coatings from particles which additionally contain B_2O_3 after the same heat treatment particles have started to melt together (figure 6). This shows that densification of the coatings can strongly be influenced by suitable sintering additives.

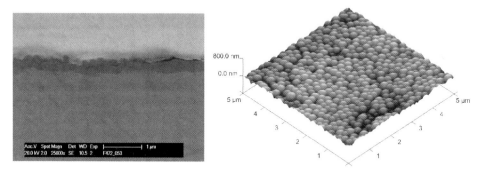

Figure 5: EPD layer on AZ31 from 220 nm particles composed of 93.4 % SiO_2 and 6.6 % P_2O_5; sintered at 400 °C; left side: SEM picture of cross-section, right side: AFM measurement

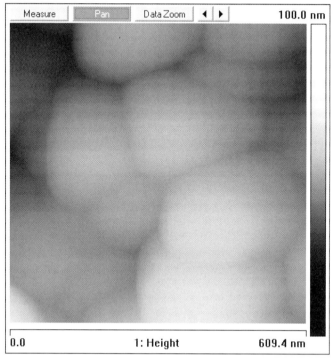

Figure 6: AFM measurement on an EPD layer on AZ31 from 160 nm particles composed of 86.7 % SiO_2, 7.5 % B_2O_3 and 5.8 % P_2O_5; sintered at 400 °C

3 Summary

By dip-coating, brushing and electrophoretic deposition thin transparent and crack-free coatings based on nanoparticles could be applied to the magnesium alloys AZ31 and AZ91. Sufficient densification of the coatings was possible at 400 °C, yet. At such temperatures the magnesium substrate should not be damaged by the heat treatment. However, the coating thickness is still quite low and should be increased to guarantee a good corrosion protection. Coating properties (mechanical stability, corrosion protection) are still under investigation.

Coatings based on commercially available nanoparticles have some advantages over coatings made from polymeric sols. The dispersions and additives to be used are much cheaper than alkoxides used for the classical sol-gel process. So the use of commercial products further increases the value of this new method especially for industrial applications.

4 Experimental Details

All chemicals and dispersions used have been obtained in the usual qualities (mainly p.a.) from the known suppliers (Sigma-Aldrich, Riedel, Fluka, Bayer, Degussa, ...) and have been used without further purification.

The substrates to be coated were first of all ground (SiC paper P4000) and degreased in caustic and acetone. For AZ31 a further acidic activation with Nital (2 % HNO_3 in ethyl alcohol, 5 sec.) resulted in the best wetting behaviour by water-based solutions. AZ91 could best be prepared by a treatment with 2 % HNO_3 in glycol/H_2O (75:25). After coating application by dipping, brushing or EPD, samples were dried at room temperature and afterwards sintered in synthetic air (heating rate: 5 K/min, holding time: 2 h).

5 Acknowledgements

This project has been funded by the German Ministry for Economics and Technology (BMWi) via the Arbeitsgemeinschaft industrieller Forschungsvereinigungen „Otto von Guericke" e.V. (AiF) with the number 14385 N.

6 References

[1] D.M. Liu, J. Mater. Sci. Lett. 1998, 17, 467–469
[2] H.Q. Nguyen, W. Fürbeth, M. Schütze, Materials and Corrosion 2002, 53, 772–778.
[3] C.J. Brinker, G. Scherer, Sol-Gel Science, Academic Press, New York, 1989.
[4] C.J. Brinker, G.C. Frye, A.J. Hurd, C.S. Ashley, Thin Solid Films 1991, 201, 97–108.
[5] M.M. Collinson, N. Moore, P.N. Deepa, M. Kanungo, Langmuir 2003, 19, 7669–7672.
[6] H. Schmidt, G. Jonschker, S. Goedicke, M. Menning, J. Sol-Gel Sci. Tech. 2000, 19, 39–51.
[7] J. Gallardo, P. Galliano, R. Moreno, A. Duran, J. Sol-Gel Sci. Tech. 2000, 19, 107–111.
[8] H. Scholze, Glas: Natur, Struktur und Eigenschaften, Springer-Verlag, Berlin, 1988.

Composite Coatings – The Newest Surface Treatment Technology For Magnesium

Dipl.-Ing. I. Ostrovsky
Palbam-Alonim AMTS, Chemetall GmbH; En-Harod, Frankfurt/Main

1 Introduction

Magnesium and magnesium alloys are specifically useful for the manufacturing of many light weight components and of many critical components for severe applications, for example for the manufacturing of secondary structural elements for aircrafts as well as of components for vehicles and electronic devices, because of their light weight and strength.

One of the significant disadvantages of magnesium and magnesium alloys is their sensitivity for corrosion. Exposure to hazardous chemical conditions causes magnesium rich surfaces to corrode quickly. Corrosion is unaesthetic and reduces strength.

A method that is often used to improve the corrosion resistance of metallic surfaces is painting. When the metallic surface is protected by a thick paint layer from the contact with corrosive agents, corrosion is prevented. However, many types of paint do not bind well to magnesium and magnesium alloy surfaces

Another problem of magnesium is low hardness of its alloys. Magnesium surface can be easy damaged during some of mechanical operations or like a result of fretting with harder material.

Additional problem of magnesium alloys is their formability. Any industrial magnesium forming process (deep drawing, forging, etc.) requires high temperature and application of special lubricants. In order to build proper lubricant layer on magnesium, it should be pretreated by a surface treatment that is stable in forming temperature and provides excellent adhesion with the lubricants.

All these problems might be solved by application of new surface treatment technology: composite coatings.

2 What is Composite Coating?

Composite coating for magnesium is newest co-development of Palbam-Alonim AMTS and Chemetall GmbH. The treatment may be classified as non-chromate conversion coating with precipitation of ceramic phase by sol-gel mechanism (Fig.1).

The result of the chemical conversion is formation of heavy soluble complex metal-fluorides layer (Fig. 2).

The sol-gel surface reaction leads to formation of silicon based ceramic phase that improve corrosion resistance and hardness of the coating.

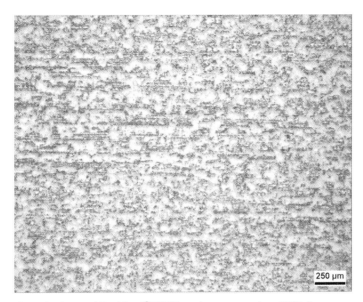

Figure 1: Picture of Gardobond® X4729 coating on magnesium AZ31 sheet.

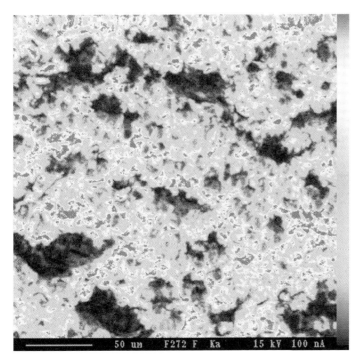

Figure 2: FKα - X-ray distribution image.

2.1 Fields of Applications

The technology of composite coatings allows obtaining of easy visible, multi-functional coating on magnesium for:

1. Powder coating, E-coat, paint and PTFE coating pretreatment;
2. Adhesive bonding pretreatment;
3. Pretreatment for magnesium forming and SPF technologies;
4. Stand-alone corrosion protection;
5. Flammability protection;
6. Electrical and thermo isolations.

The surface treatment technology intends to replace chromate conversion coatings for magnesium, some non-chromate treatments as well as anodizing process in some low cost magnesium applications.

3 Composite Coating as Paint Pretreatment

Methods based on chemical conversion of an outer metallic surface using chromate solutions are well known in the art as being useful for treating magnesium and magnesium alloy surfaces to increase corrosion resistance and paint adhesion, see for example [1] or [2]. Chromate containing coatings are mostly colored and excellent visible. However, the corrosion resistance of treated magnesium rich surfaces is typically very low – quite different from other metallic substrates coated with a chromate coating – and the environmental unfriendliness as well as the dangers for living beings of chromate solutions are definite disadvantages of these methods.

Several methods of metal surface treatment using non-chromate conversion coatings have been disclosed, for example in [3–6]. Silane solutions are environmentally friendly and lend excellent corrosion resistance to treated metal surfaces. Silane from the solution binds to a treated metallic surface forming a layer to which commonly used polymers such as paint or adhesive may be further applied, see [4].

Surface treatment disclosed in [7] teaches to provide a silane pretreatment on magnesium and magnesium alloys. Although the disclosed treatment offers excellent paint adhesion and corrosion protection, the coating is transparent and requires special on-line control methods.

Many of the present non-chromate treatment technologies are based on Group IV metals of the Periodical Table of Chemical Elements such as titanium, zirconium or hafnium, a source of fluoride ion and a mineral acid for a pH adjustment. For example, [8] discloses the use of zirconium, fluoride, nitric acid and boron to produce a uniform, colorless and clear conversion coating. Reference [9] refers to a coating comprising Group IV metals such as zirconium in combination with one or more non-fluoro anions while fluorides are specifically excluded from the processes and compositions above certain levels. The main lack of these conversion coatings is again the lack of a color and visibility, as the coatings are all clear and colorless or mostly colorless.

Recently developed well visible non-chromate conversion coatings include additionally to the metals of Group IV of the Periodic Table of Chemical Elements and to fluorides also any a special color providing component, such as an alizarine dye in [10] and such as permanganic acid and its water soluble salts in [11].

Permanganic acid is not preferred as its coloring effect is too strong and as its impurities are difficult to avoid and to remove. But the main lack of compositions containing permanganic acid or any of its salts is a low stability in contact with a magnesium rich surface so that it requires an addition of at least one sequestering agent and an extended use of chemicals.

The addition of organic dyes to process solutions usually leads to higher coating costs, to complicate compositions and to difficulties to control the process solution by optical methods like photometry

Additionally, one critical disadvantage of non-chromate conversion coatings based on Group IV metals of the Periodic Table of Chemical Elements is the very low adhesion of the formed conversion coating to fluoropolymer coatings. Anodizing coatings or phosphate coatings are usually used as pretreatment coatings for magnesium surfaces, often prior to a PTFE coating.

Anodizing coatings as well as most of the phosphate coatings are well visible on magnesium surfaces. However, as it is well known for one skilled in the art, thick crystalline phosphate conversion coatings often fail to form layers on magnesium surfaces showing sufficient corrosion resistance and paint adhesion. Providing an anodizing technology for magnesium surfaces requires complicate and expensive equipment.

Composite coatings were developed in order to fix mentioned above problems. The technology of application of these coating is simple and generally requires three technological stages completed by rinse:
1. Cleaning;
2. Pickling;
3. Composite coating.

Any special equipment is not necessary.

The coating layer obtained by the process is easy visible without addition of any dyes and/or coloring compounds (Fig. 3).

The visibility allows easy to control treatment quality on the coating line.

Figure 3: AZ31 (left) and ZK60 (right) specimens coated by Gardobond® X4729.

Composite coating shows excellent adhesion to the most industrial paint systems. The magnesium specimens (AZ31B, AZ80A-T5 and AZ91D) pretreated by Gardobond® X4729 and painted meet requirements of automotive industry. The results of the tests performed in accordance with DIN 50021 SS and evaluated in accordance with DIN 53209 and DIN 53210 are presented in Table 1.

Table 1: Results of corrosion test for painted specimens pretreated by composite coating

Paint system	Test duration (hours)	Result
Paint system for wheels	240	U1
Paint system for wheels	1008	R0
E-Coat	240	U<1

The astonishing benefit was obtained by application of composite coating as pretreatment prior PTFE coating. An industrial amorphous phosphate treatment for magnesium was used as a reference in the test. Die-cast specimens (3 specimens in batch) of AZ91D (3 specimens in batch) magnesium alloy were pretreated by the pretreatments and then coated by an industrial PTFE system. One of the specimen batches pretreated by composite coatings was also sealed by special OXSILAN® MG sealing. The specimens were tested in accordance with ASTM-117E and evaluated in accordance with ASTM-610D. The test was stopped when corrosion degree 9 (0.05% of surface area are corroded) was observed. Due to specimens pretreated by amorphous phosphate showed a difference in exposure time needed to reach the corrosion degree, the results for this batch were averaged. The results of the tests are presented on Fig. 4.

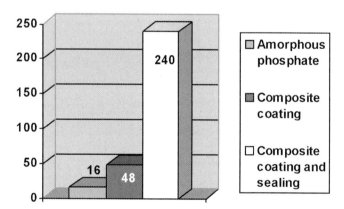

Figure 4: Results in of ASTM-117E test for pretreated specimens coated by PTFE (in hours of exposure).

The excellent adhesion of composite coatings with different paint systems and PTFE is explained by rough surface that is similar to surface morphology obtained by sand blasting or chemical blasting. However, composite coatings have clear advantages on the blasting processes:

1. Composite coatings offer to magnesium good corrosion protection prior painting. Excellent paint adhesion was observed even when the painting was done one month after the pretreatment;
2. Composite coatings do not leave impurities on magnesium surface;

3. Treatments solution does not include any component that is harmful for magnesium, such as, for example, iron and copper salts that are incorporated in some of chemical blasting processes.

3.1 Composite Coating as Pretreatment for Adhesive Bonding

Structural adhesive bonding is one of the main joining methods in aerospace industry. The rough surface of composite coatings is good substrate for industrial glues and primers that are used in adhesive bonding.

Application of composite coatings as pretreatment for aerospace adhesive bonding is now being investigated by AEROMAG project in 6th European Framework Program.

3.2 Composite Coating as Pretreatment for Magnesium Forming Processes

The coating film obtained by the presented surface treatment technology on magnesium has excellent stability in high temperature conditions. Furthermore, the rough surface of composite coatings offers excellent adhesion with lubricants used for magnesium forming.

Magnesium AZ31B sheet pretreated by Gardobond® X4729 Class L followed by application of special magnesium forming lubricants from Chemetall GmbH was successfully used by Palbam-Alonim AMTS for hot deep drawing process.

Application of composite coatings as pretreatment for other magnesium forming processes is now being investigated by MagForming project in 6th European Framework Program.

4 Application of Composite Coating for Bare Corrosion Protection of Magnesium

Composite coatings can be successfully used for bare corrosion protection of magnesium alloys in many industries. The obtained corrosion protection surpasses all chromate conversion coatings for magnesium and is comparable with many industrial anodizing processes.

In order to use composite coating as a stand-alone protection it should be post-treated in a sealing solution. The main sealing solution for composite coatings at the present time is OXSILAN® MG-0612 from Chemetall GmbH.

Coating thickness used for bare corrosion protection is usually higher than for paint pretreatment. It stays in range of 25–40 microns. Due to high rate of coating formation (3-5 microns per minute), this thickness can be obtained in acceptable by industry time.

The requirement to build higher coating thickness does not come from results of corrosion resistance obtained on thin layers (10–15 microns). The thick films of composite coatings offer to magnesium additional mechanical protection: 4–5 times improvement of the surface hardness.

The bare corrosion protection of Gardobond® X4729 Class S coating sealed by OXSILAN® MG-0612 was tested on magnesium AZ31B specimens. The results of the test are presented in Table 2.

Table 2: Results of bare corrosion resistance tests for specimens treated by Gardobond® X4729 Class S

Test Methods	Test duration (hours)	Corrosion degree (ASTM-610D)
ASTM-117E	72	9
ASTM-117E	168	7
DIN 50017 KTW	192	9–10

5   Application of Composite Coatings for Flammability Protection

Composite coatings consist of high temperature resisted compounds of fluorine and silicon. Furthermore, it is possible to build quickly very thin layers of the coating on magnesium: 40 microns, 50 microns and even more.

The both features can be used to protect magnesium against ignition. The ignition of magnesium is started when the metal has been transformed from sold to liquid state by heating.

Application of thick composite coating on magnesium allows to delay its ignition. Specimens produced from magnesium AZ31B sheet with thickness 0.7 mm and coated by Gardobond® X4729 Class S were tested for flammability resistance in accordance with FAR 25.853, Annex F, Part 25.I and met requirements of the standard.

Furthermore, the method of composite coating building allows incorporate in coating film special components that offer additional delaying of ignition.

Unfortunately, standard methods of flammability test do not allow to quantify the protection effect of composite coatings. It will be done in the near future by new methods have been developed in frame of AEROMAG project.

6   Conclusions

- Composite coatings is the newest type of surface treatment technologies that allow to build easy visible multi-functional coatings on magnesium;
- Composite coatings intend to replace chromate conversion coatings for magnesium, some non-chromate treatments as well as anodizing process in some low cost magnesium applications;
- Due to the surface treatment is very new development, its features still are being tested. The updated information will be presented on the Conference.

7   References

[1]   Ernest, J. W. US Pat. No. 2,035,380.
[2]   Bohman, W. L. US Pat. No. 3,457,124.
[3]   Van Ooij, W. J. US Pat. No. 5,292,549.
[4]   Van Ooij, W. J. US. Pat. No. 5,750,197.
[5]   Van Ooij, W. J. US Pat. No. 5,759,629.
[6]   Song, J. US Pat. No. 6,106,901.

[7] Ostrovsky, I. US Pat. No. 6,777,094.
[8] Das, N. US Pat. No. 3,964,936.
[9] Tomlinson, C. E. US Pat. No. 6,083,309.
[10] Carlson, L. R. US Pat. No. 6,464,800.
[11] Nakada, K. US Pat. No. 6,485,580.

Coating and Surface Treatment of Magnesium-Based Alloys and Composites

Volkmar Neubert[*,+], Ashraf Bakkar[*,**], and Ching-An Huang[++]
[*] Institut für Materialprüfung und Werkstofftechnik Dr. Neubert GmbH (DN), Clausthal, Germany
[+] Zentrum für Funktionswerkstoffe gGmbH, Clausthal, Germany
[**] Department of Metallurgy and Materials Engineering, Suez Canal University, Suez, Egypt
[++] Department of Mechanical Engineering, Chang Gung University, Taoyuan, Taiwan

1   Introduction

Magnesium is unable to form self-healing protective passive film and suffers high micro-galvanic corrosion when contains noble phases or impurities and macro-galvanic corrosion if being in contact with cathodic metals in wet atmosphere. Therefore, the challenge to be met in order to grantee save performance of magnesium-based materials is overcoming the high electrochemical activity of magnesium, and the save application of magnesium materials necessitates improving their corrosion resistance through suitable protection methods.

Corrosion and protection of magnesium alloys and their metal matrix composites (MMCs) are an important part of DN[*] research programs. The first step for corrosion control begins in the alloy-designing stage through alloying magnesium with compatible elements in corrosion of view point. We described, in some details, the effect of several alloying elements on the corrosion behaviour of the resultant magnesium alloys in the terms of microstructure features [1-4]. In this contribution, we present some of our recent work in coating and surface treatment technologies applied for magnesium materials, which include laser cladding, hydride coating by electrochemical ion reduction (EIR) or plasma immersion ion implantation (PIII), and electrochemical plating.

2   Laser Cladding

Laser cladding onto AS41-Mg alloy and its composite reinforced with C-short fibers was investigated using the two step method, first plasma spraying of the Al-12 Si powder alloy and then subsequently CO_2-laser remelting. The plasma sprayed coating layer has numerous pores. The density of porosity was measured by point count according to "ASTM 2002, E562-01" and found to be 18 ± 7.8 %. The as-sprayed thickness, with a very rough surface topography, varies between 343 ± 58 μm and 150 ± 29 μm, Fig. 1. The laser beam with area of about 22 × 2 mm was moved for only one track at various speeds along a length of 40 mm to yield a laser cladded specimen with area of 40 × 22 mm. The laser beam was shielded by argon gas to avoid burning of the molten pool. A more detailed experimental procedure was explained by the authors elsewhere. [5,6]

During the laser remelting, there was a tendency to void formation in both the clad layer and the substrate. High power input causes local boiling of Mg substrate (boiling point of Mg = 1110 °C) and, thus, leads to formation of pores which come up to the surface and are trapped in the clad layer, see Fig. 2. The formed pores are more numerous in the C/Mg MMC substrate than that in the Mg alloy substrate. The C-short fibers, having a thermal conductivity of about 1/30 of that of the Mg matrix alloy, have a marked influence on the temperature distribution in the substrate.

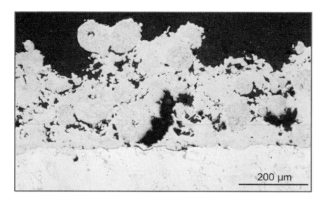

Figure 1: Cross-sectional micrograph of plasma sprayed coating.

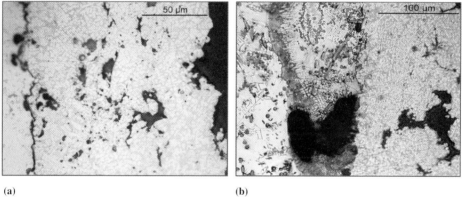

(a) (b)

Figure 2: Cross-sectional micrograph of laser clad samples at relatively high power input of power = 2.5 kW and speed =3 00 mm/min: **(a)** AS41-Mg alloy substrate, and **(b)** C/AS41-Mg MMC.

Low power inputs (low laser power and/or high scanning speed) lead to incomplete melting of the clad layer, which then retains much of the porosity in the as-sprayed layer. The most appropriate parameters – laser power and speed – were found to be in a quite narrow range.

Fig. 3 shows the SEM cross-sectional micrograph of C/Mg MMC specimen laser cladded at optimum laser power conditions by using a steady increasing displacement speed to avoid the heat accumulation. The clads have minimum porosity and the pores have no deleterious effect, because they, if found, are isolated and few in number. X-ray diffraction pattern undertaken at the substrate/clad interface indicates that the predominant interaction phases present in the interfacial layer are $Mg_{17}Al_{12}$ and Mg_2Si. [6]

Corrosion behavior investigations by polarization in 3 % NaCl solution show that the clad surfaces have corrosion rate of about two orders of magnitude lower than that of the Mg substrates, Fig. 4.

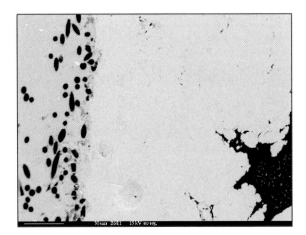

Figure 3: SEM Cross-sectional micrograph of laser clad coating onto C/AS41-Mg MMC at optimum power input (power=2.5 kW and increasing speed from 400 mm/min to 500 mm/min in seven equal steps).

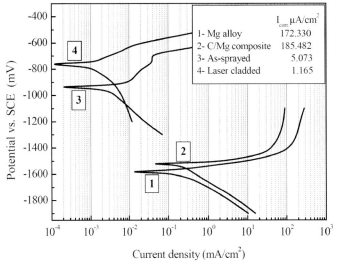

Figure 4: Potentiodynamic polarization diagrams of as-polished AS41-Mg alloy, and C/AS41-Mg composite, and of as-sprayed and laser cladded C/AS41-Mg MMC in 3 % NaCl solution.

3 Surface Modification with Hydrogen (Hydride Coating)

Two methods have been provided for hydrogen surface modification of hp Mg, AZ91 and RE-Mg alloys. One of these methods was Electrochemical Ion Reduction (EIR) of hydrogen from an alkaline electrolyte on such Mg-based cathode. The other was Plasma Immersion Ion Implantation (PIII or PI^3) into Mg-based substrate. [7]

Electrochemical ions reduction of H$^+$ (H-EIR), as a new technique for producing a magnesium hydride coating on Mg and its alloys by electrochemical means, involves the connection of the Mg specimen as a cathode in an electrolyte containing NaOH and Na_2SO_4. A pulsed cathodic current of 50 mA/cm² was applied for 7.2 ks using a Potentio-Galvanoscan controlled by a PC.

Plasma immersion ion implantation (PIII or PI³) produces a modified surface by embedding suitable atoms. The process involves immersion of the substrate in a plasma, and by applying a pulsed negative high voltage, ions from the plasma are extracted, accelerated and implanted into the substrate. The source of such hydrogen ions was H_2 gas, and the dose number of hydrogen was $27.73 \cdot 10^{17}$ ions/cm².

Secondary ion mass spectrometry (SIMS) analysis was used to monitor the secondary ion count rate of hydrogen in the modified surfaces. Fig.5 shows the SIMS depth profile of implanted and electrochemically absorbed hydrogen into AZ91 alloy. The presence of hydrogen in the bulk alloy is clearly approved and the high intensity of hydrogen ions within about 200 nm is distinctly recognized.

Fig. 6 shows the typical effects of H-modification treatments on the polarization curves of AZ91 Mg alloy in alkaline solution, pH = 12, where the Mg-based alloys are passive. The pitting potential was widely shifted to noble values by H-modification treatments. The polarization in neutral solution, Fig. 7, illustrates that the H-modification treatments reduced the corrosion rate of new RE-Mg alloy, and a pseudo-passive behavior associated with the H-modified specimen was observed in the anodic region.

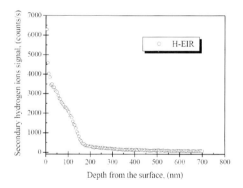

Figure 5: SIMS hydrogen depth profile through AZ91-Mg alloy: **(a)** H-modified by electrochemical ion reduction (H-EIR), **(b)** H- modified by plasma immersion ion implantation (H-PI³).

4 Electroplating

We have been actively investigating the electroplating of Mg alloys from new ionic liquid solutions used as electrolytes to dissolve metal salts at ambient temperatures. [8] Electrodeposition of Zn was successfully carried out at a constant current density of –5 mA/cm². Coatings of adherent, compact and sealed metallic zinc layers were obtained, see Fig. 8. The quality of deposition depends strongly on the chemical composition of Mg alloy substrate.

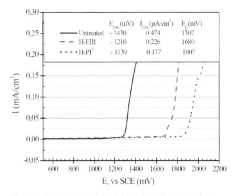 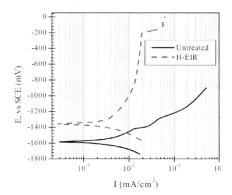

Figure 6: Segments of polarization diagrams for untreated, H-modified by electrochemical ion reduction (H-EIR), and H-modified by plasma immersion ion implantation (H-PI³) AZ91-Mg specimens in 100 PPM NaCl solution, pH = 12.

Figure 7: Potentiodynamic polarization diagrams of untreated and H-modified Mg-RE alloy (Mg Sc 1 % Y 4 % Mn 1 %) specimens, by electrochemical ion reduction (H-EIR), in 300 PPM NaCl solution, pH = 7.

Potentiodynamic polarization tests were conducted on all successful Zn plated specimens and Mg alloys substrates, and compared with Zn standard sheet (99.99 % Zn). The results showed that some deposits exhibited corrosion behavior similar to the pure Zn sheet, see Fig. 9. Nevertheless, Some deposits have very fine defects that permit the chloride solution to reach to the Mg substrate and lead to arising of galvanic corrosion between the Mg and Zn at the interface. Further work should be directed to correlate the corrosion resistance in immersion tests to the electrodeposition conditions. However, the Zn coated specimens are expected to be resistant to atmospheric corrosion.

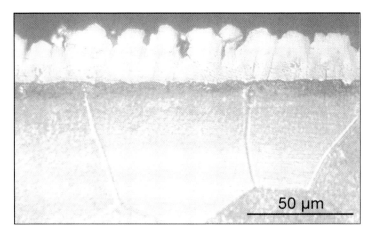

Figure 8: Cross-sectional micrograph of electrodeposited Zn on WE43-T6 Mg alloy.

Direct electrodeposition of Cu/Ni layers on Mg and its alloys, according to a new invented method, is being also studied actively. Fig. 10 shows a recent result of adherent electrodeposited Cu/Ni layers on pure Mg. Uniform and adherent coatings were obtained on hp Mg and AZ31, AZ61, AZ91, MEZ, MRI201, WE43, ZC63, and ZM21 Mg alloys.

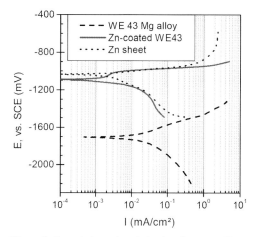

Figure 9: Potentiodynamic polarization diagrams of Mg alloy substrate, Zn coated Mg alloy, and Zn sheet in 0.1 M NaCl solution.

Electrochemical corrosion behavior of Cu/Ni coatings has been investigated in 1 % NaCl solution, and it has been found that the coatings have excellent corrosion resistance and reflect the Ni behavior in see water, Fig. 11.

Figure 10: Cross-sectional micrograph of electrodeposited Cu/Ni layers on AZ61 Mg alloy.

5 Conclusions

Our research work on the corrosion protection of Mg alloys and composites has elucidated the following points:
- A successful laser cladding of Al-Si eutectic alloy onto AS41-Mg alloy and its composite with C-short fibers could be produced by applying optimum power with accelerated speed of the scanning laser beam. The corrosion rate of the clad coating is about 100 times lower than that of the Mg substrates.
- Hydrogen surface modification of Mg alloys, both by electrochemical reduction and ion implantation methods, leads to higher corrosion resistance with no dimensional variations.
- Successful electrodeposition of Zn on RE-Mg alloys from new air and water-stable ionic liquids has been obtained. The Mg alloy composition affects strongly on the deposition quality.

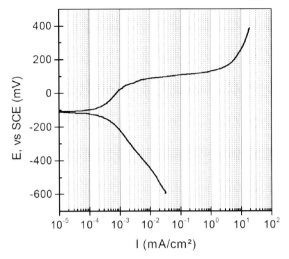

Figure 11: Potentiodynamic polarization diagram of Cu/Ni coated hp Mg in 1 % NaCl solution.

- Cu-Ni plating on Mg alloys has been shown to have extremely high corrosion resistance in sodium chloride solutions.

6   Acknowledgements

The support by the EU project IONMET within the sixth framework program, for electro-deposition from ionic liquids, is gratefully acknowledged.

7   References

[1] V. Neubert, A. Bakkar, Magnesium, Ed.: K.U. Kainer, Wiely-VCH, Weinheim, 2004, pp. 638–645.
[2] V. Neubert, I. Stulikova, B. Smola, A. Bakkar, B. L. Mordike, Metallic Materials 2004, 42, 31–41.
[3] V. Neubert, I. Stulikova, B. Smola, B. L. Mordike, M. Vlach, A. Bakkar, accepted in Materials Science and Engineering 2006.
[4] A. Bakkar, V. Neubert, accepted in Corrosion Science 2006.
[5] A. Bakkar, R. Galun, V. Neubert, Lasers in Engineering 2005, 15, 63–76.
[6] A. Bakkar, R. Galun, V. Neubert, Materials Science and Technology 2006, 22, 353–362.
[7] A. Bakkar, V. Neubert, Corrosion Science 2005, 47, 1211–1225.
[8] A.P. Abbott, G. Capper, D.L. Davies, R. Rasheed, V. Tambyrajah, U.K. Patent PCT/GB01/04300, 2000.

Anodizing of Magnesium Alloys

A. Berkani[a,1], P. Skeldon[a], G.E. Thompson[a], L.E. Fratila-Apachitei[b], I. Apachitei[b], J. Duszczyk[b]

[a]Corrosion and protection centre, The university of Manchester, Manchester, U.K.
[b]Department of materials science and engineering, Delft university of technology, Delft, The Netherlands

[1]present address: Department of materials science and engineering, Delft university of technology, Delft, The Netherlands

1 Introduction

Anodizing of magnesium alloys is the subject of more than 80 patents. Of these proprietary methods, few enjoyed commercial success and no method was considered satisfactory prior to the introduction of the HAE [1] and DOW 17 [2] processes in the 1950s. These processes use electrolytic oxidation with sparking in alkaline or slightly acidic electrolytes of complex composition. More recently, sparking as well as non-sparking anodizing methods have been developed. Among these, the spark anodizing processes MAGOXID [3], TAGNITE [4] and KERONITE [5] are commercially significant. Thus, spark anodizing appears to be the method of choice in the anodic treatment of magnesium alloys.

A preliminary investigation showed that simple ammonia-based one-component electrolytes containing aluminate, silicate, tungstate or fluoride anions can be used to form relatively uniform coatings on magnesium alloys by spark anodizing. These anions are the most frequently found in the formulation of the baths used for the surface treatment of magnesium alloys, including conversion coatings. Common to three out of four of the most recent anodizing baths, phosphate, in the form of tri-ammonium orthophosphate, has been chosen for the present work.

In previous studies by the authors [6,7], the behaviour of alloying elements during the anodizing of dilute magnesium alloys has been studied using metastable sputter-deposited alloys. A limitation of these studies, however, is that the relatively thin alloy layer limits the maximum forming voltage to about 300 V. Consequently, the results cannot be extrapolated easily to industrial practice. These limitations are overcome in the present investigation where an extended range of commercial magnesium alloys, together with a magnesium matrix composite, have been spark anodized to a maximum voltage of about 500 V, with high purity magnesium used as a reference material. This paper concentrates on the influence of the substrate composition on the sparking behaviour and the voltage-time responses at constant current density and on the composition of the resultant anodic coating.

2 Experimental Procedures

Alloys examined include: WE54, AZ91D, ZC71, WE43, and a ZC71-SiC-based metal matrix composite (MMC). High purity magnesium (99.99 %) was the reference material. Prior to anodizing, the specimens were mechanically polished. The electrolyte used was a 3 M ammonia solution with 0.05 M tri-ammonium orthophosphate. The anodizing cell was a 1-litre double jacketted glass cell where the temperature of the electyrolyte was maintained at 20 °C and agitation provided with a magnetic stirrer. A 304 stainless steel plate served as the counter-electrode. The voltage and the current were recorded using a data acquisition system. The range of current densities was 1–10 Adm^{-2}, with a maximum forming voltage of about 510 V. The coatings were characterized electronoptically and their composition determined by energy dispersive X-ray analysis (EDX) and electron probe microanalysis (EPMA).

3 Results and Discussion

3.1 Phenomenology

Common to all the alloys, the voltage initially increased rapidly. At a voltage of about 240 V, which is considered as a reasonable estimate of the sparking voltage, the first sparks could be seen on the anode surface. At about 320 V, the surface was covered with very fine sparks and small gas bubbles could be seen in its vicinity. The size of the sparks, their colour and their surface population density varied during the course of anodizing. As the voltage increased, their number decreased, their size increased and their colour changed progressively from blue to yellow. Simultaneously, the rate of gas evolution increased. At about 450 V, for all the alloys, acoustic emission was generated from local sites where persistent large sparks were observed. At this stage, gas evolution was intense with large bubbles developing on the anode prior to their detachment. Subsequently, the rate of voltage increase fell sharply and voltage fluctuations were observed. The final coatings were grey in colour and appeared uniform at all current densities for most of the substrates. The colour of the coatings formed on the ZC71 alloy changed to light brown for forming voltages of 500 V and above. In the case of the AZ91D alloy, whitish protuberances, resembling fused spots, appeared on the coatings formed at 1 Adm^{-2}, but were not evident on coatings obtained at current densities higher than 3 Adm^{-2}. The behaviour of the MMC was notably different, with the rate of voltage increase significantly lower than that of the other substrates. In this case, gas evolution occurred from the commencement of anodizing, well before the onset of sparking. Above about 420 V, the voltage was almost constant, and only a few sparking sites were present on the surface. The uniformity and esthetic appearance of the resultant coating were favored by a low current density. The anodizing behaviour of high purity magnesium was similar to that of the alloys. However, the sparking was less evident at low voltages. Up to about 480 V, the rate of gas evolution was lower and the acoustic emission at 450 V was less intense. A change occurred at about 480 V with the onset of more violent sparking together with an increase of gas evolution and the intensity of acoustic emission. Typical voltage-time responses obtained at 3 Adm^{-2} are shown in figure 1.

Generally, the curves showed three stages : an initial non-linear phase, a second linear phase and a final phase of very slow rate of voltage increase. The linear phase was not seen with the MMC. The voltage at which each transition occurred depended on the material, but it was roughly 250 to 300 V for the transition to the linear stage and 450 V for the transition to the ‚plateau' region (400 V for the MMC). As revealed in figure 1, the voltage-time curves for the WE alloys and pure magnesium are closely similar in the pre-sparking stage. The rate of voltage increase is slower for the AZ91D and is the slowest for the ZC71 alloy. The total charge density consumed in anodizing the MMC at the same current density was about 7.5 times higher than for the other substrates.

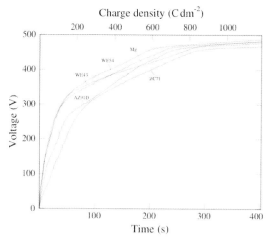

Figure 1: Voltage-time behaviour of high purity magnesium and commercial Mg alloys anodized at 3 Adm^{-2}

3.2 Relationships between Thickness, Voltage and Charge Density

The thickness data for high purity magnesium and the magnesium alloys are plotted against the forming voltage on a log-linear scale in figure 2.

Using the estimated value of the sparking voltage $U_0 = 240$ V and the corresponding thickness $d_0 = 16$ μm, the relationship between the average thickness d of the coating and the voltage U can be written:

$$d = d_0 \text{Exp}[k_1(U - U_0)] \tag{1}$$

where
$$k_1 = 9.82 \; 10^{-3} \; V^{-1}.$$

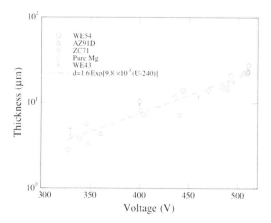

Figure 2: Thickness data against forming voltage for high purity magnesium and different Mg alloys anodized at 1 Adm^{-2} (3 Adm^{-2} for ZC71)

The relationship between the specific charge consumed q and the thickness d is of the form:

$$q = q_0 \left(\frac{d}{d_0}\right)^{k_2}, \tag{2}$$

where q_0 is the specific charge consumed at the onset of sparking and k_2 a constant which depends on the substrate. As an example, q is plotted against d in a log-log scale for the WE54 alloy in figure 3.

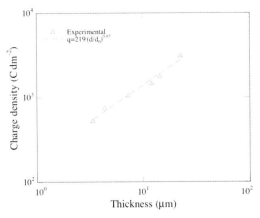

Figure 3: Specific charge against coating thickness for AZ91D anodized at 1 Adm^{-2}

A minimum current density was required to grow the films, which depended upon the substrate. Anodizing of the WE54 alloy at 0.2 Adm^{-2}, for example, lead to severe pitting without growth of the film. The ZC71 alloy and the MMC required a higher minimum current density than the other materials. Anodizing these substrates at 1 and 2 Adm^{-2} resulted in corrosion and pitting. The influence of the current density was assessed by plotting the voltage against the specific charge for different current densities. An example of such a plot is given for the WE54 alloy in figure 4.

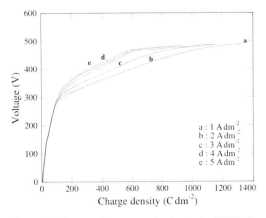

Figure 4: Influence of the current density for the WE54 alloy

It shows that in the first stage of the process, below about 280 V, the specific charge q consumed for a given forming voltage U was independent of the current density. In the sparking stage however, q decreased when the current density increased. For each material, the difference $\Delta q = q_{iMin}(U) - q_{iMax}(U)$ has been calculated for the minimum and maximum current densities used and the results plotted against the forming voltage in figure 5.

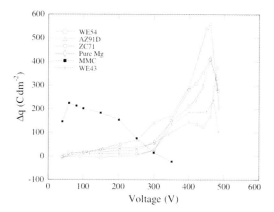

Figure 5: Influence of the current density on the specific charge q consumed for a forming voltage U. Δq is defined as $q_{iMin}(U) - q_{iMax}(U)$, where iMin and iMax are the minimum and maximum current densities used for the given substrate respectively

The influence of the current density in the sparking stage was similar for high purity magnesium and the magnesium alloys. Above about 300 V, Δq increased with the forming voltage, reaching a maximum at about 450 V after which it decreased but remained positive. In the case of the MMC, the beneficial effect of an increase in the current density diminished when the voltage increased in the pre-sparking stage. Δq values were negative above about 300 V.

3.3 Coating Composition

EDX analysis of samples anodized to different voltages showed that the coatings contained magnesium, oxygen and phosphorus, and, in the case of the alloys, minor amounts of the alloying elements. Coatings formed on the MMC contained also silicon. Phosphorus was incorporated in the coating since the commencement of anodizing and its concentration increased with anodizing time in the pre-sparking stage. The level of phosphorus in the film then remained constant throughout the sparking stage for all the substrates, with a phosphorus to magnesium atomic ratio, determined by EDX and EPMA quantitative analysis, achieving a value of about 0.7. The respective major alloying elements were present in the film with element to magnesium atomic ratios of 0.04 for Y in WE54 and WE43, 0.04 for Al in AZ91D and 0.02 for zinc in ZC71. Less silicon was detected in the films formed on the MMC in the „plateau" region of the voltage-time curve than towards the beginning of the sparking stage at 350 V. The zinc to magnesium atomic ratio in the film was the same in the ZC71-MMC and in the ZC71 alloy. EPMA depth profiling analysis of the final coatings are shown in figure 6 for WE54, AZ91D and ZC71 alloys. Phosphorus and magnesium were uniformly distributed across the thickness of the film for the three alloys, as was yttrium for the WE54 alloy. Aluminium and zinc, for the

AZ91D and ZC71 alloys respectively, were found in a few microns thick inner part of the film next to the metal/oxide interface.

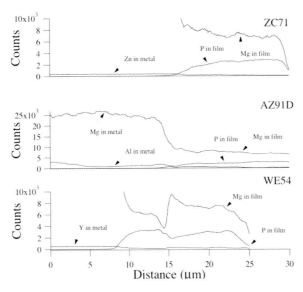

Figure 6: EPMA elemental profiles from cross sections of the final coatings on WE54, AZ91D and ZC71 alloys anodized at 1 Adm^{-2} (WE54 and AZ91D) and 3 Adm^{-2} (ZC71)

4 Conclusion

The voltage-time responses of commercial magnesium alloys anodized in electrolytes containing 3 M ammonia and 0.05 M tri-ammonium orthophosphate depend on the substrate. The curves show three distinct regions: an initial rapid increase of the voltage with time in the pre-sparking stage of the process, followed by a second approximately linear phase and a plateau-like region during the sparking stage. The sparking voltage was about 240 V, independent of the substrate. The current density does not appear to affect the pre-sparking stage of the process. In the sparking stage, however, increasing the current density decreases the charge density consumed for a given forming voltage in the case of high purity magnesium and the commercial magnesium alloys. This suggests that current efficiency for coating formation increases with the current density. This effect is maximum for a forming voltage of about 450 V for current densities of 1–10 Adm^{-2}. Anodizing of the ZC71-SiC-based matrix composite proceeds with a much lower current efficiency, compared to the other materials. The coatings, about 10–30 μm thick, contain phosphorus, incorporated from the phosphate electrolyte, and alloying elements incorporated from the substrate, and exhibit a relatively uniform composition across their thickness.

5 References

[1] Evangelides, H.A., US patent 2723952, **1955**.
[2] Anonymous, G.B. patent 762195, **1956**.

[3] Schmeling, E.L.; Roschenbleck, B.; Weidemann M.H., EP patent 333049, **1989**.
[4] Bartak, D.E.; Lemieux, B.E.; Woolsey, E.R., US patent 5240589, **1993**.
[5] Shatrov A.S., WO patent 9931303, **1999**.
[6] Bonilla, F.A.; Berkani, A.; Liu, Y.; Skeldon, P.; Thompson, G.E.; Habazaki, H.; Shimizu, K.; John, C.; Stevens, K., Journal of the Electrochemical Society, 149(1), **2002**, p. B4–b13.
[7] Abulsain, M.; Berkani, A.; Bonilla F.A.; Liu, Y.; Arenas, M.A.; Skeldon, P.; Thompson, G.E.; Bailey, P.; Noakes, T.C.Q.; Shimizu, K.; Habazaki, H., Electrochimica acta, 49(6), **2004**, p. 899–904.

Corrosion Resistance of Plasma-Anodized AZ91D Alloy: Effect of Additive

C.-E. Barchiche, C. Juers, E. Rocca, J. Hazan, J. Steinmetz

Université Henri Poincaré - NANCY I, Laboratoire de Chimie du Solide Minéral UMR CNRS 7555, BP239 – 54506 Vandoeuvre Les Nancy - FRANCE

1 Introduction

Due to its low density and especially its high strength to weight ratio, magnesium alloys are attractive for the automotive and aerospace industries, which requires the use of light materials. Indeed, this property is one of the main way to make airplanes and cars more fuel-efficient and to fulfil the economic and environmental requirements of transport. Unfortunately, magnesium alloys exhibit generally a poor corrosion resistance due to their high chemical reactivity and have to be used with coatings. Before painting, chemical conversion coatings as phosphatation [1–2], chromatation [3–4] can be applied with success and are less expensive than other kinds of coating, but produce generally thin protective layer. Anodization usually provides thick and protective inorganic layer and can be applied efficiently on Mg alloys as a precoating for paint. On magnesium alloys, various anodized coatings were developed such as HAE, DOW17, ANOMAG [5]. In these processes, an oxide layer is formed on the alloy surface by oxidation at high voltage in an alkaline bath based on potassium hydroxide and phosphate fluoride compounds. The protective layer formed during sparking caused by the electric breakdown can be modified by silicate [6–8] or aluminate [9–11] addition in the electrolyte, which leads to the formation of the ceramic $MgSiO_4$ or $MgAl_2O_4$ [12–13]. This kind of anodization is also called micro-arc anodization (MAO) or plama electrolytic anodization (PEO).

The present study, realized on AZ91D alloy, is devoted to the effect of stannate ions addition on the properties of anodic layer formed by PEO for short anodization time.

2 Materials and Experimental Methods

The samples of AZ91D magnesium alloy tested were plates of 1.2 cm thickness, prepared by high pressure die casting. Anodization experiments were carried out with a Generator Micronics Systems (Type MX 300V-10A / Model Microlab) in two-electrode cell. Two titanium plates were used as cathodes. The anodization process was performed in 3 M KOH + 0.5 M KF + 0.25 M $Na_3PO_4.12\ H_2O$ solution on AZ91D alloy (32 cm^2) with and without additions of 0.1, 0.5 M stannate. Before anodization, the surface was polished with SiC paper (grit 1200) and rinsed with ethanol. Then, the AZ91D magnesium alloy was etched in 650 g/l phosphoric acid H_3PO_4 for 30 s, rinsed in deionised water for 30 s, followed by a second immersion for 30 s in 100 g/l sodium hydroxide NaOH and finally rinsed in deionised water for 30 s. The morphology of the films was studied by scanning electron microscope (SEM HITACHI S-2500), coupled to energy dispersive X-ray spectroscopy (EDX). The electrochemical behaviour of anodized samples was studied in an aerated reference corrosive solution (D1384-87 ASTM water standard: Na_2SO_4,

148 mg/l ; NaHCO$_3$, 138 mg/l ; NaCl, 165 mg/l [14], noted ASTM water) using a three electrode electrochemical cell. The circular working electrode surface (3 cm^2) is horizontal, facing the Pt-disk used as the counter electrode and the reference electrode was a KCl-saturated calomel electrode (Hg/Hg$_2$Cl$_2$, $E = +0.242$ V/SHE).

3 Results and Discussion

3.1 Process Analysis

Figure 1 shows the evolution of the voltage during anodization in 3 M KOH + 0.5 M KF + 0.25 M Na$_3$PO$_4$.12 H$_2$O electrolyte to which we have added 0.1 M or 0.5 M K$_2$SnO$_3$.3H$_2$O. Initially, the voltage increases linearly and reaches a value of 50 V after about 70 s for the two stannate concentrations used. In this step of anodic film formation, oxygen evolution can be observed on the AZ91D alloy. During the first 100 s of anodization, the voltage increases more rapidly in presence and for a higher amount of stannate. When the voltage reached 50 V, sparking phenomenon appears on the alloy surface and the sparks increase in number with time and occur for all the duration of the anodization process. It is interesting to note that the breakdown voltage reaches slightly higher value in presence of stannate in the processing electrolyte. Thus, the insertion of tin in the anodized oxide seems to induce a higher permittivity of the oxide, which probably provokes an increase of the breakdown voltage of the oxide layer.

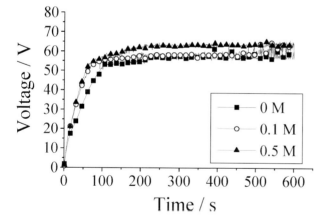

Figure 1: Voltage transient during anodization for different stannate concentrations at a constant current density (10 mA/cm^2) for 10 min.

SEM observations (Figure 2) reveal that the layer of anodized AZ91D alloy for different stannate concentrations is well covering and adherent, but presents porosities in all cases. Nevertheless, the deposit obtained without stannate is more cracked and large pores can be observed, whereas the anodic films obtained with 0.1 M and 0.5 M stannate are more homogeneous and with small spherical nodules.

The EDX analysis of the oxide layer shows the presence of tin, oxygen, potassium, sodium, phosphorus, magnesium and aluminium (Figure 3). The two later elements come directly from the oxidation of substrate (AZ91D magnesium alloy).

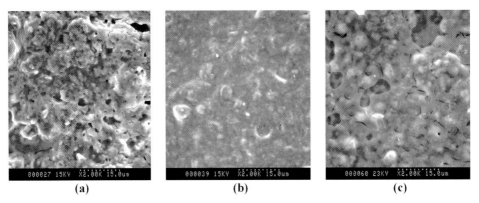

Figure 2: SEM observations of anodized AZ91D Mg alloy for different stannate concentrations: (a) 0 M, (b) 0.1 M and (c) 0.5 M (10 mA/cm^2 – 10 min).

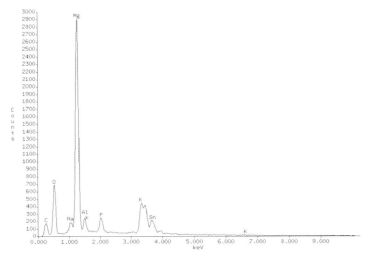

Figure 3: Typical EDX spectra of anodized layer on AZ91D.

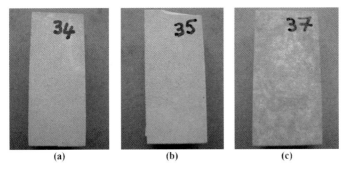

Figure 4: Appearance of some anodized AZ91D alloy for different stannate concentrations: (a) 0 M, (b) 0.1 M and (c) 0.5 M.

The macroscopic aspect shown in figure 4 reveals that the anodized layers are slightly rough and ceramic-like. In absence of tin, the oxide layer pigmentation is more grey compared to the layer with tin content.

3.2 Corrosion Behaviour

Figure 5 shows the E_{corr} evolution in ASTM standard water for the AZ91D untreated or anodized in 3 M KOH + 0.5 M KF + 0.25 M $Na_3PO_4.12\ H_2O$ electrolyte containing different stannate concentrations. After 15 hours of immersion, the corrosion potential of anodized samples stabilizes around –1.55 to –1.6 V/SCE, whereas the one of the untreated sample remains at a higher value, –1.4 V/SCE (see figure 5). So, the presence of the anodic oxide moves the E_{corr} of AZ91D to more negative potentials values, which indicates that this coating mainly inhibits the cathodic reaction of water reduction on magnesium alloy.

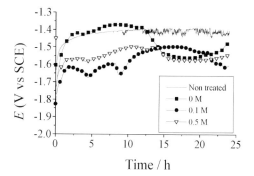

Figure 5: Corrosion potential evolution versus immersion time in ASTM water for untreated and anodized AZ91D in solution containing or not stannates (10 mA/cm^2 – 10 min).

After 24 hours of immersion, the recording of potentiodynamic curves allows to estimate the pitting behaviour of anodized samples compared to the untreated one (figure 6). The untreated alloy shows a high anodic current density with the potential increase, which is characteristic of an active state and the dissolution of the alloy, indicating that the bare alloy cannot resist to a localised attack. For the anodized alloy without stannate, a passive behaviour is observed with a low passive current density (less than 1 µA/cm^2) but becomes higher for potential values that exceed –1.3 V/SCE. Nevertheless, the passivation plateau is very short, and the coating undergoes rapidly a pitting phenomenon above –1.3 V/SCE. The anodized layers with stannates exhibit also a passivation behaviour after 24 hours of immersion in ASTM water. The passive current density on these layers is higher than the one recorded on the stannate-free anodized layer, but the passive behaviour extends on very large potential region between –1.5 and –0.9 V/SCE. So the anodized layers with stannates are stable and protective on a large domain of potential, and have better pitting corrosion characteristics.

In order to accelerate the pitting phenomenon of the anodized AZ91D alloy and to evaluate its resistance to a galvanic attack, the samples were oxidized at two potential values (E_{corr} + 100 mV and E_{corr} + 200 mV) to simulate more corrosive conditions as it is under galvanic coupling. The chronoamperometric curves for the untreated and the anodized alloy are reported in figure 7.

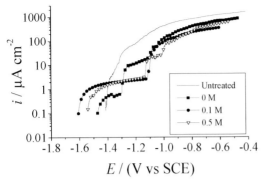

Figure 6: Potentiodynamic curves performed in ASTM water for untreated and anodized AZ91D in solution containing or not stannates (10 mA/cm² – 10 min).

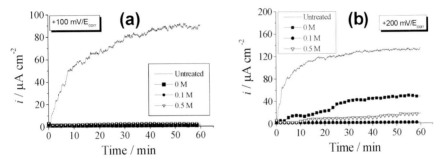

Figure 7: Chronoamperometric curves performed in ASTM water for untreated and anodized AZ91D alloy in solutions containing or not stannates at (a) $E = E_{corr} + 100$ mV and (b) $E = E_{corr} + 200$ mV (10 mA/cm² – 10 min).

For the untreated alloy (figure 7a), the anodic current density, after a sharp increase, tends to stabilise at a high value around 80 µA/cm², which demonstrates an active pitting process. In contrast, for the anodized alloy (figure 7a), the current density remains low and reaches 2 µA/cm² for anodic film obtained in the solution containing 0.1 M $K_2SnO_3.3H_2O$ as additive and 3 µA/cm² when the additive content was 0.5 M $K_2SnO_3.3H_2O$. For a higher applied overpotential (E_{corr} + 200 mV), the current density of the untreated AZ91D stabilises around 120 µA/cm², after an initial sharp increase (see figure 7b). When the alloy is anodized without stannates, the current density remains relatively high at 40 µA.cm². The best behaviour in relation to pitting resistance is obtained with a stannate-content of 0.1 M.

4 Conclusions

For short time by plasma electrolytic anodization (10 min), the addition of stannates to a 3 M KOH electrolyte allows to form a more stable and protective anodic layer. The film contains all the elements present in the electrolyte and the alloy, but remains slightly porous. Nevertheless, the anodized layer formed in stannate solutions mainly promotes an efficient inhibition of the cathodic reaction of water reduction on metal and the presence of a large passivation plateau on anodic polarization curves. This stable behaviour is interesting in the case of galvanic coupling with aluminium or iron-based materials in assembling parts. This study demonstrates also that the morphology of the performed anodic coatings plays a major role in the protection of the magnesium alloy by plasma electrolytic anodization process. Further studies will be devoted to the effect of other additives and electrical regimes on the anodization of AZ91D alloy in order to improve its corrosion behaviour.

5   Acknowledgements

This work was carried out under the postdoctoral stay of C.-E. Barchiche, funded by the European FP6 IDEA project (n° FP6-503806). The authors are grateful to the EC for support of this work.

6   References

[1]   L. Kouisni, M. Azzi, M. Zertoubi, F. Dalard, S. Maximovitch, Surface & Coatings Technology, 2004, 185, 58–67.
[2]   K. Z. Chong, T. S. Shih, Materials Chemistry and Physics, 2003, 80, 191–200.
[3]   H. Umehara, M. Takaya, T. Itoh, Aluminium International Journal, 1999, 75, 634–641.
[4]   H. Umehara, S. Terauchi, M. Takaya, Materials Science Forum, 2000, 350–351, 273–282.
[5]   E. Ghali, Uhlig's Corrosion Handbook, 2nd Edition, Ed. R. Winston Review (2000).
[6]   Y. Zhang, C. Yan, Surface & Coatings Technology (2006), in press.
[7]   H. Fukuda, Y. Matsumoto, Corrosion Science, 2004, 46, 2135–2142.
[8]   Y. Wang, J. Wang, J. Zhang, Z. Zhang, Materials and Corrosion, 2005, 56, 88–92.
[9]   O. Khaselev, J. Yahalom, Corrosion Science, 1998, 40, 1149–1160.
[10] O. Khaselev, D. Weiss, J. Yahalom, Corrosion Science, 2001, 43, 1295–1307.
[11] H.-Y. Hsiao, W.-T. Tsai, Surface & Coatings Technology, 2005, 190, 299–308.
[12] H.-Y. Hsiao, H.-C. Tsung, W.-T. Tsai, Surface & Coatings Technology, 2005, 199, 127–134.
[13] Y. Ma, X. Nie, D. O. Northwood, H. Hu, Thin Solid Films, 2006, 494, 296–301.
[14] Annual book of ASTM standards, copyright ASTM, Pa 19103 (1986).

Electroplating Processes with New Magnesium Alloys

A. Dietz, G. Klumpp
Fraunhofer Institute for Surface Engineering and Thin Films, Braunschweig

1 Abstract

The effect of the electrodeposition of zinc on rare earth magnesium alloys for the corrosion properties has been investigated. The new developed Mg-alloy MRI 207 and the commercial alloy ZE41 were coated with various processes of zinc plating. A major role plays the pretreatment with fluoride containing etchants. The quality of the zinc coating depends from the state of the magnesium surface which was influenced by the type of the different production processes like sand casting, investment casting or gravity die casting.

2 Introduction

Pure magnesium metal has a good corrosion resistance due to the passivation layer of magnesium oxide. For technical applications like in automotive or aviation industry magnesium has to be alloyed to improve the mechanical and thermal properties. However, alloying compounds like Mn, Zn or rare earth metals can cause corrosion problems and thus, an effective corrosion protection is demanded. A lot of corrosion resistant coatings and processes are described, however, principally with the very pure AZ 91 or AZ 31 alloy [1–4].

The EC-funded Project „*Integrated Design and Product Development for the Eco-efficient Production of Low-weight Aeroplane Equipment*" *(IDEA)* develops new Mg-alloys and casting processes for the application in the aviation industry. Part of the project is the development of innovative corrosion resistant systems of these new alloys.

State-of-the-art for the metal deposition on magnesium is a starting layer from an autocatalytic electroless nickel (EN) with or without a zincate pretreatment [5]. The coating will be reinforced by another electroless/electrodeposited coating and/or by a varnish. A drawback is the large electrochemical potential difference between nickel and magnesium which can result in severe corrosion damages.

The present work specifies three ways to attain the goal:

- Substitution of the nickel layer for a zinc layer to diminish the electrochemical potential difference between magnesium and the metallic layer
- Removal of the alloying elements from the surface by chemical etching to improve the adhesion of the zinc layer
- Leveling of the rough surface by chemical polishing to achieve a very dense, faultless layer

The new corrosion system is based on a zinc layer without any nickel. The metallization starts with a cementation process (Displacement deposition), followed by the electrodeposition of zinc. The quality of the first zinc coating is strictly controlled by the use of special organic and inorganic additives. These additives influence the rate of the zinc deposition as well as the hydrogen evolution caused by the magnesium dissolution.

The electrodeposited zinc coating will be additionally coated with a chromium^{VI} free passivation followed by an anti-corrosion varnish.

3 Experimental Details

The chemical composition of the magnesium alloys MRI 207 and ZE41 was determined by SEM-EDX spectroscopy and is shown in table 1.

Table 1: Chemical composition of MRI 207 and ZE 41

Composition	Zn	Zr	Nd	Y	Gd	La	Ce
MRI 207 [wt%]	0.3	0.4	3.0	1.0	5.0		
ZE 41 [wt%]	2.21	0.18				0.11	0.16

The experimental procedure is divided in five segments (see table 2): The mechanical pretreatment to remove the casting skin from magnesia was performed by a blasting process with glass pearls as blasting abrasive.

Table 2: Typical cycle of the Zn-electroplating on magnesium

	Process	Bath composition/ Product name	
Mechanical Pretreatment	Sand (Glass) Blasting		
Chemical Pretreatment	Degreasing		
		Etching (1)	Tartaric acid, 20 g/l
		Etching (2)	Hydrofluoric acid
		Activation	Ammoniumbifluoride, 40 g/l
(Polishing)	Chemical Polishing	Chromic acid, 18 wt.%, Calcium nitrate, 40 g/l	
		Magnesium fluoride, 2.5 g/l	
Plating	Electroless Zinc	Zn-sulfate, -pyrophosphate	
		Electrodeposited Zinc	Slotocyn 10 (cyanic); J = 1A/dm2
Passivation	Cr(III)-layer	Slotoplass Z21 blau	
		Sealing	

The chemical pretreatment was necessary to achieve a clean, metallic surface which was stable for the following plating process. The cleaning process involves a commercial hot degreasing, the chemical etching with tartaric acid and the activation process with ammoniumbifluoride. The etching process with hydrofluoric acid was introduced to remove the alloying elements from the surface.

A critical point for the successful coating of magnesium was the surface quality which was influenced very strongly by the casting process. Fig. 1 shows light optical microscope images of the magnesium surface made by a sand casting process (a), investment casting process (b) and gravity die casting process (c). Due to the very rough surface, the sand casting samples have been treated with a chemical polishing process to smooth the surface.

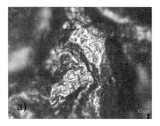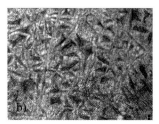

Figure 1: Light optical microscope image of the surface morphology of magnesium; a) sand casting process; b) investment casting process; c) gravity die casting

The corrosion tests of the samples have been carried out according to the ASTM B 117 Standard (Salt Spray Test) with a duration of 100 h. The mass lost has been determined. The uncoated samples have been etched with a mixture of chromic acid and silver nitrate to remove the residues. The coated samples have been cleaned by ultrasonic treatment in demineralised water.

4   Results and Discussion

4.1   Pretreatment

Figure 2 shows a SEM-image of a glass blasted, uncoated magnesium ZE 41 surface (a). The structure is caused by the impact of the glass pearls during the glass blasting. The light-coloured spots are precipitations of the alloying metals zinc, zirconium, lanthanum and cerium. However, after dipping the surface in a 20 vol. % hydrofluoric acid for some minutes, the alloying elements are dissolved (b). The EDX quantification in figure 3 confirms the result for the ZE 41 alloy: The concentration of zinc, zirconium, lanthanum and cerium is dras- tically reduced compared to the initial state.

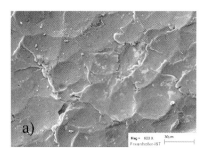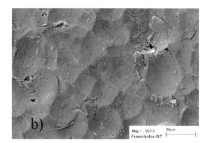

Figure 2: SEM-Images of uncoated Magnesium ZE41: a) Glass blasted; b) Glass blasted and intensely HF-etched

The removal of the alloying elements seems to be an approach to improve the quality of the coating on the magnesium alloy due to the fact that these imperfections can cause pores in the layer during the deposition.

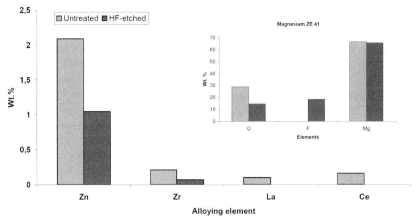

Figure 3: EDX-quantification of the element composition of Magnesium ZE 41

4.2 Chemical Polishing

Due to the demand of a faultless coating for an effective corrosion protection on magnesium, a process to level the rough surface seems to be necessary. An electropolishing process has been described in phosphorous acid [6], however, the result was an anodized surface and not a pure metallic surface.

The effect of the chemical polishing was shown in figure 4. The light optical microscope image (a) shows the very rough surface of the MRI 207 alloy, produced by the sand casting process. Image (b) shows the surface after 15 minutes etching in an etchant of chromic acid, calcium nitrate and magnesium fluoride [7]. The roughness r_a, measured by a Taylor-Hobson Form Talysurf machine decreased drastically from 13.5 µm (as-casted-state) to 3.7 µm (chemical polished).

The leveled surface increases the quality of the coating, however, due to the high dissolution rate it may be critical to thin-walled samples.

4.3 Plating

Despite the pretreatment process with fluoride containing agents which creates a thin layer of magnesium fluoride, MgF_2, a metallic interlayer is necessary to prevent a rapid dissolution of magnesium during the zinc electrodeposition (cementation process) which results in a very granular layer with a bad adhesion. The plating process of the electroless zinc layer is based on an US patent which describes the use of an electroless zinc layer as a preliminary stage to the electrodeposition of zinc [8, 9]. Figure 5 shows samples of zinc plated magnesium alloy MRI 207. The direct electroplating without the interlayer of electroless zinc results in a rough coating with a lot of blister which are the initial points of corrosion defects (a). The intermediate coating of electroless zinc is the base for a dense and smooth electrodeposited zinc layer (b).

Jiang et al. [10] used for the plating on AZ 91 magnesium alloy a zinc immersion electrolyte of zinc sulphate and phosphate with an additionally interlayer of Zn-Cu. The top coating was a Zn-Ni alloy which was deposited by a pulsed electrodeposition.

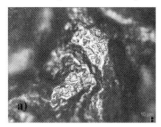 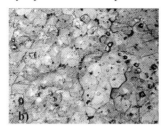

Figure 4: Light optical microscope image of the surface morphology of sand casted magnesium MRI 207 shows the effect of chemical polishing: a) as casted; b) after 15 minutes chemical polishing

 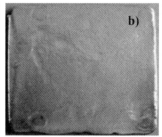

Figure 5: Photo-Images of electrodeposited zinc on Magnesium MRI 207 without (a) and with electroless zinc (b) as an interlayer

4.4 Corrosion Tests

The zinc plated samples of the MRI 207 alloy have been tested in a salt spray corrosion test equipment according ASTM B 117 for 100 h and compared with uncoated species as well as with commercial alloys AM 50 and AZ 91D. Expectedly, the very pure commercial alloys show better corrosion results, but the effect of the zinc plating of MRI 207 with the following passivation and sealing is clear. However, it was a surprise that the samples made by the sand casting process showed lower corrosion rates than the samples made by investment casting or gravity die casting (see table 3).

Table 3: Mass lost [%] of Mg alloy samples after 100 h corrosion test according ASTM B 117

Sealing	Uncoated	Zinc/Passivation
MRI 207 Sand Casting	1.52	1.39
MRI 207 Investment Casting	4.54	3.22
MRI 207 Gravity Die Casting	2.62	1.86
Commercial AM 50 HPDC	0.75	0.2
Commercial AZ 91D HPDC		0.65

5 Conclusions

The magnesium alloys ZE 41 and MRI 207 with rare earth metals have been coated with an electrodeposited zinc layer. To achieve a dense, well adhesive zinc layer, following processes have been investigated:

- Removal of the alloying elements from the surface
- Levelling the surface by chemical polishing
- Electroless deposition of a zinc interlayer

Further works will focus on the modification of the polishing process to achieve a smooth surface with a decrease of the dissolution rate.

6 Acknowledgement

The financial support of the work in the EC-funded project IDEA (contract FP6-503826) was gratefully acknowledged.

7 References

[1] J.-Y. Uan, B.-L. Yu; Magnesium Technology 2005 ; Edited by N.R. Neelameggham, H.I. Kaplan, B.R. Powell; TMS 2005, 475
[2] C. Blawert et al. ; Magnesium Technology 2005 ; Edited by N.R. Neelameggham, H.I. Kaplan, B.R. Powell; TMS 2005; 447
[3] T. Yamaguchi et al.; Magnesium Technology 2005 ; Edited by N.R. Neelameggham, H.I. Kaplan, B.R. Powell; TMS 2005; 491
[4] J.I. Skar; Materials and Corrosion 1999, 50, 2–6
[5] F. Leyendecker in Magnesium – Alloys and Technologies (Ed.: K. U. Kainer)Wiley-VCH Co. KGaA; Weinheim; 2003; 242
[6] Zhao Qun et al.; Magnesium Technology 2005; Edited by N.R. Neelameggham, H.I.
[7] Rafael Rituper; Beizen von Metallen; Eugen G. Leuze Verlag; Saulgau; 1993; p. 324
[8] Process of zinc coating magnesium articles, US Patent US 2,730,490; 1956
[9] T. W. Jelinek in Praktische Galvanotechnik; 4th ed.; Eugen G. Leuze Verlag; Saulgau; p. 185
[10] Y.F. Jiang et al. ; Thin Solid Films 2005, 484, 232–237

Post Processing

Effects of Strain Rate and Filler Alloy on the Properties of Laser Welded AZ31 Alloy Sheet

Y. Durandet[1], W. Song[1], P. Cordini[2], M. Brandt[1] and A. Ostendorf[3]
[1] CAST CRC - Industrial Research Institute Swinburne (IRIS), Melbourne, Australia
[2] Alstom LHB GmbH, Salzgitter, Germany (formally at LZH)
[3] Laser Zentrum Hannover E.V. (LZH), Hannover, Germany

1 Introduction

There are many developments taking place world wide to exploit the light weight features of magnesium such as its high specific strength or impact absorption properties. These developments rely on understanding the metallurgical mechanisms controlling its anisotropic deformation behaviour or strain rate sensitivity [1,2]. They also depend on the availability of enabling technologies such as coating [3,4] or joining [5,6] to overcome potential problems with galvanic corrosion and to integrate magnesium in the final product application.

The laser beam welding process is one possible industrial joining solution because it allows high processing speed with low and localised energy input. Laser welding of magnesium with filler alloy is attractive to relax the fit-up tolerances, as well as to control the weld metallurgy. For laser butt joining, the gap must be small enough that the beam cannot pass straight through the joint, i.e. the gap should be smaller than half the beam diameter. Wire feeding is considered to be the most practical and flexible method because it has the advantages of low contaminant pick-up, high productivity and a wide range of filler materials are usually available as MIG welding wire. The four most common filler alloys for MIG welding of magnesium are ER AZ61A, ER AZ101A, ER AZ92A and ER EZ33A and the available wire diameter is usually 1.6mm minimum. Although magnesium alloys have a low melting temperature and latent heat of fusion, a 1.6mm wire is too thick for laser welding applications where tight positioning tolerances are required due to the small interaction zone between the laser beam and material. Unlike steel or aluminium, the availability of magnesium welding wires of smaller diameter has been limited by the high costs of wire production due to the poor workability of magnesium inherent to its hexagonal lattice structure.

Smaller diameter drawn magnesium wires have been developed for robot-assisted MIG or laser welding systems [7] and trialled at LZH as part of the European InMaK project [8]. Despite the high fixed price of the magnesium wire, the high initial investment and operating costs of a laser system could partly be offset by the low usage of filler wire because laser welding uses about ten times less filler wire per meter of welded seam than MIG welding [9]. The use of 1.2mm filler wire was found to significantly increase the laser welding process stability as a result of higher permissible positioning tolerances and there was little influence of the wire alloy on the process reliability. The influence of the wire alloy on the mechanical properties of the laser weld parts was considered to be low compared to steel and aluminium.

This paper deals with the effects of filler alloy and strain rate on the tensile properties of AZ31 sheet laser welded with 1.2mm diameter filler wire of three different magnesium alloys.

2   Materials and Procedures

Samples of the parent 1.3mm thick AZ31 rolled and tempered metal sheet were supplied by LZH, along with bead-on-plate (BoP) welds produced using a 4kW Nd:YAG laser, 1.2mm diameter wire of three different magnesium alloys and two sets of welding parameters, as summarised in Table 1. BoP welds without filler alloy were produced using the 2.3kW Nd:YAG laser at IRIS and similar welding parameters as LZH.

Table 1: Details of Nd:YAG laser welded joints produced at LZH and IRIS

	Sample Set	Filler Alloy	Laser Power (kW)	Strip Feed Rate (m/min)	Wire Feed Rate (m/min)
LZH	A	AZ31	1	2.5	1.5
	B	- id -	1.3	4.5	2.5
	C	AM50	1	2.5	1.5
	D	- id -	1.3	4.5	2.5
	E	AZ92	1	2.5	1.5
	F	- id -	1.3	4.5	2.5
IRIS	LW10-10 to 24	n/a	1	2.5	n/a
	LW10-25 to 39	n/a	1.3	4.5	n/a

Tensile testing was performed at Swinburne University of Technology using a hydraulic driven MTS machine at three different speeds: 0.08, 1 and 100mm/s, corresponding to nominal strain rates of 0.0017, 0.02 and 2 s^{-1} respectively. The tensile specimens were machined from the parent AZ31 alloy sheet and from the welded samples with the loading axis transverse to the sheet rolling direction, but parallel to the seam length.

3   Results and Discussion

The effects of the filler alloy and strain rate on the properties of the laser joints for the two sets of laser welding conditions are summarized in Figure 1 (a) and (b) where the line energy or heat input per unit length of welded seam is given by the ratio of the laser power to the strip feed rate. Results overall confirmed the beneficial effects of using a filler wire.

The tensile properties of wrought AZ31 sheet are known to be influenced by texture, temperature and strain rate [2]. Without filler wire, the effects of strain rate were more significant at high line energy : the yield and tensile strength increased with strain rate. Elongation values were higher, suggesting that some annealing or strain relieving occurred in the HAZ.

At low line energy, the effects of filler alloy and strain rate on the tensile properties were not significant. However, the tensile strength and elongation of the laser joints were better with filler wire than without while the yield strength was similar, as shown in Figure 1 (a). The tensile strength and elongation of the laser joints were improved by using a filler wire, possibly because the cross section of the joints with filler wire was larger than that of the parent metal and the shape of the weld resulted in less notches in the top- and under-bead. Over the range of tensile testing speeds, the laser joints' efficiency was 93 % to 124 % of the strength of the parent metal with AZ92 and AM50 filler alloys, and 85 % to 125 % with AZ31.

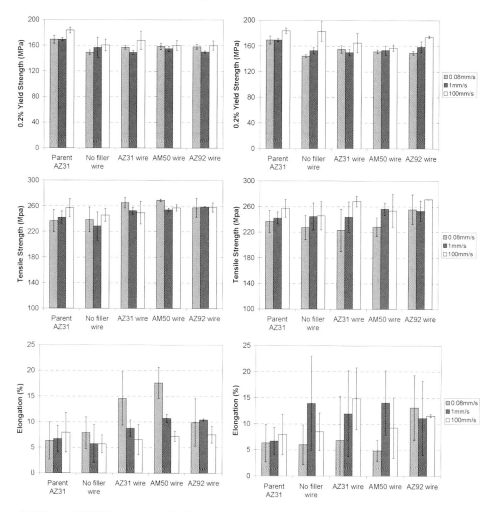

(a) 17 J/mm – 1300 W laser power, 4.5 m/min feed rate (b) 24 J/mm – 1000W laser power, 2.5 m/min feed rate

Figure 1: Effects of wire alloying and tensile testing speed on the yield strength, tensile strength and elongation at rupture of laser joints produced at (a) low and (b) high line energy.

The effects of filler alloy and strain rate on the tensile properties were more significant at high line energy, as shown in Figure 1 (b). At high line energy, the yield and tensile strengths of the laser joints with filler wire increased with strain rate, as observed for the parent metal and welds without filler. As evaporative losses and weld drop-through would have been greater at higher line energy or heat input, the compensating effects of using a filler wire were more pronounced at high rather than at low line energy. Over the range of tensile testing speeds, the laser joints' efficiency was 91–127 % of the strength of the parent metal with AZ92 filler alloy, compared to 84–115 % with AM50, and 75–117 % with AZ31.

Within the small dwell times and interaction zones between the laser beam and material, limited chemical diffusion occurred. Small changes in Al and less obvious changes in Zn concentration profiles were detected, but only over a very short diffusion length and in samples with the higher initial solution force, i.e. laser joints with AM50 and AZ92 filler alloys. On the other hand, micro-hardness profiles showed increases in hardness from 70HK in the parent metal up to 80–90 HK in the welds with AM50 and AZ92 filler alloy, and revealed some softening in the base metal near the fusion zone boundary, as shown in Figure 2. This softening was consistent with the microstructure in the HAZ (Figure 3) displaying a polygonal grain structure with high angle boundaries, slightly larger grain size and less twinning compared to the parent AZ31 metal (Figure 4). Since the filler wire provides additional absorption of the laser power, the temperature distribution in the strip during welding appeared sufficient to locally anneal the base metal and produce joints with higher elongation. At high line energy, the increased strength of the laser joints with increasing aluminum content in the filler wire can thus be related to the increased strength of the weld seam due to the presence of $Mg_{17}Al_{12}$ intermetallic phase, causing higher deformation in the base material. This was evinced by the necking and shear deformation near the weld seam of the AZ31 as shown in Figure 5. For the AZ31 filler alloy, the weld hardness was expectedly lower than the parent metal as the solidified AZ31 alloy structure did not undergo the hardening treatment imparted by the strip fabrication process to the parent AZ31 metal.

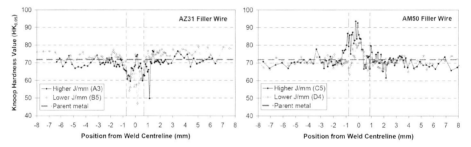

Figure 2: Example of micro-hardness profiles (Knoop indenter, 50 g load) for laser joints produced at low and high line energy with AZ31 (left) and AM50 (right) wire alloy.

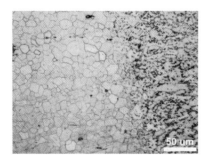
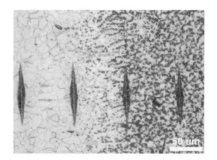

Figure 3: Example of microstructure in HAZ for laser joints produced with AZ31 wire at high line energy (left) and AZ92 wire at at low line energy (right).

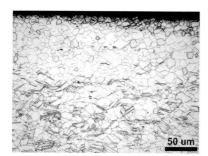
Figure 4: Microstructure of the parent metal away from the HAZ

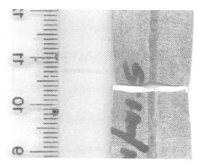
Figure 5: Example of necking of laser joint with filler wire (AM50 wire alloy, low line energy, 0.0017 s^{-1}).

Fracture surfaces of tensile specimens with low and high elongation were examined in the SEM to understand the large standard deviations in elongations and tensile strengths. Laser joint specimens with higher elongation displayed a uniform fracture surface consisting of mixed dimpled and cleaved grains, whereas specimens with low elongation displayed away from the weld area a band of smooth appearance at the surface of the parent metal strip. Similar observations were made on the fracture surface of samples of the parent (un-welded) AZ31 metal with lower elongation. A qualitative XRay-EDS analysis showed the band to be an oxidised layer (Figure 6). The surface of the rolled sheet revealed numerous micro-cracks, probably related to the strip fabrication process, which would increase the notch sensitivity of the base metal transverse to the sheet rolling direction.

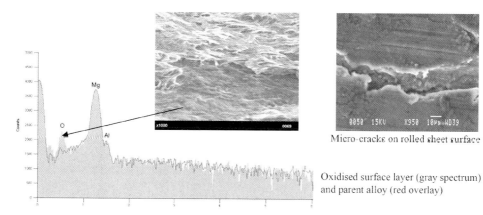

Micro-cracks on rolled sheet surface

Oxidised surface layer (gray spectrum) and parent alloy (red overlay)

Figure 6: Defects in low elongation tensile specimen of un-welded parent AZ31 alloy sheet.

4 Summary and Concluding Remarks

The effects of filler wire alloy and strain rate on the static mechanical properties of laser joints of AZ31 sheet were investigated. The tensile loading axis was aligned with the weld seam and transverse to the sheet rolling direction. While the yield strength of the laser joints was lower than the parent alloy, the tensile strength and elongation were generally comparable, and better with filler material than without. The scatter of elongmeion data was attributed to some notch sensitivity of the parent metal transverse to the sheet rolling direction. The integrity of the laser welds was not a problem.

The effects of wire alloying and strain rate on properties were most apparent at higher line energy. The deformation of the laser joints was akin to that of two-phase plastic-plastic aggregates consisting of a thin composite sheet (the matrix phase) with a single aligned continuous „fibre" (the second phase) parallel to the tensile loading axis. The „reinforcing" effects of the weld „fibre" were due to the Al content of the filler alloy strengthening the weld seam. Elongation values were higher for joints produced at higher heat input, as partial annealing in the HAZ during welding enabled more shear deformation to occur at the interface between the two phases.

At low line energy, the effects of filler wire alloy on the yield and tensile strengths at the various strain rates were not significant. This makes welding easier as readily available magnesium filler alloys can be used.

5 Acknowledgments

The CAST CRC was established and is partly supported by the Australian Government's Cooperative Research Centres (CRC) Program. The authors acknowledge the support provided by Mr Brian Dempster in machining and supplying magnesium samples for laser welding and tensile testing, and Dr Jim Harris with the operation of the Nd:YAG laser at IRIS.

6 References

[1] T. Abbott, M. Easton, R. Schmidt in Magnesium Technology 2003 (Ed.: H.I. Kaplan), TMS Annual Meeting, San Diego, California, USA, 2003, 227–230.
[2] S.R. Agnew, O. Duygulu, Materials Science Forum 2003, 419–422, 177–188.
[3] E. Ghali, Materials Science Forum 2000, 350–351, 261–272.
[4] J.E. Gray, B. Luan, Journal of Alloys and Compounds 2002, 336, 88–113.
[5] G. Kohn, A. Stern, A. Munitz in ATTCE 2001 Proceedings Volume 3: Manufacturing, Society of Automotive Engineers, Inc. (SAE), Barcelona, Spain, 2001, 201–205.
[6] L.K. France, R. Freeman in ATTCE 2001 Proceedings Volume 3: Manufacturing, Society of Automotive Engineers, Inc. (SAE), Barcelona, Spain, 2001, 207–211.
[7] Peter Baumgart (Sp) in Magnesium (6th Int. Conf. Magnesium Alloys and Their Applications) (Ed.: K.U. Kainer), Wiley-VCH, Wolfsburg, Germany, 2003, 968–973.
[8] A.M. J.-F. Lass, M. Wappelhorst, E. Hombergsmeier, H. Lanzerath, P. Baumgart, P. Cordini, in Magnesium (6th Int. Conf. Magnesium Alloys and Their Applications) (Ed.: K.U. Kainer), Wiley-VCH, Wolfsburg, Germany, 2003, 1026–1033.
[9] H. Haferkamp, M. Goede, A. Bormann, P. Cordini in Proceedings of the 3rd LANE 2001, Meisenbach-Verlag Bamberg, Erlangen, Germany, 2001, August 28–31, 333–338.

Microstructure Features of Hot Rolled AZ31 Magnesium Alloy for Friction Stir Welding

L. Commin[a], J.-E. Masse[a], M. Dumont[b], L. Barrallier[a]
a Mécasurf Team, ENSAM, Aix en Provence
b TECSEN, Université Cézanne, Marseille

1 Introduction

As the lightest structural material available so far, magnesium alloys are driving more and more interest due to their high specific properties. But welding magnesium alloys still faces many challenges. Indeed, conventional processes have exhibited some disadvantages such as a large heat affected zone, porosity, evaporative loss of alloying elements, high residual stress and distortion [1, 2].

So, Friction Stir Welding (FSW) process can be an interesting alternative as it is a solid state joining process and then it produces pore-free joints. The properties of the weld depend on the microstructure and then, its evolution during FSW has to be determined to fully understand the mechanisms involved in this process. During FSW process, the material is affected both thermally and mechanically. So, prior to the investigation of the FSW weld, a first step was to consider only the effect of the temperature parameter.

2 Experimental

In this study, 2mm thick plates of wrought commercial AZ31 alloy were used. The chemical composition of the hot rolled AZ31 alloy studied is detailed in Table 1.

Table 1: Chemical composition of AZ31 alloy (Salzgitter data)

Element	Al	Zn	Mn	Si	Cu	Ni	Fe	Ca	Other	Mg
wt.%	2.5-3.5	0.6-1.4	0.2-0.6	<0.05	<0.008	<0.002	<0.005	<0.02	<0.3	balance

The hot rolling process has been done at a 300 °C temperature. The magnesium melting point is 650 °C and the FSW process is performed a little below this temperature to keep the matter in solid-state. So, samples of wrought alloy have been submitted to a further heating cycle in order to investigate the evolution of the microstructure between the room temperature and the melting point. The wrought alloy samples will be from now designated as "as received samples". The samples, which have undergone a further heat treatment, will be denominated by the temperature at which they have been heated to.

Polished samples were etched with an acetopicral solution (0.4g picric acid, 13 ml ethanol, 3ml glacier acetic acid and 3ml boiled water [3]). They were observed using a Leitz Aristomet optical microscope and a JEOL JMS 6400 SEM.

TEM samples were made from 2 mm thick plates. One square centimeter samples were ground down to a thickness between 100 and 200 µm using several silicon carbide polishing papers. They were then electrochemically polished in a Struers Tenupol-3 jet polisher at a temperature below 10 °C with a polishing voltage of 14 V. The polishing solution was 10 pct HCl, 90 pct butoxy-2-ethanol by volume. TEM observations were performed using a FEI TECNAI G^2.

XRD spectra were determined using a SEIFERT MZ6TS diffractometer. DSC measurements were done using a SETARAM 131 at 5 °C min^{-1}.

3   Results and Discussion

The observations of the as received material and of samples heated at 380 °C and at 550 °C are shown in Figure 1. AZ31 alloy consists in only an α-Mg phase and then there was not the $Mg_{17}Al_{12}$ eutectic structure usually present in Mg alloys. The "as received" material revealed a fine equiaxed grains structure, with very few deformation twins. In the 380 °C samples, the grains size is similar, but a substantial grain growth is observed at 550 °C.

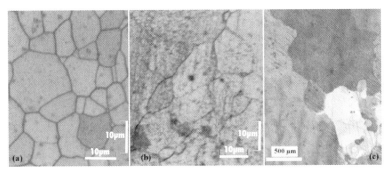

Figure 1: Optical micrograph of (a) "as received", (b) 380 °C and (c) 550 °C AZ31 alloy

SEM analysis were performed on "as received" samples and showed the presence of very bright particles (Figure 2). These particles were forming small amounts and they were distributed all along the sample with no apparent preferred localization. EDS analysis showed that theses inclusions were Al and Mn-rich phases and that there was a loss in Mg content in these areas (see Figure 3).

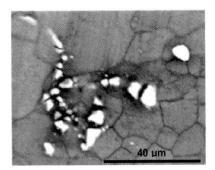

Figure 2: Secondary electrons SEM image of the "as received" AZ31 alloy

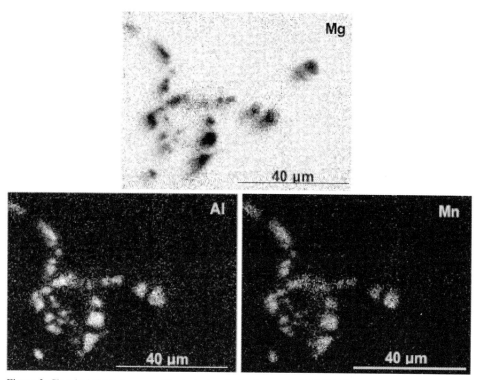

Figure 3: Chemical elements maps of the inclusions in the "as received" AZ31 alloy

SEM analysis was also performed on the 550 °C sample to check if the Al-Mn particles were still visible. Figure 4 shows that the Al-Mn phase was not dissolved at 550°C.

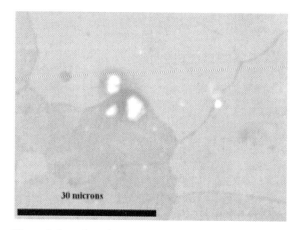

Figure 4: Secondary electrons SEM image of the 550 °C heat treated sample

The equilibrium phases at room temperature in magnesium alloys with 2 to 9 wt.% aluminium and less than 1 wt.% manganese are α-Mg solid solution, $Mg_{17}Al_{12}$ and Al_8Mn_5 [4].

The $Mg_{17}Al_{12}$ and the Al_8Mn_5 could not be identified from the XRD spectra performed on the "as received" sample. So DSC measurements were made and revealed the presence of 3 phases in the temperature range studied (see Figure 5).

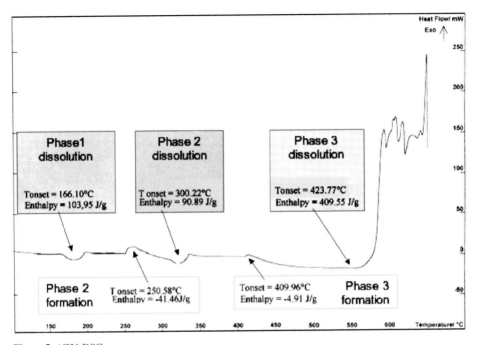

Figure 5: AZ31 DSC curve

TEM analysis performed on the as received sample, the 380 °C sample and the 550 °C sample (see Figures 6, 7 and 8). Intragranular nanoscale features can be observed in each sample ranging from about 600 nm down to 20 nm. On Figure 6 it can be seen that they are uniformly distributed into the α-Mg matrix; whereas, on Figure 7 we can see that the 600 nm needle shaped particles are concentrated in some areas and the 50 nm particles are located in different areas.

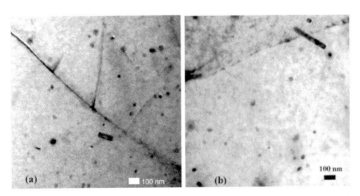

Figure 6: TEM image of the "as received" sample (a) and the 380 °C heat treated sample (b)

EDS analysis revealed that the precipitates observed in each sample were Al-Mn compounds. No $Mg_{17}Al_{12}$ phase could be identified from EDS analysis. A further observation using high resolution is needed to positively determine the presence of this phase. Figure 8 shows the SAED patterns taken along [010] axis obtained for the particles of the 380 °C heat treated samples. From the reflections indexation, the lattice parameters can be determined. It leads to $a = 0.6394$ nm and $b = 0.7724$ nm, which is consistent with the lattice parameters values given in the literature [5] for the Al_6Mn orthorhombic crystal structure.

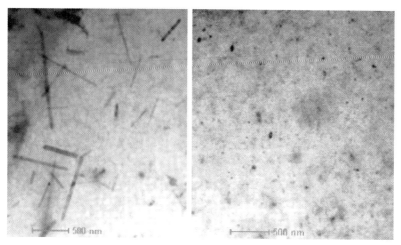

Figure 7: TEM image of the 550 °C heat treated sample.

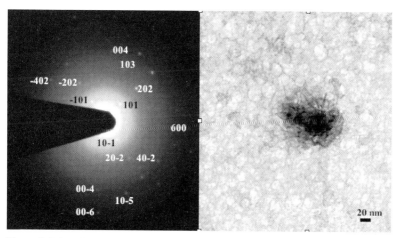

Figure 8: Bright field image of a 380°C sample particle and the corresponding SAED.

4 Conclusions

Nanoscale precipitates of different sizes have been identified as AlMn compounds within each sample studied. These precipitates are undergoing structural transformations with increasing temperature. So, the effect of temperature in friction stir welding process is likely to cause some modifications in the intragranular precipitation of AZ31 magnesium alloy.

5 Acknowledgments

This work was performed in the frame of the Aeromag Project, financed by the EC. The authors want to thank all the partners involved in this project for their support and interest.

6 References

[1] Haferkamp H. et al., Laser and electron beam welding of magnesium materials, Welding & Cutting **2000**, vol. 52, 8, p. 178–180.
[2] Munitz A. et al., Electron beam welding of magnesium AZ91D plates, Welding Journal **2000**, vol. 79, 7, p. 202–208.
[3] Nagasawa T. et al., Structure and mechanical properties of friction stir weld joints of magnesium alloys AZ31. Magnesium Technology **2000**, p. 383–387.
[4] Gertsman V.Y. et al., Microstructure and second phase particles in low and high pressure die-cast magnesium alloy AM50, Metallurgical and materials transaction A, **2005**, 36A, p. 1989–1997.
[5] Kong B.O. et al., Investigation of growth mechanism and orientation relationship of Mn-dispersoid in an Al-Zn-Mg-Mn alloy, Materials Letters,**1996**, 28, p. 385–391.

Rational Friction Welding of High Creep Resistance Magnesium Alloy

G. A. Pinheiro, C. G. Pankiewicz, J. Fernandez dos Santos, K.-U. Kainer

Institute for Research Material, GKSS-Forschungszentrum Geesthacht, Germany

Abstract

The aim of the present work was to investigate rotational friction welding of an Aluminum-Rare Earth based high creep resistance Magnesium alloy, AE 42HP. Cylindrical studs with Ø20mm were friction welded under different conditions. Further analyses of the influence of the welding parameters on the joint formation as well as temperature measurements were carried out within this context. Furthermore metallographic examination using a light electron microscope with additional grain size measurements, as well as tensile tests were carried out to obtain a better understanding and to evaluate the feasibility of Friction Welding to join magnesium alloys with continuous drive method.

The achieved results showed a good reproducibility and reliability of the friction welding system. Moreover it was proved that with such process sound welds can be produced with a refined microstructure at the bonding line. The average ASTM-grain size was in some cases 11.5, which generated a joint with a superior tensile strength if compared to that of the base material, where the average ASTM-grain size was something around -3. Tensile tests attested also the suitability of the process in performing high strength joints.

Keywords: friction welding, magnesium alloy, AE42

1. Introduction

The Mg alloys currently used in the automotive industry are high pressure die cast and have either high strength (AZ91D) or high ductility (AM60 and AM50) at room temperature. However, power train applications such as automatic transmission cases and engine blocks are limited for these alloys since their mechanical properties decrease rapidly with temperature. Mg-Si alloys, like AS21, AS41A and AS41B were exploited on a large scale in the various generation of the famous Volkswagen Beetle engine in the 70s. Nowadays no Mg engine blocks are manufactured due to high operating temperature requirements of modern engines and the cost of the alloys [1,2,3].

A die cast alloy developed by Dow Chemical Company, AE42, has shown superior mechanical properties in applications at high temperatures. This alloy was developed from a non-aluminum magnesium chemistry in which rare earths, under die-casting conditions, were shown to increase creep resistance by forming fine Mg_9RE precipitates along the grain boundaries. The presence of alloying elements results in significant advantages in comparison to Mg-Zn and Mg-Mn alloys. Table 1 shows a brief comparison between some materials industrially used.

Table 1. Mechanical properties of pure Mg in comparison with some Mg, Al and steel alloys.

Mechanical Properties	AZ91	AE42	AA2024	AA6082	AISI1022	AISI1040
Hardness (HV)	80	60HB	142	95	143	211
UTS (MPa)	275	234	427	310	474	620
Yield Strength (MPa)	145	145	345	260	330	415
Elongation at Break (%)	6	11	5	10	34	25
Modulus of Elasticity (Gpa)	44.8	45	72.4	n. a.	200	200
Shear Modulus (GPa)	17	17	27	n. a.	80	80

Friction welding is a relatively unexplored area within Mg research and has showed some advantages as short welding times, good repeatability and suitability to perform sound welds specially in dissimilar configuration (i.e. in combination with other metals), if compared with conventional fusion welding processes. In the basic and most used variation one of the workpiece rotates while the other is held stationary. Both welding surfaces come in contact under a defined pressure during a preset time. The material at the faying surfaces becomes plasticized and the parts are forged together at the end of the process. Figure 1a shows the basic configuration of this variation.

The friction weldability of Mg alloys was previously considered to be difficult, but actually many publications [4,5,6,7,8] including the AWS Handbook describe it as possible. Many friction welding studies involving Mg similar and dissimilar joints with different alloys have been carried out aiming at an optimization of process parameters, leading to sound and reliable joint formation.

Bowles et. al. [4] investigated AE42 – AZ91 friction welded joints without significant loss of hardness across the bonding line. The AZ91 side of the weld showed more significant microstructural changes related to dynamic recrystallisation taking place along an app. 500 µm wide stripe starting from the bonding line. On the other hand very few changes in the microstructure are evident on the AE42 side of the weld. Only the mechanical fragmentation of the intermetallic particles has been observed. Such a process was used to improve locally mechanical properties of an AZ91 cast component by joining AE42 studs/inserts.

Ogawa et al [5] have shown, through several experiments in a wide variety of friction welding conditions for the similar joint in AZ31, that the deformation caused by the heat input during the upset stage and upset loss can be used to evaluate the joint performance in terms of tensile strength. Friction time, rotational speed and upset pressure must have expressive values to the welds to lie in the sound joints range.

Similar AZ31 alloy joints were also studied by Kato and Tokisue [6]. According to the authors, the macrostructures in the vicinity of the weld interface are symmetrical in relation to the weld interface and joint axis regardless of the applied friction time. The hardness in the weld interface has a value similar to that of the base material and even in different parts of the joint has shown no significant difference. The tensile strength and elongation of the investigated friction welded joints [6] tend to be improved with an increase in the friction pressure and friction time. The highest impact value observed was app. 60% of the base material value due to the disappearance of the fibrous structure in the weld interface. This has also been observed in carbon steels [9] and aluminum alloys [10,11].

2. Experimental

An AE42-HP (high-purity) alloy has been selected for the present study. The material was delivered as cast ingots and spark-eroded. Table 2 lists the chemical composition while Table 3 lists both its mechanical properties according to tests carried out at GKSS and to the literature [12]. It should be mentioned that the base material properties were slightly different from those mentioned in the literature.

Table 2. Chemical composition of the AE42HP used in the present study.

El	Al	Zn	Mn	Ni	Si	Be	Ce	La	Nd	Pr	Th	Mg
%	3,9	0,003	0,41	<0,001	0,02	0,001	1,2	0,6	0,4	0,1	0,2	92

Table 3. Mechanical properties of the base material.

Material	Tensile Strength (MPa)	Elongation (%)	Hardness
AE42HP (tested)	125 (114 – 143)	6 (4 – 7)	40 (36 – 44), Hv. 0,2
AE42 [12]	230	11	60, HBS 1/5

Conventional cylindrical geometry was chosen for this work with 20mm diameter and 100mm length studs. The rotating stud was tightened to the weld head while the other was remained stationary. This geometry was suitable to fit in the friction welding machine and is commonly used to perform conventional friction welding trials. Additionally, 1,2mm holes were drilled into the studs to insert thermocouples at 5 and 10mm from the welding interface. Figure 1 shows in a) a schematic of the process and in b) the geometry of both the rotary (A) and non-rotary stud, where the thermocouples were inserted.

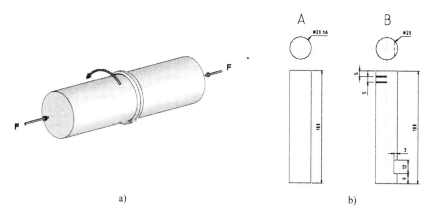

Figure 1. a) Geometry of the rotary (A) and non-rotary (B) welding studs and b) a schematic of the classic configuration used on this work.

A HSM 3000 machine has been used to perform the welds. This is a portable hydraulic powered friction welding system, designed and built by Circle Technical Services Ltd. Before the beginning with the welding operation itself some procedures were followed in order to ensure the reproducibility of the welding programme and avoid as much as possible external influences. The studs were cleaned with acetone to eliminate all possible contaminations forms that could influence the weld results. The oil temperature of the

order to ensure the reproducibility of the welding programme and avoid as much as possible external influences. The studs were cleaned with acetone to eliminate all possible contaminations forms that could influence the weld results. The oil temperature of the machine was kept at 20+-2°C prior to the beginning of the weld in order to avoid significant changes in the power delivered by the welding machine. Table 4 lists the set of 7 individual friction welding conditions investigated. Two different weld series were proposed in varying the forging force (MgMg-series) and the rotational speed and the welding pressure (Mg-series). Each set of parameters, i.e. each weld of the weldability matrix was performed five times. Four specimens of each weld were designated to tensile tests and the remaining one to metallographic analysis. Thermal cycles were recorded in all cases with using thermocouples as well as with an infrared camera.

Table 4. Friction welding conditions.

Sample label	FP (MPa)	RS (rpm)	FF (MPa)	BO (mm)
MgMg-01	1.03	4000	1.38	5.0
MgMg-02	1.03	4000	1.72	5.0
MgMg-03	1.03	4000	2.07	5.0
Mg-01	1.03	3000	1.72	5.0
Mg-02	1.03	6000	1.72	5.0
Mg-03	1.72	3000	2.41	5.0
Mg-04	1.72	6000	2.41	5.0

FP: Friction Pressure set in the welding machine; RS: Rotational Speed; FF: Forging Force; BO: Burn-off;

Welded specimens were cut (longitudinally, in the center of the studs) and etched with a Picric Acid based etchant to allow a complete examination of the joint. Finally tensile testing has been carried out on specimens with the bonding line positioned in the middle of gauge length.

3. Results and Discussion

3.1 Process Stability

Friction time, welding pressure, rotational speed and burn-off were monitored during the welding by an external data acquisition system integrated with the control unit. The acquisition system recorded each individual weld and allowed later an accurately evaluation of each weld. All groups of welds have shown very low standard deviation on important parameters. This indicates that the process was generally stable and that the HMS 3000 presented a high level of reproducibility. Table 5 shows the average value of welding time and burn-off with their respective standard deviation to give an indication of the repeatability of process itself and of the welding machine.

Table 5. Average and standard deviation of welding time and burn-off.

Samples	Welding Time (s)	Standard Deviation (s)	Real Burn-off (mm)	Standard Deviation
MgMg-01	3.03	0.27	8.70	0.19
MgMg-02	3.20	0.62	10.84	0.84
MgMg-03	3.14	0.08	13.86	1.71
Mg-01	2.11	0.03	8.54	0.51
Mg-02	2.43	0.03	9.92	0.26
Mg-03	1.50	0.06	10.32	0.44
Mg-04	1.50	0.04	11.58	0.53

In the MgMg-series the influence of the forging force has been evaluated for a welding pressure of 1.03MPa. Table 5 shows that as a result approximately the same welding time has been obtained for the three welding conditions (i.e. MgMg-01, 02 and 03).

On the other hand it was observed that the real burn-off increased significantly with forging pressure. On an average the burn-off during the stopping time was 2.42, 4.22 and 6.64mm as the forging pressure was increased, playing an important role if the final length of the pair is considered. Figure 2 shows the real burn-off versus the forging pressure.

Table 5 also shows that in the Mg-series a higher rotational speed cause a slightly increase in the welding time when the pressure was set at 1.03MPa (groups Mg-01 and Mg-02). This effect was not observed when the pressure was set as 1.72 MPa (groups Mg-03 and Mg-04) where the average welding time was exactly the same. Another characteristic that can be observed in Table 5 is the decrease on welding time when the welding pressure was increased. Burn-off was also increased with rotational speed since the stopping time increase with rotational speed due to the inertia of the system.

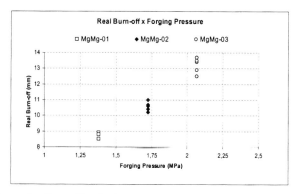

Figure 2. Dependence of real burn-off on forging pressure.

3.2 Thermal Cycle Measurements

The minimum and maximum temperatures reached during the welding process itself were recorded with a frequency of 10Hz. Therefore maximum temperatures, as presented in Table 6 for MgMg-series, could be evaluated with a precision of one tenth of a second.

Table 6. Average maximum temperature achieved during the weld.

Samples	Thermocamera (°C)	Thermocouples 5mm	Thermocouples 10mm	Welding Time (s)
MgMg-01	487,78	551.91 °C	420.82 °C	3.03
MgMg-02	476,79	550.52 °C	428.45 °C	3.20
MgMg-03	480,31	562.26 °C	418.04 °C	3.14

T1: Time between the beginning of the weld and the maximum peak of temperature.

For the MgMg-series no significant differences could be observed since those parameters are more related to welding pressure, which was constant for all three parameter sets. Forging force and welding time cannot be evaluated together with maximum temperature since forging force is applied when the welding time/burn-off is achieved. Therefore the values found for this group of specimens seemed to be close to each other and to vary randomly.

On the other hand changes on the temperature parameters can be easily observed for the Mg-series. When the pressure was fixed at 1.03MPa the total input of energy increased with the rotational speed, since the welding time was clearly longer. The above mentioned result can be seen on Table 6 comparing the groups Mg-01 and Mg-02, welded at 3000 and 6000rpm respectively. The longer welding time and consequently the higher average maximum temperature (out of 5 samples) reached by the welds made within group Mg-02 suggests a higher total energy input on those samples. In contrast to the previous analysis, if groups Mg-03 and 04, welded at 3000 and 6000 rpm respectively are compared, it could be observed that the temperature and therefore the total energy input at a pressure of 1.72 MPa decreases with rotational speed. Based on the results presented above, it can be concluded that the temperature varies randomly with rotational speed within the analyzed range of parameters.

When analysing the effect of the welding pressure on the total energy input the results were clearer. For both pairs of weld groups (Mg-01 and 03, welded at 3000 and Mg-02 and 04, welded at 6000rpm) the temperature and consequently the total energy input decreased with the welding pressure, as shown on Table 6. This tendency was observed since in the second case the preset burn-off is achieved faster owing to the higher welding pressure. Therefore the welding time is shorter and less heat is generated during the process. Figure 3 shows that the longer the welding time is, the higher the temperature tends to be.

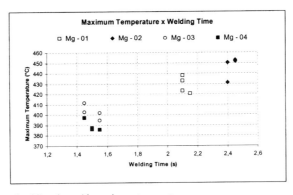

Figure 3. Variation of welding time with maximum temperature.

3.3 Metallographic Examination

Analysis of the macrographs of the group MgMg showed that sound welds with no porosity and without lack of bonding could be produced using friction welding. Bonding line is visible in the borders of the stud but not in the center, where a higher amount of plasticized material is concentrated. This effect can be explained by the lower radial forces and restricted material flow conditions closer to the rotational centre obstruct the hydro-extraction effect of both surface contamination particles and plasticized material generated in the welding interface centre. In the outer regions of the weld zone the material is easily pressed out of the friction surfaces, due to the high plastification and low deformation constrain. As the temperature of the material and the height of the plasticized zone increases, the resistance to the axial force reduces and the material is pressed out of the friction area forming the flash. These combined effects lead to an accumulation of plasticized material at the joint centre in one hand and to a defined bonding line with less plasticized material in the outer regions. Figure 4 shows an overview of sample MgMg-01.

For the sample MgMg-03, particularly, the bonding line can barely be seen even in the borders, i.e. the width of the plasticized layer is very thin in comparison to the weld carried out with lower forging forces (MgMg-01 and 02). On the other hand no significant difference can be noticed between samples MgMg-01 and 02, welded with a forging force of 1.72 and 2.07MPa respectively.

Figure 4. Overview of sample MgMg-01.

On the other hand a clear bonding line could not be seen in all four welded samples of the Mg-series duo to the intimate contact between the workpieces after the welding process. In this region no discontinuities like porosity or lack of bonding could be found through the entire diameter of the weld. As observed also within the MgMg-series, the weld interface shows the pattern of a typical friction welded joint. The microstructure at the bonding line is characterized by a thin layer of dynamically recrystalized grains without any special particles or phases precipitated in the grain boundaries (verified by EDS). Analysis of macrographs of all samples showed that the height of the plasticised layer in the center of the stud was always bigger when compared with the width of the layer at the borders. In order to evaluate the influence of the welding parameters on the shape of the bonding line, the width of the plasticised layer, i.e. the area where dynamically recristallization has taken place, was measured through the whole welding interface. According to the measurements carried out on the light microscope it was also clear that the rotational speed plays an important role on the amount of heat generation. In welds made with lower rotational speeds it was observed a noticeably higher amount of plasticised material if compared with welds made with higher rotational speeds. On the other hand it was not possible to observe a significant change in the welds made with 40 and 60MPa. Table 7 lists the results of the measurements carried out on the welded samples. Figure 5 presents a sketch representing the measured width and extension of the plasticized zone.

Table 7. Height of the plasticised layer along the welding intereface.

Position	Height of the plasticized zone (μm)			
	1	2	3	4
-10.0	580	410	385	290
-7.5	760	465	270	415
-5.0	840	410	210	665
-2.5	1015	375	825	1025
0.0 (Center)	1570	740	1020	1110
2.5	1130	850	740	865
5.0	870	420	210	575
7.5	680	280	200	445
10	405	365	230	435

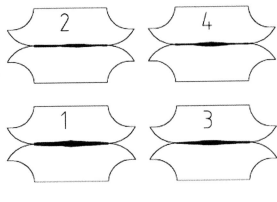

Figure 5. Bonding line shape x welding speed.

Since the relative speed and therefore the heat generation is noticeably higher in the outer regions of the diameter, it was found, as expected, that average grain size was always smaller in the center than on the periphery of the weld (see Table 8). Although welding time and maximum temperature have shown different values on the samples welded at low pressures, grain size was very similar in both cases. The major difference in grain size measurement was observed between specimens welded at high pressures. According to these results, rotational speed seems not to have a significant influence on the grain size of the welding zone. On the other hand it was observed that the pressure plays an important hole on the development of the microstructure in these alloys. In both cases the higher the welding pressure, the smaller the average grain size. Comparing the groups welded under different conditions, Table 8 shows that generally samples with larger grain size reached the highest temperatures, i.e. the higher the welding temperature is, the bigger the grain size will be.

Table 8. Grain size measurements.

Sample	ASTM – Grain Size								
	Center			Average	Outer region				Average
	1	2	3		1	2	3	4	
Mg01	11.0	11.7	11.2	**11.3**	10.4	10.7	10.6	9.5	**10.3**
Mg02	11.2	10.9	11.9	**11.3**	9.5	10.4	10.4	10.9	**10.3**
Mg03	11.8	12.6	12.2	**12.2**	10.6	10.9	11.1	11.2	**11.0**
Mg04	11.2	11.9	12.1	**11,7**	10.7	10.4	10.4	10.5	**10.5**

3.4 Mechanical Testing

Tensile tests were carried out to confirm the suitability of friction weld for performing sound welds in the selected base material. Figure 6a shows joints made under different welding conditions in having mechanical properties similar to that of the base material since no sample failed at the bonding line and no joint faced the problem of ductility loss (at 100% joint efficiency). Tensile strength and elongation were found to be equivalent while the yield strength was in all the cases noticeably superior to that of the base material.

Although the influence of the forging force in friction welding of Mg alloys is reported to be very relevant in relation to joint formation [13], samples of MgMg-series have shown that the forging force did not play a significant role in influencing the tensile strength among the specimens. As already discussed in previous sections, welding parameter, thermal cycle and therefore resultant microstructure were very similar for these three samples. As it would be consequently expected, mechanical properties within the group were also very similar.

Similarly to the behavior previously observed, samples within Mg-group presented also an adequate tensile strength with a joint efficiency of 100% and loss of ductility under some welding conditions. As mentioned above, joints have shown similar tensile strength and elongation values with a clearly higher yield strength. Specimen Mg-04, where the lowest maximum temperatures were measured, presented the best result within this group, with clear higher tensile and yield strength values in relation to base material. Figure 6b presents three tested samples showing clearly that the failure has occurred in the base material far away from the welding region and from the HAZ, positioned in the middle of the sample.

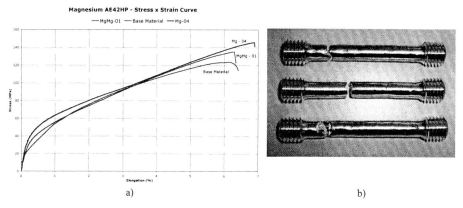

a) b)

Figure 6. In a) the Stress x Strain curves for MgMg and Mg-series in comparison with base material and in b) tensile samples with the failure positioned far away from welding region.

Although rotational speed is considered in having a lower influence on the formation of the weld interface [13], within this series, tensile strength increases clearly with rotational speed and has an undefined relation with axial pressure. Welds performed with lower rotational speed (Mg-01 and Mg-03) presented an efficiency of 98 and 90% with inferior elongation if compared with the base metal, while Mg-02 and Mg-04 presented a 100% joint efficiency. At lower speed an increase in welding pressure reduces the strength, while at higher rotation an increase in it causes a gain in tensile properties. Additionally and contrarily to the results reported on the literature [13], fully satisfactory bonded joints from the perspective of tensile strength were obtained with short welding times (<3s).

4. Summary

After the conclusion of this work, it was possible to conclude:

❖ The pair welding process/welding equipment presented a high reproducibility among the set of parameters investigated. The process itself was proved to be suitable and reliable in performing similar welds within the same set of parameters. Additionally it was observed that both the welding and forging pressure plays an important role in the course of the process and consequently in the formation and properties of the joint. On the other hand the rotational speed seemed not to influence significantly the welding process;

❖ Welding temperature varies randomly with rotational speed within the analyzed range of parameters. The effect of the welding pressure on the total energy input showed that the temperature and consequently the total energy input decreased with the welding pressure;

❖ Analysis of the macrographs of the group MgMg showed that sound welds with no porosity and without lack of bonding could be produced using friction welding. Weld interface in all cases shows the pattern of a typical friction welded joint. The microstructure at the bonding line is characterized by a thin layer of dynamically recrystalized grains without any special particles or phases precipitated in the grain boundaries. Analysis of macrographs of all samples showed that the height of the plasticized layer in the center of the stud was always bigger when compared with the height of the layer in the borders;

❖ Average grain size was always smaller in the center than on the periphery of the weld. Also rotational speed seems not to have a significant influence on the grain size of the welding zone. On the other hand the higher the welding pressure, the smaller the average grain size. Samples with larger grains reached the highest temperatures;

❖ Within the range of parameters investigated fully satisfactory bonded joints from the perspective of tensile strength were obtained. No sample failed at the bonding line and no joint faced the problem of ductility loss (at 100% joint efficiency). Tensile strength in this case increases clearly with rotational speed and was found to have an undefined relation with axial pressure.

5. Acknowledgements

The authors would like to thank both Mr. Hort for the fruitful discussions as well as Mr. Kree for the metallographic support. The research of the first author was partially sponsored by the Brazilian Council for the Development of the Research, CNPq.

6. References

[1] Bob R. Powell, Vadim Rezhets, Michael P. Balogh and Richard A. Waldo, JOM August 2002, 34-38.
[2] I.P. Moreno, T.K. Nandy, J.W. Jones, J.E. Allison and T.M. Pollock, Scripta Materialia 2003, 48, 1029-1034.
[3] H. Dieringa, A. Bowles, Norbert Hort and K.-U. Kainer, Materials Science Forum 2005, 482, 271-274.

[4] A. Bowles, N. Hort, A. Meyer, J.F. dos Santos and K.-U. Kainer, 6th International Conference on Magnesium and their Applications, September 2003, 917-923.
[5] K. Ogawa, H. Yamaguchi, H. Ochi, T. Sawai, Y Suga and Y Oki, Welding International, 2003, 17, 879-885.
[6] K. Kato and H. Tokisue, Welding International, 1994, 08, 452-457.
[7] U. Draugelates, A. Schram, B. Bouaifi and Chr. Kettler, 1998, Institute of Welding and Machining (ISAF), TU-Clausthal, Germany.
[8] K. Kato and H. Tokisue, Welding International, 2004, 18, 861-867.
[9] T. Shioya, S. Yamada and Y. Kurumatani, 1996, J Jpn Weld Soc 35(1), 65-71.
[10] H. Tokisue and K. Kato, 1978, J Jpn Weld Soc 28(9), 450-454.
[11] K. Kato and H. Tokisue, 1990, J Jpn Weld Soc 45(5), 351-355.
[12] M. M. Avedesian, H. Baker, ASM Specialty Handbook – Magnesium and Magnesium Alloys, 1999, p. 67.
[13] U. Draugelates and A. Schram, 2000, Institut für Schweißtechnik und Trennende Fertigungsverfahren, TU-Clausthal, Germany.

Development of Face Milling Process for Mg-hybrid (Mg-Al, Mg-sintered Steel) Materials

P. Ozsváth, A. Szmejkál, J. Takács
Budapest University of Technology and Economics (BME), Budapest
M. Eidenhammer, F. Obermair
Profactor Produktionsforschungs GmbH, Steyr

1 Introduction

In the automotive industry a general desire is to reduce the current energy consumption. One way to achieve this is the use of light weight metals like Mg and its' alloys. Application of magnesium in commercial engine blocks has been also extended especially in hybrid material constructions (Mg-Al and Mg-Sintered steel). [3]

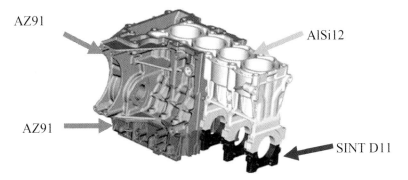

Figure 1: The scheme of a Mg-hybrid engine block and the experimental modelling materials [1]

The hybrid material is advantageous due to its low weight combined with high strength and wear characteristics. Increased difficulties in machining are the main disadvantage of hybrids, particularly Mg–Sint due to the high temperatures of sinter metal chips, which can ignite the flammable Mg chips. Also the different cutting forces between the materials call for detailed investigations of the cutting tools, their cutting edges and coatings, stable machine tools are also a necessity. In frame of an EU6 joint research program (ECOHYB) authors and further project partners targeted to develop an efficient, economical, safe and environmentally friendly machining process of magnesium-based hybrid materials. Specially designed tools, machine with safety concept and lubricants have been developed.

2 Research Tasks

The major technological tasks of the project can be grouped around the next topics. Cutting tool optimization for simultaneous face milling of the different hybrid couples was the first fundamental step. An experimental machine prototype which is capable of safe machining of Mg-hybrid parts belonged to the tasks as well as the optimization of cutting technology and lubricants. Important deliverable was the formulation of guidelines of worker's protection rules and environmentally friendliness.

In the present paper tool optimization, development of machine prototype and cutting technology tasks and achievements will be presented. Lubrication technology has been developed by the Spanish Brugarolas S.A. and Tekniker Fundation.

3 Tool Development

Determination of most suitable insert materials for both experimental Mg-hybrids, then the optimization of cutting geometry concerning cutting forces, chip temperature, surface roughness and chip formation was performed via several milling experiments. Design and manufacturing of milling head prototype was done according to the experiences of technological tests.

3.1 Determination of Cutting Edge Material

In general AZ 91 is well machinable with Al cutting geometry but at low cutting speed range build-up edge can develop as well as on the flank and the rake surface. Due to this the lowest possible cutting speed is limited. [2]

AlSi12 is also easy to cut but since the high silicon content the wearing of the tool is significant. This means that cutting speed is limited from top values. [4, 6]

The machinability of SD11 sintered steel is five times worse than magnesium. Wear is significant on the cutting tool that means upper limit of cutting speed. [5]

In commerce available inserts were tested in the milling experiments in order to choose the most suitable ones. AZ91+AlSi12 hybrid specimens were tested with 12 different insert types:

- uncoated cemented carbide inserts ($\gamma = +25°$)
- uncoated, polished cemented carbide ($\gamma = +25°$)
- coated cemented carbide inserts with Al geometry ($\gamma = +25°$)
    - conventional diamond coating
    - nano diamond coating
    - TiAlN coating
- PCD insert ($\gamma = 0°$ and $\gamma = +5°$)
- Thick diamond film coated insert ($\gamma = 0°$)
- In case of AZ91+SD11 hybrid the number of was lower and 6 different types were tested:
- uncoated cemented carbide
- coated cemented carbide ($\gamma = +7°$ and $\gamma = +25°$)

The most suitable cutting material (insert) was selected according to experiment series, where the measured data and objectives of investigation were cutting force, surface roughness of milled (hybrid) specimens, temperature of the SD11 chips; chip formation was also evaluated.

Measuring equipment:
- Kistler force measuring system (F_x, F_z, F_y) (Figure 2)

- Data acquisition with Test Point software, data evaluation with a special own program
- Mitutoyo Surftest 301
- Agema THV® 880 LWB IR camera

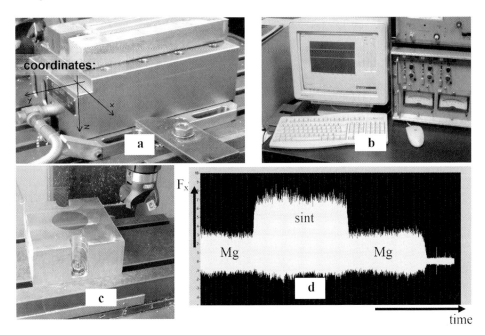

Figure 2: Kistler force measuring system (a/ and b/) AZ91+SD11 specimen (c/), typical force diagram (d/)

General principles of the face milling experiments:
- Machining with one insert
- Symmetrical positioning of milling head, $a_e = 2/3 \cdot d$ or $a_e = 1 \cdot d$
- Fixed cutting depth: $a_p = 1$ mm
- Fixed cutting speed e.g.: $v = 330$ m/min or 134 m/min
- Feed/tooth was the altering cutting parameter: $f_z = 0{,}05 \; 0{,}1 \; 0{,}2$ mm/tooth
- Tests was carried out on: AZ91, AlSi12, SD11, AZ91+AlSi12- and AZ91+SD11 hybrid

Because of high flammability of magnesium the temperature of sintered steel chips should not exceed 250°C. The chip temperature was measured with IR camera in the plane of the cut surface with line scanning mode (Figure 3), developed by BME. The cutting temperature remains under the limit when TiAlN coated cemented carbide insert with Al geometry is applied. (Finally this insert was considered to be most suitable for AZ91+SD11 hybrid.)

Figure 4 shows some kind of chip formation possibilities when different inserts are applied for milling. Favourable chip shape is fundamental for reliable chip suction which ensures safe machining conditions. In Figure 4 is visible that short and compact chip formation of AZ91 and SD11 was achieved with LC610 insert.

As result of experiments the mostly recommended cutting materials for AZ91 + AlSi12 hybrid are uncoated cemented carbide insert ($\gamma = +25°$); CVD diamond coated insert ($\gamma = 0°$) or with nano diamond coated cemented carbide insert.

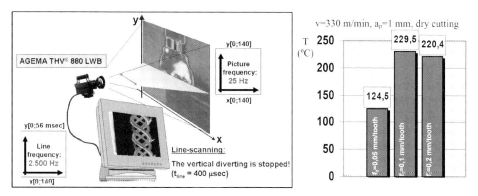

Figure 3: Scheme of line scanning mode of IR camera (left), chip temperature of SD11 with a suitable insert

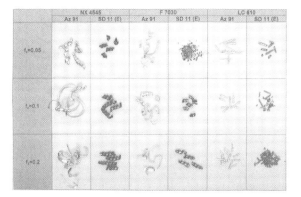

$v = 134$ m/min
$a_p = 1$ mm

Figure 4: Comparison table of chip formation when different inserts are used for dry milling of Mg+Sint

TiAlN coated cemented carbide insert ($\gamma = +25°$) proved to be the most suitable for milling of AZ91 + SD11 steel hybrid.

3.2 Cutting Geometry

The radial and axial rake angles were optimized for recommended cutting inserts according to further milling tests, detailed in point 3.1. For geometry optimization a special experimental milling head was developed which made possible to realize several cutting geometries. This milling head had four different seats with various axial rake angles (γ_p), and three different build-in-tools were developed with various axial and radial angles (γ_p / γ_f). The cutting geometry of face milling was determined by the insert–, the build-in-tool– and the seat geometry of milling head.

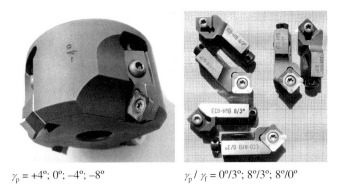

$\gamma_p = +4°; 0°; -4°; -8°$ $\gamma_p / \gamma_f = 0°/3°; 8°/3°; 8°/0°$

Figure 5: Special experimental milling head and build-in-tools from project partner LOSONCZI Ltd.

The optimal rake angles of the seat were determined separately according to the evaluation of experimental results (F, R_a, temperature, chip formation) for milling of AZ 91+AlSi12 with uncoated HW insert and AZ 91+SD11 when using TiAlN coated HW insert.

4   Machine and Tool Prototype for Machining Magnesium Hybrid Parts

The design of the tool was done at the Hungarian Losonczi Ltd. on the basis of optimization tests of cutting geometry. The focus of conception was: light metal body with adjustable build in tools and torx clamped inserts, wear resistant chip deflector (to protect the aluminium body), MQL cooling using central hole. The tool joints to the machine with HSK holding system.

The basis of the new milling machine prototype was an Anger HCP machine. It was modified in the frame of the project to reach requirements of safety and increased mechanical stability. [9]

As mentioned previously primary desire of Mg machining is the reliable chip removing system. The modified machine is equipped with a special suction device, which is suitable for removing not only chips but even more hazardous magnesium dust also.

Blowing-off system was designed with swirl nozzles for chips on flat areas, adapted to requirements of HCP machine.

Figure 6: Milling head prototype and modified machine for milling of Mg-hybrid materials

Fire detection and extinguishing system is responsible to prevent the machine if magnesium ignition occurs. The system consists of pressure relief shutters, flame arresters, infrared and thermal fire detection sensors and usage gas for extinguishing the fire.

5 Cutting Technology

On the basis of the technological tests of developed tool and machine the following results arose. Figure 7 and 8 well display the difference of technological conditions between the hybrid components. The results of the roughness measurements are very good. With magnesium and aluminium as well as magnesium and sintered steel the values especially for finishing are very low what means R_a 0,6 μm.

Also the measured cutting force in Z-direction confirmed the former gained results. Due to the optimized cutting material and geometry the tool life is absolutely reasonable if experimented technological data are applied for the milling.

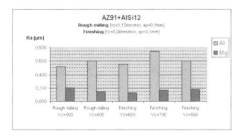 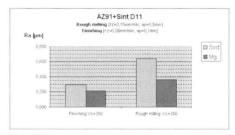

Figure 7: Surface roughness results after rough milling at the two Mg-hybrid types

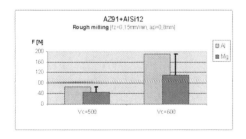 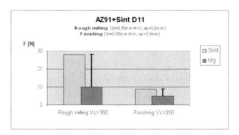

Figure 8: Cutting force (Fz) during rough milling at the two Mg-hybrid types

6 Conclusion

The pre-determined aims of development of Mg-hybrid milling have been reached. As a result of the research optimized tool, modified machine and optimized technology is available for economical and ecological machining of Mg-Al and Mg-SD11 hybrid materials.

Acknowledgement: authors are grateful for the support of EU6 CRAFT Framework Programme, project no. COOP-CT-2003-508452.

7   References

[1] J.-M. Segeud (BMW), in *Giesserei* **2004**, *9*, pp. 102–104.
[2] K. Weinert et.al., in WILEY-VCH Verlag **2003**, pp. 130–152.
[3] Z. Pál, A. Szmejkál, J. Takacs, in *22nd International Colloquium on "Advanced Manufacturing and Repair Technologies in Vehicle Industry"*, **2005**, pp. 79–84.
[4] D- Szablewski et. al., in *Annals of the CIRP*, Vol 53/1/**2004**, pp. 69–72.
[5] Benner, Diss. WZL Aachen, **2003**
[6] J. V. Grams, Diss. WZL Aachen, **2004**
[7] J. Wassenhoven, http:/home.wtal.de/wasserhoven
[8] F. Obermair, G. Klammer, Ecological end efficient High-Speed-Machining of Magnesium Lumbarda, CIM **2003**
[9] M. Eidenhammer, in *13. Österreichische HSC-Tagung* **2005**

Arc Brazing of Magnesium Alloy AZ31 with Steel (DX 53 Z)

G. Garg, J. Zschetzsche, E. Simmchen, M. Schaper, U. Fuessel, S. Pandey[*]
Technical University of Dresden, D-Dresden
*Indian Institute of Technology Delhi, India

1 Introduction

Magnesium is the lightest of the structural materials. It is about 1/4th the density of steel and 2/3 the density of aluminium. Due to its light weight and favourable strength it finds its use in the automobile industry, aerospace, defence, domestic & outdoor appliances, electronics, processing equipments, sports and other industries, but the scale of application remains limited [3,7].

Magnesium industry has been making great efforts to educate the automotive industry on the benefits of utilizing magnesium to reduce vehicle weight, cost, and complexity. According to the data it has been observed that with about 10% saving in the weight, there can be fuel savings of 20 to 30 %, without bringing about any major changes in the design [4,14]. With about 150 gm/km of exhaust produced by a new car, this exhaust can be brought down to 100–120 gm/km with the use of magnesium technology. Considering the large number of vehicles in the world, this weight saving could lead to a significant reduction in the amount of carbon dioxide released in the atmosphere. This would help in reducing the impact on global environment, in agreement with Kyoto protocol.

Many of the applications envisaged for magnesium alloys in automobiles, for example seat and door frames, will require the magnesium component to be joined to dissimilar metal sheets [12]. The most preferred metal joining technique remains brazing as it can be easily adapted in the manufacturing units with existing welding robots. But brazing of magnesium alloys is a far less explored process in the field of metal joining technology and very less information is available on this topic. Also since the alloys find extensive use in military products and aerospace, most of the research remains classified.

The primary objective of this work is to braze magnesium alloy AZ31 with steel. To achieve this objective it would be required to investigate for a suitable solder. For successful brazing of the joint, it is also necessary to evaluate the suitable parameters for the MIG (Metal Inert Gas) process. While defining the objectives it was felt necessary to conduct FEM simulation of the process in order to determine the brazing process parameters. Also, microscopic analysis and preliminary mechanical testing of the joints would be done to characterize the joints formed.

2 Experiments

It was decided to braze AZ31 magnesium alloy sheet with DX53Z steel sheet in lap joint configuration. Dimensions of the test material and their properties are as follows

Table 1: Material dimensions and properties for magnesium alloy and steel sheets

Dimensions	AZ31 Mg Alloy	DX 53 Z Steel
Length (mm)	200	200
Width (mm)	100	100
Thickness (mm)	1.6	2
Physical Properties		
Specific gravity	1.8	8.0
Melting point (°C)	565–635	1530
Thermal conductivity (W/m-°C)	96	16
Thermal Expansion (°K^{-1})	26	17
Specific Heat (J/Kg-K)	960	520
Heat Capacity (J/m^3-K)	$1.9 \cdot 10^6$	$4.0 \cdot 10^6$
Yield Strength, $R_{p0,2}$ (MPa)		260 (max)
Tensile Strength, R_m (MPa)		380 (max)

2.1 FEM Simulation

In the process of brazing, to avoid the precipitation of brittle intermetallics and the evaporation of zinc in AZ31, it is very important to control the maximum temperature reached in AZ31 sheets. This is possible only when the temperature distribution in the sheets for a given set of brazing parameters (wire feed rate, voltage, peak current, base current, pulse frequency, pulse duration, brazing speed and wire diameter) could be known. With the help of FEM simulation software ANSYS, the process of brazing was simulated for a given set of parameters and the temperature fields in the magnesium and steel plates were calculated. A 3D model was constructed and meshed using solid elements SOLID90. Figure 1 shows a meshed model of AZ31 and steel sheets brazed in lap joint.

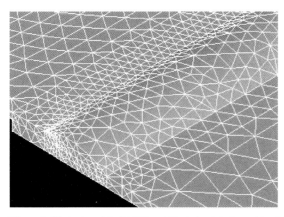

Figure 1: Meshed model of AZ31 and Steel sheets in brazed lap-joint

2.2 Brazing

Brazing would involve identification of suitable brazing solder amongst the available ones viz. AlSi5, AlMg5, Al (99.9%) and AZ61. Pulse MIG brazing process and continuous MIG brazing process would be used for the experiments. Brazing parameters estimation was done with the help of FEM calculations. However, some minor adjustments in the parameters was often needed. The brazing of magnesium alloy AZ31 with steel was carried out in the welding laboratory using MIG process. Bosch RS 60/S robot was used with the power source Cloos GLC 553 MC3. The fixture was clamped to the bed. The brazing torch was clamped to a programmable robotic arm which could be moved in all directions.

2.3 Microscopy and Mechanical Testing

Microscopic analysis of the joint was done with the help of scanning electron microscopy (SEM). It was aimed at investigating the primary cause of failure or success of the joint. Test specimens were prepared to determine the strength of the brazed joint under tension. The length of the specimens was 100 mm. The cross section area of the magnesium strip was measured to be 16 mm^2. The area of the lap joint was 100 mm^2. The specimen were mechanically clamped in the universal testing machine and the loading speed was set at 0.5 mm/min. Initial load was set to 10 N.

3 Results

3.1 FEM Simulation of the Brazing Process

FEM simulation of the brazing process was done for the evaluation of brazing parameters. This objective was accomplished by building a geometric model, meshing the model differentially, loading the model for a chosen set of brazing parameters and running the solution. The results were then viewed in the form of temperature fields illustrated by colored contours and time-temperature curves (figure 2, 3).

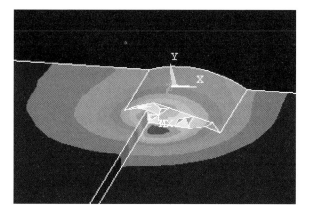

Figure 2: Temperature contours during brazing process

Temperature fields are highly sensitive to the rate of heat input per unit area. It is known that heat distribution under an electric arc is not uniform and it follows a normal distribution. Therefore, a heat distribution function was incorporated in the analysis which is expressed as

$$q(x,y) = \frac{3q}{\pi r_{0.05}^2} \exp\left(-\frac{3(x^2 + y^2)}{r_{0.05}^2}\right)$$

$$r_{0.05} = \sigma\sqrt{\frac{3}{2}}$$

$$3\sigma = r$$

$$q = \eta \times V \times I$$

where,
$q(x,y)$ is the heat distribution function in xy plane,
q is the total heat input from the source,
$r_{0.05}$ is that radius of the arc within which 95 % of the total heat input is concentrated,
σ is the standard deviation of the heat distribution,
η is the arc efficiency factor,
V is the arc voltage and
I is the arc current.

Time temperature curves were obtained by picking an element from the model and calculating the temperature change over time. The result is illustrated in figure 3. The curve shows the change of temperature in an element with time. The temperature at the start of the process is equal to room temperature and it starts rising as the brazing torch comes closer. The temperature reaches the peak when the torch is closest to the element. There after the element starts losing temperature exponentially with time. It was found that the brazing parameters for a maximum allowable temperature of 680 °C, were optimal at voltage $V = 24.3$ volts, base current $I_b = 15$ A, peak current $I_p = 30$ A, pulse frequency $f = 25$ Hz and pulse duration t of 1.1 ms.

3.2 Brazing

Brazing with solder AlSi5, AlMg5 and Al (99.9%) was not successful. Formation of brittle intermetallics with aluminium was observed which led to immediate joint failure. Scanning electron microscopy images (SEM) illustrate the braze-joint cross-section with above mentioned solders. In the joint cross-section with AlSi5 solder (figure 4a), large pores along with fragmented parts of the bead, indicative of brittle behavior can be observed. The same can be observed with commercially pure Aluminium with cracks running through whole of the cross-section (figure 4c). With AlMg5 solder (figure 4b), cracks running through a heterogeneous distribution of phases can be seen with the solder failing to bond to the base metal.

With pulse MIG process, satisfactory deposition of AZ61 could not be achieved. The problem was due to vaporization of zinc. During the process the globular metal transfer was observed and these globules were getting overheating with the pulse effect. This resulted in zinc vaporization and molten AZ61 globules often exploded leading to spatter. However, strong bonding of AZ61 solder with steel was observed. Pulse MIG parameters were changed. Wire feed rate was decreased from 2.3 m/min to 1.6 m/min. Pulse frequency and pulse duration were decreased from 60 to 40 Hz and 1.5 to 1.22 ms respectively. No major improvement was observed. Another set of parameters was tried to bring down heat input and prevent spatter but no success could be achieved.

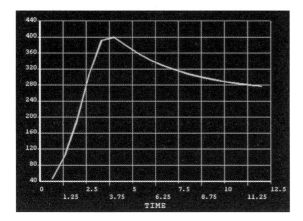

Figure 3: Time-Temperature curve for a brazing process

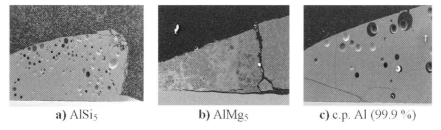

a) AlSi$_5$ b) AlMg$_5$ c) c.p. Al (99.9 %)

Figure 4: SEM images of braze joint cross sections with different solders

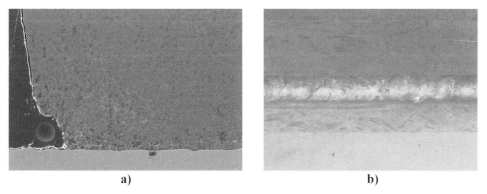

a) b)

Figure 5: a) SEM image of the braze joint with AZ61 solder; b) AZ61 braze joint illustration

The process was changed to continuous MIG. The effect was immediate as the metal transfer mode changed from globular transfer to short circuit transfer. Zinc didn't vaporize due to immediate cooling effect of the contact with the steel sheet and this allowed uniform and smooth solder deposition (figure 5b). Due to thicker diameter of the feed wire and high feed rate, a higher current of 100 A had to be used. The braze velocity was kept the same at 800 mm/min. The SEM image shows a fine bonding between magnesium alloy and steel plates without any visible signs of cracking or porosity (figure 5a). A uniform braze width indicative of stable parameters and process can be seen.

3.3 Mechanical Testing

Tensile test of the specimens from AZ61 solder joint was done on universal testing machine. Specimens of identical dimensions were prepared for the test. The lap area of the braze joint was very large as compared to the cross sectional area at the magnesium solder interface. The chances of failure of the specimen in tension were far more than in shear along the lap joint. Hence it was more logical to consider the tensile test for the specimen rather than lap shear test.

Figure 6: Specimen in fracture after tensile test

The specimen sustained a maximum load of 203.75 MPa before undergoing fracture in the magnesium strip (figure 6). The braze joint did not suffer any fracture. The extension of the specimen under this load was measured to be 0.52 mm. With clamping of some of the specimens, the magnesium strip was squeezed outwards resulting in high compressive stress along the length of the strip. This led to buckling and those specimens failed without taking any load in tension.

4 Discussion and Conclusions

FEM simulation of brazing process was successful. For a maximum allowable temperature of 680°C, optimal brazing parameters could be calculated. These parameters were used for the brazing process. However, these parameters were varied during brazing to stabilize the process, but the amount of heat input was always kept the same.

Experiments show that the brazing of magnesium alloys with aluminium based solders is not possible. The literature survey also confirmed that aluminium based solders reacts immediately with magnesium to form brittle intermetallics [1,2,9]. Although exact phases could not be determined but in case of aluminium based solders, presence of gas pores, micro and macro cracks, large grain sizes and improper bonding to steel surface was a common observation. The brazing parameters were varied to stabilize the process, but no success could be achieved.

Brazing with AZ61 solder was successful. Continuous MIG process was used instead of pulse MIG process. Pulse MIG parameters were found to be unsuitable for the process using

AZ61 solder. It was seen that in continuous MIG process the amount of current used is very high (100 A) compared to the pulse MIG but, it is to be observed that in short circuit transfer a heavy amount of current is transferred through short circuiting and not through the arc [8]. The arc remains more or less short and intermittent, so the overall heat transfer to the plates is comparatively very less. At the solder-steel interface it was seen that the contact angle is more than 90 degrees. This could be because of improper wetting of the surface [8]. The wetting ability of magnesium alloy AZ61 could be improved by addition of gallium.

The mechanical testing of test specimens show high bonding strength between steel and the solder. Despite the breaking of the test specimen because of buckling on improper loading, other specimens broke through the magnesium strip and not through the solder joint. This shows that the actual strength of the joint is far more than 203.75 MPa and more comprehensive tests need to be done.

5   Acknowledgements

This work is a collaboration of Technical University of Dresden (TUD), Germany and Indian Institute of Technology Delhi (IITD), India, organized by the German Students Exchange Service (DAAD). We thank DAAD for providing the opportunity its support.

6   References

[1]   Beck Adolf, *Magnesium und sein Legierungen*. Berlin, Verlag von Julius Springer, 1939
[2]   Busk Robert S., *Magnesium Products Design*. New York, Sponsored by the International Magnesium Association, Mercel Dekker Inc, 1987
[3]   Colasanti E., *Design and development of a light weight seat frame using magnesium extrusions and stampings*. Proc. Conf. Attributes of Magnesium for Automobile Design, Detroit, 28 Feb–3 Mar. 1994
[4]   Eliezer D.; Aghion E.; Froes F. H., *Magnesium Science, Technology and Application*. Advance Performance Materials 5, 201–212, 1998
[5]   Hassen Peter, *Physical Metallurgy*. Cambrige University Press, 1997
[6]   Houldcraft Peter; John Robert, *Welding and Cutting, A guide to fusion welding and associated cutting processes*. London, Published in association with ESAB groups, Woodhead-Faulkner, 1988
[7]   Kainer K. U., *Magnesium, Eigenschaften, Anwendungen, Potenziale*. Weinheim, Wiley-VCH, 2000
[8]   Lancaster J. F., *Metallurgy of Welding*. London, Allen & Unwin, 1987
[9]   Lashko N.; Lashko S., *Brazing and soldering of metals*. Moscow, Mir Publishers, 1979
[10] Mordike B. L.; Ebert T., *Magnesium, Properties, Applications and Potential*. Material Science and Engineering A302, 37–45, 2001
[11] Mukai Toshiji; Yamanoi Masashi; Watanabe Hiroyuki, *Ductility enhancement in AZ31 magnesium alloy by controlling its grain structure*. Scripta Materialia 45, 89–94, 2001
[12] Schaper M.; Füssel U., *Innovative Fügetechnologien für Werkstoffsysteme mit Magnesium Legierungen (Mg-Al, Mg-Fe)*. Gemeinschaftsantrag im Rahmen des Schwerepunktsprogrammes 1168 "Erweiterung des Einsatzgrenzen von Magnesiumlegierungen", 2003

[13] Schubert E.; Klassen M.; Zerner I.; Walz C.; Sepold G., *Light-weight structures produced by laser beam joining for future applications in automobile and aerospace industry.* Journal of Material Processing Technology 115, 2–8, 2001

[14] Smith L. S.; Gittos M. F., *The Joining of Magnesium Alloys to Dissimilar Materials for Automotive Applications.* TWI Proposal, GP/MAT/1129-REV.1, Febraury 1998

Electromagnetic Compression of Magnesium Tubes and Process-related Improvement of Wrought Alloys by Micro-alloying

V. Psyk[1], A. Brosius[1], C. Broer[2], M. Bosse[2], M. Schaper[2], M. Kleiner[1], Fr.-W. Bach[2]
[1] University of Dortmund, Dortmund, Germany
[2] Leibniz Universität Hannover, Hanover, Germany

1 Introduction

Over the recent years an increasing interest in applying magnesium alloys in lightweight construction concepts can be observed, because of the low density of these materials combined with their high strength. Especially for the use as structural components in automotive and rail vehicle applications this is an important advantage. On the other hand, the limited possibilities in the forming and joining technologies are still an obstacle regarding the widespread implementation of magnesium alloys into industrial productions.

In this context the electromagnetic compression (EMC) represents an interesting technique. EMC is a non-contact high-speed forming process using the energy density of pulsed magnetic fields to form workpieces made of electrically highly conductive materials. The process principle is described in detail in [1] Typical strain rates achievable by this technology are in the range of 10^3 s^{-1}. The potential to form and join aluminum tubes and hollow profiles has been investigated thoroughly [1–4]. The interdisciplinary work of a basic research project within the Priority Program 1168 is dedicated to developing adapted magnesium wrought alloys for electromagnetic joining. Thereby, the process chain "Casting of billets – Extrusion of tubular components – Electromagnetic compression" is regarded. In this project EMC of tubes made of standard magnesium wrought alloys is analyzed and on this basis process relevant material properties are defined. Micro-alloying techniques are used to improve these properties and the developed materials are characterized and compared to the standard magnesium alloys, on the one hand, and to reference aluminum alloys, already successfully processed by means of EMC, on the other hand.

2 Electromagnetic Compression Process and Relevant Material Parameters

In order to investigate to what extends the knowledge about the EMC of aluminum tubes can be transferred to magnesium, comparing experiments using the magnesium alloy AZ31 and the aluminum alloy AA5754 were carried out.

2.1 Analysis of the Free Electromagnetic Compression Pprocess

For the analysis of the EMC especially the deformation of tubular profiles (Ø40 × 2 mm) without a form defining tool (free forming operation) was considered. Thereby, apart from the measured coil current, especially the deformation of the tube's smallest cross section was regarded. Here, the displacement can be recorded during the process using an optical measurement method that is based on a shadowing principle (compare [1]). By differentiating the measured displacement over time curve, the forming velocity of the tube's relevant cross section can be determined. **Figure 1** shows the most important differences between compression processes performed on magnesium and aluminum tubes. Experiments applying the same charging energy are considered as well as tests leading to the same compressed diameter. In all cases the deformation shows a principle course well known from investigations on aluminum tubes [1]. In the beginning the current and consequently the pressure rises, but no workpiece deformation occurs, until, shortly before the current maximum is achieved, the acting magnetic pressure overcomes the resistance against deformation due to material strength, geometrical stiffness and mass inertia. At this point of time, the plastic deformation process, characterized by a rampant section of the curve, starts. Subsequently, a radial oscillation of the tube can be observed.

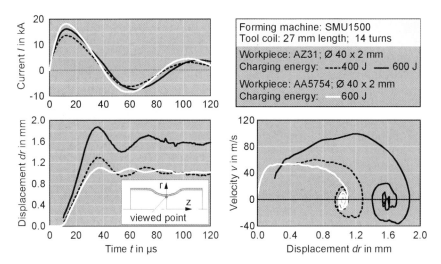

Figure 1: Current and displacement courses of compression processes on aluminum and magnesium tubes

Comparing the experiments applying the same charging energy, the higher electrical conductivity of AA5754 (σ 17 MS/m) leads to a lower common inductivity of tool coil and workpiece than in case of AZ31 (σ 12 MS/m). This correlates to a faster rise and higher amplitude of the current and, thus, of the pressure. Nevertheless, the deformation process starts earlier and the achieved forming velocity as well as the final displacement is higher in case of AZ31. Since the stiffness of the workpieces is equal and the strength of AZ31 is slightly higher, only the significantly lower density of AZ31 can be the reason for this behavior (material parameters determined by quasistatic tensile tests on tubular specimens: AA5754: yield stress $R_{p0,2} \approx 135$ MPa, tensile strength $R_m \approx 250$ MPa; AZ31: yield stress $R_{p0,2} \approx 165$ MPa, tensile strength $R_m \approx 260$ MPa).

Regarding the EMF processes leading to nearly the same compressed diameter, the comparison of the displacement curves and deformation velocities shows similar values during the plastic deformation phase, but a remarkable difference can be observed in the radial oscillation period. Here, all investigated magnesium profiles feature a noteworthy high amplitude and low frequency, which changes over the time since the oscillation is damped.

2.2 Conclusions Considering Joining Applications

The described distinctive oscillation behavior will supposably cause difficulties regarding joining operations, because the principle of force fit joints bases on an elastic (-plastic) bracing of tube and mandrel [3]. The joining partners are positioned and a radial pressure is applied so that the tube aligns to the mandrel and subsequently both joining partners are deformed together up to a maximum radial displacement. Then the applied pressure is released so that the elastic deformation of both partners is reversed. Only if the reversal of the mandrel's elastic deformation is partly avoided by the tube, an interference fit assembly is created.

The observed high amplitude of the radial oscillation in case of free EMC on magnesium tubes indicates a distinctive elastic behavior of these materials after the deformation. Thus, a force fit joint is only possible if the material of the mandrel allows extreme elastic deformations. Joining experiments have validated this expectation, showing that force fit joints could only by realized using elastomer mandrels and that even in these cases the achieved release forces were extremely low. In contrast to this, experiments using mandrels equipped with differently shaped grooves in order to produce form-fit joints have shown much more promising results. During the compression process the tube material flows into the grooves and forms an undercut, so that higher release forces can be achieved for elastomer as well as metallic mandrels. But the locally required formability is much higher compared to force-fit joints so that in many cases the formability of AZ31 will not be sufficient. Therefore, a purposeful optimization of the mandrel geometry requires the availability of magnesium alloys especially optimized with regard to the formability. Hence, the electrical conductivity is required to be as high as possible in order to achieve a good process efficiency [8].

3 Optimization of Material Properties by Micro-alloying

As described in the previous section, the most important aspects for improving magnesium alloys for EMC are the electrical conductivity and the formability. The formability can be increased by adding grain refining alloying elements e.g. calcium [6], but higher concentrations of alloying elements cause a noticeable decrease of the conductivity [7]. Thus, the conductivity of pure Magnesium (σ 22 MS/m) represents an upper limit for the achievable values of all magnesium alloys. Starting from base alloys AZ31 and lowly alloyed ZEK100, an optimum concentration of alloying elements has to be found considering the contradicting demands of formability and electrical conductivity.

3.1 Electrical Conductivity

Table 1 shows electrical conductivities measured by eddy current method (DIN EN 2004-1) at 20 °C for different magnesium alloys. The high concentration of alloying elements (especially aluminum) results in a relative low conductivity of AZ31 compared to pure magnesium. In contrast to this, the lowly alloyed ZEK100 shows a high value absolutely comparable to the aluminum alloy AA5754 (σ 17 MS/m) which was successfully processed by EMC [2]. The addition of calcium up to a concentration of 0.6 wt% in both cases has no significant influence on the electrical conductivity.

Table 1: Electrical conductivity of several Mg-alloys

Magnesium alloy	Electrical conductivity
Pure Mg	σ 22 MS/m
AZ31	σ 12 MS/m
AZ31 + 0.2 % Ca	σ 11 MS/m
AZ31 + 0.6 % Ca	σ 12 MS/m
ZEK100	σ 18 MS/m
ZEK100 + 0.2 % Ca	σ 18 MS/m
ZEK100 + 0.6 % Ca	σ 18 MS/m
ZME111	σ 16 MS/m
WE43	σ 6 MS/m

3.2 Microstructural Investigations

A significant grain refinement after the extrusion of tubes of AZ31 with 0.2 wt% Ca was observed in [6]. Due to the limited suitability of AZ31 and the results of measured electrial conductivity the base alloy ZEK100 was chosen for further microstructural investigations. Hence the influence of nucleation on the grain size by adding calcium to ZEK100 alloy was analyzed. In labor scale cast billets were extruded (extrusion ratio $R = 7.6:1$) and samples taken from extruded rods. Besides the typical tensile and compression anisotropy the mechanical properties show an improvement of the strain rate at rupture in compression tests from approximately 10 % to 15 % for ZEK100 by addition of 0.8 wt% calcium. In the next step the extrusion of hollow profiles of various alloys was carried out to observe the extrusion quality. Therefore ZEK100 billets were cast (Ø 100 mm) and tubes extruded (Ø 40 × 2 mm).

In **Figure 2** the microstructure of the ZEK100 alloy without and with addition of calcium can be seen in lengthwise extrusion direction. In case of the non-modified alloy a large amount of grains were stretched along the extrusion direction. The inhomogeneous grain size varies from 5 to 50 µm. The modification by calcium effects a significant grain refinement (3–25 µm) and a primarily globular grain geometry witch is show in **Figure 2** right. The reason for this effect is the cast constitution whereas calcium addition causes nucleation, so that the extrusion process starts with a diminished average grain size. The dynamic recrystallization proceeds homogeneously on the rim and in the center of the tube in both cases. The reduced mean grain size will improve the results of transformation by EMC and subsequent heat treating.

3.3 Formability under Quasistatic and Dynamic Conditions

The forming potential of semi-finished parts depends on the production history of the parts including all previous forming steps and on the current load case that the material is exposed to. Regarding the load case, especially the direction and the course of the acting forces are relevant. In case of EMC a highly dynamic pressure pulse is applied to the tubes causing mainly tangential compression stress. In order to get an impression of the formability of the different alloys, extruded tubes have been tested in quasistatic tensile tests and hydraulic burst tests. Thereby, the strain at failure under axial and tangential tensile stress was determined. Furthermore, several series of free electromagnetic compression processes have been performed, during which the charging energy and thus, the radial deformation was increased until material failure occurred. Subsequently, the compression of the smallest cross-section geometry was quantified using a coordinate measurement machine and on this basis the strain at failure was determined. The results of these different testing methods are summarized in **Figure 3**.

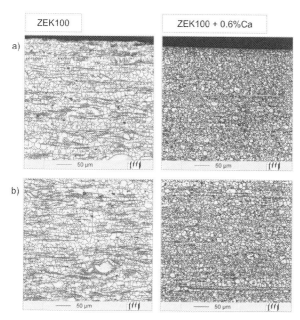

Figure 2: Microstructure of directly extruded tubes: a) outer rim, b) center

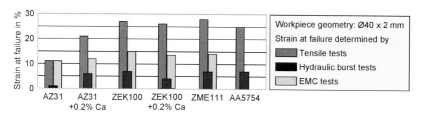

Figure 3. Formability of the different alloys determined by different testing methods

The comparison of the strains at failure determined, by tensile tests and burst tests, clearly indicates a strongly anisotropic material behavior for axial and tangential tensile load, which is typical for extruded semi-finished parts. The achievable maximum strain in case of EMC is relatively high compared to the burst tests although, due to twin-formation, the formability of magnesium alloys typically is lower in case of compressive stress than in case of tensile stress [5]. The increased forming potential can be explained by the high strain rates, which reduce localizing effects. Furthermore, the failure criterion, limiting the formability by EMC, marks another difference between the forming of magnesium and aluminum tubes. Regarding all investigated magnesium tubes, material failure occurred in the form of cracking, while in case of aluminum an increasing tangential strain is related to a stronger wrinkling effect and the forming limit is set by the allowable roundness tolerance [2]. So, no value is given for AA5754 in **Figure 3**, but as shown in [2], the material could be compressed to a tangential strain of approx. 20% with a roundness tolerance of ±0.5 mm above the radius.

Regarding the different alloys it could be shown that the forming potential of AZ31 is relatively low, but a significant improvement could be achieved by adding 0.2 wt% calcium. In case of ZEK100 and ZME111 the formability is even higher and absolutely comparable with that of AA5754. However, in case of ZEK100 the addition of 0.2 wt% calcium caused no further improvement but a slight deterioration of the formability. Therefore higher concentrations of calcium up to 1 wt% have to be investigated.

4   Summary and Outlook

To overcome problems in the forming and joining of extruded magnesium tubes, the application of electromagnetic compression for the processing of these materials was investigated. The analysis of compression processes without a form defining tool has shown that in order to achieve the same compressed diameter the lower density of magnesium tubes requires less charging energy compared to aluminum alloys. In both cases the forming behavior during the plastic deformation and the achieved forming velocities are quite similar, but the subsequent radial oscillation shows higher amplitude and lower frequency in case of magnesium profiles. This distinctive oscillation behavior causes difficulties regarding force-fit connections, so that form-fit connections seem to be more promising for the electromagnetic joining of magnesium profiles. In order to carry out detailed studies on this subject, improved magnesium alloys regarding the formability and electrical conductivity are necessary. The formability can be increased by adding alloying elements as calcium, but this causes a deterioration of the conductivity. Hence an optimum concentration regarding the contradicting demands of formability and electrical conductivity has to be found. Considering these aspects different extruded magnesium alloys have been analyzed and the potential of micro-alloying strategies to improve the material parameters could be shown. On this basis, a further optimization of the concentration of alloying elements e.g. calcium and zirconium for different basis alloys as AZ31, ZEK100 and ZME111 is required.

Another approach to adapt the properties of magnesium profiles to the demands of the electromagnetic compression process could be a purposeful reduction of the electrical conductivity up to a certain limit. In this case the resistive heating effect during the compression process will be increased which might lead to a thermal softening of the material and thus to an according increase of the formability.

5   Acknowledgements

The present study was performed by the authors within the priority programme 1168 "InnoMagTec" Project-No. Kl 619/24-1 and Project-No. Ba 851/59-1. The authors gratefully acknowledge the financial support by the German Research Foundation (DFG).

6   References

[1] Beerwald, C.: Fundamentals for Process Dimensioning and Design of Electromagnetic Forming (in German), PhD.-Thesis, Institute of Forming Technology and Lightweight Construction, University of Dortmund, 2004.

[2] Psyk, V.; Beerwald, C.; Homberg, W.; Kleiner, M.: Electromagnetic Compression as Preforming Operation for Tubular Hydroforming Parts, ICHSF 2004 - 1st Int. Conf. on High Speed Forming, Dortmund, Germany, 2004, Proceedings pp. 171–180, ISBN 3-00-012970-7.

[3] Kleiner, M.; Marré, M.; Beerwald, C.; Homberg, W.; Löhe, D.; Barreiro, P.; Schulze, V.: Investigation of force-fit joints produced by electromagnetic tube compression, Annals of the German Acad. Soc. for Prod. Eng., WGP, 2006, XIII/1.

[4] Eguia, I.; Zhang, P.; Daehn, G.: Improved Crimp-Joining of Aluminum Tubes onto Mandrels with Undulating Surfaces, ICHSF 2004– 1st Int. Conf. on High Speed Forming, Dortmund, Germany, 2004, Proceedings, pp.161 -170, ISBN 3-00-012970-7.

[5] Kammer, C.: Aluminium Taschenbuch – Band 1: Grundlagen und Werkstoffe, 15. ed., Aluminium-Verlag, Düsseldorf, 1995.

[6] Bach, Fr.-W.; Bosse, M.; Schaper, M.: Micro-alloying of Magnesium Wrought Alloys for Improved Electro-Magnetic Joining of Extrudes Hollow Profiles, TMS Annual Meeting; Magnesium Technology 2006, San Antonio, USA, pp. 265–269.

[7] Adolf Beck, Magnesium und seine Legierungen, 2. ed., Springer, Berlin, 2001.

[8] Psyk, V.; Beerwald, C.; Klaus, A.; Kleiner, M.: Characterisation of extruded magnesium profiles for electro-magnetical joining, Journal of Mat. Processing Tech., Vol. 177, Issues 1-3, 2006, pp. 266–269.

Efficient and Ecological Machining of Magnesium Hybrid Parts

C. Sanz, E. Fuentes
Tekniker, Eibar, Spain
F. Obermair
Profactor, Steyr, Austria
L. Muntada
Brugarolas, Rubí, Spain

1 Abstract

Historically, mechanical components are usually manufactured through the joint of different parts of similar or different metallic materials, offering the requested properties to the whole piece. However, new casting processes have the possibility to include other materials directly in the mould parts, for example aluminium (Al) into the final magnesium (Mg) structure. Another mixture is steel with magnesium, which also demands the machining of two different materials in one operation with different cooling-lubricant and machinability requirements.

Moreover safety problems arisen with magnesium during its processing limit widely the cutting parameters when combined with another materials. Therefore, safety equipment like explosion discharge flap, flame detectors or extinguishing systems have to be built into the machine tool.

This paper tackles the machining study from different points of view (cooling, lubrication, safety…) of the magnesium alloy "AZ91D" together with the aluminium alloy "AlSi12" and the sintered metal "Sint D11" respectively. The investigated cooling lubricant strategy is minimum quantity lubrication.

2 Introduction

In the last decade, some sectors, that manufacture structural components, are aimed at reducing energy consumption by replacing traditional materials like steel and cast iron with thermoplastics and light metals. Among the light metals commonly used in heavy manufacturing of structural components are Magnesium and Aluminium. Apart from their low weight, magnesium and their alloys have other important properties which qualify them as the optimal materials for many applications [1]. However, their corrosion and wear behaviour are not at the required level, so a recommended solution is to combine them with coatings or other materials (giving hybrid structures) in order to solve these deficiencies.

Hybrid parts arise as the result of the combination of materials of different nature [2] with the objective of achieving the optimum technical characteristics and reducing weight and costs. For this reason, in the last years hybrid parts based on light alloys are appearing, like for example magnesium-aluminium (Mg-Al) and Magnesium-Sintered metal (Mg-Sint D11) hybrid constructions. An example is a Mg-Sint D11 arrangement for the embedded main bearing into a body (Figure 1). This kind of structures is attracting a great interest because shows low weight combined with good wear and fatigue properties. However, facing machining of these hybrid parts involves a major challenge in terms of accuracy, quality and safety [3], that has to be considered in future studies since current market is starting to ask for their machining.

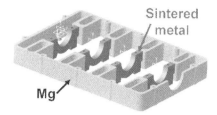

Figure 1: Magnesium hybrid part concept: embedded main bearing (Mg / Sintered Metal)

In general, machining of magnesium and their alloys presents an important fire and explosion risk due to their ease of ignition (self-sustained combustion) that depends mainly on their size and shape. Magnesium chips and dust are highly combustible substances, with a high surface/volume ratio, that are prone to oxidize, heat and ignite spontaneously [4].

Furthermore, aluminium and sintered metal alloys, used for the manufacturing of light components in automotive industry, are wear resistant materials, whose abrasive constituents induce a severe, fast wear of tool edge.

As well as these inherent problems, machining of magnesium-based hybrid components presents additional difficulties. In Mg-Al combination, both materials show different machining qualities and in order to avoid quality discontinuity, the machine-tool parameters need to be adjusted when moving from one to the other. On the other hand, fire risk is more marked when machining Mg-Sint because of the high temperatures reached by sintered metal chips and the high flammability of magnesium powder.

In order to achieve a safe machining it is absolutely necessary to use machine-tools with integrated safety concepts as well as suitable cooling lubricants and tools.

3 Safety Measures

Because of the high flammability of magnesium, safety issues related to process and workplace have to be taken into account. For this reason, suitable milling machines especially adapted to reach the requirements of safe machining of magnesium alloys have been used.

Concerning face milling tests performed at Tekniker, the used milling machine has been previously equipped with a system of suction and filtration of mist and vapour to prevent a critical concentration of mist in the working area (Figure 2a). The suction system consists of three steps: a pre-filter, an aspiro-filter and a post-filter and must not be electromagnetic or produce any spark that can provoke the magnesium deflagration. The pre-filter has been specially introduced for MQL and dry conditions, where a high formation of oil mist mixed with powder and metallic particles can be produced.

As protective measures, a security roof with anti-explosion clappers and bellows has been placed near the extractor system for gas release in case of explosion (Figure 2b). Also, the machine has been equipped with a detection and fire extinguishing system. Thermo-velocimeter sensors were placed as detectors and an Argon bottle (Figure 2c) as fire extinguishing system.

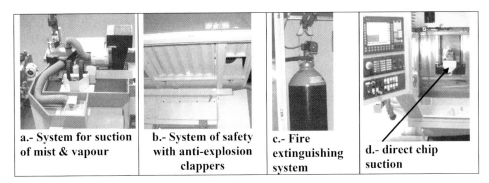

| a.- System for suction of mist & vapour | b.- System of safety with anti-explosion clappers | c.- Fire extinguishing system | d.- direct chip suction |

Figure 2: Safety devices

Otherwise, it is necessary to avoid accumulation of Mg chips in the cutting zone, due to the high probability of its ignition as consequence of possible sparks generated in the cutting zone. Concerning chip removal, the machine developed at the partner ANGER has included a suction nozzle directly under the cutting area (Figure 2d).

Anyway, before beginning the machining of magnesium hybrid parts, it is necessary to set-up all these safety systems and workers must be aware of their work environment, recognize potential hazards and take corrective actions to mitigate hazards.

4   Cooling/Lubrication Strategy

Water-based fluids are not recommended for the machining of magnesium-based hybrid parts because of the high reactivity between magnesium and water being the use of neat oils the best alternative. These neat oils are non-soluble cooling/lubricants that must fulfil additional technical requirements for the machining of these materials.

A low evaporation tendency of oil is particularly important for machining Mg-based hybrid parts because volatile compounds are responsible for fires. Besides, the flash point of the lubricant should be kept as high as possible being recommended values above 150 °C, that means a low tendency of oil to form a flammable mixture with air. Concerning viscosity, in general low viscosity oils can dissipate the heat better and evaporate more easily, which can be favourable from an environmental point of view to reduce waste on chips. However, it is advantageous to use in this case higher viscosity fluids to reduce oil vapour and oil mist formation which presents lower flammable point than Magnesium. For this reason, the viscosity at 40 °C of the lubricants should be in a range between 30 and 50 cSt. The lack of reactivity of cutting fluids with Mg is also an important factor when machining this material. Their low acid number gives an approximate notion about the low corrosiveness of the oil and it usually corresponds to results obtained with immersion tests.

Because of environmental and health hazards of lubricants, it is becoming absolutely necessary on the one hand, to develop more environmentally & healthy friendly lubricants, minimizing or avoiding, the toxic compounds involved in their formulations and on the other, to minimize their use. In this sense, MQL systems represent an interesting alternative because combine the functionality of the cutting fluid with a low consumption of lubricant and they are considered as total loss systems based on spraying the lubricant near the cutting area. Due to the atomised fluid is dispersed between tool and workpiece or is finally carried out by the chips, there is no fluid waste generated in the process. Two alternatives: Dry and MQL techniques have been analysed for the face milling of Mg-based hybrid parts.

Taking into account all these considerations, a neat oil of synthetic nature based on Polyol Ester developed by Brugarolas has been used in tests for the safe, efficient and environmentally friendly machining of magnesium hybrid components.

5   Face Milling of Hybrid Parts

Face milling tests have been carried out in a comparative way under dry and MQL conditions on AZ91D/Sint D11 and AZ91D/AlSi12 combined plates using commercial coated and uncoated cemented carbide inserts, as well as a commercial toolholder (Figure 3).

Figure 3: Cutting tools

For these milling tests, Mg-Sint combinations have been prepared by infix cylinders of Sint D11 in AZ91D magnesium alloy plates and aluminium flanges into magnesium parts, as it is visualized in Figure 4.

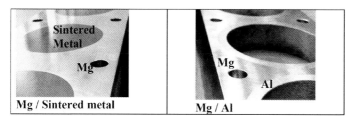

Figure 4: Hybrid parts

These workpieces serve for simulating partly the machining operations on real parts.

For *AZ91D/Sint D11 combination*, initial machining conditions considered for this kind of operation have been a Cutting speed (v_c) of 200 m/min, a feed per tooth (f_z) of 0.25 mm/z, an axial depth of cut (a_p) of 0.3 mm and a radial depth of cut (a_e) of 50 mm.

A MQL system based on micro pumps and supplying the oil-air mixture through two nozzles has been used in testing. The total consumption of lubricant is 22 ml/h.

As main outputs to control the process have been considered surface roughness (Ra), tool wear (V_B) and spark generation.

For *AZ91D/AlSi12 combination*, cutting parameters are for rough cutting: v_c: 500–600 m/min, f_z: 0.15 mm/z, a_p: 0.8mm, a_e: 50 mm and for finishing: v_c: 600–800 m/min, f_z: 0.08 mm/z, a_p: 0.1 mm, a_e: 50 mm. The MQL strategy was also used here.

4.1 Results

4.1.1 Magnesium/Sintered Metal

At the first tested cutting conditions (v_c: 200m/min, f_z: 0.25 mm/z, a_p: 0.3 mm, a_e: 50 mm), tests only progressed for 3.51 m² of machined area in dry and MQL conditions for the detection of possible differences related to surface roughness and flank wear. The evolution of surface roughness (Ra) and wear for Mg & Sint D11 with the machined area is represented in Figure 5.

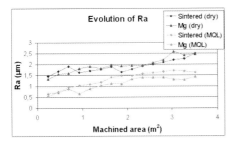
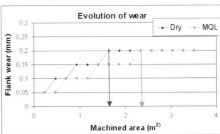

Figure 5: Ra and wear evolution in the machining of Mg-Sint

In these cutting tests, MQL has offered better surface quality for magnesium and Sintered Metal, with values of Ra lower than 1.46 and 1.83 µm, respectively. Besides, MQL has showed a slight improvement concerning flank wear, delaying the obtaining of a flank wear of 0.2 mm in 0.7 m².

Increasing the cutting speed up to 250 and 300 m/min, for a_p of 0.3 mm, sparks were detected both in dry and MQL conditions with the assumed risk of fire, and tests in these conditions were stopped. In view of that the maximum cutting speed for not producing sparks is around 200 m/min, it was decided to keep this cutting speed and try an increase of a_p up to 0.5 mm. But, in these conditions, sparks were detected once again at dry and with MQL, meaning that higher a_p involve reduction in speed, occurring new suitable conditions at v_c: 150m/min, f_z: 0.25 mm/z, a_p: 0.5 mm and a_e: 50 mm.

Chips were slightly smaller with MQL as consequence of decreasing the plastic deformation, reduction in friction and increasing shear angle.

4.1.2 Magnesium/Aluminium

In the tested conditions, a quite significant difference concerning Ra values was observed for both roughing and finishing operations. In finishing, the Mg/Al combination offered Ra 0.2 µm in magnesium and Ra 0.8 µm in aluminium area, giving this last material the limiting value. This good surface quality could be reached only with a specially designed toolholder and well fitted cutting inserts.

The height difference at the boundary between the two materials is a very important aspect, keeping always in mind the need of smooth borders for further assembling of parts.

As a result of the used parameters and strategy for milling, no significant step was detected after machining. The measurements showed height differences lower than 3µm. This is a very good result in comparison to the obtained in previous cutting processes where steps up to 20µm occurred, that was a problem for screwing parts together. In Figure 6 the boundary between magnesium and aluminium can be seen at a microscope level.

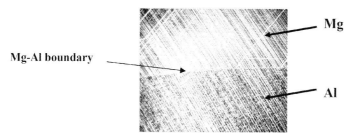

Figure 6: Mg–Al boundary

5 Conclusions

Machining of Mg-based hybrid parts with the minimum quantity lubrication strategy needs a well balanced combination of the machine tool concept, the lubrication, the tools and the cutting parameter. By considering this and the safety guidelines for machining of magnesium, an economical and environmentally friendly machining of this material could be made.

6 Acknowledgements

Authors would like to thank European Commission for its financial assistance in the development of CRAFT project "ECOHYB" within the 6th Framework Programme as well as to the consortium for supporting this work.

7 References

[1] Hydro Magnesium "Machining Magnesium"
[2] 132.248.76.38/posgrado/materiales.html
[3] sme.cordis.lu/docs/cooperative_projects_Vol1_web.pdf
[4] http://tis.eh.doe.gov/techstd/standard/hdbk1081/hbk1081c.html

Experiences with the Machining of Magnesium

Dr. René Schwerin, Stefan Joksch
Oemeta Chemische Werke GmbH, Uetersen

1   Introduction

The role of water miscible metal working fluids in the machining of magnesium alloys gets increasingly more important. The machining process with defined or undefined cutting edge does not represent a great challenge. The entire peripheral treatment process must be adjusted to the special characteristics of magnesium.

The cutting fluid stability during high salt load, an appropriate corrosion protection and the reduction of the reactivity of the magnesium have to be considered. Further conditions for the successful processing are an optimal cleaning of the dirty emulsion to decrease the salt load and the constant monitoring of all important parameters during the process.

On the basis of best practice with all its challenges, the machining of magnesium with water miscible metal working fluids will be presented below.

2   Magnesium Alloys as new Materials

In the beginning of the 19th century, when magnesium was first available as a raw material, it fascinated people, because of its unique characteristics as light metal with high mechanical strength. Combined with the simplicity of the machining process, the utilization of magnesium alloys increased until the early 20th century, when it was used for example in the Volkswagen "Käfer" automobile. 21 kg of magnesium was integrated in this vehicle.

After the second world war aluminum and its alloys replaced magnesium as a raw material, because of its better corrosion protection characteristics. Especially tests under salt water conditions presented a great challenge for magnesium and its alloys.

After the development of better resistant magnesium alloys the trend to use this light metal once again became stronger. These new alloys have corrosion protection scores of a hundredfold better compared to the older ones, because of far less critical amounts of impurities.

Nowadays new foundry technology allows for parts in a variety of applications. From car body parts, engine and power train components to computer, communication and consumer electronics. The trend to use the weight saving opportunities of magnesium continues.

3   Machining of Magnesium with Water Miscible Cutting Fluids

3.1   Characteristics of Magnesium in a Water Based System

There is no extraordinary requirement in the mechanical cutting process. The entire peripheral treatment must be adjusted to its special characteristics.

The application of neat oils or a dry process has a variety of disadvantages. Special fire protection requirements have to be taken. The removal of swarf combined with its explosive properties when using oils or dangerous magnesium dust when no cutting fluid is used have put the water miscible system back into consideration, after the usage of water miscible cutting fluids again was allowed in Germany in 1995.

Magnesium as an element of the 2nd group is reactive to water. Like aluminum it is forming an oxide layer, which protects the metal. Nevertheless it is possible, that magnesium reacts with water:

$$Mg + 2 H_2O \rightarrow Mg^{2+} + 2 OH^- + H_2\uparrow \qquad (1)$$

The reactivity of magnesium has to be decreased by formulating the cutting fluid only with suitable raw material. Furthermore the emulsion needs good stability even at high water hardness (formation of Mg^{2+}) and the pH has to be buffered to reduce its increase (formation of OH^-). The German government safety organization (BG) has compiled recommendations, how to work with magnesium.[1]

3.2   Water Miscible Cutting Fluids for Magnesium Machining

The chemical reaction (1) is the main focus, while formulating a water miscible cutting fluid. The formation of reactive hydrogen has to be minimized. Hydrogen can form an explosive mixture with air (from 4 Vol.% to 77 Vol.%). The nascent hydrogen is a strong reducing agent, that reacts with other elements of the alloy. Together with phosphorous, the formation of phosphane (PH_3) is possible. It has a strong garlic-like smell and is toxic (limit at workplace: 0,1 mg/m³).

If the coolant is aggressive on magnesium, the reaction (1) is boosted. Another effect is a possible change to the surface condition of the magnesium part. The metallic surface can be stained. The more reactions take place, the worse the staining gets. (Figure 1)

Figure 1: Staining of magnesium chips (left) and test strips (right) after immersing in cutting fluid

To reduce the reactivity of the coolant special inhibitors are used, which have an affinity to the surface of magnesium. They act like an extra oxide layer and protect the surface. Nevertheless on the long run the hardness of the emulsion will rise during its sump life, so the emulsifiers, that are used, have to be very resistant. Nonionic emulsifiers are able to secure stability up to very high salt loads (1200 mg/l, Figure 2).

Figure 2: Separated emulsion (left) and intact emulsion (right) stressed with magnesium chips.

The pH-value of an emulsion is an important indicator for its condition. When machining magnesium the alkaline buffer capacity is important for two reasons. First of all a high pH-value is a concern for the health and safety of employees when contact with the emulsion is unavoidable. On the other hand the solubility of magnesium is pH-dependant. If the value exceeds a certain limit, it is possible to get salt residues within the machines. Certain alkanole amines, which have a buffer capacity and a good magnesium compatibility, are applicable and used for cooling fluids for the magnesium machining. The differences in magnesium compatibility of several formulations is presented in Figure 3.

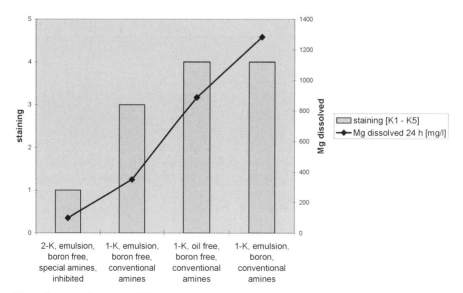

Figure 3: Magnesium compatibility of four different formulations

3.3 Maintenance of Emulsions in Magnesium Machining

To secure a safe and economic process, the maintenance of the emulsion is an important part in magnesium machining. The supplier of the coolant and the operator should work closely together, when running and controlling the systems.

The instant removal of swarf and magnesium chips is another way to minimize the reaction between metal surface and water by reducing the contact time. Therefore the wetting and flushing properties of the emulsion should be optimized to take the debris as quickly as possible to the filtration system.

The filtration system should consist of highly efficient 3-dimensional filter fleece and/or centrifugal separators to remove even the smallest particles, that are formed during the machining process. The smaller the particle size is, the higher is its relative reactive surface.

Even if all peripheral maintenance parameters described here are fulfilled and the cutting fluid is optimized for magnesium machining, the reaction between magnesium and water is never totally avoidable. If the pH-value increases and the water hardness is drawing closer to critical values, there is a chemical treatment available, which can extend the sump life of the emulsion. The desalination filtration system removes molecules up to a certain size. Magnesium ions are filtered so the water hardness will drop down. If using a two component product only certain chemicals get lost in the filtrate. They can be topped up again accordingly with one of the concentrates. The sump life of the emulsion can be extended and costs are reduced.

To ease handling and allow recycling, the chips are pressed to briquettes. This allows for a safe and economic use of the magnesium containing debris.

4 Conclusion

The most important facts of magnesium machining with water miscible cutting fluids are:
- reactivity of magnesium with water has to be reduced
- ingredients of the concentrate have to be magnesium compatible
- stability of the emulsion even at high hardness levels
- good wetting and flushing properties are essential
- maintenance and control of the emulsion is of high priority
- desalination filtration system is available as maintenance tool
- recycling of magnesium chips as briquettes

5 References

[1] BGFE, Fachausschuss Leichtmetall der BGZ, Umgang mit Magnesium, aktualisiert **2005**

Magnesium Matrix Composites

In Mould Heating of Continuous Carbon Fibre Preforms for Reinforced High Pressure Die Casting Magnesium Parts

C. Oberschelp[1], G. Klaus[1], A. Bührig-Polaczek[1], Jens Werner[2], A. Langkamp[2], W. Hufenbach[2], N. Hort[3], H. Dieringa[3], K. Kainer[3]

[1]RWTH Aachen University, GI – Giesserei-Institut, D-52056 Aachen, Germany
[2]Technische Universität Dresden, ILK Institut für Leichtbau und Kunststofftechnik, D-01062 Dresden, Germany
[3]GKSS Forschungszentrum Geesthacht GmbH, MagIC – Magnesium Innovations Center, D-21502 Geesthacht, Germany

1   Introduction

There is an increasing demand for lightweight structural parts which are capable of bearing increased high temperature properties combined with advanced abrasion resistance. The aim of this work is to reinforce magnesium castings locally with carbon fibres. In addition to improve wear resistance, it is also possible to improve the creep resistance. With the help of the very productive high pressure die casting process, carbon fibre preforms are put into a special die, subsequently heated and infiltrated with the magnesium alloy AE42. The preforms are made of T300J carbon continuous fibres and are specially stitched to achieve higher dimensional stability. The casting alloy AE42 is particularly known for its high corrosion resistance and for its good mechanical properties at elevated temperatures. The content of aluminium promotes the infiltration of carbon fibres but is restricted to avoid a possible deterioration of the fibre surface by chemical reactions.

2   Experimental

Within the infiltration trials a cold-chamber die casting machine and a specially designed mould were utilised. The production of continuous carbon fibre reinforced castings was conducted on a BÜHLER® horizontal cold-chamber die casting machine featuring real time control and a die locking force of 725 t. According to the special requirements of magnesium in high pressure die casting, a vacuum-assisted mould filling, a heated casting chamber charged with protective gas, a tempered mould and injection plunger were utilised.

A mould featuring a newly designed clamping device [4], [6] was developed for the production of continuous carbon fibre reinforced magnesium parts. The cavity size is 210×290 mm^2. The preform measuring 140×210 mm^2 is located in the division plane and fixed by an appropriate clamping device. The thickness of the specimen can be varied by up to 15 mm with a maximum infiltration thickness of 5 mm. For the production of the preforms, multiple layers of woven fabric were stitched together with carbon yarn at the ILK.

In order to achieve a high infiltration rate, the carbon preforms have been preheated to a temperature close to the liquidus temperature of the alloy. A heating device was developed, which makes use of the electrical properties of the carbon fibre preforms, as described in Ref [4]. Within a few seconds, the heater generates with ease temperatures higher than 500 °C.

The high pressure die casting trials were conducted with canvas type woven fabrics. It was discovered that this type of woven fabric featured the highest dimensional stability for the trials conducted. The stability can be further increased by stitching multiple layers of the preform applying a special sewing technique. As described in a preliminary work [4], three different kinds of stitching techniques were tested for their stiffness (Figure 1 a–c). It turned out that a rhombic alignment of the sewing prevents a dislocation of the fibres which can occur as a result of the melt pressure during casting. A further benefit of the rhombic-shaped stitched preform is the elimination of fibre splicing on the mechanically stressed side of the preform and inflating by the melt.

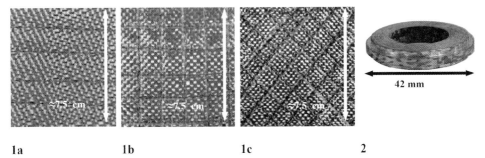

1a　　　　　　　　　1b　　　　　　　　　1c　　　　　　　　　2

Figure 1: a) Longitudinally stitched perform, b) Longitudinally and horizontally stitched preform, c) Rhombically shaped stitched preform　　**Figure 2:** machined insert

The inserts (figure 2) were made of Toray T300 canvas coiled to a hollow cylinder and infiltrated by the DGPI method (**D**ifferential **G**as **P**ressure **I**nfiltration [5]) with the matrix metal. The matrix metal consists of pure magnesium alloyed with 0.2 wt.-% Al, which promotes the fibre wetting. The cylindrical inserts were machined to 7 mm thick disks with a 3.0 mm shoulder on the outside. The function of the shoulder is to support the mechanical and chemical contact between the insert with the surrounding metal (AE42) after the recasting process.

For the infiltration of the samples, the magnesium alloy AE42 was used. Magnesium alloy with aluminium and rare earth content possess good mechanical properties at room temperature (similar to AZ 91), good corrosion resistance and cast ability [7]. Furthermore, rare earth containing alloys features high deformation and creep resistance at higher temperatures. The aluminium promotes the infiltration of carbon fibres but should be restricted because it may harm the fibre surface by chemical reactions.

3 Results and Discussion

The quality of the chemical bond of preforms and inserts with the cast alloy was intensively examined by means of metallographic, x-ray and EDX-analysis. Furthermore, the abrasion of the coated mould inserts during casting was analysed.

Initially, there was a need to find shot parameters which avoid inadequate cavity filling and wetting of the preforms and secondly, there was a need to find parameters which avoid displacement and damage to the fibres. Therefore, the melt velocity had to be diminished in front of the preform. To avoid pre-solidification of the melt and to fully infiltrate the fibres, the pressure had to be subsequently increased after the braking phase. Figure 3 shows a test specimen (210 × 290 mm) with a 1 mm thick infiltrated preform featuring the described rhombic sewing. Figure 3 shows the x-ray-picture of this (right) and of another specimen (left) featuring a longitudinally stitched preform. The progress achieved through the improved stitching technique is clearly visible. The preforms which are longitudinally stitched suffer upon initial melt contact and are badly bent. The perform featuring rhombic stitching is able to resist the force of the incoming melt. Oxides or porosity can not be recognised in the x-ray-picture on the right.

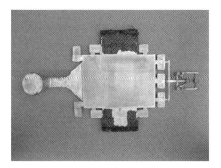
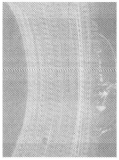
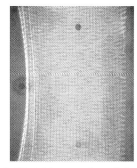

Figure 3: Test specimen with ingot and overflows

Figure 4: X-ray-picture of an infiltrated 1 mm preform. (left longitudinally stitched, right rhombic sewing)

In Figure 5 the quality of infiltration is shown. The fibres are all well infiltrated due to the fibre heating of 500 °C and the applied injection pressure. The chemical composition at several analysed positions was identified by EDX-analysis (Figure 6). The results (Table 1) show that local chemical composition differs from the basic composition between the fibre spacing and at the inter-phases of carbon fibre and matrix. The individually analysed positions show the appearance of rare earth rich phases directly at the fibres.

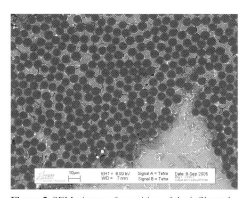
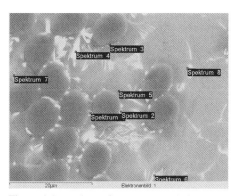

Figure 5. SEM-picture of a position of the infiltrated preform. Magnification: 500 x

Figure 6. SEM-picture for an EDX-Analysis

The inserts were examined before and after high pressure die casting to receive information whether the die casting process affected the insert quality. The result is that a preheating of the inserts may cause thick oxide layers. A 15 min. period of exposure at 450 °C in an environmental atmosphere may lead to 15 μm thick oxide layer around the part. When the inserts are preheated in the closed die, only very thin or no oxide layers appear.

Table 1: Local chemical composition in the fibre spacing

Pos.	Mg [wt.-%]	Al [wt.-%]	Mn [wt.-%]	La [wt.-%]	Ce [wt.-%]	Nd [wt.-%]	Others [wt.-%]	Σ
1	37.31	28.25	16.81	3.60	10.77	3.79		100.0
2	53.66	22.50	9.84	2.81	8.17	2.77	0.24	100.0
3	52.32	18.76	0.83	5.16	17.49	6.25	0.21	100.0
4	34.35	33.73	1.11	7.62	18.50	4.70	0.20	100.0
5	96.14	2.67	0.40	0.32	0.34	0.20	–	100.0
6	84.04	8.69	0.30	1.54	5.00	0.51	0,23	100.0
7	70.17	26.68	0.22	0.27	0.52	0.49	1.39	100.0
8	83.98	7.86	0.28	2.08	4.43	1.07	–	100.0

Metallographic micrographs (Figure 7 and Figure 8) confirm that a good chemical and mechanical bond between insert and magnesium casting exists. Especially at the mixed interface pure Mg/carbon fibres/AE42 at the chemical bond between insert and surrounding metal is promoted.

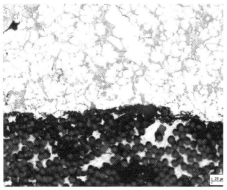 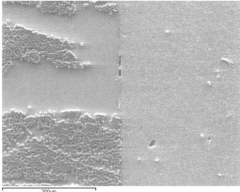

Figure 7: Metallographic analysis of the insert boundary, 500x magnification

Figure 8: EDX analysis of the insert boundary

Figure 9 shows a diagram which reveals the dependency of the heating on power and on the thickness of a preform. To gain a temperature of 700 °C for a preform with a thickness of 3.0 mm, a power of 12.4 kW and heating-time of 19 sec. are required. To heat up a preform with a thickness of 5.0 mm, a power of 13.6 kW and heating-time of 25 sec. result in a temperature of 700 °C.

Because of heating up the preforms electrically in the closed die halves, there was the need to insulate the dies to avoid the discharge of the electricity. Therefore, an Al_2O_3-coating with a thickness of 20 µm for the mould inserts was chosen to ensure reliable electric insulation. In addition to that, the coating had to be resistant enough to face the process temperatures and inert for chemical reactions with the liquid magnesium.

The experiments show that an Al_2O_3-coating is a good insulator in fact, but could not fulfil the requirements concerning durability. Within the period of just a few casting procedures, the coating showed defects on its surface (Figure 10). It is assumed that the different thermal expansions of hot working steel and coating lead to this degradation combined with the high melt pressure and a sticking of the cast part due to a chemical reaction. As a further effect of the radiation heating of the red glowing preform, the coating is badly stressed by quickly rising temperature at the surface. To avoid such defects caused by abrasion, other types of coating should be used in future. Most of the coatings used in the hpd-casting industry exhibit conductivity and are therefore inappropriate. Therefore, further trials will be conducted with thicker zirconium dioxide coating on a tie layer.

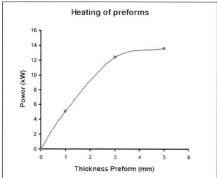

Figure 9: Power versus preform-thickness. Heating of preforms up to 700 °C

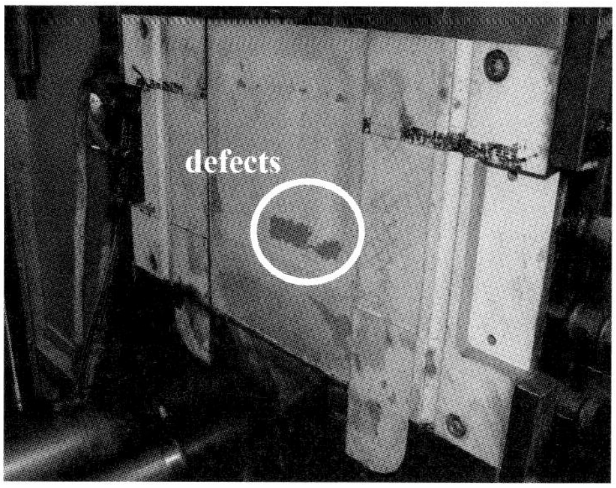

Figure 10: Defects on the surface of the coating

4 Conclusions

The results show that an infiltration of carbon fibres in high pressure die casting is possible. Especially, the importance of a preheating of the fibres is emphasised because it is known that complete perform infiltration depends on temperatures which are close to the liquidus temperature of the melt. With the implementation of the fibre heating device the demand for higher fibre temperatures could be met.

The results also show that it is possible to reinforce castings locally with inserts. The inserts utilised are composed of pure Mg and carbon fibres and build up a satisfactory interface together with the surrounding matrix of the AE42 casting. Especially at the interface Mg/carbon fibre/AE42 matrix, good bonding was observed.

As presented in this work, the high pressure die casting method combined with continuous carbon fibre reinforcements is a promising technology for future trends. With the new electrical fibre heating system, a better process parameter control can be achieved. In the future, additional trials to infiltrate preforms thicker than 1.0 mm will follow.

5 Acknowledgements

The authors would like to extend their gratitude to the Helmholtz Association for the financial support required to conduct the study. Furthermore, the partners in the project „Key Technologies for Advanced Engineering Materials/Schlüsselwerkstoffe für den Leichtbau" would like to thank NORSK HYDRO for supplying the alloy AE42 and the INSTITUT FOR SURFACE TECHNOLOGY for coating the die inserts.

6 References

[1] Internet page Think Ceramics – technische Keramik; www.keramverband.de
[2] Internet page *Edelstahl Witten-Krefeld*; www.edelstahl-witten-krefeld.de
[3] R&G Faserverbundwerkstoffe 2003; brochure
[4] Klaus, G., Oberschelp, C., Fehlbier, M., Bührig-Placzek, A., Werner, J., Hufenbach W.: Fabrication of Carbon Long Fibre Reinforced Magnesium Parts in High Pressure Die Casting. *TMS 2006, 135th Annual Meeting and Exhibition*
[5] Hufenbach, W.: Hochfeste und extrem leichte Bauteile aus langfaserverstärktem Magnesium. *Dresdner Transferbrief* 9 (2001) 4, S. 16
[6] Hufenbach, W., Bührig-Polaczek, A., Klaus, G., Fehlbier, M., Langkamp, A.: Fabrication of fibre reinforced magnesium alloys. *Ustron, April 22, 2005*
[7] Kainer, K.U.: Magnesium – Eigenschaften, Anwendungen, Potenziale; WILEY-VCH Verlag GmbH; ISBN 3-527-29979-3
[8] *Magnesium Taschenbuch*; 1.Auflage 2000; Hrsg.: Aluminium-Zentrale Düsseldorf; ISBN 3-87017-264-9

Mechanical Behaviour of an Mg-8Li Alloy Reinforced With SiC Particles

Z. Száraz, Z. Trojanová

Department of Metal Physics, Charles University Prague, Ke Karlovu 5, CZ-121 16 Praha 2, Czech Republic

1 Abstract

The paper reports results of the investigation of the microstructure and the deformation behaviour of a Mg-8Li composite reinforced with 8 % volume fraction of silicon carbide particulates. The composite was prepared using a powder metallurgical technique. The matrix alloy is a composition of two phases: hcp and bcc . Compressive deformation properties of the composite were investigated in the temperature range from room temperature up to 300 °C. A significant work hardening has been observed at temperatures up to 100 °C. Softening occurs at higher temperatures. The stress-strain curves obtained at temperatures 250 °C and 300 °C have a flat character. The yield stress and the maximum stress decrease with increasing temperature. The reinforcing effect of particles decreases with increasing temperature. Various strengthening mechanisms originating from the matrix and the reinforcing particles are discussed. As the main strengthening mechanism in the Mg-8Li/SiC composite has been found the Hall-Petch strengthening. A contribution due to an increase of the dislocation density in the composite is also significant.

Light and transition electron microscopy were used for study of the microstructure of the composite.

2 Introduction

The use of extremely light materials plays a dominant role for different structural applications. The demand for further weight reduction e.g. in automotive industry makes the magnesium alloys attractive. In the transportation industry there is an increased interest in application for improving fuel efficiency through vehicle weight reduction. Aerospace applications underline the importance of density reduction [1]. The Mg-based alloys exhibit a high specific strength. The strength decreases rapidly with increasing temperature. On the other hand, Mg and α-Mg alloys possess a poor ductility and formability because of its hexagonal closed packed structure. To make up this shortcoming and further reduce weight can achieve by alloying magnesium with lithium of extremely low density (0.58 g/cm^3). Lithium is a suitable alloying element that decreases density from 1.8 g/cm^3 to 1.35 1.8 g/cm^3 and increases ductility of magnesium alloys [2]. The room-temperature elastic modulus of Mg-Li alloys is 42 GPa, which is almost as high as the value for conventional Mg alloys, 45 GPa [3].

The Mg-Li phase diagram shows that Lithium has relatively high solid solubility in hcp -phase up to 5.5 wt.% (17 at.%) [4]. Moreover, Mg with Li content greater then 11 wt% crystallize in β phase, which has bcc crystal structure, thereby improving machinability and weldability of alloy. Mechanical properties of the hcp -phase are better in comparison with the bcc alloys, while the ductility is worse. Disadvantages of Mg-Li alloys with bcc structure are a high chemical activity and poor corrosion resistivity.

According to the Mg–Li phase diagram, with Li content between 5.5 to 11 wt.%, bcc - phase of Li solid solution will co-exist with the hcp -phase of Mg solid solution at room temperature. Some compromise would be an alloy with 8 wt% of Li (a mixture of phases +), which might exhibit both improved mechanical properties as well as a good corrosion resistance.

However, magnesium-lithium alloys have low strength, creep resistance, and poor corrosion resistance. Conventionally, Mg–Li alloys can be strengthened by the addition of a third element for forming ternary Mg–Li–X (usually X=Al, Zn, Si) alloys. Reinforcing the alloy with the ceramic fibres or particles is the other feasible way to increase strength and to prevent mechanical properties degradation of Mg–Li alloys [5]. The fiber or particle reinforcement can essentially improve strength and stiffness of Mg–Li alloys thus giving metallic materials with very high specific strength and high specific stiffness.

The aim of the present work is to evaluate the effect of addition of SiC particulates as reinforcement on the microstructures and mechanical properties of Mg-8Li alloy, to study the influence of temperature on compressive properties of composite and discuss possible hardening and softening mechanisms.

3   Experimental Procedure

The material used in this study was Mg-8Li magnesium matrix composite fabricated by a powder metallurgy method. Mixing of the matrix powders with SiC particles was carried out in an asymmetrically moved mixer with subsequent milling in a ball mill. The powder was capsulated in magnesium containers and extruded at 400 °C using a 400 t horizontal extrusion press. The volume fraction of SiC particles was about 8 %. The composite samples were not thermally treated.

Compression tests were performed in temperature range from room temperature to 300 °C using an INSTRON testing machine. Cylindrical specimens of 5 mm diameter and 9 mm height were deformed with a constant crosshead speed corresponding to nominal (initial) strain rate of $1.85 \cdot 10^{-4}$ s^{-1}. The temperature in the furnace was controlled to within ±1°C. Protective argon-atmosphere was used at elevated temperatures. Test data were collected by a computer data-acquisition system, from which deformation properties such as yield stress, maximum strength and elongation to fracture were determined. The microstructure and substructure of composites was observed by the light microscope Olympus and transmission electron microscopy (TEM).

4   Experimental Results

Optical micrograph of the Mg8Li composite prepared by powder metallurgy is shown in Fig. 1. It can be seen that the distribution of SiC particulates in the matrix alloy is not homogenous. There is obvious agglomeration of reinforcement in composite. The mean SiC particle size was found to be about 9 m. Lighter (α) phase and Li richer (β) phase are visible. A mean grain size has been estimated to be 5±2 μm, which is less than the particles size.

The compressive true stress-strain curves at strain rate of $1.85 \cdot 10^{-4}$ s^{-1} and different temperatures of the Mg-8Li composite are shown in Fig. 2. The temperature significantly affects the shape of the deformation curves. Compressive flow stress decreases with the increase of temperature.

Figure 1: Optical micrograph of the as-prepared composite

Samples were deformed up to fracture at RT and 50 °C. Deformation at higher temperatures was interrupted at about true strain of 0.4. Significant work hardening was estimated for temperature from room temperature up 100 °C. Curves obtained at temperatures higher have a flat character. It indicates that some recovery process(-es) take(s) place.

Characteristic stresses, the yield stress σ_{02} and the maximum stress σ_{max}, were estimated from stress-strain curves and are introduced in Fig. 3. Observed difference between the yield stress and maximum stress is at room temperature about 206 MPa and it decreases rapidly with increasing temperature. A very small or no difference between the σ_{max} and the σ_{02} at higher temperatures is apparent from Fig. 3 and reflects the flat character of the stress strain curves.

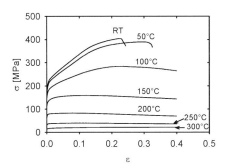
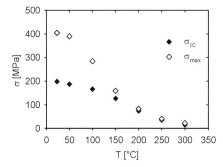

Figure 2: True stress-strain curves obtained at different temperatures in compression

Figure 3: Temperature dependence of the yield stress σ_{02} and the maximum stress σ_{max}

TEM microstructure of the non-deformed composite is introduced in Fig 4. TEM investigation revealed the presence of high density of dislocation in as-prepared samples. Dislocations are generated during the hot extrusion and subsequent cooling. Dislocation tangles are formed in the vicinity of grain boundaries and SiC particles.

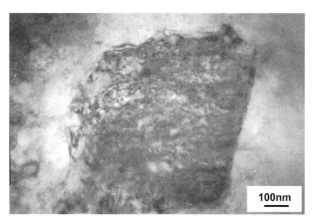

Figure 4: Transmission electron micrograph of the as-prepared sample

5 Discussion

Mechanical properties of composites reinforced by particles are determinate not only by properties of matrix and the reinforcing phase, but also with their interaction. The volume fraction of reinforcement, the distribution and the properties of the reinforcing phase influence dislocation behavior and the mechanical properties of composites. Particles may also influence other material properties as the wear resistance or damping capacity. Another important factor is the preparation method.

It is not straightforward to propose theoretical model including all deformation mechanisms predicting mechanical properties of given composite. In the literature several theoretical models were developed to explain the strengthening in composites [6–11]. Individual models take into account only some mechanisms and influence of other mechanisms is neglected. However, we can estimate contribution of various factors and to establish the most important mechanism(-s).

Trojanova et. al [12] investigated the deformation behaviour of Mg-8Li alloy with the grain size of 180 μm. They estimated the yield stress value at room temperature of $\sigma_{02alloy}$ = 85 MPa. This value was used as σ_m for calculation in the Table 1. Various strengthening mechanisms which can operate in composites are listed in Table 1. Contributions to the composite strengthening were calculated according to formulae listed in the Table, shear modulus G = 17 GPa, α_1 constant has been taken according to Lavrentev [13] α_1 = 0.35, K_y = 0.28 MNm$^{-3/2}$ [14]. As it follows form the Table 1, the load transfer and the Orowan mechanism play only marginal role for the composite strengthening. In case of equiaxed particles the load transfer from the matrix to the particles is not effective. Furthermore, the mean grain size of the composite, prepared by powder metallurgical method is smaller than the reinforcing particles. In this case of hot extruded Mg-8Li with SiC particles the main hardening mechanism seems to be the reinforcing effect of the grain refinement. Smaller matrix grain size also reduces twinning which is important deformation mode in the Mg alloys. A higher dislocation density in the composite material yields a higher level of internal stress. Enhanced dislocation density may also significantly increase the composite yield stress. Thermal stresses arising due to a big difference between the thermal expansion coefficients may be partially accommodated by dislocation generation in the matrix. Tensile residual stresses exist also at ambient temperature and they relax very slowly.

Table 1: Possible strengthening contributions in Mg-8Li composite

Mechanism			Contribution
Load transfer [6]	$\sigma_{LT} = \sigma_m 0.5 f$	f volume fraction σ_m matrix stress	3 MPa
Thermal dislocations [7]	$\rho_T = \dfrac{B f \Delta \alpha \Delta T}{b(1-f)} \dfrac{1}{t}$	b Burgers vector $B=12$ for particles $\Delta u \Delta T$ is the thermal strain	$1.78 \cdot 10^{12}$ m^{-2}
Dislocation geometrically necessary [9]	$\rho_G = \dfrac{f 8 \varepsilon_p}{bt}$	ε_p plastic strain t is the particle diameter	$3.49 \cdot 10^{11}$ m^{-2}
Enhanced dislocation density	$\Delta \sigma_D = \alpha_1 \psi G b (\rho_T + \rho_T)^{1/2}$	α_1 constant ψ Taylor factor G shear modulus	17 MPa
Orowan strengthening [9]	$\Delta \sigma_{OR} = \left(\dfrac{Gb}{\Lambda} + \dfrac{5}{2\pi} G f \varepsilon_p \right)$	Λ distance between fibres	0 MPa
Grain size refinement [10]	$\Delta \sigma_{GS} = K_y \left(d_2^{-1/2} - d_1^{-1/2} \right)$	K_y constant d_1, d_2 grain size	101 MPa
Average residual stress is matrix [11]	$\langle \sigma_m \rangle_{max} = \dfrac{2}{3} \sigma_y \ln\left(\dfrac{1}{f} \right) \dfrac{f}{1-f}$	σ_y yield stress in the matrix	13 MPa
σ_{total} [MPa]			219 MPa
$\sigma_{02 experimental}$ [MPa]			198 MPa

Small grain size gives the possibility for grain boundary sliding, which is probably the important deformation mechanism at elevated temperatures.

6 Conclusions

Mg-8Li based composite was prepared by powder metallurgy technique. The matrix alloy exhibits dual phase structure (+). Compression tests were carried out in temperature interval from room temperature to 300 °C. The stress-strain curves show high work hardening at lower temperatures. The Hall-Petch strengthening has been found to be probably the main strengthening mechanism. The influence of the increased dislocation density is also important. Other possible mechanisms play in strengthening only a marginal role. Rapid decrease of the yield stress and maximum stress at elevated temperatures in Mg-8Li composite indicates some grain boundary deformation mechanisms.

7 Acknowledgements

This work received a support from the Research Project 1M 2560471601 „Eco-centre for Applied Research of Non-ferrous Metals" that is financed by the Ministry of Education, Youth and Sports of the Czech Republic. Additional support was provided by the Grant Agency of the Czech Republic under Grant No. 103/06/0708.

8 References

[1] B. L. Mordike, and K. U. Kainer, (ed.), *Magnesium Alloys and Their Applications*. Werkstoff-Informationsgesellschaft, Frankfurt **1998**.
[2] F.W. Bach, M. Schaper, C. Jaschik, Mater. Sci. Forum **2003**, 1037, 419–422.
[3] M.M. Advesian, H. Baker, (Eds.), Magnesium and Magnesium Alloys, ASM International, Materials Park, **1999**, 19.
[4] Massalski, T. B., *Binary Alloy Phase Diagrams*. ASM, Metals Park, OH **1986**, 1487.
[5] S. Kúdela, Int. J Mater Prod Technol. **2003**, 18, 91.
[6] Jr. R.M. Aikin, L. Christodoulou, Scripta Metal. Mater. 1991, 25, 9–14.
[7] R.J. Arsenault, N. Shi, Mater. Sci. Engn. **1986**, 81, 175–187.
[8] M.F. Ashby, Phil Mag. **1970**, 21, 399–424.
[9] H. Lilholt, Mater. Sci. Eng. **1991**, A 135, 161–171.
[10] E. Carreño-Morelli, S.E. Urreta and Schiller, *Proceedings of the International Conference on Fatigue of Composites* (Ed.: S. Degallaix, C. Bathias, R. Fougères), Société Française de Métallurgie et de Matériaux, Paris, France, **1997**, p. 112–19.
[11] F. Delannay in Comprehensive Composite Materials (Ed.: T.W. Clyne), Vol. 3, Elsevier, Amsterdam; **2000**. 341–369.
[12] Z. Trojanová, P. Luká
[13] , J. Mater. Process. Techn. **2005**, 162-163, 416–421.
[14] F.F. Lavrentev, Y.A. Pokhil, I.N. Zolotukhina, Mater. Sci. Engn. **1978**, 32, 113–119.
[15] M. Mabuchi and K.Higashi, Acta Mater. **1996**, 44, 4611–4618.

Production of AZ91D/CNF by Compocasting/Ultrasonic Agitation Conjugated Process

Renato Galvao da Silveira Mussi, Tetsuichi Motegi, Fumi Tanabe, Hideyuki Kawamura
Chiba Institute of Technology
Kazuo Anzai, Daisuke Shiba, Masashi Suganuma
Nissei Plastic Industrial Co., Ltd.

1 Introduction

Magnesium alloys have recently attracted significant interest of researchers due to their high strength-weight ratio, which makes of them potential options for replacing heavier materials in some automobile parts [1] and in structural parts in electronic devices, such mobile phone external parts [2]. It is also expected that Magnesium will play an important role in lightweight construction in the near future [3], due not only to its reduced weight, but also because it can be found in abundance in Earth's crust. However, further utilization of Mg alloys is limited by its mechanical properties, which have high specific values but are insufficient at more requiring applications. The number of researches aiming improvement of Mg alloys mechanical properties has increased significantly, resulting in a variety of new and adapted production methods and Mg-alloy-based materials for many different applications, as it can be seen in the case of Japan. [2]

Among different new materials based on Mg alloys, metal matrix composites (MMCs) reinforced with ceramic materials have been proving to be a good options, as addition of ceramic short-fibers or particulates to magnesium alloys are reported in literature to have improved significantly their mechanical and tribological properties. [4-6] Considering such properties to be strongly dependent on various parameters, such as quantity, size and properties of the reinforcement, utilization of carbon nano-tubes (CNTs), a recently discovered allotrope of carbon with extremely high tensile strength and elastic modulus [7], has produced plastic matrix composites with improved mechanical and electrical properties.[8]

Although successful production of Mg alloy reinforced with CNTs has been previously reported [9–11], homogeneously dispersing them either in solid or liquid metal is a difficult task. Among various production processes, traditional compocasting by mechanical stirring produces good quality composites at low cost, providing better wetting and dispersion of reinforcement [12] as the semi-solid slurry keeps it from floating or settling [13, 14]. In previous works [15,16], utilization of a modified compocasting process obtained, for the best conditions, slurries containing 10μm-agglomerates of carbon nano-fibers (CNFs) that, although homogeneously distributed through Mg and Al alloys, were not completely wetted by the alloys. Among various possible solutions to improve dispersion of CNF, cavitation by ultrasonic agitation appeared to be the best option.

Initially used in liquid metals for flow velocity measurements [17,18], ultrasonic vibrations started to be used to induce cavitation into molten metals, as it was shown to produce refined microstructure [19–21]. As for MMC production [22–24], ultrasonic agitation improved wetting of ceramic particles by the metallic matrix, resulting in sound composites. Thus, ultrasonic agitation was applied to slurries containing CNF and semi-solid Mg alloy, resulting in complete dispersion of the CNF agglomerates into the matrix. In the present work, the process is described and the influence of different experimental conditions on the quality of the composites obtained discussed.

2 Compocasting/Ultrasonic Agitation Conjugated Process

2.1 Matrix and Reinforcement

Commercial AZ91D, an Mg-9%Al-1%Zn alloy was used as matrix due to its good castability, corrosion resistance and mechanical properties. Another advantage of this Mg alloy is in its wide solidification interval of over 150 °C, which allows better control on the solid fraction of the slurry during process. The material used to reinforce AZ91D was CNF, which is an extremely thin fiber 10 m long and 100–200 nm in diameter, with density of 2.0 g/cm^3. CNFs are believed to have similar properties to those of carbon nano-tubes, i.e., high tensile strength and electrical conductivity, although such values are still to be determined.

2.2 Modified Compocasting Process

In order to produce the slurry containing semi-solid AZ91D and CNF that would be ultrasonically agitated, a modified compocasting process was employed. In steady of starting to stir metal at liquid state and lower its temperature until semi-solid state is obtained, in this modified compocasting semi-solid metal is produced before being poured in the mould where stirring will be performed, allowing the process to be carried out at lower temperatures and in shorter times. The modification consists in pouring molten metal over an inclined cooling plate that is located over the mould as, according to crystal separation theory [25], by doing so, the metal flowing on the plate causes the solidification nuclei generated on the cold surface to detached and flow into the mould kept in a furnace at constant temperature.

2.2.1 Production of Semi-solid AZ91D using an Inclined Cooling Plate

Inclined cooling plates have been used in various researches to produce semi-solid slurries of different kinds of metals [26–28], and parameters such as plate geometry and pouring temperature were reported to strongly affect the quality of the material obtained. Considering that semi-solid metal with refined globular primary phase results in higher fluidity [29] and, consequently in better stirring conditions to disperse the reinforcement, before performing experiments involving mechanical stirring, the best conditions to produce refined semi-solid AZ91D were determined. Results reported in previous works [27] were used to set some of the parameters as constants, and experiments were carried out varying the lengths of the cooling plate. The values of the parameters kept constant during the experiments are 605 °C, 595 °C and 60° for pouring temperature, tundish furnace temperature and cooling plate inclination, respectively.

The experiments consisted in pouring molten AZ91D on the inclined cooling plate and after holding the metal for 10 seconds in the mould, in order to allow the solidification nuclei generated on the cooling plate to grow and stabilize in the tundish, it was rapidly solidified in water. AZ91D with the most refined microstructure (Figure 1), consisting of primary crystals average diameter of 37m, was obtained for an inclined cooling plate of 160mm in length. Both shorter and longer plates produced coarser primary globules. Shorter cooling lengths generate insufficient number of solidification nuclei, which have enough space to grow during solidification. Longer cooling plates, on the other hand, generate excessive number of solidification nuclei, which promptly interact with each other, allowing coalescence, and consequently coarsening to occur. The results obtained for the best conditions for production of semi-solid AZ91D were used in all posterior experiments involving mechanical stirring and ultrasonic agitation.

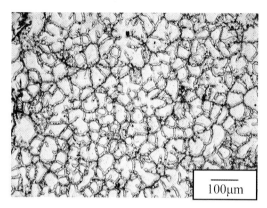

Figure 1: Microstructure of AZ91D obtained using an inclined cooling plate with 160 mm in length

2.2.2 Production of Semi-solid AZ91D/CNF Slurry by Mechanical Stirring

In mechanical stirring process, the most important parameters are considered to be speed, temperature and time during which stirring is performed. [30] Once more, values reported in previous works [15] were used to set stirring temperature at 588 °C–593 °C, and stirring speed at 1750 rpm, as different value for these parameter could result in inadequate fluidity of the slurry while stirred, possibly resulting in poor inclusion or dispersion of CNFs. In the present experiment, AZ91D/CNF slurry was produced by mechanically agitating semi-solid AZ91D with the aid of an edged turbine, while CNF is added into vortex generated around the paddle. After 1 vol% of CNF were completely included, mechanical stirring proceeds at constant speed and temperature for periods that varied from 1 to 25 minutes., at the end of which the paddle, still in movement, was pulled out from the slurry. The mould containing the slurry was then rapidly solidified in water. Argon gas was continuously blown on the surface of the slurry to prevent Mg from react with oxygen during isothermal stirring.

Semi-solid AZ91D/CNF slurries produced before ultrasonic agitation presented the most homogeneous distribution of reinforcement along the matrix after being mechanically stirred for 20 minutes. Although when observed by SEM, these composites produced at the best conditions of the CNF presented some areas with well dispersed CNFs, as in Figure 2a, most of the CNFs were found as small agglomerates (Figure 2b) homogeneously distributed along the surface analyzed. Magnesium primary phase obtained after stirring were twice as coarser as those shown in Figure 1, which could be due to mechanisms of coalescence during the long stirring times in regions in the mould of reduced turbulence.

Composites stirred for less than 20 minutes did not exhibit the same coarsening in microstructure, but the shorter the stirring times the bigger were the CNF agglomerates found along the matrix, and the worse their distribution. Further stirring the slurry, however, caused only further coarsening, but have not resulted in better dispersion of the reinforcement, which means that the limit for dispersion of CNF in semi-solid AZ91D by this modified compocasting method was reached after 20 minutes of mechanical stirring at 1750 rpm and semi-solid metal with 10 to 20% volume fraction.

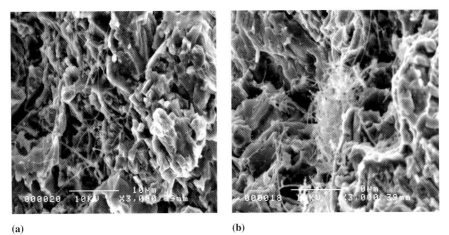

(a)   (b)

Figure 2: SEM photographs of (a) dispersed and (b) agglomerated CNF found in AZ91D stirred for 20 minutes

2.3 Ultrasonic Agitation Process

Ultrasonic agitation of the slurry obtained by modified compocasting was performed with the aid of an ultrasonic welding equipment (USG) adapted for the present experiment. The parameters tested in this phase were ultrasonic wave amplitude (11 to 20 m), application time (1 to 20 min.) and diameter of the waveguide responsible for conduce the ultrasonic waves from the USG into the slurry (2 to 4 cm). Argon gas was also used along the process and the composite slurry rapidly solidified in water at the end of the agitation. Schematic illustration of the whole compocasting/ultrasonic agitation process is shown in Figure 3.

CNF agglomerates present in the semi-solid slurry act as bubbles, and besides lowering the cavitation threshold, the high temperature spots, high-pressure pulses and cumulative jets produced when they collapse during ultrasonic agitation seem to increase the bath temperature [23], which is not enough to completely melt the solid crystals, but facilitate them to split, leading to an extremely refined microstructure. Such intense turbulence promoted by cavitation seems also to break the dendrite arms being formed and smash the unstable solid crystals present in the slurry. Cavitation threshold for this process could not be reached for wave amplitudes shorter than 13m whatever the agitation time used. The higher the amplitudes used the shorter the times needed to completely disperse CNF in semi-solid AZ91D, and for over 16m complete dispersion was achieved in less than 5 minutes.

As an option to reduce time of ultrasonic agitation, discontinuous application of ultrasonic waves proved to be more effective than uninterrupted treatment, as 16 and 18m vibration for fixed times followed by 15sec. resting intervals, when dendrites slightly grow to be broken during the following vibration period, adequately dispersing 1 vol% CNF through the Mg alloy in about 3minutes. The microstructure produced at such conditions consisted in a refined primary phase with about 10m average diameter (Figure 4). The highest amplitude produced by the equipment used in this experiment had 20m, but it was found to cause overheating in the transducer, and considered unnecessary, as similar results were obtaining at less extreme conditions. Use of different waveguide diameters has not significantly improved the results, meaning that the most important parameters in this process were ultrasonic wave amplitudes and agitation time. The average Vickers hardness of the composites was 63.5HV, while the as-received AZ91D and the material produced using the inclined cooling plate exhibited hardness of 55HV and 58HV, respectively. This improvement in hardness is probably explained by the refined microstructure obtained, as similar values were obtained for unreinforced AZ91D with the same microstructure.

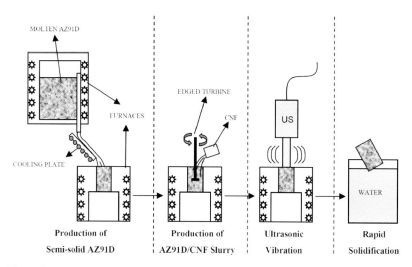

Figure 3: Schematic illustration of compocasting/ultrasonic agitation conjugated process

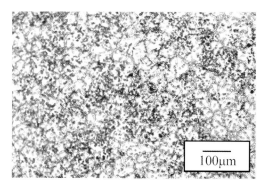

Figure 4: Typical microstructure of AZ91D/1vol%CNF composite produced by compocasting/ultrasonic agitation conjugated process (wave amplitude 16m and 3-minute discontinuous application)

3 Summary

Compocasting/ultrasonic agitation conjugated process proved to be an excellent option to produce magnesium alloy matrix composites reinforced with 1 vol% carbon nano-fibers, providing complete dispersion of reinforcement. Composites produced by this method showed refined primary phase and improved hardness In order to produce sound composites, many process parameters should be controlled, among which are ultrasonic wave amplitude, ultrasonic agitation and mechanical stirring time, and length of the inclined cooling plate.

4 References

[1] B. L. Mordike, T. Ebert, Mater. Sci. Eng. A, 2001, 302(1), 37–45.
[2] Y. Kojima, S. Kamado, Mater. Sci. F. 2005, 488–489, 9–16.
[3] C. Jaschik, H. Haferkamp, M. Niemeyer, in Magnesium Alloys and Their Application, (Ed.: K. U. Kainer), Wiley-VCH, Munich, Germany, 2000, 41–46.
[4] C. Fritze, in Magnesium Alloys and Technology (Ed.: K. U. Kainer), Wiley-VCH Verlag, Weinheim, Germany, 2003, Chapter 11.
[5] F. Moll and K. U. Kainer in Magnesium Alloys and Technology (Ed.: K. U. Kainer), Wiley-VCH Verlag, Weinheim, Germany, 2003, Chapter 12.
[6] J. Lan, Y. Yang, X. Li, Mater. Sci. Eng.A, 2004, 386, 284–290.
[7] M. Yu, O. Lourie, M. J. Dyer et al., Science, 2000, 287 (5453), 637–640.
[8] R. Andrews, D. Jacques, A. Rao et al.: Ap. Phys. Letters 1999, 75(9), 1329–1331.
[9] Z. Youshou, X. Yiyu, L. Sinian, et al., Mater. Sci. F. 2005, 488-489, 897–900.
[10] L. Sinian, S. Souzhi, Y. Tianqin, et al., Mater. Sci. F. 2005, 488-489, 893–896.
[11] Y. Morisada, H. Fujii, T. Nagaoka, M. Fukusumi, Mater. Sci. Eng. A, 2006, 419, 344–348.
[12] J. W. McCoy, C. Jones, F. E. Wawner, SAMPE Quart. 1988, 19(2), 37–50.
[13] R. Mehrabian, R. G. Riek, M. C. Flemings, Metal. Trans. 1974, 5, 1899.
[14] E. F. Fascetta, R. G. Riek, R. Mehrabian, M. C. Flemings, Trans. AFS. 1973, 81, 81.
[15] R. Mussi, T. Motegi, F. Tanabe et al., in Proceedings: MP3. 2005, 4, 211–215.
[16] R. Mussi, T. Motegi, F. Tanabe et al., Mater. Sci. F. 2006, 519-521, 609–614.
[17] S. Eckert, B. Willers, G. Gerbeth, Metal. Mater. Trans. A 2005, 36A, 267–270.
[18] S. Eckert, G. Gerbeth, V. I. Melnikov, Experiments in Fluids 2003, 35, 381–388.
[19] G.I. Eskin, Advanced Performance Materials 1997, 4(2), 223–232.
[20] C. Vives, JOM 1998, 50(2).
[21] Radjai, K. Miwa, Metal. Mater. Trans. A 2002, 33A, 3025–3030.
[22] Y. Genma, Y. Tsunekawa, M. Okumiya et al., JIM 1997, 38(3), 232–239.
[23] G. I. Eskin, D.G. Eskin, Ultrasonics Sonochemistry 2003, 10(4-5), 297–301.
[24] Y. Tsunekawa, H. Suzuki, Y. Genma, Materials & Design 2001, 22(6), 467–472.
[25] Ohno, T. Motegi, H. Soda, Iron and Steel Inst. Japan 1971, 18–23.
[26] F. Tanabe, T. Motegi, E. Sugiura, J. Japan Inst. Metals, 2003, 67(6), 291–294.
[27] E. Yano, N. Wada, N. Nishikawa, T. Motegi, J. Japan Inst. Metals, 2002, 66, 1131–1134.
[28] T. Motegi, E. Yano, N. Wada, Y. Tamura, Mater. Sci. F. 2003, 419–422, 605–610.
[29] H.K. Moon, PhD Thesis, MIT, EUA, 1990, 104–115.
[30] V. Laurent, P. Jarry, G. Regazzoni, and D. Apelian, J. Mater. Sci. 1992, 27, 4447–4459.

Thresholds in Creep Behaviour of Magnesium Alloy Matrix Composites with Short Alumina Fibres

K. Milika, F. Dobeš
Institute of Physics of Materials AS CR, Brno

1 Introduction

Many of contemporary models of creep behavior of metal matrix composites (MMCs) are based on threshold stress concept. According to this concept, the dependence of the minimum creep rate $\dot{\varepsilon}_{min}$ on the applied stress σ is described as

$$\dot{\varepsilon}_{min} = f(\sigma - \sigma_{th}) , \tag{1}$$

where σ_{th} is the stress under which the material investigated does not creep at a given temperature. The function f has usually a power form, so that

$$\dot{\varepsilon}_{min} \propto (\sigma - \sigma_{th})^{n_{th}} , \tag{2}$$

where n_{th} is the threshold exponent. Both parameters σ_{th} and n_{th} are crucial from the point of the applicability of various models. In principle, a prescription or definition of one of these parameters can strongly influence the value of the second parameter in the description of given data set.

Li and Langdon [1] summarized common features of creep behavior of MMCs and analyzed creep behavior of an AZ91 magnesium alloy reinforced with short alumina fibers at temperatures from 473 to 673 K. A marked threshold behavior was observed in creep of this composite consisting in a drastic decrease of the creep rate $\dot{\varepsilon}_{min}$ when the stress decreases to a certain value. However, such marked threshold behavior is not quite typical for MMCs reinforced with short fibers. In contradiction with Li and Langdon, other authors [2-4] investigating MMCs Al alloy and Mg alloy matrices reinforced with short fibers observed apparent Norton power dependence

$$\dot{\varepsilon}_{min} \propto \sigma^{n_c} \tag{3}$$

with a constant parameter n_c at a given temperature in wide experimental interval of stresses.

In the present contribution, the threshold description is applied and verified on creep data for (i) magnesium alloy AS21 reinforced by 27 vol.% of randomly planar distributed short Saffil fibers and (ii) alloy ZE41 based composites with 15 and 20 vol.% of Saffil fibers with very similar dimension and distribution parameters. Both matrix alloys and pure magnesium were also tested for referential purposes.

2 Experimental

Metal matrix composites with two different matrix magnesium alloys were used for experiments. The first matrix was a magnesium alloy AS21 of the chemical composition (in wt.%) 2.2 Al, 1.0 Si, 0.1 Mn and Mg balanced. Composite with the volume fraction $f = 27$ vol.% of Saffil fibers were prepared by squeeze casting in the Zentrum für Funktionswerkstoffe, Clausthal-Zellerfeld, Germany. The cast blocks had rather parallelepiped shape with dimensions approximately $50 \times 90 \times 90$ mm. Fiber pre-form had a shape of rectangular parallelepiped $25 \times 25 \times 70$ mm and was situated at the centre of the bottom of the blocks. Fiber pre-form consisted of planar randomly distributed short δ-alumina fibers. The fibers contained 97 % Al_2O_3 and 3 % SiO_2 and their average diameter and length were $d_r = 3$ μm and up to $l_r = 150$ μm, respectively. Thus, the maximum aspect ratio l_r / d_r of used distributed fibers reached the value high value 50. The reference monolith alloy AS21 was also prepared by squeeze casting. Cast blocks of the alloy and the composite were not subjected to a further heat treatment procedure before creep testing. The average grain size of the alloy was approximately 0.3 mm.

The alloy ZE41 and both their composites were also prepared by squeeze casting. The nominal chemical composition of the alloy is: (in wt.%) 4.2 Zn, 0.7 Zr, 1.2 RE (rare-earth element) and Mg balanced. Two different Saffil pre-forms with nominal volume fractions $f = 15$ and 20 vol.% of fibers were used. Other parameters of the fiber distribution were very similar in all three investigated composites. All three ZE41 matrix materials were heat treated according T5 and ageing at 450 K for 16 hours.

Magnesium of purity approximately 99.95 % was also prepared by squeeze casting in order to eliminate possible influences of preparation.

Parallelepiped creep specimens with dimensions $6 \times 6 \times 12$ mm were cut from the cast blocks of all composites so that their longitudinal axis was parallel with planes of preferential fiber positions.

Compressive creep tests under constant stress σ were performed using special IPM creep machine [5]. Identical temperature regime was kept before the test loading. Tests were performed in a protective argon atmosphere.

All procedures for the optimization of parameters in equations for description of obtained creep data were based on the least square method.

3 Results and Discussions

Three usual creep stages, i.e., primary, secondary and tertiary creep were observed in creep curves of all materials investigated. As a rule, the secondary creep has not been established and, therefore, the minimum creep rates $\dot{\varepsilon}_{min}$ could only evaluated from the curves. This rate was taken as the characteristic of the creep resistance in following considerations.

Applied stress dependences of the minimum creep rate $\dot{\varepsilon}_{min}$ for the alloy AS21 and corresponding dependences for magnesium are compared in Fig. 1. As the first step, the Norton creep relationship (3) with constant exponent n_c was fitted for a description of all dependences. It can be seen that drawn full straight lines fit well the data in all cases. The obtained value $n_c = 7$ resulting for Mg agrees well with those reported for coarse grained Mg in literature [6]. Therefore, this value is taken as the basic referential value of the stress exponent in the following analyses of threshold creep behavior. In fact, such an approach can be considered rather phenomenological without detailed knowledge of mechanisms controlling creep of pure magnesium.

In accord with this approach, possible threshold behavior described with $n_{th} = 7$ was proved for a description of $\dot{\varepsilon}_{min} = \dot{\varepsilon}_{min}(\sigma)$ dependences of the alloy AS21. Note that the enhanced creep resistance of this alloy in comparison with Mg is given by solid solution hardening (probably minor part) as well as by precipitation hardening (major contribution). Result of the fitting the AS21 data by Eq. (2) with $n_{th} = 7$ is illustrated in Fig. 2 by dashed lines. Apparently, such a description can be considered justifiable in the whole experimental stress range. The obtained threshold stresses characterize phenomenologically the influence of the simultaneous integral solid solution and precipitation hardening in creep of the alloy. However, their direct physical explanation is not simple.

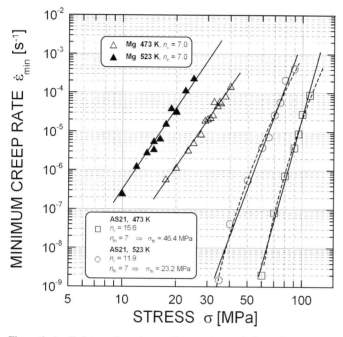

Figure 1: Applied stress dependences of magnesium and alloy AS21

The applied stress dependences of the rate $\dot{\varepsilon}_{min}$ of the composite AS21 with Saffil fibers are plotted in Fig. 2. The statistically quite acceptable Norton law description (Eq. 3) – see full straight lines – yielded similar exponents $n_c = 14.9$ and 14.1 for both temperatures 473 and 523 K, respectively. As the further step, an attempt to describe these dependences by means of the common relationship

$$\dot{\varepsilon}_{min} \propto (\sigma - \sigma_{th,exp})^{n_{th,exp}} \qquad (4)$$

with non-prescripted („free") threshold stress exponent $n_{th,exp}$ yielded quite different values of this exponent for both temperatures, as the corresponding fitting – see dotted lines – is apparently good from statistical point of view. Therefore, the description with a „free" exponent $n_{th,exp}$ seems to be questionable.

Data fitting by means of the pre-scripted exponent $n_{th} = 7$ is plotted by dashed lines. The fitting is surprisingly good – statistically at least comparable with the Norton law fitting. Optimum values of the stress σ_{th}, i.e., $\sigma_{th} = 83$ MPa for 473 K and $\sigma_{th} = 61.7$ MPa for 523 K, were obtained.

The applied stress dependences of the minimum creep rate $\dot{\varepsilon}_{min}$ for the alloy ZE41 and its both composites investigated are plotted in Fig. 3. The shapes of all dependences clearly disqualify the description by means of the Norton law (Eq. 3) with a constant exponent n_c; the shapes correspond to a threshold behavior. Two possibilities of the descriptions of these dependences by threshold relationship were proved.

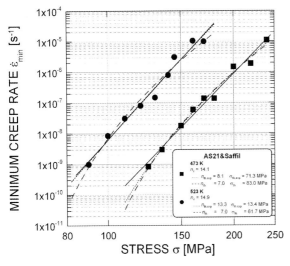

Figure 2: Applied stress dependences of the AS21 based composite

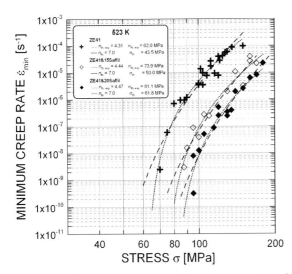

Figure 3: Applied stress dependences of the minimum creep rate for the alloy ZE41 and both its composites

First, the common threshold stress description with a non-prescript („free") exponent $n_{th,\,exp}$, Eq. (4) was fitted - see dotted lines. Practically identical values of the exponent $n_{th,\,exp} \approx 4$ were obtained in all three materials. Such exponents correspond well to the description of creep behavior of dispersion strengthened alloys suggested originally by Lagneborg [6] in 1968. Threshold stresses 62.0, 73.9 and 81.1 MPa were calculated for the alloy ZE41, composites ZE41&15Saffil and ZE41&20Saffil, respectively.

Second, a description with the pre-scripted exponent $n_{th} = 7$ corresponding to pure magnesium was proved – see dashed lines. Statistically, such description is not quite suitable in comparison with the description by means of the „free" exponent $n_{th,\,exp} \approx 4$. The fitting yielded optimum values σ_{th} = 43.5, 53.0 and 61.8 MPa for the alloy ZE41, composites ZE41&15Saffil and ZE41&20Saffil, respectively. All these values are substantially lower than the relevant values of the stress $\sigma_{th,exp}$. From the course of the fitted lines, it is obviously seen that values for $n_{th} = 7$ are apparently undervalued. In contrast to the composite with AS21 matrix materials, the description of the data of all three ZE41 alloy based materials by means of the „free" exponent $n_{th,\,exp}$ is much better in comparison with the description by means of $n_{th} = 7$.

The threshold description with $n_{th} = 7$ was tested on the data of Li and Langdon [1], Fig. 4, for the composite with AZ91 alloy prepared identically as our experimental materials. In contrast to our testing, these authors used double-shear creep tests. Li and Langdon assumed that at lower applied stresses σ the creep behavior can be described by threshold stress exponent $n_{th} = 3$ which corresponds to viscous glide process. At higher stresses the exponent n_{th} increases, $n_{th} > 3$, which the authors attribute to a breakaway of dislocation from their solute atom atmospheres. The estimated threshold stresses then arise from an attractive interaction between dislocations and precipitates in the matrix.

The description of the data based on $n_{th} = 7$ is compared with the description used by mentioned authors in Fig. 4. It is apparent from the figure that the former description (dotted lines) seems statistically better than the latter fitting beyond a great scatter of the data.

The description using threshold exponent $n_{th} = 7$ can be simply interpreted as phenomenological. However, such description can be also explained as an approach based on an assumption that the basic creep mechanisms in threshold behavior are identical with the mechanism governing the creep of pure magnesium. Then, the experimental threshold stresses should characterize integral influence of all phenomena in structure contributing to threshold behavior. However, the behavior of the ZE41 alloy and its composites does not satisfy such conception. One of the possible explanations of different behavior of these materials could consist in the fact that used testing temperature 523 K is greater than temperature of the last step of the heat treatment T5 applied on all ZE41 materials, 450 K for 16 hours. Simultaneous influence of creep strain and temperature resulting in eventual changes in parameters of dispersed phases could be the cause of different behavior. However, such explanation is purely hypothetical and its verification needs detailed structure investigation.

4 Acknowledgements

The work was supported by Grant Agency of the Czech Republic under contract No. 106/06/1354.

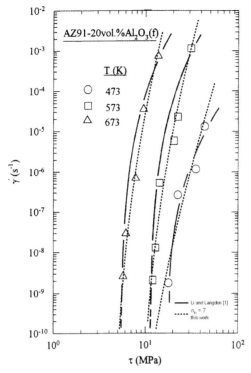

Figure 4: Results of an application of the threshold description with the exponent $n_{th} = 7$ on the data of Li and Langdon [1]

5 References

[1] Li, Y.; Langdon T. G., Met. Mat. Trans A **1999**, *30A*, 2059–2066.
[2] Yawny, A.; Kausträter, G.; Skrotzki, B.; Eggeler, G., Scripta Mater. **2002**, *46*, 837–842.
[3] Kuchaová, K.; Horkel, T.; Dlouhý, A., In: *Creep Behaviour of Advanced Materials for the 21st Century.* - Warrendale, Minerals, Metals and Materials Society 1999. p. 127–136.
[4] Milika, K.; Dobeš, F. Journal of Alloys and Compounds **2004**, *378*, 167–171.
[5] Dobeš, F.; Zvina, O.; adek, J., J. Test. Eval. **1986**, *14*, 271–273.
[6] Frost, H. J.; Ashby, M. F., DEFORMATION-MECHANISM MAPS - The Plasticity and Creep of Metals and Ceramics - Pergamon Press **1982**, p. 49.
[7] Lagneborg, R., In: *Proc. 3rd Int. Conf. on Strength of Metals and Alloys.* Cambridge, England, 20-25 Aug. **1973**. Vol. 1, 316–320.

Anisotropy of Dilatation Characteristics in Mg and Mg4Li Fibre Composites

A. Rudajevová, P. Lukáč
Charles University, Prague

1   Introduction

It is well known that materials composed of two or more components with different coefficients of thermal expansion (CTE) may be thermally deformed at room temperature after fabrication. This deformation and its development with increasing temperature may influence a number of physical properties, which may be used to study this phenomenon. The thermal strains were studied, for example, by neutron diffraction [1] or by dilatometer [2]. In both cases, it can be seen that the measured parameters, i.e. the lattice constant and the relative elongation, exhibit hysteresis after heating and cooling takes place. The thermal stresses can result in asymmetric stress-strain responses under tensile and compressive loading [3].

In the literature, we can find numerous papers dealing with the study of fibre composites with a longitudinal orientation. There are only a few investigations on thermal expansion of composites with other fibre-orientations. The most recent papers dealing with this problem were published by Kumar et al. [3, 4].

The aim of this work is to investigate the thermal expansion behaviour of Mg-10 vol.% Saffil fibre and Mg4Li-10 vol.% Saffil fibre composites with either a parallel or a longitudinal orientation of 2D-randon fibre array planes with respect to the specimen axes. The results will be analyzed using the elastic-plastic model in the temperature range from room temperature up to 380 °C.

2   Experimental Details

The specimens of Mg-10 vol.% Saffil fibre and Mg4Li-10 vol.% Saffil fibre composites were prepared by pressure infiltration of Saffil fibre pre-forms in a laboratory autoclave under an argon pressure of up to 6 MPa at temperatures of 888–908 K in the Institute of Materials and Machine Mechanics, in Bratislava, Slovakia.

The linear thermal expansion of the composite specimens was measured in an argon atmosphere using a Netzsch 402E dilatometer over the temperature range from room temperature to 380 °C at heating and cooling rates of 5 K/min. The specimens studied were 6 mm in diameter and 50 mm in length. The thermal expansion curves for composites were measured during three heating and cooling cycles. The results obtained in the second thermal cycle were the same as those in the third cycle. Therefore, in the following, dilatation characteristics obtained in the first and second thermal cycles will be presented and discussed. The composites were investigated in the as-prepared state.

The Saffil fibres remain predominantly in the planar-random arrangement. Two types of composite specimen were used: a) planes of planar randomly distributed short fibres lay parallel to the longitudinal axis – LD composites; and b) planes of short fibres were perpendicular to the specimen axis – TD composites.

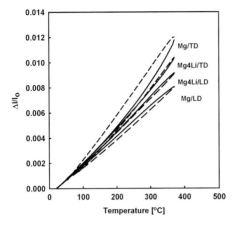
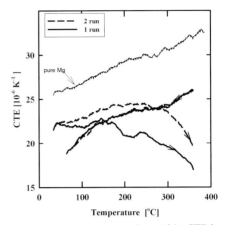

Figure 1: Temperature dependence of the relative elongation for the LD and TD Mg-10 vol.% Saffil fibre and Mg4Li-10 vol.% Saffil fibre composites

Figure 2: Temperature dependence of the CTE for the LD Mg-10 vol.% Saffil fibre composite

3 Results

Figure1 shows the temperature dependence of the relative elongation of Mg-10 vol.% Saffil fibre and Mg4Li-10 vol.% Saffil fibre composites for LD and TD ordering of the fibres. Results were obtained in the second thermal cycle. During heating the relative elongation increases faster for TD composites than for LD composites. Hysteresis is perceptible on all temperature dependences of the relative elongation. The temperature dependence of the CTE for pure Mg and Mg-10 vol.% Saffil fibre composite with the LD orientation of the fibres is plotted in Figure 2. The CTE during heating of the as-cast composite has different character than the CTE in the second and further thermal cycles. The CTE for cooling is the same for all thermal cycles. The CTE decreases from 250 °C during heating. However the CTE increases with increasing temperature during heating for TD composites (Figure 3) in the whole temperature range studied. The CTE during heating is higher for TD composites than for LD composites. Figure 4 shows temperature dependence of the thermal and residual strain for LD and TD Mg-10 vol.% Saffil fibre composites. The relative elongation measured of the composite in the dilatometer $(\Delta l / l_0)_{meas}$ may be expressed as

$$\left(\frac{\Delta l}{l_o}\right)_{meas} = \frac{\Delta l}{l_o} + \left(\frac{\Delta d}{l_o}\right)_{RS} + \left(\frac{\Delta d}{l_o}\right)_{TS} \qquad (1)$$

where $(\Delta l / l_0)$ is the relative elongation due to thermal vibrations, $(\Delta d / l_0)_{RS}$ is the residual strain and $(\Delta d / l_0)_{TS}$ is the thermal strain. $(\Delta l / l_0)$ can be determined by the rule of mixtures and it increases with increasing temperature. While the temperature dependence of the residual and thermal strain is presented for the first thermal cycle only the thermal strain exist in the composite in further thermal cycles. The residual strain is removed during the first thermal cycle.

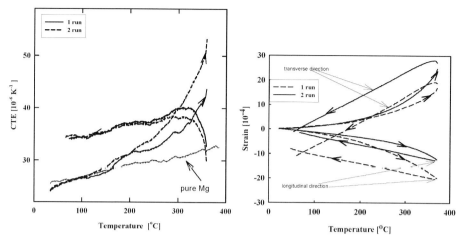

Figure 3: Temperature dependence of the CTE for the TD Mg-10 vol.% Saffil fibre composite

Figure 4: Temperature dependence of the residual and thermal strain for the LD and TD Mg-10 vol.% Saffil fibre composite (1 run - the residual strain and thermal strain, 2 run - the thermal strain)

The similar results as for Mg-10 vol.% Saffil fibre composites were obtained for Mg4Li-10 vol.% Saffil fibre composites. The temperature dependences of the CTE only for heating and the second thermal cycle are shown in Figure 5. While the CTE decreases from 300 °C for LD composite, that increases from 300 °C for TD composite.

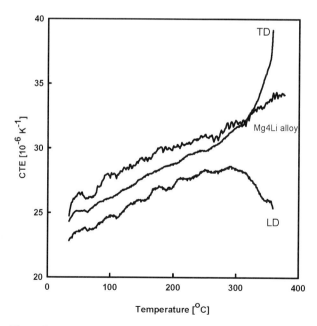

Figure 5: Temperature dependence of the CTE for the LD and TD Mg4Li-10 vol.% Saffil fibre composite (heating in the second thermal cycle)

4 Discussion

If the thermal and elastic properties of the composite components are different, then the internal thermal stresses and with them the associated thermal strains are present in the composite. We differentiate, in principle, two types of deformation, residual strain and thermal strain. The residual strain depends on the thermal and mechanical history of the material and it may have various origins, mechanical, thermal or both. It may be removed by heating, which is associated with the permanent change of the specimen length. Thermal strain is the strain due only to temperature changes during thermal cycling. It does not depend on the thermal history of the composite. It is consequence of the difference of the CTE of the composite components and it depends on the elastic modulus and the matrix yield stress.

While the residual strain is connected with a permanent length changes after the thermal cycle, the thermal strain influences the expansion characteristics during heating and cooling so that no permanent strain occurs after thermal cycle. The different course of the thermal strain during heating and cooling causes hysteresis. Figure 1 shows the influence of the thermal strain on course of the temperature dependence of the relative elongation for all composite investigated in this work. Smaller influence of the thermal strain on the relative elongation was found for Mg4Li-10 vol.% Saffil fibre composite than for Mg-10 vol.% Saffil fibre composite. This difference is a consequence of different chemical composition of the interface between matrix and reinforcement [5].

Appreciable difference of the dilatation characteristics of the LD and TD Mg-10 vol.% Saffil fibre composites is perceptible from the figures 2 and 3. While absolute values of the CTE of the TD composites are higher than for the pure Mg, that one are lower for the LD composites. Decrease of the CTE for the LD composites and increase of the CTE for TD composites are a consequence of the stresses in the matrix. The tensile stress and with them connected tensile strain exist in the matrix in direction of the fibres at room temperature. During heating the tensile stress and strain decreases (to reach the free stress state) in matrix and macroscopic reduction of the sample length occurs. Macroscopic elongation of the sample occurs during heating in the TD composites because the compression strain in the matrix decrease. Release of the residual and thermal strain during the thermal cycle arises only during heating as that is shown in Figure 4. Release of the both kinds of strains is elastic in the whole temperature range [6]. Above a certain temperature when the yield stress of matrix is reached the plastic strain in matrix near the interface between matrix and reinforcement influences this process and release of the thermal strain is faster.

Figure 5 shows that the CTE of Mg4Li alloy is nearly the same as for pure Mg. The pure Mg and Mg4Li alloy have also nearly the same mechanical properties [7]. Character of the temperature dependence of the CTE for LD and TD for the Mg4Li-10 vol.% Saffil fibre composite is nearly the same as that for Mg-10 vol.% Saffil fibre composite. A difference is found in the transition temperature between elastic and elastic-plastic mechanism of release of the thermal strain. While this transition temperature is about 250 °C in the LD-Mg based composite that is about 300 °C for the LD-Mg4Li based composite. The temperature dependence of the CTE for the TD-Mg based composite does not show practically any linear part that means that the elastic-plastic mechanism of release of the thermal strain occurs from beginning of the thermal cycle. The temperature dependence of the CTE for TD-Mg4Li composite is linear up to 300 °C. The presence of the Li in the interface between matrix and reinforcement is evidently principal reason of the different behavior of the both composites.

5 Conclusions

Hysteresis was found in the temperature dependences of the relative elongation of the Mg-10 vol.% Saffil fibre and Mg4Li-10 vol.% Saffil fibre composites with the LD and TD orientation of the fibres over the temperature range from room temperature up to 380 °C. This effect is higher for Mg-10 vol.% Saffil fibre composites. The relative elongation of the LD composites is lower than the TD composites in the whole temperature range studied. Release of the thermal strain with increasing temperature is cause of these expansion characteristics. At low temperatures the release of the thermal strain occurs by elastic mechanism. Above a certain temperature when yield stress is reached plastic strain influences the release of the thermal strain that decreases faster. The transition temperatures between elastic and elastic-plastic mechanism are higher for Mg4Li-10 vol.% Saffil fibre composites. This is evidently effect of the Li in the interface between matrix and reinforcement.

6 Acknowledgements

This work is a part of the research program MSM 0021620834 that is financed by the Ministry of Education of the Czech Republic. The authors are also grateful for the support offered by the Grant Agency of the Czech Republic under Grant 106/06/1354.

7 References

[1] C. M. Warwick, T. W. Clyne: Fundamental Relationships Between Metal-Matrix Composites, Liaw PK, Gungor MN. editors. Indianopolis, TMS; **1990** 209.
[2] D. Mari, A. D. Krawitz, J. W. Richardson, W. Benoit, Mater. Sci. Eng. **1996**, A209, 197.
[3] S. Kumar, A. K. Mondal, H. Dieringa, K. U. Kainer, Comp. Sci. Technol. **2003**, 63, 1805.
[4] S. Kumar, S. Ingole, H. Dieringa, K. U. Kainer, Comp. Sci. Technol. **2004**, 64, 1179.
[5] S. Kúdela in: Fibre-Matrix Interaction in Mg Composites, PhD Thesis, Institute of Materials and Machine Mechanics, Slovak Academy of Sciences, Bratislava, **2002**.
[6] A. Rudajevová, P. Lukác,. Acta Mater. **2003**, 51, 5579.
[7] M. E. Dric, F. M. Elkin, I. I. Gurjev, B. I. Bondarev, F. V. Trochova, V. F. Sergejevskaja, T. N. Osokina : Magnijevo- litijevije splavy, „Metalurgija", Moskva, **1980**, 22.

Magnesium-Hydroxyapatite Composites as an Approach to Degradable Biomaterials

F. Witte*, J. Fischer**, P. Maier**, C. Blawert**, M. Störmer**, N. Hort**
*Hannover Medical School, Laboratory for Biomechanics and Biomaterials
**GKSS Research Centre Geesthacht, Institute for Materials Science, Magnesium Innovation Centre

1 Introduction

Previous in vivo studies have shown that magnesium alloys are suitable as biomaterials for degradable implants. Current research indicates that there is still a high demand to create magnesium alloys with adjustable in-vivo corrosion rates while the mechanical properties are not compromised. A novel approach to this challenge might be the application of metal matrix composites (MMC) based on magnesium alloys. MMCs have gained increasing interest in the last decade based on their higher specific stiffness, strength, creep resistance and minimized sensibility to galvanic corrosion [1]. The advantage to use MMCs as biomaterials are the adjustable mechanical properties (Young's modulus, strength) as well as the adjustable corrosion properties by choosing the right combination of components. An appropriate way to control magnesium corrosion is to elevate the surrounding pH of the magnesium alloy to stabilize the corrosion products in the protective corrosion layer. Especially calcium has been known for years to reduce the susceptibility of magnesium to corrode when added in amounts of a few tenths of weight percents [2, 3]. As a natural bone composite, hydroxyapatite (HA) is known to possess a low solubility in body environment [4, 5]. Therefore, HA particles seem to be suitable as reinforcements in magnesium based MMCs. In previous studies the magnesium alloy AZ91D has revealed local corrosion in vitro and in vivo [6, 7]. In this study a MMC made of AZ91D as a matrix and HA particles as reinforcements has been investigated in vitro for mechanical and corrosive properties.

2 Experimental

The MMCs were produced by mixing gas atomised AZ91D powder with 20 wt% of HA powder into an AZ31 can of 70 mm diameter. Subsequently, this packet was extruded at 400 °C to a diameter of 18 mm. The initial powder particle size was approximately 30 µm. Material samples were machined from the extruded rod.

Vickers hardness measurements ($HV_{0.1}$) and nanoindentation (load of 1000 µN) have been carried out for mechanical characterization of the material. Nanoindentation, using a modified scanning force microscope as a triboscope (Hysitron Triboscope™), has been employed. Both methods have a strong positive relationship to Young's modulus, yield and ultimate stress [8, 9]. Optical microscopy has been performed to obtain an overview of distribution and size of components within the material. Scanning electron microscopy, in detail EDX analyses, has been carried out to evaluate the composition of defined areas in the microstructure. The samples were prepared as reported by Kree et al. [10].

For corrosion experiments, cylindrical specimens of 10 mm diameter and 5 mm thickness were machined from the extruded rods. The two flat cylinder surfaces were grinded with 1200 grid SiC paper and cleaned in ethanol prior to corrosion tests. For the immersion tests a 1 mm diameter hole was drilled into the specimens providing a fixture for the specimens during testing. Immersion testing was performed for 24 and 72 hours in artificial sea water (DIN 50905), modified eagle medium (cell solution) and modified eagle solution with proteins (10 % fetal calf serum) at 37 °C. Three specimens for each condition were immersed in 200 ml of solution, without touching the container wall or each other. The solution temperature was controlled by a flow of heated water in a water bath in which the container were placed (37 ± 0.2 °C). The electrolyte was not stirred during the experiments and saturated with atmospheric oxygen. The corrosion rates were determined by weight loss measurements using chromic acid for removing the corrosion products.

The electrochemical tests were performed in artificial sea water on flat cylindrical surfaces (perpendicular to the extrusion direction) using an ACM Gill AC potentiostat. A sequence of tests, starting with 30 min measurement of the rest potential, followed by a single potentiodynamic polarisation and 72 hours single electrochemical impedance spectroscopy measurements (EIS) was used with 1 hour rest between each measurement. The electrolyte temperature (37 ± 0.2 °C) was controlled by a flow of heated water in the double wall electrolytic cell with an electrolyte volume of 333 ml. The electrolyte was stirred during the experiments and saturated with atmospheric oxygen. A three electrode set-up with an Ag/AgCl reference and a platinum counter electrode was used. After 30 min recording of the free corrosion potential, the polarization scan was started from –200 mV relative to the free corrosion potential with a scan rate of 0.2 mV/s. The test was terminated when a corrosion current density of 0.1 mA/cm^2 was exceeded. From the cathodic branch of the polarization curve the corrosion rate was determined using the Tafel slope. The EIS measurements were carried out over a frequency range from 30 kHz to 0.1 Hz. The amplitude of the sinusoidal signals was 10 mV. The resistance values (charge transfer resistance) obtained at 0° phase shift were used as an indicator of the corrosion resistance of the specimens.

For phase analysis by means of XRD the samples were investigated in parallel beam geometry, using Cu-K$_1$ radiation. The X-ray diffractometer (Bruker D8 Advance) with a line focus was equipped with a Göbel mirror and a 1 mm slit on the primary side. On the secondary side there are a 0.6 mm backscattering slit and a 0.2 mm detector slit. The diffraction patterns were measured from 2θ (8–92°) for each sample. The increment was 0.04° with a step time of 16s. For the qualitative phase identification, the database PDF-2 (Release 2002) from the International Center for Diffraction (ICDD) was used [11].

The SRμCT experimental set-up at the Hamburger Synchrotronstrahlungslabor (HASYLAB) at Deutsches Elektronensynchrotron (DESY) was used at beamline BW2 at a photon energy of 24 keV. To perform in-situ corrosion measurements a cuvette was made of a radiation resistant polymer (PEEK) to contain the corrosive solution and the magnesium sample during the tomographic measurements. Artificial sea water with an initial pH of 7.4 and a temperature of 37 °C was used as a corrosive solution. The corrosive solution in the cuvette was constantly exchanged (50ml/min) by a peristaltic pump from a reservoir with constant temperature and pH. Cylindrical samples (MMC-HA: 3.02 mm diameter, 5.0 mm height, AZ91D: 3.13 mm diameter, 5.0 mm height) for in-situ corrosion measurements were immersed in the cuvette and attenuation images were acquired every 45 minutes during a period of 20 hours. The initial sample volumes used to observe the corrosion were slightly different due to limited machining abilities (MMC-HA: 12.93 mm³; AZ91D: 13.81 mm³). The data transfer from the CCD was accelerated by double binning of the image data. Furthermore, a step-size of 0.5° rotation angle was chosen to decrease the overall time for scanning a single tomogram to 30 minutes for a rotational range from 0° to 180°. A voxel edge length of 4.7 µm was obtained. Changes in surface morphology and remaining metallic sample volume were observed. The corrosion rate was calculated according to [7].

3 Results and Discussion

The HA formations appeared to be very brittle and were difficult to indent without damaging them. The average $HV_{0.1}$ of the HA conglomerates was 111, while the average $HV_{0.1}$ of the MMC-HA was 73. However, single hardness values of MMC-HA varied from 60 to 95 depending on the contribution and size of apatite conglomerates. Low hardness ($HV_{0.1}$ 60) was observed with large and inhomogenously distributed conglomerates, while finer and more homogenously distributed conglomerates lead to higher hardness ($HV_{0.1}$ 95). This shows that size and distribution of the HA reinforcements can change the overall hardness of the composite material compared to standard magnesium alloys (up to $HV_{0.1}$ 80) [12, 13].

A limitation of the nano-indentation measurement was that the diamond tip could slip off the apatite formation, or even shifting them, resulting in wrong hardness values. Due to the brittleness and lose contact of the HA conglomerates in the magnesium matrix, the tensile and compressive strength of the composite is not expected to be significantly higher than of non-reinforced magnesium alloys. Since the microhardness of cortical bone is reported be to 49.8 [14], the tested MMC-HA seemed to be in closer agreement to the characteristics of the natural bone, which compressive yield strength is 130–180 MPa [15]. The Young's modulus of the matrix has been measured with nano-indentation at 40 GPa, which is typical for Mg alloys.

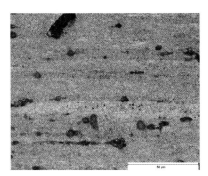

Figure 1: Microstructure of the longitudinal direction of MMC-HA. The extrusion direction is orientated horizontally

The microstructure of the sample in longitudinal direction revealed that the grains in the extrusion direction appeared as horizontal elongated grains (Fig. 1). The magnesium matrix microstructure consisted of small grains of a size of a few micrometers (1-5 µm). Furthermore, bands of larger grains were observed, due to incomplete recrystallization. HA conglomerations were inhomogeneously spread over the sample, seen as black formations of various sizes. The size differed from less than a micrometer up to 20 µm and even more if agglomerated. However, a few HA formations showed possible facets of crystallized areas.

The EDX analysis of the matrix material revealed a typical composition for an AZ91D magnesium alloy, except the slightly enriched calcium, which could be attributed to the HA additions. EDX analysis of HA conglomerate revealed amounts of phosphorous and calcium which are typical for calcium phosphate or rather HA. The representative areas of magnesium matrix and HA formation were clearly identified by EDX analysis (Table 1).

Table 1: Composition in wt.%

Position	Composition in wt.%					
	Al	Zn	Si	Ca	P	Mg
Matrix	8.31	1.09				bal.
Apatite conglomerates	1.14		1.77	61.21	31.12	4.75

After 24 and 72 hours of immersion in artificial sea water and cell solutions with and without protein addition (Fig. 2) the weight change was measured. The specimens in artificial sea water exhibit high corrosion attack which is consistent with a large weight gain and a rough surface covered with various corrosion products. The specimens immersed in cell solutions with and without protein revealed glassy surface layers with black underlayers on the specimens. The protein addition to the cell solution resulted in a visible gas evolution at the beginning of the immersion test, which stopped after a couple of hours. After the removal of the corrosion products with chromic acid the weight loss indicated that there was corrosion attack on all specimens. It is most likely that especially the strong localised corrosion attacks responsible for the major weight loss in the immersion tests required some initiation time.

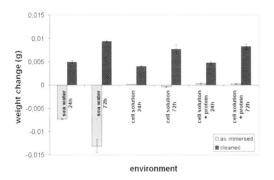

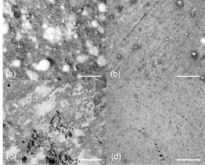

Figure 2: Weight changes of immersed MMC-HA samples in artificial sea water, cell solutions with and without proteins after 24 and 72 hours

Figure 3: Surface appearance after 72 hours of immersion in a) artificial sea water, b) cell solution with protein, c) artificial sea water (cleaned) and d) cell solution with protein (cleaned)

The specimens in sea water (Fig. 3 a, c) suffered from strong localised corrosion attack, while all the specimens in cell solution showed a more uniform corrosion attack (Fig. 3 b, d). The conversion of the weight losses into corrosion rates revealed that in all solutions the corrosion rate is decreasing with increasing immersion time. The lower chloride concentration in the cell solution was most likely the reason for the different corrosion morphologies.

Addition of proteins to the cell solution increased the corrosion rate of the MMC-HA slightly, but did not change the corrosion morphology. The protective layer on the specimens immersed in cell solution was more uniform and the whole surface was covered with corrosion products. Even though the corrosion rates in cell solution and artificial sea water did not differ very much, the more uniform corrosion in cell solution should be much more suitable for mechanically loaded implants. In cell solutions a uniform amorphous surface layer formed which protected successfully the surfaces. Even in the presence of chloride ions in cell solution the MMC-HA specimens did not suffer from localised corrosion. However, in sea water the protective layer was less uniform and pitting occurred. In contrast to many other Mg based MMCs, the corrosion rate of the AZ91D matrix was not enhanced by HA reinforcements. From the corrosion point of view the HA particles act neutral and might be integrated into the corrosion layer stabilizing the corrosion products. Further studies will concentrate on the role of HA particles in the passive layer formation.

The short term polarisation measurements of MMC-HA resulted in a corrosion rate of 1.25 ± 0.16 mm/year which were lower than the corrosion rates determined in the immersion test (2.0–3.2 mm/yr). In the EIS measurements there was a general trend that the charge transfer resistance was increasing from about 1000 Ωcm^2 to 1300 Ωcm^2 with longer immersion times. However, a larger scatter was found ranging from 750 to 1600 Ωcm^2 occurring randomly over the whole test period. The scatter in electrochemical measurements was interpreted as break down and reformation of the protective surface layer. The break down was responsible for the observed localised corrosion. Break down of the protective films occurred randomly over the whole period of the test.

To characterize the raw material by means of XRD, pure HA powder was investigated first. The diffraction pattern showed the reflections expected for HA $Ca_{10}(PO_4)_6(OH)_2$ with a hexagonal unit cell ($a = 9.424$ Å and $c = 6.879$ Å) [16]. The determination of the lattice parameter has shown a precision of better than 0.01 Å. The strong lines of HA were in the range of $2\theta = 25°$ to $55°$. The diffraction pattern of the extruded MMC-HA sample exhibited dominant hcp-Mg reflections. In addition, there were reflections from the $Mg_{17}Al_{12}$-phase, which originated from the AZ91D matrix and clear reflections from HA could be observed.

XRD measurements for specimens immersed to artificial sea water showed again reflections from hcp-Mg, $Mg_{17}Al_{12}$, HA, $CaCO_3$ and $Mg(OH)_2$. The following lattice parameters were determined: $a = 3.19$ Å and $c = 5.20$ Å for hcp-Mg and $a = 10.59$ Å for bcc-$Mg_{17}Al_{12}$. The diffraction pattern of HA matches again quite well to the pattern 74–565 from the PDF-2 database [16]. The positions of the reflections from the HA particles have been found to match nearly those of the powdered sample. However, some new reflections at $2\theta = 26.1°$, $27.1°$, $33.0°$ and $45.8°$, belonging to a new orthorhombic phase were found. The lattice parameters were determined at $a = 4.97$ Å, $b = 8.00$ Å and $c = 5.76$ Å. Based on the components of the artificial sea water, calcium carbonate was identified by means of phase determination. The carbonate originates from the artificial sea water and therefore the weight of the samples increased after immersion testing (Fig. 3). Due to the low solubility of HA, it is most likely that the calcium of the precipitated calcium carbonate originates from calcium chloride, which is dissolved in the sea water.

After immersion testing for 24 and 72 hours in cell solutions without and with protein, the diffraction patterns were quite similar that of the non-immersed MMC-HA sample. The reflections can again clearly be assigned to hcp-Mg, bcc-$Mg_{17}Al_{12}$, and HA. No reflections from $CaCO_3$ were found after immersion in cell solution with and without protein addition. At lower 2Θ angles, the background was increased and the peaks were broad which might be due to the fact that the respective top layers were amorphous. The comparison of the sample surfaces after 24 and 72 hours of immersion supports the later assumption, because the intensity of the reflections of $CaCO_3$ increases form 24 to 72 hours of immersion. On the other hand, the peak intensities of hcp-Mg decrease with increasing immersion time, which might be also affected by an increase in surface roughness. In summary, we have found clear evidence of the formation of calcium carbonate after immersion in artificial sea water. This explains the formation of white crystals on the sample surfaces. The phase identification showed that all samples contained hcp-Mg, $Mg_{17}Al_{12}$, and HA before and after immersion. The observation of a broad peak indicated that the top layer was covered by amorphous phases. After immersion of MMC-HA samples in cell solutions with and without proteins, we did not find any hints for the formation of $CaCO_3$.

In general, low corrosion attack occurred on the MMC-HA sample. Initial pit formation was observed 2-3 hours after immersion in artificial sea water in SRµCT. The small pits continuously grew until 9 hours after immersion, while higher X-ray absorbent precipitations were observed homogenously distributed on the whole MMC-HA surface. These precipitations continuously accumulated on the surface until the end of the experiment. The pitting formations were fully covered by these precipitations after 9 hours and the growth rate of the pits decreased nearly completely. In contrast, the AZ91D sample exhibited severe pitting after immersion in artificial sea water. The pits grew faster and deeper than in the MMC-HA sample. The precipitations were found on uncorroded sample surface as well as a cover on some shallow pits, acting as a protective layer. XRD analyses revealed that $CaCO_3$ was found in these precipitations. These precipitations seemed to reduce the local corrosion attack and led to more uniform corrosion morphology, especially with MMC-HA, while uncovered areas of AZ91D exhibited severe deep pitting and high corrosion rates. The observed sample volume of AZ91D decreased about 2.31 % (0.32 mm^3) of its metal volume after 20 hours of immersion, while the changes in the samples volume of MMC-HA were to low for a reliable determination. Artefacts from small high absorbing areas occurred during the reconstruction procedure and interfered with the thin corrosion layer of the MMC-HA sample. Therefore, the number of voxels changing their grey values according to corrosion products in the MMC-HA sample was to low to calculate a reliable change in the sample volume. In contrast, the prominent corrosion of AZ91D in artificial sea water facilitated the calculation of the corrosion rate (7.94 mm/yr) according to [7].

4 Conclusion

In summary, our results clearly show that magnesium-based metal-matrix composites are novel biomaterials with adjustable mechanical and corrosive properties. The distribution and size of the hydroxyapatite reinforcements are of major importance for mechanical and corrosive properties. In contrast to previous studies, the addition of hydroxyapatite reinforcements enhanced the corrosion resistance of the magnesium matrix in artificial sea water and cell solutions. Further studies will focus on combinations of different magnesium alloys as a matrix material and on the selection of different biocompatible reinforcements.

5 References

[1] Clyne TW. Vol.3 of Comprehensive Composite Materials. Pergamon, 2000.
[2] Beck A. Magnesium und seine Legierungen. Berlin: Springer, 1939.

[3] Emley EF. Principles of Magnesium Technology. Oxford: Pergamon Press, 1966.
[4] Fulmer MT, Ison IC, Hankermayer CR, Constantz BR, Ross J. Biomaterials 2002, 23(3), 751–755.
[5] Tadic D, Epple M. Biomaterials 2004, 25(6), 987–994.
[6] Witte F, Kaese V, Haferkamp H, Switzer E, Meyer-Lindenberg A, Wirth CJ et al. Biomaterials 2005, 26(17), 3557–3563.
[7] Witte F, Fischer J, Nellesen J, Crostack HA, Kaese V, Pisch A et al. Biomaterials 2006, 27(7), 1013–1018.
[8] Currey J, Brear K. J. Mater. Sci. Mater. Med. 1990, 1, 14–20.
[9] Evans P, Behiri J, Currey J, Bonfield W. J. Mater. Sci. Mater. Med. 1990, 1, 38–43.
[10] Kree V, Bohlen J, Letzig D, Kainer KU. Practical Metallography 2004, 41[5], 233–246.
[11] The Powder Diffraction File^{TM}. International Center for Diffraction Data (ICDD), PDF Release 2002, http://www.icdd.com 2002.
[12] Caceres C, Poole W, Bowles A, Davidson C. Materials Science and Engineering A 2005, 402, 269–277.
[13] Tabor D. The Hardness of Metals. London: Oxford University Press, 1951.
[14] Hodgskinson R, Currey JD, Evans GP. J. Orthop. Res. 1989, 7(5), 754–758.
[15] Staiger MP, Pietak AM, Huadmai J, Dias G. Biomaterials 2006, 27(9), 1728–1734.
[16] Sudarsanan K, Young RA. Acta Cryst. 1969, B25, 1534–1543.

Applications

AM-lite®: an Innovative New Alloy Opens Fresh Opportunities for Magnesium Applications

C.Kettler*, T.Fuest**, G.Stoesser**, T. Abbott*, M. Murray* and G.Dunlop*
* Advanced Magnesium Technologies, Heidelberg, Germany and Brisbane, Australia
** MacDermid Industrial Solutions, Forst, Germany

1   Abstract

The new magnesium alloy, AM-lite®, has been specifically developed to satisfy demand for a light weight diecastable alloy that is highly suitable for decorative surface finishing applications. The key technical attributes of AM-lite® are its light weight, metallic feel, excellent diecastability, oxidation resistance and suitability for decorative surface finishing operations such as electroplating and painting. Electroplating, using the Bondal® Mg pretreatment, results in excellent coating performance. The alloy offers cost savings compared to zinc, aluminium, magnesium alloy AZ91D and also injection moulded plastics, in the manufacture of surface finished articles.

2   Introduction

Diecasting provides the biggest market for magnesium alloys. However, at ~170 000 tpa, the current size of this market is very small compared with markets for zinc and aluminium diecasting alloys and also much smaller than that for injection moulded plastics. While markets for magnesium diecasting alloys continue to grow, it is clear that there would be considerably greater opportunities for magnesium if alloys could be developed that have more attractive properties compared with competitor materials.

AM-lite® has been specifically developed for a wide range of diecasting applications where diecastability and decorative surface finishes are of prime importance [1,2]. Currently AZ91D is the major magnesium diecasting alloy that competes against zinc, plastics, and to a lesser extent aluminium, for general diecast and moulded parts. This total market is very large and AZ91D only manages to capture a small portion of it. The new alloy, AM-lite®, complements AZ91D by addressing some of its deficiencies and is opening up new opportunities in areas where other materials have previously been favoured.

AM-lite® and its accompanying surface finishing technology is the product of intensive research and development by AMT, its research partner CAST, and MacDermid. The new alloy has considerable potential to expand markets for magnesium alloys into areas that have previously been the domain of zinc and aluminium diecasting alloys and injection moulded plastics.

The decorative diecasting market requires an alloy that has the light weight attributes of magnesium combined with the diecastability and surface finishing characteristics of zinc. Over the past two decades zinc has lost much of the automotive decorative market, primarily to plastics but also to aluminium and magnesium AZ91D, because of its weight. More recently the rapid rise in zinc prices has made it increasingly favourable to substitute zinc diecastings with other materials that can fulfil the requirements of decorative applications. More than three times as many parts per unit weight of metal can be produced from magnesium alloys compared to zinc. Thus, based on current metal prices, the cost of metal in diecast parts is ~70 % cheaper for AM-lite® than for zinc.

While magnesium AZ91D has many admirable attributes and has served as a work horse alloy for the magnesium industry for many years, it cannot fulfil all of the requirements required for a wide range of decorative applications. Most importantly, it is difficult and expensive to electroplate AZ91D. Moreover the diecastability and as-cast surface quality of AZ91D is considerably inferior to zinc so that diecast AZ91D generally requires extensive pre-work (polishing, buffing, filling) as part of surface finishing operations. AZ91D also has the typical disadvantages of magnesium alloys relating to melt losses in the foundry, due to dross and sludge formation, and the need for external recycling of casting returns.

Diecast AM-lite® has surface characteristics that are very suitable for surface finishing operations and MacDermid has developed specific chemical pre-treatments, Bondal® Mg for electroplating and Chemidize Mag Prime for painting and powder coating, that enable high quality durable electroplated and painted surface finishes to be applied.

3   General Features of AM-lite®

AM-lite® is an Mg-Zn-Al alloy that makes use of its special composition to provide unique properties. It is a highly diecastable light weight alloy that has excellent surface finishing characteristics. The alloy is oxidation resistant in the temperature range of solidification and below. This results in less oxide on the surface of castings and gives them a longer shelf life prior to surface finishing operations. The mechanical strength of AM-lite® is somewhat greater than AZ91D and its elevated creep strength is considerably higher.

3.1   Diecastability and As-cast Surface Quality

AM-lite® exhibits considerably improved fluidity and diecastability which in many ways is similar to that of zinc diecasting alloys. The usual difficulties experienced with diecasting of AZ91D, such as cracking along flow lines, cold cracks and hot cracks, are significantly reduced with AM-lite®. Because of the higher alloy fluidity the dimensions of runners and volume of overflows can usually be reduced.

As might be expected, the optimum diecasting conditions for AM-lite® are different from those for AZ91D. In fact AM-lite®'s operating window, for the production of good quality castings, is wider and therefore the alloy provides greater stability of operation in mass production situations and allows for easier automation. This is especially true as AM-lite® does not adhere to, or react with, the die (soldering) as does AZ91D. Hence, die maintenance is significantly reduced with AM-lite®, thus reducing down-time and improving automation. Very thin sections are more easily achieved with AM-lite® than for AZ91D and the alloy will reproduce very fine detail from the die.

In general terms, AM-lite® can be cast into thin sections that are 10–20 % thinner than can be cast with AZ91D. This is of particular value for intricate thin walled parts for portable electronics such as mobile phones and laptop computer housings.

The excellent diecastability of AM-lite® stretches to an ability to produce an as-cast mirror surface finish. Production of such a finish requires a polished die surface, minimization of the amount of die lubricant used and appropriate adjustment of diecasting parameters. An example of a mirror as-cast surface is shown in Fig. 1. As-cast surfaces such as this provide considerable advantages in surface finishing.

Figure 1: Reflected image in the as-cast surface of an AM-lite® diecasting produced under conditions that provide a mirror finish

3.2 Mechanical Properties of AM-lite Diecastings

The mechanical properties of AM-lite® diecastings are summarised and compared with AZ91D in Table 1. The tensile properties in this table were obtained on 2 mm thick diecast plates. Both the yield strength (0.2 % offset proof stress) and ultimate tensile strength of AM-lite are higher than for AZ91D. The most important difference in mechanical properties can be seen with reference to Fig 2. AM-lite® maintains its linear elastic behaviour to higher stresses than AZ91D. The limit of linear elastic behaviour for AZ91D is ~40 MPa while for AM-lite® this limit is ~100 MPa. This means that, for stiffness critical parts, the design strength of AM-lite® is considerably higher than that for AZ91D.

AM-lite® also has better creep strength than AZ91D and zinc alloys. Zinc alloys creep significantly even at room temperature under relatively low loads. As shown in Fig 3, the creep strength of diecast AM-lite® at temperatures around 100 °C is about 65 % greater than AZ91D and 600 % greater than a typical zinc diecasting alloy.

Table 1: Mechanical properties of AM-lite® diecastings compared to AZ91D

Property	AM-lite	AZ91D
Yield stress (0.2 % proof), MPa	160–170	120–150
Ultimate tensile strength, MPa	230–250	180–205
Young's modulus, GPa	45	44
Elastic limit, MPa	~100	~40
Tensile elongation, %	3–4	3–4

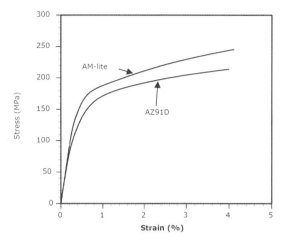

Figure 2: Tensile curves for AM-lite® and AZ91D. 2mm thick diecast plates

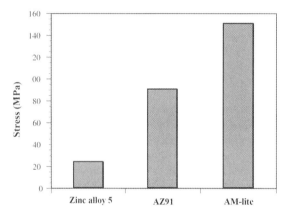

Figure 3: Comparison of creep strength of AM-lite®, AZ91D and zinc Alloy 5. Stress required for 0.5 % strain after 100 h at 93 °C (200 °F).

3.3 Oxidation Resistance

The strong resistance of AM-lite® to oxidation during solidification is illustrated in Fig 4 which shows samples of AZ91D and AM-lite that have been allowed to solidify in air in open muffin-cup moulds. Under such circumstances AZ91D oxidises extensively and eventually burns but AM-lite freezes with a shiny surface.

The oxidation behaviour of AM-lite® in the molten state and during solidification is responsible for many of the alloy's surface properties. For example, during diecasting, the surface film on a liquid metal influences its flow characteristics and in particular the ability for opposing flows to bond metallurgically during filling of the cavity. The high resistance to oxidation of AM-lite® enables such opposing flows to bond effectively during solidification contributing to a much lower defect density in these regions. It also allows molten metal to flow smoothly under pressure to fill small interstices created through shrinkage and, in particular, contributes to the smooth shiny as-cast surface of AM-lite® die castings whose oxidation resistance persists after casting.

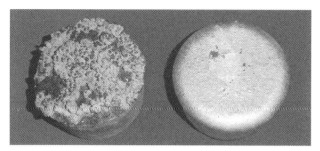

Figure 4: Comparison of oxidation of AZ91D (left) and AM-lite® (right) solidified in air

The oxidation resistance of AM-lite® has other important effects. While use of a cover gas is still required in melt handling, the amount of dross formed is significantly reduced. This may be helped by a furnace temperature that is typically 20 °C lower than that for AZ91D. Experience from industrial trials has shown that melt losses due to dross formation are ~75 % less for AM-lite® than those usually experienced for AZ91D under similar circumstances. Sludge formation is virtually negligible. Reduced melt losses are an important cost saving for diecasters. Another virtue is that there is a significant reduction of the amount of burning of magnesium adhering to tools when they are removed from the furnace, or when removing dross, leading to less fuming and a cleaner foundry environment.

4   Surface Finishing of AM-lite®

AM-lite® is suitable for surface finishing by a wide range of surface finishing processes. The high quality as-cast surfaces that can be achieved during diecasting reduce the need for polishing, buffing and filling. The alloy can be surface finished by a wide range of processes including:
- electroplating
- powder coating
- painting
- electrophoretic coating
- anodising
- physical vapour deposition.

4.1   Electroplating

The as-cast surface of AM-lite® provides an improved substrate for electroplating because of two main reasons. Firstly, the high as-cast surface definition reduces the amount of buffing, or polishing, that is necessary to produce a prepared polished surface for coating. Secondly, the surface chemistry of AM-lite® is such that good adhesion of deposited layers is obtained.

Key differences in procedures for electroplating of different materials arise in the pretreatment sequences prior to electrodeposition of subsequent layers of copper, nickel and chrome (or what ever top coat is desired). For electroplating of zinc, pretreatment only involves cleaning and activation of the surface. However for aluminium and magnesium AZ91D steps of etching, deoxidation and zincating must be added. Even then the quality of adhesion of electroplated layers on aluminium and AZ91D is significantly lower than for zinc.

Pretreatment of AM-lite with the Bondal Mg process developed by MacDermid reduces the separate etching and activation steps that are necessary for aluminium and AZ91D to a single step and utilizes a newly developed zincating process. The process is simple and provides better adhesion than is possible on AZ91D. The quality of electroplating should be similar to that expected of electroplated zinc diecastings.

The Bondal® Mg pretreatment process for AM-lite® allows the full range of possibilities for deposition of decorative electroplated layers and top coats that are available to other materials. Electroplated AM-lite® displays excellent coating performance in standard testing regimes. For example, no chipping or adhesion loss is experienced in standard saw cut tests. Accelerated CASS corrosion tests in accordance with ASTM and DIN standards have shown the absence of corrosion for up to 72 hours which is equivalent to more than 500 hours in standard neutral salt spray testing.

Figure 5: Diecast AM-lite® automobile parts electroplated using the Bondal® Mg process

4.2 Conversion Coating for Painting and Powder Coating

Because the surface chemistry of AM-lite® is very different to conventional magnesium alloys, such as AZ91D, AM50 and AM-60, the passivating conversion coatings that are commonly used for painted and powder coated surfaces of these alloys are generally not suitable for AM-lite. MacDermid has developed a conversion coating process, Chemidize Mag Prime, that provides excellent corrosion protection to AM-lite® while at the same time providing a highly suitable undercoat for paint and powder coat finishes. Chemidize Mag Prime is very environmentally friendly as it does not utilise any Cr compounds.

Powder coated and painted samples, that had been pretreated with Chemidize Mag Prime, have been shown to survive 240h neutral salt spray testing in accordance with DIN 50021 SS and DIN EN ISO 6270-2 CH.

5 Comparison with other Materials

Table 2 is a score card that has been drawn up to show the relative attributes of competing materials for diecast or moulded parts. The scores are in the range 1–5 with 5 being best and 1 being worst. While such a quick comparison does not allow a detailed comparison of the nuances of each material, it can be easily understood that, when considered over a range of properties, AM-lite® is consistently better than its competitors for many applications.

Table 2: Score card for attributes of competing materials for diecast and moulded parts

Attribute	AM-lite®	Zn	Al	AZ91D	Plastics
Diecastability / mouldability	4	5	2	3	5
As-cast surface quality	4	5	2	3	5
Electroplating	4	5	2	1	2
Painting/Powder coating	4	3	5	3	3
Melt loss	4	4	4	1	5
Recycling	4	5	5	3	1
Productivity	4	4	3	2	5
Mechanical properties	4	3	5	4	1
Conductivity (metallic feel)	4	4	5	4	1
Density (lightness)	5	1	4	5	5
Cost of finished product (5 is cheapest)	5	3	4	3	3

5.1 Comparison with Zinc

Zn diecasting alloys are highly diecastable and their electroplating characteristics are excellent. The main disadvantages of zinc are its high density and the high current cost of the base metal. The density of AM-lite® is less than 1/3 of the density of Zn and therefore more than 3 times as many parts can be made with AM-lite® for the same weight of material. Based on current metal prices, this results in the cost of metal per AM-lite® part being only about 1/4 of the cost of zinc alloy for castings of the same volume.

The electroplating characteristics of AM-lite® using MacDermid's Bondal® Mg technology allow the same standard of electroplated surfaces to be produced as can be formed on Zn diecastings with only a small increase in the cost of electroplating.

The high density of zinc makes zinc diecastings unfavourable in applications, such as automotive and personal electronics, where light weight is important. Painted and powder coated zinc parts are also vulnerable to the formation of surface blisters during baking due to the expansion of gases trapped in the metal during diecasting. AM-lite® is resistant to the formation of blister defects during baking because the alloy has an inherent creep resistance that stops the growth of gas bubbles. This much improved creep resistance of AM-lite® also enables the design of thinner bosses for bolted joints compared to what is necessary for zinc.

5.2 Comparison with Aluminium

Aluminium alloys are not as diecastable as zinc or magnesium alloys. Therefore they are not used for thin walled or intricate diecastings. Aluminium is more favoured for thick walled robust parts such as automotive transmission housings and other structural parts. Aluminium also has a 50 % higher density than magnesium alloys so the latter have a distinct advantage when weight is at a premium. Current high prices for aluminium combined with the larger number of parts that can be made from the same weight of AM-lite® mean that there is a cost saving of about 14 % on the input metal for AM-lite® castings of the same volume. In die casting, aluminium is significantly more aggressive on dies than AM-lite® and therefore the cost per part due to die replacement is much more expensive for aluminium.

For these reasons, AM-lite® is being considered by a number of manufacturers for current aluminium applications where the advantages of light weight, improved diecastability, longer die life, ease of electroplating and lower cost make compelling arguments.

5.3 Comparison with Magnesium Alloy AZ91D

As can be seen in Table 2, AM-lite® is superior to AZ91D in many respects. Most importantly, AM-lite® is more diecastable, can be electroplated and saves substantial costs for surface finished components. The improved diecastability enables sections that are 10 % thinner to be readily diecast and also allows casting yields to be improved through the use of thinner runners and smaller overflows. The as-cast surface is improved and the oxidation resistance of AM-lite® provides a longer shelf life for unfinished castings.

In melt handling, costly melt losses due to dross and sludge are reduced, from an industry average of 4 % for AZ91D to about 1 % for AM-lite®, through a virtual elimination of sludge and significantly less dross formation.

AM-lite® has a higher design strength than AZ91D and this allows for the incorporation of even thinner sections and thinner bosses in bolted sections. The greater creep resistance of AM-lite® also makes it an option for components, such as valve covers, that experience raised temperatures and require maintenance of tight bolted joints over long periods.

The most important advantage of AM-lite® over AZ91D is however its excellent surface finishing characteristics and, in particular, its ability to be electroplated.

5.4 Comparison with Plastics

Electroplated plastics might look like metals but, because the underlying plastic substrate is thermally non-conductive, they do not feel like metals. This simple matter of aesthetics often disappoints consumers when it is realized, when touched, that an outwardly metallic looking item clearly isn't what it was thought to be. Because it is a conductive metal, AM-lite® does have a "metallic feel" and so is more attractive to customers.

AM-lite® has a much higher stiffness, yield strength and creep strength than unreinforced plastics thus allowing improved thin section light weight designs. The alloy's electrical and thermal conductivity, and EMS shielding, is also important for the design of consumer electronics such as mobile phones and laptop computers.

6 Applications of AM-lite®

From the above it can be understood that AM-lite® is highly suitable for a wide range of products that previously may have used zinc, aluminium, magnesium AZ91D and plastics. Thus applications are being uncovered in many areas where these materials have been or are currently being used. These include:

- building and furniture hardware, such as handles, knobs and cover plates
- light fittings
- sanitary ware, such as tap handles, kitchen and bathroom fittings
- hand tool housings
- enclosures for portable electronics such as mobile phones and laptop computers
- motor cycle parts such as cover plates, badges and filler caps
- automotive decorative components and trim such as door handles, console panels and gear shift paddles.

7 Conclusions

1. AM-lite® is a new general purpose magnesium diecasting alloy especially suited for decorative applications.
2. AM-lite® is highly competitive with zinc, aluminium, magnesium AZ91D and plastics for most decorative applications.
3. Key attributes for AM-lite® include:
    - light weight
    - high diecastability
    - improved as-cast surface for all surface finishing operations
    - can be readily electroplated
    - resistant to oxidation
    - potential for in-cell recycling
    - higher design strength than zinc, magnesium AZ91D and plastics
    - cost savings in both diecasting and surface finishing
4. AM-lite® can be readily electroplated using the Bondal® Mg electroplating technology.
5. Conversion coating with Chemidize Mag Prime provides an excellent passivating base for painted and powder coated surfaces
6. Cost savings in both diecasting and surface finishing

8 References

[1] Abbott, T.; Murray, M.; Dunlop G., in *Magnesium Technology 2006*, San Antonio, TMS 2006, p. 481–485
[2] Dunlop, G.; Abbott, T.; Murray, M.; Bettles, C.; Gibson, M.; in *Proceedings International Magnesium Association World Conference*, Beijing, IMA 2006, p. 24–33

Aluminum Bolts in Magnesium Engine Components

Ingemar Bertilsson[1], Andreas Fischersworring-Bunk[2], Thomas Marx[3], Filip Bergman[1], Martin Kunst[2], Jan Thiele[2], Zheng Tan[1]
1. Volvo, Göteborg, Sweden, 2. BMW Group, Munich, Germany, 3. Adam Opel GmbH, Rüsselsheim, Germany

1　Abstract

Eight partners from Europe and one from North America have joined efforts in an EU-supported project to find new ways for sustainable production of Mg-based engine blocks for cars. One of the main specific technical objectives was the development of improved bolt fastening system for bolted assemblies in Mg-components exposed to high temperature and a model for the retained clamping forces in such systems. This includes the fatigue strength characterisation of aluminum bolts, galvanic corrosion performance testing, assembly friction testing, bolt load retention loss testing and the numerical analysis using FE techniques. The magnesium alloys under investigation are the Mg-Al-RE (AE44, AE35) and the Mg-Al-Sr (AJ52A, AJ62A) alloys.

2　Introduction

Using steel bolts when assembling magnesium engine components will results in problems with loss in clamping force and galvanic corrosion. Aluminum bolts will significantly reduce these problems. In this paper the joining of magnesium engine components with aluminum bots is discussed. The results are based on an EC supported project; Mg-Engine.

Mg-Engine was running from 2002–2005 with participants from five automotive companies (BMW, Fiat, Opel, Renault, and Volvo), two magnesium alloy suppliers (Hydro Magnesium and Noranda Magnesium), one die caster (Honsel) and one Research institute (Risö National Labororatory). Kamax was an associated partner for the joining work package. The overall objective of Mg-Engine was to generate knowledge about how to design and produce an automotive engine in magnesium, and thus save weight and decrease the environmental impact from cars. An overall description of the project and the general outcome of the project can be found in the paper by Schröder Pedersen et al. (1).

Four different magnesium alloys were investigated. The improved high temperature properties were obtained by Rare Earth additions (AE 35 and AE 44 from Hydro Magnesium), or strontium additions (AJ52 and AJ62 from Noranda Magnesium). More information about the alloys can be found in (2) for the AE alloys, and in (3, 4) for the AJ alloys.

3　Selection of Aluminum Bolt Alloy

Three different aluminum alloys were initially screened as possible candidates for the production of aluminum bolts within the project. It was Al 6056, Al 6062 and Al 7075. These alloys are commercially available as wire and can be manufactured in a large scale. Al 7075 had the best static high temperature properties, but the worst galvanic corrosion properties vs. magnesium due to the high copper content. Al 6082 had the best galvanic corrosion properties but to low strength, Al 6056 was found to be the best compromise between the different properties and it was used in most of the investigations.

4 Mechanical Properties of Al bolts and Joined Assemblies

The bolt samples for the evaluation of the strength properties and the fatigue properties were provided by Kamax in the dimensions M8, M10 and M12. The samples were manufactured in two alternative production process sequences: "Final Heat Treatment" and "Final Thread Rolling" (see Table 1). The precipitation heat treatment condition of the wrought alloy EN AW-6056 (EN 573-3) was T6.

Table 1. Strength properties of the bolt material

Dimension	Bolt Head Type	Production Process Sequence	0,2%-Proportional Limit $R_{p0.2\ RT}$	Ultimate Tensile Strength UTS/R_m
M8x90	External Hexalobular E12 with flange ⌀ 16,5 mm	Final Heat Treatment (FHT)	365 N/mm²	413 N/mm²
M8x90	External Hexalobular E12 with flange ⌀ 16,5 mm	Final Thread Rolling (FTR)	368 N/mm²	413 N/mm²
M10x65	External Hexalobular E12 with flange ⌀ 19,5 mm	Final Heat Treatment (FHT)	375 N/mm²	412 N/mm²
M10x65	External Hexalobular E12 with flange ⌀ 19,5 mm	Final Thread Rolling (FTR)	375 N/mm²	418 N/mm²
M12x85	External Hexalobular E18 with flange ⌀ 21,0 mm	Final Heat Treatment (FHT)	382 N/mm²	423 N/mm²

The strength properties were tested on machined bolts (sample size $n = 5$) i.e. the values represent material values. Assuming a minimum achievable yield strength $R_{p0.2min}$ of 350 N/mm², the initial assembly clamping force of the aluminum bolts is given in Table 2 (compared to a steel bolt grade 8.8). The direct weight saving using Al bolts instead of steel bolts for the same clamping force is about 1/3.

Table 2: Comparison of the clamping force bandwidth: steel versus aluminum bolt

Mat. Dia.	Steel 8.8 ($R_{p0,2min}$ = 640 N/mm²)	Alu 6056 ($R_{p0,2min}$ = 350 N/mm²)
M8	12–18 kN	6,5–10 kN
M10	19–28,5 kN	10,5–15,5 kN
M12	2–41,5 kN	15–23 kN

Utilisation: $v = 90\ \%\ R_{p0,2min}$; Friction Coefficient: $\mu = 0,10$–$0,16$

The fatigue properties of the aluminum bolts were tested according to ISO 3800/DIN969 ("Threaded fasteners – Axial load fatigue testing") with the stair case method (Limit cycle number $N_G = 10^7$). The purpose was to survey the influence of selected parameters on the fatigue strength $\sigma_{a50\%}$ of the bolts and to establish S/N-curves (Wöhler-diagrams) for the variants (see Table 3).

Table 3: Test Parameters Fatigue Testing of Bolts

Parameter	Parameter Values
Nominal Diameter of the Bolt	M8, M10, M12
Production Sequence	Final Heat Treatment (FHT), Final Thread Rolling (FTR)
Condition	as delivered (T6), aged for 2,200 hrs (3 months) at 150 °C
Mean Load σ_m	50 % $R_{p0,2\ RT}$, 70 % $R_{p0,2\ RT}$, 90 % $R_{p0,2\ RT}$
Assembly Method	Preassembled over yield point (Torque-Angle controlled tightening)
Test Temperature	–40 °C, RT, 80 °C, 120 °C, 150 °C

The results of the performed fatigue test series are summarized in the following diagrams. The first diagram (Figure 1.) contains the acquired values for the fatigue strength of the test series O1 to O16, which were mainly run at RT. The second diagram (Figure 2.) shows the outcome of the test series with varied test temperature (–40 °C, RT, 80 °C, 120 °C and 150 °C). For comparison: the fatigue strength of a steel bolt is at $\sigma_{a50\%}$ = 40 to 60 N/mm².

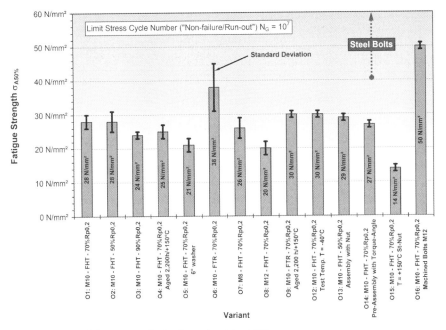

Figure 1: Results fatigue tests series (performed mainly at RT)

The parameters respectively influence factors with the highest impact on the fatigue strength are:

- the **production sequence**: final heat treatment (FHT) versus final thread rolling (FTR),
- the **test temperature** (–40 °C, RT, 80 °C, 120 °C, 150 °C).

The difference between the fatigue strength of the bolts M10 FHT ($\sigma_{a50\%}$ = 28 N/mm²) and the bolts M10 FTR ($\sigma_{a50\%}$ = 38 N/mm²) is nearly at 40 % (exact: 36 %). The difference of the fatigue strength of the bolts M10 FHT tested at RT and at 150 °C is at 50 %.

The **bolt dimension** (nominal diameter) seems to have also an impact. The difference between the fatigue strength of the bolts M12 FHT and the bolts M10 FHT is 30 %. However it was not finally clarified, if this result is also influenced by the microstructure condition of the respective material.

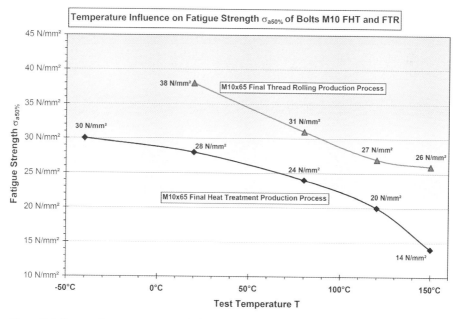

Figure 2: Influence of test temperature on fatigue strength $\sigma_{a50\%}$

Ageing (2,200 hrs @ 150 °C) of the bolts M10 with final thread rolling production sequence (FTR) also reduces the fatigue strength significantly from $\sigma_{a50\%}$ = 38 N/mm² to $\sigma_{a50\%}$ = 30 N/mm². This equals to a drop of –21 %. This result can be explained by residual stresses induced in the thread root by the work hardening effect of the final thread rolling process. The induced residual stresses increase the fatigue strength of the aluminum bolts. This effect is also known with steel bolts. The ageing at 150 °C causes a reduction of the residual stresses which in consequence leads to a drop of the fatigue strength. Also after ageing the finally thread rolled bolts have higher fatigue strength than the finally heat treated bolts.

The **variation of the mean load F_m** ($\Rightarrow$ mean stress σ_m) showed only a minor influence on the fatigue strength of the tested bolts. At a value of σ_m = 90 %$R_{p0,2}$, the fatigue strength drops slightly from $\sigma_{a50\%}$ = 28 N/mm² to $\sigma_{a50\%}$ = 24 N/mm² (–14 %).

5 Galvanic Corrosion of Magnesium – Aluminum Bolt Joints

The general corrosion properties are good. The galvanic corrosion of magnesium can be very fast in contact with noble elements and an electrolyte. The engine compartment is not the worst affected part of the car, but after some time salt and dirt will accumulate and when humid an electrolyte will be available.

In order to simulate galvanic corrosion in a car a SCAB outdoor corrosion test was preformed. In the SCAB test, the corrosion process is accelerated by twice a week spraying the test specimens with a 5 % sodium chloride solution during a six months outdoor exposure. By this way, corrosion similar to the one arising in a salt environment can be simulated. In this test aluminum bolts with and without a commercial lubricants were combined with magnesium. The weight loss and the pit depth of magnesium were measured after the test.

The weight loss for the magnesium alloys in combination Al 6056 bolts lubricated with OKS 1700 can be seen in Figure 3. The weight loss is considered to be low. The AJ alloys, with strontium additions, are slightly better than the AE alloys with Rare Earth additions.

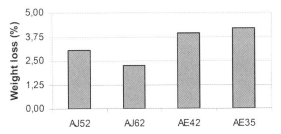

Figure 3: Avreage weight loss during SCAB testing for the different magnesium alloys combined with Al bolts in Al 6056 lubricated with OKS 1700

In case of galvanic corrosion, pitting corrosion is more dangerous than general attack, as perforation is not allowed. Therefore, analysis of galvanic corrosion is often characterized by the pit depths measured. The average pit depths are given in Figure 4. The error bars in the figures indicate the range of maximum and minimum pit depths of eight sections on three samples. The pit corrosions show almost the same tendency as weight losses for the corresponding combinations. The different aluminum bolt alloys combined with AJ 62 are shown in Figure 4. It can be seen that Al 6082 gives the lowest pit depth, while the other differences are lost in the scatter.

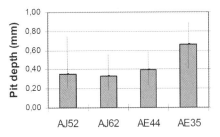 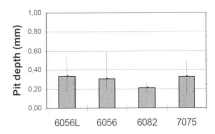

Figure 4: Pit depth measured after SCAB testing. In the left figure the pit depth for the different investigated magnesium alloys combined with Al bolts in alloy 6056 lubricated with OKS 1700 are shown. In the right figure the pit depth in magnesium AJ62 combined with different aluminum bolt alloys are shown. 6056L denotes Al 6056 lubricated with OKS 1700.

6   Assembly Friction

The assembly friction is an important parameter in the design and assembly of bolted components. Thus a large experimental investigation was made in order to characterize the assembling friction properties of aluminum screws in the different magnesium and aluminum alloys. Two commercial lubricants (OKS 1700 and Gleitmo-605) with two different thicknesses were tested. Different combinations of material and different geometries as well as repeated assembling were studied (1).

In Figure 5 the friction coefficient for magnesium and the reference aluminum alloy is shown using different lubricants and thicknesses of the lubricant. It can be seen that magnesium has a slightly higher friction coefficient than aluminum. By a proper selection of lubricant and thickness of the lubricant it is possible to obtain a friction coefficient suitable to the normal design practice used. The different automotive companies in this project aim for slightly different friction coefficient due differences in design practice and history. All these values can be obtained for aluminum bolts in magnesium. No significant differences in friction coefficient were found between the different magnesium alloys.

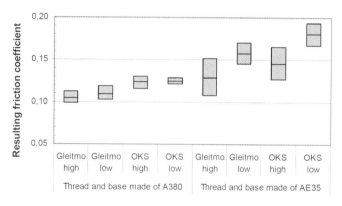

Figure 5: Diagram showing mean value (marked line) of assembly friction coefficient and the maximum spread (grey area). Maximum and minimum spreads are indicated using the lowest and highest measured values. High and low after the name of the lubricants corresponds to the concentration (thickness) of the lubricant.

7   Design and Simulation of Bolted Joints

A key functional question related to the design of a bolted joint in magnesium is on how to ensure that the loss in retention load remains within a given design limit for the foreseen operational lifetime. The loss in retention load is dominated by the compressive relaxation in the magnesium sleeve/thread material and tentatively by the tensile creep effects in the bolt material, in particular in the case of an aluminum bolt. To assess the expected loss in retention capability with regard to the design question, a FEA based sensitivity study was performed to identify the key design parameters and their order of relative importance using a Pareto analysis of the bolted joint predictions.

7.1 Simulation Results

The numerical sensitivity analysis uses a parametric axis-symmetric FE model and a visco-plastic material deformation model for the Mg alloys (5) and both an elasto-plastic and a visco-plastic model for the Al6056 bolt material. The creep material model is using Norton's model in it's time hardening formulation available in ABAQUS (6). The numerical study was performed with variations in clamping length, thread engagement length, bolt head diameter, bolt material (aluminum, steel), thread material (Al, Mg), sleeve material (Al, Mg) and temperature (RT, 100 °C, 150 °C) for an arbitrary dwell time of 250 hrs viz. 1000 hrs with a total of 15 combinations. The metric M8 bolt with an initial clamping load of 11 kN was used as reference configuration for the Pareto analysis. The overall performance is governed by the load rebalance between the Al bolt and the Mg flange viz. thread during the relaxation process. An example for a FE analysis result is the given in Figure 6 which shows for the M8 aluminum bolt and both sleeve and thread in magnesium AJ52 material the distributed stress and creep strain after 250hrs dwell time at 150 °C.

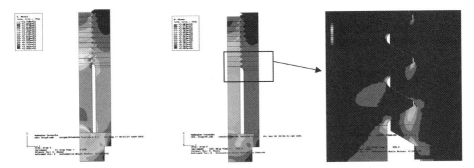

Figure 6: Stress redistribution prior / after 250 hrs @ 150 °C for AJ52 time hardening model with creep of the Al-bolt and creep strain distribution in the thread bolt engagement area

A comparison for the predicted relative loss in retention load for the M8 bolt is given in Figure 7 for a dwell time of 250 hrs. The variation in temperature is identified with Var1-RT, Var2-100 °C, Var3-150 °C for the Al-thread/Mg-sleeve combination. The Mg-thread/Mg-sleeve combinations at 150 °C are identified as Var9 for the Al bolt and as Var15 for a reference steel bolt with the largest loss in retention load.

The comparison shows, that it is of paramount importance to know the creep performance of the Al bolt, since it contributes significantly to the overall loss in retention load and changes the relative order of importance of the different geometrical design parameters and the impact of the Mg alloy performance. As an example for the impact of the creep effect in the aluminum bolt is given in Figure 8: Influence of geometric parameters on the pretension force loss (Var9) by geometrical parameter with (left) and without (right) creep in the Al bolt8 and presenting the influence of the Mg alloy and geometrical design parameter variations on the absolute loss in retention force for the Mg-thread/Mg-sleeve material combination. The variation for the sensitivity analysis is expressed in terms of the bolt diameter. In the direct comparison the variables of importance are changed from the clamping length to the bolt head diameter.

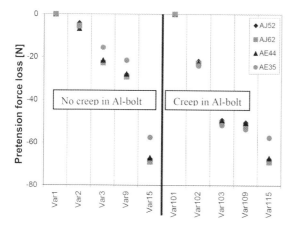

Figure 7: Predicted relative load retention loss for the reference bolted joint geometry

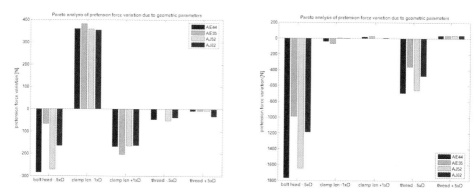

Figure 8: Influence of geometric parameters on the pretension force loss (Var9) by geometrical parameter with (left) and without (right) creep in the Al bolt

7.2 Simulation Conclusions

From all the predictions for the bolted joint assembly operated at 150 °C, it can be concluded the design should be for approximately 60 % loss in retention load when using standard bolt torque loadings applicable to Al bolts as recommended by the bolt manufacturers. With this loss in retention load however it is possible to operate a component. In the design of the bolted joint in general this will mean that the bolt diameter needs to be increased to provide a similar load carrying capacity when compared to a steel bolt joint in aluminum.

For a bolted joint affected by the creep of the Al bolt, the most important parameter is the clamping length followed by the bolt head diameter. The bolt clamping length has the largest impact, with a decrease resulting in an increase of retained load. However the impact from geometry and alloy combination is significantly smaller compared to the prediction without creep in the Al bolt. Future R&D activities should be focused on the development of Al-bolt material alternatives with good tensile creep performance, maintaining the demonstrated good contact corrosion properties of the Al6056 material. Since the overall performance of the bolted joint is governed by the non-linear load redistribution between the Al bolt due to tensile creep and the compressive relaxation in the flange/thread material, additional R&D activities should be directed to establish an optimum torque load to minimize the retention loss.

8 Conclusions

In this paper different aspects of the use of aluminum bolts in magnesium are discussed and it is showed that aluminum bolts is a very interesting alternative to steel bolts in magnesium at elevated temperatures. Most of the investigations were made with bolts in aluminum alloy 6056. The simulation of loss of retention load showed that a significant portion of the relaxation took place in the bolt itself, thus further R & D work is needed in order to develop a bolt alloy with improved creep properties combined with good galvanic corrosion behavior.

Finally thread rolled bolts have better fatigue properties than finally heat treated bolts. Although long time ageing at 150 °C decrease the fatigue properties for both types of bolts, the finally thread rolled bolts remains the best.

Combinations of different magnesium and aluminum alloys were tested for galvanic corrosion in a SCAB test. Generally the properties were good; the best bolt alloy was Al 6082. The strontium based alloys showed slightly lower galvanic corrosion than the rare earth based alloys. For mounting purposes it has been showed that a suitable friction coefficient can be achieved between aluminum bolts and magnesium by using a commercial lubricant.

9 Acknowledgment

The authors would like to acknowledge the financial support of the project by the European Commission under the grant GRD1-2000-25580, contract no. G3RD-CT2001-00304. The authors also wish to thank the respective partner organizations for the permission to publish this paper.

10 References

[1] A. Schröder Pedersen, A Fischersworring-Bunk, M. Kunst, I. Bertilsson, I. De Lima, M. Smith, m. Wappelhorst, P. Bakke, S. Serine, M. Durando, T. Marx, "Light Weight Engine Construction through Extended and Sustainable Use of Mg-Allous", SAE **2006**-01-0068, (Detroit, MI: SAE2006 World Congress)
[2] P Bakke; H Westengen, Proc. 2nd Intl. Light Metals Technology Conference, 2005, June 8-10, St. Wolfgang, Austria, Ed: H. Kaufmann, ISBN-3-902092-03-3, **2005**, p. 57–62
[3] E. Baril, P. Labelle and M.O. Pekguleryuz, "Elevated Temperature Mg-Al-Sr: Creep Resistance, Mechanical Properties, and Microstructure", JOM, Vol. 55, **2003**, 34–39.
[4] ASTM B93 SB93/B93M-05, "Standard Specification for Magnesium Alloys in Ingot Form for Sand Castings, Permanent Mold Castings, and Die Castings"
[5] P. Baake, A Fischersworring-Bunk, I. de Lima, H. Lillholt, I. Bertilsson, P. Labelle, 'The EU Mg-Engine project – generation of material property data for four die cast Mg-alloys', SAE **2006**-01-0070, (Detroit, MI: SAE2006 World Congress)
[6] ABAQUS Users' manual, Version 6.4

Experimental Warm Hydroforming System for Mg AZ31 Alloy Sheet Using a Low Melting Point Alloy as Forming Medium

M. Steffensen, J. Danckert
Department of Production, Aalborg University, Denmark

1 Introduction

Sheet hydroforming has due to its advantages such as high limiting forming ratio, capability of forming complicated shapes and good surface quality gained much attention and use. Magnesium (Mg) alloy sheet material has in the recent years also gained much attention due to its lower density compared to that of steel and aluminum, damping properties, electromagnetic shielding and favorable part quality compared to that of die casted Mg parts. Given the advantages of the hydroforming process and the Mg alloy sheet material it is highly interesting to do hydroforming of Mg alloy sheet, both from an academic and an industrial point of view.

In order to do successful hydroforming of Mg alloy sheet the forming needs to be done within a temperature range of about 170–250 °C [1] due to the hexagonal structure of Mg material. The high forming temperature makes the hydroforming process substantially more difficult than at room temperature. The hydroforming tools need to be equipped with heating devices and a control system to ensure an accurate and uniform temperature of the tool. Furthermore, the forming medium needs to be able to withstand the high temperature which can make the it very expensive, problematic in regard to thermal stability and a number of safety issues have do be dealt with.

A number of researchers have already dealt with the topic of hydroforming Mg alloy sheet material. [2] deals with the design of a warm hydromechanical deep drawing tool and compares the limit forming ratios from experiments with the tool to that of conventional warm deep drawing of the same cylindrical part in AZ31 alloy sheet. In [3] hydromechanical deep drawing experiments have been conducted in AZ31 alloy sheet at elevated temperatures to investigate the effect of the counter pressure and punch temperature on the achievable strains at the bottom of the formed part. [4] presents results from hydroforming a hang-on part for a license plate in AZ31 alloy sheet at both room temperature and elevated temperature to investigate the geometrical accuracy among other things. In [5] investigations have been made of warm hydroforming a cover for a mobile phone in AZ31 alloy sheet and identifying the right pre-bulging pressure and counter pressure to successfully forming the part. [6] describes equipment for warm hydroforming and the use of it for identifying the process window when forming a model bonnet in AZ31 alloy sheet.

One common feature of all the mentioned research activities is that when hydroforming Mg alloys sheet the used forming medium is always oil. However the use of oil is problematic with regard to thermal stability and safety issues. An alternative to oil could be a low melting point alloy.

The article describes the design of an experimental warm hydroforming system for Mg alloy sheet material in which a low melting point alloy is used as forming medium.

2   System Design

In order to investigate warm hydroforming of Mg alloy sheet material in which a low melting point alloy is used as forming medium a system has been designed and built to accomplish this. When mounted in the hydraulic press the system looks as shown in Figure 1.

Figure 1: Warm hydroforming system mounted in the hydraulic press

The working principal behind the system is pretty similar to that of a regular sheet hydroforming system used for steel or aluminum at room temperature. This means that a punch which equals the inner geometry of the part to form is used and by drawing the punch into the Mg alloy sheet and further into a fluid filled cavity the displacement of the fluid will lead to a pressure build-up. The pressure forces will press the Mg alloy sheet up against the punch and the sheet will as the punch moves down take the shape of the punch without the use of a draw die. The system does however differ in 3 ways compared to that of a regular hydroforming system. Firstly, the cavity, die and blank hold are heated. Secondly, the forming medium in the cavity is heated and lastly, a separation is made between the warm forming medium in the cavity and the hydraulic system that controls the cavity pressure.

2.1   Design Criteria

During the design a number of design criteria have been set up for the system, an extract of these are:

- The system should have a working temperature range of 170–250 °C.
- The system warm-up time to 200 °C should not exceed 1 hour in order for this not to be too time consuming.
- The system is to be equipped with a temperature control so that a constant temperature of ±1 °C within the specified range can be kept during used. Furthermore, the temperature of the

die, blank holder and forming medium should be controlled separately so that different temperatures can be set on the respective parts.
- The system should be built in a way such that different forming media like Lead-Bi alloy can be tested without having to make changes to the system. Furthermore, the forming medium should be functional within the given working temperature range and must not harm the Mg alloy sheet material.
- Safety measures must be taken to prevent any warm medium to reach the user if the system should malfunction.

3 System Parts

From the assembly drawing shown in Figure 2 an overview is given of the upper (part 1–6) and lower (part 7–16) tool parts. The upper parts are assembled together and when mounted in the press these are connected to the main cylinder. The punch (part 6) is connected to the triple cylinder on the press which makes it possible to move the punch up and down independently of the rest of the upper part. When the upper part moves down, part 2 and part 8 will close the upper and lower tool parts together by the pressure supplied from the main cylinder. The lower parts 7–12 and 15–16 are all connected to the cavity (part 13) which is fixed on the press table.

3.1 Heating System and Control

The warm hydroforming system has 3 heating units; 1 in the upper part of the tool for heating the blank holder (part 4 on Figure 2), and 2 in the lower part of the tool for heating partly the die (part 10) and partly the cavity (part 13) and thereby the forming medium. Each of these heating units consists of 6 cylindrical heating cartridges, though of different sizes. To minimize the heat conduction between the warm and cold tool parts insulation plates are placed in both the upper and lower tool parts (parts 3, 9, 11, 15 on Figure 2). In order to comply with the design criteria of being able to heat the tool to 200 °C within 1 hour a total of approximately 10 kW are required. To control, i.e. running the tool up to the desired temperature within the range of 170–250 °C and retaining it, the temperature of the blank holder, die and cavity is controlled via separate PID controllers with matching temperature sensors.

3.2 Cavity Pressure and Control

In order to control the cavity pressure, i.e. the pressure on the forming medium during forming, a separation is made between the forming medium inside the cavity and a traditional hydraulic system that controls the pressure. The physical separation is made by the cavity punch (part 14 on Figure 2) which is connected via a piston rod to a hydraulic cylinder placed below the press table. The hydraulic system consists of a pump station and two proportional valves which enables control of the counter pressure and pre-bulge pressure. The current hydraulic system enables the counter pressure to be controlled within the range of 0–700 bar and the pre-bulge pressure within range of 0–200 bar.

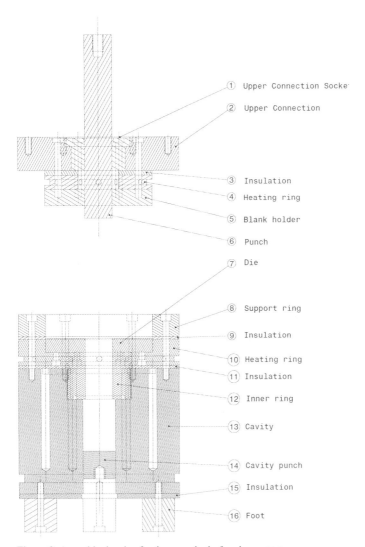

Figure 2: Assembly drawing for the warm hydroforming system

3.3 Forming Medium

For the warm hydroforming system is has been chosen to experiment with low melting point alloys as forming medium due to high prices of oils for this type of application but also due to the fact that low melting point alloys have not been tried in warm hydroforming before. The initial low melting point alloy used as forming medium is one made by MCP/HEK in Germany. The name of the specific alloy is MCP 138 and is a tin-bismuth alloy. Selected properties are listed in Table 1.

Table 1: Properties of the MCP 138 low melting point alloy used as forming medium

Property	Value
Melting temperature [°C]	138
Specific melting energy [J/kg]	44,800
Specific heat capacity, solid [J/kg K]	167
Specific heat capacity, fluid [J/kg K]	201
Heat conductivity [W/m K]	18.5
Specific weight [g/cm^3]	8.58
Price [Euro/kg]	~ 13

In order to get the low melting point alloy into the tool cavity, pieces of the slab material are cut off and melted in an oven and then poured into the cavity. After ending an experiment session and turning off the tool heating the MCP 138 will solidify inside the cavity but when the tool heating is turned on again the MCP 138 will melt and be usable for forming again.

3.3.1   Shielding and Safety

Dealing with warm forming and especially the warm forming medium necessitates some safety measures when using the system. The warm forming medium can be dangerous if for example a leakage occurs and the medium splashes out and possibly hits the user(s). For this reason a safety shield consisting of thick transparent polycarbonate plates has been put up covering all sides of the press table work area. By this the tool is completely sealed off during forming but still it is possible to see the tool from all sides.

4   Conclusion

The initial experiments of using the warm hydroforming system to form soft steel and aluminum has proven that the system works.

Compared to using oil as forming medium the use of a low melting point alloy sets at least one limitation which is that this type of forming medium also has a lower temperature bound to when it can be used, i.e. the melting temperature. For example the current forming medium used in the system has a melting temperature of around 140 °C and an upper temperature bound of around 200 °C at which the main alloying elements separate. This means that working temperature range for conducting experiments is 140–200 °C but this can however be changed by changing to a different type of low melting point alloy.

Having worked with the low melting point alloy forming medium it seems easy to handle which mostly comes from the fact that workpieces are easy to clean after forming by just using a clean cloth to wipe the workpiece while they are is still warm. Furthermore, any forming medium that might come out of the warm tool either during or after forming will solidify. This makes the forming medium easy to collect and reuse or to clean up by using an industrial vacuum cleaner.

Soon coming experiments of using the hydroforming system for forming cylindrical 50 mm cups in Mg AZ31 alloy sheet will show if the use of a low melting point alloy will be a good alternative to that of using oil as forming medium.

5 Acknowledgements

The authors would like to thank the "Danish Agency for Technology and Production" for funding this project – "Warm Hydroforming of Sheet Metal Parts Made from Magnesium Alloys" (26-03-0200).

6 References

[1] Steffensen, M., State of Art within Magnesium Forming – a literature study, Internal report at Department of Production, Aalborg University, 2006.
[2] Kurz, G., Magnesium Technology 2004, Minerals, Metals and Materials Society, 2004, p. 67–71.
[3] Siegert, K., Jager, S., in Proceedings Hydro/Sheet/Gas Forming Technology and Modeling, SAE 2004 World Congress, SAE International, 2004, p. 57–62.
[4] Geiger, M., Merklein, M., Hecht, J., in Proceedings 8th ICTP 2005, International Conference on Technology of Plasticity, 2005.
[5] Zhang, S., Ren, L., Zhou, L., Xu, Y., Yuan, A,. Ruan, L., in Proceedings IDDRG 2006, International Deep Drawing Research Group, 2006, p. 319–324.
[6] Neugebauer, R., Sterzing, A., Seifert, M., Kurka, P., in Proceedings 8th ICTP 2005, International Conference on Technology of Plasticity, 2005.

Magnesium with Magnetic Properties for Sensory Use

Fr.-W. Bach, M. Schaper, M. Rodman, M. Nowak
Leibniz University of Hannover, Hannover, Germany

1 Abstract

Magnesium with its density of 1,74 g/cm^3 shows outstanding weight specific materials properties and is the lightest construction metal. It is therefore used especially for lightweight constructions and strongly accelerated parts. By adding magnetic properties to magnesium alloys these special material properties and thus its fields of application will be extended [1]. Aim of this research is the development of magnesium alloys with ferromagnetic properties.

Magnetic properties can be obtained by either addition of magnetic powders to the magnesium melt or the subsequent creation of intermetallic phases on the grain boundaries. Future parts featuring magnetic properties will provide the possibility to store data in the material itself. These parts can also obtain sensory qualities which can be used for the magnetic determination of position and movement. By usage of the Villari-Effect it is possible to detect an elongation of the material.

This article reports the results of experiments to place magnetic powders for data storage in different magnesium alloys. Different kind of magnetic powders are used. Hard magnetic powders are used for sensory qualities and soft magnetic powders are used for the data memory usage. The dissolution behavior of these different magnetic powders in magnesium melts is analyzed by electron probe micro analysis. Also the influence of different casting methods on the embedding of magnetic powders in the magnesium matrix is being examined. Further the dispersion of the magnetic particles in various castings is determined by the usage of X-rays analysis and computed tomography. The article further reports first experimental results on the creation of intermetallic magnetic phases by alloying magnesium with rare earth elements (e.g. RExCoy) and the use of the Villari-Effect to detect an elongation of the material and to determine mechanical stresses.

2 Introduction

Within the scope of the Research Center 653 „Gentelligent parts in their lifecycle" methods are developed to cancel the physical seperation of part and belonging information. These new parts shall use biological principles in the manner to pass the information gained in a parts lifecycle to the next generation of parts. Further these parts can use their sensory abilities to detect mechanical loads and store these informations in the material inherent data memory. In combination with hall-probes it is possible to provide non-contact sensors for the determination of velocity and position of magnetic parts. Taking into account the dissolution behavior of the different magnetic powders in magnesium melts a part shall be developed containing a consistent and technically utilizable allocation of the magnetic intermetallic phases or magnetic particles.

3 Experiments and Discussion

3.1 Particle Dispersion in Cast Parts

The classification of magnetic materials is done by their coercive field strength. Soft magnetic materials have a coercive field strength ranging from 10^{-1}–10^{3} A/m whereas hard magnetic materials' coercive field strength ranges from 10^{4}–10^{7} A/m. Their flux density B_t remains after release of the external magnetic field. For economic reasons oxide ceramics (e.g. Ba- or Sr-hexaferrites) are most often used. If higher energy products are required magnetic materials of rare earth elements are used. These are especially used for the realization of small-sized magnetic systems. For sensory technics the demand for a high flux density are lower and ferritic materials are used. By the combination of magnets as transducers and hall probes as detectors it is possible to replace many mechanical sensors by non-contact sensors in automotive applications [3]. By coding magnetic zones in the surface zone of parts it is possible to determine for example rotating rate and the angle of rotation of rotating parts [4].

Different magnetic powders have been examined to determine their applicability as magnetic materials with regard to their use in magnesium melts. The hard magnetic materials AlNiCo, SmCo and NeFeB consisting of rare earth elements have been examined. Table 1 displays the properties of these magnetic materials.

Table 1: Properties of magnetic materials

	Energy product BH_{max} [kJ/m³]	Coercive field strength H_c [kA/m]	Retentivity B_r [mT]
Ba- Sr-hexaferrites	30	250	400
NdFeB	330	950	1270
SmCo	240	780	1100
AlNiCo	75	150	1300

For the sensory use of parts it inevitable that inserted magnetic particles are allocated homogenously in the part. Therefore magnetic particles have been inserted in magnesium melts and cast into a cylindrical mold. The cast cylinder has been examined by computed tomography to determine the allocation of the particles in the cylinder. AlNiCo particles were used with a concentration of 20 wt.-% in the magnesium melt. To get most accurate results the cylinder has been examined lengthwise and along the cross sectional area. The results of the CT analysis is shown in figures 1 and 2. These figures displays the particle allocation and the inserted particles allocated homogenously in the cylinder.

3.2 Dissolution Behavior of Magnetic Materials in Magnesium Melts

To use the magnesium alloys with magnetic properties as sensors their capability to be magnetized has to be granted. Therefore experimental casts have been done with dissimilar magnetic powders and a varying content of magnetic powders to examine the dissolution behavior of the magnetic materials in magnesium melts. AlNiCo, SmCo and NdFeB powders have been used. The content of powders varied between 10–40 wt.-%. The samples have been analyzed by electron probe micro analysis (EPMA). Figures 3 and 4 show micrographs of magnesium containing SmCo.

Figure 1: View of CT analysis, lengthwise

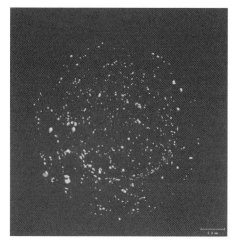

Figure 2: View of CT analysis, cross section

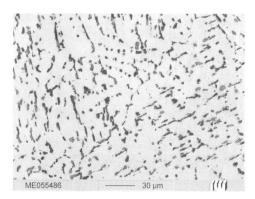

Figure 3: Micrograph Mg + SmCo (10 wt.-%)

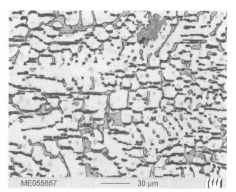

Figure 4: Micrograph Mg + SmCo (40 wt.-%)

The SmCo phases are allocated homogenously in the cast part. The results of the electron probe micro analysis for a content of 40 wt.-% of SmCo can be seen in figures 5 and 6. The results of the EPMA show in evidence that the SmCo powder did not dissolve and magnetic phases exist.

Figures 7 and 8 show micrographs of magnesium containing magnetic NdFeB powder. The inserted material is, as for SmCo powder too, homogenously allocated and precipitated on the grain boundaries. Figures 9 and 10 show the results of the EPMA for magnesium with a content of 40 wt.-% NdFeB powder.

For magnesium with NdFeB powder, only Neodymium and minor amounts of boron could be proved by the EPMA and was carried out in line scans. Iron could as well be proved but only in minimal amounts. These elements dissolved in the magnesium melt. The same experiments have been carried out for AlNiCo powder. Here the EPMA and the line scans state the same reults as for NdFeB. To obtain magnetic phases NdFeB respectively AlNiCo phases are inevitable so that these castings do not have any magnetic properties.

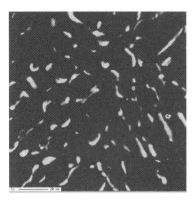 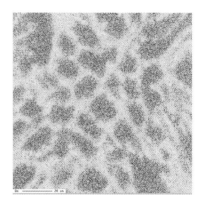

Figure 5: Cobalt mapping (40 wt.-% SmCo)   **Figure 6:** Samarium mapping (40 wt.-% SmCo)

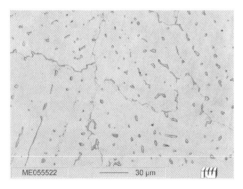 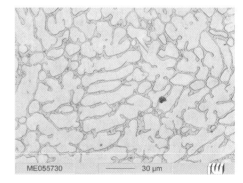

Figure 7: Micrograph Mg + NdFeB (10 wt.-%)   **Figure 8:** Micrograph Mg + NdFeB (40 wt.-%)

As the magnetic powders of NdFeB and AlNiCo dissolve in magnesium melts they cannot be used in combination with mold casting for the cooling rates are too low. As an alternative a casting process with a high cooling rate is inevitable. Die-casting could be used as an alternative manufacturing process as it offers high cooling rates and is commonly used in the industry for the production of magnesium cast parts.

3.3 Capability to Magnetize Specimens of Magnesium with Magnetic Properties

To qualify the produced magnesium alloys as sensor materials other experiments have been carried out to examine the capability to magnetize the produced magnesium alloys. As the magnetic materials NdFeB and AlNiCo cannot not be considered due to their high solubility in magnesium melts only magnesium alloys containing SmCo have been analyzed.

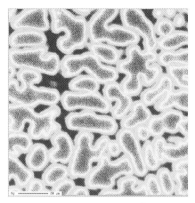 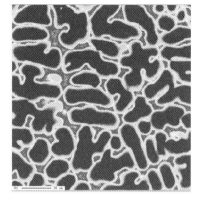

Figure 9: Magnesium mapping (40 wt.-% NdFeB)   **Figure 10:** Neodymium mapping (40 wt.-% NdFeB)

Cylindrical samples with 30 mm in diameter and 70 mm in legth have been made. These samples have been magnetized in the direction from 0° to 180°. The resulting field strength in the sample after the magnetization has been measured peripherically by using a hall sensor with an eddy-current technology. The results are pictured in figure 11. The magnetized samples show a significantly higher field strength compared to the field strength before the magnetization. Hence it is possible to magnetize magnesium alloys with magnetic SmCo phases.

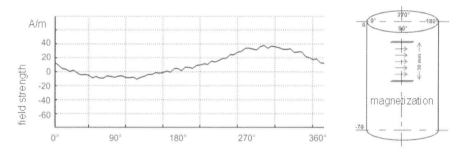

Figure 11: Field strength of magnetized sample

4 Summary

Different hard magnetic powders have been examined due to their dissolution behavior in magnesium melts. Only SmCo powder did not dissolve and magnetic phases of SmCo have been obtained. NdFeB and AlNiCo powders did dissolve in the magnesium melt and no magnetic phases could be obtained in the magnesium castings. So these powders cannot be considered for the use as magnetic powders for the depicted aim. In combination with different casting methods offering higher cooling rates, e.g. die-casting, these powders could be used.

The carried out computed tomography analysis of the cast samples showed a homogenous allocation of the particles in the magnesium casting. Samples of magnesium containing SmCo have been succesfully magnetized and the measured field strength increased significantly. This means that after the release of external magnetic field the magnetic field in the magnetic phases in the casting remains. Furthermore it could be observed that the remaining field strength after the magnetization in the magnesium casting does not increase proportionally with the amount of magnetic powders.

5   Outlook

By use of the Villari-effect it is possible to determine an elongation of a ferromagnetic part as the field strength changes with any elongation. It is possible to detect an actual mechanical strain of a mechnical loaded specimen by usage of a developed eddy-current method. It is as well possible to detect changes within the specimen if it has been mechanically loaded and unloaded again. That will make it possible to analyze the mechanical strains that a part has undergone in its life cycle. On that basis it will be possible to improve this part on real mechanical loadings that had been effective on the part in its life cycle.

Continuative work will be the test and improvement of the manufacturing processes for magnesium alloys with magnetic properties. In this regard different casting methods, like die-casting, meltspinning, lost-foam casting, centrifugal casting for the production of magnesium alloys with magnetic properties are to be tested. Standard magnesium alloys shall be extended with magnetic properties. In this context these alloys are to be tested for an adequate level of the mechanical properties. Some magnetic materials affect the corrosion behavior of magnesium due to their chemical composition. Different combinations of magnesium alloys and magnetic powders shall be tested and evaluated regarding their corrosion resistance.

6   References

[1]   Michalowsky, L.: Magnettechnik, Grundlagen und Anwendungen, 2. Auflage, Fachbuchverlag, Leipzig **1995**
[2]   Hornbogen, E.; Warlimont, H., in Metallkunde, Magnetische Werkstoffe, Springer Verlag, Berlin **1996**
[3]   Schmidtbauer, A., in Kunststoffgebundene Dauermagnete – Werkstoffe, Fertigungsverfahren und Eigenschaften (Ed:. Ehrenstein, G.W.; Drummer, D.), Springer-VDI-Verlag, Düsseldorf **2004**

Research Programs

The MAGFORGE Project: European Community Research on Forging of Magnesium Alloys

W.H. Sillekens*, D. Letzig**
* TNO Science and Industry, Eindhoven
** GKSS Research Center, Geesthacht

1 Introduction

Forging is an industrial manufacturing method in which metal parts/components are shaped from feedstock (slugs) by applying compressive forces through various tools and dies. Where closed-die forging is the most common variant – see Figure 1 – this metalworking category also includes such processes as swaging and impact extrusion. Part shapes can be intricate, and as such it is a competitive process to casting. Forging may be done at room temperature (cold forging) or at elevated temperatures (warm/hot forging), each with its distinct assets and drawbacks. Although forging yields (near to) net-shape products, most forgings require some kind of finishing such as machining for reasons of accuracy and surface quality, heat treating to modify mechanical properties, and/or coating to enhance appearance or resistance to corrosion and wear.

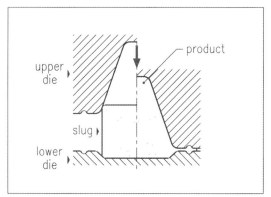

Figure 1: Principle of closed-die forging (left) and some typical products (right)

During processing, metal flow and grain structure can be controlled, so forged components generally have good strength and toughness as compared to cast counterparts and no porosity. Hence, they are often used reliably for highly stressed and critical applications. Well-known product examples are hand tools such as spanners and wrenches, but the product range includes a wide variety of components, including for motor and drive-train applications such as gear wheels and axles. Most common product material is steel, but forging of aluminum is also established practice.

The main market for forged components is automotive; the forging industry is thus faced with some particular trends that relate to developments within this sector [1]. The general platform strategy as well as the increase in diesel-engine powered cars and four-wheel driven sports-utility vehicles are implying volume increase for forged components. On the other hand, there is a continuing price pressure: where costs (for materials, labor, energy and the like) go up, the main customers and their system suppliers do not value that these are passed on to deliveries. Further, the automotive industry has committed itself to substantially reduce fuel consumption and exhaust emissions (amongst others CO_2), for which weight saving at all levels is crucial. Priorities in lightweight structure design are the un-sprung mass (wheels and their suspension), the front end before the front axle and the mass between the front axle and the instrument panel [2]. All these are typical areas where forged components are used.

In this respect, it is increasingly recognized that aluminum (with a density of 2,700 kg/m^3) and magnesium (1,800 kg/m^3) are attractive alternatives for steel (7,800 kg/m^3). To further specify this aspect, Figure 2 gives an overview of the intrinsic weight-saving potential for some magnesium wrought alloys as related to an aluminum alloy which is in common use for forgings. The graph distinguishes between some distinct modes of loading a beam, taking into account the relevant material properties: modulus of elasticity E, yield stress YS, and density ρ (for other geometries and modes of loading, other material indices apply). Although these data depend somewhat on the specific assumptions as well as on the actual property values, this basic approach clearly demonstrates that benefits are anticipated for *strength-related* and in particular for *bending-relevant* parts.

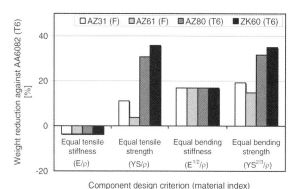

Figure 2: Weight-saving potential of magnesium over aluminum for some typical loading situations (beam)

A consortium of industrial associations, companies and research organizations has recently started to work on magnesium forging within the European research project „Magnesium Forged Components for Structural Lightweight Transport Applications (MAGFORGE)". The project is presented in this paper. Its overall goal is to provide tailored and cost-effective technologies for the industrial manufacturing of magnesium forged components, which is to enable the European forging industry to innovate and enhance its manufacturing capabilities for structural lightweight components in high-volume applications, particularly in view of automotive end use.

2 State-of-the-Art

For forging, steel is the traditional material of choice and still dominating the sector. Rooted in the tradition of the blacksmith, forging workshops in Europe have developed over the last decades into rationalized manufacturing enterprises with modern facilities for process planning, as well as for forging and finishing. Although the annual world production of non-ferrous forgings (aluminum, brass,…) is only a few percent of the overall market, its volume has increased strongly as of late. Noteworthy, the forging industry in the Far East seems to have adopted aluminum more extensively than in Europe.

Forging of magnesium in Europe and beyond is still restricted to a few companies that operate for niche markets such as specialty wheels and aerospace parts [3]. Technical issues that relate to this are underdeveloped mechanical properties of available materials (including large variations in and between batches) and lacking analytical capabilities on process and product performance (based on handcraft rather than on scientific insights). A further issue has to do with the unfamiliarity with the material throughout the forging sector and at the end users, so that the threshold for implementation is considerable. As a result, the size of scale is still small: magnesium accounts only for a minor share of the non-ferrous forging volume.

To further illustrate the state-of-the-art for magnesium forging, Table 1 compiles some technical data for the commonly used alloys [4–5]. These go back to two alloy families (AZ and ZK), where the former is more for general-purpose use and the latter for specialty applications. It should be noted that the mechanical properties apply to longitudinal direction, which as a rule gives the best values. One particular concern is that the strength under compressive loading is substantially lower than under tensile loading. While the tensile properties of the high-strength magnesium alloys are in the same range as those of the aluminum forging alloy AA6082 in T6 temper condition (TYS = 260 MPa, UTS = 310 MPa and Elong. = 6 % [6]), it is notably this aspect of anisotropy which is due for improvement.

Table 1: Current magnesium alloys for forging

Alloy (temper)[a]	Forging temperature [°C]		Typical mechanical properties[b]				Other[c]	
	Product	Die	TYS [MPa]	UTS [MPa]	Elong. [%]	CYS [MPa]	Weldability	Corr. Resist.
AZ31 (F)	290–345	260–315	195	260	9	85	E	G
AZ61 (F)	315–370	290–345	180	295	12	115	G	G
AZ80 (T6)	290–400	205–290	250	345	5	185	G	G
ZK60 (T6)	290–385	205–290	270	325	11	170	NR	F

a) Temper condition affects mechanical properties
b) Longitudinal direction – TYS = tensile yield stress; UTS = ultimate tensile strength; Elong. = tensile elongation; CYS = compressive yield stress
c) Relative ratings: E = excellent; G = good; F = fair; NR = not recommended

Although such general material property data are documented in handbooks, specific information on forging technology for magnesium like on tool design, process parameters and the control over mechanical properties of the part is only scarcely available and typically not at the disposal of the sector. Basically, many of the processing aspects need still to be explored and while the technical potential is promising, the current price level is an important factor impeding a more widespread use [7].

3 Project Outline

3.1 Objectives and Work Contents

The MAGFORGE project treats the subject of magnesium forging along the manufacturing chain rather than as an isolated technological issue. The main areas as well as the overall project objectives are gathered in Table 2. Targeted scientific and technological advancements are to enable high-volume applications of magnesium forgings as they will decisively contribute to pushing down cost price and providing weight-saving benefits over conventional solutions to the end user.

Table 2: Project objectives

Area	Objectives
Scientific and Technological Aspects	
Materials	Magnesium alloys and preparation methods, yielding feedstock with forge-ability (manufacturability, productivity) similar to aluminum and improved as-forged properties (tensile yield stress 200–300 MPa, ultimate tensile strength 300–350 MPa, elongation 15–25 %; compressive versus tensile yield strength ratio 0.8; fatigue: endurance strength >150 MPa at 10^7–10^8 cycles)
Processes	Forging technology with a technically satisfying level of predictability (success rate for „first time right" 80 %) on working conditions (sequence, heating, temperature, speed, lubrication,…) and tool design (geometrical preferences, coatings) Machining technology for finishing of magnesium forgings with high productivity (material-removal rate $\geq$ 20 % higher than for aluminum) and acceptable surface quality (similar to aluminum), while complying with industrial safety and ecology standards (oil/lubricant content in chips $\leq$4 wt%)
Components	Magnesium demonstrator components for typical application fields with considerable lower weight than current aluminum versions (reduction 15–25 %) and proper functional performance (depending on the aimed application)
Economical Aspect	
Manufacturing costs	Forging and finishing operations in the workshop with a cost-price level similar to aluminum (5–10 €/kg, or equivalently 9–18 €/dm^3, of product)

To achieve these objectives, the work focuses on the next aspects.
- *Materials*: chemical modification and grain refining in order to come to a substantial widening and improved consistency of property profiles (formability, strength, isotropy).
- *Forging processes*: modeling and simulation in order to obtain scientifically based FEM design tools regarding metal flow and product properties, as well as tooling and operation in order to define appropriate tool designs (including tool coatings) and process windows.
- *Machining processes*: tooling and operation in order to define appropriate cutting tools, strategies and parameters for high-speed, safe and ecological finishing of magnesium forgings.
- *Components*: product engineering of demonstrators in order to validate a new generation of structural components for high-volume applications with substantially lower weight.

Dissemination and training will be an integral part of the program so as to ensure that the results reach the different stakeholder categories. The anchoring of knowledge and skills from the project includes the origination of a „handbook" and training opportunities that treat the relevant aspects relating to the use of magnesium in forging practice, with a particular emphasis on introducing newcomers to the field.

Potential use of magnesium forged components in vehicles includes parts from the chassis (wheel hubs, suspension arms, shock-absorber housings,...), the interior (seat components, pedal brackets, airbag housings,...), the engine/transmission (axles, fuel-tight appendages,...), and the body (space-frame nodes, brackets,...). The ambition level for these components ranges from moderate to high. Other markets are sports and leisure (bicycle parts, backpack frames,…), electronics (portables, hand tools,…), machinery (acceleration parts), and medical applications (wheel-chair parts,…). Although volumes are smaller, margins may be better so that such components may serve as initial applications, particularly where the use of novel materials is used as a marketing instrument.

3.2 Co-operation

The MAGFORGE project is conducted within the Sixth Framework Program of the European Commission (EC). Being a so-called Collective Research project, it addresses the needs of a sector primarily consisting of small- and medium-sized companies, in this case that of the forging and associated industries.

There are three types of participants: Industrial Associations/Groupings (IAGs), Small/Medium-size Enterprises (SMEs), and RTD Performers (RTDs). Where the RTDs conduct the research and technological development work, the SMEs play an important role in securing industrial relevance and validating results, while the IAGs' main part is in dissemination and training. The participants are listed in Table 3. The IAGs are national forging, tooling and automotive suppliers' associations; the SMEs include forging companies, tool manufacturers (including coating) and a manufacturer of milling equipment. The duration of the project is 36 months, starting in July 2006. With an overall budget of 2.9 M (EC contribution 1.7 M), the total manpower effort adds up to 27 person-years.

3.3 State of Affairs

Upon drafting this paper, the project is still in its early phase with the participants being in the process of leveling their expertise and initiating the activities on material, process and component development. Milestones are scheduled not before the end of the first project year, so that scientific/technological results cannot be reported yet.

From the onset of the project, however, it is considered important to inform the different parties within the forging industry, the magnesium industry as well as possible end users on the proceedings so as to provide for a receptive breeding ground for the technical outcomes of the project. More in particular, the members of the involved IAGs are encouraged at this stage to already express their interest and become engaged in especially the knowledge-transfer activities (such as workshops) that will be organized in due course. For other parties there will also be on-line information available from an internet homepage.

4 Acknowledgement

MAGFORGE is being conducted for the European Commission within the framework of the specific research and technological development program „Integrating and strengthening the European Research Area" under contract number COLL-CT-2006-030208. The financial support is gratefully acknowledged.

Table 3: The consortium

#	Name	Type	Location (Country)
1	Nederlandse Organisatie voor Toegepast Natuurwetenschappelijk Onderzoek (TNO)	RTD[a]	Eindhoven (NL)
2	Association Française de Forge	IAG	Courbevoie (F)
3	Deutsche Gesellschaft für Materialkunde e.V.	IAG	Frankfurt (D)
4	Confederation of British Metalforming	IAG	West Bromwich (UK)
5	Svaz Kovaren eske Republiky	IAG	Ostrava-Pustkovec (CZ)
6	Societatea de Forja din Romania	IAG	Bucharest (RO)
7	Razvojni Center Orodjarstva Slovenije	IAG	Celje (SI)
8	Asociación Española de Fabricantes de Equipos y Componentes para Automoción	IAG	Madrid (E)
9	Gospodarsko Interesno Zdruenja ACS, Slovenski Avtomobilski Grozd	IAG	Ljubljana (SI)
10	Nonferro Metaalnijverheid N.V.	SME	Mechelen (B)
11	Forgialluminio 3 s.r.l.	SME	Pedavena (I)
12	Leiber Poland Sp. Z o.o.	SME	Ruda Slaska (PL)
13	SC Bimetal SA Bucharest	SME	Bucharest (RO)
14	Kovinar D.O.O	SME	Vitanje (SI)
15	Pressmetall AB	SME	Köping (S)
16	Duroc Tooling i Robertsfors AB	SME	Robertsfors (S)
17	MAT PlasMATec GmbH	SME	Dresden (D)
18	Soraluce Sociedad Cooperativa	SME	Bergara (E)
19	GKSS Forschungszentrum Geesthacht GmbH	RTD	Geesthacht (D)
20	Univerza v Ljubljani	RTD	Ljubljana (SI)
21	IDEKO Sociedad Cooperativa	RTD	Elgoibar (E)
22	Universitatea POLITEHNICA din Bucureti	RTD	Bucharest (RO)
23	Svensk Verktygsteknik AB	RTD	Luleå/Göteborg (S)

a) Project coordinator

4 References

[1] www.euroforge.com.
[2] H. Hagen, B. Rüttimann, Aluminium 2004, 80, 124–132.
[3] J. Becker, G. Fischer, Proceedings of the 11th Magnesium Automotive and End User Seminar 2003.
[4] M.M. Avedesian, H. Baker, Magnesium and Magnesium Alloys, ASM International, Ohio USA, 1999.
[5] R.S. Busk, Magnesium Products Design, Marcel Dekker, New York, 1987.
[6] European standard EN 586-2, European Committee for Standardization, Brussels, 1994.
[7] G. Fischer, J. Becker, A. Stich, Materialwissenschaft und Werkstofftechnik 2000: 993–999 (*in German*).

Reports of the Priority Programme 1168 of the Deutsche Forschungsgemeinschaft (German Research counsil) InnoMagTec

Detecting the Solid Fraction of Commercial Mg Alloys by Heat-Transfer Modeling: Comparison of DTA and DSC Experimental Basis Data

Djordje Mirkovi; Rainer Schmid-Fetzer
Clausthal University of Technology, Clausthal-Zellerfeld

1 Introduction

The solidification curve of an alloy, i.e. the local solid fraction vs. temperature during solidification, provides important information for the control of casting processes and is also a key input parameter for solidification simulation software. In this study a comparison of improved in-situ methods for determination of solidification curves is presented. Various other methods had been applied to obtain or calculate solidification curves. The detailed overview of these methods regarding their advantages and limitations is given in [1] or will be published [2].

A mathematical heat-transfer model (HTM) based on the Tian linear first-order differential equation is applied to the measured DTA (differential thermal analysis) and DSC (differential scanning calorimetry) curves of the following magnesium alloys: AZ31, AM50, AZ61, AZ62 and AZ91, thus providing their corresponding solidification curves. The selected alloys cover wide ranges of typical applications such as rolling (AZ31), extrusion (AZ61), sand casting (AZ62) and die casting (AM50 and AZ91).

Compared to earlier work, a better simulation of the measured DTA and DSC signals is attained through an independent measurement of the time constant as function of temperature for the applied equipment. A further improvement is achieved through a more impartial interpretation of the measured DTA/DSC curves, where the termination of solidification is not being artificially defined by operator. Both improvements enable a better desmearing of the DTA/DSC signal and reduce the error induced by the operator. The comparison of different applied measurement techniques (DTA and DSC) and their respective limitations to obtain solidification curves is in the focus of this presentation.

Reproducibility and reliability of obtained DSC-HTM solidification curves are also tested. For this reason independent samples of AZ62 and AZ91 alloys are measured by DSC at the group of Prof. Ferro at the University of Genova applying entirely different experimental set up. Both DSC experimental data recorded in Clausthal and Genova are processed by the improved heat-transfer model and resulting solidification curves are compared.

2 DTA and DSC Experiments and Corresponding Heat-Transfer Model (DTA-HTM and DSC-HTM)

2.1 Experimental Procedure

The alloys studied in this work, AZ31, AM50, AZ61, AZ62 and AZ91, were prepared from master alloys (AM50, AZ61 and AZ91 by Norsk Hydro, Bottrop; AZ31 by Dead Sea Magnesium Ltd., Israel). AZ62 was prepared by adding high purity Zn to the AZ61 master alloy. High oxygen affinity, vapor pressure of the measured Mg alloys, as well high reactivity of alloying element aluminum significantly complicate searching of the proper sample crucible for DTA/ DSC measurements. The commercially available crucible solutions have been studied, but rejected owing to various limitations. Details on selection criteria why tantalum has been chosen as crucible (capsule) material are published elsewhere [1]. The cross-sections of applied Ta-capsules are shown in Figure 1. They differ for DTA and DSC. The capsules are sealed by electric arc welding under argon with 1.5 bar total pressure.

Figure 1: Cross-section of the used DTA and DSC Ta-capsules. The capsules are shown with Mg alloy sample after being measured by DTA and DSC, respectively

The developed Ta-capsule is particularly adapted to the methodic requirements of both applied methods of thermal analysis. The height-width ratio of the DTA capsule is adapted to the bottom mounted thermocouple and bottom was made from as thin as possible Ta sheet material. The design of DSC capsule suits very well applied cylinder-type DSC equipment. Aspects of reliability and handling are also considered during the development of the Ta-capsules.

DTA apparatus in Clausthal and DSC in Clausthal and Genova have been used to measure AZ62 and AZ91 alloys. All other alloys are measured only by DSC in Clausthal. For each alloy composition two independent samples were studied and two or three heating/cooling cycles were performed for each sample. For DTA overall uncertainty is estimated to be ± 3 K and for both DSC systems ± 3 K and ± 0.3 kJ/mol by temperature determination and caloric measurements, respectively.

DTA experiments in Clausthal are performed using a Netzsch DTA 404S (NETZSCH GmbH, Selb, Germany) apparatus. DSC experiments in Clausthal are performed by Setaram MHTC 96 heat-flow DSC equipment (SETARAM Instrumentation, Caluire, France). This is cylinder-type measuring system with thermally-coupled sample containers which were helium flushed during the measurement. DSC experiments in Genova are performed using a heat-flux twin cylindrical Calvet-type calorimetric system Setaram TG-DSC 111. Also cylindrical Ta-capsules (not shown in Figure 1) with 6 mm inner diameter and 13 mm height were used.

2.2 Heat-transfer Model (HTM)

The modeling of the heat flow between DTA/DSC cells and the furnace applied in this work is based on the so-called Tian equation [3, 4] for the heat flow Φ_r [J/s] produced through reaction or transition inside the sample. The equations 1a and 1b are given for DTA and DSC, respectively:

DTA: $R_1 \Phi_r = \Delta T_M + R_1(C_S + C_C)\dfrac{\mathrm{d}(\Delta T_M)}{\mathrm{d}t}$ [K] (1a)

DSC: $\Phi_r = \Phi_M + R_1(C_S + C_C)\dfrac{\mathrm{d}\Phi_M}{\mathrm{d}t}$ [J] (1b)

where t is the time and for DSC $\Phi_M = (\Phi_{FR} - \Phi_{FS})$ is the heat-flow difference between sample and reference which is directly measured during the DSC experiment. Correspondingly, for DTA, $T_M = (T_{MS} - T_{MR})$ stays for the directly measured temperature difference between sample and reference cells. It is worth noting that both directly measured quantities Φ_M and T_M are not directly equivalent to the actual heat flow (the rate of the reaction) in the sample due to the thermal lag (smearing) between event and corresponding thermocouple response [5]. The term $R_1(C_S + C_C)$ is commonly marked as τ_1 and named the first time constant. The τ_1 is defined by heat flow resistance between furnace and the sample thermocouple (R_1) and the heat capacity of the sample cell (sample and crucible together). Related desmearing details for DTA have been discussed in [1], for DSC it will be given separately [2]. A schematic diagram of the both DTA and DSC methods and quantities relevant for the heat-transfer model are shown in **Figure 2**.

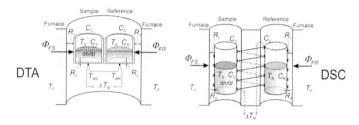

Figure 2: Schematic view of the DTA and DSC, with quantities relevant for the heat transfer model

It is not straightforward to establish an analytical correlation between released enthalpy of solidification and liquid (or solid) phase fraction using thermal analysis techniques. Among other authors [6, 7], Chen at al. assumed a linear dependence of heat evolution during solidification on the liquid phase fraction f_l, in their application of the Tian equation to DSC experiment [8] and with a similar heat-transfer model to DTA experiment [9, 10] in order to determine solidification curves. Assuming this linearization, the heat flow generated by exothermic or endothermic reaction inside the sample, Φ_r can be expressed by the heat evolution $H\,\mathrm{d}(1-f_l)$ which occurs in the sample:

$$\Phi_r = \dfrac{H\,\mathrm{d}(1 - f_l)}{\mathrm{d}t} \qquad (2)$$

where H is the total latent heat of solidification, assumed to be a constant for each alloy.

This linear dependence is often blamed to be a very rough simplifying approximation. Some authors [11, 12] claimed that the accurate liquid fraction cannot be obtained directly using enthalpy measurements. However, if we calculate enthalpy as function of f_l for the studied alloys using our thermodynamic Mg-database [14] and Pandat program [15] for Scheil solidification, Figure 3 is obtained. Calculated line for equilibrium case is not shown, it produces nearly a straight line. Even for the enthalpy release calculated under Scheil's condition, the assumption of linear heat release might be accepted as a first approximation. The specific heat contribution, included in Figure 3, is also essentially linear and does not change this statement. A more detailed discussion about this very important subject relating enthalpy released during solidification and liquid phase fraction will be given separately [2].

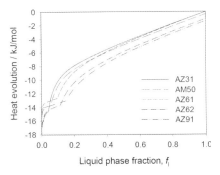

Figure 3: Dependence of heat evolution (total enthalpy of alloy relative to its value at liquidus temperature) on liquid phase fraction during the solidification; calculated according to Scheil conditions for the studied alloys

After combining Eqs.(1a/1b) and (2), the finite-difference form of final equation can be used for calculation of solidification curves. For this calculation, the knowledge of time constant τ_l, as characteristic for the applied DTA/DSC equipment and experimental conditions, is necessary. In this study τ_l has been determined independently using the cooling curves of pure metals measured at the same experimental conditions as the actual magnesium alloy samples (same Ta-capsule setup, scanning rates etc.). This is in contrast to the previous work [8, 9, 10], where for the time constant a determination from the exponential tail of the last peak of the DTA/DSC curve was applied. This becomes questionable in the case when an unknown alloy sample with unknown solidification path or potential solid state reactions could corrupt the results. Detailed discussion why the peak decay of an alloy cannot be used for the determination of the time constants will be given by [2].

The next important point necessary to be discussed is the way how the experimental DTA/DSC cooling curve is being processed to the solidification curve of investigated alloy. It is not necessary and may be even wrong to artificially proclaim one point on the curve, such as maximum of the last peak on the cooling curve, as the end of solidification, as suggested by Chen et al. [8]. The total heat released should be related to the entire area including the exponential tail [15]. More details on the novel evaluation method for treating the measured DSC signal will be published soon [2].

3   Results and Discussion

Exemplary, experimental cooling curves and solidification curves are shown for AZ62 and AZ91 alloys. Figure 4a shows the original experimental DTA and both DSC cooling curves and Figure 4b the liquid phase fraction curves derived from them using the present DTA-HTM/DSC-HTM for alloy AZ62. It is obvious that f_l curves related to DSC curves show very good agreement for all experimental conditions. Even if the DSC curves are associated with two independent alloy samples and are recorded applying two completely different experimental set ups, the derived f_l curves virtually coincide in the graph. For comparison, the solidification curves calculated by limiting solidification models (equilibrium solidification and Scheil model) are also given. They essentially bracket the DSC-HTM experimental curves.

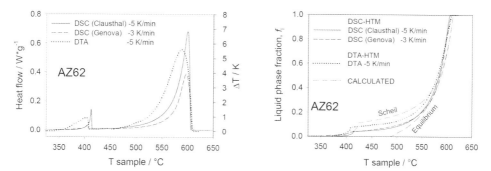

Figure 4: a) The DTA/DSC signals of AZ62 alloy, b) **DTA-HTM**/DSC-HTM calculated solidification curves of AZ62 alloy. Scheil and equilibrium calculations are also shown.

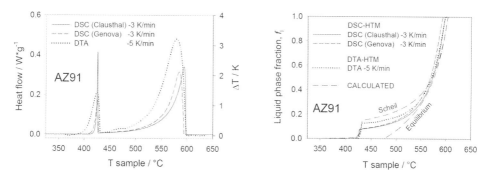

Figure 5: a) The DTA/DSC signals of AZ91 alloy, b) **DTA-HTM**/DSC-HTM calculated solidification curves of AZ91 alloy. Scheil and equilibrium calculations are also shown.

Figures 5a and 5b show the analogous data for the alloy AZ62. It is noticeable for both alloys that the solidification curves obtain by DTA-HTM exhibit higher amount of the phase (quantified by integration of the secondary signal), than those defined by DSC-HTM. This disagreement might be explained by the worse base line stability of DTA. DSC, which is superior compared to DTA with respect to quantitative heat flow measurement and base line stability, is considered being capable to provide more reliable experimental basis data which are further processed to solidification curves. For both DTA and DSC, R_l decrease implies resolution improvement, but also proportional peak area decrease for DTA.

The solidification curves calculated according to Scheil conditions show a qualitatively similar path compared to those given by DSC-HTM in Figures 5b and 6b. This is in accordance with the metallographic analysis of DTA-cooled samples [1] where even under the slow cooling rate of 1 K/min the non-equilibrium particles are precipitated, resembling a Scheil solidification mode, rather than equilibrium.

The Scheil and equilibrium solidification models simulate two limiting cases of the solidification process. Even these two simple models require a complete thermodynamic description of the alloy system under consideration. Based on that information the enthalpy-temperature relation may be calculated. A higher accuracy may be obtained by using solute-diffusion micromodels as shown by Boettinger and Kattner [16]. Dong et al. [11] also claim the necessity to use microsegregation models to obtain solidification curves on the basis of heat evolution measurements. We are in complete agreement with this approach to use micromodels for a detailed description *if* all the thermodynamic *and* kinetic data are available.

In this work a method of desmearing is introduced which involves independent determination of a time constant as function of temperature for the applied equipment. A further improvement is achieved through a more impartial interpretation of the measured DTA/DSC curves, where the termination of solidification is not being artificially defined by operator. The solidification curves obtained by DTA/HTM and DSC/HTM are compared where the base line construction is emphasized as one of the most important parameters influencing the reproducibility and reliability of solidification curves obtained from DTA or DSC data.

4   Acknowledgements

This study is supported by the German Research Foundation (DFG) in the Priority Programme "DFG-SPP 1168: InnoMagTec" under grant no. Schm 588/27(29). The use of the DSC equipment by courtesy of Prof. Riccardo Ferro, Università di Genova, is gratefully acknowledged.

5   References

[1]   Mirkovi D., Schmid-Fetzer R.; Z. Metallkd. **2006**, *97*, 2, 119–129.
[2]   Mirkovi D., Schmid-Fetzer R.; to be published
[3]   Hemminger W., Cammenga H.; Methoden der Thermischen Analyse, Springer-Verlag Berlin, Heidelberg, 1989.
[4]   Höhne G.W.H., Hemminger W., Flammersheim H.-J.; Differential scanning Calorimetry: an introduction for practitioners, Springer-Verlag Berlin, Heidelberg, 1996.
[5]   Loeblich K.R.; Thermochim. Acta **1985**, *83*, 99–106.
[6]   Fedriksson H., Rogberg B.; Metal Science **1979**, *3*, 685–690.
[7]   Bäckerud L., Krol E., Tamminen J.; Solidification characteristics of aluminium alloys; Vol. 1: Wrought Alloys, Skan Aluminium, Oslo, **1986**, 65.
[8]   Chen S.W., Lin C.C., Chen C.M.; Metall. Trans. A **1998**, *29A*, 1965–1972.
[9]   Chen S.W., Huang C.C.; Acta mater. **1996**, *44*, 5, 1955–1965.
[10] Chen S.W., Jeng S.C.; Metall. Trans. A **1997**, *28A*, 503–504.
[11] Dong H.B., Shin M.R.M., Kurum E.C., Cama H., Hunt J.D.; Fluid phase equilibria, **2003**, *212*, 1, 199–208.
[12] Larouche D., Laroche C., Bouchard M.; Acta mat. **2003**, *51*, 2161–2170.
[13] Schmid-Fetzer R., Gröbner J.; Adv. Eng. Mater., **2001**, *3*, 12, 947–961.
[14] Chen S.L., Daniel S., Zhang F., Chang Y.A., Oates W.A., Schmid-Fetzer R.; J. Phase Equilibria **2001**, *22*, 373–378.
[15] Gray A.P., in: Analytical Calorimetry, (Ed.: Porter R.S., Johnson J.F.), Plenum Press, New York, USA **1968**, 209–218.
[16] Boettinger W.J., Kattner U.R.; Metall. Trans. A **2002**, *33A*, 1779–1794.

Thermodynamic Simulation of Blending Mg-Alloys by Thixomolding

A. Kozlov, M. B. Djurdjevic, R. Schmid-Fetzer
Clausthal University of Technology, Clausthal-Zellerfeld, Germany

1 Introduction

Thixomolding offers a simple way of producing new Mg-alloys by mixing different feedstock chips at room temperature. This way of blending or mixing two or more commercially available standard Mg alloys will produce a new alloy with different melting range, solidification path, phase distribution and metallurgical and mechanical properties. This concept, originally developed by Young et al. [1] by premixing different types of powder, has been recently experimentally applied by Nandy et al. [2] and Czerwinski [3] for designing new magnesium alloys by thixomolding processing.

The most suitable metallic materials for thixoforming processes are based on Al and Mg alloys. Most commercial magnesium alloys are AZ based on Mg-Al-Zn-Mn and AM based on Mg-Al-Mn systems, offering good castability, corrosion resistance and mechanical properties. These alloys have appreciable melting range and their microstructure becomes globular under applied shear force. Aluminum as a major alloying element in these alloys is added to improve strength, hardness and to widen the freezing range of solidification. Zinc as a second important alloying element in this alloy is used to improve the room temperature strength, fluidity and resistance to corrosion. Manganese is mainly used for improving the seawater corrosion resistance, by removing iron and possibly other heavy metal elements into relatively harmless intermetallic compounds [4, 5].

The aim of this work is to demonstrate the applicability of thermodynamic modeling to generate detailed information about phase formation in this process. This knowledge is crucial since in addition to the other process parameters one has to consider the mixing ratios of the feedstock alloys. Thus, the design of the best slurry produced by blending two or even more magnesium alloys may be supported by such simulations. For this purpose the process is pictured by appropriate thermodynamic calculations using a multicomponent Mg alloy database. Important calculated results are fraction solid, process temperature, thermal sensitivity, freezing ranges and the detailed constitution of the slurry produced and the solidified new alloy. Based on this information promising alloys and process parameters may be identified, thus focusing the necessary, though costly, thixomolding experiments to a smaller number. A waste of effort and time in less promising conditions may be avoided.

2 Thermodynamic Calculation as a Tool to Analyze a Blending of Mg-alloys by Thixomolding

2.1 Outline of Thermodynamic Approach

Thixomolding is high-speed, net-shape, semisolid, magnesium injection molding process. In this semi-continuous process room-temperature magnesium chips are heated to semi-solid slurry inside a barrel and screw and die-cast into precision-molded components. The whole process can be divided in two stages.

The first stage consists of introducing blended Mg alloys feedstock (in our case AZ91D and AM60B alloys in the ration 2:1) in the form of metal granules chips or pellets at room temperature to a heated barrel and screw system. The barrel is continuously purged using a counter flow of argon gas to prevent oxidation of alloys. The blended Mg alloys feedstock is heated up to a preset thixoforming temperature T_{TXM} between liquidus and solidus ($T_{LIQ} > T_{TXM} > T_{SOL}$), under high shear rate mixing. The T_{TXM} temperature is one of the critical parameters of this thixoforming process that beside alloy composition controls the ratio between solid and liquid phases prior to casting.

In a previous work one limiting case of this process was simulated by assuming complete equilibrium to be attained in the first stage while the feedstock is heated to the T_{TXM} temperature [6]. As already noted in that work [6], these conditions may not prevail in high-throughput thixomolding. In the present report the other limiting case is treated, that is the first stage is also treated as a limiting non-equilibrium process by applying the Scheil conditions, just like the second stage of rapid solidification. The metallurgical basis for these assumptions will be elaborated in more detail in a separate paper [7].

For the temperature range between T_{LIQ} and T_{TXM} the non-equilibrium primary magnesium solid solutions particles (Mg^{AZ}) and (Mg^{AM}) will be formed, the superscript AZ or AM is denoting the origin from the AZ91D and AM60B alloys, respectively. They are shaped into desired globules by the high shear force. The amount of these primary magnesium particles (Mg^{AZ}) and (Mg^{AM}) can be calculated using thermodynamic non-equilibrium calculation [7].

During the second stage of thixoforming process, the screw quickly injects solid particles suspended in liquid matrix, so called slurry, into preheated metal mold. The semisolid slurry, strictly speaking the residual liquid, will solidify in the less than a second in the mold under non-equilibrium conditions. Here the Scheil conditions are generally a safe assumption, given the very short solidification time in the permanent mold in thixomolding or other thixoforming processes. Under this model with its blocked diffusion in the solid state also the amount of primary (Mg^{AZ}) and (Mg^{AM}) present at T_{TXM} will be frozen-in into the final microstructure.

2.2 Examples of Blending Two Alloys with Ratio 2AZ91D:1AM60B

In this work we present preliminary results of alloys with mixing composition 67% of AZ91D+33% of AM60B.

It was mentioned above that blending of Mg-alloys by thixomolding has been recently experimentally applied by Nandy et al. [2] and Czerwinski [3] for designing a new magnesium alloy by thixomolding processing. Czerwinski studied the alloys with mixing composition 50% AZ91D + 50% AM60B. The results of our calculation for alloys with this composition

(AZ91D:AM60B) will be given later [7]. For thermodynamic simulation of blending alloys with ratio 2AZ91D:1AM60B we apply the chemical composition of the AZ91D and AM60B magnesium alloy (Table 1) and thixomolding temperature 600 °C according with [3]. Alloys with mixing composition 33%AZ91D+67%AM60B and 67%AZ91D+33%AM60B were studied by Nandy et al. [2]. They used for experimental investigation base alloys AM60B and AZ91D with nominal composition of Al 6 and 9 wt.%, respectively, without giving the full chemical analysis. During thixomolding, barrel temperature was set between 590–620 °C and volume fractions of the solid particles in the "high solid fraction alloys" vary between 20–30 %, those in the "low solid fraction range" from 5–10 %. In that work [2] more attention was given to investigations of chemistry composition of primary and secondary crystallized magnesium and properties of the new alloys.

Table 1: Chemical compositions of the AZ91D and AM60B precursors and the alloy obtained by their mixing in ratio 2:1 (2AZ91D:1AM60B)]

	Al [2]	Al [3]	Zn	Mn	Si	Cu	Fe	Ni	Mg
AM60B	6	5.82a	<0.01a	0.31a	0.03a	naa	<0.01a	<0.01a	Balance
2AZ91D:1AM60B	8	7.74	0.44	0.30	-	-	-	-	Balance
AZ91D	9	8.69a	0.66a	0.29a	0.02a	<0.01a	<0.01a	<0.01a	Balance

[a] – complete analysis only given in ref. [3]

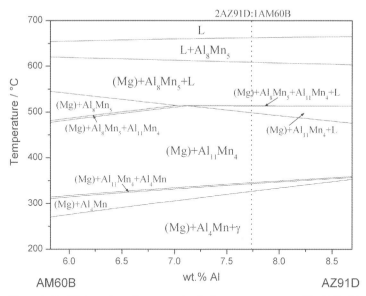

Figure 1: Equilibrium phase diagram section in the quaternary system Mg-Al-Zn-Mn along the composition section from AM60B to AZ91D (Mg-Al5.82Mn0.31 to Mg-Al8.69Zn0.66Mn0.29). Dotted line represents the blended alloy composition with ratio 2AZ91D:1AM60B.

For specified chemical composition of the AM60B and AZ91D magnesium alloy (see Table 1) the phase diagram section has been calculated using Pandat software package [8] together with a magnesium database developed in our group, the basis of which is given elsewhere [9].

The equilibrium phase diagram section is shown in Figure 1. The dotted line represents the blended alloy composition with ratio 2AZ91D:1AM60B. The composition of this alloy is listed in the second line of Table 1, the trace elements Si, Cu, Fe and Ni were not used for the thermodynamic calculation. Even though the actual process is off-equilibrium, this specific phase diagram section gives an overall view of the phase relations between AM60B, 2AZ91D:1AM60B, and AZ91D. It also shows the phases to be expected after a high temperature equilibration treatment. It is noted that under equilibrium conditions the γ phase (γ-AlMg or γ-Al$_{12}$Mg$_{17}$) is formed not during solidification but in a solid state transformation below 270 to 353°C. This is in contrast to the subsequent non-equilibrium calculations.

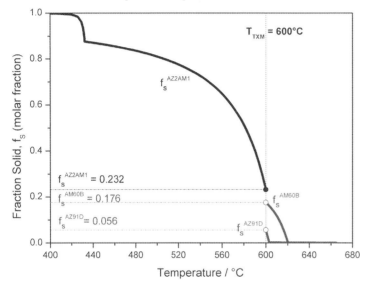

Figure 2: Calculated fraction solid for AZ2AM1 Mg-alloy (formed by mixing 2 parts of AZ91D and 1 part of AM60B alloys at room temperature): The non-equilibrium (Scheil) fraction solid of the AM60 and AZ91D alloys are presented for $T > T_{TXM}$ temperature stages. For the preset working temperature $T_{TXM} = 600$ °C, the amounts of fraction solid of precursors AZ91D and AM60B alloys (open circle) as well as the fraction solid of the mixed AZ2AM1 alloy (filled circle) are given.

For the simulation of the actual thixomolding process with non-equilibrium (fast) solidification another type of calculation, the Scheil model was used. Figure 2 shows calculated fraction solid curves of blended alloy with ratio 2AZ91D:1AM60B, denoted as AZ2AM1. The non-equilibrium solidification in the second stage is marked as bold curve, f_S^{AZ2AM1}. The preset value of $T_{TXM} = 600$ °C defines the fractions of individual solids, $f_S^{AZ91D} = 0.056$ and $f_S^{AM60B} = 0.176$ for AZ91D and AM60B respectively. They add up to the starting value for the blended alloy AZ2AM1 with $f_S^{AZ2AM1} = 0.232$ at 600 °C. These values of fractions solid reasonably conform to experimental values from [2], where it was observed that coarse globular solid particles in the "high solid fraction alloys" [2] vary between 20–30 %. No further details were reported.

Most multicomponent alloys solidify to more than one single phase. In fact each bend in the fraction solid curve of Fig. 2 signals the appearance of a new phase. The detailed information about all phases after solidification of the blended alloy AZ2AM1 is given in Fig. 3 and is a key factor in understanding the complex microstructure. Figure 3 shows the amount of each individual phase in molar fractions and the temperature range over which it is formed. The superscript

# denotes phases generated in the second stage of the process, the rapid solidification from 600 °C. A logarithmic scale is used for a quantitative representation of the minority phases. The cutoff limit was set to a phase amount of 0.01 %. There are three trace phases forming with an even smaller amount. The box in the left part of the diagram summarizes the final fraction of each phase after complete solidification.

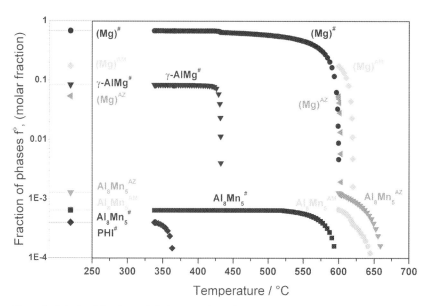

Figure 3: Calculated fractions of all phases vs. formation temperature for the AM60B, AZ91D and AZ2AM1 alloys plotted in logarithmic scale. We distinguish between the primary phases $Al_8Mn_5^{AM}$, $(Mg)^{AM}$, $Al_8Mn_5^{AZ}$ and $(Mg)^{AZ}$ formed during the first stage of the process, non-equilibrium melting of the AM60B and AZ91D alloys, the range between their liquidus and T_{TXM} temperatures is shown in light grey for orientation. All other phases marked by hash symbol (#), are formed during fast solidification of the residual melt produced by mixing residual liquids of two alloys (AM60B and AZ91D) at the process point T_{TXM} = 600 °C. The box in the left part of the diagram displays the final fraction of phases in the solidified microstructure.

According to Fig. 3, first the minute amounts of Al_8Mn_5 phase are formed. These calculated, very small molar fractions of primary $Al_8Mn_5^{AM}$ and $Al_8Mn_5^{AZ}$ phases are shown between liquidus and T_{TXM} temperature. The calculated fraction of primary magnesium phases $(Mg)^{AM}$ and $(Mg)^{AZ}$ are also shown between liquidus and T_{TXM} temperature; their formation temperature is below those of the Al_8Mn_5 phases. In the second process step, after casting the slurry into the die, both phases $Al_8Mn_5^{\#}$ and $(Mg)^{\#}$ are rapidly solidified jointly from the mixed residual liquid of blended alloy AZ2AM1. Magnesium phase $(Mg)^{\#}$ will be easily recognized as a fine microstructure. Even though $(Mg)^{AM}$, $(Mg)^{AZ}$ and $(Mg)^{\#}$ are the same *phase* (*hcp* solid solution), they are distinctly different in their morphological appearance in the microstructure. Following the complete solidification path of this alloy, after $Al_8Mn_5^{\#}$ and $(Mg)^{\#}$, the next phases γ-AlMg$^{\#}$, PHI$^{\#}$ φ-AlMgZn) and the trace phases $Al_{11}Mn_4^{\#}$, τ-AlMgZn$^{\#}$, MgZn$^{\#}$ are precipitated. The latter three are formed in miniscule amounts below 0.01 %. The final microstructure is composed of seven different phases frozen-in this non-equilibrium state, and two of them appear in three distinct morphological appearances. Finally, this calculated microstructure is in general agreement with experimental observation [2]. Nandy et al. observed many of the primary magnesium particles surrounded by a continuous matrix of fine "secondary" magnesium and γ-AlMg phases. Moreover, additionally large irregular particles were occasionally observed which were rich in Mn and Al and these particles were present in all the alloys, however, their volume fraction was extremely low. As indicated by the present calculations, the Al-Mn intermetallic compounds, such as Al_8Mn_5 (and some $Al_{11}Mn_4$) precipitate during solidification and the phase fraction of these particles is very small. According with [5] the actual precipitate typically contains some iron, $Al_8(Mn,Fe)_5$ which is not yet incorporated in the present calculations.

3   Acknowledgement

This study is supported by the German Research Foundation (DFG) in the Priority Programme "DFG-SPP 1168: InnoMagTec" under grant no. Schm 588/26.

4   References

[1]   Young, R.M.K., Clyne, T.W., J. Mater. Sci., **1986**, *21*, 1057–1059.
[2]   Nandy, T.K., Wayne, Jones J., Pollock, T.M., Walukas, D.M., Decker, R.F., Magnesium Technology TMS, **2002**, 215–220.
[3]   Czerwinski, F., Acta Mater., **2004**, *52*, 5057–5069.
[4]   Petterson, G., H´øier, R., Lohne, O., Westengen, H., Mater. Sci. Eng. A, **1996**, *207*, 115–120.
[5]   Kwon, S.-L., Kim, S.-J., Byun, J.-Y., Ha, N.-F., Shim, J.-D., J. Mater. Sci. Lett., **2003**, *22*, 199–200
[6]   Djurdjevic, M.B., Schmid-Fetzer, R., Mater. Sci. Eng. A, **2006**, *417*, 24–33.
[7]   Kozlov, A., Djurdjevic; M. B., Schmid-Fetzer, R., to be published.
[8]   Chen, S.-L., Daniel, S., Zhang, F., Chang, Y. A., Yan, X.-Y., Xie, F.-Y., Schmid-Fetzer, R., Oates, W. A., Calphad, **2002**, *26*, 175–188.
[9]   Schmid-Fetzer, R., Gröbner, J., Adv. Eng. Mater., **2001**, *3*, 947–961.

Quantitative Understanding and Modeling of Microstructural Formation by Solidification Technique of Magnesium Alloys

G. Klaus (Sp), B. Böttger, J. Eiken, A. Bührig-Polaczek, I. Steinbach
RWTH Aachen (Germany)
M. Ohno, J. Gröbner, R. Schmid-Fetzer
Clausthal University, Clausthal-Zellerfeld (Germany)

1 Abstract

This work reports on progress in an ongoing joint research project on the quantitative acquisition and modeling of microstructure formation during the solidification of commercial magnesium alloys. It aims at the development of a microstructure quality criteria to assist the computational design of advanced magnesium based cast alloys.

Thermodynamic chemistry, casting experiments and phase field simulations are combined tio reach this goal. Evaluating published thermodynamic data and experimental data, measured by thermal analysis, it was found, that liquidus and solidus temperatures of AZ and AM alloys for a wide range substantially differ from the corresponding temperatures in casting experiments. This discrepancy can be resolved using thermodynamic calculations. Casting experiments were performed to reveal the influence of the chemical composition and the cooling conditions. Using the thermodynamic data and the experimental process parameters as an input, a PhaseField model is calibrated, that can be used to tailor the alloy to the demands of different casting processes.

This study is supported by the German Research Foundation (DFG) in the Priority Programme "DFG-SPP 1168: InnoMagTec".

2 Introduction

Boundary conditions in industrial casting process are many and depend on the particular process. As a result the microstructure can change from columnar to equiaxed morpholgy. In bridgman type experiments the directional solidification of AZ31 was investigated [1], [2] and input parameters for the quantitative phase field simulation for microstructure evaluation were adjusted. Later on, the experimental and simulated results have been compared and found to be in good agreement.

In the present paper the equiaxed solidification of AZ31 was investigated and input parameters were calibrated for quantitative phase field simulation. Therefore an experimental setup was developed at the Foundry Institute of RWTH Aachen University, which allows for well controlled equiaxed solidification conditions with a wide range of cooling conditions.

3 Casting Experiment

During the experiments cooling rates were measured for different processes and cast parts to gain knowledge about the boundary conditions for solidification. For a magnesium sand casting (Figure 1) with 7 mm sections and thick walled junctions 0.7–1.0 K/s have been determined. For a cylindrical shaped tensile test specimen with 16 [mm] in diameter the cooling rate was about 12.5 K/s. For a high pressure die casting part (Figure 2) 20.0 K/s were determined at a 15 mm section and 45.5 K/s at a 8 mm section. The casting of all parts was performed with AZ31 at 740 °C, the given cooling rates were determined between 620 °C and 520 °C. The cooling conditions of the high pressure die casting were evaluated using commercial numerical simulation software.

Many parts cast for automotive or air craft application can only be produced by sand casting because of their complex shape. Therfore it is necessary to develop an alloy, which solidifies with a fine grain structure even at low cooling rates, typical for sand casting. The experiment described below is designed to simulate these conditions.

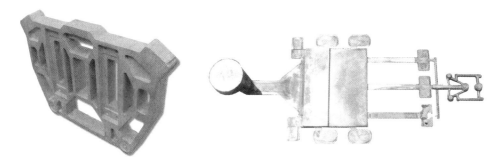

Figure 1: Sand casting   **Figure 2:** High pressure die casting part

4 Design of an Experimental Setup for Equiaxed Solidification

A reproducible evaluation of the microstructure evolution during solidification needs an undisturbed equal heat transfer in all three directions of the coordinate system. From geometric point of view a sphere shaped casting would perfectly fulfill this requirement. Apart from the influence of gravity, solidification would start at the same time on the surface and progress in the direction of the center. Due to the shrinkage of metals a cavity would occur in the center of the casting, making it impossible to conduct a temperature/time measurement in the thermal centre of the sample. It was assumed, that a cylindrically shaped casting with a geometric ratio of height/diameter = 1 would fulfill the conflictive requirements concerning an equal heat transfer and feeding, that has to take place in vertical direction.

The temperature/time measurement should be conducted in the centre and at the edge of the casting in order to evaluate the local progress of solidification and to determine parameters for simulation. Furthermore the nucleation and growth of phases [3] can be determined using thermal analyse. The temperature measurement requireres a long solidification time in order to detect phases with small amounts of latent heat, which cause just low peaks in the temperature/time curve. The sand mold has three cavities placed in a 120° angle around the central ingate.

The cylindrical cavities all feature a cubic geometric ratio of height/diameter = 1 and have got the sizes 30, 60 and 90 mm. Due to the similar conditions only the solidification rate varies as a function of the modulus and the influence of the cooling conditions on phase formation and the solidification morphology can be observed.

5   Numerical Simulation of the Mold Filling and Solidification Process

Before constructing the pattern (Figure 4) a numerical simulation was conducted to proof the conditions for a quasi-undirectional solidification. The simulated temperature profil is illustrated in figure 3 and shows a quasi spherical temperature distribution in each cylinder. Furthermore the ingate and runners were designed for a turbulence free mold filling process. The calm melt flow is required to the entrainement of oxides and the subsequent precipitation of gas, which might influence the experimental results. The single cylinders were placed in an adequate distance of each other to avoid a mutual thermal influence.

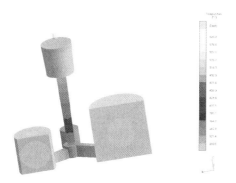

Figure 3: Temperature profil during solidification

Figure 4: Al pattern with the position of the thermocouples

Figure 5 shows a 90 mm specimen of pure aluminium (99.999 wt% Al), which was cast to gain basic knowledge about the experiment concerning the microstructure evaluation. The horizontal greyish colored stripes at the surface of the casting indicate, that the mold filling was calm and without turbulences. The macro etched cross section of the sample (Figure 6) shows the centre oriented growth of the single cristals, which confirms the simulated results for the temperature distribution. A further measurement of the temperature profile is in preparation.

Figure 5: Casted specimen

Figure 6: macro etched 90 mm Al sample

6 Description of the Casting Experiment

The experimental setup consisted of an resistive heated oven with a mild steel crucible featuring a capacity of 4 kg. The melt was charged with a mixture based on Argon and 2.0 vol % SF6. The sand molds were made of furan resin hardend sand with an addition of the inhibitors sulfur and boric acid in powder form. Casting was performend using AZ31 at a casting temperature of 760 °C. The temperature was choosen to prevent premature solidification and to melt up phases like Al8Mn5. The filling of the casting took five seconds. The temperature was measured for about 1400 seconds untill 400 °C were reached. Below this temperature no valuable information could be gained. The total weight of a casting (Figure 5) was about 1.4 kg .

7 Metallographic Preparation and Evaluation

Each cylinder was cut in transversal direction and metallographically prepared. For the evaluation of the microstructure the samples were color etched and digitally photographed and examined by a digital picture analyze. The micrographs utilized for the comparison between phase field simulation and experiment were taken in the centre of the samples close to the thermocouple. Figure 7 shows the temperature measurement of the three cylindrical samples which were used to define the boundary conditions for the phase field simulation. The 90 mm sample solidifies with a speed of 0.23 K/s the thinner samples with 0.46 and 1.34 K/s. Figure 8 shows the microstructure corresponding to the pictured cooling curves. As expected the grain size decreases with increasing cooling rate.

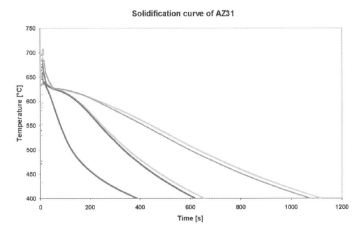

Figure 7: solidification curves of AZ31

8 Thermodynamic Assessment of Mg-Al-Mn-Zn System

The Calphad method, based on well established thermodynamic data is able to predict the phase equilibria in multi-component system with quite a high accuracy [4]. These reliable data are needed for the phase field simulation for microstructure evolution. Based on the thermodynamic descriptions of the sub-ternary systems Mg-Al-Mn, Mg-Al-Zn and Mg-Mn-Zn , the quaternary system Mg-Al-Mn-Zn was calculated. A detailed comparison between calculated and experimental results, reported in Ref [5], demonstrates the relability of the calculated data, which are in good agreement with experimental results. Furthermore it was found, that the widely accepted liquidus temperature of AZ and AM alloys do not represent equilibrium conditions. Even at extremly slow cooling rates below 1 K/min the alloy AZ91 solidfies far from equilibrium rates. This discrepancy has to be taken into acount, when data from thermal analyse is evaluated.

9   Microstructure Simulation using the Phase-field Method

A phase-field model for multicomponent alloys [6, 7], has been applied to simulate the microstructure evolution for the model castings. Thermodynamic data of the alloy AZ31 were taken from the newly assessed Calphad database by Schmid-Fetzer et al. [2]. to meet the cooling conditions of the experiment, one heat extraction rate was deduced for each measured cooling curve, that was used as an input for the simulation. Heterogeneous nucleation was assumed on small nucleants particles with different radii distributed randomly within the calculation area with a given nucleant distribution.

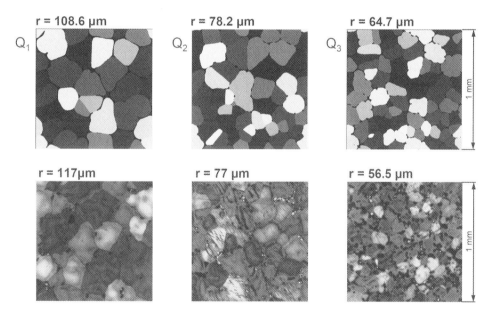

Figure 8: Comparison of simulated results with the experiment

The particles are not resolved in the simulation, but define locally the critical nucleation temperature required for locally nucleation. After nucleation the new grain releases latent heat and redistributes solute, which hinders further nucleation in the direct neighborhood of the grain. If the release of latent heat by all grains balances the heat extraction, maximum undercooling is reached and the nucleation process stops. Only nucleation sites with smaller critical undercooling than the maximum reached undercooling are activated, while the others do not become active and will be overgrown (see also [8]).

Figure 8 shows the simulated grain structure in comparison with the experimental data. The mean grain radii have been evaluated by dividing the calculation domain (respectively the metallographic observation domain) by the number of grains and taking the radius of an equivalent circle. The simulated results reproduce the clear tendency of an improved grain refinement with increasing cooling rate and are also in excellent quantitative agreement with the experimental results, proving the validity of the model assumptions and the numerical implementation. Simulations with varying alloy content and varying nucleation densities are presented elsewhere [9]

10 Conclusion

Casting experiments and numerical simulations were performed to establish a tool for the prediction of grain density in Magnesium casings dependent on the cooling rate. The casting experiments were designed to ensure conditions of equiaxed dendritic growth and to vary the cooling rate by two orders of magnitude. Numerical process simulation was used to reconstruct the local conditions within the model casting. Phase-Field simulation was used to simulate the nucleation process and the equiaxed dendritic growth, the competition of which determines the final grain structure. Using a thermodynamic database for the Mg alloy system will allow for the variation of the alloy content and a specific alloy optimization for different casting process windows.

11 Acknowledgement

We thank the German Research Foundation (DFG) for funding within the Priority Programme SPP-1168.

12 References

[1] B. Böttger, J.Eiken, M. Ohno, G. Klaus, M. Fehlbier; R.Schmid-Fetzer, I.Steinbach, A. Bührig-Polaczek "Controlling Microstructure in Magnesium Alloys: A Combined Thermodynamic, Experimental and Simulation Approach"
[2] J. Eiken, B. Böttger, I. Steinbach "Simulation of micrstucture evolution during solidification of magnesium-based alloys"
[3] Bäckerud, L.; Guocai, C.; Tamminen, J. "solidification characteristics of aluminium alloys – volume 2. AFS/SKANALUMINIUM (1990)"
[4] Y. Austin Chang, Shuanglin Chen, Fan Zhang, Xinyan Yan, Fanyou Xie, Rainer Schmid-Fetzer, W. Alan Oates, "Phase Diagram Calculation: Past, Present and Future" Prog. Mater. Sci., 49, 2004 313–345
[5] M. Ohno, D. Mirkovic and R. Schmid-Fetzer, "Liquidus and Solidus Temperatures of Mg rich Mg-Al-Mn-Zn Alloys" Acta Mater., (submitted).
[6] J. Eiken, B. Böttger, I. Steinbach, Physical Review E 73 (2006), p066122
[7] B. Böttger et al., Advanced Engineering Materials 8, (2006), p241
[8] T. E. Quested, A. L. Greer, Acta Materialica 52 (2004), p3859
[9] J.Eiken, B.Böttger, I.Steinbach, this conference

Production of a Fine-grained Mg Alloy AZ31 with SiC Particles

Gabriele Vidrich[1], Andreas Schiffl[2], Franka Pravdic[2], Hans Ferkel[3], Yuri Estrin[1], Helmut Kaufmann[2], Norbert Hort[4]

[1] Technische Universität Clausthal, 38678 Clausthal-Zellerfeld, Germany
[2] ARC Leichtmetallkompetenzzentrum Ranshofen GmbH (LKR), 5282 Ranshofen, Austria
[3] now: Volkswagen AG, Konzernforschung Werkstoffe, 38436 Wolfsburg, Germany
[4] GKSS Forschungszentrum Geesthacht, 21502 Geesthacht, Germany

1   Introduction

Due to an increasing demand for lightweight materials the production of magnesium materials grows by about 15 % per year [1–4]. Since Mg alloys are mostly used in the as-cast form, numerous efforts to improve the properties of cast materials have been undertaken. Further processing of cast alloys, for example via thixoforming, requires a globular structure of the precursor material.

There are also increasing efforts to develop wrought alloys. In order to achieve better ductility, fine-grained Mg-based materials are advantageous, and a possible avenue for that is to use grain-refining additives. For mold and continuous casting grain-refined magnesium alloys show better mechanical properties under static and dynamic loading. In fine-grained Mg alloys twinning is largely hindered, therefore the mechanical anisotropy can be reduced and deformation at moderate temperature is easier.

Most commercially used magnesium alloys contain aluminum, and therefore zirconium, an effective grain refining agent [5–7], cannot be employed because of its toxic effect. According to literature [8–13], SiC can act as a grain refining agent for Mg-Al alloys, but at present there is no grain refining system available commercially.

In the present work, a method allowing grain refinement of commercial AZ31 alloys via continuous casting from Mg melts was developed based on the use of ceramic particles.

In order to disperse the ceramic particles in the melt, precursor material highly loaded with SiC particles was added to the melt followed by casting. The particles act as heterogeneous nucleation sites in the material. The precursor material consists of inert particles, which were coated with AZ91 by means of mechanical milling of the SiC powder and the matrix metal micro powder.

2   Experimental

The micro scaled SiC particles with a median diameter before ball milling of about 2 µm had a diamond structure with a density of 3.22 g/cm^3 and a melting point of 2700 °C. The AZ91 micro powder employed, produced by gas atomization of the corresponding melt under Argon containing 1 % Oxygen for powder passivation, was of spherical shape with a median particle diameter of about 70µm. Mg alloy and SiC powder were milled for 12 hours at 250 rpm in a planetary ball mill (Retsch, PM400) in a sealed Argon atmosphere. The milling balls (diameter 10 mm) and milling vial (500 mL volume) were made of corundum. The weight ratio ball-to-powder was 1:10. 15 vol. % of SiC powder was dispersed via this milling technique in the master alloy. The composite was encapsulated in an evacuated aluminum container (70 mm in diameter), degassed at about 350 °C and extruded by a preheated (350 °C) 400 t horizontal extrusion press (outlet 16 mm).

Before extrusion the precursor material was investigated by X-ray diffraction and differential thermal analysis (DTA), Netzsch STA 409. DTA measurements were carried out between room temperature and 800°C with a ramp rate of 5 K/min.

For casting, the fully automated continuous casting line MAGNUMCAST® was used. Pretreatment of the melt was performed on an electrically heated holding furnace with a capacity of 600 kg. A mixture of HCF and N_2 was used as cover gas for the molten alloy. The SiC master alloy was added in small chunks that were preheated and dissolved on a casting spoon just below the surface of the melt. The melt was stirred gently for a homogeneous distribution of the dissolving master alloy throughout the melt. Billets with a diameter of 250 mm were cast with a casting speed of 1.5 mm/s and a temperature of the melt of 700 °C. As a reference, one casting without a grain refiner was also done under the same conditions.

The amount of Si in the SiC containing AZ31 alloy was below 0.1 wt. %. Samples cut out of the cast billet were polished and etched for light microscopy investigations. Tensile tests as well as XRD and DTA measurements were carried out. The cast material was extruded and investigated as well.

3   Results and Discussion

After milling the size of the ceramic particles in the master alloy was determined. It was found that the median particle diameter decreased.

To ensure the stability of the SiC particles in the Mg melt, the behavior of these particles was investigated by X-ray diffraction and thermal analysis. Figure 1 shows X-ray diffraction spectra of the SiC micro powder, pure AZ91 micro powder and the milled master alloy.

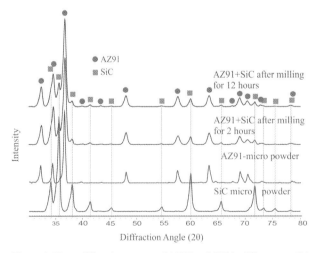

Figure 1: X-ray diffraction spectra of AZ91 and SiC in different conditions

It can be seen that after 12 h milling, the AZ91 diffraction pattern are broadened. This indicates that during milling heavy deformation of the AZ91 particles takes place. The deformation causes micro particle hardening, particle cleavage and cold welding of the micro particle fragments, which finally results in a fine dispersion of the ceramic SiC particles in the material.

DTA spectra of pure AZ91 and the AZ91 precursor material give no indication of new phases or reactions of SiC with the matrix.

Light microscope investigations show a very inhomogeneous grain size distribution with large grain sizes for the as-cast AZ31. In comparison, AZ31 containing SiC particles exhibits a much finer grain with a homogeneous size distribution from the rim to the center of the cast billet. The grain sizes found in pure AZ31 vary from 930 µm (10 mm from the rim) to 620 µm in the center of the billet. By comparison, the SiC particle-containing alloy shows grain sizes from 190 µm near the rim to 320 µm in the center. It was found that with increasing residence time of the ceramic particles in the melt before casting (residence time from 10 minutes up to 3 hours) the grain sizes decreases.

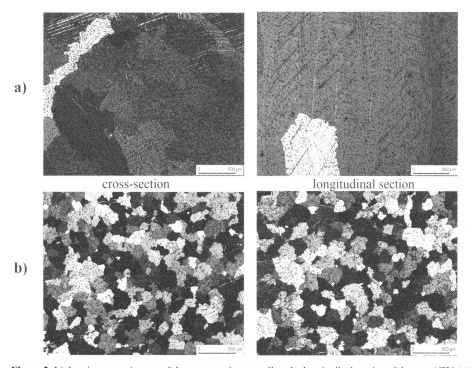

Figure 2: Light microscope images of the cross-section as well as the longitudinal section of the cast AZ31 (a) and the grain-refined AZ31 (b)

According to Fig. 2 the alloy containing SiC exhibits no significant difference in grain size and grain shape between the cross-section and a longitudinal section. In both directions the grain is finer and more globular compared to pure AZ31 alloy.

Obviously a small amount of the SiC master alloy leads to a significant grain refining effect.

Segregation analyses of the elements revealed a homogeneous distribution across the cross-section of a strand.

Figure 3 shows representative results of the tensile tests carried out at room temperature on the as-cast material for a strain rate of 10^{-3} s^{-1}. It can be seen that the alloy containing SiC particles exhibits elongation to failure twice as big as that of the initial AZ31. The tensile elongation increases from 5 % to 11.7 % in case of the grain refined cast material. It was found that grain refining leads to a significant increase in ductility. While there is a decrease in yield strength, the ultimate tensile strength rises appreciably, from 170 MPa to 205 MPa, thus exhibiting an enhanced hardening potential.

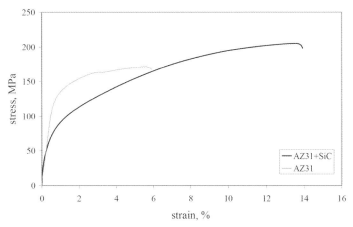

Figure 3: Representative tensile tests for the as-cast materials carried out at room temperature, strain rate 10^{-3} s^{-1}

After extrusion, the behavior of the as-extruded material under tensile deformation was investigated. Figure 4 shows the stress-strain curves for both alloys at room temperature (RT), 100 °C and 150 °C for a strain rate of 10^{-3} s^{-1}.

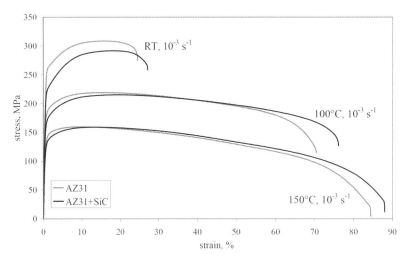

Figure 4: Stress vs. strain curves for the alloys in as-extruded condition at different temperatures

It can be seen that in case of pure AZ31 smaller tensile elongation can be achieved as compared to the grain-refined alloy. Although the pure alloy shows higher values for the yield strength, only at room temperature does it also exhibits a higher ultimate tensile strength. At elevated temperatures there is no difference in the UTS between both materials. It can be seen that the SiC containing alloy reaches the UTS at higher strains. This behavior was observed for all temperatures.

4   Conclusion

It was demonstrated that addition of small amounts of the SiC-containing master alloy leads to a significant grain refinement effect in AZ31 produced by continuous casting. The material generated in this way shows a fine grain with a homogeneous grain size distribution. There is no difference in grain size and grain shape between radial and longitudinal directions of the cast billet.

The cast material exhibits a much higher tensile elongation in the case of the grain refined alloy. After extrusion, a small increase in tensile elongation over that of the pure AZ31 alloy is still retained. Although the grain-refined as-cast material shows decreased yield strength, higher values of UTS were recorded.

5   Acknowledgment

This study was supported by the German Research Foundation (DFG) in the Priority Programme "DFG-SPP 1168: InnoMagTec" under grants Nos. ES 74/21, KA 2255/2 and HO 3234/1.

6   References

[1]   B.L. Mordike and K.U. Kainer, Magnesium Alloys and their Applications, Werkstoffe-Informationsgesellschaft, Frankfurt, Germany (1998).
[2]   G. Neite et al., in R.W. Cahn, P.Haasen, E.J. Kramer (Eds.), VCH-Verlag, Mater. Sci. Technol. 8 (1996) 114.
[3]   D.J. Lloyd, Intern. Mater. Rev. 39 (1994) 1.
[4]   R. Oakley, R.F. Cochrane, R. Stevens, Key Eng. Mater. 104 (1995) 387.
[5]   F. Sauerwald, Über Walzlegierungen mit Zirkonium, Zeitschrift f. Metallkunde, 41, (1950) 81–87.
[6]   D. Wenwen et al., Microstructure and mechanical properties of Mg-Al based alloy with calcium and rare earth additions, Mat. Sci. Eng. A356 (2003) 1–7.
[7]   G. Schmidt, Beitrag zur kornfeinenden Wirkung des Zirkoniums in Magnesium-Legierungen, PhD Thesis, TU Berlin, Germany (1975).
[8]   M.-X. Zhang, et al., Crystallography of grain refinement in Mg-Al based alloys, Acta Mater. 53 (2005) 3261–3270.
[9]   Q. Jin, et al., Grain refining mechanism of carbon addition method in a Mg-Al magensium alloy, Scripta Mater. 49 (2003) 1129–1132.

[10] M. Qian, P. Cao, Discussions on grain refinement of magnesium alloys by carbon inoculation, Scripta Mater. 52 (2005) 415–419.
[11] Q. Jin, J.-P. Eom, S.-G. Lim, Reply to comments on "Grain refining mechanism of a carbon addition method in a Mg-Al magnesium alloy", Scripta Mater. 52 (2005) 421–423.
[12] Y. Lee, Grain refinement of Magnesium, PhD Thesis, University of Queensland, Australia (2002).
[13] Y. Liu, X. Liu, B. Xiufang, Grain refinement of Mg-Al alloys with Al4C3-SiC/Al master alloy, Mater. Lett. 58 (2004) 1282–1287.

Grain Refinement of AZ31 by (SiC)$_P$: Theoretical Calculation and Experiment

R. Günther[1], Ch. Hartig[1], R. Bormann[1]
[1]Hamburg University of Technology, Inst. of Materials Science and Technology, D-21073 Hamburg, Germany

1 Introduction

Thermomechanical treatment of wrought Magnesium alloys should preferably start with improved feedstock from cast materials already having small and homogeneous grain sizes. However, due to slow cooling rates involved in the processing of the feedstock, constitutional undercooling cannot be effectively utilized to obtain a homogeneous and fine microstructure. Consequently, grain refinement has to be achieved via in-situ refinement by primary precipitated metallic or intermetallic phases [1], or by inoculation of the melt via e.g. ceramic particles that remain stable in the melt due to their high thermodynamic stability. Ceramic inoculants such as SiC and Al_4C_3 are known to act as grain refiner even in Al-containing magnesium alloys, [2], however the microscopic mechanisms are not well understood. In addition, it is still not clear what are the ideal volume fraction, particle size and size distribution for given alloy composition and cooling rate. For this, in the present work a theoretical nucleation and growth model is developed in order to quantify the grain refining potential of SiC-inoculants in AZ31.

The simulation method for the heterogeneous nucleation is based on the hemispherical cap model proposed by Greer et al. [3]. It allows a prediction of the grain size as a function of the particle size distribution, the content of ceramic inoculants, the cooling rate and the alloy constitution. In order to test the model assumptions, experiments with SiC-inoculation of AZ31 alloy were accomplished. Further, this model will be generalized in order to predict the efficiency of inoculant particles in common Mg-alloys.

2 Experimental

For preparation of the SiC master alloy powder blends containing pure Mg- and SiC-powders were milled in a planetary ball mill (type P5, Fritsch, Idar-Oberstein, Germany). Subsequently, the AZ31 alloy was melted in a steel-made crucible using a resistance furnace. The SiC master alloy was added to the molten AZ31 in the crucible at 760 °C–780 °C. Casting was performed under a protective atmosphere of argon and SF_6. Prior to casting, the melt was stirred mechanically for 10 minutes. The alloy was cast in a cylindrical steel crucible of 8 cm in diameter and 6 cm height as well as 10 cm in diameter and 41 cm height, resulting in ingots of about 0.6 kg and 8 kg, respectively. For comparison of grain sizes with those of conventional AZ31, castings under the same condition without refiner were performed. For the 0.6 kg ingot, cooling rates were determined by four thermocouples mounted at different heights and radial positions with respect to the centre of the cylindrical steel crucible.

For optical microscopy, all specimens were sectioned, mounted, polished and then etched in a solution of picric and acetic acid. Microstructures were investigated using an optical microscope (LEICA DMLM). Grain sizes of the ingots were measured at the outer surface, in the centre of transverse sections and in between (D/4), applying the line-intersect method.

3 Modeling

The nucleation and growth of grains from the melt was modeled using the hemispherical cap model of Greer et al. [3]. In this model a barrier for heterogeneous nucleation is assumed that depends on the radius of the inoculant particle only. The required "curvature undercooling" ΔT_C at the onset of free growth then is related with the diameter d of a nucleant particle by

$$\Delta T_c = \frac{4\sigma}{\Delta S_V \cdot d} \tag{1}$$

The free growth-velocity dr/dt of successful inoculant particles is controlled by the diffusion balance of the solutes between melt and solid. It is approximately proportional to the reciprocal of the so-called growth-restriction factor Q [3, 4]:

$$\frac{dr}{dt} \propto \frac{D_S \Delta T_S}{r \cdot Q} \tag{2}$$

$$\Delta T_S = T_m - T - \Delta T_C \tag{3}$$

where D_s is the diffusion constant of the solutes in the melt and ΔT_s is the solutal undercooling, the difference between the observed undercooling $T_m - T$ and the curvature undercooling ΔT_c. The particle sizes of inoculant particles at the beginning of the cooling process are given as a distribution function $\varphi(r, t = 0)$. The further evolution of this function then follows from a conservation law for the particle number which is coupled with an energy balance describing the crystallization enthalpy [3, 4]:

$$\frac{\partial \varphi}{\partial t} + \frac{\partial}{\partial r}\left(\frac{dr}{dt}\varphi\right) = 0; \quad \varphi(r,t_0) = \varphi_0 \tag{4}$$

$$\frac{dT}{dt} = -\dot{T}_{ext} + \frac{4\pi}{3}\frac{H_V}{C_P}\frac{d}{dt}\int_0^\infty \varphi(r,t)r^3 dr \tag{5}$$

where H_v, C_p are the crystallization enthalpy per volume and the specific heat per volume.

The differential equations (4) and (5) can be solved numerically using appropriate discretization schemes [3, 4]. In order to calculate the final grain size in a completely solidified volume, the number of particles N_s per volume with a positive growth-velocity V at $T = T - T_{min} = 0.2$ K was taken as the number of successful nuclei and the grain size D was calculated from [3]:

$$D = \sqrt[3]{\frac{1}{2N_S}} \tag{6}$$

The physical parameters for AZ31 required for the solution of eq. (4) and eq. (5) are shown in table 1.

Table 1. The value k is a linear superposition of k(Mg-Al) and k(Mg-Zn) calculated from parameters in Ref. [5]. The growth-restriction value Q is linear superposed of Q(Al) and Q(Zn) with respect to AZ31 composition.

Quantity	Symbol	Units	Value	Reference
Diffusivity in melt	D_S	m^2/s	2,7·10^8	[6]
Entropy of fusion per unit volume	ΔS_V	J/K m^3	6,508·10^5	[5]
Enthalpy of fusion per unit volume	ΔH_V	J/m^3	6,07·10^6	[5]
Heat capacity of melt per unit volume	C_{pV}	J/K m^3	2,456·10^6	[5]
Solid-Liquid interfacial energy	s	mJ/m^2	115,0	[7]
Growth-restriction factor	Q		18,2	
Equilibrium partition coefficient	K		0,308	

4 Results and Discussion

4.1 Simulation

Based on reasonable physical data, the simulation method allows the prediction of the grain size as a function of the mean diameter and size distribution of the refiner particles, the alloy elements and the cooling rate. Figure 1 shows the predicted grain sizes as a function of the cooling rate for typical AZ-alloys AZ31, AZ61 and AZ91. As experimentally observed for solidification in general, faster cooling gives a finer grain size. Quantitatively, this is reflected correctly by the model. As shown, there is a strong dependency of the grain size at cooling rates lower than approx. 5 K/s, tending to saturate at high cooling rates. Typical casting processes are in a regime where the grain size could be significantly affected by the cooling rate. Furthermore, the tendency to smaller grain sizes with increasing aluminium content [8] is predicted by the model. This can be described by the growth-restriction factor Q, which is a function of the concentration of the solutes. The derivative of solid fraction with temperature at the onset of solidification is inversely proportional to Q [9]. Thus, increasing Q implies a slower onset of solidification, slower latent heat release and therefore a higher undercooling before recalescence, leading to initiation of free growth on more particles.

In order to minimize potential negative effects of high refiner contents on the mechanical properties, the knowledge of the minimum content needed for sufficient grain refinement is indispensable. Therefore, the grain size was simulated as a function of the refiner content for AZ31 and SiC inoculants based on an experimental particle size distribution and a cooling rate of 5 K/s (fig. 2). With increasing refiner content the grain size decreases rapidly and saturates at higher contents. For the underlying simulation conditions, a sufficient grain size reduction should be achieved at a refiner level of approximately 0.01 wt%.

Figure 3 shows the grain size dependency on the mean particle diameter of SiC in AZ31 with a refiner content of 0.04 wt%. The prediction of the most effective diameter is in the same range as predicted by Greer et al. [3] for aluminium alloys. Further evidence for the predicted order of magnitude of the effective inoculant size is given by experimental investigations of a magnesium-zirconium alloy that have shown that almost all effective zirconium particles had a size between 1 and 5 μm in diameter [10].

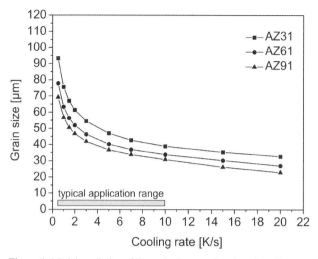

Figure 1: Model prediction of the grain size as a function of cooling rate and alloy constitution

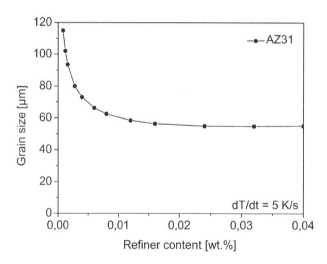

Figure 2: Effect of the addition level of the refiner on the resulting grain size calculated for AZ31 inoculated with SiC at a cooling rate of 5 K/s

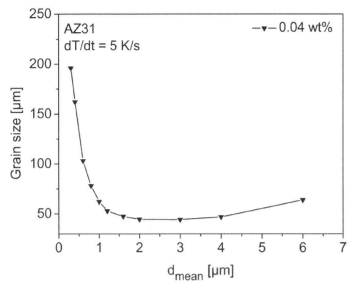

Figure 3: Calculated dependency of the mean particle size of SiC with a constant weight fraction of 0.04% on the resulting grain size for AZ31 with a cooling rate of 5 K/s

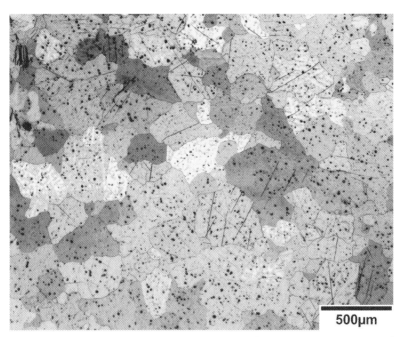

Figure 4: Microstructure of a conventional AZ31 alloy, 0.6 kg ingot. Black spots are caused by etching

4.2 Experiments

In order to test the model assumptions, experiments based on SiC-inoculation of AZ31 have been conducted. For this, a master alloy was prepared as described in chapter 2. After milling, the mean diameter of the SiC particles has been estimated to approx. 2.7 µm by metallographic observations. The master alloy was added to the AZ31 melt with a corresponding content of 0.4 wt% of SiC particles, which subsequently was cast in ingots of 0.6 kg and 8 kg weight. Figure 4 and figure 5 show the microstructure of the conventional and the corresponding refined AZ31 alloy of the 0.6 kg ingot, respectively. While the conventional alloy exhibits a relatively coarse and inhomogeneous microstructure with an overall mean grain size of 145 µm, the refined alloy reveals a substantially homogeneous microstructure with a significant decrease in grain size of a factor of 3 down to 49 µm (fig. 6). For the 8 kg ingots, a reduction in grain size of a factor 2.8 is reached, however, the overall grain sizes are larger due to the lower cooling rate, i.e. 369 µm of the non-refined alloy compared to 134 µm for the refined one (fig. 7).

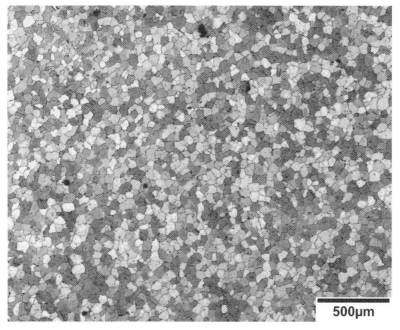

Figure 5: Microstructure of a conventional AZ31 alloy refined by SiC inoculation, 0.6 kg ingot

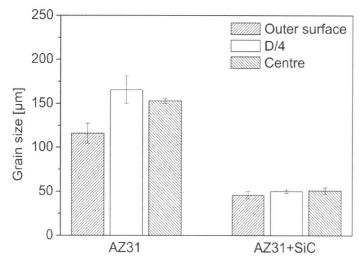

Figure 6: 0.6 kg ingots: Mean grain size at outer surface, mid-radius (D/4) and central region of the conventional AZ31 alloy in comparison to the refined AZ31 alloy. Error bars indicate the scatter of mean grain sizes in different fields

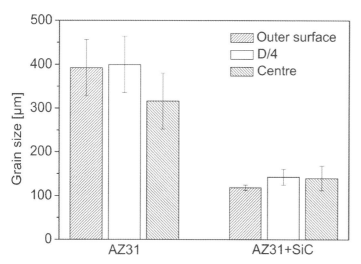

Figure 7: 8 kg ingots: Mean grain size at outer surface, mid-radius (D/4) and central region of the conventional AZ31 alloy in comparison to the refined AZ31 alloy. Error bars indicate the scatter of mean grain sizes in different fields

For comparison, the grain size as a function of cooling rate based on the accurate experimental parameters (mean particle size, particle size distribution, content of refiner, alloy composition) has been modeled quantitatively. As shown in figure 8, the experiments based on SiC-inoculation of AZ31 verify the model assumptions.

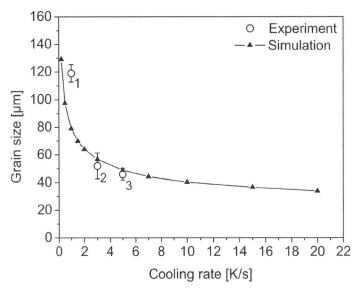

Figure 8: Comparison of the model predictions with experimental data. Data point 1 corresponds to the 8 kg ingot measured 10 mm from the bottom, data point 2 and 3 correspond to the 0.6 kg ingots, grain sizes have been measured at the middle vertical section and in 10 mm height from the bottom of the ingot. Error bars indicate the scatter of mean grain sizes in different fields

The comparison of the experimental data with the theoretical calculation of the grain size as function of the cooling rate exhibits a close match between experiment and theory: data point 1 corresponds to the 8 kg ingot measured 10mm from the bottom, whereas data point 2 and 3 correspond to the 0.6 kg ingots, where the grain sizes have been measured at the middle vertical section and in 10 mm height from the bottom of the ingot. Cooling rates of the small ingot were measured to 5 K/s at the bottom, and 3 K/s at mid-height. The corresponding cooling rate of the 8 kg ingot has been estimated to about 1 K/s.

5 Conclusions

The grain refinement efficiency of $(SiC)_p$ in Mg-alloy AZ31 for two different ingot sizes has been examined experimentally and compared with theoretical calculations. It is found that addition of $(SiC)_p$ results in significant grain refinement of a factor of about 3, regardless of the ingot dimensions. However, the ingot with 0.6 kg weight leads to smaller grain sizes of 50 µm due to the higher cooling rate. The results demonstrate that the theoretical model may assist as a potent tool to develop universal grain refiners for Mg-alloys, however, more comprehensive experiments have to be done to optimize the model assumptions, i.e. consideration of coalescence of the inoculants.

6   Acknowledgements

The authors would like to particularly thank the valuable discussions with L. A. Greer and T. E. Quested. In addition, the authors thank Dr.-Ing. N. Hort and M. Nürnberg, GKSS Research Centre, Institute for Materials Research, for support of the casting experiments. The financial support of the Deutsche Forschungsgemeinschaft (DFG) within the SPP Nr. 1168 programme "InnoMagTec" is gratefully acknowledged.

7   References

[1]   T. Laser, M.R. Nürnberg, A. Janz, C. Hartig, D. Letzig, R. Schmid-Fetzer, R. Bormann, Acta Mater **2006**, *54*, 3033-3041.
[2]   Y.H. Liu, X.F. Liu, X.F. Bian, Materials Letters **2004**, *58*, 1282-1287.
[3]   A.L. Greer, A.M. Bunn, A. Tronche, P.V. Evans, D.J. Bristow, Acta Mater **2000**, *48*, 2823-2835.
[4]   R. Günther, C. Hartig, R. Bormann, Acta Mater **2006**. Submitted
[5]   A.A. Nayeb-Hashemi, J.B. Clark. Phase Diagrams of Binary Magnesium Alloys. ASM International, Metals Park, Ohio, **1988**.
[6]   personal communication with G. Frohberg
[7]   A.R. Miedema, F.J.A. den Broeder, Zeitschrift der Metallkunde **1979**, *70*, 14-20.
[8]   D.H. StJohn, M. Qian, M.A. Easton, P. Cao, Z. Hildebrand, Metall Mater Trans A **2005**, *36A*, 1669-1679.
[9]   M.A. Easton, D.H. StJohn, Acta Mater **2001**, *49*, 1867-1878.
[10] M. Qian, D.H. StJohn, M.T. Frost, Scripta Mater **2004**, *50*, 1115-1119.

Creep Resistance of Highly Calcium-alloyed AZ91

D. Amberger[1], P. Eisenlohr[2], M. Göken[1]
[1] Institut für Werkstoffwissenschaften, LS 1, Universität Erlangen-Nürnberg, Martensstraße 5, 91058 Erlangen, Germany,
[2] Max-Planck-Institut für Eisenforschung, Max-Planck-Straße 1, 40237 Düsseldorf, Germany

1 Abstract

In this work, the influence of Calcium on the creep resistance of AZ91 magnesium alloy has been investigated. Different amounts of Ca (varying from 1 to 5 wt%) have been added to AZ 91. Creep tests have been performed in compression at different temperatures varying from 150 °C to 200 °C and stress levels between 100 MPa and 170 MPa. It can be seen that with an increasing amount of Ca the creep resistance improves for all tested temperatures and stress levels.

2 Introduction

Primarily owing to their very low density, Mg-base alloys are attractive candidate materials for light-weight construction. Aluminum as the most widely employed alloying element provides corrosion resistance and excellent castability such that even thin-walled structures can be produced in a cost-efficient manner. For the most common alloy AZ91, the microstructure after pressure die-casting consists of dendritic grains of primarily solidified alpha-phase (solid-solution) which is surrounded by a mantle of beta-phase (intermetallic $Mg_{17}Al_{12}$). The plastic deformation behavior of such a composite microstructure has been successfully modeled [1] under the assumption of a soft phase (alpha) being embedded in a hard phase (beta). The assumption of a relatively stronger beta-phase recently gained support from deformation tests on bulk $Mg_{17}Al_{12}$.

The low creep resistance even at moderate temperatures is the main disadvantage of AZ91 which precludes its use in many structural applications, *e.g.*, within the automotive powertrain, and which has lead to a continuing development of more creep-resistant alloys [2].

The present study aims to explore one of the promising roads to improve the creep resistance of AZ91 by using Calcium as an additional alloying element. This low-cost element does not significantly increase the alloy density and is known to suppress the formation of $Mg_{17}Al_{12}$ in favor of Al_2Ca (and/or Mg_2Ca depending on the Ca/Al-ratio) [3]. We will report deformation data from uniaxial compression tests and interpret the results in terms of the microstructural features observed by optical microscopy.

3 Experimental

AZ91 base material was alloyed with 1, 3, and 5 wt% Ca to yield AXZ911, AXZ931, and AXZ951, respectively. These alloys were produced by thixomolding a semi-solid slurry at

605 °C, corresponding to a volume fraction of primary solid phase of less than 5 %, into the shape of a plate of 6 mm thickness (process details can be found in, *e.g.*, [4]).

Samples of dimension 4.5 × 4.5 × 5.5 mm³ were machined from the centre of the plate. Loading faces were ground to plane-parallelism within 10 μm. Deformation tests in simple compression at constant stress and temperature have been performed in the range of 100–200 MPa and 150–200 °C.

Microstructural observations with an optical light microscope (Leica DmRm 200) were performed and concentrated on the center of the deformed compression samples which were sectioned, ground, mechanically polished and finally etched in a solution of 1 ml HNO_3 and 25 ml Ethanol.

4 Results and Discussion

The typical evolution of the creep resistance with strain for AZ91 with increasing Ca-content c_{Ca} is shown in Figure 1a) in the convenient form of strain rate $\dot{\varepsilon}$ *vs.* strain ε. With increasing strain, all tested materials exhibit a decreasing strain rate (primary creep) which, after reaching a minimum in $\dot{\varepsilon}$ (maximum creep resistance), increases again by less than an order of magnitude in the strain range investigated.

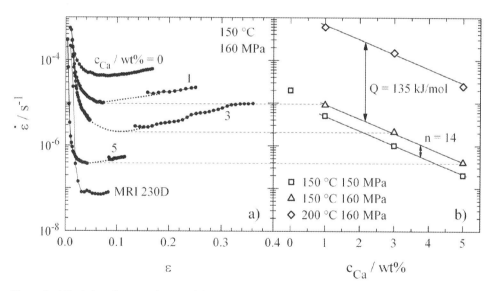

Figure 1: a) Evolution of creep resistance with strain for various additions of calcium to AZ91. Highly creep-resistant commercial alloy MRI 230D [5] shown for comparison; b) minimum creep rates at different testing conditions as function of Ca-content.

4.1 Minimum Creep Rate

The systematic decrease in the minimum creep rate $\dot{\varepsilon}_{min}$ with increasing c_{Ca} is illustrated in Figure 1b) which includes two additional deformation conditions of either increased temperature T or decreased stress σ. Within the present experimental window of deformation conditions a strengthening effect of $d\ln\dot{\varepsilon}_{min}/dc_{Ca} = -2.2/\text{wt\%}$, which is rather independent from σ and T, is found. Furthermore, no saturation of this beneficial effect of additional Ca is discernible up to the maximum of 5 wt% Ca used here.

Also rather independent from the Ca-content a stress exponent $n = d\ln\dot{\varepsilon}_{min}/d\sigma \approx 14$ and apparent activation energy $Q = -R d\ln\dot{\varepsilon}_{min}/d(1/T) \approx 135 \text{kJ/mol}$ follows from the data plotted in Figure 1b) with R being the gas constant. As for AZ91, the value found for Q can be correlated to self-diffusion of Mg (135 kJ/mol) [6].

A broader summary of minimum creep rates is given in Figure 2 in a temperature-normalized manner (with k_B: Boltzmann constant, D: coefficient of self-diffusion, G: shear modulus and b: magnitude of Burgers vector, taken from [6]). All alloys show a rather large n at the upper limit of the investigated stress range. However, comparing AZ91 to AXZ951, the stress exponents in the technologically relevant range of low creep rates differ by approximately a factor of 2, since AZ91 exhibits a significant decrease of n with decreasing $\dot{\varepsilon}$ or increasing T, i.e., losing the potential to substantially increase creep life by moderate reductions in stress. Whether a similarly strong decrease in n with σ is also present for AXZ951 (at much lower absolute $\dot{\varepsilon}$ compared to AZ91) is to be investigated in longer-term experiments than presently accomplished.

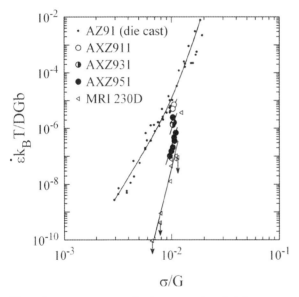

Figure 2: Temperature normalized representation of minimum creep rates for various additions of calcium to AZ91 as function of stress (normalized by shear modulus G). Highly creep-resistant commercial alloy MRI 230D [5] shown for comparison.

Figure 3 presents a sequence of light-optical microscope images of the as-thixomolded microstructure for increasing Ca-content. Three major transitions can be observed from left to right: First, the intermetallic phase (IP) which precipitates in the interdendritic regions from the supersaturated residual melt during casting changes from being exclusively $Mg_{17}Al_{12}$ in AZ91 over a mixture of $Mg_{17}Al_{12}$ and Al_2Ca to being predominantly Al_2Ca at the largest c_{Ca}. Whether additional Mg_2Ca is present has not been investigated yet, but the maximum Ca/Al-ratio of 5/9 = 0.55 is still below the reported threshold value (= 0.8 [1]) for the formation of Mg_2Ca [7]. Second, in contrast to AZ91 where the IP is discontinuous, a rather continuous skeleton encloses the alpha-phase in the case of AXZ951. Third, the volume fraction of IP, which was derived from the area fraction visible in the micrographs of Figure 3, increases with increasing c_{Ca} from 0.1 to 0.3.

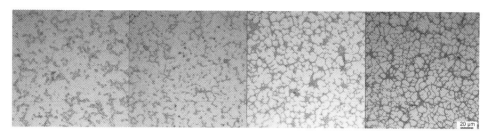

Figure 3: Light-optical micrographs in as-thixomolded condition. From left to right; AZ91, AXZ911, AXZ931 and AXZ951.

The observed increase in creep resistance (Figure 2) is easily interpreted when the microstructure (Figure 3) is modeled as a composite consisting of soft interiors with hard shells: Assuming Al_2Ca to be at least as difficult to deform and thermally stable as $Mg_{17}Al_{12}$, both the increase in its total volume fraction as well as a growing connectivity renders the composite microstructure more creep resistant. Should Al_2Ca be even stronger than $Mg_{17}Al_{12}$, the change in the IP alone would account for part of the increase in creep resistance.

In order to develop a quantitative model for AXZ-type alloys the individual deformation resistance of both composite constituents needs to be assessed at the elevated temperatures of interest here. The relative degree of solid solution strengthening within the interior alpha-phase as a function of Ca-content requires the analysis of its chemical composition. High-resolution transmission electron microscopy can determine whether fine precipitates contribute to interior strength as was found for the highly creep-resistant alloy MRI 230D [8]. Investigations along these lines are planned for future research.

4.2 Evolution of Deformation Resistance

Figure 4a) illustrates the evolution of the deformation resistance with strain for the case of AXZ951 (containing 5 wt% Ca) under different stress levels at a temperature of 200 °C. It is seen that the strain necessary to reach the maximum deformation resistance (open circles in Figure 4a) decreases with decreasing stress. Furthermore, the relative increase in strain rate with strain at the beginning of the tertiary stage is larger for smaller stress and seems to reduce to a common level at strains of about 0.25 for all stress levels investigated.

During the primary stage the dislocation structure in the alpha-phase builds up and causes workhardening. In addition, invoking above mentioned composite approach, a certain strain is required until the load is distributed among soft and hard constituents to assure equal rates of straining in them. Both these effects entail a lower strain hardening rate with increasing stress as observed (Figure 4a).

[1.]This threshold approximately corresponds to the atomic weight ratio of Al_2Ca, *i.e.*, only after the available aluminum has been locked into Al_2Ca the excess calcium precipitates as Mg_2Ca.

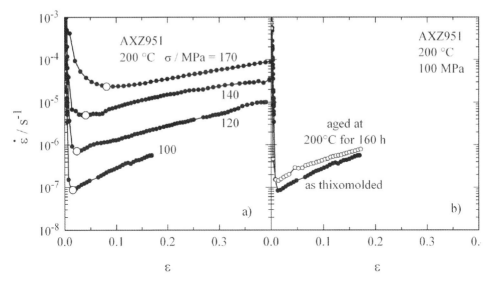

Figure 4: a) Evolution of creep resistance with strain for AXZ951 at different stress levels; b) influence of aging on maximum deformation resistance and the subsequent softening.

The softening in the tertiary stage is connected to a change in microstructure since plastic instabilities or damage can be largely excluded in compression. At present, micrographs of the microstructural evolution have not been prepared yet. However, two general causes of the softening are imaginable: A prolonged exposure to the high testing temperatures may lead to changes in the precipitate structure thus weakening the material, or the composite microstructure is destroyed by straining it. The larger rate of softening $d\dot\varepsilon/d\varepsilon$ at smaller stress indicates that strain cannot be the exclusive cause of softening. On the other hand, the congruence of the rates of softening at large strains, i. e., at relatively high strain rates suggests that the effect of time becomes negligible compared to strain-induced softening at sufficiently high creep rates.

To substantiate this, the creep responses of two samples are compared in Figure 4b). Both samples were deformed under identical conditions of 100 MPa at 200 °C. However, while the first sample was tested in the as-thixomolded state, the second sample was prior aged at 200 °C for 160 h, which is the time the first sample took to reach a strain of 0.17. The minimum creep rate of the aged sample falls in between the minimum creep rate and the rate reached at $\varepsilon = 0.17$ of the as-thixomolded one. Hence a mixture of time-dependent and strain-induced softening occurs in the strain rate range around 10^{-7} s^{-1}. Probably, the time-dependent reaction is the dissolution of the skeleton consisting of Al_2Ca. Such dissolution has been observed during heat-treatments of Mg-Al alloys containing different levels of Ca [9]. Since Liu et al. aged at 415 °C their dissolution rates are significantly larger than expected in this work. It is planned to investigate the microstructure of deformed samples to clarify the origin of the tertiary stage.

5 Acknowledgement

The authors gratefully acknowledge the financial support of the Deutsche Forschungsgemeinschaft under contract number GO 741/9-2. Thanks are due to Prof. R.F. Singer, Dr. A. Lohmüller and H. Eibisch from Neue Materialien Fürth GmbH, Germany, for generously supplying the thixomolded Mg-alloys.

6 References

[1] Blum W, Li YJ, Zeng XH, Zhang P, von Großmann B, Haberling C; Metall Mat Trans **2005** *36A*, 1721–1728.
[2] Luo A; Int Mater Reviews **2004** *49*, 13–30.
[3] Pekguleryuz MO, Kaya AA in *Magnesium Alloys and their Applications (Ed Kainer KU) Wiley VCH Verlag* **2003** 74–93.
[4] Lohmüller A, Scharrer M, Hilbinger M, Jenning R, Hartmann M, Singer RF; Giessereiforschung **2005** *57*, 2–9.
[5] US patent 6,139,651 **2000**.
[6] Frost HJ, Ashby MF; *Deformation Mechanism Maps*, Pergamon Press, Oxford **1982**.
[7] Ninomiya T, Ojiro T, Kubota K; Acta metall mater **1995** *43*, 669–674.
[8] Backes B, Durst K, Blum W, Göken M in *Praktische Metallographie Sonderband 37, Metallographietagung Erlangen 2005 (Ed. Göken M) Werkstoffinformationsgesellschaft mbH Frankfurt* **2005**, 59–64.
[9] Liu MP, Wang QD, Zeng XQ, Wei YH, Zhu YP, Lu C, Z Metallkd **2003** *94*, 886–891.

Experimental and Numerical Investigation of Texture Development During Hot Rolling of Magnesium Alloys

Rudolf Kawalla[1], Christian Schmidt[1], Hermann Riedel[2], Aruna Prakash[2]
[1] Freiberg University of Mining and Technology, Freiberg; [2] Fraunhofer Institute for Mechanics of Materials (IWM), Freiburg

1 Introduction

Due to the deformation mechanisms of magnesium and the typical basal texture rolled magnesium sheets show a significant asymmetry of flow stress in tension and compression which is disadvantageous for subsequent processing and leads to anisotropic properties. In order to avoid this undesired behavior it is necessary to achieve non-basal texture during rolling, or at least, to reduce the intensity of the basal texture component. The reduction of the anisotropy caused by the basal texture is very important for subsequent forming processes. The knowledge of microstructure and texture development during forming opens up possibilities to influence those processes by thermo mechanical treatment and by utilizing recrystallization phenomena and promises significant progress in terms of production and processability.

This work aims at optimizing the hot rolling process with special consideration of texture effects by means of numerical simulation. The development of the model is carried out in close cooperation with the experimental work. Therefore, the experiments described in the following shall be seen as a contribution to the clarification of the forming mechanisms and the accompanying specifics and effects of texture development. At the same time those results are used for the development and verification of a simulation model, which shall enable the user in the future to predict and influence the texture development during forming of magnesium alloys.

2 Experimental Procedure

The experiments comprised compression tests with conventionally produced feedstock as well as rolling tests with cast-rolled material (TRC – twin roll casting).

The samples (Ø = 10mm, length = 18mm) for the compression tests were made from a multistage caliber rolled bar of continuous casting AZ31. Subsequent to a 10 min heat treatment at 520 °C the compression test was carried out on a servohydraulic press in rolling direction at seven different temperatures between 280 °C und 500 °C with a logarithmic deformation degree of $\varphi = 1$ and a strain rate of $10~\text{s}^{-1}$. In a second test series samples of the same kind were compressed with small deformation degree of $\varphi = 0.025$ to $\varphi = 0.15$ in order to investigate the distinctive behavior of the flow curve in its slope.

The strips (thickness 6mm) used as feedstock for the rolling tests were manufactured with the cast-rolling technology (TRC) by the MgF Magnesium Flachprodukte GmbH Freiberg [1]. Their chemical composition is in accordance with AZ31 with slightly reduced Al content. The results for the development of microstructure and texture dwelled on in 3.3 were obtained in rolling trials which were carried out on the finishing line of the semi-continuous hot rolling mill at the Institute of Metal Forming at the Freiberg University of Mining and Technology. The rolling trails consisted of tests in which two different initial rolling speeds (IRS: 1m/s, 5m/s) at three different initial rolling temperatures (IRT: 300 °C, 400 °C, and 500 °C – all strips preheated to 500 °C) were applied on a rolling process with three passes and an overall deformation degree of $\varphi = 1.3$. Rolling was performed in cast-rolling direction.

All samples were quenched in water right after deformation to conserve the microstructure after deformation which is not only of interest for the model development. The texture and microstructure investigations were carried out in two different metallographic sections, at 1/2 and at 4/5 of the specimen height, with the objective of detecting an existing gradient of properties.

3 Results

3.1 Compression Tests

After compression tests the specimens' shape was more oval than round which is an indication for anisotropic properties already after caliber rolling. The texture related cause of the oval shape of the specimen assumed after macroscopic examination is confirmed by the examination of the initial texture [2]. It showed that after caliber-rolling the majority of crystals were oriented with its (0002)-crystallographic direction perpendicular to the rolling direction and towards the direction of the preferred widening. In this case, the phenomena of twinning which clearly acts as an additional forming mechanism can be seen as cause of that behavior. The polfigures of the deformed specimens show a texture with a basal character (Fig 1). Deformation accomplished only by main slip systems would lead to a strong basal texture. Asserting a different result it is assumed that secondary slip systems must have been involved in the forming process [3]. The reduction of the basal texture component at higher temperatures supports that supposition. However, the basal component of the texture remains predominantly and decreases only slightly at higher temperatures. Hence, the texture development can not sufficiently be affected only by a variation in forming temperature.

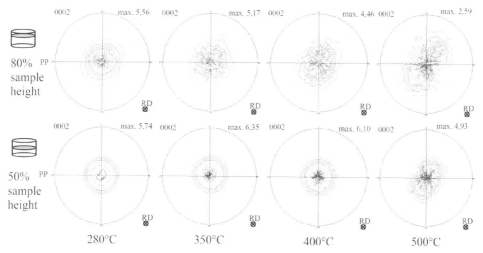

Figure 1: Texture development subject to temperature

Analyzing the flow curves of those tests an anomaly appearing at all temperatures could be found in the range of $\varphi = 0...0.2$. That anomaly is reflected in several inflection points in the slope of the flow curve at small deformation degrees [2]. So far it oftentimes was suspected that with increasing temperature and the thereby activation of additional slip systems the addiction to mechanical twinning would decrease in favor of the new activated slip systems and not occur at high temperatures anymore. In order to ascertain whether mechanical twinning or recrystallization phenomena are responsible for that behavior, additional compression tests at 280 °C and 400 °C with small deformation degrees were conducted thereupon (Fig. 2).

Those additional experiments showed a good compliance between the development of twins in the microstructure and the change in flow stress in the flow curve. Therefore, it could be asserted that mechanical twinning appears even at high temperatures and it was found responsible for the flow curve behavior [2]. It is therefore of relevance when evaluating deformation processes in a wide range of temperatures. This, however, holds only true if the initial orientation of grains is pleasant for twinning.

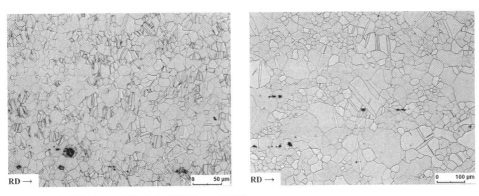

Figure 2: left: 280 °C, $\varphi = 0.05$, right: 400 °C, $\varphi = 0.025$

3.2 Simulation

In the first approach a recrystallization model has been combined with a visco-plastic self-consisted (VPSC) model [4–6]. The entire model [7] calculates the texture development as a result of deformation and recrystallization, and the resulting anisotropic plastic properties. Deformation is produced by crystallographic slip on various slip systems as well as by mechanical twinning. The influence of twinning on texture development is described by a predominant twin reorientation (PTR) model [8]. The recrystallization model is based on the fact that no new texture components are generated by recrystallization. Only the distribution of the individual components originating form the deformation is changing.

The above mentioned model was adapted in a first step to the flow curves of the compression tests where two different combinations of deformation mechanisms were tested [2]. The first combination comprised slip on basal planes and second order pyramidal planes as well as mechanical twinning. In the second combination slip on prismatic systems was additionally permitted but did not improve the adaptation. The consideration of recrystallization, however, seems to be essential. Even though the calculated flow curves do not match exactly with the experimental ones yet, they nevertheless describe the run already quiet well.

In the next step the polfigures of the compression tests were calculated and compared with the experimental texture (Fig. 3). The comparison shows that the experimental texture is reached in the simulation already after 30 % of strain. It intensifies significantly as the deformation simulation continues [2]. For this reason further experiments and improvements in terms of adaptation will be necessary.

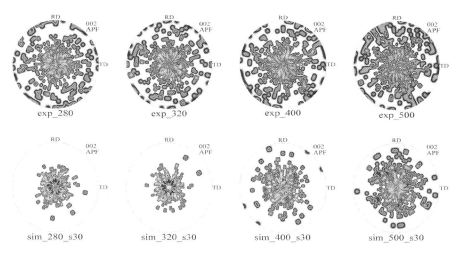

Figure 3: Experimental polfigures (first row) after compression tests at 280 °C, 320 °C, 400 °C, 500 °C; beneath: simulated polfigures at 30 % of deformation

3.3 Rolling Tests with Cast-rolled Feedstock

The typical structure of a cast-rolled Mg strips is a result of the superposed solidification and transportation process during cast-rolling. The melt starts to solidify at the surface of the rollers and the solidification continuous toward the middle of the strip, which results in a structure with about 45° tilted segregation sections between equiangular dendrites (Fig. 4). The solidification is finished before the strip reaches the center line of the roller whereby the strip experiences a partly rolling. That preformed, crushed casting structure possesses a remarkable formability, that rolling with 75 % reduction per pass can be possible [1].

Starting with that structure, which did not significantly change during preheating up to 500 °C and cooling down to the initial rolling temperatures (IRT), the microstructure during the first pass at 1m/s IRS at 300 °C IRT changed into a microstructure with deformed grains superposed by dynamic recrystallization. The same holds true for the second pass whereas in the third pass due to the lower forming temperature not only no recrystallized structure could be observed anymore but also twinning was now clearly apparent. With rising initial temperature the share of recrystallized grains increased especially after the first and second pass. After the third pass at 400 °C (IRT), however, a deformation structure without twins or recrystallization was found, whereas at 500 °C (IRT) dynamic recrystallization had already started to alter the microstructure slightly.

In contrast to that behavior the samples rolled with 5 m/s IRS possessed already after the first pass at 300 °C IRT a strongly dynamically recrystallized microstructure with a mean grain size of about 4 μm. The share of deformation structure was significantly lower compared to 1 m/s

IRS. After the second pass the recrystallized share of microstructure and its mean grain size increased until only recrystallized structure was found after the third pass. A similar behavior was found for the higher IRT, however, the mean grain size of the recrystallized grains increased with rising temperatures [9]. Twinning could only be observed insignificantly after the first pass at 300 °C and 400 °C IRT.

The previously assumed texture related reason for the good forming ability of cast-rolled magnesium is confirmed by the polfigures of the cast-rolled condition (Fig. 5 left). In that connection the special structure (Fig. 4) originating from the cast-rolling process plays an important roll. Even thought the mid layer of the strip possess already a quiet homogeneous texture with only some small peeks, the surrounding layer, however, is fare more responsible for the good formability. Its texture shows that most of the crystals in that layer are orientated with its basal planes tilted by about 45° to the sheet normal (Fig. 5 top left). Especially those crystals are extremely advantageously orientated for dislocation movement in a first forming process as their preferentially activated basal planes have the highest Schmid factor in that condition.

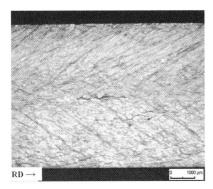

Figure 4: Cast-rolled strip, macro etching

This is also indicated by the resulting texture after the first rolling pass. Even though a basal character develops, its intensity however, is weaker than after compression tests. Additionally, the distribution of intensities scatters considerably in and against rolling direction. In the surrounding layer the basal texture is clearly less developed. As rolling continuous the basal texture component increases in the mid layer but only slightly in the surrounding layer. The scatter around rolling direction remains as Fig. 5 (right hand side) exemplifies. Furthermore, unlike the compression tests showed, a temperature dependency of the basal texture development could be observed in those experiments.

4 Summary

The present research work has been concentrated on the clarification of selected problems of deformation mechanisms for the aspired deformation-recrystallization model. The model at the current state of development is already able to describe many experimental results at least quantitatively. Among other things the experimental results showed that twinning occurs even at high temperatures and is responsible for the anomaly in the slope of the flow curve in the range of small deformation degrees. During compression tests which were mainly required for the adaptation of the model a temperature dependency of the basal texture development could barely be observed. This, however, does not apply to the rolling trails with cast-rolled feedstock. In addition, the reasons for the remarkable forming ability of cast-rolled magnesium strips could be revealed by examining the distinctive initial texture and tracing its development during rolling.

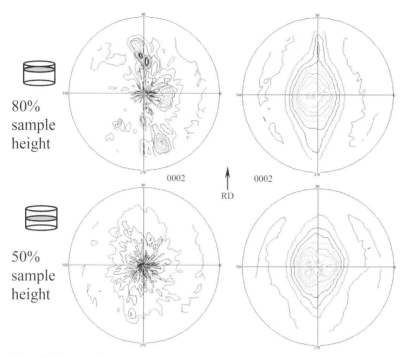

Figure 5: left: cast-rolled Mg strip right: after second pass at 400°C IRT (rolling temperature of pass 355°C), 60% overall reduction

Subject of further work will be the detection and utilization of recrystallization phenomena. The cognitions of the experiments will be taken into account for the model development in order to reproduce all experimentally found mechanisms.

5 Acknowledgements

The financial support of the project by the German Research Foundation (Deutsche Forschungsgemeinschaft DFG, KA 1591/8-1 and RI 329/27-1) in the framework of the priority program SPP 1168 is gratefully acknowledged.

6 References

[1] C. Schmidt, R. Kawalla, B. Engl, N.D. Cuong: An innovative method to produce magnesium strip. Magnesium – broad horizons, Moscow, **2005**, proceedings

[2] C. Schmidt, R. Kawalla, T. Walde, H. Riedel, A. Prakash: Experimental and numerical investigation of texture development during hot rolling of magnesium alloy AZ31. Thermec **2006**, Vancouver, proceedings
[3] S.R. Agnew, M.H. Yoo, C.N. Tomé: Acta mater. 49 (**2001**), p. 4277– 4289
[4] R.A. Lebensohn and C.N. Tomé: Acta Metall. Mater. 41 (**1993**), p. 2611–2624
[5] T. Walde: Dissertation at Karlsruhe University, **2004**, (carried out at Fraunhofer-IWM)
[6] T. Walde, H. Riedel: Materials Science Forum 426-432 (**2003**) p. 3679–3684
[7] T. Walde, H. Riedel: Solid State Phenomena 105 (**2005**), p. 285–290.
[8] C.N. Tomé, R.A. Lebensohn, U.F. Kocks: Acta metall. mater. 39 (**1991**), p. 2667–2680
[9] R. Kawalla, M. Ullmann, M. Oswald, C. Schmidt: Properties of strips and sheets of magnesium alloy produced by cast-rolling technology. 7th International Conference on Magnesium Alloys and their Applications, Dresden, **2006**, proceedings

The Influence of Calcium and Cerium Mischmetal on the Deformation Behavior of Mg-3Al-1Zn

T. Laser[a], M. R. Nürnberg[b], Ch. Hartig[a], D. Letzig[b], R. Bormann[a]

[a] Hamburg University of Technology, Inst. of Matierials Science and Technology, Hamburg, Germany
[b] GKSS Research Centre Geesthacht, Magnesium Innovation Centre - MagIC, Geesthacht, Germany

1 Introduction

The use of magnesium wrought alloys in industrial applications is still limited due to their lower strength and formability compared to other lightweight materials such as aluminum. It is well known that micro alloying elements may improve the thermomechanical processing ability as well as the mechanical properties of a material. This investigation is aimed at enhancing the limits of the standard alloy Mg-3Al-1Zn (AZ31) by producing aluminum-rich intermetallic phases via micro alloying with calcium and cerium mischmetal. The presence of these second-phase particles can further stimulate recrystallization during a thermomechanical treatment, supporting the adjustment of a homogeneous fine-grain-sized microstructure. In order to clarify the potential effect of calcium and cerium mischmetal alloying on the grain refinement during recrystallization, the alloys in this study were hot-rolled and characterized with respect to the evolution of microstructure, texture and some basic plastic properties.

2 Material and Experimental Procedure

The AZ31-based alloys were prepared using pure magnesium (99.95 wt.%), pure aluminum (99.95 wt.%), pure zinc (99.8 wt.%) and pure calcium granulate (99.9 wt.%). Mn was added in form of a master alloy (M2) containing Mg with 1.7 wt.% Mn. Cerium mischmetal was added in form of small commercially produced tablets.

Four different alloys based on conventional AZ31 (alloy I) containing additions of 0.4 wt.% cerium mischmetal (alloy II), 0.7 wt.% calcium (alloy III) and a combination of 0.6 wt.% cerium mischmetal and 0.8 wt.% calcium (alloy IV) were cast. The AZ31 alloys were melted in a 14-litre steel-crucible using a resistance furnace. The master alloy M2, cerium mischmetal and the alloying elements Al, Zn and calcium granulate were added to the molten Mg in the crucible at 760 °C. Casting was performed under a protective atmosphere of argon and SF_6. The melt was held at 760 °C for 60 min to make sure that the alloying elements were homogeneously dissolved. The alloys were cast into a cylindrical steel container of 10 cm in diameter and 41 cm in length. The weight of the cast billets was about 8 kg. The chemical compositions of the alloys were analyzed using the ICP-MS Agilent 7500 technique and are shown in Tab. 1. As high purity materials were used, the contents of Si, Fe, Ni, and Cu were below the maximum levels allowed in the ASTM B275 specification.

For hot-rolling, blocks with a thickness of 6 mm were cut out of each cast ingot with the rolling direction parallel to the cylindrical axis. All blocks were homogenized at 400°C for 20 h and then air-cooled to room temperature. Before hot-rolling, the blocks were reheated to 450 °C for

30 min and then rolled on a two-high mill with a roll diameter of 200 mm. No heating of the mill rolls was applied. Rolling was carried out up to a total strain of $\varphi = 1.6$ achieved over three runs. Each run comprised several reduction steps that were followed by 3 min reheating at 450 °C. A run was followed by subsequent annealing for 30 min at 400 °C. Tensile samples parallel (RD) and perpendicular (TD) to the rolling direction were machined out of the resulting sheet material (sheet dimensions: ~1.2 mm × 200 mm × 300 mm). Room temperature tensile tests were performed on a Zwick universal testing machine at a constant strain rate of $1 \cdot 10^{-4}$ s^{-1}.

Table 1: Chemical compositions (wt.%) of the studied alloys

Alloy	Al	Zn	Mn	Cer-MM	Ca	Si	Fe	Ni	Cu	Others	Mg
I	3.1	1.0	0.28	–	<0.01	<0.01	0.006	<0.001	<0.001	<0.1	Balance
II	3.3	1.0	0.23	0.38	0.015	0.01	0.026	<0.001	0.001	<0.15	Balance
III	3.3	1.3	0.22	–	0.66	0.01	0.016	<0.001	0.001	<0.1	Balance
IV	3.1	1.0	0.22	0.57	0.8	0.015	0.025	<0.001	<0.001	<0.2	Balance

All specimens for optical microscopy were sectioned, cold-mounted, polished and then etched in a solution of picric and acetic acid [1]. Microstructures were observed by optical (LEICA DMLM) and scanning electron microscopy (ZEISS DSM 962) using an accelerating voltage of 20 keV. The grain sizes were measured using the linear intercept method [2].

The textures of the rolled sheets were measured on a D8 DISCOVER X-ray diffractometer (BRUKER AXS Inc.) using reflection geometry and Cu-K$_\alpha$ radiation. The orientation distribution function (ODF) and complete pole figures were calculated using the harmonic method of Bunge [3].

3 Results and Discussion

The microstructures of the four AZ31 based alloys are shown in Fig. 1. Grain size measurements (Fig. 2) using optical microscopy, according to ASTM E 112-96, indicate that the average grain size within the lower part of the as-cast AZ31 alloy cylinder is around 380 µm ± 40 µm. The addition of 0.4 wt.% cerium mischmetal to the AZ31 based alloy (II) leads to an increase in the mean grain size to 700 µm ± 70 µm. Modification with 0.7 wt.% Ca (III) leads to a mean grain size of 440 µm ± 30 µm, while the alloy containing 0.6wt.% cerium mischmetal and 0.8wt.% Ca (IV) has a mean grain size of 270 µm ± 100 µm, which is the smallest measured of the four alloys.

Hot-rolling of sheets resulted in fine-grained and equiaxed microstructures that are presented in Fig. 3. Deformation twins are not observed in all the sheets. (Twins observed in alloy II were generated during metallographic preparation.) The heat treatment after the last rolling pass led to a fully recrystallized microstructure for all alloys. Second phase particles of Al-rich phases are distributed preferably at grain boundaries, but also found in the interior of the grains. In addition, the rolling process produces a fragmentation of some second phase precipitates. This can very well be observed in the alloy IV that contains the highest amount of Al-rich particles (Fig. 3d).

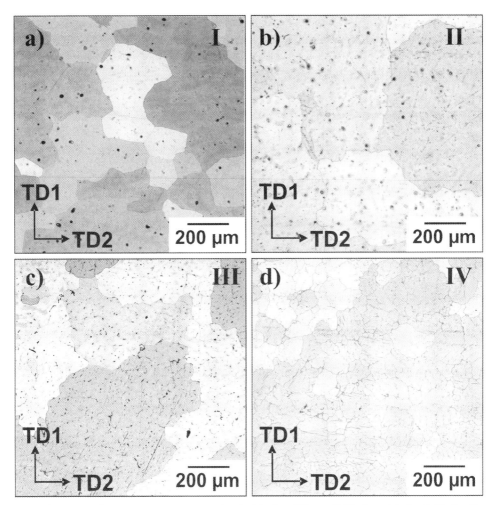

Figure 1: As-cast microstructures (transverse sections) of the four different AZ31 alloys: (a) alloy I, (b) alloy II, (c) alloy III and (d) alloy IV

The mean grain sizes of the rolled sheets are presented in Tab. 2. The conventional AZ31 alloy I exhibits the smallest mean grain size. The Ca containing alloys III and IV show a mean grain size of 15.3 μm. The modification with cerium mischmetal (alloy II) exhibits the largest mean grain size, approximately 20 % larger than the conventional alloy I. However taking into account the relatively large scatter in the grain sizes of each alloy, this difference is insignificant.

In Fig. 4, the (0002) (basal) pole figures of the sheets are presented. All alloy modifications show a more or less pronounced basal texture, which means that most of the grains have their c-axis perpendicular to the rolling plane. The texture for the conventional AZ31 alloy (compare Fig. 4a) is commonly observed for multi-pass rolling with annealing between two passes [4]. The basal plane pole figure exhibits a maximum intensity of 9.15. This texture is more pronounced than the texture of the cerium mischmetal modified AZ31 sheet (alloy II), which shows a maximum intensity of 7.89. The calcium containing alloys show significantly weaker textures with maximum intensities of 4.79 for alloy III and 3.79 for alloy IV. Obviously, the addition of the alloying elements reduces the strong basal fiber texture of AZ31. The results also show that the addition of cerium mischmetal leads to a reduction in the ovality, whereas the addition of calcium results in a transverse expansion of the basal poles.

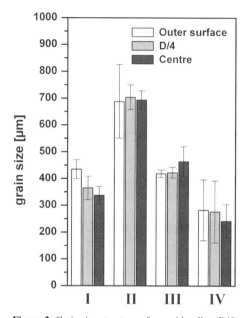

Figure 2: Grain size at outer surface, mid-radius (D/4) and central region of alloys I, II, III and IV. Error bars indicate the scatter of mean grain sizes in different fields

Table 2: Grain sizes of hot-rolled sheets, Gaussian standard deviations calculated from 800 grains

Alloy	Mean grain size / μm	Standard deviation / μm
I	14.0	8.3
II	16.9	7.9
III	15.3	7.7
IV	15.3	8.3

Tensile tests were performed at room temperature. The yield strengths (YS) and the ultimate tensile strengths (UTS) are presented in Fig. 5. The absolute values of the yield strength in rolling direction are around 150 MPa. This order of magnitude is commonly observed for fine-grained and non work-hardened AZ31 alloys [5]. The modification of AZ31 with cerium mischmetal results in a slight reduction of the yield strength. As a large part of the aluminum is bound in rare earth intermetallic phases, this reduction may be caused by the lower aluminum content of the magnesium solid solution of alloy II, thus resulting in a lower solid solution hardening. The addition of calcium in alloys III and IV raises the yield strength compared to alloy I.

Here, small particles of the fragmented intermetallic phases might cause a significant hardening effect, thus compensating the lower solid solution hardening due to the reduced Al content in the Mg matrix.

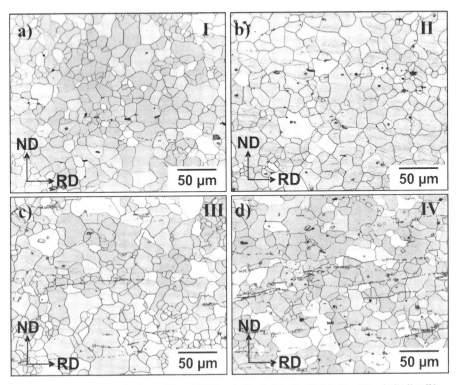

Figure 3: Microstructures after hot-rolling of AZ31: (a) alloy I, (b) alloy II, (c) alloy III and (d) alloy IV

As can been seen in Fig. 5a, the yield stress of AZ31 (alloy I) in the transverse direction (TD) is greater than in the rolling direction (RD). Alloy I exhibits a yield strength anisotropy of $YS_{RD}/YS_{TD} = 0.88$. Micro-alloying AZ31 with 0.38 wt.% cerium mischmetal (alloy II) leads to isotropic yield strength behavior $YS_{RD}/YS_{TD} = 0.99$. The addition of calcium reverses the yield strength anisotropy. For alloy III (AZ31+Ca) the anisotropy results in $YS_{RD}/YS_{TD} = 1.09$. For alloy IV the biggest difference in yield strength of transverse and rolling direction can be found: $YS_{RD}/YS_{TD} = 1.17$. The change of the yield anisotropy can be understood as a consequence of the different textures. With increasing extension of the basal poles in the transverse direction the yield strength in TD decreases. An optimal ratio for the yield strength is reached with the addition of 0.38 wt.% cerium mischmetal.

For all alloys, the values of the ultimate tensile stress (UTS) and the elongation to fracture in the rolling direction were higher than those measured in transverse direction. For all TD specimens, the transverse direction is much less ductile. This might be caused by micro-pores in the as-cast microstructure. Due to the rolling process these pores are elongated in the rolling direction and may thus reach a critical length for crack nucleation under transverse loading. The plastic elongations (A5) measured in the rolling direction are in the range 12%-22%. High values of the UTS are only achieved when a large plastic elongation is reached.

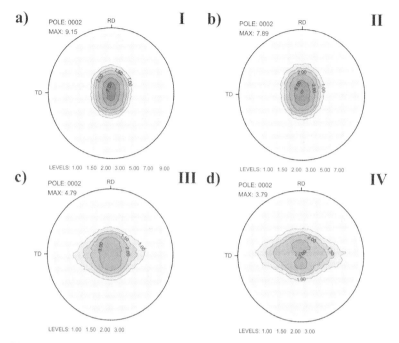

Figure 4: (0002) pole figures of hot-rolled AZ31 sheets: (a) alloy I, (b) alloy II, (c) alloy III and (d) alloy IV

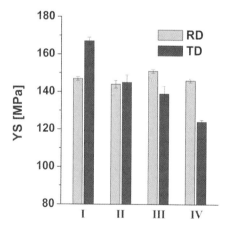

Figure 5: Mechanical properties at RT of hot-rolled AZ31 sheets: (a) yield strength and (b) ultimate tensile strength. The error bars indicate the scatter of three to five tensile tests per alloy.

4 Conclusions

It can be stated that a small amount of extra micro-alloying elements leads to significant changes in the properties of AZ31. Changes in (a) the as-cast microstructure, (b) the development of texture and (c) the mechanical properties were observed. The alloying elements reduce the strong texture and modify the yield anisotropy. The strength of the recrystallized microstructure can be improved by the addition of calcium.

5   Acknowledgements

The financial support by the German Research Foundation (DFG) in the priority program SPP 1168 "InnoMagTec" (BO 691/8 & LE 1395/2) is gratefully acknowledged.

6   References

[1] V. Kree, J. Bohlen, D. Letzig, K. U. Kainer, Praktische Metallographie 41 (2004) 233–246.
[2] H. E. Exner, H. P. Hougardy, Einführung in die quantitative Gefügeanalyse, Dt. Ges. für Metallkunde, Oberursel, 1986.
[3] H.-J. Bunge, Texture analysis in materials science: mathematical methods, Cuvillier, Göttingen, 1993.
[4] S. R. Agnew in: H. I. E. Kaplan HIE, editor. TMS annual meeting: TMS The Minerals Metals & Materials Society, 2002. p.169.
[5] C. Kammer, Aluminium-Zentrale, Magnesium-Taschenbuch, Aluminium-Verl., Düsseldorf, 2000.

Corrosion Fatigue Behaviour of Newly Developed Mg Wrought Alloys

C. Fleck[1], A. Schildknecht[1], K.A. Weidenmann[2], A. Wanner[2], D. Löhe[2]

[1] Technische Universität Berlin, Germany
[2] Universität Karlsruhe, Germany

1 Introduction

Recently, Mg-alloys found new interest especially for the weight-saving realisation of future transport systems [1–3]. During service, vehicle components are exposed to stochastic cyclic loads with relatively high numbers of cycles [4, 5]. The cyclic deformation behaviour of Mg-alloys has not been thoroughly investigated up to now [3]. Newer investigations focused on the influence of microstructural parameters, for example the texture, different surface states and treatments, and of the loading conditions [e.g. 4–7]. In comparison, a quite low number of investigations concerning the cylic deformation behaviour and the corresponding microstructural processes has been published. These examinations were mainly performed in laboratory air and rather seldom in corrosive media. Depending on the production process and the heat treatment, cyclic softening [8, 9], mainly cyclic hardening [10, 11] or hardening following an initial short softening [12] were observed. Due to the direction dependent formation of deformation twins the mechanical properties in laboratory air are strongly influenced by the different deformation behaviour under tensile and compressive loading [13,14]. In corrosive media, transcrystalline crack growth was observed in rotating bending loaded AZ91-, AE42- and AM50-specimens [15]. The cyclic deformation experiments were partially supplemented by microstructural examinations showing slip bands and twin formation [9,12] or characterising crack formation and growth in dependence of microstructural parameters [12,16–20].

2 Materials and Methods

2.1 Material States

2.2.1 Compositions and Microstructures

The investigation comprised different wrought Mg alloys (ZEK 100, LAE 442) in two heat treatment states, and the standard extrusion alloy AZ31 as a reference. The alloys were prepared, cast and extruded at the Institut für Werkstoffkunde, Universität Hannover. For the production of the rods, the materials were extruded at 300 °C after preheating to 340 °C. The rods were then cooled in air (state "0") or air-cooled and tempered at 300 °C for 30 min (state "1"). The compositions of the materials are presented in Table 1.

Figure 1 shows the microstructure of longitudinal sections. The alloy AZ 31-0 shows a typical recrystallised microstructure developed during extrusion. It contains a low level of precipitates which are composed of Al and Mn according to EDX analyses. The slightly inhomogeneous grain size over the cross-section due to different plastic deformation seems to homogenise during the heat treatment. The ZEK and LAE alloys also show precipitates composed of rare earth elements as well as zinc and aluminium, respectively. They are more pronounced for the ZEK alloy. In the case of ZEK 100, the heat treatment leads to a slightly more homogeneous microstructure with smaller and more finely dispersed precipitates. In contrast, the heat treated LAE 442-1 exhibits a coarsening of the precipitates.

Table 1: Compositions of the investigated alloys

Material	Composition [weight-%]
AZ 31	96.4 Mg 2.4 Al 0.8 Zn 0.4 Mn 0.1 Si
LAE 442	89.6 Mg 4.0 Li 3.9 Al 2.2 REa 0.2 Mn $1.1 \cdot 10^{-3}$ Be $2.9 \cdot 10^{-3}$ Fe (REa: 54,0 Ce 22,3 La 17,8 Nd 5,8 Pr)
ZEK 100	98.4 Mg 1.4 Zn 0.1 Zr 0.1 REb (REb: 49,1 Ce 35,9 La 11,0 Nd 4,0 Pr)

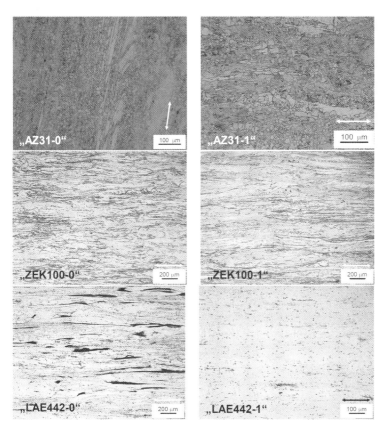

Figure 1: Microstructures of the investigated alloys (arrows = extrusion direction)

2.2.2   Texture and Residual Stress Measurements

The crystallographic texture of the alloys was measured by X-ray diffractometry with CuKα-radiation (60 kV, 80 mA, 10 s/measuring point). The 2θ-angles were $34,4° \pm 0,5°$ for the (0002)- und $36,6° \pm 0,5°$ for the (10-11)-pole figures. The tilting angle α was varied in steps of 5° between 0° and 90°, the rotation angle β in steps of 10° between 0° and 360°. The measurements show pronounced differences between the single alloys. AZ31 and ZEK100 show a distinct (0001)-texture, while the one for the LAE alloy is very weak. The texture of the ZEK alloy is slightly weaker after the heat treatment.

Residual stresses were also measured by X-ray diffraction mainly on extruded rods by the $\sin^2\psi$-method. The measurements were performed on (11-22)-diffraction reflexes under 13 distance angles ($-60° \leq \psi \leq +60°$). The alloys AZ31 and ZEK100 exhibit compressive residual surface stresses which cannot be neglected as compared to the yield strengths of the alloys. The normal stress components are much higher than the shear stress components. It is remarkable that the LAE alloy does not exhibit significant residual stresses.

2.2   Fatigue Tests

For the fatigue tests, a servohydraulic testing machine with a digital control system was used. The specimens were turned from the extruded rods and mechanically polished in the gauge length with diamond suspension (grain size 6 μm). Load controlled load increase tests ($\sigma_{a,min}$ = 60 MPa, ΔN_{step} = 2 · 10^4, $\Delta\sigma_a$ = 10 MPa in ΔN = 2 · 10^3) were performed at a frequency of 5 Hz in air and in 0,5 M (2,6 wt.-%) NaCl-solution. Exact alignment of the hydraulic grips ensured loading of the specimens nearly without bending moments. To characterise the cyclic deformation behaviour, stress-strain hysteresis measurements were performed. The specimen and fracture surfaces were examined by light and electron microscopy. Furthermore, microstructural changes were investigated by transmission electron microscopy.

3   Results

3.1   Cyclic Deformation Behaviour

The cyclic deformation behaviour was characterised by stress-strain-hysteresis measurements. The dependence of the plastic strain amplitude on the number of cycles is shown in Fig. 2. In this and all following graphs the development of the stress amplitude is also shown. In the non-heat treated state, the alloys AZ31 and ZEK100 show a continuous softening. After heat treatment, both alloys exhibit a strong softening when the load is increased followed by a pronounced hardening during the individual loading steps for AZ31 and by saturation with a nearly constant plastic strain amplitude for ZEK100. This behaviour is observed until failure takes place. Qualitatively, the behaviour of the alloy LAE442 is similar to that of AZ31-1. However, the plastic deformation is lower and constant on a quite low level up to a loading step of 90 MPa ("0") and 110 MPa ("1").

As expected, loading in corrosive environment strongly influences the cyclic deformation behaviour. As shown in Fig. 3, the behaviour is much more brittle with a very low continuous softening at the most. Only the alloy AZ31 in the non-heat-treated state exhibits a slight softening when the load is increased and a saturation state during the loading steps. In the heat-treated state, this behaviour is only observed in the last loading step before specimen failure.

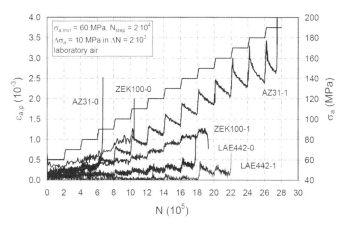

Figure 2: Plastic strain amplitude in dependence of the cycle number for loading in air

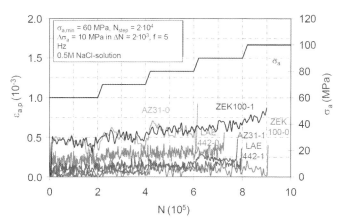

Figure 3: Plastic strain amplitude in dependence of the cycle number for loading in 0,5M NaCl-solution

The strong texture of the materials leads to an anisotropic deformation behaviour under tensile and compressive loading. This leads to pronounced cyclic creep in compressive direction (cf. Figs. 4, 5). As all tests were performed without mean stresses ($|\sigma_m| \leq 2$ MPa), the development of the total and the plastic mean strain are equivalent. When loaded in air, all alloys and all heat-treatment states besides AZ31-1 show these pronounced cyclic creep effects with a continuous increase in the total mean strain. For higher stress amplitudes, there is a jumplike increase when the load is raised with a constant course during the following loading step, for example for the alloy ZEK 100-0. As was observed for the cyclic deformation behaviour, the alloy AZ31-1 behaves strongly different concerning cyclic creep. Up to a stress amplitude of 120 MPa, the specimen length shows a slight but continuous increase. During the following loading steps creep in compressive direction occurs, first jumpwise up to a stress amplitude of 160 MPa, then countinuously.

Like the cyclic deformation behaviour, the cyclic creep behaviour is strongly influenced by the environmental conditions (Fig. 5). All alloys show a continuous decrease in the total mean strain.

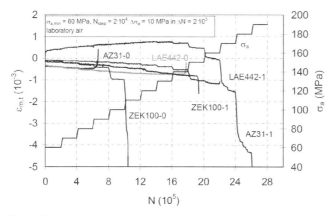

Figure 4: Total mean strain as a function of the cycle number for loading in air

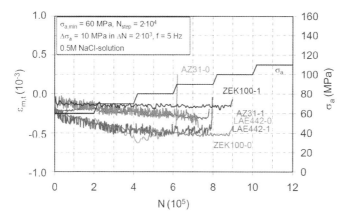

Figure 5: Total mean strain as a function of cycle number for loading in 0,5 M NaCl-solution

3.2 Fractography

To identify the microstructural processes during fatigue laoding the specimen surfaces and the fracture surfaces were examined by scanning electron microscopy. Loading in air leads to a typical fatigue fracture with a smooth fatigue induced and a coarser final fracture surface (Fig. 6). In the transition zone fatigue striations as well as dimples which are characteristic for a ductile fracture are visible. Loading in 0.5 M NaCl-solution also leads to a fracture surface on which a relatively smooth fatigue fracture and the rougher final fracture areas are clearly distinguishable (Figs. 7, 8). In comparison with loading in air, the fatigue fracture areas are smoother for loading in saline solution. Generally, the failure starts at a low number of cracks on the specimen surface; with increasing crack length, the roughness increases as expected. The LAE- and ZEK-alloys showed a high number of secondary cracks in the area of the crack origin which partially may be due to hydrogen embrittlement (Fig. 8).

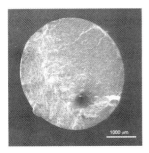 

Figure 6: SEM: Survey of the fracture surface of a ZEK100-0 specimen loaded in air (left) and magnification of the fatigue fracture surface (right)

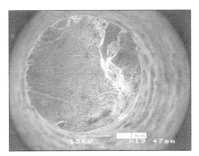 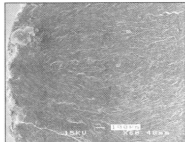

Figure 7: SEM: Survey of the fracture surface of an AZ31-0 alloy loaded in 0.5 M NaCl-solution (left) and magnification of the area of the crack origin (right)

 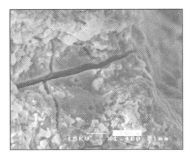

Figure 8: SEM: Survey of the fracture surface of a ZEK100-0 alloy loaded in 0.5 M NaCl-solution (left) and magnification of the area of the crack origin (right)

4 Summary and Conclusions

For loading in laboratory air, the alloys investigated differ quite strongly in their cyclic deformation behaviour. Further, a positive influence of the heat treatment was observed. This effect is probably due to precipitation hardening for the ZEK- and LAE-alloys and to the homogenisation of the grain size for the AZ-alloy. Besides LAE, the extruded bars exhibit relatively high compressive surfaces stresses not negligible in comparison with the compressive yield strength. From a certain stress amplitude, the yield strength is reached during the compressive half cycle with a pronounced influence on the deformation behaviour.

5 References

[1] Kainer, K.U. in: *Werkstoffwoche '98, Band VI, Metalle / Simulation Metalle,* Wiley VCH Weinheim **1999**, p. 439–448.
[2] Burger, W.; Nientit, G., in *Werkstofftag '94, Duisburg, 09./10.03.1994,* VDI Berichte Nr. 1080, VDI, Düsseldorf **1994**, p. 17–23.
[3] Zenner, H.; Renner, F., Mat.-wiss. u. Werkstofftech. **2001**, *32*, 68–75.
[4] Mayer, H.; Papakyriacou, M; Stanzl-Tschegg, S.; Tschegg, E.; Zettl, B.; Lipowsky, H.; Rösch, R.; Stich, A., Materials and Corrosion **1999**, *50*, 81–89.
[5] Beste, A.; Lipowsky, Hj.; Mayer, H.; Tschegg, S.; Kotte, K.-L.; Pannkoke, K., in *Magnesium Alloys and their Applications,* Werkstoff-Informationsgesellschaft mbH Frankfurt **1998**, p. 253–258.
[6] Ogarevic, V. V.; Stephens, R. I., Annual Review of Materials Science **1990**, *20*, 141–177.
[7] Sonsino, C. M.; Dieterich, K.; Wenk, L.; Till, A., in *Magnesium Alloys and their Applications,* Wiley-VCH Weinheim **2000**, p. 304–311.
[8] Srivatsan, T. S.; Wei, L.; Chang, C. F., Engineering Fracture Mechanics **1997**, *56(6)*, 735–758.
[9] Altenberger, I.; Jägg, S.; Scholtes, B., in *Tagung Werkstoffprüfung, Bad Nauheim, 03./ 04.12.1998,* DVM, Berlin **1998**, p. 55–65.
[10] Zheng, L.; Zhongguang, W.; Yue, W.; Zhongyang, L., Journal of Materials Science Letters **1999**, *18*, 1567–1569.
[11] Liu, Z.; Wang, Z. G.; Wang, Y.; Chen, L. H.; Zhao, H. J.; Klein, F., Materials Science and Technology **2001**, *17(3)*, 264–268.
[12] Potzies, C., Zum Ermüdungsverhalten der Magnesium-Gußlegierung G-MgAl9Zn1 (AZ91), Dissertation TU Berlin, GCA-Verlag Herdecke **2002**.
[13] Noster, U.; Scholtes, B., Materials Science Forum **2003**, *419–422 I*, 103–108.
[14] Noster, U.; Scholtes B., Zeitschrift für Metallkunde **2003**, *94(5)*, 559–563.
[15] Berger, C.; Troßmann, T.; Gugau, M., Materialwissenschaft und Werkstofftechnik, **2003**, *34(9)*, 812–832
[16] Eisenmeier, G.; Mughrabi, H.; Holzwarth, B.; Höppel, H. W.; Ding, H. Z., in *Mechanismusorientierte Lebensdauervorhersage für zyklisch beanspruchte metallische Werkstoffe, Kolloquium im Schwerpunktprogramm der Deutschen Forschungsgemeinschaft, 24./25.02.2000,* DVM Berlin **2000**, p. 153–164.
[17] Eisenmeier, G.; Holzwarth, B.; Hoppel, H.-W.; Mughrabi, H., Materials Science and Engineering A **2001**, *319–321*, 578–582.
[18] Eisenmeier, G.; Höppel, H. W.; Ding, H. Z.; Mughrabi, H., Materialwissenschaft und Werkstofftechnik **2002**, *33(5)*, 238–243.
[19] Horstemeyer, M. F.; Yang, N.; Gall, K.; McDowell, D.; Fan, J.; Gullett, P., Fatigue and Fracture of Engineering Materials and Structures **2002**, *25(11)*, 1045–1056.
[20] Bag, A.; Zhou, W., Journal of Materials Science Letters **2001**, *20(5)*, 457–459.

Comparison of Dieless Clinching and Dieless Rivet Clinching of Magnesium

R. Neugebauer, S. Dietrich, C. Kraus
Fraunhofer Institute for Machine Tools and Forming Technology IWU, Chemnitz

1 Introduction

Joining by forming of magnesium alloys is restricted by the limited forming capability of magnesium at room temperature. For this reason heating of the parts to temperatures of 220 °C or more is required in state-of-the-art joining methods using a contoured die (like clinching or self-pierce riveting) to form joints without cracks. Researches on these joining methods have shown that a minimum heating time of 3 to 6 seconds is needed to achieve joints of acceptable quality [1, 2]. Dieless clinching and dieless rivet clinching are new joining by forming methods, where a flat anvil is used as a counter tool instead of a contoured die. In this methods the crack inducing tensile stresses in the parts are reduced considerably during joining [3, 4, 5]. Although crack-free joining without heating of the magnesium parts has not been successful so far, a heating of the magnesium parts to temperatures of about 150 °C often is sufficient in the dieless joining methods. Moreover the heat transfer from a flat anvil to the parts can be much faster than from a contoured die of the same diameter. The combination of both effects make it possible to decrease the heating time to only one second or less [4, 5]. Dieless clinching and dieless rivet clinching, both methods were simulated using the Finite Elements Method (FEM) to analyze the influence of the geometrical parameters of the punch in dieless clinching and of the rivet in dieless rivet clinching. In this paper new results of investigations on the two processes are introduced.

2 Principle of Dieless Clinching and Dieless Rivet Clinching

As to be seen in Fig. 1 in both methods an upper part (3) will be joined with a bottom part (4). At the beginning of the joining process (Fig. 1, a) these two parts to be joined (3, 4) are lying partially overlapped on a flat anvil (5). For forming the connection the clamp (2) and the punch (1) are moved towards the parts to be joined (3, 4). First the clamp (2) gets into contact to the upper part (3) and a limited pressure is applied on the parts (3, 4) without deforming them. Then in dieless clinching (Fig. 1, top) the punch (1) is pressed directly into the parts (3, 4) with a high force. In the case of dieless rivet clinching (Fig. 1, bottom) the punch (1) is pressing the rivet (6) into the parts (3, 4). The material of the parts (3, 4) is displaced partially and flows in the opposite direction to the movement of the punch (1), thereby pushing the clamp (2) upwards. Thus an elevation is formed on the downside of the bottom sheet (it is important that the clamp pressure is sufficient to avoid warping of the parts but still allows for the formation of the elevation). The size of this elevation is increasing the further the punch (1) or the rivet (6) are pressed into the parts (3, 4). As soon as the material of the upper part (3) has come into contact with the shoulder of the punch (1) the material flow against the movement of the punch (1) is stopped

(Fig. 1,b). When the punch (1) is pressed even further into the parts (3, 4) the elevation at the downside of the bottom part is flattened and the material displaced by the punch (1) or the rivet (6) is forced to flow in radial direction, thus forming an interlock between the parts (Fig. 1,c). The connection has been formed in a single step. The material volume that the punch (1) in dieless clinching and the rivet (6) in dieless rivet clinching displaces is entirely directed towards the formation of the elevation [6,7]. In dieless clinching the punch (1) is pulled out of the connection after the end of the joining process. In dieless rivet clinching the rivet (6) stays in the connection and thereby stabilizes it, thus substantially increasing the shear strength of the connection. The shape of the rivet and its deformation in the joining process can support the formation of the interlock [6]. If heating of the parts is necessary, heating cartridges can be attached to the flat anvil. The heat transfer takes place, as soon as the parts are fixed between anvil and clamp. Thus it is possible to join parts made of magnesium or to join parts made of magnesium to parts made of other materials, such as aluminum or steel [4, 5].

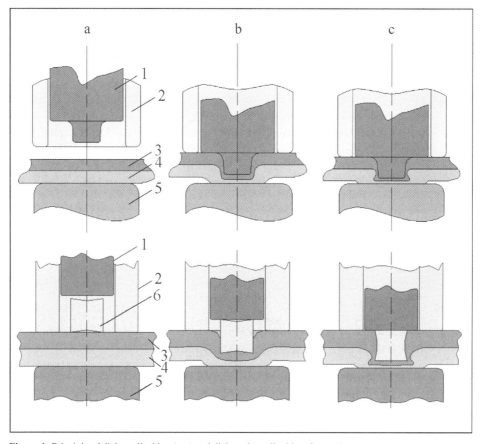

Figure 1: Principle of dieless clinching (top) and dieless rivet clinching (bottom)

3   Advantages of Dieless Joining by Forming in Joining Magnesium

Besides of the general advantages of dieless joining by forming methods, such as the reduction of tools (one single flat anvil for all joining tasks instead of extensive sets of dies), the high process reliability (almost no wear of the flat anvil, no rim that can be damaged, no cavity and therefore no space for deposits etc.) [8] there are further advantages as the high misalignment tolerance and the low heating times that make the dieless joining methods especially suitable for joining magnesium. The main reason for the low heating times when applying dieless methods is – besides the near absence of tensile stresses during the joining process reducing the required temperature of the magnesium – that there is a good heat transfer from the heated anvil to the magnesium part. A costly cooling system for the prevention of misalignments between the joining tools, due to heat induced deformations of the joining equipment – as it is needed in thermal assisted joining by forming with a contoured die – is not required [3].

4   Influence of the Rivet Geometry In Dieless Rivet Clinching in Comparison to the Punch Geometry in Dieless Clinching

4.1   Influence of the Shape of the Rivet

Instead of the punch in dieless clinching, which is pulled out of the connection after forming it, the rivet clinched connection is formed with an auxiliary part. This part – the rivet – stays in the connection. As there is no need to pull the rivet out of the connection again it is possible to use shapes that have an undercut, which is not possible for the punch in dieless clinching. Moreover, for every rivet is only used once, plastic deformation of the rivet is not only acceptable but can even help in the formation of the interlock between the parts to be joined.

Various rivet shapes have been investigated. Some of these shapes are to be seen in the middle column of Fig. 2. In the top row a cylindrical rivet is shown. This rivet shape closely resembles the punch geometry in dieless clinching. In the middle row of Fig. 2 there is an undercut-type rivet with a distinct waistline and in the bottom row there is a rivet with contoured ends. In the right column of Fig. 2 the formation of the connections for these rivet shapes is to be seen.

It has been shown in the investigations, that all these types of rivet shapes are adequate for the formation of connections with a substantial interlock.

However although using the cylindrical rivet results in good connections there always remains a risk for the rivet to loosen and fall out of the connection later as there is no interlock between the rivet itself and the parts. The other two rivet types both show an interlock between the rivet and the upper part. In case of the undercut-type rivet the material of the upper part is flowing into the waist of the rivet and in case of the rivet with the contoured ends a deformation of the rivet takes place in such a way, that the rivet is flattened at the end of the process and an interlock between the rivet and the upper part results (compare Fig. 2). As there is an interlock between the rivets and the upper part there is in both cases no risk for the rivet to loosen and fall out of the connection later. The connections made using rivets with contoured ends show in many cases an increased interlock compared to connections made with rivets having no contoured ends.

Attention has to be paid to the limitations of using these rivet shapes. Connections with the undercut-type rivets tend to have small cavities in the waist area if the waist is too tight. It could be proven, that the optimum waist diameter is about 0,2 mm less than the rivet diameter for rivet diameters in the range of 3–5 mm. The connections made using the rivet with contoured ends always show reduced neck thicknesses. The reduction in the neck thickness of the connection increases the more distinct the contour of the ends is getting.

Rivet type	Geometry	Formed connection (FEM)
Cylindrical	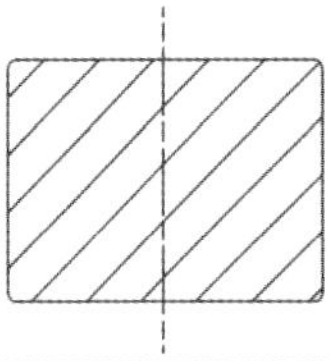	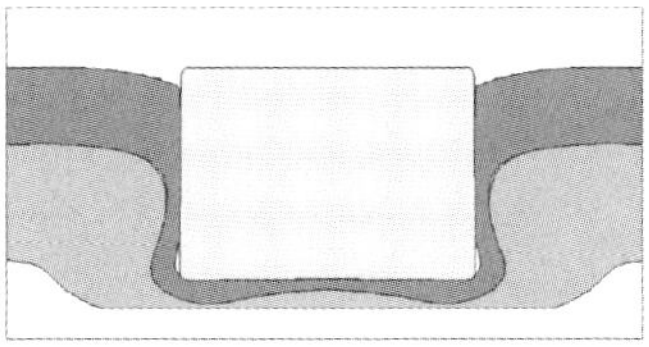
Undercut-type	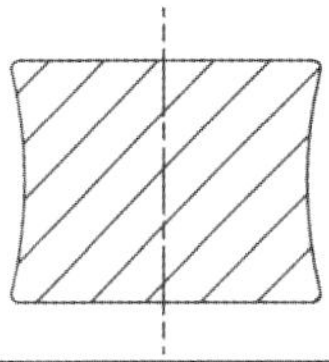	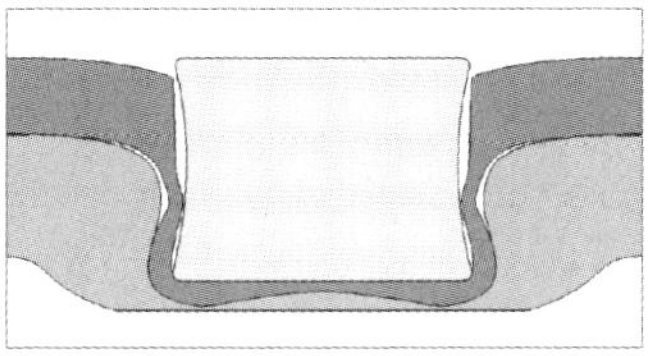
Countoured ends	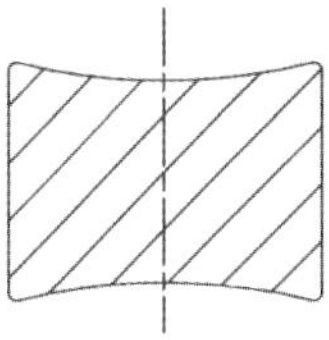	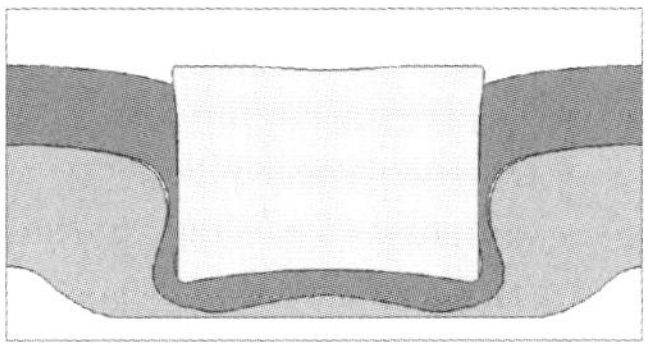

Figure 2: Connections formed with rivets of different shapes

4.2 Influence of the Diameter of the Rivet in Dieless Rivet Clinching

It was to investigate, whether the diameter of the rivet in dieless rivet clinching has the same influence on the formation of the connection as the diameter of the punch in dieless clinching. As it has been shown in previous investigations an increased punch diameter in dieless clinching always leads to an increased neck thickness of the formed connection. The obtainable interlock of the formed connection has an maximum at a certain punch diameter, the size of which is depending on the joining task. At punch diameters above or below this diameter the interlock is decreasing [4, 5].

The investigations on dieless rivet clinching showed, that regardless of the rivet shape the neck thickness has the same dependency on the rivet diameter as in dieless clinching on the punch diameter. An increasing of the rivet diameter results in an increasing of the neck thickness. The size of the interlock has also shown an increase for increasing rivet diameters. However, contrary to the investigations in dieless clinching there was no rivet diameter to be found in the investigated range of rivet diameters were the interlock would show a maximum (compare Fig. 3). The reason for this is probably, that the maximum will only be reached at even bigger diameters, that haven't been investigated due to a lack of practical relevance.

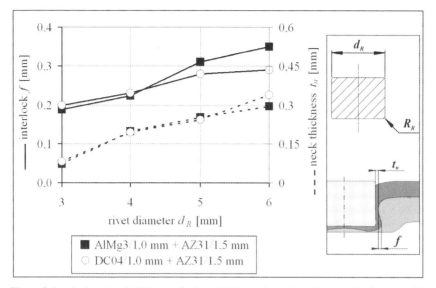

Figure 3: Interlock and neck thickness of selected joining tasks in dependency on the diameter of the rivet

4.3 Influence of Edge Radius in Dieless Rivet Clinching

The edge radius of the rivet is a geometrical parameter in dieless rivet clinching that resembles the punch radius in dieless clinching. In previous investigations it had been shown, that the edge radius of the punch has a major influence on the formation of the connections in dieless clinching. An increased edge radius at the punch in dieless clinching leads to a decreasing interlock and an increasing neck thickness [4, 5].

It could be shown in investigations of the dieless rivet clinching process, that the edge radius of the rivet has an influence of similar importance. Regardless of the shape of the rivet and regardless of the joining task an increasing edge radius of the rivet in dieless rivet clinching leads always to a decreasing interlock and an increasing neck thickness. The reason for this is that the material flow is forced in radial direction to form the interlock when the edge is sharp. As soon as the radius is getting bigger the material can flow better into the neck of the connection, thus increasing the neck thickness and reducing the available material for the formation of the interlock. In Fig. 4 the dependency of the interlock and the neck thickness of the formed connections on the edge radius of the rivet is shown for selected joining tasks.

5 Influence of the Hardness of the Rivet

The rivet can have a high or a low hardness. If it has a high hardness then it will only be slightly deformed in the joining process. If it has a low hardness then it will be deformed much more. In the investigations the influence of the hardness of the rivet on the formation of the connection was to be determined. The hardness was divided into seven levels as it is known from self-pierce riveting – level 6 being the highest hardness and level 0 the lowest.

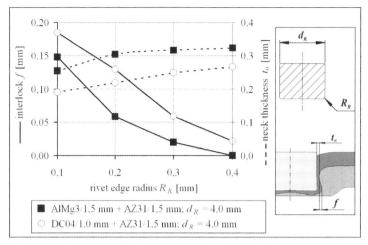

Figure 4: Interlock and neck thickness of selected joining tasks in dependency on the edge radius of the rivet in dieless rivet clinching

In Fig. 5 the dependency of the interlock and the neck thickness of the formed connection on the hardness level of the rivet is to be seen. If the hardness is too low the rivet is flattened completely in an early stage of the process and no interlock is to be obtained, for the material between the rivet and the anvil is not forced to flow in radial direction. Moreover the neck of the connection can be destroyed by the strong radial flowing of the material of the rivet. For this reason even when dieless rivet clinching soft magnesium alloys a rivet having at least a hardness level of 2 is required. An even higher hardness results in some cases in a higher neck thickness (compare Fig 5), but experience shows, that at the highest hardness levels the rivets tend to get cracks during the joining process. The optimum hardness level for the dieless rivet clinching of magnesium alloys is between 2 and 4.

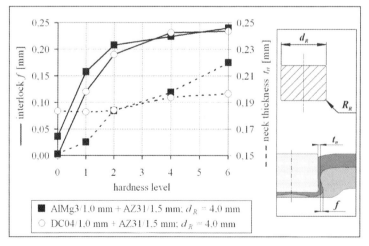

Figure 5: Interlock and neck thickness of selected joining tasks in dependency on the hardness level of the rivet in dieless rivet clinching

6 Conclusions

The dieless rivet clinching process was simulated with the Finite Elements Method to analyze the influences of geometrical parameters of the rivet and the influence of the hardness of the rivet on the formation of the connection. The investigations of different rivet shapes have shown, that a cylindrical rivet shape is adequate for forming good connections with a substantial interlock. A rivet shape with a slight waist or a slight contour in the ends of the rivet allows for the formation of connections with similar neck thickness and interlock between the parts, that additionally have an interlock between the rivet and the upper part, thus avoiding that the rivet falls out of the connection later on.

The influences of other geometrical parameters of the rivet in dieless rivet clinching such as edge radius or rivet diameter show a strong resemblance to the influence of the geometrical parameters of the punch in dieless clinching. Investigations on the hardness of the rivet have shown, that a minimum hardness is required to get connections with a substantial interlock and a sufficient neck thickness.

Further research will include experiments to verify the FEM based analysis about the influences of the geometrical parameters of the rivet and the influence of the hardness of the rivet on the formation of the connection as well as experiments to ascertain the strength of dieless clinched and dieless rivet clinched connections.

7 Acknowledgment

The authors are grateful to the Deutsche Forschungsgemeinschaft DFG (German Research Foundation) for generously supporting our research work in the priority program 1168 „Extending the range of applications of magnesium alloys".

8 References

[1] S. Hübner, Clinchen moderner Blechwerkstoffe, Dissertation, University Hannover, Germany, **2005**
[2] Doege, E.; Hübner, S; Thoms, V.; Bräunling, S., Thermisch unterstütztes Clinchen von Blechen und Bauteilen aus Magnesium-Knetlegierungen, Europäische Forschungsgesellschaft für Blechverarbeitung e. V., Forschungsbericht Nr. 203, Hannover, Germany, **2003**
[3] Neugebauer, R.; Mauermann, R.; Dietrich, S., in *11th International Conference on Sheet Metal SheMet 05*, Erlangen, Germany, **2005**, p. 203–209
[4] Neugebauer, R.; Dietrich, S.; Kraus, C., in *12th Paderborner Symposium Fügetechnik*, Paderborn, Germany, **2005**, p. 100–110.
[5] Neugebauer, R.; Dietrich, S.; Kraus, C., in *THERMEC '2006*, Vancouver, Canada, **2006**
[6] Neugebauer, R.; Mauermann, R.; Dietrich, S., in *International Conference on Technology of Plasticity,* Yokohama, Japan, 28. October–1. November, **2002**, p. 1291–1296
[7] Neugebauer, R.; Mauermann, R.; Dietrich, S., Journal of the JSTP, 507 (Volume 44), **2003**, p. 52–53 (p.370–371)
[8] Dietrich, S.; Mauermann, R.; Voelkner, W., Blech InForm 4/2003, Carl Hanser Verlag, München, Germany, **2003**, p. 50–52

Influence of Cutting and Non-Cutting Processes on the Corrosion Behavior and the Mechanical Properties of Magnesium Alloys

Friedrich-Wilhelm Bach*, Berend Denkena*, Klaus Weinert**; Patrick Alpers*, Martin Bosse*, Niels Hammer**,
* Leibniz University Hannover, Hannover; Germany
** University Dortmund, Dortmund; Germany

1 Introduction

An important aim for the automotive industry is to reduce vehicle weight in order to decrease fuel consumption and emission of CO_2. The use of light metals as construction materials is generally considered of key importance for the future. Magnesium alloys fulfill the demands of a low specific weight with excellent machining properties and high recycling potential.

However, along with the excellent properties, there are some disadvantages of magnesium alloys. The lower corrosion resistance in comparison with other light materials (e.g. aluminium alloys) is one of the reasons for the rare application as a construction material. Therefore the essential aim of the presented joint research project within the priority program 1168 of the German Research Foundation (DFG) is to influence the mechanical properties and corrosion resistance by cutting and non-cutting processes.

2 Experimental Procedure

The surface of the workpiece is the most loaded area at technical applications [1]. To obtain functional surfaces milling, drilling and tapping operations are essential machining technologies. However these common cutting processes often lead to changes of the microstructure. One of these consequences can be changes in the micro hardness in the subsurface zone. Previous experimental investigations showed a correlation between cutting and non-cutting processes and the corrosion behavior of different magnesium alloys [2]. Milling tests were carried out with the aim of obtaining detailed knowledge about the corrosion behavior of differently machined workpiece surfaces.

Besides of cutting operations, non-cutting mechanical treatments of the workpiece can effect the surface and subsurface properties of the material. The surface roughness, the induced residual stress, the strain hardening and the hardness can be influenced. Proven methods for non-cutting surface treatments are rolling and shot peening. The deep rolling process is a surface finishing process for metal which cold forms the top layer of the material instead of removing it. The task of this mechanical process is to improve the fatigue strength of dynamically loaded components. This post process treatment was established in the industry for different kinds of material (e.g. high strength steel, titanium, aluminum) and applications (e.g. general machine construction, vehicle construction, power plants, engines).

In order to increase the use of magnesium as a structural material in lightweight constructions it is necessary to improve the corrosion resistance of the workpiece surface. The influence of surface treatment, different alloy compositions and varying heat treatments were analyzed in salt spray corrosion tests according to DIN 50021 SS. Electrochemical noise measurements were carried to determine the influence of different surface treatment by cutting and non-cutting processes. The measurements were made to characterize the corrosion attack on the surface after a short time in dependence of the roughness and subsurface properties.

3 Results and Discussing

3.1 Analysis of Surface and Subsurface Effects of Milling Operations

As testing materials the two sand cast, high strength magnesium alloys EQ21 and WE43 were investigated. For the milling processes, a face milling cutter $d = 80$ mm with 4 inserts was used under dry cutting conditions. Uncoated, cemented carbide (HW) inserts and inserts out of poly crystalline diamond (PCD) with the same shape specification SPGW1204EDR were applied under variable cutting conditions. Furthermore, metallographic examinations as well as analyses of the thermo mechanical loads acting on the machined surfaces during the cutting process were carried out. For the metallographic examinations, adapted strategies of specimen preparation for these magnesium alloys had to be developed on the basis of already existing strategies [3].

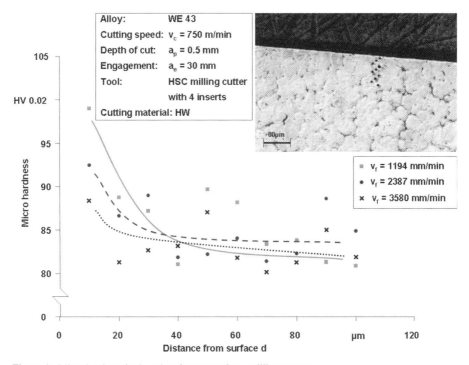

Figure 1: Micro hardness in the subsurface zone after a milling process

Micro hardness measurements in the subsurface zone were performed to determine the interactions between the machining parameters and the increase of hardness. Figure 1 exemplarily depicts a micro hardness depth series after a milling operation. It can be observed, that a low feed speed leads to a more significant hardness increase close to the surface. An increase of hardness up to 100 HV0.02 can be detected, because of the thermal influence when machining with low feed speeds. The core hardness of the material can be measured at approximately 80–85 HV0.02. Due to the mechanical loads and an increased duration of the thermal exposure at the surface the workpiece will be heated up, so that changes in the micro structure are the result. Besides the cutting parameter related thermal and mechanical influences on the micro structure the cutting material and the boundary conditions of the process are also influencing factors for the surface properties.

Figure 2 shows the average surface roughness for constant feed speeds of v_f = 2550 mm/min on different magnesium alloys and after application of different cutting tool materials. A comparison of varying cutting conditions which lead to different surface qualities can be seen. Good surface qualities can be achieved by the combination of high cutting speeds and low feed. This leads to a moderate feed speed and a thermally induced reduction of the material strength on the shear plane.

A comparison of the two investigated alloys WE43 and EQ21 shows no significant differences with respect to the surfaces roughness. Related to the thermal material behaviour it can be noticed, that EQ21 shows a high stability for a micro hardness increase [2]. As its thermal conductivity is twice as high as the thermal conductivity of WE43, the generated heat is distributed within the complete workpiece, so that significant micro hardness changes hardly occur.

The cutting tools out of cemented carbide (HW) lead to the lowest surface roughness. This can be explained by an improved chip formation. However, PCD tools are commonly applied in cutting operations of magnesium alloys due to their favorable wear behavior.

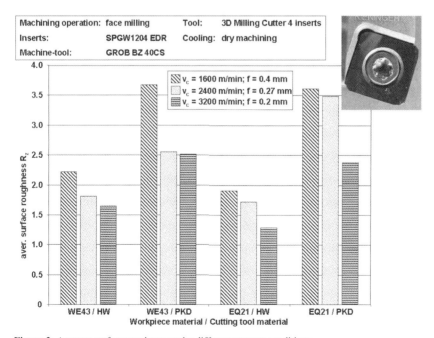

Figure 2: Average surface roughness under different process conditions

3.2 Corrosion Behavior of Milled Magnesium Alloys

The corrosion rate of the magnesium sand cast material WE43 (4 m% Yttrium, 3m% Rare Earth) is analyzed in the salt spray test according to DIN 50021 SS. The WE43 materials is known as high strength casting alloy with high temperature creep resistance up to 300 °C. In order to ensure a homogeneous microstructure and to achieve the same conditions in the corrosion trials all parts were solution heat treated and age-hardened (T6) [4]. The tests determine the corrosion behavior by measuring the weight difference of the samples before and after the corrosion test. After 3–7 day of cyclic testing in 5 w% NaCl salt solution the weight loss is measured by dissolving the corrosion products in chromic solution. The surface of the samples is processed in a milling process under dry cutting condition. Uncoated cemented carbide (HM) and polycrystalline diamond carbide (PCD) cutting inserts were tested under variable cutting conditions.

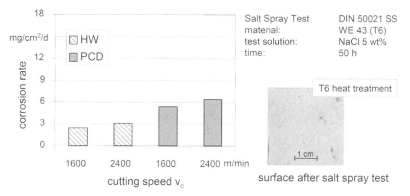

Figure 3: Salt spray corrosion test of WE43

The gravimetric studies of WE43 after 50 h of testing in NaCl-spray are shown in figure 3. The corrosion rate of the samples machined with polycrystalline diamond tools (6 mg/cm^2/d) is significant higher than the corrosion rate of the uncoated cemented carbide tool (3 mg/cm^0/d). It can be observed that increasing cutting speed intensifies the rate of weight loss per unit area after on day. However, the corrosion test with 5 wt% salt solution is too aggressive and indicates only a tendency [5]. One influencing factor leading to a lower weight loss due to the corrosion could be the lower surface roughness produced with cemented carbide tools.

The electrochemical noise measurement was used to characterize the corrosion behaviour of different surface conditions which were generated in a milling process under variation of the process parameter. In comparison to conventional noise measuring methods the correlated noise measuring technique (CorrELNoise) shows a more sensitive detecting behaviour [6].

This measuring technique was applied to analyse the initial corrosion events on the influenced test surfaces. The corrosion cell is built up with a two electrode setup of identical test samples and an test area of 0.5 cm². Figure 4 illustrates the measuring setup. Two WE43 samples were processed with the same cutting speed and tool material. Afterwards these sample were tested in low concentrated NaCl solution (0.05 wt%). The correlated noise response of the current and the potential was recorded.

In figure 4 the comparison of the current noise over a period of 800 seconds of two WE43 specimen cut with PCD insets is shown. An increase of corrosion after 400 sec is indicated by the rising current curve for the sample which is machined with higher cutting speed (2400 m min⁻¹). This points out a stronger corrosion attack at the surface which was already assumed after the salt spray corrosion test. The corrElNoise-test of samples cut with HW inserts the current curves in figure 5 indicates a decrease of corrosion current after 600 sec. However it tends to diminish the current noise which correlates to the results of gravimetric studies. A significant influence of different cutting speed on the corrosion resistance could not be determined (figure 4 and 5). Further adjustment of this methodology is necessary to detect the influence of the cutting speed on corrosion behaviour of processed surfaces.

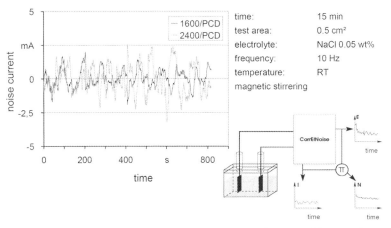

Figure 4: Noise vs. time curves of WE43 machined with PCD inserts at different cutting speeds v_c

In figure 4 the comparison of the current noise over a period of 800 seconds of two WE43 specimen cut with PCD insets is shown. An increase of corrosion after 400 sec is indicated by the rising current curve for the sample which is machined with higher cutting speed (2400 m min⁻¹). This points out a stronger corrosion attack at the surface which was already assumed after the salt spray corrosion test. The corrElNoise-test of samples cut with HW inserts the current curves in figure 5 indicates a decrease of corrosion current after 600 sec. However it tends to diminish the current noise which correlates to the results of gravimetric studies. A significant influence of different cutting speed on the corrosion resistance could not be determined (figure 4 and 5). Further adjustment of this methodology is necessary to detect the influence of the cutting speed on corrosion behaviour of processed surfaces.

3.3   Analysis of Surface and Subsurface Effects of Non-cutting Operations

The wrought alloy AZ31 and the cast alloy AM50 are prepared by deep rolling with the hydrostatic tool HG6 of Ecoroll. The results were carried out at constant parameters of velocity and feed rate. The pressure is varied from 4 MPa to 10 MPa in steps of 2 MPa. Using a Kistler dynamometer the average axial force is detected to $F_{N,1} = 79$ N (4 MPa), $F_{N,2} = 138$ N (6 MPa), $F_{N,3} = 184$ N (8 MPa) and $F_{N,4} = 229$ N (10 MPa). This axial force plastically forms the top layer. Micrographs show the formation of twins close to the surface caused by the poor cold workability of the hexagonal lattice structure. The depth effect becomes more significant by an increasing axial force, so that the maximal depth of detected twins is about 800 μm (p_W = 10 MPa). The plastically form of the top layer was also shown in the better surface roughness, which gets from R_a = 1.18 μm (turning) to R_a = 0.1 μm (deep rolling). This lower surface roughness of R_a = 0.1 μm is independent of the hydrostatic pressure. A small pressure of 4 MPa is enough to smooth the surface [2].

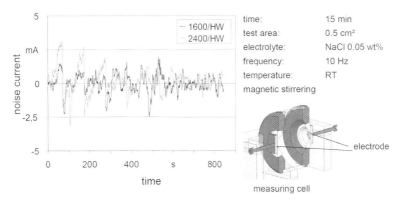

Figure 5: Noise vs. time curves of WE43 machined with HW inserts at different cutting speeds v_c

Another effect of the deep rolling is the increasing hardness caused by the strain hardening of the material. Tests show that the average surface hardness is directly proportional to the deep rolling pressure. Figure 6 shows the characteristics regarding the micro hardness. The base hardness of the extruded AZ31 alloy is about 60 HV0.025. A turning operation leads to rise up the micro hardness close to the surface. An additional deep rolling process increases the hardness and the area of influence. This effect becomes more significant as the rolling pressure is increased.

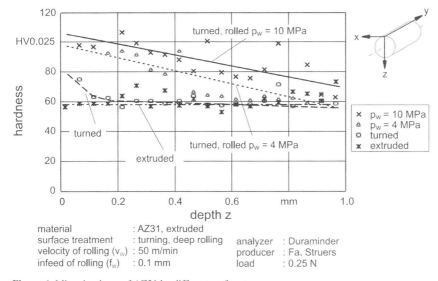

Figure 6: Micro hardness of AZ31 by different surface treatments

The $\sin^2$-ψ-method was used to detect the stress. A qualitative measurement shows compressive stress in the surface zone in all deep rolled workpieces (Figure 7). The highest compressive stress is detected at rolling pressure of 4 MPa. Further analyses show a migration of the maximum of the compressive stress into the workpiece while the pressure of rolling is increased Winkler, J., in Herstellung rotationssymmetrischer Funktionsflächen aus Magnesiumwerkstoffen durch Drehen und Festwalzen, VDI Verlag GmbH, Düsseldorf, Germany, 2000. This effect depends on the Hertzian stress (Hertzsche Pressung).

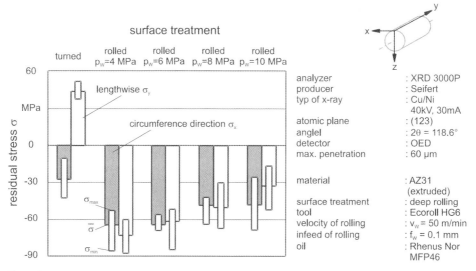

Figure 7: Residual stress of deep rolled AZ31

Rotating bending tests are used to characterize the fatigue strength. Experiments with the AM50 alloy show, that the number of cycles N at a load of 80 MPa grows up from nearly $N = 1 \cdot 10^5$ (turned) to $N = 9 \cdot 10^5$ ($p_W = 4$ MPa). A similar effect is observed with the extruded AZ31 alloy.

To separate the influence of surface roughness and compressive stress the high pressure water blasting is an alternative surface after treatment. Furthermore it has to be taken into account that a peening process does not import placement of impurities into the material. Tests with different pressures from $p_S = 32$ MPa to $p_S = 72$ MPa have shown, that in opposite to the rolling process the surface roughness increases by a rising pressure. Abrasive metal removal is detected by blasting with a pressure of 52 MPa. These tests also demonstrate, that it is possible to rise up the hardness and to induce compressive stress with a water blasting process. To reduce the surface roughness without increasing the inherent compressive stress, a polishing process was also analyzed.

Figure 8 illustrates the average, maximal and minimal number of loading cycles N for a rotating bending test with different finishing operations at AZ31. The compressive stress is the main influence to rise up the fatigue strength.

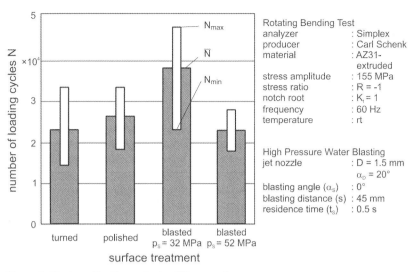

Figure 8: Number of loading cycles by different surface treatment

4   Summary and Outlook

To draw a conclusion concerning cutting process related effects on the surface integrity cutting parameters as well as the cutting material can be identified as main influencing factors. Especially the intensity and the duration of thermal loads lead to changes in the micro structure of the magnesium alloys. Further investigations will be carried out to show the process conditions which lead to improved corrosion properties.

Besides the investigations concerning the process related effects on the surface integrity when milling, adapted cutting technologies for an improved use of magnesium alloys will also be developed. Former investigations showed that conventional tapping processes lead to significant deformations in the subsurface zone of magnesium workpieces when manufacturing threads [8]. In this context the innovative cutting and forming technology for threads is a promising concept to produce threads in magnesium with increased strength properties. Furthermore, the combination of these high strength threads for nuts and bolts made from magnesium offers the opportunity to avoid the contact corrosion by using identical materials.

Different corrosion tests show the influence of surface treatment by cutting processes on the corrosion behavior of magnesium alloys. A correlation of different milling parameters and corrosion rate using the salt spray tests is complicated by the typical scattering of this test method. Therefore, the correlated electrochemical noise measurement is used to localize corrosion effects on the machined surfaces especially in the beginning of the corrosion process. Further investigation will concentrate on the corrosion analysis of coated and surface treated magnesium components. Adhesive tests will be done under corrosive attack in the salt-spray test to characterize the protecting effect of the coating system. Electrochemical noise measurement will be carried out to characterize defects in the surface coating under corrosive load.

Due to thermal and mechanical loads in cutting operations the properties of the surface and the subsurface zone are effected. To increase the influenced zone and to adjust compressive residual stress one possibility is the use of a deep rolling process.

The next step of the research activities is to increase the area of influenced subsurface by a combination of different effects while cutting. In this way, no additional rolling tool is needed to obtain the favorable effects of a surface treatment process.

Furthermore, a combination of surface coating and a following deep rolling process will be analyzed. The aim of this process is to increase the mechanical properties and to reduce the corrosion behavior of magnesium components. In first tests AZ31 sheets were coated with a MAG-PASS-COAT® (free from chromium). Afterwards the surface was finished by a rolling process. Analyzing the behavior, this combination demonstrated a positive influence on corrosion properties.

5   Acknowledgement

The studies about the Influence of cutting and non-cutting processes on the corrosion behavior and the mechanical properties of magnesium alloys were carried out within the priority program 1168 „InnoMagTec", supported by the German Research Foundation (DFG). The research work is being conducted in the joint project of the Institute of Production Engineering and Machine Tools and the Institute of Materials Science, Leibniz University Hannover, and the Department of Machining Technology, University of Dortmund.

6   References

[1] Scholtes, B.; Voehringer, O., in Mechanische Oberflächenbehandlung, (Ed.: E. Broszeit, H. Steindorf), DGM Informationsgesellschaft, **1996**, Oberursel, Germany, p. 3–20
[2] Denkena, B., Alpers, P., Weinert, K., Hammer, N., in Annals of the German Academic Society for Production Engineering, XIII, **2006**
[3] Kree, V., Bohlen, J., Letzig, D., Kainer, K. U., in Praktische Metallographie, **2004**, 41, München, Germany, p. 233–245
[4] Stevenson, A., in ASM Handbook, **1991**, Heat Treading Volume 4, p. 899–906
[5] Bach, Fr.-W., Haferkamp, H., Kaese, V., Niemeyer, M., Tai, P., Wilk, O., Magnesium Taschenbuch, 1st ed., Aluminium-Verlag, Düsseldorf, Germany, **2000**, p. 285–298
[6] Rösler, B., Schiller A.C., in Materials and Corrosion **2001**, 52, Weinheim, Germany, p. 413–417
[7] Winkler, J., in Herstellung rotationssymmetrischer Funktionsflächen aus Magnesiumwerkstoffen durch Drehen und Festwalzen, VDI Verlag GmbH, Düsseldorf, Germany, **2000**
[8] Hammer, N., Weinert K., in Magnesiumbearbeitung- die thermischen Effekte des Gewindebohrens, in WB Werkstatt und Betrieb, **2005**, 138, p. 27–30

Development of Innovative Technologies for Joining Magnesium Alloys and Dissimilar Materials Related to Components

Silke Mücklich, Bernhard Wielage
Lehrstuhl für Verbundwerkstoffe, TU Chemnitz, Germany
Magnus Horstmann, Ortwin Hahn
Laboratorium für Werkstoff- und Fügetechnik, Univ. Paderborn, Germany

1 Introduction

Welding technologies offer a broad range of joining parameters (for instance [Schra01]), but they are in most of the cases not appropriated for joining of high alloyed materials as well as for joining of dissimilar materials. Only beam welding and friction welding can be used for joining of dissimilar materials (e.g. [Bor05]). The main drawback of welding processes is the high temperature load of the base materials. Thus alternative heat-poor joining technologies for joining magnesium alloys both among each other as well as to aluminum alloys and steel are examined in the present study. Magnesium alloy AZ31 which is the most common sheet material was used in the present investigations as a base material for joining. To obtain dissimilar joints AlMgSi alloys as well as unalloyed steels have been used as the second base material.

2 Soldering

2.1 Experimental

Soldering was carried out at 350 °C. Filler materials used were tin based in the case of AZ31/steel joints and Zn-Mg-Al based for joining of magnesium and aluminum alloys. Zn-Mg-Al filler materials were newly developed and are described in the following chapter. Ultrasonic soldering was used to avoid the use of fluxes during soldering. The samples were heated to the soldering temperature. Afterwards ultrasound was introduced to the samples for 5 s. Immediately after soldering the samples were cooled down at air. The examinations described here are based on investigations at soldered joints of magnesium cast and wrought alloys which have been studied in [Mue05]. The microstructures of the samples were investigated using scanning electron microscopy and the mechanical behaviour was measured during lap shear tests.

2.2 Filler Development

There are several requirements for solders used to magnesium base materials: The first point is that the temperature range for soldering has to fit to the base materials. Aluminum alloyed magnesium alloys posses a eutectic temperature at 437 °C. Additional alloying elements can possibly decrease this temperature. To avoid remelting in the base material sample the soldering temperature has to fall far below this eutectic temperature. The second main requirement is that the difference in the electrochemical potentials between the filler and the base materials should be as low as possible. Furthermore the filler materials should be easy to manage during sample preparation. Since there are no appropriated filler materials available new filler materials were investigated and developed based on the ternary system Al-Mg-Zn. The ternary diagram shows a low melting eutectic at 339 °C. Filler materials with compositions corresponding to this eutectic were obtained by rapid solidification in form of 0,1 mm thick tapes. The filler metal used in this study consists of 40 Mg, 50 Zn, and 10 Al (in mass-%).

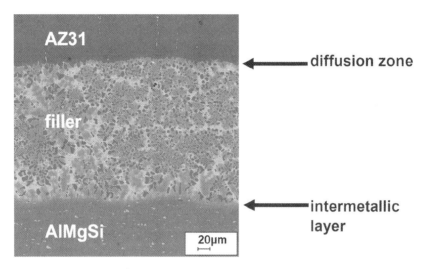

Figure 1: Microstructure of an soldered seam between AZ31 and AlMgSi base materials

2.3 Microstructure of Soldered Joints

Soldered joints which are obtained using dissimilar base materials form microstructures like it is shown in Figure 1. The soldered seam consists of magnesium solid solution, which appears dark in contrast, and several intermetallic phases showing lighter gray contrast. At the interface to the magnesium base material a diffusion zone can be observed containing higher amounts of zinc than AZ31. An Mg17Al12 layer containing zinc on substitutional lattice sites can be observed at the interface to the aluminium base material. During the soldering process the magnesium as well as the aluminum content of the filler metal are increasing while the zinc content is decreasing. The overall composition after soldering within the soldered seam was measured to be 59 Mg, 21Al, and 20 Zn (in mass-%). The influence of the filler material composition has been described in [Wie05]. Additional details of the microstructure of joints with dissimilar base materials have been discussed in [Wie05a].

2.4 Mechanical Properties of Joints

The highest shear strength observed at AZ31/AZ31 soldered joints is higher than at joints of dissimilar materials. This fact is probably due to the formation of an intermetallic $Mg_{17}Al_{12}$ layer forming at the interface to the aluminum base material.

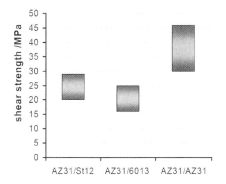

Figure 2: Comparison of the shear strength of AZ31/AZ31 soldered joints and joints with dissimilar materials

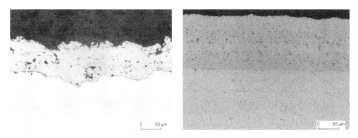

Figure 3: Zn coating on AZ31 sheet (left: thermal arc spraying, right: Galvanic Zn coating)

3 Adhesive Bonding

3.1 Experimental

Up to now there is no extensive knowledge in adhesive bonding of magnesium components. For instance it is not considerably clarified, to which extent pre-treating systems, that lead to a good joint strength with aluminium alloys, are adequately applicable for the pre-treatment of magnesium alloys. Therefore the strength as well as the aging behaviour of different adhesive joints on magnesium substrates were researched under variation of the pre-treatment operation. Different chemical, mechanical and physical treatments were applied. Also coatings were researched as a basis for glueability e.g. a Zn coat which was brought to AZ31 sheet material by using thermal arc spraying and electrochemical methods (Figure 3). For the assays two commercially available adhesives were investigated, a 1-component as well as a 2-component epoxy. In addition to the pure bonding of magnesium specimen, an optimization of adhesives was accomplished for the application in a hem flange bonding process.

3.2 Mechanical Properties of Joints

The significant differences in the shear strengths result from a varied failure behaviour of the joints. For the pre-treatment with ScotchBrite appears a substrate-close cohesive failure of the adhesive, while the specimens which are coated via thermal arc spraying, fail in the coating itself. The minor adhesion of the thermally sprayed coat thus led to a minor strength level.

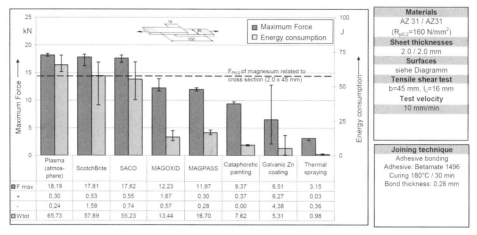

Figure 4: Comparison of the shear strength of AZ31/AZ31 bonded joints with different surface pre-treatments

3.3 Optimization of Adhesive for Hem Flange Bonding

With an appropriate heating of the substrates, the hemming process, where two components are joined together by folding the edge of one over the other, can be process-sure realized for magnesium. The process itself was executed in several hemming steps by using inductive heating. Due to the thermally supported process to deform the magnesium substrate, up to now a combination of roller hemming and adhesive bonding wasn't possible according to the state of knowledge, because the adhesive already shows a chemical reaction in the first pre-hemming step, when the material is heated till 230 °C. This leads to an insufficient deformation and wetting ability of the adhesive at the final hemming step, since the common curing temperatures of hot-curing systems in the automotive industry only face a value up to 180 °C. If the adhesive is exposed to such a high temperature, this can lead to an abrupt curing, which has a hurtful effect to the properties of the adhesive. At this point the development of a process technology was necessary to achieve optimal characteristics.

With the help of a calculated formal reaction kinetics model the reaction process for different heating rates and different adhesives could be determined. Using an advanced adhesive, the research has shown that it doesn't cure under the considered thermally supported heating process.

4 Mechanical Joining

4.1 Experimental

Different mechanical joining techniques were researched. Blind riveting, screwdriving, hemming, clinching, clinch riveting and self piercing riveting with semi tubular rivet as well as with solid rivet came into operation. For the joining processes which require a deformation of the magnesium substrate, a process modification is necessary, by that the deformation process is supported by an inductive heating of the magnesium. The joining process was conducted at a temperature of approx. 280 °C. At this temperature magnesium shows an increased deformation ability, resulting from the thermal activation of additional slip systems in the crystal structure. The procedure was introduced in [Hah04]. The rivets and screws used in this project consisted of steel and aluminium to compare the corrosion resistance of the connections after ageing tests. To some extent the rivets were coated with different layers.

4.2 Mechanical Properties of Joints

The load bearing capacity of the different connections was compared in quasi-static tensile shear tests. The results of these tests for the thermally supported joining techniques are shown in Figure 5. The different maximum forces result from the different failure behaviour of the joining techniques. While a neck-crack of the joint leads to fracture for the clinch elements, the insertion of a clinch rivet reinforces the connection in the direction of the shearing strain. In this case it appears an unbuttoning of the clinch rivet due to the tensile shear test. The connections realized with self piercing riveting with semi-tubular rivet fail in the magnesium base material by building an elongated hole. Here the strength of the magnesium substrates is utilized best. To take a look at the results of the different combinations of materials, it appears that the die sided configuration of AZ31 is to the best advantage for clinching and clinch riveting. On the other hand the punch sided disposition of AZ31 achieves better outcomes for self piercing riveting.

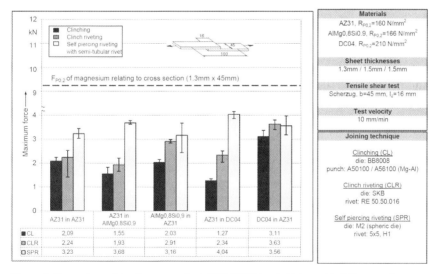

Figure 5: Comparison of the shear strength of AZ31/AZ31 bonded joints with different surface pre-treatments

4.3 Mechanical Joining by Forming at Room Temperature

In a continuative research it was ascertained that an increased forming velocity causes a better flow behaviour of the magnesium substrates at room temperature. It is necessary to utilize this effect for the mechanical joining by forming with an accelerated setting process up to 100m/s. Therefore different methods are researched to achieve an accelerated setting velocity. Among other things this includes the joining via expansion of a gas volume. The research can be based on results of successful works in the area of impulse joining with one-sided accessibility [Drah05]. Initial tests for the self piercing riveting with semi-tubular rivet have shown that good joining element characteristics are producible even at room temperature (Fig. 6). Important influence factors on the joining process are the inserted energy or rather the joining velocity, sheet thickness, the hardness of the rivet as well as the geometry of the die.

Figure 6: Impulse joined connection at room temperature with a common die (self piercing riveting with semi tubular rivet, AZ31/Al and Al/Mg, steel rivet)

5 Conclusions and Acknowledgements

Previous research has shown that it is possible to produce high-quality joints for magnesium alloys as well as in material mix by using the joining techniques soldering, adhesive bonding and also forming techniques. For this purpose systematic optimizations of the joining systems are necessary. For soldering this means particularly the development of the used filler metals. In consideration of an optimal adhesive joint it deserves an appropriate surface pre-treatment of the substrates. Attempts to improve mechanical joining of magnesium are targeted on a better material deformation by using conventional joining techniques on the one hand. On the other hand innovative mechanical joining concepts are observed which consider the comparatively poor deformability of magnesium at room temperature. The shown results represent an essential research step for a later industrial application. At this point it is necessary to gain detailed information about the realized joints in a component related application. At present appropriate research is carried out within the DFG-priority programme 1168 (InnoMagTec).

6 Acknowledgments

The authors are very grateful to the German Research Foundation (DFG - Deutsche Forschungsgemeinschaft) for funding this study within the scope of priority programme 1168 (reference numbers WI688/68-2 and HA1005/14-2).

7 References

[Bor05]   Borrisutthekul, R.; Miyashita, Y.; Mutoh, Y.: Dissimilar material laser welding between magnesium alloy AZ31B and aluminum alloy A5052-O, Science and Technology of Advanced Materials. 6 (2005) 2, 199–204

[Dra05]   Draht, T.: Entwicklung des Bolzensetzens für Blech-Profil-Verbindungen im Fahrzeugbau, Dissertation, Laboratorium für Werkstoff- und Fügetechnik, Universität Paderborn (2005)

[Hah04]   Hahn, O.; Tan, Y.; Schroeder, M.; Horstmann, M.: Thermally Supported Mechanical Joining of Magnesium Components, Proceedings "International Conference on Magnesium", 20. - 24.09.2004, Beijing, P. R. China (2004)

[Mue05]   Mücklich, S.: Beitrag zum flussmittelfreien Löten von Magnesiumwerkstoffen mit angepassten Lotwerkstoffen. Deutsche Dissertation. Werkstoffe und Werkstofftechnische Anwendungen. Hrsg.: TU Chemnitz, B. Wielage, vol. 21 (2005)

[Schra01] Schram, A.; Kettler, C.: Stoffschlüssiges Fügen von Magnesiumlegierungen. Magnesium - Eigenschaften, Anwendungen, Potenziale. K.U. Kainer (Hrsg.) Wiley-VCH Verlag GmbH, Weinheim - New York - Chichester - Brisbane - Singapore - Toronto. 2000, 161 ff.

[Wie05]   Wielage, B.; Mücklich, S.: Einfluss der Lotzusammensetzung auf die Gefügeentwicklung in Magnesium-Lötverbindungen. Prakt. Metallogr. **42** (2005) 10, 513–527

[Wie05a] Wielage, B.; Mücklich, S.: Gefügeuntersuchungen an gelöteten Mischverbindungen zwischen Magnesium und Aluminium bzw. Stahl. Fortschritte in der Metallographie. Hrsg.: G. Petzow, M. Göken, Werkstoffinformationsgesellschaft mbH, Band 37 (2005), 47–52

Monotonic Properties And Cyclic Deformation Behavior of Friction Stir Welded Mg/Mg-, Mg/Al- And Al/Al-Joints

Guntram Wagner, Markus Gutensohn, Dietmar Eifler
Institute of Materials Science and Engineering, University of Kaiserslautern, Germany

1 Introduction

Friction stir welds (FSW) are based on an intensive plastic deformation of the joining partners in the welding zone. Joints with a yield strength close to the initial state of the original materials can be realized by this low temperature welding technique. In the paper, investigations of monotonic and cyclic properties of Mg/Mg, Mg/Al and Al/Al light weight joints are presented. Plastic strain amplitude and electrical resistance measurements during cyclic loading were applied to get detailed information about microstructural changes in the welding zone of the joints. Furthermore microscopical investigations of the welding areas were carried out.

2 Welding Equipment

The Institute of Materials Science and Engineering, University of Kaiserslautern, uses a milling machine from Deckel Maho, DMG Germany, to perform the friction stir weldings. The spindle of the machine has an output power of 15 kW with a maximum rotational speed of 12 000 rpm. The maximum dimensions of the plates to be welded are 800×600 mm^2. The milling machine is able to perform friction stir weldings of light weight material sheets up to a thickness of 10 mm. The tilt angle of the machine head can be electronically adjusted. The machine is controlled by a CNC-operating system. Extensive integrated measurement equipments allow an on-line detection of the temperatures in the welding zone and the welding forces in x-, y- and z-direction.

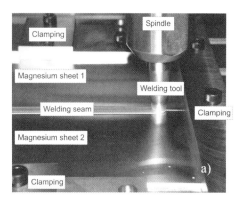 

Figure 1: a) Friction stir welding of magnesium sheets; **b)** FSW-tool for welding Mg-alloys

In comparison to a special FSW machine the use of a conventional milling machine has several advantages. The price for these machines is lower and the integrated tool pick-up system enables the alternative use of welding and milling tools.

To perform the FSW-welds, different clamping units have been constructed and tested. Figure 1. a) shows the spindle with one of the welding tools used and the clamping during a welding process. The welding tool was made of a hot work tool steel with a high temperature strength, a high wear resistance and a good machinability. In contrast to conventional FSW-welds of Al-alloys a special modification in the design of the shoulder was carried out for the welding of Mg-alloys. To provide a better material flow towards the centre of the tool three channels were milled into the shoulder of the welding tool (Figure 1. b)).

3   Materials and Specimens

For the welding experiments the Mg-alloy AZ31 and the Al-alloy AA5454 were used. The chemical composition of the alloys are shown in **Table 1.**

Table 1: a) Chemical composition of AZ31 (MgAl3Zn1)

elements	Si	Cu	Al	Zn	Fe	Ni	Mn	others.
ma.-%	0.1	0.1	3	1	0.03	0.005	<0.4	0.1

Table 1: b) Chemical composition of AA5454 (AlMg3Mn)

elements	Si	Fe	Cu	Mn	Mg	Cr	Ti	Zn	others.
ma.-%	0.25	0.4	0.1	1	3	<0.2	0.2	0.25	0.15

Due to his hexagonal lattice with an elongation after fraction of 7 % (AlMg3Mn: 9 %) the ductility of magnesium is somewhat lower then the one of aluminium. This property has also detrimental effects on the weldability of the Mg-alloy. In tensile tests the yield strength could be determined in a range of 190 to 200 MPa. The tensile strength was about 290 MPa.

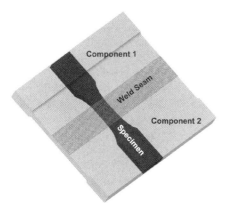

Figure 2: Position of a specimen in a FSW-joint

For the butt welding experiments cutted AZ31 and AlMg3Mn sheets in the dimension of 125 × 300 mm² were used. After cleaning with ethyl alcohol the sheets were fixed on the machine table by a special clamping device.

To provide an exact movement of the tool pin along the contact area of the sheets, the sheets are measured in width and thickness at several points. These data are the basis for the CNC-welding program. The welding process starts with the penetration of the tool pin. Then a dwell time of 10 s follows before the transversal feed starts. The welding seam has a length of 250 mm. After welding the sheets are cutted into pieces of 20 × 250 mm² for the monotonic and cyclic investigations as shown in Figure 2.

4 Results

4.1 Force and Temperature Measurements during the FSW Process

With the aim to optimize the welding process a thermographic system was positioned above the welding zone. In addition thermocouples were implemented along the seam to describe the temperature distribution during the welding (Figure 3). The welding forces during the FSW-process were measured with load cells integrated in the anvil.

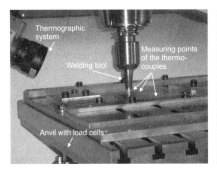

Figure 3: Set-up of the force, thermographic and thermometric measuring systems during the FSW-process

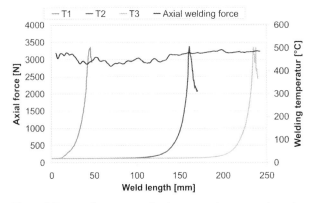

Figure 4: Force and temperature development at three measuring points during FSW of an AZ31/AZ31-joint

In the following characteristic forces and temperatures developing during the FSW-process are shown exemplarily for an AZ31/AZ31-joint (Figure 4). The Mg-alloy sheets had a thickness of 3.5 mm and the joint was welded with a velocity of 120 mm/min.

The resulting initial welding force of about 3000 N is nearly constant during the whole welding process. The temperature rise and the maximum temperature of about 500 °C for each measuring point are nearly identical over the whole seam.

4.2 Monotonic and Cyclic Loading

With 260 MPa the tensile strength of the AZ31/AZ31-FSW-joint reaches about 90% of the tensile strength of the initial material [4]. The elongation at fracture of 7% corresponds to the original material. Moreover Figure 5 shows, that for optimized process parameters the obtainable mechanical properties are independent from the individual position within the weld seam.

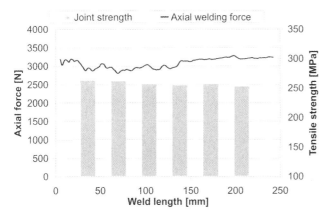

Figure 5: Axial welding force and tensile strength along a FSW AZ31/AZ31-seam

A cross section of a AZ31/AZ31-FSW-joint (Figure 6, Figure 7 A, B, C in detail) shows characteristic microstructural changes.

In the middle of the butt weld the FSW-nugget is visible (C). This area is followed by the thermally affected (B) and the heat affected (A) zone. The microstructure clearly changes from the middle of the weld to the non-welded original material.

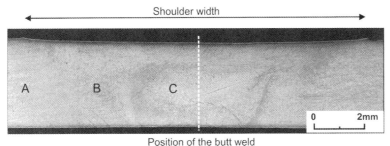

Figure 6: Cross section of an AZ31/AZ31-FSW-joint

For AlMg3Mn/AlMg3Mn-joints the influence of the welding velocity on the developing micro hardness profile was investigated (Figure 8). With an increasing welding velocity the thermally and mechanically influenced volume gets smaller, because of the lower energy induced locally in the welding seam. A larger volume with a lower micro hardness leads to a higher elongation at fracture with nearly constant tensile strength of the joints.

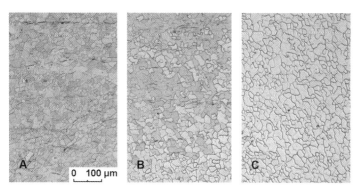

Figure 7: Cross section of the heat affected zone **A**, the thermo mechanical affected zone **B** and the weld nugget **C** of an AZ31/AZ31-FSW-joint

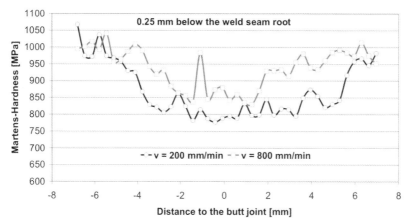

Figure 8: Micro hardness profiles of AlMg3Mn/AlMg3Mn-FSW-joints with different welding velocities

For dissimilar friction stir welds exemplarily AZ31/AlMgSi0,5-joints were investigated. In the cross-section in Figure. 9 a very high plastic deformation is visible, which induces an intensive stirring of the AZ31/AlMgSi0,5-sheets.

At the Institute of Materials Science and Engineering comprehensive methods to characterize the cyclic deformation behavior of metallic materials are available. To get information about the microstructural changes in the materials and their fatigue behavior, the plastic strain amplitude and the electrical resistance are measured during cyclic loading. This technique was successfully applied at FSW-specimens of AlMg3/AlMg3-joints (Figure 10.). Until approximately $130 \cdot 10^3$ cycles the plastic strain amplitude indicates a very small cyclic softening. After about

$135 \cdot 10^3$ cycles the plastic strain amplitude strongly increases due to microcrack growth short before final failure.

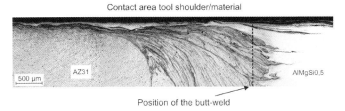

Figure 9: Cross-section of an AZ31/AlMgSi0,5-FSW-joint

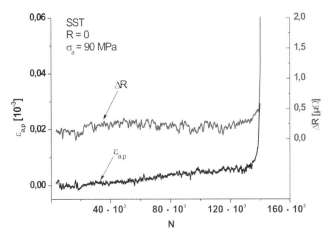

Figure 10: Plastic strain amplitude and electrical resistance measured during cyclic loading of an AlMg3Mn/AlMg3Mn-FSW-joint

The electrical resistance seems to be constant for the first $120 \cdot 10^3$ cycles. Afterwards the resistance values increase. This is an indication for a developing fatigue damage in the joint. For about $140 \cdot 10^3$ cycles an accelerated increase of the electrical resistance occurs as an indication of instable crack growth. These investigations show that the measured physical values change in a characteristic manner as a function of cyclic load and time. The electrical resistance seems to be more sensitive to identify the fatigue damage. In the future these measurements should be used for lifetime predictions of friction stir welded components.

5 Conclusions

The results demonstrate that friction stir welding is an effective method to join similar and dissimilar light weight materials like Mg-alloys and Al-alloys. The achieved tensile strength of the Mg-alloy/Mg-alloy-joints was nearly 90 % of the tensile strength of the original material. The fatigue behavior of friction stir welded light weight materials joints can be characterized by an on-line measurement of the plastic strain amplitude and the electrical resistance.

6 Acknowledgements

The investigations are supported by the DFG Priority Program 1168: „Erweiterung der Einsatzgrenzen von Magnesiumlegierungen"

7 References

[1] K. U. Kainer, Wiley-VCH, Weinheim, 2000
[2] R. Zettler, A. da Silva, S. Rodrigues, A. Blanco, J. dos Santos, AEM, 2006, 8, 415–421
[3] B. Ebel-Wolf, F. Walther, D. Eifler, DVM-Bericht 611, 2005, 240–244
[4] L. Sunggon, K. Sangshik, L. Cang-Gil, D. Chang, J. K. Sung, Metallurgical and Materials Transactions, Vol. 36A, 2005, 1609–1012

Authors

Aarstad, K.   207
Abaturov, I.S.   240
Abbott, T.   26, 967
Abramovich, H.   601
Abu Leil, T.   37; 49
Abu-Farha, Fadi   399
Adamiec, J.   506
Adeva, P.   699
Alexopoulos, N.D.   743
Alpers, P.   1076
Al-Samman, T.   553
Amberger, D.   1042
Amira, S.   775
Anada, H.   419
Ando, Y.   281
Andreatta, F.   749
Anzai, K.   941
Apachitei, I.   749; 849
Arruebarrena, G.   165; 228, 473
Astafiev, V.V.   240
Atrens, A.   715
Avraham, S.   473

Bach, F.-W.   215; 432, 547, 789, 909, 991, 1076
Bakkar, A.   842
Bakke, P.   55; 93
Bamberger, M.   473; 486
Barchiche, C.-E.   782; 856, 862
Baril, E.   498
Barnett, M.R.   571, 588
Barrallier, L.   877
Barucca, G.   466
Ben Hassen, S.   809
Bender, S.   721
Ben-Dov, A.   14, 165, 228
Ben-Hamu, G.   727, 769
Berçot, P.   809
Berger, C.   658, 734, 802
Bergman, F.   976
Berkani, A.   749, 849
Bertilsson, I.   976
Bettles, C.J.   480
Blandin, J.J.   393, 633

Blawert, C.   37, 49, 262, 344, 715, 958
Bober, J.   344
Bobrovnitchii, G.S.   131
Boese, E.   721
Bohlen, J.   158
Bohlen, J.   607
Böhme, W.   645
Boissière, R.   393
Boomsma, J.P.   324
Bormann, R.   158, 1033, 1055
Bosse, M.   909, 1076
Böttger, B.   151, 1021
Bousselmi, L.   809
Bowles, A.L.   55, 93
Brandt, M.   871
Brocks, W.   145
Brodova, I.G.   240
Broer, C.   909
Brokmeier, H.-G.   357, 529
Bronfin, B.   14, 336, 473
Brokmeier, H.-G.   357, 529
Brosius, A.   909
Bührig-Polaczek, A.   929, 1021
Buršík, J.   37, 49
Buršíková, V.   37, 49

Camin, B.   687
Cao, P.   8, 189
Chamos, A.N.   743
Chang Dong Yim   221, 763
Chang, Y.A.   74
Ching-An Huang   842
Chmelík, F.   607
Cibis, R.   81, 87, 100
Cizek, J.   517
Commin, L.   877
Cordini, P.   871
Cottam, R.   318
Cross, C.E.   727
Cwajna, J.   506

Danckert, J.   985
David, P.   658
Davis, B.   318
Dehm, G.   486

Denkena, B.   1076
Deveneyi, S.   1, 228
Di Cristoforo, A.   466
Dieringa, H.   929
Dietrich, S.   1069
Dietz, A.   862
Dietze, G.   639
Dietzel, W.   37, 49, 666, 715
Dilthey, U.   734
Djurdjevic, M.B.   1015
Dobatkin, S.V.   137
Dobatkina, T.V.   137
Dobeš, F.   699, 947
Dobro, P.   607
Doglione, R.   466
Drevenstedt, A.   297
Drozd, Z.   594
Dubé, D.   775
Dumont, M.   877
Dunlop, G.   967
Durandet, Y.   871
Düring, M.   377, 560
Duszczyk, J.   119, 324, 413, 541, 749, 849
Dzwonczyk, J.   119, 541, 749
Easton, M.   26, 189
Ebeling, T.   158
Ebel-Wolf, B.   673
Eibisch, H.   106, 248
Eidenhammer, M.   894
Eifler, D.   673, 1092
Eiken, J.   151, 1021
Eisenlohr, P.   1042
Eliezer, A.   822, 727, 769
Ellermeier, J.   734
Engh, T. A.   207
Eppel, K.   734, 802
Esderts, A.   658
Essadiqui , E.   535
Estrin, Y.   1027

Fabián, T.   512
Fan, Z.   256
Fechner, D.   262
Feil, Florian   828
Ferkel, H.   1027
Fernandez dos Santos, J.   883
Ferragut, R.   466

Fischer, J.   958
Fischersworring-Bunk, A.   498, 976
Fleck, C.   1062
Fratila-Apachitei, L. E.   749
Fratila-Apachitei, L. E.   849
Fuentes, E.   916
Fuessel, U.   901
Fuest, T.   967
Fujimoto, H.   330, 581
Funami, K.   704
Fürbeth, W.   828
Furui, M.   419
Fusheng Pan   297

Gable, B.M.   447
Galvani , C.   535
Galvao da Silveira Mussi, R.   941
Garces, G.   699
Garg, G.   901
Geiger, M.   386
Ghali, E.   775
Gharghouri, M.A.   680
Gibson, M.A.   480
Gill, L.R.   441
Gjestland, H.   175
Glatzel, U.   498
Goellner, J.   721
Göken, M.   1042
Gökena, J.   620
Golovko, A.N.   432
Gorny, A.   486
Gorsse, S.   480
Gottstein, G.   553
Gradinger, R.   297
Graff, S.   145
Grimm, J.   658
Gröbner, J.   32, 74, 1021
Gugau, M.   658, 802
Günther, R.   1033
Gutensohn, M.   1092
Gutman, E.M.   709

Hahn, O.   1085
Hama, T.   330
Hammer, N.   1076
Hampl, M.   32
Han, E.H.   666

Hanada, K.   125
Hartig, Ch.   158, 1033, 1055
Hartmann, H.   377
Hartmann, M.   248
Hassel, T.   432, 789
Hatsukano, K.   125
Hazan, J.   782, 856, 862
Heaney, S.   20
Hecht, J.   386
Helis, L.   581
Hepke, M.   215, 547
Heyn, A.   721
Hilbinger, R.M.   248
Hintze, W.   344
Hiraga, K.   523
Hnilica, F.   693
Honsel, Ch.   165
Horstmann, M.   1085
Hort, N.   37, 49, 262, 620, 929, 958, 1027
Huang, Y.   49
Hufenbach, W.   929
Huichang, T.   406
Hurtado, I.   165, 228
Hutchinson, C.R.   480

Iozzia, A.   466
Islamgaliev, R.K.   517

Janz, A.   74
Jenning, R.   248
Ji, S.   256
Jirásková, Y.   37, 49
Joksch, S.   922
Juers, C.   782, 856, 862
Kainer, K.-U.   37, 49, 262, 336, 344, 529, 607, 666, 715, 883, 929
Kaneko, J.   627
Karabet, A.   377
Karger, A.   215
Katgerman, L.   324
Kato, M.   125
Kaufmann, H.   1027
Kawabata, K.   113
Kawalla, R.   274, 364, 1048
Kawamura, H.   941
Ke, W.   666
Keckés, J.   486

Keshavarz, Z.   588
Kettler, C.   967
Kettner, M.   297
Khan, S.A.   651
Khosroshahi, R.A.   796
Khraisheh, M.   399
Kiebus, A.   81, 87, 100, , 492, 506
Kilian, H.   297
Kitamura, H.   419
Klaus, G.   929, 1021
Kleiner, M.   909
Klumpp, G.   862
Koch, R.   645
Koh, H.   370
Koike, T.   651
Kondoh, K.   8, 113
Konno, K.   523
Köster, U.   757
Kotler, A.   601
Kozlov, A.   32, 1015
Kraus, C.   789, 1069
Kühlein, W.   297
Kulyasova, O.   517
Kun, Wu   529
Kunst, M.   498, 976
Kužel, R.   512
Kwang Seon Shin   43

L'Esperance, G.   498
Labelle, P.   498
Ladstaetter, E.   297
Landkof, B.   601
Langdon, T.G.   419
Langkamp, A.   929
Lasera, T.   1055
Lavrenko, V. A.   816
Leeflang, M.A.   119, 413
Letzig, D.   158, 344, 607, 999, 1055
Levi, G.   486
Leyens, C.   613
Li, J.   535
Li Peijie   240
Liebscher, C.   498
Lieven, E.   645
Lindemann, J.   613
Liu, G.   256
Liu, P.   535

Löffler, A. 336
Löhe, D. 1062
Lohmüller, A. 106, 248
Long, S. 297
Lorimer, G. 318, 441
Luká, P. 575, 607, 953
Lussana, D. 466
Lutz, A. 297
Lyon, P. 20, 441

Ma, Qian 8, 189, 200
Mahmood, S. 14, 165, 228
Maier, P. 262, 958
Marx, T. 976
Masny, H. 734
Massazza, M. 466
Masse, J.-E. 877
Matsuura, M. 523
Matsuzaki, K. 125
Medraj, M. 74
Mendi, C.L. 480
Mengucci, P. 466
Merklein, M. 386
Milika, K. 699, 947
Minamiguchi, S. 370
Mirkovi, D. 74, 1009
Mitsugu, M. 268, 281
Mitsuhiko, Y. 523, 651
Mizuno, S. 704
Moscovitch, N. 14
Motegi, T. 941
Mücklich, S. 1085
Muddle, B.C. 447
Mueller, K. 406, 425
Mueller, S. 406, 425, 687
Muntada, L. 916
Murakoshi, Y. 125
Murray, M. 967
Mutoh, Y. 651

Neubert, V. 842
Neugebauer, R. 1069
Newiak, F. 377
Nie, J.F. 26, 447
Nikitina, N.I. 137
Nishida, S. 268, 281
Nishijama, M. 523

Noster, U. 297
Nowak, M. 991
Nürnberg, M.R. 1055
Obermair, F. 894, 916
Oberschelp, C. 929
Oenášek, V. 620, 693
Oginuma, H. 113
Ohno, M. 1021
Okayasu, K. 581
Ostendorf, A. 871
Ostrovsky, I. 834
Oswald, M. 364
Ovsyannikov, B. V. 352
Ozsváth, P. 894

Palkowski, H. 357
Pandey, S. 901
Pankiewicz, C.G. 883
Pantelakis, S.G. 743
Parvini Ahmadi, N. 796
Pako, J. 506
Pelcová, J. 67, 453
Perez, P. 699
Pieper, C. 757
Pieper, G. 377
Pinheiro, G. A. 883
Pinkernelle, A. 165, 228
Plamondon, P. 498
Popel, P.S. 240
Popov, M.V. 137
Prakash, A. 274, 1048
Pravdic, F. 1027
Prochazka, I. 517
Psyk, V. 909
Puech, S. 633
Rao, K.P. 49
Rashkova, B. 486
Razrazi, M. 809
Regener, D. 639
Reimers, W. 406, 425, 687
Richter, U. 377
Riedel, H. 274, 1048
Riontino, G. 466
Robson, J. 318
Rocca, E. 782, 856, 862
Rodman, M. 547, 991
Rokhlin, L.L. 137

Rossberg, A.   547
Rucki, P. M.   357
Rudajevová, A.   953
Rzycho, T.   87, 100

Sadayappan, M.   234
Sahoo, M.   234, 680
Sakai, T.   370
Sakurada, Y.   330
Sanz, C.   916
Schaeffer, L.   560
Schaper, M.   215, 901, 909, 991
Scharrer, M.   106, 248
Schiffl, A.   1027
Schildknecht, A.   1062
Schmid-Fetzer, R.   32, 74, 1009, 1015, 1021
Schmidt, C.   274, 364, 1048
Schreyer, A.   529
Schröder, A.   165, 228
Schumann, S.   336
Schütze, M.   828
Schwarz, S.   645
Schwebke, B.   529
Schwerin, R.   922
Shiba, D.   941
Shimizu, T.   125
Shin, K.S.   769
Shvets, V. A.   816
Sideris, A.J.   131
Sillekens, W. H.   999
Simmchen, E.   901
Singer, R.F.   106, 248
Skeldona, P.   849
Slooff, F.   324, 541
Smola, B.   67, 453, 517, 620, 693
Song, G.   715
Song, W.   871
Soubeyroux, J.L.   633
Stadler, F.   297
Steffensen, M.   985
Steglich, D.   145
Steinbach, I.   151, 1021
Steinhoff, K.   620
Steinmetz, J.   782, 856, 862
Stich, A.   386
StJohn, D.H.   8, 189
Stoesser, G.   967

Störmer, M.   958
Stulíková, I.   67, 453, 517, 620, 693
Su Geun Lim   763
Su Jang R.   763
Sugamata, M.   627, 941
Sun You, B.   221, 763
Swiostek, J.   344
Syed, I.   20
Syvertsen, M.   207
Száraz, Z.   935
Szmejkál, A.   894

Tada, K.   288
Takács, J.   894
Takano, Y.   281
Takaya, M.   704
Takeshi, Y.   281
Takuda, H.   330
Talash, V.N.   816
Tamehiro,   704
Tamura, Y.   704
Tanabe, F.   941
Tang Aitao   297
Tatiana, K.V.   816
Thiele, J.   976
Thompson, G.E.   849
Thomson, J.P.   234
Timofeev, V. N.   137
Torabadi, A.   796
Townsend, J.   14, 165, 228
Tranell, G.   207
Tremblay, R.   775
Triki, E.   809
Trojanová, Z.   512, 575, 594, 935
Troßmann, T.   658, 734, 802

Ullmann, M.   364
Unigovski, Y.B.   709
Utsunomiya, H.   370

Vainola, J.   14
Väinölä, J.   165, 228
Vidrich, G.   1027
Viehweger, B.   377, 560
Vlach, M.   517

Wagner, G.   1092
Walther, F.   673
Wang, Y.   256
Wanner, A.   1062
Weidenmann, K.A.   1062
Weimin, Gan   529
Weinert, K.   1076
Weiss, D.J.   234
Weiss, K.   165, 228
Wendt, A.   165, 228
Werner, J.   215, 929
Westengen, H.   55, 93, 175
Wielage, B.   1085
Winzer, N   715
Witte, F.   822, 958
Woeste, K.   734
Wolter, P.   406
Woo Chul Cho   43
Wu, G.H.   221

Xiaoguang, Qiao   529
Xiaoshi, Hu   529
Xu, P.   727
Xu, S.   529, 680

Yang, M.   297
Yeon, Jun Chung   43
Yi, S.-B.   357
Young, Min Kim   763
Yuchang, Su   447
Yue, S.   535
Yukutake, E.   627

Zábranský, K.   49
Žaludová, N.   67
Zander, D.   757
Zarandi, F.   535
Zeng, R.C.   666
Zhang, P.   613
Zheng, M.   529
Zheng Tan   976
Zhou, J.   119, 324, 413, 541
Zhu, J.   74
Zschetzsche, J.   901

Subject Index

A

acoustic emission   607
active environment impact   709
additives
– calcium   535
– carbon   8
– corrosion resistance   856
AE44   55
AE44   680
aerospace
– alloy   20
– applications   473
– investment-casting process   228
age-hardenable   137, 480
ageing   575
aircraft industry
– gravity casting   14
– wrought alloys   305
AJ62A   680
alkaline media, corrosion behavior   775
alloys
– aircraft industry   14
– anodizing   849
– arc brazing   901
– AZ   see Az alloys
– binary   789
– biomaterials   822
– blending   1015
– casting-rolling technology   364
– coating   834
– coating   842
– comparative corrosion   816
– composition   789
– containing Al   26, 43, 74, 87, 93, 100, 119, 125, 498, 541, 588, 666, 749, 763, 894, 1055, 1092
– containing Ca   49, 119, 125, 432, 535, 789, 1042
– containing Cu   523, 816
– containing Gd   441, 523
– containing Li   419, 594, 935, 953
– containing Mn   693, 769
– containing Nd   441, 517, 693
– containing Ni   131, 816
– containing rare earth   67
– containing Sc   693
– containing Si   529
– containing Sm   137
– containing Sn   43, 49, 119, 480
– containing Sr   74, 87, 453, 498
– containing Tb   517
– containing Ti   699
– containing Y   318, 453, 523, 693
– containing Zn   26, 93, 125, 441, 453, 486, 541, 588, 666, 749, 769, 1055
– containing Zr   453
– corrosion   743, 757, 775, 782, 796, 1062, 1076
– crack propagation   666
– creep properties   687
– cyclic deformation   673
– deformation behavior   1055
– development   1–142
– die casting   81
– dilatation characteristics   953
– electroplating   862
– elektron 21   459
– elevated temperature   55
– fillers   871
– finely dispersed particles   704
– formation process   393
– fracture toughness behavior   639
– grain refinement   8, 1033
– heat-treatment   49
– hot rolling   274
– joining   1085
– melt filtration   221
– microstructural formation   1021
– microstructure   37
– modification   32
– molten   281
– new material   3
– plasma-anodized   856

– reinforcement   941
– relaxation oscillations   601
– SiC particles   1027
– strip   268
– structure   3, 506
– texture development   1048
– ultra fine grained (ufg)   517
– warm hydroforming system   985
– wrought   295–438
alumina fibres, creep behaviour   947
aluminum bolts   976
AM50   262, 620, 680, 757
AM60, fatigue behavior   651
am-lite ®   967
amorphous precipitates   523
angular pressing, Mg-8%Li Alloy   419
anisotropy   953
– AZ31   158, 607
anodized layers   651
– AM60   651
anodization   849
– AZ91D   856
aqueous chloride media   809
arc brazing   901
as-cast alloys   37
automobile wheels   352
AZ alloys   687
– extrusion   406
AZ31   113, 158, 535, 553, 727, 763, 877, 1027
– acoustic emission   607
– arc brazing   901
– deformation   581
– dent resistance   627
– ductile failure   571
– formability   330
– formation process   399
– grain refinement   1033
– hydroforming system   985
– microstructural evolution   560
– plastic deformation   425
– roll-compaction   113
– salt spray corrosion   743
– strain rate   871
AZ61
– deformation   581
– formability   330

AZ80   613
– microstructural evolution   560
– forming properties   336
AZ91   106, 757
– creep resistance   1042
AZ91D   856
– compocasting   935
– corrosion behavior   775
– viscosity   240

B

billet temperature   413
binary alloys, corrosion behaviour   789
biomaterials
– degradable   958
– magnesium alloys   822
biomedical science, stent   432
blending, thixomolding   1015
bolts   976
brazing   901
bulk metallic glass   633

C

calcium   1055
– additive   240, 535
– alloy modification   32
– texture evolution   158
– wrought alloys   769
cantilever beam   601
carbon additives   8
carbon fibre preforms   929
carbon steel, corrosion resistance   809
cast alloy, rolling   620
casting   173–294
casting
– corrosion   757
– gravity   14
– high pressure   175
– thin-walled   165
casting-rolling technology   364
catalyst   131
cathodic polarization   816
ceramic foam filtration   221
cerium   1055
– alloy modification   32
China, wrought alloys   297

chloride
– aqueous media 809
– concentrations 789
– corrosion 775, 789, 796
civil aircraft application, wrought alloys 305
clinching 1069
coating 828, 842
– composite 834
commercial alloys
– die forging 344
– solid fraction 1009
commercially pure Mg 318
comparative corrosion 816
compocasting, AZ91D/CNF 935
composite coatings 834
composites 953
– coating 842
– magnesium matrix 927–964
composition 393
– binary alloys 789
– rare earth-Mn alloys 67
composition control, Mg-Al-Zn 93
compression
– electromagnetic 909
– uniaxial 581
– wrought alloys 324
constitutive analysis 324
contacting condition 281
continuous carbon fibre preforms 929
continuous casting, horizontal 288
continuously wrought-alloys, microstructural evolution 560
coring structure, WE54 492
corrosion 707–868
– chloridic 796
– comparative 816
– heat-treatment 49
– microstructure 757
– salt spray 743
– welded magnesium joints 734
corrosion behavior 775, 782, 1076
– AZ31-xCa 763
– binary alloys 789
– fatigue 1062
– wrought alloys 727
corrosion interfaces, extruded alloys 749

corrosion protection 828
corrosion resistance 856
– magnesium layer 809
corrosion testing, material specific 721
cover gases, ecologically harmful 215
crack propagation 666
creep, Mg-10 vol.% Ti PM-Composite 699
creep asymmetry 680
creep behavior 947
– Mg4Y2Nd1Sc1Mn 693
– Rare Earth-Mn alloys 67
creep properties 687
creep resistance 1042
– friction welding 883
creep strength, AZ91 106
crystallization, AZ91D 240
current-potential-curves, passivation behavior 802
cutting processes, corrosion 1076
cyclic current-potential-curves, passivation behavior 802
cyclic deformation 1092
– characterization 673

D

damping capacity 601
deep-drawing applications, magnesium sheets 547
deformation 137, 594
– behavior 1055
– high strain-rate 541
– high temperature 581
– temperature 512
degradable biomaterials 958
dent resistance 627
Deutsche Forschungsgemeinschaft 1007–1098
diamond synthesis 131
die cast alloys 256
die cast components, durability 658
die casting 81, 100, 175, 639, 673, 680
–high pressure 929
–MgAl6Mn 645
–Mg-Al-Sr (AJ) 498
die forging 344
– disks 352
dieless clinching 1069

dilatation characteristics, Mg4Li 953
disks, die forging 352
dislocation structure, evolution 512
dispersed particles 704
drag process, contacting condition 281
draught rolling 370
ductile failure, AZ31 571
durability 658
DX 53Z 901
dynamic dent resistance 627
dynamic recrystallization 318
dynamic strain ageing 575

E
ECAP 529
ecological machining 916
ecologically harmful cover gases, reduction 215
elaboration, bulk metallic glass 633
electrochemical properties 789
– corrosion 816
electrodeposited magnesium layer 809
electromagnetic compression 909
electroplating 862
Elektron 21 alloy 20, 459
elevated temperature
– alloys 55
– formability 330
engine components 976
environment 822
– impact 709
equal-channel angular pressing 419
european community research 999
EV31 466
experimental basis data, heat-transfer modeling 1009
extrusion 406
– AZ31 113, 413
– conditions 419
– corrosion 749
– Mg-Al-Ca-Sn alloy 119
– wrought-alloys 560

F
face milling 894
failure, ductile 571

fatigue
– AZ31 727
– behavior 651
– crack propagation 666
feedstock, semi-continuously casted 377
fibre composites 953
filler alloy 871
filtration 221
fine graines 1027
fine magnesium crystals, creation 200
finely dispersed particles 704
flow instabilities 541
fluorine solubility 207
foam filtration 221
forging, MagForge 999
formability 386
– change 330
formation process
– AZ31 399
– superplastic 393
– magnesium sheet-metal 377
forming properties, wrought alloys 336
fracture toughness 639
friction welding 883
– AZ31 877
– joints 1092

G
gadolinium 441
German Research Council 1007–1098
glass, elaboration 633
grain refinement
– analytical methodology 189
– AZ31 1033
– magnesium alloys 8
grains, thermal stability 517
gravity casting 14
grignard reagents, corrosion resistance 809
growth instabilities, control 200

H
harmful cover gases, reduction 215
heat treatment 49, 453
– AZ alloys 687
– carbon fibre preforms 929
– Elektron 21 alloy 459

– Mg4Y2Nd1Sc1Mn 693
– Mg-Al 506
– WE54 492
heat-transfer modeling, commercial alloys 1009
high creep resistance, friction welding 883
high pressure
– diamond synthesis 131
– die casting 26, 175, 929
– torsion 517
high speed rolling 370
high strain-rate deformation 541
high temperature 43, 131
– deformation 581
highly calcium alloys 1042
horizontal continuous casting 288
hot chamber, die casting 81
hot rolling 274, 877
– texture development 1048
hot tearing susceptibility 234
hot working 560
hybrid materials 894
hybrid parts, ecological machining 916
hydroforming system, AZ31 985
hydroxyapatite 958

I
impact loading 639
improved creep strength, AZ91 106
injection molding 248
INNOMagTec 1007–1098
inorganic coatings 828
interfaces, corrosion 749
investment casting 228
– elektron 21 20

J
joining, different alloys 1085
joints 1092
– corrosion 734

L
large draught rolling 370
laser welding, AZ31 871
layer, electrodeposited 809
LIST, evaluation of Mg SCC 715
local mechanical properties 645

long period stacking structures 523
low melting point alloy 985

M
machining 922
– ecological 916
MagForge 999
magnesium
– aircraft industry 20
– crystals 200
– hot chamber die-casting 81
– injection molding 248
– molten 207, 215
– recycling alloy 262
– slab 288
magnesium based alloys 782
magnesium matrix composites 927–964
– creep behaviour 947
magnesium sheets
– metal 377
– tribology 386
magnesium-hydroxyapatite composites 958
mechanical anisotropy, AZ31 607
mechanical behaviour, bulk metallic glass 633
mechanical properties 569–708
– binary alloys 789
– local 645
melt drag process 268
melting 173–294
melt-quenched Mg98Cu1R1 523
metallic glass 633
Mg-Al alloy structure 506
Mg-Al-Ca-Sn alloy 119
Mg-4Al-4Re, die casting 100
Mg-Al-Sn alloys 43
Mg-Al-Sr alloys 74, 87, 498
Mg-Al-Zn family 93, 541
Mg-3Al-1Zn
– deformation behavior 1055
– sheet 588
MgCa0,8, stent 432
Mg98Cu1R1, stacking structures 523
Mg-4Li
– deformation behavior 594
– fibre composites 953

Mg-8Li   935
Mg-8%Li, equal-channel angular pressing   419
Mg-Nd-Gd alloys   441
Mg-Ni alloys   131
Mg-Si alloys   529
Mg-Sm alloys   137
Mg-Sn-phase formation   32
Mg-Sn-Ca alloys   37
Mg-3Sn-2Ca alloy   49
Mg-Sn-Mn alloys   37
Mg SCC   715
Mg-10 vol.% Ti Pm-composite   699
Mg-4Tb-2Nd alloy   517
Mg-Y alloys   318
Mg4Y2Nd1Sc1Mn, creep behavior   693
Mg-Y-Nd-Zn-Zr alloy   453
Mg-Zn based alloys   486
Mg-Zn-Al, microstructure   26
Mg-Zn-Mn based alloys   769
microalloying   480, 909
microhardness, heat-treatment   49
microstructural characterization, Mg-Al-Sr (AJ)   498
microstructural evolution   439–568
– AZ31   535
– Mg-Al-Ca-Sn alloy   119
– Mg-Zn alloys   486
microstructural formation   1021
microstructural stability   473
microstructure   37, 43, 165, 256, 459, 594, 645, 673
– AZ31   113, 877
– corrosion   757
– effects   393
– formation   151
– heat-treatment   49
– magnesium sheets   547
– Mg-4Tb-2Nd   517
– Mg-Al-Sr alloy   87
– Mg-Zn-Al alloys   26, 125
– strip cast magnesium   357
milling, hybrid materials   894
mischmetal, deformation behavior   1055
modelling   143–172
modified magnesium alloys, die forging   344

molten magnesium   281
– cover gases   215
– protection   207
monotonic properties   1092
mould heating   929
MRI 153M   658, 673
– corrosion behavior   775
MRI 301F, forming properties   336
MRI 230D   673

N

nanostructure   3
nanoparticle, corrosion protection   828
neutron diffraction   529
non-cutting processes, corrosion   1076
nucleation, control   200

O

orientation, Mg-10 vol.% Ti PM-composite   699

P

particles, finely dispersed   704
passivation behavior   802
performance gains   55
permanent molds, tearing susceptibility   234
phase composition, rare earth-Mn alloys   67
phase equilibria, Mg-Al-Sr alloys   74
phase formation
– Mg-Sn alloys   32
– Mg-Zn alloys   486
phase-field simulation   151
plasma, anodizing   856
plastic anisotropy, AZ31   158
plastic deformation   137, 425
polarization, cathodic   816
post processing   869–926
precipitate distributions, refinement   480
precipitation   453
– secondary   447
– sequence   441
preforms, carbon fibre   929
pressure die cast   639
priority programme 1168–1007

process chain approach, civil aircraft application 305
production technology 453

Q
quench sensitivity, Mg-Nd-Gd alloys 441

R
ram speed 413
rare earth-Mn alloys 67
rational friction welding 883
recrystallization, commercially pure Mg 318
recycling alloy 262
recycling melting 173–294
refinement
– grain 189
– AZ31 1033
reinforcement
– magnesium parts 929
– SiC particles 935
relaxation oscillations 601
research programs 997–1098
rheo-diecast Mg-Alloys 256
rivet clinching 1069
roll-compaction, AZ31 alloy 113
rolling
– AM50 620
– AZ80 613
– contacting condition 281
– hot 877
– magnesium sheets 547
– single pass 370
– texture development 1048

S
salt spray corrosion 743
sand casting
– Elektron 21 alloy 20
– Mg-Zn-Al 26
– amorphous precipitates 523
second phase 763
secondary precipitation, WE54 447
semi-continuously casted feedstock 377
sensory use 991
severe plastic deformation, Mg-Sm alloys 137

sheet-metal, tribological properties 377
sheets, casting-rolling technology 364
short alumina fibres, creep behaviour 947
silicium, wrought alloys 769
SiC particles
– alloy reinforcement 935
– AZ31 1027
(SiC)P, grain refinement 1033
simulations 143–172
– blending 1015
– microstructure 165, 645
– phase-field 151
– texture development 274, 1048
single crystal, magnesium 145
single pass large draught rolling 370
sintered steel, hybrid materials 894
slab, magnesium alloys 288
slip, tension testing 588
solidification technique 1021
solubility, fluorine 207
specific damping capacity 601
specimen orientation 699
spherical magnesium crystals, creation 200
squeeze cast 693
SSRT, evaluation of Mg SCC 715
stacking structures 523
static dent resistance 627
static loading 639
steel, arc brazing 901
stent 432
stir welding 877
strain 575
– path 553
– rate, AZ31 871
strip cast alloys 125
strip cast magnesium, microstructure 357
strips, casting-rolling technology 364
structural behaviour, magnesium 145
structural durability, die cast components 658
structural evolution, EV31 466
structure
– Mg-Sm alloys 137
– nano/sub-micron 3
structure formation, AZ91D 240
sub-micron structure 3

superplastic forming   393, 399
superplastic properties   419
surface rolling, AZ80   613
surface treatment   707–868
– composite coatings   834
tearing susceptibility   234

T

technologies, innovative   1085
temperatures
– AZ91D   240
– billet   413
– elevated   55, 330
– high   43, 131, 581
– viscosity   240, 709
tensile behavior   743
tensile-compressive creep
    asymmetry   680
tension testing, Mg-3Al-1Zn   588
texture development   529, 581, 1048
– AZ31   553
– calcium   158
– numerical investigation   274
thermal stability, Mg-4Tb-2Nd   517
thermal treatment
– AZ91D   240
– EV31   466
thermodynamic simulation, blending   1015
thermomechanical processing,
    AZ31   535
thin-walled magnesium castings   165
Thixomolding   106, 248, 1015
torsion, Mg-4Tb-2Nd   517
toughness   639
tribology   386
– magnesium sheet-metal   377
tubes   432
– electromagnetic compression   909
twin roll caster   125
twinning, ductile failure   571

U

ultra fine grained (UFG)   512
– Mg-4Tb-2Nd   517
ultrasonic agitation, AZ91D/CNF   935
uniaxial compression   581

V

viscosity
– environment impact   709
– temperature dependence   240

W

warm forming   386
warm hydroforming, AZ31   985
WE43   796
WE54   447, 92, 796
welded magnesium joints, corrosion   734
welding   883
– friction   877, 883, 1092
– laser   871
– stir   877
weldments, AZ31   727
wrought alloys   295–438, 560, 743, 769
– AZ31   727
– compression tests   324
– corrosion fatigue behaviour   1062
– development   297
– improvement   909
– surface rolling   613

X

x-ray diffraction, UFG Magnesium   512

Y

yielding, magnesium   145

Z

zinc   441
– additives   67
zirconium, WE54   492